Textures of Materials
ICOTOM-10

Proceedings of the

10th International Conference
on
Textures of Materials
Part 2

ICOTOM-10

Clausthal, Germany, 20-24 September 1993

Edited by

H.J. Bunge

TRANS TECH PUBLICATIONS
Switzerland - Germany - UK - USA

ISBN 0-87849-681-5

Volumes 157-162 of
Materials Science Forum
ISSN 0255-5476

***Distributed** in the Americas by*

Trans Tech Publications Ltd
LPS Distribution Center
52 LaBombard Rd. North
Lebanon, NH 03766
USA
Phone: (603) 448 0037
Fax: (603) 448 2576

and worldwide by

Trans Tech Publications Ltd
Hardstrasse 13
CH-4714 Aedermannsdorf
Switzerland
Fax: (++41) 62 74 10 58

International Committee

R. Penelle (Chairman)
E. Aernoudt
G. Gottstein
H. Hu
W.B. Hutchinson
K. Lücke
W. Truszkowski
H. Abe†
H.J. Bunge
M. Hatherly
J.F. Humphreys
J.S. Kallend
S. Nagashima

Local Committee

H.J. Bunge
H.G. Brokmeier
F. Haessner
I. Nielsen
H. Quade
M. Wehrhahn
W. Böcker
H. Bunge
H. Klein
N.J. Park
A. Schulze
S. Zaefferer
J. Brinkmann
E. Haessner
U. Köhler
G. Quade
R. Schwarzer

OVERVIEW

Part 1

Plenary Lectures 1

1. Experimental Methods of Pole Figure and Individual Grain Orientation Analysis 57

2. Calculation and Representation of ODF, MODF and Other Textural Quantities 257

3. Experimental Studies of Texture Formation 499

3.1 Deformation Textures 583

3.2 Recrystallization Textures 885

Part 2

3.3 Local Textures (Shear Bands, EBSP, Misorientation) 1117

3.4 Textures of Special Materials 1325

4. Texture and Directional Properties of Materials 1521

5. Mathematical Modelling of Texture Formation and Materials Properties 1701

6. Technological Applications of Texture Studies 1915

7. Implication of Texture in Powder Diffraction Methods: Internal Stress Analysis, Phase Analysis, Structure Analysis 2049

Author Index 2083

Subject Index 2089

Abbreviations 2101

CONTENTS - Part 1

Plenary Lectures

A Historical Survey of Texture Development between ICOTOM-1 and ICOTOM-10 3
J. Grewen

Statistical Crystallography of the Polycrystal 13
H.J. Bunge

Orientation Imaging Microscopy: New Possibilities for Microstructural Investigations Using Automated BKD Analysis 31
B.L. Adams D.J. Dingley K. Kunze S.I. Wright

Textures in Ceramic Materials 43
K.J. Bowman

1. EXPERIMENTAL METHODS OF POLE FIGURE AND INDIVIDUAL GRAIN ORIENTATION ANALYSIS

Invited Lecture:

Texture Analysis by Neutron Diffraction **59**
H.-G. Brokmeier

Panel Discussion:

Advanced Experimental Techniques in X-Ray Texture Analysis **71**
H.J. Bunge R. Großterlinden A. Haase R. Ortega J.A. Szpunar P. Van Houtte

Quantitative Texture Analysis of Thin Polycrystalline Layers 97
G. Bermig J. Tobisch K. Helming

An Experimental Apparatus for Quantitative On-Line Texture Analysis 103
P. Blandford J.A. Szpunar

Investigation of Inhomogeneous Textures of Coatings and Near-Surface-Layers 111
J.T. Bonarski L. Wcislak H.J. Bunge

A New Method for Texture Measurement Based on an X-Ray Imaging Plate System 119
H.-G. Brokmeier M. Ermrich

X-Ray Method for Continuous Tracking of Grain Boundary Motion: Investigation of Grain Boundary Migration in Al-Bicrystals 125
U. Czubayko D.A. Molodov B.-C. Petersen G. Gottstein L.S. Shvindlerman

A High-Pressure Device for In-Situ Measurements in a Neutron Beam 131
J. Heinitz N.N. Isakov A.N. Nikitin W.A. Sukhoparov K. Ullemeyer K. Walther

Electronmicrodiffraction (EBSP) in the Scanning Electron Microscope (SEM) - Further Hardware Development to Improve Pattern Quality 137
J. Hjelen A.H. Qvale Ø. Gomo

A High Resolution Electron Diffraction Method for On-Line Texture Analysis 143
R. Høier J. Bentdal O. Daaland E. Nes

Automatic Recognition of Deformed and Recrystallized Regions in Partly Recrystallized Samples Using Electron Back Scattering Patterns 149
N.C. Krieger Lassen D. Juul Jensen K. Conradsen

Optimization of the Texture Determination of Thin Films from X-Ray Diffraction Measurements 159
B. Moreau F. Wagner H. Göbel

Determination of Texture in CuZnAl Shape Memory Alloys in the High Temperature Austenitic Phase 167
N.J. Park H.J. Bunge

Orientation Connectivity in Polycrystals 175
V. Randle

High Temperature Texture Goniometer for the Measurement of Transformation Textures 181
F.R. Reher W. Hänel U. Czubayko G. Gottstein

A CCD Camera System for the Acquisition of Backscatter Kikuchi Patterns on an SEM 187
R.A. Schwarzer

An Inexpensive CCD Camera System for the Recording and On-Line Interpretation of TEM Kikuchi Patterns 189
R.A. Schwarzer S. Zaefferer

On-line Interpretation of SAD Channeling Patterns 195
R.A. Schwarzer S. Zaefferer

Preparation of High-Resistance or Sensitive Samples for Grain Orientation Measurement with Electron Microscopes 201
R.A. Schwarzer

Equipment for Texture Measurement in Thin Films 207
J.A. Szpunar P. Blandford

Application of the Method for Non-Destructive Evaluation of Texture Heterogeneity 213
J. Tarasiuk K. Wierzbanowski A. Baczmanski

Influence of Texture on Backscattered Ultrasonic Noise 221
R.B. Thompson K.Y. Han I. Yalda-Mooshabad J.H. Rose F.J. Margetan

Characterisation of Multilayers Crystallographic Texture 227
A. Tizliouine J. Bessières J.J. Heizmann J.F. Bobo

Determination of the Texture of a TiNi Shape Memory Alloy by Treatment of a Set of Overlapping Rays 235
A. Vadon J.J. Heizmann M. Chaoui

On-Line Determination of Deformation Systems for Cubic and Hexagonal Materials with the TEM 241
S. Zaefferer R.A. Schwarzer

On-line Interpretation of Spot and Kikuchi Patterns 247
S. Zaefferer R.A. Schwarzer

Determination of Magnetic Textures in Ferromagnetics by Means of Neutron Diffraction 251
U. Zink H.-G. Brokmeier H.J. Bunge

2. CALCULATION AND REPRESENTATION OF ODF, MODF AND OTHER TEXTURAL QUANTITIES

Invited Lectures:

Statistics, Evaluation and Representation of Single Grain Orientation Measurements 259
O. Engler G. Gottstein J. Pospiech

Experimental Errors in Quantitative Texture Analysis from Diffraction Pole Figures 275
A. Mücklich P. Klimaneck

Coordinate Free Tensorial Representation of *N*-Point Correlation Functions for Microstructure by Harmonic Polynomials 287
B.L. Adams P.I. Etingof D. Sam

Ultrasonic Characterization of the Texture of Aluminum Alloys 295
A. Anderson R.B. Thompson R. Bolingbroke

Some Comments about Texture Analysis: Comparison between Harmonic and Vector Methods 301
B. Bacroix Th. Chauveau P. Gargano A.A. Pochettino

Texture of High Temperature Deformed $\gamma-\alpha_2$ Titanium Aluminides 309
G. Bermig J. Tobisch C. Scheffzük C. Oertel

Practical Considerations in the Calculation of Orientation Distribution Functions from Electron Back-Scattered Diffraction Patterns 315
A.W. Bowen

Approximation of Orientation Distribution of Grains in Polycrystalline Samples by Means of Gaussians 323
T.I. Bucharova T.M. Ivanova D.I. Nikolayev T.I. Savyolova

Automatic Quantitative Analysis of Texture Components from CODF Data 327
M.J. Cai W.B. Lee

ODF Analysis and Physical Properties on Personal Computer Basis 333
E. Dahlem-Klein H. Klein N.J. Park H.J. Bunge

Series Expansion and Positivity 341
M. Dahms

Variation Width, Modelling Assumptions, and Interpretation of ODFs from Diffraction PDFs 349
J.J. Fundenberger H. Schaeben

Texture Analysis from Diffraction Pole Figures of Structurally Inhomogeneous Polycrystals 357
I. Girlich P. Klimanek

Some Applications of the Texture Component Model 363
K. Helming

Orientation Representation on the 4-D Unit Sphere, the Analogon of the 3-D Pole Sphere 369
G. Ibe

Estimate of the Grain Number for True ODF Determination by Individual Orientation Detecting Method 375
Y.S. Liu R. Penelle F. Wang J.Z. Xu Z.D. Liang

Absorption Correction for Non-Standard Geometry for Pole Figure Measurements 381
D.I. Nikolayev K. Walther

Approximation of the ODF by Gaussians for Sharp Textures 387
D.I. Nikolayev T.I. Savyolova

Numerical Economization of the Series Method 393
D.I. Nikolayev

The ODF Calculation from Pole Figures for Different Types of Crystal and Sample Symmetries 401
K. Pawlik

Statistical Analysis of Single Grain Orientation Data Generated from Model Textures 407
J. Pospiech J. Jura G. Gottstein

Analysis of the ADC Method for Direct ODF Calculation by Use of
Gauss Models and Standard Functions 413
D. Raabe K. Lücke

Inverse Formulas for Orientation Distribution Function 419
T.I. Savyolova

Analogy and Duality of Quantitative Texture Analysis by Harmonic or
Indicator Functions 423
H. Schaeben

Exploratory Orientation Data Analysis: Kernel Density Estimation and Clustering 431
H. Schaeben

Methods for Determining Small Differences between Measured Textures 439
R. Schouwenaars P. Van Houtte

Linear Regression for Texture Comparison 447
J. Tarasiuk K. Wierzbanowski

Smoothing Pole Figures Using Tensor Products of Trigonometric and
Polynomial Splines 453
C. Traas H. Siemes H. Schaeben

Direct Correspondence between the Volumic Fractions Found by Texture Analysis
with the Vector Method and with the Harmonic Method and Inverse Correspondence 459
A. Vadon J.J. Heizmann T. Baudin R. Penelle

Quantitative Texture Analysis with the Vector Method, Using Pole Figures of Rays
and a Set of Overlapping Rays 465
A. Vadon J.J. Heizmann

Texture Analysis from Diffraction Spectra 473
H.-R. Wenk S. Matthies L. Lutterotti

Display of the ODF by an Image Analyzer 481
L.F. Wu Y. Wang W.P. Liu H.J. Bunge

PC Program for the Calculation of ODF from SAD Pole Figures or
from Individual Grain Orientations 487
W. Xia R.A. Schwarzer

A Computer Program for Determining Volume Fractions of Texture
Components in Cubic Materials 493
L. Zuo J. Muller C. Esling

3. EXPERIMENTAL STUDIES OF TEXTURE FORMATION

Development of Microtextures in Cold-Rolled Iron-Oligocrystals 501
J. Boeslau D. Raabe

Texture Analysis in the Monoclinic Martensitic Phase of a CuZnAl Shape Memory Alloy 507
M. Dahms N.J. Park H.J. Bunge

Texture and Phase Development in Al-40 at% Ni Composites 515
H. Gertel-Kloos H.-G. Brokmeier

Oxidation of Titanium, Zirconium and their Alloys, Texture of their Oxide Scales 521
I. Halley-Demoulin C. Valot D. Ciosmak M.Lallemant J.J. Heizmann C. Laruelle

Texture Analysis of an Amphibolite using the Component Method 529
K. Helming R. Kruse S. Siegesmund

Changes in Texture by Irradiation with Ar^{++} of Iridium and Platinum Layers on Oriented Sapphire Substrates 535
V. Jung W. Schneider

Cube Texture in Aluminum Single Crystal of a (118)[711] Orientation 541
T. Kamijo K. Obata M. Ishizuki H. Fukutomi

The Texture Competition in Solidification Process 547
D.Y. Li J.A. Szpunar

Texture of Al-7%Si Ingots 555
D.Y. Li C. Rocaniere S. Das J.A. Szpunar

Texture Transformation due to Martensitic Phase Transformation in CuZnAl Shape Memory Alloys 563
N.J. Park H.J. Bunge

Effect of Phosphorous on the Development of Textures in Ti-Stabilized Steel Sheets 571
Y.B. Park S.K. Chang D. Raabe K.Lücke

Describing the Rolling Texture in Very Low Stacking Fault Energy Alloys 577
W. Truszkowski J. Pospiech J. Jura J.Gryziecki K. Pawlik

3.1 DEFORMATION TEXTURES

Invited Lectures:

Deformation Texture Formation from Single Crystals 585
J.H. Driver

Rolling and Anealing Textures of BCC Metals **597**
D. Raabe K. Lücke

Study of the Texture of Steelcord during the Wet Drawing Process
- Influence of the Patenting and the Friction on the Dies 611
A. Abdelaoui T. Montesin J.J. Heizmann J.B. Pelletier

The Effects of the Cold Rolling Process on the Texture of Zircaloy-4 Sheets 617
J.L. Béchade B. Bacroix R. Guillen

Texture and Microstructure of Channel-Die Deformed Aluminum Bicrystals
with S Orientations 627
M. Blicharski R. Becker H. Hu

Pb Texture of Pb-Al Composites Determined by the Iterative Series
Expansion Method and the Component Method 633
H.-G. Brokmeier K. Helming T. Eschner

Plane Strain Compression of Titanium Crystals 639
N. Cheneau-Späth J.H. Driver

Development of Orientation and Damage during the Plastic Deformation
of Semicrystalline Polyoxymethylene 645
A. Dahoun C. G'sell M.J. Philippe G.R. Canova A. Molinari

Correlation between Texture and Dislocation Structure in Deformed Metals 653
V.N. Dnieprenko

On the Inhomogeneous Rolling Texture Transition 659
B.J. Duggan C.S. Lee R.E. Smallman

The Development of Deformation Textures as a Function of the Phase
Fractions in Two-Phase Titanium-Based Alloys 665
D. Dunst R. Dendievel H. Mecking

Influence of the Rolling Temperature on the Texture Gradient in an Al-Mg-Si Alloy 673
O. Engler J. Hirsch K. Karhausen G. Gottstein

Rolling Texture Development in Cu-Mn-Alloys 679
O. Engler C. Pithan K. Lücke

Texture Development in Extruded Al-Cu Composites 685
H. Gertel-Kloos H.-G. Brokmeier H.J. Bunge

Textural and Microstructural Evolution during Cold-Rolling of Pure Nickel 693
N. Hansen D. Juul Jensen D.A. Hughes

Circular Texture in Thin Wires 701
J.J. Heizmann T. Montesin A. Vadon

Textures of Rolled and Wire Drawn Cu-20%Nb 709
F. Heringhaus D. Raabe U. Hangen G. Gottstein

Fiber Texture Formation and Mechanical Properties in Drawn Fine Copper Wires 715
N. Inakazu Y. Kaneno H. Inoue

Isothermal Forging Textures in γ-TiAl Base Alloys 721
H. Inoue Y. Yoshida N. Inakazu

Crystallographic Texture Changes of Oriented Ethylene-Vinylalcohol Copolymer Films on Rolling 727
T. Ishida M. Oka Y. Kimura K. Sekine

Examples of Neutron Diffraction Texture Analysis on One and the Same Chalcopyrite Sample before and after Experimental Deformation 733
E.M. Jansen H.-G. Brokmeier H. Siemes

Microstructural Analysis of an Experimentally Deformed Chalcopyrite Grain by Orientation Imaging Microscopy 739
E.M. Jansen K. Kunze

Texture Development in Al 3003 during Hot Plane Strain Compression 745
D. Juul Jensen H. Shi R.K. Bolingbroke

Ridging Phenomena in Textured Ferritic Stainless Steel Sheets 753
H.M. Kim J.A. Szpunar

Effect of Copper Alloying on the Development of Cold Rolling Textures in Extra Low Carbon Steels 761
J.H. Kim H.-C. Lee Y.B. Park K. Lücke

Texture Inhomogeneities in Rolled Profiles 767
H. Klein H.J. Bunge H. Schneider

Texture and Yield Locus Evolution in 70/30 Brass under Different Deformation Paths 783
R.A. Lebensohn C. Vial Edwards A.A. Pochettino C.N. Tomé

Quantitative Texture Analysis of Naturally Deformed Polyphase Dolomite Marbles and its Kinematic Significance 789
B. Leiss K. Helming S. Siegesmund K. Weber

Examination of Texture Discontinuity at Sheet Surface due to Cold Rolling 795
J. Li S. Saimoto H. Sang

Strain and Temperature Dependence of Deformation Banding 801
E. Lim G.W. Greene B.J. Duggan

Hot Rolling Textures of Aluminium 807
Cl. Maurice J.H. Driver

Textures in Two Phase Intermetallic TiAl Alloys after Compression and Extrusion 813
H. Mecking J. Seeger Ch. Hartig G. Frommeyer

Preferred Orientation of a Naturally and Experimentally Deformed Pyrrhotite Ore by X-Ray and Neutron Diffraction Texture Analysis 821
E. Niederschlag H.-G. Brokmeier H. Siemes

Texture Development of Cold Drawn Wires in (α+ß) CuZnAl Shape Memory Alloys 827
N.J. Park C.Q. Wang H.J. Bunge

Temperature Effects on Rolling Texture Formation in Zirconium Alloys 835
A.A. Pochettino P. Sánchez R. Lebensohn C.N. Tomé

Textures of Cold Rolled and Annealed Tantalum 841
D. Raabe B. Mülders G. Gottstein K. Lücke

Texture Evolution during the Drawing of an Interstitial-Free Low Carbon Steel 847
C. Schuman M.J. Philippe C. Esling M. Jallon M. Hergesheimer A. Lefort

Texture and Microstructure Development in a Cold-Rolled Duplex Stainless Steel Annealed at 800 °C 853
C.H. Shek G.J. Shen J.K.L. Lai B.J. Duggan

About Fair Defining the Texture of Sheet Zinc 859
M. Staszewski

Influence of Grain Boundary Orientation on Deformation and Texture Development in Rolled {112}<111> Cu-Bicrystals 865
P. Wagner N. Akdut K. Lücke G. Gottstein

Formation of Biaxial-Stretching Textures in FCC Sheet Metals 873
Y. Zhou K.W. Neale

Influence of Initial Texture on the Development of Deep Drawing Textures in FCC Metals 879
Y. Zhou J. Savoie J.J. Jonas

3.2 RECRYSTALLIZATION TEXTURES

Invited Lectures:

Formation of the Recrystallization Texture in Deformed Single Crystals 887
P. Haasen P.-J. Wilbrandt

Texture Evolution by Recrystallization and Grain Growth in Cold Rolled and Annealed (100)[001] Single Crystal of Fe-3%Si 899
J. Harase Y. Ushigami N. Takahashi

Texture Development during Secondary Recrystallization **905**
V. Yu. Novikov

Rolling and Annealing Texture in Twin Roll Cast Commercial Purity Aluminium 913
S. Benum O. Engler E. Nes

Deformation and Recrystallisation Textures in Al-SiC Metal-Matrix Composites 919
A.W. Bowen M.G. Ardakani F.J. Humphreys

Preferred Grain Boundary Orientations Formed during Secondary Recrystallization of a Cu-0.5 at%-Mn Alloy 927
B. Dörner P.-J. Wilbrandt P. Haasen

Behavior of Statistical Texture Parameters Applied to Single Grain Orientation Measurements in Recrystallized Al-Mn 933
O. Engler P. Yang G. Gottstein J. Jura J. Pospiech

Formation of Recrystallization Textures and Plastic Anisotropy in Al-Mg-Si Alloys 939
O. Engler A. Chavooshi J. Hirsch G. Gottstein

Investigation into the Nucleation Process of Recrystallization in Cold Rolled Boron Doped Ni_3Al 945
C. Escher G. Gottstein

The Use of Σ Operators in Investigating Secondary Recrystallization of 3% Silicon Steel 953
P. Gangli J.A. Szpunar

Control of the BCC Metals Texture by Pre-Recrystallization Annealing during Deformation 959
I.V. Gervasyeva V.V. Gubernatorov

Analysis of Misorientation Characteristics in Annealed and Deformed Copper 965
F. Heidelbach J. Pospiech H.-R. Wenk

Growth Selection Experiments on Tensile Deformed <100>-Oriented Aluminium(99.998%) Single Crystals 971
M. Heinrich P.-J. Wilbrandt P. Haasen

The Effect of Intermetallic Particles and Fibers on Recrystallization Textures of Cold Rolled Aluminum Alloys 977
Y. Kaneno H. Inoue N. Inakazu

Effect of Substitutional Alloying Elements on the Development of Recrystallization Textures in Extra Low Carbon Steel 983
J.H. Kim H.-C. Lee Y.B. Park K. Lücke

Texture Development during Secondary Recrystallization of Grain Oriented Silicon Steel 989
K.T. Lee J.A. Szpunar

Texture Evolution during Recrystallisation of Ultra-Low Carbon Steel 997
E. Lindh B. Hutchinson P. Bate

Texture Development of Amorphous FeBSi Alloy during Crystallization 1003
J. Lu J.A. Szpunar

Recrystallization Texture in Fe-28Al-5Cr Alloy Sheet 1009
W. Mao Z. Sun

Pancake Grain Structure and Recrystallization Texture Formation in Ultra-Low Carbon Al-Killed Sheet Steels 1015
N. Mizui

Cube Texture Generation Dependence on Deformation Textures in Cold Rolled OFE Copper 1021
C.T. Necker R.D. Doherty A.D. Rollett

Recrystallization Texture in Fe 3% Si Alloys Obtained by Direct Casting: Characterization by EBSD and X-Ray Diffraction 1027
P. Paillard T. Baudin R. Penelle

Effects of Precipitations on the Annealing Textures of Ferritic Stainless Steels 1033
D. Raabe G. Brückner K. Lücke

Texture Development of Strip Cast Ferritic Stainless Steel 1039
D. Raabe M. Hölscher M. Dubke F. Reher K. Lücke

Deformation and Recrystallization Textures of Cu-Si Alloys 1045
W. Ratuszek A. Bunsch K. Chrusciel

Influence of α Region Hot Rolling on the Rolling and Recrystallization Texture of Cold Rolled Ti-Bearing Extra Low Carbon Steel Sheets 1051
T. Senuma

Grain Boundary Misorientation Changes during Grain Growth in Pure Aluminium 1057
L.S. Shvindlerman V.G. Sursaeva V.P. Yashnikov R.G. Faulkner

Grain Structure Evolution during Grain Growth in 2-D Aluminum Foils 1063
L.S. Shvindlerman V.G. Sursaeva V.Y. Novikov R.G. Faulkner

In situ Observations of the Initial Stage of Recrystallization of Highly Rolled Phosphorous Copper 1069
K. Sztwiertnia F. Haessner

Recrystallization Texture of Cross-Rolled Aluminum Alloy 1075
S. To W.B. Lee B. Ralph

Dynamic Observation of the Growth of Secondary Recrystallized Grains
of Fe-3% Si Alloy Utilizing Synchrotron X-Ray Topography 1081
Y. Ushigami K. Kawasaki T. Nakayama Y. Suga J. Harase N. Takahashi

On the Formation of Cube Texture in Aluminium 1087
H.E. Vatne O. Daaland E. Nes

The Effect of Hot-Rolling Parameters on Texture Development
and Microstructure in Aluminium AA5182 1095
R.F. Visser I.M. Wolff B.D. Harty

Texture and Grain Boundary Character Distribution (GBCD) in
B-Free Ductile Polycrystalline Ni_3Al 1103
T. Watanabe T. Hirano T. Ochiai H. Oikawa

Evolution of the Microtexture during Recrystallization of High Purity Al-Mn 1109
P. Yang O. Engler G. Gottstein

Author Index (Parts 1 and 2) xxxvii

Subject Index (Parts 1 and 2) xliii

Abbreviations lv

CONTENTS - Part 2

3.3 LOCAL TEXTURES (SHEAR BANDS, EBSP, MISORIENTATION)

Invited Lecture:

Substructure Analysis in Textured Metallic Materials **1119**
P. Klimanek

Individual Grain Orientation Relations after High-Speed Hot Rolling of Steel Rods 1131
M. Barthel D. Gerth R.A. Schwarzer P. Klimanek U. Messerschmidt

Local Orientation Investigation on the Ridging Phenomenon in Fe 17% Cr Steel 1137
K. Bethke M. Hölscher K. Lücke

The Influence of Initial Microstructure on the Recrystallization Textures of Aluminium Alloys after Hot Deformation by Laboratory Simulation 1145
R.K. Bolingbroke G.J. Marshall R.A. Ricks

Influence of the Initial Texture on the Microstructure- and Microtexture Development during High Temperature Low Cycle Fatigue 1153
S. Brodesser G. Brückner G. Gottstein

EBSP and SIMS Studies of Oxygen Tracer Diffusion in the High Temperature Superconductor $La_{2-x}Sr_xCuO_4$ 1161
J. Claus G. Borchardt S. Weber S. Scherrer

Deformation Banding and Its Influence on High SFE FCC Rolling Texture Development 1167
B.J. Duggan C.S. Lee R.E. Smallman

Evidence for the Existence of a Special Class of Crystallographic Misorientations 1175
D.P. Field

The Dependence of Dislocation Density and Cell Size on Crystallographic Orientation in Aluminum 1181
D.P. Field H. Weiland

Microtexture Development during High Temperature Deformation of Nimonic 80A Single Crystals 1189
J. Fischer-Bühner D. Ponge G. Gottstein

The Influence of the Extrusion Speed on Texture in the Surface Layer of Alunimium Profiles Investigated by the EBSP Technique 1197
T. Furu Ø. Sødahl E. Nes L. Hanssen O. Lohne

Stress-Induced Grain Growth in Thin Al-1%Si Layers on Si/SiO_2 Substrates 1205
D. Gerth S. Zaefferer R.A. Schwarzer

Grain Subdivision during Deformation of Polycrystalline Aluminium 1211
N. Hansen D. Juul Jensen

Texture Estimate from Minimum Ranges of SAD Pole Figures 1219
K. Helming R.A. Schwarzer

Evaluation of the High Temperature ß Texture of a Sample of TA6V from Individual Orientation Measurements of Plates of the α Phase at Room Temperature 1225
M. Humbert H. Moustahfid F. Wagner M.J. Philippe

Influence of Shear Banding on the Texture in Rolled and Channel-Die Compressed Polycrystalline Copper 1231
Z. Jasienski J. Pospiech A. Piatkowski J. Kusnierz A. Litwora K. Pawlik H. Paul

Local Orientation Changes during Shear Banding of Copper Single Crystals Compressed in Channel-Die 1237
Z. Jasienski T. Baudin R. Penelle A. Piatkowski

Local Microstructural Investigations in Recrystallized Quartzite Using Orientation Imaging Microscopy 1243
K. Kunze B.L. Adams F. Heidelbach H.-R. Wenk

Effect of Texture on Strain Localization in Deformation Processing 1251
W.B. Lee

A Local Texture Analysis of a Hot Deformed Al-Li-Zr-Alloy 1257
J. Mizera J.H. Driver

About the Texture Influence on the Frequency of CSL Boundaries in Steels 1263
A. Morawiec J.A. Szpunar

Misorientation Distributions and the Transition to Continuous Recrystallisation in a Strip Cast Aluminium Alloy 1271
A. Oscarsson B. Hutchinson B. Nicol P. Bate H.-E. Ekström

Direct Observation of Orientation Changes Following Channel Die Compression of Polycrystalline Aluminum: Comparison between Experiment and Model 1277
S. Panchanadeeswaran R. Becker R.D. Doherty K. Kunze

Microtexture Evolution of Necklace Structures during Hot Working 1283
D. Ponge E. Brünger G. Gottstein

Orientational Aspects of the Morphological Elements of the Microstructure in Highly Cold Rolled Pure Copper and Phosphorous Copper 1291
K. Sztwiertnia F. Haessner

Microtexture Determination of As-Drawn Tungsten Wires by Backscatter Kikuchi Diffraction in the Scanning Electron Microscope 1299
K.Z. Troost M.H.J. Slangen E. Gerritsen

Micro-Texture Analysis of Diffusion Bonded Interfaces in Al-Li Alloy Sheet 1305
H.S. Ubhi A.W. Bowen

Microtextural Characterization of Annealed and Deformed Copper 1313
S.I. Wright F. Heidelbach

Determination of Local Texture and Deformation Systems in TiAl6V4 and T40 1319
S. Zaefferer D. Gerth R.A. Schwarzer

3.4 TEXTURES OF SPECIAL MATERIALS

Invited Lectures:

Textures in Thin Films **1327**
D.B. Knorr

Texture Formation in Hexagonal Materials **1337**
M.-J. Philippe

The Effects of Impurity Elements on the Preferred Orientation during Melt Spinning of Dilute Zn-Alloys 1351
M.V. Akdeniz J.V. Wood

Relations between Texture, Superplastic Deformation and Mechanical Properties of Thin TA6V Slabs 1357
O. Benay A.S. Lucas S. Obadia A. Vadon

Texture in Zircaloy-4: Effect of Thermomechanical Treatments 1365
L.P.M. Brandão C.S. Da Costa Viana

Texture of Electroplated Coatings of Copper and Bismuth Telluride 1371
H. Chaouni P. Magri J. Bessières C. Boulanger J.J. Heizmann

X-Ray Texture Analysis in Films by the Reflection Method: Principal Aspects and Applications 1379
D. Chateigner P. Germi M. Pernet

Experimental Studies of Texture Development in Co-Cr Films 1387
L. Cheng-Zhang J.C. Lodder J.A. Szpunar

Electric Field Induced Domain Formation in Surface Stabilized Ferroelectric Liquid Crystal Cells 1393
I. Dierking F. Gießelmann J. Schacht P. Zugenmaier

Epitaxy in Texture Formation of Electrodeposited Cu-Coatings 1405
I. Handreg P. Klimanek

Texture Formation in Al_2O_3 Substrates 1411
J. Huber W. Krahn J. Ernst A. Böcker H.J. Bunge

Preferred Orientations in Polycrystalline C_{60} 1417
K. Ito Y. Takayama K. Suenaga T. Katoh Y. Ishida

Inhomogeneous Textures in Tantalum Sheets 1423
H. Klein C. Heubeck H.J. Bunge

Texture and Grain Structure Effects on the Reliability of Microelectronic Interconnects 1435
D.B. Knorr K.P. Rodbell

Textures of Aluminum and Copper Thin Films 1443
D.B. Knorr D.P. Tracy

Controlled Texture Development in Hot-Forged Lithium Fluoride 1449
K.L. Kruger K.J. Bowman

Texture Analysis in Zircaloy Cladding Tube Material for Nuclear Fuel 1455
S.D. Le Roux D.J. Van der Merwe

Natural Fibre Textures in a Naturally Textured Material 1463
A. Oscarsson U. Sahlberg

Macrotextures of Stainless Fe-Cr Steels 1469
D. Raabe K. Lücke

Neutron Diffraction Measurements of Texture in Silicon Carbide Whisker-Reinforced Aluminum Composites 1475
J.H. Root H.J. Rack

Formation of Oblique Shape and Lattice Preferred Orientation in a Quartz Band of a Gneissic Mylonite 1481
W. Skrotzki J. Dornbusch F. Heinicke K. Ullemeyer

Texture of Aluminium and Iron Thin Films and Texture of Bilayers Al/Fe Deposited on Silicon 1487
A. Tizliouine J. Bessières J.J. Heizmann J.F. Bobo

X-Ray Diffraction Study of Texture Evolution in Electrodeposited Zinc Layers 1495
I. Tomov

The Effect of Precipitates on Texture Development 1501
H.E. Vatne O. Engler E. Nes

On the Influence of a Weak Preferred Orientation on the Strength of Aluminium Oxide Ceramic 1507
W. Winter H.-G. Brokmeier H. Siemes

Influence of Segregation of Phosphorous on Texture Development in Cold-Rolled Fe-P Alloys during Annealing 1513
L. Zhang L. Xiong H. Ning D. Ye B.J. Duggan

4. TEXTURE AND DIRECTIONAL PROPERTIES OF MATERIALS

Invited Lecture:

Anisotropic Properties of Minerals and Rocks **1523**
H. Kern

The Influence of Microstructure and Texture on Fracture Toughness in Titanium Alloy CORONA-5 1537
S. Benhaddad C. Quesne R. Penelle

Orientation Dependence of the Permeability in Textured Soft Magnetic Materials 1543
M. Birsan J.A. Szpunar

Anomalies of Young's Modulus in Fe-Cu Composites after High Degrees of Deformation 1551
W. Böcker H.J. Bunge T. Reinert

Stress Birefringence in Textured Silver Chloride 1559
P. Dietz H. Gieleßen

Limited Fibre Components in Texture Analysis: Texture Contribution in Anisotropy of Physical Properties 1565
V.N. Dnieprenko S.V. Divinskii

Local Mechanical Properties in Thin Aluminium Layers on Silicon Substrates Calculated from Measured Grain Orientations 1571
D. Gerth R.A. Schwarzer

Toughness Anisotropy in Brittle Materials 1577
M.D. Grah K.J. Bowman

Numerical Analysis of Stiffness in Sheet Products Based on Crystal Anisotropy 1585
S. Hiwatashi T. Hatakeyama K. Ushioda M. Usuda

The Dependence of the Anisotropy and Texture Properties of Injection Moulded Liquid Crystal Polymer Parts on Moulding Parameters 1593
W. Hufenbach M. Lepper

Use of Irreducible Spherical Tensors in the Calculation of the Mean Elastic Properties of Polycrystals 1599
M. Humbert L. Zuo J. Muller C. Esling

Derivation of Yield Criteria of Cubic Metals from Schmid's Law 1603
H.-T. Jeong D.N. Lee K.H. Oh

Calculation of Average Elastic Moduli of Polycrystalline Materials Including $BaTiO_3$ and High-T_c Superconductors 1609
H. Kiewel L. Fritsche

Textures and Plastic Strain Ratios of Planar Isotropic Sheet Metals 1623
I.S. Kim

Texture and Formability of Aluminum Sheets 1629
S. Kohara

Strain Localization and Fracture in Anisotropic FCC Metal Sheets: Shear Bands 1635
J. Kusnierz

On Some Methodical Developments Concerning Calculations Performed Directly in the Orientation Space 1641
S. Matthies G.W. Vinel

On the Geometric Mean of Physical Tensors Using Orientation Distributions 1647
S. Matthies M. Humbert

The Effect of Texture and Strain on the r-Value of Heavy Gauge Tantalum Plate 1653
Ch. Michaluk J. Bingert C.S. Choi

Determination of Young's Modulus in Textured CuZnAl Shape Memory Alloys 1663
N.J. Park H.J. Bunge

Effect of Texture and Microstructure on the Mechanical Properties of Zn Alloys 1671
M.J. Philippe I. Beaujean E. Bouzy M. Diot J. Wegria C. Esling

Diamagnetic Anisotropy of Precambrian Quartzites (Moeda Formation, Taquaral Valley, Minas Gerais, Brazil) 1675
H. Quade T. Reinert D. Schmidt

Magnetic Anisotropy and Texture of Banded Hematite Ores 1681
H. Quade T. Reinert

The Work Hardening of Pearlite during Wire Drawing 1689
P. Watté P. Van Houtte E. Aernoudt J. Gil Sevillano W. Van Raemdonck I. Lefever

Texture Gradient Effects in Tantalum 1695
S.I. Wright A.J. Beaudoin G.T. Gray III

5. MATHEMATICAL MODELLING OF TEXTURE FORMATION AND MATERIALS PROPERTIES

Invited Lectures:

Finite Element Modelling of Polycrystalline Solids **1703**
P.R. Dawson A.J. Beaudoin K.K. Mathur

Modelling the Effects of Static and Dynamic Recrystallization on Texture Development **1713**
J.J. Jonas L.S. Tóth T. Urabe

Theory of Grain Boundary Structure Effects on Mechanical Behaviour 1731
B.L. Adams T.A. Mason T. Olson D.D. Sam

Simulation of Rolling and Deep Drawing Textures in Ferritic Steels: Application to Ear Profiles Calculation in Deep Drawing 1739
D. Ceccaldi F. Yala T. Baudin R. Penelle F. Royer

An Equilibrium-Based Model for Anisotropic Deformations of Polycrystalline Materials 1747
Y.B. Chastel P.R. Dawson

Interconnection of Texture Development and Alternative Slipping of Different Types of Slip Systems: Computer Simulation 1753
S.V. Divinskii V.N. Dnieprenko

A New Theory of the FCC Rolling Texture Transition 1759
B.J. Duggan C.S. Lee R.E. Smallman

Effect of Cube Nucleus Distribution on Cube Texture 1765
B.J. Duggan C.Y. Chung

Texture Development and Simulation of Inhomogeneous Deformation of FeCr during Hot Rolling 1771
A.I. Fedosseev D. Raabe G. Gottstein

On the Effect of Grain Orientation on Deformation Texture 1777
J. Hirsch

Experimental and Theoretical Study of the Recrystallization Texture of a Low Carbon Steel Sheet 1783
L. Kestens U. Köhler P. Van Houtte E. Aernoudt H.J. Bunge

Modelling Cyclic Deformation Textures with Orientation Flow Fields 1791
H. Klein H.J. Bunge

Failures to Model the Development of a Cube Texture during the High Temperature Compression of Al-Mg Alloys 1797
U.F. Kocks S.R. Chen P.R. Dawson

Calculation of the Recrystallization Textures of Cubic Metals 1803
U. Köhler H.J. Bunge

Comparison of a Self-Consistent Approach and a Pure Kinematical Model for Plastic Deformation and Texture Development 1809
R.A. Lebensohn R.E. Bolmaro

Lattice Rotation during Plastic Deformation with Grain Subdivision 1815
T. Leffers

Texture of Microstructures in BCC Metals for Various Loading Paths 1821
X. Lemoine M. Berveiller D. Muller

Modelling of the Texture Formation in Electrodeposition Process 1827
D.Y. Li J.A. Szpunar

Simulation of Recrystallization Texture in Copper 1839
Y.S. Liu F. Wang J.Z. Xu Z.D. Liang

Grain Growth Simulation by Monte Carlo Method in a HiB Fe 3% Si Alloy 1847
P. Paillard R. Penelle T. Baudin

Effect of Deep Drawing on Texture Development in Extra Low Carbon Steel Sheets 1855
J. Savoie D. Daniel J.J. Jonas

Contribution of EBSP to the Determination of the Rotation Flow Field 1861
M. Serghat M.J. Philippe C. Esling B. Bouzy

A Modified Self Consistent Viscoplastic Model Based on Finite Element Results 1869
L.S. Tóth A. Molinari

Prediction of Forming Limits of Titanium Sheets Using the Perturbation Analysis with Texture Development 1875
L.S. Tóth D. Dudzinski A. Molinari

Taylor Simulation of Cyclic Textures at the Surface of Drawn Wires Using a Simple Flow Field Model 1881
P. Van Houtte P. Watté E. Aernoudt J. Gil Sevillano I. Lefever W. Van Raemdonck

Modelling Microstructural Evolution of Multiple Texture Components during Recrystallization 1887
R.A. Vandermeer D. Juul Jensen

On the Theory of Compromise Texture 1895
H.E. Vatne T.O. Saetre E. Nes

Anisotropic Finite-Element Prediction of Texture Evolution in Material Forming 1901
N. Wang F.R. Hall I. Pillinger P. Hartley C.E.N. Sturgess P. De Wever A. Van Bael J. Winters P. Van Houtte

Finite-Element Prediction of Heterogeneous Material Flow during Tensile Testing of Anisotropic Material 1909
J. Winters A. Van Bael P. Van Houtte N. Wang I. Pillinger P. Hartley C.E.N. Sturgess

6. TECHNOLOGICAL APPLICATIONS OF TEXTURE STUDIES

Invited Lectures:

Practical Aspects of Texture Control in Low Carbon Steels 1917
B. Hutchinson

Inspection and Control by On-Line Texture Measurement 1929
H.J. Kopineck

The Influence of Texture on the Magnetic Properties of CoCr Films 1941
L. Cheng-Zhang J.C. Lodder J.A. Szpunar

Effect of Strain Path Change on Anisotropy of Yield Stresses of Cubic Structure Sheet Metals 1947
J.H. Chung D.N. Lee

A Quantitative Analysis of Earing during Deep Drawing of Tin Plate Steel and Aluminum 1953
A.P. Clarke P. Van Houtte S. Saimoto

Earing Prediction From Experimental and Texture Data 1961
C.S. Da Costa Viana N.V.V. De Avila

Earing in Single Crystal Sheet Metals 1967
C.S. Da Costa Viana

The Influence of Texture and Microstructure on Corrosion-Fatigue in Ti-6Al-4V 1971
J.K. Gregory H.-G. Brokmeier

Texture Evolution during Deep Drawing in Aluminium Sheet 1979
J. Hirsch T.J. Rickert

Texture Inhomogeneities in High Tensile Strength Steels Processed with Low Temperature Controlled Rolling 1985
H. Inagaki K. Inoue

Annealing Textures in Aluminium Deformed by Hot Plane Strain Compression 1991
D. Juul Jensen R.K. Bolingbroke H. Shi R. Shahani T. Furu

The Effect of Grain Boundary Structure on the Intergranular Corrosion of Stainless Steel 1997
H.M. Kim J.A. Szpunar

Study on Factors Affecting r-Value of Cu Precipitation-Hardening Cold-Rolled Steel Sheet 2005
M. Morita Y. Hosoya

Importance of Process Parameters for Texture and Properties of Microalloyed Deep Drawing Steels 2011
C.-P. Reip W. Bleck R. Großterlinden U. Lotter

Earing and Textures in Austenitic Stainless Steel Type 305 2017
T.J. Rickert

Direct Oberservation of the Nucleation and Growth Rates of Cube and Non-Cube Grains in Warm Plane-Strain Extruded Commercial Purity Aluminum 2025
I. Samajdar R.D. Doherty S. Panchanadeeswaran K. Kunze

Global Texture Development in Cold Work of Copper and Brass by Pass Rolling 2031
H. Schneider P. Klimanek H.-G. Brokmeier

Texture Development in Ferrous Materials during Deformation by Pass Rolling or Compression 2037
H. Schneider P. Klimanek

Quantitative Correlation of Texture and Earing in Al-Alloys 2043
P. Wagner K. Lücke

7. IMPLICATION OF TEXTURE IN POWDER DIFFRACTION METHODS: INTERNAL STRESS ANALYSIS, PHASE ANALYSIS, STRUCTURE ANALYSIS

Determination of Residual Stresses in Plastically Deformed Polycrystalline Material 2051
A. Baczmanski K. Wierzbanowski P. Lipinski

X-Ray Residual Macrostress of Texturized Polycrystals from Pole Figures 2059
A.R. Gokhman

A New Method for Crystal Structure Analysis of Textured Powder Samples 2067
R. Hedel H.J. Bunge G. Reck

Wire Geometry Correction of Textures and Stress Measurements on Wires 2075
K. Van Acker P. Van Houtte E. Aernoudt

Author Index (Parts 1 and 2) 2083

Subject Index (Parts 1 and 2) 2089

Abbreviations 2101

3.3 Local Textures: Shear Bands, EBSP, Misorientation

Materials Science Forum Vol. 157-162 (1994) pp. 1119-1130

SUBSTRUCTURE ANALYSIS IN TEXTURED METALLIC MATERIALS

P. Klimanek

Institut für Metallkunde der Berg-Akademie,
Gustav-Zeuner-Str. 5, D-09596 Freiberg, Germany

Keywords: Substructure Analysis, TEM, X-Ray Diffraction, Neutron Diffraction, Diffraction-Multiplet, Texture-Substructure Interrelations

ABSTRACT

Because microstructure formation in polycrystalline materials leads frequently to preferred orientations or orientation changes, a texture-related methodology of both local substructure analysis by TEM and integrated substructure analysis from the line broadening of X-ray or neutron diffraction peaks becomes necessary. The present paper is mainly concerned with the procedures of the second type, the methodology and application of which can be based on the fact that the observable X-ray or neutron reflections of a textured polycrystal are, usually, weighted and orientation - dependent sums (diffraction multiplets) of the partial reflections due to the texture components being existent.

INTRODUCTION

It is generally accepted that both preferred orientations and orientation changes occurring in polycrystalline materials during processes as plastic deformation, recovery and recrystallization, but also in crystal growth (e.g. in thin layers) or phase transformations, are closely related to the operating mechanisms of microstructure formation. Information about the actually existing interrelations is needed

- in texture analysis for the explanation of the mechanisms and for modeling (simulation) of texture development, and in
- microstructure analysis for the identification of mechanisms of microstructure formation (structure-process correlation) and for a proper quantitative description or prediction of material properties (structure-property correlation).

The present paper is concerned particularly with the connections between the texture and the substructure (i.e. the density and the spatial arrangement of lattice defects as dislocations, stacking faults, twins etc. causing lattice disorder of the 2nd kind [1,2]) in cold-worked polycrystalline materials. In order to get reliable knowledge of them, a texture-related methodolology of both

* l o c a l substructure analysis by means of transmission electron microscopy (TEM) and
* i n t e g r a t e d substructure analysis from the line broadening of X-ray (XDA) or neutron (NDA) diffraction peaks

is necessary. Procedures of the first type, which are based on the observation of defect contrasts in combination with local (Kikuchi lines, CBED, cp. [3,4,5]) and global (XDA or NDA) texture analysis, are well established and have been used during the last decade particularly for the investigation of f.c.c. materials. ([6-26], for instance). In contrast, the methodology and application of line - broadening analysis in textured materials is less well developed. Such an analysis, which will be the main topic of the following considerations, has to take into account that the observable intensity distributions $I(x,y)$ ($x = 2\Delta\sin\theta/\lambda$ - change of the diffraction vector; y - direction of the sample-related axis system) of X-ray or neutron reflections of polycrystals with various preferred orientations g_j are usually weighted and, consequently, orientation - dependent averages of partial reflections $I_j(x-x_j, y)$ due to the texture components g_j [30-32]], which have to be separated. Important advantages of the line-broadening analysis are that it can be applied in materials with high defect densities, and that the results, especially in the case of NDA, are statistically representative for the crystallite ensemble.

TEXTURE-SUBSTRUCTURE INTERRELATIONS IN COLD-ROLLED MATERIALS FROM TEM INVESTIGATIONS

Interrelations between the texture and the substructure are of interest in view of

- the role of the operating deformation mechanisms (e.g. dislocation slip, twinning, shear banding),
- the types (e.g. cell or subgrain structures, spatial arrangements of stacking faults and twins) of the formed defect structures, and
- the defect content (e.g. dislocation densities, cell or subgrain size, stacking fault and/or twin concentrations)

in crystallites with different orientations. TEM studies which took into account especially the first two aspects, were performed for cold-rolled cubic metals and alloys by numerous authors. The results for f.c.c. materials ([6-26], for instance) are summarized in Table 1. Materials with b.c.c. lattice were less intensively studied, but it can be supposed that the substructure development is (qualitatively) comparable to that of f.c.c. structures with high stacking-fault energy (SFE).
From the TEM follows that in materials with low SFE the components of the rolling texture (Table 2) can be associated with different deformation modes and have different substructures (An overwiev was recently be given in [21]). The Cu orientation {112}<111> should be caused essentially by multiple and cross slip, the Bs orientation {110}<112> can be related to normal slip, slip of partial dislocations and also to shear banding, and the formation of the Goss texture {011}<100> can be explained by twinning of the Cu (or the S) component and subsequent slip via [28]

{112}<111> -> {255}<511> -> {011}<100>,

but above all by shear banding. However, because of the superposi-

Table 1: Substructure development in the plastic deformation of f.c.c. metals and alloys [6 - 26]

Strain	SFE High Al Ni	SFE Medium Cu	SFE Low Brass CrNi steels
		dislocation slip	
$\epsilon > 0.1$	- equiaxial cell structures - deformation bands	- equiaxial cell structures - microbands	- deformation bands - SF packages - twin bundles
$\epsilon > 0.3$	- cell refinement	- clustering of microbands	
$\epsilon > 0.6$	- self organization of the dislocation structure	- shear bands	- alignment of twins - replacement of twins by shear bands - shear bands
$\epsilon > 0.9$	- shear bands		- resumption of slip in shear bands - nonoctahedral slip - formation of recovery twins

Table 2: Rolling textures in cubic materials

FCC metals and alloys [13]	BCC metalls and alloys [27]
α-fibre {011} ‖ ND Gs Bs {011}<100> -> {011}<211>	α-fibre <110> ‖ RD {001}<110> -> {111}<110>
β-fibre {011}60° RD Bs Cu {011}<211> -> {112}<111> via S : {123}<634>	γ-fibre {111} ‖ ND {111}<110> -> {111}<112>
	η-fibre {001}<100> -> {011}<110>

tion of mechanisms of different structural level (e.g. dislocation slip, twinning, shear banding) with similar effect on the orientation changes, an detailed interpretation of the TEM results is difficult and, in part, controversially. For this reason, but also because the microstructure and texture development is strongly influenced by small and in practice often not sufficiently known changes of the conditions of a deformation process ([24,29], for instance) an unambiguous and reliable prediction of the texture-substructure interrelations seems very difficult and further work is necessary.

INTEGRATED SUBSTRUCTURE ANALYSIS FROM X-RAY OR NEUTRON DIFFRACTION-LINE BROADENING

Methodological Background

From the TEM observations it is evident that the development of deformation textures in polycrystals leads, in general, to structural inhomogeneity. That means, the grain ensemble is divided into crystallite groups N_j and volume fractions w_j, respectively, with signicantly different substructures (e.g. dislocation densities) and preferred orientations (texture components).
Because the volume fractions w_j form the components of a lattice disorder statistics (LDS; Fig. 1) the X-ray or neutron diffraction peaks of textured polycrystals measured for an arbitrary direction y of the sample-related axis system are usually diffraction multiplets (i.e. weighted sums of partial reflections due to the components of the LDS or the texture components, respectively; Fig. 1):

$$I(g-h^*,y) = \sum_j w_j P_j(h,y) I_j(g-h_j^*) = \sum_j W_j(h,y) I_j(g-h_j^*) \quad (1)$$

with

$g = 2\sin\theta/\lambda$ - amount of the diffraction vector $\vec{g} = (\vec{s}-\vec{s}_0)/\lambda$

$h^* \approx h$ - vector of the reciprocal lattice related to the reflecting lattice planes h = {hkl}

$P_{h,j}(y)$ - pole density of the lattice planes h = {hkl} of the j-th crystallite group

$W_j(h,y)$ - texture-weighted volume fraction of the j-th crystallite group

$I_j(g-h_j^*)$ - normalized ($w_j = 1$) intensity distribution of a partial reflection

The characteristics of diffraction multiplets were discussed in detail in [30-32]. Their main parameters are defined by the following equations :

- Integrated intensity

$$I(h,y) = \sum_j w_j P_j(h,y) I_j(h) = \sum_j W_j(h,y) I_j(h) \quad (2)$$

- Integrated line width

$$1/B(h,y) = I(0,y)/ I(g-h^*,y)\, d(g-h^*)$$

$$= \sum_j w_j P_j(h,y) I_j(h) \alpha_j(h,y)/B_j(h)$$

$$= \sum_j \omega_j(y) \alpha_j(h,y)/B_j(h) \quad (3)$$

with $\alpha_j = I_j(x_j-h_j^*)/I_j(0)$; $x_j = h^*-h_j^*$

and $\omega_j = w_j P_j(h,y) I_j(h)/\sum_j w_j P_j(h,y)I_j(h)$

- Fourier coefficients of the line shape

$$C_h(L,y) = A_h(L,y) \pm iB_h(L,y)$$
$$= \sum_j w_j P_j(h,y) I_{j(h)} \exp[2\pi i x_j L] C_{h,j}(L)$$
$$= \sum_j \omega_j \exp[2\pi i x_j L] \; C_{h,j}(L) \qquad (4)$$

with $L = nd_h$ - measuring length.

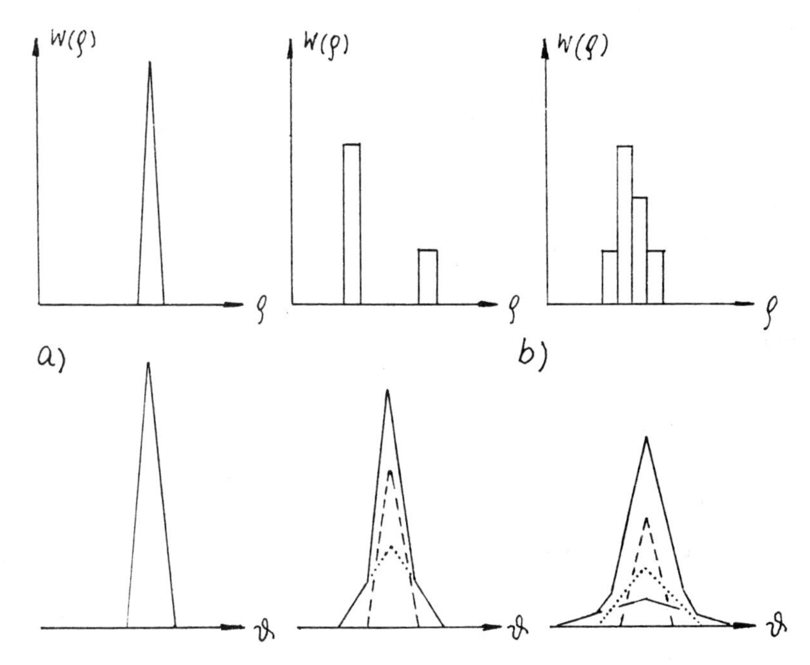

Fig. 1 : Types of lattice-disorder statistics and diffraction peaks in real polycrystalline materials (a - structurally homogeneous materials with diffraction singlets, b - structurally inhomogeneous materials with diffraction multiplets)

In order to describe the orientation-dependence of the diffraction (and microstructure characteristics, respectively) of structurally inhomogeneous polycrystals, the well-known conception of pole figures can be used. In this connection the following relationships are valid:

- Experimental orientation (diffraction) pole figures $P_h^*(y)$ [33] are related to the observable intensity pole figures $I_h(y)$ by the expression [34,35] (w_j^* - volume fractions, $P_j^*(h,y)$ - appearent texture components)

$$P_h(y) = K(h,y) I_h(y) = \sum w_j^* \; P_j^*(h,y)$$

- Strain pole figures $S_h(y) = (\Delta d_h/d_h)_y$ [36,37] describing the orientation dependence of macrostresses are obtained from line - position pole figures $\Theta_h(y)$ via the Bragg equation.

- Line-broadening (or line-width, respectively) pole figures $B_h(y)$ can be measured directly [38] or derived from the Fourier coefficients via the relation

$$B_h(y) = 1/\left(\sum_j \omega_j(y)\alpha_j(h,y)/B_j\right) = 1/\left(\sum_L A_h(L,y)\right)$$

- Pole figures of substructure parameters (e.g. size or strain pole figures $L_h(y)$ and $\epsilon_h(y)$ [39,40]) may be derived from the Fourier coefficients $C_h(L,y)$ of the diffraction-line shapes measured for different orientations y.

However, because the X-ray or neutron eflections associated with arbitrary directions y of the sample-related axis system are, in general, diffraction multiplets with a complicated meaning, an adequate interpretation of the line-broadening or substructure pol figures just mentioned is very difficult. Reliable integrated substructure analysis in textured materials should therefore be based on diffraction-line shapes related to special sample orientations, which can be considered as diffraction singlets or interpreted in a relatively simple manner, respectively. In this connection the following cases of equation (1) are of interest:

- The substructure is comparable for all grain orientations and

$$I(g-h^*,y) \approx I(g-h^*) \sum_j w_j P_{h,j}(y) \approx I(g-h^*)\ P_h^*(y) \qquad (5)$$

can be treated as a diffraction singlet.

- The LDS of the sample material is determined by the texture components. In this case the partial reflections of the LDS can be separated, because the intensity distribution (1) is reduced to a diffraction singlet

$$I(g-h^*,y) \approx w_1\ P_1(h,y)\ I_1(g-h_1^*) \qquad (6)$$

if the orientation y is the maximum position of a texture component and/or fulfills the condition $P_{h,j}(y) \approx 0$ or, at least, $P_{h,j}(y) \ll P_{h,1}(y)$ for $j \neq 1$.

- The texture is weak or similar for all components of the LDS and equation (1) gets the form

$$I(g-h^*,y) \approx P_h(y) \sum_j w_j(y)\ I_j(g-h_j^*) \qquad (7)$$

In this case the meaning of $I(g-h^*,y)$ is, in general, complicated and a separation of components of the LDS is possible only in special microstructures [30-32].

Experimental Results

* Dislocation Densities in Deformed Steel X5CrTi17

As a first example of substructure analysis the estimation of dislocation densities in ferritic steel X5CrTi17 may be considered [41,42]. The material was plastically deformed at room temperature by tension or compression and experimentally investigated by X-ray scattering and high-resolution neutron diffraction. For the estimation of the defect densities the Krivoglaz-Wilkens theory [43-46] of dislocation-induced diffraction line broadening was used; the

procedure is described in [41,42], for instance.
Fig. 2 shows the dislocation densities N_d(hkl) obtained from the Fourier coefficients of the line shape of various X-ray reflections {hkl} after tension [41]. Because the sample surface was perpendicular to the tensile axis and the axis of the (weak) <110> fibre texture of the specimens, respectively, the observed peak profiles are (approximately) diffraction singlets and the different defect densities N_d(hkl) indicate an orientation dependence of the substructure development.

Dislocation densities N_d(hkl) estimated after compression of particularly fine-grained (D ≈ 20 μm) cylindrical specimens by both XDA and high-resolution NDA [42] are presented in Fig.3. The agreement of the results is, especially for the {211} reflection, very good. But because in this case the sample surface was parallel to the deformation direction (and, consequently, to the axis of the compression texture <111>+<100>), the reflections have the meaning of diffraction multiplets and the observation of a similar relationship $N_d(211)>N_d(110)$ as for the tensile specimens is unexpected.

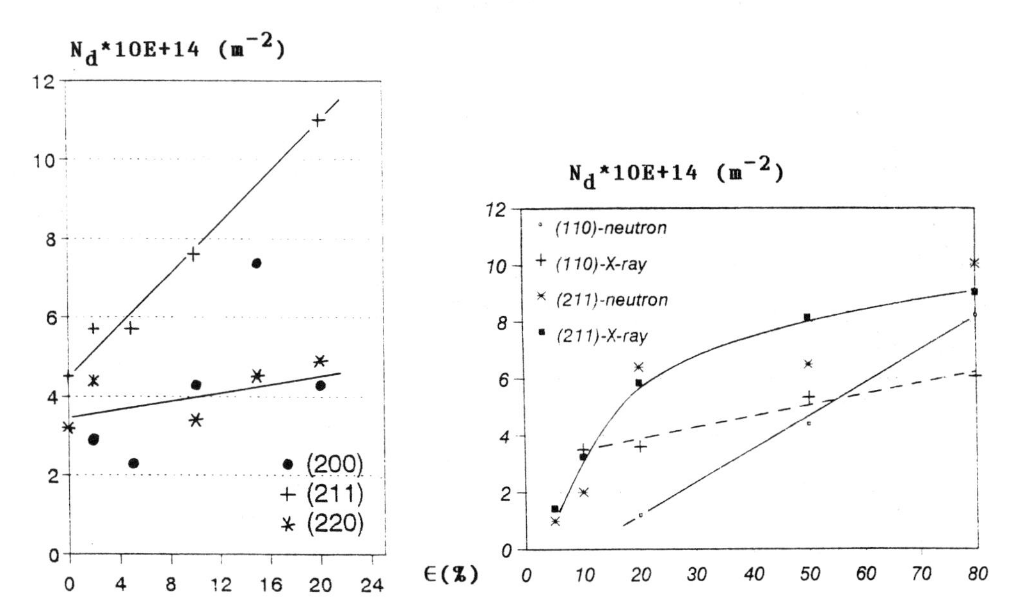

Fig.2: Dislocation densities N_d after tension (XDA)

Fig.3: Dislocation densities N_d in compression (XDA, NDA)

* Texture-related Diffraction Line Broadening in α-Brass CuZn20

From the TEM investigations [6-26] summarized in Table 1 follows that in cold-rolled α-brass the substructures of crystallites with various preferred orientation are significantly different and that the LDS of the material should be defined by these texture components. Therefore, according to equation (6) a proper integrated substructure analysis requires the separation of the diffraction - line broadening related to the individual texture components by measurement of the X-ray or neutron reflections for selected sample orientations y. Possibilities of a suitable choice of such orientations y for the reflections {111}, {200}, and {220} due to the main components {hkl}<uvw> occurring in the rolling texture of f.c.c. materials (Table 2) are demonstrated in Fig. 4.

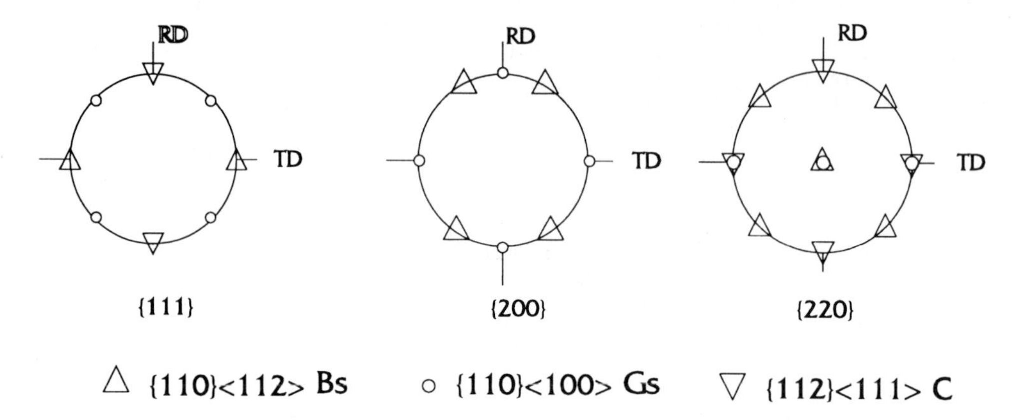

Fig. 4 : Choice of special directions y for the substructure analysis in cold-rolled α brass

In order to measure the orientation dependence of the diffraction - line broadening experimentally, cube samples with the dimensions 20x20x20 mm^3 were constructed by oriented stacking of cold-rolled sheets of equal thickness reduction ∈ and cut in such a manner that the desired direction y was perpendicular to the intersection plane. A more detailed description of the experimental work, which is concerned with the evaluation of both the line widths and the Fourier coefficients of the physical diffraction line-shape, can be found in [47]; some results are presented in Table 3.

Table 3: Physical line widths $\beta_{2\theta}$ (°) determined from Stokes-corrected Fourier coefficients

TC	∈(%)	y	111	y	200	y	220
Cu	30	RD	0.096				
	60		0.206				
	90		0.300				
Bs	30	TD	0.174	34°	0.333	45°	0.367
	60		0.244		0.471		0.542
	90		0.459		0.657		0.772
Gs	30	45°	0.177	TD	0.412		
	60		0.254		0.546		
	90		0.416		0.826		
Bs/Gs	30					ND	0.431
	60						0.550
	90						0.759

TC - texture component

From the data it becomes evidently that the line broadening of the {111} reflection related to the Cu orientation {112}<111> is much smaller than that of the {111} peaks due to the Bs and the Gs orientations {011}<211> and {011}<100>, respectively. The possible reasons of the phenomenon , a proper interpretation of which requires further experimental work and a specification of the substructure models underlying the procedures of data evaluation [47], are
- different mean lattice strains due to internal stress of the 2nd kind and the dislocation content (density and arrangement), and
- various geometrical anisotropy (i.e. an orientation - dependent particle size effect) resulting from the density and the spatial distribution of dislocation walls or subboundaries with high desorientation, twins and stacking faults

within the crystallites of the texture components.

* Substructure Analysis in Hot-Worked Austenite X50Ni24

Hot-working of austenitic (f.c.c.) steels gives often rise to dynamic recrystallization. The microstructure of the material is then a mixture of deformed (d) and recrystallized (r) crystallites defining the components of a binary LDS, and its X-ray reflections become diffraction doublets with the Fourier coefficients

$$A(L) = \omega_r A_r(L) + \omega_d A_d(L) \approx \omega_r + \omega_d A_d(L) \qquad (8)$$

having the weight factors

$$\omega_r = \frac{w_r P_{h,r}(y)}{(w_r P_{h,r}(y) + w_d P_{h,d}(y))}, \qquad \omega_d = \frac{w_d P_{h,d}(y)}{(w_r P_{h,r}(y) + w_d P_{h,d}(y))}$$

Equation (8) shows that, because of $A_r(L) \approx 1$ and $A_d(L) \rightarrow 0$ at sufficiently large L, from the asymptotic behaviour of a plot A(L) the quantity ω_r can be determined and, in addition, the separation of the Fourier coefficients $A_d(L)$ describing the deformed microstructure component becomes possible.
Application of the procedure may be demonstrated by the results of X-ray diffraction experiments with an alloy X50Ni24 after hot-rolling at 950°C [48]. Dynamic recrystallization took place at thickness reductions $\epsilon > 30\%$. As demonstrated in Fig. 4 for the reflections {111} and {222}, the behaviour of the Stokes-corrected Fourier coefficients is then clearly in agreement with the predictions of the equation (9).
Microstructure parameters determined from the Fourier coefficients of the X-ray reflections [48] and by means of TEM [49] and optical microscopy (volume fraction of the recrystallized grains) in samples with different rolling degrees are summarized in the Table 4. The results may be commented as follows:
* After a thickness reduction $\epsilon = 10\%$, which does no lead to dynamic recrystallization, the agreement of the substructure parameters estimated by both methods is very good.
* The weight factors ω_r determined from the asymptotic behaviour of the Fourier coefficients of partially recrystallized samples correspond well with the volume fractions w_r measured by optical microscopy. The differences are within the limits of error of the methods, but it is of interest that they can also be explained by different ($P_d > P_r$) orientation densities $P_{h,j}(y)$ of the microstructure components.

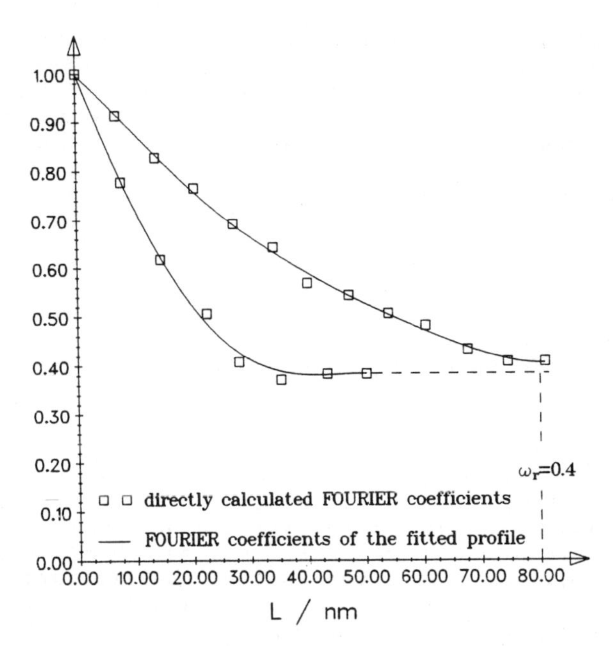

Fig. 5 : Stokes-corrected Fourier coefficients of the reflections (111) and (222) of the material X50Ni24 after hot-rolling at 950°C (ϵ = 40%)

Table 4: Substructure characteristics in hot-rolled steel X50Ni24 as determined by X-ray diffraction and TEM

Rolling degree ϵ	10 %	40 %	60 %
X-RAY ANALYSIS			
Volume fraction ω_r	0	0.4	0.6
Total dislocation density N_d 10^{-14} m^{-2}	3.1	4.0	12.0
Subgrain size	0.2	> 0.5	> 0.5
TEM / OPTICAL MICROSCOPY			
Volume fraction w_r	0	0.3	0.5
Dislocation density N_d 10^{-14} m^{-2} within subgrains	3.3	1.5	2,0
Subgrain size	0.5	0.9	0.6

* The differences of the dislocation densities N_d estimated by XDA and TEM after deformation degrees of 40 and 60% are due to the fact that in XDA the total dislocation content of the non-recrystallized microstructure fraction is obtained, while the TEM data are typical for the interior of the subgrains.

* In agreement with the TEM observations the subgrain (particle) size of the non-recrystallized microstructure fraction cannot be estimated quantitatively by XDA after deformation of 40% and 60%, because it does not cause significant line broadening. Another X-ray diffraction procedure, which could be used in this case, was described in [50].

CONCLUSIONS

From the investigations presented here the following conclusions can be drawn:

- TEM investigations of f.c.c. metallic materials have clearly shown that the substructures of cold - worked polycrystalline materials must, usually, be interpreted in terms of a lattice-disorder statistics associated with the components of the deformation texture. Accordingly, integrated substructure analysis by XDA or NDA has to take into account that diffraction peaks due to such materials are diffraction multiplets.
- Using the kinematical approach of X-ray and neutron scattering by structurally inhomogeneous polycrystals presented in [30-32] a suitable methodology for the analysis of diffraction-multiplets of textured materials can be obtained. However, in order to get reliable quantitative values of substructure parameters, the development of more specified, texture-related substructure models is necessary.

ACKNOWLEDGEMENTS

The experimental results presented here were obtained in cooperation with Ms. A.Philipp, Ms. A.Riedel, and Mr. T.Kschidock. In addition it is grateful acknowledged that a part of the investigations was financially supported by the Federal Ministry for Research and Technology (BMFT) of the FRG.

REFERENCES

[1] Krivoglaz, M.A.: Fiz. Metallov Metalloved. 1961,
[2] Krivoglaz, M.A.: Scattering theory of X-rays and thermal neutrons by imperfect crystals (In Russian). Nauka, Moscow 1967
[3] Schwarzer, R.A., Weiland, H. in: Experimental Techniques of Texture Analysis (Ed. H.J. Bunge). DGM, Oberursel 1986, 287
[4] Humphreys ,F.J.: Proc. ICOTOM 8 (Ed.: J.Kallend, G.Gottstein) The Metallurgical Society, 1988, 171
[5] Adams, B.L., Wright, S.I., Kunze, K.: Metall.Trans. 1993, A 24, 819
[6] Duggan,B.J., Hatherly,M., Hutchinson, W.B., Wakefield, P.T.: Metal Sci. 1978, 12, 343
[7] Hutchinson, W.B., Duggan, B.J., Hatherly, M.: Metals Technol. 1979, 6, 398
[8] Gil Sevillano, J., Van Houtte, P., Aernoudt, E. : Progr. Mater. Sci. 1980, 25, 71
[9] Plege, B., Noda, T., Bunge, H.J.: Z. Metallkde. 1981, 72, 641
[10] Scherke, R., Klimanek, P. Bergner, D.: Freiberger Forschungsheft 1982, B225, 65
[11] Yeung, W.Y., Duggan, B.J.: Acta Metall. 1986, 34, 653

[12] Yeung, W.Y., Duggan, B.J.: Mater.Sci.Technol. 1986, 2, 552
[13] Hirsch, J., Lücke, K.: Acta Metall. 1988, 36, 2863 and 2883
[14] Hirsch,J., Lücke,K., Hatherly,M.: Acta Metall. 1988, 36, 2905
[15] Köhlhoff, G.D., Malin, A.S., Lücke, K., Hatherly, M.: Acta Metall. 1988, 36, 2841
[16] Bay, B., Hansen, N., Kuhlmann-Wilsdorf, D.: Mater. Sci. Eng. 1989, A113, 585
[17] Donadille, C., Valle, R., Dervin, P., Penelle, R.: Acta Metall. 1989, 37, 1547
[18] Yeung, W. Y.: Acta Metall. Mater. 1990, 38, 1109
[19] Leffers, T., Bilde-Sørensen, J.B.: Acta Metall. Mat. 1990, 38, 1917
[20] Leffers, T., Ananthan, V.S.: Textures & Microstructures, 1991 ,14-18, 971
[21] Leffers, T., Juul Jensen, D.: Textures and Microstructures, 14-18 (1991) 933
[22] Hughes, D.A., Hansen, N.: Mater. Sci. Technol. 1991, 7, 544
[23] Bay, B., Hansen, N ., Hughes, D.A., Kuhlmann-Wilsdorf, D.: Acta Metall. Mater. 1992, 40, 205
[24] Leffers, T., Juul Jansen, D.: Scripta Metall. Mater. 1992, 27
[25] Hughes, D.A., Hansen, N.: Metall. Trans. 1993, A24, 2021
[26] Hansen, N., Juul Jensen, D., Hughes, D.A. : This issue
[27] Lücke, K., Hölscher, M.: Textures & Microstructures 1991,14-18, 585
[28] Wassermann, G.: Z. Metallkde. 1963, 54, 61,
[29] Juul Jensen, D., Hansen, N., Humphreys, F.J.: Proc. ICOTOM 8 (Ed. J.S.Kallend, G. Gottstein). The Metallurgical Society, Warrendale/Pennsylvania 1988, 431
[30] Klimanek, P.: Freiberger Forschungsheft 1988, B 265, 76
[31] Klimanek, P.: X-Ray and Neutron Structure Analysis in Materials Science (Ed. J.Hasek). Plenum Press, New York 1989, 125
[32] Klimanek, P.: Proc.EPDIC 1, Mat.Sci.Forum 1991, 79-82/Pt.1, 73
[33] Wever, F.: Z. Physik 1924, 28, 69
[34] Mücklich, A., Klimanek, P.: this issue
[35] Girlich, I., Klimanek, P.: this issue
[36] Hoffmann, J., Neff, H., Scholtes, B., Macherauch, E. Härtereitechn. Mitteilg. 1983, 38, 180
[37] Hauk, V., Vaessen, G., Weber, G.: Härtereitechn. Mitteilg. 1985, 40, 122
[38] Schubert, A., Michel, B.: phys.stat.sol.(a) 1989, 111 K137
[39] Tidu & Heizmann : Textures & Microstructures 1991, 14-18, 79
[40] Perlovich, Yu. : this issue
[41] Klimanek, P.: Proc. EUROMAT 3 (1993), J. de Physique: in press
[42] Klimanek, P. Kschidock, T., Mikula, P., Lukas, P., Vrana, P.: Proc. EUROMAT 3 (1993), J. de Physique: in press
[43] Krivoglaz, M.A., Martynenko, O.V., Ryaboshapka, K.P.: Fizika Metallov Metalloved. 1983, 55, 5
[44] Krivoglaz, M.A.: Diffraction theory of X-rays and neutrons by non-ideal crystals (in Russian). Naukova Dumka, Kiev 1983
[45] Wilkens, M. in: Fundamental Aspects of Dislocation Theory (Ed. J.A. Simmons et al.) Vol.II, 1195, N.S.B. Special Publ., Washington 1970
[46] Wilkens, M.: phys.stat.sol.(a) 1970, 2, 359
[47] Klimanek, P., Riedel, A.: Submitted to Z. Metallkde.
[48] Philipp, A. : Thesis. Freiberg Academy of Mining, 1988
[49] Martin, U., : Thesis. Freiberg Academy of Mining, 1983
[50] Koch, M., Oettel, H., Klimanek, P., Ohser, J.: Z. Metallkde. 1987, 78, 310

Materials Science Forum Vol. 157-162 (1994) pp. 1131-1136

INDIVIDUAL GRAIN ORIENTATION RELATIONS AFTER HIGH-SPEED HOT ROLLING OF STEEL RODS

M. Barthel [1], D. Gerth [2], R.A. Schwarzer [2], P. Klimanek [3] and U. Messerschmidt [1]

[1] Max-Planck-Institut für Mikrostrukturphysik, Am Weinberg 2, D-06120 Halle, Germany

[2] Institut für Metallkunde und Metallphysik, TU Clausthal, Grosser Bruch 23, D-38678 Clausthal-Zellerfeld, Germany

[3] TU Bergakademie Freiberg, Institut für Metallkunde, Gustav-Zeuner-Str. 5, D-09596 Freiberg, Germany

Keywords: Hot Deformation, Pass Rolling, Local Texture, Recovery, Recrystallization, TEM, Kikuchi Pattern

ABSTRACT

High-speed hot working of ferrite X8CrTi17 performed by pass-rolling results in a very complex microstructure. Previous transmission electron microscopy (TEM) investigations combined with stereological methods delivered qualitative and quantitative information about mean grain and subgrain size and size distribution. Here we report local texture determination by TEM Kikuchi diagram measurements. An increasing misorientation between individual grains with increasing number of passes is observed, which is accompanied by the onset of dynamic and postdynamic recrystallization at higher deformation rates.

INTRODUCTION

High-speed hot rolling of steel rods is a complicated deformation process leading to inhomogeneous microstructures. An essential characteristic of the procedure is significant adiabatic overheating, which in ferritic high-alloy steel X8CrTi17 induces dynamic and postdynamic recrystallization [3-6]. In order to understand the microstructure formation in some detail, its systematic investigation at consecutive stages of the deformation process is necessary. Because all structure changes in the material occur at the micrometer scale, TEM must be used for this purpose. The main capabilities of TEM microstructure analysis are illustrated in Fig.1. Results of investigations concerning the development of the mean grain and subgrain size and the corresponding size distribution by stereological methods [1,2] were reported in [3,4,6].
This paper presents results of local orientation analysis by means of Kikuchi patterns, which was performed after pass-rolling in a round-oval groove sequence at strain rates between 10^2 und 10^3 s^{-1}.

EXPERIMENTAL PROCEDURE

The sample material of the present work is characterized in [3]. The deformation experiments were performed with steel bars of a circular initial cross section (d=10 mm), using a research pass-rolling mill with a four step groove sequence of the type round-oval [7]. The deformation temperature

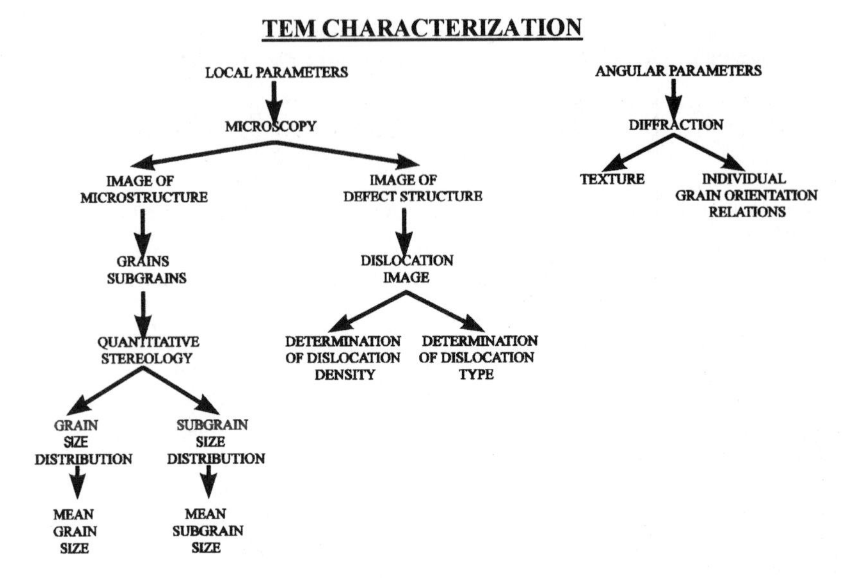

Fig. 1: TEM characterization of polycrystalline materials

was 1073 K, the deformation rates were in the range of 70-1500 s^{-1}. In order to obtain the microstructure corresponding to the deformation states after the individual steps of the groove sequence, the wires were directly introduced into a water bath. The quenching time was <1 s.
The microstructure after hot rolling was studied by both a high voltage transmission electron microscope (JEM 1000) with an accelerating voltage of 1 MeV and a medium voltage transmission electron microscope (Philips EM 430) with an accelerating voltage of 300 kV. Discs for TEM were prepared perpendicular to the wire axis, i.e. the beam direction was parallel to the rolling direction. The single grain orientations were measured by means of Kikuchi patterns [8]. The computer program [9] available for this purpose permits the determination of both crystallite orientations and misorientations.
Experimental results of the individual grain orientation measurements are usually represented by plotting the orientations for two reference directions in the workpiece (rolling direction and a normal direction on it) point by point on inverse pole figures as well as by the corresponding pseudo-colour orientation images of the grains and subgrains. In this way correlations between grain structure and grain orientation can be recognized easily.

EXPERIMENTAL RESULTS AND DISCUSSION

- DEFORMATION IN THE 1st PASS

Fig. 2 shows the TEM micrograph of the microstructure of the ferritic steel X8CrTi17 after hot rolling in the first pass at a deformation rate of 160 s^{-1}. The microstructure is homogeneous. It consists of dislocation cells or subgrains, respectively, but significant differences in the size of these particles are not observed. Both the inverse pole figure of the rolling direction and the pseudo-colour orientation image of the microstructure (Fig. 3) indicate a very sharp texture of the subcrystallites. The orientation differences (misorientations) between the neighbouring grains are smaller than 3.5°, i.e. all boundaries are cell walls or small-angle grain boundaries, respectively, and this result confirms the conclusion [6] that, under the conditions of the present work, the only restoration process during the first step of pass-rolling is dynamic recovery.

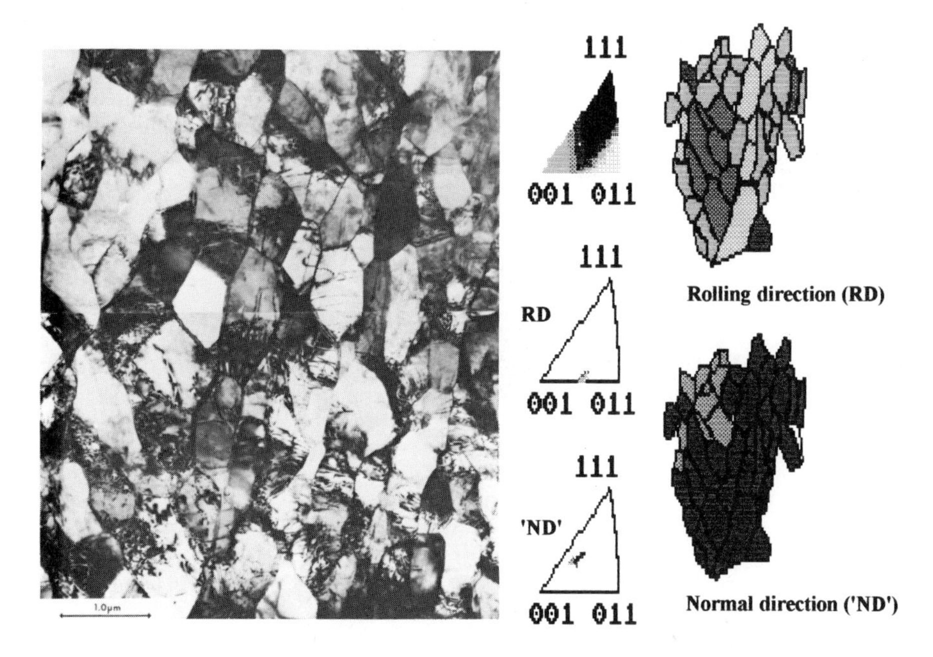

Fig. 2: TEM micrograph

Fig. 3: Inverse pole figure, pseudo-colour representation of the grain and subgrain orientation

- DEFORMATION IN THE 2nd PASS

During hot rolling in the second pass of the groove sequence, which took place at a deformation rate of 450 s^{-1}, the microstructure becomes inhomogeneous and significant differences in both the defect content and in the particle size are visible (Fig. 4).

Fig. 4: Microstructure after the 2nd pass

The local texture analysis (Fig. 5) indicates that both low angle (misorientation <15°) and high angle (misorientation >15°) grain boundaries are present. Moreover, from Fig. 5 it follows that, in general, neighbouring subgrains or grains have nearly the same orientation whereas the orientation differences of regions with greater distances frequently correspond to high-angle boundaries. It is suggested that the occurrence of such high-angle boundaries (which are different from those existing before the hot working) indicates the beginning of recrystallization.

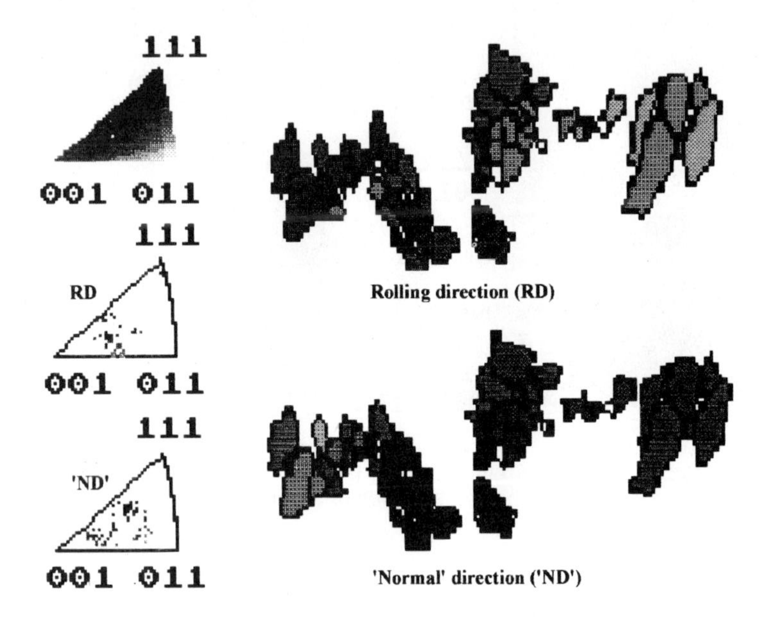

Fig. 5: Inverse pole figure, pseudo-colour representation of the grain and subgrain orientation

- DEFORMATION IN THE 4th PASS

During pass-rolling in the 4th pass of the groove sequence, which was performed at a deformation rate of 870 s^{-1} the inhomogeneity of the microstructure is significantly increased (Fig. 6). In the TEM micrographs three types of domains can be distinguished:

- small subgrains (Fig. 6; $D_S \leq 2$ μm with misorientations <15° corresponding to low-angle boundaries),
- nearly dislocation-free small grains (Fig. 7; $D_S \approx 5$ μm with misorientations >15° corresponding to high-angle boundaries), and
- relatively large grains (Fig. 8; $D_S \approx 15$ μm with misorientations >15° corresponding to high-angle boundaries) with significant dislocation density.

The results confirm [3,4,6] that dynamic recrystallization (DRX) and postdynamic recrystallization (PDRX) took place. In this connection it is assumed that the large grains with relatively high dislocation density (Fig. 8) result from DRX and the smaller ones (Fig. 7) from PDRX. The inverse pole figure (Fig. 9) of the rolling direction indicates the existence following microtextures: $\{hkl\}\ \langle 100 \rangle$ and $\{hkl\}\ \langle 110 \rangle$. Neighbouring large grains have different orientations.

Fig. 6: Microstructure after the 4th pass

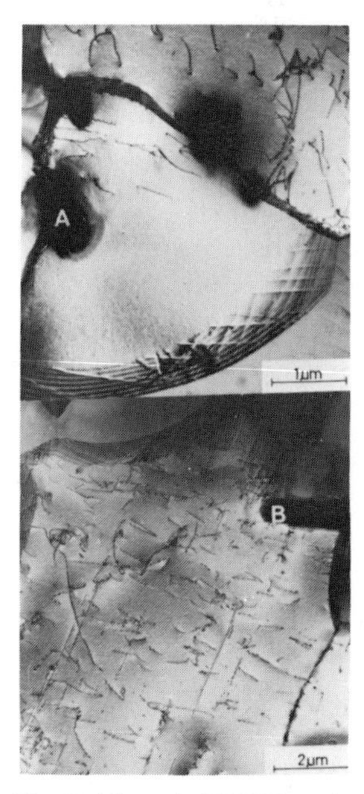

Fig. 7 (above): PDRX grain
Fig. 8 (below): DRX grain

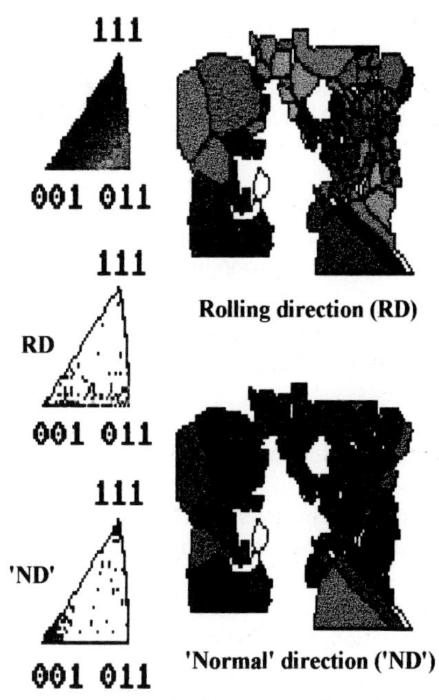

Fig. 9: Inverse pole figure, pseudo-colour representation

CONCLUSIONS

The orientation measurements presented here confirm the conclusions drawn from the stereological analysis of the microstructure occurring in high-speed pass rolling of the steel X8CrTi17. Owing to significant local adiabatic overheating, recrystallization processes are observed which are unusual for the material at low deformation rates [10]. From the local texture analysis it also follows that the microstructure development within the other non-recrystallized volume fraction is, especially during the first step of the groove sequence, in agreement with the model proposed in [11,12] for f.c.c. metals with high stacking fault energy. Qualitatively this may be expected, but a more detailed discussion must take into account that the dislocation character and the rearrangement of the dislocations in b.c.c. and f.c.c. materials is different, and that the development of the deformation pattern should be modified by the temperature.

REFERENCES

1) Barthel, M., Baither,D.:
Prakt. Met. 1987, 24, 391-397

2) Barthel, M., Baither, D.:
Prakt. Met. 1988, 26, 261-271

3) Klimanek, P., Hensger, K.-E. , Schubert, A., Barthel, M.:
Czech. J. Phys. 1988, B38, 373-376

4) Barthel, M., Klimanek, P., Hensger, K.-E.:
Czech. J. Phys. 1988, B38, 377-379

5) Trinks, U., Hensger, K.-E., Klimanek, P., Martin, U., Cyrener, K.:
In: Electron Microscopy in Plasticity and Fracture Research of Materials
Akademie-Verlag 1990, 149-155

6) Barthel, M., Klimanek, P., Hensger, K.-E.:
In: Electron Microscopy in Plasticity and Fracture Research of Materials
Akademie-Verlag 1990, 465-471

7) Hensel, A., Oelstöter, G., Lietzmann, K.-D.:
Neue Hütte 1982, 1, 37

8) Schwarzer, R. A.:
Textures and Microstructures 1990, 13, 15-30

9) Gerth, D., Schwarzer, R. A.:
Textures and Microstructures 1993, 21, 177-193

10) Schubert, A.
Dissertation, Bergakademie Freiberg 1985

11) Hansen, N., Bay, B.B., Hughes, D.A., Kuhlmann-Wilsdorf, D.:
Acta Metall. Mater. 1992, 40, 205

12) Hansen, N., Jensen, J., Hughes, D.A.: this issue

Materials Science Forum Vol. 157-162 (1994) pp. 1137-1144

LOCAL ORIENTATION INVESTIGATION ON THE RIDGING PHENOMENON IN Fe 17% Cr STEEL

K. Bethke [1], M. Hölscher [1,2] and K. Lücke [1]

[1] Institut für Metallkunde und Metallphysik,
RWTH Aachen, Kopernikusstr. 14, D-52074 Aachen, Germany

[2] Now with Krupp Stahl AG, Hildener Str. 80, D-40597 Düsseldorf, Germany

Keywords: EBSP, ODDF, Misorientation Correlation Index MCI, Rodrigues Vectors, Ferritic Stainless Steel, Ridging Phenomenon, Stochastic Process

Abstract

Fe 17% Cr ferritic stainless steel sheets, after cold rolling and annealing, exhibits ridging along the rolling direction during subsequent tensile deformation. The resulting deterioration in surface quality is a severe and long known problem in industrial processing. Here, direct local orientation measurements were performed by means of the EBSP (**E**lectron **B**ack **S**cattering **P**attern) method in the form of line or area scans in all three main planes of the sheet. The connection between microscopic and macroscopic features was revealed by space resolved orientation determination correlated with calculations of shape changes of individual cross- sections. As a main result, the corrugation character of the sheet and thus also ridging emerged. Finally, the ridging profile was simulated on the mathematical basis of stochastic processes.

Introduction

The ridging problem in ferritic stainless steel sheets is well known and unsolved for more than two decades. It consists of the conspicious attitude that a sheet, being flat after cold rolling and recrystallization, changes into a corrugated sheet after a subsequent tensile deformation as it is normally applied in case of industrial transformation processes. The ropes or ridges, which possess a depth of about 10- 20 μm and can be seen by bare eyes, have an extension over the whole sheet parallel to the former rolling direction whereas in the transverse direction the width of the ridges reaches 1 to 2 mm superimposed by other ripples on a much smaller scale. It is of importance to recognize that this phenomenon is not a surface effect or caused by changes of thickness of the sheet, but, as shown first by Appel [1] consists in a true corrugation with both faces of the sheet exhibiting the same profile, and, as shown in [2], the ridging effect becomes even more pronounced when they are removed (in contrast to [3]).

In literature, by Appel and Lücke [4] and also by other authors (e. g. [5- 8]), this problem has been related to the texture of the sheet by assuming that differently oriented grains exhibit different deformations, preferably shear deformations [4], [6], and thus experience different shape changes. The main problem in understanding this phenomenon arises, however, if one considers that the mean grain size is about 10 μm. This means that, because of the factor of 100 between the magnitudes of ridging width and grain diameter, this effect cannot be attributed to the deformation of individual grains. In order to explain this striking difference one could imagine that large groups of grains with nearly equal orientation generate the effect as assumed e.g. by [8]. In the present paper, however, we will show on the basis of microscopic local orientation determinations that even this idea is not sufficient since the groups of neighboured equally oriented grains are not large enough. Instead, the macroscopic deformation of the cross- section plane obtained as sum of the (microscopic) deformations of the individual grains of even very different orientation can be shown to be responsible for the ridging which thus obtains the character of a stochastic phenomenon.

Experimental

The sample was an 80% cold rolled Fe 17% Cr ferritic stainless steel sheet of a thickness of 0.5 mm which was recrystallized and subsequently tensile deformed parallel to the rolling direction by an elongation of 7%. The grain size was about 10 μm in all three directions.

The orientation measurements were performed by a scanning electron microscope attached to a semi automatic EBSD facility [9- 12] (**E**lectron **B**ack **S**cattering **D**iffraction). For the present investigations the precision for the angular resolution of orientation determination is about 1° and for the spatial resolution about 1 μm. The experiments include line and area scans in the RD/TD (rolling) plane as well as in the ND/TD (transverse) and ND/RD (longitudinal) section with a total number of more than 7000 investigated orientations. The measurements have not been taken grain by grain, but at fixed equal distances with the size of these scan steps adapted to the size of the mean grain diameter of about 10 μm. In this paper the following experiments are presented:

1) Two line scans, one parallel to TD and one parallel to RD with a scan width of 10 μm and 900 measuring points for each scan leading to a total length of 9 mm. The scanning lines were chosen to lie in the center plane of the sheet, i. e. just in the middle between the two rolling surfaces (refering to Figs. 1, 3, 4, 5. 6, 7).
2) One area scan of 1000 measuring points situated in a transverse section (⊥ RD) around the middle of the sheet thickness with a distance between measuring points of 5 μm. Along ND 25 points corresponding to 0.125 mm (1 / 4 of the sheet thickness) and along TD 40 points corresponding to 0.20 mm were measured (refering to Fig. 2).

The samples were electropolished before measurement in order to remove deformation zones which is necessary for obtaining good scattering patterns, and, finally, were slightly etched.

Results of the Orientation Measurements

Figs. 1 exhibit the difference between the orientations at the various measuring points on the one hand and a reference orientation on the other hand for pieces of the line scans || TD (Fig. 1a) and || RD (Fig.1b) in the rolling plane. Here the reference orientation is arbitrarily chosen to be the (001)[1-10] direction. These orientation differences are represented by Rodrigues vectors the length of which describes the angle and the direction of which the rotation axis in the Rodrigues space [13]. Supposed that neighbouring measuring points correspond to neighbouring grains, the difference between neighbouring vectors would express the misorientation of these grains (see also [14]). In Fig. 1a the vectors and thus the orientations strongly differ in direction and length from measuring point to measuring point along TD. The vectors in Fig. 1b, in contrast, mainly occupy two prefered directions which can be identified to be about (111)<1 -1 0> and (1 1 -2)<1 -1 0>. Moreover, they are grouped in bundles of vectors with similar direction and length indicating groups of similar orientations.

The greatest length of such coherent orientation groups is seen to be in the order of 50 grains, i. e. of 0.5 mm.

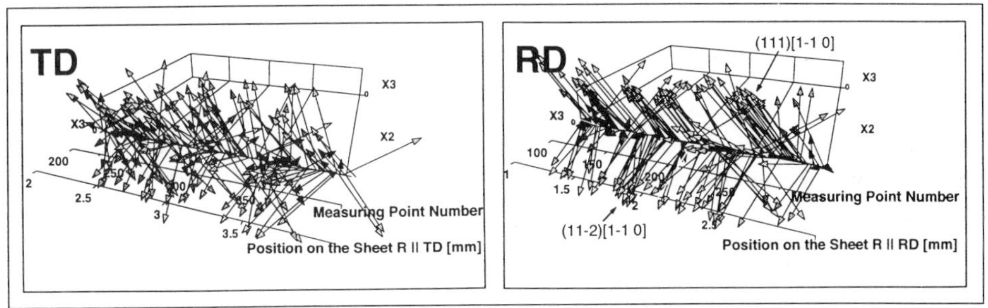

Fig. 1: Orientation difference between the orientation at a given measuring point and the (001)[1-1 0] direction, represented by Rodrigues Vectors; a) for a piece of the TD scan, b) for a piece of the RD scan.

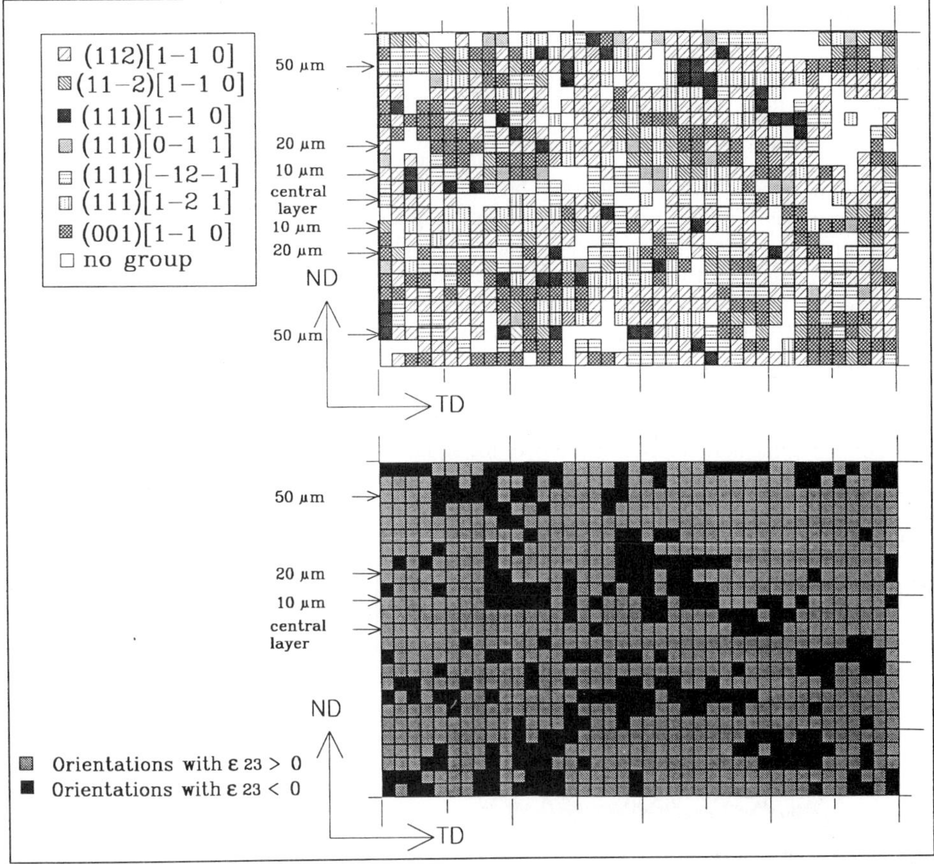

Fig. 2: a) Area scan in the plane ⊥ RD, 1000 orientations: Measured orientations collected into 7 classes defined as neighbourhoods of 7 ideal orientations

b) The same orientations as in Fig. 2a collected into two classes distinguished by the shear component $\varepsilon_{23} > 0$ and $\varepsilon_{23} < 0$

Furthermore, in the transverse section (plane ⊥ RD) an area with 1000 measuring points, 25 along ND and 40 along TD was scanned. Their distances between the points was chosen to be only 5 μm which is slightly smaller than the mean grain diameter. This area was again placed around the central layer of the sheet and within a region of strong corrugations seen at the surface by tilting the sample in the SEM. The results for scanning along ND were found to be similar to those of scanning along TD (Fig.1a). This can be concluded from Fig. 2a., where all orientations are collected in classes within a misorientation angle of 10° relative to the seven main ideal orientations of the X- ray texture. These seven orientations were found, in agreement with the present measurements, to all be situated on the α and γ fibre and to consist of {001}<110>, {112}<110>, {111}<110> and {111}<112> including the three symmetrically equivalent orientations. Also here no large clustering of similar orientations in the cross- section plane is found but rather a randomly looking distribution.

Evaluation of the Measurements

The above measurements allow some important conclusions. First, one clearly has an anisotropic orientation arrangement in the rolling plane: When proceeding along the transverse direction one finds strong orientation changes practically from grain to grain whereas along the tensile direction larger ranges of similar orientations up to about 50 grain diameters (0.5 mm) occur. Second, the basic idea that differently oriented single grains which underlie different deformations are responsible for the formation of ridges cannot be correct: For a ridge width of 1 mm and a sheet thickness of 0.5 mm the cross- section involved in forming one ridge contains about 100 x 50 = 5000 grains all of different orientation. Third, one can see from Fig. 1a that also the concept of grains being small compared to the ridge dimensions but arranged in big clusters of nearly equal orientations cannot be true; According to Fig. 2a the length of clusters of equal orientations consists of only a few grain sizes per cross- section and thus is at least 500 times smaller than the observed extension of a ridge (which in RD has a length of many centimetres).

The latter conclusion is confirmed by calculating the misorientation correlation index MCI [15]. This quantity is the analogon to the texture index of an orientation distribution function (ODF) of an orientation difference distribution function (ODDF) and describes its sharpness. It is defined as the normalized integral over the square of the ODDF or, equivalently, the sum over the squares of the coefficients of the series expansion of the ODDF weighted by $1 / (2\ell +1)$ with ℓ being the respective expansion degree. A purely random ODDF would result in the constant value of 1 for the MCI.

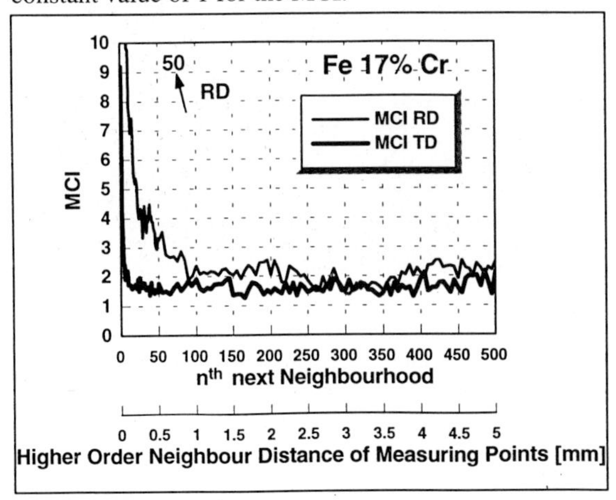

Fig. 3 Correlation Index of the Orientation Difference Distribution Function (Misorientation Correlation Function) MCI as function of higher order neighbour distances between measuring points (n = number of nearest neighbour distances).

In Fig. 3 the MCI is shown for the distribution of orientation differences between the nearest, the next nearest, the overnext nearest, etc. measuring points up to the 500th neighbourhood. Firstly one sees that for the scan along TD only for a range of one to two measuring points MCI > 1, i.e. a strong orientation difference correlation is found. The range for the RD scan is larger, but also only extends over 50 to 100 measuring points corresponding to 0.5 to 1 mm. Secondly one sees that the MCI values, i.e. the strength of correlation, for the nearest neighbour orientation differences is higher by a factor of 5 for the RD than for the TD scan which agrees well with Fig.1. Third, a more thorough analysis additionally reveals that the maximum values in the misorientation distribution are situated at the misorientation angle zero, i.e. that they are caused by clustering of similar orientations .

Simulations and Calculations

On the basis of the above results in the following the response of the differently oriented individual grains with respect to their plastic deformation will be regarded. As will be shown elsewhere [16], the consideration only of the main slip systems of the variously oriented grains would lead with respect to ridging to a preference of the shear ε_{23} on the transverse section plane ($\perp$ RD) in ND direction. For two symmetrically equivalent orientations this shear will have the same magnitude but different signs. Thus the contributions of e.g. (11-2)[1-10] and (111)[1-10] will have one sign (here called +) whereas (112)[1-10] and (111)[0-11] will have the other sign (here -). As EBSD distinguishes between symmetrically equivalent orientations, the spatial distribution of such shears taking into account also the signs can be determined. This is done in Fig. 2b for the orientations of Fig. 2a, but for simplicity reasons it is only distinguished between shears $\varepsilon_{23} > 0$ and < 0. The results show first that the occupation of these two classes are of different magnitude revealing a preponderance of one sense of shear (about 3:1), and second, the coherent areas of orientation classes exhibiting only one sign become somewhat larger than the classes for single orientations shown in Fig. 2a.

For obtaining a more quantitative understanding for shear stresses in individual grains causing macroscopic effects like corrugations, such shears must be assumed to be really carried out to some extent. This is in contrast to the Taylor theory which requires the complete suppression of shears, but in a certain agreement with more modern concepts (e. g. [17], [18]). For a linear row of grains the different signs of shears would then lead to a zig-zag line of the grains with the ups and downs depending on the sign of the shear and being given for the nth grain by $\Delta h_n = 2 \cdot \varepsilon_{23}(n) \cdot 10\ \mu m$ (Fig. 4). If one has thin sheets consisting of several layers of grains, it will be assumed that, as first approximation, the effects of the various grains lying on top of each other superimpose leading, as a kind of average deformation, to a bending of the sheet which will be observable as corrugation.

For a very rough quantitative estimate we will first assume that each grain will produce the same amount of shear Δh, but that this shear and thus the contribution to the corrugation profile can be positive or negative with equal probability. Then the resulting profile consisting of smaller and larger hills and valleys is the result of a stochastic process and thus analogous to e. g. the results of coin tossing experiments. In Fig. 5 such results for three

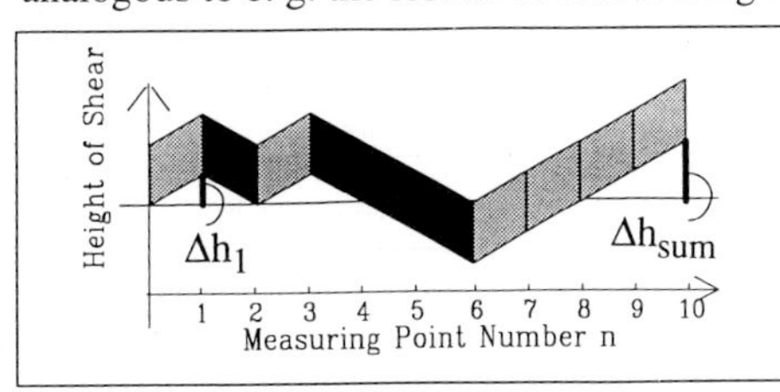

Fig. 4: Shearings $\varepsilon_{23}(n)$ in the transverse section plane of the sheet (schematic)

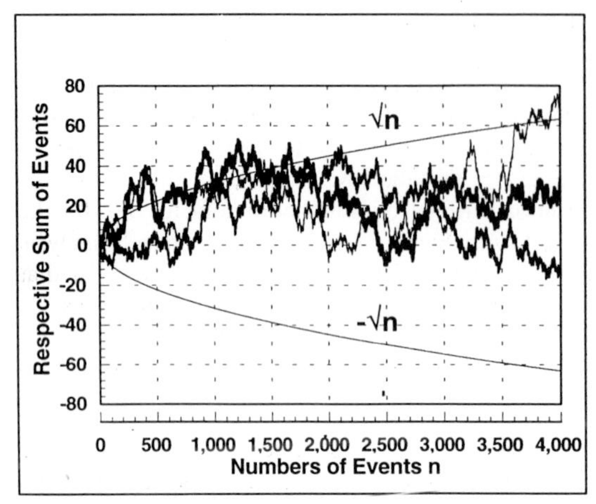

Fig.5: Results of 3 stochastic coin tossing experiments

series of coin tossings obtained by a computer simulation setting $\varepsilon_{23} = \pm 1$ are shown. On the abcissa the number n of the tossing events and on the ordinate the sum s_n of the results of the tossings up to the corresponding number n is plotted. The mean value of the sum of all possible sequences, of course, will be zero for each n, and the mean deviation can be shown to be $+\sqrt{n}$ or $-\sqrt{n}$. The individual curves, however, form hills and valleys of different depths and wavelengths as is also observed for the ridging profiles. The curves of Fig. 5 are calculated for a total number of 4000 events which corresponds to the whole width of 40 mm of the sheet with a mean grain diameter of 10 μm.

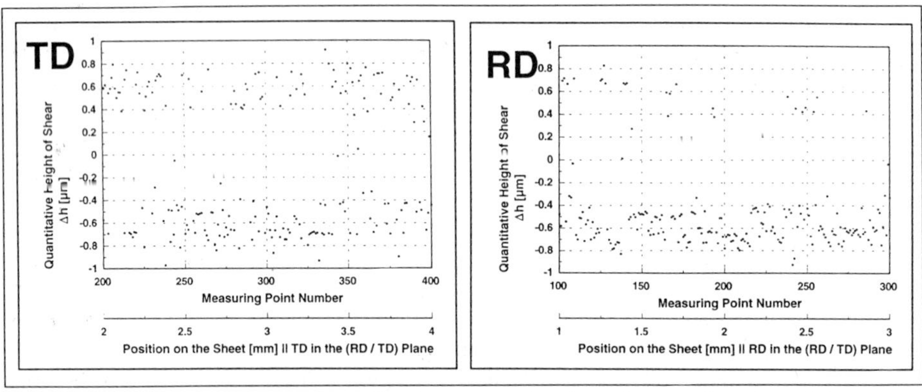

Fig. 6: Quantitative height of shear Δh of every single orientation as a function of the position on the sheet (measuring point number n) a) for TD and b) for RD scan

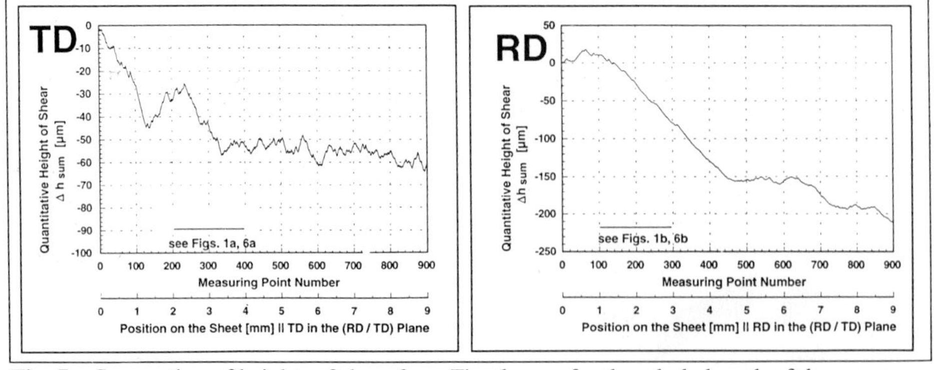

Fig. 7: Summation of heights of shear from Fig. 6; now for the whole length of the scans; a) for TD; b) for RD

This qualitative picture can now be improved to a more quantitative one by taking into account that differently oriented grains produce different shears and thus also different Δh. For this purpose for those parts of the scans applied in Fig. 1 the shear ε_{23} is calculated for each measuring point as that of a 7% tensile deformed single crystal. The corresponding increase or decrease in height for the nth grain given by $\Delta h_n = h_0 \cdot 2\, \varepsilon_{23}(n)$ (h_0 = grain size) is then plotted in Fig. 6. Fig. 7 shows the sum s_n of the individual Δh_n values from Fig. 6, now for the entire lengths of scan. One can see that for the TD scan (Fig. 6a) the positive and negative values nearly cancel which is due to the small coherence lengths of only one grain diameter. For the RD scan, in contrast, the negative Δh values strongly outweigh the positive ones (Fig. 6b) which is due to a longer coherence length. This is in agreement with the above observations. A more thorough quantitative evaluation and comparison will be given in a subsequent paper, [16].

Summary

For the first time large numbers of local orientation measurements were applied for clarifying the reason of the rippling of rolled and recrystallized ferritic stainless steels after tensile deformation. These EBSD measurements were accompanied by shear profile calculations and stochastic model simulations to make understandable the corrugation of the sheet. The results show that the ridges (width in TD ≈ 1 mm) do not represent the deformation of individual grains (diameter ≈ 10 μm) and that also larger groups of equally oriented grains do not exist. The ridging must be caused by a superposition of the deformation of the great number of grains involved in forming a ridge of 1 mm, particularly of the shear deformations ε_{23}. These are caused by activating the main slip systems by the applied tensile stress which is in contrast to the Taylor theory requiring that all shears are zero. In this superposition the shears of the various grains largely cancel each other and it is only the remaining rest which occurs as rippling. Its magnitude can be estimated by stochastic model calculations. One sees that this model yields both corrugations and the also observed superimposed small ripples. The appearance of ripples in form of stripes is due to the larger length of coherence in RD.

The present investigations also ruled out some models proposed in literature. For instance, the model of Chao [5] proposed an anisotropic flow mechanism which, however, does not lead to a corrugated sheet with both surfaces exhibiting the same profile. The plastic buckling mechanism of Wright [7] in which bands are buckling in a matrix, requires individual orientations arranged in groups forming such a kind of matrix as not being found here. In the model of Takechi et al. [6] the disregard of the existing annealing texture components and the lack of sufficient accuracy of the etch pit method rendered the model insufficient.

The authors like to thank Mr. J. A. Salsgiver, Pittsburgh, for interesting discussions.

References

[1] Appel, H. G., Becker, H., Z. Metallkunde, 54, 1963, 724
[2] Hölscher, M., Lücke, K., to be published, and Hölscher, M., Diss., RWTH Aachen, FRG, 1987
[3] Salsgiver, J. A., Larsen, J. M., Bornemann, P. R., Recrystallization ´90 ed. by Chandra, T., The Minerals, Metals and Materials Society, 1990
[4] in Appel, H. G., Diss., RWTH Aachen, 1971
[5] Chao, H-.C., ASM Trans. Quart., 60, 1967, 37
[6] Takechi, H., Kato, H., Sunami, T., Nakayama, T., Trans. Jap. Inst. Metals, 8, 1967, 233
[7] Wright, R.N., Met. Trans., 3, 1972, 83
[8] Defilippi, J. D., Chao, H-. C., Met. Trans., 1971, 3209
[9] Venables, J.A., Harland, C.J., Phil. Mag., 27, 1973a, 1193
[10] Hjelen, J., HØjer, R., Nes, E., in Proc. XI Int. Conf. on Electron Microscopy, Kyoto, Japan, 1, 1986, 751
[11] Dingley, D. J., Scanning Electron Microscopy II, 1984, 569
[12] Engler, O., Gottstein, G., Steel Research, 63, 1992, 413
[13] Frank, F. C., Int. Conf. on Textures of Materials (ICOTOM 8), ed. by Kallend, J. S. and Gottstein, G., The Metallurgical Society, 1988, p. 3
[14] Pospiech, J., Lücke, K., Sztwiertnia, K., Acta Met., 41,1, 1993, 305
[15] Weiland, H., Acta. Met., 40, 5,1992, 1083
[16] Bethke, K. and Lücke, K., in publication
[17] Honneff, H. and Mecking, H., in Textures of Materials, eds. Gottstein, G. and Lücke, K., Springer, 1978, p. 265
[18] Fortunier, R., Hirsch, J., in Theoretical Methods of Texture Analysis, 1987, ed. by H. J. Bunge, p. 231
Schmitter, U., Dipl. Thesis, RWTH Aachen, FRG, 1991

Materials Science Forum Vol. 157-162 (1994) pp. 1145-1152

THE INFLUENCE OF INITIAL MICROSTRUCTURE ON THE RECRYSTALLIZATION TEXTURES OF ALUMINIUM ALLOYS AFTER HOT DEFORMATION BY LABORATORY SIMULATION

R.K. Bolingbroke, G.J. Marshall and R.A. Ricks

Alcan International Limited, Southam Road, Banbury, Oxon, England

Keywords: Aluminium Alloy, AA1050, Cube Texture, Hot Deformation, Recrystallization

ABSTRACT

The recrystallization textures produced after annealing following commercial hot rolling of Al alloys are known to be important in controlling anisotropy of properties in the final product. Of particular interest is the generation of cube orientated crystallites, which compete with those grains nucleated elsewhere in the deformed matrix, in determining the texture balance in the recrystallized microstructure. These annealed textures are believed to be influenced by both the hot deformation conditions and the microstructure prior to deformation. Both have been assessed in this study with the initial microstructure altered in terms of grain size and texture. Samples have been deformed by Plane Strain Compression (PSC) and immediately annealed or water quenched to preserve the deformed structure. The textures are discussed with regard to the possible mechanisms of the nucleation of recrystallized grains with these particular crystallographic orientations.

INTRODUCTION

The control of crystallographic texture in commercial rolling operations is of considerable importance to provide the customer with a product that meets his requirements. For the majority of aluminium alloys this involves the control of the cube texture component during recrystallization annealing and the balance with the rolling deformation textures developed during cold rolling. Although the latter is fairly well documented[1,2] the former, especially following hot rolling, is less clearly understood.

Work has been reported in which hot deformed microstructures have been studied in detail to establish the origins of the cube component using both TEM and EBSP techniques[3,4]. Both have identified recovered cube orientated subgrains in the deformed microstructure as possible nucleation sites for cube grains even though the cube component has been widely reported to be unstable during deformation and, in particular, cold rolling[1]. More recently, this concept of a stable cube orientation during high temperature deformation has been experimentally verified using Channel Die Compression (CDC) single crystal experiments[5,6]; however, the exact mechanism has still to be substantiated. Taken together, these studies support an orientated nucleation theory based on the preservation of cube orientations during hot rolling. The observation of 'preserved' cube sites [3], however, was made on commercially hot rolled samples with the problems of 'quenching-in' the deformed microstructure and the single crystal studies[5,6] were performed at very low, non-commercial strain rates and need extrapolating to polycrystals.

The ideas proposed above depend on the strength of the cube orientation in the deformed microstructure as well as the relative deformed grain and subgrain sizes. Both of these are clearly dependent on the initial, starting microstructure and the deformation conditions. It is these microstructural and thermomechanical parameters that have been examined in the present study with the aim of identifying which are the most important in controlling the cube component in the annealed microstructure. The results should also provide interesting data to further examine the latest theories of cube grain formation.

EXPERIMENTAL PROCEDURE

The study has been based on the hot deformation of AA1050 by Plane Strain Compression (PSC) testing and the measurement of bulk textures in the as quenched and fully annealed conditions. Prior to deformation a number of different microstructures were developed by laboratory thermomechanical processing to obtain a range of materials with varying recrystallized grain sizes and textures. These microstructural parameters are given in Table 1. The microstructures were designed so that each of the microstructural parameters could be studied independently. For example, where the initial cube texture strength was to be assessed the grain sizes of each of the relevant materials was kept approximately constant (materials #2 and #5). Where possible material from the same DC cast ingot has been used to ensure identical size distributions of large intermetallic particles after laboratory rolling and, hence, the effects of differences in nucleation around these particles will have been minimised.

material	grain size (μm)	volume fraction (%)				
		cube	Goss	brass	Cu	S
#1	80	5.6	4.0	7.6	8.1	14.9
#2	330	9.4	4.6	8.7	4.9	12.1
#4	40	5.3	4.3	9.9	5.4	14.3
#5	250	25.1	4.0	5.9	3.8	12.2

Table 1. Microstructural information for the materials used in the study.

PSC testing was carried out on a specially designed machine capable of accurate control of the deformation parameters strain, strain rate and temperature. A matrix of deformation conditions and initial, predeformed microstructures were studied to assess these effects on deformed and recrystallized textures and these will be explained in the following section. Following deformation samples were water quenched within 2 seconds of the end of the test to preserve the deformed microstructure. Where necessary, samples were annealed in a fluidised bed at the deformation temperature for times long enough to allow full recrystallization to occur.

Deformed and recrystallized textures from the ½ thickness position were quantified by a bulk x-ray technique using Mo Kα x-rays. Four incomplete pole figures ((111), (200), (220) and (113)) were measured from which the ODF was derived using the series expansion method up to L=22[re]. The volume fractions of each of the major crystallographic components found in aluminium alloys were calculated allowing for a 15° spread around the ideal orientations.

RESULTS

Figure 1 shows data relating to the influence of the strain or total reduction on the deformation textures of two of the materials, #2 (initial cube component (Rc^0) 9.4%) and #5 (Rc^0 = 25.1%), deformed at 400°C at a strain rate of 2.5/s. For both materials the texture

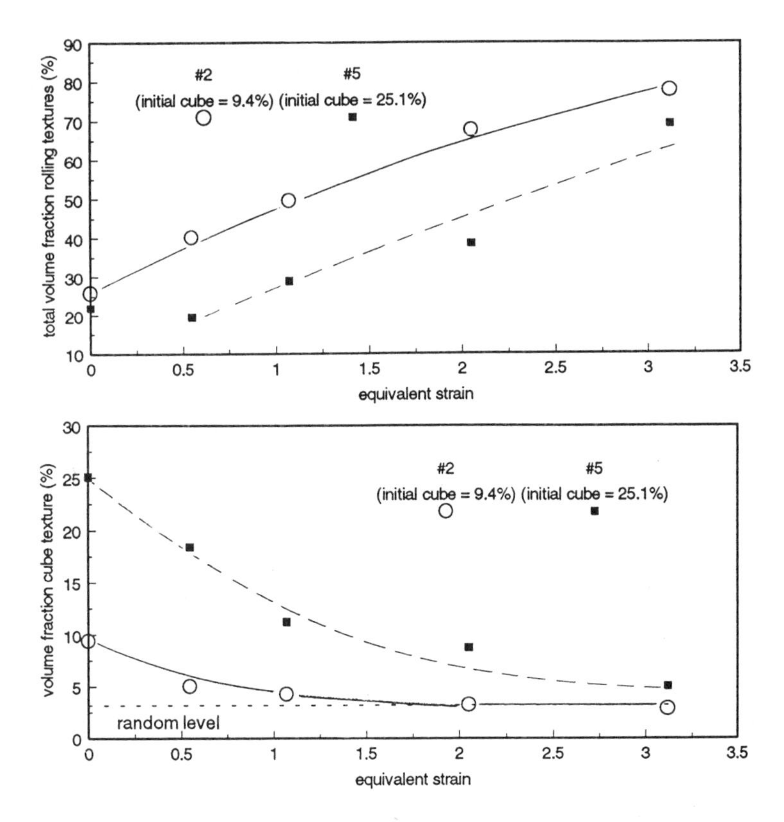

Figure 1. The effect of strain and initial cube strength on the development of (a) the rolling textures (bs + S + Cu) and (b) the cube texture after deformation (temperature = 400°C, strain rate = 2.5/s).

consists of a mixture of the rolling components, which gradually strengthen with increasing strain (Figure 1a), and the cube component (Figure 1b). The cube diminishes, initially fairly rapidly, though thereafter the rate of decline is markedly reduced and even at the largest strain examined (3.1) the strength of the cube orientation in material #5 remains above the random level. The disparity between the cube strength in the two starting microstructures persists across the range of strains examined. Upon annealing (400°C /1 hour) this variation in the remaining cube is mirrored in the fully recrystallized texture with material #5 consistently displaying a higher concentration of cube compared to #2 at equivalent strains (Figure 2). However, the more prominent effect is the increase in the cube component with increasing strain. Over the range of reductions studied a 3-fold increase in the cube strength was measured.

The other deformation parameters also appear to influence the generation of the cube component in the fully annealed condition. Figure 3 shows the effect in terms of the Zener-Holloman parameter (Z), incorporating both the strain rate and the deformation temperature. All the samples in this study were deformed to the same equivalent strain of 2.0. Generally an increase in Z (higher strain rate or lower temperature) is reflected in a weakening of the cube orientation, however, the data suggest that Z may not be an appropriate parameter to describe the behaviour in this particular case. At $Z=10^{14}/s$ the strength of the cube component can change significantly depending on the combination of deformation temperature and strain rate, with a higher temperature and higher strain rate resulting in a more pronounced cube orientation in the recrystallized microstructure. This behaviour is reflected, to some degree, in the as deformed texture which is shown for material #5 in Figure 4. Considerably more

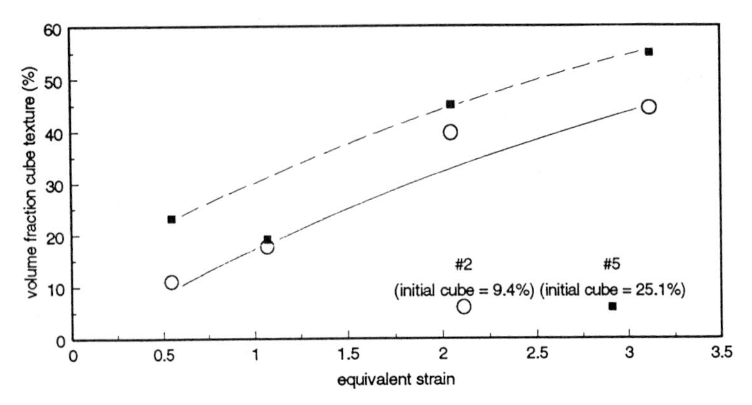

Figure 2. The effect of strain and initial cube strength on the development of the cube texture after deformation and annealing (400°C for 1 hour). Defomation temperature = 400°C, strain rate = 2.5/s.

cube remains in the deformed state after the higher temperature/lower strain rate deformation.

Finally, the influence of initial grain size in the undeformed microstructure has been examined by comparing materials #4, #1 and #2, which reflect an increase in the recrystallized grain size from 40μm to 330μm. From Table 1 it can be appreciated that each of these materials have approximately the same initial texture though that for material #2 is slightly stronger. The texture data from samples deformed at 400°C at 2.5/s to an equivalent strain of 2.0 followed by a recrystallization anneal are displayed in Figure 5. For the deformation conditions used in the study a negligible difference is observed.

DISCUSSION

The experimental data clearly highlight the degree of change possible in the strength of the cube component after high temperature deformation and annealing by control of the microstructure and thermomechanical processing. The results demonstrating the effect of strain (Figure 2) support the two important parameters believed necessary for cube nucleation following deformation, that of the presence of some cube orientated subgrains and the development of high angle grain boundaries adjacent to them. Recently, the former has been proposed as a mechanism for nucleation of grains with, or close to, the cube orientation during subsequent annealing[3-6]. It is postulated that cube subgrains that can 'survive' high temperature deformation are able to recover faster, either statically or dynamically, than subgrains of other orientations due to the limited dislocation types present after deformation.

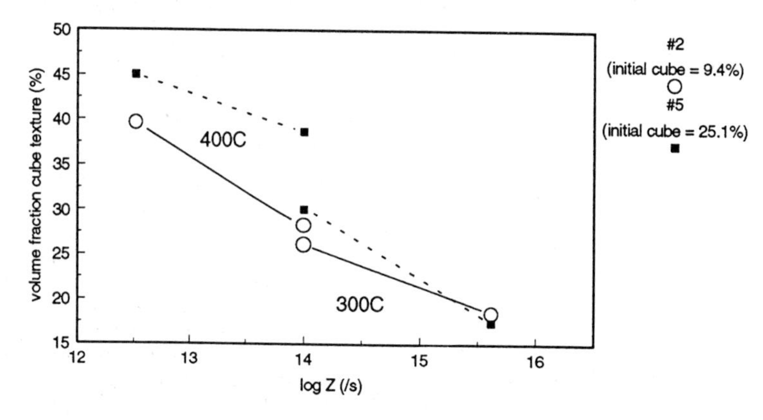

Figure 3. The influence of Zener-Holloman parameter on the strength of the cube texture after deformation (eq. strain = 2.0) and annealing.

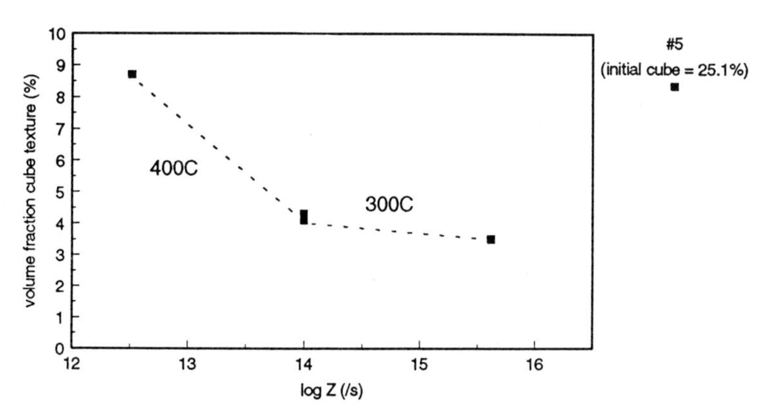

Figure 4. The influence of Zener-Holloman parameter on the strength of the cube texture after deformation (eq. strain = 2.0).

Ridha and Hutchinson[7] have shown in pure copper that only two sets of orthogonal dislocations need be active to accomodate plane strain in a cube orientated crystallite. This orthogonality inhibits interaction between the two types of dislocations and, hence, annihilation leading to recovery is enhanced. Effectively this provides the recovered cube orientated subgrains with a nucleation advantage over other potential sites and this is reflected in the texture of the fully recrystallized structure. Observations that the cube orientation can be stable in single crystals under specific, ideal deformation conditions has been attributed to slip on planes other than {111}, most notably on those parallel to {110}[5,6]. Whilst the present study can not comment on the operative slip systems in polycrystalline AA1050, the data in Figures 1b and 2 show that the presence of cube orientated matrix after deformation is clearly an advantage in obtaining high cube volume fractions after annealing. The retained cube component after deformation is shown in the phi 2 = 0° sections of the ODF's for materials #2 and #5 deformed to a strain of 2.0 at 400°C and a strain rate of 2.5/s (Figure 6). In both, though especially for material #5, a distinct cube component can be discerned with little rotation around the rolling direction (RD) as has been reported for cold rolled aluminium[2,8]. An alternative mechanism to account for the emergence of the cube orientation during annealing is that of generation of new cube orientated nucleation sites in the deformed microstructure. Certainly at low temperatures it has been demonstrated that cube orientated grains can nucleate at transition bands in a similar alloy[9], though no evidence of such features has been reported after high temperature deformation. Such a mechanism is required, however, to account for more recent observations showing the development of substantial levels of cube texture after annealing in materials that contained only a random cube component prior to hot deformation[10,11]. The latter demonstrated that cube grains nucleated

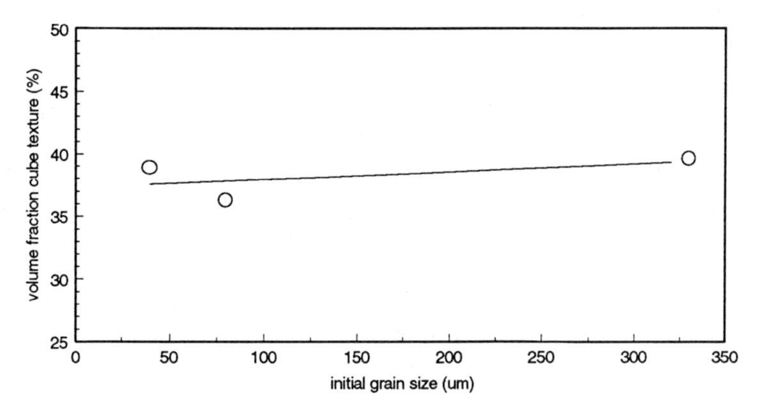

Figure 5. The effect of initial annealed grain size on the strength of the cube texture after deformation (temperature = 400°C, strain rate = 2.5/s, eq. strain = 2.0) and annealing (40°C for 1 hour).

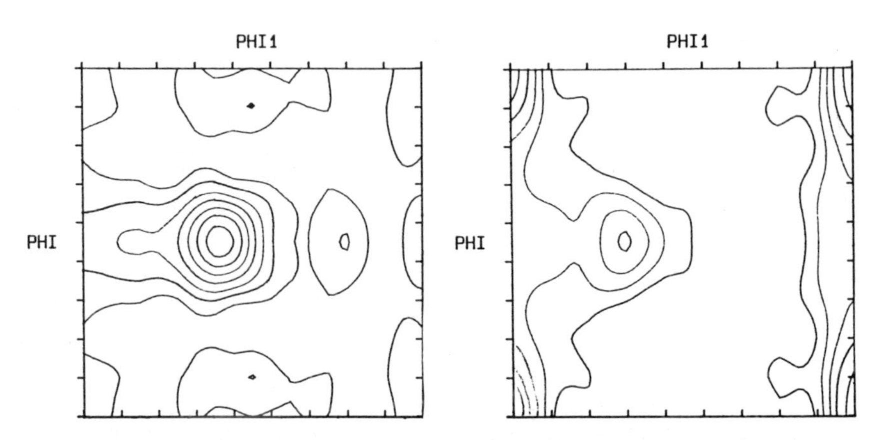

Figure 6. ODF phi2 = 0° sections of (a) material #2 and (b) material #5 after deformation (temperature = 400°C, strain rate = 2.5/s, eq. strain = 2.0)

at deformed grain boundaries during the initial stages of recrystallization.

Although the increased presence of the cube component after deformation is shown to be important the more potent controlling parameter is that of strain (Figure 2), presumably through the development of a fine subgrain structure and the increased misorientation between neighbouring subgrains and deformed grains[3,12]. The enhanced mobility of the latter would further improve the growth potential of cube recrystallized grains nucleated in the substructure, however, the present data do not indicate that this necessarily depends on the initial, predeformed grain size as proposed by Daaland et al[3]. They argued that the cube subgrain nucleation sites were more potent when the original deformed grain size approached that of the subgrains contained within them. The data in Figure 5 show no such effect over a wide range of initial grain sizes from 40μm to 330μm even though the initial textures were similar. After deformation to a strain of 2.0, corresponding to an 82% reduction, the deformed grain dimensions perpendicular to the rolling direction have been measured to be approximately 10μm and 36μm for materials #4 and #2 respectively, both of which are still greater than the subgrain size measured in a similar microstructure of around 3.5μm[12]. The data in Figure 2, demonstrating the pronounced effect of strain, suggest that it is an increased grain boundary area and maximised cube nucleation sites after hot deformation that are necessary to optimise the cube component in the recrystallized microstructure. Solely increasing the reduction may develop more high angle boundaries though at the expense of the number of potential nucleation sites, unless the initial microstructure contains more cube or the deformation conditions increase the propensity for other cube nucleation sites.

The data shown in Figure 3, showing the effect of Z, indicate that the degree of reduction alone appears to be insufficient to predict the level of cube produced after annealing. At higher values of Z one might expect a greater driving force for the recrystallization reaction due to a refinement in the subgrain size[12]. However, as highlighted in Figure 4, and as already observed by Driver et al[5,6] a lowering of the deformation temperature and an enhancement in the strain rate leads to a reduction in the amount of cube present in the deformed microstructure. Furthermore, at lower temperatures an augmented contribution to the nucleation rate from sites around coarse intermetallic particles would be expected[13], further weakening the cube component in the fully recrystallized microstructure[14]. Although particle effects could be incorporated into future micromechanical models to predict cube formation during hot deformation, the use of Z alone must be considered doubtful. The dissimilar behaviour of samples deformed at the same Z, though at different temperatures and strain rates, indicates that these parameters may have to be considered separately, or at least in a way so that they describe both the number of cube nucleation sites and the development of the substructure in terms of the subgrain size and misorientation between and within deformed grains.

CONCLUSIONS

1. The development of the cube component in AA1050 during annealing after hot deformation is highly dependent on the thermomechanical conditions, especially the degree of reduction. An increase in the strain and deformation temperature and a reduction in the strain rate all augment the level of cube texture after a recrystallization anneal. It is suggested that the Zener-Holloman parameter is inappropriate to describe the contribution of the latter two parameters to the deformation texture.

2. The texture prior to deformation is also considered important as this, along with the deformation conditions, affects the level of the cube component in the as deformed microstructure which, in turn, can be reflected in the fully annealed texture. This, however, can not solely account for all the observations of cube nucleation following hot deformation and further mechanisms need to be invoked before the arguement of the origin of the cube orientation can be settled. These need to include the role of the deformed microstructure in generating potential cube nucleation sites.

ACKNOWLEDGEMENTS

This work was carried out as part of a BRITE/EURAM programme BREU-CT91-0399

REFERENCES

1) Hansen, N., Bay, B., Juul Jensen, D., Leffers, T.: Strength of Metals and Alloys, Montreal, Canada, 12-16 August 1985, Pergamon Press, Oxford, 1, 317
2) Hollinshead, P.A., Sheppard, T.: Met Trans A, 1989, 20A, 1495
3) Daaland, O., Maurice, C., Driver, J., Raynaud, G-M., Lequeu, P., Strid, J., Nes, E., ICAA3, Trondheim, Norway, 22-26 June 1992, 297
4) Weiland, H., Hirsch, J.R.: ICOTOM9, Textures and Microstructures, 1991, 14-18, 647
5) Maurice, Cl., Driver, J.H.: Acta Metall. Mater., 1993, 41, 6, 1653
6) Akef, A., Driver, J.H.: Recrystallization '92, Materials Science Forum, 1993, 113-115, 103
7) Ridha, A.A, Hutchinson, W.B.: Acta Metall, 1982, 30, 1929
8) Hirsch, J.R.: Recrystallization '90, ed. Chandra, T., TMS, 1990, 759
9) Hjelen, J., Ørsund, R., Nes, E.: Acta Metall Mater., 1991, 39, 7, 1377
10) Alcan International unpublished work, 1993
11) Doherty, R.D., Kashyap, K., Panchanadeeswaran, S.: Acta Metall Mater., 1993, 41, 3029
12) Furu T.: SINTEF unpublished work, 1993
13) Kalu, P.N., Humphreys, F.J.: Aluminium Technology '86, ed:Sheppard, T., Institute of Metals, 1986, 197
14) Ørsund, R.., Nes, E.: Scripta Metallugica, 1988, 22, 665

Materials Science Forum Vol. 157-162 (1994) pp. 1153-1160

INFLUENCE OF THE INITIAL TEXTURE ON THE MICROSTRUCTURE- AND MICROTEXTURE DEVELOPMENT DURING HIGH TEMPERATURE LOW CYCLE FATIGUE

S. Brodesser, G. Brückner and G. Gottstein

Institut für Metallkunde und Metallphysik, RWTH Aachen, D-52056 Aachen, Germany

Keywords: EBSP, High Temperature Low Cycle Fatigue, Grain Boundary Migration, Microtexture, Misorientation Distribution, Rodrigues Vector

ABSTRACT

The microstructure and microtexture evolution during high temperature low cycle fatigue (HT-LCF) was investigated in pure nickel. At elevated temperatures grain boundary motion, grain growth and grain boundary alignment with the maximum shear stress take place. These phenomena are connected with an obvious change of microtexture and mesotexture, observed in "quasi in situ" experiments. The orientations of the individual grains were determined by means of **E**lectron **B**ack **S**catter **D**iffraction (**EBSD**) in a scanning electron microscope. The current investigations will be discussed with regard to the physical processes that control the microstructural changes.

INTRODUCTION

A variety of metallic materials undergo conspicuous microstructural changes during cyclic deformation at elevated temperatures due to migration of the grain boundaries and their reorientation under 45° with respect to the stress axis. These facts have been often reported in literature (Yavari and Langdon [1], Snowden [2]), but there is still no satisfactory interpretation of the physical mechanisms that control the microstructural evolution. The grain boundary arrangement with the maximum shear stress causes a damage of the material, because grain boundary sliding becomes intensified. Therefore, it is important to investigate it to avoid the premature failure of structures under HT-LCF. This kind of damage was first detected in high purity metals used as cables or wires and having a very special initial microstructure and -texture. So, samples with the typical structure and texture of cables were investigated first. Further, samples of the same material and purity but various initial textures were studied.

Additionally to the statistical investigation of the problem individual grains and grain boundaries were observed. The microstructure and -texture were determined before and after deformation using the same samples. These "quasi in-situ" measurements can give exact information about the reaction of the individual grains or grain boundaries deformed under defined experimental conditions.

EXPERIMENTAL

To investigate the influence of the initial textures on the microstructural evolution specimens of high purity nickel (99.99%) were used. From rods of extrusion processed material LCF samples with cable texture were machined. Cathodically refined metal was used to create samples with nearly randomly distributed orientations and a third initial texture was created by cross rolling and

recrystallization. Prior to deformation all samples were annealed for 5 to 40 h (depending on the grain size) at 900°C in high vacuum in order to establish a microstructure which was essentially stable against recrystallization and normal grain growth at deformation temperature. The specimens were deformed by tension/compression fatigue in high vacuum with a total (elastic and plastic) strain amplitude of 0.5% at a cyclic frequency of $13 \cdot 10^{-3}$Hz. The differently treated specimens were deformed to 300 cycles at 600°C. After cooling down to room temperature (no quenching) the interior of the gauge section of the specimens was subjected to metallographic and microtexture evaluation [3].
For the "quasi in-situ" experiments the same cathodically refined metal, annealed in the same way, was used. In contrast to the previous specimens a part of the gauge section of each sample was polished and prepared for microstructure and -texture evaluation. After this evaluation the specimens were deformed under the conditions described before with a strain amplitude of 1% up to 100 cycles. Without any new preparation the microstructure and -texture were measured again after deformation. The orientations of individual grains were determined by the **e**lectron **b**ackscatter **d**iffraction technique (EBSD) [4] in a **s**canning **e**lectron **m**icroscope (SEM). For each sample the orientations of more than 200 grains were determined and the orientation relationships between neighbouring grains (between 350 and 500) were calculated. For the "quasi in-situ" experiments in total the orientations of more than 500 grains for each processing step containing 1500 misorientations between contiguous grains were measured.

RESULTS

During cyclic deformation an obvious change of microstructure takes place for the extrusion processed samples (figure 1). The initial microstructure with very large grains and many small twins inside changes into a structure with a relatively homogeneous grain size distribution. Furthermore, the alignment of a high percentage of grain boundaries under 45° with respect to the stress axis is remarkable. These microstructural changes are correlated with an evolution of the microtexture. Figure 2 shows the corresponding ODFs calculated from the measured individual grain orientations by associating with each orientation a Gauss-type scatter of 5° about the exact orientation.

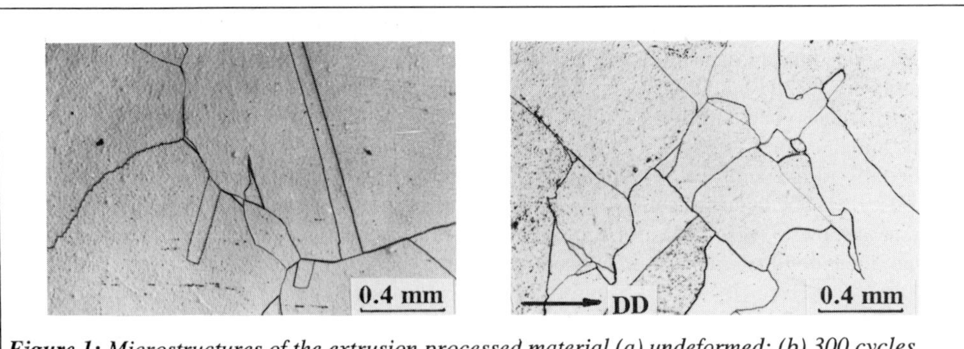

Figure 1: Microstructures of the extrusion processed material (a) undeformed; (b) 300 cycles

For microtexture evaluation the grain size remained unconsidered with the exception that for the very inhomogeneous microstructure of the initial sample the grain size of the very large grains was taken into account. The macrotexture changes with progressing deformation from an incomplete <111> fibre texture to an incomplete <100> fibre texture with a maximum near the cube orientation. From the individual grain orientations the orientation relationships between next neighbour grains were calculated and represented in figure 3 in Rodrigues space [5,6], where a misorientation is represented by a vector **R** defined as $\mathbf{R} = \mathbf{n} \cdot \tan \omega/2$. The direction of the vector **n** is parallel to the rotation axis and the length of the vector is proportional to the angle of rotation. Each misorientation was associated with a Gauss type scatter of 2° about the exact orientation relationship.

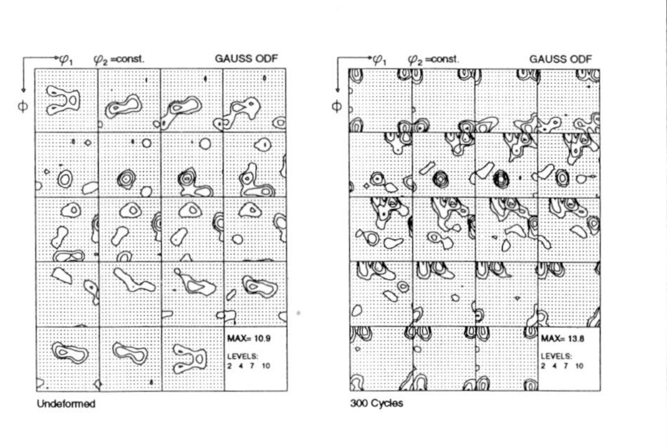

Figure 2: ODFs of the extrusion processed material

Comparing the two misorientation distribution functions (MODFs) (figure 3) there are characteristic changes. The undeformed specimen shows conspicuously high misorientation densities only close to the origin, corresponding to a frequent occurrence of small angle boundaries. Furthermore first and second order twin orientation relationships marked with Σ3 and Σ9 occurred. After 300 deformation cycles new orientation relationships with <111> rotation axes and rotation angles between 30° and 45° emerge. They can be associated with low Σ coincidence site orientation relationships (CSL), especially Σ7, Σ13 and Σ19. With regard to the question whether these characteristic changes are directly correlated with this special microstructure and -texture, various initial conditions were created.

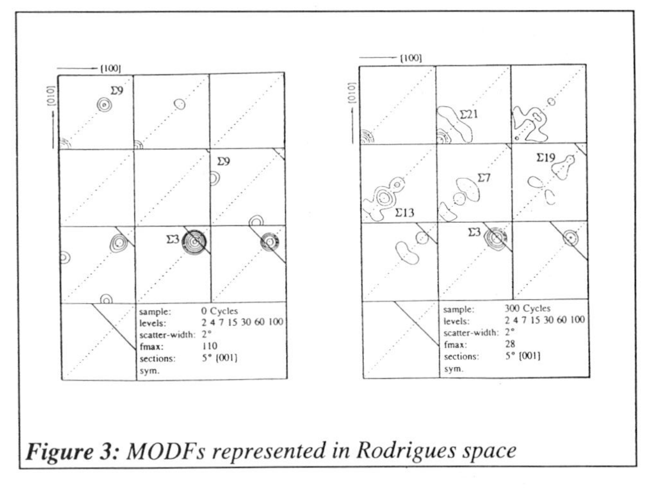

Figure 3: MODFs represented in Rodrigues space

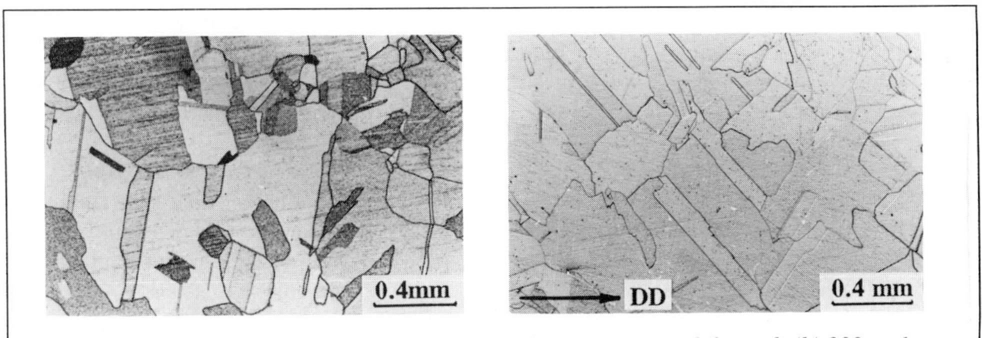

Figure 4: Microstructures of the cathodically refined material (a) undeformed; (b) 300 cycles

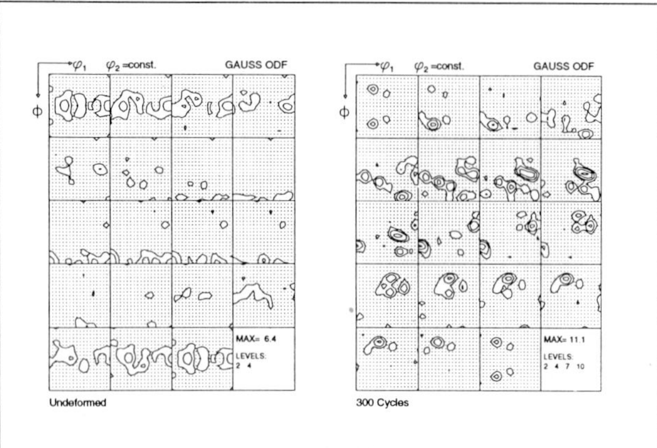

***Figure 5:** ODFs of the cathodically refined material*

Figure 4 shows the microstructure evolution of the cathodically refined material and figure 5 the corresponding texture. During deformation grain growth and grain boundary migration take place. The alignment of the grain boundaries with the direction of maximum shear stress is also obvious. The initial texture can be described as a weak and incomplete α-fibre, which disappears with progressing deformation but without forming a pronounced new component. The following results were found for the cross rolled and recrystallized samples. The fine grained and homogeneous microstructure (figure 6) shows an increase of the grain size and the reorientation of a lot of grains under 45° with respect to the stress axis. The initial texture, a sharp and symmetric α-fibre, remained nearly stable. The texture maximum decreases insignificantly, and there is some more orientation scatter around the transverse direction (figure 7).

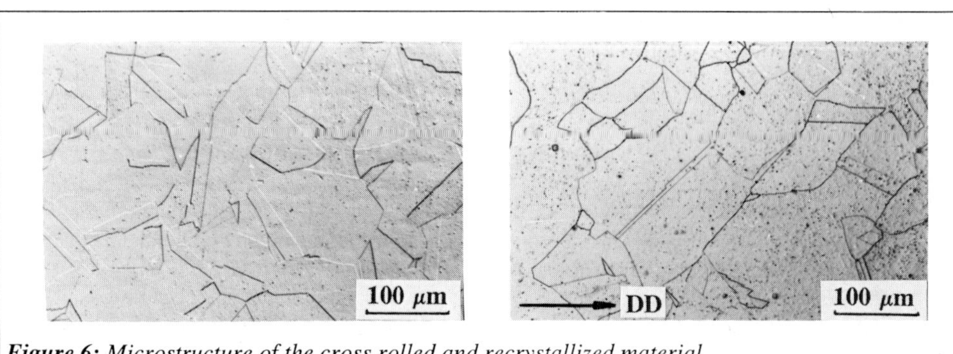

***Figure 6:** Microstructure of the cross rolled and recrystallized material*

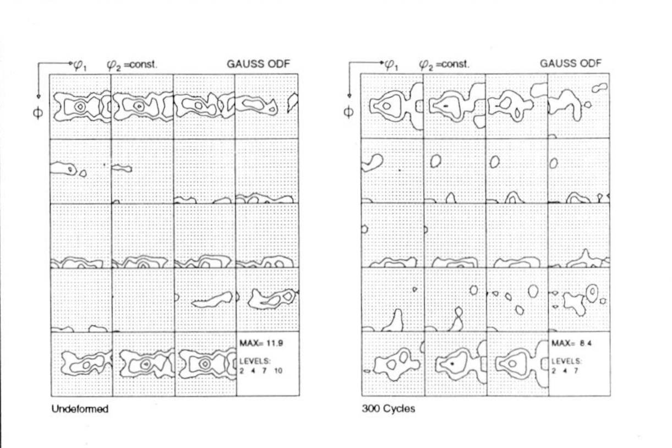

***Figure 7:** ODFs of the cross rolled and recrystallized material*

The corresponding orientation relationships represented in Rodrigues space (figure 8) were very similar for the two different initial conditions and also after deformation. The dominant orientation relationships were twins of first generation ($\Sigma 3$). Twins of second generation and low angle grain boundaries also show a noticeable occurrence. There are only a few other orientation relationships marked in the figure with $\Sigma 7$, $\Sigma 13$, $\Sigma 19$ and

Σ21, but their number and occurrence is not markedly different neither between the different samples, nor before and after deformation.

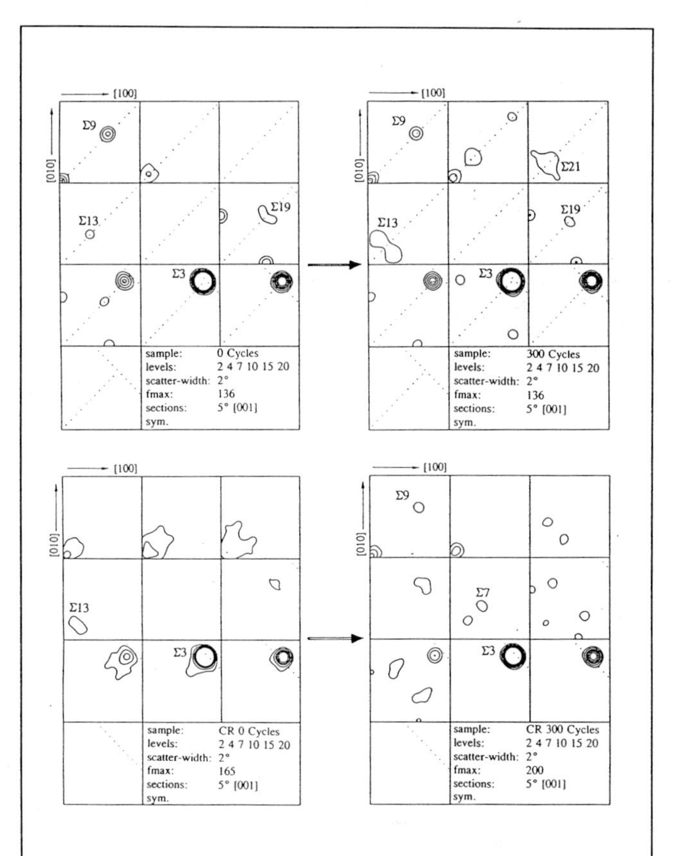

***Figure 8:** Rodrigues distribution of the **(a)** cathodically refined and **(b)** cross rolled and recrystallized material*

Discussion

Influence of the initial texture

From this study and also from previous investigations it is evident that the microstructure changes significantly during HT-LCF. All samples show an increase of the grain size and the, for cyclic deformation, typical grain boundary alignment with the maximum shear stress. But comparing the extrusion processed material with the samples processed differently, there is a much higher percentage of grain boundaries that reached the 45° position in the extrusion processed samples (30%) than in the other specimens (only 15%). With regard to the texture evolution it is obvious, that in general a change of the texture takes place. But it depends on the initial texture, how strong this change can be. If there is a sharp and symmetric texture in a fine grained material, like in the cross rolled and recrystallized material, or nearly randomly distributed orientations like in the cathodically refined material, only a weak texture change without formation of significant new orientations takes place. The same results were found regarding at the MODFs which did not show any significant change of the mesotexture for these specimens. Only for the samples prepared from the extrusion processed material a noticeable and systematical evolution of micro- and mesotexture was observed. But, from the results of this study, it cannot be decided whether microstructure, texture of the material or grain boundary structure play an essential role in this complex process. Furthermore, the physical mechanisms and driving forces for the observed phenomena remained to be clarified. Therefore, the "quasi in-situ" experiments were chosen to obtain a deeper understanding of this important problem.

Quasi in-situ experiments

Figure 9 shows the same area of microstructure after 10 and after 100 cycles, respectively. In this micrograph a variety of processes that take place during the microstructural evolution under HT-LCF are summarized.

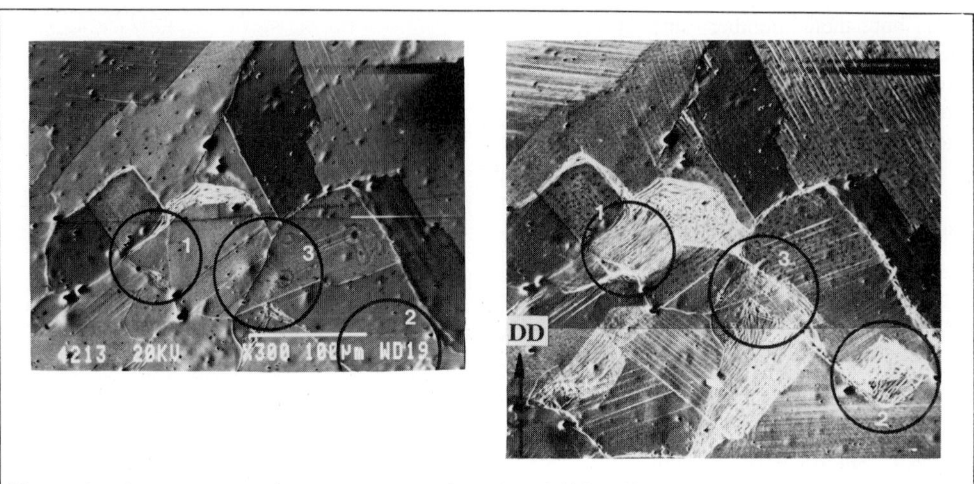

Figure 9: *The same area of microstructure after 10 and 100 cycles*

Comparing the grains at the two stages of deformation, there are grains that remain undeformed, others are shrinking or growing. The shrinking and disappearing of a whole grain is demonstrated in circled area (indicated by a number 1 in the figure). The occurrence of a new grain due to grain boundary migration can be seen in the 2nd circle. The advantage of this kind of microstructural evaluation obviously is, that all grains or grain boundaries can be classified exactly. Therefore, the single grain orientations can be subdivided into several groups. Figure 10 shows the ODFs calculated from the individual grain orientations before and after deformation, figure 11 for the corresponding groups of grains. These figures clarify, that the texture of all grains is composed of the texture of the different groups of grains. During deformation a microtextural evolution is obvious. A tendency of the orientations to rotate into <100> positions and the formation of <100> orientations is apparent, while other orientations are diminishing or rotating away. The orientation distributions for the different groups of grains (figure 11) show, that the components near the cube component with its scatter around the <100> direction arise from the growing and the newly formed grains. The texture of the growing grains is controlled by a sharp cube orientation and a lot of the new grains also arrange in a near cube orientation with a maximum at (5°,0°,0°) in Euler space. Regarding the ODF of the shrinking grains, no orientation densities in the cube position (0°,0°,0°) were found, but most of the shrinking grains belonged to the α fibre. Therefore it is obvious, that the microtextural evolution is controlled by the generation and growth of grains with special orientations.

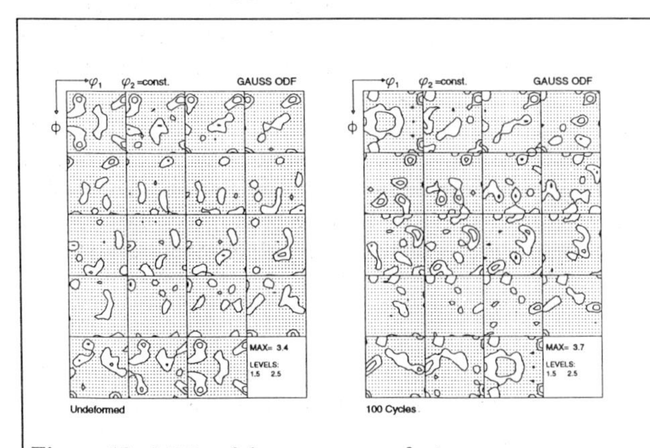

Figure 10: *ODFs of the same area of microstructure*

The microstructural changes can also be examined with regard to the grain boundaries (figure 9). There are grain boundaries which remain absolutely stable during deformation, some are migrating more or less and some are migrating into

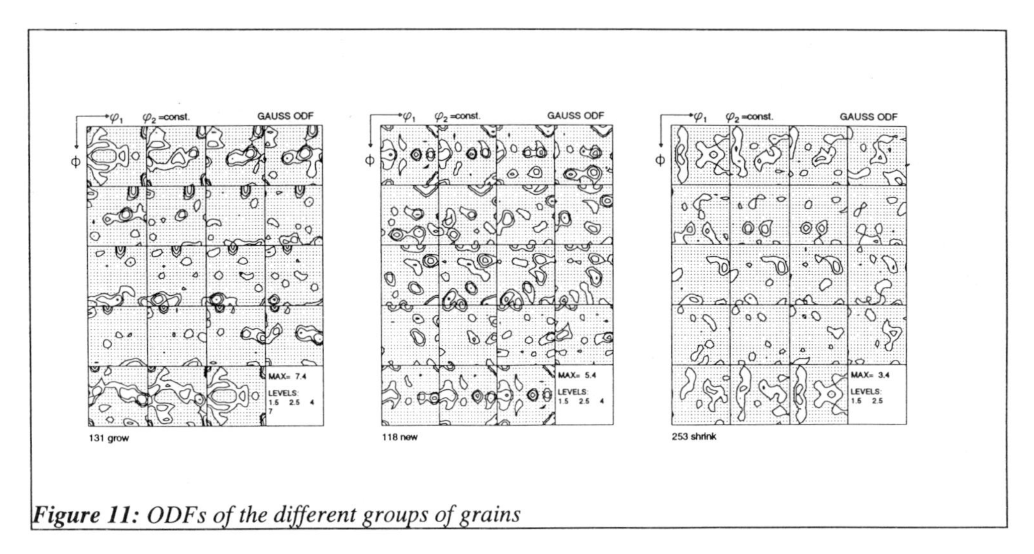

Figure 11: ODFs of the different groups of grains

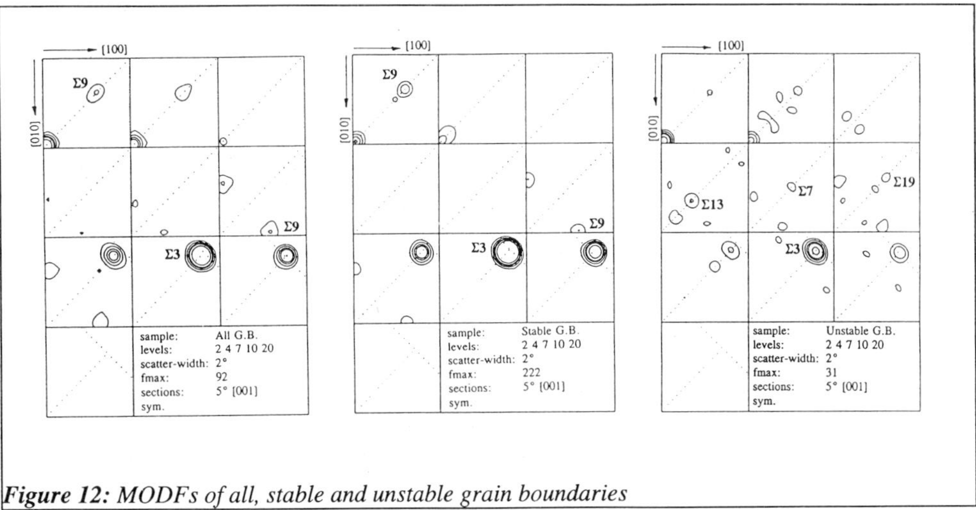

Figure 12: MODFs of all, stable and unstable grain boundaries

the often described 45° position with respect to the stress axis, as can be seen in the 3rd circle. The orientation relationships between neighbouring grains are shown in figures 12 and 13. Figure 12 describes the misorientation distribution for all next neighbour grains and the corresponding misorientations of the groups of the stable and unstable grain boundaries, respectively. The Rodrigues distribution of all grain boundaries is dominated by low angle boundaries and twins, mainly of first generation ($\Sigma 3$) but also of second generation ($\Sigma 9$). Only the twins and some of the low angle boundaries remain stable during cyclic deformation, as obvious from the very high density of $\Sigma 3$. The character of the unstable grain boundaries is either random, or of high mobility ($\Sigma 13$, $\Sigma 7$, $\Sigma 19$). Those orientation relationships have a <111> axis in common with angles of rotation between 30 and 45 degrees. Also a relatively high number of twin orientation relationships was observed to be unstable, but most of them are incoherent twins, easy recognizable from the appearance of the

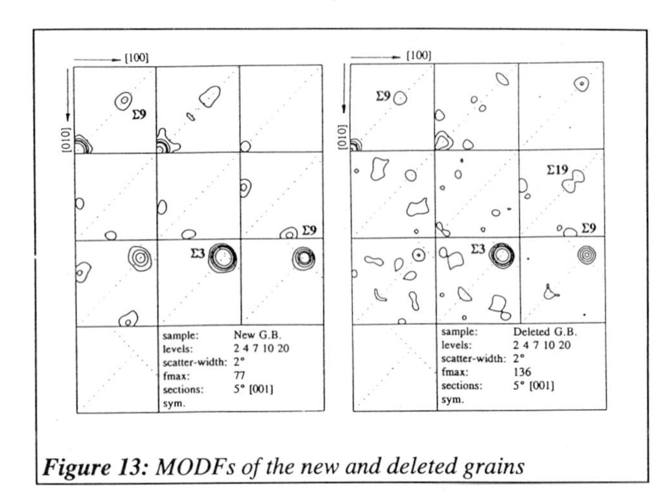

Figure 13: MODFs of the new and deleted grains

boundary in the microstructure. The MODFs of new and deleted boundaries are given in figure 13. The Rodrigues distribution of the new grain boundaries clarifies the nucleation mechanisms of the new grains. The occurrence of exclusively twins and low angle boundaries can be interpreted as nucleation by twinning and by the formation of a subgrain structure. The disappearing grain boundaries are of non special character except for some low angle boundaries and twins, which were mostly annihilated in the course of migration of other not twin character boundaries.

CONCLUSIONS

The microstructural and microtextural evolution during HT-LCF was investigated in high purity nickel. The following results were obtained:

- Microstructural changes like grain growth, grain boundary migration and boundary reorientation with respect to the stress axis were observed.
- Microstructure and microtexture change, but with different rates depending on the initial condition.
- The texture change can be related to the microtextural evolution of several groups of grains.
- The stable, unstable, new and deleted grain boundaries can be associated with the orientation relationships.
- Three different processes were found to control the evolution of microstructure and microtexture during HT-LCF namely:
  i) the formation of a substructure,
  ii) the twinning mechanism and
  iii) the high mobility grain boundaries.

REFERENCES

[1] Yavari, P. and Langdon, T.G.: Acta metall., 1983, 31, 1595
[2] Snowden, K.U.: Phil.Mag., 1960, 6, 321
[3] Brodesser, S. and Gottstein, G.: Text. & Microstr., 1993, 20, 179
[4] Engler, O. and Gottstein, G.: Steel Research, 1992, 63, 413
[5] Frank, F.C.: ICOTOM 8 ,1988 (eds. Kallend, J.S. and Gottstein, G.), TMS Warrendale, 3
[6] Neumann, P.: Text. & Microstr., 1991, 14-18, 53

Materials Science Forum Vol. 157-162 (1994) pp. 1161-1166

EBSP AND SIMS STUDIES OF OXYGEN TRACER DIFFUSION IN THE HIGH TEMPERATURE SUPERCONDUCTOR $La_{2-x}Sr_xCuO_4$

J. Claus [1], G. Borchardt [1], S. Weber [2] and S. Scherrer [2]

[1] AG Elektronische Materialien, TU Clausthal,
Robert-Koch-Str. 42, D-38678 Clausthal-Zellerfeld, Germany

[2] Laboratoire de Métallurgie Physique et Sciences des Matériaux,
Ecole des Mines de Nancy, F-54042 Nancy Cédex, France

Keywords: High Temperature Superconductors, $La_{2-x}Sr_xCuO_4$, Oxygen Tracer Diffusion, Electron Back Scattering Pattern (EBSP) Technique, Secondary Ion Mass Spectrometry (SIMS)

ABSTRACT

Oxygen diffusion in the ceramic high temperature superconductor $La_{2-x}Sr_xCuO_4$ (x = 0 and 0.15) was investigated as a function of temperature and crystal orientation. In polycrystalline samples the crystal orientation of single grains was determined by EBSP measurements with a modified SEM. Oxygen-18 diffusion was investigated in the temperature range between 600 °C and 900 °C. From the isotope depth profiles measured by SIMS tracer diffusivities of oxygen were determined as a function of crystal orientation. $La_{2-x}Sr_xCuO_4$ showed a strong anisotropy of oxygen diffusivity for all dopant concentrations.

INTRODUCTION

$La_{2-x}Sr_xCuO_4$ is a ceramic high temperature superconductor (figure 1) with a transition temperature $T_C \cong 37$ K (x = 0.15). Oxygen ion vacancies are believed to play a critical role in the superconducting properties of $La_{2-x}Sr_xCuO_4$. These oxygen vacancies are introduced for charge compensation as Sr^{2+} ions are substituted for La^{3+}. Indeed, T_C depends on the strontium concentration [1], [2]. Therefore, an understanding of oxygen diffusion in terms of the oxygen defect structure is important for the accurate determination of annealing times and temperatures required to achieve the desired properties. Oxygen diffusivities were measured in polycrystalline $La_{2-x}Sr_xCuO_4$ (x = 0.1, 0.15, 0.20) [3] and in $La_{2-x}Sr_xCuO_4$ single crystals (x = 0, 0.07, 0.09, 0.12) [4] at lower temperatures (250 °C - 500 °C) using an ^{18}O exchange method followed by SIMS analysis. This technique in principle allows the direct determination of the oxygen tracer diffusion coefficient since the ^{18}O diffusion profile is obtained under conditions where the chemical potential of oxygen is constant. As large high quality $La_{2-x}Sr_xCuO_4$ single crystals with smooth as-grown surfaces are difficult to obtain the combination of EBSP and SIMS was used in order to correlate the oxygen diffusivities measured by SIMS on single grains in a polycrystalline sample with their crystallographic orientation determined by EBSP. For samples with large grains (> 100 μm) this combined analytical approach allows to obtain the same information on transport processes as from single crystals from polycrystalline material which is easier to prepare. The influence of grain boundary diffusion can be avoided by analyzing large grains (grain size much larger than the analyzed SIMS area) and by applying short diffusion anneals to minimize the depth of the isotope diffusion profiles.

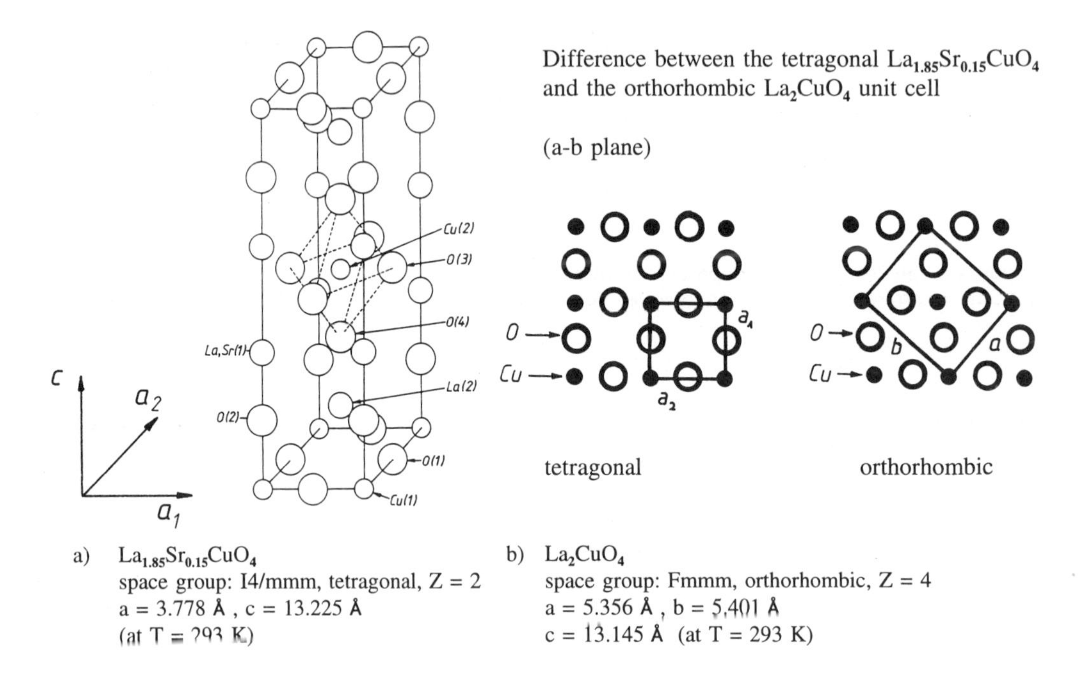

a) $La_{1.85}Sr_{0.15}CuO_4$
space group: I4/mmm, tetragonal, Z = 2
a = 3.778 Å , c = 13.225 Å
(at T = 293 K)

b) La_2CuO_4
space group: Fmmm, orthorhombic, Z = 4
a = 5.356 Å , b = 5.401 Å
c = 13.145 Å (at T = 293 K)

Fig. 1 Crystal structures of $La_{1.85}Sr_{0.15}CuO_4$ and La_2CuO_4 after [2]

EXPERIMENTAL PROCEDURES

A. Sample preparation and EBSP measurements

The polycrystalline $La_{2-x}Sr_xCuO_4$ samples were prepared from high purity (> 99.9 %) La_2O_3, $SrCO_3$ and CuO powders by standard powder ceramic techniques with a grain size up to about 200 μm. The diffusion samples were predominantly single phase superconductors with a density up to 97 % of the theoretical density. The very fine porosity was closed and more widely spaced than the diffusion distances. The $La_{2-x}Sr_xCuO_4$ samples were polished to a surface roughness of less than 50 nm and were annealed at 800 °C to remove surface damage.
The crystallographic orientation of single grains in the polycrystalline samples was determined by EBSP measurements with a modified SEM (JEOL JSM 820). Since the total energy loss of the back scattered electrons is rather small compared to the primary beam energy a phosphor screen was placed at a distance of some centimetres to the sample surface which was tilted by an angle of 30 degrees against the primary electron beam. The electron back scattering patterns on the screen were detected through a lead glass window by a CCD Camera (HAMAMATSU) coupled with a

camera control unit and an image processor. The resolution of the orientation measurements was about 1 µm (I = 1 nA and U_{prim} = 30 kV). The evaluation of the electron back scattering patterns was practiced by a personal computer supported on-line method. Figure 2 shows the experimental EBSP arrangement. Since this method allows the determination of single grain orientations of a polycrystalline bulk sample no large-scale sample preparation is required and standard preparation methods can be used. After the orientation measurements the samples were annealed in a pure ^{16}O atmosphere for times much longer than the ^{18}O exchange time at the same temperature and pressure like the exchange conditions to ensure a constant oxygen chemical potential to depths greater than the ^{18}O profile.

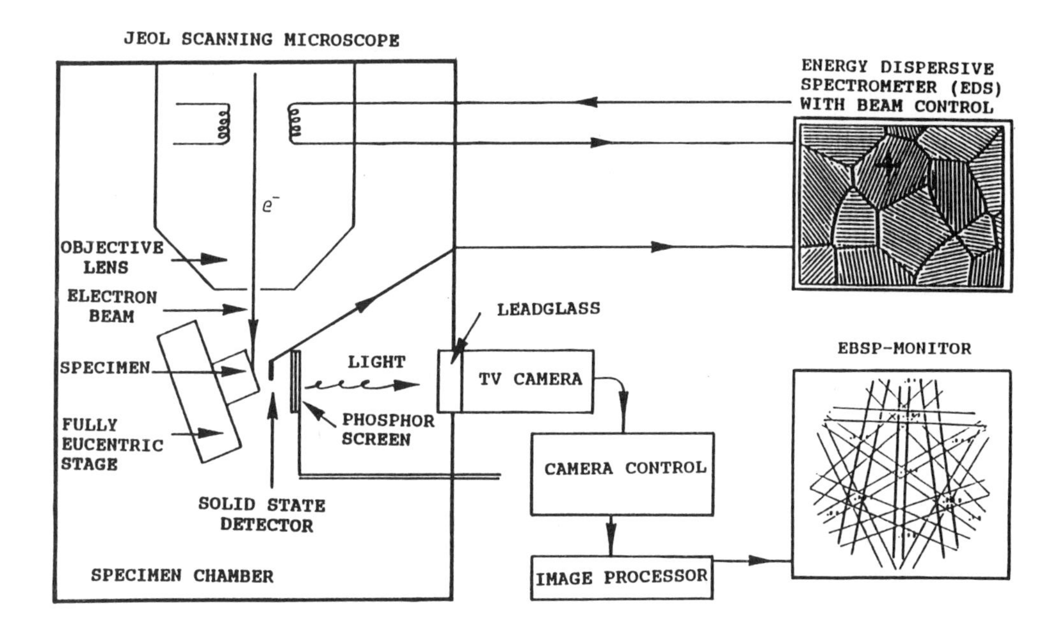

Fig. 2 Experimental arrangement of the EBSP equipment after [5]

B. SIMS and data analysis

The ^{18}O diffusion profiles were measured with a VG SIMSLAB using a Ga^+ source and a 1-5 nA current. The sputtered area was either 40 µm by 25 µm or 25 µm by 15 µm. The total secondary ion signal was electronically gated to the central 20 % of the analyzed area. No sample charging occurred. The sputter depth z was calculated from profiles measured with a profilometer (TENCOR). The oxygen intensities were used to calculate the ^{18}O concentration C(z,t) as follows: $C(z,t) = I(^{18}O) / [I(^{16}O) + I(^{18}O)]$. Surface exchange was found to limit the surface enrichment of ^{18}O at lower diffusion temperatures. Although the gas was 98 % ^{18}O, surface enrichments of 25 % at 600 °C and 75 % at 750 °C in $La_{1.85}Sr_{0.15}CuO_4$ were found. In La_2CuO_4 the ^{18}O surface concentration was only as high as 2.5 % at 700 °C and 8 % at 850 °C. So the data were fitted to a solution of the diffusion equation for an exchange reaction controlled by first-order kinetics (equation 1) with rate constant K for the surface exchange reaction [6] and C_O for the initial ^{18}O concentration in the sample.

$$(1)\qquad \frac{C(z,t) - C_0}{C_G - C_0} = \mathrm{erfc}\left(\frac{z}{2\sqrt{Dt}}\right) - \exp\left(\frac{K}{D}\cdot z + \frac{K^2}{D}\cdot t\right)\cdot \mathrm{erfc}\left(\frac{z}{2\sqrt{Dt}} + \frac{K}{D}\cdot\sqrt{Dt}\right)$$

The boundary condition at the surface (z = 0) is given by equation 2:

$$(2)\qquad - D \cdot (\delta C/\delta z)_{z=0} = K \cdot (C_G - C_S)$$

C_G = equilibrium concentration with the vapour pressure in the atmosphere
C_S = actual concentration in the surface at any time

In anisotropic crystals the diffusion is characterized by the diffusion coefficients parallel to the principal crystallographic axes. In the orthorhombic structure the diffusion coefficient in an arbitrary direction ϕ is given by equation 3 where α, β and γ are the angles between the diffusion direction and the three principal axes and D_a, D_b and D_c are the diffusion coefficients parallel to these axes:

$$(3)\qquad D_\phi = D_a \cdot \cos^2\alpha + D_b \cdot \cos^2\beta + D_c \cdot \cos^2\gamma$$

In the tetragonal structure $D_a = D_b = D_{ab}$ and D_ϕ is then given by equation 4 where γ is the angle between the diffusion direction and the c axis.

$$(4)\qquad D_\phi = D_{ab} \cdot \sin^2\gamma + D_c \cdot \cos^2\gamma$$

EXPERIMENTAL RESULTS

In the undoped orthorhombic La_2CuO_4 even shortest diffusion anneals (5 min) at 850 °C resulted in a penetration profile which was flat over several micrometres with a low $^{18}O/^{16}O$ ratio but significantly higher than the natural isotopic abundance. This indicates a very rapid diffusion and surface exchange reaction controlled kinetics. Oxygen diffusivities for La_2CuO_4 and $La_{1.85}Sr_{0.15}CuO_4$ are plotted as a function of the reciprocal temperature in figure 3 from ref. [3] and [4] and in figure 4 from this work.

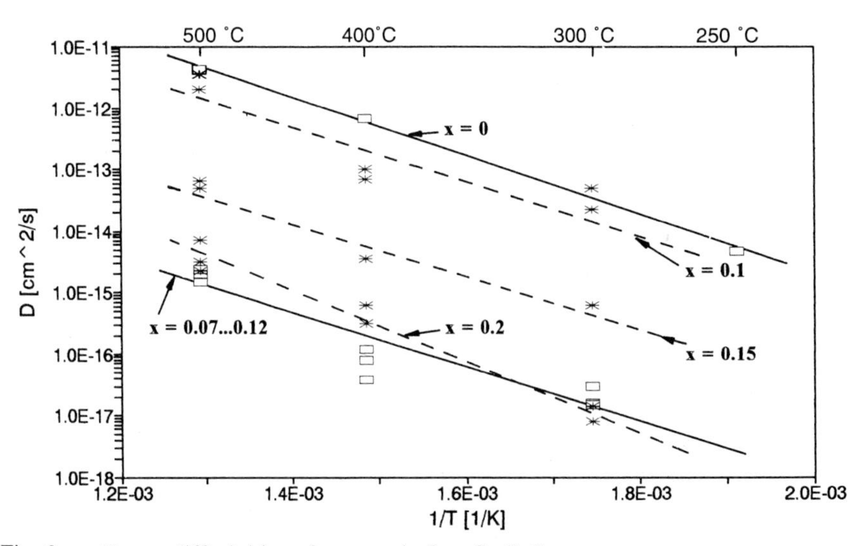

Fig. 3 Tracer diffusivities of oxygen in $La_{2-x}Sr_xCuO_4$
✳ = ref. [3] polycrystal; □ = ref. [4] single crystal, c direction

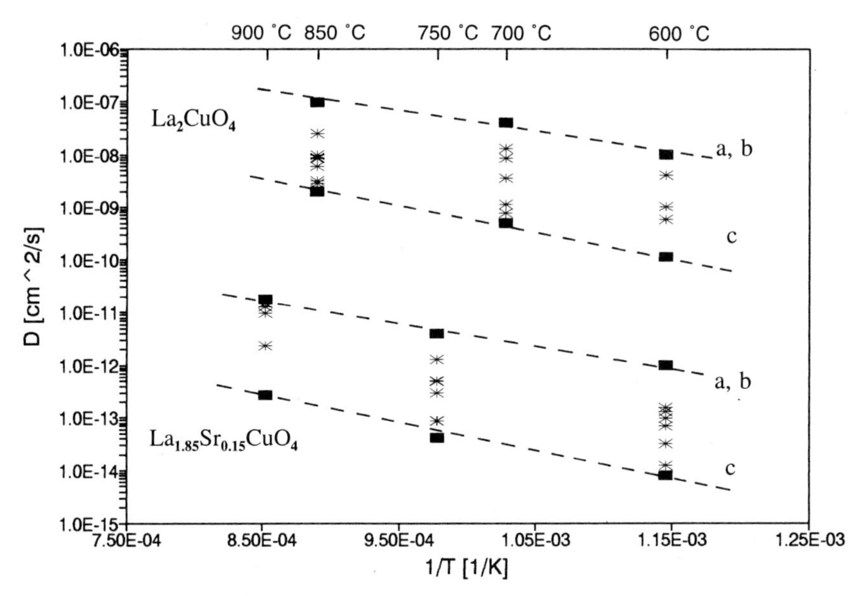

Fig. 4 Tracer diffusivities of oxygen in La_2CuO_4 and $La_{1.85}Sr_{0.15}CuO_4$ as a function of temperature and crystallographic orientation (this work).
■ = calculated diffusivities in the a-b plane and in the c direction
✱ = each point represents a measured diffusivity in a specific crystallographic direction

The diffusivities in the tetragonal $La_{1.85}Sr_{0.15}CuO_4$ phase are significantly smaller (about 4 orders of magnitude) than in La_2CuO_4 whereas D is decreasing with increasing strontium content. The oxygen diffusivities are about 2 orders of magnitude smaller in the c direction than in the a-b plane for all strontium concentrations. The values of D are reproducible within a factor of 2. The activation energies Q_i calculated for oxygen diffusion in the a-b plane and in the c direction of La_2CuO_4 and $La_{1.85}Sr_{0.15}CuO_4$ and the values of the preexponential factor D_O in the equation for the diffusivity $D = D_O \cdot \exp(-Q_i/RT)$ are listed in table 1 (600 °C $\leq T \leq$ 900 °C).

type of sample	Q_i [eV]	D_O [cm²/s]
La_2CuO_4	0.76 ± 0.13 (a-b plane) 0.97 ± 0.06 (c direction)	$2.7 \cdot 10^{-4}$ (a-b plane) $4.8 \cdot 10^{-5}$ (c direction)
$La_{1.85}Sr_{0.15}CuO_4$	0.84 ± 0.10 (a-b plane) 1.02 ± 0.12 (c direction)	$6.5 \cdot 10^{-8}$ (a-b plane) $5.7 \cdot 10^{-9}$ (c direction)

Table 1 Activation energy Q_i and preexponential factor D_O for La_2CuO_4 and $La_{1.85}Sr_{0.15}CuO_4$

DISCUSSION

The principal result that the oxygen diffusivity D decreases with increasing strontium additions appears to be inconsistent with thermogravimetric measurements [7] which show that the replacement of La^{3+} by Sr^{2+} results in an increase in the oxygen vacancy concentration. This behaviour can be explained by the incorporation of strontium according to equations 5 and 6:

(5) $2\ SrO + 2\ La_{La} + 1/2\ O_2 \Rightarrow La_2O_3 + 2\ Sr'_{La} + 2\ h^{\bullet}$

(6) $2\ SrO + 2\ La_{La} + O_O \Rightarrow La_2O_3 + 2\ Sr'_{La} + V_O^{\bullet\bullet}$

As a result both $h^{\bullet}$ and $V_O^{\bullet\bullet}$ concentrations are increasing in agreement with experiment [8]. Therefore one expects an increase in D rather than the observed decrease which was also found by Rothman [3] and Opila [4]. Within the scatter of the data, the activation energy is nearly unchanged as strontium is added to levels up to $x \cong 0.15$ (see also [4]). This indicates that the diffusion mechanism is unchanged with low strontium additions. For $x = 0.2$ the activation energy increases according to Routbort's data [3]. A qualitative explanation can be given by a combination of two models [3], [4]: A decrease in D_O with increasing x ($0 \leq x \leq 0.15$) would be consistent with oxygen interstitials being the more mobile oxygen species. Thermogravimetric and chemical analysis [9] showed a slight oxygen excess in La_2CuO_{4+y} in pure oxygen atmospheres. In addition, at higher defect concentrations the observed strontium dependence can be explained by an extended defect model based on the ordering of oxygen vacancies in strontium doped material [3]. Positron annihilation results [10] show that oxygen vacancies may be bound to strontium ion clusters being immobilized (increasing activation energy), where a cluster is defined as two or more strontium ions on nearest neighbour lanthanum sites. This association reaction as well as the oxidation of Cu^{2+} to Cu^{3+} compensates the effective negative charge of the Sr'_{La} ions (model of local charge balance). This model is quite successful in predicting the dependence of the oxygen vacancy concentration on the strontium content, the dependence of T_C on composition which was assumed to be related to the Cu^{3+} concentration and in explaining both the diffusion and the positron annihilation results. It can be assumed that at lower strontium concentrations interstitial diffusion may dominate while at higher strontium concentrations vacancy diffusion may become dominant.
Furthermore, a strong anisotropy of oxygen tracer diffusion for all dopant concentrations with diffusivities in c direction being almost two orders of magnitude lower than in the a-b plane was found. Diffusion along the c direction is slower due to the higher number of equivalent nearest neighbour sites (either vacancies or interstitials) in the high mobility a-b plane. Because of the lower activation energy of 0.76 - 0.84 eV in the a-b plane (c direction 0.97 - 1.02 eV) the anisotropy effect decreases with increasing temperature.

ACKNOWLEDGMENTS

We thank Dr. H. Fischer, Universität Karlsruhe, for the sample material and Prof. M.- J. Philippe, Université de Metz, and Mr. J.- M. Hiver, Ecole des Mines de Nancy, for their help with the EBSP equipment.

REFERENCES

[1] Van Dover, R. B., Cava, R. J., Batlogg, B., Rietman, E. A.: Phys. Rev. B, 1987, 35, 5337
[2] Cava, R. J., Santoro, A., Johnson, D. W. Jr., Rhodes, W. W.: Phys. Rev. B, 1987, 35, 6716
[3] Routbort, J. L., Rothman, S. J., et al.: J. Mater. Res., 1988, 3, 116
[4] Opila, E. J.,Tuller, H. L., Wuensch, B. J., Maier, J.: Mater. Res. Soc. Symp. Proc., 1991, 209, 795
[5] Beaujean, I., Université de Metz, personal communication
[6] Crank, J.: The Mathematics of Diffusion, Oxford, University Press, London, 1956, p. 34
[7] Nguyen, N., Choisnet, J., Hervieu, M., Raveau, B.: J. Solid State Chem., 1981, 39,120
[8] Opila, E. J., Pfundtner, G., Maier, J., Tuller, H. L., Wuensch, B. J.: Mater. Sci. Eng., 1992, B13, 165
[9] Maier, J., Pfundtner, G., Opila, E. J., Tuller, H. L., Wuensch, B. J.: in: Freyhart, Flükiger, Penckert (eds.) High Temperature Superconductors-Materials Aspects, DGM-Verlag, Oberursel, 1991, p. 853
[10] Smedskjaer, L. C., Routbort, J. L., Flandermeyer, B. K., Rothman, S. J., Legnini, D. G.: Phys. Rev. B, 1987, 36, 3903

Materials Science Forum Vol. 157-162 (1994) pp. 1167-1174

DEFORMATION BANDING AND ITS INFLUENCE ON HIGH SFE FCC ROLLING TEXTURE DEVELOPMENT

B.J. Duggan [1], C.S. Lee [2], and R.E. Smallman [2]

[1] Department of Mechanical Engineering, The University of Hong Kong, Hong Kong

[2] School of Metallurgy and Materials, University of Birmingham, UK

Keywords: Deformation Banding Stability, Modelling, FCC Textures

ABSTRACT

Deformation banding is a deformation mode which subdivides crystals into lath-like volumes in three dimensions. Its inclusion into modelling in a Taylor framework produces predictions closer to experiment that CS and B are seen to coexist without special manipulation. {110}<112> is formed progressively in the new model and {123}<634> should be considered as an important rolling texture component in both high and low SFE materials.

INTRODUCTION

Deformation banding is a process in which different regions of a deforming crystal gradually rotate to different orientations as deformation proceeds. Its importance to the deformation process has been repeatedly emphasized by Barrett et al. [1 - 3] and Chin [4]. However, only a limited amount of work has been done [5-8] mainly on single crystals or for recrystallisation studies. The purpose of the present work has been to investigate the influence of this deformation mode in polycrystalline materials.

EXPERIMENTAL RESULTS AND DISCUSSION

High purity copper (99.99%) of two different grain sizes, 40 and 3000 μm, with weak initial textures were cold rolled to various strains up to 92% reduction. The behaviour of the two materials has been confirmed to be typical by various standard techniques including X-ray diffraction, optical and transmission electron microscopy. The results shown in this paper were obtained by a crystallographic etching technique recently developed by Köhlhoff et al. [9]. Edge sections reveal that coarse grained copper has many layers of distinct orientation. The most remarkable feature is the unexpected thickness of the layered structure. The thickness of the layers each with different orientations ranges from about 1 to 50 μm and has an average value of 17.2 μm (measured by a linear intercept method at 4 different locations in two specimens containing over 100 layers). This value is much smaller than the average grain thickness after 85% reduction, which is about 3000 x 0.15 or 450 μm. It therefore follows that each grain deforms to give on the average about 26 (=450/17.2) layers. The banding is three dimensional.

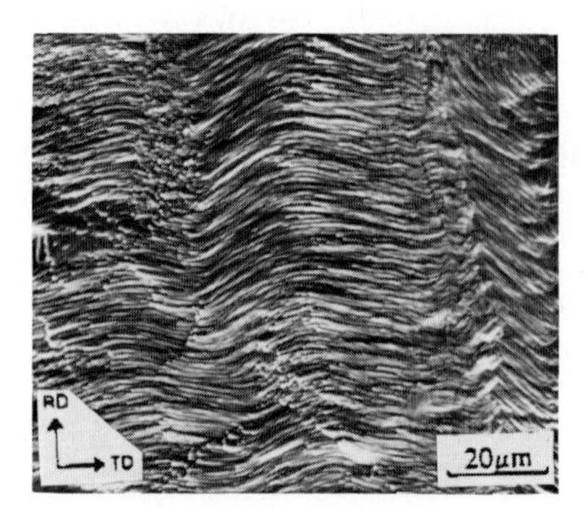

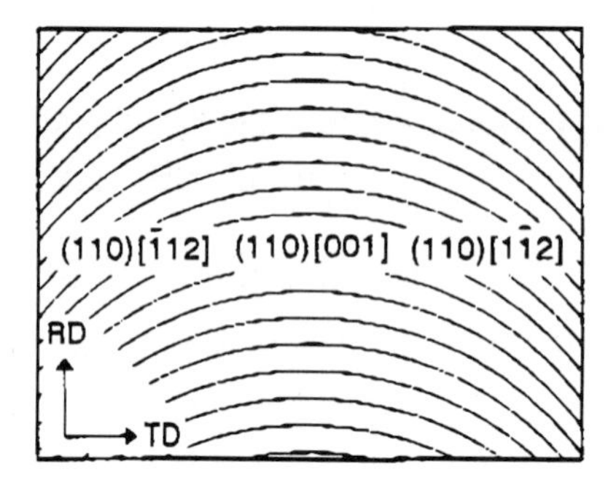

Fig. 1. a) SEM micrograph of crystallographically etched rolling plane section of coarse grained copper after 70% reduction. The grain splits into alternative bands of complementary B orientations (i.e. (110)[1$\bar{1}$2] and (110)[$\bar{1}$12]); b) Schematic diagram of the etching patterns of (110)[1$\bar{1}$2], (110)[$\bar{1}$12] and (110)[001] when observed in the rolling plane.

Figure 1a is a micrograph taken from the rolling plane section of the coarse grained copper after 75% reduction. It clearly shows that a grain is divided into bands of two symmetrically related {110}<112> orientations (i.e. (110)[1$\bar{1}$2] and (110)[$\bar{1}$12]) with (110)[001] in between, fig. 1b). The average band width in the rolling plane section is measured to be 120 μm. This is again much narrower than the original grain width of 3000 μm, suggesting that on the average each grain is divided into 25 (=3000/12) regions by banding with boundaries parallel to the longitudinal planes. It therefore appears that banding of at least two different geometries is operating simultaneously, namely banding with planes parallel to the rolling and longitudinal planes respectively. As a consequence, grains are three dimensionally subdivided and on the average, each grain will give over 600 (25 x 26, assuming that the two banding geometries are independent) bands in three dimensions.

Deformation banding was also observed in the fine grained copper but was not as pronounced as in the coarse grained copper. From the measurement of the longitudinal section, each grain gives an average of 2.4 layers. This value together with the corresponding one from the coarse grained copper agrees reasonably well with a recent prediction by Lee et al. [10] that the average number of bands (in one dimension) formed in a grain is proportional to the square root of the grain size. No measurements have been carried out on the rolling plane of the fine grained copper because satisfactory etching cannot be achieved due to the small layer thickness in the normal direction.

Because of these important microstructural changes, it is likely that deformation banding also significantly affects the development of the deformation texture. It has been believed for many years that copper deformed at room temperature does so mainly by homogeneous slip processes. Deformation models based on homogeneous slip have indeed predicted deformation textures similar to those observed experimentally [11-12]. However, these predicted textures have major differences compared with the experimentally determined textures. For example, the {110}<112> orientation is one of the major stable rolling texture components in copper but is predicted by neither the Full Constraint (FC) nor the Relaxed Constraint (RC) Taylor models [13]. Calculation with these models shows that {110}<112> is stable only if shear in the rolling plane is allowed. However, relaxing this shear constraint in the deformation not only produces a severe incompatible shape change with the surrounding, but also destroys other experimentally observed stable texture components, namely $\{123\}<634>$ and $\{112\}<111>$. An earlier analysis by Aernoudt and Stüwe [14] affords a possible solution in suggesting that if grains of $(110)[1\bar{1}\bar{2}]$ and $(110)[\bar{1}12]$ orientations are put side by side, the opposite shears will then be cancelled out. While they made no further suggestion on how the special microstructural arrangement forms, it is clear that this structure which stabilises the $\{110\}<112>$ orientation could be produced by deformation banding.

MODELLING OF DEFORMATION BANDING

It is clear from the above discussion that the deficiency of the current deformation models is mainly due to their failure to address the significance of deformation banding. In order to study its influence on the deformation processes, a texture simulation model (DB model) is developed by incorporating deformation banding into the Taylor model. This is done by assuming that all grains behave macroscopically as "Taylor grains", whether deformation banding occurs or not. In the original Taylor model, it is assumed that the strain in each grain equals the macroscopic external strain. If deformation banding occurs, this assumption becomes: *The average strain over different regions of a banded grain should equal the macroscopic external strain.* Thus, the microscopic deformation within each segment can be different from the microscopic one, so long as their average strain is compatible to themselves and their surroundings.

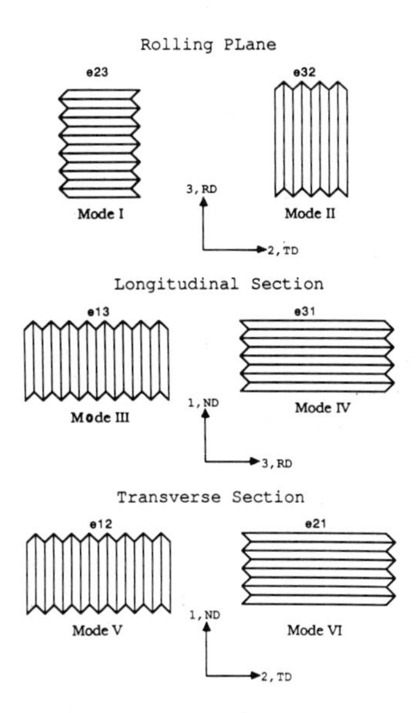

Fig. 2. A schematic diagram showing six geometrically possible modes of deformation banding in rolled grain.

From a purely geometrical point of view, there can be six different modes of deformation banding (fig. 2). However, only three banding modes II, IV and VI, which are consistent with the banding geometries observed experimentally, need further consideration.

With the three banding modes operating simultaneously, the three shear strain components (ε_{31}, ε_{21} and ε_{32}) of grains are no longer constrained to zero. They are partly relaxed. The term partly is used because the relaxation is not as free as it is assumed in the RC model. In the RC model, the shear strain ε_{31} and ε_{21} are relaxed without any further restriction such as the geometric constraint from the neighbouring grains. In real materials when deformation banding take place, these shear strains are in fact partly relaxed with constraints from the neighbouring bands. There are two types of constraints, the first requires that the average shear strains over the grain are zero; the second type requires continuity of displacement across band boundaries. Within these restrictions, the shear strains can be any value. Roughly speaking, the material's behaviour is in-between the description by the FC and the RC model.

However, modelling of the simultaneous operating of the three banding modes is still not possible due to the mathematical complexity. Instead, a one dimensional model, in which only one banding mode is considered, is used to estimate the possible effect of banding. The influence of the remaining two banding modes are approximated by some simplification processes. As the main effect of the banding is to allow a partial relaxation of the shear, it is natural to approximate its effect by either fully relaxed or fully constrainted shears. These correspond to the RC and the FC model respectively. Either of the approximation is acceptable as both the RC and the FC models predict textures similar to the experimentally obtained copper type texture. However, as the quality of the textures predicted by the RC model are generally considered better, the former approximation was hence used.

The remaining question is which of the two modes should be approximated. A sensible choice is to consider the two modes which by freely relaxing the corresponding shears produce least error in the predicted texture. From the results of the classical RC model, the two modes corresponding to the relaxation of ε_{31} and ε_{21} should be used. It is because freely relaxing ε_{32} not only produces a larger incompatible deformation, but also destroys the other stable texture component (S and C) in copper. Thus only mode II banding is modelled and mode IV and VI are approximated by freely relaxing the corresponding shear ε_{31} and ε_{21} as in the classical RC model.

With these approximations, the one dimensional model would suffer from some strain compatibility problems as in the RC model, e.g. ε_{31} and ε_{21} might not be continuous across the band boundaries. In the RC model, it is suggested that these incompatibilities do not cause severe error so long as the grain thickness is small compared to the other two dimensions. Concerning this, it should be pointed out that the individual band in a banded grain has in fact a more favourable geometry for relaxation than a non-banded grain in the RC model. As the compatibility problem does not cause significant error in the predictions of the RC model, it is unlikely that it will in the DB model. Perhaps the best justification of ignoring this problem in the present one dimensional deformation banding model is that the simulated textures are closer to the experimental textures.

A one dimensional mode based on mode II banding is then formulated with the following rules:

I. The average strain over a grain equals the macroscopic strain:

$$\varepsilon_{32}^{(1)}x^{(1)} + \varepsilon_{32}^{(2)}x^{(2)} = 0 \quad \text{with} \quad x^{(1)} + x^{(2)} = 1; \quad x = \text{volume fraction;}$$

$\varepsilon_{11}^{(j)} = e_{11}$; $\varepsilon_{22}^{(j)} = 0$; $\varepsilon_{33}^{(j)} = -e_{11}$; and $\varepsilon_{21}^{(j)}$ and $\varepsilon_{31}^{(j)}$ relaxed;

where the bracketed superscripts refer to the deformation band number.

II. Total internal work done, W_{DB}, is minimum:

i.e. $W_{DB} = x^{(1)} \sum \gamma_i^{(1)}\tau_i + x^{(2)} \sum \gamma_i^{(2)}\tau_i$ is minimized by selecting suitable slip systems which can accommodate the strain in I; where γ_i and τ_i are respectively the shear strain and CRSS associated with the ith slip system.

γ_i and crystal rotations can then be solved with the above rules.

A random array of 100 grains was used as the computer material and texture computed after 30, 54, 79 and 92% equivalent rolling reduction, the strain in each step being 0.05. It is unrealistically assumed that all grains deform by deformation banding. The results are shown in fig. 3. For comparative purpose, fig. 4 shows the measured texture of copper rolled 75% obtained by Hirsch and Lücke (5) using the Euler space projection, superimposed on the computed textures based on deformation banding.

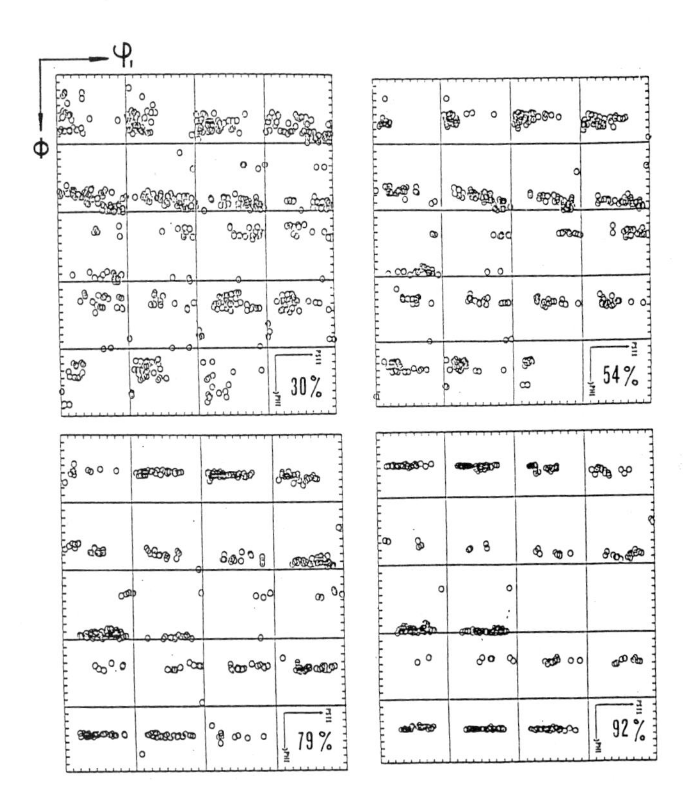

Fig. 3 Simulated ODFs based on the DB model for different rolling strains (a) 30%, (b) 54%, (c) 79% and (d) 92%.

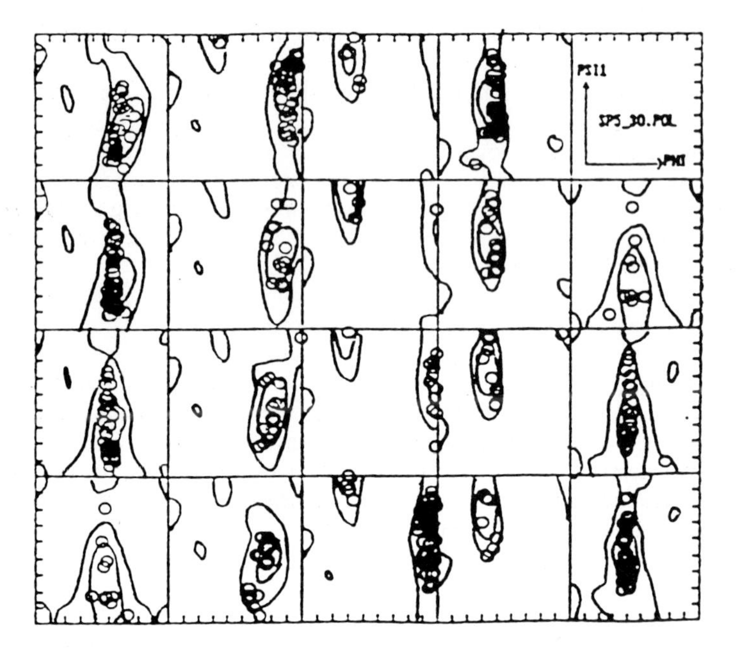

Fig. 4 DB simulation after 79% deformation superimposed on the experimental contours published by Hirsch and Lücke (13)

The orientation dependence of deformation banding is a natural consequence of the model in a subtle sense. For example the S orientation is forced to divide by deformation banding in the computer model but the orientations are rather similar and remain so until 87% reduction, figure 5a. Only after 92% does the orientation spread, figure 5b. The behaviour of {100}<001> and {111}<110> oriented grains are typical of another set of orientations, in which deformation banding produces divergent rotation paths upon further deformation, figure 5c and d. The practical consequences of these different kinds of behaviour on the modelling lies in an overestimate of the effect of deformation banding in orientations such as S. In reality, grains of S orientation would deform by conventional relaxed constraints processes rather than by deformation banding and thus would be stable. Deformation banding in the computer makes S unstable at high strains and hence underestimates its contribution to the final texture.

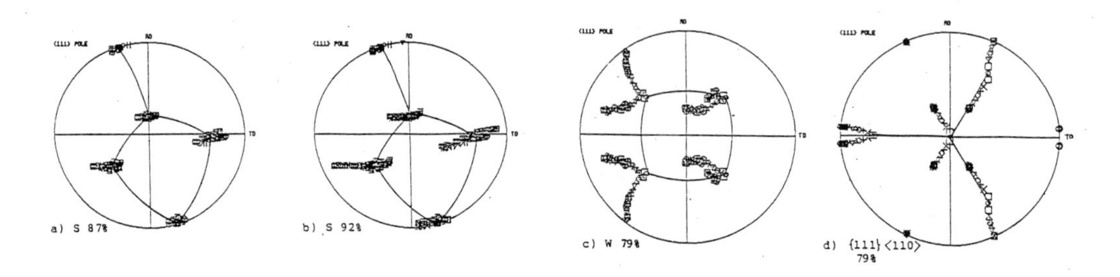

Fig. 5 Orientation dependence of DB by computer modelling (a) S, 87% (b) S, 92% (c) {100}<001> 79% (d) {111}<110> 79%

The predictions are fairly accurate and superior to the older models. (1) All the components are predicted which is not true of the FC and RC Taylor models. (2) The predicted components are closer in orientation to the experimental peaks, in particular the component {4,4,11}<11 11 8> is sparsely populated. In the other models special techniques have to be used to rid the simulation of this component. (3) The predicted texture has wider spread than other model textures and in this respect is closer to experiment. (4) All other models use more slip systems than are observed to operate. The deformation banding model requires fewer, and so is closer to experiment.

CONCLUSIONS

Deformation banding is shown to be an important deformation mode in copper. It is more pronounced in coarse-grained than fine-grained materials. In the coarse-grained material, different banding geometries form simultaneously, three dimensionally dividing grains into a large number of bands. The influence of deformation banding on texture development is estimated by incorporating deformation banding into the Taylor model. The simulated rolling textures from the deformation banding model are closer to the experimental textures than those from the Relaxed Constraint model. In particular, the {110}<112> texture component in rolled copper is shown to be formed by deformation banding using both microstructure observations and texture simulations.

ACKNOWLEDGEMENT

C.S. Lee thanks the Croucher Foundation for its support in the form of a Studentship and Fellowship.

REFERENCES

1. Barrett, C.S.: Trans. AIME, 1939, 135, 296.
2. Barrett, C.S., Levenson, L.H.: Trans. AIME., 1940, 137, 112.
3. Barrett, C.S.: The Structure of Metals (1st ed., McGraw-Hill, N.Y.), 1943, 381-419.
4. Chin G.Y.: Textures in Research and Practice, Berlin, 1969, 236.
5. Bellier, S.P. and Doherty, R.D.: Acta Metall., 1977, 25, 521.
6. Inokuti, Y. and Doherty, R.D.: Acta Metall., 1978, 26, 61.
7. Higashida, K., Takamura, J. and Narita, N.: Mat. Sci. Eng., 1986, 81, 239.
8. Akef, A. and Driver, J.H.: Mat. Sci. Eng., 1991, A132, 245.
9. Köhlhoff, G.D., Sun, X. and Lücke, K.: Proc. 8th Int. Conf. on Textures of Materials, 1988, 183.
10. Lee, C.S., Duggan, B.J. and Smallman, R.E.: Acta Metall. Mater. (accepted).
11. Kallend, J.S. and Davies, G.J.: Phil. Mag., 1972, 25, 471.
12. Dillamore, I.L. and Katoh, H.: Met. Sci., 1974, 8, 21.
13. Hirsch, J. and Lücke, K.: Acta Metall. 1988, 36, 2883.
14. Aernoudt, E., Stüwe, H.P. and Metallk, Z.: 1970, 61, 128.
15. Lee, C.S. and Duggan, B.J.: Acta Metall. Mater. (accepted).
16. Lee, C.S., Smallman, R.E, and Duggan, B.J. to be published.
17. Lee, C.S. and Duggan, B.J.: Scripta Metall. Mater., 1992, 28, 121.

Materials Science Forum Vol. 157-162 (1994) pp. 1175-1180

EVIDENCE FOR THE EXISTENCE OF A SPECIAL CLASS OF CRYSTALLOGRAPHIC MISORIENTATIONS

D.P. Field

Fabricating Technology Division, Alcoa Technical Center,
Alcoa Center, PA 15069-0001, USA

Keywords: Grain Boundary Specialness, CSL, Intergranular Damage

ABSTRACT

Grain boundary specialness is most often discussed in terms of the coincident site lattice (CSL) theory. Recent contributions by Adams and co-workers claim that a certain class of misorientations, defined in group theory as having a multiplicity greater than one, respond differently than random boundaries. The present work critically examines this claim and offers a physical description of the mathematical boundary on which these misorientations lie. Careful review of the available literature offers little support for CSL theory as a means of defining grain boundary specialness. Three-dimensional descriptions of the crystallographic structure of grain boundaries are essential in defining specialness. Data from damaged materials are presented which offer qualified support of Adams' claims. Detailed investigations are required to further restrict the specialness criteria and reduce the domain of boundaries classified as special.

INTRODUCTION

It is widely observed that several phenomena occur heterogeneously at grain boundaries in polycrystalline materials. These include intergranular void growth, fracture by fatigue or embrittlement, interfacial segregation, corrosion, and nucleation of new crystallites among others. Crystallite interfaces of similar plane normal orientation often behave differently under identical external conditions. These types of observations have led to the development of theories which explain "specialness" of interfaces according to geometric criteria based upon the crystallographic misorientation across the boundary. While the crystallographic structure of a grain boundary possesses five degrees of freedom, it has long been theorized that special types of interfaces can be identified by their misorientation alone. The most simple criterion is that of misorientation angle. Low angle boundaries are in low energy configurations, and hence, most stable. The most common description of special boundaries comes from the coincident site lattice (CSL) theory. CSL theory characterizes lattice misorientations by the fraction of lattice sites which lie at identical positions in two interpenetrating lattices [1-3]. The inverse of this fraction is the sigma (Σ) value of the boundary. Low Σ boundaries, which possess a high fraction of coincident sites, are claimed to be damage resistant and generally desirable boundaries. As Σ increases, the specialness of the boundaries decreases until high values which behave as "random, high-angle boundaries." A less restrictive theory is that of plane-matching (PM) [4,5], or the equivalent coincident axial direction (CAD) theory [6]. This describes misorientations based upon planes or axes aligning in adjoining crystallites. For cubic lattices, rotations about the <100>, <110>, and <111> type axes are considered to have special properties. As discussed by Warrington and Boon [6], these criteria include several of the misorientations which define various CSL boundaries. The PM and CAD theories describe lines through orientation space, while the CSL theory defines points in this space. DeChamps has used this type of comparison to conclude that CSL is the preferred theory because it is more restrictive [7]. This conclusion assumes that specialness is accurately described by the CSL theory.

A concise mathematical description of the asymmetric domain of crystallographic misorientation between two lattices possessing cubic symmetry has been defined in the space of Euler angles [8,9]. It has been shown that each of the CSL misorientations up to $\Sigma = 49$ ($\Sigma \neq 39b$) can be mapped onto the boundary of the asymmetric domain making CSL interfaces a subset of the set of special boundaries defined by Zhao and Adams [9]. In addition, the special PM and CAD boundaries lie on the edges of this asymmetric domain. It was first postulated in a 1990 paper that grain boundaries whose crystallographic misorientations lie upon the boundary of the asymmetric domain are a special class of interfaces which preferentially damage [10]. This conclusion was supported in later studies [11,12]. These grain boundaries have a group multiplicity greater than one in misorientation space meaning that different, but equivalent, misorientations map to the same point in the space of Euler angles.

This paper compares experimental data describing the effects of three different grain boundary damage mechanisms; intergranular void growth during creep, grain boundary embrittlement, and fatigue crack growth. Crystallographic misorientations were measured in each instance and examined in the space of the asymmetric domain. These data are used to offer support for the theory that misorientations which lie on the boundary of the asymmetric domain are preferentially damaged.

DESCRIPTION OF AN ASYMMETRIC DOMAIN OF CUBIC-CUBIC MISORIENTATIONS IN EULER SPACE

The parameterization of misorientations between two lattices can be formulated by various descriptors. Common among these are Euler angles. This paper will focus upon Euler angles as defined by Bunge [13], and given in his notation as the ordered triplet $\Delta g = \Delta g(\varphi_1, \Phi, \varphi_2)$. The misorientation, Δg, is defined as $g_A = \Delta g \cdot g_B$, where g_A and g_B represent the orientations of adjoining crystallites A and B. There exist 24 elements of the point group O(432), for cubic lattices, which map the representation of an orientation to a physically indistinguishable position. For misorientations there are 24 squared, or 576 elements of the relevant point group. This can be described by

$$\Delta g_e = h_i \, \Delta g \, h_j \qquad (1)$$

where Δg_e is an equivalent (or physically indistinguishable) misorientation and h_i and h_j are elements of the point group O(432). As indicated in the previous section Zhao and Adams have given a description of an asymmetric domain in this space [8,9]. The interior of the domain contains the set of all possible misorientations such that each point represents a physically distinguishable misorientation, meaning that only the identity element satisfies equation (1). Misorientations for which more than the identity element survives to satisfy equation (1) lie on the surface of the asymmetric domain. These are said to have group multiplicity greater than one. Multiplicity in this sense defines the number of physically indistinguishable positions in misorientation space. A group multiplicity of two reduces by half the number of physically indistinguishable parameterizations of a misorientation. The maximum multiplicity possible is 48 for crystallites of identical orientation, and those with no particular symmetry possess a multiplicity of one.

Figure 1 schematically describes an asymmetric domain of cubic-cubic misorientations in Euler space as defined by Zhao and Adams [8,9]. Discussion of the derivation of this and other domains is given explicitly in reference [14] and will not be discussed here. The physical significance of the boundary of this domain has not previously been described. This can most easily be envisioned by using an axis and angle representation of the misorientation. All pairs of lattices have an axis in common around which a rotation angle can be defined to describe the misorientation between the two. This axis, described crystallographically, is that which is coincident in the two crystallites. Physical interpretation of the boundary shown in Figure 1 can be described in these terms as given in Table 1. Misorientations represented by rotations about:

- <100> axes start at C with ω, the angle of rotation, equal to zero and travel along the line CD until $\omega=\pi/2$, which maps back to C.
- <110> begin at C ($\omega=0$) and travel along the line CA until $\omega = \arccos(1/3)$, which is A. As ω increases about <110> the misorientation travels along the line AB until $\omega = \pi/2$. Past $\pi/2$, the misorientation backs along the same path until $\omega = \pi$ which is the identity at C again.
- <111> begin at D and travel along the line DA, $\pi/3$ radians to A. Past $\pi/3$, the misorientation backs along the same path to $\omega = 2\pi/3$, which is the identity, and the process is repeated.
- *<hhk>* with $|h|>|k|$, such as <332> lie on the planar surfaces defined by ABC or ABD.
- *<hhk>* with $|k|>|h|$, such as <322> lie on the curved surface defined by ACD.
- *<hkl>* as a general rule do not lie on the boundary of the asymmetric domain.

Axes such as <99 99 98> and <99 98 98>, which satisfy the final two conditions listed above, must lie near, and on either side of, the line AD which is <111>. Similar comparisons can be made for near <100> and near <110> axes. The final two conditions are dependent upon rotation angle. For small angles the results described are correct, but as the angle increases above a given value for each axis, the misorientation may be represented on a surface other than that indicated. However, it always lies on the boundary of the asymmetric domain.

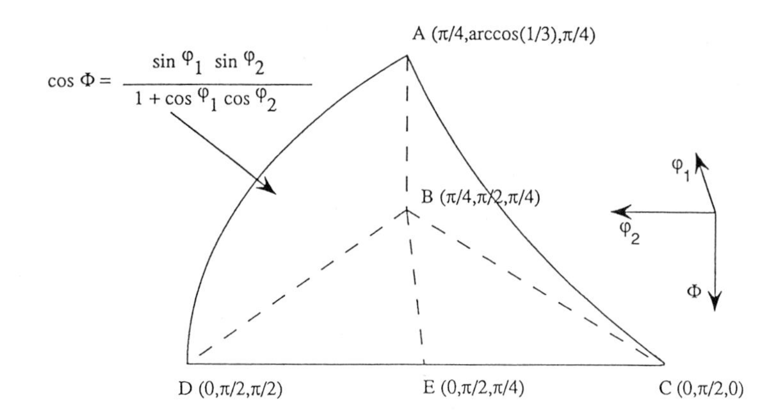

Figure 1 - Schematic of an asymmetric domain of cubic-cubic misorientations in Euler space.

Table 1 - Group multiplicities and physical significance of positions on the surface of the asymmetric domain.

Points	Multiplicity	Physical Significance
A	12	60 degree rotation about <111> (Σ3)
B	8	90 degree rotation about <110>
C	48	Identity
D	48	Identity
E	16	45 degree rotation about <100>
C and D are equivalent points		
Lines	**Multiplicity**	**Physical Significance**
AB	4	110 axes common and arccos(1/3)<ω<π/2
AC	4	110 axes common and ω<arccos(1/3)
AD	6	111 axes common
BE	2	110 aligned with 100 axis in adjoining grain
CD	8	100 axes common
BC and BD are equivalent lines		
Surfaces	**Multiplicity**	**Physical Significance**
ABC	2	*hhk* axes common (|*h*|>|*k*|)
ABD	2	*hhk* axes common (|*h*|>|*k*|)
ACD	2	*hhk* axes common (|*h*|<|*k*|)

RESULTS

Experimental data of damage on grain boundaries for three different materials under differing conditions have been examined. The effects of intergranular void growth during creep of commercial purity copper, grain boundary embrittlement of an Fe-1%Si alloy, and fatigue crack growth of an Al-Zn-Mg-Cu alloy were investigated. Crystallographic misorientations were measured in each instance using BKD analyses and examined in the space of the asymmetric domain.

Reference [12] discusses commercial purity copper (99.99%) which was subjected to creep in plane strain tension at a temperature of 540°C (0.6 T_m) for 18 hours (0.6 of the fracture life). Diffusional void growth occured heterogeneously on grain boundaries. Distributions of grain boundary geometry and crystallography were measured with information residing in an eight-dimensional space consisting of the lattice orientation of each grain and the orientation of the grain boundary plane. It was observed that damage occured preferentially on boundaries normal to the axis of maximum tensile stress. The space was reduced to two-dimensional projections by fixing the boundary normal orientation at the maximum value and examining only the misorientation information in the asymmetric domain. These are shown in Figure 2. While grain boundary normal information is contained in the data, it is not mentioned here as the other data sets examined do not presently contain this information. The distribution of damaged boundaries in this space divided by the total distribution gives the fraction of damaged boundaries of a given character. The peaks in the function generally lie on the boundary of the asymmetric domain. The data show that peaks in the damage function occur near positions of Σ11 (33.68°,79.53°,33.68°) and Σ19b (33.69°,71.59°,56.31°) in CSL theory while avoiding the Σ3 (45°,70.53°,45°) misorientation. The structure was heavily twinned and consisted of a high fraction of Σ3 boundaries. Damage was never observed on the coherent twin plane of this material, but was readily observed on off twin boundaries for the Σ3 misorientation. Figure 3 shows a micrograph of a twin boundary with no damage on the coherent twin interface, but severe damage on the off twin portion.

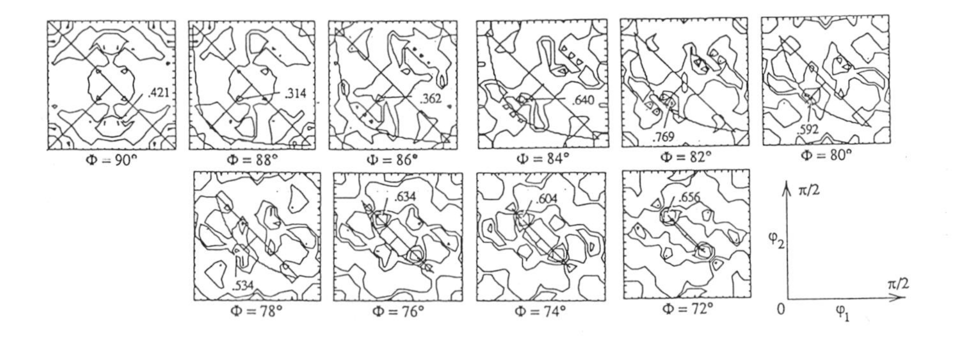

Figure 2 - Fraction of damaged boundary as a function of misorientation for Cu crept at 0.6 T_m.

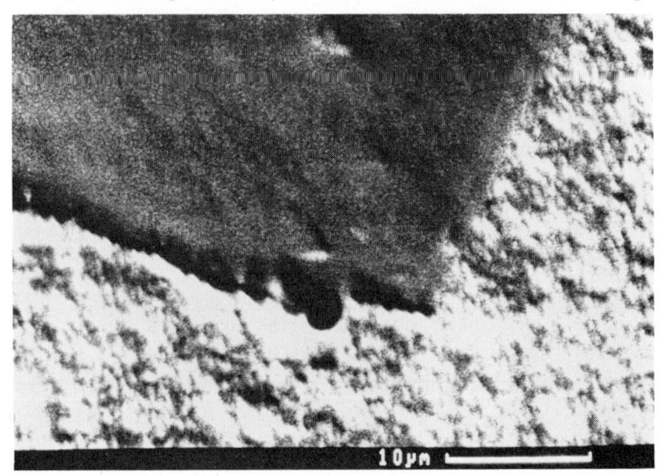

Figure 3 - Micrograph of a Σ3 twin boundary damaged only on the off-twin plane.

Data were obtained from H. Weiland and R. Burns of Alcoa Technical Center for a fatigue cracked Al-Zn-Mg-Cu alloy. A fatigue crack had propagated through several grains in the structure. A polished section through the crack was prepared and orientations were measured along lines normal to it in the plane. This resulted in data describing the misorientation distribution of the material as a whole, in addition to that for boundaries through which the crack passed. Figures 4.a and 4.b show positions of the total and cracked misorientations respectively. The limited data available show that the crack preferentially avoided low angle boundaries and generally chose misorientations of multiplicity greater than one. Out of 21 cracked boundaries measured, 18 were near the boundary of the asymmetric domain. Eight of these satisfied Brandon's criterion [15] for CSL classification; three as $\Sigma 3$ boundaries, one as $\Sigma 5$ (0°,90°,36.86°), and the remainder as higher Σ values.

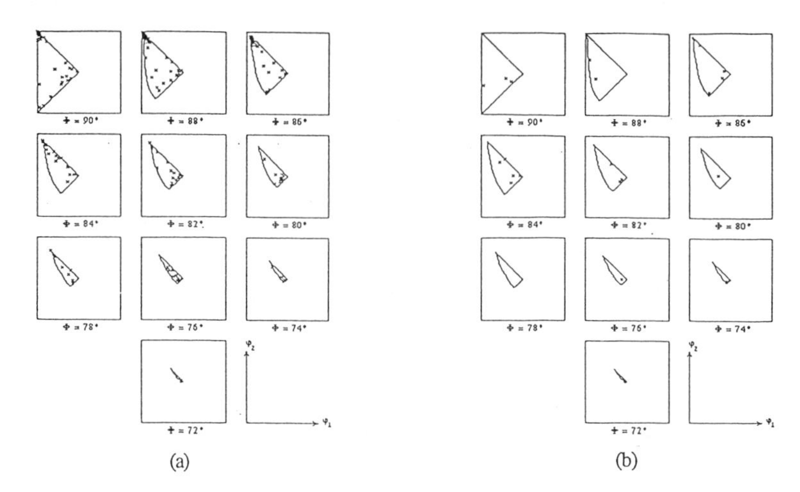

Figure 4 - Positions of a) the total, and b) the fatigue cracked misorientations in an Al-Zn-Mg-Cu alloy.

A recent investigation of an Fe - 1%Si alloy including a concentration of 180 ppm P to embrittle the grain boundaries before being lightly deformed at liquid nitrogen temperatures was performed by S. Wardle and J. Hack of Yale University [16]. This resulted in selected grain boundary cracking. Only the boundaries which were most prone to damage under these conditions were assumed to have separated during deformation. Figure 5 shows the distribution of cracked boundaries in misorientation space. Peaks in this function occur again on the boundary of the asymmetric domain, some of which satisfy Brandon's criterion. CSL boundaries which are preferentially damaged in this data set include $\Sigma 7$ (26.56°,73.4°,63.44°) and $\Sigma 19b$.

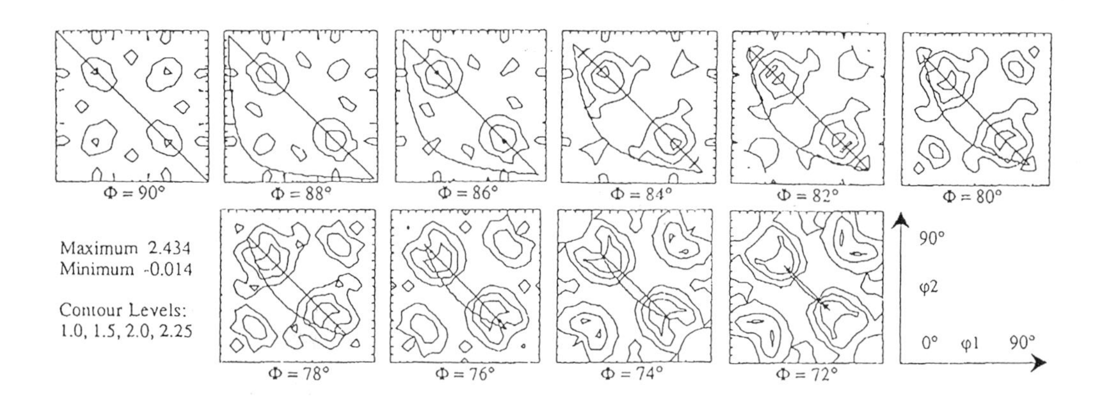

Figure 5 - The distribution of cracked boundaries in an Fe + 1%Si alloy embrittled by 180 ppm P.

DISCUSSION

The three sets of data which were presented in the preceding section contain a single common trend. Namely, damage preferentially occurs on boundaries with misorientations of group multiplicity greater than one. No evidence exists to support CSL theory without consideration of grain boundary plane orientation. Recent studies which include plane normal orientation in conjunction with CSL theory as a criteria for specialness identify Σ3 as a special boundary which preferentially avoids damage, but offer little evidence to further support the theory [17,18]. This contrasts findings of various researchers who strongly support the theory that CSL boundaries are damage resistant and offer other desirable properties contrasted with random, high-angle boundaries [cf. 19]. Very few investigations of grain boundary specialness in polycrystals have included plane normal orientation in the analyses. Insufficient data exist at this point to conclude that low sigma boundaries, in general, do not possess desirable properties, but data do exist which demand that the grain boundary plane is an important part of the specialness criteria.

Grain boundaries which have misorientations of group multiplicity greater than one are shown in the three investigations described herein to damage preferentially. This includes some CSL and some non-CSL boundaries. Certainly, boundary plane orientation, which received little discussion in this paper, is of great importance in determining specialness. It is also obvious that the condition of multiplicity (in conjunction with plane normal orientation) is insufficient in distinguishing between interfaces with desirable properties and those without. Certain boundaries with high group multiplicity are damage resistant if the boundary plane is in the preferential position (for example, Σ1 and Σ3). This amounts to support of an expanded CAD theory which includes all misorientations defined by rotations about <*hhk*> type axes. One final consideration is that boundaries, in general, tend to prefer symmetric positions. Evidence exists that damage occurs preferentially on these boundaries. However, if damage occured randomly in a structure with only symmetric boundaries, peaks in the damage function would appear on the boundary of the asymmetric domain. It has been observed in each of the investigations described above that the distribution of damaged boundaries differs significantly from distributions of all boundaries. This indicates that peaks in the damage function properly identify boundaries which are preferentially damaged.

CONCLUSIONS

The identification of grain boundary structures which promote or deter various mechanical or chemical processes at interfaces may lead to the development of polycrystalline materials with superior properties. Analyses of copper, aluminum and ferrous alloys by electron back-scatter diffraction method in the SEM have shown that some grain boundaries possess special properties. Intergranular void growth, fracture by embrittlement, and fatigue cracking have been shown to be dependent upon grain boundary structure. CSL theory appears generally to offer inaccurate criteria for grain boundary specialness (except, perhaps, for the coherent Σ3 twin) while the less restrictive criteria of high group multiplicity is more accurate, but insufficient. Three-dimensional grain morphology information including boundary plane orientation is necessary in determining the character of boundaries with special properties.

REFERENCES

1. Bollmann, W.: Crystal Defects and Crystalline Interfaces, Springer, New York, 1970.
2. Warrington, D.H. and Bufalini, P.: Scripta Metall., 1971, 5, 771.
3. Grimmer, H., Bollmann, W., and Warrington, D.H.: Acta Cryst., 1974, A30, 197.
4. Pumphrey, P.H.: Scripta Metall., 1972, 6, 107.
5. Pumphrey, P.H.: Scripta Metall., 1973, 7, 893.
6. Warrington, D.H. and Boon, M.: Acta Metall, 1975, 23, 599.
7. Dechamps, M., Yazidi, S., and Barbier, F.: Science of Ceramics 14, ed. D. Taylor, The Institute of Ceramics, Shelton, Stoke-on-Trent, UK, 1988, 569.
8. Zhao, J. and Adams, B.L.: Acta Cryst., 1988, A44, 326.
9. Adams, B.L., Zhao, J., and Grimmer, H.: Acta Cryst., 1990, A46, 620.
10. Zhao, J., Adams, B.L., and O'Hara, D.: Acta metall. mater.,1990, 38, 953.
11. Field, D.P. and Adams, B.L.: Acta metall. mater., 1992, 40, 1145.
12. Field, D.P. and Adams, B.L.: Metall. Trans., 1992, 23A, 2515.
13. Bunge, H.J.: Texture Analysis in Materials Science, 1982, Butterworths, London.
14. Zhao, J.: Analysis of Misorientation in Cubic Polycrystals, 1988, PhD. Dissertation, Brigham Young University.
15. Brandon, D.G.: Acta metall., 1966, 14, 1479.
16. Wardle, S. and Hack, J.: Personal Communication, 1993.
17. Liu, W., Bayerlein, M., Mughrabi, H., Day, A., and Quested, P.N.: Acta metall. mater., 1992, 40, 1763.
18. Lin, H. and Pope, D.P.: Acta metall. mater., 1993, 41, 553.
19. Watanabe, T.: Matls. Sci. Forum, 1989, 46, 25.

Materials Science Forum Vol. 157-162 (1994) pp. 1181-1188

THE DEPENDENCE OF DISLOCATION DENSITY AND CELL SIZE ON CRYSTALLOGRAPHIC ORIENTATION IN ALUMINUM

D.P. Field and H. Weiland

Alcoa Technical Center, Alcoa Center, PA 15069-0001, USA

Keywords: Dislocation Structures, Aluminum, Constitutive Modeling

ABSTRACT

The functional dependence of dislocation density and cell size on crystallographic orientation in commercial purity aluminum is examined by direct observation of dislocation structures in the TEM. As a crystalline material plastically deforms and subsequently recovers, dislocations generated during deformation seek positions of low energy. This results in the formation of various dislocation structures and densities. It is observed that these structures are dependent upon crystallographic orientation. A framework for incorporating these measures into constitutive models is discussed. Wide scatter in the data may indicate that higher order statistics should be included in the analysis to describe the dependence of dislocation structures upon neighboring orientations and other features of the microstructure.

INTRODUCTION

Modern state variable material models are typically based upon observed phenomenology of a material. From experimental data, parameters are fit to selected equational forms representing constitutive behavior and evolution of structural variables. These state variables, while often founded on principles of sound phenomenology, do not directly represent any physical features of the microstructure. Rather, they include the effects of a large number of features and mechanisms which control deformation behavior. Because of this the equations representing the behavior can have no foundation in physical principles, but are an attempt to mathematically represent the observed response of the material. This type of modeling is generally successful in representing large sets of experimental data and in interpolating between data sets. Extrapolation to deformation conditions which are not experimentally investigated is less successful. The utility of such models in predicting material behavior and optimizing processing parameters is limited to the range of conditions spanned by the experimental matrix, and it is unrealistic to attempt using such a model for alloy design. Using the information obtained from the state variables for predictions of subsequent recovery and recrystallization (or other metallurgical processes dependent upon thermomechanical history) is difficult as the values do not relate to physical entities and are not experimentally measurable.

In some deformation ranges and for some materials, the evolution of dislocation structures may control deformation behavior. These structures are directly observable using transmission electron microscopy (TEM). It is apparent that these structures evolve with deformation and certain features may serve as state variables. This would enable the development of a constitutive model based on the evolution of physically measurable entities. A qualitative description of the evolution of dislocation structures has been given by various authors [1-3]. Dislocation density, cell size, and the magnitude of crystallographic misorientation between neighboring cells are common parameters evaluated quantitatively. While several attempts have been made at describing microstructural evolution with strain, and a few widely accepted empirical equations have been developed, there has never been conclusive evidence of dominant microstructural features controlling deformation behavior. The ultimate goal of this research is to identify those features which control deformation of commercial purity aluminum in a given domain of interest, namely that of small strain deformation, and use these in modeling the material behavior.

As a crystalline material plastically deforms and subsequently recovers, dislocations generated during deformation seek positions of low energy. This results in the formation of various dislocation structures and densities.

Equations relating these parameters to flow stress have been derived from a wide range of experiments. Common among these are

$$\sigma - \sigma_o = \alpha_1 \mu b \rho_t^{1/2} \text{ and } \sigma - \sigma_o = \alpha_2 \mu \left(\frac{b}{\lambda} \right)^m \quad (1)$$

where ρ, λ, μ, and b are the dislocation density, average cell size, shear modulus, and Burgers vector respectively, and m is a constant. It is evident in polycrystals (and even single crystals) that dislocations are not homogeneously distributed throughout the structure, especially for small imposed strains. Grains oriented favorably for deformation suffer larger strains than those with high Taylor factors, oriented so as to geometrically oppose deformation. Preliminary studies indicate that this is reflected in the observed dislocation structures. Figure 1 shows the structures of two grains from the same specimen compressed to two percent strain at 250°C. One grain was oriented such that a <111> axis was parallel to the axis of compression, and the second crystallite was oriented with a <100> axis in this position. The heterogeneity of dislocation motion requires that the dislocation density and cell size in each crystallite be considered individually, and subsequently homogenized for relation to macroscopic behavior.

Equations (1) can be modified to predict deformation behavior of polycrystals by including a description of the distribution of crystallite orientations in the material,

$$\sigma - \sigma_o = \alpha_1(\dot{\varepsilon}, T)\mu(T) b \left[\frac{1}{8\pi^2} \oint_{g \in SO(3)} f(g) \rho_t(g) dg \right]^{1/2}$$

and,

$$\sigma - \sigma_o = \alpha_2(\dot{\varepsilon}, T)\mu(T) b^m \left[\frac{1}{8\pi^2} \oint_{g \in SO(3)} \left(\frac{f(g)}{\lambda(g)} \right) dg \right]^m \quad (2)$$

$f(g)$ is the normalized crystallite orientation distribution function (ODF) which serves as a weighting factor for the integration of $\rho(g)$ and $\lambda(g)$, respectively the dislocation density and cell size (or mean free path for dislocation motion) as functions of orientation, g. The integrations are performed over the special orthogonal group of three-dimensional rotations, $SO(3)$, and normalized by the factor immediately preceding the integral. α_1 and α_2 are assumed to be functions of strain rate and temperature and will not be further discussed in this treatment. Orientational dependence of μ is neglected, but could be included in the integrand without much further complication. This would not appreciably affect the result for aluminum alloys, but may be significant for some materials.

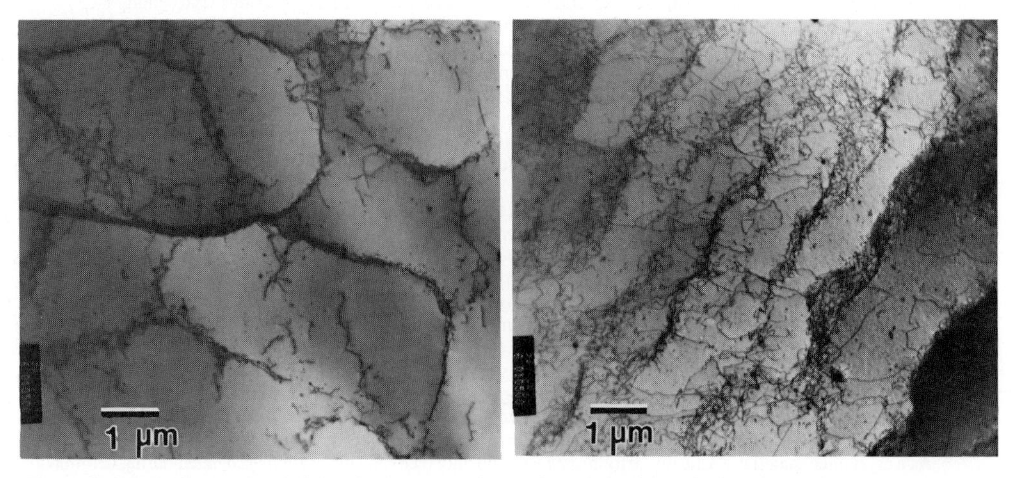

Figure 1 - TEM micrographs of dislocation structures formed in commercial purity aluminum after two percent uniaxial compressive strain at 250°C showing orientation dependence of the structures. The crystallite on the left is oriented with a <100> axis aligned with the deformation axis while that on the right is a <111> type orientation.

As a first step in formulating constitutive relations such as those given in Equations (2), the orientational dependence of dislocation density and mean free path must be established. A combination of experimental observations and theoretical developments are necessary in developing the proper relations. Various theories exist on the evolution of dislocation structures. These have recently been reviewed by Kubin [4]. In addition, it is rapidly becoming possible to numerically track the evolution of a large number of dislocations in a three dimensional lattice. These types of developments will be crucial in formulating the necessary evolution equations for developing state variable based constitutive models. The present treatise will focus solely on the experimentally observed dislocation structures and not attempt to develop the necessary equations.

EXPERIMENTAL DETAILS

Commercial purity aluminum having an equiaxed grain size on the order of 100 μm was deformed in uniaxial compression at temperatures ranging from room temperature to 450°C and a strain rate of $2.0 \times 10^{-4}\ s^{-1}$. The maximum imposed strain was 0.02. All specimens were observed using channeling contrast in the SEM to investigate subgrain development. There were no subgrains observed in any of the specimens, and all further analyses were constrained to TEM observations. After deformation, and just prior to analysis, TEM foils were prepared whose normal was oriented parallel with the axis of compression. Foils were prepared for specimens deformed at room temperature, 250°C, and 450°C. Undeformed specimens, and those with 0.5 and 2.0 percent strain were investigated. Since an orientation dependence of dislocation structure formation was presupposed, grains of similar crystallographic orientation were investigated for each specimen. This was accomplished by preparing several foils from each specimen and measuring several orientations from each foil. Dislocation structures were quantified for those locations exhibiting the proper orientations. The orientations concentrated upon were those having <100>, <111>, and <922> axes aligned with the axis of compression. These correspond to Taylor factors of 2.45, 3.67, and 2.56 respectively.

Foils were oriented in the TEM so as to remain as normal to the electron beam as possible. Four poles which satisfied the diffraction condition were chosen for imaging the dislocations. These were observed to yield essentially identical images of the dislocation structure. Whichever of these was closest to the desired position for the foil to remain normal to the beam was chosen to view the dislocation structure. Foil thicknesses were measured and were in the range of 150 to 300 nm.

QUANTIFICATION OF DISLOCATION DENSITY AND CELL SIZE

Images of dislocation structures seen in the TEM are two-dimensional projections of a three-dimensional space. Dislocation lines running through the foil, and satisfying the proper diffraction criteria, are projected onto a surface for imaging. A proper measure of dislocation density may be obtained from these sections assuming the foil thickness is accurately known and stereographic projections are taken to obtain three-dimensional information, or appropriate stereological procedures are followed [5]. For this analysis, the two-dimensional image is viewed as a true representation of the dislocation structure on a plane section and the principle of Delesse (area fraction equivalent to volume fraction) is employed in extending the analysis to three dimensions. Another critical assumption is that all dislocations of consequence are viewed in the image.

The algorithm developed to analyze the TEM images defines dislocations as all pixels in the image which appear darker than an input threshold value between 0 and 255. The user can interactively change the threshold value to dictate the positions of the dislocations. A window of observation is also identified by the user to avoid portions of the image which may be unsuitable for automatic analysis. On screen display of the original image as well as the image with the areas observed shown in black and white pixels offers the user assurance that proper measurements are made. The simplicity of the algorithm necessitates the use of this feature to ensure that the dislocations are properly identified.

Line scans are made through the image as it is read into memory. The individual lengths between dislocations are identified and written into a file. If the dislocations are very close to each other they are considered to be part of a dense dislocation wall and are not considered separately in the analysis of lengths between dislocations. As the cell interiors are largely free from dislocations, the average length between dislocations is taken as a measure of the dislocation cell size or mean free path for dislocation motion, λ. This is the inverse of the stereological measure P_L, intersections of a dislocation per unit length of test line, with an intersection defined as either a single dislocation or a dislocation cluster. A measure of the total dislocation density is given by the area fraction of dislocation lines projected onto the plane surface given by the ratio of "dark" to overall number of pixels. Assuming a foil thickness on the order of 2000 Angstroms, a relative dislocation density of 0.1 as given by this measure corresponds to a density of approximately $10^{11}\ m^{-2}$. Continuing with our assumption of a simple two-dimensional analysis, the line length per unit area of test section, L_A, is given by the well-known relation

$$L_A = \pi/2\ P_L \tag{3}$$

Since the area fraction of dislocations, A_A, is known from a simple ratio, and the length per area is known from equation (3), the average thickness, h, of dislocation clusters (including dislocation cell walls, or bands) is given by

$$h = A_A/L_A \tag{4}$$

In addition to the dislocation structure information the crystallographic orientation of each grain analyzed is measured from Kikuchi patterns obtained in the TEM.

RESULTS OF ANALYSES

Over 60 grains from several deformed specimens were analyzed to obtain preliminary results. The lattice orientation of each grain analyzed was consistent with one of the three specified in the above section on experimental details. Namely, <100>, <111>, and <922> axes aligned with the axis of compression. These were observed to exhibit a wide range of dislocation cells, clusters and walls. In addition, several foils were averaged to obtain the initial dislocation density, cell size, and wall thickness. All data points shown in the following figures for the undeformed condition are the result of this average. Scatter in the data for the undeformed specimens was minimal, indicating that the initial dislocation density and cell size for the annealed material was homogeneous throughout the structure and independent of orientation.

The data was first examined as a function of deformation parameters without regarding lattice orientations. A typical set of results is seen in Figure 2 which shows the evolution of relative dislocation density with strain for a deformation temperature of 250°C. Since this is a two-dimensional analysis, no inferences of actual densities are obtained but the data is useful in defining trends. Figure 2 supports general knowledge that dislocation density increases with strain and shows that it can be measured even for high temperature deformations and very small strains. There is a wide scatter in the data of Figure 2, and the scatter increases with strain in this range. Figure 3 plots the mean free path for dislocation motion for the same set of foils examined in Figure 2. Again, a general trend is observed that mean free path decreases with increasing strain, but there is wide scatter in the data which increases with strain. Cell wall thickness increases with strain in this domain with similar scatter.

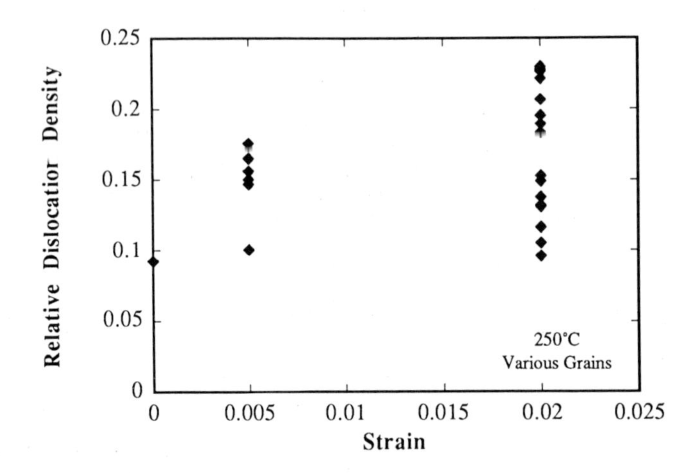

Figure 2 - Evolution of relative dislocation density with strain for a deformation temperature of 250°C.

Examining grains of similar orientation significantly reduces the amount of scatter in the data. Figure 4 shows the relative dislocation density of grains oriented such that a <111> axis is aligned with the compression direction. These grains have the highest Taylor factor of all those investigated, requiring the maximum amount of dislocation motion for unit strain. Comparison of data from <111> grains against others indicates that dislocation density increases more rapidly and mean free path decreases more rapidly in these grains than for those having a lower Taylor factor. Figure 5 contains the mean free path data from <922> type grains compared against that for <111> grains, showing the more rapid evolution of dislocation structure for <111> grains. Similar results were generally observed for all conditions investigated.

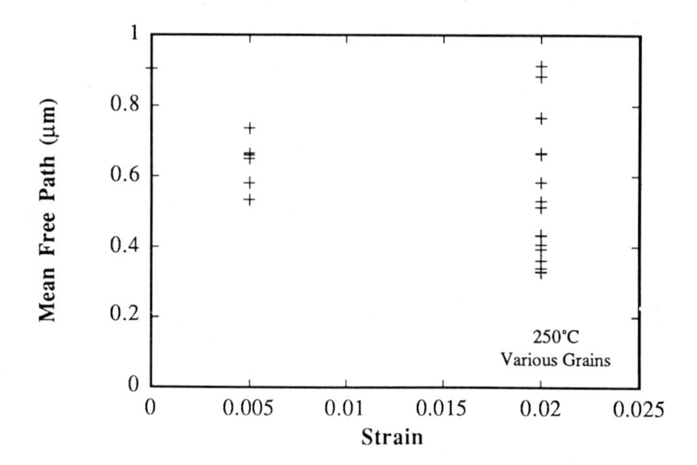

Figure 3 - Mean free path for dislocation motion measured for specimens deformed at 250°C.

The data was also examined for dependence upon deformation temperature. Figure 6 shows the mean free path as a function of temperature for <111> type grains. The trend seen here remains consistent throughout all data sets; mean free path for dislocation motion increases with increasing deformation temperature. Surprisingly, there is no apparent trend in the data for the temperature dependence of dislocation density.

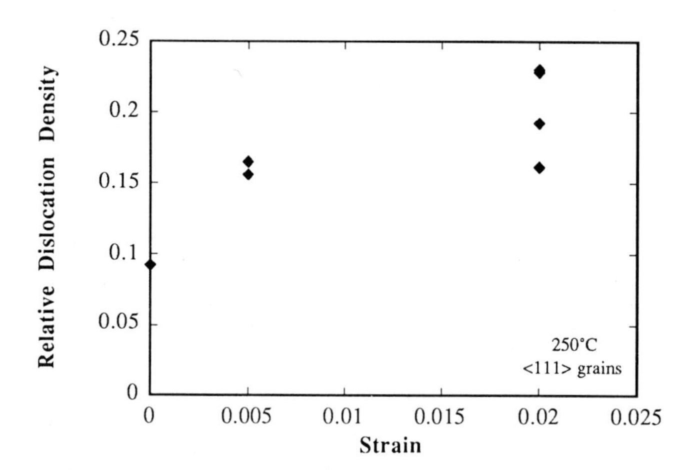

Figure 4 - Relative dislocation density of <111> type grains in specimens deformed at 250°C.

DISCUSSION

The ultimate goal of this investigation is to determine the feasibility of correlating dislocation structure evolution with macroscopic flow behavior of aluminum during small strain deformation. From the results presented in the previous section it is apparent that, in this deformation range, the dislocation structures are dependent upon lattice orientation. Wide scatter exists in the data showing dislocation density and mean free path without regard to lattice orientation. This is reduced significantly by examining like orientations independently, indicating the orientation dependence of these features. Even when orientation dependence is considered it can be reasonably argued that considerable scatter remains in the data. This is likely an indication of the effects of neighboring orientations on deformation behavior. Non-local effects probably play a significant role in the development of dislocation structures since these often form as a consequence of strain compatibility requirements between adjoining crystallites.

The experimental techniques for dislocation structure analysis employed in this investigation yield only rough approximations of the actual structure. The assumption that the TEM images are simple two-dimensional views of the dislocation structure rather than projections of a three-dimensional space onto plane sections introduces significant errors into the analysis. The information obtained from this type of analysis is only useful because a consistent procedure was followed. Anisotropy may exist in the structures which introduces errors if foils from sections of differing orientations are treated equally. This problem was avoided entirely in this analysis as anisotropy was not considered, and all foils were of similar normal orientation. Proper characterization of the dislocation structures may be performed by analyzing various sections and considering the intersections of the dislocation lines with the test section. A procedure for performing this analysis to accurately determine dislocation density whether the foil thicknesses are known or not is given in reference [5]. The technique involves analyzing sections cut at various angles to a reference direction and using stereological relations to back out the line length per unit volume. Other features of the dislocation structures can be determined using similar procedures [6].

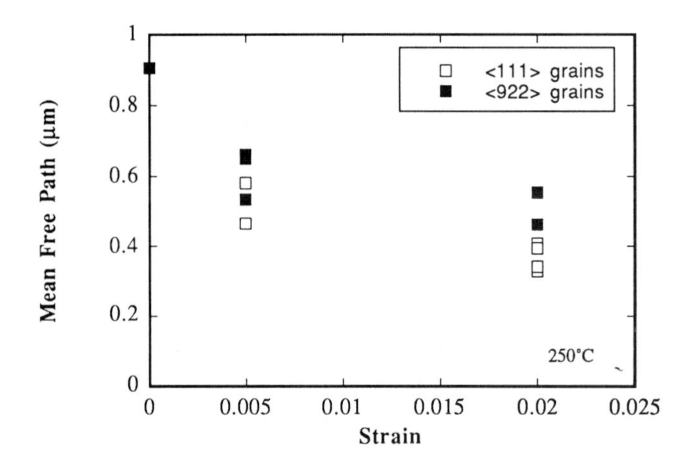

Figure 5 - Comparison of mean free path from <922> type grains against that for <111> grains.

To determine the magnitude of the effect neighboring orientations play on the development of dislocation structures in a polycrystal, investigations using single crystals should be performed using identical deformation conditions. Statistical analyses can then be performed to determine whether the scatter in the data is much greater in polycrystalline materials than for single crystal deformations. Assuming non-local effects are important, equations (2) can be modified to include the necessary higher order statistics:

$$\sigma - \sigma_o = \alpha_1(\dot{\varepsilon},T)\mu(T)b\left[\iiint_{g,r,g'} h(g'|g,\mathbf{r})\rho_t(g'|g,\mathbf{r})\,dg\,d\mathbf{r}\,dg'\right]^{1/2}$$

$$\sigma - \sigma_o = \alpha_2(\dot{\varepsilon},T)\mu(T)b^m\left[\iiint_{g,r,g'} \left(\frac{h(g'|g,\mathbf{r})}{\lambda(g'|g,\mathbf{r})}\right)dg\,d\mathbf{r}\,dg'\right]^m \tag{5}$$

where $h(g'|g,r)$ is the 2-point conditional probability density function of an orientation, g', lying at the head of a vector, $\boldsymbol{r}$, whose tail lies in a grain of orientation g.

In addition to higher order statistics, it may be necessary to include more detailed information of the dislocation structures for proper flow stress modeling. This might include characterization of the anisotropic character of the important microstructural features. Using functions such as this, which reside in many-dimension spaces, it becomes impossible to empirically derive the necessary evolution equations. Model development must be guided by theory and verified by experiment for potential success. Dislocation structure evolution theory has recently been reviewed by Kubin [4]. In addition, large scale numerical simulations of dislocation structure formation are rapidly becoming feasible.

CONCLUSIONS

While this investigation does not attempt to correlate the evolution of dislocation structures with flow stress of the material, it shows that their evolution with deformation can be quantified using simple techniques, even for very small strains. This implies that it is reasonable to use physically measurable features of the microstructure as state variables in small strain constitutive models as long as orientation dependence is specified. Dislocation density and mean free path for dislocation motion are functions of lattice orientation in lightly deformed aluminum, and probably functions of non-local characteristics of the microstructure as well.

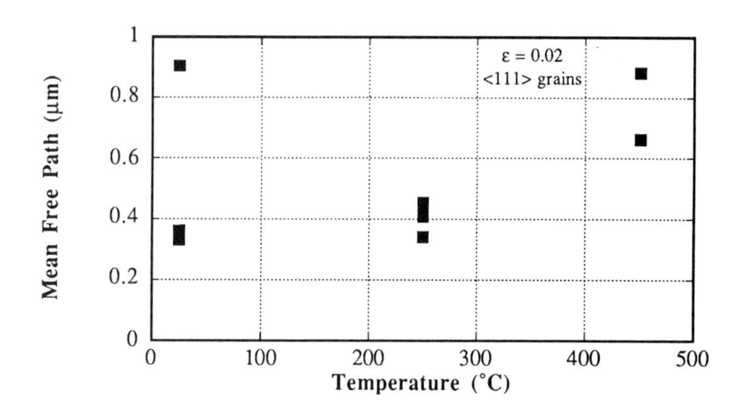

Figure 6 - Mean free path as a function of temperature for <111> type grains.

ACKNOWLEDGEMENTS

Helpful discussions of this project with P.T. Wang, V.M. Sample, and R.C. Becker, all of ATC, are gratefully acknowledged. The deformation experiments were performed by G. Petrikovic and C. Husack aided in the TEM work.

REFERENCES

1. Bay, B., Hansen, N., and Kuhlmann-Wilsdorf, D.: Mat. Sci. Eng., 1989, A113, 385.
2. Hughes, D.A.: Scripta Metall. Mater., 1992, 27, 969.
3. Thompson, A.W.: Metall. Trans., 1977, 8A, 833.
4. Kubin, L.P.: Materials Science and Technology, 1992, 6.
5. Gokhale, A.M.: Journal of Microscopy, 1991, 167, 1-8.
6. Adams, B.L. and Field, D.P.: Metall. Trans., 1992, 23A, 2501.

Materials Science Forum Vol. 157-162 (1994) pp. 1189-1196

MICROTEXTURE DEVELOPMENT DURING HIGH TEMPERATURE DEFORMATION OF NIMONIC 80A SINGLE CRYSTALS

J. Fischer-Bühner, D. Ponge and G. Gottstein

Institut für Metallkunde und Metallphysik, RWTH Aachen, D-52056 Aachen, Germany

Keywords: Subgrain Structure, Misorientation Distribution, Nucleation, Dynamic Recrystallization

ABSTRACT

The evolution of microtexture and misorientation distribution during high temperature deformation of Nimonic 80A single crystals was investigated by means of single grain orientation measurements. Tensile tests on <111>-oriented specimens were carried out at 1200°C and three different strain rates. The orientations of individual subgrains and dynamically recrystallized grains were measured by means of **E**lectron **B**ack **S**catter **D**iffraction (**EBSD**) in a scanning electron microscope. It was found that strong subgrain rotations only occured in highly stressed regions at inclusions. The maximum misorientation Θ_{max} between neighbouring subgrains partly exceeded 15° at these inhomogeneities, while Θ_{max} remained low ($\approx$6-7°) in homogeneously deformed regions. It is proposed that these instabilities of the subgrain structure lead to nucleation of dynamic recrystallization.

INTRODUCTION

In metals with low or medium high stacking fault energy, deformation at elevated temperature can lead to dynamic recrystallization (DRX), even in single crystals [1]. The use of single crystals turned out to be particularly suitable for investigations of fundamental features of DRX, mainly for the following reasons: the onset of DRX can easily be detected from the macroscopic stress-strain behaviour; dynamically recrystallized grains can unambiguously be distinguished from the deformed matrix, so that orientation relationships between the new grains and the matrix can be determined. Previous investigations on copper [2,4], silver [3], nickel [4] and Nimonic 80A [6,7] single crystals have shown, that macroscopically measurable parameters like the stress or strain at the onset of DRX cannot be unambiguously related with critical nucleation conditions. It was concluded that these parameter are not sufficiently sensitive to microstructural details [2]. Additionally it was found, that the subgrain size distribution does not provide a suitable criterion for the initiation of DRX [5].

In contrast to polycrystals, the formation and growth of new grains in deformed single crystals requires a supercritical misorientation to develop between subgrains in order to generate a mobile boundary. It was the aim of the present study to determine, if such instabilities of the subgrain structure form during high temperature deformation. The method of local orientation determination by means of Electron Back Scatter Diffraction (**EBSD**) [8,9] offers the possibility to determine the evolution of microtexture and misorientation distribution. As testing material the nickelbase superalloy Nimonic 80A (Cr 20, Al 2.5, Ti 2.5, C 0.02, in wt%; T_{sol}=1290°C) was chosen. It was found especially suitable for this study, because the dislocation or subgrain structure can be easily revealed by decoration during precipitation at 750°C [6,7].

EXPERIMENTAL

Cylindrical tensile specimens with <111>-orientation of the tensile axis and a gauge section of 6mm in length and 2mm in diameter were used in this study. The orientation of <111>, which causes multiple slip, was chosen in order to obtain results which are relevant also to polycrystalline material. Deformation was carried out in vacuum at 1200°C and three different strain rates $\dot{\varepsilon}_1 = 4 \cdot 10^{-5}/s$, $\dot{\varepsilon}_2 = 2 \cdot 10^{-4}/s$, $\dot{\varepsilon}_3 = 1 \cdot 10^{-3}/s$. Before testing the specimens were electropolished to remove surface damage introduced during machining and finally, solution treated by annealing for 1 hour at 1200°C . Therefore the material was single phase at testing temperature so that deformation was not affected by precipitates. However, metallographic observation revealed coarse non-metallic inclusions in almost all specimens, obviously originating from the manufacturing process, which indeed had great influence on the microstructure and microtexture development.
Immediately after terminating deformation, quenching by cold helium gas was used to freeze the microstructure at deformation temperature. Subsequently, the deformed specimens were annealed for 8 hours at 750°C, which is far below the recrystallization temperature due to the high solute content. During annealing $Cr_{23}C_6$ precipitated preferentially at dislocations and grain boundaries. This way, the dislocation or subgrain structure was revealed and conveniently investigated by means of scanning electron microscopy and even light microscopy. For this purpose the deformed specimens were sectioned parallel to the tensile axis, metallographically prepared and finally etched for 20sec in a reagent of 5g $CuSO_4$ + 20ml H_2O + 20ml HCl.
The single grain orientation measurements were carried out by means of **EBSD** in a scanning electron microscope. For this, the incident electron beam is focused onto a certain sample location of interest , in this case a subgrain or a dynamically recrystallized grain. The generated diffraction pattern is intercepted on a phosphorous screen and displayed on a monitor using a low-light TV-camera. Computer-aided analysis of the pattern finally yields the orientation of the specific location within an accuracy of 1°. A detailed description of this technique and experimental device is given elsewhere [9].

In order to obtain relevant results at least 200 orientations per specimen were determined in characteristic regions. For illustration pole figures were calculated from the single grain orientation data of each specimen. Furthermore, the misorientations between neighbouring subgrains were calculated and evaluated with respect to the evolution of misorientation distribution with increasing deformation degree. Of most importance was the maximum misorientation between next neighbour subgrains, especially close to the onset of dynamic recrystallization. Also, the orientations of dynamically recrystallized grains were investigated.

RESULTS AND DISCUSSION

In figure 1 typical hardening curves obtained at the three different strain rates are shown. The shear strain to the onset of DRX grows with increasing strain rate and the sudden drop in flow stress due to DRX is more pronounced with decreasing strain rate. The deformation curve at the highest strain rate $\dot{\varepsilon}_3 = 1\cdot10^{-3}$/s unexpectedly appears to be more flat than the curves obtained at the lower strain rates. But it was found by means of diameter measurements that deformation with $\dot{\varepsilon}_3$ led to slightly inhomogeneous deformation over the gauge length which is consistent with previous experiments on Nimonic 80A single crystals at higher strain rates and lower temperatures [10]. However, deformation at $\dot{\varepsilon}_1$ and $\dot{\varepsilon}_2$ was homogeneous over the gauge length. Recalculation of the shear stress at the onset of DRX in the specimen deformed with $\dot{\varepsilon}_3$ (based on the minimum diameter measured in the corresponding specimen) resulted in τ_{max}=15 MPa instead of 13.4 MPa so that the true curve is considerably steeper.

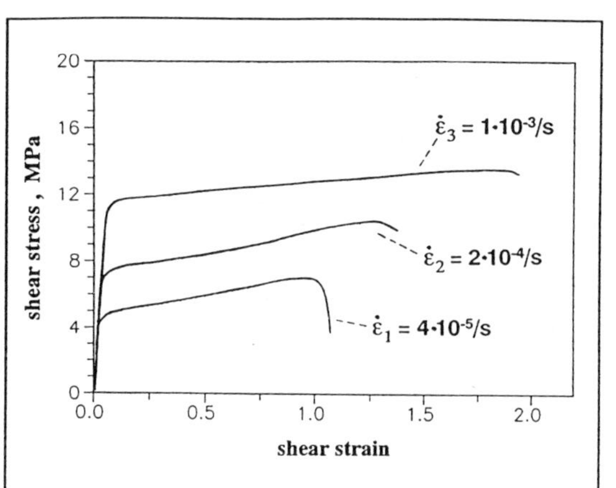

Figure 1: Hardening curves of <111>-oriented Nimonic 80A single crystals at 1200°C and different strain rates.

Despite the high deformation temperature in no case a stationary deformation state, characterised by constant flow stress with increasing strain, was reached before the onset of DRX . This is in contrast to previous experiments on copper-, nickel- and silver single crystals [2,3,4], obviously due to the high content of alloying elements.

Figure 2 shows the microstructure evolution at constant strain rate ($\dot{\varepsilon}_2$). With increasing shear strain the subgrain structure, consisting of equiaxed subgrains, becomes more pronounced and homogeneous (figures 2a-c). No kink bands were observed. But inhomogeneities of the subgrain structure can be seen at the inclusions, which appear as coarse dark spots in the micrographs. New grains only were detected in specimen, which were deformed up to the drop in flow stress. Figure 2d shows new grains obviously originating from dynamic recrystallization because of the pronounced subgrain structure within the grains. For comparison figure 2e shows new grains in a specimen, which was not quenched after the deformation test. Because of the lack of a deformation structure within the grains they undoubtedly formed after deformation, i.e. they result from static recrystallization.

The microtexture development at constant strain rate ($\dot{\varepsilon}_2$) is shown in figure 3. In figures 3a-c the dashed lines represent the initial orientation of the single crystals. The amount of subgrain rotation during deformation, which is illustrated by the dotted lines, grows with increasing shear strain. Most strongly rotated subgrains were detected in the inhomogeneities at inclusions. Figure 3d demonstrates that the orientations of the dynamically recrystallized grains belong to a single twin chain. It extends over 4 generations and its origin is supposed to be in the outermost part of the subgrain spread because the orientation marked with dashed lines (figure 3d) nearly is in twin relation to these subgrain orientations. However, no dynamically recrystallized grain with an orientation of such a strongly rotated subgrain was detected in the investigated part of this sample.
The evolution of misorientation distribution is presented in figure 4. First of all only subgrain orientations measured in homogeneously deformed regions were considered. Figure 4a shows that the maximum misorientation between next neighbour subgrains approaches a plateau value of Θ_{max}=6° with increasing degree of deformation at constant strain rate ($\dot{\varepsilon}_2$). Therefore, no generation of mobile (high angle) grain boundaries between subgrains was found in these regions, and

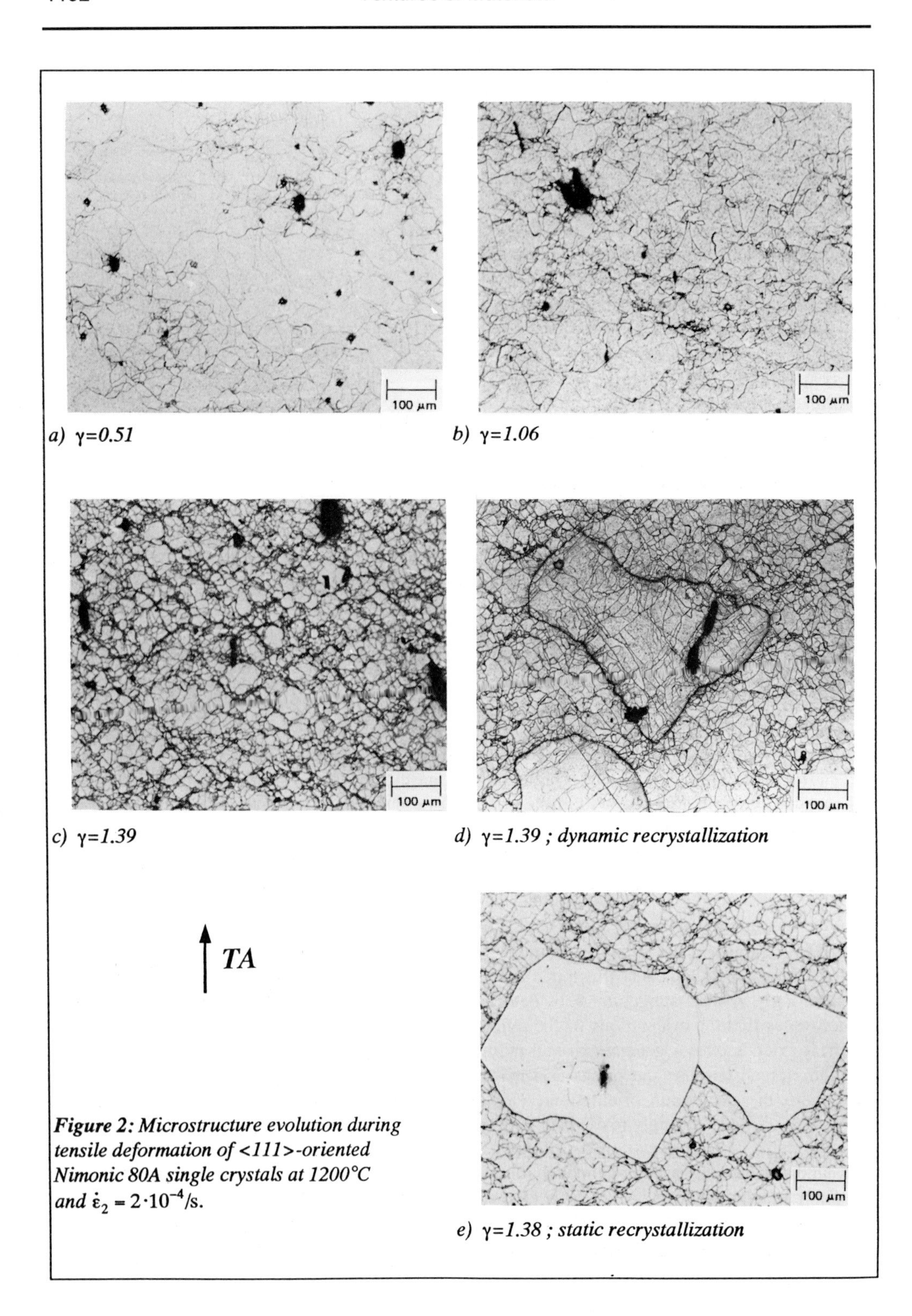

a) γ=*0.51*

b) γ=*1.06*

c) γ=*1.39*

d) γ=*1.39 ; dynamic recrystallization*

e) γ=*1.38 ; static recrystallization*

Figure 2: *Microstructure evolution during tensile deformation of <111>-oriented Nimonic 80A single crystals at 1200°C and* $\dot{\varepsilon}_2 = 2 \cdot 10^{-4}$/s.

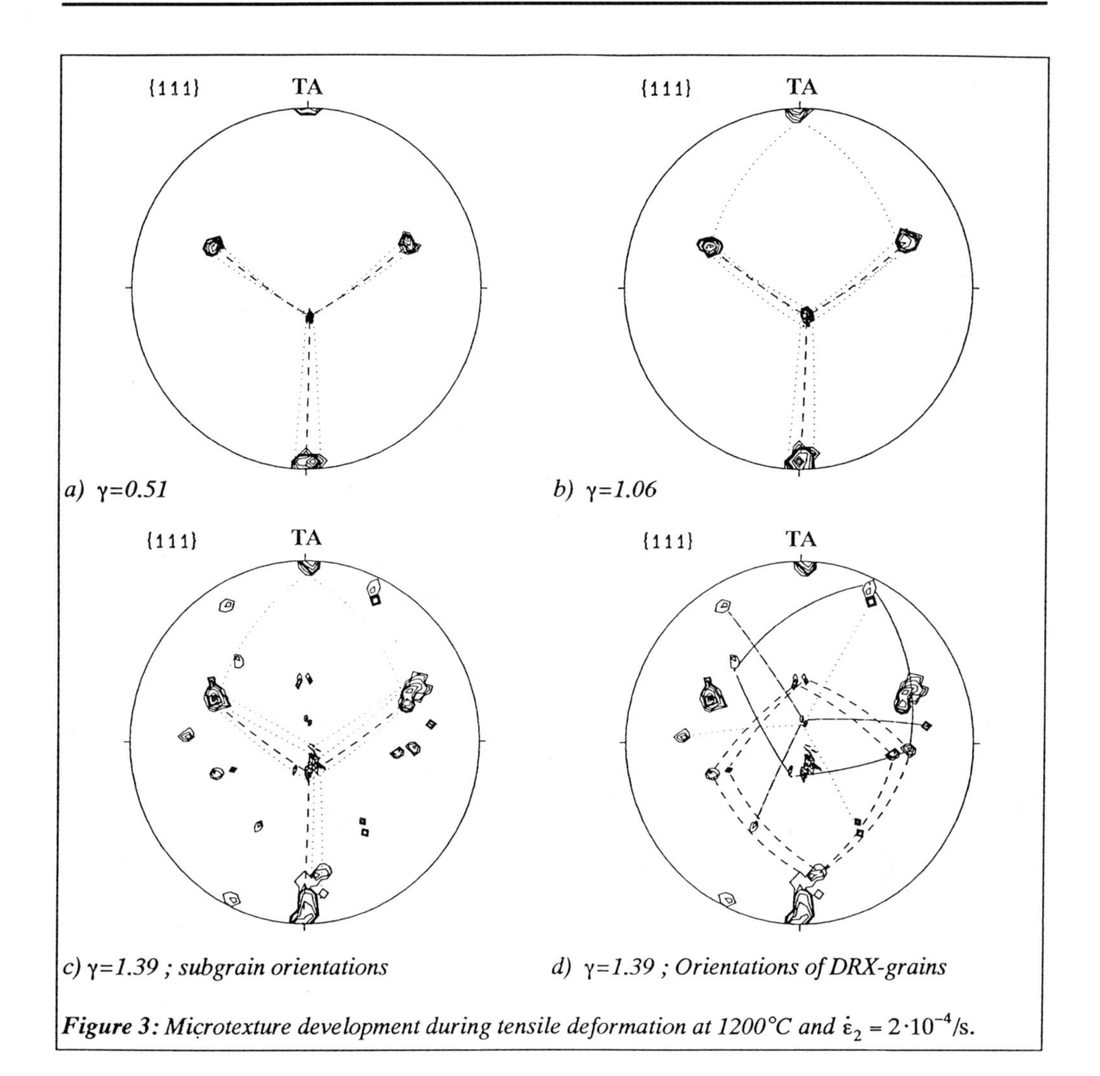

a) γ*=0.51* *b)* γ*=1.06*

c) γ*=1.39 ; subgrain orientations* *d)* γ*=1.39 ; Orientations of DRX-grains*

Figure 3: *Microtexture development during tensile deformation at 1200°C and* $\dot{\varepsilon}_2 = 2\cdot10^{-4}$/s.

this observation holds for the other strain rates, too. Nevertheless dynamic recrystallization was observed. But at all strain rates the dynamically recrystallized specimens showed considerably larger misorientations in the inhomogeneously deformed regions at inclusions (figure 4b). At the highest strain rate ($\dot{\varepsilon}_3$) the generation of high angle grain boundaries (Θ_{max} >15°) between subgrains was directly observed. This specific location is shown in figure 5a together with the microtexture of the sample. The marked orientations in the pole figure correspond to the marked subgrains in the micrograph. They belong to the outermost part of the subgrain spread, which was found in this sample. In addition to the high misorientation between these subgrains and the neighbouring subgrains, small twin lamellae were detected within them. A comparison of the orientations of dynamically recrystallized grains (figure 5b) with subgrain orientations yields that the orientations of dynamically recrystallized grains correspond to the orientations of strongly rotated subgrains (grains marked with 1 in figure 5b) or to twins of them (grains 2 and 3 : 1st order twins; grain 4 : 2nd order twin; figure 5b).

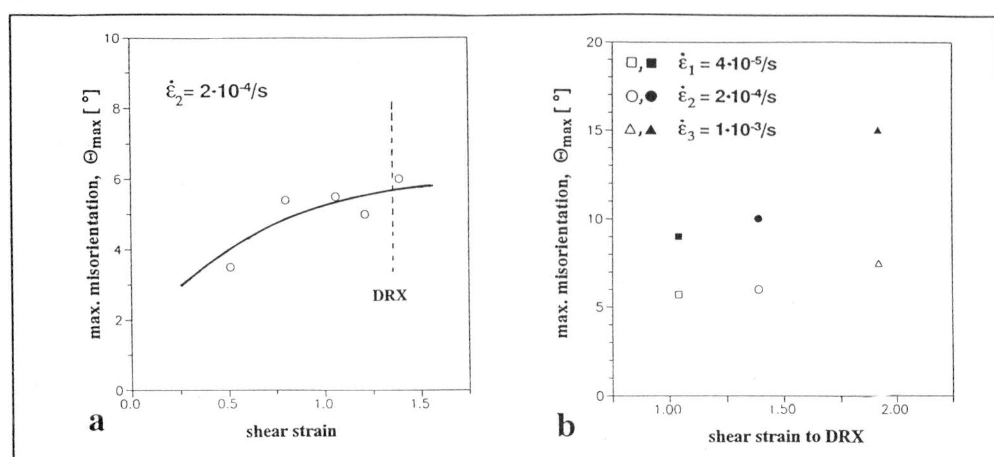

***Figure 4:** Maximum misorientation between next neighbour subgrains; a) at constant strain rate $\dot{\varepsilon}_2 = 2 \cdot 10^{-4}$/s in homogeneous regions; b) for different strain rates at the onset of DRX with (filled symbols) and without (open symbols) consideration of inhomogeneities at inclusions.*

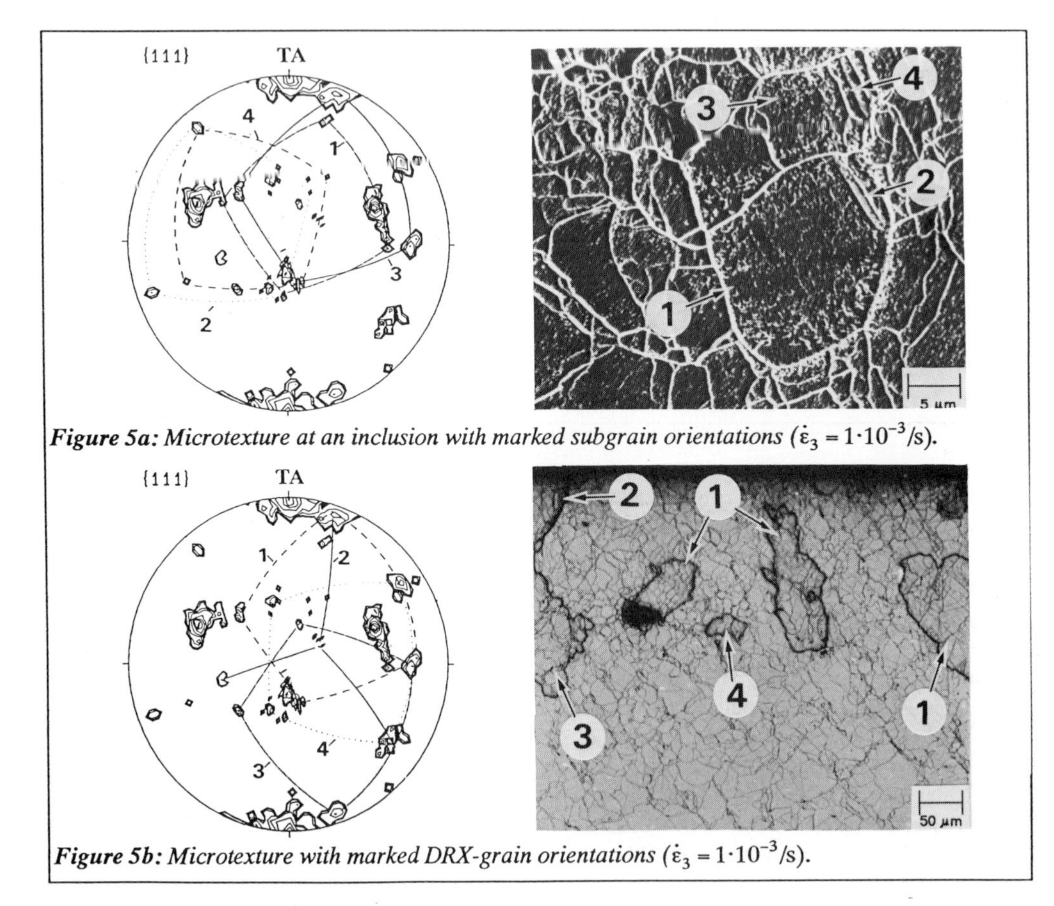

***Figure 5a:** Microtexture at an inclusion with marked subgrain orientations ($\dot{\varepsilon}_3 = 1 \cdot 10^{-3}$/s).*

***Figure 5b:** Microtexture with marked DRX-grain orientations ($\dot{\varepsilon}_3 = 1 \cdot 10^{-3}$/s).*

These results strongly suggest that dynamic recrystallization was triggered by instabilities of the subgrain structure which formed in highly stressed regions at inclusions. However, the critical step during the nucleation process cannot be concluded with certainty. There are some indications that the generation of a mobile (high angle) boundary between subgrains is the triggering step, leading to discontinuous subgrain growth, because all orientations of DRX-grains originated in regions of the subgrain structure, where the largest misorientations were found. In no case they could be attributed to subgrain orientations of homogeneously deformed regions, where misorientations never exceeded 6-7°. But discontinuously grown subgrains only were found at strain rate $\dot{\varepsilon}_3$ (figure 6), which supports the opinion that twinning is the triggering step in nucleation of DRX. Finally these results reveal a characteristic feature of DRX, namely that it is nucleated at locations, which are rather untypical for the whole sample. This feature explains the difficulty for a direct observation of the nucleation process.

Summary And Conclusions

1. In <111>-oriented Nimonic 80A single crystals tensile deformation at 1200°C leads to dynamic recrystallization.
2. The microtexture evolution reveals increasing subgrain rotations with growing shear strain, with the most strongly rotated subgrains appearing in inhomogeneities at inclusions.
3. The maximum misorientation between subgrains remains low in homogeneous deformation zones and is considerably larger in highly stressed regions at inclusions, partly exceeding 15°.
4. The orientations of dynamically recrystallized grains originate in the outermost spread of subgrain orientations; both discontinuously grown subgrains and twins of these subgrains are present.
5. Nucleation of DRX in Nimonic 80A single crystals is triggered in inhomogeneities at inclusions by instabilities of the subgrain structure. These instabilities consist of strongly rotated subgrains with mobile high angle grain boundaries, which were generated due to the strong rotation during deformation or due to twinning.

Acknowledgements

The authors are grateful to Prof. V.I. Levit from the Institute of Metal Physics in Jekaterinburg/Russia for support and supply of the single crystalline specimens

References

1) Gottstein, G., Mecking, H.,: Recrystallization of metallic materials (ed. by F. Haessner), 1978, 2nd edn, 195
2) Gottstein, G., Zabardjadi, D., Mecking, H.: Met. Sci., 1979, 13, 223
3) Stuitje, P.J.T.. Gottstein, G.: Z. Metallk., 1980, 71, 279
4) Gottstein, G., Kocks, U.F.: Acta metall. et mater., 1983, 35, 1261
5) Wantzen, A., Karduck, P., Gottstein, G.: ICSMA 5, 1979, 1, 517
6) Levit, V.I. et al.: Phys. Met. Metall., 1982, 54, 135
7) Levit, V.I., Bakhteyeva, N.D.: Phys. Met. Metall., 1983, 55/2, 124
8) Venables, J.A., Harland, C.J.: Phil. Mag., 1973, 27, 1193
9) Engler, O., Gottstein, G.: steel research, 1992, 63, 413
10) Levit, V.I., priv. communication

Materials Science Forum Vol. 157-162 (1994) pp. 1197-1204

THE INFLUENCE OF THE EXTRUSION SPEED ON TEXTURE IN THE SURFACE LAYER OF ALUMINIUM PROFILES INVESTIGATED BY THE EBSP TECHNIQUE

T. Furu [1], Ø. Sødahl [3], E. Nes [3], L. Hanssen [2] and O. Lohne [1]

[1] SINTEF Div. of Metallurgy, N-7034 Trondheim, Norway

[2] SINTEF Structures and Concrete, N-7034 Trondheim, Norway

[3] Dept. of Metallurgy, The Norwegian Institute of Technology, N-7034 Trondheim, Norway

Keywords: Aluminium, Extrusion, Surface Layer Texture, EBSP-Technique

ABSTRACT

Billets of the industrial AlMgSi alloy AA6082 have been extruded to flat profiles at different extrusion speeds followed by water quenching. In the centre of the profile the deformation mode is plane strain. Except for a thin surface layer, the material is not recrystallized. The recrystallized grains in the surface layer show a random texture and the grain size decreases as the ram speed increases. These observations have been related to particle stimulated nucleation (PSN) mechanisms.

1. INTRODUCTION

During hot extrusion of aluminium the cast microstructure is changed. In some alloys the new microstructure consists of recrystallized grains, in other alloys the recrystallization may be suppressed or partly suppressed e.g. by the presence of dispersoids. In heat treatable aluminium alloys a non recrystallized grain structure gives better mechanical properties in the extrusion direction than recrystallized material, the so-called "press effect".

The distribution of grain size and texture in the surface layer of extruded profiles depend on strain, strain rate and temperature in the deformation zone in the emerging extrusion and on the cooling rate afterwards. Structural differences may therefore appear, also across the same profile. Anodized profiles are given a pretreatment of chemical etching which removes a surface layer of typically 20-30 μm thickness. The light reflecting properties of the anodized profiles depend among other parameters upon the microstructure and the roughness of the surface. If the grain structure or texture vary, constant surface properties are no longer obtained. Such differences may be looked upon as cosmetic defects. The aim of this work is to obtain a better quantitative knowledge of the relationships which govern the surface structure in order to optimize billet microstructure and industrial practice to obtain products with a more consistent surface appearance.

2. MATERIAL AND EXPERIMENTAL PROCEDURE

Material
Billets of an AA6082 alloy (0.65wt%Mg, 1.0wt%Si, 0.2wt%Fe, 0.55wt%Mn and balance Al) were industrially DC-cast. The billets were machined to avoid material from the surface inverse segregation zone to enter the profiles during extrusion, homogenized at 575 °C for 3 hours and water quenched. Metallographical examinations showed that α-Al(FeMn)Si primary particles and AlMnSi dispersoids were present after the homogenization heat treatment.

Experimental
The cylindrical billets with a diameter of 97 mm and a length of 300 mm were preheated to about 490 °C, loaded into an extrusion container with a diameter of 100 mm and extruded at various ram speeds to rectangular shaped profiles of 4.0 x 78.5 mm^2, corresponding to an extrusion ratio of 25. The temperature of the container and the die was 470 °C when each extrusion trial started. The surface temperature of the profile at each pass was recorded with a thermocouple situated in the die bearing surface about 2.2 mm from the entrance of the die orifice and touching the profile surface. The die had a bearing length of 5 mm and a relief angle of 0° in the unloaded condition. At a distance of about 200 mm from the die a water spray quenching device was situated. The extrusion ram speed being constant for each billet, was varied from 0.5 mm/s to 25 mm/s corresponding to profile speeds ranging from 0.75 m/min to 38 m/min, respectively. The time from the material left the die to entering the water quenching device therefore varied from about 15 seconds to 0.3 seconds.

In this investigation material from the profile surface and from a cross section taken from the central part of the extrusion at half the total length was used for metallographical documentation. The profile surface was just electropolished (shock polishing for 2-3 sec. at 40 V in A2). The cross-section was ground and electropolished. The grain sizes were measured in an optical microscope and texture by SEM equipped with the SINTEF EBSP-system.

The variations in temperature T and strain rate ε were calculated at the midline positions where the material entered the die orifice and where a two-dimensional metal flow is expected. This was done by using a two-dimensional finite element program ALMA 2D [1]. This program is based on a two-dimensional Eulerian description with thermo-mechanical coupling in the time domain. A modified Zener-Hollomon relationship is used for the material behaviour. This has been established by hot torsion testing. The calculations of temperature were compared with the measurements of the extrusion surface temperature.

3. EXPERIMENTAL RESULTS

In the Figures 1(b), 2(b) and 3(b) micrographs from both the profile surface and the longitudinal transverse section for the different materials are presented. As indicated the materials are denoted as material 1 to material 3 where material 1 has the lowest and material 3 the highest ram speed. Texture measurements belonging to these microstructures obtained by the EBSP-technique from different regions of the specimens are also presented. The calculated values of T, $\dot{\varepsilon}$ and Z are shown in the Figures 1a-3a. In the surface layer the materials 1 and 2 are only partly recrystallized while material 3 is completely recrystallized with a mean grain size of 18 μm. The fact that the first two materials are only partly recrystallized gives the possibility to study the texture in both recrystallized and non-recrystallized grains in the surface of the extrusion.

As can be seen from Figure 1(b) the non-recrystallized structure in the centre region shows a texture

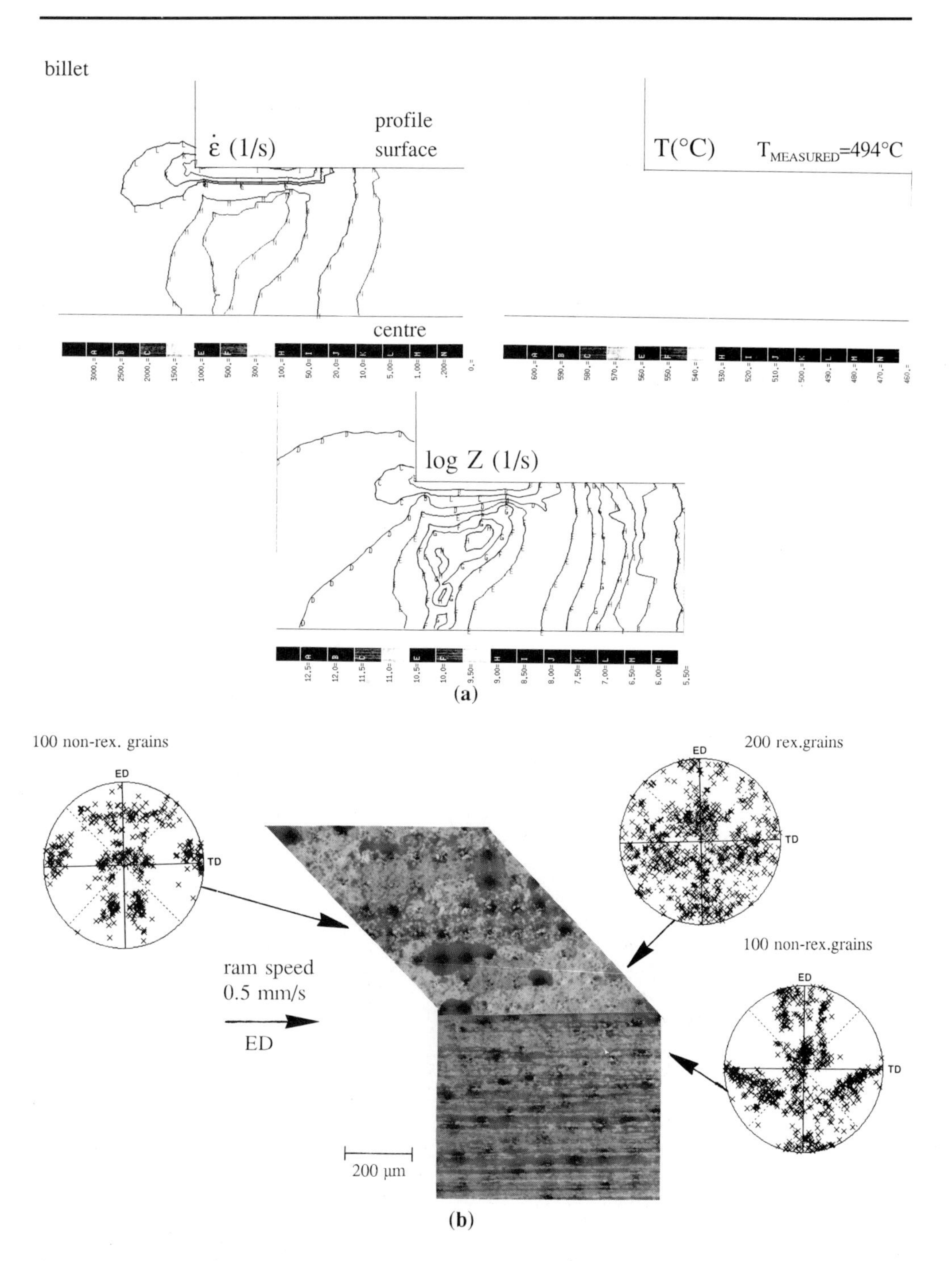

Figure 1. **Material 1:** (**a**) FEM-calculations of strain rate $\dot{\varepsilon}$, temperature T and Zener-Hollomon parameter Z across the extrusion thickness. (**b**) Microstructure in longitudinal and extrusion plane sections and (111) pole figures showing the texture at different positions of the profile. Surface layer about 50% recrystallized.

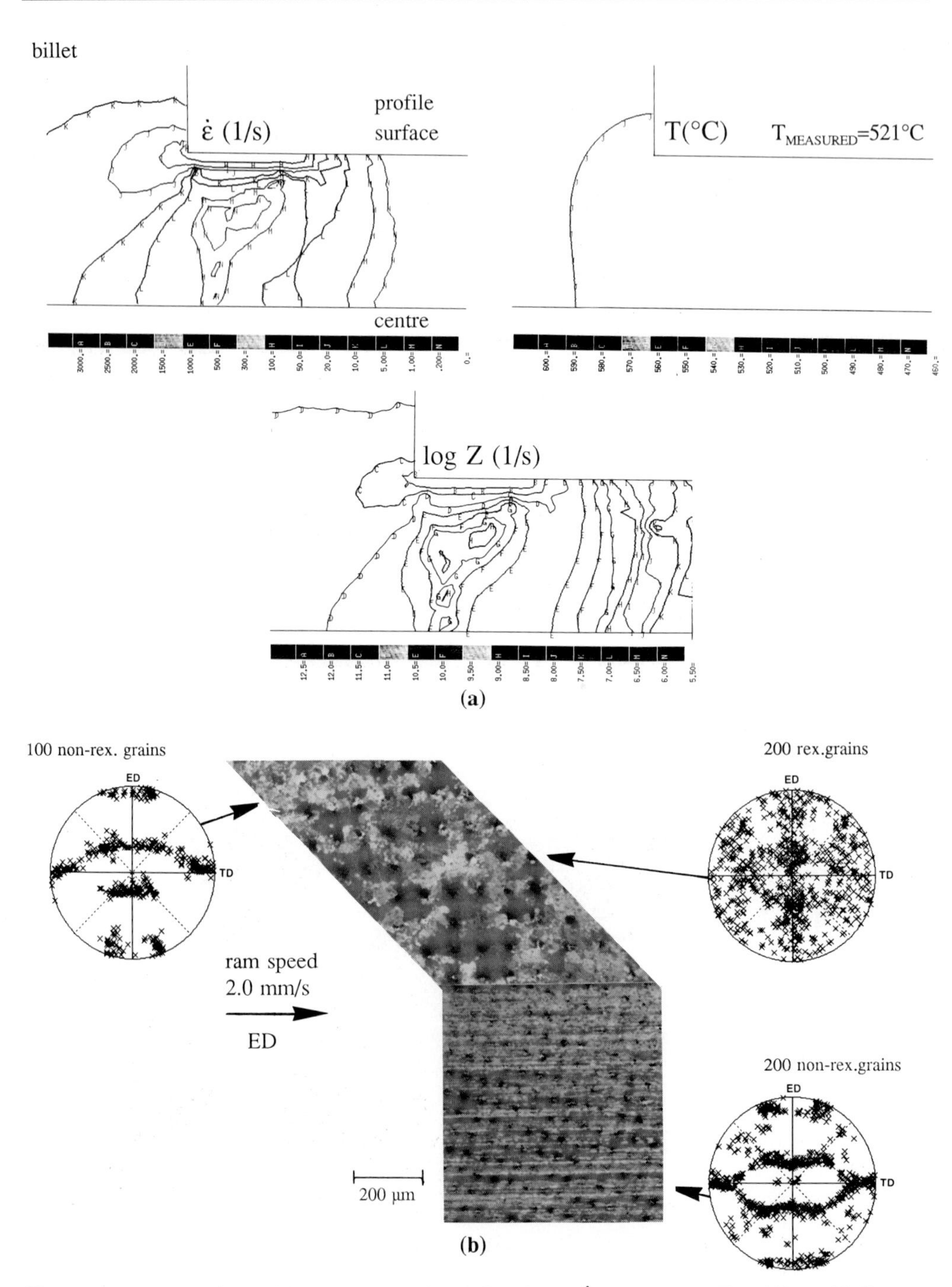

Figure 2. **Material 2:** (**a**) FEM-calculations of strain rate $\dot{\varepsilon}$, temperature T and Zener-Hollomon parameter across the extrusion thickness. (**b**) Microstructure in longitudinal and extrusion plane sections and (111) pole figures showing the texture at different positions of the profile. Surface layer about 70% recrystallized.

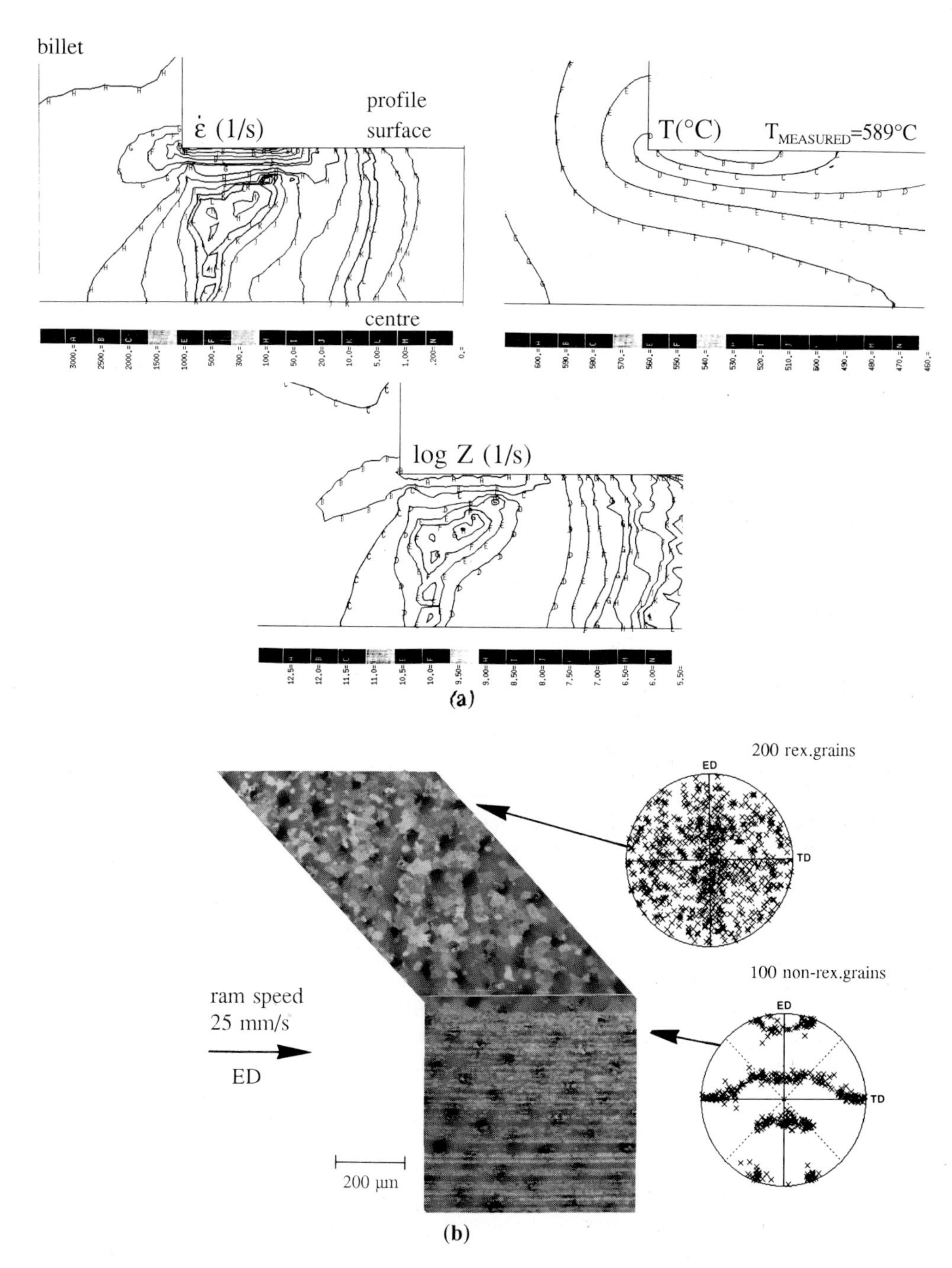

Figure 3. **Material 3:** (**a**) FEM-calculations of strain rate $\dot{\varepsilon}$, temperature T and Zener-Hollomon parameter across the extrusion thickness. (**b**) Microstructure in longitudinal and extrusion plane sections and (111) pole figures showing the texture at different positions of the profile. Surface layer: 100% recrystallized.

which is typical for plane strain deformation conditions, e.g. rolled sheet. This is as expected for material taken far from the edges of the profile.

The non-recrystallized structures in the surface layer of materials 1 and 2 reveal both well known deformation textures components. In the case of the lowest ram speed a rather sharp (211)<011> texture is observed, whilst the material deformed at 2 mm/s exhibit a texture incorporating the Cu, Bs and the S texture components. When studying the deformation texture in the region just below the surface layer of material 3 it is seen that this texture is almost identical to the texture in the non-recrystallized surface layer of material 2. Also the texture just below the surface layer of material 1 is dominated by the deformation texture components Cu, Bs and S.

The EBSP results from the recrystallized grains in the surface layer indicate that the texture is more random at high ram speeds than at low ram speeds. The deformation texture components Cu,S and Bs are more prominent at 0.5 and 2 mm/s than at 25 mm/s.

4. DISCUSSION

Although the number of grains analysed in each pole figure is limited the experimental texture observations may be classified as follows:

(i) In the midplane the non recrystallized grain structure shows a symmetrical texture pattern as expected for plane strain deformation.

(ii) Near the surface the texture of the non-recrystallized material differ from the texture in the mid-plane. This is as expected for a material flow having a shear strain component in the surface.

(iii) The recrystallized grains in the surface layer show nearly a random texture, especially at higher extrusion speeds.

The experimental texture observations described above may be discussed in terms of the FEM-calculations presented in the Figures 1-3, which demonstrate quite substantial variations in strain rate and temperature across the extrusion thickness. As can be seen from Figure 3 the variation in strain rate and temperature are, as expected, largest at the highest extrusion speeds. At the highest extrusion speed the strain rate varies between just above 0 in the middle section to higher than 3500 s^{-1} at the surface, whilst it varies only between 0 and 60 s^{-1} for the lowest extrusion speed.

While the deformation temperature is seen to be almost constant throughout the extrudate for the lowest extrusion speed, a variation between 530°C and 590°C is obtained at the highest speed. It should be noticed that the measured temperatures in the surface corresponds quite well with the FEM-calculated temperatures for all extrusion speeds. The variation in the Zener-Hollomon parameter across the extrusion thickness is also presented ($Z=\dot{\varepsilon}\exp(Q/RT)$, where Q is the activation energy for deformation and set to 156 kJ/mole [2]). For all extrusion speeds the Z-value is seen to vary sharply across the extrusion thickness. However, as the extrusion ram speed varies from 0.5 mm/s to 25 mm/s, which means a factor of 50, the variation in Z is less than a factor of 4.

As shown in the Figures 1-3 the recrystallization textures in the surface layer becomes more random the higher the extrusion speed. In addition, as can be seen from Table 1. the surface layer grain size decreases as the ram speed increases. At the highest speed the grain size becomes very small taken into account the rather low Z-value. The grain size is seen to vary quite significant over a relative small variation in the Z-value. It is tempting to associate these observations to particle stimulated nucleation mechanisms (PSN) and that the particles are more effective as nucleation positions the higher the ram speed. Further work on this aspect is in progress.

Table 1. Maximum values of the strain rate ε, profile surface temperature T and the Zener-Hollomon parameter obtained by FEM-calculations and the surface layer grain size for the different materials.

Material	$\dot{\varepsilon}_{max}$ (s^{-1})	$T_{measured}$ (°C)	$T^{max}_{calculated}$ (°C)	Z (s^{-1})	Grain size(μm)
1	67	494	498	$2.5 \cdot 10^{12}$	172 (‖ED) 96 (⊥ED)
2	270	521	528	$4.1 \cdot 10^{12}$	31
3	3480	589	590	$9.7 \cdot 10^{12}$	18

5. CONCLUSIONS

- The EBSP-TECHNIQUE is a powerful tool in measuring texture variations in the surface layer (and in other local areas).
- The texture of the recrystallized grains in the surface layer is nearly random.

ACKNOWLEDGEMENT

The authors express their thanks to NFR, Hydro Aluminium and Raufoss A/S for financial support and Raufoss A/S for supplying the alloy which has been investigated.

REFERENCES

1) Grasmo, G., Holthe, K., Støren, S., Valberg, H., Flatval, R., Hanssen, L., Lefstad, M., Lohne, O., Welo, T., Ørsund, R. and Herberg, J., Proc. Fifth Int. Al. Extrusion Techn. Seminar (ET92), Chicago, 1992, 2, 367
2) Wong, W.A. and Jonas, J.J., Trans. Met. Soc. A.I.M.E., 1968, 242,2271

Materials Science Forum Vol. 157-162 (1994) pp. 1205-1210

STRESS-INDUCED GRAIN GROWTH IN THIN Al-1%Si LAYERS ON Si/SiO_2 SUBSTRATES

D. Gerth, S. Zaefferer and R.A. Schwarzer

Institut für Metallkunde und Metallphysik der TU, D-38678 Clausthal-Zellerfeld, Germany

Keywords: Grain Growth, Thin Films, Stress, Strain, Local Texture, Al-1%Si, Electron Diffraction, Kikuchi Patterns, TEM

ABSTRACT

Grain growth has been encountered during thermal cycling of thin Al-1%Si layers on oxidised silicon substrates. Individual grain orientations are measured on-line by the Kikuchi pattern method. Stereological parameters are obtained from digitised TEM micrographs. Local texture is illustrated by orientation images, and the orientation of individual grains are represented by points in inverse pole figures, in the Euler and in the Rodrigues space. The ordinary <111> fibre texture is weakened by fast heating, since grains with the <111> direction deviating from the layer normal grow in preference under biaxial mechanical stress. Some potential driving forces are calculated on the basis of the measured stereological and orientation parameters. Minimisation of grain boundary energy is the main driving force. In addition the minimisation of deformation energy is important at lower temperatures, while the minimisation of surface energy is important at higher temperatures.

I. INTRODUCTION

Unexpected grain growth has been encountered during thermal cycling of thin aluminium-alloy layers on oxidised silicon substrates. During temperature changes a mechanical stress is induced in thin layers on silicon substrates due to the mismatch between the thermal expansion coefficients of layer and substrate. Some possible driving forces for grain growth may depend on the local stress-strain state. For example, the deformation energy depends on the dislocation density which in turn is influenced by the morphology of a grain and its neighbours as well as by the crystallographic orientation of the glide planes with respect to the acting stress in the grain. Load on grain boundaries, as a result of inhomogeneous vertical strain, may affect the mobility as well as the diffusivity of atoms. Transmission electron microscopy (TEM) is a means to study the morphology and crystallographic orientation on a grain-specific scale. From TEM investigations statistical values and local differences of elastic constants can be computed as a function of grain orientation during the heating process.

II. EXPERIMENTAL

Investigations were carried out with 800 nm thick Al-1%Si layers on thermally oxidised silicon substrates (SiO_2 thickness 100 nm). The films were sputter deposited at room temperature. The samples were heated in the vacuum at various temperatures for different times. Two heating regimes were used:

- direct heating to a fixed temperature with a further annealing for 30 minutes (named isothermal annealing),
- stepwise heating up to 550 °C in steps of 50 °C and 6 minutes.

For transmission electron microscopy, samples were prepared by chemical polishing. The back of a sample was thinned to a hole by dipping the silicon disk in a mixture of HNO_3 and HF (6.5:1). The aluminium layer was protected from etch by a droplet of glycerine under a plastic lid. The morphology and the crystallographic orientation of the grains in the layer were investigated with a TEM at an accelerating voltage of 300 kV (PHILIPS EM 430). The crystal orientations were measured on-line using Kikuchi patterns [1]. The micrographs of the areas with the measured grains were digitised in order to obtain the grain boundary coordinates.

III. GRAIN GROWTH

Figure 1 shows a TEM micrograph of the grain structure of the heated films. The films have a columnar structure which is attained at about 200 °C. The mean values of diameter, grain boundary curvature and number of corners are given in table 1.

The mean grain diameter increases from about 100 nm after deposition (grainy structure) to 1.1 µm after annealing at 400 °C and to 3.3 µm after annealing at 550 °C. The layers attain a columnar structure by grain growth at about 200 °C. The pronounced grain growth above about 400 °C was also found in earlier investigations [2]. After stepwise heating up to 550 °C the average grain diameter is larger (3.7 µm) than after isothermal annealing at 550 °C. In the layer region on substrate larger grain growth was found [2] (cf. figure 2). In the following it is shown that the increased grain growth is affected by stress.

Fig. 1 TEM micrographs of 800 nm Al-1%Si layers heated on 100 nm SiO_2/Si substrates at different temperatures.

Table 1 Morphological parameters of 800 nm Al-1%Si layers on oxidised silicon substrates after different heating regimes.

Heating regime	Average grain diameter *d* [µm]	Average grain boundary curvature *c* [10^3/m]	Average number of corners *e*
isothermal 200 °C	1.07	328	5.79
isothermal 400 °C	1.50	258	6.14
isothermal 550 °C	3.30	100	5.97
stepwise 550 °C	3.72	91	5.87

Fig. 2 TEM micrograph of an *in situ* heated 800 nm thick Al-1%Si.

left: free-standing layer
right: layer on SiO_2 substrate

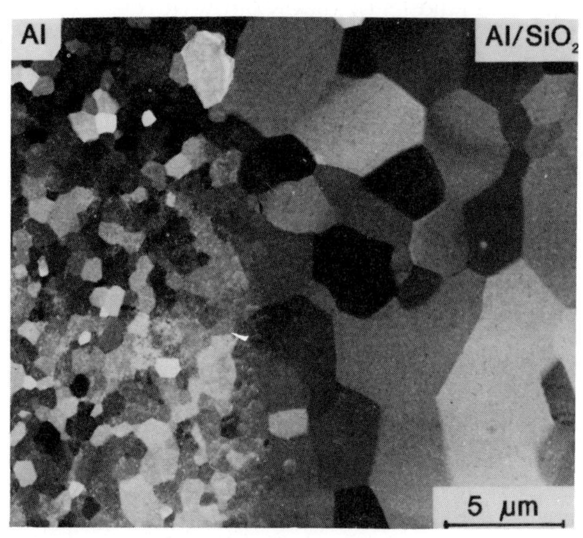

IV. GRAIN SPECIFIC CRYSTAL ORIENTATIONS

Figure 3 shows the local orientation distributions of the layers by inverse pole figures for the plane-normal direction and for an arbitrary reference direction in the layer plane. It turns out that the majority of the grains are oriented closely along a <111> fibre texture. On the as-deposited samples a <111> fibre texture was confirmed by x-ray measurements. The fibre texture is weakened during grain growth at higher temperatures. The texture is more pronounced after stepwise heating up to 550 °C than it is after isothermal annealing at 550 °C.

The spatial distribution of orientations may be illustrated by false-colour maps [3, 4, 5] which allow to recognise local orientation inhomogeneities and orientation differences (figure 4) with great ease. The most appropriate representation depends on the given texture and on the actual physical problem [5]. An inverse pole figure may often be adequate for a fibre texture using the Miller indices of the fibre direction as parameters for a false-colour map.

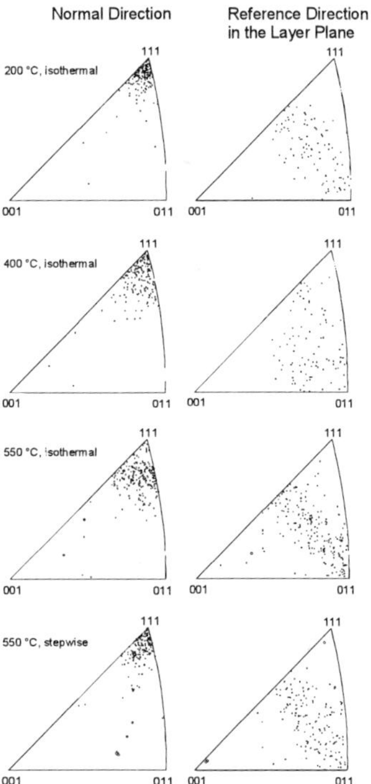

Fig. 3 Inverse pole figures of 800 nm Al-1%Si layers heated on SiO_2/Si substrates at different temperatures.

V. DRIVING FORCES OF GRAIN GROWTH

The prevalent driving forces of grain growth depend on temperature and heating modes. Some driving forces have been discussed in literature. In thin aluminium films grain growth is often an instantaneous process of grain coalescence [2, 6].

The elastic energy, p_{Ela}, is expressed as follows:

$$p_{Ela} = \Delta\left(\frac{u}{\varepsilon^2}\right)\varepsilon^2 \quad . \qquad (1)$$

u is the calculated elastic energy per volume and strain, and ε the measured strain. The difference of elastic energy between abutting grains which are oriented with <100> or <111>, respectively, in the normal direction of the sample was assumed as an upper limit. The strain was estimated from stress measurements [2]. Grain boundary energy, γ_{Gb} , decreases by grain coalescence. Assuming six-sided grains, the decrease of energy by consumption of a grain boundary is [7]:

$$p_{Coal} = \sqrt{\frac{8}{3\sqrt{3}\,\pi}}\ \frac{\gamma_{Gb}}{d} \approx 0.7\ \frac{\gamma_{Gb}}{d} \qquad (2)$$

where d is the average grain diameter. Differences in surface energy may be calculated from [7, 8]:

$$p_{Surf} = \frac{0{,}06\,\gamma_{Gb}}{h} \quad . \qquad (4)$$

where h is the layer thickness.

The driving force from the deformation energy is [9]:

$$p_{Def} = \Delta\rho\ G\ b^2 \quad , \qquad (5)$$

where G is the shear modulus, **b** the Burgers vector and $\Delta\varrho$ the difference of dislocation density between neighbouring grains. For a qualitative calculation, the difference of the number of dislocations between neighbouring grains, ΔN_{Disl} , must be known. It was estimated from TEM micrographs.

The backdriving force was calculated for one-sided and two-sided grooving after [7, 10]:

$$p_{groov1} = -\frac{\gamma_{Gb}}{18\ h}\ , \qquad p_{groov2} = -\frac{\gamma_{Gb}}{9\ h} \quad . \qquad (6)$$

Table 2 shows the orders of magnitude of the driving forces in dependence of temperature. The backdriving force from Zener drag was not calculated. It is assumed to have roughly the same value as the difference from driving and backdriving forces when grain growth is finished.

VI. CONCLUSION

The main driving force for grain growth seems to be due to the minimisation of grain boundary energy. In addition deformation energy is high at low temperatures and may further contribute to grain growth. At higher temperatures the minimisation of surface energy becomes increasingly important. Enhanced grain growth may be induced by an excess of deformation energy in layer regions sticking to the substrate. The difference in grain growth is not supposed to be due to the backdriving force, since the differences between

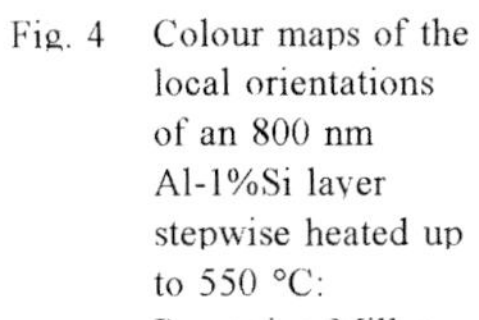

Fig. 4 Colour maps of the local orientations of an 800 nm Al-1%Si layer stepwise heated up to 550 °C:

I) using Miller indices and inverse pole figures,

II) using Euler angles,

III) using Rodrigues vectors.

I)

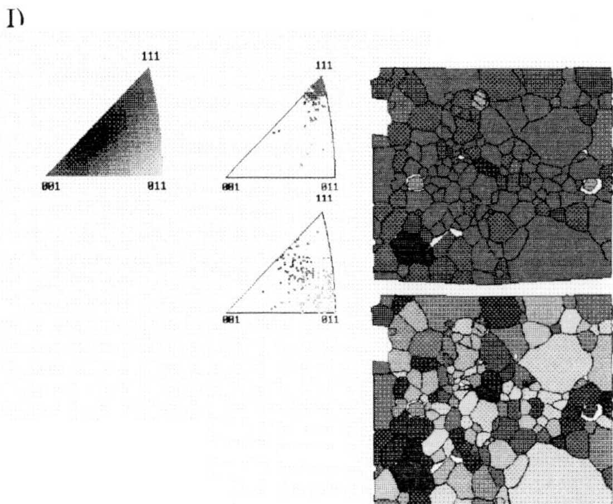

II)

III)

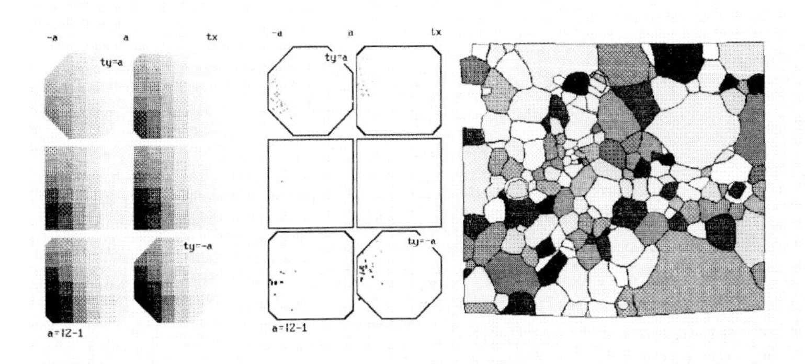

one-sided and two-sided grooving are small (table 2). As a further mechanism instantaneous grain coalescence may contribute to increased grain growth in regions with high stress.

Table 2 Calculated values of driving forces of grain growth at different temperatures.

Shear modulus G [10^9 Pa]				24
Elastic energy per volume and strain squared u/ε^2 [10^9 Pa]				112
Difference of elastic energy per volume and strain squared $\Delta(u/\varepsilon^2)$ [10^9 Pa]				10
Burgers vector $\|b\|$ [10^{-10} m]				2.86
Layer thickness h [10^{-6} m]				0.8
T [°C]		130	300	500
Average grain diameter d [10^{-6} m]		0.9	1.3	2.6
Elastic stress σ_{el} [10^6 Pa]		80	20	5
Strain ε [10^{-3}]		0.74	0.18	0.05
Specific grain boundary energy γ_{Gb} [10^{-3} J/m^2]		200...600	150..400	100..300
Difference of total number of dislocations, ΔN_{Disl}, between neighbouring grains		30...70	5...20	1...10
Difference of dislocation density, $\Delta\rho$ [10^{12} m^{-2}], between neighbouring grains		50...120	6...25	1...6
Deformation energy	p_{Def} [10^3 Pa]	98...235	12..49	2..12
Grain boundary energy	p_{Coal} [10^3 Pa]	194...581	100...268	33...100
Elastic energy	p_{Ela} [10^3 Pa]	5.5	0.3	0.03
Surface energy	p_{Surf} [10^3 Pa]	15...45	11...30	7...22
Grooving	p_{groov2} [10^3 Pa]	-(28...84)	-(20...56)	-(14...42)
	p_{groov1} [10^3 Pa]	-(14...42)	-(10...28)	-(7...21)

ACKNOWLEDGMENT

The authors are indepted to the Deutsche Forschungsgemeinschaft for financial support (DFG project BU 374/22-1).

REFERENCES

[1] Schwarzer, R. A.: Textures and Microstructures, 1990, 13, 15
[2] Gerth, D., Katzer, D., and Krohn, M.: Thin Solid Films, 1992, 208, 67
[3] Inokuti, Y., Maeda, C., and Ito, Y.: Trans. Iron Steel Inst. Japan, 1987, 27, 302
[4] Dingley, D. J., Day, A., and Bewick, A.: Textures and Microstructures, 1991,14-18,91
[5] Gerth, D., and Schwarzer, R. A.: Textures and Microstructures, 1993, 21, 177
[6] Thon, A., and Brokman, A.: J. Appl. Phys. 1988, 63, 5331
[7] Gerth, D., Katzer, D., and Schwarzer, R. A.: to be published
[8] Gangulee, A.: J. Appl. Phys., 1974, 45, 3749
[9] Lücke, K., and Rixen, R.: Z. Metallkunde, 1968, 59, 321
[10] Chaudhari, P.: J. Vac. Sci. Technol., 1971, 9, 520

Materials Science Forum Vol. 157-162 (1994) pp. 1211-1218

GRAIN SUBDIVISION DURING DEFORMATION OF POLYCRYSTALLINE ALUMINIUM

N. Hansen and D. Juul Jensen

Materials Department, Risø National Laboratory, DK-4000 Roskilde, Denmark

Keywords: Aluminium, Cold-Rolling, Local Orientation, EBSP-Technique, Grain Subdivision

ABSTRACT

High purity aluminium (99.996%) with a grain size of about 200 μm has been deformed by cold-rolling from 5 to 30% reductions. The local crystallographic orientations at three strain levels have been characterized using the electron back scattering pattern (EBSP) technique. The deformation leads to grain subdivision on a scale which is small compared to the grain size and no deformation bands have been found. The observed pattern of crystallographic changes across the grains supplements TEM observations and is in accordance with a general framework for the microstructural evolution.

1. INTRODUCTION

Deformation microstructures in high purity aluminium have been described within an evolutionary framework common to medium or high stacking fault energy FCC polycrystalline metals (1 - 4). Within this framework deforming grains are subdivided into rotated volume elements on different levels for all deformation modes (4). The reason is that the number and selection of simultaneously acting slip systems differ between neighbouring volume elements in a grain and that in every volume the number of slip systems operating is in general smaller than the five suggested in the Taylor model (5). This deformation pattern leads to a reduction in flow stress as the number of dislocation intersections is reduced. However, the strain accommodation will be less perfect (1,2). This pattern for plastic deformation, that individual volume elements are characterized by their own combination of slip systems will lead to differences in lattice rotations between neighbouring elements when the material is strained.

The microstructural evolution leading to grain subdivision and lattice rotations has been observed by many different techniques, from TEM (2 - 4) on a fine scale compared to the grain size to etch pit techniques (6) and Kossel technique (7) on a relatively coarse scale. For studies on an

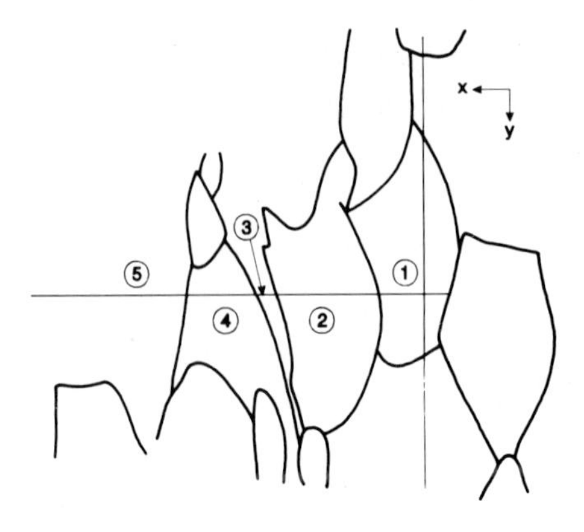

Fig. 1. Schematic drawing of the microstructure and the scanning directions.

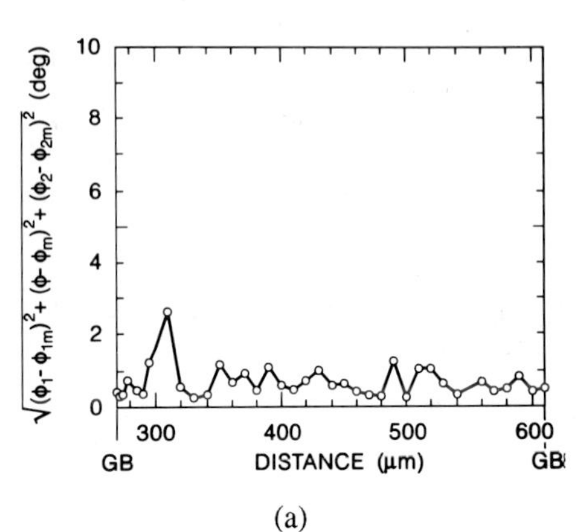

Fig. 2. EBSP analyses of rotation in the rolling plane. A grain is scanned stepwise from grain boundary to grain boundary, see text. 5% reduction (a) 20% reduction (b) and 30% reduction (c)

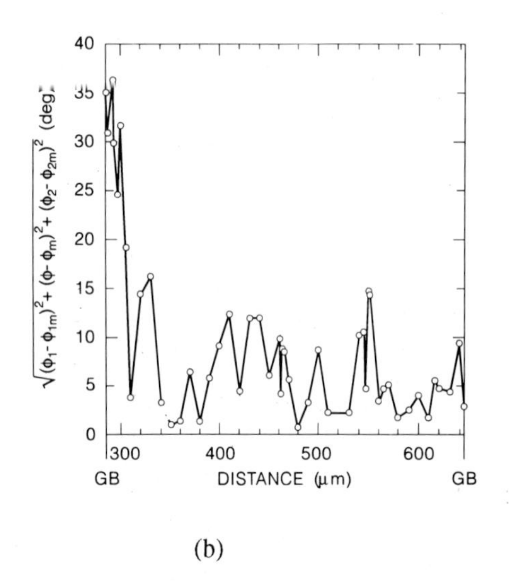

(b)

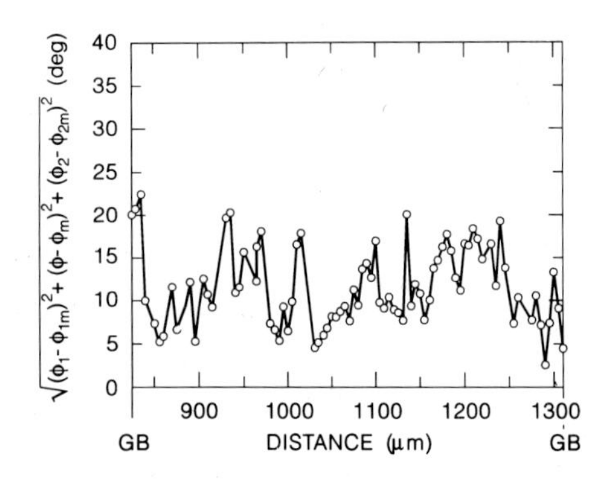

(c)

intermediate scale the electron back scattering pattern (EBSP) technique offers a new possibility and studies of deformed materials have started to appear (8). The EBSP technique has been chosen in the present work in order to supplement many detailed TEM investigations of deformation microstructures in pure aluminium (4, 9).

2. EXPERIMENTAL

High purity aluminium (99.996%) with a grain size of about 200 μm has been deformed by cold-rolling to 5, 20 and 30% reduction (ε_{VM} = 0.05, 0.22 and 0.36).

The textures of ~1 cm^3 bulk specimens have been determined by neutron diffraction and the three dimensional orientation distribution functions calculated using the series expansion method. In the undeformed state the material has a fairly strong cube {100} <001> texture. In total ~50 vol.% of the material has orientations within 15° of the ideal cube orientation. During deformation the initial cube texture weakens and the typical rolling texture ({110} <112> + {123} <634> + {112} <110>) develops.

The local crystallographic orientations within the deformed grains have been determined in the rolling plane by the electron back scattering pattern (EBSP) technique. The spatial resolution of this technique is about 1 μm. Scans have been made along RD and TD through several grains in steps of typically 2 - 5 μm (see Fig. 1). At each step the crystallographic orientation is determined from the EBSP. In this way maps of orientations along the selected scan lines are obtained. The accuracy of the indexing in a recrystallized sample is 0.5 - 1° (10).

3. DATA REPRESENTATION AND RESULTS

Two methods have been used to represent the local orientation measurement:

Method 1: Orientations in the form of Euler angles (φ_1, Φ, φ_2) are related to the average orientation of the grain (φ_{1m}, Φ_m, φ_{2m}) as

$$D = \sqrt{(\varphi_1 - \varphi_{1m})^2 + (\phi - \phi_m)^2 + (\varphi_2 - \varphi_{2m})^2} \qquad 1)$$

There appears to be no significant difference between the scans in the RD and TD directions and the results are plotted in Fig. 2 for one direction of the scan. These observations show that relatively large differences in lattice rotation exist within the individual grains and that these rotations increase with the degree of deformation. In good agreement with Ref. 6, it has been found that the range of orientations increases from 0 - 3° at 5% reduction to 0 - 15° at 20% and 5 - 20° at 30%. The length scale of the variation in lattice orientation is relatively small and shows that the grain has subdivided into regions which are one to two orders of magnitude smaller than the grain size. The lattice orientation has also been studied in the grain boundary region and it has been found that in many cases there is no significant difference in orientations between the grain boundary region and the grain interior. However, approaching some boundaries a large change in lattice orientation can be observed as shown in Fig. 2b.

Method 2: The local misorientation can be plotted with respect to a reference orientation or with respect to a neighbouring volume. The latter representation is chosen here. In both cases the representation is based on calculations of angle/axis pairs. However, it is problematic to plot the angle (θ_m) and the axes (l_m) for hundreds of orientations within a grain (with the specimen position

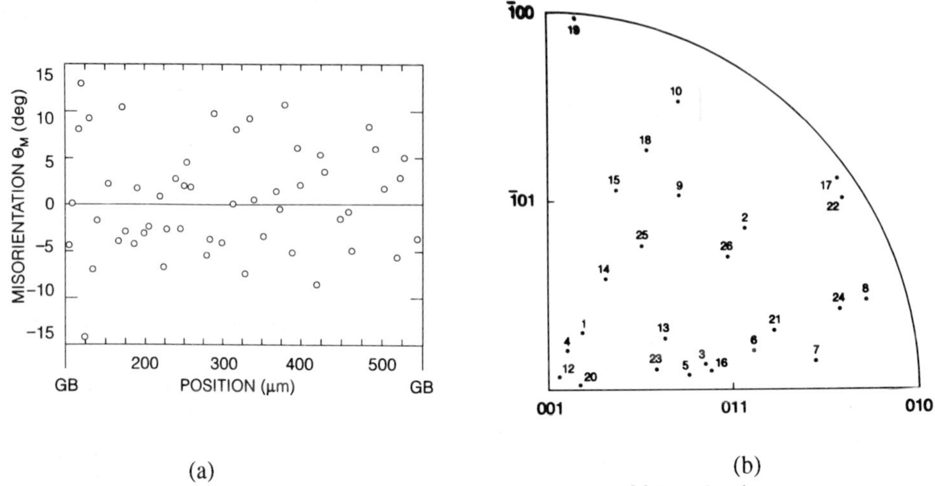

(a) (b)

Fig. 3. Angles, θ_m (a) and axis, l_m (b) for a specimen rolled to 30% reduction, see text.

of each measurement being identifiable). Here it has been chosen to plot φ_M versus position in the grain and l_m in a stereographic projection using a number to identify the position.

A standard calculation of θ_m and l_m has been used (11) based on the following equation for the calculation of θ_m.

$$\cos\theta_m = \tfrac{1}{2}\,(a_{11} + a_{22} + a_{33} - 1) \tag{2}$$

where a_{11}, a_{22} and a_{33} are the diagonal elements of the misorientation matrix. θ_m is therefore always positive. In order to obtain information about the spatial orientation changes in a subdivided grain (for example whether they are cumulative or alternating), it has been chosen to calculate θ_m with a sign based on the position of l_m in the stereographic projection: θ_m is taken to be positive if l_m is in right hand triangle of the stereographic projection and to be negative when l_m is in the left hand triangle.

For a specimen cold-rolled 30%, θ_m and l_m are plotted in Fig. 3. Fig. 3a shows that θ_m alternates between positive and negative values with an average near to zero. This figure, in accordance with Fig. 2, also shows that cumulative orientation changes can be found, but only over quite small distances. Fig. 3b shows larger variations in the orientation of l_m with apparently no systematic changes. A similar behaviour to that shown in Fig. 3 has been found in all the examined grains deformed both at 30% and at lower reductions.

4. DISCUSSION

Grain subdivision during deformation of polycrystalline metals takes place both on a macroscopic and microscopic scale. A macroscopic subdivision had been found in the early work by Barrett and Levenson (6) studying coarse-grained aluminium deformed in compression. It was observed that individual grains did not rotate as units but deformed inhomogeneously with a range of orientations. Optically it was found that well defined narrow bands formed across the grains in parallel sets and that the misorientation across such bands increased with the degree of deformation. These bands were called deformation bands and separated volumes of material of a common orientation which deform uniformly with different combinations of slip systems. These volumes separating the deformation bands were later termed matrix bands (12).

Deformation bands are a typical phenomenon in unstable single crystals which split into diverging texture components during deformation. These bands have a width of a few micrometres and an extensive length. Occasionally deformation bands have also been observed in polycrystalline metals most frequently in coarse-grained specimens (13). Examples are deformation bands identified by transmission Kossel diffraction in aluminium having a grain size of 800 μm (8) and by the EBSP technique in aluminium in the as-cast condition (14). In both cases specimens have been deformed 40% in compression. The spacing between the deformation bands is of the order of 50 to 100 micrometres and their length is hundreds of micrometres.

Deformation bands on a macroscopic scale as described above have not been observed in the present work. Deformation bands separated by matrix bands should show up as a characteristic change in D_m and θ_m (Fig. 2 and Fig. 3) when scanning across a grain. These figures should be rather constant when intersecting a matrix band but should change abruptly over short distances when a deformation band is passed. Instead of this behaviour an alternating change in D_m and θ_m is observed when scanning across the grains and this behaviour is observed at all the strains examined. This crystallographic information is reflected in the microstructure as observed by TEM which will be shown in the following.

A typical microstructure observed at medium strains is shown in a schematic drawing (Fig. 4). This figure shows a characteristic subdivision into volume elements each characterized by its own combination of slip systems. Such volume elements have been termed cell blocks (4) and they are composed of ordinary cells. Between the blocks dislocation boundaries form which are much longer and typically have larger misorientation across them than ordinary cell walls. These boundaries are geometrically necessary boundaries (16) and have been termed dense dislocation walls and microbands (4). There is a clear tendency that rotations across neighbouring boundaries are of similar magnitude but alternate in sign (9). The lattice rotation determined by microdiffraction across dense dislocation walls and microbands increases to about 10 - 15° at a rolling reduction of 30% (9).

Another characteristic feature is that dense dislocation walls and microbands appear to have a fairly constant macroscopic orientation with respect to the specimen axes within a particular strain regime. For example in rolling at small to intermediate strains (0.1 - 0.5) the majority of the boundaries is arranged almost parallel to the transverse direction and plus and minus 40° to the rolling plane, and a smaller fraction has an orientation about 15° to the rolling plane (17). The mean distance between the boundaries is approximately 5 μm (9) and smaller in areas where two families intersect each other.

The orientation changes observed by EBSP cannot directly be related to the TEM-observations as the structural origin of the EBSP pattern is unknown. However, based on the frequency, macroscopic orientation and misorientations of the dense dislocation walls and microbands observed by TEM it is highly probable that it is the crystallographic characteristics of this structure which are reflected in the local crystallographic orientations determined by EBSP.

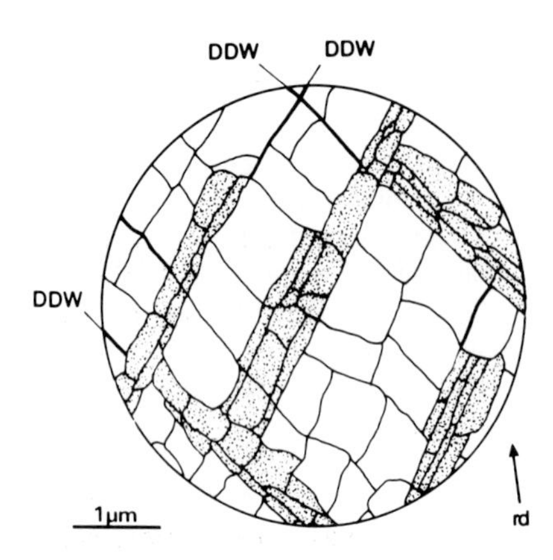

Fig. 4. Schematic drawing (longitudinal section) of grain subdivision showing the formation of cell blocks bounded by geometrically necessary boundaries in the form of (i) dense dislocation walls (DDWs) which are drawn as thick lines and (b) microbands which are dotted. These boundaries delineate the cell blocks which contain ordinary dislocation cells (15).

5. CONCLUSIONS

For polycrystalline high purity aluminium (99.996%) cold rolled from 5 to 30% reduction local crystallographic orientation measurements by the EBSP technique have led to the following conclusions:

- The grain subdivides during deformation and the misorientations within the grains increase with strain to quite high values.

- The crystallographic orientations show large scatter across the grains. Cumulative misorientations are found over short distances (~10 μm) whereas deformation bands on a macroscopic scale are not observed.

- The observed pattern of crystallographic changes supplements TEM observations and is in accordance with a general framework for the microstructural evolution with increasing strain common to medium and high stacking fault energy FCC metals.

ACKNOWLEDGEMENTS

The authors thank E. Sørensen for preparing the manuscript, and T.E. Skov, H. Nilsson and P.B. Olesen for making the figures.

REFERENCES

1) Kuhlmann-Wilsdorf, D.: Mat. Sci. Eng., 1989, A113, 1.

2) Hansen, N.: Mat. Sci. Tech., 1990, 6, 1990.

3) Hughes, D.A. and Hansen, N.: Mat. Sci. Techn., 1991, 7, 544.

4) Bay, B., Hansen, N., Hughes, D.A. and Kuhlmann-Wilsdorf, D.: Acta Metall. Mater., 1992, 40, 205.

5) Taylor, G.J.: J. Inst. Metals, 1938, 62, 307.

6) Barrett, C.S. and Levenson, L.H.: Trans. AIME, 1940, 137, 112.

7) Bellier, S.P. and Doherty, R.D.: Acta Metall. 1977, 25, 521.

8) Microscale Textures of Materials, Textures and Microstructures, 1993, 1-4, 242 pp.

9) Bay, B., Hansen, N. and Kuhlmann-Wilsdorf, D.: Mat. Sci. Eng., 1989, A113, 385.

10) Schmidt, N.H., Bilde-Sørensen, J.B. and Juul Jensen, D.: Scan. Microscopy, 1991, 5, 637.

11) Randle, V.: Microtexture Determination and Its Applications, The Institute of Metals, London, 1992, 169 pp.

12) Inohuti, Y. and Doherty, R.D.: Acta Metall., 1978, 26, 61.

13) Hansen, N.: Scripta Metall. Mater., 1991, 25, 1557.

14) Doherty, R.D., Kashyap, K and Panchanadeeswaran, S.: Acta Metall. Mater., 1993, 41, 3029.

15) Bay, B., Hansen, N. and Kuhlmann-Wilsdorf, D.: Mat. Sci. Eng., 1992, A 158, 139.

16) Kuhlmann-Wilsdorf, D. and Hansen, N.: Scripta Metall. Mater., 1991, 25, 1557.

17) Juul Jensen, D. and Hansen, N.: Acta Metall. Mater., 1990, 38, 1369.

Materials Science Forum Vol. 157-162 (1994) pp. 1219-1224

TEXTURE ESTIMATE FROM MINIMUM RANGES OF SAD POLE FIGURES

K. Helming and R.A. Schwarzer

Institut für Metallkunde und Metallphysik der TU, D-38678 Clausthal-Zellerfeld, Germany

Keywords: ODF Analysis, Minimal Pole Figure Ranges, Texture Components, Shear Bands, TEM Measurement

ABSTRACT

Only a small number of incomplete pole figures are required for texture estimates using the Component Method. Applications to TEM measurements of shear bands in brass and titanium are given.

INTRODUCTION

Pole figures can be measured from very small sample regions down to about 1 μm in diameter with the TEM in selected area diffraction mode (SAD) [1]. The high sensitivity of electron diffraction favors the study of extremely small sample volumes of fine grain material and high degrees of deformation. Promising fields of application are the analysis of local texture in thin films, multiphase textures, and shear bands after cold rolling. Correction procedures have been developed to interpret quantitatively the diffracted intensities [2]. Allowance is made for the increase in diffracting volume and absorption with increasing sample tilt. The measured pole figures are incomplete. The blind areas are biangular sections on the reference sphere. Hence standard programs cannot be used which have been developed for the calculation of the ODF from x-ray pole figures. An advantage is to apply an iteration method for pole figure inversion [3,4] whereby a constant but positive value is given to the immeasurable poles ("positivity method" [5]). Materials of low crystal symmetry, however, require increasingly more experimental pole figures to provide an accurate harmonic expansion of the ODF [6], while the diffraction rings are closer spaced or even overlap in electron diffraction patterns with decreasing crystal symmetry, and less SAD pole figures can be measured. Furthermore, statistics of the measured data may be poor, since only a small number of grains may fall in the measured field. For these cases the component method can be used to advantage to obtain texture estimates from incomplete pole figures of any crystal symmetry [7,10,11].

MINIMAL POLE FIGURE RANGES (MPR)

A necessary condition for the texture determination from pole figures is that the number of available pole figures and the measured ranges must be large enough to allow the unique determination of any crystal orientation. The minimum number and the range of pole figures only depend on crystal symmetry and the indices of the pole figures [8]. The minimum radius of a contiguous spherical segment, measured on the reference sphere, can be calculated easily using the MPR program [8]. For the (100), (110) and (111) reflections measured in the cubic system this radius is 45°, and for the (10.0) and (10.2) reflections in the hexagonal system (c/a = 1.587) the radius is 56°.

DETERMINATION OF TEXTURE COMPONENTS

Texture components are described by their orientation g, the half-width b and the intensity $\mathcal{I}$, which is related to the volume fraction, of the simulated Gauss peaks in the orientation space. The peaks, visible in diffraction pole figures, are checked successively for consistency in crystal direction and intensity in all experimental pole figures to find first estimates for the component parameters g and b [11]. A refinement of all component parameters and the determination of the unknown normalization factors of the pole figures can be done by using non-linear optimization methods. Manual operations of the component search and numerical refinement of the component parameters have to performed until residuals in the difference pole figures (difference between experimental and recalculated pole figures) are small.

Main benefits of the Texture Component Model are:

- If the necessary condition mentioned above is fulfilled, a texture estimate can be obtained from incomplete pole figures with uncommon measured areas and for any crystal symmetry. The pole figures may be superpositions of diffraction signals from different lattice planes and different phases [11].

- As a rule the use of texture components reduces the number of necessary data for the quantitative texture description [10]. Processes and properties correlated with texture are understood more readily due to the high compression of texture information. Pole figures can be calculated from the Component ODF which cannot be measured, if the diffraction rings overlap in the pattern or the structure factor is low.

EXAMPLES

Textures of shear bands were determined from SAD pole figures for brass and titanium (Figure 2, 3). To analyse the texture of shear bands and deformation bands, local pole figures of deformed Ti were measured (Figure 4). Several experimental pole figures measured along the bands have been added up before the calculation of the texture estimate was carried out in order to improve statistics. The calculated component description of the three ODF requires no more than 19 components in each case. Complete theoretical pole figures are calculated from the ODF estimate and displayed side by side along with the measured pole figures for a visual comparison (cf. Figures 2, 3, 4).

REFERENCES:

1) Schwarzer R.: Beitr. elektronenmikr. Direktabb. Oberfl., 1985, **18**, 61-68
2) Schwarzer R.: Beitr. elektronenmikr. Direktabb. Oberfl., 1983, **16**, 131-134
3) Weiland H. and S. Panchanadeeswaran: Textures and Microstr., 1993, **20**, 7-86
4) Xia Wei and R. Schwarzer : ICOTOM 10
5) Dahms M. and H.J. Bunge: Textures and Microstructures, 1988, **10**, 21-35
6) Bunge H.-J.: Texture Analysis in Materials Science. Butterworths 1982
7) Helming K. and T. Eschner: Cryst. Res. Technol., 1990, **20**, K 203
8) Helming K.: Textures and Microstructures, 1992, **19**, 45
10) Helming K.: ICOTOM 10, 1993,
11) Helming K., K. Kruse and S. Siegesmund: ICOTOM 10

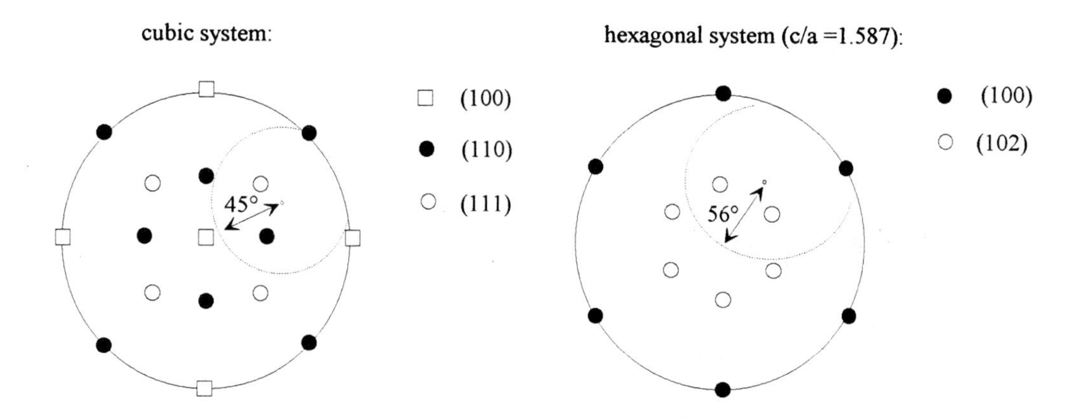

Figure 1: Minimal pole figure ranges for cubic crystal symmetry (O_h), detectable reflections (100),(110) and (111), and hexagonal symmetry with (100) and (102).

	Brass shear					Ti shear					Ti deformation				
c	$\mathcal{I}^c$	α^c	β^c	γ^c	b^c	$\mathcal{I}^c$	α^c	β^c	γ^c	b^c	$\mathcal{I}^c$	α^c	β^c	γ^c	b^c
1	34	37	38	87	8	21	92	38	30	12	52	274	23	19	24
2	30	97	36	37	14	18	67	20	52	20	27	91	30	58	22
3	26	94	21	42	13	13	289	29	9	15	25	83	38	55	13
4	23	135	29	27	13	12	84	37	8	12	23	270	35	5	16
5	21	38	31	80	9	10	323	24	46	13	18	272	37	9	21
6	20	6	40	83	13	10	268	41	43	8	13	295	23	31	13
7	18	171	44	8	7	7	104	22	44	8	8	208	16	10	9
8	17	86	23	45	11	7	252	19	49	8	8	317	21	36	13
9	15	21	35	83	9	6	340	18	24	8	8	89	55	59	8
10	15	170	44	7	9	5	14	23	28	6	6	52	19	15	10
11	15	276	44	15	12	5	160	72	36	8	6	86	66	57	9
12	13	145	22	23	7	5	267	89	4	8	5	277	47	2	7
13	12	30	41	8	7	4	310	37	40	6	4	96	36	39	6
14	9	21	45	60	8	4	68	2	48	8	4	130	18	4	10
15	9	356	44	79	7	3	23	19	30	4	4	45	21	21	10
16	6	32	31	79	6	2	326	47	24	5	4	89	68	48	10
17											4	322	22	31	7
18											3	89	51	55	4
19											3	269	55	3	7

Table 1: Component description of shear/deformation bands in brass and titanium with volume fraction $\mathcal{I}^c$ [$^o/_{oo}$], preferred orientation $g^c = \{\alpha^c, \beta^c, \gamma^c\}$ and half-width b^c [^{o}]. The orientation are described by three Euler angles α, β, γ rotating the sample coordinate system into the crystal coordinate system by three successive rotations on the topical $\vec{Z}, \vec{Y}$ and $\vec{Z}$ axis.

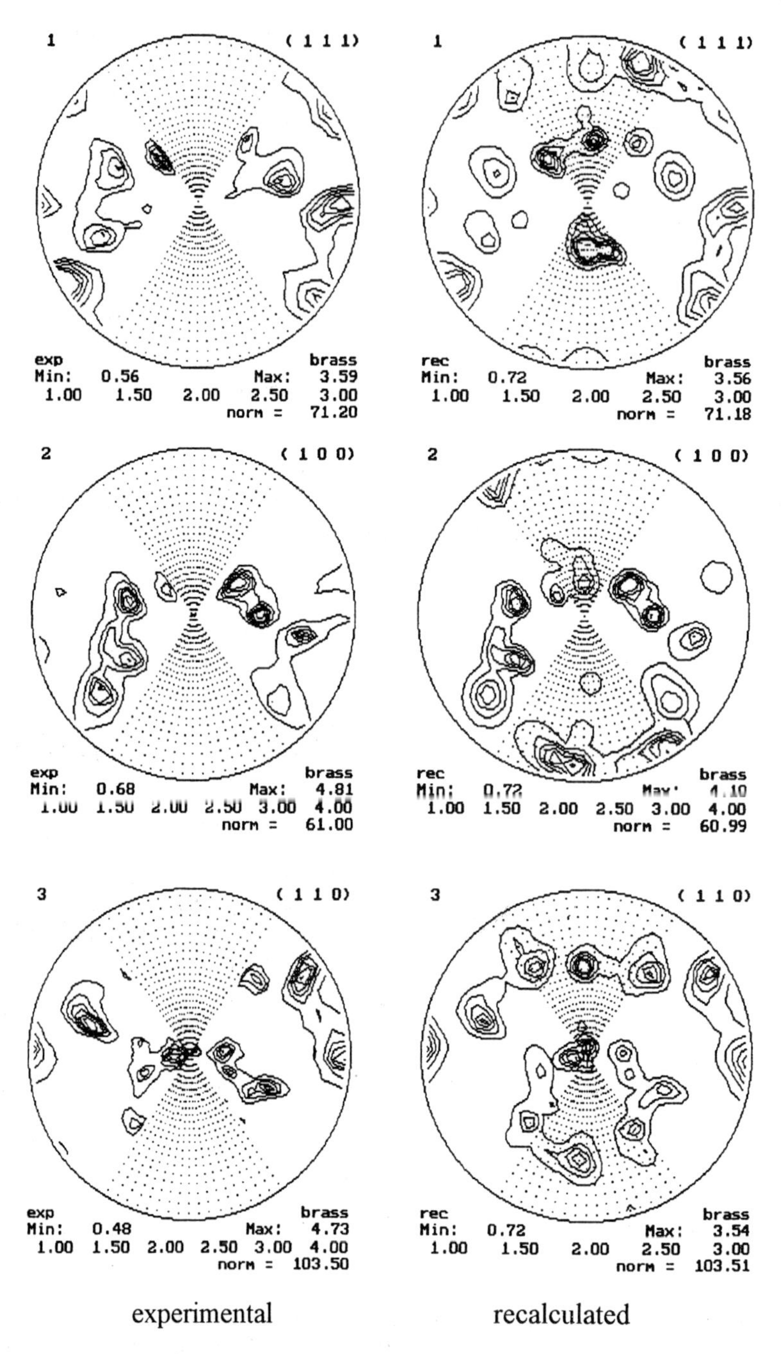

Figure 2: Incomplete experimental pole figures (111), (100) and (110) of a shear band in brass in comparison with the complete pole figures recalculated from the Component Model.

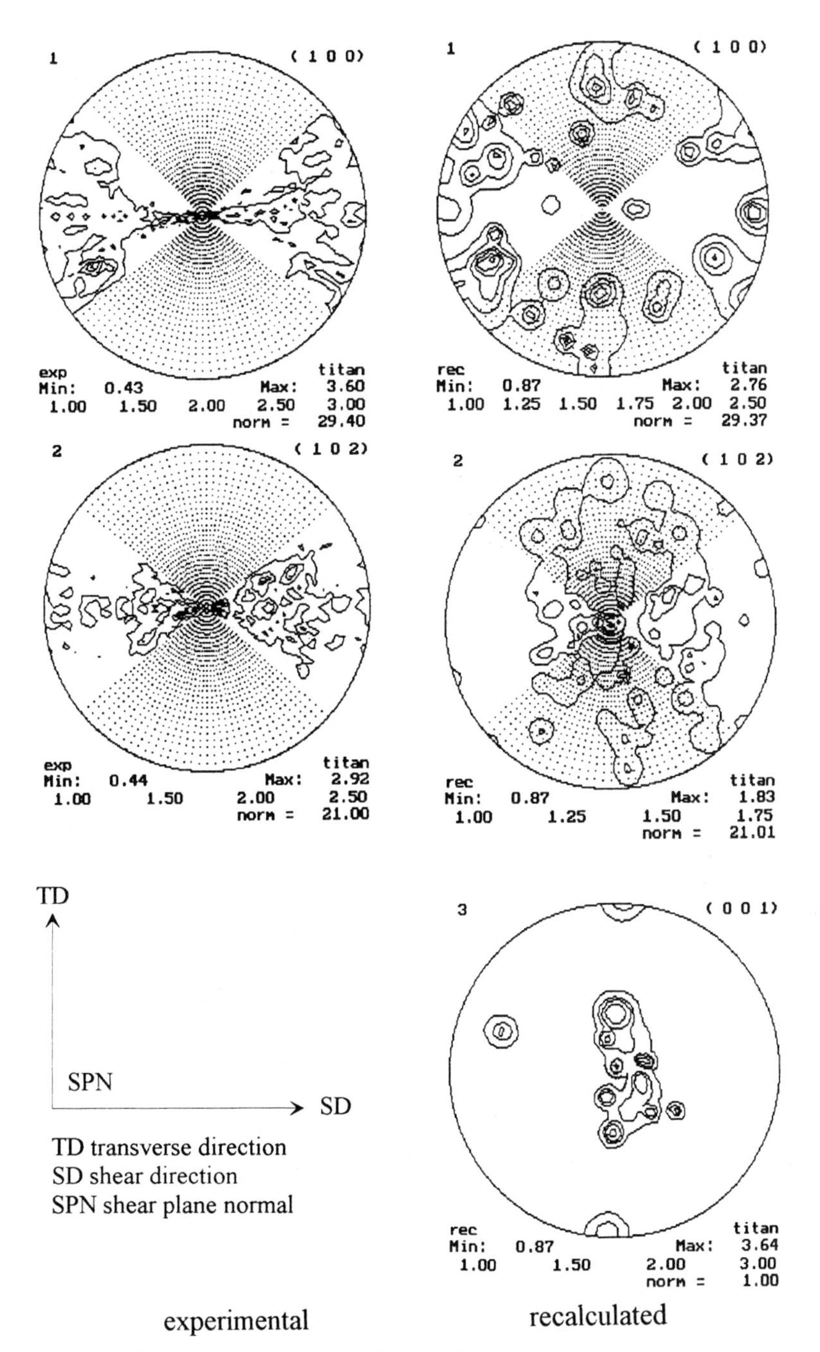

Figure 3: Incomplete experimental pole figures (10.0) and (10.2) of a shear band in titanium in comparison with the complete pole figures (10.0); (10.2) and (00.1) recalculated from the Component Model. Unmeasured regions are shaded.

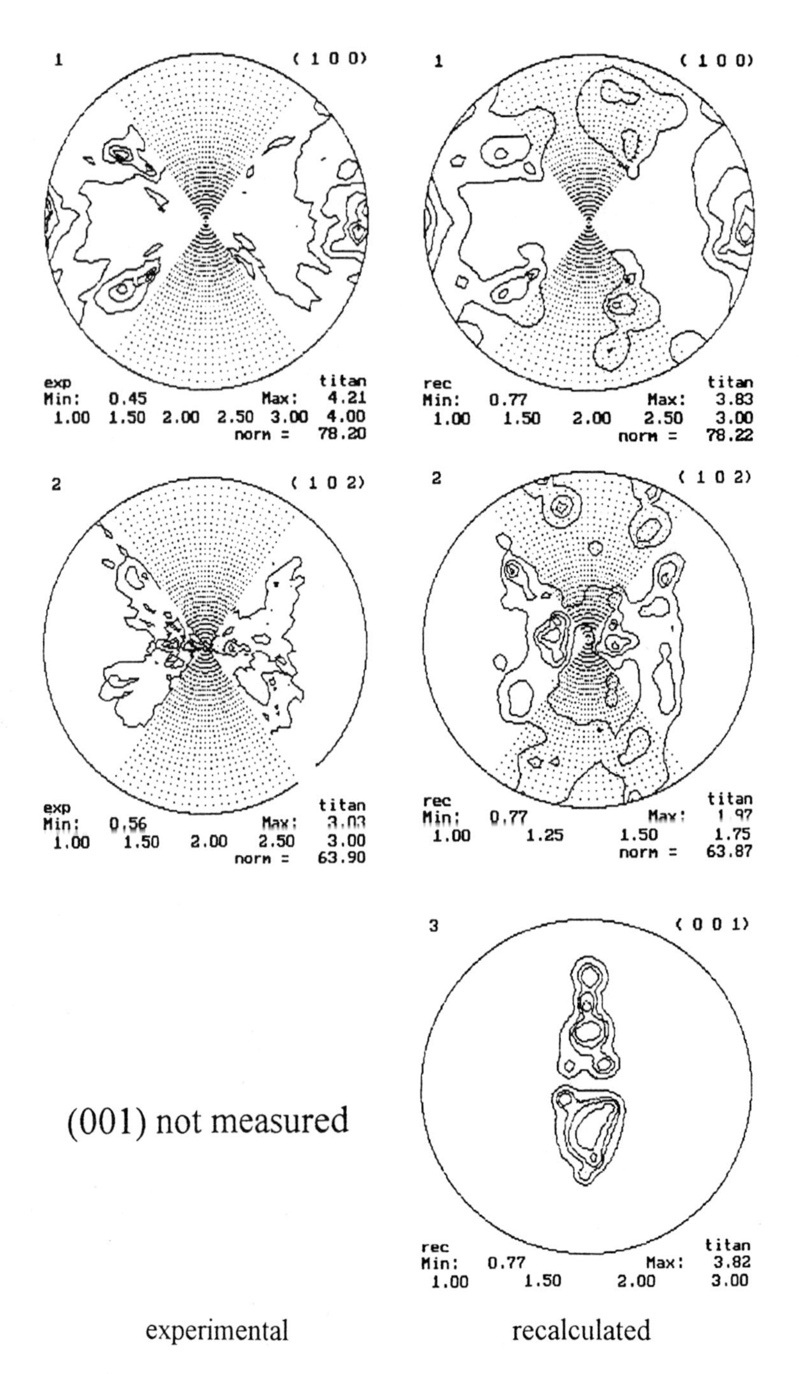

Figure 4: Incomplete experimental pole figures (10.0) and (10.2) of a deformation band in titanium in comparison with the complete pole figures (10.0), (10.2) and (00.1) recalculated from the Component Model. Unmeasured regions are shaded.

Materials Science Forum Vol. 157-162 (1994) pp. 1225-1230

EVALUATION OF THE HIGH TEMPERATURE β TEXTURE OF A SAMPLE OF TA6V FROM INDIVIDUAL ORIENTATION MEASUREMENTS OF PLATES OF THE α PHASE AT ROOM TEMPERATURE

M. Humbert, H. Moustahfid, F. Wagner and M.J. Philippe

LM2P, ISGMP UFR MIM-Sciences, Ile du Saulcy, F-57045 Metz Cedex 01, France

Keywords: High Temperature Texture, Single Orientation Determination, Phase Transformation

ABSTRACT

This contribution describes a method, which allows us to determine indirectly the O.D.F. of the high temperature β phase of a TA6V sample from the orientations of α plates at room temperature. In this method, it is indispensable to recognize, after phase transformation, the boundaries of the former parent β grains and to correlate, for each former parent β grain, several corresponding α plates inherited by phase transformation. The results obtained in the case of a TA6V specimen are compared with the texture of the residual β phase.

INTRODUCTION

The improvement of materials like Ti, Zr, Zy, TA6V destined to given applications often requires an adaptation of their microstructures and textures. This can be done by performing high temperature treatements such as rolling and forging which can modify the mechanical conditions of phase transformations occuring during the cooling down to room temperature. Of course, knowledge of the high temperature texture constitutes an important information for such a control. The aim of this work was to evaluate the high temperature ß texture of a TA6V sample. Unfortunately, the direct determination of the texture of the high temperature phase is not easy because it requires diffraction experiments through high temperature furnaces. Moreover, the determination of the high temperature ß O.D.F. ($f(g)_ß$) cannot be obtained from the inherited room temperature α O.D.F. ($f(g)_\alpha$). A possibility would be to consider the O.D.F. of the retained ß phase as a good representation of the high temperature ß O.D.F.. But the amount of the retained ß phase can be very small for satisfactory pole figure measurements. One must also notice, in particular cases, the superposition of the diffraction peaks of $(00.2)_\alpha$ and $(110)_ß$, which makes the O.D.F. determination uneasy.

An indirect method to evaluate the ß O.D.F, which has not yet been explored, is presented in this contribution. It consists in determining the orientation and the volume of a large number of the former ß grains which are statistically representative of the whole sample. The orientation of a former ß grain is obtained by correlating the orientations of several inherited α plates, resulting from its phase tranformation. This procedure is only possible if the α plates can be clearly localized in a former ß grain and if a relatively strict orientation relation between the α and β lattices exists. The orientations of the inherited α plates can be determined from E.B.S.P.

SAMPLE ELABORATION

The TA6V sample was obtained according to the following thermal treatment: the material was kept at 1050 C° for 30 minutes after heating up at a rate of 1400 C° per hour. At the end of this stage, the whole material was in ß phase. Then, a cooling down to room temperature followed at a rate of 500 C°per hour. So we obtained a material with a microstructure formed by very thin α plates occuring during the phase transformation of the former ß grains. An important fact is that the boundaries of the ß grains were easily recognisable due to the presence of very thin edges of α material. One observed that the ß grains were almost equiaxed with a size varying from 400 to 600 μm. The thin α plates were nevertheless large enough (greater than 1μm) to determine their orientations from E.B.S.P.. Moreover, this thermal treatment allowed us to retain approximately 20 percent of the ß phase at room temperature. This retained β material surrounded the boundaries of the α plates, as shown on micrographs of figures 1 and 2. The α edges make the boundaries of the former ß parent grains distinguishable, that is imperative in our method. Three families of α plates can be easily observed in each former ß grain. The mean size of the plates is 25 μm long and 6 μm wide.

PRINCIPLE OF THE DETERMINATION OF THE ORIENTATION OF A PARENT β GRAIN

The orientation of a given parent ß grain is represented by the rotation g_0^β which puts the macroscopical reference frame into concidence with the crystal reference frame.

The symmetrically equivalent rotations $\left\{S_i^\beta . g_0^\beta\right\}$ are also representative of this orientation. $\left\{S_i^\beta ; i=1,2 ...., 24\right\}$ are the 24 rotation elements of the symmetry group G_β of the initial cubic lattice. During the cooling down, a phase transformation occurs and each ß grain transforms into α plates of different orientations (variants), which are linked to the orientation of the former ß grain. Theoretically, 12 orientations (variants) can be found according to the symmetries of the phase before the transformation (cubic) and after the transformation (hexagonal). The ith orientation (variant) among the twelve is represented by the rotation g_i^α which is related to the initial rotation g_0^β by the relation [1]:

$$g_i^\alpha = \Delta g . S_i^\beta g_0^\beta \qquad \text{for a given } S_i^\beta \tag{1}$$

In relation (1), Δg represents the orientation relationship between the initial lattice ß and the final one α. In the case of the transformation from BCC (ß) to HCP (α), we used Burgers's relation [2] :

$$(110)_{BCC} \, // \, (0001)_{HCP}$$
$$[\bar{1}1\bar{1}]_{BCC} \, // \, [2\bar{1}\bar{1}0]_{HCP}$$

to express Δg in terms of Euler angles ($\Delta g = \{135°, 90°, 354°74\}$).

The set of rotations $\left\{S_j^\alpha . g_i^\alpha\right\}$ is also representative of this orientation (variant). In this expression, $\left\{S_j^\alpha ; i = 1, 2 ...., 12\right\}$ are the 12 rotation elements of the symmetry group of the final hexagonal lattice G_α. The orientation of one α inherited plate is described by one rotation, say g_1^α. The potential parent orientations of the variant g_1^α are defined by the 12 rotations corresponding to the 12 rotation elements S_j^α:

$$\Delta g^{-1} . S_j^\alpha g_1^\alpha \qquad (2)$$

These rotations can be classified into 6 groups of 2 rotations which define the same orientation. This fact is due to the commutation properties of Δg with S_j^α and S_i^β [1] . By assuming that the elements S_j^α have been arranged in such a way that S_j^α ($j = 1..6$) and S_{j+6}^α give the same potential parent orientation, the 6 potential parent orientations are defined by 6 sets of 24 rotations. A set corresponds to a fixed value of j ($1 \leq j \leq 6$)[1]:

$$\left\{S_i^\beta . \Delta g^{-1} . S_j^\alpha . g_1^\alpha\right\} \text{ for j fixed and } S_i^\beta \in G_\beta \qquad (3)$$

The unknown orientation of the parent β grain is one of these 6 orientations. These 6 orientations can be considered as a set A_1 of 6 elements. The rotation g_2^α, representing the orientation of a second α plate within the considered parent β grain allows us to find the corresponding set A_2 of 6 elements which also contains the solution. We proceeded in the same way with the α plates of different orientations belonging to the same β grain. The orientation of the parent β grain is the unique common element of all the set A_i. Therefore, the orientation of the parent β grain is found when the intersection of different sets A_i reduces to a unique element. We performed numerical tests which showed that the intersections of 3 sets obtained by considering 3 any variants taken among the 12 possible variants reduce to 1 unique element for 216 combinations out of the 220 possible combinations. The use of an additional fourth variant allows us to obtain the solution every time. In our investigation, we decided to use only 3 variants for such determinations. This was justified by the small frequency of the failing cases. Practically, in this study 5 determinations failed out of 120.

RESULTS

The orientations of at least 3 differently oriented α plates per β grain were deduced from the corresponding Electron Backscattering Patterns forming Kikuchi lines [3] on a phosphor screen inserted in a Scannig Electron Microscope (JEOL 820 SEM). EBSP's were indexed by a code provided by N. Schmidt [4]. Using the method described in the previous section, we succeeded in determining the orientations g_i^{β} of 115 parent β grains from 3*115 variants. Figure 3 shows the projection of the c-axes of 3*115 α plates. In this study the coordinate system (A, B, C) is chosen; B corresponds to the rolling direction. The micrographs show that the parent β grains had ,in average, the same volume. So we built up a continuous O.D.F. by superposition [5-6] of 115 Gaussian peaks of the same weigth $\omega_i = 1/N$, characterized by a fixed width Φ_i of 8° degrees.

The Phi1 sections of the β O.D.F. are shown in figure 4. One distinguishes 2 main orientations characterized by rotations (90°,40°,45°) and (270°,40°,45°) and their symmetrically equivalent ones in the Euler space. These 2 orientations can also be recognized in the corresponding (200) pole figure (figure 5). One observes a very good agreement between this figure and the experimental (200) pole figure (figure 6) of the β phase retained after cooling down to room temperature. Although this figure does not fully determine the texture of the β phase at room temperature, it appears that, in this particular case, the texture of the retained β phase gives a good indication of the texture of the high temperature β phase. Knowledge of the high temperature β texture also allows us to simulate the α texture, inherited during a phase transformation occuring without variant selection. By comparing this simulated α texture and the real one, it becomes possible to conclude about the type of transformation really occuring and further about the variant selection. This interesting point will be detailed elsewhere [7]. For the present investigation it is possible to conclude that the transformation occured without variant selection. This is corroborated by the comparison of figure 3 and figure 7, which shows the (00.2) pole figure of the α phase at room temperature, calculated from the β O.D.F. by assuming no variant selection.

CONCLUSION

Individual orientations (variants) of several α plates determined by EBSP can be used for the determination of the orientation of their parent β grain provided that a strict orientation relation between α and β lattices exists and that the boundaries of the parent β grain can be easily recognized. The additional knowledge of the volume of β grains makes the evaluation of the high temperature β texture possible. This method is particularly interesting when no high temperature diffraction equipment is available and if available, when the grain size is too large for a statistically reliable texture determination. Knowledge of the high temperature phase texture is indispensable in studying the variant selection, occurring during the $\beta-\alpha$ phase transformation of materials.

REFERENCES

1) M. Humbert, F. Wagner and C. Esling: *J. Appl. Cryst.*, 1992, 25, 724.
2) W.G. Burgers: *Physica*, 1934, 1, 561.
3) D.J. Dingley: *Proceedings of ICOTOM 8*, TMS., 1988, Warrendale .
4) N.H. Schmidt , J.B. Bilde-Sorensen and D. Juul Jensen: *Scanning Microscopy*, 1991, 3, 637.
5) H.J. Bunge: *Texture Analysis in Materials Science*, Butterworths,1982, London.
6) F. Wagner: in *Experimental Techniques of Texture Analysis*, , DGM,1986, Informationgesellschaft -Verlag.
7) M. Humbert, H. Moustahfid, F. Wagner and M.J. Philippe to be published

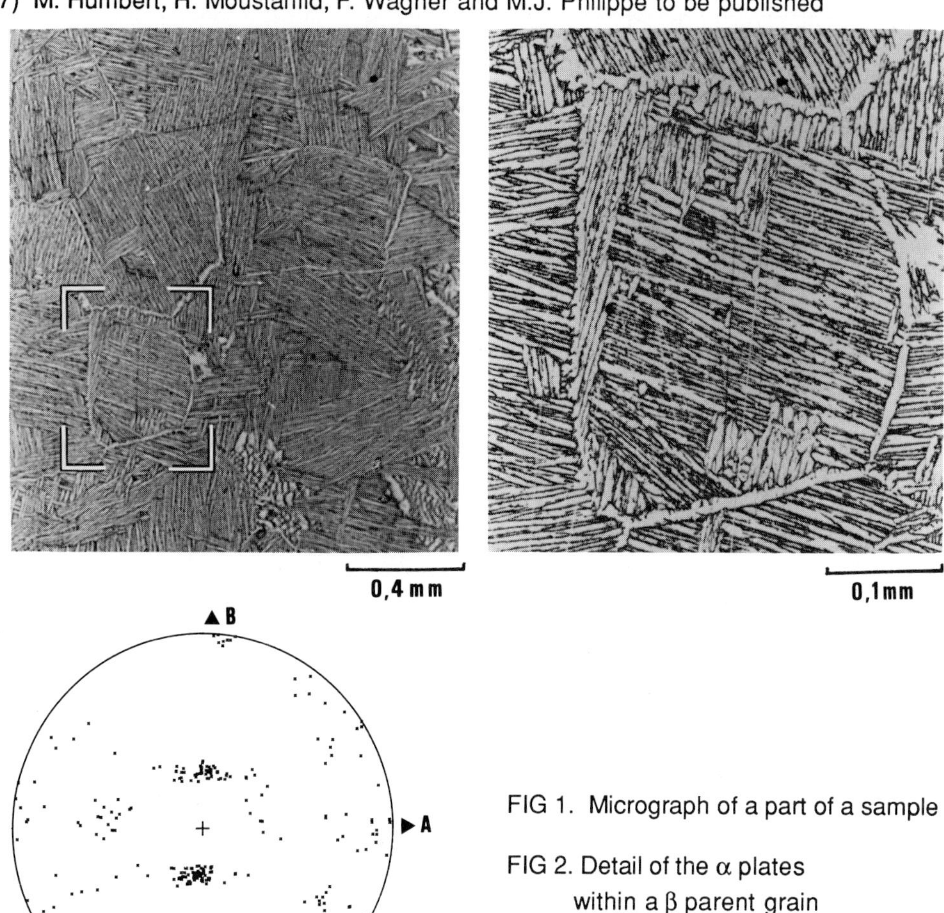

FIG 1. Micrograph of a part of a sample

FIG 2. Detail of the α plates within a β parent grain

FIG 3. Projections of the c axes of the 345 measured α plates

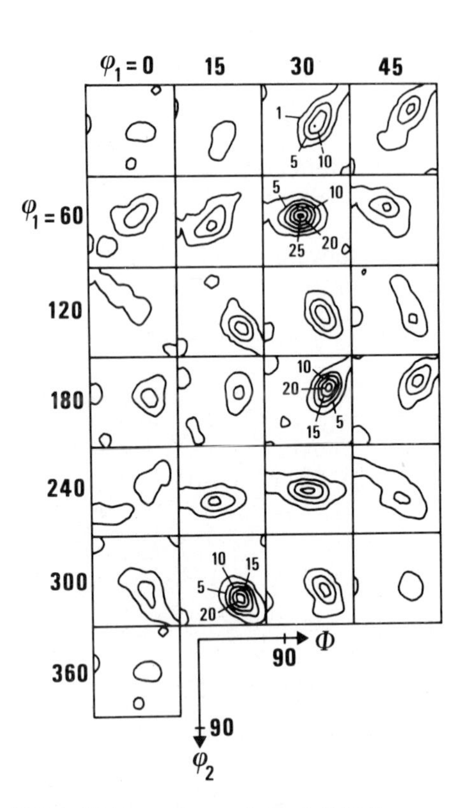

FIG 4. Phi1 sections of the β O.D.F. evaluated from the individual orientations of sets of 3 variants.

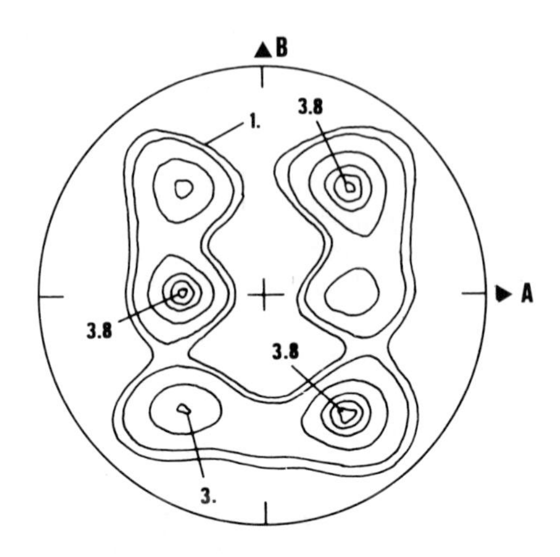

FIG 5. Calculated (200) pole figure of the β phase.

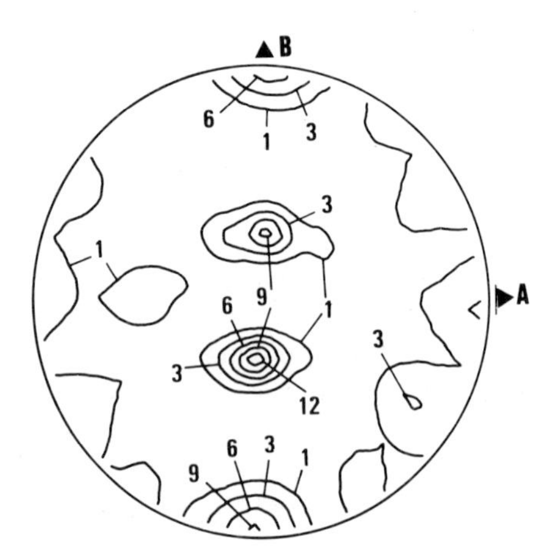

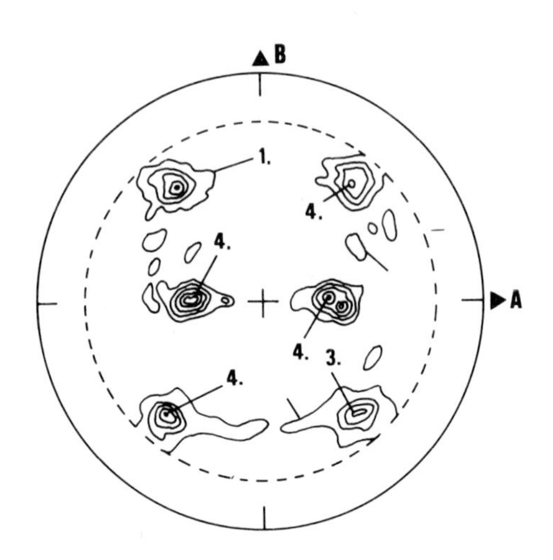

FIG 6. Experimental (200) pole figure of the residual β phase.

FIG 7. c pole figure corresponding to the recalculated α O.D.F.

Materials Science Forum Vol. 157-162 (1994) pp. 1231-1236

INFLUENCE OF SHEAR BANDING ON THE TEXTURE IN ROLLED AND CHANNEL-DIE COMPRESSED POLYCRYSTALLINE COPPER

Z. Jasienski, J. Pospiech, A. Piatkowski, J. Kusnierz, A. Litwora, K. Pawlik and H. Paul

Aleksander Krupkowski Institute for Metal Research,
Polish Academy of Sciences, ul. Reymonta 25, PL-30-059 Kraków, Poland

Keywords: Deformation Texture, Shear Bands, Channnel-Die Compression, Rolling

ABSTRACT

The development of shear bands and their crystallographic effects have been investigated in fine-grained and coarse-grained high purity copper in a wide range of deformation by rolling or channel-die. The results have shown that the scattering of the C-component appears on the one side in the {114}<221> position, and further on in the position {001}<110> and on the other side it appears in the position {111}<112>. A special attention is also given to the differences in the formation of textures produced by two deformation modes.

INTRODUCTION

The subject of investigations in the present paper is the influence of conditions imposed by channel-die (CD) or rolling (R) on the formation of textures. Differences are expected in the formation of texture components in connection with the differences in material flow [1,2,3]. Simultanously, the local influences on texture are considered, those which may become revealed through the geometry of macroscopic shear bands (MSBs) formation. Due to controlled CD compression, conditions are created for the generation of MSBs with relatively great local concentrations. The MSBs formed in rolled samples (especially in the middle layer) are rather randomly distributed and their volume fraction is relatively low. Investigations performed on single crystals indicate that the MSBs are often of crystallographic nature and may have a definite effect on texture [4,5]. The orientation distribution function (ODF) calculated from pole figures by the ADC method and employed in this paper creates the possibilities of identifying the crystallographic effects of the phenomenon of MSBs formation [6,7].

DEFORMATION GEOMETRY

The differences in material flow procured by these two deformation modes: CD and R should produce different texture characteristics. This is to be expected since the walls of the channel prevent the widening of the strips. In a frictionless CD compression (taking x, y, z parallel to the compression plane normal, channel wall normal and channel axis, respectively) the lateral spreading is not allowed $d\epsilon_{yy} = 0$, and because of volume constancy $d\epsilon_{zz} = -d\epsilon_{xx}$. The channel walls maintain a rectangular cross-section on the z-face, therefore $d\epsilon_{xy} = 0$. The other shears, $d\epsilon_{yz}$ and $d\epsilon_{xz}$, are both allowed. Thus it can be assumed that the rotation about the channel axis z is hindered, whereas, due to the constraint ϵ_{yz} the rotation about the compression direction x is the most preferred. In rolling the situation is more complex; here the friction plays an essential role. As a consequence of the existing differences in the geometry of deformation one should expect various contributions of the characteristic components C, S and B.

Investigations conducted on single crystals have shown that formation of MSBs is favoured in grains with the orientation C–{112}< 111 > [4,5], and that these bands are of crystallographic nature. The formation of MSBs through a coplanar slip in the octahedric planes is associated with the appearance of the L – {114}< 221 >

component characterizing the local texture of the MSB inclined at 35° angle to the compression plane. The appearance of the $N = \{111\}< 211 >$ orientation, characterizing the local orientation of a non-active MSB, corresponds to a change of the inclination of the MSB (rotation around TD) into a position parallel to the compression plane.

EXPERIMENTAL PROCEDURE

High purity (99.99%) copper samples of the grain size of 30 μm and 150 μm were used in CD compression or in rolling. Samples for CD compression of the dimensions $10 \times 10 \times 10$ mm^3 were wrapped in teflon foil. Those intended for rolling without lubrication had the dimensions $10 \times 10 \times 30$ mm^3 . For each sample obtained after a successive deformation stage (25, 40, 50, 60, 75, 80, 90, 95%) by means of unidirectional rolling and CD compression the metallographic observations of the MSBs development were performed and using x-ray reflection technique the four incomplete pole figures {111}, {200}, {220} and {113} were measured. These observations as well as the measurements were conducted for the middle layer and on the surface of each sample. The ODF as the basis for texture analysis was calculated from the measured pole figures by the ADC method [6] in the space of Euler angles $(\varphi_1, \phi, \varphi_2)$.

RESULTS

The calculation results of the ODF for each of the deformed samples were characterized here by the orientation densities (ODF-values) along the skeleton line running through typical positions of the rolling texture of copper: $C \cong \{112\}< 111 >$, $S \cong \{213\}< 364 >$ and $B \cong \{011\}< 211 >$. They are shown in Figs 1a and 1b for CD compressed or rolled fine-grained samples, respectively.

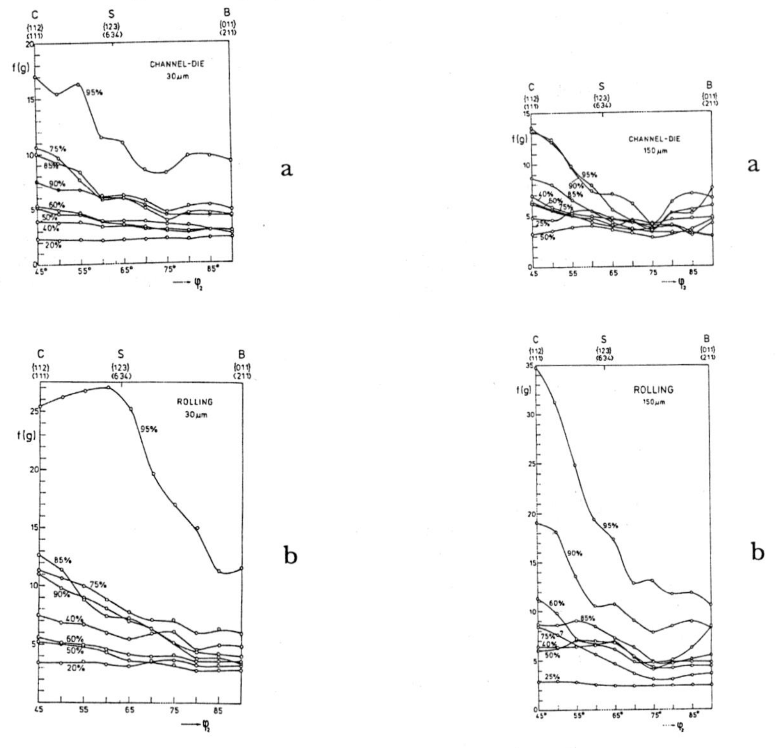

Fig. 1: Orientation density $f(g)$ along the β-fibre after various degrees of reduction by (a) CD compression and (b) cold-rolling. Fine-grained samples of 30 μm grain size.

Fig. 2: Orientation density $f(g)$ along the β-fibre after various degrees of reduction by (a) CD compression and (b) cold-rolling. Coarse-grained samples of 150 μm grain size.

From the ODF analysis it follows that there are no essential differences in texture formation in the fine-grained samples deformed up to 60% reduction in either of the applied deformation modes. A similar situation is observed in the coarse-grained samples deformed up to 50% of reduction. The distribution densities, concentrating along the same skeleton line for CD and rolling, increase weakly towards the C-orientation, attaining the maximum in the position ≅ (225)[$\bar{5}\bar{5}$4] for the fine-grained samples only. The differences in textures formed by CD compression or rolling can be observed mainly for fine-grained samples, in the interval of higher reductions, starting from 75%.

Generally in CD compressed samples the increase of the orientation density up to 95 % reduction is the highest in the surroundings of the C-position, whereas in rolled samples one observes a tendency for orientation concentration in the surroundings of C and S in such a way that at 95 % reduction the S-component with wide scattering comprising the C-position becomes prominent. The density in C-position in the textures of CD compressed and rolled samples does not increase in a monotonic way with the reduction. As it follows from Figs 1a and 3a, the density of the C-component in CD compressed fine-grained samples decreases at 85 % and 90% of reduction, and subsequently increases at 95 % reduction. Also in the rolling texture a diminution of the density in C-position (but on a less scale) is observed after 90% reduction (Fig. 1b), similarly as it has been shown in [8]. The decrease of the density in the C-position for the coarse-grained samples is observed first at 50 % of reduction and further at 75 % of reduction by either of the applied deformation mode (Figs 2 and 3).

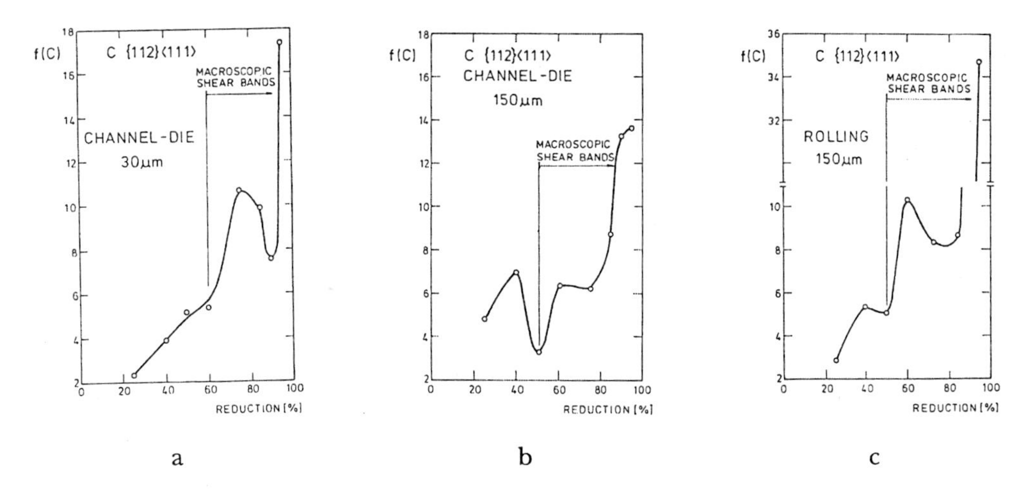

Fig.3: Density $f(C)$ of the C = {112}< 111 > position for various degrees of reduction: (a) CD compression. Samples of 30 μm grain size. (b) CD compression. Samples of 150 μm grain size. (c) Cold-rolling. Samples of 150 μm grain size.

It is remarkable that together with the density decrease in the C-position there appear specific changes in its surroundings, observed on the ODF cross-sections in Figs 4 and 5. In the case of CD compression in the interval of 75 % - 90 % reduction there occurs a splitting of the C-component, more clearly visible in the surface texture of both the fine-grained (Fig. 4) and the coarse-grained (Fig. 5) samples. The splitting area comprises, on the one hand, the position L = {114}< 221 >, and on the other the position N = {111}< 112 >. Traces of this splitting are still visible at 95 % reduction. Also starting from 75 % to 90 % reduction in the rolling texture of these samples a scattering of the C-component appears toward the L–position, with the tendency to comprise the positions between P = {001}< 110 > and {100}< 001 >, whereas toward the N-position only a weak scattering can be recognized.

It is important to note that in the surface texture of the rolled samples the scattering toward the N–position is not observed, on the other hand, the density in the P-position is stronger than in the middle layers of the samples. The described changes in the surroundings of the C-component become revealed in the deformation interval which is characterized by the development of MSBs both in the CD compressed and in the rolled samples (in fine-grained samples at 60 % reduction and in coarse-grained at 50 % reduction). As an example, the optical micrographs of MSBs traces on the lateral face (perpendicular to the transverse direction) of CD compressed fine-grained samples at 84.0 % and 85.5 % reduction are shown in Figs 6a and 6b. It can be observed (Fig. 6a), that the first MSB (A) intersecting one of the platen surfaces is reflected from that platen in a symmetric

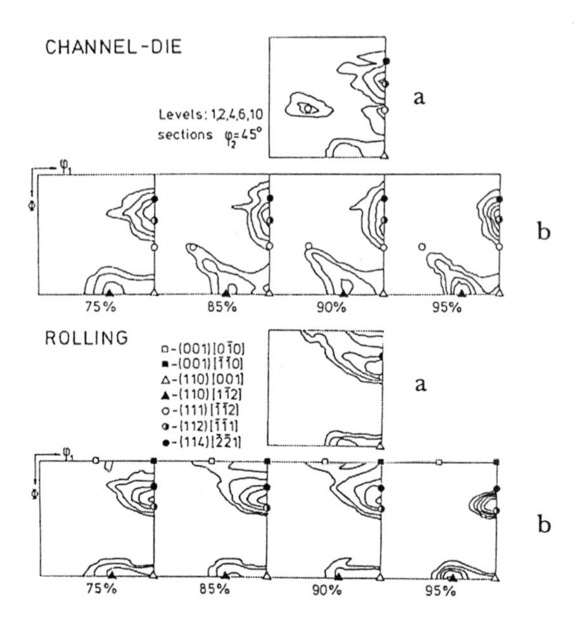

Fig. 4: ODFs representation in $\varphi_2 = 45°$ section for samples of 30 μm grain size, channel-die compressed or cold-rolled to 75 %, 85 %, 90 %, 95 %: (a) surface layers, (b) middle layers.

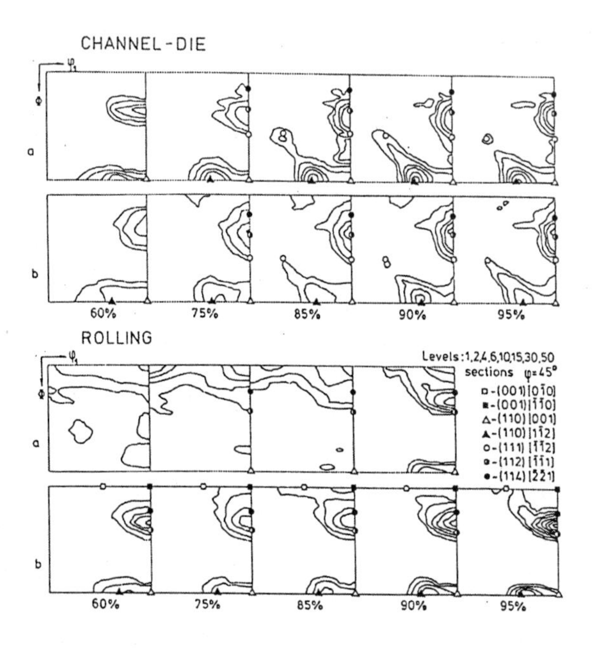

Fig. 5: ODFs representation in $\varphi_2 = 45°$ section for samples of 150 μm grain size, channel-die compressed or cold-rolled to 60 %, 75 %, 85 %, 90 %, 95 %: (a) surface layers, (b) middle layers.

kink-band like a macroscopic band. Inside this band, the second, newly formed MSB (B) appears, which tends to turn the non-active band A toward the compression plane. As the deformation proceeds (Fig. 6b) a subsequent set of MSBs is formed within the large kink band, so that the crossings of the bands are concentrated at the sample surface and in its middle part. The MSBs formed in rolled samples are rather randomly distributed and straight, especially in the middle layer (Fig.7). In this layer the volume fraction of the MSBs is relatively low.

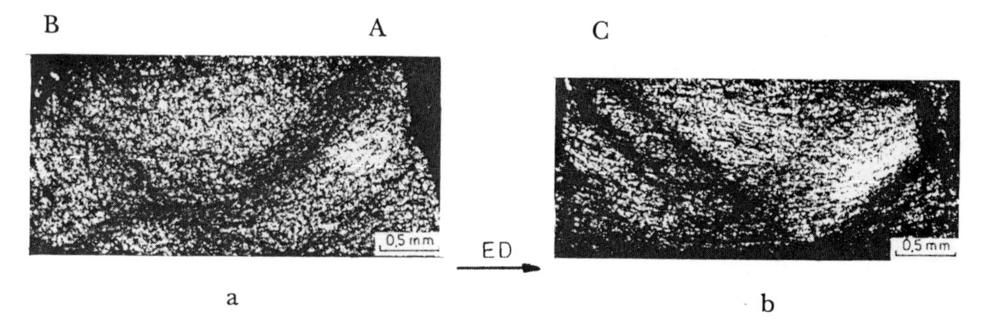

Fig. 6: Optical micrograph of channel-die compressed sample showing MSBs traces on the lateral face: (a) at 84.0 %, (b) at 85.5 %.

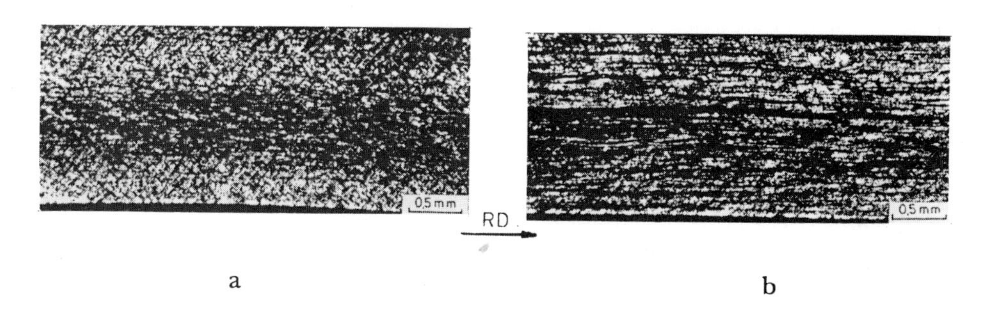

Fig. 7: Optical micrograph of cold-rolled samples at 85 % reduction showing MSBs traces on the face perpendicular to the transverse direction: (a) grain size of 30 μm, (b) grain size of 150 μm.

DISCUSSION AND CONCLUSIONS

In the textures of samples deformed by CD compression and by rolling the global difference is characterized by a smaller contribution of the S-component in the texture of CD compressed fine-grained samples, especially for 95 % reductions. The difference, already announced in literature [3] should be associated with the limitations imposed by the the channel geometry. Subtle changes of the density in the surroundings of C-orientation have not been an object of interest so far, as the C-component is generally assumed to be stable in the copper-type rolling or the CD texture. Simultaneously, it is an orientation which favours the occuring of localized shear instabilities in fcc single crystals. The result of the present study show that in CD compressed or rolled polycrystalline samples the decrease of the C-orientation density at large strains was preceded by the formation of MSBs at angles between 35° and 45° (55°) to the compression (or rolling) plane. The specific density changes in the surroundings of the C-orientation found in the texture of our samples are similar to those observed previously [4,5] for {211}<111> single crystals compressed in CD, where the MSBs formation was strongly dependent on the slip geometry. A favorable slip geometry is related with two coplanar {111}<110> slip systems or an equivalent system {111}<112> (perpendicular to the transverse direction, the <112> direction is the vector addition of two coplanar <110> directions). For C-orientation, the approach of the octahedric slip plane (inclined at an angle about 20° to the compression plane) to a position parallel to the MSB plane requires a rotation by the angle 15°–25° around the transverse direction (parallel to <110> direction). This rotation of the lattice localized inside MSBs increases the resolved shear stress on the MSB underlying the {111} slip plane and corresponds to the observed tendency for scattering of the C-component to the position L = {114}<221> and further on towards the P-position. The density increase in the L-position is the natural outcome of the concentration of the newly formed MSBs, increasing with deformation, and therefore it represents a common feature of both the CD and rolling textures.

The second feature, but not a common one to both methods is the tendency for scattering of the C-component into the N={111}<112> position. It is mainly observed in the CD compression texture at 85 % and 90 % reduction. This observation seems to be coupled with the tendency for MSBs reorientation, visualized by bending of the non-active MSBs (Fig. 6). There is a tendency for the band to align with the compression plane, which corresponds to the rotation of the MSB underlying {111} slip plane into a position parallel to the compression plane. It means that the non-active MSBs actually rotate in an opposite direction than does the lattice inside the MSBs. In samples compressed in CD the occurence of the N-component in the middle part of the sample should be associated with strong concentration of curvatures at the MSBs crossing locations. Observation shows that in the middle part of the rolled samples the MSBs remain generally straight and the density of the crossing locations is low, thus appearance of N-orientation cannot be expected. Instead in the N–orientation, the density in the P–orientation increases at 90 %. As it is known from the investigations of copper single crystals [3,4], the orientations N and P are not stable during rolling and CD compression. With increasing deformation the orientation density in the P-position diminishes in favour of the density in {211}<111> position, and the density in N–position falls apart into the {011}<100> and {211}<111> components. The results are consistent with the situation observed at 95 % reduction. The occurence of the cube component in rolled samples can be explained as being due to high rolling rate, which affects the formation of this component.

The obtained results indicate that the MSBs within polycrystals behave in a similar manner as those within single crystals and that the intaragranular crystallographic configuration controls the origin and the propagation of MSBs through the grain boundaries. The presented differentiation of the texture of samples subjected to rolling or CD compression is independent of the grain size which affects only the deformation degree necessary to initiate the macroscopic pracess of the shear bands formation.

REFERENCES

1) Wonsiewicz, B.C. , Chin,G.Y. , Met. Trans. 1970, **1**, 271.
2) Kocks,V.F., Chandra,H., Acta metall. 1982, **30**, 595.
3) Hirsch,J., Lücke,K., Acta metall. 1988, **36**, 2883.
4) Jasieński , Z. , Piątkowski , A. , Proceed. of the 8th ICSMA, Tampere, ed. P.O.Kettunen et al., Pergamon Press, 1988, **1**, 367.
5) Jasieński , Z. , Piątkowski , A. , Proceed. of the 9th ICSMA, Haifa, ed. D.G.Brandon et al., Freund Publ., 1991, **2**, 1025.
6) Pawlik, K.: Phys.Stat.Sol.(b). 1986. **134**, 477
7) Pawlik,K., Pospiech,J., Lücke,K., Proceed. ICOTOM9, Avignon 1990, 25.
8) Leffers,T., Hansen,N., Proceed. of the 13th Riso Int. Symp. on Materials Science, ed. S.I.Andersen et al., Riso Nat. Lab. 1992, 57.

Materials Science Forum Vol. 157-162 (1994) pp. 1237-1242

LOCAL ORIENTATION CHANGES DURING SHEAR BANDING OF COPPER SINGLE CRYSTALS COMPRESSED IN CHANNEL-DIE

Z. Jasienski [1], T. Baudin [2], R. Penelle [2] and A. Piatkowski [1]

[1] Aleksander Krupkowski Institute for Metal Research, Polish Academy of Sciences, ul. Reymonta 25, PL-30-059 Kraków, Polen

[2] Laboratoire de Métallurgie Structurale, Université Paris-Sud, F-91405 Orsay, France

Keywords: Single Crystals, Deformation Textures, Channel-Die Compression

ABSTRACT

Experimental results concerning the appearence of the local orientation changes inside a macroscopic shear band formed in channel-die compressed $(112)[\bar{1}\bar{1}1]$ copper single crystals are presented.

INTRODUCTION

The opinion that the macroscopic shear bands (MSB) are of non-crystallographic nature is widespread in literature [1,2]. It is expected that these bands with no specifed crystallographic habit plane do not produce any orientation change in the crystal matrix. This suggestion was used to account for the stability of the $(112)[\bar{1}\bar{1}1]$ orientation in the channel-die compressed copper single crystals (or the C-component in copper rolling texture), since this orientation favours the occuring of localized MSB.

In the author's earlier works [3,4] a hypothesis was suggested that the scatterings of the $(112)[\bar{1}\bar{1}1]$ orientation observed at large strains are mainly related to the local orientation changes inside the macroscopic shear band and that these bands are of crystallographic nature [5,6].

In the present paper some experimental results supporting this hypothesis are presented.

EXPERIMENTAL METHODS

Copper single crystals of the $(112)[\bar{1}\bar{1}1]$ orientation were used in the experiments. The samples 10 x 10 x 10 mm^3 were channel-die compressed (Fig.1) up to 27% and 75% of reduction.

The neutron diffraction was used to determine the global texture in the volume of the deformed crystals, including the expected dispersion of the orientations within a macroscopic shear band.

To determine directly the local dispersion of orientations within the shear band and to define the orientation of the shear plane, the X-ray measurements of the {111} pole figures were carried out in the plane of sections made along the middle plane of a macroscopic shear band (Fig.1a, section 1) and along a plane outside the band (Fig.1a, section 2) in the bordering zone of the crystal matrix.

Using the electron back-scaterring patterns (EBSP), the dispersion of the orientations was determined in the deformed matrix and within the macroscopic shear band. Patterns of Kikuchi lines were obtained from the lateral face of the crystal.

The measurements by neutron diffraction and the EBSP method were performed at the Laboratoire de Métallurgie Structurale of the University Paris-Sud.

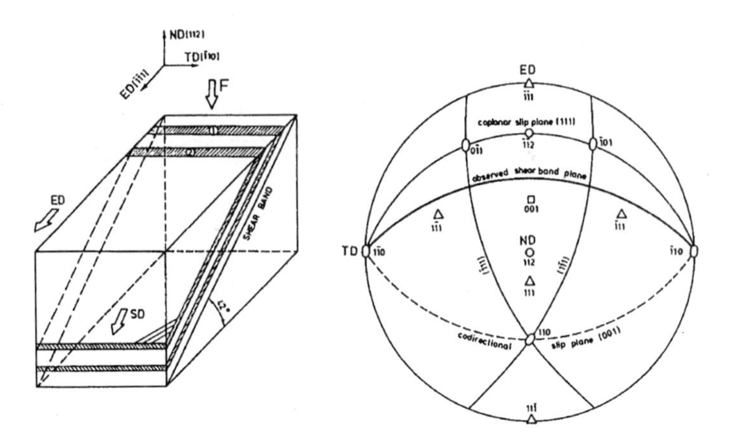

Fig.1. Macroscopic shear band in a channel-die compressed single crystal of the (112)[$\bar{1}\bar{1}1$] orientation.

RESULTS

Optical microscopy. In the channel-die compressed (112)[$\bar{1}\bar{1}1$] single crystals the operation of a single, perfectly straight macroscopic shear bands is observed at 27% of reduction (ε=0.32). The transverse direction (TD) is parallel to the MSB's material plane and this plane is inclined at an angle of about 42° to the extension direction (ED), as sketched in Fig.1a. The MSB, about 600 μm thick is composed of a set of narrow shear bands which are non-regularly spaced and are inclined to the extension direction at various angles from 35° to 50° (Fig.2). The apparent non-crystallographic nature of the MSB's material plane is visualized on a stereographic projection in Fig.1b. The trace of the MSB is misorientated with respect to the octahedric plane (111) of a coplanar slip by a 22° rotation around the transverse direction toward the compression axis.

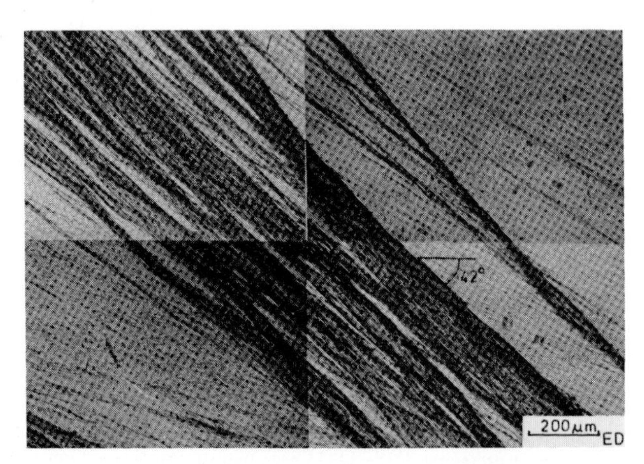

Fig.2. Optical micrograph of the MSB trace on the lateral face of the (112)[$\bar{1}\bar{1}1$] single crystal of copper at 27% of reduction. Note the misorientation between the narrow shear bands inside the MSB.

Neutron diffraction. The figures of the poles {111} and {200} determined by the neutron diffraction measurements (Fig.3) show that past the onset of MSB localization, the spreading of the (112)[$\bar{1}\bar{1}1$] matrix orientation increases with deformation by channel-die compression. These are two tendencies in the lattice rotations, the rotations being essentially around the transverse direction (parallel to [$\bar{1}$10]), but each of them is in opposite sense.

At the 27% of reduction (Fig. 3a and 3b) the scattering of the (112)[$\bar{1}\bar{1}1$] matrix orientation appears toward the (114)[$\bar{2}\bar{2}1$] position with the tendency comprising the (116)[$\bar{3}\bar{3}1$] and next the (001)[110] positions at the limits of the scattering area (Fig.3b). According to this tendency, the angle between the octahedric plane (111) of the coplanar slip and the compression plane increases from 35° at (114)[$\bar{2}\bar{2}$1] position, through about 42° at (116)[$\bar{3}\bar{3}$1] one, to about 50° near the (001)[110] position. Note that in the matrix orientation this angle is of 19.5° only. The described misorientation of the crystallographic slip plane (111) is in good agreement with those of the nawrrow shear bands material planes (35°-50°) as observed within the MSB (Fig.2). Thus, the first tendency of the scattering of (112)[$\bar{1}\bar{1}$1] matrix orientation will be associated with the local orientation changes inside the newly formed MSB.

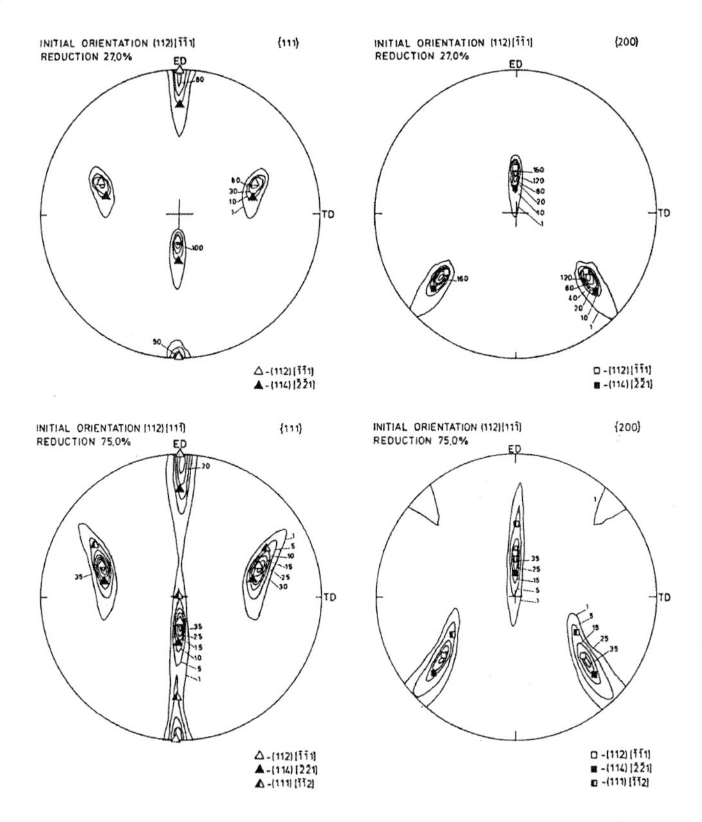

Fig.3. The {111} and {200} pole figures obtained by the neutron diffraction showing the scatterings of the (112)[$\bar{1}\bar{1}$1] matrix orientation in channel-die compressed single crystals of copper: a) and b) At 27% of reduction (ε=0.32); c) and d) At 75% of reduction (ε=1.39).

At the 75% of reduction (ε=1.39), the increase of scattering of the (112)[$\bar{1}\bar{1}$1] matrix orientation is mainly connected with the second tendency of the scattering toward the (111)[$\bar{1}\bar{1}$2] position as shown in Fig.3c and Fig.3d. This new tendency appears together with the first one which represents the opposite sense of the crystal lattice rotation around the transverse direction. Since the first tendency of scattering has not disappeared with increasing deformation, the opposite one seems to be related with the bending of the MSB as a whole, leading to an S-shaped band. In this case the near platen parts of the MSB turn into a position nearly parallel to the compression plane.

X-ray diffraction. The {111} pole figures are represented in the plane of sections parallel to the material plane of the newly formed MSB in the crystal compressed up to 27% of reduction. The {111} pole figure (Fig.4a) from section 1 situated inside the MSB shows that the positions of the MSB's material plane and the shear direction (SD) are defined by the (111)[$\bar{1}\bar{1}$2] orientation. Accordingly, the (116)[$\bar{3}\bar{3}$1] orientation (as marked by the positions of the normal (ND) and the extension (ED) directions) is the main component of the local texture of this MSB inclined at the angle of 42° to the extension direction. In the matrix zone bordering the MSB,

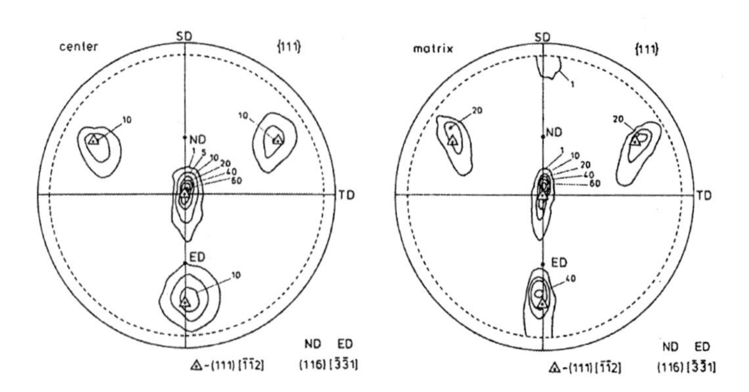

Fig.4. The {111} pole figures obtained by the X-ray diffraction in the plane of the sections: a) Inside the MSB formed at 27% of reduction; b) Outside the MSB in the matrix zone 2, bordering the band.

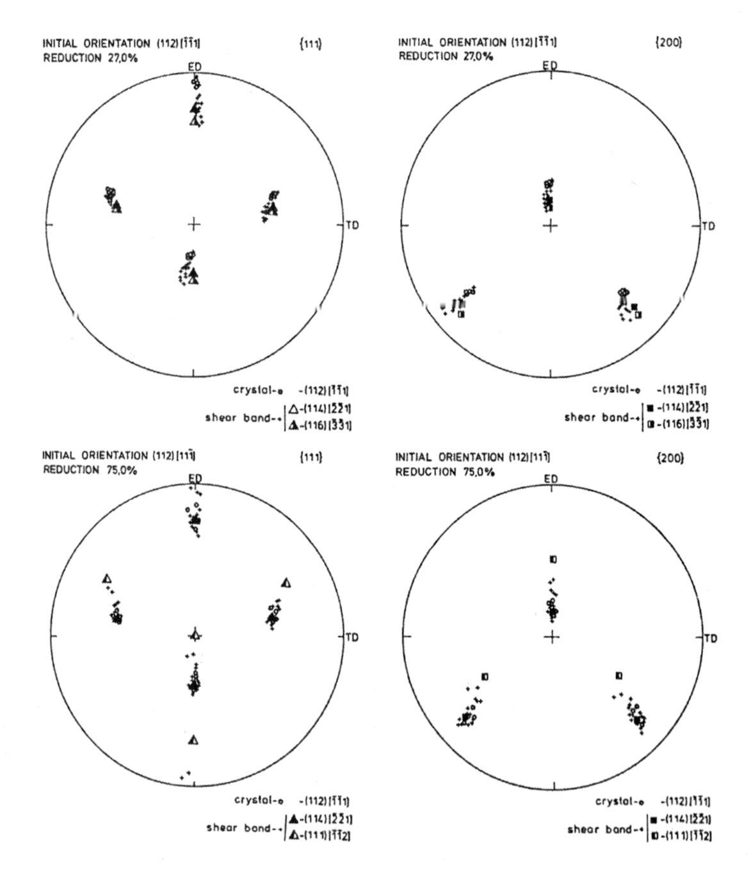

Fig.5. The {111} and {200} pole figures obtained from the EBSP measurements on the lateral face of the (112)[$\bar{1}\bar{1}$1] single crystal of copper: a) and b) At 27% of reduction (ε=0.32); c) and d) At 75% of reduction (ε=1.39).

as shown in Fig.4b, the maximum of the density of poles {111} is shifted by about 10° from the center of the pole figure toward the normal direction (ND) which indicates that the initial matrix orientation is approached. These results confirm directly that the first tendency of scattering of the (112)[$\bar{1}\bar{1}$1] matrix orientation is related only with the local orientation changes inside the newly formed MSB. On the other side, taking into account the observed lattice rotation inside the MSB, the crystallographic nature of the (111)[$\bar{1}\bar{1}$2] shearing seems to be well evidenced. Under plane strain conditions, a combination of two coplanar and equally stressed slip systems (a <211> direction is the resultant of two <110> slip directions) operates inside the MSB.

EBSP measurements. The orientations determined in the matrix of the deformed crystal and inside the MSBs are represented separately in the {111} and {200} pole figures (Fig.5).

At 27% of reduction, the initial (112)[$\bar{1}\bar{1}$1] orientation remained essentially unchanged in the matrix, as observed in Fig.5a and Fig.5b. On the other hand, the local orientation changes are concentrated inside the MSB, as it is expected from the neutron diffraction pole figures and can be expressed by the same (114)[$\bar{2}\bar{2}$1] and the (116)[$\bar{3}\bar{3}$1] positions.

At 75% of reduction (Fig.5c and 5d) the new tendency of scattering toward the (111)[2$\bar{1}\bar{1}$] position is cleary observed inside the MSB. It is worth noticing that at this large strain, the orientation of the matrix is shifted by about 10° from its initial (112)[$\bar{1}\bar{1}$1] position. It can be related to the overall shape change of the deformed crystal.

CONCLUSIONS

1. The scattering of the (112)[$\bar{1}\bar{1}$1] matrix orientation at large strain is the natural outcome of the local lattice reorientation inside the newly formed macroscopic shear band.
2. The material plane of the MSB and the shear direction are nearly parallel to a crystallographic {111} plane and a <211> direction, respectively.
3. In a channel–die compressed (112)[$\bar{1}\bar{1}$1] single crystal, the formation of the MSBs through a combination of two coplanar slip systems is related with the appearance of the (114)[$\bar{2}\bar{2}$1] and (116)[$\bar{3}\bar{3}$1] components of texture, characterizing the local orientation changes inside the MSB.

REFERENCES

1) Lee,W.B., Chan,K.C., Scripta metall.mater., 1990, **24**, 997.
2) Ananthan,V.S., Leffers,T., Hansen,N., Scripta metall.mater., 1991, **25**, 137.
3) Jasieński,Z., Piątkowski,A., Proceed. of the 8th ICSMA, Tampere, ed. P.O.Kettunen et al., Pergamon Press, 1988, **1**, 367.
4) Jasieński,Z., Piątkowski,A., Proceed. of the 9th ICSMA, Haifa, ed. D.G.Brandon et al., Freund Publ., 1991, **2**, 1025.
5) Donadille,C., Valle,R., Dervin,P., Penelle,R., Acta metall.,1989, **37**, 1547.
6) Harren,S.V., Dève,H.E., Asaro,R.J., Acta metall., 1988, **36**, 2435.

Materials Science Forum Vol. 157-162 (1994) pp. 1243-1250

LOCAL MICROSTRUCTURAL INVESTIGATIONS IN RECRYSTALLIZED QUARTZITE USING ORIENTATION IMAGING MICROSCOPY

K. Kunze [1,3], B.L. Adams [1], F. Heidelbach [2] and H.-R. Wenk [2]

[1] Dept. of Manufacturing Engineering,
Brigham Young University, 435 CTB, Provo, UT 84602, USA

[2] Dept. of Geology and Geophysics,
University of California, Berkeley, CA 94720-4767, USA

[3] Present address: Institut für Metallkunde,
Grosser Bruch 23, D-38678 Clausthal-Zellerfeld, Germany

Keywords: Orientation Imaging, Quartz, Misorientation Distribution

ABSTRACT

The crystallographic texture of a recrystallized quartz mylonite has been measured by electron backscatter Kikuchi diffraction using an automated indexing system (orientation imaging microscope). The results compare excellently with neutron diffraction data from the same sample and prove the reliability of the method for trigonal crystal symmetry. The ability of the technique to identify Dauphiné twins is demonstrated. The misorientation distribution function for nearest neighbors was derived from the single orientation measurements. Dauphiné twins (60° or 180° rotation around [0001]) form the dominant maximum in the misorientation distribution.

INTRODUCTION

The development of preferred orientations during recrystallization depends strongly on the generation and mobility of high angle grain boundaries [1]. A grain boundary is characterized by the misorientation between the two adjacent crystals as well as the shape and orientation of the grain boundary itself. The misorientation between two grains can be derived only from single orientation measurements.

The approach of single grain measurements is well-known to structural geologists. Back in 1950 Sander [2] introduced the "axis distribution analysis" (AVA) which correlates local orientations with the microstructure. Later such measurements were used to obtain statistical information about neighbor misorientations [3]. These early studies relied on orientation measurements with the optical microscope. Particularly for quartz they were restricted to c-axis data. More recently geologists have been using the technique of electron channeling in the SEM for orientation mapping of quartz polycrystals [4,5].

The technique of orientation imaging microscopy (OIM) [6,7] uses the automated analysis of electron backscatter Kikuchi diffraction (BKD) patterns in the SEM. It provides an excellent tool to obtain complete and statistically reliable data about local crystal orientations. In this paper the OIM method has been applied to a recrystallized quartz mylonite. The main question was, how far the strong texture of this mylonite is accompanied by misorientation correlations, i.e., whether the grain boundaries show preferred misorientations, which could be used to interpret the recrystallization behaviour of quartzite.

SAMPLE DESCRIPTION AND EXPERIMENTAL DETAILS

The sample BRG420 from the Bergell Alps is a fully recrystallized quartz mylonite, which contains besides quartz only a few muscovite grains that define the macroscopic foliation. The sample has also a well defined lineation. The microstructure is characterized by coarse grains (average diameter about 500 μm) with little undulatory extinction and sharp grain boundaries. Neutron diffraction experiments performed on a cube of 8 cm^3 from the same sample are described in more detail elsewhere [8].

A slab of the sample cut parallel to the lineation and perpendicular to the foliation was mechanically ground down to a particle size of 0.25 μm and then polished with SYTON silica solution for about 12 hours in order to remove surface damage. The sample surface was not carbon coated because this would have degenerated the backscattered diffraction patterns. Charging effects in the SEM were reduced by using a low accelerating voltage (8 kV). The OIM measurements covered an area of about 14 x 14 mm^2, which was scanned on a regular hexagonal grid with a stepsize of 50 μm.

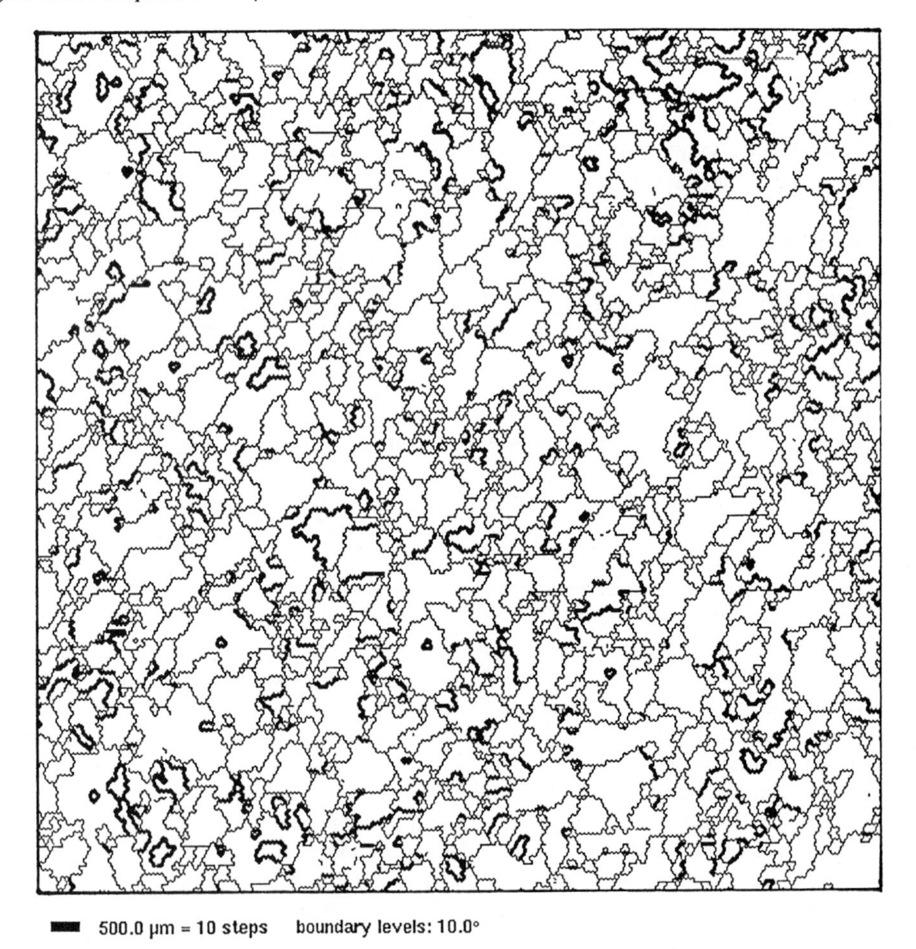

Figure 1: Orientation imaging micrograph of the Quartz sample; thin lines: grain boundaries with misorientation angle greater than 10°; thick lines: Dauphiné twin boundaries.

METHOD

Orientation imaging microscopy [6,7] is a fully automated technique to rapidly identify electron backscatter Kikuchi diagrams. Here the method has been adapted to investigate materials of low crystal symmetry, in particular quartz (trigonal symmetry). Compared to BKD patterns from cubic materials the patterns from low symmetry crystal structures are composed of a much larger number of different reflections, each of them having only a small symmetry-induced multiplicity. Based on structure factor calculations and direct observation of the BKD patterns a table of the main reflections has been created as a data base for the indexing procedure (table 1). As required for an unique indexing, this list contains mostly true trigonal (hkil)-families, with their counterparts (khil) having negligible scattering intensities. The positive and negative rhombohedral forms $(10\bar{1}1)$ and $(01\bar{1}1)$ can be distinguished by the different intensity of their Kikuchi bands. Thus it is possible to determine the complete crystal orientation. However, lefthanded and righthanded quartz crystals are not distinguished. The ability of the technique to correctly index BKD patterns of any possible orientation has been tested by numerical pattern simulation for all orientations on a 5° grid in one fundamental region of orientation space.

Figure 2: Reflected light micrograph of the same area as in figure 1.

Table 1: Main reflections in BKD patterns from Quartz (ordered by intensity)

Reflection Family (hkil)	Lattice Spacing d [Å]	Structure Amplitude [relative units]	Effective Symmetry
$10\bar{1}1$	3.37	100.0	trigonal
$20\bar{2}2$	1.68	18.8	
$01\bar{1}1$	3.37	41.9	trigonal
$02\bar{2}3$	1.39	26.7	trigonal
$30\bar{3}1$	1.37	22.9	trigonal
$10\bar{1}0$	4.26	20.6	hexagonal
$20\bar{2}0$	2.13	12.5	
$11\bar{2}2$	1.83	20.6	hexagonal
$01\bar{1}2$	2.31	14.7	trigonal
$11\bar{2}0$	2.46	13.9	hexagonal
$21\bar{3}1$	1.54	10.8	trigonal

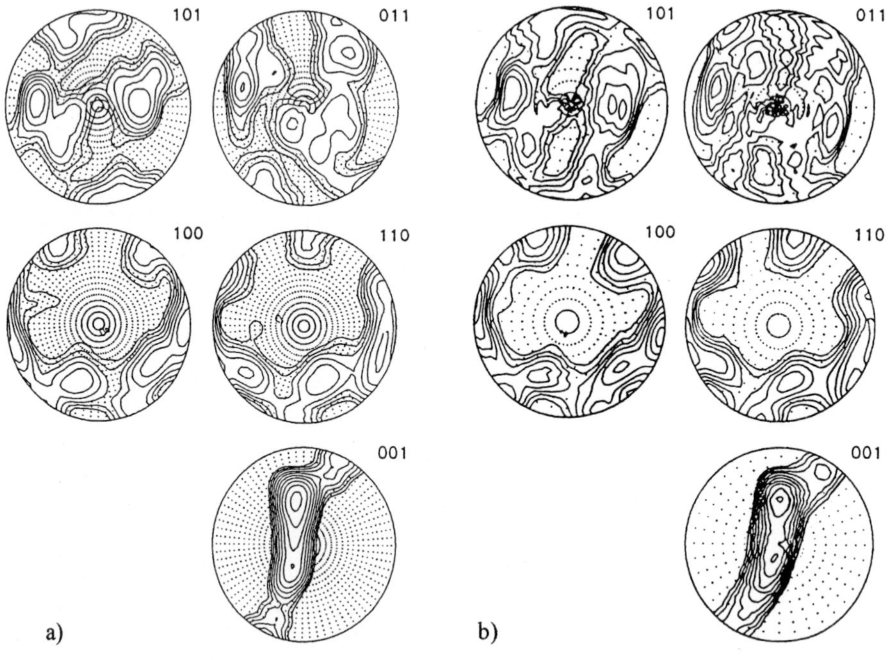

Figure 3: Pole figures obtained from OIM measurements (a) and neutron diffraction (b); equal area projection perpendicular to foliation and parallel to lineation, trace of foliation is horizontal; contours are 0.5, 0.71, 1.0, 1.4, 2.0, 2.8, 4, 5.6, 8 and 11.2 m.r.d., dotted below 1.0 m.r.d.

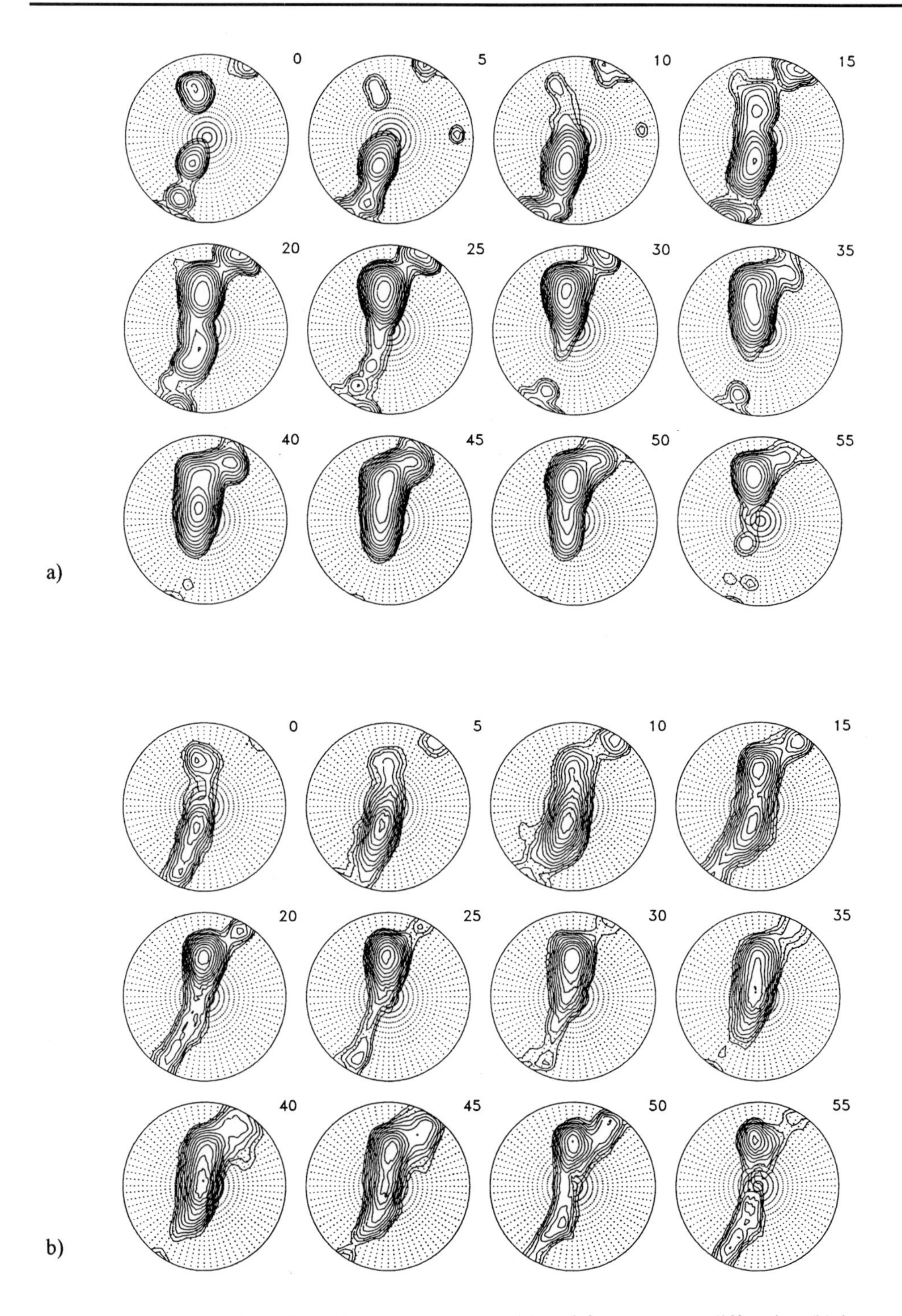

Figure 4: ODFs obtained from OIM measurements (a) and from neutron diffraction (b) in σ-sections; contours as in figure 3.

CALCULATION OF DISTRIBUTION FUNCTIONS

Continuous distribution functions like ODF, MDF and pole figures were calculated from individual grain orientations using convolution methods with Gaussians of 15° scattering width. The presented results obtained by harmonic expansions compare well with calculations by discrete methods [9,10] (not shown here). For each nearest neighbor pair on the measurement grid the misorientation with the smallest possible rotation angle was derived. To eliminate neighbor pairs within the same grain, only misorientations with a minimum rotation angle greater than 10° have been considered for the MDF calculation. According to the trigonal crystal symmetry the fundamental region of distinguishable misorientations does not contain rotations with rotation angles greater than 104.5° [11]. The misorientation space displayed here still contains a threefold redundancy due to the threefold rotation symmetry about the c-axis [0001]. The additional mirror symmetry of the MDF inside of these basic regions (120°-sectors) results from the assumption, that any misorientation and its inverse were considered to be equivalent.

RESULTS

Grain maps were formed by analysis of the misorientations of nearest neighbor pairs on the measurement grid. Misorientations with rotation angles exceeding 10° are represented by a thin line segment between the respective points. Misorientations close to the twin relationship of a 180°(=60°) rotation about [0001] are highlighted by a thick line. The grain map (figure 1) compares well with a reflected light micrograph of the measured area (figure 2). Particularly the twinned areas (thick boundaries in grain map) can easily be identified in the image of the etched surface structure of the sample.

The ODFs and pole figures calculated from the OIM measurements compare excellently with those derived from neutron diffraction (figures 3,4). Considering that the results were obtained by completely independent measurements and data processing both the peak positions and amplitudes give good agreement (ODF maximum is 35.1 from neutrons and 31.4 from OIM). The texture of the sample is best described by the asymmetric girdle of [0001] axes which is roughly perpendicular to the foliation and lineation. The c-axes form an elongated maximum within the girdle with the highest concentration in the foliation and perpendicular to the lineation. The a-axis orientations parallel to the lineation are preferred. This type of texture is typical for quartz mylonite which underwent ductile deformation.

The misorientation distribution (figure 5) is dominated by a single maximum in the position of the Dauphiné twin (a 180°(=60°) rotation around [0001]).

DISCUSSION AND CONCLUSIONS

The results presented here demonstrate that orientation imaging microscopy is feasible for ceramic samples with trigonal crystal symmetry. It is possible to obtain good BKD patterns if proper care is taken with preparing the sample surface. For this quartzite sample the statistics of the measurements were sufficient despite the fact, that the OIM measurements covered about 1,000 grains compared to some 65,000 grains in the neutron experiment.

The results indicate that Dauphiné twinning dominates the grain boundary misorientation distribution. Dauphiné twins have been documented as mechanical twins in experimentally deformed quartzites [12]. All the MDF-maxima with rotation axes near [0001] are closely related to the strong c-axis texture (If the c-axes form a sharp maximum the only possible differences in crystal orientations are rotations around [0001]). A more complete analysis of preferred misorientations should therefore account for the texture related part of the MDF [13].

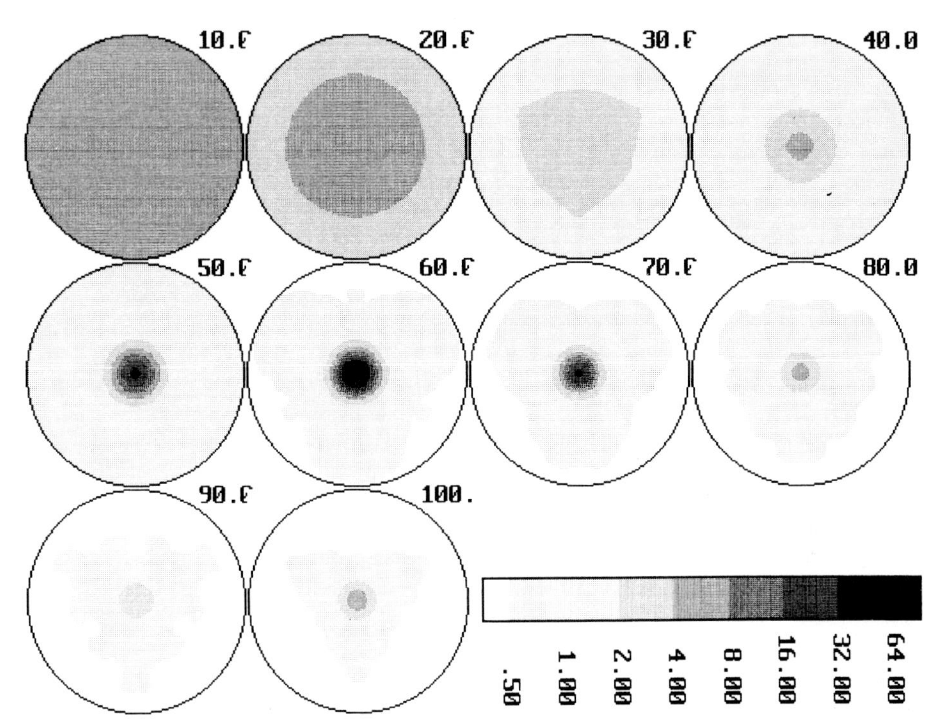

Figure 5: MDF of all nearest neighbor pairs with misorientation angle $\omega > 10°$ represented in ω - sections of axis/angle space.

ACKNOWLEDGEMENTS

We are grateful to S. Matthies and J. Pospiech for providing their programs to calculate distribution functions from single orientation data. We also appreciate the expertise from N.H. Schmidt and N.O. Olesen with the sample preparation. Support through NSF Materials Research Group Award DMR-9001278 (KK and BLA) as well as NSF grant EAR 9017237 and IGPP-LANL (HRW and FH) is gratefully acknowledged.

REFERENCES

1) Gottstein, G.: Rekristallisation metallischer Werkstoffe. DGM Oberursel, 1984, p. 276.
2) Sander, B.: Einfuehrung in die Gefuegekunde der geologischen Koerper.
Band II: Die Korngefuege. Springer Berlin, 1950, p. 399.
3) Wenk, H.-R.: Schweiz. Min. Petr. Mitt., 1965, 45, 467.
4) Lloyd, G.E., Olesen, N.O. and Schmidt, N.H.: Scanning, 1988, 10, 163.
5) Lloyd, G.E., Law, R.D., Mainprice, D. and Wheeler, J.: J. Struct. Geol., 1992, 14, 1074.
6) Adams, B.L., Wright, S.I. and Kunze, K.: Metall.. Trans., 1993, 24A, 819.
7) Adams, B.L., Kunze, K., Wright, S.I. and Dingley, D.J.: this volume.
8) Schäfer, W., Jansen, E., Merz, P., Will, G. and Wenk, H.-R.: Physica, 1992, B180-181, 1035.
9) Matthies, S.: this volume.
10) Heidelbach, F., Pospiech, J. and Wenk, H.-R.: this volume.
11) Neumann, P.: Textures and Microstructures, 1991, 14-18, 53.
12) Tullis, J. and Tullis, T.E.: Geophys. Monogr., 1972, 16, 67.
13) Pospiech, J., Haessner, F. and Sztwiertnia, K.: Acta. Met., 1992, 41, 305.

Materials Science Forum Vol. 157-162 (1994) pp. 1251-1256

EFFECT OF TEXTURE ON STRAIN LOCALIZATION IN DEFORMATION PROCESSING

W.B. Lee

Department of Manufacturing Engineering, Hong Kong Polytechnic,
Hung Hom, Kowloon, Hong Kong

Keywords: Strain Localization, Shear Band Fracture, Localized Necking, Shear Angle, Workability, Rolling, Sheet Metal Forming, Machining

ABSTRACT: Strain localization in material deformation processes often takes the form of shear fracture or localized necking. Texture is an important material variable in the mesoplasticity modelling of the strain localization phenomenon. It is possible to predict quantitatively the effect of crystallographic texture on the occurrence and magnitude of these strain localization in a wide range of manufacturing processes from rolling, sheet metal forming to machining. Such information will be needed in the industrial control of texture to meet particular workability requirement.

1. INTRODUCTION

In deformation processing, workpiece materials are confined in tools and dies and forced to flow in certain directions to give the required shape of the finished products. The term "workability" refers to the capacity of a material to undergo desired shape changes. Failures in a wide spectrum of forming processes are often associated with the inability of the material to distribute plastic strain uniformly. Plastic deformation often appears to be macroscopically homogeneous, is intrinsically heterogenous when viewed from a micro or meso scale. The discrete nature of dislocation slip trigger non-uniform plastic deformation at a stress level far below the theoretical shear strength, and leads to complex structural instabilities.

The useful ductility of a material is often terminated by its tendency to strain localization. A detailed review of the strain localization problem and workability of metals can be found in the work of Semiatain and Jonas [1], and a review of shear localization by Bai and Dodd [2]. The above works, however, put more emphasizes on the macroscopic aspects of of the problem. In this paper, it is shown that the development of strain localization in the form of *shear banding* or *localized necking* in a range of important deformation processes from rolling, sheet metal forming to machining (figure 1) can be related to the crystallographic textures of a workpiece material, and the onset and magnitude of such strain inhomogeneities can be quantitatively predicted from a given known texture data.

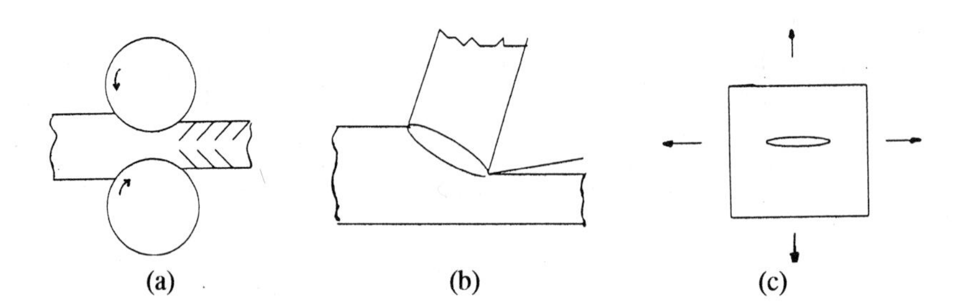

Figure 1. Strain localization in common deformation processing of materials (a) shear band fracture in rolling (b) shear zone formation in machining and (c) localized necking in sheet metals.

2. SCALES FOR PLASTICITY INVESTIGATION

The strain localization problem in deformation processing can be explored from different scales. Two approaches have been used in the past. These are (1) the macroplasticity approach and (2) the microplasticity approach. The major framework of continuum mechanics is developed at a time modern analytical techniques was not available for the characterization of material structures. A full quantitative characterization of *meso-structures* (i.e, the geometry, shape, size of various phases and crystallographic textures, etc.) becomes now possible due to advances in modern data acquisition techniques such as image processing and quantitative texture analysis. The mesoplasticity approach [3] represents an important connection between the continuum-based macroplasticity and the atomistic physical theory of microplasticity, a connection in both length scales and the method of investigation. It is the meso-structures and not the dislocation scale structures that can be readily determined and controlled in an industrial environment. Among the material variables, texture is one of the most important structural parameter at the meso scale.

3. EXAMPLES OF STRAIN LOCALIZATION

3.1 Grain Orientation Dependency of Shear Band Cracks in Rolling

Strain localization in the form of shear bands is an important part of the ductile fracture mechanism as fracture eventually occurs along these shear bands. Shear banding can limit the ductility available for the required deformation. Extensive shears on these bands have been associated with failure during rolling of some aluminium and copper alloys [4]. Although the shear bands are macroscopic in nature, they are very sensitive to the crystallographic orientations of the micro-volume of the material in which they are formed [5]. Figure 2(a) shows a shear band crack in an aged-hardened 2024 aluminium alloy cold rolled to a rolling strain of 0.92. Not all grains formed shear bands. It is interesting to note that grains which did not form shear bands themselves behaved differently to the 'invasion of shear band cracks from their neighbouring grains. Some grains acted as 'crack-stoppers' to the invading grains. Figure 2(b) shows two shear band cracks advancing from opposite directions failing to unite together at a third intervening grain, and the crack branched off in opposite directions. The crack often follows the general directions of the sample-scale shear bands but not those of the grain-scale shear bands. The critical strain at which a shear band will appear depends on the work hardening exponent of the slip planes and the texture softening factor. It has been shown by Lee and Chan [6] that the most likely shear band angle is given by the one which has the same minimum texture softening factor among the ones which has the same minimum number of slip systems and shear strength. The theoretical sample-scale shear band angles calculated for various local grain orientations indicate that the shear band propagation angle is sensitive to local variation in the texture proportion [5]. Other things being equal, polycrystals with weak crystallographic textures should be preferred to sharp textures in the stopping of shear-band crack propagation. Further work will be needed to explore the textural effect of shear band cracks on the fracture toughness of textured polycrystals.

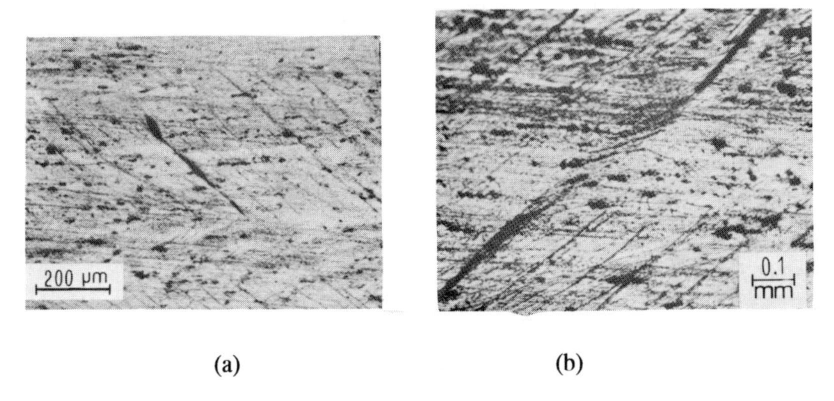

(a) (b)

Fig.2 (a) shear band crack in an aged-hardened Al-Cu alloy (b) Stopping of shear band cracks at a third intervening grain.

3.2 Formation of Looper Lines and Localized Necking in Sheet Metals

Looper lines are parabolic markings or streaks that are encountered in the deep drawing and spinning of a wide range of aluminium sheet components. An example of looper lines in a spin-formed commercially pure aluminium sheet is shown in figure 3 . A closer inspection of these looper lines will reveal that they are shallow surface undulations. When these undulations are deep, they become preferred sites for the occurrence of localized necking, especially in thin gauge materials. These surface markings was thought to be caused by the presence of long columnar grains in the original cast ingot [6]. The origin of these surface undulations has been proposed by Lee and co-workers [7] to arise from the differential thinning as a result of the non-compliance of surface grains with the macroscopic imposed shape change. The differential thinning is caused by a difference in the strain paths taken between regions of texture segregation and their immediate neighbouring grains as shown in figure 4. Suppose the region (B) with texture segregation tends to deform in plane strain taking the strain path OB, while the bulk material (A) is deforming along the path OB, the thickness at B will decrease more quickly than the thickness at A. This type of deformation behaviour is possible for surface grains, as there are much less constraints near the free surface. The problem of predicting the strain path for

Figure 3. Looper lines and loaclized necking in a spin-formed C.P. aluminium sheet.

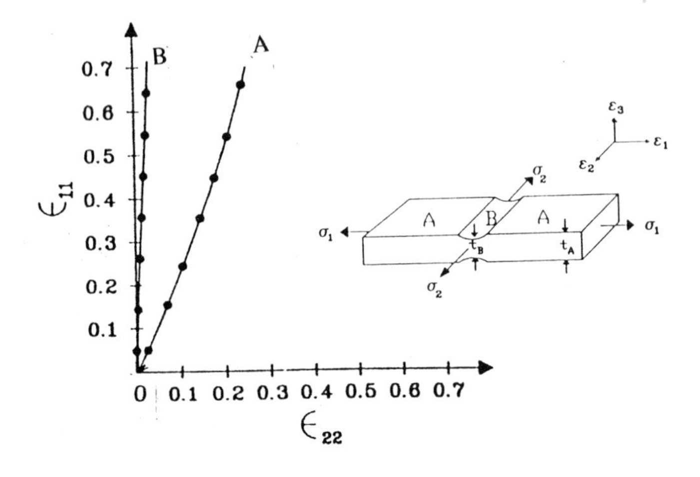

Figure 4. Different strain paths taken by the bulk material (A) and the texture colony (B) causing differential thinning.

a textured region and a given imposed stress state is equivalent to that of determining the strain tensor of the region for each incremental step of deformation, taking into account the grain rotation after each successive step [9].

An numerical example showing the effect of local texture variation on the magnitude of the relative thinning has been reported by Lee and Chan [8] in an aluminium sheet. The above analysis shows that a groove of sufficient large size can be developed without assuming any initial defects or "equivalent defects" in a smooth and damage-free sheet. Furthermore, it is shown that the growth of the groove is sensitive not only to the texture variations abut also to the alignment of the texture colony

3.3 Textural Effect on Shear Zone in Metal Cutting

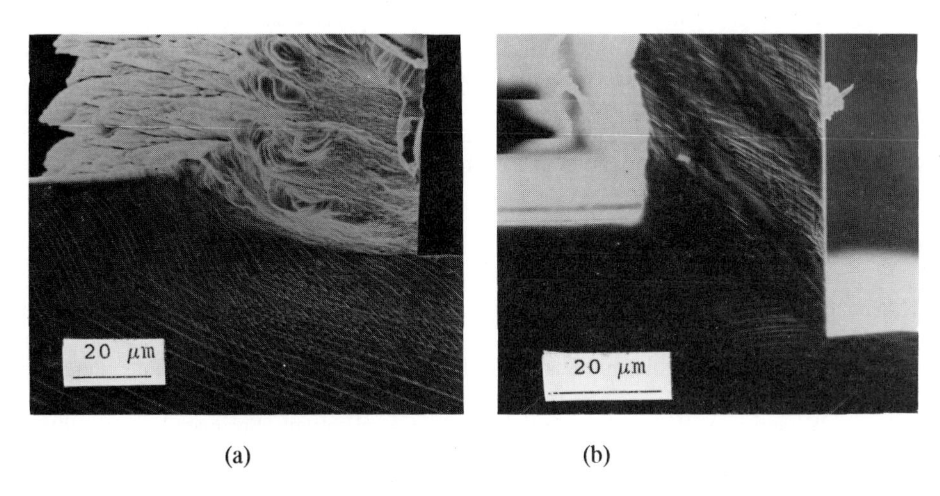

(a) (b)

Figure 5. SEM micrograph of shear zone formation in a copper single crystal when cutting is done along (a) [320] and (b) [324] on a (230) plane.

In machining ductile metals and alloys, the strain is highly localized in a shear zone which separates the chip from the workpiece material (figure 1b). The magnitude of the shear angle (or the idealized orientation of the shear zone) indicates the machinability of the work materials. A large shear angle is associated with continuous chip formation, good surface finish and low cutting forces. The shear zone has been found to vary with the tool geometry, the cutting conditions and the workpiece materials. Various theoretical shear angle equations have been derived in the past but met with partial success only. Most studies of the cutting mechanism are based on the assumption that the material is isotropic and is a homogeneous continuum. The effect of material anisotropy is often not included in the theoretical analysis. One important source of material anisotropy lies in the crystallographic nature of the metallic substrate. During machining, the zone of workpiece material in contact with the tool tip acts as strong source of dislocations. Fine cracks are produced near the vicinity of the tool tip and trigger the primary shearing process. An example of the effect of crystallography on the shear zone angle in cutting a copper single crystal is shown in figure 5. The shear zone angle changed from 20° to 33° when cutting was done along different crystallographic directions.

A predictive model has been proposed by Lee [11,12] to analyze the orientation of the shear zone in machining textured materials and in single crystal cutting. The finding of a direction at which shearing will occur will be similar to the analysis of the shear banding problem [11]. It should be noted that the shear plane may not be parallel to a particular crystallographic slip plane of the crystal, and are not dislocation glide planes as implied by other workers. The shear in the band has to be accomplished by homogeneously distributed slip, i.e., all operative slip systems co-operate in the development of the shear zone.

4. CONCLUSIONS

The above three examples illustrate the importance of texture studies in the understanding of strain localization problem in many technologically important deformation processes from rolling, sheet metal forming to machining. The wide application of texture analysis and the incorporation of crystallographic orientation effect in material based constitutive equations make it possible to study the effect of texture on strain localization problem in a more quantitative manner. This will enable the right type of textures to be engineered into a material to control its flow behaviour during deformation processing.

REFERENCES

[1] Semiatain, S.L and Jonas, J.J:Formability and Workability of Metals, American Society of Metals, 1984
[2] Bai, Y and Dodd, B.:Adiabatic Shear Localization, Theories and Applications, Pergamon, 1993
[3] Yang, W. and Lee, W.B.:Mesoplasticity and its Applications, Springer Verlag, 1993
[4] Brown, K.:Journal of the Institute of Metals, 1972,100,341
[5] Lee, W.B. and Chan, K.C.:Int.J.Fracture, 1991,52,207.
[6] Lee, W.B. and K.C.Chan:Acta Metallurgica and Materialia,1991,39,411
[7] Jackson, C.M. and Welch, E.H.:Sheet Metal Industries, January,1975,34
[8] Lee, W.B., K.Y.Yu and B.Ralph, " Proceedings of the 17th International Deep Drawing Research Group (IDDRG) Biennial Congress, 11-13 June, 1992, Shengyeng, 422
[9] Chan, K.C. and Lee, W.B.:Int.J.Mech.Sci., 1990,32,497
[10] Lee, W.B. and Chan, K.C.:Textures and Microstructures, 1991,14-18,1221
[11] Lee, W.B.: Proceedings of the 4th International Conference in Metal Cutting, Non-conventional Machining and their Automation, 25-27 April, 1989, Beijing Institute of Technology, China.
[12] Lee, W.B. and Zhou, M.:Int.J.Machine Tool & Manuf., 1993,33,439

Materials Science Forum Vol. 157-162 (1994) pp. 1257-1262

A LOCAL TEXTURE ANALYSIS OF A HOT DEFORMED Al-Li-Zr ALLOY

J. Mizera and J.H. Driver

Materials Department, Ecole des Mines de St-Etienne,
158, Cours Fauriel, F-42023 Saint-Etienne Cédex 2, France

Keywords: Al-Li-Zr Alloy, Hot Deformation, Local Textures, Orientation Gradients

ABSTRACT

The microtextures of a model Al-2.2% Li-0.1% Zr alloy after hot extrusion and hot rolling have been characterized by microdiffraction techniques to determine the spatial distribution of the grain orientations e.g. the decomposition model of Goss {011}<100> to Brass {011}<112> components.

The alloy has been examined in two deformed states: (I)extruded to ε=2 and (II) hot rolled to ε=3.6. In the extruded state, containing ≈18% Goss and 3% Bs, many of the grains lying in the rolling plane exhibit a continuous orientation gradient from Goss to Bs. A significant proportion of the grains (measured in the RD/ND plane) also possess orientations along the classical β fibre (Bs↔S↔Cu). After further hot rolling, which increases the Bs component to 11%, a larger number of paired Bs^+ and Bs^- grains are found in the RD/TD plane in agreement with the ND rot. decomposition model of Goss to Bs^+ and Bs^-. Another remarkable feature of the microstructure is the high proportion (≈50%) of grains which exhibit continuous orientation gradients between the usual hot rolling texture components: i.e. Cu↔S by rotations about the <111> pole closest to ND and S↔Bs by RD rotations.

1. INTRODUCTION

The development of the "Brass" texture component {011}<211> in hot rolled FCC metals is now generally recognised but the physical mechanisms that lead to this component have not yet been completely elucidated. For example, the Bs component is only predicted by relaxed constraints Taylor-type grain deformation models when the ε_{TR} shear is relaxed. However, it is widely accepted that large ε_{TR} shears in rolled grains are unlikely because of the resulting strain incompatibilities with the surrounding material.

AERNOUDT and STÜWE [1] have pointed out that the ε_{TR} shear incompatibility in a rolled grain can be compensated by a similar shear of opposite sign in a neighbouring grain to maintain a zero macroscopic ε_{TR} shear. As suggested by these authors, a spatial configuration of symmetrical Bs oriented grain pairs can be developed by the decomposition of Goss, {011}<100> oriented grains, by rotations of opposite sign about ND (figure 1). This Goss decomposition model has received some indirect confirmation from macroscopic X-ray texture measurements: e. g. HIRSCH and LÜCKE [2, 3] have shown that a strong Bs texture component develops during cold rolling of aluminium with an initially strong Goss component. More direct evidence would be provided by the presence of grain pairs of symmetry-related Bs orientations in the RD/TD plane; new microtexture techniques such as ECP and EBSD should be able to detect these particular spatial configurations. The purpose of this study is to investigate, using electron microtexture measurements, the local grain orientations and spatial arrangements in a hot rolled aluminium alloy which develops a Bs texture component.

2. ALLOY AND EXPERIMENTAL TECHNIQUES

Two electron microdiffraction techniques have been employed, in a suitably modified SEM, to characterize the orientations of individual grains and subgrains: ECP (Electron Channeling Patterns) and EBKD (Electron Back Scattered Kikuchi Diffraction). These microtexture techniques are based on electron channelling phenomena in a crystal lattice which produce Kikuchi-type lines, as described for example by JOY et al. [4] and RANDLE [5].

The present study has been performed on a model Al-Li alloy of composition Al-2.2% Li-0.1% Zr. This relatively simple alloy exhibits most of the typical features of texture formation in more complex commercial Al-Li alloys and also contains a sufficiently large Zr content to completely inhibit recrystallization during and after hot rolling. However it should be noted that the absence of nucleating agents lead to a very coarse dendritic grain structure after casting.

The textures presented here were measured after two deformation reductions: (i) hot extruded to 13 mm plate ($\varepsilon \approx 2.4$) and (ii) further hot rolled to 4 mm sheet corresponding to an additional strain of 1.2 and a total strain of 3.6. Both the rolling plane and long-transverse sections have been analysed. As described below, many of the grains exhibit a very wide orientation spread: the grain orientations have therefore been measured at the subgrain level by analysing at least 10 subgrains per grain. These subgrains, of dimensions 5 to 15 μm according to the section, are indicated in the micrographs.

3. RESULTS

3.1. Hot Extruded Plate (ε=2.4)

A long transverse plane section taken from the plate centre was examined by EBKD over 4 areas between the surface and mid-thickness (1650 subgrains analysed). The overall texture obtained from these measurements indicated a β-fibre hot rolling texture but with a particularly high density of S {123}<634> oriented grains. Previous X-ray ODF analysis [6] had revealed heterogeneous texture distributions both from the plate centre edge and through the thickness. Obviously care should be exercised in generalizing from local to macroscopic textures in such heterogeneous material. In this work, the local texture analyses were always first compared with the macroscopic textures obtained by systematic X-ray analysis to ensure a certain degree of representativity.

The most typical feature of the EBKD analysis of the transverse plane was the usual random orientation spread of the subgrains about the grain orientation (±10°). Note that all 4 variants of the S orientation were found at this stage. Some Bs and Goss grains were observed but there was no indication of any Bs pairing or Goss splitting in this plane. One rather surprising feature, however, was the presence of a few grains with continuous orientation gradients along ND from S to Bs and S to Cu.

In the extrusion (rolling) plane 5 areas were examined by ECP (≈600 subgrains over a total length of ≈12 mm along TD). The macroscopic (X-rays) and average microscopic textures in this plane section revealed the presence of strong texture components around Goss and Bs. The volume fractions of these components as determined by ECP (i.e. the ratio of the area of the grains of a component to the total area examined) coincide with those obtained by X-ray ODF analysis; Goss: 21% (ECP) and 18% (X-ray), Bs: 5% (ECP) and 3% (X-rays).

Figure 2 illustrates a typical area together with the {111} pole figures of the individual grains, most of which are Goss or Bs orientations. Here again, many of the larger grains exhibit an orientation gradient from one component to another. For example, the orientation of grain 2 in figure 2(a) changes progressively from Goss to Bs along the transverse direction. This is shown in greater detail in figure 2(b) where the Euler angles of subgrain A to G are plotted as a function of distance along TD. The angle ϕ_1 changes by ≈25° over the grain from -80 to -55°. This is interpreted as a Goss grain breaking up into a near -Bs orientation by rotations about ND. This grain has a Goss neighbour (N° 1) on one side and, on the Bs side, a S orientation (N° 3), the latter being followed by 3 Bs grains.

The same type of continuous orientation gradient between Bs and Goss was observed in other grains together with occasional pairs of complementary Bs orientations. Some of the S orientated grains also exhibited a continuous rotation by ≈30° about the normal direction.

3.2. Hot Rolled Sheet (ε=3.6)

In the long transverse plane a complete EBKD analysis was performed on all the grains along a line from surface to mid-thickness (2 mm in length, 720 subgrains). The overall texture taken from this line is made up of β-fibre components with S and Bs as the strongest components. For example, between the surface and 3/4 thickness two variants of S and Bs predominate whereas the opposing variants of S and Bs are found near the centre. Minor amounts of Goss, Cu and Goss/Cube orientations are also present.

At this higher deformation, a higher proportion of grains have orientation gradients. All of these occur along the β-fibre, i.e. Bs↔S↔Cu but no complementary Bs pairs were found through the thickness.

In the rolling plane however large numbers of these pairs were observed. In this section ≈600 subgrains were analysed by ECP in 3 areas of total width 5.6 mm. The overall texture of these areas contained about 18% near-Bs components in good agreement with the X-ray measurements. One of these areas is shown in figure 3 with the associated {111} pole figures. Grains numbered 4 to 8 exhibit a particularly interesting orientation distribution: Bs^+, Bs^-, S, Bs^+, Bs^-. This type of configuration was frequently found in the rolling plane of the hot rolled sheet indicating that the Bs^+, Bs^- pairs are often surrounded by near-S oriented grains. Note also in figure 3 the large orientation gradients that occur in some of the grains (eg. grains 2, 3 and 11). The orientation gradients in these types of grain were typically:

(i) between Cu and S components by rotations about the <111> pole closest to ND and,
(ii) between S and Bs by RD rotations.

These orientation gradients are described in greater detail by MIZERA [7].

4. DISCUSSION AND CONCLUSIONS

The microtexture analysis clearly shows that during hot rolling complementary Bs (+-) pairs develop in the rolling plane from Goss texture components. The presence of the pairs in the rolling plane and their absence from the transverse plane are consistent with the AERNOUDT and STÜWE [1] hypothesis of Goss grains splitting in the RD/TD plane to Bs pairs by rotations of opposite sense about ND. However figure 1 is undoubtedly a very schematic view of this decomposition which generally occurs by the creation of continuous orientation gradients within the grain (e. g. the Goss→Bs transition in figure 2), and probably as a consequence of inhomogeneous reaction stresses between grains. Thus most rolling experiments on Goss oriented single crystals (e. g. LATAS et al. [8]) indicate that the orientation is stable up to very high strains. The observation of Goss to Bs transformations in polycrystals at strains of ≈2 would suggest that the presence of adjacent grains, eg. S oriented grains, could favour localized slip systems that develop Goss to Bs rotations.

The conclusions of the present microtexture study of hot rolled Al-Li-Zr alloy can be summarized as follows:
- a strong β-fibre texture with ≈20% Bs component is developed, in agreement with more complex industrial alloys,
- large orientation gradients develop within a significant proportion of the grains,
- in the extruded state many grains exhibit a continuous orientation spread from Goss to Bs in the rolling plane,
- after further hot rolling many pairs of complementary Bs grains are found in the rolling plane.

The above observations confirm that Bs oriented grains can develop by orientation splitting of Goss grains in accordance with the model of AERNOUDT and STÜWE [1].

REFERENCES

1) Aernoudt, E.; Stüwe P.: Z. Metalkd., 1970, 61 (H2),128.
2) Hirsch, J.; Lücke, K.: in "Proc. Inter. Conf. Aluminium Alloys-Physical and Mechanical Properties", eds. Starke A. et Sanders T. H., Chameleon Press, London, 1986, 1725.
3) Hirsch, J.; Lücke, K.: Acta Metall., 1988, 36, 2863, 2883 and 2905.
4) Joy, D. C.; Newbury, D. E.; Davidson, D. L.: J. Appl. Phys., 1982, 53, R81.
5) Randle, V.: "Microtexture Determination and its Application", The Institute of Materials, 1992.
6) Engler, O.; Mizera, J.; Driver, J.; Lücke, K.: Textures and Microstr., 1991, 14-18, 1153.
7) Mizera, J.: Doctoral Thesis, EMSE-INPG, 1993.
8) Latas, W.; Wrobel, M.; Gorczyca, S.: Archives of Metallurgy, 1987, 32, 530.

ACKNOWLEDGEMENTS

The authors wish to thank CRV (Pechiney) for the provision of alloys. This work has been supported by the EEC through a Brite-Euram contract.

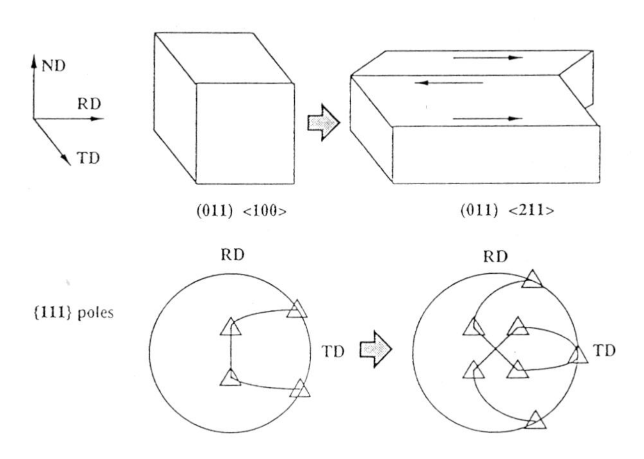

Figure 1. Decomposition model of Goss to Brass components.

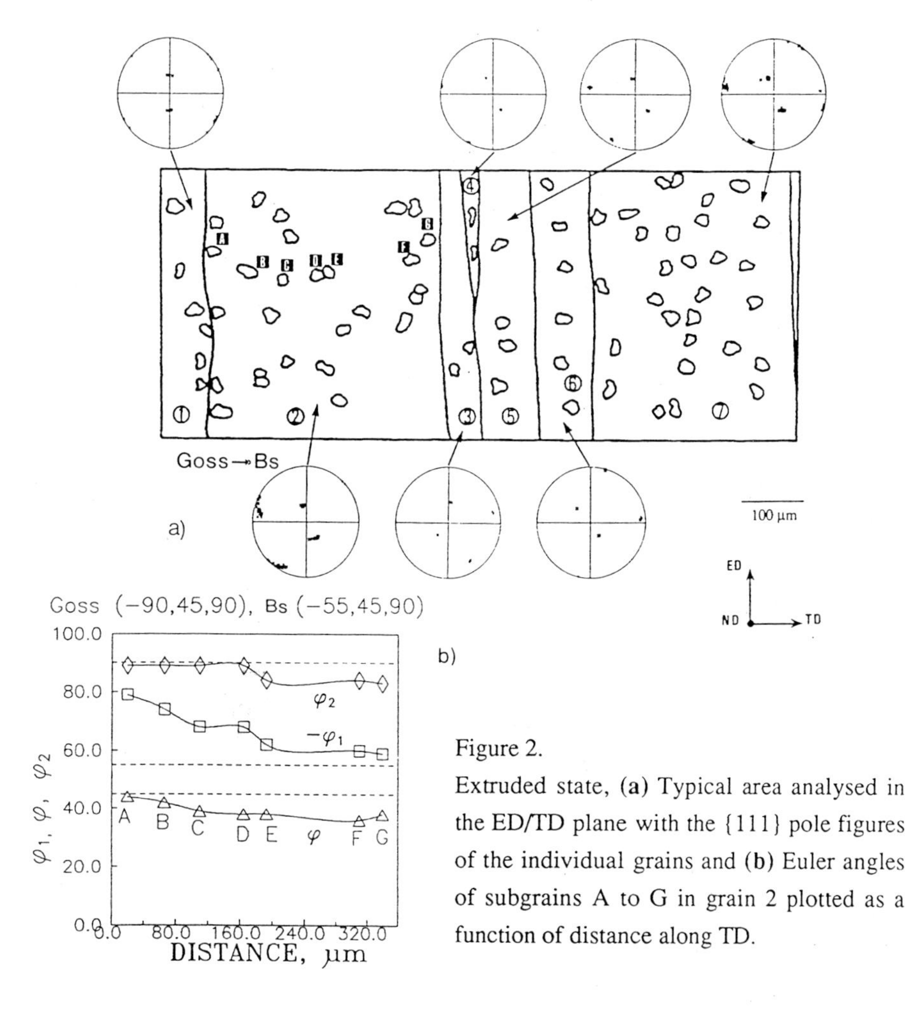

Figure 2.
Extruded state, (**a**) Typical area analysed in the ED/TD plane with the {111} pole figures of the individual grains and (**b**) Euler angles of subgrains A to G in grain 2 plotted as a function of distance along TD.

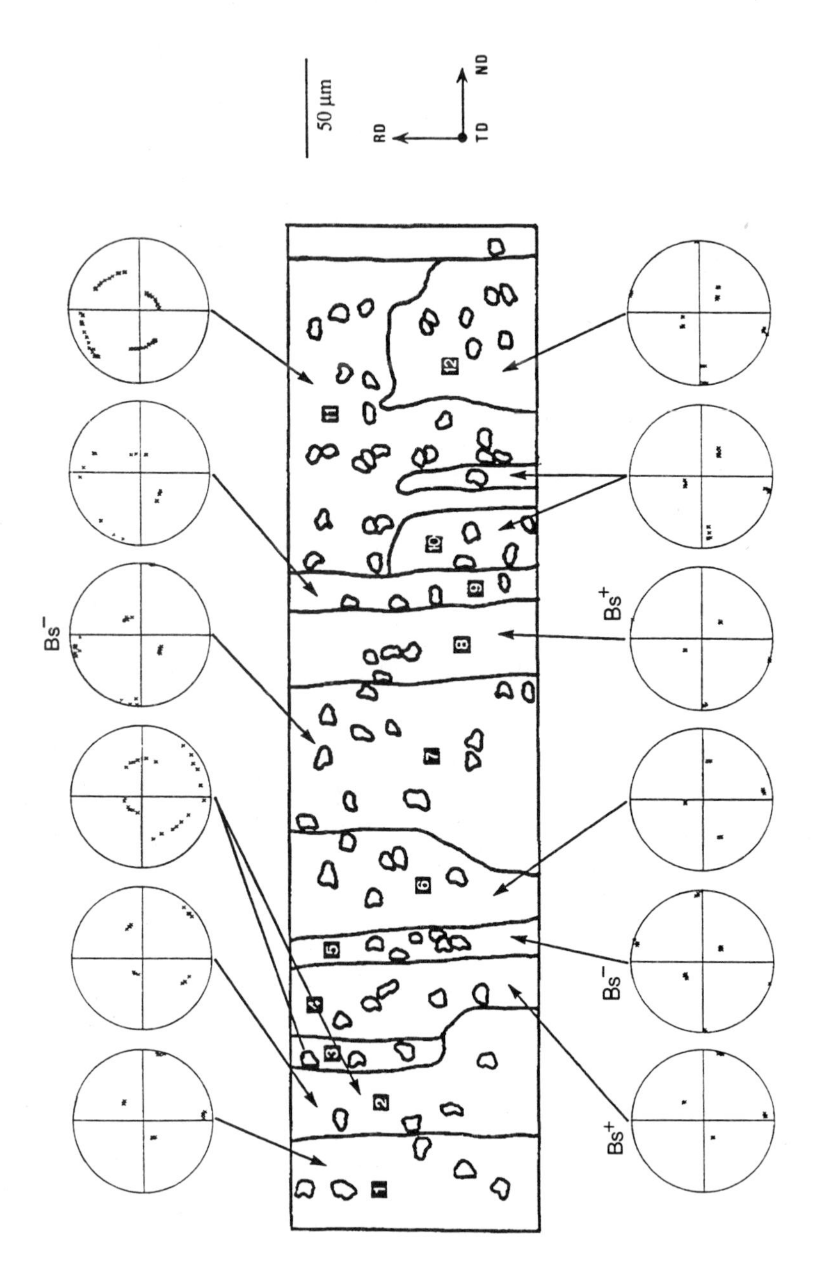

Figure 3. Hot rolled state, an area of the 12 grains analysed in the ND/RD section with the associated {111} pole figures

Materials Science Forum Vol. 157-162 (1994) pp. 1263-1270

ABOUT THE TEXTURE INFLUENCE ON THE FREQUENCY OF CSL BOUNDARIES IN STEELS

A. Morawiec and J.A. Szpunar

Department of Mining and Metallurgical Engineering,
McGill University, 3450 University Street, Montreal, Quebec H3A 2A7, Canada

Keywords: CSL Boundaries, MODF, ODF

Abstract

The influence of texture on the frequencies of CSL boundaries and mis-orientation between grains was investigated. The method used assumes that the material is spatially disordered. Following that assumption examples illustrating the influence of texture strength on the frequency of grain mis-orientation and the probability of CSL mis-orientation are presented.

The method described can also be used to determine the CSL statistics of orientation differences with respect to any chosen grain orientation.

Introduction

The crystal orientation distribution function (ODF) is a quantitative description of the frequency of the distribution of grain orientations in Euler Angle space. It is possible to extract from the ODF different types of information characterizing the probability of mis-orientation between grains. Garbacz and Grabski [1] demonstrated that for simple fibre textures, the texture data can be used to obtain the frequency of CSL boundaries in materials. A most general approach which applies to any ODF, was then proposed by A. Morawiec, J.A. Szpunar and D. Hinz [2]. The method assumes that there is no correlation between the location and orientation of the grains. Such an assumption cannot be justified if grain orientation clustering is observed or when grain to grain orientation differences are determined in addition to texture by other factors, i.e. twinning.

In the first step or calculation, the statistical mis-orientation function is derived. The next step involves calculations of grains with orientation differences corresponding to CSL boundaries (assuming Brandon's accuracy criterion). This allows calculation of the frequency of various Sigma boundaries. Realization of these steps is made using computer simulation. Pairs of orientations are generated such that the total distribution agrees with the known ODF.

This information obtained from the ODF facilitates the understanding of various grain boundary related properties of polycrystalline materials. Low CSL boundaries have a lower energy, higher electrical resistivity, a greater resistance to intergranular fracture and cavitation, a higher corrosion resistance as well as many other special properties [3].

Several authors [4,5] consider low CSL boundaries as being responsible for the selection process during anomalous grain growth. In this type of research, information about the statistical mis-orientation between growing grains and the matrix grains is important. Such differences in orientation between grains can also be obtained using the proposed method of ODF analysis.

In the next chapter certain examples illustrating the application of such ODF analysis will be presented.

Example of Texture Influence of Frequency of CSL Boundaries

Garbacz and Grabski [1] and Gertsman et al. [6] concluded their investigation on texture and CSL frequencies relationship by saying that texture strongly influences these frequencies. However, the hypothetical textures they used were extraordinarily sharp.

To illustrate the relationship between the texture sharpness and CSL frequencies we will analyse an exemplary texture of rolled steel given in Fig. 1. Full statistics of the CSL distribution are given in Fig. 2 [2]. None of the densities of CSL boundaries in the specimen investigated exceed 4%. In Fig. 3, the CSL statistics were calculated with respect to exemplary grain orientation given in the Euler's Angle space by g = 10, 55, 10 and g = 90, 60, 45 (Fig. 3). Given statistics also show a rather low CSL density, which does not differ significantly from the CSL statistics for a random, non-textured specimen. Data obtained for randomly oriented grains are marked on Fig. 3 by black squares. Similar statistics calculated with respect to various other grain orientation show that CSL probability does not differ significantly from that obtained for the random specimen.

The influence of texture strength on CSL statistics might be illustrated by using an example of hypothetical texture shown in Fig. 4a. We will assume that the ODF maxima will always be in the same place and that the strength of these maxima will increase from 8 to 800 random units. Figs. 4 and 5, demonstrate the changes in density calculatled as a function of the disorientation angle and the CSL statistics for two different strengths of the ODF maxima. Such analysis shows that in the case of a well defined texture certain CSL relationships, CSL 5 in this case, are well represented, and that their percentage increases strongly with the texture strength. To illustrate how the sharpness of the texture influences the CSL frequency, one can use two variables. One, D, characterizes the difference in frequencies of CSL-mis-orientation relationships between random and non-random specimens and is defined as a sum over CSL boundaries up to CSL 27 and another is the texture variance, so called index [2].

The texturies used in the calculation of the relationship between the D factor and the variance are similar to those presented in Fig. 1. They consist of the so called γ-fibre whose strength was changed to obtain a variation in texture. Results presented in Fig. 6 illustrate a typical relationship between two selected parameters. A high percentage of CSL type boundaries can only be obtained if strong texturing is present.

Such strong texture is often manufactured in industrial processes. A classical example is a Goss type texture in Fe-Si oriented steel. An exemplary CSL distribution illustrating a high percentge of low angle boundaries for sharp (500 in random units) Goss texture is given in Fig.7

Summing up this discussion we would like to stress that the presented interpretation of the ODF will allow us to extract from it new information about the mis-orientation between grains and about CSL grain boundaries. Such information can be used in research on recrystallization of metals and as well in correlating the frequency distribution of CSL boundaries and various mechanical and physical properties of materials.

[1] A. Garbacz and N.W. Grabski, Scripta Metall. 23 (1989), 1369.
[2] A. Morawiec, J.A. Szpunar and D. Hinz, Acta Metall. Mater. (to be published).
[3] K.T. Aust and G. Palumbo, Structure and Property Relationship for Interfaces, ed. by J.L. Walter et al., ASM 1991.

[4] J. Harase, R. Shimuzu and D.J. Dingly, Acta Met., 39 (1991) 763-770.
[5] R. Penelle, T. Baudin, P. Pallard and L. Mora, Textures and Microstructures 14-18 (1991) 597-611.
[6] V. Yu. Gertsman, A.P. Zhilyaev, A.I. Pshenichnyuk and R.Z. Valiev, Acta Metall. Mater. 40 (1992) 1433

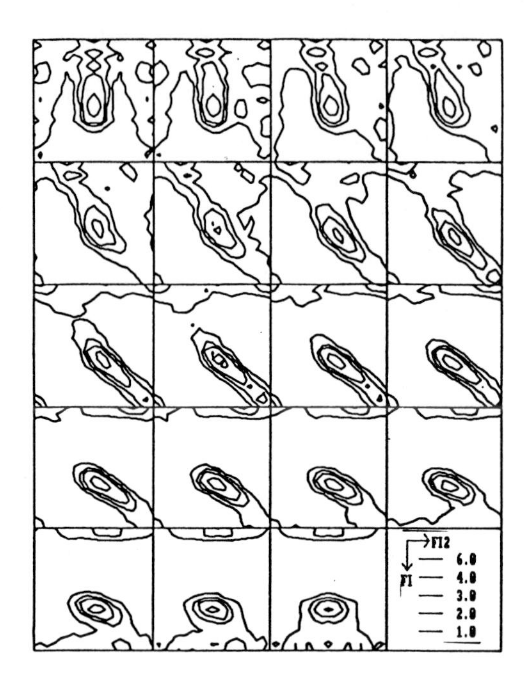

Fig. 1. ODF for cold-rolled steel.

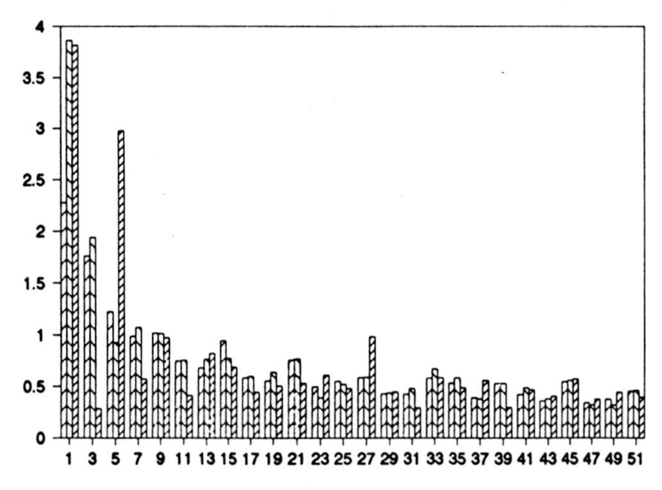

Fig. 2. The frequencies of CSL-mis-orientations. First bar represents results obtained for a random orientation distribution; second bar to texture Fig. 1; third bar mis-orientation calculated with respect to Goss orientation [2].

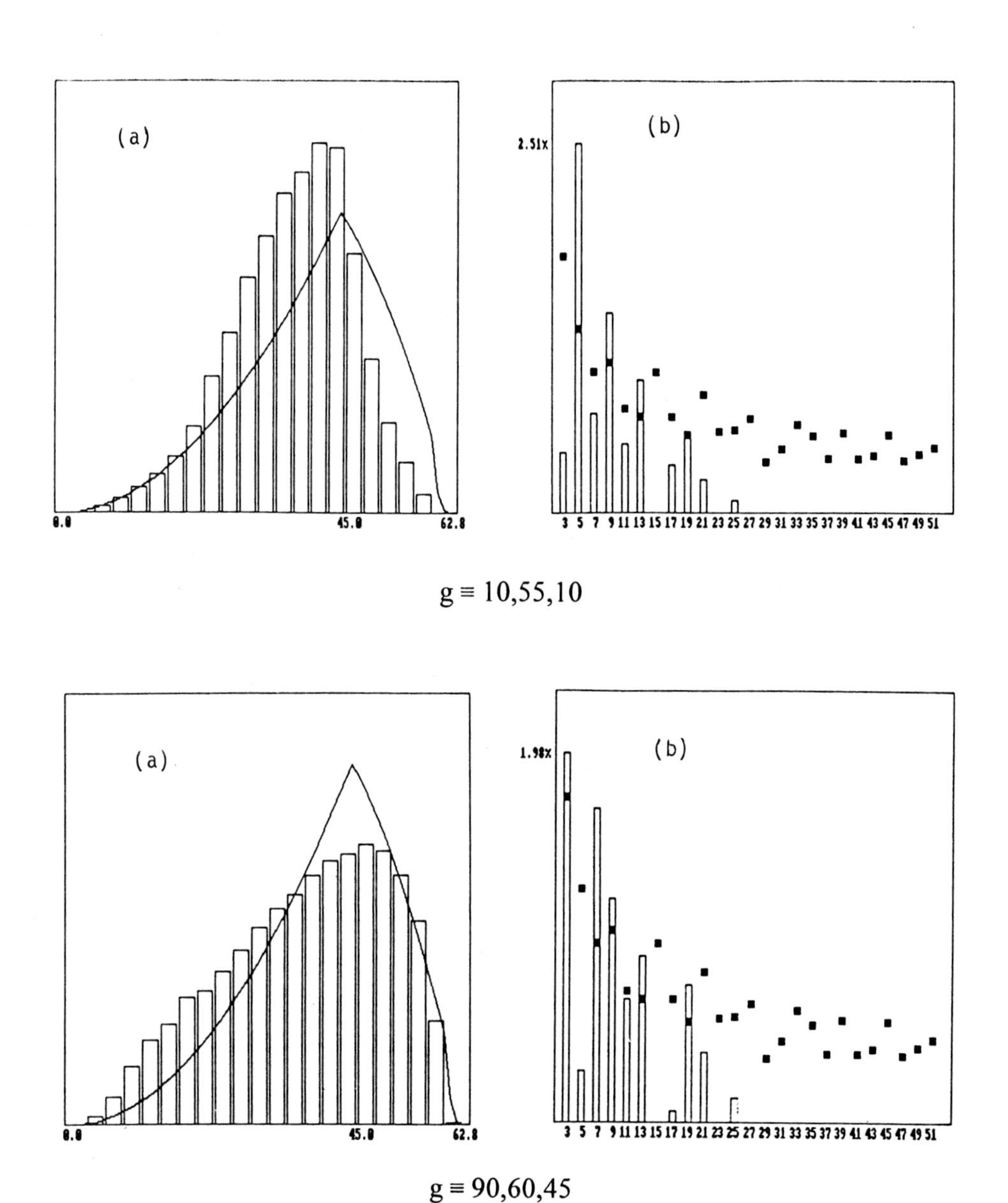

Fig. 3. (a) Mis-orientation distribution for two given grain orientation, (b) CSL statistics for two given grain orientation. Texture described by ODF in Fig. 1.

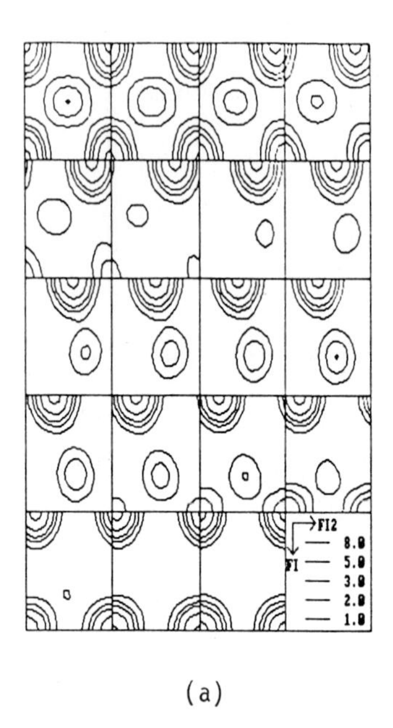

(a)

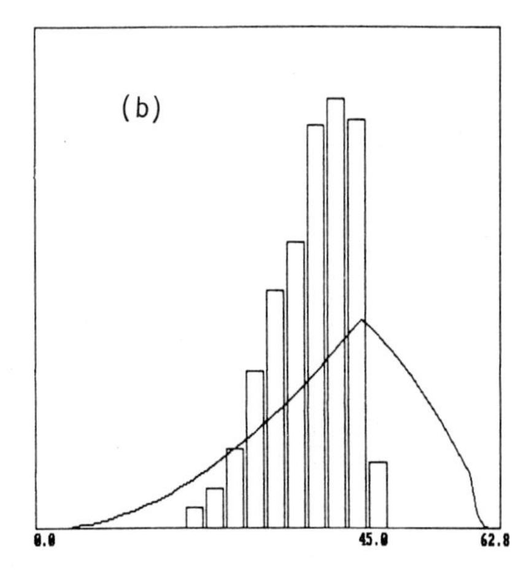

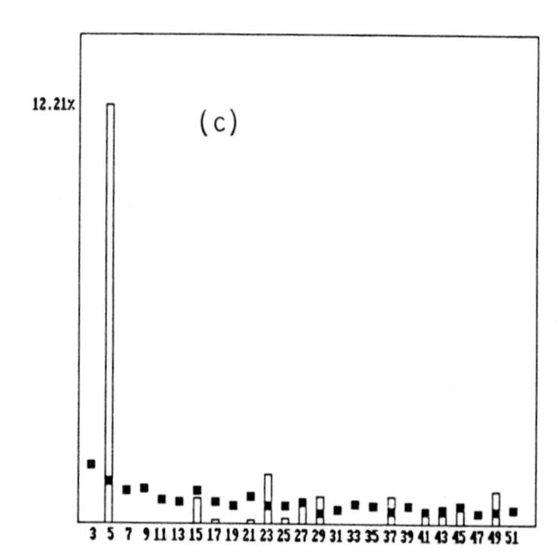

Fig. 4. (a) Hypothetical texture, ODF maxima 8 units.
(b) Misorientation angle statistics.
(c) CSL frequencies.

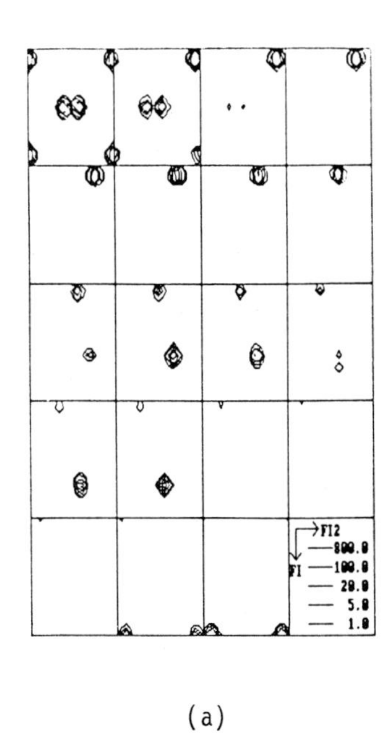

(a)

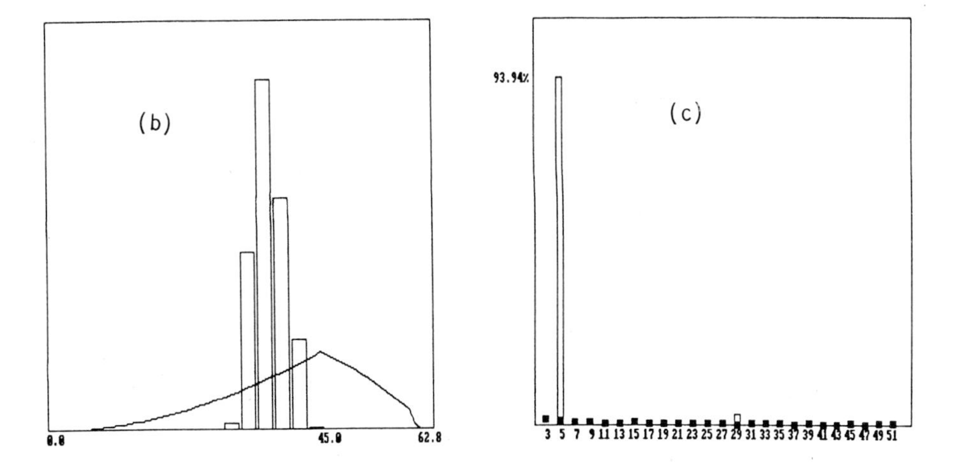

Fig. 5. (a) Hypothetical texture, ODF maxima 800 units.
(b) Mis-orientation angle statistics.
(c) CSL frequencies.

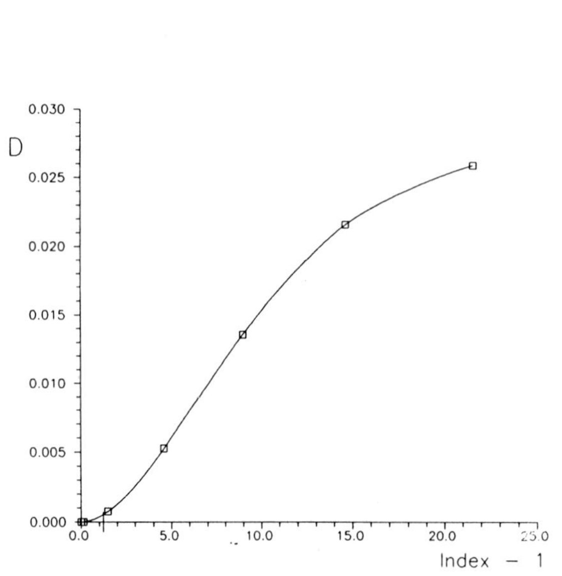

Fig. 6. Dependence of the frequencies of CSL-mis-orientation relationship on texture sharpness.

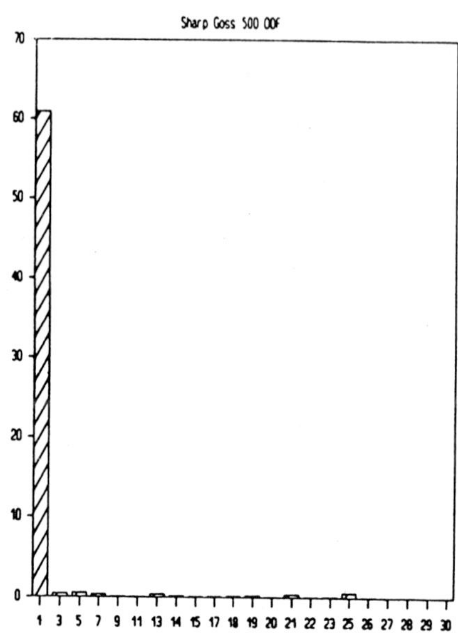

Fig. 7. CSL frequencies for grain oriented Fe-Si steel.

Materials Science Forum Vol. 157-162 (1994) pp. 1271-1276

MISORIENTATION DISTRIBUTIONS AND THE TRANSITION TO CONTINUOUS RECRYSTALLISATION IN A STRIP CAST ALUMINIUM ALLOY

A. Oscarsson [1], B. Hutchinson [1], B. Nicol [2], P. Bate [3] and H.-E. Ekström [4]

[1] Department of Mechanical Metallurgy, Institutet för Metallforskning, Drottning Kristinas Väg 48, S-114 28 Stockholm, Sweden

[2] Comalco Research Centre, 15 Edgars Rd, Thomastown Victoria 3074, Australia

[3] Interdisciplinary Research Centre, University of Birmingham, Edgbaston, Birmingham B15 2TT, England

[4] Gränges Technology Centre, S-612 81 Finspång, Sweden

Keywords: Strip Casting, Al-Alloy, Orientation, Misorientation, Continuous and Discontinuous Recrystallization

ABSTRACT

The orientation and misorientation distributions of a strip cast Al-Fe-Si (AA1145) alloy have been studied in detail. This material exhibits a transition from discontinuous recrystallisation to a continuous recrystallisation process during annealing for cold reductions exceeding about 95%.

Results show that the proportion of high angle boundaries increased with increasing cold strain, and the distribution of misorientations was found to approach that predicted from ODF data for random spatial distribution of the grains. Effects of different initial distributions of grain boundary misorientations during annealing are probed using a computer model of microstructure evolution.

INTRODUCTION

The present study is a part of a larger investigation into the textures, microstructures and properties of aluminium sheet rolled to heavy reductions and annealed in the course of foil manufacture. During the final annealing of aluminium foil products it is important to avoid the occurrence of large recrystallised grains which severely diminish formability. Experience has shown that different alloys, to a greater or lesser extent, undergo a gradual coarsening of the substructure on annealing, sometimes called continuous recrystallisation, which permits a favourable combination of strength and formability to be achieved. Under other conditions, recrystallisation occurs in a classical discontinuous manner whereby individual grains nucleate and grow to a large size, a situation which is prejudicial to the ductility of the thin sheet. The present work aims at characterising better the continuous recrystallisation process and trying to understand the reasons for its occurrence.

EXPERIMENTAL

The material used was commercially processed AA1145 alloy containing 0.35 % Fe and 0.16 % Si. It was strip-cast to a thickness of 6.15 mm then cold rolled in a number of passes to 0.08 mm. Cold rolled samples were taken at final and intermediate gauges and these were annealed at various temperatures using a simulated batch annealing process. Specimens were examined by mechanical testing, optical and scanning electron microscopy and X-ray texture (ODF) analysis. Electron back-scattering pattern (EBSP) measurements were made to determine the distributions of boundary misorientations present after different degrees of cold rolling. These were compared with non-correlated misorientation distributions obtained from the ODF measurements. Computer modelling of structure coarsening in 2-dimensional analogue structures has also been applied for comparison with the observed behaviour during annealing.

RESULTS

Figure 1 summarises strength data for the aluminium sheets with different combinations of cold rolling reduction and annealing temperature. For the lower rolling reductions, up to ~90 %, there is a relatively abrupt loss of strength over a narrow range of temperature and this 'softening temperature' decreases with increasing cold deformation in the conventional manner. At higher reductions, the softening process becomes extended over a much wider range of temperature, and the later stages actually become retarded with increasing cold deformation.

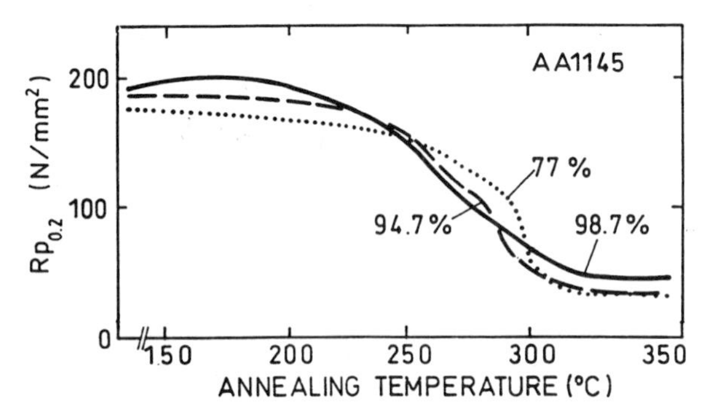

Figure 1
Variation of yield stress with annealing temperature for different cold rolling reductions.

Microstructural examination of annealed sheets shows that recrystallisation occurs in a classical discontinuous manner following the lower cold rolling reductions, with a quite distinct difference in appearance between the large growing grains and the fine deformed substructure. This is exemplified in figure 2. For higher cold rolling deformations, and particularly at 98.7 % reduction,

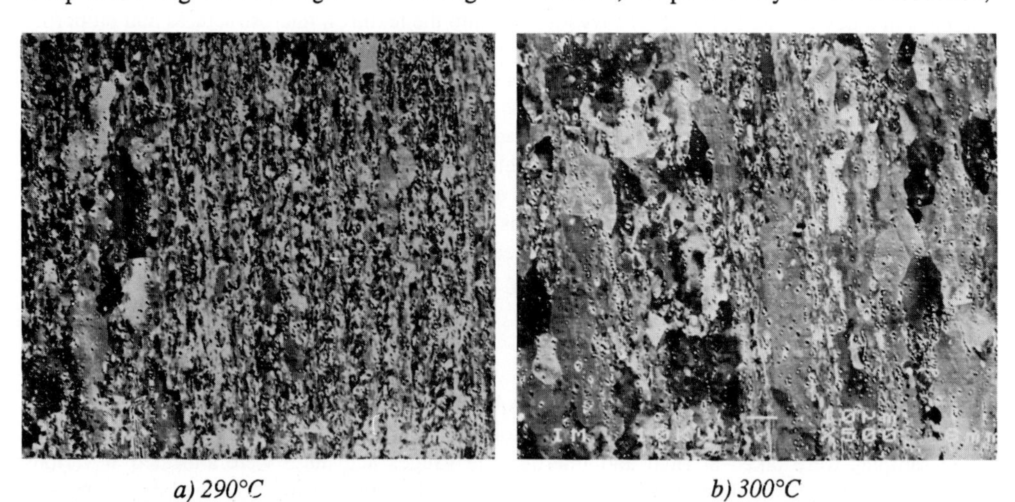

a) 290°C *b) 300°C*

Figure 2 SEM/BSE micrographs showing discontinuous recrystallisation after 77.6% cold rolling reduction.

there is a change towards a general coarsening process of the type which has frequently been termed 'continuous recrystallisation' or sometimes 'extended recovery'. In the present alloy, the growth is not perfectly continuous in the sense that the relative size distribution of crystallites is always the same, but it is evident that the majority of crystallites are in the process of either growing or shrinking, figure 3.

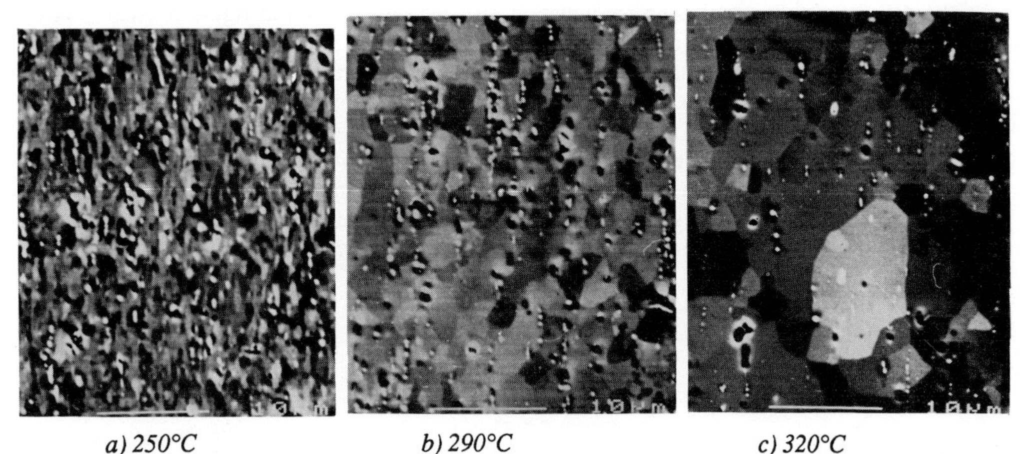

a) 250°C *b) 290°C* *c) 320°C*

Figure 3 *SEM/BSE micrographs showing continuous recrystallisation after 98.7% cold rolling reduction.*

The textures of the rolled sheets are all of the usual fcc rolling type. The β-fibre intensities plotted in figure 4 are already well developed following strip-casting which involves a significant degree of plastic deformation. Sharpening of the texture occurs during the early stages of cold rolling, followed by a static period and then a further development of the Cu and R-components for cold rolling above 97 % reduction, although with no further strengthening of the brass component.

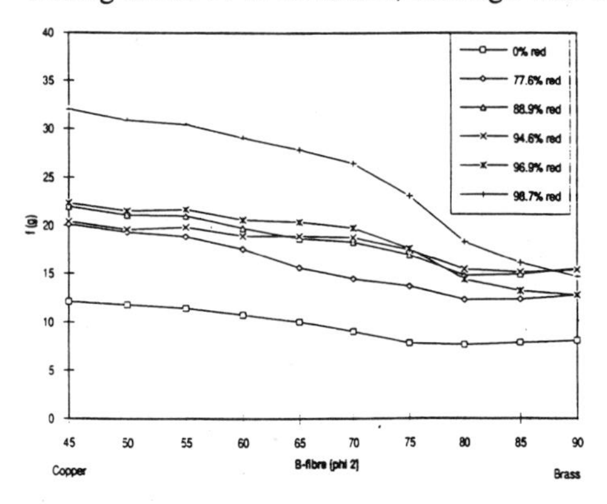

Figure 4 *Orientation densities along the β-fibre after various stages of cold rolling reduction.*

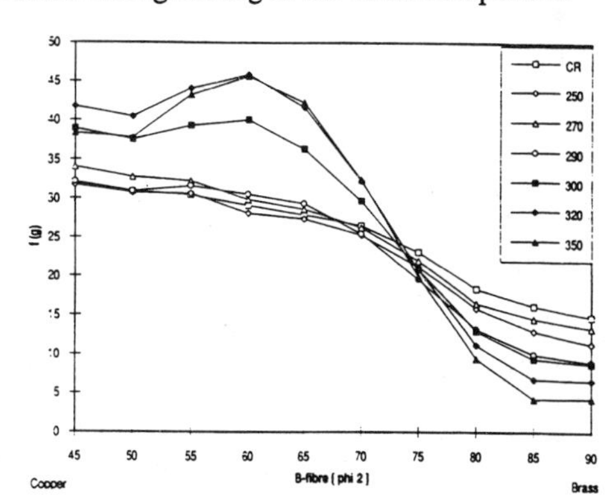

Figure 5 *Orientation densities along the β-fibre after 98.7% cold rolling reduction and annealing at different temperatures.*

Annealing textures in the fully softened condition are markedly dependent on the prior rolling reduction. For low reductions the cube texture dominates but the Cu and especially the R-component become prominent after 90 % reduction or more. The texture is, however, not invariant during annealing of the 98,7 % rolled sheet where continuous recrystallisation dominates, figure 5. There is seen to be a strengthening of the Cu and R-components and an associated decrease near the brass components.

The structure of the cold rolled aluminium consists of a contiguous network of grain boundaries and subboundaries which can seldom be distinguished simply on the basis of their appearance. The EBSP method was therefore applied to determine the distribution of (minimum) misorientation angles for the boundaries. Measurements were made on a central section parallel to the rolling plane along tracks parallel to RD and TD, corresponding to 200 boundaries per specimen. A simplified representation of these misorientation distributions is given in figure 6. Also shown in figure 6 are the misorientation distributions calculated from the X-ray ODF results for the same samples, where it is assumed that no spatial correlation of misorientations exists. With increasing rolling reduction the uncorrelated misorientation distributions show a decrease at high angles and an increase at low angles which is simply explained by sharpening of the texture. The experimental (EBSP) misorientation distributions reveal, however, an opposite trend with an increase in the proportion of high angle misorientations following heavier deformation, i.e. fewer subboundaries and more high angle boundaries. There is, however, still some degree of spatial correlation even after 98.7 % reduction since the two distributions are not coincident.

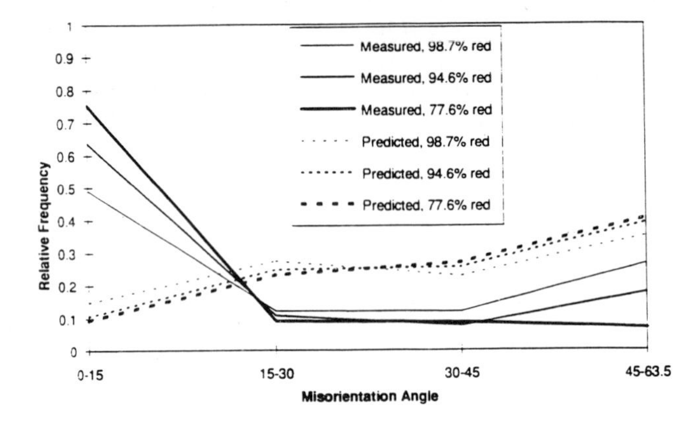

Figure 6 Misorientation distributions measured by EBSP (solid lines) and predicted from X-ray ODF results (dotted lines) after various stages of cold rolling reduction.

DISCUSSION

The mechanical property and microstructure observations demonstrate that a change occurs in the nature of the recrystallisation process as the degree of rolling is increased. At the same time, the annealing texture changes from being predominantly cube to being quite similar to the deformation texture although with a weakening of the brass component. Such textural behaviour has often been attributed in the past to a change from discontinuous to continuous recrystallisation [e.g. 1]. However, care is necessary when making deductions on the basis of textures alone since the R-texture is frequently a result of discontinuous recrystallisation [2] and, as shown here, there is a change in the texture even when recrystallisation occurs continuously.

Most descriptions of the continuous recrystallisation phenomenon [e.g. 3] associate it with pinning of subboundaries by second phases and their gradual release due to particle coarsening. Although the present alloy contains second phases as well as a significant content of iron in solid solution, there is a good reason to doubt the particle coarsening model. Most significantly, it cannot explain why the level of strain is so important to the occurrence of discontinuous/continuous behaviour. A well developed subgrain structure is present even after the lowest cold reduction of ~77 %.

An alternative viewpoint [4] is that the nature of the recrystallisation process depends on the relative proportions of mobile and immobile boundaries which are present in the network of boundaries which constitutes the deformed (and recovered) substructure. It was calculated that the original grains which are ~ 12 µm across become flattened to the thickness of a single crystallite at approximately 95 % reduction, implying a large proportion of high angle boundaries at about this stage. The present measurements of misorientation are qualitatively in agreement with such a viewpoint. It is known that low angle boundaries have very low mobility compared to high angle ones and that the transition in behaviour occurs in the vicinity of 15°. After 77.6 % rolling reduction the proportion of boundaries which are mobile is 25 % according to Figure 6 whereas after 98.7 % reduction this proportion increases to 51 %. Thus, a much greater fraction of the boundary network is expected to be mobile following heavy reductions.

The effect of having different proportions of low angle and high angle boundaries with different mobilities has been examined by computer simulation using a 2-dimensional network in the manner described by Frost et al. (5) and Humphreys (6). Investigations are in progress whereby the texture topology of the rolled sheet can be used as the initial state for the computation, but the present results are limited to a simplified model consisting only of different proportions of low angle (5° Gaussian spread) boundaries and random high angle boundaries. The assumed energies and mobilities as functions of misorientation are shown in figure 7. No account is taken of second phase particles although these certainly play a role in reality.

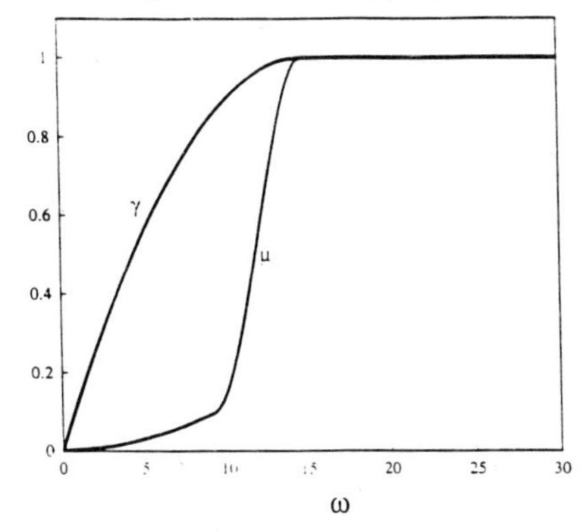

Figure 7
Assumed specific boundary energy (γ) and mobility (μ) as functions of misorientation angle (ω).

Representative simulated microstructures are shown in figure 8 and the average grain sizes are plotted as a function of time in figure 9. Classical parabolic growth rates are predicted for the cases

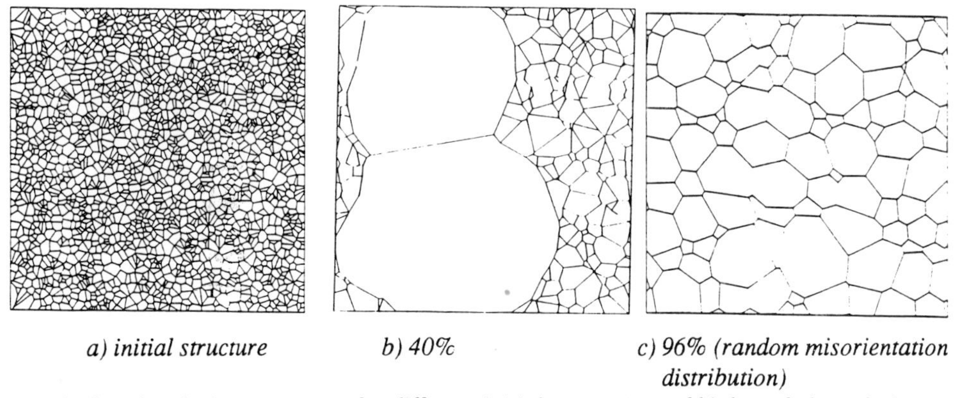

a) initial structure *b) 40%* *c) 96% (random misorientation distribution)*

Figure 8 Simulated microstructures for different initial proportions of high angle boundaries

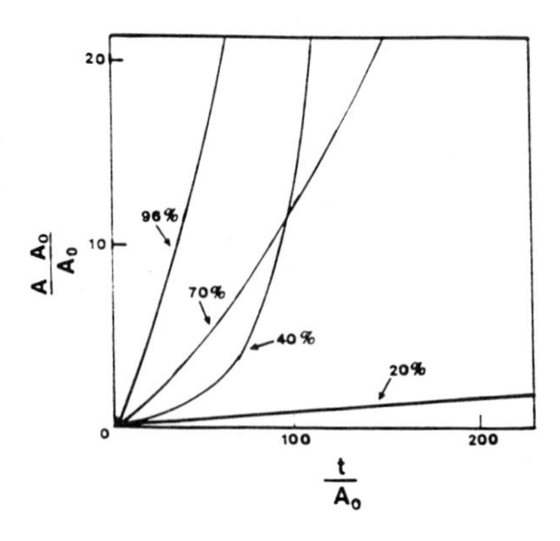

Figure 9
Average grain sizes $(A-A_0)/A_0$) plotted as a function of time for simulated microstructures having different proportions of boundaries with different misorientations.

with solely low or high angle boundaries although with widely different growth rates. For mixed structures containing a small proportion of high angle boundaries the growth process becomes highly abnormal, displaying the characteristics of discontinuous recrystallisation. The average size in such cases may become larger than for structures comprised only of high angle boundaries. As the proportion of high angle boundaries increases the process becomes progressively more normal with more rapid initial rates of coarsening but a slower development later on. Such behaviour is in excellent qualitative agreement with the changes in softening behaviour shown in figure 1 and the grain boundary misorientation data for different rolling reductions summarised in figure 6. The proportion of high angle boundaries necessary to produce a transition from discontinuous to continuous recrystallisation lies in the region of 70 % which is greater than that deduced experimentally from EBSP measurements (~ 35 % assuming a constant high mobility for misorientations greater than 15°). However, the EBSP experiments did not include measurements in the thickness direction of the sheet which is expected to show a higher frequency of high angle boundaries.

In summary it may be concluded that continuous recrystallisation in this alloy can be explained qualitatively and even semi-quantitatively in terms of the proliferation of mobile high angle boundaries within the deformed substructure at very high levels of plastic strain.

ACKNOWLEDGEMENT

This work has been sponsored by the Swedish National Board for Technical Development, Gränges AB and Gränges Luxembourg Aluminium. The authors wish to thank Richard Davis, Philippe Charlier and co-workers of Gränges Luxembourg Aluminium for their careful choice of the industrially processed materials.

REFERENCES

1. Grewen, J., Z. Metallkunde., 1968, 59, 236
2. Hirsch, J. and Lücke, K., Acta Met., 1985, 33, 1927.
3. Hornbogen, E., Met. Trans., 1979, 10A, 947.
4. Oscarsson, A., Ekström, H-E. and Hutchinson, W.B., Recrystallisation '92, Materials Science Forum, 1993, 113-115, 177.
5. Frost, H.J. et al., Scripta Met., 1988, 22, 65.
6. Humphreys, F.J., Recrystallisation '92, Materials Science Forum, 1993, 113-115, 329.

Materials Science Forum Vol. 157-162 (1994) pp. 1277-1282

DIRECT OBSERVATION OF ORIENTATION CHANGES FOLLOWING CHANNEL DIE COMPRESSION OF POLYCRYSTALLINE ALUMINUM: COMPARISON BETWEEN EXPERIMENT AND MODEL

S. Panchanadeeswaran [1], R. Becker [1], R.D. Doherty [2] and K. Kunze [3]

[1] Fabricating Technical Division, Alcoa Laboratories, 100 Technical Drive, Alcoa Center, PA 15069-0001, USA

[2] Drexel University, Philadelphia, PA 19104, USA

[3] Department of Manufacturing Engineering, BYU, Provo Utah, USA

Keywords: Backscattered Kikuchi Diffraction, Split Sample, Orientation Change on Deformation, Internal Grains, Inhomogeneous Strain, Taylor Model. Finite Element Model

ABSTRACT By use of a split channel die sample, the grain rotations in about 100 internal grains produced by near plane-strain compression of 40% at 375°C have been directly measured in polycrystalline aluminum. This type of experiment first carried out by Barrett and Levenson [1] in 1940 for uniaxial compression has been here, successfully repeated for rolling-type strains using a modern local orientation technique - that of Back-scattered Kikuchi Diffraction (BKD). The grains rotate, as predicted by a Taylor type model, to the standard β fibers of rolled fcc metals ODF textures. However the model failed to predict either the large misorientations seen in each grain or the observed positions along the β fibers. A two dimensional grain based finite element model correctly predicted large in grain misorientations but only about the, stress independent, axis normal to the two dimensional grain array.

INTRODUCTION Although there are numerous studies of the development of deformation texture, there have been very few direct observations of deformation induced grain rotation. In a unique study in 1940, Barrett and Levenson[1] reported grain rotations of internal grains in very coarse grained aluminum uniaxially compressed in a split sample. The orientation of the interface between the two parts of the split sample appears to have been normal to the compression axis. They reported firstly that individual grains fragmented into regions with steadily increasing misorientations. The misorientation "spread 7 to 10° at 10% reduction in thickness, two to three times this much at 30% and four to five times this at 60%. This spread appeared to be greater in grains on the inner faces than in grains at the surface in contact with the lubricated compression plates, but is considerable in all grains large and small." For small strains, they were able to follow the grain rotations and found that only half of the grains rotated as predicted by Taylor's homogeneous strain model[2]. They also reported that the internal faces became roughened by deformation and remarked that "The heterogeneous flow of which this is proof is important to an understanding of the nature of lattice rotations in aggregates, for it means that each grain and each fragment of a grain deforms in a manner influenced by the everchanging flow of its neighbors." After more than 50 years it appears time that this important experiment be extended to other deformation modes, to a normal grain-sized material and with the crystal rotations studied using modern high resolution techniques. The present study does this using warm channel die compression of DC-cast, commercial-purity, aluminum. The paper extends the initial observations[3] and compares the results with those predicted by a Taylor type model[2,4] and by a finite element (FE) model[5] in which modeled grain interactions replace the assumption of homogeneous strain.

EXPERIMENTAL METHODS DC cast Al-0.17% Fe-0.07%Si, with an equiaxed grain size of about 200μm, was machined to produce two samples that fitted together in a channel die compression holder. The split interface was parallel to the compression axis and the free or "rolling direction", figure 1. One of the surfaces was electropolished and anodised. A region 2mm by 2 mm was marked by microhardness indentations and observed under polarized light. After stripping the oxide film, the orientations of all the grains in the marked region were measured manually by BKD.

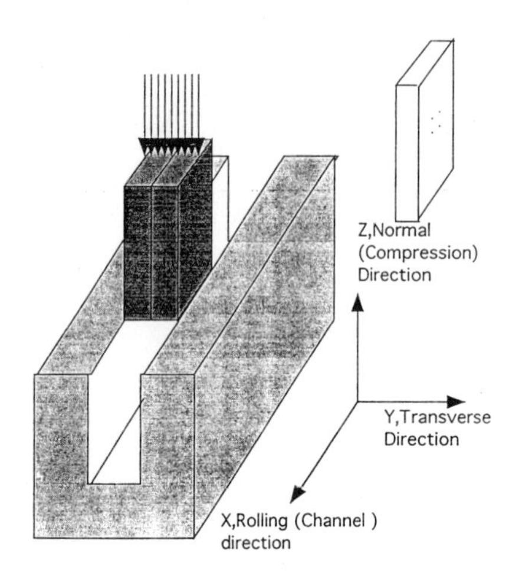

Figure 1 Split sample assembly for channel die compression.

The split sample was reassembled with the marked face internal to the assembly and warm plane strain compressed 40% at 375°C at a strain rate of $0.05s^{-1}$. After deformation the grains were reanalysed using both manual BKD and by the recent automatic BKD technique of Orientation Imaging Micrsocopy (OIM)[6]. The OIM studies were carried out both on the initial "internal" surface and after electro-polishing about 10μm from this surface. Standard X-ray textures were determined (i) from the interior of an unsplit sample, (ii) from the surface of the mating sample used in the BKD studies and (iii) from the interior of the mating sample. The details of the modeling methods have been previously described[3,5]. They need not be repeated here except to point out that both models used the same mechanical input obtained by experiment on the alloy used[3]. The FE model[5] used a two dimensional grain model, here with the measured orientations on the plane normal to the transverse direction, Figs. 1 and 2.

In the experiment, and thus in the model, there is no knowledge of the orientations facing the grains in the mating sample. The model assumes that the grains continue into the mating surface giving columnar grains. In the previous use of the model[5] the hexagonal grains faced the compression direction.

EXPERIMENTAL AND MODELLING RESULTS:

i) OVERALL TEXTURE. As previously observed[3] from the displacement of the microhardness indentations, the split sample underwent, in addition to the plane strain compression of 0.58, a small shear of 0.19 parallel to the "rolling plane" and in the "rolling direction". The surface of the sample was roughened on a grain scale as in the early study[1]. In the present study it was noticed that the hardness indentations were also visible on the mating sample. That is, metal had partially flowed into the indentations from the mating sample, producing small flattened extrusions on the surface of the mating sample. The general texture was observed and recorded as ODFs using both BKD methods (a 5° Gaussian spread being imposed on each observation) and from X-ray pole figures. Qualitatively the samples all showed equivalent textures along the β fiber from brass through S to copper, usual for rolled Al.[7]. As might be expected, the peak intensity in the ODF was higher in measurements from smaller areas than from larger areas. However the intensities along the ODF fiber were equivalent between average X-ray measurements and from local BKD, but at different areas observed. From an area of $1mm^2$, by BKD, the highest intensity was found at 12 times random from S to copper.The peak intensity fell to 7 times in the same positions when BKD measurements were obtained from the full area of $4mm^2$. X-ray generated ODFs from the surface of the mating sample, using an area of $1cm^2$, gave a clear β fiber with a peak intensity of 4.7 times random at brass and 4 times at copper. When the area was reduced to the $4mm^2$ area used with BKD, although the peak intensity rose to 6.3 at brass, other, non-standard texture components began to be visible. The X-ray generated ODF from the center of the mating sample, using a full area of $1cm^2$, was identical to that seen by BKD at an area of $4mm^2$. These had clean β fibers with the peak intensity of 6 times random located from S to copper. In all these ODFs, orthorhombic sample symmetry was imposed by averaging in the usual way[7]. An unsplit sample deformed to the same reduction in height, at the same temperature and strain rate, but without any additional rolling plane shear also had a clean β fiber with the peak intensity located from S to copper. However the peak intensity was slightly higher - at 9 times random. Finally the three sets of BKD measurements, two on the initial surface and the third after polishing 10μm from the surface, all gave essentially identical ODFs from the same marked region of the sample, at $4mm^2$. These general texture studies clearly suggest that the deformation at the interior surface of the split sample appeared to be essentially unchanged from that of a true interior region in the polycrystalline aluminum studied. However it is clearly desirable for this result to be checked by further studies particularly with BKD and X-ray ODFs determined from identical samples and with both studies repeated for deformation on split and unsplit samples.

The Taylor model was used to predict the overall texture in several ways. Using the BKD measured 100 grains, based on 234 individually measured initial orientations, and imposing both the compression strain and the additional rolling plane shear, the Taylor model[4] predicted a standard β fiber ODF. The ODF had a peak intensity at S to copper of 11.6 times random, significantly higher than the BKD observed results from the same deformed grains. A model without the additional shear increased the peak intensity to 13.3 times random but did not otherwise change the predicted texture. As in the experimental studies, a 5°Gausian spread was imposed on each orientation. Simulations were also run both with and without shear using 1000 randomly oriented grains to provide a comparison with the X-ray generated ODFs. Here the same form of ODF was seen: a β fiber with a peak intensity from S to copper of 8.6 times random with the rolling plane shear and of 10 times without the shear. The present results match very well the usual observation [see for example 8] that Taylor type models based on homogeneous deformation match overall fcc rolling textures rather well, though with intensities along the β fiber predicted to be higher than those observed.

INDIVIDUAL GRAIN ROTATIONS. The observations previously reported[3] for 25 grains have now been extended to over 60 grains with new observations made using OIM to give many more observations per grain. The OIM studies were then repeated after 10μm had been electropolished from the initial surface. First of all it was found that the electropolishing showed no significant change in the measured orientations. So the previous concern[3] about the measurements being made so close to the internal "grain boundary" provided by the mating surface, was found to be unwarranted. The second point of importance is that the use of the OIM technique, figure 2b, allowed a nearly perfect grain to grain match with the undeformed grain structure, figure 2a. This microstructural match removed any residual uncertainty about which undeformed grain the measured deformed orientations had come from.

In the initial report of this work[3] the measured in-grain rotations showed large spreads of orientation within each grain, in contrast to the unique grain rotation predicted by the homogeneous strain assumption of the Taylor model. In addition, the Taylor model, in almost every case, failed to match the magnitude or direction of the observed average grain rotation. These initial results have been fully confirmed by the present studies. This is illustrated in figure 3 for grains, A and J', previously shown [3], and for two other grains, H and K. The 4 grains are from the region seen in figure 2.

For each grain in figure 3, the initial orientation is shown as an open square, the unique Taylor predicted rotated orientation is shown as an open diamond, the FE predicted range of orientations are shown as crosses, and the measured orientations are shown as short vertical lines. (There is a slightly different Taylor model prediction shown with the finite element calculations - this arose from slightly different boundary conditions used for the smaller area modeled in the FE study). For grain A, most of the measurements show only a small rotation, though a few large rotations are also seen. Both the Taylor and the FE models correctly predicted, in this case, the small average rotation seen. As in the previous FE model simulation[5], a spread of deformed orientations was correctly predicted. In the present case the simulation spreads are about the axis normal to the grain structure, the transverse axis in this model. In the previous simulation[5] the orientation spread was predicted to occur about the compression axis which had there been set normal to the grain array. Repeated tests of the FE model always show lattice rotation primarily about the axis nomal to the two dimensional grain structure. This indicates that the FE predicted grain spreading appears to be due to the grain interactions. As seen in figure 3 for grains J', H and K, the experimental average grain rotations were much larger than those predicted, at least by the Taylor model, and were in completely different directions from the rotations predicted by either model. A general summary of the local observations in comparison with the Taylor model predictions indicates that:

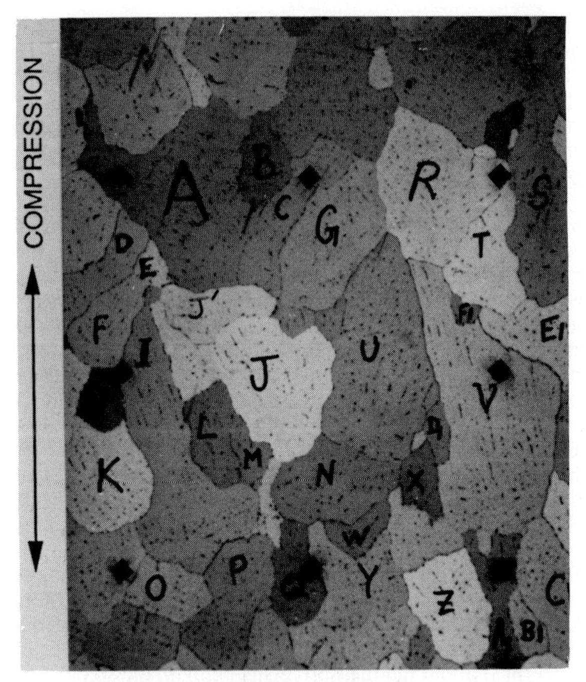

Figure 2a Optical micrograph of part of the undeformed sample.

1) In every grain, a wide range of misorientations was found, often upto or more than 25°.

(2) In 8 out of 62 grains where the initial orientation had been close to one of the rolling texture components (e.g. grain A), the predicted small average rotations were found. Even in these grains however, the small rotations seen for the center of the observed distributions, were still not correctly predicted.

(3) For one other grain, L, which was close to the Goss orientation the lack of any major rotation of the center of the observed spread was correctly predicted.

(4) In all other grains the model failed to predict either the magnitude and/or the direction of rotation.

5) In 24 cases, including grains H and K of Figure 3, the grains moved to orientations close to one of the rolling texture components, but not to the texture component predicted by the model. 10 of these 24 grains, again including grains H and K, were initially close to one rolling texture component but they rotated to a different rolling component.

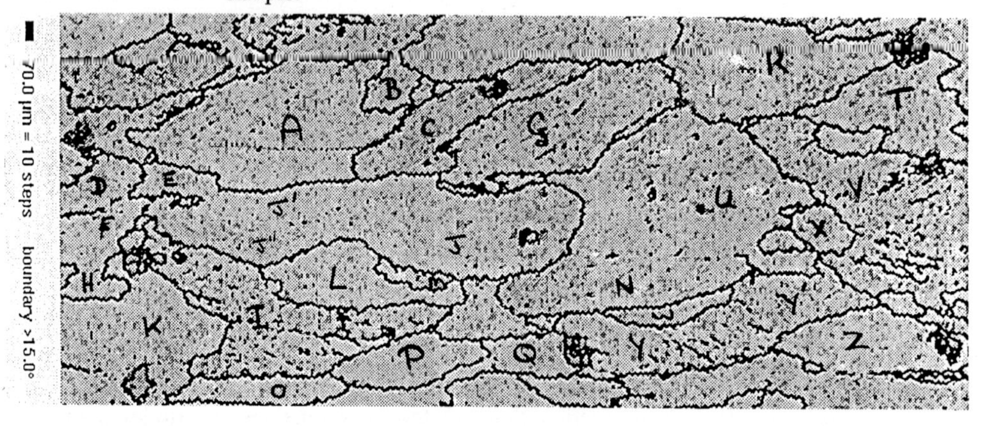

Figure 2b OIM of part of the undeformed sample after deformation. Misorientations of 15° are shown. The microhardness indatations are visible as dark regions of low quality BD patterns.

It is clear from the summary given above that the Taylor type model failed almost completely to predict the orientation changes seen in the 62 grains analysed. The model fails to predict the large orientation spreads seen in each grain, a result previously reported by Barrett and Levenson [1] and frequently reported for unsplit samples studied on sectioning after deformation (see for example [9, 10]) . In addition, apart from the 9 grains noted in points (2) and (3) above, the model fails to match, in any way, the observed general grain rotations. The FE model that allowed intergrain interactions is clearly superior to the Taylor model in that it successfully predicts the inhomogeneous strain as indicated by the in-grain misorientations. For other details of the FE model's prediction of strain inhomogeneity, the original report by Becker [5]should be consulted.

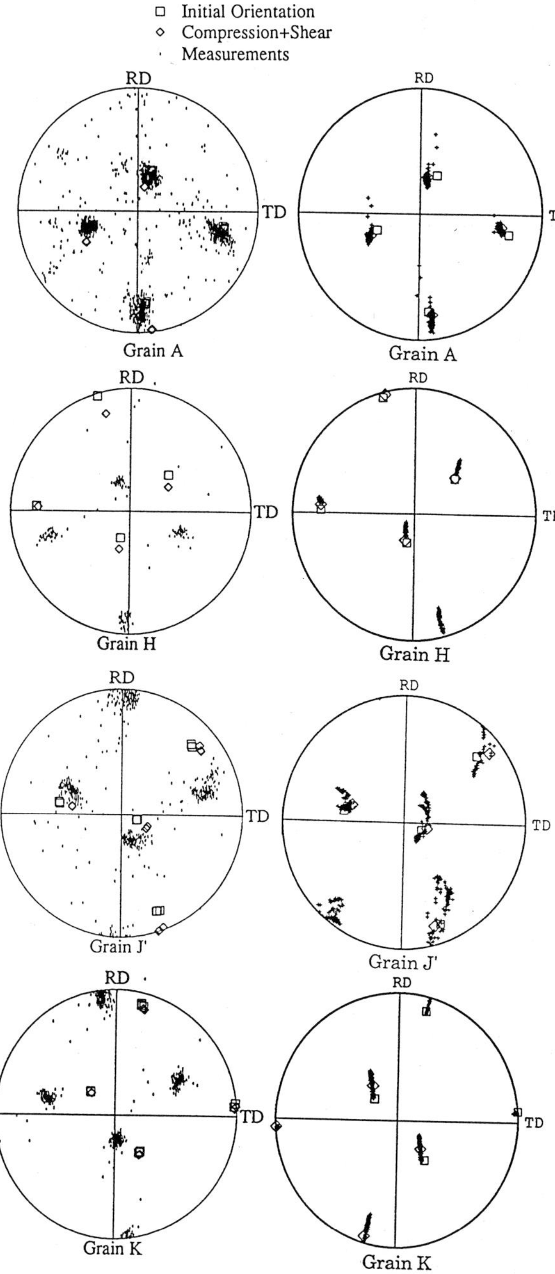

Figure 3. {111} pole figures of measured and predicted grain rotations after 40% compression. The FE predicted post - strain orientations are marked by +.

The FE model is still incomplete, however, since the type of spread predicted is controlled by the orientation of the plane in which the two dimensional array of grains is modeled. Given this failure, it is not surprising that the FE model was hardly any more successful than the Taylor model in predicting the direction of grain rotation during deformation. The grain interactions as seen by the FE model produce large orientation spreads bout only about amodel dependent axis. So it is then perhaps not surprising that the FE model fails to capture the true average grain rotation. The failure occurs since the FE model is unaware of the closest grains - those mating the studied grain array across the interface. A future split sample experiment in which the measured polycrystalline grain array abuts a single crystal of known orientation should supply some of the missing information needed to stimulate a 2 grain thick 3 dimensional FE model.

A final question naturally arises as to why there can be this general disagreement for individual grain rotations when the Taylor model provided such a good general qualitative agreement with the measured average texture. This question arises since the Taylor based model was successful in predicting the average texture not only in this study but in many previous studies, as seen in many papers in recent ICOTOM proceedings. That is, the model appears qualitatively successful at the global level, but it fails here almost completely at the level of individual grains. An initial and partial answer to this question is provided by the common observation that the observed orientations of the grains after deformation were indeed on the predicted rolling texture fibers, thus accounting for the match with the average texture predictions. Although the individual grains were found on the β fibers, they were not found in the predicted position along the β fibers. That is, the obseved rotation of most grains was to a predicted rolling texture orientation but not the rolling texture orientation predicted for that grain. The observation that many grains are correctly located at or close to the rolling texture fibers but not in their predicted positions along the fiber is then compatable with the global success but local failure of the model. This is, as yet, only a partial answer to the question since the present authors have no satisfactory explanation for the mechanistic origin of this discrepancy or at least have no tractable model that can make successful predictions of the grain rotations.

To explore the idea of occupation of rolling texture fibers but at unpredicted locations along the fiber, the ODFs were recalculated, for both the experimental and the predicted orientation results, but without enforcing sample orthorhombic symmetry. ODFs of this type are generally known as cubic-triclinic ODFs, Bunge [11]. The first term refers to the cubic crystal symmetry, the second to the triclinic sample symmetry. When this was done, the previously reported close agreement between the predicted and the BKD measured textures from the same set of starting grains was then lost. With the cubic-triclinic ODF the peak intensities vary in each ODF between the 4 different S components and the two different brass and copper components. Critically, however, the particular components showing peak intensity in the predicted ODF were not the ones showing peak intensity in the measured ODF. That is, the previous agreement between prediction and measured ODF was an artifact of inforcing orthorhombic symmetry. Such symmetry appears invalid when only 100 grains are observed.

Unfortunately lack of space prevents any comparison of results of this work with other studies of this topic, particularly those of Driver et al. [12, 13].

CONCLUSIONS The first conclusion is that deformation is heterogeneous on a grain to grain level, giving rise to large in-grain misorientations. Although the homogeneous Taylor model did predict the average texture produced in plane strain compression rather well, it failed totally at the level of individual grains. The failures are the previously observed wide misorientation within each grain and, although the model succcessfully predicts the ODF β fibers, it rotates most grains to different positions on the fibers from those observed. The current two dimensional grain FE model which successfully captures the large in-grain misorientations does not yet match the experimental results in other ways. A full three dimensional FE grain model, using measured starting grain orientations, is clearly the next step required to predict deformation microstructure at the level required, for example, for recrystallization purposes [10].

ACKNOWLEDGEMENTS Dr. O. Richmond and Dr. L.A. Lalli for suggesting this problem, Dr. J. Liu for discussions on the BKD technique, Dr. V.Sample for the channel die compression tests, Dr. U.F.Kocks and Dr. A.D. Rollett and others for stimulating discussions of this topic, Prof B. Adams for use of the Automatic BKD facility (OIM) and the financial support of Alcoa and NSF (DMR 9001378).

REFERENCES
1) Barrett, C.S. and Levenson, L.H.: TMS-AIME., 1940, 137, 112.
2) Taylor, G.I.: J.Inst.Metals, 1938, 62, 307.
3) Panchanadeeswaran, S. and Doherty, R.D.: Scripta Metal. and Mater. 1993, 28, 213.
4) Asaro, R.J. and Needleman, A.: Acta Metal. 1985, 33, 93.
5) Becker, R.: Acta Metal. and Mater. 1991, 39, 1211.
6) Adams, B.L., Wright, S..I. and Kunze, K.:Metall.Trans. 1993, 24A, 819.
7) Panchanadeeswaran, S., Sample, V.M, Rollett, A.D., Doherty, R.D., Fricke, W.G.Jr., and Lalli, L.A.: ICOTOM 8, Eds Kallend, J.S and Gottstein ,G., Met. Soc. AIME. (1988) p.485.
8) Kocks, U.F.: ICOTOM 8, p 285.
9) Bellier, R.D. and Doherty, R.D.: Acta Metal. 1977, 25, 521.
10) Doherty, R.D, Panchanadeeswaran. S. and Kashyap, K.: Acta Metal. and Mater. 1993, 41, In press.
11) Bunge, H.J. " Texture Analysis in Materials Science", Butterworths, London, 1982.
12) Skalli, A., Fortunier, R., Fillet, R and Driver, J.H.: Acta Metal. 1985, 33, 997.
13 Fortunier, R. and Driver, J.H.: Acta Metal. 1987, 35, 1355.

Materials Science Forum Vol. 157-162 (1994) pp. 1283-1290

MICROTEXTURE EVOLUTION OF NECKLACE STRUCTURES DURING HOT WORKING

D. Ponge, E. Brünger and G. Gottstein

Institut für Metallkunde und Metallphysik,
RWTH Aachen, Kopernikusstr. 14, D-52056 Aachen, Germany

Keywords: Dynamic Recrystallization, Necklace Structure, EBSP, Cu, Ni_3Al, Single Grain Orientation Measurements

ABSTRACT

Single grain orientation measurements by means of EBSD were utilized to investigate the onset and the evolution of necklace structures during dynamic recrystallization (DRX) in Cu and Ni_3Al. The two different metallic materials were investigated to elucidate the role of dislocation substructure on DRX, since the dislocations in the intermetallic compound Ni_3Al are randomly distributed, whereas Cu develops a well defined cell structure. In both materials DRX starts with bulging of the pre-existing grain boundaries, but in Cu indications for additional nucleation sites within the grain interior were found.

INTRODUCTION

Many materials undergo dynamic recrystallization (DRX) during hot working. Besides a reduction of flow stress, DRX strongly affects microstructural evolution during high temperature deformation. Initially coarse grained material suffers conspicuous grain refinement under hot working conditions in particular at high strain rates and moderately high temperatures. Grain refinement is obtained by the formation and expansion of a fine grained layer of dynamically recrystallized grains along prior grain boundaries, a so called "necklace-structure". The mechanism, by which the first layer of new grains forms, is well established. Previous investigations suggest that pre-existing grain boundaries constitute the principal nucleation sites for DRX [1,2]. Localized strain-induced grain boundary migration leads to bulging of the boundaries in the manner first analysed quantitatively by Bailey and Hirsch [3].
The mechanisms, by which subsequent layers of dynamically recrystallized grains are formed, are virtually unknown. When in the course of DRX the pre-existing grain boundaries are entirely covered with new grains (site saturation), bulging would have to proceed from the small recrystallized grains which would require a very high boundary curvature. This makes further nucleation by bulging unlikely, because the very high driving force necessary to offset the high surface energy of the bulge is not available in hot deformed microstructures. Roberts et al. [2] propose from metallographic observations, that when the nucleation sites at the original grain boundary are exhausted, the reaction should then proceed via regular nucleation at the interface between recrystallized and unrecrystallized material. This needs to be confirmed, since additional nucleation mechanisms at these interfaces are unknown so far.

The current study is concerned with the evolution and progression of a necklace structure during high-temperature deformation. The inhomogenity of driving force distribution may affect the nucleation of new grains. It is observed that even pure metals, which do form cells and subgrains throughout the crystals during deformation, develop a necklace structure by DRX [4]. To study the influence of the formation of a substructure on DRX, two different materials were chosen: The intermetallic compound Ni_3Al reveals a comparatively uniform dislocation distribution with no cell structure formation during hot working, whereas Cu reveals a well defined cell structure.
Moreover, Ni_3Al exhibits a low grain boundary mobility. This facilitates the investigation since only furnace cooling is required to freeze the dynamically recrystallized microstructure. In the case of Cu, water quenching becomes necessary, otherwise a statically recrystallized structure is obtained.

EXPERIMENTAL PROCEDURE

Ni_3Al was produced from pure constituents by vacuum induction melting with the composition $Ni_{76}Al_{24}$+0.24 at%B. The as cast alloy was annealed in a sequence 1050°C/6h, 900°C/8h and 750°C/8h in vacuum. The material was cold compressed to a strain of 30% and subsequently recrystallized at 1050°C/0.5h to produce a mean grain size of 100μm. The mechanical tests were carried out in compression under high vacuum (<10^{-3}Pa). The specimens had cylindrical shape with 6.3mm diameter and 11mm length. BN-powder was used as lubricant between specimen and compression dies. Deformation was conducted at 950°C to various strains with a constant displacement rate corresponding to an initial strain rate of $2 \cdot 10^{-4} / s$. Subsequent to deformation the samples were quenched by spraying cold nitrogen gas onto the specimen with a quench rate of approximately 50°C/s. After dismounting from the testing machine, the samples were sectioned parallel to the compression axis, and finally mechanically and electrochemically polished.
Electrolytic tough pitch (ETP) copper of technical purity (≥99.90%, 0.005-0.040%O_2) was annealed at 1020°C/1.5h to provide a uniform grain size of 150μm and torsion tested at 400°C with a strain rate of $2 \cdot 10^{-1} / s$. The samples were water quenched in 1~2 seconds immediately after deformation. After sectioning parallel to the torsion axis, the samples were mechanically and electrochemically polished.
The Ni_3Al and Cu samples were mounted in a SEM equipped with an integrated EBSD-unit for single grain orientation measurements [5].

RESULTS

The stress strain curves of both Ni_3Al and Cu revealed a shape with a maximum which is typical for the occurrence of DRX. The strain corresponding to the maximum stress was 4.9% for Ni_3Al and 190% for Cu. In both cases the test was terminated before a steady state was attained. The orientations of the necklace grains and their parent grains were determined and from this the misorientation across the respective grain boundaries was calculated.

i) Ni_3Al
Figure 2 shows a low magnification micrograph of a typical necklace structure in Ni_3Al. TEM investigations revealed that the new grains contain dislocations so that these grains can be considered as dynamically recrystallized grains. Some of the parent grains contain microbands. They run parallel to the traces of {100} planes, which are the slip planes of the operating slip system {100}<011> in Ni_3Al above the peak temperature (≅600°C)[6].
Figure 3 shows a typical example for the bulge mechanism at low deformations. Two twins have formed at the bulging front and are labelled as Σ3.

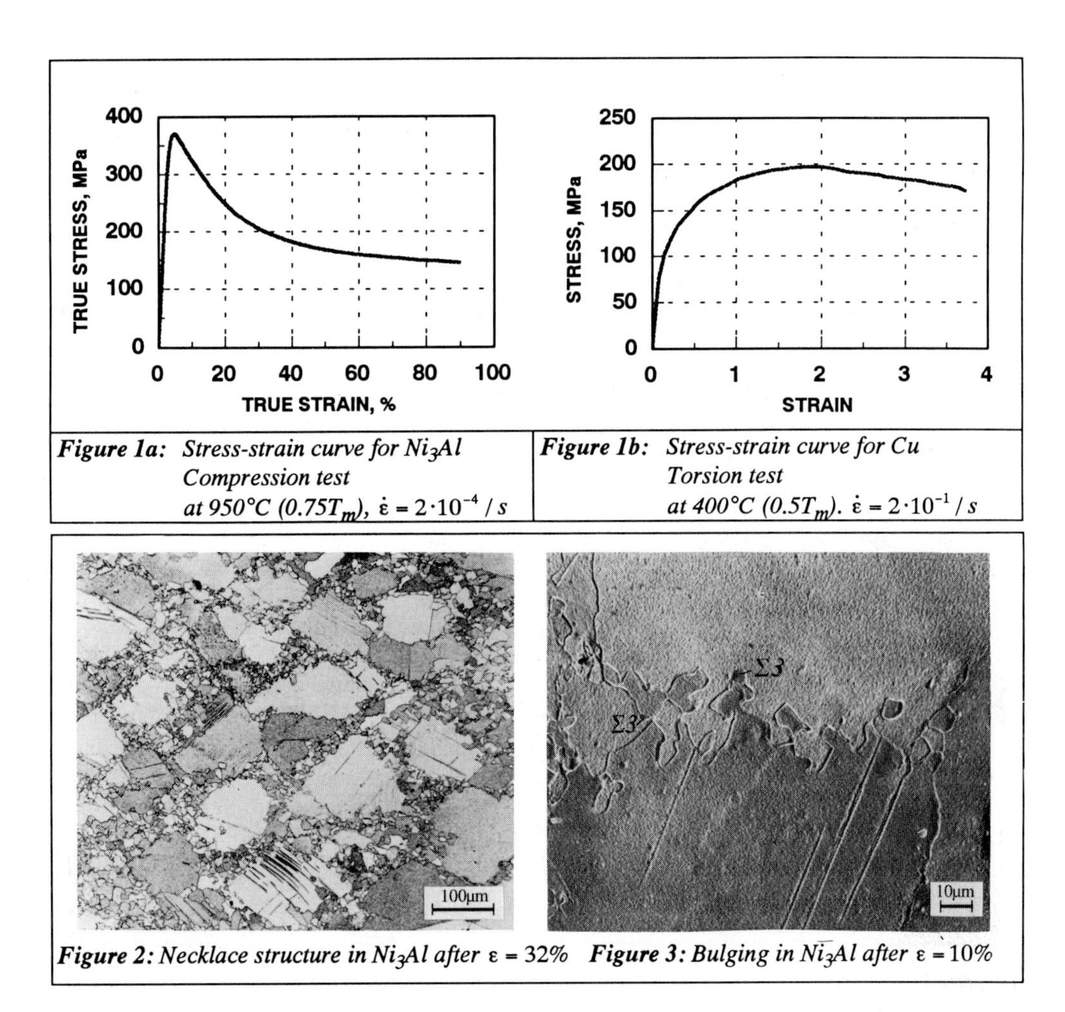

Figure 1a: *Stress-strain curve for Ni_3Al Compression test at 950°C ($0.75T_m$), $\dot{\varepsilon} = 2 \cdot 10^{-4}/s$*

Figure 1b: *Stress-strain curve for Cu Torsion test at 400°C ($0.5T_m$). $\dot{\varepsilon} = 2 \cdot 10^{-1}/s$*

Figure 2: Necklace structure in Ni_3Al after $\varepsilon = 32\%$ *Figure 3: Bulging in Ni_3Al after* $\varepsilon = 10\%$

In the course of DRX the deformed original grains are consumed by the spreading of layers of dynamically recrystallized grains. In figure 4 and 5 a layer of new grains between two parent grains is shown for ε=25% and ε=100%. The microtexture is represented by different shadings of the grains. The angle of rotation is related to the misorientation between the new grain and one of the parent grains (Only the smaller angle of rotation was considered). Unshaded grains have misorientations $\geq$22.5° to both of the parent grains. Small angle boundaries (angle of misorientation $\leq$15°) are labelled as 1, twins of first, second and third generation as 3, 9 and 27, respectively. All other boundaries are not labelled.
The fraction of special boundaries (other than small angle boundaries and twins) is not a function of strain and remains virtually constant at 5%.Twin boundaries occur with a frequency of $\cong$6.5%, independent of strain.
To handle the large amount of orientation data, a simplified classification was established: For every new grain located between two parent grains the angles of misorientation between this grain and the two parent grains were determined and only the orientation relationship with the smaller angle of rotation was considered.

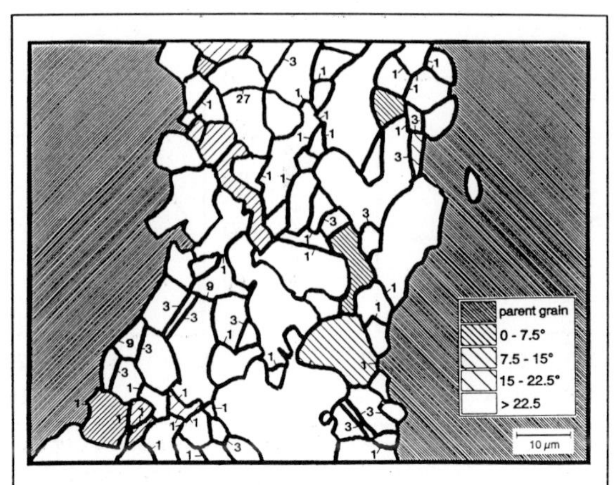

Figure 4: Necklace structure in Ni_3Al after $\varepsilon = 25\%$

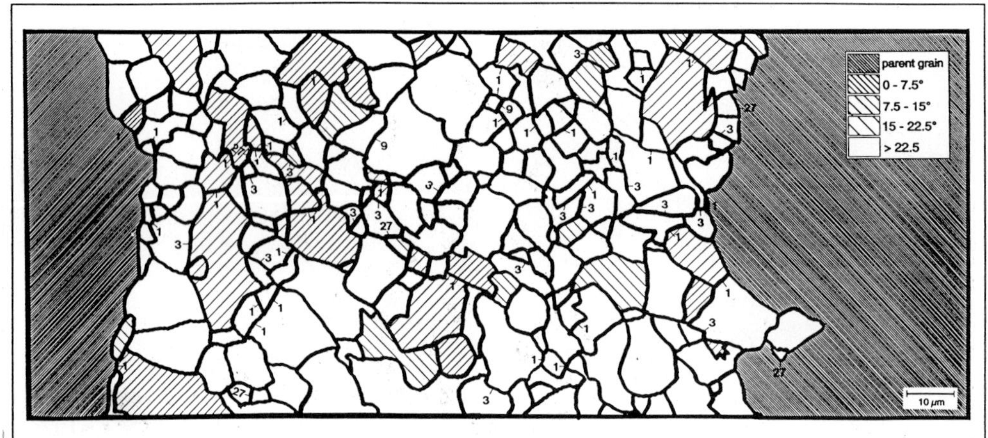

Figure 5: Necklace structure in Ni_3Al after $\varepsilon = 100\%$

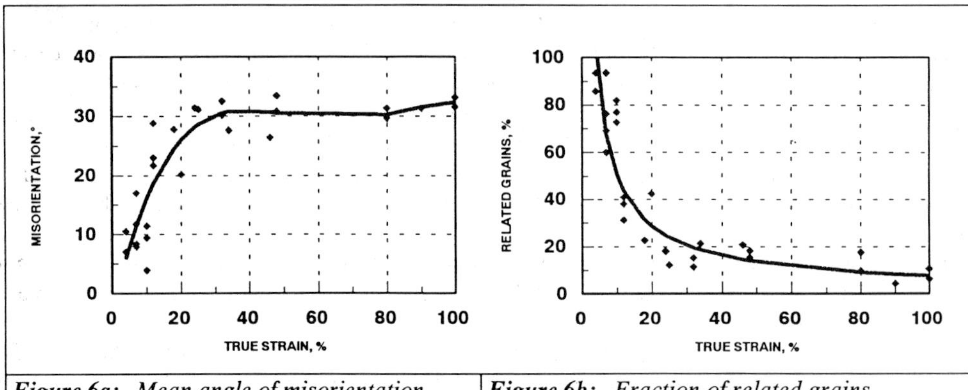

Figure 6a: *Mean angle of misorientation versus strain for Ni_3Al*

Figure 6b: *Fraction of related grains versus strain for Ni_3Al*

If the angular misorientation between a new grain and one of the parent grains was less than 15°, this new grain was classified as "related grain". For each evaluated specimen area the mean angle of misorientation and the fraction of related grains was determined.
The results are plotted in figure 6 as a function of strain. At low strains, where the first layer of new grains are formed by bulging, the mean angle of misorientation starts with values of 5° (figure 6a). This value rises with increasing strain and reaches a steady state at 25% strain with a value of approximately 30°. At small strains almost all new grains (≅90%) are related grains (figure 6b). This value decreases with increasing strain to values less than 10%.

ii) Cu
At low strain levels the grain boundaries in Cu become curved and bulging of pre-existing grain boundaries occurs (figure 7). Newly formed grains which are not situated at an original grain boundaries were found as well (figure 8). These grains contain dislocations and can be considered as dynamically recrystallized grains. At higher strain levels small grains with new orientations appear at the original grain boundaries, predominately at sites, where three or more parent grains meet. In figure 9 there are altogether five parent grains. Two of them are separated by new grains. The orientations of most of these new grains are close to one of the parent grain orientations. At very high strain levels (figure 10) wide regions are covered with small grains which are separated by high angle boundaries. The parent grains developed a pronounced cell structure. In most of the parent grains the EBSD-patterns were very obscure. It seems that the electron beam illuminated more then a single dislocation cell which develop high misorientations of up to 15° within a individual parent grain.
Hence there is a large orientation spread within a single parent grain. This fact makes it difficult to establish an orientation relationship between a new grain and its parent grain.
Also high angle grain boundaries are not always visible in the SEM image. All these problems makes interpretation of the orientation measurements in Cu more difficult. Therefore, the grain classification introduced for Ni_3Al was omitted.

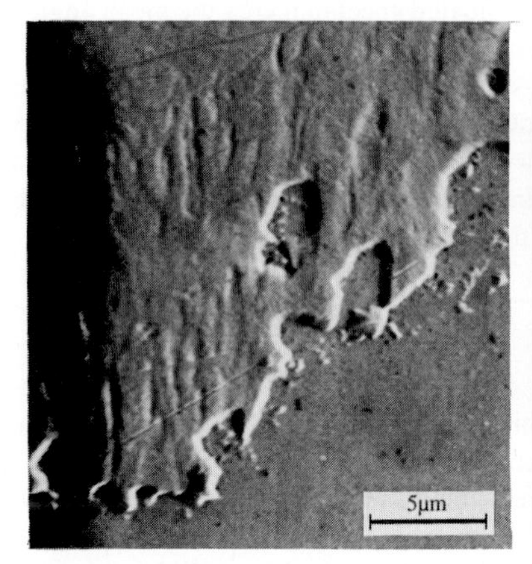

Figure 7: Bulging in Cu after ε = 60%

Figure 8: Cells and a new grain within a parent grain

Figure 9: New grains at pre-existing grain boundaries in Cu after ε = 120%

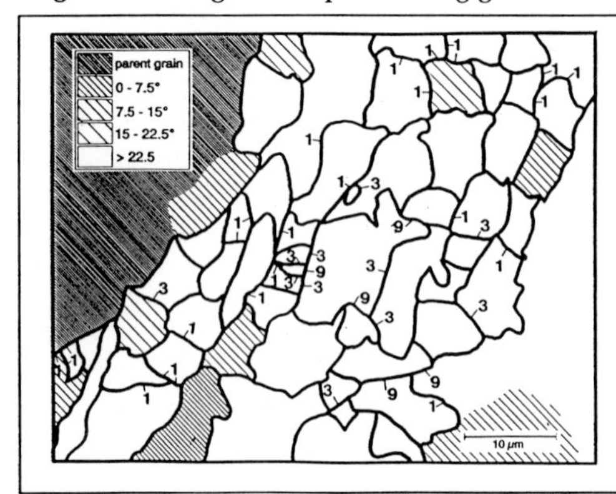

Figure 10: Cu, ε = 200%

DISCUSSION

i) Ni_3Al
From previous investigations on materials other than Ni_3Al it seemed to be established that the first recrystallized grains occur at prior grain boundaries and tend to completely decorate these pre-existing grain boundaries, which is referred to as a necklace structure. The Bailey-Hirsch mechanism [1] was proposed to be the principal nucleation mechanism. The current study confirms these results for Ni_3Al. If the first layer of new grains is formed by this mechanism, the new grains should reveal orientations similar to their parent grains, which indeed is observed. At low strain levels the mean angle of misorientation is in the order of 5-10°.
A more difficult problem, however, poses the expansion of the necklace structure. From the current study it is evident that with progressing strain an increasing number of grains cannot be associated to one of the parent grains in terms of a similar orientation. The question arises how the new orientations come about. Actually, there may be two reasons for the evolution of large misorientations and the loss of coherence between DRX grain and parent grain: grain rotation during superplastic flow or annealing twin formation and evolution of twin chains.
In fact, at the chosen deformation temperature of $0.75T_m$ the observed dynamically recrystallized grain size of $10\mu m$ is sufficiently small to undergo superplastic flow in the fine grained areas. During superplastic deformation also grain boundary sliding and/or - diffusion contribute to the plastic flow and correspondingly, to a macroscopic rotation of the grain. As a matter of fact, randomization of the crystallographic texture is observed during superplastic flow and also a progressing loss of orientation correlation was observed during DRX of Ni_3Al. As reported elsewhere [7], the strain rate sensitivity m in Ni_3Al increases with progressing DRX to values $m>0.4$. This is a strong indication for the occurrence of superplastic flow during DRX.
New orientations may also be generated by the formation of annealing twins of first or higher order during grain boundary migration. Twin formation and multiple twinning was reported before for nucleation of DRX in single crystals of Cu, Ag, Ni and Cu-alloys. In an investigation of DRX in polycrystalline austenitic Fe-Ni alloys [8], also twins were found at the bulging front. The strong tendency for annealing twins during static recrystallization of Ni_3Al has been evidenced [9] and the current results show that annealing twins occur in Ni_3Al as well during DRX (figure 3). The EBSD-data reveal a frequency of $\approx$6.5% twin boundaries within the necklace layers. However it would be difficult to track twin chains in the observed complex grain arrangement of the necklace structure.
In the course of DRX in Ni_3Al the deformed original grains were only consumed by the progression of the necklace layers. No indications for nucleation inside the deformed parent grains remote from the pre-existing grain boundaries were found. Since the original grain boundaries provide the only nucleation sites, the evolution of DRX in Ni_3Al strongly depends on the initial grain size D_0.

ii) Cu
In contrast to Ni_3Al the tendency to distinct necklace formation was much less pronounced and defined in Cu. Actually, no gradual consumption of the original grains by the expansion of distinct necklace layers was observed. Rather, at a few locations new grains with new orientations appeared at the prior grain boundaries. These structures are similar to necklace layers at an early stage, but they never completely enclosed the original grains. Instead, extended recrystallized areas spread out from these nucleation sites. Especially at tripelpoints these structures were apparent.
A major difference between Cu and Ni_3Al is that Cu develops a well defined cell structure during deformation and the EBSD-measurements confirmed a well developed substructure with orientation gradients up to 10° within the original grains. At grain boundary junctions the substructure development seemed to be accelerated and the DRX microstructure in cell forming materials may

not be generated by expansion of necklaces rather than by patches or clusters of DRX grains. For these reasons it is at least questionable to call these structures "necklace-structures".
From the current investigations it is evident that nucleation in Cu is not confined to the pre-existing grain boundaries. Rather additional nucleation takes place at deformation bands. Jonas et al. [10] have pointed out that especially in coarse grained Cu with an initial grain size larger than 100μm deformation bands are the preferred nucleation sites. In fact, we found indications for nucleation inside the parent grains remote from the original grain boundaries. As an example figure 9 reveals a dynamically recrystallized nucleus (as evident by its dislocation content) which has formed inside the substructure of an original grain.

CONCLUSIONS

The evolution of DRX in initially coarse grained Ni_3Al and Cu was investigated by means of single grain orientation measurement with the EBSD-method. The following results were obtained:

- In Ni_3Al DRX nucleation always started at pre-existing grain boundaries. Bulging of these boundaries leads to the first layer of new grains with orientations similar to the parent grains.
- DRX proceeds by gradual expansion of these layers into the original grains. Nucleation remote from original grain boundaries was not observed.
- At high strain levels most of the dynamically recrystallized grains reveal orientations quite different from those of the parent grains. We propose to attribute this finding to the formation of twin chains during boundary migration or rotation of new grains by superplastic flow.

- In Cu distinct necklace layers which would enclose the original grains were not observed. The DRX microstructure in this material seems to be generated in expanding patches of DRX grains which may be nucleated at prior grain boundaries or inside the grains. Evidence for both nucleation at the original grain boundaries and inside the grains was found in this study.
- Further investigations are necessary to elucidate the evolution of microstructure and texture in this material.

ACKNOWLEDGEMENTS

The authors are grateful to Professor J. J. Jonas and Dr. D. Bai for performing the torsion tests on copper.

REFERENCES

1) Roberts, W.; Ahlblom, B.: Acta Met, 1978, 26, 801
2) Roberts, W.; Bodén, H.; Ahlblom, B.: Met. Sci., 1979, 13, 195
3) Bailey, A.R.; Hirsch, P.B.: Proc. R. Soc., 1962, A267, 11
4) Jonas, J.J.; Sakei, T.; Deformation, Processing and Structure, ASM, 1984, 185
5) Engler, O.; Gottstein, G.: Steel Research 1992, 63, 413
6) Copley, S.M.; Kear, B.H.: Trans. Metal. Soc. AIME, 1967, 239, 977
7) Ponge, D.; Grummon, D.S.;Lee, C.S.;Gottstein, G.: Mat. Sci. Forum, 1993, 113-115, 605
8) Furubayashi, E.; Nakamura, M.: Proc. ICOTOM-6, 1981, 619
9) Escher, C.; Gottstein, G.: same conference
10) Blaz, L.; Sakai, T.; Jonas, J.J.: Metal Science, 1983, 17, 609

Materials Science Forum Vol. 157-162 (1994) pp. 1291-1298

ORIENTATIONAL ASPECTS OF THE MORPHOLOGICAL ELEMENTS OF THE MICROSTRUCTURE IN HIGHLY COLD ROLLED PURE COPPER AND PHOSPHORUS COPPER

K. Sztwiertnia and F. Haessner

Institut für Werkstoffe, Technische Universität Braunschweig,
Langer Kamp 8, D-38106 Braunschweig, Germany

Keywords: Local Textures, Shear Bands, Microbands, Orientations, Misorientations, Copper, Phosphorus Copper

ABSTRACT

High-purity copper (99.9998%) and alloy containing .96 wt-% phosphorus after 95% cold-rolling were used as examples of f.c.c. metals and alloys for investigating the orientation aspects of microstructural inhomogeneities in these classes of materials. Space elements of the microstructure such as cells or microbands were studied by the TEM. The local textures of these elements were determined. The main components of the microstructure as cells in pure copper or microbands in phosphorus copper reproduce the global textures of these materials. The local textures of the minor microstructural elements can be quite different from this global texture.

INTRODUCTION

A considerable amount of data is available in literature on the deformation behaviour of rolling f.c.c. metals and alloys [see f.i. 1,2,3,4]. The main parameter governing this behaviour is the stacking fault energy (SFE). Pure copper and brass have often been used here as prototype materials of high / medium and medium / low SFE respectively [e.g. 1,3]. In these papers most attention was paid to morphological features of the heterogenous microstructures. In this contribution which is the continuation of an earlier study [5], we have analysed the local orientation distribution assigned to characteristic inhomogeneities of highly rolled pure copper and phosphorus copper (Cu+0.96%P) as examples of medium / low SFE material.

EXPERIMENTAL PROCEDURE AND DATA EVALUATION

Using samples of the same materials as in [5] i.e.: pure copper (purity 99.9998%) and phosphorus copper (Cu+0.96%P) cold rolled to 95%, thin foils were prepared perpendicular to the transverse direction (the longitudinal section). In the TEM, at an acceleration voltage of 120kV, bright field images and Kikuchi-patterns (convergent nanobeam electron diffraction) were obtained. The orientations were measured along lines crossing the particular space elements of the microstructure such as microbands. Thereby it was possible to obtain the orientations and consistently the misorientations of neighbouring elements (see Fig. 1).

From the measured individual orientations, partial distribution functions (asigned to the particular microstructural features) of the orientations (ODF) and of the misorientations between adjacent regions (MODF) were calculated in the same manner as in [5]. The partial ODFs are plotted in Euler angles space (φ_1, Φ, φ_2). The MODFs are plotted in the space formed by axis of rotation and rotation angle coordinates (**a**, ω), more convenient here for interpretation.

RESULTS

Pure copper

Highly rolled pure copper shows a layered structure composed of sharp walled cell blocks (CB's) and microbands (MB's). The boundaries of CB's and MB's (clusters of MB's) lie nearly parallel to the rolling plane. This structure is fairly often crossed by shear bands (SB's) (see f.i. [1]). The morphology of this arrangement is shown schematically in Fig.1.

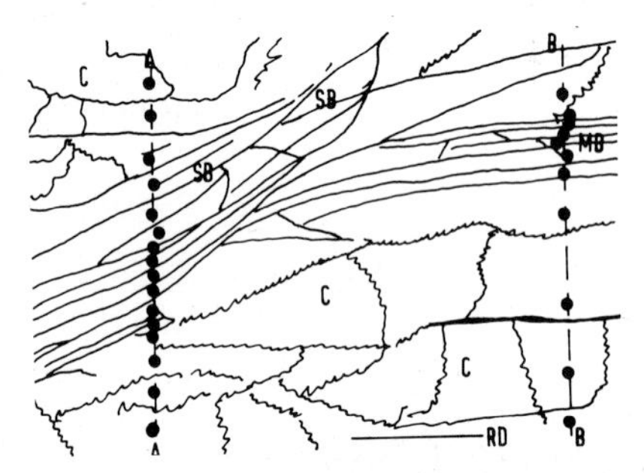

Fig.1: Schematic diagram of the microstructure in highly rolled pure copper: C-cells, MB-microbands, SB-shear bands. Orientations were measured along the lines A-A and B-B. The points on the lines indicate positions where a new orientation can be detected.

Cells

Most of the material consists of cell blocks elongated in rolling direction which are subdivided into cells. The orientation distribution of the cells in the CB's shows no significant differences compared with the typical components of the global texture of rolled copper as S {123} <634>, C {112}<111> and B {011}<211>, see Fig.3a and quantitatively Table I.

The boundaries between CB's, "dense dislocation walls" (DDW's) after [6,7], were sharp and characterized by rather high disorientations. Typical examples for DDW's are shown in Fig.2 between points 38-37 and 26-27. The disorientation angles (ω_d) were 55° and 54° respectively. The distribution of orientation differences across the DDW's between adjacent CB's is shown in Fig.4a. Four orientation relationships dominate: 60°<111>, 50°<101>, ~25°<101> and ~45°<111>. All of these misorientations correspond to coincidence-site lattic (CSL) orientation relationships with Σ equal to 3, 11, 19a and 19b respectively. Where diffuse walls or dislocation tangles were located between neighbouring cells, the orientation differences were small (as for instance between points 25-37, 23-25 and 23-15 in Fig.2: 4°, 9° and 9° respectively). The disorientation angle ranged between 0° and 10° having a maximum at 4°.

Microbands

In [6, 7,8] the authors distinguish between the 1st and the 2nd generation of microbands. In highly rolled material, as in the present case, this differentiation is rather difficult to make out since their appearance can be very similar. We call all laminar subgrains which separate the layers of CB's, microbands in accordance with Malin and Hatherly [1]. These bands, their thickness ranged from 0.1-0.2 µm, run parallel or nearly parallel to the rolling plane. Their orientation distribution (Fig.3b) can be described by three components: ~{011}<511> (φ_1=~15°, Φ=~45°, φ_2=0°), ~{013}<100> (0°, ~20°, 0°) and C. It must be noted that the first component is located in the scatter around the global texture components B and Goss. Thus if the texture of the microbands is described in terms of the four global texture components, as is done in Table I by reason of comparison, the shares of the B and Goss components are overestimated. In fact about 43% of the orientations in MB's can be identified as {011}<511>, thus the real portion of each of the B- and Goss-components does not exceed 10%.

The boundaries between the MB's were quite sharp. About 50% of the disorientation angles (ω_d) were smaller than 15°. They range between 0° and 10°, having the highest frequencies around 3° -6°. In the MODF (Fig.4b) calculated for ω_d>15° one can find maxima at: 60°<111> (Σ= 3), ~50°<110> (Σ=11), ~35°<211> (Σ=35a) and 45°<211> (Σ=21b).

Shear bands

The shear bands cross the array of the aligned MB's and CB's at an angle of 25-35° to the rolling plane. They consist of narrow subgrains, elongated in the direction of the SB. The band thickness ranged from 0.5-1.0 µm and the subgrains within the bands varied in width from 0.01-0.2 µm.

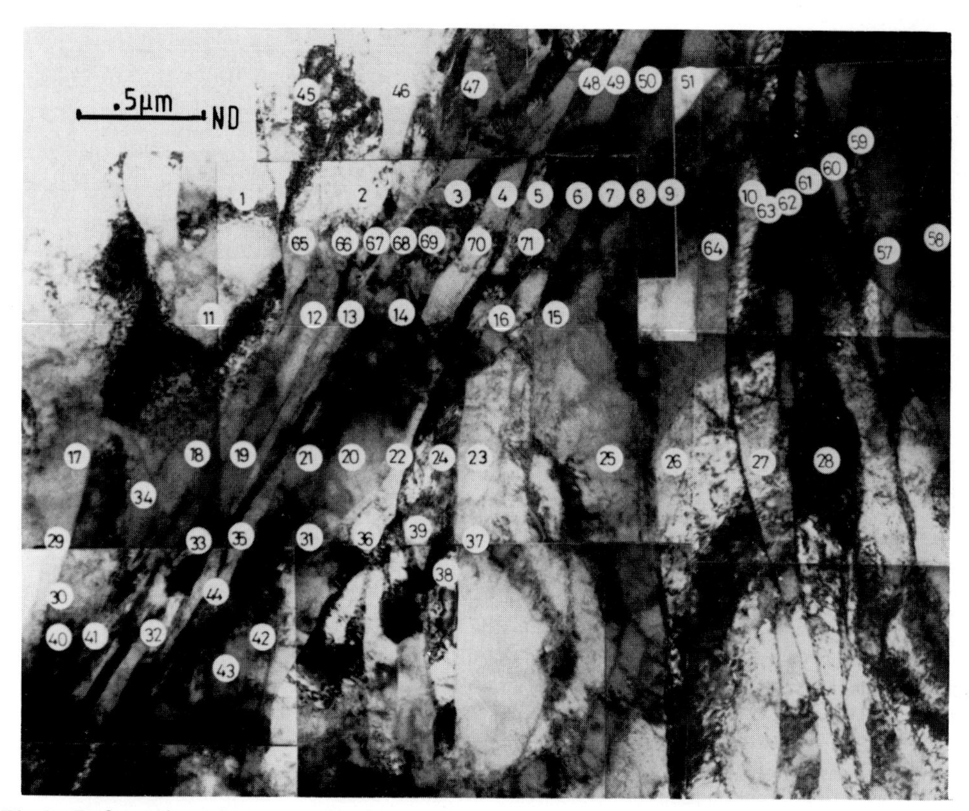

Fig.2: Deformation microstructure of 95% cold-rolled pure copper (longitudinal section).

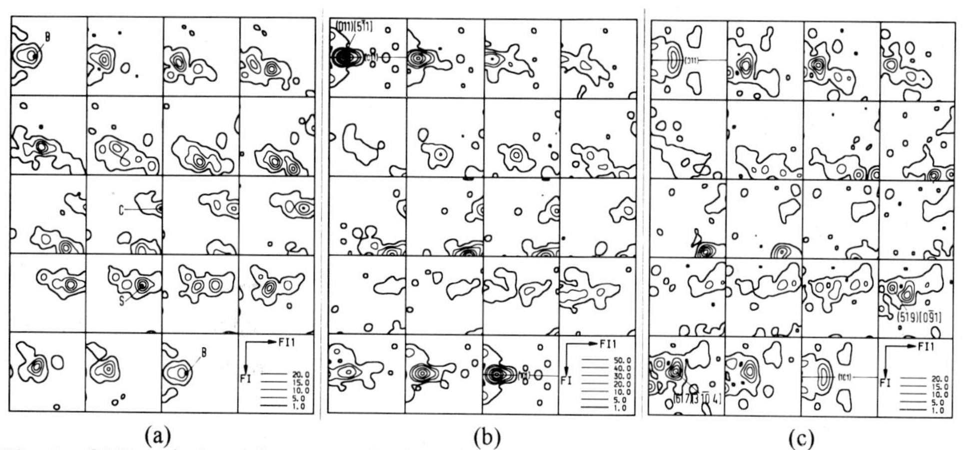

(a) (b) (c)

Fig. 3: ODFs calculated from sets of orientation measurements on the morphological elements of the microstructure of 95% cold-rolled pure copper:
a) ODF of cells (133 orientations),
b) ODF of microbands (110 orientations),
c) ODF of shear bands (141 orientations).

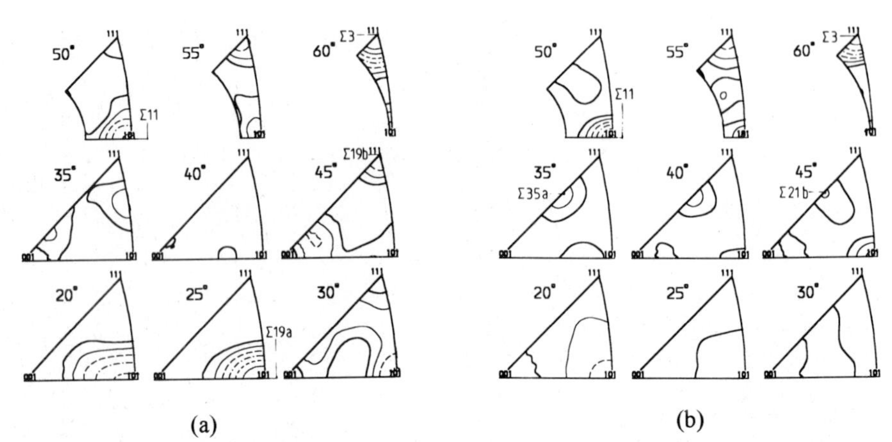

Fig.4: MODFs calculated from sets of misorientations between adjacent morphological elements of the microstructure of 95% cold-rolled pure copper:
a) between cell blocks (43 misorientations, ω_d>15°, levels at 1., 2., 3., ...),
b) between microbands (38 misorientations, ω_d>15°, levels at 1., 2., 3., ...).
CSL positions are marked with the Σ values.

The orientation distribution in the SB's shows a preferential spread of orientations with {110} near the rolling plane (Fig.3c). The maxima of the distribution are located near the positions {617}<3 10 4> (~30°, ~40°, ~80°) and {519}<091> (~15°, ~30°, 80°). Both orientations are included in the scatter around the global texture components B and Goss respectively, Thus analogous to the corresponding case of the MB's, the shares of B- and Goss-components listed in Table I are overestimated.

The boundaries between neighbouring subgrains in the SB's are usually sharp and the disorientation angles (ω_d) across the boundaries are either relatively small (about 60% of them varied between 0° and 7°, with the maximum about 6°) or very high (between 45° and 62° with the maximum at 60°<111>). This maximum corresponds to the stacking of thin twin plates lying parallel to the band which were observed in few locations in the SB's.

The disorientation angles between the subgrais in the SB's and the cells resp. subgrains in the matrix vary between 3-9° (over 60% of the misorientations) or attain high values (>25°). The rotations axes show a wide scattering around the <210> direction for $\omega_d \in$(20-25°), around the <111> direction for $\omega_d \in$(35-60°) having the maximum at 60°<111> and around the <110> direction for ω_d=60°. Thus here the CSL relations are not as distinctly favoured as in the case of the MB-MB or CB-CB misorientations.

Phosporus Copper

Highly rolled phosphorus copper exhibits, rather typical for metals/alloys with a low stacking fault energy, a wavy microband/shear band structure. In this structure we find rarely dispersed rhombic domains of twin lamellae (for a comparison see f.i. [2, 3, 4]).

Microbands
Most of the material shows a microband structure. The band thickness varies in width from 0.02-0.1 µm, thus the MB's in phosphorus copper are about 10x thinner than the MB's in pure copper. They lie nearly parallel to the rolling plane often forming large, wave-shaped curvatures (Fig.5). The orientation distribution in the MB's, Fig.6a, looks like a typical brass rolling texture. Compared to the global texture of phosphorus copper [5] the share of the Goss component is considerably lower (see Table I).

The boundaries between neighbouring bands were usualy sharp and the misorientation across the boundaries were either small or high: About half of the disorientation angles (ω_d) are randomly distributed in the range 0-12°, the other half show ω_d-values >20°. For the higher ω_d values some preference for four rotation axes is observed: 45°<111> (Σ19a), 50-60°<110> (Σ11 and Σ33b), ~50°<331> (Σ25b) and 35° ~<311> (Fig.7).

Fig.5: Deformation microstructure of 95% cold-rolled phosporus copper (longitudinal section). MB-microbands, SB-shear bands, R-rhombic domain of twin lamellae

Shear Bands

The second importand component of the structure are shear bands. They are frequently observed in grains (regions) which are characterized by a layered structure consisting of twin lammellae roughly parallel to the rolling plane (see the next section of this paper). These bands are inclined at 30-40° to the rolling plane and delimit the rhomb (lens) -shaped twinned areas. The substructure within the SB's consists of small, elongated crystallites. The bands thickness ranged from 0.1-0.3 µm and the individual crystallites within the bands varied in width from 0.05-0.1 µm. The orientation distribution of these crystallites shows a widely scattered range but is not random (Fig.6b). There is, as in pure copper, a preference for sheet plane orientations near {110} with a maximum about {110}<611> (~15°, 45°, 0°) and some spread from this maximum towards {012}<100> (0°, ~25°, 0°).

Rhombic Domains of Twin Lamellae

Beside the morphological features such as MB's or SB's a minor component (less than 10% of the total volume) was observed: Lens or rhomb-shaped regions of 1-2 µm in size characterized by a layered structure consisting of twin lamellae roughly parallel to the rolling plane. These domains are more or less completely preserved in the structure. We have observed rhombic domains with a nearly perfect structure of very thin (0.01-0.04 µm) twins which were about {111}<112> oriented and others with wide scattered orientations where only traces of {111}<uvw> orientations can be found and where the thickness of the lamellae varied on a distinct larger scale (up to 0.1 µm) as in "perfectly" twinned domains. All of these domains were bounded by the sets of SB's.

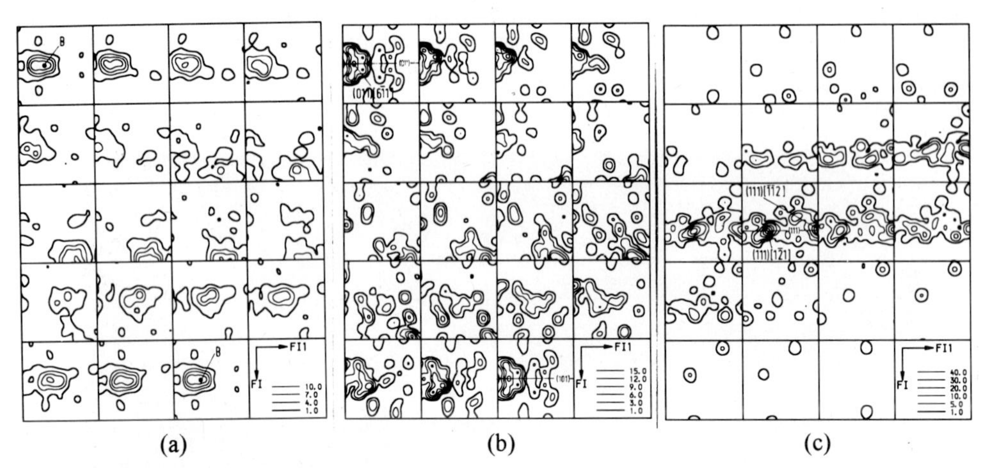

Fig.6.: ODFs calculated from sets of orientation measurements on the morphological elements of the microstructure of 95% cold-rolled phosphorus copper:
a) ODF of microbands (191 orientations),
b) ODF of shear bands (51 orientations),
c) ODF of twin lamellae in rhombic domains (28 orientations).

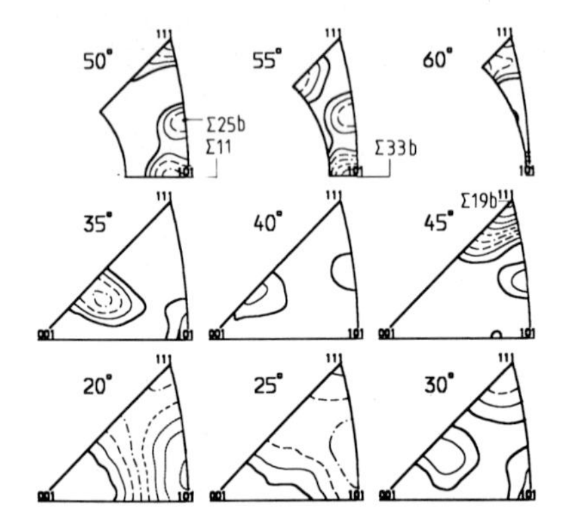

Fig.7: MODF calculated from set of 50 misorientations (ω_d>15°) between microbands in 95% cold-rolled phosphorus copper, levels at 1., 1.5, 2., 2.5, ...).
CSL positions are marked with the Σ values.

DISCUSSION

It is not surprising that the orientation distribution of the main morphological components as cells in pure copper or microbands in phosphorus copper reproduce the global texture. However, the local textures of the minor microstructural elements can be quite different from the global texture. In particular cases some microstructural elements show very specific texture components as e.g. the rhombic domains of twin lamellae the {111}<112> or {111}<uvw> orientations. In other cases the particular element can contribute to the scattering around the main components of the global texture or the background. These contributions are usually imperceptible in the global representation. An example for this case is the orientation distribution in SB's in both materials. Both distributions show maxima lying in the scatter around the components of the global texture, other regularities of the widely scattered orientation distributions in SB's are not visible.

According to Williams [10] the main shear textures in f.c.c. metals are {100}<011>, {111}<110> and {111}<112> the last of which exists in two different variants with respect to the shear axes. In the reference system of the rolled sample these three types of orientations become: {112}<111>,

~{021}<512>, {114}<221> and {110}<001>. The traces of all these have been found in the present study. The share of the latter in SB's is higher than in the global texture in pure copper and higher than in the MB's in phosporus copper (see the Table I). Therefore, it is shown that the SB's contribute an important share to the Goss component. It must, however, be noted that the contributions of the Goss component in SB's of both materials, though significant, never exceeds ten to twenty precent. The relatively high number of near Goss-oriented crystallites in SB's in phosphorus copper is probably related to the fact that in this material twinned areas exist with dominating orientation {111}<112>. When shear banding appears in such regions, the mechanism proposed by Yeung and Duggan [9] or Donadille et al. [11] seems probable, leading to shear bands of ~{110}<001> orientation which corresponds to a 35° rotation about TD.

Table I.

	pure copper				**phosphorus copper**		
	cells	SB's	MB's	global texture [after 5]	MB's	SB's	global texture [after 5]
S	41%	16%	13%	44%	19%	10%	20%
C	17%	5%	10%	20%	-	-	-
B	12%	19%	13%	12%	24%	8%	28%
Goss	4%	6%	21%	2%	4%	12%	13%

The shares of main texture components in particular space elements of the microstructure. Orientations having a disorientation angle ω_d relative to S, C, B or Goss smaller than 15° were counted under S, C, B or Goss respectively.

ACKNOWLEDGMENTS

This work has been supported by the Deutsche Forschungsgemeinschaft. The authors wish to thank Ch.Grusewski for the preparation of foils and photographs.

REFERENCES

1) Malin, A.S. and Hatherly, M.: Metal Sci., 1979, 13, 463.
2) Blicharski, M. and Gorczyca, S.: Metal Sci., 1978, 12, 303.
3) Duggan, B.J., Hatherly, M., Hutchinson, W.B. and Wakefield, P.T.: Metal Sci., 1978, 12, 343.
4) Plege, B., Noda, T. und Bunge, H.J.: Z.Metallkde., 1981, 72, 641.
5) Sztwiertnia, K. and Haeßner, F.: Textures and Microstructures, Proc. of the Symp. "Microscale Textures of Materials" Cincinnati 1991, ed. H.J.Bunge, Gordon and Breach Science Publishers 1993, 87.
6) Hansen, N.: Scripta met., 1992, 27, 1447.
7) Bay, B., Hansen, N., Hughes D.A. and Kuhlmann-Wilsdorf, D.: Acta metall. mater., 1992, 40, 205.
8) Ananthan, V.S., Leffers T. and Hansen, N.: Scripta met., 1991, 25, 137.
9) Yeung, W.Y. and Duggan, B.J.: Acta metall., 1987 35, 541.
10) Williams, R.O.: Trans. A.I.M.E., 1962, 224, 129
11) Donadille, C.Valle, R.Dervin P. and Penelle, R.: Acta metall., 1989, 37, 1547.

Materials Science Forum Vol. 157-162 (1994) pp. 1299-1304

MICROTEXTURE DETERMINATION OF AS-DRAWN TUNGSTEN WIRES BY BACKSCATTER KIKUCHI DIFFRACTION IN THE SCANNING ELECTRON MICROSCOPE

K.Z. Troost [1], M.H.J. Slangen [1] and E. Gerritsen [2]

[1] Philips Research Laboratories,
Postbus 80 000, NL-5600 JA Eindhoven, The Netherlands

[2] Philips GmbH Forschungslaboratorien Aachen,
Postfach 1980, D-5100 Aachen, Germany

Keywords: Backscatter Kikuchi Diffraction (BKD), Electron Backscattering Patterns (EBSP), Scanning Electron Microscopy (SEM), Microscale Texture Determination, ODF Determination, Thin Wires, As-Drawn Tungsten Wires, Fiber Textures, Cylindrical Textures

ABSTRACT
Backscatter Kikuchi diffraction in the scanning electron microscope is applied to the microscale texture determination of 215-μm as-drawn tungsten wires. Wire sections were electrochemically etched to a tip with an apex angle of about 20° and the individual orientations of about 200 grains in 20x20 μm^2 areas on the conical tip surface, situated at about half the wire radius from the wire axis, were determined. A distribution of [110]($11k$) and [110]($hh1$) cylindrical texture components was found of which the relative strength varied with the position within the wire and between wires. Frequently, only one of the equivalent ($11k$) and ($11\bar{k}$) or ($hh1$) and ($hh\bar{1}$) planes locally showed a preference to align parallelly to the wire circumference, corresponding to a "chirality" of the local cylindrical texture.

INTRODUCTION
The study of the texture of tungsten (W) wires has a history that goes back to the first applications of X-ray diffraction (XRD) to the determination of textures. Detailed knowledge of the texture of W wires is of importance since the texture may have an influence on their high temperature properties in lamps. Already in 1921, the well-known (110)-fiber texture of as-drawn tungsten wires was established in an XRD experiment by Burger within Philips Research [1]. In a later XRD study [2], Jeffries reported that crystallites in as-drawn W wires, apart from arranging with their [110] direction along the wire axis, show a preference to have a set of (100) lattice planes aligned parallel to the wire circumference. However, Rieck [3], instead of finding the "cylindrical" texture proposed by Jeffries, observed a "classical" [110] fiber texture, which means that the [110] direction coincides with the wire axis but that the orientation in the plane perpendicular to the wire axis is completely random. Still later, Leber reported a broad cylindrical texture with contributions roughly between [110](111) and [111](211) which he attributes to textures

characteristic to the swaged tungsten rods, the starting material for tungsten wire drawing, persisting to some extent throughout the wire drawing process. Also in more recent XRD experiments, cylindrical textures were observed, like a [110](110) cylindrical texture in 1-mm steel wires and a [111](211) texture in 98-μm aluminum wires [5]. In Ref. 5, also the use of electron back-scattering patterns (EBSP), now mostly called backscatter Kikuchi diffraction (BKD), in the scanning electron microscope (SEM) was proposed for studying the microtexture of wires.

For the determination of textures in polycrystalline materials, XRD is routinely used because of the simplicity of the involved measurement procedure. The main disadvantage of XRD is that, by the nature of X-rays, these cannot be focussed by electric or magnetic fields. For studies requiring a high lateral resolution one has to resort to beams of charged particles, like ions or electrons, which can readily be focussed. Finely focussed electron beams in the transmission electron microscope (TEM) and the scanning electron microscope (SEM) [6] are especially suited for highly local diffraction studies. A key advantage of diffraction techniques in TEM and SEM over XRD is that the exact orientation of individual crystallites of sub-micron size can directly be determined to generate the complete local orientation distribution function (ODF). Direct measurement of the misorientations between neighbouring grains also enables the measurement of the orientation difference distribution fuction (ODDF), something which is fundamentally impossible when irradiating many grains at a time, as is the case with XRD. A further peculiarity of XRD is that the orientation distribution function (ODF) has to be reconstructed from multiple data sets, whereas the diffraction methods in TEM and SEM are direct. Admittedly, it is quite laborious to obtain a statistically sufficient amount of diffraction data in TEM and SEM, since several hundreds of individual orientations are needed [7]. However, efforts for full automation of the collection and interpretation of data by BKD in the SEM are in progress [8,9]. The thin W wires investigated here are a good example of a specimen type of which the true texture determination requires the high spatial resolution offered by BKD in the SEM. In all of the earlier texture experiments on thin wires of tungsten and other metals, XRD was used and the lack of spatial resolution was circumvented by mounting a large number of wires parallelly on a frame or a glass plate to obtain a specimen of sufficient area [3-5]. We here note that because of the random rotational mounting angle of the wire sections and the uncertain mounting direction of the wire sections either parallel or antiparallel to the drawing direction, local differences in the wire texture may have been averaged out.

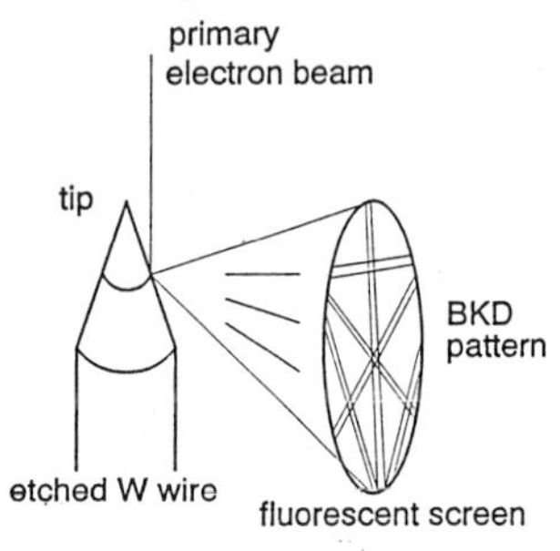

Figure 1: Experimental geometry for the recording of BKD patterns from tungsten wire tips.

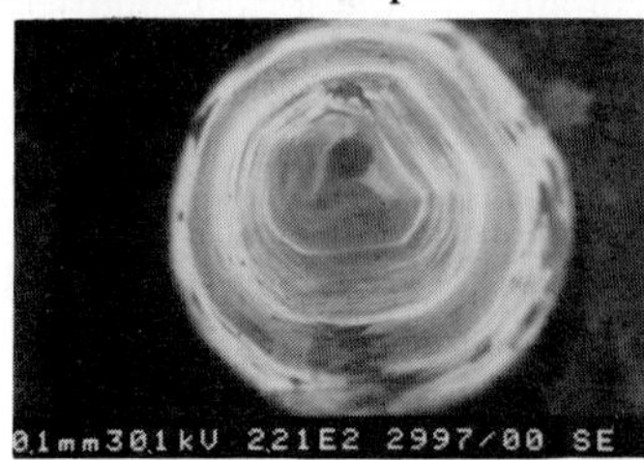

Figure 2: Top view of a section of a cold-drawn 200-μm thick W wire electrochemically etched to a point with an apex angle of about 20°.

BACKGROUND

We have briefly summarized the mechanism of the formation of BKD patterns in the SEM in an earlier publication [10]. In summary, BKD patterns consist of Kikuchi bands, which are quite similar to X-ray Kossel rings, but with a much smaller diffraction angle $\theta_{hkl} \approx \lambda/2d_{hkl}$ of order 1°, since for a typical energy of 10 kV, the electron wavelength of about 10^{-2} Å is about hundred times smaller than that for X-rays. From the positions of Kikuchi bands in BKD patterns the orientations of the individual crystallites can be determined using special software [11].

RESULTS

We have performed texture measurements on 215-μm as-drawn W wires by BKD in the SEM. The wires were obtained from swaged tungsten rod in a sequence of drawing operations. Wire sections of about 1-cm length were electrochemically etched in a NaOH solution to a point with an apex angle of about 20° and subsequently mounted in a Philips SEM 525 M with the wire axis directed vertically and the wire tip pointing upwards. In this way, the electron beam, running vertically, hits the conical tip surface at about 10°, which ensures the recording of high-quality BKD patterns. A fluorescent screen for the observation of the BKD patterns is mounted vertically next to the impact point of the electron beam on the wire tip (figure 1). We call the direction pointing from the impact point perpendicularly to the fluorescent screen the viewing direction.

Figure 3: SEM micrograph of part of the conical tip of a cold-drawn 200-μm thick W wire showing elongated grains of a few μm length and of a few tenths of a μm width.

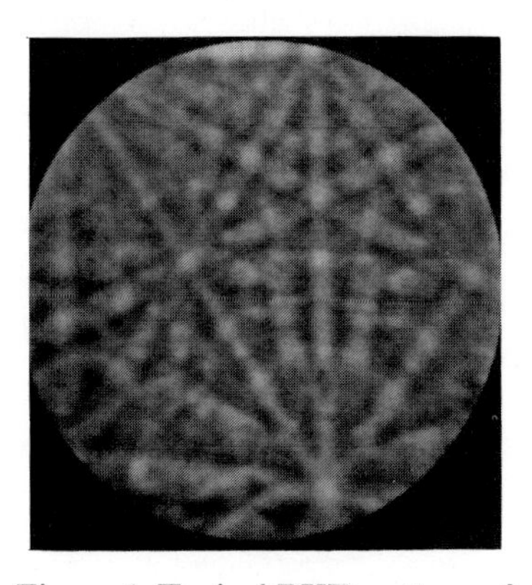

Figure 4: Typical BKD pattern of an individual W grain after numerical subtraction of the uneven background.

Figure 2 shows a top view of an etched wire in the SEM. The etched tip is not entirely rotationally symmetric due to preferential etching of the wire in specific directions, possibly caused by local differences in the wire texture. An SEM micrograph of part of the conical tip surface (figure 3) clearly shows grains which are elongated along the wire axis and which are a few μm long and about 1 μm wide. The contrast between different grains is caused by channeling of electrons along lattice planes, which effect also underlies the formation of BKD patterns.

For a number of wires, about 200 BKD patterns from individual crystallites within a 20x20 μm^2 area, situated at about half the wire diameter from the wire axis, were recorded. In figure 4, a typical BKD pattern of a W grain after background subtraction is given. Clearly, a large number of Kikuchi bands crossing in poles is observed. In figure 5 the individual grain orientations obtained in a typical

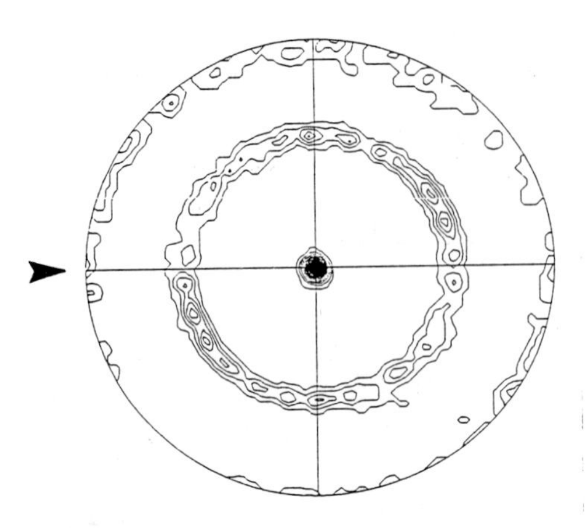

Figure 5: (110) Pole figure by BKD in SEM consisting of about 200 individual grain orientations from a 215-μm thick as-drawn W wire as. A clear (110) fiber texture is observed.

experiment are presented as a contour plot of the (110) pole figure in stereographic projection on the plane perpendicular to the wire axis. The viewing direction is indicated by the large arrow head. A sharp (110) fiber texture is concluded from the pronounced maximum in the center of the pole figure. The intensity in the ring at 60° from the origin is due to equivalent <110> directions, like [101], [011], $[10\bar{1}]$ and $[01\bar{1}]$, and the intensity on the ring at 90° from the origin is due to the $[1\bar{1}0]$ direction which is perpendicular to [110]. It is evident that the intensity on the ring at 60° from the center is not evenly distributed. This points to preferential orientation of grains within the plane perpendicular to the wire axis while having their [110] direction fixed along the wire axis. However, when considering if the intensity variation is significant, one must realize that several effects tend to smear out maxima and minima in the intensity distribution on the ring: (i) the intensity contributions of equivalent <110> directions interfere, (ii) the 20-μm width of the sampling area causes a smearing of about 10°, and (iii) the limited number of grains may cause statistical variations.

In order to reduce the number of interfering contributions we have plotted a (111) pole figure of the same data in figure 6. Hereby, the number of equivalent contributions is reduced from four to two. Again, the viewing direction is indicated by the big arrow head. With the [110] direction fixed parallel to the wire axis, the $[\bar{1}11]$ and the $[\bar{1}1\bar{1}]$ directions both lie in the plane perpendicular to the wire axis, with $[\bar{1}1\bar{1}]$ lying at 70.5° from $[\bar{1}11]$ anticlockwise around [110]. These directions contribute to intensity on the outermost ring of the pole figure, at 90° from the origin. The two other equivalent [111] and $[11\bar{1}]$ directions contribute on a ring at 35.3° from the origin. As can be concluded from the intensity maximum in the viewing direction and the related intensity maxima due to equivalent <111>

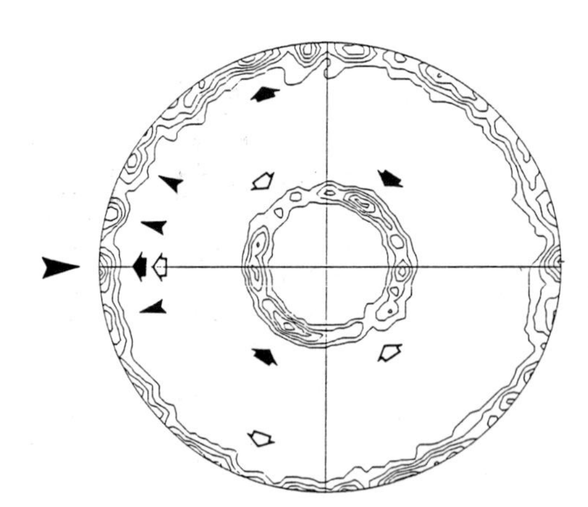

Figure 6: (111) Pole figure by BKD in SEM (same data as in figure 5) showing an anisotropic distribution of orientations in the plane perpendicular to the wire axis.

directions indicated by solid arrows, there is a distinct preference for **only** directions near $[\bar{1}11]$ to be pointing perpendicularly to the wire circumference. Directions near the $[\bar{1}1\bar{1}]$ direction in fact show a tendency **not** to lie perpendicularly to the wire circumference as can be concluded from the intensity minima indicated by the open arrows. This corresponds to a local cylindrical texture with a "chirality". Apart from the strong maximum of $[\bar{1}11]$ in the viewing direction, several less pronounced maxima, marked by small arrows heads, are also observed. These may be attributed to other $[110](11k)$ and $[110](hh1)$ cylindrical texture components.
From a series of measurements on different positions on the wire tip and on different wires it was concluded that the relative strength of the different cylindrical texture components varies with the position on the wire tip and from wire to wire. We have not yet investigated the dependence of macroscopic wire properties on details in the wire texture.

CONCLUSIONS

We conclude that by BKD in the SEM a coupling between diffraction and spatial information from individual micron-sized grains can be made. This enables the direct determination of the ODF as well as the ODDF, both on a scale of microns, something which is not possible by using XRD. BKD in the SEM is especially suited for the mictrotexture determination of small specimens, like the 215-μm as-drawn W wires studied here. In these wires, we have observed a distribution of $[110](11k)$ and $[110](hh1)$ cylindrical texture components, often with a local tendency for only one of the two equivalent $(11k)$ and $(11\bar{k})$ or $(hh1)$ and $(hh\bar{1})$ planes to align parallelly to the wire circumference, corresponding to a "chirality" of the local texture.

REFERENCES

1) Burger, H.C.: Physica, 1921, 1, 214
2) Jeffries, Z.: Trans. AIME, 1924, 70, 303
3) Rieck, G.D.: Philips Research Reports, 1957, 12, 432
4) Leber, S., 1965, Trans. Met. Soc. AIME, 233, 953
5) Montesin, T., Heizman, J. J.: J. Appl. Cryst., 1992, 25, 665
6) Harland, C.J., Akhter, P., Venables, J.A.: Phys. E, 1981, 14, 175
7) Baudin, Th., Paillard, P. and Penelle, R.: J. Appl. Cryst., 1992, 25, 400
8) Lassen, N. Chr.: Micron Microscopica Acta, 1992, 23, 191
9) Wright, S.I., Zhao, J.W., Adams, B.L.: Text. Micros., 1992, 13, 123
10) Troost, K.Z., van der Sluis, P., Gravesteijn, D.J.: Appl. Phys. Lett., 1993, 62, 1110
11) N.-H. Schmidt Scientific Software, Randers, Denmark.

Materials Science Forum Vol. 157-162 (1994) pp. 1305-1312

MICRO-TEXTURE ANALYSIS OF DIFFUSION BONDED INTERFACES IN Al-Li ALLOY SHEET

H.S. Ubhi and A.W. Bowen

Materials and Structures Department, DRA Farnborough, Hants GU14 6TD, UK

Keywords: Diffusion Bonding, Solid State Bonding, Transient Liquid Phase Bonding, Al-Li Alloys, Inerfaces, Micro-Texture, Electron Back-Scattered Diffraction

ABSTRACT

The results of experiments to induce boundary migration to improve the microstructural continuity across the interface in diffusion bonded 8090 Al-Li alloy sheet, by joining surfaces with large grain mis-orientations and by deformation and heat treatment, are reported. The micro-textures at the bond interfaces were determined by the Electron Back-Scattered Diffraction (EBSD) technique and it has been found that significant boundary migration could be induced only by plane strain compression at superplastic deformation rates.

INTRODUCTION

Extensive research into the diffusion bonding of 8090 Al-Li alloy sheet using solid state and transient liquid phase bonding techniques [1] has resulted in the production of bonds with shear strengths that approach those of the parent metal ie the bond shear strengths are ~190MPa, as opposed to 200 MPa for the parent metal [2,3].

The interfaces of both types of bond are planar; even in the case of the transient liquid phase bond there is still a layer of much larger, weaker grains at the interface. In these situations, when failure occurs, planar interfaces may offer an extensive, preferential, path for inter-granular fracture. To improve the microstructural continuity across the interface, and hence remove any preference for fracture to be confined to such regions, investigations have been carried out into the micro-texture of the interface region and on means of inducing boundary migration that would break up the planar interface. The results of the initial phase of this investigation are contained in this paper.

EXPERIMENTAL

Samples of 4mm thick Al-Li alloy 8090 sheet (composition 2.5 Li, 0.6 Cu, 0.8Mg, 0.12Zr, 0.1Fe, 0.05Si in wt%) were bonded via solid state and transient liquid phase bonding processes [1]. Sections through the bond were prepared by electrolytic polishing, followed by chemical etching. Micro-textures were determined from EBSD patterns, the hardware for which was attached to a Philips SEM505 scanning electron microscope. The diffraction patterns were analysed on-line using software developed by Dingley [4] and Schmidt [5].

RESULTS AND DISCUSSION

Typical microstructures at, and near, the bond interfaces of both the transient liquid phase and solid state bonds are shown in Figures 1 and 2 respectively. The corresponding micro-texture information is also included in these figures.

The interface of the transient liquid phase bond consisted of large equiaxed grains (Figure 1a) that have the rotated cube orientation ({100}<011>, Figure 1 (b)). The matrix orientations close to the interface, on the other hand, consist predominantly of the copper ({112}<111>) and some cube ({100}<001>) orientations (Figure 1 (c).

The solid state bond interface is clearly very planar (Figure 2a). The grains in the top sheet interface are predominantly of the rotated cube ({100}<011>) orientation (Figure 2b), whereas in the bottom sheet their orientations are the copper type ({112}<111>) (Figure 2c). Whilst these results indicate that a large misorientation could exist across the bond interface before bonding, this was not found to be the usual case, since the cube oriented grains extend only a few grains deep and are usually removed during surface preparation. Typically, therefore, in surface-to-surface bonds the orientations on both sides of the bond are likely to be the copper type.

Although transient liquid phase bonding clearly removes the original sheet surface, it is not a favoured procedure because it also changes the local chemistry and produces a planar array of large grains. A reduction in thickness of the inter-layer leads to finer grains but not to a corresponding increase in strength [6]. Continued reductions in inter-layer thickness would ultimately lead to a solid state bond. All effort has thus been directed at providing conditions for boundary migration in *solid state* bonds.

Since no evidence was found for the pinning of the interface by oxide particles or that lithium loss or porosity influenced bond stability, and that the deformation occurring during the diffusion bonding process seemed to be below the critical strain [7], the thermal stability of grain boundaries in the unrecrystallised sheet was taken to be due to the pinning effects of Al_3Zr particles, the AlLi δ phase and to the solute present, particularly lithium.

The options available for inducing boundary migration at the bond interface therefore included:

a) the deliberate introduction of different grain orientations on either side of interface.

This was achieved by making use of the marked texture gradients seen in this type of alloy. An example of this gradient is shown in Figure 3, where data for contiguous grain orientations from the sheet surface through to the centre are presented. It is clear that to a depth of about one third thickness the copper orientation dominates, with some cube component (Figure 3b), and beyond

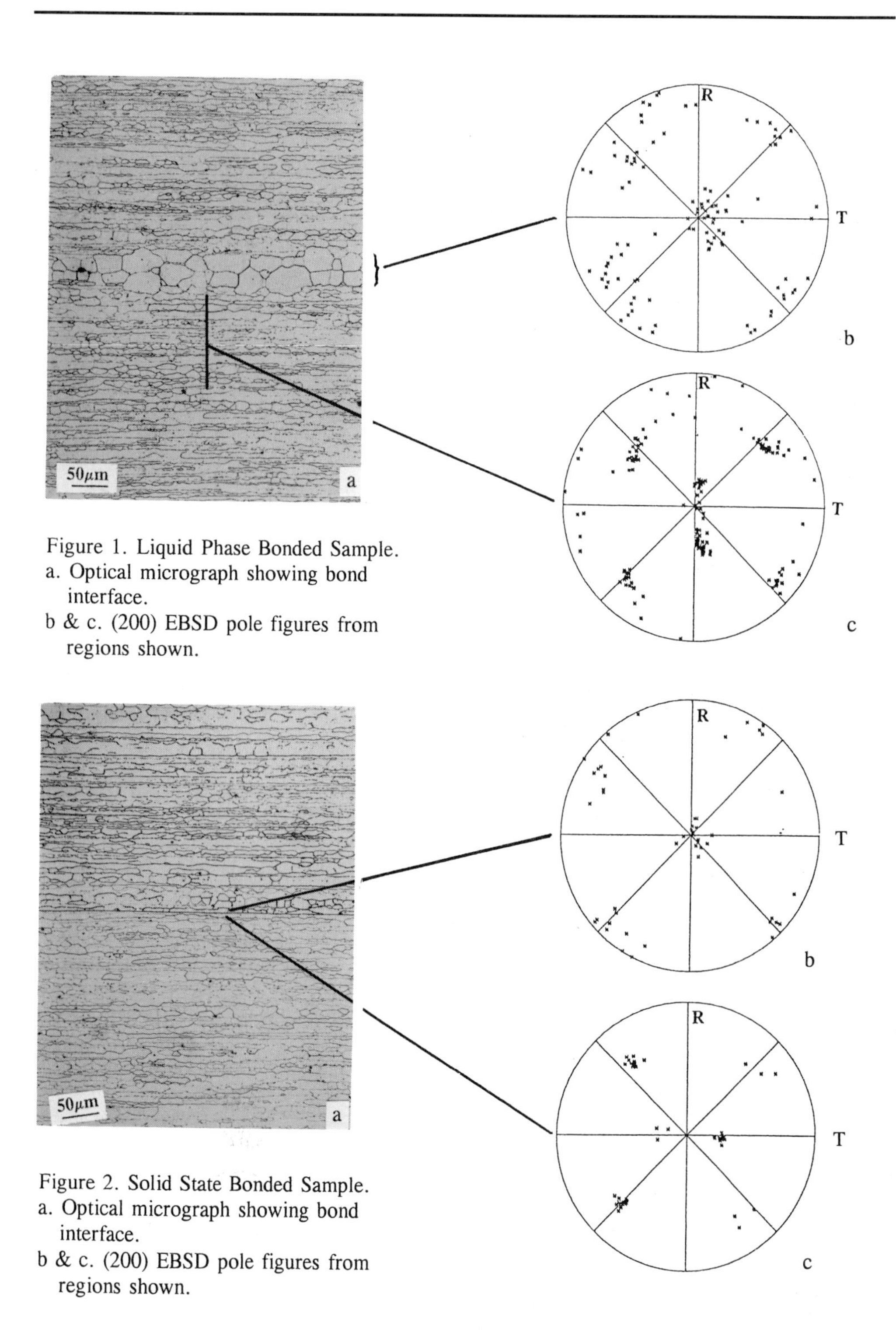

Figure 1. Liquid Phase Bonded Sample.
a. Optical micrograph showing bond interface.
b & c. (200) EBSD pole figures from regions shown.

Figure 2. Solid State Bonded Sample.
a. Optical micrograph showing bond interface.
b & c. (200) EBSD pole figures from regions shown.

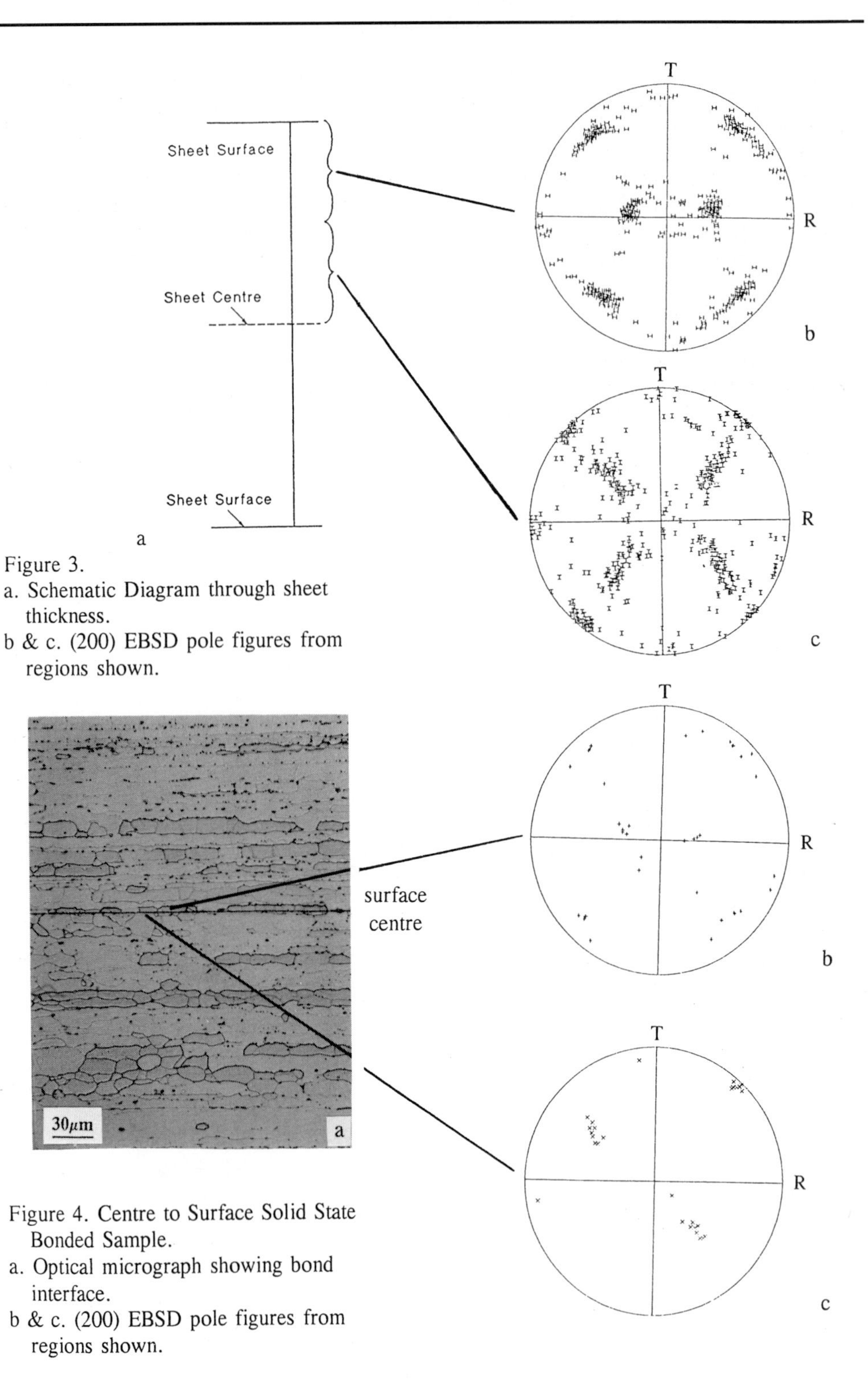

Figure 3.
a. Schematic Diagram through sheet thickness.
b & c. (200) EBSD pole figures from regions shown.

Figure 4. Centre to Surface Solid State Bonded Sample.
a. Optical micrograph showing bond interface.
b & c. (200) EBSD pole figures from regions shown.

that this is replaced by the brass orientation ({110}<112>), again with some cube component (Figure 3c). Bonded samples were therefore prepared where one sheet had been deliberately machined to the mid-plane, giving surface-to-centre bonds.

Figure 4 shows the optical microstructure and micro-textures of such a centre-to-surface bond. It is clear that the bond interface is still planar with no evidence for boundary migration. In addition, the population distributions of grains on either side of the bond have not changed (cf Figures 3b and 4b, and 3c and 4c), indicating that there is a marked reluctance for the interface to migrate.

b) deformation of a centre-to-surface bonded sample followed by an anneal.

Deformation at room temperature was introduced into a centre-to-surface bonded joint by plane strain compression. The sample fractured at about 5% strain. Figure 5a shows the optical microstructure of the failed sample, with evidence of considerable deformation at the fracture surface and the bending of the bond interface. Clearly, this method of introducing cold deformation is not viable routinely. Nevertheless, to see the effect of heat treatment on the bond interface the fractured sample was heat treated at 520° C for 20 minutes in a salt bath followed by water quenching. The optical microstructure after this treatment is shown in Figure 5b. While grain growth has occurred at the fracture surface, and can be attributed to lithium loss [8], there is no evidence of preferential recrystallisation at the bond interface itself. This was confirmed by micro-texture measurements. It can be concluded therefore that, in this high Zr 8090 Al-Li alloy, cold deformation followed by annealing does not induce boundary migration, even across boundaries of large misorientations.

c) combined deformation and annealing.

This was simulated by superplastic deformation, which has been shown to lead to grain growth in this type of high Zr alloy [9]. Figure 6a shows the optical microstructure of a centre-to-surface bond after plane strain compression to about 10% total strain at 560° C at a strain rate of $3x10^{-4}$ s^{-1} followed by air cooling. Shown in Figures 6b and 6c are the micro-textures at the interface which indicate little change from the initial texture (shown in Figures 3 b&c). However, on closer inspection of optical micrographs, there was isolated evidence of some boundary migration. This was confirmed by the examination of samples that had undergone extensive superplastic strain (0.7 strain) (Figure 7). Clearly, therefore, combined deformation and annealing can induce extensive interface migration even in the presence of grain boundary pinning particles. It has been argued that this is due to stress assisted grain boundary migration rather than dynamic recrystallisation [9].

CONCLUSIONS

Micro-texture studies of diffusion bonded interfaces in 8090 Al-Li alloy sheet, using the EBSD technique, have shown that:

1. Interfaces formed during solid state diffusion bonding of high Zr 8090 Al-Li alloy sheet are very stable even when mis-orientations between grains across the joining surfaces are high.

2. Interfaces were also resistant to post bonding deformation and annealing treatments.

3. Migration of the interface could be induced by deformation under superplastic deformation conditions.

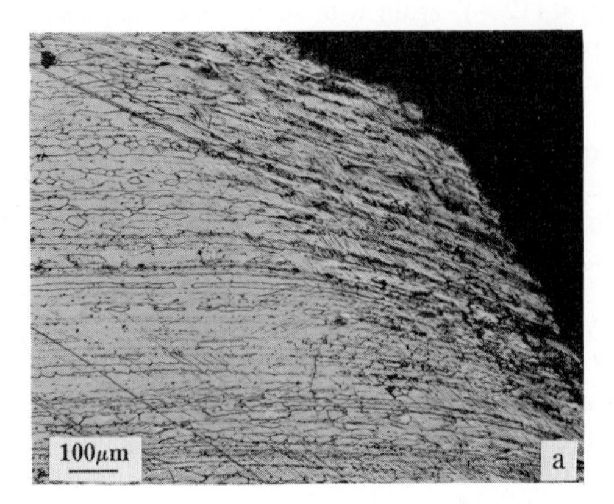

Figure 5.
a. Optical micrograph showing a section through the centre-to-surface bond fractured during plane strain compression at room temperature.
b. As above, but heat treated at 520° C for 20 minutes in a salt bath followed by water quenching.

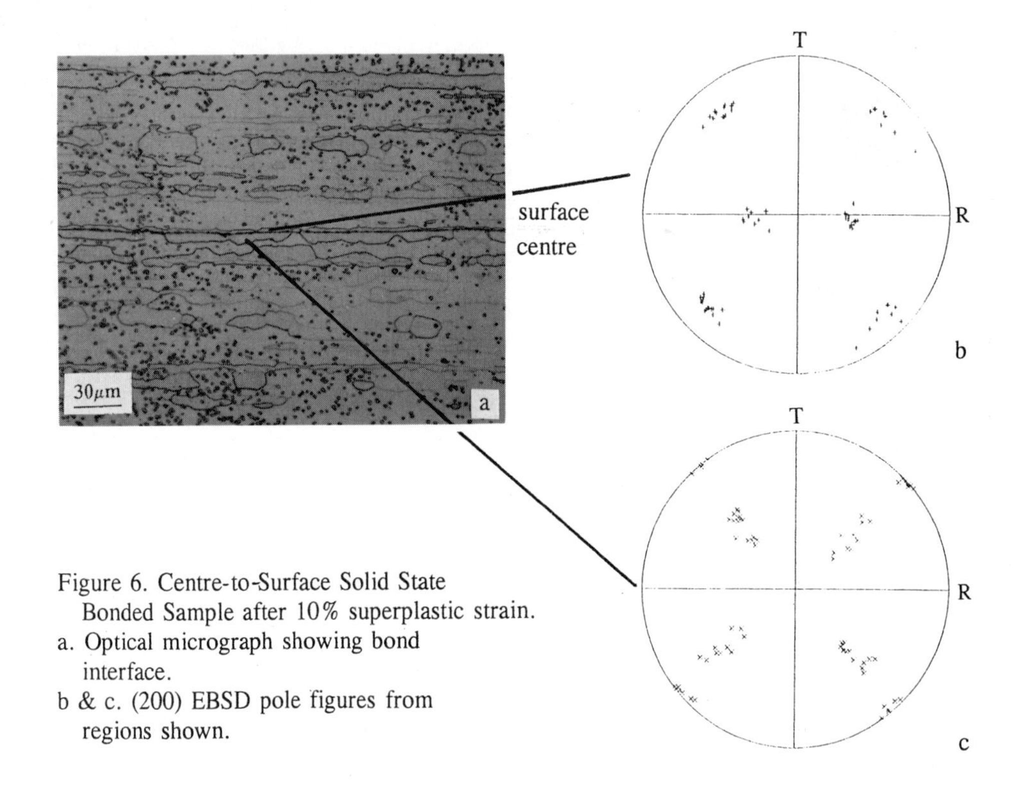

Figure 6. Centre-to-Surface Solid State Bonded Sample after 10% superplastic strain.
a. Optical micrograph showing bond interface.
b & c. (200) EBSD pole figures from regions shown.

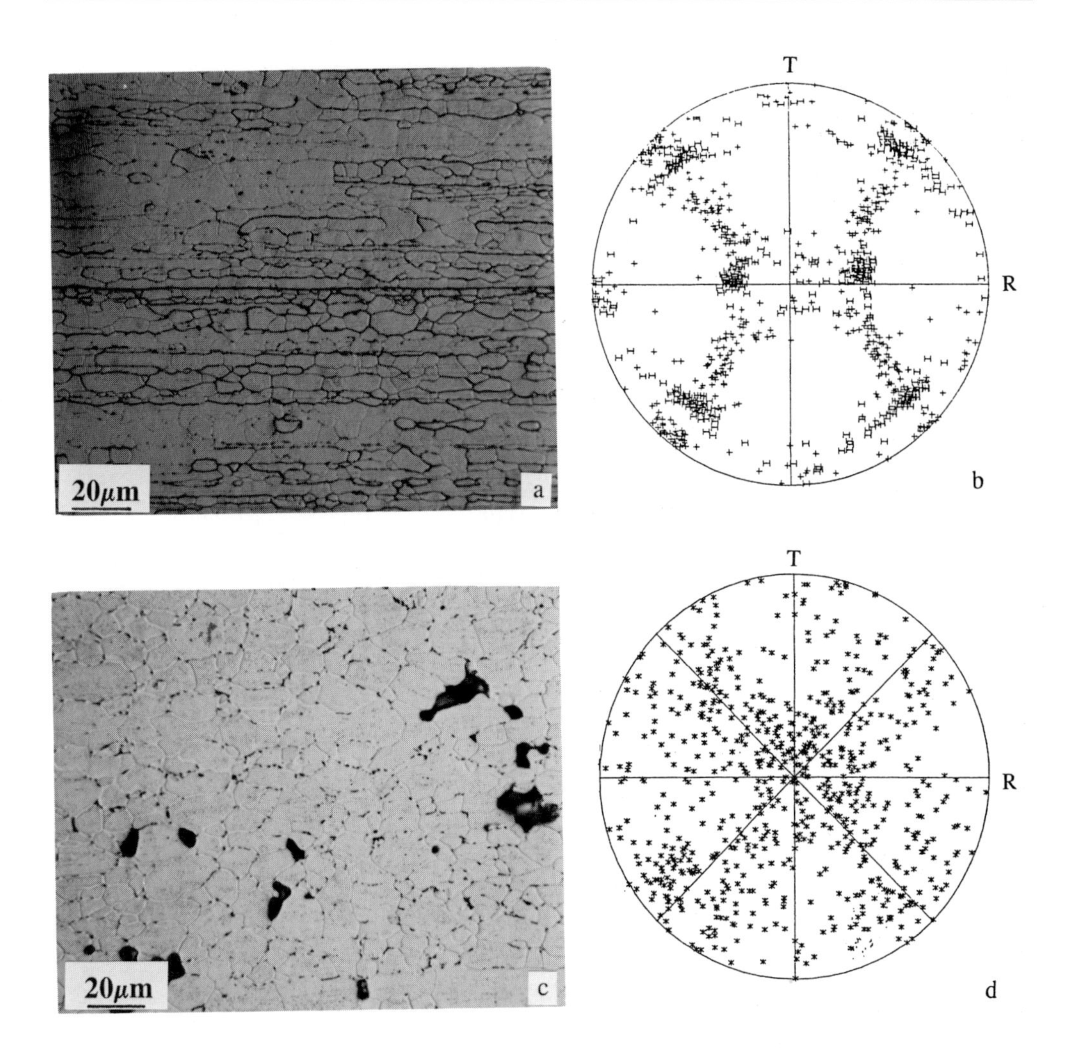

Figure 7.
a. Optical micrograph of surface-to-surface solid state bond.
b. (200) EBSD pole figure across sheet thickness and through the bond interface before deformation.
c. Optical micrograph of surface-to-surface solid state bond after superplastic strain of 0.7. Note that cavitation has occurred because deformation did not involve a differential pressure arrangement.
d. (200) EBSD pole figure across sheet thickness and through a region that contained the bond interface after deformation.

ACKNOWLEDGEMENTS

The authors wish to thank Drs D V Dunford, for supplying the specimens, and A Wisby, for assistance with the plane strain compression tests. The financial support of MoD, through AD Sci (Air) 1, and DTI, through the CARAD programmes, is acknowledged.

REFERENCES

1) Dunford, D.V. and Partridge, P.G.: Mater. Sci. Tech., 1992, 8, 385.
2) Dunford, D.V. and Partridge, P.G.: J. Mater. Sci., 1990, 25, 4957.
3) Dunford, D.V. and Partridge, P.G.: J. Mater. Sci., 1990, 26, 2625.
4) Dingley, D.J.: Proc. 8th Int. Conf. on Textures of Materials (eds J S Kallend & G Gottstein) AIME Warrendale, Pa USA pp 189.
5) Schmidt, N-H.: Bøsbrovej 21, Randers, Denmark.
6) Dunford, D. V. et al: Proc 6th Int. Conf. on Al-Li Alloys DGM (1992) pp 1057.
7) Gilmore, C.J. et al: J. Mater. Sci., 1991, 26, 3119.
8) Partridge, P.G.: Int. Met. Rev., 1990, 35, 37.
9) Bowen, A. W.: Textures and Microstructures, 1988, 8&9, 233.

Materials Science Forum Vol. 157-162 (1994) pp. 1313-1318

MICROTEXTURAL CHARACTERIZATION OF ANNEALED AND DEFORMED COPPER

S.I. Wright [1] and F. Heidelbach [2]

[1] Materials Technology: Metallurgy, Los Alamos National Laboratory, Mail Stop K762, Los Alamos, NM 87545, USA

[2] Department of Geology and Geophysics, University of California, Berkeley, CA 94720, USA

Keywords: Microtexture, Misorientation Distribution, Electron Backscatter Diffraction

ABSTRACT

Grain boundary textures of copper in two different states are examined using backscatter electron Kikuchi diffraction patterns (BEKPs) in the scanning electron microscope. Local orientation relationships across a former recrystallization twin boundary after deformation are explored in detail using automatic BEKP analysis. The misorientation distribution of the annealed sample is dominated by twin relationships (primary and secondary). The deformed sample has a misorientation distribution close to random and the twin relationships are lost. The twin boundary is transformed into a "normal" boundary.

INTRODUCTION

Whereas the development of preferred orientation in copper during deformation is well investigated and also reproduced in models of polycrystalline plasticity, the development of the misorientation texture, which is an important part of the analysis of grain boundaries, has received far less attention. Advances in the measurement of microtextures using Backscattered Electron Kikuchi diffraction Patterns (BEKPs) in the scanning electron microscope (see review by Dingley and Randle [1]) allow for texture studies of misorientations between grains. The recent introduction of automated analysis of BEKPs (see recent works by Adams *et al.* [2] and Wright [3]) enables quantitative studies of misorientation distributions to be performed. In this study, these techniques are applied to the investigation of the misorientation structure of annealed (undeformed) and deformed copper.

EXPERIMENTAL DETAILS

The starting material was high purity copper (99.9999%) which was annealed at 600°C for 30 minutes. A sample of the starting material was rolled to 50% height reduction. X-ray pole figures were obtained from both the annealed and deformed samples and used to determine orientation distributions using the WIMV [4] algorithm as implemented in *popLA* [5], a texture analysis software package. The annealed sample exhibited a random texture, whereas, the deformed sample exhibited a <110> fiber texture with a maximum at 5.14 times random. The general rolling texture components (copper, brass and Goss) were present; however a complete rolling texture was not developed because the sample was too small for general rolling conditions to exist. There was significant strain in the transverse direction.

Grain maps covering about 1mm^2 of sample surface were prepared for both samples. Using BEKPs, the orientations of 333 and 200 single grains were measured in the annealed and deformed samples. The grain maps were coupled with the measured orientations to determine the lattice misorientations across 881 (annealed state) and 504 (deformed state) next neighbor boundaries. Large spreads (up to 15°) in orientation within single grains were observed. Thus, the measured orientation associated with a single grain orientation measurement only represents the average orientations for the whole grain. Pole figures derived from the BEKP measurements were in good agreement with those measured with X-rays.

Approximately fifteen thousand automatic BEKP measurements were made in the neighborhood of what appeared to be a former annealing twin in the deformed sample. The orientation measurements were made on a regular hexagonal grid with 1 μm spacing between measurements. At each point on the grid the orientation (in Euler angles according to Bunge's[6] formalism), the position of the electron beam on the sample and a parameter describing the quality of the corresponding diffraction pattern was recorded.

MISORIENTATION DISTRIBUTIONS

The distribution of minimum misorientation angle for grain boundaries measured in the annealed and deformed samples is compared with the random distribution in figure 1. The "random" distribution is based on the crystallographic texture. It was generated by forming all possible orientation pairs from the single grain measurements in each set and then calculating the misorientation angle for each pair. The two maxima in the annealed sample correspond to primary and secondary annealing twins ($\Sigma3$ and $\Sigma9$ boundaries). The axis–angle pairs describing these twins are a 60° rotation about the <111> axis (the primary annealing twin) and a 38° rotation about the <110> axis. The secondary twin is produced when two different primary twins of the same grain share a common boundary.

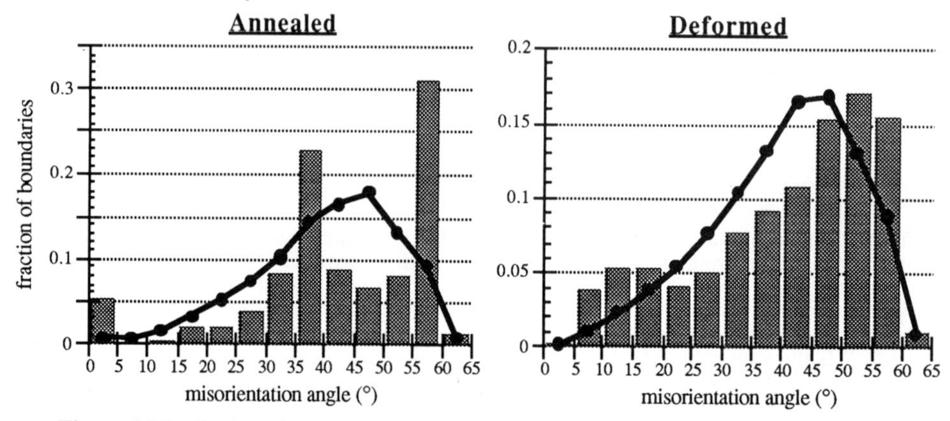

Figure 1 Distribution of minimum misorientation angle in annealed and deformed copper.

Misorientation distribution functions (MDFs) (see [7] for example) were calculated for each sample from the single orientation data. The generalized spherical harmonic series expansion method of Bunge [6] was followed. The MDF calculations were carried out to an order of l = 20. Each misorientation was represented by a Gaussian with a 5° half–width [8]. The results for the MDF calculations are shown in figure 2.

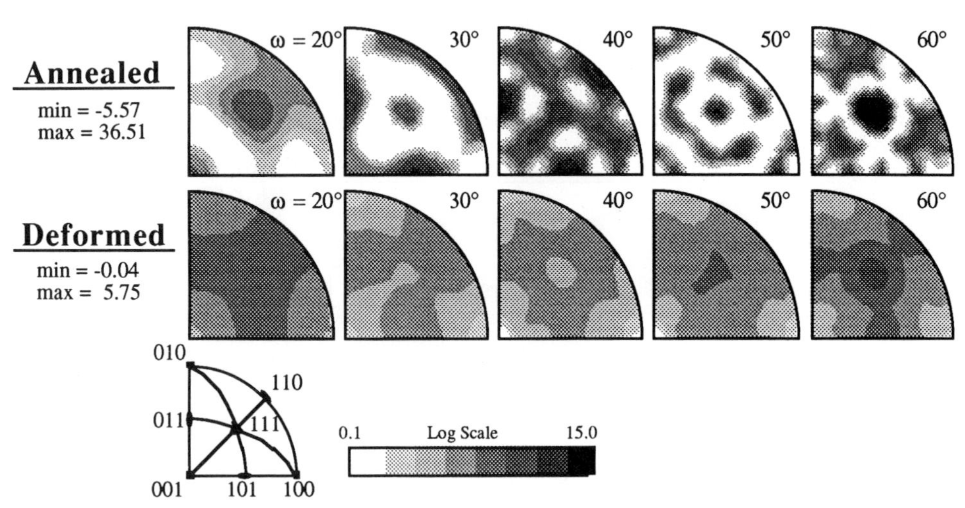

Figure 2 MDFs of annealed and deformed copper plotted in axis–angle space.

It appears in figures 1 and 2 that the imposed deformation breaks the twin structure down into boundaries of random misorientation. There appears to be some remnant of the original annealing twins in the deformed sample. However, the misorientation texture in the deformed sample is considerably weaker than in the annealed sample. It is interesting to note that a strong texture does not imply a strong misorientation texture. In this example, the opposite is true. The deformed sample has a stronger texture than the annealed sample but it exhibits a weaker misorientation texture.

TWIN BOUNDARY CHARACTERIZATION

In order to investigate the break down of the twin relationship more closely, a grain of the deformed sample containing remnant recrystallization twins was examined using automated BEKP analysis. Figure 3 shows a reconstruction of the grain boundary map from the measurements.

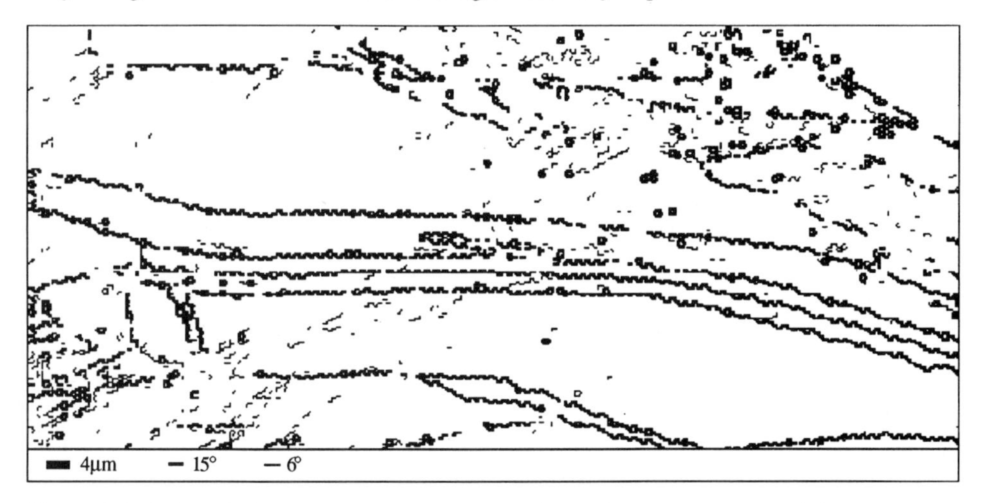

Figure 3 Grain boundaries reconstructed from automatic BEKP measurements on deformed copper.

In figure 3 the thick lines separate neighboring measurement points with misorientation angles exceeding 15° and thin lines separate neighboring points with misorientation angles greater than 6° and less than 15°.

To aid in the analysis of the automatic BEKP orientation measurements on the deformed copper a schematic of the measured region is given in figure 4. The grains labeled **A** and **B** appear to be former recrystallization twins.

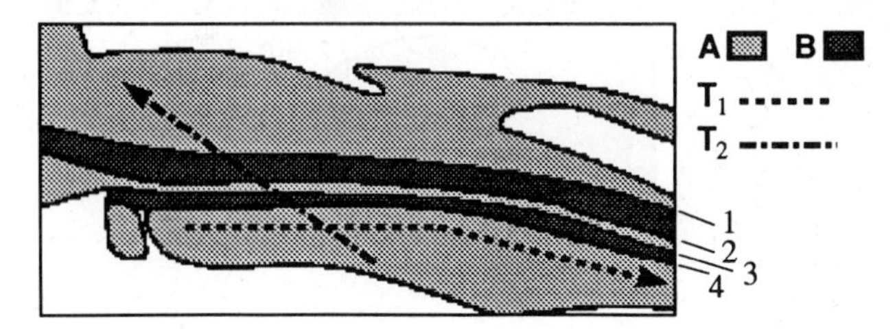

Figure 4 Schematic of figure 3 showing twins, twin boundaries and trace lines.

An intergranular trace (denoted by $\mathbf{T}_1$ in figure 4) and a transgranular trace (denoted by $\mathbf{T}_2$) were formed by collecting all measured orientations within 1.5μm of prescribed lines through the measurement space. The misorientation angles between neighboring points on the two traces as well as the misorientation angles between the first point on the two traces and each successive point are both displayed in figure 5. It should be noted that there is a fairly large spread in orientation within the grains as can be observed in the first trace.

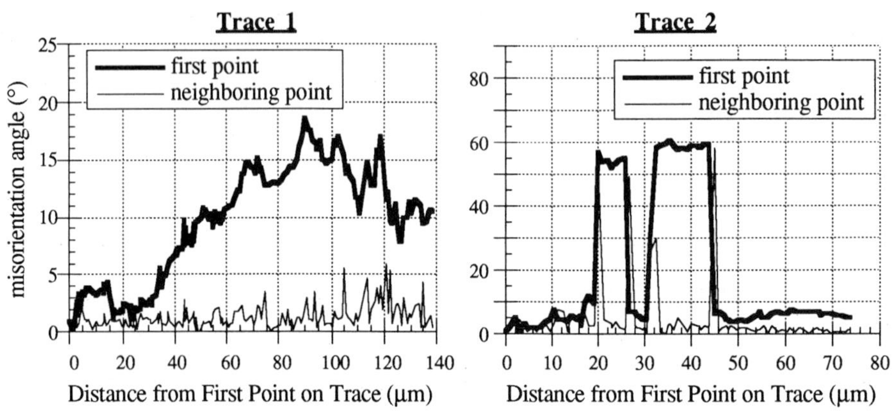

Figure 5 Misorientations between neighboring measurements and between each measurement and the starting measurement along trace lines in deformed copper

In order to examine more closely the spatial distribution of orientation within the individual grains, the average orientations for both the **A** and the **B** grains were determined by calculating an orientation distribution function (ODF) from the automatic orientation measurements using the harmonic analysis described previously (l = 16). The locations in Euler space of the two highest peaks were found and used as reference orientations. The deviations of the measured orientations from these reference orientations were mapped onto a gray scale as shown in figure 6. In this image dark areas correspond to areas very near the reference orientations; whereas, white areas have orientations greater than 15° from the reference orientations. A similar spread of orientation within one grain was also found in channel die deformed aluminum[9].

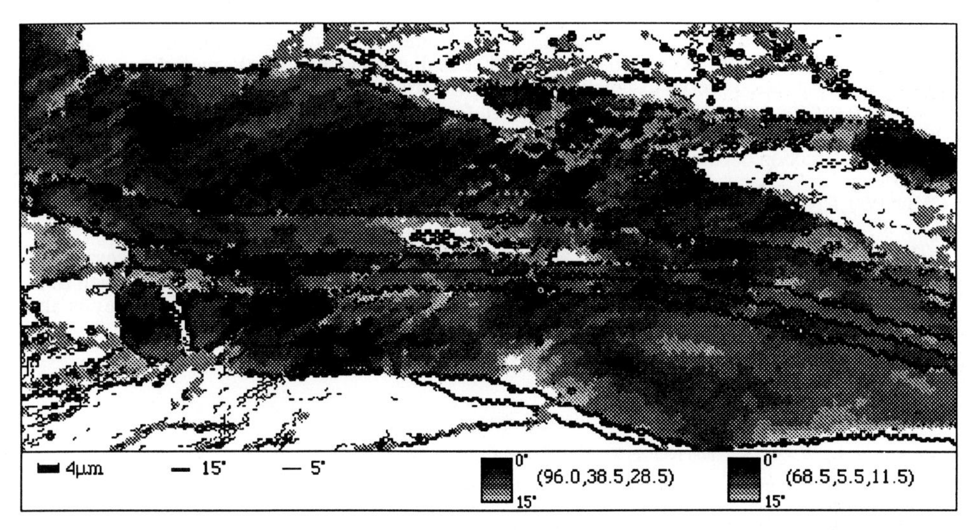

Figure 6 Gray scale image formed by deviation of measured orientations from reference orientations.

Misorientations along the boundaries denoted by the labels 1, 2, 3 and 4 in figure 4 were collected and examined in detail. The misorientations greater than 5° across boundary 1 are plotted in axis-angle space in figure 7. The results for the other three boundaries look very similar. The boundary is clearly no longer a classic <111> twin. It appears to have broken up into <110> and approximate <112> components. Only one "twin" boundary is examined here and the general break down of all twin boundaries into specific components is unlikely since the MDF calculated for the deformed sample shows no evidence of preferred misorientation around the components occurring in the case detailed here.

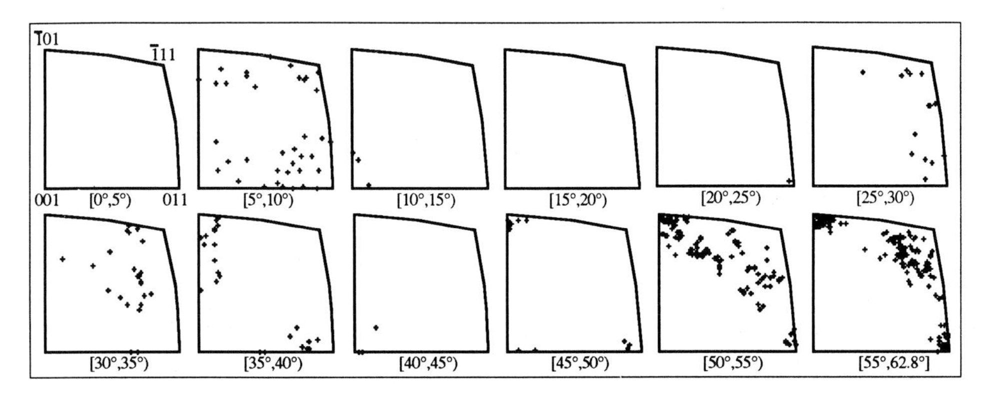

Figure 7 Misorientations along grain boundary 1 in the deformed copper sample.

The spatial distribution of misorientation along the grain boundary was examined by partitioning the boundary into four segments containing an equal number of measured misorientations. No evidence for a systematic spatial distribution of misorientations along the boundary was observed.

DISCUSSION AND CONCLUSIONS

Even though deformation produces material with preferred orientation it appears to randomize the grain boundary texture in copper. It appears that as dislocations move through the material the resulting lattice rotations serve to break down the twin structure in a stochastic manner. An effort was made to use the automatic BEKP data to search for dislocation cell wall structures. However, no evidence was found for such structures. This may be due, in part, to the resolution of the BEKP technique. Diffraction occurs from an area with mean diameter of approximately 0.5μm.

ACKNOWLEDGMENTS

The authors acknowledge K. Kunze and B. L. Adams for allowing the authors access to the automatic BEKP analysis system at Brigham Young University. The help of U. F. Kocks and S. R. Chen of Los Alamos National Laboratory as well as that of H.–R. Wenk at U C Berkeley is also gratefully acknowledged.

REFERENCES

1) Dingley, D. J. and Randle, V.: J. Mat. Sci., 1992, 27, 4545.

2) Adams, B. L., Wright, S. I. and Kunze, K.: Metall. Trans., 1993, 24A, 819.

3) Wright, S. I.: J. Computer Assisted Microscopy, 1993, 5, 207.

4) Matthies, S. and Vinel, G. W.: phys. stat. sol. (b), 1982, 112, K111.

5) Kallend, J. S., Kocks, U. F., Rollett, A. D. and Wenk, H.-R.: Mat. Sci. Eng., 1991, A132, 1.

6) Bunge, H.-J., *Texture Analysis in Materials Science. Mathematical Methods*., Butterworths, London (1982).

7) Pospiech, J., Sztwiertnia, K. and Haessner, F.: Textures Microstruct., 1986, 6, 201.

8) Wagner, F., Wenk, H. R., Esling, C. and Bunge, H. J.: phys. stat. sol. (b), 1981, 67, 269.

9) Weiland, H.: Acta Metall. Mat., 1992, 40, 1083.

Materials Science Forum Vol. 157-162 (1994) pp. 1319-1324

DETERMINATION OF LOCAL TEXTURE AND DEFORMATION SYSTEMS IN TiAl6V4 AND T40

S. Zaefferer, D. Gerth and R.A. Schwarzer

Institut für Metallkunde und Metallphysik der TU,
Grosser Bruch 23, D-38678 Clausthal-Zellerfeld, Germany

Keywords: Titanium, Deformation Systems, Recrystallization, Electron Diffraction, Kikuchi Patterns

ABSTRACT

The spatial orientation distribution and the deformation systems of titanium alloys were investigated. No clear correlation between the degree of local deformation and single grain orientation was found in partially recrystallized TiAl6V4. Fully recrystallized sheets of T40 and TiAl6V4 were deformed up to 10 % by biaxial expansion. The developing dislocation structure was examined. A correlation between the type of dislocation and single grain orientation was found.

I. INTRODUCTION

Within the scope of a fundamental research project of the European Community on texture control of hexagonal metals the microstructure, especially the occurrence of deformation systems, and the spatial orientation distribution in titanium alloys was investigated by transmission electron microscopy. In hexagonal materials, unlike in cubic crystals, several glide and twinning systems may act in dependence of, for example, crystal orientation and degree of local deformation. The knowledge of these correlations is important for the mechanical behaviour and the modelization of texture evolution during plastic deformation.

II. MATERIAL

TiAl6V4 is a dual phase alloy consisting of the hcp α-phase and about 5 % bcc ß-phase. T40 is a single α-phase alloy of titanium containing 900 ppm oxygen. TiAl6V4 sheets, partially recrystallized after 20 % reduction by cold rolling, as well as TiAl6V4 sheets with a fully recrystallized microstructure were studied. From T40 only fully recrystallized material was investigated. The fully recrystallized material of both alloys was deformed 5 % and 10 % by biaxial expansion.

III. EXPERIMENTAL

The investigations were carried out on a Philips TEM EM 430. The individual grain orientations were determined by on-line interpretation of Kikuchi patterns [1]. This method features a high precision in orientation measurement and a high spatial resolution. The measurement of single grain orientations is possible even in small recrystallized grains or in highly deformed grains in the partially recrystallized microstructure.
The investigation of deformation systems was performed with a computer program which facilitates the on-line determination of Burgers vectors, dislocation line directions, glide planes and twinning systems [2]. The determination of Burgers vectors and line directions, however, is limited to a degree of deformation smaller than 10 %, since individual dislocations must be recognized after successive tilts of the sample about large angles. The investigation of twin systems is limited to twin widths larger than 0.2 μm.

IV. RESULTS

The individual grain orientations are represented in inverse pole figure plots as well as in false-colour orientation images [3]. Individual grains in the microstructure are attributed distinct colours which are characteristic of the positions of their *crystallographic directions* on the standard orientation triangle, i.e. the orientation triangle is superimposed by a colour triangle to make an orientation-to-colour scale. Since the representation of an *orientation* requires *two* orientation triangles (one for each reference direction), two false-colour maps are required for the representation of the full spatial orientation distribution. Correlations between grain structure, grain orientation and active deformation systems can be recognized easily from the false-colour images.

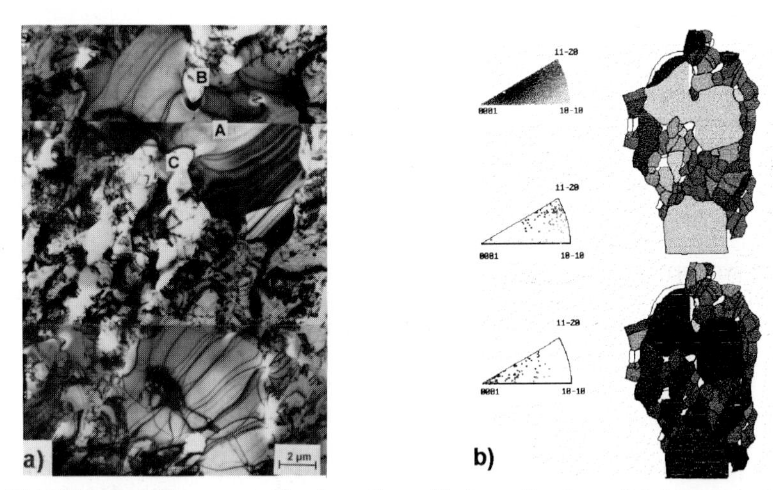

Fig. 1: TiAl6V4, partially recrystallized, section with large fraction of deformed grains.
a) TEM micrograph. b) Orientation distribution in false-colour images.

TiAl6V4, partially recrystallized

The distribution of grain orientations was measured in sections containing deformed as well as recrystallized microstructure. TEM micrographs, orientation distribution images and inverse pole figures are given in figures 1 and 2. In the recrystallized sections different states of grain growth can be seen. Some grains are already in a stable state (figure 2), exhibiting straight grain boundaries and an equally distributed grain size. Other grains (figure 1) are still growing, indicated by curved grain boundaries between deformed grains. The moving grain boundary of grain A is pinned by a grain, B, of special orientation and a cubic crystal, C.

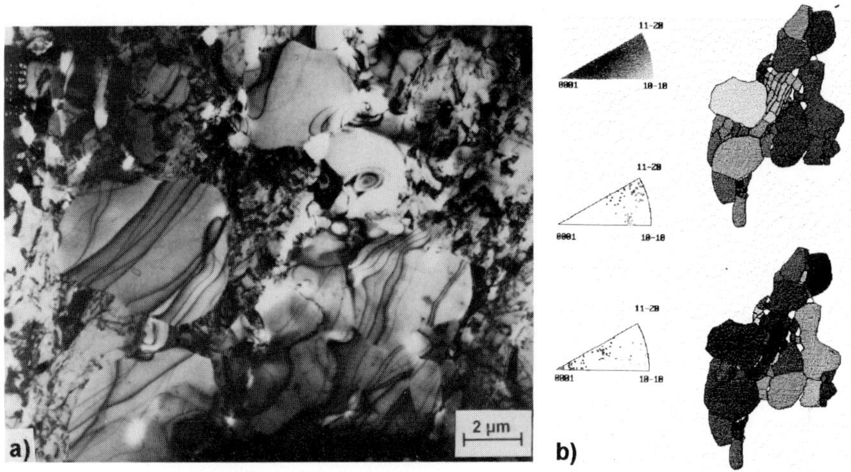

Fig. 2: TiAl6V4, partially recrystallized, section with mainly recrystallized grains
a) TEM micrograph. b) Orientation distribution in false-colour images.

TiAl6V4, recrystallized and deformed 10% by biaxial expansion

The spatial orientation distribution and microstructure is given in figure 3. The crystallographic c-axes of the grains are preferentially oriented in sheet normal direction. Only very few grains show a larger deviation from this orientation. The investigation of deformation systems (Burgers vectors and slip planes of dislocations) yielded the following results:

- Almost all dislocations are of a-type ($<1120>$ Burgers vectors).
- A large fraction (≈70 %) of these dislocations have screw character. Since screw dislocations in this alloy often form slip bands, the slip planes could be determined to be of {0001}.
- Dislocations of edge or mixed character were found to glide in preference on pyramidal planes $\{10\bar{1}1\}$.
- With the exception of dislocations in one grain which is indicated by an arrow in figure 3, c+a-dislocations ($<11\bar{2}3>$ Burgers vectors) are scarcely found.

T40, recrystallized and deformed 5 % by biaxial expansion

The measured orientations are given in inverse pole figures by number in figure 5a. The sample exhibits a weak texture. Grain size (average diameter 30 µm) is large as compared to the TiAl6V4 samples (7 µm). After 5 % of deformation the development of twins can be observed (figure 4a). Since the twins are very small, the twin system could not be determined. Depending on the individual grain orientation, different slip systems were found in the grains (figures 4b, 5b, 5c). The diffraction condition in figure 4b is set such that only a-dislocations are visible. In figure 5b and 5c only c+a-dislocations are in diffraction contrast condition. Grains with the c-axis close to the sample normal direction (A) mainly show a-dislocations on pyramidal planes $\{10\bar{1}1\}$. However, grains with c-axes lying closer to the rolling or transverse direction (B) contain a high fraction of c+a-dislocations on pyramidal planes $\{10\bar{1}1\}$.

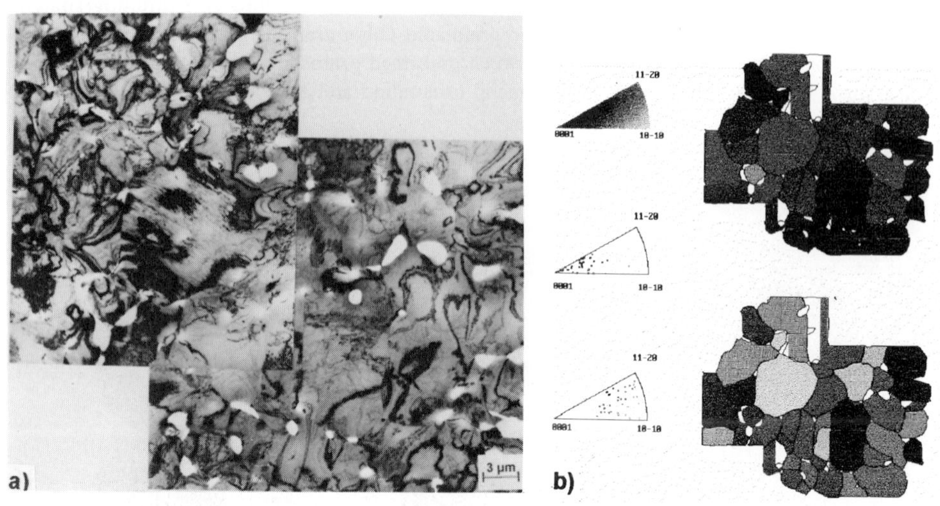

Fig. 3: TiAl6V4, recrystallized and deformed 10 % in biaxial expansion.
a) TEM micrograph. b) Orientation distribution in false-colour images.

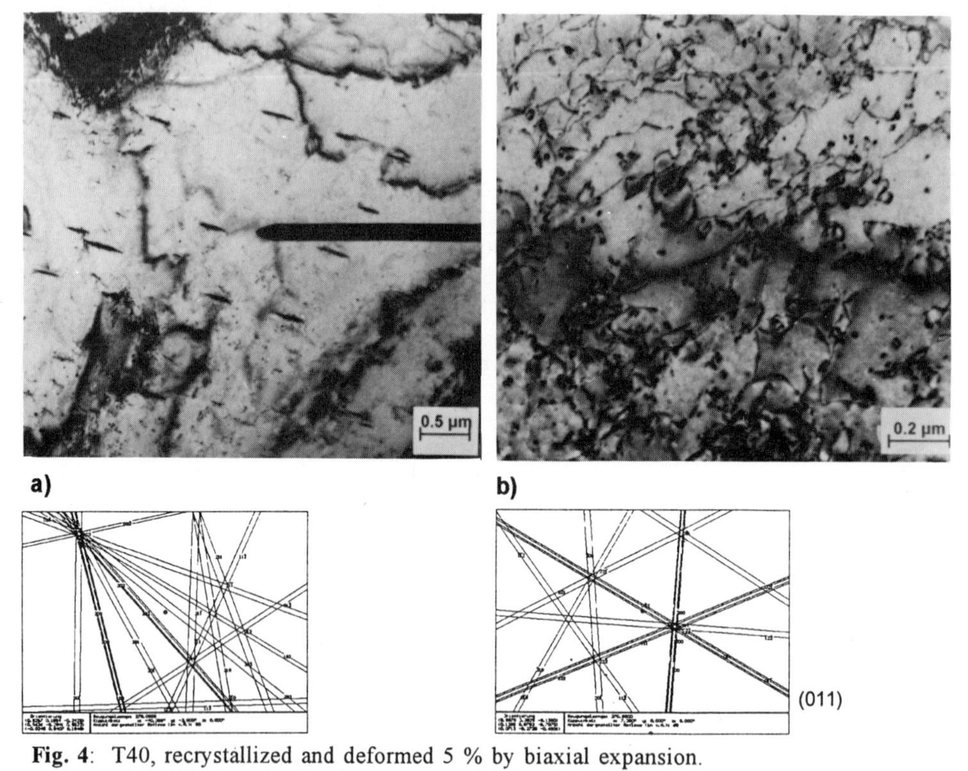

Fig. 4: T40, recrystallized and deformed 5 % by biaxial expansion.
a) TEM micrograph of developing twins and Kikuchi pattern of grain orientation.
b) TEM micrograph of $\langle 11\bar{2}0 \rangle$ dislocations, $(01\bar{1}1)$ reflection excited.

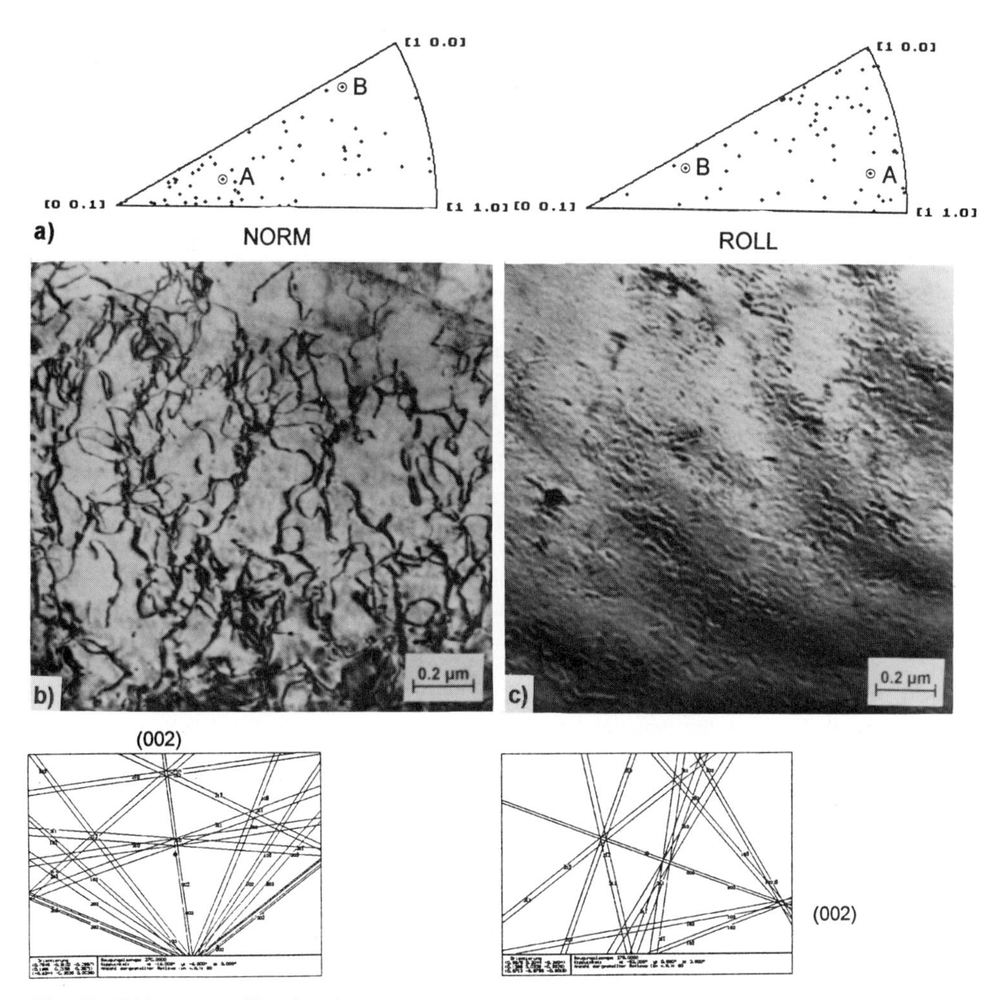

Fig. 5: T40, recrystallized and 5 % deformed by biaxial expansion.
a) Inverse pole figures, two different orientations (**A, B**) are marked. b) TEM micrograph of <11$\bar{2}$3>-dislocations in a grain of orientation **B**, (0002) reflection excited, high dislocation density. c) TEM micrograph of <11$\bar{2}$3>-dislocations in a grain of orientation **A**, (0002) reflection excited, low dislocation density.

V. DISCUSSION

Partially recrystallized TiAl6V4

In a first investigation [4] a correlation between grain orientation and degree of local deformation, measured as dislocation density, was found: Crystallites are the closer oriented to (11$\bar{2}$0)[10$\bar{1}$0] the more they are deformed, i.e. the higher their dislocation density is. In contrast to these measurements this correlation could not be verified in more recent investigations. The false-colour orientation images (figures 1 and 2) do not indicate a significant difference in the orientation distribution of deformed and recrystallized sections.

In figure 2 a large grain (A) is growing in the deformed structure. Its grain boundary is pinned at several sites by grains. At C a grain of cubic crystal structure is pinning the grain boundary. In fully recrystallized material the cubic grains are often found along grain boundaries from which is concluded that precipitations of ß-phase may act as a barrier for grain growth. Grain B shows a considerably deviating orientation which obviously impedes grain growth.

Fully recrystallized TiAl6V4, deformed 10 % by biaxial expansion

The comparison of the orientation distribution of the partially (figures 1 and 2) and fully recrystallized material (figure 3) indicates a marked difference between both. The change in texture after two cycles of rolling and recrystallization is verified by x-ray investigations [4]. The preferred appearance of dislocations with $\langle 11\bar{2}0 \rangle$ Burgers vectors can be explained from the grain orientations: Most of the crystals are oriented such that their c-axes are almost parallel to the normal direction of the sheet. The forces during biaxial expansion are acting perpendicular to the sheet normal, and the $\langle 11\bar{2}0 \rangle$ Burgers vectors are normal to the c-axis of the crystal. This means that the main tension directions and the main deformation directions are coplanar to each other, and dislocations with c+a-Burgers vectors are not necessary. Only grains with deviating orientations, such as the grain marked in figure 3, show a higher density of dislocations with $\langle 11\bar{2}3 \rangle$ Burgers vectors. Deformation of these grains parallel to the sheet surface requires a deformation component in $\langle 0001 \rangle$-direction, hence the presence of c+a-dislocations. A large amount of screw dislocations with $\langle 11\bar{2}0 \rangle$ Burgers vectors were found to glide on (0001) planes. This type of deformation is uncommon for this alloy and cannot be explained yet.

T40, recrystallized and deformed 5 % by biaxial expansion

In contrast to TiAl6V4, this material deforms not only by dislocation glide but by twinning as well. In the 5 % deformed sample the development of twins is observed (figure 5a). The twin systems, however, cannot be determined, since the twin width is still too small to give rise to clear Kikuchi patterns.

The sample shows a similar dislocation distribution as observed in TiAl6V4. Grains with c-axes closely oriented in the sheet normal direction contain almost only $\langle 11\bar{2}0 \rangle$ Burgers vectors (figure 5b), whereas grains with deviating orientations show a high density of $\langle 11\bar{2}3 \rangle$ Burgers vectors (figure 4b). The same explanation is valid here as was given for the TiAl6V4 sample. The distribution of dislocations with $\langle 11\bar{2}3 \rangle$ Burgers vectors is not homogeneous across a grain. Often a concentration of dislocations is observed close to grain triple junctions. This finding is supposed to be due to the necessary accommodation of deformation between neighbouring grains.

ACKNOWLEDGMENT

Financial and personnel support is gratefully acknowledged to the European Community (BRITE EURAM project BREU 0117 C (EBD)). The specimen material has been supplied by M.J. Philippe, Université de Metz.

REFERENCES

1) Zaefferer, S., and Schwarzer, R. A.: On-line interpretation of spot and Kikuchi patterns. Materials Science Forum (1994) / Proc. ICOTOM 10, 1993 (this volume)
2) Zaefferer, S., and Schwarzer, R. A.: On-line determination of deformations systems for cubic and hexagonal materials with the TEM.
Materials Science Forum (1994) / Proc. ICOTOM 10, 1993 (this volume)
3) Gerth, D., and Schwarzer, R. A.: Textures and Microstructures 1993, 21, 177-193
4) Philippe, M. J. et al.: Proc. 7th World Conf. on Titanium, San Diego 1992 in print

3.4 Textures of Special Materials

Materials Science Forum Vol. 157-162 (1994) pp. 1327-1336

TEXTURES IN THIN FILMS

D.B. Knorr

Materials Engineering Department and Center for Integrated Electonics,
Rensselaer Polytechnic Institute, Troy, NY 12180-2590, USA

Keywords: Interfacial Energy, Nucleation, Coalescence, Growth, Granular Epitaxy, Structure Zone Model, Columnar Grains

ABSTRACT

The formation of polycrystalline films by the Volmer-Weber process is considered. Film formation is divided into two stages: 1) nucleation and coalescence and 2) growth. The development of texture cannot be viewed in isolation so its understanding must be coupled to the kinetic processes that control nucleation, coalescence, and growth. Texture during the nucleation and coalescence stage is influenced by stable cluster configurations, surface energy anisotropy, and surface diffusivity, D_s. Impurities can substantially alter these quantities. Growth is considered in the context of the Structure Zone Model. The important processes are both the interfacial energy and D_s, as above, and the grain boundary mobility and lattice diffusivity; the relative contribution of each changes with T/T_m. The influence of all parameters on all stages of film formation are considered from the standpoint of texture development.

INTRODUCTION

No universal definition of a "thin film" exists because the field of films and coating is extraordinarily broad covering a multitude of functional applications, of materials, and of deposition technologies. Several criteria are used to distinguish thin films from thick films and coatings: 1) the film properties, application, or behavior are defined by the thickness; 2) the layer is physically thin such that the thickness does not exceed a low multiple of the dominant microstructural feature such as grain size; and 3) the ratio of interfacial area to volume is large so interfacial transport and/or interfacial energy impact both the microstructural development and the properties. An additional defining aspect is the three step film formation process: 1) synthesis of the specie(s) to be deposited, 2) transport to the substrate, and 3) incorporation into the film resulting in growth normal to the substrate. The precursor material/specie has a different physical and, often, chemical form than in the final film.

A wide variety of deposition methods creates films intended for functional applications [1] such as optical, electrical, mechanical, chemical (catalysis and corrosion protection), and decorative. The most extensive discussion will concern thin film formation by physical vapor deposition (PVD) [2,3] in the context of microelectronic applications. The energy carried by the vapor atoms/ions to the growing deposit is especially important because adatom surface mobility is affected. The energy ranges of various processes cover from 0.025-1 eV for evaporation, 1-100 eV for sputtering, and 3-1000 eV for ion assisted deposition. The source of the energy that affects adatom and, possibly, cluster mobility is:

- evaporation: thermal ejection of atoms;
- sputtering: reflected neutral working gas atoms [4], atoms ejected from the target, accelerated ions (special case of bias sputtering);
- ion-assisted: accelerated ions from an ion source or from self-ions.

The nucleation and early growth of films proceeds by any of three growth modes [5]. The Volmer - Weber (VW) process [6], which is of interest here, involves the nucleation and growth of discreet islands. The ability for an assemblage of atoms to wet the substrate surface depends on the relative interfacial free energies [7] so the VW growth mode dominates when

$$\gamma_{sv} \leq \gamma_{fs} + \gamma_{fv} \tag{1}$$

where γ_{sv}, γ_{fs}, and γ_{fv} are interfacial free energies for the substrate-vapor, film-substrate, film-vapor, respectively. Metallic materials are considered most extensively here with more limited discussion on ceramic and semiconductor films.

The formation of crystallographically oriented deposits and their modification by annealing must be understood within the context of thin film microstructure. Texture is one aspect that includes grain structure and morphology, defect structure (point, dislocations, grain boundaries, etc.), and coarsening (in-situ and post-deposition annealing). The formation of thin films is carried out far from equilibrium so specific details or predictions of microstructural development are very difficult. The deposition processes depend on a variety of variables, both controlled and uncontrolled. Broad trends can be identified using the Structure Zone Model (SZM) which is introduced in the next section. Previous treatments of thin films approached the subject from the standpoint of grain structure/morphology, internal stress, or deposition technology. This paper considers these phenomena using texture as the common theme. The coupled texture/microstructure problem is developed by considering the thermodynamic and kinetic parameters that control the development of morphological and/or crystallographic anisotropy.

MICROSTRUCTURAL DEVELOPMENT

Nucleation and Coalescence

The earliest stages of film formation are treated as kinetic processes involving the nucleation of discreet islands, growth of those islands by addition of adatoms and by island coarsening, coalescence of the individual islands to form an interconnected network but not continuous film, and channel growth by progressive coverage of the remaining substrate area until the film is continuous [8]. Figure 1 demonstrates the important kinetic processes that dissipate the deposition energy in the initial nucleation of islands [5]. Condensation occurs readily, if the vapor is highly supersaturated, to form a critical cluster with one to several atoms [9]. The independent variables are the deposition rate [10] and the effective energy acquired by the adatoms, which depends on the substrate temperature and on the energy per atom or ion in the impinging vapor. The coalescence of discreet islands has been described as "liquid-like" where substantial local mass transport in the form of island rotation, island flattening, etc., occurs [11]. An important point is that grain boundaries form in the film for the first time during the coalescence process.

Two interrelated factors influence the orientation of critical nuclei: 1) the stability of small atomic assemblages and 2) the interfacial energy where both substrate/nucleus and nucleus/vapor interfaces must be considered. The qualitative model of Walton [12] proposes that the structure of the critical nucleus determines the orientation. High vapor supersaturation (low substrate temperature and/or high deposition rate) is predicted to minimize epitaxy on single crystal substrates [13], potentially leading to randomly oriented nuclei [12], in the absence of either strong nucleus adsorption or lattice matching. Stable nuclei with the fewest number of atoms in FCC are predicted in the order (111), (100), and (110) as supersaturation decreases. The substrate might influence which configuration is more strongly adsorbed through the substrate/cluster interfacial energy or by suppressing growth of particular orientations. The concept of "epitaxy temperature" for films on single crystal substrates is derived from such arguments.

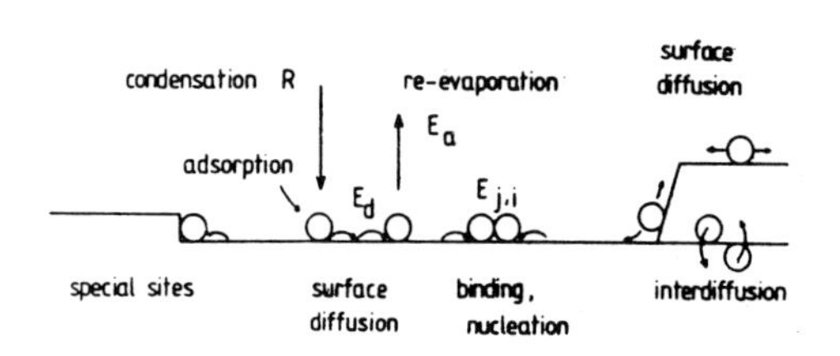

Fig. 1. Kinetic processes and energies involved in nucleation and growth [5].

The concepts of "nucleus adsorption" and "wetting" are manifestations of the concept of interfacial free energy. Since the interfacial area to volume ratio is very large for discreet nuclei, minimization of interfacial free energy is an obvious driving force for texture formation. This is particularly the case for the interface exposed to the ambient and the interface in contact with the substrate. The densest atomic plane tends to be parallel to the substrate if there is high atomic mobility while little orientation results from low mobility [14]. For example, the surface energy in face centered cubic crystals [15] ranks as $\gamma_{(111)}<\gamma_{(100)}<\gamma_{(110)}$ so an (111) fiber texture might be expected in the absence of a substrate effect; this situation occurs for minimal wetting on a single crystal substrate and for an amorphous substrate. If the substrate is crystalline, lattice matching can reduce the interfacial energy further [13] whereby the elastic misfit energy arbitrates which orientation forms, e.g., cubic oriented films on an (001) substrate [16]. Orientations other than (111) are usually predicted in these situations if the substrate is other than (111) oriented. Lattice matching might not necessarily occur when a cluster reaches critical size because island nucleation occurs at defects on a single crystal substrate (heterogeneous) rather than on the undisturbed crystal lattice (homogeneous) as seen on figure 1. A polycrystalline film nucleating on a polycrystalline underlayer might develop local preferred orientations of nuclei on individual grains in a process referred to as "granular epitaxy".

The coalescence stage presents an opportunity for texture evolution. Coalescence does not occur merely by discreet islands impinging but by an extensive rearrangement of material that has been described as "liquid-like" [11,17,18] or as "sintering" [8]. For gold on single crystal graphite [19], the texture changes from an (111) fiber to orientations with some in-plane orientation influenced by the graphite upon coalescence. In another case [20], the coalescence of large (111) grains and small (200) grains is followed by the rotation of the (200) grains into (111) twin orientations; this process minimizes the interfacial energy of the system due to the very low twin boundary energy compared to the grain boundary energy. For Pd on NaCl(001) [21], island coalescence into an epitaxial film is accompanied by dislocations and microtwins. Boundary migration also accompanies the rearrangement of material [18,22].

Impurities have an important, but often overlooked, influence on all aspects of film formation. Contamination can originate from several sources such as chemisorbed or physically absorbed species: 1) adsorption of oxygen, water vapor, or hydrocarbons from the ambient or from desorption of species from the vacuum components, 2) reaction of surfaces with contaminants resulting in a surface layer, or 3) residues of pretreatments for cleaning or for processing. Potential impurity effects that relate to texture development are best summarized by Barna [23]:

Nucleation - impurity atoms/molecules preferentially bind to nucleation sites;

Diffusion - alter, usually decrease, surface diffusivity by occupying surface sites;

Surface energy - change magnitude and, often, reduce anisotropy between different crystallographic planes; preferential segregation to faces with particular orientations;

Coalescence - accumulation and/or segregation to grain boundaries which can inhibit in-situ or subsequent grain growth [24];

Subsequent film thickening - impurity coating on growing film can alter the nucleation and growth mechanisms by changing mode of film thickening, e.g., inducing renucleation on an existing deposit.

Early work [25] on several fcc materials demonstrates that impurities from a relatively high deposition ambient pressure prevent epitaxy on a single crystal substrate and produce randomly oriented deposits. An intermediate pressure gives an oriented fiber texture while ultrahigh vacuum conditions produce epitaxy. Similarly, Nb deposition on GaAs(111) is polycrystalline after high vacuum deposition [22] but epitaxial for deposition in an UHV environment [26]. This effect is attributed to the greater impurity level in the high vacuum system [22]. Several studies on the deposition of aluminum elaborate on various aspects of these effects. At high oxygen impurity content, oxygen adsorption depends on the crystal face orientation [27,28]. Contaminants also reduce the grain size [29], presumably due to impurity boundary pinning. A careful study of oxygen [30] and water vapor [31] effects on film microstructure show that texture depends strongly on the level of contamination. Figure 2 demonstrates that a strong (111) texture at low contaminant levels changes to a nearly random texture at moderate levels. Higher impurity levels also induce substantial changes in the surface morphology of the film [27,28,30,31].

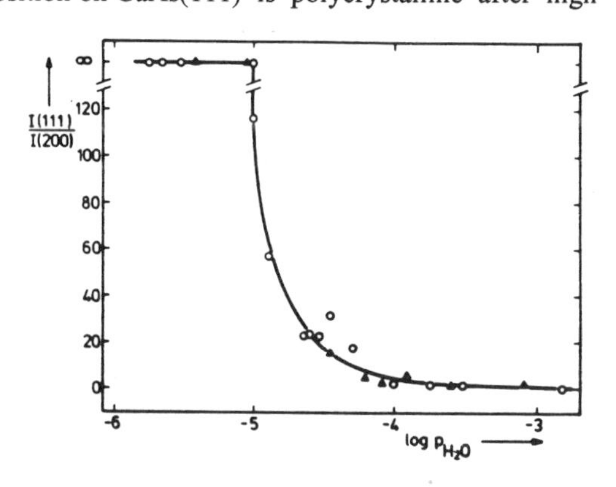

Fig. 2. The effect of water vapor impurity on the texture of evaporated aluminum. Texture index is the ratio of intensities from a Bragg-Brentano X-ray diffraction scan.

Growth

The development of microstructure as deposition proceeds is best described by the Structure Zone Model (SZM). The microstructural attributes are described as a function of homologous temperature and deposition conditions related to the energy incident on the growing film. The first description of such microstructures was by Movchan and Demchishan (MD) [32] for evaporated materials. The SZM for evaporated coatings is shown in figure 3(a). Subsequently, Thornton [33,34] extended the SZM to include magnetron sputtered films by an additional axis to account for the effects of working gas pressure; his SZM is given in figure 3(b). Since sputter deposition is more energetic at lower working gas pressure, each zone develops at a lower substrate temperature. Although separate zones are depicted in both models, there is a continuous transition between zones as temperature increases. The major distinction between the two SZM schemes is the variation with working gas pressure and the presence of Zone T in sputtered coatings. The structures and homologous temperature ranges are closest to the MD model at higher working gas pressures where the sputtered atoms/reflected neutrals are substantially thermalized to energies characteristic evaporation [35]. The homologous temperature ranges vary for different materials, and some zones are missing depending on the material, e.g., Zone T is often missing in pure metals [2]. An important point is that both the Movchan and Demchishan and the Thornton SZMs are based on *thick* deposits where the coating thickness in Zone 2 or lower is > 10 times the grain size or the coating is

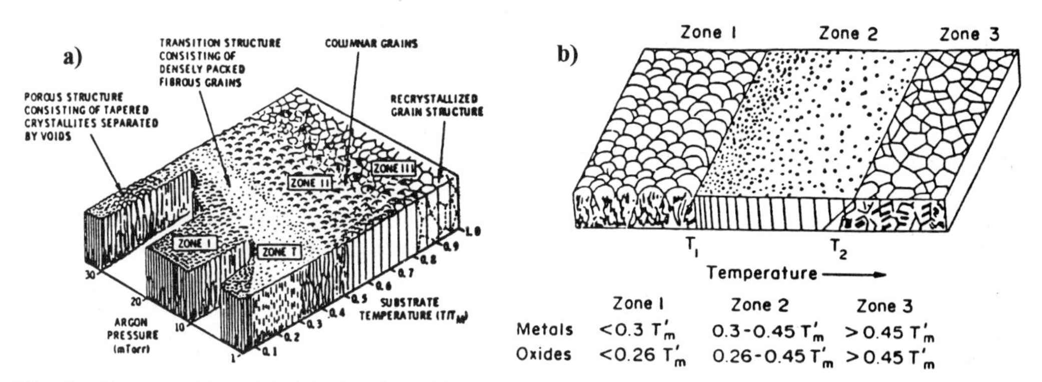

Fig. 3. Structure Zone Models for deposition by: a) evaporation due to Movchan and Demchishan [32], and b) sputtering due to Thornton [33].

physically thick. A general description of the grain structure and properties of each zone follows:

Zone 1: The low temperature zone consists of tapered, narrow crystals separated by voided boundaries. Crystal diameter increases with temperature. The internal substructure is difficult to resolve, the dislocation density is high, and the mechanical properties are poor.
Zone T: The transition zone has a dense array of poorly defined, fibrous grains without the gross voiding found in Zone 1. The substructure is similar to Zone 1, and mechanical properties improve due to the absence of gross voiding, but are still marginal.
Zone 2: A fully dense columnar morphology with significantly less substructure within the grains is present. Mechanical properties are characterized by the best ductility.
Zone 3: A change from a columnar grain morphology is caused either by recrystallization, in the case where dislocation density or stored strain energy are high, or by in-situ grain growth.

Several groups have refined the SZM to apply to ***thin*** films and to reflect a more complete understanding of the substructure in Zones 1, T, and 2 deposits. Messier and co-workers [36,37] modify the interpretation of the Zones 1 and T regions by concluding that Zone T is a sub-zone of Zone 1 based on microstructural similarities. They further subdivide Zone 1 as a function of film thickness to account for development of substructure. Similar to the effect of working gas pressure on the extent of Zones 1 and T, energy from atom bombardment expands Zone T. Hentzell et al. [38] show that the columnar grains in Zone 1 are actually stacks of similarly oriented small grains. The role of grain boundary mobility and grain growth is incorporated into their SZM. Finally, Thornton [39] modifies some interpretations of his SZM to reflect observations in thin films, particularly the suppression of voiding with energetic particle bombardment. The revised SZM in figure 4 will be used to discuss the kinetic processes that contribute to microstructural development and, particularly, to the evolution of texture. Four

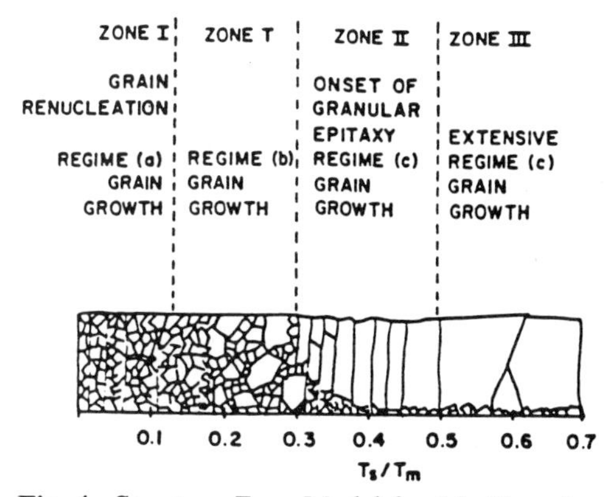

Fig. 4. Structure Zone Model for thin films due to Hentzsell et al. [38].

parameters are important: 1) interfacial energy, γ_i, 2) surface diffusivity, D_s, 3) grain boundary mobility, M_{bdry}, and 4) lattice diffusivity, D_l. The relative importance of each parameter varies depending on the Zone under consideration.

Zone 1 microstructures are determined by limitations on adatom mobility, otherwise expressed in terms of D_s. Geometric shadowing is very important where high points on the surface of the growing film receive a greater atomic flux. Adjacent "valleys" become increasingly deficient in material which is the origin of voids. Substrate roughness stabilizes the Zone 1 structure by suppressing coalescence and by encouraging columnar growth. Strong textures are observed at low temperatures [2,40] where the closest packed plane is often parallel to the substrate. The absence of any appreciable surface transport and of boundary mobility leads to the conclusion that any texture originates with nucleation. Zone T is characterized by evolution in these trends. Adatom mobility increases to the extent that shadowing is no longer important, which is responsible for the reduced voiding and for the stronger columnar grain boundaries. A few high mobility boundaries of special orientation are capable of creating a limited number of large grains in a matrix where no grain growth has otherwise occurred. As temperature increases, more boundaries become mobile. The mechanisms of texture development are expected to be similar to Zone 1 with the possible exception of large grains which might have distinct orientations.

The dense columnar grain structure in Zone 2 is caused by both the high adatom mobility and the boundary mobility. A variety of terms have been used to describe the growth process in Zone 2: "surface recrystallization" [32], "evolutionary selection" [41], and "granular epitaxy" [40]. These terms embody the processes that dominate: 1) grains with specific orientations have a growth rate advantage, and 2) high surface diffusivity permits mass redistribution so no shadowing occurs as in Zones 1 and T. Renucleation does not occur so adatoms are incorporated directly into the growing grains which gives rise to the "granular epitaxy" term. The activation energy for growth approximates the activation energy for surface diffusion [32], although the magnitude of the activation energy for boundary diffusion is comparable [40], so the latter transport path is also possible.

Schematic representations of columnar growth structures are shown in figure 5 [34]. The preferred Zone 2 growth of specific orientations is best illustrated in figure 5(c). This structure is most representative of thick films where substantial increases in grain size accompany thickening [42]. In thin films, the selection processes remain active, but the tapered grains do not have the opportunity to fully develop due to restrictions on the lateral grain growth relative to limited film thickening and due to adatom incorporation on existing columnar grains which suppresses the renucleation processes depicted in figure 5(d). Modeling and computer simulation have successfully quantified the complementary and competing processes involved in columnar growth from Zone 1 through Zone 2. A growth instability leads to the surface topography responsible for the geometric shadowing characteristic of Zone 1 [43,44]. The effects of surface diffusion and of deposition rate on the transition from Zone 1 to Zone 2 structures is

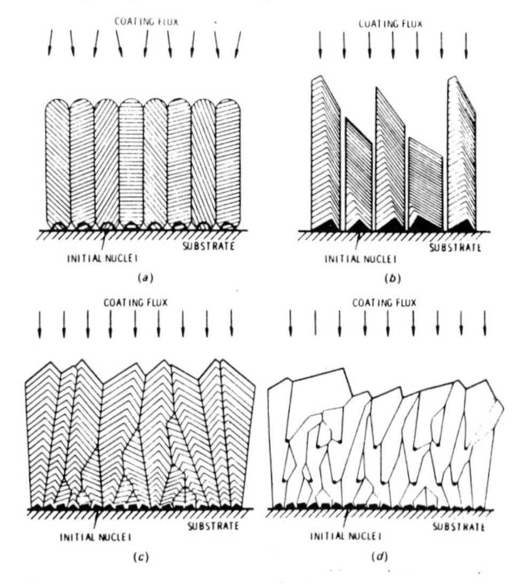

Fig. 5. Schematic Zone 2 growth structures, which depend on adatom mobility: a) $D_s \sim 0$, b) $D_s \sim 0$, texture effect on condensation, c) fast D_s, and d) fast D_s with renucleation.

predicted as well [45,46]. Growth selection responsible for the tapered crystals in Zone 2 is also considered [42]. Several models consider texture development with columnar growth of the tapered grains. The emergence of a specific orientation is attributed to surface energy anisotropy where growth of low surface energy grains is preferred [45]. A model by Lee [47,48] also predicts that low surface energy interfaces will be exposed to the vapor flux, but only at low temperature and/or low deposition rate. Higher energy interfaces are preferred when the concentration of vapor is high, which enhances the stability of interfaces better able to accommodate the lower surface diffusion distances. Muller [46] also considers the influence of deposition rate on the zone transition; an higher deposition rate is predicted to shift the transition temperature to higher temperature.

Zone 3 growth structures are established above T/T_m by bulk diffusion [34] and grain boundary mobility [40] so extensive in-situ grain growth occurs. Activation energy is higher than for the case where surface diffusion dominates and the magnitude is comparable to bulk diffusion [49]. The ready grain boundary mobility is indicative of *in-situ* grain growth during deposition. Grain size increases as substrate temperate increases in this regime [40]. In thin films, the grain size to thickness ratio can exceed unity limited by the grain growth stagnation phenomenon [50]. Texture development in thin films is due to the grain growth which can have an important surface energy contribution. As thickening proceeds, grains with low interfacial energies are preferred. Grain sizes many times the film thickness are possible [51].

Other Processes

Two additional deposition-related phenomena are ion assisted deposition and oblique deposition. Both affect the microstructure of the film and create unique textures. An understanding of the influence of deposition energy and of the microstructure in the context of the SZM are necessary to fully define the texture formation process. Ion assisted deposition (also called ion plating [52]) involves the bombardment of the substrate with charged particles during film growth. The type of ion, its flux, and its energy can be carefully controlled, unlike sputtering where these parameters are usually not well known or controlled. Thus, ion assisted deposition has been widely used to study deposition processes. The ions can be either inert gas or self- ions, i.e., the same specie as the film. Several excellent references on ion assisted deposition are available [52-55]. Texture is generally enhanced by ion deposition and sometimes changes from one orientation to another [53]. Ion bombardment effectively cleans the substrate there by minimizing the effects of adsorbed impurities as represented by figure 2. Ion assisted deposition is an energetic process so surface mobility is enhanced resulting in larger and more widely spaced clusters [56]. During the growth stage preferential resputtering due to ion bombardment favors certain orientations [57]. The extent of geometric shadowing is reduced because ion bombardment erodes the peaks and fills the valleys [58]; the impact of these processes are obviously most important in Zone 1 and Zone T microstructures. Yet another mechanism for ion induced texture formation is preferential channelling. Dobrev [59] shows that ion bombardment changes a (111) fiber texture in Ag to (110) and attributes the change to channeling; the orientation is traced back to the cluster stage. In summary, ion-assisted deposition contributes to and modifies texture by several mechanism: surface cleaning, enhanced surface diffusivity, preferential resputtering, and preferential channeling.

Oblique deposition effects are most evident where columnar grains form. Growth is normal to the incident beam resulting in oblique grains due to geometric shadowing and limited surface diffusion. This effect can be superimposed on the effects of ion assisted deposition. The formation of annular textures [60] is evident in Fe [61] and Nb [62,63].

SUMMARY

Texture development occurs during all phases of film formation: nucleation, coalescence, and growth. Interfacial energy and diffusive transport are important in all of these stages. Surface diffusivity controls adatom mobility while volume diffusion contributes to in-situ coarsening for elevated temperature deposition. The Structure Zone Model is an useful tool to understand the development of grain structure; texture must be interpreted in terms of the grain structure and its evolution. Finally, ion-assisted and oblique deposition results are discussed in terms of these and additional processes which promote texture.

REFERENCES

1. Bunshah, R.F.: in **Deposition Technologies for Films and Coatings**, R.F. Bunshah et al., eds., 1982, (Noyes, Park Ridge, NJ), 1.
2. Bunshah, R.F.: in **Deposition Technologies for Films and Coatings**, R.F. Bunshah et al., eds., 1982, (Noyes, Park Ridge, NJ), 83.
3. Thornton, J.A.: in **Deposition Technologies for Films and Coatings**, R.F. Bunshah et al., eds. 1982, (Noyes, Park Ridge, NJ), 170.
4. Winters, H.F., Coufal, H.J. and Eckstein, W.: J. Vac. Sci. Tech. A, 1993, 11, 657.
5. Venables, J.A.: in **Proceedings, 9th International Vacuum Congress and 5th International Conference on Solid Surfaces**, 1983, J.L. Segovia, ed., (ASEVA, Madrid), 26.
6. Volmer, M. and Weber, A.: Z. Phys. Chem., 1926, 119, 227.
7. Gilmer, G.H. and Grabow, M.H.: J. Metals, June 1987, 19.
8. Lewis, B. and Anderson, J.C.: **Nucleation and Growth of Thin Films**, 1978 (Academic Press, London).
9. Zinsmeister, G.: Thin Solid Films, 1968, 2, 497.
10. Fridrich, J. and Kohout, J.: Thin Solid Films, 1971, 7, R49.
11. Chopra, K.L., and Randlett, M.R.: J. Appl. Phys., 1968, 39, 1874.
12. Walton, D.: , Phil. Mag., 1962, 14, 1671.
13. Singh, H.P. and Murr, L.E. Met. Trans., 1972, 3, 983.
14. Evans, D.M. and Wilman, H.: Acta Cryst., 1952, 5, 731.
15. Sundquist, B.E.: Acta Metall., 1964, 12, 67.
16. Kato, M., Wada, M., Sato, A., and Mori, T.: Acta Metall., 1989, 37 , 749.
17. Andersson, T. and Granqvist, C.G.: J. Appl. Phys., 1977, 48 , 1673.
18. Jacobs, M.H., Pashley, D.W., and Stowell, M.J.: Phil. Mag., 1966, 13, 129.
19. Darby, T.P. and Wayman, C.M.: J. Cryst. Growth, 1975, 28, 41.
20. Matthews, J.W.: J. Vac. Sci. Tech., 1966, 3, 133.
21. Murr, L.E.: Thin Solid Films, 1971, 7, 101.
22. Fisher, I.M. and Smith, D.A.: Textures and Microstructures, 1991, 13, 91.
23. Barna, P.B.: in **Proceedings, 9th International Vacuum Congress and 5th International Conference on Solid Surfaces**, 1983 J.L. Segovia, ed., (ASEVA, Madrid), 383.
24. Kehrer, H.P.:Thin Solid Films, 1971, 7, R43.
25. Murr, L.E. and Inman, M.C.:Phil. Mag., 1966, 14, 135.
26. Eisenberg, M., Smith, D.A., Heilblum, M., and Segmuller, A.P.: Appl. Phys. Lett., 1986, 49, 422.
27. Barna, A., Barna, P.B., Radnoczi, G., Reicha, F.M., and Toth, L.: Phys. Stat. Sol. (a), 1979, 55, 427.
28. Barna, P.B., Reicha, F.M., Barcza, G., Gosztola, L., and Koltai, F.: Vacuum, 1983, 33, 25.
29. Reimer, J.D.: J. Vac. Sci. Tech. A, 1984, 2, 242.
30. Verkerk, M.J. and van der Kolk, G.J.: J. Vac. Sci. Tech. A, 1986, 4, 3101.
31. Verkerk, M.J. and Brankaert, W.A.M.C.: Thin Solid Films, 139, 77.
32. Movchan, B.A. and Demchishin, A.V.: Phys. Metal. Metallogr., 1969, 28, 83.
33. Thornton, J.A.: J. Vac. Sci. Technol., 1974, 11, 666.
34. Thornton, J.A.: Ann. Rev. Mater. Sci., 1977, 7, 239.
35. Thornton, J.A.: in **Deposition Technologies for Films and Coatings**, R.F. Bunshah et al., eds.

1982, (Noyes, Park Ridge, NJ), 170.
36. Messier, R. and Ross, R.C.: J. Appl. Phys., 1982, 53, 6220; J. Appl. Phys., 1981, 52, 5329; J. Appl. Phys., 1983, 54, 5744; J. Appl. Phys., 1984, 56, 347.
37. Messier, R., Giri, A.P., and Roy, R.A.: J. Vac. Sci. Technol. A, 1984, 2, 500.
38. Hentzell, H.T.G., Andersson, B., and Karlsson, S.-E.: Acta Metall., 1983, 31, 2103.
39. Thornton, J.A.: J. Vac. Sci. Technol. A, 1986, 4, 3059.
40. Grovenor, C.R.M., Hertzell, H.T.G., and Smith, D.A.: Acta Metall., 1984, 32, 773.
41. van der Drift, A.: Philips Res. Rep., 1967, 22, 267.
42. Dammers, A.J. and Radelaar, S.: Textures and Microstructures, 1991, 14-18, 757.
43. Karunasiri, R.P.U., Bruinsma, R., and Rudnick, J.: Phys. Rev. Lett., 1989, 62, 788.
44. Lichter, S. and Chen, J.: Phys. Rev. Lett., 1986, 56, 1396.
45. Mazor, A., Srolovitz, D.J., Hagan, P.S., and Bukiet, B.G.: Phys. Rev. Lett., 1988, 60, 424.
46. Muller, K.-H.: J. Appl. Phys., 1985, 58, 2573.
47. Lee, D.N.: J. Mater. Sci., 1989, 24, 4375.
48. Lee, D.N.: Textures and Microstructures, 1991, 14-18, 763.
49. Movchan, B.A., Demchishin, A.V., and Kooluck, L.D.: J. Vac. Sci. Technol., 1974, 11, 869.
50. Mullins, W.W.: Acta Metall., 1958, 6, 414.
51. Gangulee, A.: J. Appl. Phys., 1974, 45, 3749.
52. Mattox, D.M.: in **Deposition Technologies for Films and Coatings**, 1982, R.F. Bunshah et al., eds., (Noyes, Park Ridge, NJ), 244.
53. Harper, J.M.E.: in **Plasma-Surface Interactions and Processing of Materials**, 1990, O. Auciello et al., eds., (Kluwer, New York), 251.
54. Randhawa, H.: Thin Solid Films, 1991, 196, 329.
55. Takagi, T.: Thin Solid Films, 1982, 92, 1.
56. Kay, E. and Rossnagel, S.M.: in **Handbook of Ion Beam Processing Technology**, 1989, J.J. Cuomo, S.M. Rossnagel, and H.R. Kaufman, eds., (Noyes, Park Ridge, NJ), 170.
57. Bradley, R.M., Harper, J.M.E., and Smith, D.A.: J. Appl. Phys., 1986, 60, 4160.
58. Bland, R.D., Kominiak, G.J., and Mattox, D.M.: J. Vac. Sci. Tech., 1974, 4, 671.
59. Dobrev, D.: Thin Solid Films, 1982, 92, 41.
60. Knorr, D.B.: in **Materials Reliability Issues in Microelectronics III**, 1993, K.P. Rodbell, W.F. Filter, H.J. Frost, and P.S. Ho, eds., Vol. 309 (MRS, Pittsburgh, PA), 75.
61. Hashimoto, T., Okamoto, K., Hara, K., Kamiya, M., and Fujiwara, H.: Thin Solid Films, 1982, 91, 143.
62. Bradley, R.M.: in **Handbook of Ion Beam Processing Technology**, 1989, J.J. Cuomo, S.M. Rossnagel, and H.R. Kaufman, eds., (Noyes, Park Ridge, NJ), 300.
63. Yu, L.S., Harper, J.M.E., Cuomo, J.J., and Smith, D.A.: J. Vac. Sci. Tech. A, 1986, 4A, 443.

Materials Science Forum Vol. 157-162 (1994) pp. 1337-1350

TEXTURE FORMATION IN HEXAGONAL MATERIALS

M.J. Philippe

LM2P/ISGMP, UFR MIM-Sciences, Université de Metz,
Ile du Saulcy, F-57045 Metz Cédex 1, France

Keywords: Texture Formation, Hexagonal Materials, Rolling Texture, Recrystallization, Modelling

ABSTRACT

This presentation deals with the texture evolution in the most current hexagonal materials, i.e. zinc, magnesium, titanium and zirconium and for the most current mechanical treatments, i.e. hot rolling, cold rolling, and recrystallization. For all these alloys, the texture evolutions during hot rolling will be commented on, including the phase transformations when they exist. In the same way, the texture evolutions observed during cold rolling will be explained by means of the evolutions of the microstructure and deformation mechanisms. The results of the modelling using plasticity models will be compared with the experimental evolutions. The texture evolutions during recrystallization depend on the alloy and the initial texture.

INTRODUCTION

This presentation deals with the texture evolution in the most current hexagonal materials, i.e. zinc, magnesium, titanium and zirconium and for the most current mechanical treatments, i.e. hot rolling, cold rolling, and recrystallization.

For this purpose, we decided to group the Zn and Mg alloys, both alloys having a melting temperature less than about 650 °C, showing no phase transformation, and having a compactness ratio c/a higher than 1.63, having similar cold deformation mechanisms and thus rather similar texture evolutions [1, 2, 3]. However, we will comment on the specificities of each of them. For the same reasons, we group the Ti and Zr alloys which have a melting temperature higher than 1700 °C, a phase transformation BCC->HCP, a compactness ratio c/a less than 1.63, deformation mechanisms and texture evolution almost identical [2, 3, 4, 5,]. The specificities will also be commented on.

MAGNESIUM AND ZINC ALLOYS

- Experimental results

The magnesium and zinc alloys are either single-phased or biphased. The second phase is generally harder than the matrix and in small proportion, as precipitates or platelets.
Hot rolling (at temperatures higher than about 200 °C for Mg and higher than 50 °C for Zn) causes the c axes to group around the ND of the sheet plane, leading to a maximum in the center or tilted at a few degrees to ND in the ND-RD of the 00.2 PF. (Fig. 1) [6]. (For this presentation to be homogeneous, we will present essentially the pole figures, the ODF not being available for all the presented examples.)

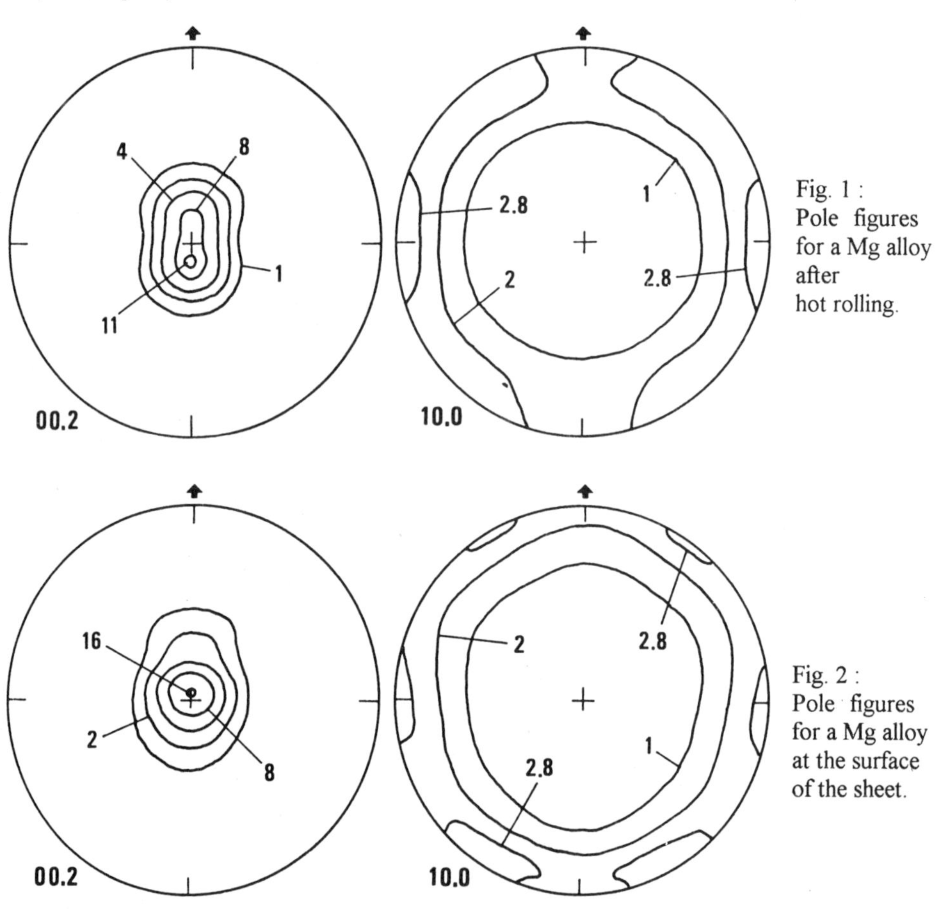

Fig. 1 : Pole figures for a Mg alloy after hot rolling.

Fig. 2 : Pole figures for a Mg alloy at the surface of the sheet.

On the 10.0 pole figure, the intensity strengthens in the TD direction, but only weakly. The texture with central pole is always present in the sheet surface where the sheet undergoes maximal shear (Fig. 2).

During cold rolling and in the particular case of zinc, it is necessary to cool down below 0°C in order to avoid that the heating due to rolling leads to a recrystallization, even if partial. The c axes go away from the center of the 00.2 PF and a maximum appears at 10-20° to ND which increases with the deformation degree (20° for Zn alloys and 10° for Mg alloys) [7, 8]. The component having the 10.0 directions in TD also increases (Fig. 3). These components are all the more pronounced that the grain size is small and that there is a second hard phase, especially in the form of platelets (as in ZnCuTi alloys, Fig 4).

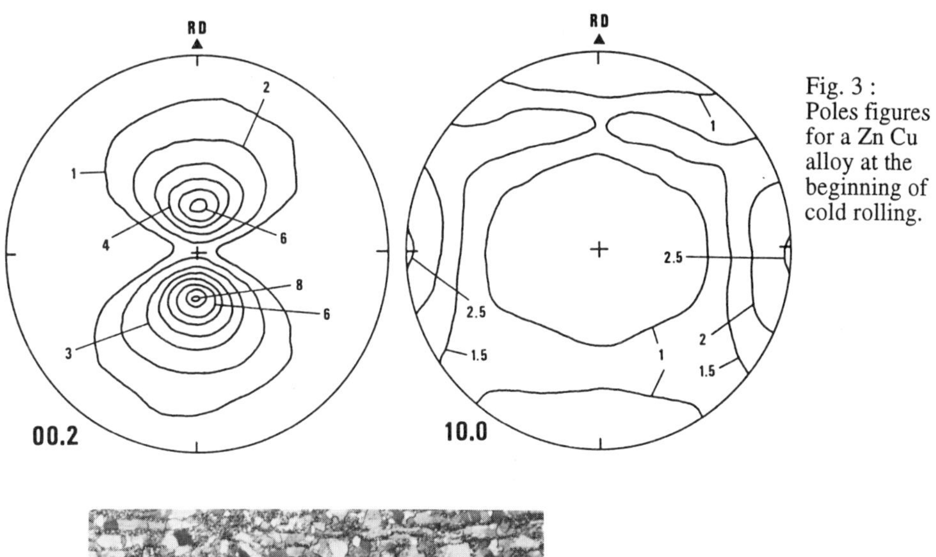

Fig. 3 : Poles figures for a Zn Cu alloy at the beginning of cold rolling.

20 μm

Fig. 4 : Micrography of a Zn Cu Ti alloy.

In the case of zinc alloys, a second component appears during cold rolling with c axes in RD [9]. The intensity of this component increases with the deformation degree (Fig 5). This component does not exist in cold rolled magnesium. This is certainly due to {10.2} twinning (Fig. 6). For Zn, {10.2} twinning is activated by compression of the c axis whereas for the other materials (Mg, Ti, and Zr), it is activated by a tension of the c axis [10] (Fig. 7). The component with c axes in RD is due to <c+a > gliding and to {10.2} compression twinning in Zn. For the other materials (Mg, Ti, and Zr), this component due to <c+a> gliding is suppressed during cold rolling by {10.2} tension twinning (tension of the c axis). This decreases the component with c axes in RD or even suppresses it and brings the c axes back in ND.

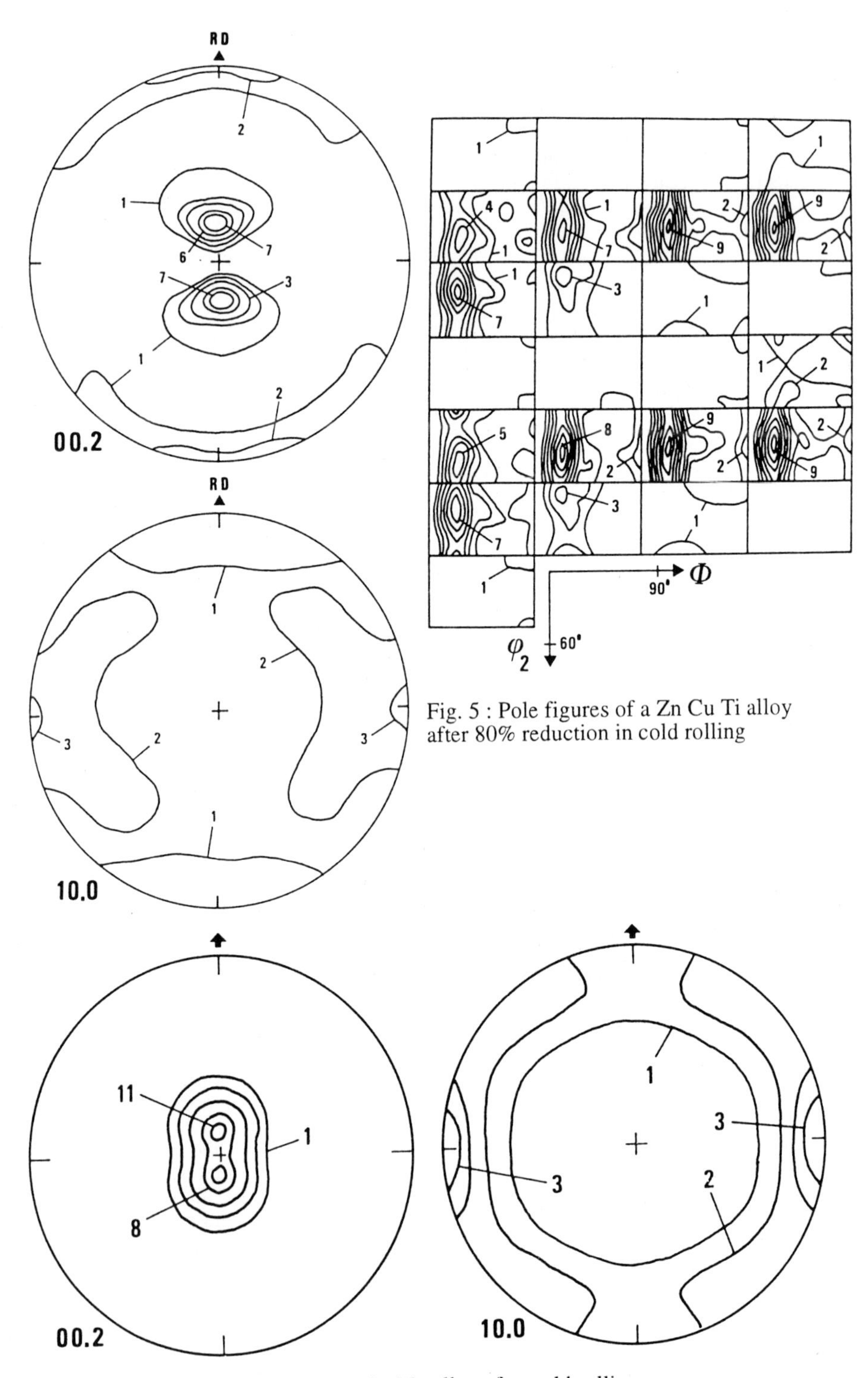

Fig. 5 : Pole figures of a Zn Cu Ti alloy after 80% reduction in cold rolling

Fig. 6 : Pole figures of a Mg alloy after cold rolling.

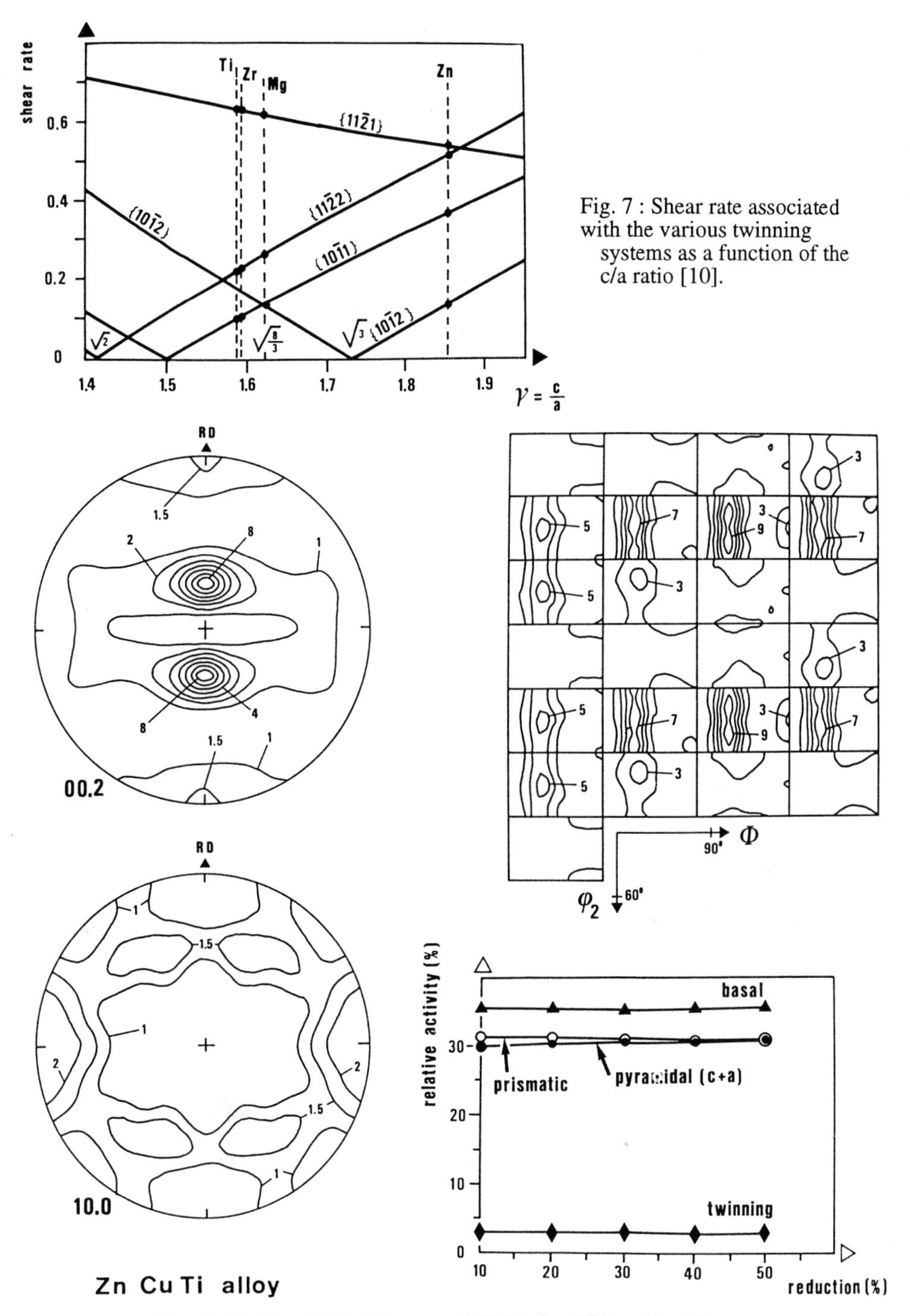

Fig. 7 : Shear rate associated with the various twinning systems as a function of the c/a ratio [10].

Fig. 8 : Predicted Pole Figures and ODF after 80% cold rolling

- Modelling

Different authors have modelized the texture evolution during cold rolling [11, 12, 13] and very good results have been obtained for the ZnCuTi alloys which deformation is homogeneous because of the presence of TiZn15 platelets [13].
The use of the Taylor model in a Full Constrained version and starting with the true hot rolling texture as initial texture made it possible to obtain the results presented here. We can notice a good agreement for all the PFs and the ODF (Fig. 8). In this modelling, basal gliding and prismatic <a> and pyramidal <c+a> gliding are introduced. The introduction of {10.2} twinning increases the component in RD. For Mg this component does not exist. With the same mechanisms the maxima on 00.2 PFs are at 20° from ND for Zn and at 10° from ND for Mg, the difference being due to the c/a ratio.

- Recrystallization

Although only a few theoretical studies concern the static recrystallization of hexagonal materials, a descriptive approach leads to the following observations.
Recrystallization causes for zinc the decreasing then the disappearance of c axes parallel to RD whereas the components at 10 to 20° to ND decrease. If recrystallization goes far, we almost obtain a central pole. In the 10.0 PFs the intensities of the peak in TD decrease, the 10.0 poles rotate around c axes [6, 9].

TITANIUM AND ZIRCONIUM ALLOYS

- Experimental results

Titanium and zirconium alloys present a phase transformation from the β BCC high temperature phase to the HCP low temperature phase. According to the annealing and/or deformation temperatures, the microstructures are different.

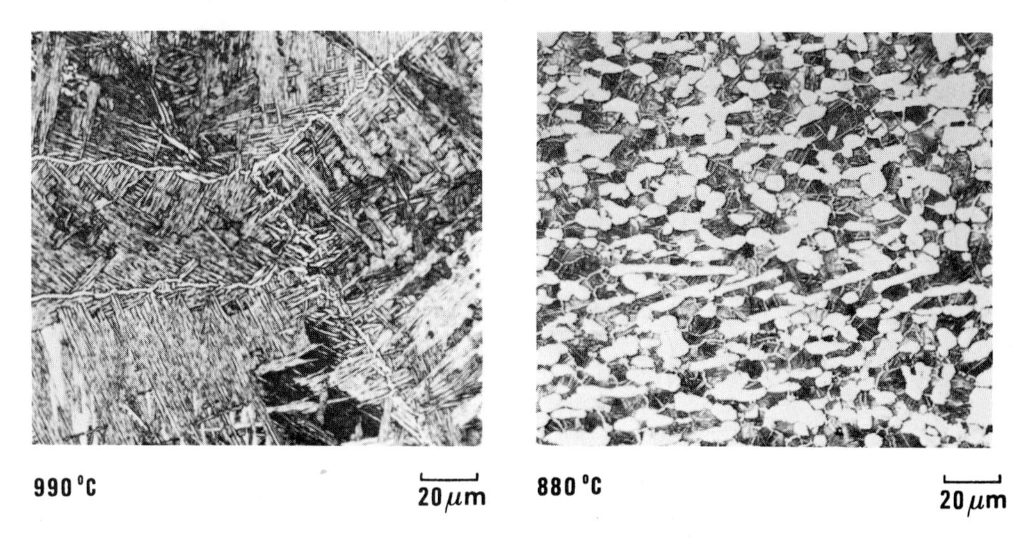

Fig. 9 : Microstructures of α/β alloys according to different thermomechanical treatments [18].

Thus it is possible to have either a completely lamellar structure (due to the transformation from β to α) or bimodal (equiaxed and lamellar) when deformation was performed in the α + β domain (Fig. 9) [14, 15, 16, 17]. So the textures determined after cooling correspond to the hot deformation textures of the primary α phase and to the trace of the deformation texture of the β phase which has been transformed in α after hot deformation. During the phase transformation variant selection occurs.

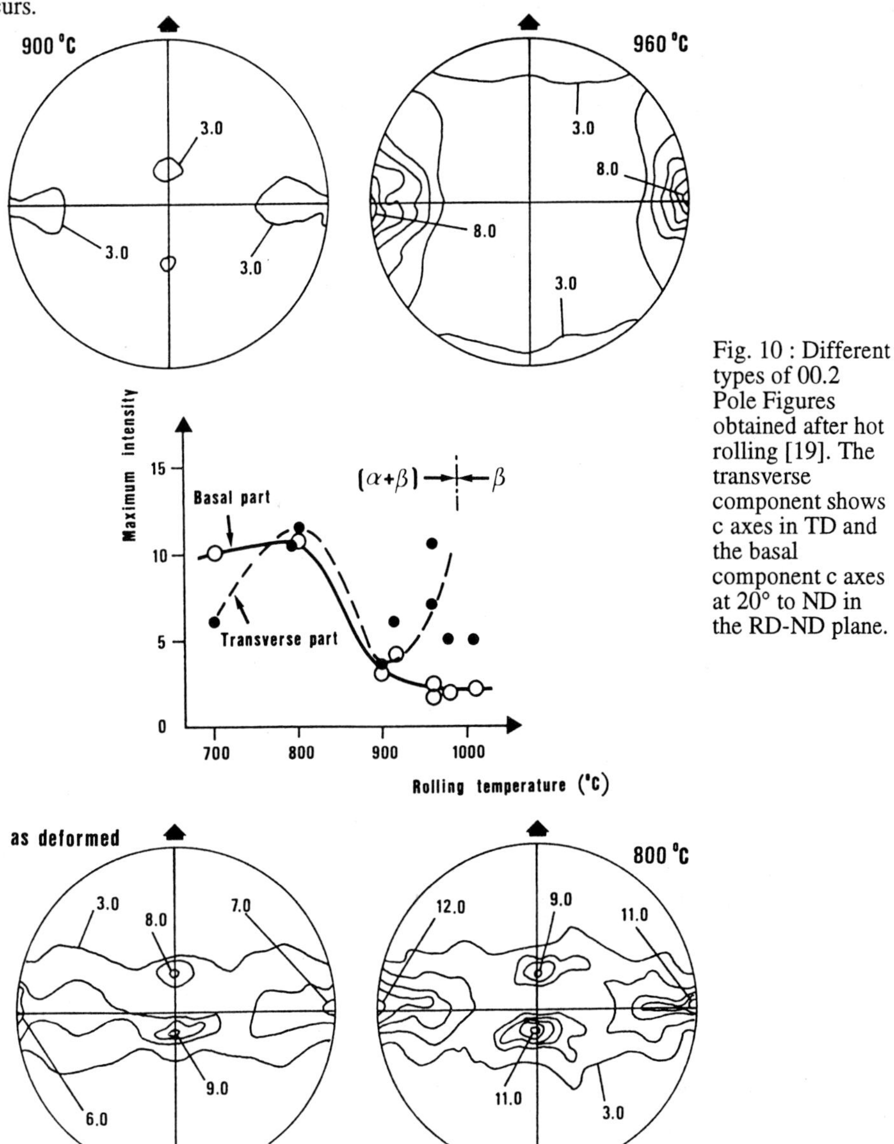

Fig. 10 : Different types of 00.2 Pole Figures obtained after hot rolling [19]. The transverse component shows c axes in TD and the basal component c axes at 20° to ND in the RD-ND plane.

For example in TA6V4 and according to the deformation temperatures different texture types are noticed (Fig. 10) [12]. Rolling a sample with prevailing β phase at a temperature near but below the

β transus leads, after phase transformation, to a transverse texture with c axes in TD. We can notice the variant selection which was active because the transformation β -> α occurred before the recrystallization of the β phase. Beyond the β transus, this phase undergoes recrystallization before transformation and we can notice in the 00.2 PF the 6 variants for a given β parent orientation.
When rolling is performed at a lower temperature (800°C for example), the transverse component is always present but so-called basal components with c axes at 20° to ND in the ND-RD plane are also present, their intensity being all the higher as the temperature is low and the proportion of the α phase during deformation is large [20]. This is valid for all α + β alloys. Concerning the <10.0> axes, even if from the rolling in β phase on, a <10.0> parallel to RD appears, this component strengthens when the rolling temperature decreases and when the deformation degree increases [21].
The cold rolled textures of the titanium and zirconium alloys are comprised between two extreme cases which are presented ; all the other alloys having an intermediary evolution [22].
The first case corresponds to practically pure titanium and zirconium alloys like the T35 and ZrNQ (i.e. Nuclear Quality) which have quite identical texture evolutions [23, 24]. After 50% cold rolling, the texture is already similar but less pronounced than at 80% reduction as we can see on this figure, where the texture index decreases up to 50% deformation and increases afterwards (Fig.11) [25].

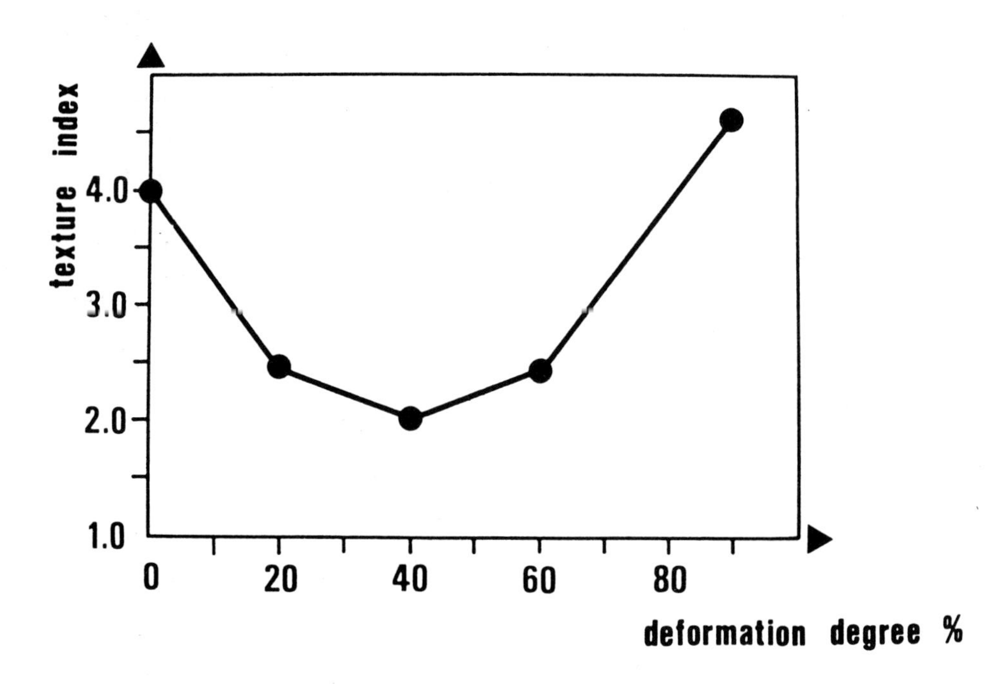

Fig. 11 : Texture index as a function of the deformation degree [25].

Even if at the beginning the deformation is not homogeneous, at 50% deformation the grains are practically all equishaped and all twinned for both alloys. The twinned volume fraction is smaller in the case of ZrNQ [26]. After 80% reduction, the texture of T35 alloys presents two main components whose c axes are at 30° to ND in the ND-TD plane and whose <10.0> directions are parallel to RD (Fig. 12). The texture of the ZrNQ alloy is similar but the c axes are tilted at 20° to ND only. These alloys have average grain sizes of about 30 to 50 mm.

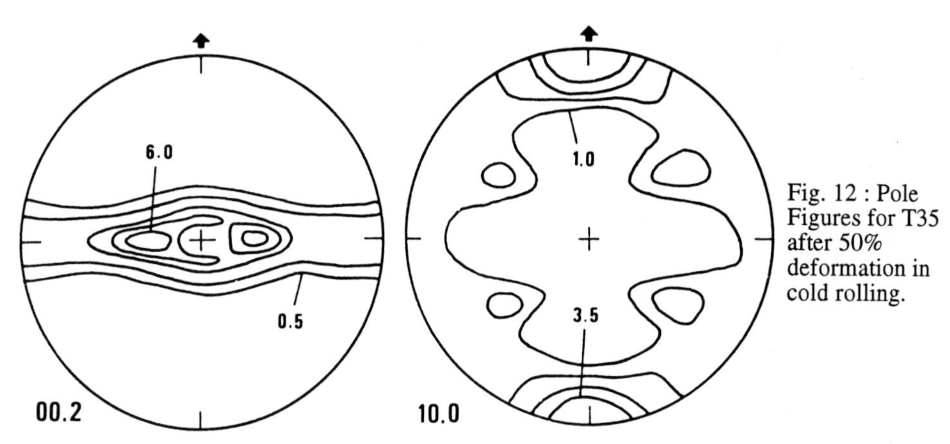

Fig. 12 : Pole Figures for T35 after 50% deformation in cold rolling.

The second extreme case corresponds to alloys like the Zy4 and the T6Al which have also quite similar texture evolutions although more pronounced for the latter (Fig. 13). At 50% deformation, the texture of Zy4 shows two main components whose c axes are at 20° to ND in the ND-RD plane and <10.0> directions are close to RD [27].

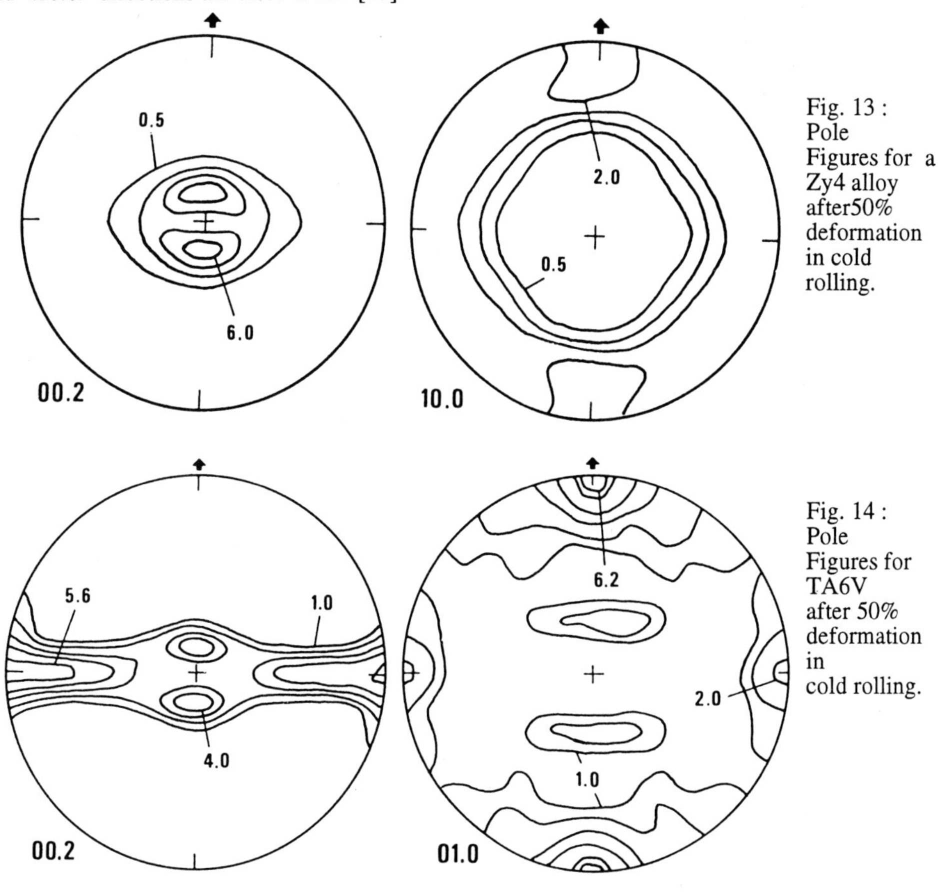

Fig. 13 : Pole Figures for a Zy4 alloy after50% deformation in cold rolling.

Fig. 14 : Pole Figures for TA6V after 50% deformation in cold rolling.

The TA6V4 texture evolution is the same and the components with c axes at 20° from ND in the ND-RD plane increase with deformation (Fig. 14). In addition, there are two components with c axes in TD, remaining from hot rolling. This component decreases during cold rolling and c axes rotate towards ND [28].
The differences between these two extreme cases are due to alloying elements. The first case concerns the non- or low-alloyed materials where the grain size is large and where twinning can be activated or materials with a second softer phase as for example residual β phase in ZrNb. This second phase accommodates the deformation at the boundaries of the α grains.
The second case concerns the alloyed materials with a second phase harder than the matrix and where the a grain size is small and where the deformation is essentially adapted by gliding. So this texture evolution can be explained as follows.

- Modelling

Many works has been done in this field [28,29,30] and we can present the most relevant results. In the first case of alloys with larger grain size, the deformation mechanisms are gliding on prismatic planes, cross slip with <a> direction and twinning. In the first stage of deformation, {10.2} and {11.2} twinning are activated up to 50% deformation. After about 50% deformation twinning is not activated any more and <c+a> pyramidal gliding has thus to be activated instead of {11.2} compression twinning. This shows that there is an anisotropic hardening on each gliding and twinning system. Moreover the texture evolution is complex and some minor intermediary components may be attributed to twinning. As seen previously, {10.2} twinning moves the 0002 poles back to the PF center and may explain the {00.1}<10.0> component which builds up at 50% deformation (components A and D on the fig. 15) [31]. Both components decrease afterwards because the c+a glide empties the PF center. Component C remains the same because these grains deform essentially by prismatic glide. For this first case the modelling of the texture evolution has to be made by stages owing to the evolution of the critical stresses of slip and twinning systems.

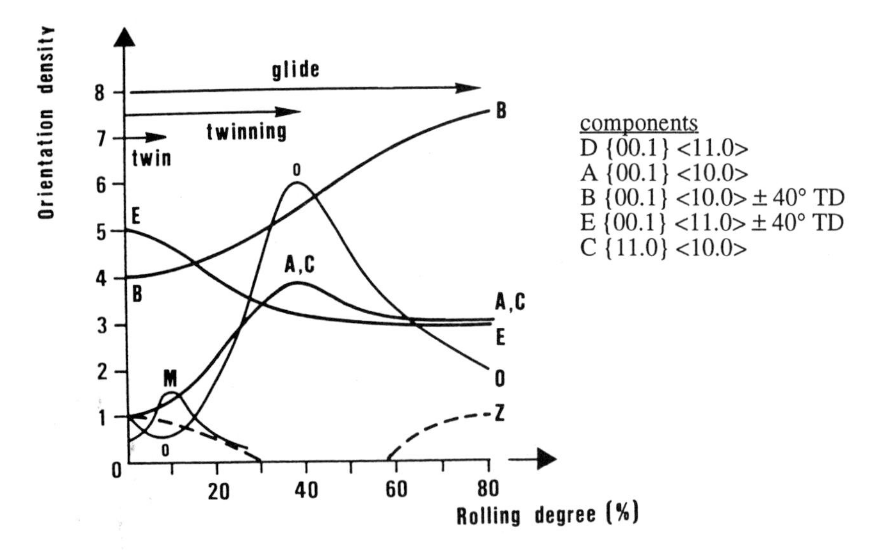

Fig. 15 : Evolution of the different texture components as a function of the cold rolling degree [31]

Then by taking the evolutions of the microstructure and of the deformation mechanisms into account the modelling allowed to obtain good results presented here on T40 (Fig 16 and 17) [32].
In the second case, the grain sizes are smaller for the TA6V and T6Al alloys. Twinning is not very active and essentially prismatic, basal and pyramidal <c+a> glidings are active (Fig. 16 and 17). In these cases the modellings also give good results both for the PFs and ODF analyses. In all the modellings mentioned here, the authors noticed the necessity to have good knowledge about the evolution of the microstructure and of the deformation mechanisms [32].

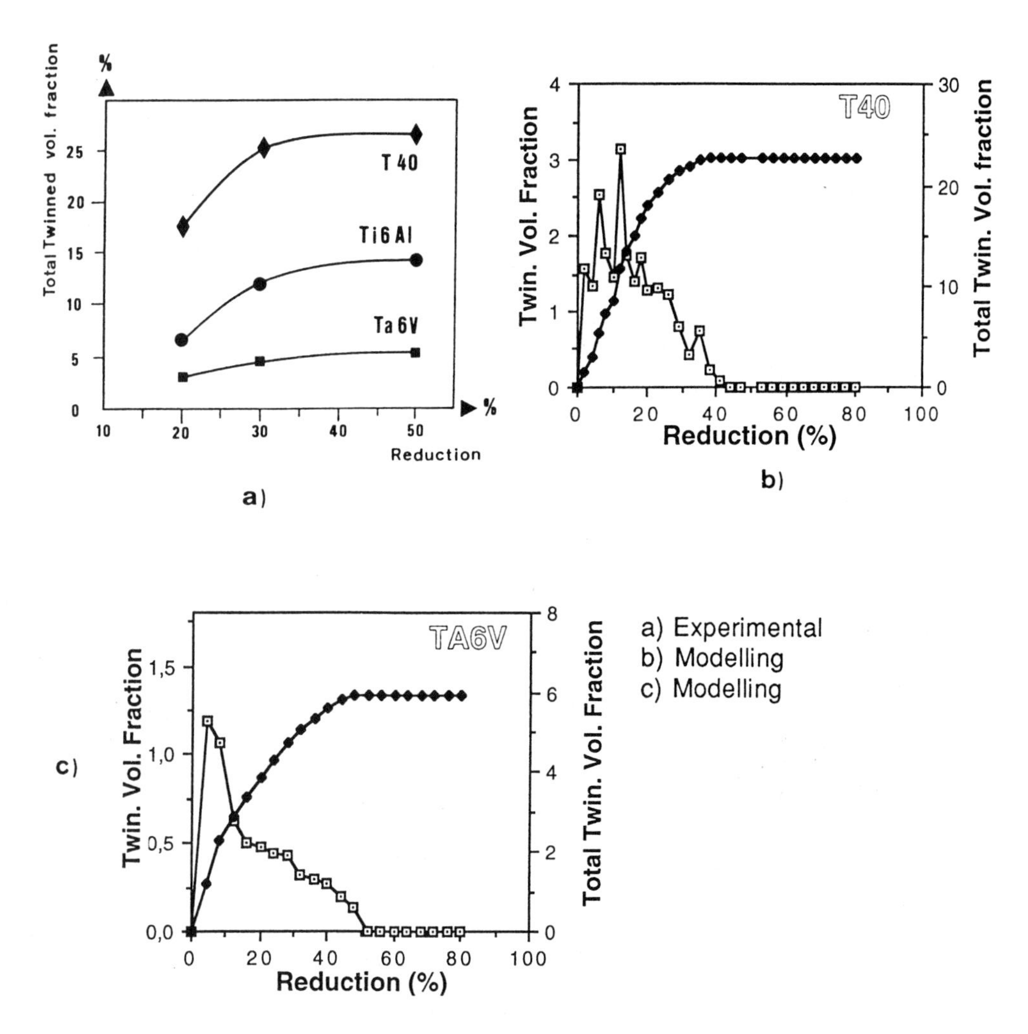

Fig. 16 : Evolution for the relative activity of twinning systems as a function of the reduction (%).

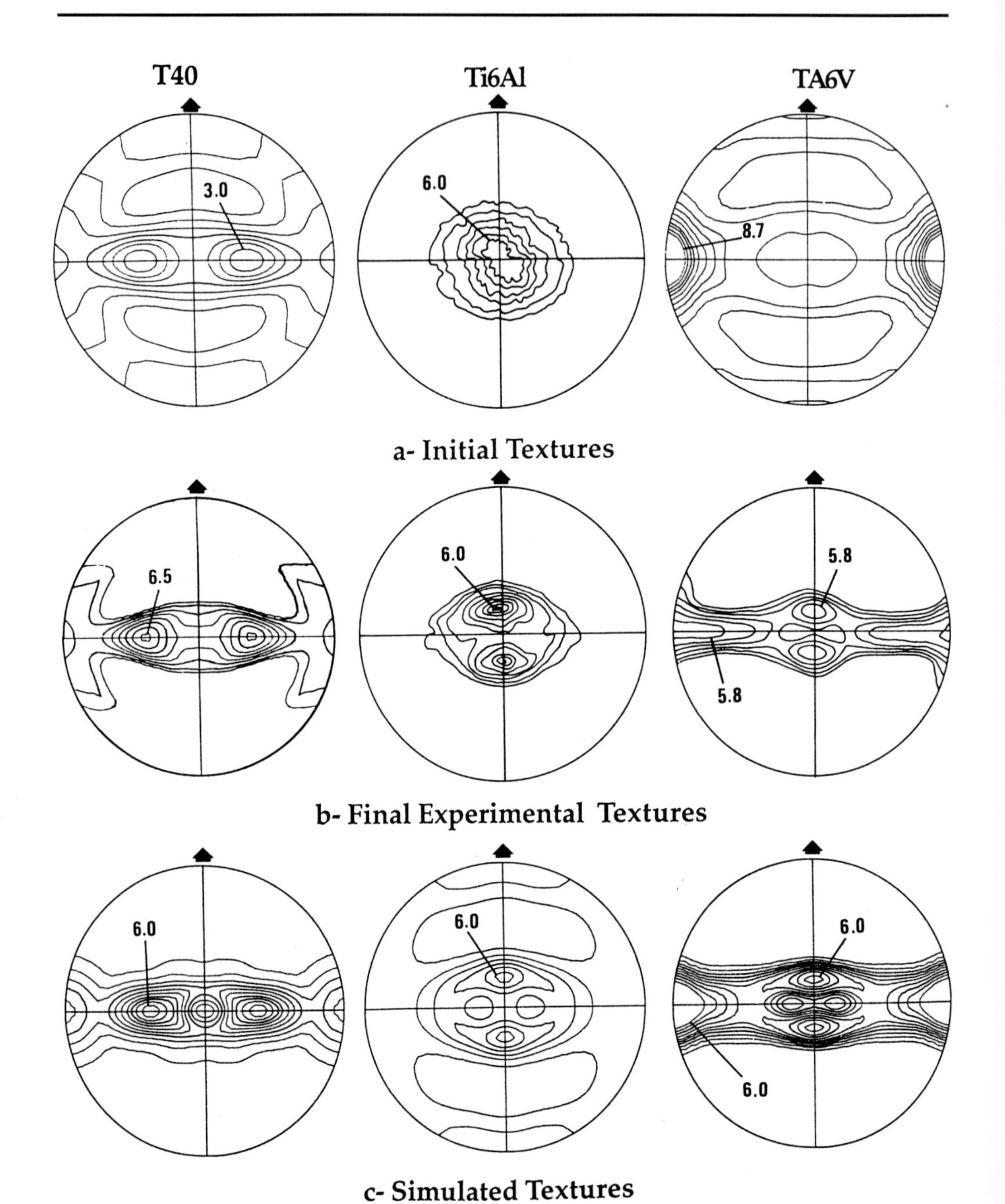

Fig. 17 : Comparison of (00.2) Pole Figures [32]
a) Initial textures
b) Final experimental textures
c) Simulated textures.

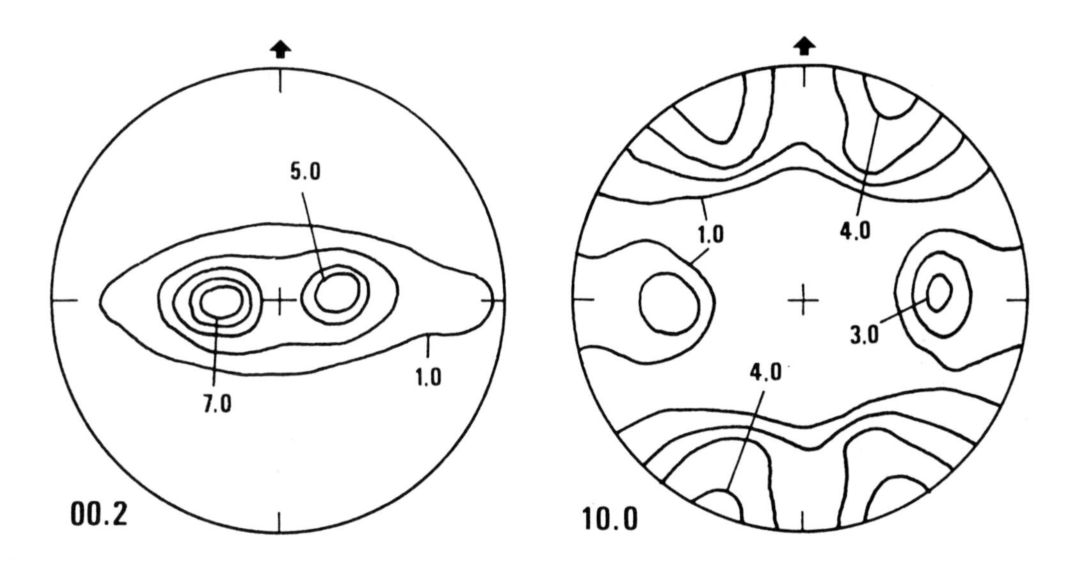

Fig. 18 : Pole Figures of T35 after recrystallization.

- Recrystallization

In the first case where c axes are in the TD-ND plane, a recrystallization leads to a 30° rotation of the <10.0> directions around the c axes and the maxima of the 00.2 PF decrease (Fig. 18) [33].

For the textures where the peaks of the 00.2 PF are at 20° to ND in the RD-ND plane, recrystallization causes the components' intensity to decrease and if recrystallization goes far enough, a central pole builds up as in the case of zinc and magnesium [34].

CONCLUSION

Texture evolution during cold rolling in hexagonal material has already been well studied. It is possible now to explain this texture evolution with the different deformation mechanisms and to modelize it. But the modelling of cold rolling for hexagonal alloys should be further ameliorated, especially by improving knowledge on twinning and gliding CRSSs and on hardening laws.
The largest lacks appear in particular for deformations at high temperatures (in β and $\alpha + \beta$). However these gaps should be bridged as it is necessary to modelize the texture evolution in the α+β domain in order to be able to treat separately the deformation textures and transformation textures.

ACKNOWLEDGMENTS

The author expresses her gratitude to Dr. Y. COMBRES (CEZUS) for his help and for kindly having provided documents and measurements on titanium alloys, to Dr. J. WEGRIA for having provided zinc alloys and to Dr. G. NUSSBAUM for magnesium alloys. My thanks go also to my colleagues of the LM2P laboratory.

REFERENCES

[1] CALNAN, E.A. and CLEWS, C.J.B.: Phil. Mag., 1951, 42, 919
[2] WASSERMANN, G. and GREWEN, J.: Texturen metallischer Werkstoffe. Springer Verlag (1962).
[3] DILLAMORE, I.L. and ROBERTS, W.T.: Metallurgical Reviews, 1965, 10, 271.
[4] WILLIAMS, D.N. and EPPELSHEIMER, D.S.: 1952, 170, 146.
[5] EPPELSHEIMER, D.S. and GOULD, D.S.: Nature, 1956, 177, 241.
[6] GELIOT, R.: Microthesis, Metz (F) (1993).
[7] VALQUCH, M.A.: Metallwissensch., 1932, 11, 165.
[8] WEGRIA, J.: Thesis, UST Lille (F) (1984).
[9] PHILIPPE, M.J., MELLAB, F.E., WAGNER, F., WEGRIA, J. and ESLING, C.: Modelling of Plastic Deformation and its Engineering Applications. Proc. 13th Risø Intern. Symposium on Materials Science, 1992, 385.
[10] YOO, M.H.: Metal. Trans., 1981, 12A, 409.
[11] WIERZBANOWSKI, K.: Atch. Hutn., 1982, 27, 189.
[12] SZTWIERTNIA, K., MUELLER, H. and HAESSNER, F.: Materials Science and Technology, 1985, 1, 380.
[13] PHILIPPE, M.J., WAGNER, F., MELLAB, F.E., ESLING, E. and WEGRIA, J.: Acta Met. and Mat., 1993 in print.
[14] FROES, F.H. and HIGHBERGER, W.T.: Journal of Metals, 1980, 32, 57.
[15] TRICOT, R.: Matériaux et Techniques, 1988, 1-2, 47.
[16] CHAMPIN, B. and DE GELAS, B.: Les Techniques de l'Ingénieur, Dunod, 1976, 1335.
[17] COMBRES, Y. and CHAMPIN, B.: Matériaux et Techniques, 1991, 5-6, 31.
[18] BENHADDAD, S.: Thesis, Orsay (F) (1992).
[19] LUETJERING, G. and PETERS, M.: Titanium Science and Technology, Ed. Kimura and Izumi, Tokyo, 925, (1980).
[20] MORII, K., HARTIG, C., MECKING, H., NAKAYAMA, Y. and LUETJERING, G.: Proc. ICOTOM 8, Eds. Kallend and Gottstein, 1988, 991.
[21] COMBRES, Y.: Private Communication
[22] PHILIPPE, M.J., WAGNER, F., and ESLING, C.: Proc. ICOTOM 8, Eds. Kallend and Gottstein, 1988, 837.
[23] DERVIN, P., MARDON, J.P., PERNOT, M., PENELLE, R. and LACOMBE, P.: J. Less. Common. Metals, 1977, 55, 25.
[24] ALSGAROV, A.H., ADAMESKU, R.A. and GELO, P.V.: Russ. Metal, 1977, 2, 116.
[25] NAUER-GERHARDT, C.U. and BUNGE, H.J.: Proc. ICOTOM 8, Eds. Kallend and Gottstein, (1988).
[26] PHILIPPE, M.J., ESLING, C. and HOCHEID, B.: Proc. ICOTOM 7, Eds. Brakman, Jongenburger and Mittemeijer, (1984) 519.
[27] PHILIPPE, M.J., WAGNER, F., MELLAB, F.E. and ESLING, C.: Proc. ICOTOM 9 / Textures and Microstructures, 1991, 14-18, 1091.
[28] THORNBURG, T.R. and PIEHLER, H.R.: Metal Trans., 1975, 6A, 1511.
[29] WIERZBANOWSKI, K.: Scripta Met., 1979, 13, 759.
[30] TOME, C.N., LEBENSOHN, R.A. and KOCKS, U.F.: Acta Met. and Mater., 1991, 39, 2667.
[31] LEE, H.P., ESLING, C. and BUNGE, H.J.: Textures and Microstructures, 1988, 7, 317.
[32] PHILIPPE, M.J., SERGHAT, M., VAN HOUTTE, P. and ESLING, C.: Acta Met. and Mater. to be published.
[33] NAKA, S., PENELLE, R., VALLE, R. and LACOMBE, P.: Proc. ICOTOM 5, Eds. Gottstein, G. and Lücke, K., 1978, 405.
[34] CHARQUET, D. and BLANC, G.: Proc. ICOTOM 7, Eds. Brakman, C.M., Jongenburger, P.N. and Mittemeijer, E.J., 1984, 485.

Materials Science Forum Vol. 157-162 (1994) pp. 1351-1356

THE EFFECTS OF IMPURITY ELEMENTS ON THE PREFERRED ORIENTATION DURING MELT SPINNING OF DILUTE Zn-ALLOYS

M.V. Akdeniz [1,2] and J.V. Wood [1,3]

[1] Materials Discipline, Faculty of Technology, The Open University, Walton Hall, Milton Keynes, MK7 6AA, UK

[2] Metallurgical Engineering Department, Middle East Technical University, TR-06531 Ankara, Turkey

[3] Department of Materials Engineering and Materials Design, The University of Nottingham, Nottingham, UK

Keywords: Solidification Texture, Preferred Orientation, Zn-Based Alloys, Melt Spinning, Rapid Solidification

ABSTRACT

The development of solidification textures in dilute zinc alloys during rapid solidification by means of melt spinning has been studied by the deliberate addition of impurity elements. Rapidly solidified High Purity (HP) zinc (99.999%) and dilute zinc alloys (Zn-Mg, Zn-Sn, Zn-Cu) develop very strong texture where the basal plane, (0001), in melt spun ribbon is coincident with the solid-liquid interface and perpendicular to the [0001] growth and heat flow directions. However, Commercial Purity (CP) zinc (99.98%), containing Pb and Bi as the main impurity elements, and Zn-0.03% Sb alloy displayed different and distinct preferred orientations at the free surface of the ribbon. The wheel side of the ribbon is textured as in the high purity (HP) sample whereas at the free surface the basal planes tend to be perpendicular to the ribbon surface. This requires reorientation of the basal planes and the instability of the solid-liquid interface during solidification. Differences in severity and texture formation in HP and dilute zinc alloys can be attributed to the change in solid-liquid interface morphology (i.e. planar to celullar) and the formation of celullar substructure in the top portion of the melt spun ribbon.

1. INTRODUCTION

In hcp metals the casting texture depends on the c/a ratio [1]; Mg with a c/a just less than the ideal value (1.63) has a <$11\bar{2}0$> fibre axis in the columnar zone whereas Cd and Zn with c/a greater than the ideal value prefer to grow in the <$10\bar{1}0$> direction. In contrast to cubic metals, where the orientation of grains in the outer chill zone is random, Zn and Cd also develop a casting texture in the outer chill zone with the basal plane (0001) parallel to the chill surface. The morphology and the formation of the substructures in dilute zinc alloys are strongly influenced by the the orientation of the hexagonal basal plane relative to the solid - liquid interface during solidification. In single crystal studies of zinc [2,3] it has been reported that when the basal plane is parallel to the solid - liquid interface regular hexagonal cells form whereas when growth takes place along a direction on the basal plane which is perpendicular to the solid - liquid interface, the regular hexagonaloid cells are greatly elongated and become lamellae parallel to the basal plane. There is an increase in the effect of crystal orientation on the direction of the substructure formation at growth rates $\sim 10^{-4}$ m/s [4] and in certain conditions cells are elongated in a <$10\bar{1}0$> direction on the (0001) basal plane if the growth rate is further increased [4,5]. Nevertheless, the stability of the planar interface which leads to the supression of the formation of substructures is important at high solidification rates, >0.01 m/s, and in undercooled melts as is encountered during rapid solidification [6-8].

The rapid solidification of hcp metals by means of piston and anvil [9,10], twin roller [11] and melt spinning [12-14] techniques exhibit very strong preferred orientations. In all cases, the basal plane was parallel to the ribbon / foil surface, i.e. chill contact surface. Texture formation in melt spinning seems to match the orientations developed during normal casting. In the case of melt spun cubic metals, [100] texture develop at the free surface as in columnar zone orientation in normal casting [15,16]. However the quench surface (wheel side) shows more randomness as compared to the free surface [17].

This paper will describe the effects of the type and the content of the impurity elements on the formation of solidification texture during melt spinning .

2. EXPERIMENTAL

Commercially pure (99.98%) and high purity (99.999%) zinc were used in this investigation. Mg, Sb, Sn, and Cu were subsequently chosen to be deliberately added as impurity elements to investigate the effect of the impurity atomic species and concentration on the development of solidification texture of rapidly solidified zinc. The chemical analysis of melt spun dilute zinc alloys are given in Table 1.

Nominal Composition	Chemical Analysis (wt%)							
	Mg	Pb	Bi	Cd	Fe	Sn	Cu	Sb
High Purity	—	0.7 ppm	—	0.1 ppm	0.3 ppm	—	—	—
Commercial Purity	—	0.02	0.022	<0.003	<0.003	—	—	—
Zn - 0.03 % Mg	0.005	<0.003	<0.001	<0.003	0.009	—	—	—
Zn - 0.03 % Sn	—	<0.003	<0.001	<0.003	0.006	0.032	—	—
Zn - 0.03 % Cu	—	<0.003	<0.001	<0.003	0.008	—	0.045	—
Zn - 0.03 % Sb	—	<0.003	<0.001	<0.003	0.003	—	—	0.03

Table 1. The chemical analysis of melt spun Zn-based alloys

The production of rapidly solidified material was achieved by standard CBMS technique. Small pieces of material cut from the as-cast rod were melted by means of RF induction heating in either quartz or stainless steel crucible. The latter were coated with a thin layer of graphite. In addition to round nozzle crucibles (orifice diameter 0.8 mm) a stainless steel crucible with a 0.75 x 10 mm slot orifice was also employed. The wheel and crucible assembly were enclose in a chamber which was initially evacuated to 10^{-2} Pa and double flushed with Ar to prevent oxidation of materials during melting and spinning. The temperature of the melt was monitored by inserting a thermocouple into the crucible. When the desired processing parameters were achieved an Ar back pressure of 0.1 MPa was applied to eject the melt onto a 254 mm diameter water cooled copper wheel. A distance of 1.5 mm between the crucible and the wheel surface was maintained for all experiments. The ribbon widths were about 3 mm and 10 mm for round and rectengular cross section nozzles respectively. The thickness of the ribbon was dependent on the wheel peripheral speed which varied from 10 to 30 m/s.

The texture measurements of samples were performed by obtaining (0001) pole figures for the surfaces in contact with the wheel (quench surface) and the free surface of the ribbon using CuK_{α} radiation at a setting of 40 kV at 20 mA. Specimens were positioned in the goniometer such that the long transverse direction of the ribbon, i.e. the melt spinning direction, was parallel to the *rolling direction* in the stereographic projection.

The Schulz back reflection method allowed determination of pole figures [18]. Pole figure data were collected from the almost equal solid angle intervals along the triangular mesh in the stereographic projection. A total of 613 data points were sampled for about 5 s each. The intensities were normalized over the pole figure to yield values relative to the intensities of random specimens. Intensity contours were then plotted at suitable intervals on the pole figures using a standart computer program.

3. RESULTS

A typical pole figure showing preferred orientation from RS (**R**apidly **S**olidified) high purity (99.999%) zinc is illustrated in figure 1 for both surfaces of the ribbon. It can be seen that RS zinc has a strong preferred orientation with the basal plane parallel to the ribbon surface, and that the degree of this texture is markedly reduced at the free surface of the ribbon. In this orientation, the basal plane (0001) in melt spun ribbon is coincident with the solid liquid interface and perpendicular to the growth and heat flow directions which are defined by the [0001] direction. Thus, this texture can be characterised by a simple [0001] fibre texture and corresponds to chill zone orientation found in normal casting. Similar solidification textures were also observed in rapidly solidified dilute Zn - 0.005% Mg, Zn - 0.032% Sn and Zn - 0.045% Cu alloys. In all cases the basal planes are parallel to the ribbon surface. The only distinction between HP zinc and impure zinc is that the severity of texture is much more pronounced in HP zinc than in impure samples for both surfaces.

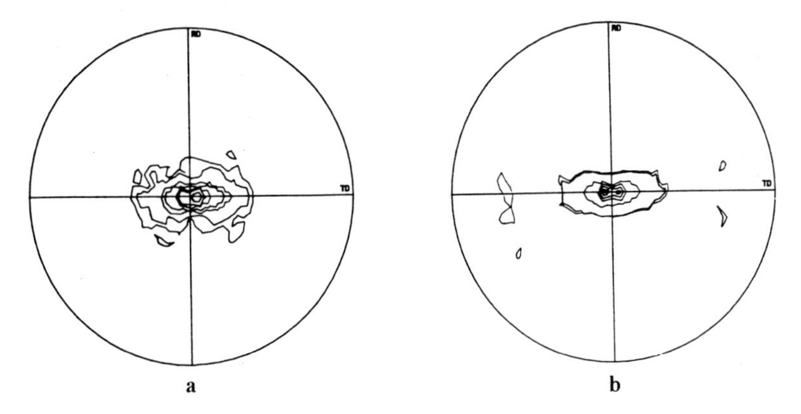

Figure 1. (0002) pole figure of melt spun high purity zinc (a) wheel side, (b) free side

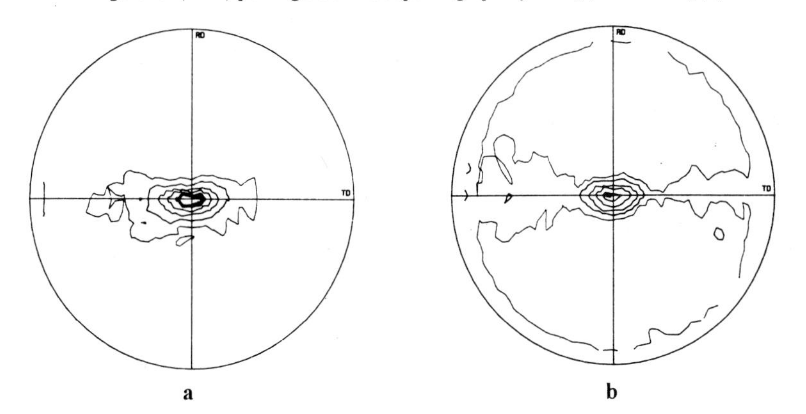

Figure 2. (0002) pole figure of melt spuncommercial purity zinc (a) wheel side, (b) free side

However CP (Commercial Purity) zinc, containing Pb and Bi as the main impurity elements, and the Zn - Sb alloy displayed different orientations at the free surface of the ribbon. The wheel side of the ribbon where the planar growth is observed is textured as in the HP sample whereas at the free surface the basal planes tend to be perpendicular to the ribbon surface. This texture is more conspicuous in CP zinc and the basal plane orientations are illustrated in figure 2.

Pole figure data coupled with the microstructural observations indicate that orientation relationship between the growth direction and formation and morphology of growing solid liquid interface depends on the type of the impurity elements present. For example, as HP zinc solidifies with a planar interface, figure 3, leading to strong preferred orientation, Bi and Sb cause deviation of this preferred orientation during growth which brings about the elongation of these cells in the direction of the growth and the formation of a eutectic like *lamellar-cellular* structure in the top portion of the ribbon, figure 4, . Sn and Mg have no effect on this preferred orientation but form regular cell structures, figure 5, whereas Cu additions seem to have no effect at all on either orientation or formation of substructures and grow with a fully planar interface as in HP zinc under the same solidification conditions, figure 6.

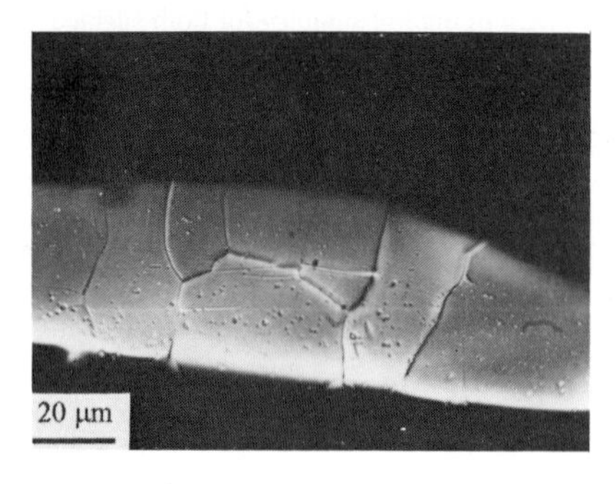

Figure 3. Optical micrograph of melt spun high purity zinc

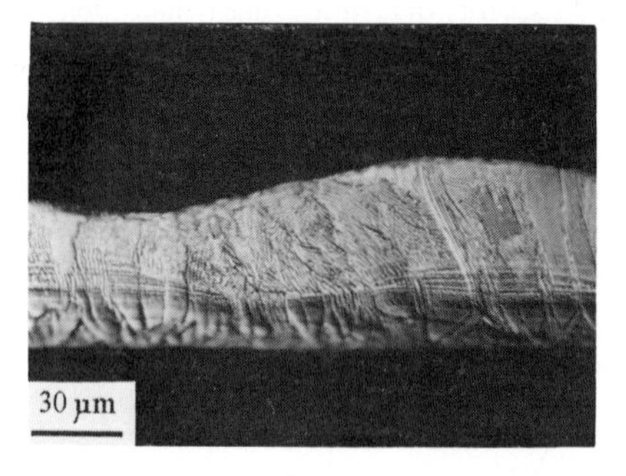

Figure 4. Optical micrograph of melt spun commercial purity zinc showing lamellar-cellular structure

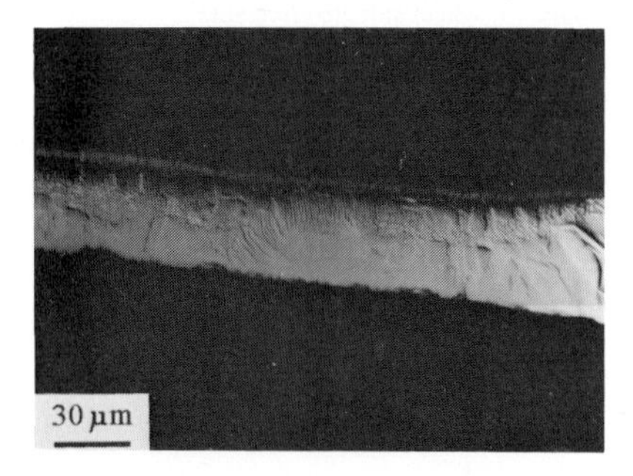

Figure 5. Optical micrograph of melt spun Zn-0.005 Mg dilute alloy

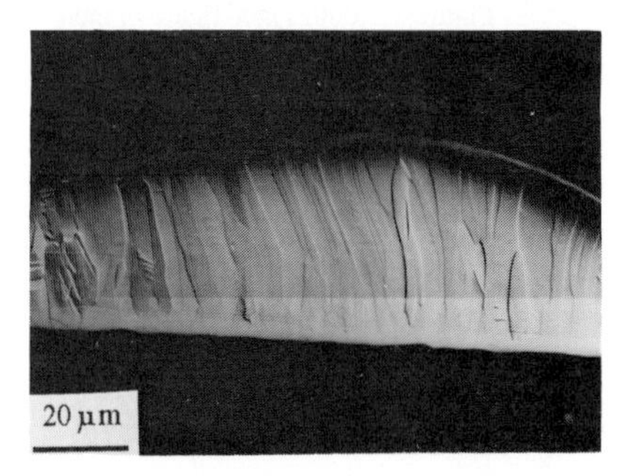

Figure 6. Optical micrograph of melt spun Zn-0.045%Cu dilute alloy

4. DISCUSSION

The preferred [0001] growth direction and the resulting strong [0001] fibre texture during planar solidification are likely to be determined by highly anisotropic nature of zinc. Grains having higher growth rates will determine the preferred orientation as a result of competitive growth. Since, in melt spinning, the heat flow direction is normal to the plane of the ribbon, the most favourable orientation for Zn is that where the basal planes coincide with the growing solid - liquid interface and are perpendicular to the heat flow direction. Such an orientation would allow the fastest rate of heat transfer to the substrate and consequently the fastest growth rate in the ribbon normal which is the [0001] direction. Differences in severity can be attributed to the formation of a cellular structure in the top portion of the ribbon as the cellular solidification of these metals does not favour any particular orientation [19]. This would tend to suggest that during rapid solidification of impure zinc the change in interface morphology, i.e. planar to cellular, gives a less severe texture than that in ribbon which solidifies with a fully planar interface.

The formation of cellular substructure or instability of the planar interface is a function of the growth rate. Thus transition from planar to cellular solidification occurs at a certain critical growth rate above which the interface will always be stable under rapid solidification conditions [6-8]. However, the actual growth rate during melt spinning takes into account both the initial high growth regime and the subsequent comparatively lower growth period during isothermal solidification. Therefore during the initial high growth regime if the actual growth rate is higher than the critical value the interface is stable and plane front solidification occurs. The actual growth rate is then progressively reduces to a constant value when the isothermal solidification conditions are achieved. The instability, which leads to to the cellular solidification, starts as soon as the actual growth rate reaches the critical value. Such structural transition across the ribbon thickness is observed in CP zinc, Zn-Mg and Zn-Sn dilute alloys in which the planar growth at the wheel side is followed by a cellular solidification at the free side of the ribbon. For the case of Zn-Cu dilute alloy the critical growth rate seems to well below the actual growth rate thus, this alloy does not display any transition or instability during rapid solidification.

Although investigations into planar interface breakdown and the formation of substructures in dilute zinc alloys for solidification rates $<10^{-4}$ ms^{-1} under directional solidification conditions agree well with constitutional supercooling theory, the stability of the planar interface has become important at high solidification rates, > 0.01 ms^{-1}, and in undercooled melts as is encountered during rapid solidifiacation [7,8]. However, in agreement with the single crystal studies, present results also show that the morphology of the cellular substructure in dilute zinc alloys is determined by the basal plane orientation relative to the solid-liquid interface if the instability conditions are met during melt spinning or under rapid solidification conditions. For example in Zn-Sn and Zn-Mg dilute alloys where the basal plane is parallel to the solid-liquid interface there is a tendency to form regular hexagonaloid cells in the top portion of the ribbon and in some regions cells are elongated parallel to this plane in the melt spinning direction, figure 5. However, the elongation of cells parallel to the basal plane is more prominent if the basal plane is perpendicular to solid liquid interface as observed in CP zinc. These cells are higly elongated in the direction of growth and become lamellae which eventually display a eutectic-like *lamellar-cellular* (LC) structures, figure 4. The cell formation in the *lamellar - cellular* region in the top portion of the ribbon demonstrates a tendency for the basal planes to be perpendicular to the solid liquid interface and requires almost 90° changing in orientation during growth. This change seems to be associated with the solidification stresses which arise from the differential contraction or expansion of the impurity elements. Although most metals and their alloys contract when they solidify, Bi and Sb are the two main exceptions that expand on freezing. Reorientation of the basal planes by these stresses during growth necessitates plastic deformation with a specific stress configuration. However, this only provides an orientation favourable for the existing cells to be elongated in the growth direction, but it does not give any indication of the initial formation of these cells. Hence, an instability of the advancing solid-liquid interface which leads to the cellular solidification is also required for this unusual structure to form.

5. CONCLUSIONS

(i) The basal plane orientation relative to the solid liquid interface determines the cellular morphology of dilute zinc alloys. Yet, certain impurities have specific effects on the solidification textures and microstructures of RS zinc under the same solidification conditions.

(ii) Samples containing Mg and Sn form regular cell structures at the free side of the ribbon whereas Cu containing samples solidify with a planar interface and do not develop any substructure. The preferred orientations in these samples are the same as in HP zinc in which basal plane parallel to the ribbon surface. Deliberate addition of impurities reduces the severity of the texture as a result of the formation of a cellular substructure in the top portion of the ribbon.

(iii) CP zinc, containing Bi, and Zn-Sb dilute alloy display different and distinct preferred orientations which are associated with the formation of lamellar-cellular structure in the top portion of the ribbon. Regions of planar growth are textured as in HP zinc. The basal planes are then reoriented during growth and tend to be perpendicular to the solid liquid interface in the region where the LC structures start to develop.

References

1 Edmuns, G.: *Trans. Metall. Soc. AIME,* 1941, 143, 183.
2 Hulme, K. F.: *Acta Metall,* 1954, 2, 810.
3 Audero, M. A. and Biloni, H.: *J. Cryst. Growth,* 1973, 18, 257.
4 Kratochvil, P., Formankova, H. and Sichova, H.: *Czech. J. Phys.,* 1961, 11, 679.
5 Bocek, M., Kratochvil P. and Valouch, M.: *Czech. J. Phys.,* 1958, 8, 557.
6 Mullins, W. W. and Sekerka, R. F.: *J. Applied Physics,* 1964, 35, 44
7 Coriell, S. R. and Sekerka, R. F.: *Rapid Solidification Processing: Principles and Technologies, II,* R. Mehrabian, B. H. Kear and M. Cohen eds., Baton Rouge, Claitor's Publ. Div., 1980, p. 35
8 Trivedi, R. and Kurz, W.: *Acta Metall,* 1986, 34, 1663
9 Laine, E. and Lahteenmaki, I.: *J. Mater. Sci.,* 1971, 6, 1421.
10 Nayar, P. K. K. : *Scripta Met.,* 1979, 13, 1115.
11 Adam, S., Babic, E. and Ocko, E.: *Fizika,* 1978, 10, 239
12 Romanova, A. V. and Bukhalenko, V. V.: *Fiz. Metal. Metalloved.,* 1973, 35, 1313.
13 Kavesh, S. : *Am. Inst. Chem. Eng.,* 1978, 74, 1.
14 Blake, N. W. and Smith, R. W.: *Can. J. Phys.,* 1982, 60, 1720.
15 Huang, S. C. Laforce, R. P., Ritter, A. M. and Goehner, R. P.: *Met. Trans. A.,*1985, 16A, 1773
16 Matsuura, M.: *Rapidly Solidified Materials*, P. W. Lee and R. S. Carbonara eds., ASM, Metals Park, Ohio,1986, p. 261
17 Tewari, S. N.: *Met. Trans. A.*, 1988, 19A, 1711
18 Bate, P. and Price, D. C.: *J. Phys. E: Sci. Instrum.,* 1987, 20, 51.
19 Hellawell, A. and Herbert, P. M.: *Proc. R. Soc. London,* 1962, 269, 560.

Materials Science Forum Vol. 157-162 (1994) pp. 1357-1364

RELATIONS BETWEEN TEXTURE, SUPERPLASTIC DEFORMATION AND MECHANICAL PROPERTIES OF THIN TA6V SLABS

O. Benay [1], A.S. Lucas [1], S. Obadia [1] and A. Vadon [2]

[1] SNECMA, Département des Matériaux et Procédés, YKOM 4, BP 81, F-91003 Evry Cédex, France

[2] ISGMP, Ile du Saulcy, F-57045 Metz Cédex 1, France

Keywords: TA6V, Superplasticity, Texture, Tensile Properties

ABSTRACT :

Superplastic tensile tests have been carried out up to elongations of 20 and 100 % at 925°C on a thin TA6V slab. To study the influence of texture, test pieces axes have been oriented with four angles from the rolling direction.
No pronounced effect of texture on superplastic behaviour is clearly enhanced by the usual stress/strain rate law.
The type of microstructural changes and the variations of the texture intensity may be correlated with the strain range. As for the starting material, tensile properties of the stretched material at room temperature are strongly dependent on texture.

1. INTRODUCTION

Superplastic forming combined with bounding diffusion of sheets is, at present time, employed in Aeronautics, in the manufacture of some weakly stressed components [1]. Reduction both in manufacturing cost and structure weight may be achieved by the use of this process compared with alternative technologies.
That is the reason why it seems now interesting to perfect it for making more complex and loaded components as Ti-6Al-4V large chord fan blade.
Therefore, it is necessary to control correctly the different steps of the thermo-mechanical cycle, especially the superplastic deformation conditions and the factors which affect the post formed characteristics.
Data on superplastic deformation and post formed properties are given by several previous works [2,3]. However,they are closely connected with starting material.
In the present work, experiments were carried out on Ti-6Al-4V slabs which are, at present time used by Snecma for perfecting the SPF-DB technology. First, the effect of superplastic strain on crystallographic texture, microstructure and tensile properties evolution is studied. Then the role of crystallographic texture on tensile properties at room temperature and superplastic behaviour is also described.

2. EXPERIMENTAL

2.1. Starting material

The studied material was a Ti-6Al-4V rolled slab 1,2 mm thick. Its composition was :

composants	Ti	Al	V	C	Fe	O2	N2	H2
% WT	base	6.08	3.87	0.02	0.14	0.113	0.01	0.003

2.2. Uniaxial superplastic straining test

All the test pieces were machined in the same slab. In order to study the influence of texture, their axes were oriented with four different angles α = 0, 30, 60 and 90° from the rolling direction.
The gauge lengths were 15 mm wide and 20 mm long. In order to minimize their deformation the test piece heads were reinforced by plates of 2 mm thick Ti-6Al-4V sheet electron beam welded.
Uniaxial tensile test were carried out up to elongations of 20 % and 100 % at 925°C in an argon atmosphere. The average strain-rate was chosen equal to 2.5 $10^{-4}s^{-1}$ and maintained constant by increasing the cross-head speed after about 50 % of the final strain required. Only the load was continuously monitored. After deformation, test pieces were cooled at about 25°C/min.
In these superplastic conditions of temperature and strain rate, the equation for stress σ versus strain rate $\dot{\epsilon}$ was supposed to be : $\sigma = K \dot{\epsilon}^m$
The strain rate sensitivity index, m, was calculated at the cross head velocity stepping by : m = Ln(P1/P2)/Ln(V1/V2).
where P is the load at the cross-head speed V.
K was determined by $\sigma_{final} = K \dot{\epsilon}^m_{final}$

2.3. Room tensile tests

Room tensile tests were performed with an INSTRON 4505 strain machine, using an MTS 63258-14 biaxial strain gauge.
First, starting material was tested. As for the 925°C deformation tests, axes of the test pieces were oriented with an angle of 0°, 30°, 60° or 90° from the rolling direction.
Post formed properties were measured on the stretched test pieces. Before testing the alpha-case layer formed during the thermo-mechanical treatment, was removed by vapor blasting and etching and the gauge lengths were machined again.

2.4. Texture analyses

$(10\bar{1}0)$, (0002), $(10\bar{1}1)$ and $(11\bar{2}0)$ pole figures were measured for the α-phase. Analyses were carried out with the CuKα radiation on a Inel texture goniometer fitted with a 120° curved position sensitive detector.
Both Harmonic and Vector methods were used to calculate the Orientation Distribution function (ODF).
Quantitative analysis has been expressed in terms of volume ratios of material taking into account a disorientation of 10° around each preferential texture component.

3. RESULTS AND DISCUSSION

3.1. Starting material

Initial microstructure contained elongated alpha grains aligned in the rolling direction and embedded in an acicular alpha/beta matrix (figure 1.a).
The texture of the starting slab was characterized by two preferential components :
- a transverse texture $(1\bar{2}10)\ \langle 10\bar{1}0\rangle$ (T orientation)
- a near-basal texture $(1\bar{1}03)\ \langle\bar{2}112\rangle$ (B orientation)

This texture is characteristic of an unidirectional rolling performed between 800 and 900°C on a both a lamelar and globular structure [4].
The main results of room temperature tensile tests are given on figure 2. The curves versus the angle of tensile test show that the Young's modulus as the fracture stress are influenced by texture.
The evolution of the fracture stress, σr could be explained with a simplified model applying the Schmid law [5,6]. Considering that all slip systems have the same CRSS, the Schmid factor μ, can be calculated for a hexagonal close packed crystal (figure 3). As a function of the location of the various slip systems to the testing direction, experimental σr is found to be inversely proportional to μ, as predicted.
As for the Young's modulus, a simplified model using the elastic properties of the single crystal [7] weighted by the ODF have been developped. Calculated values are in good agreement with the experimental results (table 1). Texture seems to be the primary factor influencing the Young's modulus for such a microstructure.

3.2. Microstructure after superplastic deformation

After 20 % superplastic strain (figure 1b) primary alpha grains remained nearly unchanged, still aligned in the rolling direction. On the other hand, the matrix has been transformed : most of the acicular alpha grains have become thicker and rounded, but smaller than the primary alpha grains. The alpha grains are surrounded by the beta phase.
After 100 % superplastic strain (figure 1c) whatever the angle between the tensile orientation and the rolling direction, the microstructure seems to be nearly equiaxed. However, the rolling direction is still reminded by the macroscopic effect of band of the alpha grains. In an other way the beta transformed/beta acicular structure is no more continuous but brought up between the alpha grains.
At elevated temperature the energy stored up by the material is probably higher during the 100 % strain test than the 20 % strain one. Then, the number of alpha nuclei in the beta phase would be higher in the first case and would lead to the finest acicular structure. In the second case, then alpha nuclei, because fewer, might grow more easily to give a coarsener structure.

3.3. Texture after superplastic deformation

After both 20 % and 100 % deformation, texture analyses have shown that the preferential crystallographic orientations, found for the starting material, have been kept on. Only variations on intensities have been noticed, irrespective of the tensile direction. As shown on figure 4, in terms of volume fraction, the

intensity of the B-orientation remained the same after 20 % of strain and was sightly weakened after 100 % strain.
In the other hand, the intensity of the T-orientation increased in both cases, comparatively with the starting material. This one is maximal after 20 % deformation.
Then previous works [3,8] reported a loss in texture intensity after higher strain range. But internal previous works have also shown that a heat treatment at 925°C, without deformation leads to an increase of the texture intensity. Superplastic deformation mechanisms and temperature would have so two opposite effects on texture.
In the present work, for 20 % strain, the influence of temperature would be dominant. For 100 % strain, thermal effect would start to be reduced by superplastic deformation mechanisms. Texture intensity might be a function of superplastic strain range with a maximum value located under 100 % strain.

3.4. Texture and superplastic behaviour

For each tensile orientation, the parameters, K and m, of the law $\sigma = K\dot{\epsilon}^m$ have been calculated from the experimental results, as explained in the paragraph 2.2.
In the experimental range of strain rate, flow stress is plotted versus strain rate after 20 % deformation (figure 5a) and 100 % deformation (figure 5b).
After the 100 % strain, at given $\dot{\epsilon}$, the evolution of the flow stress with stretching orientation could be compared with the evolution of the fracture stress at room temperature. Then, both grain-boundary sliding and slip deformation are involved in superplasticity. However, the conditions for these mechanisms to occur are not well known. In the present case, slip-deformation mode would be dominant.
After the 20 % strain, the four curves are parallel. There is no influence of the orientation on the strain rate sensitivity coefficient. At given $\dot{\epsilon}$, the arrangement of flow stresses is different from the 100 % strain one.
In fact, the coefficients of the law are determined at different between the two tests. Thus, m is time calculated from a velocity step happening at about 10 % stretching in the 100 % strain one. In the same way, K is determined from σ 20 % in the first case and σ 100 % in the other one.
Then, significative microstructural differences are shown by the micrographic study and the texture analysis. This differences are probably the result of a microstructural evolution occurred during the test connected with the range of the strain or the time during which the test pieces are kept at 925°C.
This law seems to be not sufficient to describe correctly the superplastic behaviour. It would be necessary to add a time factor, which characterize microstructure evolution, to refine it.

3.5. Tensile properties after superplastic deformation

Only the results obtained at 20 % stretching are presented on figure 2.
The evolution of the stresses with the tensile orientation is the same one as for the starting material. Plastic deformation is so induced by identical mechanisms.

In an other way a systematic decrease of the values is recorded. This fact could be explained as follows. The intensification of texture could lead a decrease of the number of possible slip systems. For each tensile orientation, one of them would be particularly favored, making the deformation accommodation more difficult. From that, rupture would occur more easily.
The evolution of the Young's modulus with the tensile orientation is also the same as for the starting material. It is still dependent on texture. However, the values are higher. This increase could be partially explained by the intensification of the T-orientation, as is shown by comparison of experimental and calculated values (table 1). But the differences observed suggest that Young's modulus is also influenced by an other secondary microstructural factor.

4. Conclusion

1. This present work has shown that there is no pronounced effect of texture on the superplastic behaviour of thin Ti6Al4V slab at 925°C. At least, this influence is not clearly enhanced by the law which has been employed.
2. Microstructural changes have been noticed after superplastic deformation. But its final aspect may be correlated with the strain range.
3. The type of starting texture remained the same after 20 and 100 % strain, irrespective of the tensile direction. But the evolution of the texture intensity could be attributed to the competitive effect of temperature and superplastic deformation.
4. Tensile properties are strongly dependent on crystallographic texture after superplastic deformation like before. Especially, calculated Young's modulus are in good agreement with experiments.
5. An increase of the Young's modulus values have been noticed after 20 % strain test. But for this same strain, there is a diminution of about 10 per cent on the fracture stress.

REFERENCES

(1) AIR & COSMOS/AVIATION MAGAZINE 1993, 1417, 26
(2) Mac Darmaid, D.S. and Partridge, P.G. : Conference "Superplasticity and Superplastic forming", 1988, 529 Edited by Hamilton, C.H. and Paton, N.E.
(3) Dunford, D.V., Wisbey A. and Partridge, P.G. Mater. Sci. and Techno, 1991, 7, 62
(4) Lütjering, G. and Wagner, L., 177. Mecking, H., Lütjering, G. and Morii, K., 239 : "Directional Properties of materials", 1988, Edited by Bunge, H.J.
(5) Schmid, E. : Z. Elektrochem, 1931, 37, 447
(6) Peters, M., Gysler, A. and Lütjering, G. : Metall. Trans., 1984, 15A, 1597
(7) Flowers Jr, J.W., O' Brien, K.C. and Mc Eleney, P.C. : J. Less. Common Metals, 1964, 7, 393
(8) Bowen, A.W., Mc Darmaid, D.S. and Partridge, P.G. : J. Mater. Sci., 1991, 26, 3457

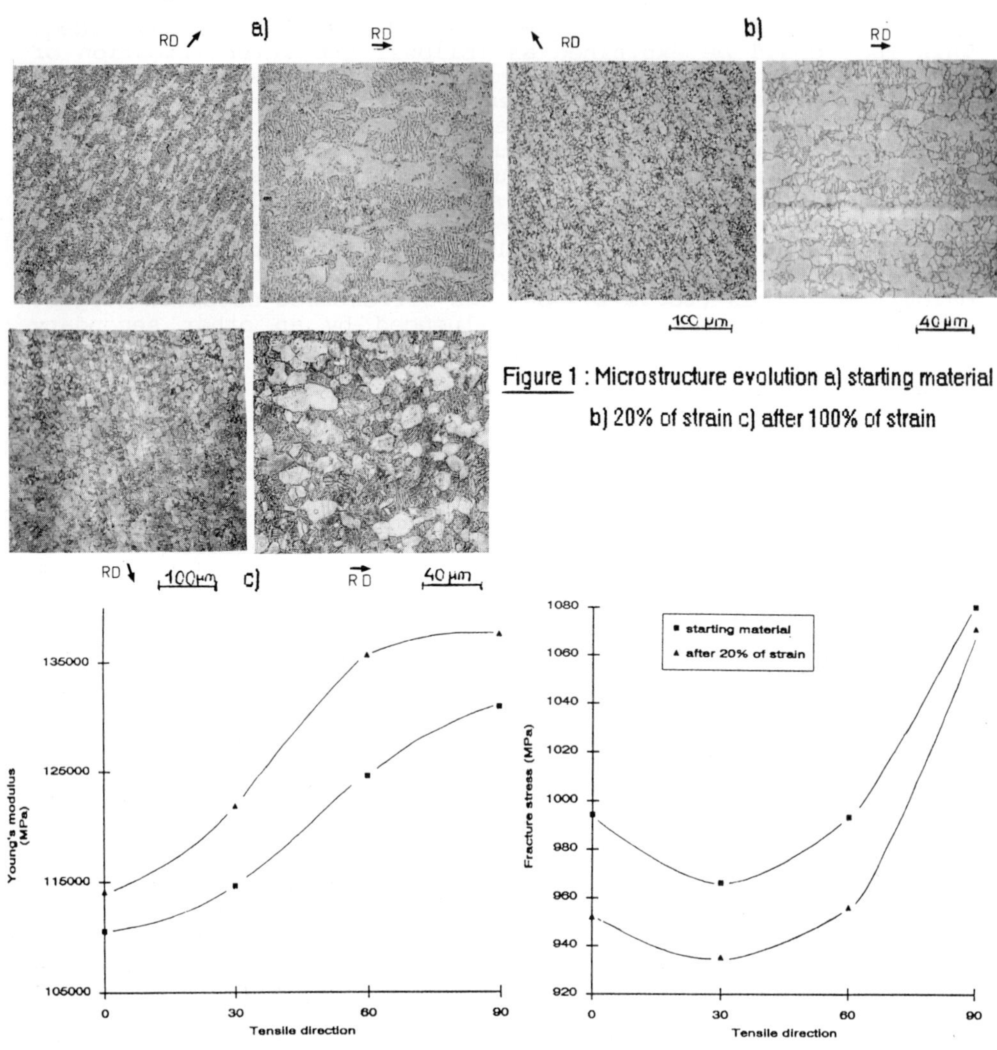

Figure 1 : Microstructure evolution a) starting material b) 20% of strain c) after 100% of strain

Figure 2 : Young's Modulus and Fracture Stress versus the tensile direction

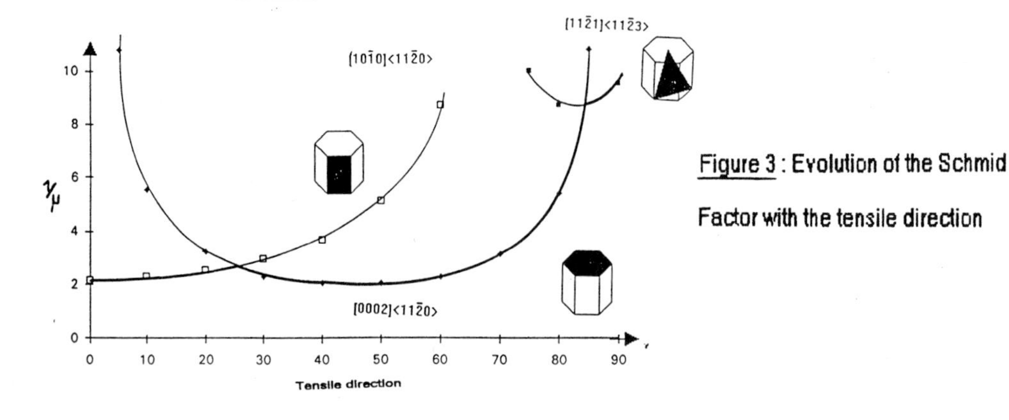

Figure 3 : Evolution of the Schmid Factor with the tensile direction

	STARTING	MATERIAL	AFTER 20% OF	DEFORMATION
	E experimental MPa	E calculated MPa	E experimental MPa	E calculated MPa
0	110460	110390	114080	109130
30	114580	113010	121880	112450
60	124560	119780	135640	121228
90	130900	130970	137580	134523

Table 1 : Comparison between experimental and calculated Young's Modulus

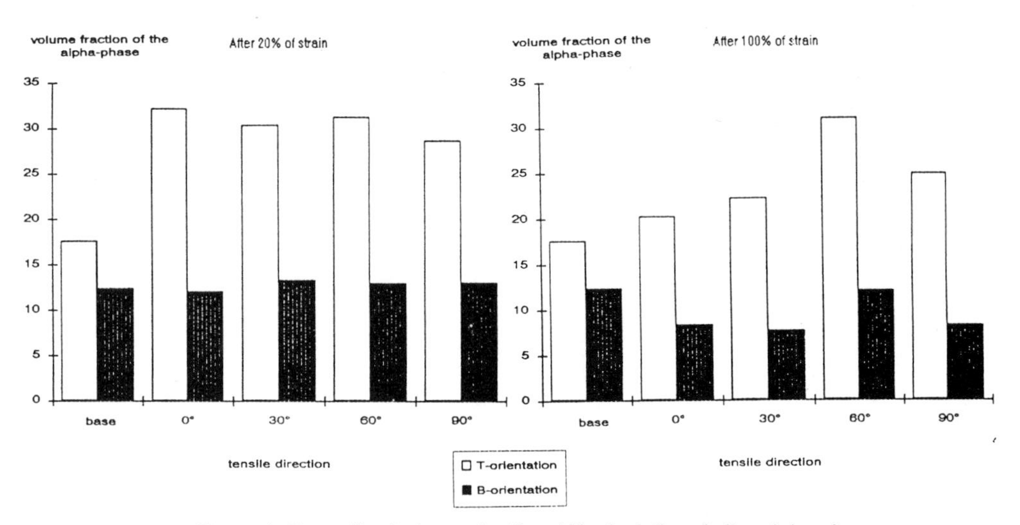

Figure 4 : Proportion between the B and T-orientations in the alpha-phase

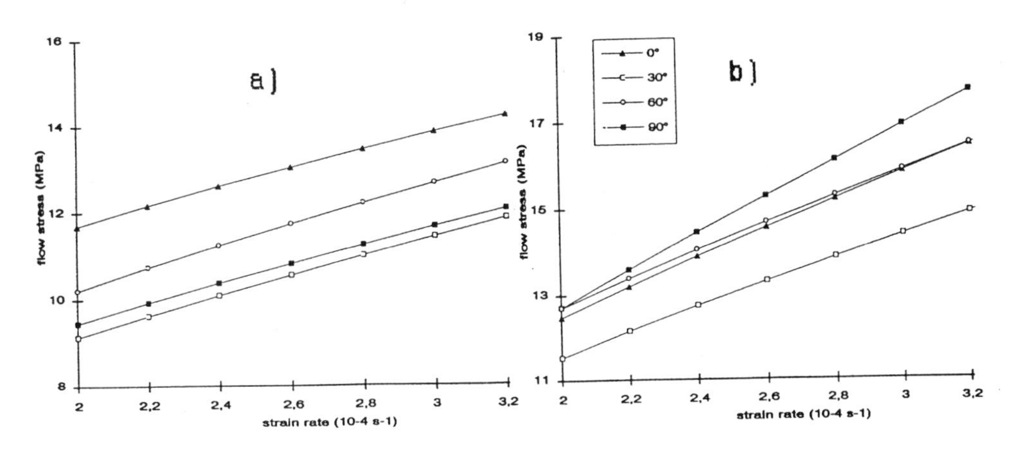

Figure 5 : Superplastic behaviour of TA6V slabs a) after 20% of strain b) after 100% of strain

Materials Science Forum Vol. 157-162 (1994) pp. 1365-1370

TEXTURE IN ZIRCALOY-4: EFFECT OF THERMOMECHANICAL TREATMENTS

L.P.M. Brandão and C.S. da Costa Viana

IME-SE/4, Instituto Militar Engenharia,
Praça Gen. Tibúrcio, 80-Urca, BR-22290-270 Rio de Janeiro - RJ, Brazil

Keywords: Zircaloy-4, Cold Rolling Texture, Annealing Texture

ABSTRACT :

The high plastic anisotropy of zircaloy-4 alloy is a function of its low slip systems options, its asymmetrical distribution and the strict crystallographic relationship for twinning. The principal deformation mechanisms are prismatic slip, pyramidal slip (c+a type) and twinning. The deformation along the c-axis is produced by the latter two mechanisms. Nowadays, one of the most important discussions is about the ease with which each mechanism occurs. All of this is important because it is well known that the plastic behavior of this alloy is associated with its crystallographic texture which is a function of its deformation mechanisms. The aim of this work was to investigate the texture development in Zircaloy-4. This work was carried out in two stages. First, the relation of cold rolling to texture was evaluated for several reduction factors varying from 30 to 87.7%. Subsequently, the cold worked material was annealed 550 and 750°C followed by air cooling.

INTRODUCTION

Texture in hexagonal materials has been extensively studied [1-4] and specifically in zirconium alloys some works are notworthy[1,2,4]. In the present work an attempt is made to explore the effect of cold rolling and annealing on the development of texture in Zircaloy-4 sheet material.

EXPERIMENTAL PROCEDURE

The Zircaloy-4 was received as a cold rolled thin sheet with 2.1 mm thickness. The as-received material was annealed at 750°C in vacuum followed by air cooling with subsequent cold rolling at room temperature. The cold rolling produced specimens with 30, 40, 50, 60, 70 ,80% and 87% reductions. Specimens from each reduction were submitted for texture analysis. The cold rolled materials were annealed in two ways. The room temperature cold rolled samples were annealed at 550°C and 750°C. The textures were quantified via the crystallite orientation distribution function (CODF) with a series expansion of l=20.

RESULTS

ROOM TEMPERATURE ROLLING

The main features of the texture of the cold rolled at room temperature material are presented in table 1. The CODFs of the 50 and 87% cold rolled materials are shown in figure 1.

Table 1. CODF of the room temperature cold rolled material.

Reduction	Main Orientations	CODF Height	TSP
30%	$(\bar{1}014)[1\bar{2}10]$	6.94	1.50
40%	$(\bar{1}014)[1\bar{2}10]$	7.11	1.53
50%	$(\bar{1}014)[1\bar{2}10]$	7.05	1.58
60%	$(\bar{1}014)[1\bar{2}10]$	7.33	1.67
70%	$(\bar{1}014)[1\bar{2}10]$	7.53	1.69
80%	$(\bar{1}014)[1\bar{2}10]$	7.97	1.72
87%	$(\bar{1}014)[1\bar{2}10]$	8.80	1.79

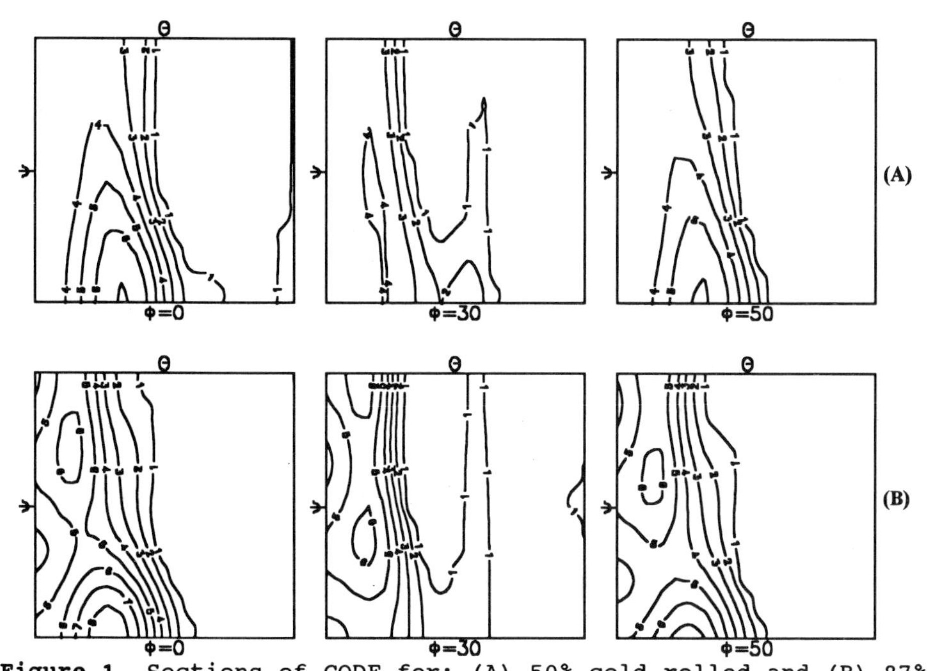

Figure 1. Sections of CODF for: (A) 50% cold rolled and (B) 87% cold rolled Zircaloy-4.

It can be observed, in table 1, that the CODF maxima occur for the same orientations for all reductions. The principal texture components define a tube of orientations that runs from $(\bar{1}014)[1\bar{2}10]$ at $\phi=0°$ to $(\bar{4}31\ 14)[\bar{1}\bar{1}20]$ at $\phi=50°$ for the 87% cold rolled material. A second tube near $(\bar{1}018)[7\bar{3}\bar{4}1]$ at $\phi=0°$ and $(\bar{1}01\ 16)[4\bar{7}31]$ at $\phi=50°$, for the 87% cold reduction, appears for reductions greater than 50%. This tube appears in the 60% cold rolled material with an intensity of 6.3 and reaches an intensity of 8.8 for the 87% reduction. Both of these tubes were near (0002), at 25° and 10° away, respectively. The texture severity parameter (TSP) increases with the cold reduction in the range examined.

ZIRCALOY-4 ANNEALED AT 550°C

The main results of the CODF analysis for the material cold rolled at room temperature and annealed at 550°C are presented in table 2. The CODF of the 80% cold rolled and annealed material is shown in figure 2.

Table 2. Zircaloy-4 cold rolled and annealed at 550°C .

Reduction	Main Orientations	CODF Height	TSP
40%	$(\bar{1}014)[1\bar{2}10]$	4.80	1.29
60%	$(\bar{1}014)[1\bar{2}10]$	5.49	1.38
70%	$(\bar{1}014)[1\bar{2}10]$	5.54	1.49
80%	$(\bar{1}014)[1\bar{2}10]$	6.02	1.58

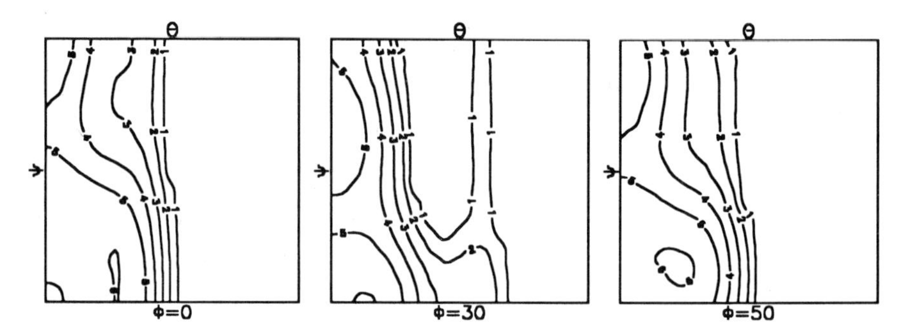

Figure 2. Sections of CODF for Zircaloy-4 80% cold rolled at room temperature and annealed at 550°C.

These results show that the CODF maxima of the 550°C annealed sample occurred for orientations similar to those of the cold rolled materials but with smaller intensities. The smaller values of the TSP obtained for the annealed materials also confirm this assertion. The decrease of these two values (maximum function height and TSP) allows the conclusion that the annealing treatment weakened the cold rolling texture.

The CODF analisys of the results showed that the principal texture components define, for 40% deformation, a tube of orientations that runs from $(\bar{1}014)[1\bar{2}10]$ at $\phi=0°$ to $(\bar{3}2115)[\bar{1}\bar{5}60]$ at $\phi=50°$. The intensities, however, are less than for the cold rolled materials. The observed second tube in the cold rolling textures for reductions greater than 50% disappeared with annealing. For the 80% reduction a new peak near $(0001)[1\bar{1}00]$ was seen to emmerge, at $\theta=0°$ in all ϕ sections of the CODF which reached an intensity value of 5.8.

ZIRCALOY-4 ANNEALED AT 750°C

The material cold rolled at room temperature between 40 and 87% was annealed at 750°C. The principal results of the texture analysis of these materials are shown in table 3. The CODF of the annealed material previously cold rolled at 80% is presented in figure 3.

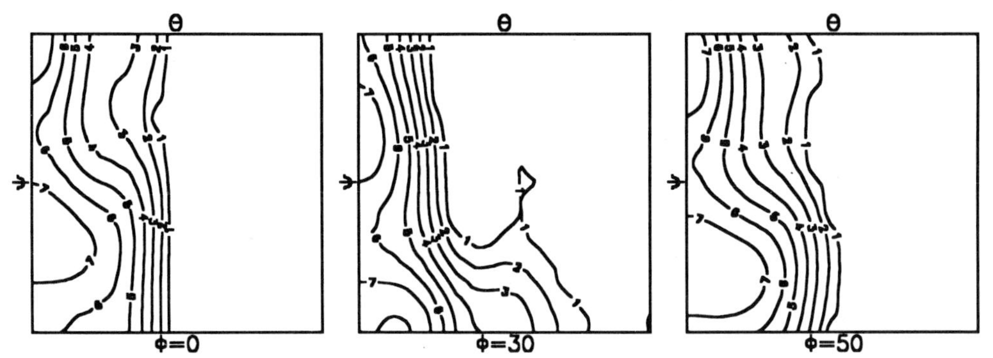

Figure 3. Sections of CODF for Zircaloy-4 80% cold rolled at room temperature and annealed at 750°C.

Table 3. Zircaloy-4 cold rolled and annealed at 750°C texture results.

Reduction	Main Orientations	CODF Height	TSP
40%	$(\bar{1}014)[1\bar{2}10]$	5.78	1.38
50%	$(\bar{1}014)[1\bar{2}10]$	5.53	1.41
60%	$(0001)[1\bar{1}00]$	6.11	1.56
70%	$(0001)[1\bar{1}00]$	6.55	1.58
80%	$(0001)[1\bar{1}00]$	7.90	1.76
87%	$(0001)[1\bar{1}00]$	8.38	1.79

These results show a change of behaviour when the amount of reduction reaches 60%. For 40% and 50% reductions the maximum CODF height occured for the $(\bar{1}014)[1\bar{2}10]$ component. The maximum CODF height for reductions equal to or greater than 60% moved to the $\{0001\}<1\bar{1}00>$ component. The TSP and the CODF peak became greater with increasing reduction. The observed cold rolled second tube also disapears with annealing

As was observed for the material annealed at 550°C, the 40% and 50% rolled samples retained the same cold rolling peaks on annealing. An additional peak at $\{0001\}<1\bar{1}00>$ can also be observed in the 750°C annealed material with intensities of 4.2 and 5.2 at 40% and 50% cold reduction, respectively.

Comparing the TSP of the annealed samples to those of the cold rolling textures shows that, for 80% and 87%, the annealing testures were sharper, i.e. no weakening effect was observed in this case.

CONCLUSIONS

- The main component of Zircaloy-4 cold rolled at room temperature in the reduction range between 30 and 87% was identified as $(\bar{1}014)[1\bar{2}10]$;
- The peak height increases with cold rolling for both cold rolling and annealing textures;
- For reductions higher than 50% a second important component, $(\bar{1}018)[7\overline{34}1]$, develops, increasing with the extent of cold rolling deformation;
- The CODF of the material cold rolled at room temperature and annealed at 550°C shows a maximum at $(\bar{1}014)[1\bar{2}10]$, being, therefore, similar to the cold rolling texture. The TSP is, however, smaller for the annealed specimen;
- Two distinct behaviours were observed for the cold rolled-and-annealed at 750°C material: (i) for reductions up to 50% the main component was observed to be $(\bar{1}014)[1\bar{2}10]$; (ii) for reductions higher than 50% the main component is $(0001)[(1\bar{1}00]$. In both cases the TSP increases with cold reduction.

REFERENCES

1) Tenckhoof,E.: Zirconium in the Nuclear Industries, ASTM STP 754, 1982, 5.

2) Kallstrom, K.: Can. Met. Quaart., 1972, 11, 185.

3)Knight, F.: PhD Theses, University of Cambridge, 1978.

4) Tenckhoof,E.: ASTM STP 966, 1988,1.

Materials Science Forum Vol. 157-162 (1994) pp. 1371-1378

TEXTURE OF ELECTROPLATED COATINGS OF COPPER AND BISMUTH TELLURIDE

H. Chaouni [1], P. Magri [2], J. Bessières [1], C. Boulanger [2] and J.-J. Heizmann [1]

[1] LMPC/ISGMP, Université de Metz, Ile du Saulcy, F-57045 Metz Cédex 1, France

[2] LEM Université de Metz, URA CNRS 158, F-57045 Metz Cédex 1, France

Keywords: Electrodeposition, Copper, Bismuth Telluride, Texture, Pole Figure

ABSTRACT

Electroplating is the process of depositing a thin layer on a surface by means of electrolysis. Its purpose is to alter the surface characteristics in order to improve apparance, resistance to abrasion, ability to withstand corrosive agents or to obtain other desired properties.

In the present work Copper and Bismuth telluride coatings are studied. For both materials the present work evidences that, the substratum nature, the various electrochemical parameters as bath compositions, surface preparation and current densities have a great influence on the texture of the deposited layers.

I INTRODUCTION

Electroplating may be defined as the production of metal coatings on substratum in order to combine desired surface properties, as lubricant in the drawing process or protective layer for instance.

Copper and Bismuth Telluride electrodeposits are suitable to demonstrate the relationship between texture and plating bath conditions.

II COPPER COATING

An electroplating system is used for copper deposition from pyrophosphate solution, rather than sulfate solution because the effiency is better and the copper layer is more adhesive. This system consists essentially of a plating bath, a source of current densities and two electrodes. Under the influence of the current, reactions occur at the electrodes, the cations migrate to the cathode where they are deposited as metallic copper. During the coating, whatever is the process (PVD, CVD or Sputtering) usually the densiest lattice planes are deposited parallely to the surface of the substratum. The formation of the coating is then governed by two processes which are the nucleation and the growth of the grains.

Experimental procedure

All the copper film are deposited on a low carbon steel with keeping the same surface preparation, including at the end an electropolishing in order to avoid small asperities on the surface

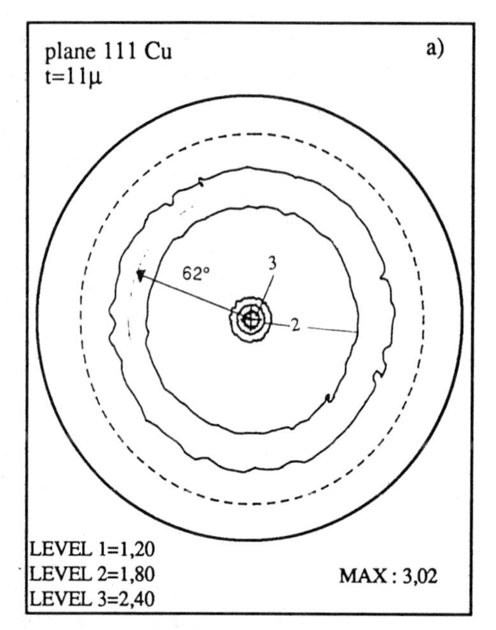

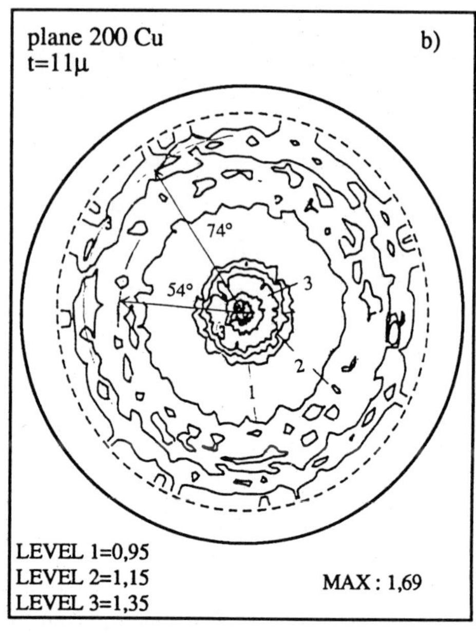

Figure1 : 11μm thick copper layer deposited on steel
a) (111) pole figure
b) (200) pole figure

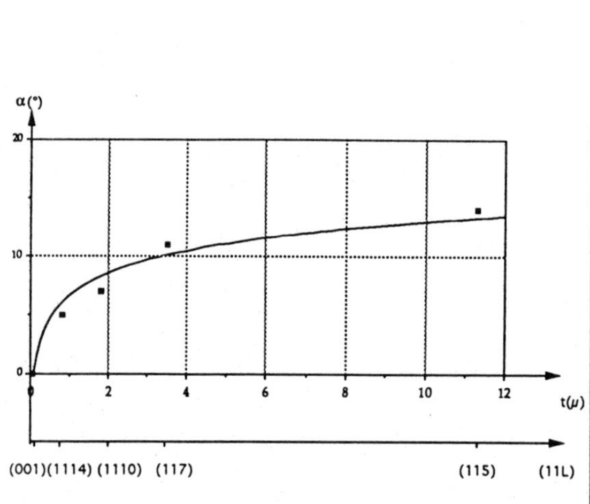

Figure 2 : evolution of angular tilt α with respect to the thickness of the layer

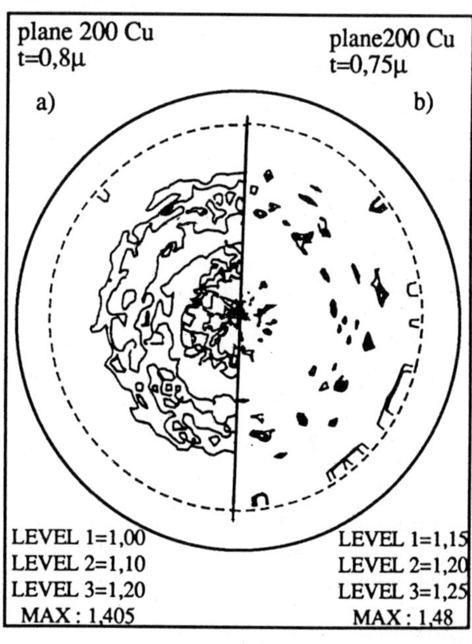

Figure 3 : (200)pole figure of 0,75μm thick copper deposited on :
a) as reseived rolled steel
b) the 50% more rolled steel

leading to irregularities of the electric field (the substratum must be thoroughly clean and free of any surface films).

Deposited thicknesses are ranged between 1 μm and 12 μm.They are measured within 5% by the increasing of mass of the sample or by the duration of the deposition time because the Faraday's law is fully verified.

General observation

(111) and (200) pole figures are measured. All the presented pole figures are in normalised scale and intensity correction has been made according to the thickness of the layer. Whatever is the thickness pole figures can be compared because the reference diffracting volume is the bulk.
Tracy and al [2] indicate that <111> fiber, <200> and <115> fibers are the most important observed texture whatever the deposition process is. The main orientation being always the <111>. Their sharpness depending on the underlayer texture.

-Influence of the thickness

Figure (1a) shows the (111) pole figure of 11 μm thick layer.The central peak belongs to the fiber<111>, but the 70° ring seems absent. So one or several other fiber textures must be present to explain the ring located at 62°. The (200) pole figure presented figure (1b) shows a (200) ring located at ψ=15°, and two other rings weaker than previous one located at 54°and at 74°. The 74°ring and the 15° one are corresponding to a <115> fibrous texture whereas the 54° belong to the previous<111> fiber.

When the thickness of the copper layer decreases, the diameter of first (200) ring decreases too.

Figure (2) gives the angular tilt α between the (100) planes and the surface of the sample, with respect to the thickness. As the thickness decreases the fibrous axis <11L> approximates the <001>one.

On very thin evaporated copper films less than 650A, deposited on NaCl, Heizmann and al [1] have observed that (100) planes are mainly (more than 75%) parallel to the surface whereas the <111> component is very weak (4%).

- Influence of current density

Preliminary studies were made with a Hull testing electroplating cell, to determine the current densities able to give a suitable layer. With current densities ranged between 3,5 to 5A/dm2 used to produce a 1,2 μm thick copper film, no important change in the texture components were observed.

- Influence of substratum

The same deposition conditions were used to prepare a copper film 0,8 μm thick on a as received cold rolled steel and on the previous steel 50% more rolled. Their surfaces were prepared in the same way (diamond polishing followed by a light electropolishing).
The main difference between the two texture observed on the (200) pole figures is: the two copper layers have the orientations seen previously nevertheless on the 50 % rolled steel the grains formed during the growth of the layer are larger than the grain formed on the as received steel (figure 3a,b).

III BISMUTH TELLURIDE COATINGS

The Bi_2Te_3 and its derived doped components are very interesting material in the field of thermoelectric behaviour at ambiant temperature.

A specific electrochemical synthesis of Bi_2Te_3 has been developped [3] to be able to produce a wanted thickness of low cost thin film on a wide suface. The electric properties of the coating will depend on the orientation and the homogeneity of the film because Bi_2Te_3 has R(-3)m structure with c/a higher than 7 (Hexagonal cell). This ratio can induce strong anisotropic effects on conductivity and resistivity.

Bi_2Te_3 electrochemical deposition is performed in acid baths according to the reaction:

$$3\ HTeO_2^{+} + 2\ BiO^{+} + 18\ e^{-} + 13\ H^{+} \rightarrow Bi_2Te_3 + 8\ H_2O.$$

The Faraday's efficiency is approximately 100 % under normal conditions.

Crystallographic arrangement can be described as a pilling up of atoms layers in the order [4] :

Te(1) - Bi - Te(2) - Bi - Te(1)

The Te(1) - Te(1) bond is insured by van der Waals forces, which create a marked anisotropy in the mechanical properties of the material like cleavage planes in a direction perpendicular to the c axis, and in the physical properties.

The knowledge of texture and electrochemical conditions are important to the developpement of efficient materials for thermoelectric systems.

Experimental procedure

A specific electroplating system is used for bismuth telluride deposition. $\theta-2\theta$ diffractogramm shows only Bi_2Te_3 and/or $Bi_{2+x}Te_{3-x}$ which have very close structure.

General observation

Fiber textures are always observed. According to the thicknesses, the substratum and the current density <00.1>, <10.10>, <11.0>, <01.8>, <01.5>, <10.0>fiber axes normal to the surface can be present. These fibers are presented on (01.5) pole figures (fig a,.....h). All the fiber orientations observed are corresponding to dense lattice planes of Bi_2Te_3 parallel to the surface.

In spite of these great number of orientations which seem appear in a hazardous way, the growth of the layer and the induced orientations can be understood by texture measurements and scanning electron microscopy observations.

Growing of the orientation

Pole figures

Keeping the same thikness of the layer (5,3 μm), poles figures show the complexity of the orientations formation.

These layers are deposited on the same substratum which is either an aluminium rod or a stainless steel rod.

The first pole figure (a,b) shows the influence of the substratum : <01.5> fiber is observed on Aluminium whereas <10.10> is observed on steel.

The following pole figure (c,d) indicates that density of current may influence the orientation. <01.8> fiber texture instead of <01.5>on Aluminium and <00.1> instead of <10.10> on steel.

Pole figure (e,f) shows the influence of the surface finish of the sample. On the well polished (1μm diamond paste) there is a <01.8>fiber whereas on a standart polishing until 3 μm diamond paste <10.10> fiber is present.

The last pole figure (g.h) measured on thick layers (20 μm) deposited either on aluminium or steel shows the same fiber orientation <11.0>. When the thickness increases the same orientation appears wathever the substratum .

At the beginning of the electroplating when the thickness is thinner than 1 μm the <01.5> fiber is observed on stainless steel.

S.E.M. observation

On figure 5 are presented, with the same magnification, the surface of the layers of Bi_2Te_3 observed during the electroplating process: pictures are taken when the layers have reached the same thickness on the two different substratums.

Pictures **a** to **d** are taken on the surface of layers which are grown on aluminium. Pictures **e** to **h** are taken on steel. At the beginning (0.3 μm) there is nucleation of Bi_2Te_3 on the substratum, the nuclei grow to give rose shaped crystals either on steel or aluminium. Nevertheless one can notice the same size of the crystals on steel whereas on aluminium large and very small crystals are observed. This could be explained by an insulating effect of Al_2O_3 thin film which inhibit the electroplating process and leads to islands of growth.

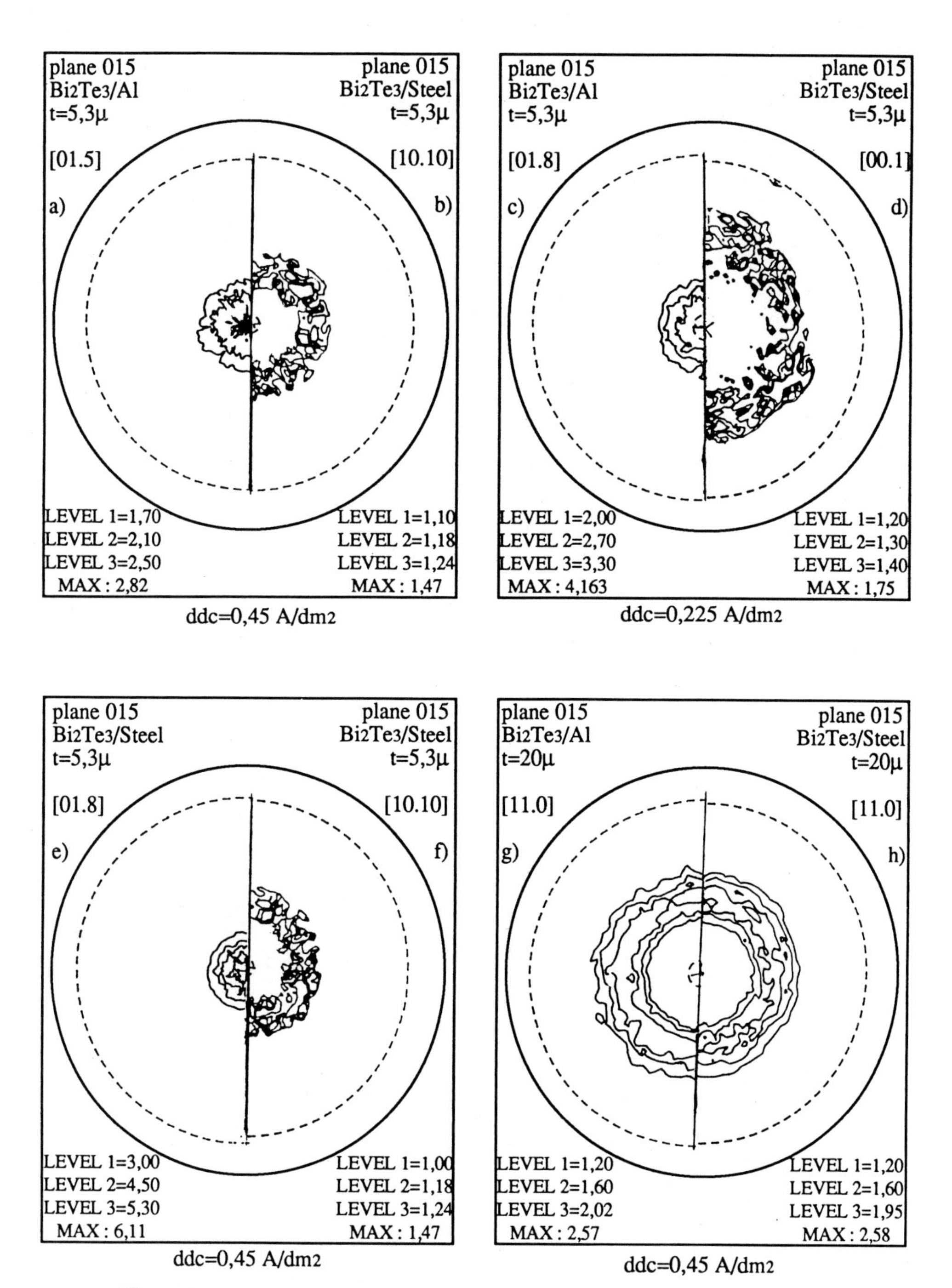

Figure 4 : (01.5) pole figures of Bi_2Te_3 layer 5μm thick (a to f) and 20μm thick (g,h) deposited on aluminium substratum (a,c,g) and stainless steel substratum (b,d,e,f,h)

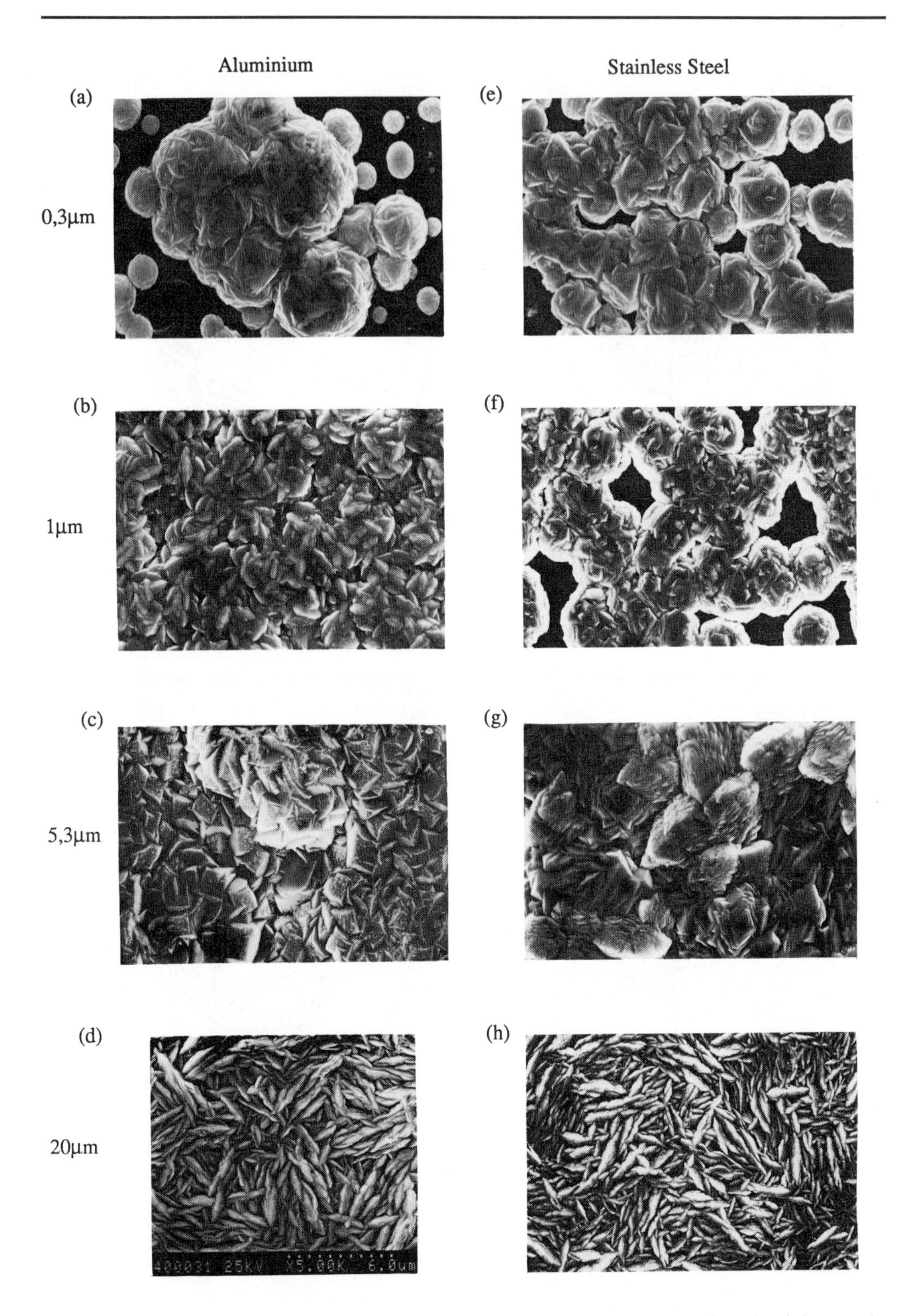

Figure 5 : Bi_2Te_3 layer deposited on aluminium substratum (a,b,c,d) and stainless steel (e,f,g,h); 0,3μm (a,e); 1μm (b,f); 5μm (c,g); 20μm (d,h) thick

When the thickness of the layer is not very important, the nucleation is depending on the substratum and on its surface finish. Nevertheless orientations found are <00.1> fiber axis or fiber axis close to <00.1> as: <01.8>,<10.10>- - - -. Three fold axis which is characteristic of the c axis, can be seen normal to the pictures (5 a,e,f).

After growing of the nuclei, there is a new type of nucleation. New nuclei are formed on the previous crystals. These new nuclei grow with their (110) plane parallel to the surface of the substratum: two fold axis of these nuclei are well seen on picture 6.

At this step of the growing (about 1µm) there is a change in the growing conditions, the new nuclei are now rather formed on the layer itself than on the substratum wich is becoming entirely coated by the electrodeposited layer. This leads to the <110> fiber texture seen picture **d** and **h.**

Conclusions

For both materials which are strongly different electrodeposition seems follow the same rules. At the beginning, orientations of the layers are greatly influenced by the nature of the substratum, its surface finish and the electroplating conditions. After few micrometer or less (depending on the layer and on the substratum properties) a new orientation appears. It appears when there is electrodeposition on the layer itself, leading then to a negligible influence of the substratum.

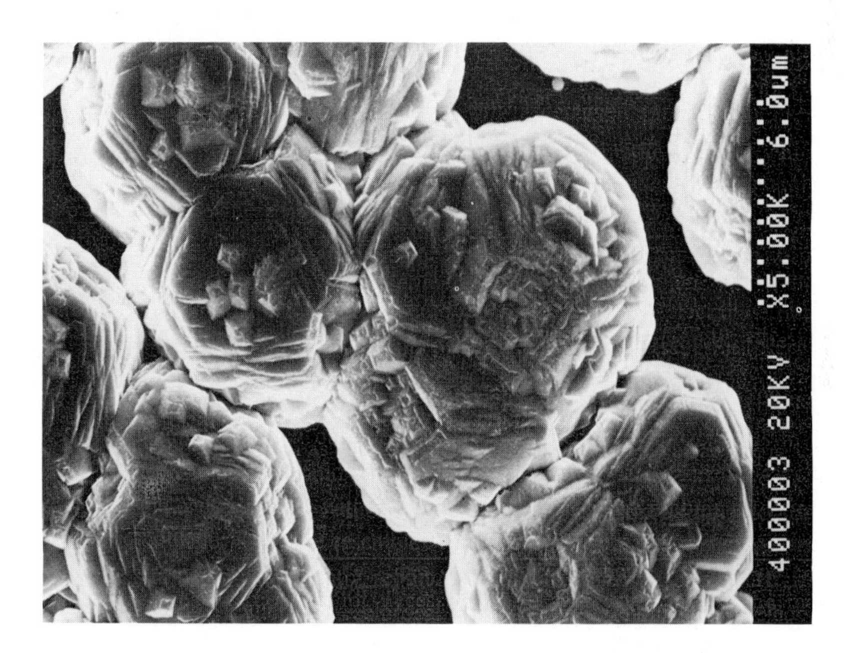

Figure 6 : 1µm thick layer of Bi_2Te_3 deposited on stainless steel

References

1)-Heizmann,J.J. Schlatter,D. Vadon,A. Baltzinger,C. Bessieres,J. : Text.and microstructure, 1991, 14-18, 18
2)-Tracy,D.P. Knorr,D.B : J.Electronic Materials, 1993, 22, 611
3)-Magri,P.: DEA, 1992, METZ
4)-Wiese,J.R. and Muldawer,L : J.Phys.chem.Solids, Pergamon Press, 1960, 15, 13

Materials Science Forum Vol. 157-162 (1994) pp. 1379-1386

X-RAY TEXTURE ANALYSIS IN FILMS BY THE REFLECTION METHOD: PRINCIPAL ASPECTS AND APPLICATIONS

D. Chateigner, P. Germi and M. Pernet

Laboratoire de Cristallographie, CNRS, BP 166, F-38042 Grenoble Cédex 09, France

Keywords: Film Texture Analysis, Defocusing Correction, High-T_c Superconductors

ABSTRACT

We review the principal characteristics of defocusing correction in thin films texture analysis using the Schulz reflection technique. Consequences are examined on $YBa_2Cu_3O_7$ films compared to classical bulk-like corrections. Principal effects are revealed in the high tilt angle region and, for highly oriented films ($YBa_2Cu_3O_7$ films on MgO), the film-like corrections offers better precision on the quantitative evaluation of the different parts of the texture than the bulk-one.

INTRODUCTION

Texture analysis of polycrystalline compounds have been proved of great interest to explain physical properties of materials, especially if these latter are highly anisotropic. A random distribution of crystallite orientations giving oftenly averaged physical properties, the texturation is needed to improve them for optimal utilizations. The characterization of the resulting textures is consequently of crucial importance. Numerous methods have been employed to improve the texturation either of powdered bulk samples (rolling, uniaxial pressure, magnetic alignment, crystalline growth ...) or thin film compounds (laser ablation, sputtering, chemical vapour deposition ...). In the case of the Schulz reflection method [1] (which we use in our experiments) these two kinds of samples have to be considered differently since the irradiated volumes do not evolve identically. Thin films texture analysis using this method provides more informations than classical experiments like rocking curves or φ-scan procedures. This technics gives information from the total thickness of the film, but the usual intensity corrections have to be modified for taking into account the enhancement of the irradiated volume during the tilt of the sample. We remind here the principle of intensity corrections for defocusing in the case of thin films in view to establish the true pole figures.

The recent development of high Tc superconductors has given rise to a great interest in their texture analysis, particularly for thin films in order to connect a preferred crystallographic orientation to physical properties. On the other hand we show how we can estimate the relative proportion of the principal textures existing in $YBa_2Cu_3O_7$/MgO superconducting samples from the operated corrections.

EXPERIMENTAL

The texture analyses have been realized with a Courbon S.A. (St. Etienne) texture sample holder mounted on a horizontal θ-2θ goniometer (GMI-Grenoble). We use the $Cu_{K\bar{\alpha}}$ radiation supplied by a Rigaku RU300E rotating anode which offers up to 60kVx300mA in power. The incident beam is monochromatised by the (002) reflection of a flat graphite (Huber-Rimsting) specimen of 0.4° mosaïc spread, which is essential for the study of thin films deposited on single crystalline substrates [2]. The beam is collimated by two crossed slits of 0.84mm large and 0.6mm high. The diffracted line is detected through a horizontal slit of 1mm by an Inel proportionnal counter. The signal is finally filtered by an SCA Ortec tension switcher before the data are collected in a Compaq 386/20 computer. The correction and pole figure layout is realized with a Compaq 486/50 computer using our own programs (Cortexg, Phiscan and Pofint) written in Turbo Pascal 5.5.

Since we currently observe strong textures in oxide films [3] the step angles choosen for the pole figure scans have to be small (0.5° or less), in view to avoid wrong interpolated intensity levels. Consequently isointensity lines are not useful and we draw direct pole figures where only experimental points are represented. The studied samples have been elaborated by MOCVD at LMGP-Grenoble (YBCO films). This technic is well described elsewhere [4].

The randomly oriented bulk sample of $YBa_2Cu_3O_7$ (YBCO) has been synthezised by classical solid state reactions starting with the 97% pure oxide powders (Y_2O_3, BaO and CuO). The random orientation of the resulting powders has been controled by comparing their θ-2θ spectrum with the theoretical ones. The agreement was satisfactory, by less than 2% of difference for the majority of the peaks below θ=40°. The defocusing curves realized on these samples were measured by integrating the diffracted intensity during one rotation of the sample around its normal, φ, for each tilt position, χ. The orthorhombic symmetry with 3.82Å x 3.88Å x 11.67Å unit cell parameters of $YBa_2Cu_3O_7$ implies a great number of neighbouring peaks. Consequently a relatively thin aperture of the detecting slit is needed and we recorded the defocusing curves with a high counting time of several minutes for each χ position to improve the statistics (in order to get precisely peak and background curves at high χ values).

THEORETICAL BACKGROUND

In the following we develop the useful expressions for the correction of pole figures in the case of Schulz reflection technics for the study of thin compounds. On our diffractometer the value χ=0° is define when the φ axis is colinear with the diffraction vector, **k**, and χ=90° when **k**⊥φ. For high enough χ values it is well known that defocusing occurs causing an intensity loss by peak broadening and irradiated surface increasing [5,6]. Usually this effect is corrected by the experimental set up of defocusing curves obtained from bulk samples whithout preferred orientation [7]. Such samples are generally feasible, and offer the possibility to correct pole figures up to χ≈80°. On the other hand, making untextured thin compounds is practically impossible, even on polycrystalline or amorphous substrates [8,9]. Such undesired textures can be promoted by interaction with the substrate (particularly with well matched interfaces) or the result of anisotropic growth speed of the crystallites. De facto, it becomes impossible to correct pole figures made on films. One solution consists in an analytical correction of the data using the experimental bulk defocusing curves. This means that the calculated curves for layers are available with the same experimental arrangement and material as for the bulk.

Films

Wenk & al. [10] proposed for the first time a formulae adapted to the correction of single layer deposited on substrates, based on the previous work of Schulz, and applied it to polycrystalline films of silicon. We take again here this formulation which gives the defocused intensity of the film, $I^f(\chi)$, from the one of the bulk, $I^b(\chi)$:

$$\frac{I^f(\chi)}{I^b(\chi)} = 1 - \exp\left(\frac{-2\mu T}{\sin\theta\cos\chi}\right) \qquad (1)$$

Here θ is the Bragg angle of the considered (hkl) reflection, T and μ are respectively the thickness and the linear absorption coefficient of the film.
Due to the exponential term the defocusing curve of the film does not content a constant region for its low χ part , like the bulk generally exhibits. In addition we can remark that the allowable domain of the pole figure is extended in the high χ region since the irradiated volume increases with χ.

Substrates

If the substrate is crystalline, it would be interesting to study the influence of its texture on that of the layer. Here also, the pole figures cannot be corrected directly like a simple bulk (uncovered) since there is an absorption of both incident an diffracted beam into the film. We proposed recently the following expression for the correction [11] :

$$\frac{I^s(\chi)}{I^{bs}(\chi)} = \exp\left(\frac{-2\mu T}{\sin\theta_s\cos\chi}\right) \qquad (2)$$

$I^s(\chi)$ and $I^{bs}(\chi)$ are respectively the defocused intensity of the covered and uncovered substrate, while θ_s is the Bragg angle for the considered peak of the substrate.
Since the absorption of the beam increases with χ, the pole figure is less extended on its periphery than for an uncovered bulk.

Multilayers

The particular physical properties of multilayer compounds had given interest in the microscopic and macroscopic comprehension of their structure, where texture analysis may play an important rule of characterization. By combining equations (1) and (2), we can refine the appropriate correction form needed in this analysis. For a multilayer, the jth layer from the top of the sample is diffracting an intensity, $I_j(\chi)$, which has to be corrected by :

$$\frac{I_j(\chi)}{I_j^b(\chi)} = \left[1 - \exp\left(\frac{-2\mu_j T_j}{\sin\theta_j\cos\chi}\right)\right]\exp\left(\frac{-2\sum_{i=1}^{j-1}\mu_i T_i}{\sin\theta_j\cos\chi}\right) \qquad (3)$$

$I_j^b(\chi)$ is the bulk defocusing curve for the material constituting the jth layer at the Bragg angle θ_j. The indice i entering the summation refers to all the layers which are deposited on top of layer j, and the summation represents the absorption along these layers. The principal requirement here is the obligation to analyse a peak which is sufficiently separated from the others.

From (3) we can deduce the correction formulae for different types of multilayers. Let us consider the most current type, which consists of two kinds of materials, A and B, deposited J times one AB unit on the other periodically stacked (means $T_A=T_B$). We call this compound $(AB)_J$ multilayer. The first deposited layer is B (interface) and the last is A (top of the sample). In such configuration, equation (3) gives :

$$\frac{I_A(\chi)}{I_A^b(\chi)} = K_A(\chi)\left(1 + \sum_{i=1}^{J-1}\exp\left[\frac{-2i(\mu_A T_A + \mu_B T_B)}{\sin\theta_A\cos\chi}\right]\right) \qquad (4)$$

and

$$\frac{I_B(\chi)}{I_B^b} = K_B(\chi)\sum_{i=1}^{J}\exp\left[\frac{-2[i\mu_A T_A + (i-1)\mu_B T_B]}{\sin\theta_B\cos\chi}\right] \quad (5)$$

$$\text{with} \qquad K_X(\chi) = 1-\exp\left(\frac{-2\mu_X T_X}{\sin\theta_X\cos\chi}\right) \quad (6)$$

Here X refers to the considered phase A or B. The b exponent denotes the bulk defocusing curve of the X phase. Pursuing for more general cases, we can obtain the correction expression for three or more different layers in each unit of the stack. If the stack unit is composed by K different layers, and deposited J times, with k in the intervall [A..K], we can write :

$$\frac{I_X(\chi)}{I_X^b(\chi)} = K_X(\chi)\left(\delta_X + \sum_{i=1}^{J-\delta_X}\exp\left[\frac{-2}{\sin\theta_X\cos\chi}\left(i\sum_{k=1}^{x-1}\mu_k T_k + (i-1+\delta_X)\sum_{k=x}^{K}\mu_k T_k\right)\right]\right) \quad (7)$$

with $\delta_X=1$ if k=1 and $\delta_X=0$ for k>1.

For sufficiently thin layers (relatively to the linear absoption coefficient) the exponential term can be developped and the $K_X(\chi)$ coefficient can be replaced by [12] :

$$K_X(\chi) = \frac{-2\mu_X T_X}{\sin\theta_X\cos\chi} \quad (8)$$

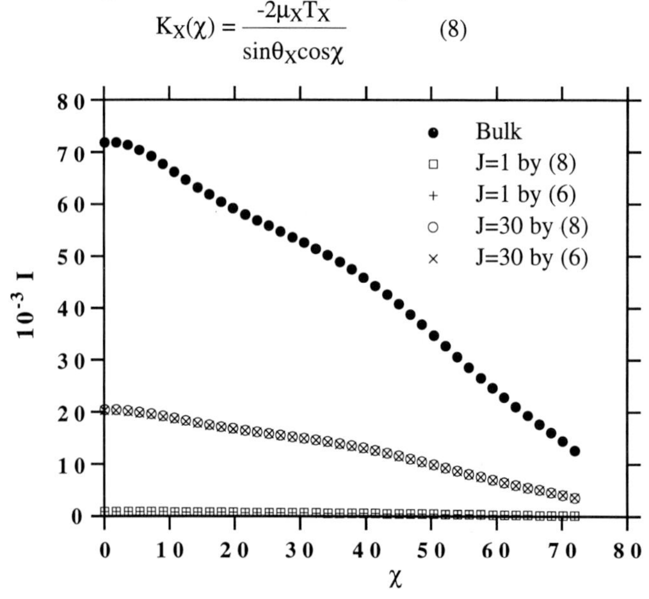

Fig.1 : Defocusing curves for the (103) reflection of the YBaCuO phase at θ=16.44° for an YBCO/YSZ multilayer sample. The layer type curves have been calculated through the equations (6) and (8) for two number of deposited AB unit, J=1 and J=30. The thicknesses of the layers are 50Å for both YBCO and YSZ phases and we calculated $\mu_{YBCO}=0.10915\mu m^{-1}$ and $\mu_{YSZ}=0.06667\mu m^{-1}$.

A comparison between the employment of (6) and (8) is made on figure 1 in the case of a YBCO/YSZ multilayer sample, for J=1 and J=30. The thickness of the two phases are 50Å and we made the calculation for the (103) reflection of YBCO. The linear absorption coefficients are $\mu_{YBCO}=0.10915\ \mu m^{-1}$ and $\mu_{YSZ}=0.06667\ \mu m^{-1}$. The two formulations (6) and (8) give defocusing curves in very good agreement in this case.

The increase in number of (AB) units, J, tends rapidly to the same defocusing as for bulk, and consequently needs an identical correction. On figure 2 are represented the correction curves,

$I(\chi)/I(0)$, of the same sample as figure 1. We clearly see that for ten AB units and more, these curves are quite identical for the experimental bulk and for the calculated one. However this effect is lowered for thicker layers and higher μ coefficients.

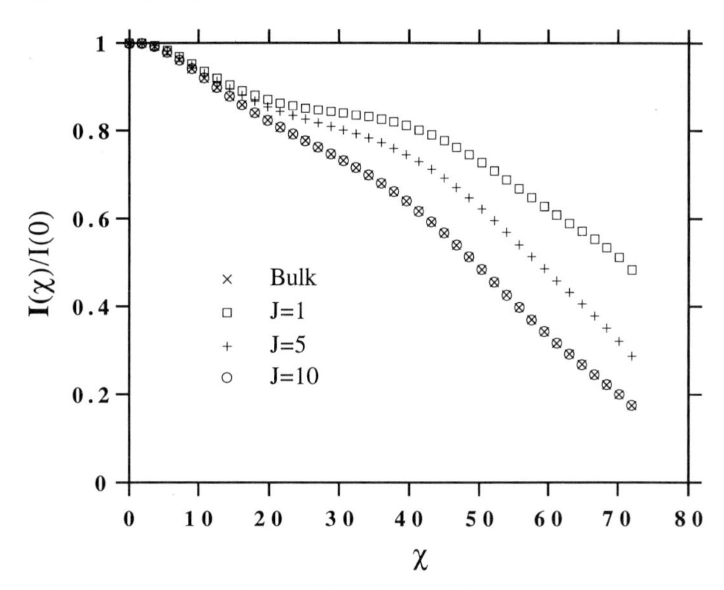

Fig.2 : Correction curves for the (103) reflection of the YBCO phase of the same sample as in figure 1. The number of deposited (AB) units is increasing. The calculated curves are given for J=1, 5 and 10 and are compared to the experimental bulk one. Notice the quite similar evolution of the bulk curve and for J=10.

EXAMPLES

Single crystalline MgO substrates take an important place in high temperature superconductor realizations because they exhibit a relatively low dielectric constant, and due to their cheapness in front of other possible single crystals. Deposited on such substrates, the YBCO phase shows physical properties comparable to the best one when the texture of the film is strong. The studied samples all exhibit critical temperatures over 85K, with transition width less than 2K. On such substrates the texture is characterized by **c** axes of the crystallites oriented perpendicularly to the substrate plane, and with the **a** and **b** parameters epitaxially aligned with MgO unit cell parameters (this texture is called $\mathbf{c}_{\perp 0}$). In such configuration one of the essential physical property, the critical current density Jc, can flow in the **ab** planes where it is about ten time greater than along the **c** axis. On the other hand Jc is greatly affected by misoriented crystallites which cause high angle grain boundaries [13,14]. Two types of misorientations can be easily produced in the case of MgO substrate. The first type, $\mathbf{a}_{\perp}$, corresponding to **a**-axis perpendicular to the surface, depends principally of the substrate temperature during the deposition. This orientation (with the $\mathbf{b}_{\perp}$ component) is well revealed by the examination of the (102/012) pole figure. The second type corresponds to $\mathbf{c}_{\perp}$ crystallites rotated by $\varphi°$ around **c** from the principal direction ($\mathbf{c}_{\perp\varphi}$) the most current occuring with φ=45°. This misorientation is explained by the formation of a near coincidence site lattice (NCSL) [15,16] between YBCO and MgO named Σ'=8 referring to the MgO lattice, and is shown by studying the (013/103) pole figure. The better statistic of the (013/103) diffraction compared to (012/102) made us choose the former for the evaluation of $\mathbf{c}_{\perp 45}$. Figure 3 is a multipole figure of the (012/102) $\mathbf{a}_\perp$ and $\mathbf{c}_\perp$ component (θ=13.90°), and (013/103) reflections (θ=16.35°) of a 2200Å thick film of YBCO on MgO, respectively delimited on the figure by the inner, outer and middle ring. Since only the (00l) and (h00) peaks appeared on the θ-2θ scan realized on this sample, the three respective pole figures were done in the angular range of interest with the following parameters : 27.45°<χ<39.15°, 50.65°<χ<62.35° and 39.15°<χ<50.85°, with Δχ=Δφ=0.45°. These zones are delimited by the green circles. For one second of integration time, we observed respectively maximum intensities of 345, 4170 and 24952 counts after correction. We integrated the total intensities of the (012/102) poles in a region of 8° in φ centered on the pole positions. After average on the four poles of each perpendicular component we found $\mathbf{a}_\perp/(\mathbf{a}_\perp+\mathbf{c}_\perp)$=10±1%, where $\mathbf{c}_\perp$ denotes the sum $(\mathbf{c}_{\perp 45}+\mathbf{c}_{\perp 0})$.

The figure 4 is a detailed view of the (013/103) pole figure of the same sample represented in a three dimensionnal φ-scan format. The horizontal and depth axes are respectively the φ and χ angles while the vertical axis represents the intensity level. This allows us to see clearly the $\mathbf{c}_{\perp 45}$ poles placed in this case far below the minimal intensity level of a classical pole figure (like in figure 3). After integration, we found $\mathbf{c}_{\perp 45}/\mathbf{c}_\perp$=0.90±0.06%. We estimated the error levels by integrating the empty parts of the corrected pole figures on the same angle range as the poles.

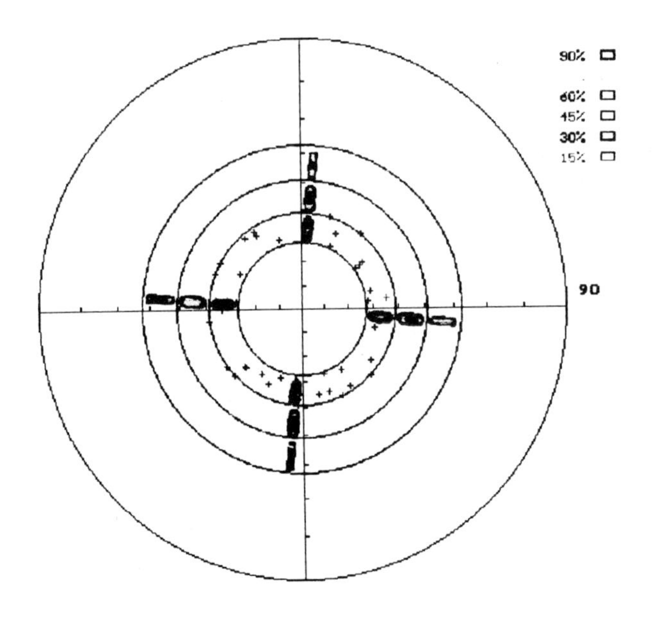

Figure 3 : Multipole figure of the (012/102)$\mathbf{a}_\perp$, (012/102)$\mathbf{c}_\perp$ and (013/103) reflections (resp. inner, outer and middle ring) of a 2200Å thick YBCO film on (100)MgO. θ=13.90° (012/102) and θ=16.35° (013/103). Note the presence of an important part of $\mathbf{a}_\perp$ oriented crystallites. (012/102)$\mathbf{a}_\perp$: Imax=345 cts, (012/102)$\mathbf{c}_\perp$: Imax=4170 cts and (013/103) : Imax=24952 cts.

It is clear that an accurate correction of the pole figures has to be carried out before to give quantitative appreciations of the relative parts of the texture. For samples with such strong textures the differences between film and bulk-like corrections are first sensitive on the maximal intensity value and consequently on the integrated ratios.

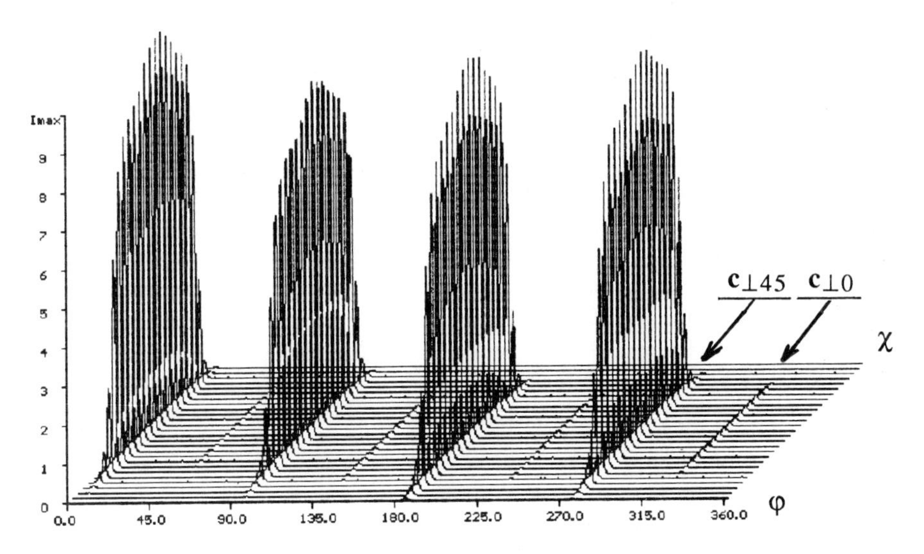

Figure 4 : Three dimensionnal φ-scan representation of the (013/103) pole figure of the fig.3. The $\mathbf{c}_{\perp 45}$ component is clearly visible. $39.15° < \chi < 50.85°$, $\Delta\chi = \Delta\varphi = 0.45°$. Green lines defines every 2.25° in χ.

For example, bulk-like corrections on the previous YBCO sample gives $\mathbf{a}_{\perp}/(\mathbf{a}_{\perp}+\mathbf{c}_{\perp})=9\pm2\%$ and $\mathbf{c}_{\perp 45}/\mathbf{c}_{\perp}=1\pm0.15\%$. While average value of the ratios are slightly shifted down and up respectively, both errors are doubled. The maximal intensities of the (012/102)$\mathbf{a}\perp$, (012/102)$\mathbf{c}_{\perp}$ and (013/103) pole figures increase from the previously found values up to 454, 6155 and 36096 cts respectively. Of course more differences between film and bulk-like corrections are observed in $I_{max.}$ for higher χ's, i.e. (012/102)$\mathbf{c}_{\perp}$ component. These effects are increasing with the thickness of the layer.

CONCLUSION

The texture studies of various thin compounds can be ruled out if particular corrections for defocusing and absorption are accomplished. We have detailed these later in some of the most current configurations of thin samples. Their application have been proved efficient for the quantitative determination of the principal components of the texture in highly textured YBCO films analyses. The effect of film-like corrections is more sensitive as the tilt angle becomes higher. In the case of multilayers we have shown that the correction to be applied becomes identical to the bulk one after some deposited unit.

ACKNOWLEDGEMENTS

The Alcatel-Alsthom industry is greatly acknowledged for its partial support in the thesis work of one of us (D. Chateigner). We also wish to thank C. Dubourdieu from the LMGP-Grenoble who provided the sample.

REFERENCES

1) Schulz, L.G.: J. of Appl. Phys., 1949, 20, 1030.
2) Wenk, H.R.: J. of Appl. Cryst., 1992, 25, 524.
3) Pernet, M. & al.: EMRS-92 proceedings, J. of All. & Comp., 1993, 195, 149.

4) Thomas, O. & al.: Int. Workshop on HTSC thin films, Rome, April 1991.
5) Chernock, W.P. & Beck, P.A.: J. of Appl. Phys., 1952, 23, 341.
6) Gale, B. & Griffiths, D.: Brit. J of Appl. Phys., 1960, 11, 96.
7) Feng, C.: J. of Appl. Phys., 1965, 36, 3432.
8) Iijima, Y. & al.: App. Phys. Lett., 1992, 60, 769.
9) Narumi, E. & al.: Japan. J. of Appl. Phys., 1991, 30, L585.
10) Wenk, H.R. & al.: J. of Appl. Phys., 1990, 67, 572.
11) Chateigner, D. & al.: J. of Appl. Cryst., 1992, 25, 766.
12) Chateigner, D. & al.: to appear in J. of Appl. Cryst.
13) Dimos, D. & al.: Phys. Rev. Lett., 1988, 61, 219.
14) Chisholm, M.F. & Pennycook, S.J.: Nature, 1991, 351, 47.
15) Gertsman, V.Y.: Scrip. Met. & Mat., 1992, 27, 291.
16) Suzuki, H. & al.: Physica. C, 1990, 190, 75.

Materials Science Forum Vol. 157-162 (1994) pp. 1387-1392

EXPERIMENTAL STUDIES OF TEXTURE DEVELOPMENT IN Co-Cr FILMS

L. Cheng-Zhang [1], J.C. Lodder [2] and J.A. Szpunar [3]

[1] On leave from The Institute of Computing Technology, Academia Sinica, PO Box 2704-6, Beijing, PR China

[2] MESA Research Institute, Faculty of Electrical Engineering, University of Twente, PO Box 217, NL-7500 AE Enschede, The Netherlands

[3] Department of Metallurgical Engineering, McGill University, 3450 University Street, Montreal, Quebec H3A 2A7, Canada

Keywords: Co-Cr Films, Texture, Structure, Internal Stresses

ABSTRACT

Specimens of Co-Cr thin films were chosen to investigate the role of the substrate, composition and thickness of the films as texture is developed. In general, the films investigated were strongly textured. The orientation ratio, ORc, was defined to describe the strength of the texture. Experimental data showed that for films 5-200nm, the ORc value increased with increasing thickness, while for another series of films 46-980nm, the texture strength decreases (starting at 130nm). Films having different compositions were investigated in the thickness range of 5-80nm. It was found that Co-Cr films with a higher Cr content exhibited a stronger texture and better magnetic properties than films with a lower Cr content. This situation is reversed for films thicknesses above 200nm. An optimal choice of substrate layer is also a key factor in determining the texture and magnetic properties of thin Co-Cr films. Our investigations demonstrated that a Ge/SiO_2/Si or Si substrate promotes the formation of the preferred (0002) texture, while Ge/Si and Si_3N_4/Si substrates were responsible for the deterioration of the (0002) texture.

INTRODUCTION

The excellent microstructure of the columnar grains and the strong C-axis texture of Co-Cr films makes them extremely useful as perpendicular recording media. Experimental data show that the texture and the magnetic properties of Co-Cr films are strongly influenced by the film thickness, the composition, the type of the substrate and/or seedlayer and the preparation methods.
In our previous papers [1,2], the texture evolution and the magnetic properties for two Series of $Co_{81}Cr_{19}$ films were studied as functions of the thickness in a range of 5-980 nm. It was demonstrated that fibre texture is observed in all tested films, and the (0002) planes are aligned parallel to the film surface.
In addition, experimental data show[1] that as the film thickness increase from 46 to 1200 nm, the crystal grains gradually change from having an equiaxial shape to a typically columnar one. However, the (0002) texture in the thinnest film is still stronger than the texture of a random distribution of the C-axis.
One factor, which affects the C-axis alignment and the surface morphology, is the structure of the substrate and seedlayer [3-7]. It was shown [3,5,6] that the C-axis of Co-Cr films can be aligned along either the direction perpendicular or parallel to the film plane, depending on the type and the thickness of the seedlayer. For instance, the use of an appropriate substrate and seedlayer, such as Ti, Au, Si and Ge [3,5,6], promotes the alignment of the C-axis along the film normal and the formation of the (0002) texture, while a choice of Cr seedlayer will promote the alignment of the C-axis in the in-plane direction and a (1120)Co ∥(100)Cr epitaxial orientation relationship [4,6]. In view of this, it is important to investigate the influence of different substrates and seedlayers on the development of the texture, the internal stress and magnetic behaviour for Co-Cr films.
Based on the these reasons, we will concentrate our attention in this report on investigating the influence of the Cr content and the substrate on the texture and magnetic properties of Co-Cr films.

EXPERIMENTAL PROCEDURE

Series A $Co_{81}Cr_{19}$ and Series B $Co_{77}Cr_{23}$ films were chosen to investigate the changes in the texture ith the thickness. These films were sputtered on $Ge/SiO_2/Si$ substrates. The thicknesses of the tested films range from 5 to 200 nm. In addition, another Series D $Co_{81}Cr_{19}$ films having the same thickness of 100 nm were selected to study the influence of the substrate and the seedlayer on the texture and magnetic properties. These films were sputtered on $Ge/SiO_2/Si$, Si_3N_4/Si, Ge(1)/Si and Si substrates . The rf voltage during sputtering deposition is 1600 V for most of the Co-Cr films , except for the $Co_{81}Cr_{19}$/Ge(1)/Si film. This last film was prepared by slowly increasing the rf voltage from 0 to 1600 V during 10 seconds. The film thickness was measured using SEM, STM (Scanning Tunnel Microscopy) and/or AFM (Atomic Force Microscopy). The magnetic properties were measured using VSM and a Torque magnetometer. The maximum applied field for measuring the hysteresis loop is about 840 kA/m.

The structure of the films was measured using Rigaku diffraction (Cu-Kα radiation). The texture of the same specimens was measured using a Siemens D-500 texture goniometer (Mo-Kα radiation), and the pole figure for hcp (0002) crystallographic planes was plotted. In addition, a more precise orientation distribution F(α) of the (0002) pole density was measured as a function of the specimen tilt angle α. During this measurement, the angular step was 0.1° and the data were collected in a 12° angular interval. The ratio ORc (α,β) will be used later as a measure of the texture strength. This ratio is defined by the following expression:

$$OR_c(\alpha,\beta) = \frac{\left(\frac{1}{\alpha_2-\alpha_1}\right)\int_{\alpha_1}^{\alpha_2}\int_0^{2\pi} F(\alpha,\beta)\, d\alpha\, d\beta}{\frac{2}{\pi}\int_0^{\frac{\pi}{2}}\int_0^{2\pi} F(\alpha,\beta)\, d\alpha\, d\beta} \quad (1)$$

Where α and β are the radial and circumferential angles of the pole figure, with α_1 and α_2 being the initial and final tilt angles for each measurement step. The difference between α_2 -α_1 is equal to 0.1°. Since we observed that for Co-Cr films the measured texture has a high degree of rotational symmetry around the direction normal to the film surface, the orientation distribution F(α) will be used to describe a (0002) pole density. For the sake of convenience, the ORc value is defined as the maximum of ORc (α,β) and will be called the orientation ratio of the crystallites. The C-axis dispersion $\Delta\alpha50$ is the full width-half maximum of the orientation distribution F(α) and will be used to characterize the dispersion of (0002) planes in Co-Cr films. The most relevant properties for three Series A, C and D of Co-Cr films are summarized in Tables 1, 2 and 3 respectively.

Table 1: Properties for Series A $Co_{81}Cr_{19}$ films sputtered on $Ge/SiO_2/Si$ substrates.

No	Thickness nm	Ms kA/m	Mr⊥ kA/m	Hc⊥ kA/m	K1 kJ/M³	Rs‖	ORm⊥	ORc	Δα50 degree
A1	5	360	354	4	122	0.08	12.5	3.4	4.75°
A2	12	465	9.3	4.3	116	0.83	0.024	9.6	3.94°
A3	25	500	30	15.4	127	0.33	0.18	22	3.25°
A4	50	485	36.8	112	112	0.15	0.73	32	2.84°
A5	100	510	60	47.2	134	0.09	1.34	40	2.93°
A6	200	500	84.5	69.9	131	0.06	2.84	44	3.03°

Table 2: Properties for Series C $Co_{77}Cr_{23}$ films sputtered on $Ge/SiO_2/Si$ substrates.

No	thickness nm	Ms kA/m	Mr⊥ kA/m	Hc⊥ kA/m	K1 KJ/m³	Rs‖	ORm⊥	ORc	Δα50 degree
C1	5	330	305	8	36.1	0.05	18.6	4.7	3.88°
C2	12	330	33	8	43.1	0.2	0.5	12	3.63°
C3	25	365	62	20	51.1	0.13	1.3	24	3.25°
C4	50	370	86	52.3	67.5	0.04	5.8	36	2.84°
C5	100	320	92	66.3	60.9	0.04	7.3	37	2.68°

Table 3: The influence of the substrate and/or seedlayer on the properties for Series D $Co_{81}Cr_{19}$ films(thickness=100 nm).

No	Substrate	Ms kA/m	Hc⊥ kA/m	Hc1⊥ kA/m	K1 kJ/m³	Rs‖	ORm⊥	ORc	Δα50 degrees
D1	$Ge/SiO_2/Si$	510	47.2	60	134	0.09	1.34	40	2.9°
D2	Si	450	107	150	133.5	0.09	2.85	38	2.5°
D3	Si_3N_4	350	72.3	115	67.6	0.56	0.34	2.8	6.6°
D4	Ge(1)/Si	478	97.7	121	49.5	0.54	0.38	2.0	7.8°

Note that Rs⊥=(Mr⊥/Ms) and Rs‖=(Mr‖/Ms) are the squareness ratios along the direction perpendicular to the plane and in-plane directions respectively. ORm⊥=(Rs⊥/Rs‖) is the so-called orientation ratio of the magnetization, reflecting the anisotropy of the remanence along the easy and hard directions. Hc⊥ and Hc1⊥ are the measured and calculated coercivities along the film normal respectively.

INFLUENCE OF THE COMPOSITION AND FILM THICKNESS ON THE TEXTURE EVOLUTION

The measured grain diameter versus thickness for $Co_{81}Cr_{19}$ and $Co_{77}Cr_{23}$ films is plotted in Fig.1. As expected, the size of the columnar grains increases with increasing thickness in the range of 5-1200 nm. The grain-size changes with the thickness in $Co_{81}Cr_{19}$ and $Co_{77}Cr_{23}$ films are similar.

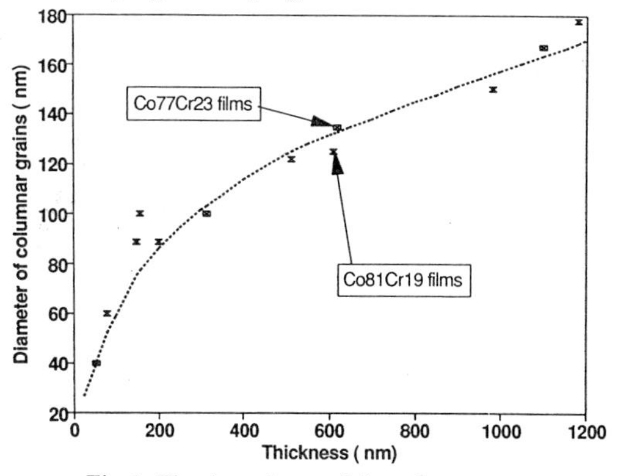

Fig.1: The dependence of the columnar diameter on the thickness for Co-Cr films

Typical results of texture measurement for $Co_{81}Cr_{19}$ and $Co_{77}Cr_{23}$ films, which have the same thickness of 12 nm, are presented in the form of a (0002) pole figure in Figs. 2a and 2b. The only difference between them is that the $Co_{77}Cr_{23}$ film has higher maxima of pole densities (≐28) than the $Co_{81}Cr_{19}$ film (=24). Experimental data show that the pole figures for the remaining specimens of the Series A and C are similar to that of the 12 nm film, except that the maximum of the pole densities increases with increasing thickness in the range of 5-100 nm.

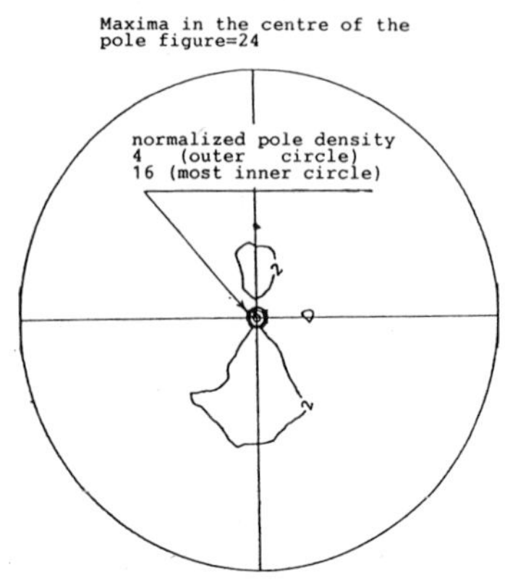

Fig.2a: The (0002) pole figure for $Co_{81}Cr_{19}$ film (12 nm)

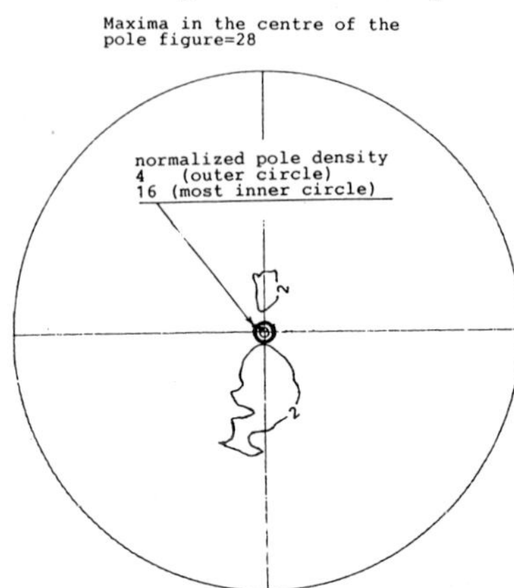

Fig.2b: The (0002) pole figure for $Co_{77}Cr_{23}$ film (12 nm)

The angular dependence of the orientation ratio ORc(α) of the crystallites as a function of the film thickness for $Co_{81}Cr_{19}$ and $Co_{77}Cr_{23}$ films is presented in Figs. 3a and 3b. It can be seen from these figures that in the range of 5-100 nm, the orientation ratio ORc(α) increases with increasing film thickness, and the orientation direction of the (0002) crystallographic planes aligns with the film normal.

In order to investigate the influence of the Cr content on the texture evolution, the dependence of the orientation ratio ORc of the crystallites on the film thickness for Series A $Co_{81}Cr_{19}$ and Series C $Co_{77}Cr_{23}$ films is plotted in Fig.4. It can be seen from this figure that for both $Co_{81}Cr_{19}$ and $Co_{77}Cr_{23}$ films, the ORc values increase with increasing thickness, but in the range of 5-80 nm the $Co_{77}Cr_{23}$ films have larger ORc values than those of $Co_{81}Cr_{19}$ films. In the case of $Co_{77}Cr_{23}$ films, there is an optimal thickness, where ORc is maximized at about 80 nm. The optimal thickness for the Series B $Co_{81}Cr_{19}$ films [1] is about 130 nm. It seems that a higher Cr content in Co-Cr films gives a lower optimum thickness.

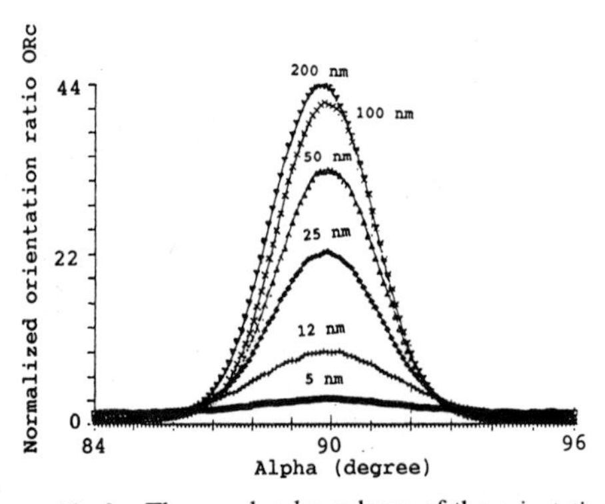

Fig.3a: The angular dependence of the orientation ratio ORc as a function of the thickness for $Co_{81}Cr_{19}$ films.

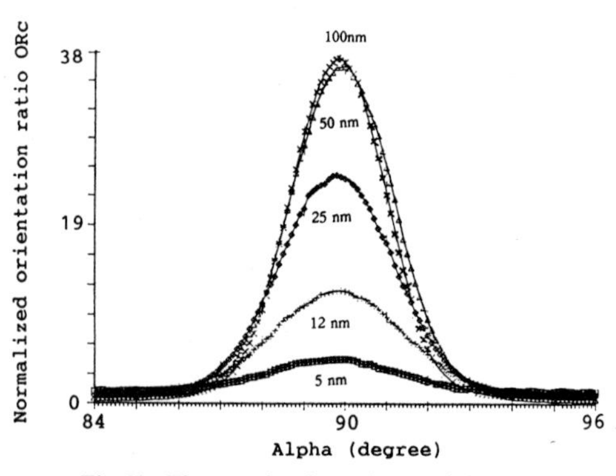

Fig.3b: The angular dependence of the orientation ratio ORc as a function of the thickness for $Co_{77}Cr_{23}$ films

The dependence of the C-axis dispersion $\Delta\alpha 50$ on the thickness is plotted in Fig. 5. There is an optimum thickness (about 50 nm) for $Co_{81}Cr_{19}$ films, that minimizes the $\Delta\alpha 50$ value. However, for $Co_{77}Cr_{23}$ films in the range of 5-100 nm, the $\Delta\alpha 50$ value decreases monotonically with increasing thickness. In our case, $Co_{77}Cr_{23}$ films have lower $\Delta\alpha 50$ values than those of $Co_{81}Cr_{19}$ films.

It can be concluded that a sputtered Co-Cr film with higher Cr content is a better choice as a perpendicular recording film, if it is thinner than about 80 nm. Such a film has a higher orientation ratio ORc of the crystallites, a higher squareness ratio $Rs\perp$, a higher coercivity $Hc\perp$ and a higher remanence $Mr\perp$ as well as lower in-plane squareness ratio $Rs\|$ and lower $\Delta\alpha 50$ value.

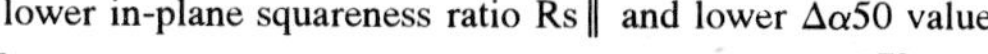

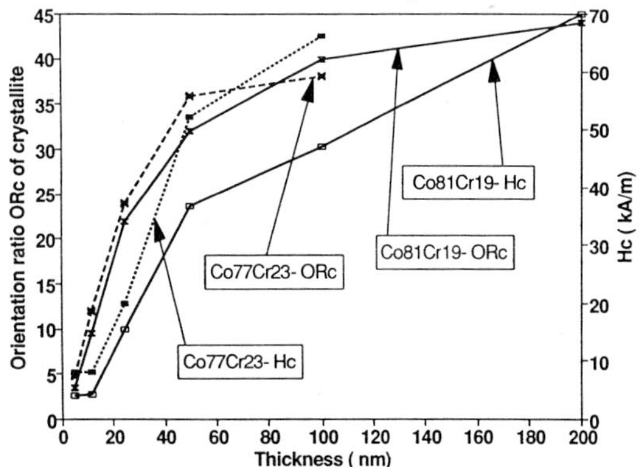

Fig.4: The comparison of the thickness dependence of the orientation ratio ORc and $Hc\perp$ for $Co_{81}Cr_{19}$ and $Co_{77}Cr_{23}$ films

Fig.5: The comparison of the dependence of $\Delta\alpha 50$ value on the thickness for $Co_{81}Cr_{19}$ and $Co_{77}Cr_{23}$ films

THE INFLUENCE OF THE SUBSTRATE AND SEEDLAYER ON THE TEXTURE

Series D $Co_{81}Cr_{19}$ films having the same thickness (100 nm) were selected to investigate the effect of the substrate and seedlayer on the texture. These films were sputtered on Si, Ge/SiO_2/Si, Si_3N_4/Si and Ge(1)/Si respectively. As shown in Table 3, the texture and magnetic properties of these tested films are closely related to the substrate and seedlayer materials. Experimental data show that the utilization of Ge/SiO_2/Si and Si promotes the formation of a strong and perpendicularly oriented hcp texture on the substrate, and that Si_3N_4/Si and Ge(1)/Si structures will greatly deteriorate a perpendicularly oriented texture. For example, $Co_{81}Cr_{19}$/SiO_2/Si and $Co_{81}Cr_{19}$/Si specimens have much larger orientation ratios ORc, which reach 40 and 38 respectively, compared to those of $Co_{81}Cr_{19}$/Si_3N_4/Si and $Co_{81}Cr_{19}$/Ge(1)/Si specimens, whose orientation ratios ORc are only 2.8 and 2.0 respectively. In addition, the first two films exhibit lower C-axis dispersions $\Delta\alpha 50$ (2.9° and 2.8°) than those of the other two films (6.6° and 7.8°). A preferred perpendicular C-axis orientation for $Co_{81}Cr_{19}$ films results in a higher magnetocrystalline anisotropy K1, higher orientation ratio $ORm\perp$ of the magnetization as well as a lower in-plane squareness ratio $Rs\|$, as shown in Table 3. For example, the magnetocrystalline anisotropy constant K1, the orientation ratio $ORm\perp$ and the squareness ratio $Rs\|$ for $Co_{81}Cr_{19}$/Ge/SiO_2/Si specimen are 134 kJ/m^3, 1.34 and 0.09 respectively, but for $Co_{81}Cr_{19}$/Si_3N_4 specimen the corresponding values are 67.6 kJ/m^3, 0.34 and 0.56 respectively. This demonstrates how important the correct choice of substrate and seedlayer is for preparing high quality perpendicular media.

Experimental data show that the utilization of a Ge/SiO_2/Si or a Si substrate promotes the formation of a preferred (0002) texture, which results in higher coercivity and higher remanence, and the adoption of a Ge(1)/Si or a Si_3N_4/Si substrate will greatly deteriorate the (0002) texture and magnetic properties in $Co_{81}Cr_{19}$ film.

ACKNOWLEDGEMENTS

The authors acknowledge the financial support of the Natural Sciences and Engineering Research Council of Canada. We would like to thank Dr. W. Geerts and Dr. P.ten Berge, University of Twente, The Netherlands for providing the tested specimens and experimental data on magnetic properties. The authors are also obliged to Mr. S. Poplawski, McGill University, Canada, for doing the measurements on X-ray diffraction and texture.

REFERENCES

1) Li Cheng-Zhang, J.C. Lodder and J.A. Szpunar, "The development of texture in Co-Cr films", to be published in IEEE Trans. on Mag. 1993

2) Li Cheng-Zhang, J.C. Lodder and J.A. Szpunar, "The influence of the textural development on magnetic properties in $Co_{81}Cr_{19}$ films" to be published in J. Magn. Magn. Mater., 1993

3) P. Berge, J.C. Lodder, S. Porthun and Th. J.A. Popma, J. Magn. Magn. Mater. Vol. 113, pp. 36-46, 1992,

4) H.-C. Tsai, B.B. Lai and A. Eltoukhy, J. Appl. Phys. Vol. 71, pp. 3579-3585, 1992,

5) M. Futamoto, Y. Honda, H. Kakibayashi and K. Yoshida, IEEE Trans. Vol. Mag-21, pp. 1426, 1985,

6) G. Pan, D.J. Mapps, M.A. Akhteo, J.C. Lodder, P.ten Berge, H.Y. Wong and J.N. Chapman, J. Magn. Magn. Mater. Vol. 113, pp.21-28, 1992,

7) M.A. Parker, K.E. Johnson, C. Hwang and A. Bermea, IEEE Trans. Vol. Mag-27, No.6, 1991,

8) J.L. Pressesky, S.Y. Lee, N. Heiman, D. Williams, T. Coughlin and D.E. Speliotis, IEEE Trans. Vol. Mag-26, no. 5, pp.1596-1598, 1990,

9) Y. Arisaka, K. Sato, Y. Yamada and K. Chiba, IEEE Trans. Vol. Mag-27, No. 6, pp.4742-4744, 1991,

10) B. Szpunar and J.A. Szpunar, J. Magn. Magn. Mater. Vol. 43, pp.317-326, 1984,

11) Li Cheng-Zhang and J.A. Szpunar, "The correlation of texture and magnetic properties in Co-Cr film", to be published in J. Electronic Materials, 1993,

Materials Science Forum Vol. 157-162 (1994) pp. 1393-1404

ELECTRIC FIELD INDUCED DOMAIN FORMATION IN SURFACE STABILIZED FERROELECTRIC LIQUID CRYSTAL CELLS

I. Dierking, F. Gießelmann, J. Schacht and P. Zugenmaier

Institut für Physikalische Chemie, TU Clausthal,
Arnold-Sommerfeld-Str. 4, D-38678 Clausthal-Zellerfeld, Germany

Keywords: Liquid Crystals, Ferroelectricity, Chirality, Smectic C*, Mesophse

ABSTRACT

Two types of domains have been observed for S_C^* ferroelectric liquid crystals in surface stabilized cells (SSFLC) by application of a high electric field with the smectic layers tilted by the amount of the chevron angle with respect to the normal of the rubbing direction in the substrate plane. The layer structure resembles that of a chevron configuration in the plane of the substrate similar to the recently reported stripe-shaped SSFLC structure. The two domain types "appear" to switch in a reciprocal fashion when applying an AC electric field and observed between crossed polarizers. The temperature dependence of this effect has been investigated and an explanation proposed analogous to a striped texture model.

INTRODUCTION

A variety of director configurations can be observed in S_C^* ferroelectric liquid crystals, depending on anchoring strength and geometry of the sample. In thick samples prepared on a glass slide without orientation layers, typical textures may appear, e.g. the well known fan-shaped texture. The fans may be modulated with equidistant disclination lines due to a helical director configuration [1,2]. This line texture may also be observed in LC cells with a small gap and orientation layers depending on the value of the S_C^* pitch [3]. If the cell gap is comparable to or smaller than the S_C^* pitch, the helix is usually supressed by surface effects and a so called surface stabilized ferroelectric liquid crystal (SSFLC) geometry is obtained [3]. In the simplest case, one observes a structure, where the layers are arranged perpendicular to the substrate plane and to the rubbing direction of the orientational layers with the molecules tilted with respect to the layer normal (bookshelf geometry). Due to anchoring effects at the substrate, one usually observes a structure with the layers tilted with respect to the substrate normal (chevron geometry [4,5]).
Recently an electric field induced layer structure has been observed, which resembles that of a chevron turned by 90°. The smectic layers are tilted with respect to the normal of the rubbing direction in the substrate plane by the amount of the chevron angle ($\triangleq$ to the director tilt angle), in contrast to the well known chevron geometry, where the smectic layers are tilted with respect to the substrate normal. Application of a high electric field to the LC cell results in the formation of two domain types, which seem to switch in an reciprocal way when applying an AC electric field and placed between crossed polarizers. The formation of the two domain types is analogous to observations by PATEL et al.[6,7], although the conditions applied are quite different. The measurements were carried out on a diarylethane-α-chloroester,

$$C_8H_{17}O-\langle\bigcirc\rangle-COO-\langle\bigcirc\rangle-CH_2-CH_2-\langle\bigcirc\rangle-OOC-\overset{*}{C}H(Cl)-CH(CH_3)-CH_3$$

derived from D-valine, which will be referred to as compound D8. Synthesis, purification, characterization and temperature dependent measurements of several physical parameters of this compound and other homologous of the series will be published elsewhere [8].

A detailed review on textures of liquid crystalline phases can be found in a book by DEMUS/RICHTER [9] and GRAY/GOODBY [10]. A recent review on properties of liquid crystals was published by CHANDRASEKHAR [11]. Properties and applications of ferroelectric liquid crystals are summarized in a book by GOODBY et al. [12].

EXPERIMENTAL

The phase sequence of D8 as studied by polarization microscopy (Olympus BH-2, equipped with a Mettler FP 52 hot stage and a Mettler FP 5 temperature controller) and DSC measurements (Perkin Elmer DSC 7) is given by:

$$\mathrm{I}\ 134.5\ \mathrm{N}^{*}\ 129.4\ \mathrm{TGB\ A}^{*}\ 129.2\ \mathrm{S}_{\mathrm{A}}^{*}\ 125.6\ \mathrm{S}_{\mathrm{C}}^{*}\ 92.9\ \mathrm{S}_{\mathrm{I}}^{*}\ 83.7\ \mathrm{S}_{\mathrm{F}}^{*}$$

All measurements presented below were performed on samples prepared in commercially available liquid crystal cells (E.H.C. Co. Ltd.) with a cell gap of 4 μm, ITO electrodes and parallel rubbed polyimide orientation layers.
Determination of the value of the optical director tilt angle and the angular position of the layer normal was performed using a method analogous to the one introduced by BAHR et al.[13]. We applied an electric square wave field with frequency f=200 Hz and an amplitude of E=1 MV/m. The cell is placed in the hot stage with the rubbing direction $\vec{x}$ coinciding with an angular position of the hot stage marking $\varphi=0°$. Recording the transmissions of the two switching states with *"polarization up"* (I_{up}) and *"polarization down"* (I_{down}) as a function of the rotation angle φ, which is defined as the rotation angle of the hot stage in the substrate plane, one can determine the director tilt angle Θ by the phase shift between the $I_{up}(\varphi)$ and $I_{down}(\varphi)$ curves and the angular position of the layer normal $\vec{k}$ with respect to the rubbing direction $\vec{x}$ from the intersection of the $I_{up}(\varphi)$ and $I_{down}(\varphi)$ curves.

EXPERIMENTAL RESULTS

The first series of measurements was obtained by slowly heating the oriented sample in SSFLC cells through the S_C^* phase and recording the $I_{up}(\varphi)$ and $I_{down}(\varphi)$ data for several different temperatures. These data are depicted in figure 1 for temperatures T=118, 120, 122, 123, 124 and 125°C.
It is clear, that the phase shift between two respective curves (i.e. the angular difference between the two minima of intensity) and, therefore the director tilt angle Θ decreases with increasing temperature, as expected. The intersection points of the two corresponding curves remain constant at an rotation angle $\varphi=0°$, which means that the position of the layer normal $\vec{k}$ is independent of temperature and encloses an angle $\alpha=0°$ with the rubbing direction $\vec{x}$, i.e. rubbing direction and layer normal coincide for all temperatures. The structure of this configuration is the well known geometry depicted in figure 2(a)-(c) as a top view of the cell. At zero electric field, two spontaneously formed domains are observed, one in the polarization "up" and one in the polarization "down" state (figure 2(a)). The angle between the respective directors $\vec{n}$ is given by 2Θ. Applying an electric field E

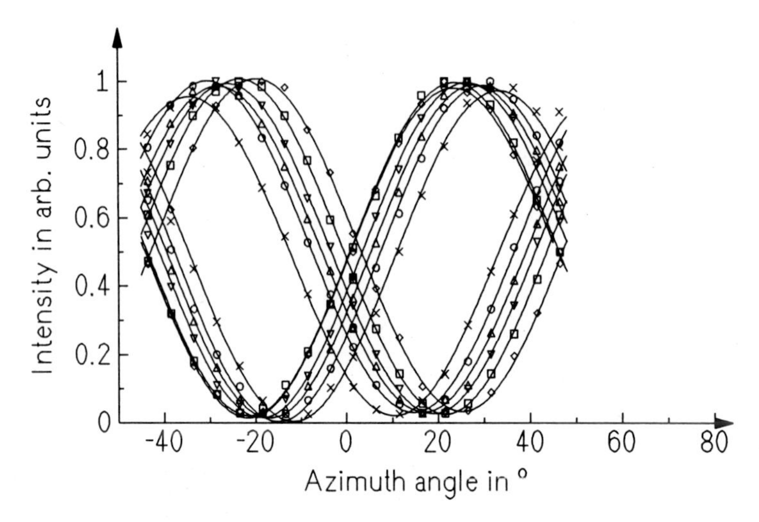

Figure 1: Intensity of the transmitted light as a function of the rotation angle. (◇) T=118°C, (□) T=120°C, (∇) T=122°C, (Δ) T=123°C, (○) T=124°C and (×) T=125°C.

perpendicular to the substrate plane in the $-\vec{z}$ direction, all molecules reorient in the polarization "down" state and the director $\vec{n}$ encloses an angle Θ with the layer normal $\vec{k}$ (figure 2(b)). Reversing the direction of the electric field, the molecules reorient by an angle 2Θ into the polarization "up" state (figure 2(c)).

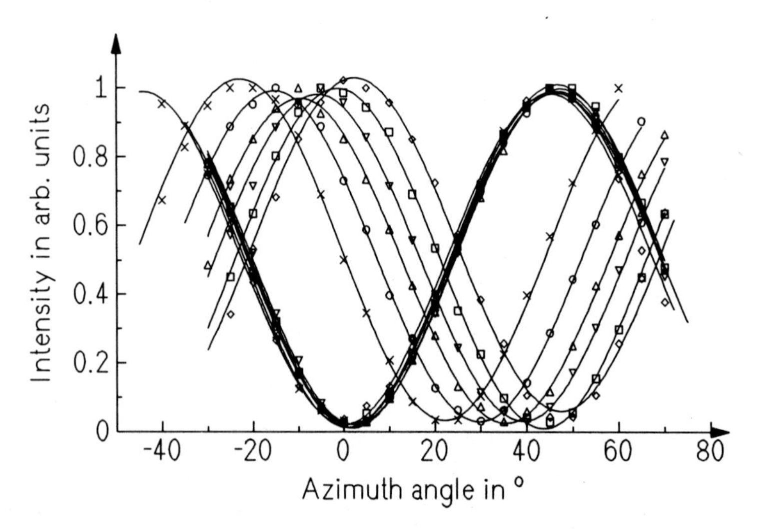

Figure 3: Second measurement series after the high field treatment (◇) T=118°C, (□) T=120°C, (∇) T=122°C, (Δ) T=123°C, (○) T=124°C and (×) T=125°C.

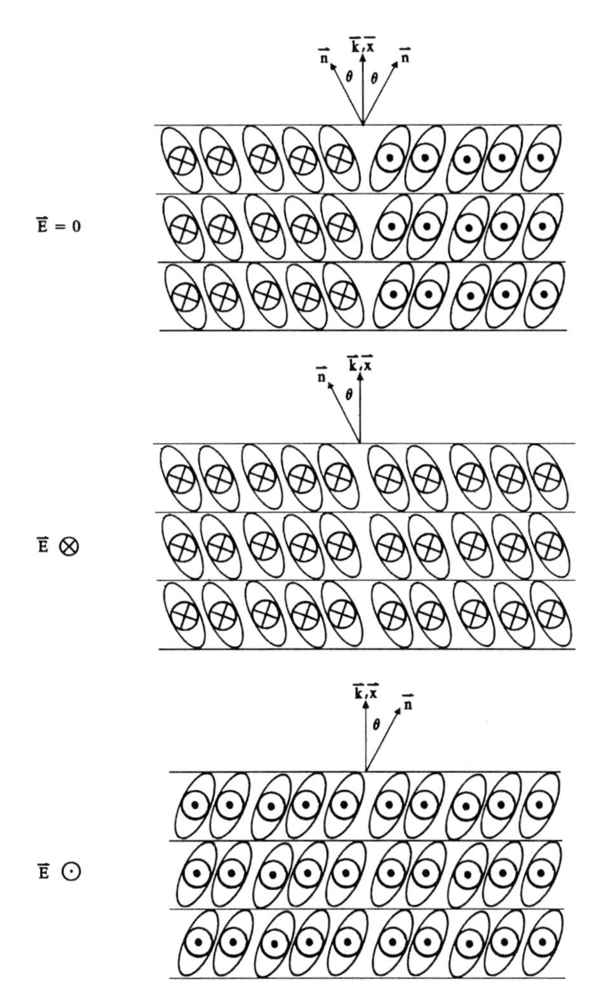

Figure 2: Top view of the SSFLC cell for different directions of the electric field E, before the high field treatment. (a) E=0, (b) E in -z direction and (c) E in +z direction.

The second series of measurements was obtained after application of a high electric field with amplitude E=10 MV/m at T=115°C for about 5 seconds. The $I_{up}(\varphi)$ and $I_{down}(\varphi)$ curves are now considerably different from the respective ones of the first series and are depicted in figure 3. The phase shift between two respective curves is still the same as before, but the angular position of the layer normal has changed and is now dependent on temperature. The director tilt angle still has the same value as in the first measurement series (figure 4), but the angle α between the rubbing direction $\vec{x}$ (φ=0°) and the layer normal $\vec{k}$ is now equal to the director tilt angle ($\alpha=\Theta$). Figure 5 depicts the dependence of the layer angle α on the director tilt angle Θ (parameter temperature). The line shows a least-square fit, which yields a slope of 1.007, leading to $\alpha(T)=\Theta(T)$, disregarding the small offset angle $\alpha=0.7°$ for $\Theta=0°$.

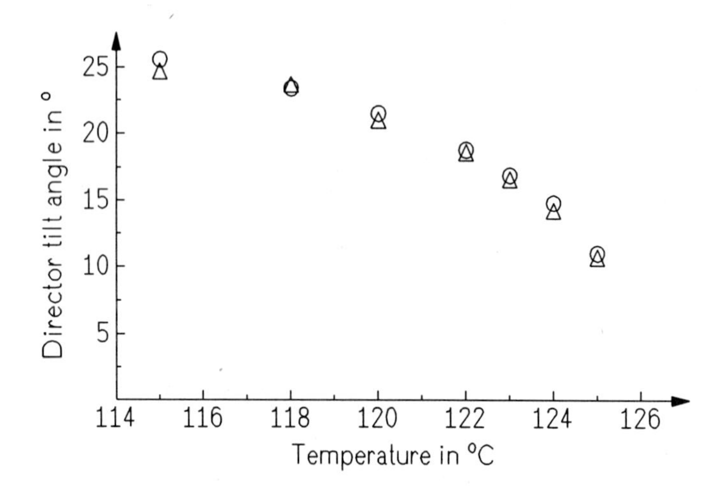

Figure 4: Director tilt angle Θ as a function of temperature: (O) first measurement series and (Δ) second measurement series.

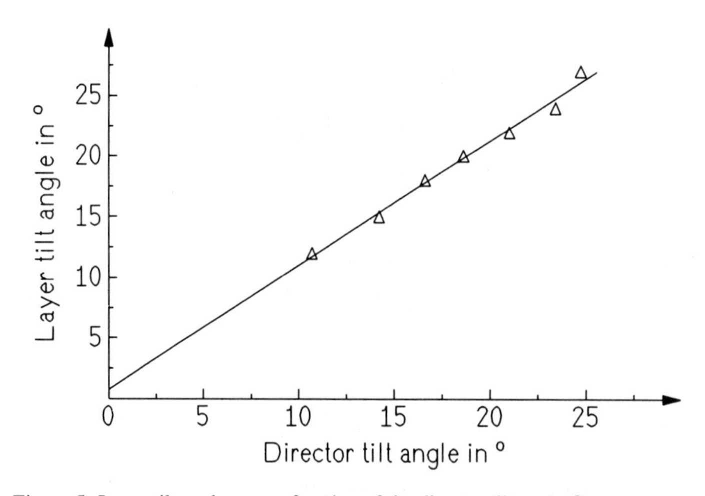

Figure 5: Layer tilt angle α as a function of the director tilt angle Θ.

A structural model which explaines this behaviour is depicted in figure 6(a)-(c) and supported by texture photographs (figure 7(a)-(c)) taken at a hot stage position $\varphi=0°$ and temperature T=110 °C. After the high field treatment the molecules are oriented, with the

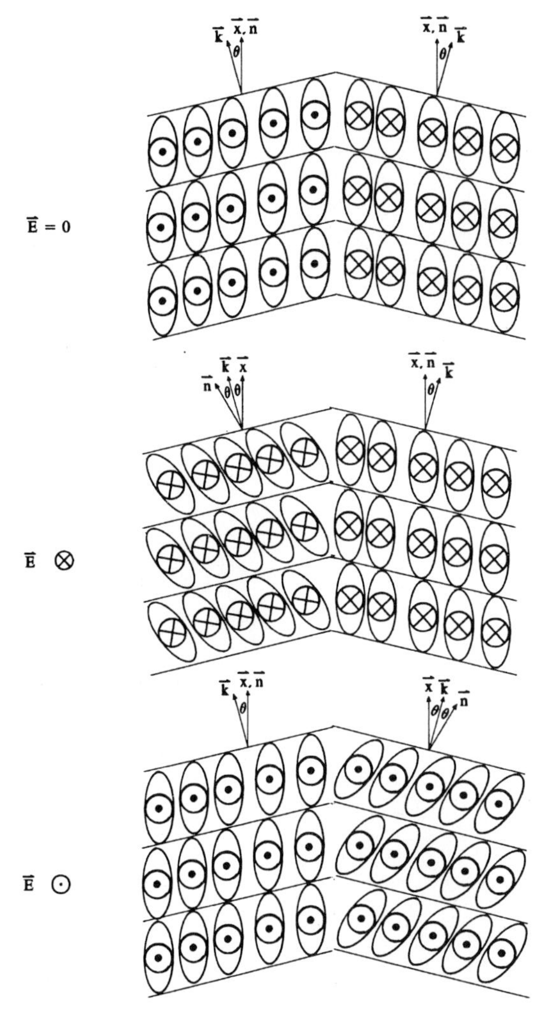

Figure 6: Top view of the cell after the high field treatment (a) E=0, (b) E in -z direction and (c) E in +z direction.

director $\vec{n}$ coinciding with the rubbing direction $\vec{x}$. This is only possible, if the layers reorient by an angle $\alpha=\Theta$, since the director tilt angle is essentially fixed for a given temperature. At zero field after the high field treatment, the two domain types "1" and "2", newly formed, have the same director orientation $\vec{n}$ (figure 6(a)) and are in the dark state between crossed polarizers (figure 7(a)). Applying an electric field perpendicular to the substrate plane in the $-\vec{z}$ direction, molecules of domain type "1", which are in the polarization "up" state, switch by an angle 2Θ along the tilt cone, while the molecules of domain type "2", which are already in the polarization "down" state, stay in their position (figure 6(b) and figure 7(b)). Reversing the direction of the electric field, all molecules reorient along the tilt cone by an

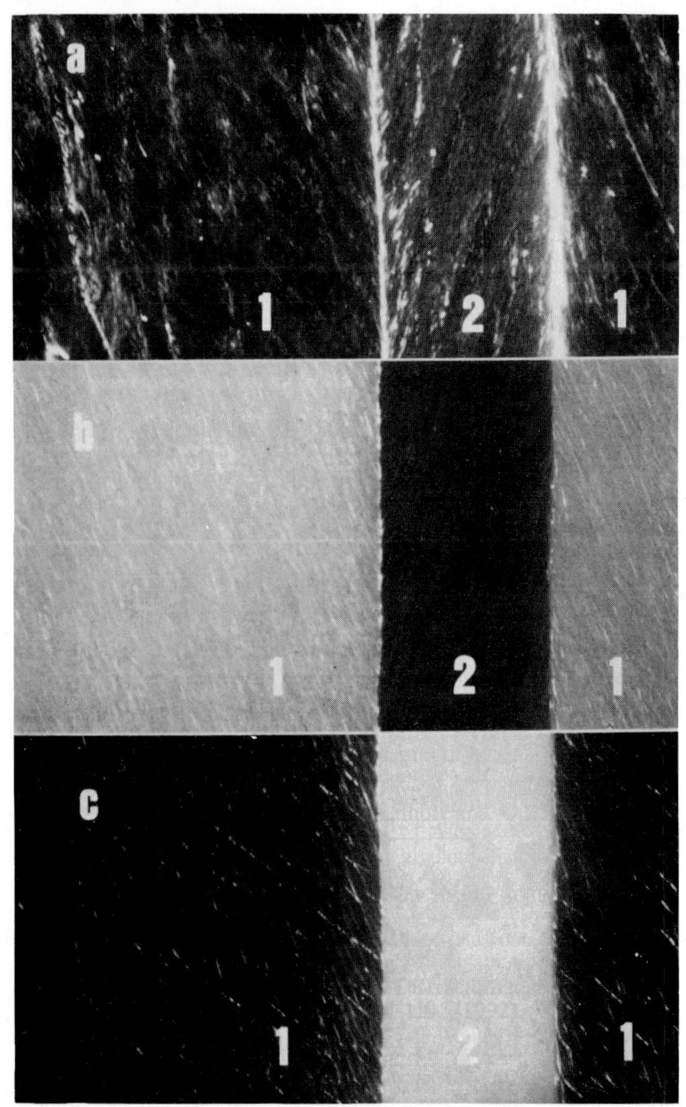

Figure 7: Textures of the field induced domains between crossed polarizers: (a) E=0 V, (b) E=-2 MV/m DC and (c) E=+2 MV/m DC. The depicted area is 1x0.5 mm^2 in each case.

angle 2Θ and domain type "1" now optically appears like domain type "2" before, and vice versa (figure 6(c) and figure 7(c)). Observing the cell between crossed polarizers while applying a low frequency square wave field, the switching between the configurations depicted in figure 6(b) and 6(c) optically "appears" as if the two domain types switch in a reciprocal fashion (figures 7(b) and 7(c)) due to the different layer orientations in both domain types.

DISCUSSION

Recently, several papers have discussed models for the stripe-shaped texture [14-18]. The field induced domain formation in SSFLC cells can be explained by a model proposed by SHAO et al.[14] for the stripe-shaped texture. It was usually found, that the width of the stripes is approximately equal to the cell gap. The formation of large field induced domains as observed in our case, which may extend across the whole electrode area of the cell, can be described by an analogous mechanism as proposed in [14].

Figure 8 (a): Model of the vertical chevron layer structure, the initial structure in SSFLC cells before the high field treatment.

The initial layer structure in the cell before the high field treatment is of the vertical chevron type [4,5] with zig-zag defects mediating regions of chevron shaped layers, tilted in opposite direction [5]. These intermediate zones are of bookshelf type with layers tilted by the amount of the chevron angle with respect to the rubbing direction [14]. This structure is depicted in the model of figure 8(a). The high field treatment forces the chevron layers to straighten up as has been demonstrated in several publications [19-21]. This means that by application of a reasonably high electric field, the bookshelf areas, previously mediating two regions of opposite chevron direction, grow at the expense of the chevron regions, until the layer structure in the cell is of bookshelf type. In this configuration the layer normal is inclined to the rubbing direction by the amount of the previous chevron angle. The intermediate zone between two bookshelf regions of opposite tilt is of the chevron type. The layer structure of the cell now resembles that of a chevron structure turned by 90° resulting in a horizontal chevron structure which is depicted in the model of figure 8(b).

Figure 8 (b): Model of the horizontal chevron structure after the high field treatment.

The here investigated field induced domain formation in SSFLC cells is the transition from a vertical chevron structure with bookshelf intermediate zones between regions of opposite chevron direction to a horizontal chevron structure with chevron intermediate zones between bookshelf areas with opposite tilt of the layer normal with respect to the rubbing direction.

CONCLUSIONS

The electric field induced domain formation in SSFLC cells has been investigated by polarization microscopy and electrooptical methods. The layer structure of the cell after the high field treatment resembles that of a horizontal chevron structure with the layer normal tilted with respect to the rubbing direction by the amount of the previous chevron angle, which in our case is equal to the director tilt angle. The process of changing the initial vertical chevron structure to the domain structure can be described in terms of a model [14] proposed for the stripe-shaped texture, although there is no relation between the size or shape of the domains and the cell gap, as it is found for stripe-shaped textures.

This work was supported by a grant from the *Deutsche Forschngsgemeinschaft*.

REFERENCES

[1] MARTINOT-LAGARDE, PH., 1976, *Jour. de Phys.*, **37**, C3-129

[2] MARTINOT-LAGARDE, PH., DUKE, R. and DURAND, G., 1981, *Mol. Cryst. Liq. Cryst.*, **75**, 249

[3] CLARK, N.A. and LAGERWALL, S.T., 1980, *Appl. Phys. Lett.*, **36**, 899

[4] RIEKER, T.P., CLARK, N.A., SMITH, G.S., PARMAR, D.S., SIROTA, E.B. and SAFINYA, C.R., 1987, *Phys. Rev. Lett.*, **59**, 2658

[5] CLARK, N.A. and RIEKER, T.P., 1988, *Phys. Rev. A*, **37**, 1053

[6] PATEL, J.S. and GOODBY, J.W., 1986, *J. Appl. Phys.*, **59**, 2355

[7] PATEL, J.S., LEE, SIN-DOO and GOODBY, J.W., 1989, *Phys. Rev. A*, **40**, 2854

[8] DIERKING, I., GIEßELMANN, F., KUßEROW, J. and ZUGENMAIER, P., *submitted to Liq. Cryst.*

[9] DEMUS, D. and RICHTER, L., *Textures of Liquid Crystals*, Verlag Chemie, Weinheim, 1978

[10] GRAY, G.W. and GOODBY, J.W., *Smectic Liquid Crystals-Textures and Structures*, Leonard Hill, Glasgow, 1984

[11] CHANDRASEKHAR, S., *Liquid Crystals*, 2. Edition, Cambridge University Press, Cambridge, 1992

[12] GOODBY, J.W. et al., *Ferroelectric Liquid Crystals-Principles, Properties and Applications*, Gordon and Breach Science Publishers, Philadelphia, 1991

[13] BAHR, CH. and HEPPKE, G., 1987, *Liq. Cryst.*, **2**, 825

[14] SHAO, R.F., WILLIS, P.C. and CLARK, N.A., 1991, *Ferroelectrics*, **121**, 127

[15] LEJCEK, L. and PIRKL, S., 1990, *Liq. Cryst.*, **8**, 871

[16] FÜNFSCHILLING, J. and SCHADT, M., 1991, *Jap. Jour. Appl. Phys.*, **30**, 741

[17] JAKLI, A. and SAUPE, A., 1992, *Phys. Rev. A*, **45**, 5674

[18] ASAO, Y. and UCHIDA, T., 1993, *Jap. Jour. Appl. Phys. Lett.*, **32**, L604

[19] JOHNO, M., CHANDANI, A.D.L., OUCHI, Y., TAKEZOE, H., FUKUDA, A., ICHIHASHI, M. and FURUKAWA, K., 1989, *Jap. Jour. Appl. Phys. Lett.*, **28**, L119

[20] ITOH, K., JOHNO, M., TAKANISHI, Y., OUCHI, Y., TAKEZOE, H. and FUKUDA, A., 1991, *Jap. Jour. Appl. Phys.*, **30**, 735

[21] OH-E, M., ISOGAI, M. and KITAMURA, T., 1992, *Liq. Cryst.*, **11**, 101

Materials Science Forum Vol. 157-162 (1994) pp. 1405-1410

EPITAXY IN TEXTURE FORMATION OF ELECTRODEPOSITED Cu-COATINGS

I. Handreg and P. Klimanek

Institut für Metallkunde der Berg-Akademie,
Gustav-Zeuner-Str. 5, D-09596 Freiberg/Sa., Germany

Keywords: Cu-Coatings, Epitaxial Growth, Electrodeposition, Texture, Microstructure

ABSTRACT

The present work is concerned with the conditions of epitaxial growth of copper coatings on polycrystalline substrates of copper (homoepitaxy), high-alloy austenitic steel (X8CrNiTi18.10) and α-iron (heteroepitaxy). The coatings were electrodeposited from acid and basic electrolytes free of additions at various bath temperatures and current densities. In order to determine the ocurrence and the degree of epitaxial relationships in dependence on the kind of electrolyte and the deposition parameters quantitative texture analysis and optical microscopy were used.

INTRODUCTION

For a long time it is well known that metallic coatings can be formed via epitaxial growth under certain conditions [1,2,3].
During the deposition process the type of growth mechanism and consequently the microstructure formation of the coating is selected by the competition of different parameters (e.g. bath temperature, current density, lattice misfit) in a manner schematized in figure 1. If only one growth mode is predominating in the progress of the deposition the coating still remain homogeneously. However in the most cases the growth mechanism changed in consequence of the modified conditions throughout the electrodeposition and results in inhomogeneous layers.

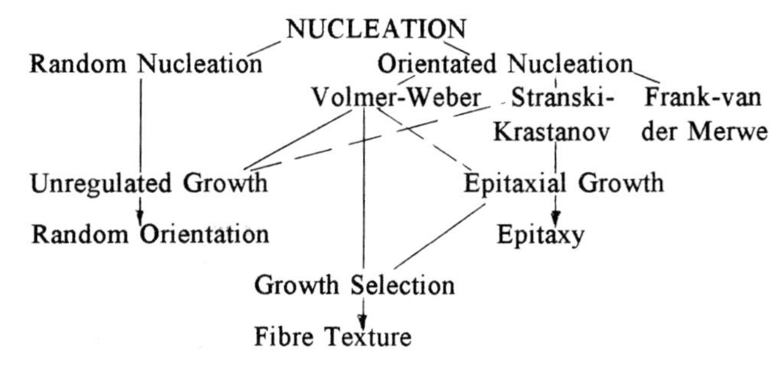

Fig. 1: Schema of texture formation in electrodeposits

The Frank-van der Merwe and the Stranski-Krastanov modes occur in systems of strong deposit-substrate interaction (weak misfit) and base on the formation of initial pseudomorphic layers [4]. In

the first mode misfit dislocations formed above a critical thickness to release from the strain accumulated in the growing deposit. In the second mode the strain energy in the layer exceeds the interfacial energy and result in the formation of nuclei [4,5].
There are different explanations of the three dimensional Volmer-Weber mode which appears in systems of weak interfacial force [4,6]. The interpretation by a post-nucleation process may be supported by many in-situ electron microscopy investigations at vapour-deposited films and the fact that good epitaxy was found in systems of large lattice misfits [4]. If the growth of the epitaxial nuclei is not inhibited, the substrate orientation is reproduced by the coating up to a thickness of 100 μm [7]. But the superposition of many factors caused by the comlex interactions of the deposition parameters and substrate characteristics results in a loss of epitaxy. In dependence on the bath conditions a direction with a attached maximum of crystallization rate will be selected [7,8,9,10]. Also there are many geometric considerations to explain epitaxial relationships in varios systems independent from the deposition methode [11,12,13,14].

EXPERIMENTAL PROCEDURES

The present work is concerned with the conditions of epitaxial growth of electrodeposited Cu-coatings on polycrystalline substrates of copper (homoepitaxy), α-iron, high-alloy austenitic steel (X8CrNiTi18.10) (heteroepitaxy) from acid or basic electrolytes like presented in table 1.

Table 1 : Conditions of electrodeposition

substrate material	electro-lyte	temperature [°C]	current d. [Adm^{-2}]	thickness [μm]
copper	acid basic	-10 to 110 20 to 90	1 to 4 1.66 to 2.5	50 20 to 50
high-alloy austenitc steel	acid basic	28 to 80 25 to 90	2 to 5 0.5 to 5	30 to 50 5 to 50
recrystallized α-iron	cyanide\acid basic	20 50 to 90	4 0.5 to 5	1 to 80 5 to 50

In order to detect the existence of orientation relationships between the crystallites of the substrates and the coatings, optical microscopy (cross-section imaging) and quantitative texture analysis of both the substrate and the coating were used. The texture characterization was based on X-ray diffraction pole figures and the interpretation of the three dimensional orientation distribution functions (ODF`s) calculated by the Bunge-formalism [15].

RESULTS AND DISCUSSION

Copper on Copper

Epitaxial growth of copper electrodeposited on copper were found both for single-crystalline [16,17] and polycrystalline substrates [7]. The results obtained under the conditions of table 1 can be summarized as follows:

* At low bath temperatures (-10°C to 5°C) fine columnar grains were formed parallel to the direction of the electric field lines with a strong <110>-fibre texture.
* With increasing temperatures the microstructure becomes more and more inhomogeneous. A transition from a fibre texture to an orientation distribution with orthorhombic symmetry takes place.
* At sufficiently high temperatures (above 20°C) the coating is completely formed by epitaxial

growth across the whole thickness (like represented in Fig. 2a).

* The addition of an organic substance (1g/l P4000) and, consequently, increasing inhibition of the growth results in a shifting of the texture transition temperature towards higher values (above 40°C).
* A decreasing of current density removed the starting temperature of epitaxial growth to distinctly lower bath temperatures [18].

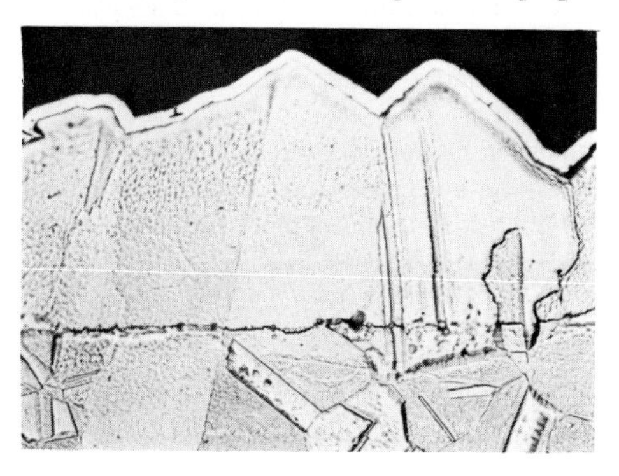

a) acid electrolyte

b) basic electrolyte

Fig. 2: Optical micrographs of copper on copper at 90°C

Figure 3 represents the orientation densities of the estimated main components in dependence on the bath temperature for deposition from an acid electrolyte without any organics.
It is evident at temperatures above 20°C the coating takes on the orientation of the substrate.
The utilization of a basic electrolyte (pyrophosphate basis) complicates the relationship between temperature and micostructure, caused by the more difficult dissociation of the complex ion. In spite of this at sufficiently high temperatures (90°C) epitaxial growth occurs. In comparison with the deposition from the acid electrolyte in the optical micrographs (Fig. 2b) and also in the pole figures an increasing disturbance of the epitaxial growth across the coating thickness becomes visible.

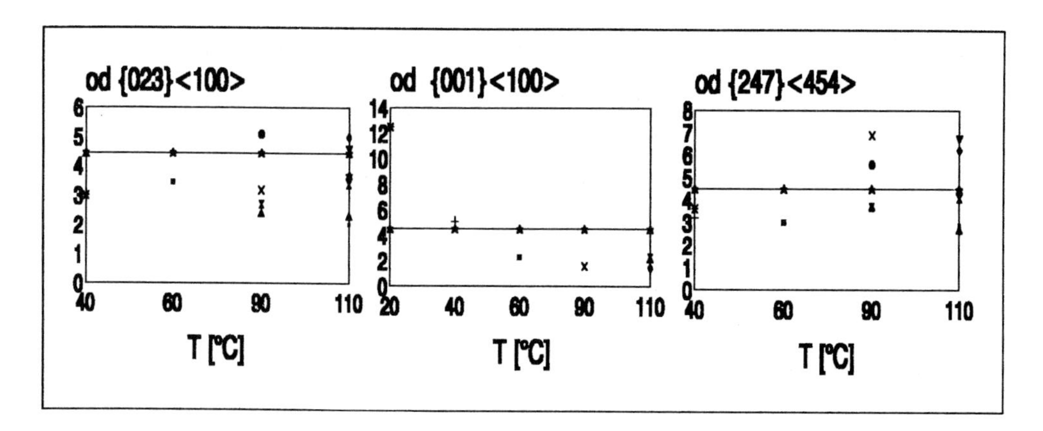

a) {023}<100>-component b) {001}<100> cube component c) {247}<454> Cu-P-position

Fig. 3: Orientation densities in dependence on the deposition temperature (---substrate level)

Copper on high-alloy steel

An austenitic steel (X8CrNiTi18.10) were choosen for investigations of heteroepitaxial growth of copper since the difference of the lattice parameters is only about 1%. Such a small misfit should favours the initial pseudomorphism and consequently the Frank-van der Merwe nucleation.
After immersion of the substrates in a $CuCl_2$-solution copper coatings were deposited from an acid electrolyte in a temperature range from 28°C up to 80°C. Simultaneously the surface state was varied by the surface pretreatment as follows : grinded with 600 paper, diamond polished and also grinded and additionally etched with aqua regia. The latter state is similar to that used in the deposition of copper on copper.
In the case of the etched substrates there is an agreement in the temperature dependence of texture formation observed for polycrystalline copper substrates : at low temperatures a <110>-fibre texture is formed and with increasing temperatures (up to 80°C) the coating shows more and more epitaxial growth (Fig. 4).

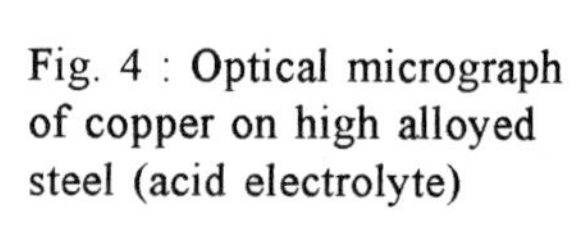

Fig. 4 : Optical micrograph of copper on high alloyed steel (acid electrolyte)

The substrate texture of the austenitic substrate can be characterized by a weak {001}<110> component (OD 2.2, at ϕ_1=45 to 0°,Φ=const.=0°,ϕ_2=0 to 90°) and a α-fibre {011}<011> - {001}<011> with a maximum in the orientation density of 3.58 at ϕ_1=55.92°, Φ=59.37°, ϕ_2=60°. Figure 5 demonstrates the skeleton lines for the main fibre (011)[01$\bar{1}$] - (100)[01$\bar{1}$] of the substrate and the coatings, respectively, in the orientation space. There is an good agreement in orientation densities as well as in the Euler angles of the maximum of the α-fibre generated via epitaxial growth.

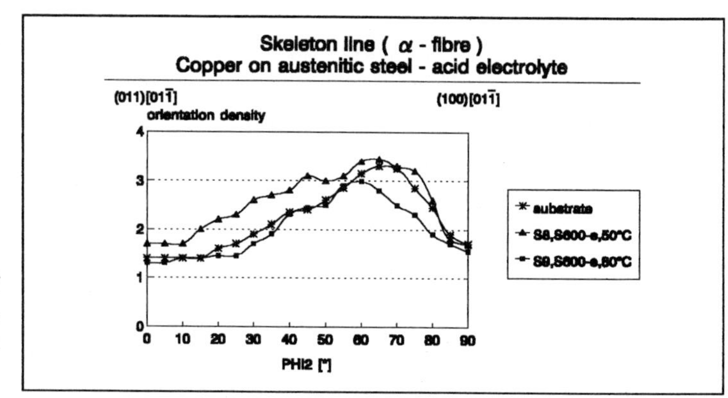

Fig. 5: Skeleton lines of copper-coatings on austenitic steel (acid electrolyte)

In the case of the basic electrolyte smaller current densities and\or higher bath temperatures are nessesary to induce epitaxial growth than in the case of the acid electrolyte . Figure 6 represents the determined texture formation in dependence on bath temperature and current density .

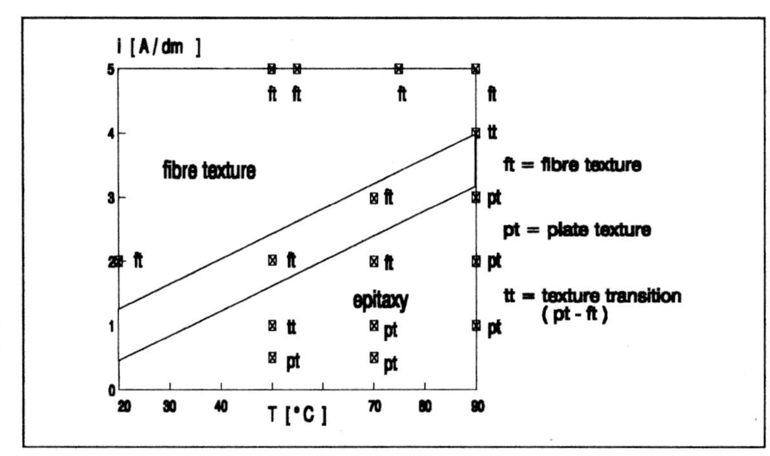

Fig. 6: Texture formation : copper on austenitic steel (basic electrolyte)

Similar to the deposition from the acid electrolyte at low temperatures and\or high current densities fibre textures are formed. A favourization of epitaxial growth by increasing bath temperatures and\or decreasing current densities results in a transition to a orientation distribution closely related to the substrate texture.
Investigations of coatings deposited onto substrates with a deranged surface layer (generated by grinding) show a "distorted" epitaxy.

Copper on α-iron

An example for a system of weak interaction is the f.c.c.-b.c.c. system of copper and α-iron [11,13]. The expected Volmer-Weber-mode were experimentally confirmed by means of TEM [15].
At investigations of copper coatings deposited onto α-iron from an acid electrolyte (after cyanide pre-copper-plating, 1μm) at room temperature no global epitaxy were observed. The variation of thickness (see at table 1) shows a inhomogeneous formation of microstructure. The texture is describable by a superposition of different fibre components like <110>, <111>, <100>, <210> and even <221> or <331> in dependence on the distance from the interface of substrate surface and coating [19].
The utilization of the basic electrolyte in the temperature range from 50°C up to 90°C at surface states comparable to the high-alloy steel sheets results in a similar texture formation (figure 7).

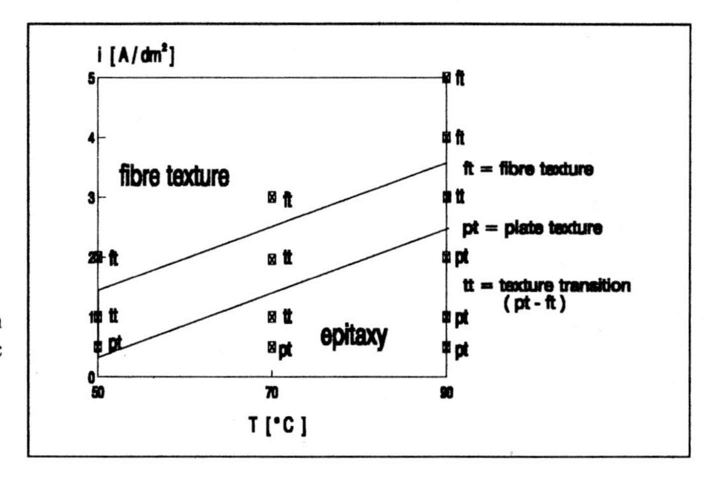

Fig. 7: Texture formation Copper on α-iron (basic electrolyte)

In this connection the main component of the iron substrate {110}<001> refers to an epitaxial growth in relationships given by Kurdjumov-Sachs.

CONCLUSIONS

The results of the investigations can be summarized as follows:
*Epitaxial growth of electrodeposited copper coatings is essentially stimulated by an increasing electrolyte temperature and\or decreasing current densities.
*The start temperature of (global) epitaxial growth depends on the type and the composition of the electrolyte. Particulary by addition of organic inhibitor substances or utilization of a basic electrolyte it is significantly shifted towards higher temperatures.
*Both homo- and heteroepitaxy of Cu-coatings are clearly related to the surface preparation of the substrates, grinding and etching favour the epitaxial growth.
*Deformed layers close to the surface of the substrates result in a distortion of epitaxy.
*Copper coatings deposited from a basic electrolyte on α-iron grown corresponding to Kurdjumov-Sachs relationship.
The experiments indicate that like to other deposition techniques (e.g. [5,6]) surface cleanless and activation and atomic mobility (bath temperatures, current densities) are nessesary suppositions for epitaxial growth.

ACKNOWLEDGEMENTS

The present work was sponsored by the Arbeitsgemeinschaft Industrieller Forschungsvereinigungen e.V. (AIF) of the FRG.

REFERENCES

[1] Huntington, A.K. : Trans. Faraday Soc. Bd.1(1905),324-325
[2] Wassermann, G.: Texturen metallischer Werkstoffe. Springer- Verlag, Berlin1939
[3] Fischer, H.: Elektr. Abscheidung und Elektrokrist. von Met., Springer Verlag,Berlin 1954
[4] Honjo, G., K. Takayanagi, K. Kobayashi, K. Yagi : Phys.stat. sol. (a), 55(1979)353-367
[5] Nix, W.D., R.F.M. Medalist : Met.Trans.,Vol.20A,9(1989)2217
[6] Thompson, C.V., Floro, J. : J. Appl. Phys., 67(1990)9, 4099
[7] Wassermann, G., J. Grewen : Texturen met. Werkstoffe, Springer-Verlag, 1962, 2. Aufl.
[8] Finch, G. I. : Zeit. f. Elektrochemie, 54(1950)6, 457
[9] Glockner, R., Knaupp, E. : Z. Phys., 76(1932)829-848
[10] Lee, D.N. : J. Mat. Science, 24(1989)4375-4378
[11] Kato, M. : Mat. Scien. and Eng.,A146(1991),pp 205-216
[12] Zur, A., T. C. Mc. Gill : J. Appl. Phys.,55(1984)2, 378-386
[13] Kato, M., T. Kubo, T. Mori : Acta. Met.,36(1988)8,2071-2081
[14] Bruce, L.A., H. Jaeger : Phil. Mag., 36(1977)6,131-1354; Phil.Mag.A,37(1978)3,337-354; Phil.Mag.A,38(1978)2,223-240
[15] Bunge, H.-J. : Quantitative Texture Analysis in Materials Science. Butterworth, London, 1984
[16] Lighty, P.E., D. Shanefield : J. Appl. Phys.34(1963)8,2233
[17] Mirzamaani, M., R. Weil : Plating and surface finishing,(1986)8, 96-100
[18] Handreg, I., P. KLimanek, H. Weidner : Conf. Euromat, Paris,6(1993), in press
[19] Klimanek, P. et.al : Schlußbericht-Projekt 29D, AIF, 1993

Materials Science Forum Vol. 157-162 (1994) pp. 1411-1416

TEXTURE FORMATION IN Al_2O_3 SUBSTRATES

J. Huber [1], W. Krahn [1], J. Ernst [1], A. Böcker [2] and H.J. Bunge [2]

[1] Hoechst CeramTec, Postfach 108, D-95601 Marktredwitz, Germany

[2] Institut für Metallkunde und Metallphysik der TU,
Grosser Bruch 23, D-38678 Clausthal-Zellerfeld, Germany

Keywords: Alumina Al_2O_3, Tape Casting, Sintering, Preferred Orientation, ODF Analysis, Grain Growth

ABSTRACT

Texture formation in Al_2O_3 substrates is studied. During green forming by tape casting preferred orientation occurs by rigid particle rotation. This process may be influenced by surface tension and shear flow near the surface. The green texture is modified by sintering in the course of grain coarsening. All textures found in this investigation were fibre textures with the basal plane parallel to the substrate plane. The final texture strength can be expressed in terms of sintering time and temperature.

INTRODUCTION

In Al_2O_3 substrates preferred orientation of the crystallites was found in a wide variation range [1, 4]. It was also found that the degree of texture considerably influences physical and technological properties of these materials [2]. On the other hand, not very much was known about how these textures are being formed during the production process and how they can be controlled with respect to desired physical properties.

EXPERIMENTAL TECHNIQUE

Texture measurements were carried out with an automatic Texture Diffractometer ATEMA-C using $Co - K_\alpha$ radiation. Pole figures were measured in steps of $\Delta\alpha = 5^o$, $\Delta\beta = 3.6^o$ up to $\alpha_{max} = 70^o$. It turned out, that the observed textures were axially symmetric fibre textures with the basal plane parallel to the substrate plane. Hence, measurement of the basal plane pole figure would be appropiate. However, the structure factors of the basal reflections are very weak. Hence, the ODF was calculated from several other pole figures which have rather low pole densities [3]. The texture could be described by the ideal orientation (0001) || subatrate plane with Gaussian spread about it figure 1. It can be characterized by the maximum pole density $P_{(0006)max} = f_{max}$ or by the spread width. These two parameters are related to each other according to figure 2.

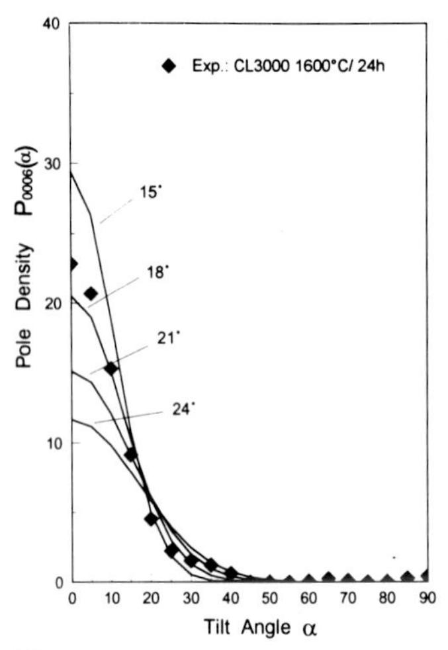

Figure 1
Pole density (0006) as a function of the angle α to the substrate normal direction
points: experimental measurements
lines: calculated with Gaussian spread of different spread width

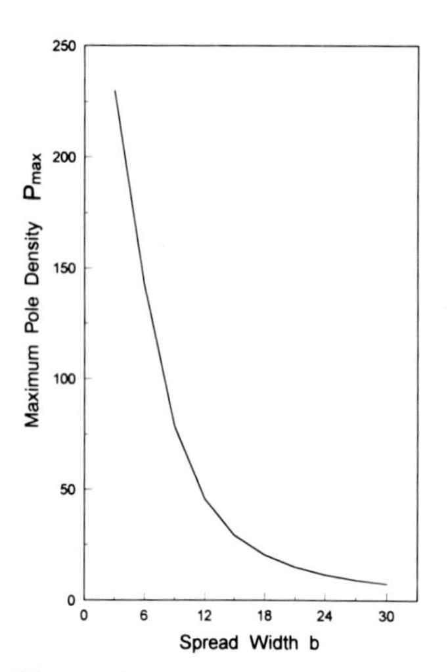

Figure 2
Relationsship
between maximum pole density and Gaussian spread width

RESULTS

The substrates were produced in an industrial production plant by tape casting and subsequent sintering. This allowed to measure the "green texture" as well as the final texture. At first, the accuracy ot texture determination as well as the reproducibility of texture formation was studied. A survey of that is given in table 1.

	Variation of texture strength
Reproducibility of texture strength in the same sample statistical error	~ 2.5%
Estimated systematical error (with constant experimental conditions)	~ 8.0%
Reproducibility within the same lot	~ 5.0%
Through–thickness inhomogeneity	≈ 14.0%

Table 1
Reproducibility of the measured textures.

	Green texture			Sintering texture		
	from	to	Band width	from	to	Band width
Kind of raw alumina	1.9	8.2	6.3	3.1	12.6	9.5
Sintering aids	4.6	6.2	1.6	8.6	10.1	1.5
Binder	6.3	7.0	0.7	10.5	11.5	1.0
Solvent	5.5	6.3	0.8	8.6	10.6	2.0
Softener	7.0	7.7	0.7	12.0	13.4	1.4
Raw material processing	5.0	5.4	0.4	10.4	14.1	3.7
Slurry drying	6.9	7.2	0.3	12.2	12.4	0.2
Slurry caster	4.6	5.2	0.6	8.8	9.9	1.1
Tape thickness	6.8	7.2	0.4	11.3	12.3	1.0
Subsequent mechanical compaction	6.3	6.7	0.4	9.4	10.0	0.6

Table 2
Band width of the influence of the green-forming parameters on the green and sintering textures

Second, the "band-width" of several influencing parameters was investigated as is shown in

table 2. From these results it can be concluded that most of the studied parameters have a (stronger or weaker) influence on the green- as well as final texture.
The technological tape casting process was modelled by manual casting in two different ways, i.e. simply pouring the slurry onto a glass plate or shear-spreading it with a blade. These two processes produced strong differences in the green textures as is seen in figure 3.

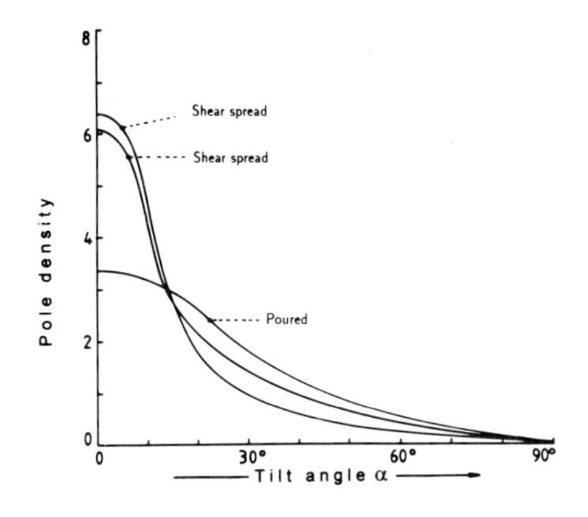

Figure 3
Pole density distribution of green-tapes prepared in different ways:
– poured onto a glass substrate
– shear spread

Furthermore, texture development during the drying process was studied by in-situ measurements using the "$\vartheta - \vartheta$"-geometry. It is seen in figure 4 that the texture remains essentially constant in this period.

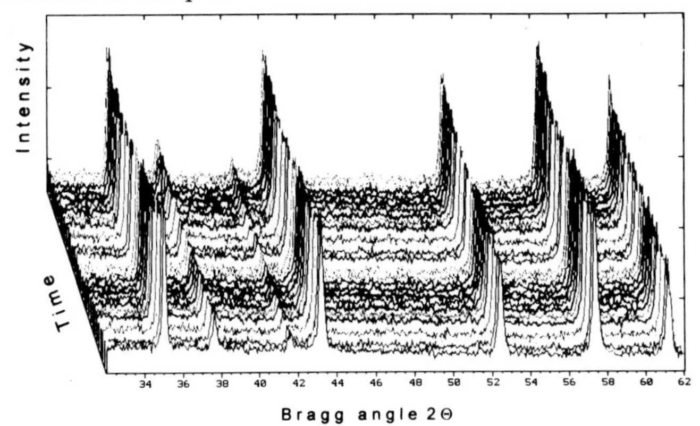

Figure 4
Part of the diffraction diagram measured in real time during drying of the slurry

During sintering the textures changed in degree but not in type particularly at high temperatures as is illustrated in figure 5. A small but nevertheless significant increase was found at the beginning followed by a temperature range with no influence on the texture and a third range with strongly increasing texture degree. In this latter region the time dependence of texture is as shown in figure 6.
It can also be conduded that the increase of texture is correlated with grain growth as is illustrated in figure 7

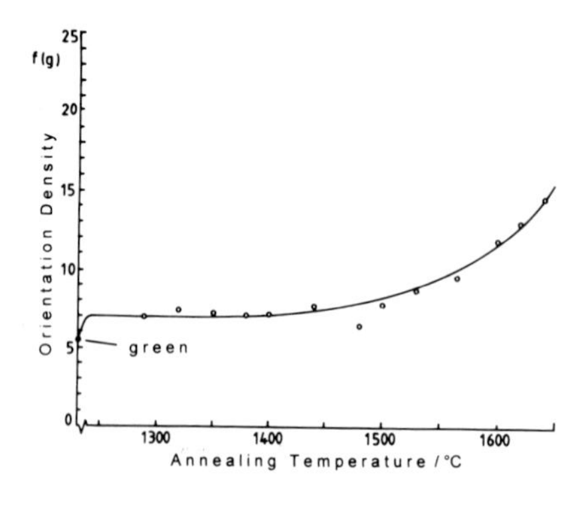

Figure 5
Maximum orientation density after sintering 12^h at different temeratures

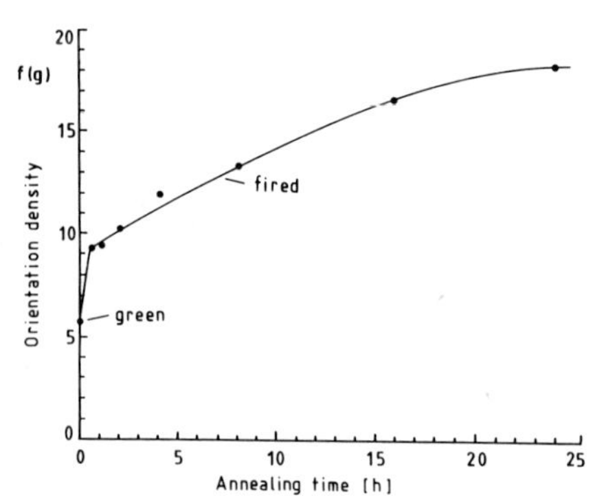

Figure 6
Maximum pole density after sintering at 1600°C as a function of holding time

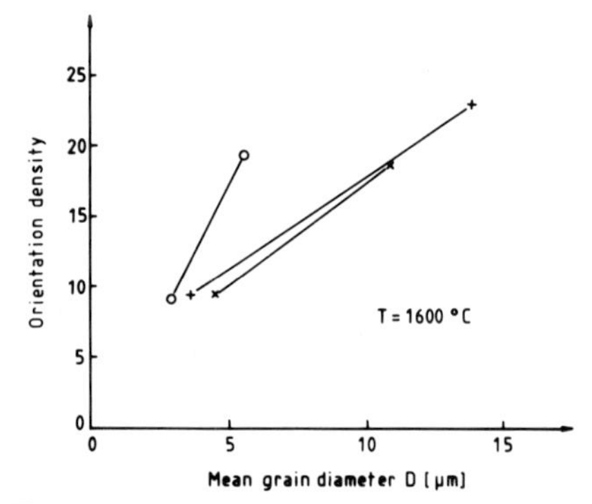

Figure 7
Increase of maximum orientation density and mean grain diameter by sintering at 1600°C

Since, in this particular case, the texture can be characterized by a single parameter an empirical formula for the degree of texture could be obtained. At first, it seems reasonable to introduce the "texture strength" parameter φ

$$\varphi = f_{max} - 1 \quad , \quad 0 \leq \varphi \leq \infty \tag{1}$$

The texture after sintering can be expressed in the form [5]

$$\varphi^{Sinter} = \varphi^{green} \cdot q \left[1 + A \cdot e^{-\frac{Q}{R \cdot T}} \cdot \sqrt{t}\right] \tag{2}$$

where T is sintering temperature and t sintering time. The factor q is near to 1 and the factor A is assumed to depend on grain boundary mobility m, specific grain boundary energy γ and initial particles size $\bar{r}_o$ in the form

$$A \sim \frac{\gamma \cdot m}{\bar{r}_o} \tag{3}$$

The time-dependence according to a square-root law is verified in figure 8, the temperature-dependence, according to a Boltzman factor is seen in figure 9.

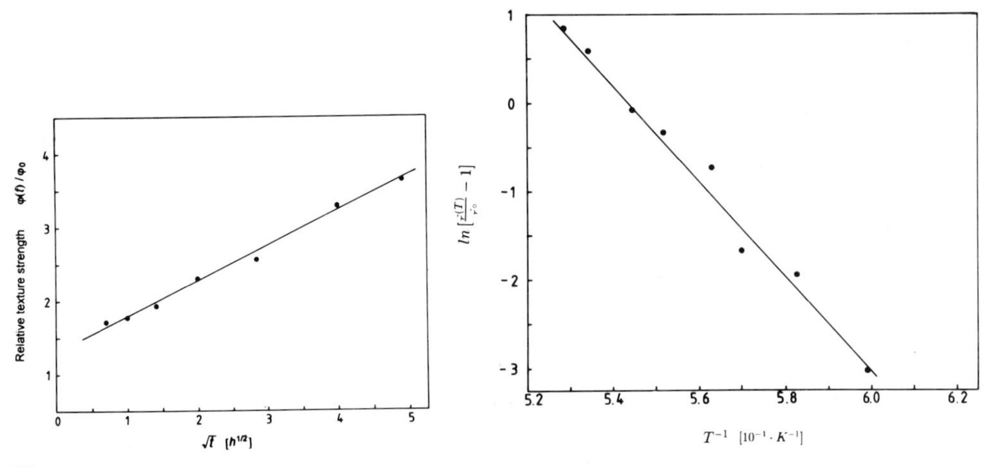

Figure 8
The relative texture strength plotted versus square-root of sintering time

Figure 9
Relative texture strength in an Arrhenius plot as a function of sintering temperature

CONCLUSIONS

According to equation 2, the final texture depends simultaneously on two factors i.e. the "green" factor and the "sintering" factor. It is

$$0 \leq \varphi^{green} \quad , \quad 1 \leq [sintering factor] \tag{4}$$

If strong textures are wanted, then both factors must be as high as possible. If random distribution is desired then the decisive factor is φ^{green} which can reach zero. Both factors depend strongly on the properties of the used alumina powders, i.e. particle size- and shape distribution as is illustrated schematically in figure 10.

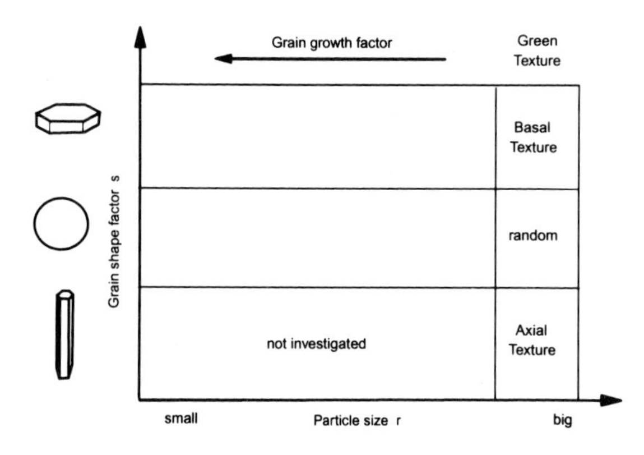

Figure 10
Schematic diagram illustrating the influence factors for texture formation in the green foil and after sintering

It is assumed that the big particles determine the grain growth texture after sintering, but a small particle fraction is needed in order to provide sufficient driving force. On the other hand it is the crystallographic "shape" of the particles that determines the green texture due to particle rotation. It is further assumed that in most cases, particle size and particle shape are statistically correlated.
Finally, it can be concluded from figure 3 that at least two different physical processes may constribute to the formation of the green texture, i.e. surface tension in the slurry and shear flow of a near surface region.

REFERENCES

1) Dimarcello, F.V., Key, P.L. and Williams, J.C.
Journ. Amer. Ceram. Soc 1972, 55, 509

2) Klein, H., Waibel, B., Martin, W. and Bunge, H.J.
In: Joining Ceramics, Glass and Metal.
Ed. W. Kraft DGM Informationsgesellschaft
Oberursel 1989, 285

3) Böcker, A., Brokmeier, H.G. and Bunge, H.J.
Journ. Europ. Ceram. Soc. 1991, 8, 187

4) Böcker, A., Bunge, H.J., Ernst, J. and Krahn, W.
In: Recrystallization '92. Ed. M. Fuentes and J. Gil Sevillano.
Trans Tech Publ. 1993, 575

5) Böcker, A., Bunge, H.J., Huber, J. and Krahn, W.
Journ. Europ. Ceram. Soc. (submitted)

Materials Science Forum Vol. 157-162 (1994) pp. 1417-1422

PREFERRED ORIENTATIONS IN POLYCRYSTALLINE C_{60}

K. Ito, Y. Takayama, K. Suenaga, T. Katoh and Y. Ishida

Department of Materials Science, Faculty of Engineering,
The University of Tokyo, 7-3-1 Hongo Bunkyo-ku, Tokyo 113, Japan

Keywords: Annealing, Carbon Sixty, Compression, Drawing, Preferred Orientation, Rolling

ABSTRACT

Following the suggestion obtained from our preliminary work that fcc carbon sixty crystals would deform through some crystallographic systems, their textures after deformation and annealing have been examined. No preferred orientations are found in a drawn specimen, nor in a drawn and annealed one. The rolling deformation has formed weak preferred orientations somewhat dissimilar to fcc metals, while a texture made by the compression is found to be similar to that of fcc metals. The preferred orientations developed by rolling and compression as well as microstructures have been changed by annealing.

1. INTRODUCTION

A molecule of carbon sixty consists of 60 carbon atoms, making a football-shaped cage 0.71nm in diameter. The molecules crystallize at room temperature into the cubic closed packed structure with a lattice constant of 1.42nm, whose (2n 0 0) X-ray diffractions are lacking[1]. Traces of {111} planes are observed on surface of monocrystals subjected to loading, suggesting that the crystals deform plastically at room temperature through some crystallographic systems. This plastic deformability is noteworthy, because most of carbon materials cannot be deformed plastically.

In the present work, preferred orientations are examined in polycrystalline aggregates of this new material after large plastic deformation as well as after subsequent annealing.

2. EXPERIMENTAL PROCEDURE

Raw material was produced by a conventional method of contact arc vapourization of a graphite rod in a 50 Torr atmosphere of helium, followed by extraction of the resultant graphite soot with toluene. Carbon sixty molecules were further separated by column chromatography on alumina with toluene + hexane. They were crystallized by annealing at 150 deg Celsius under a dynamic vacuum of about 0.001 Torr.

Crystals were compacted at room temperature in aluminium block under a pressure of 85 MN per square m, as shown by figure 1 (a). The materials were deformed together with the embedding block, as shown by figure 1(b), either by compression, or unidirectional rolling , or by drawing to about 90% reduction in height, or in thickness, or in area. After compression as well as rolling, the aluminium containers were stripped off and small specimens about 0.5mm in thickness, 1mm in width and 1mm in length were cut out from the deformed materials. In the case of drawing, small specimens were cut out together with the sheath. The cut out specimens were annealed for 1h at 300 deg Celsius in an evacuated capsule.

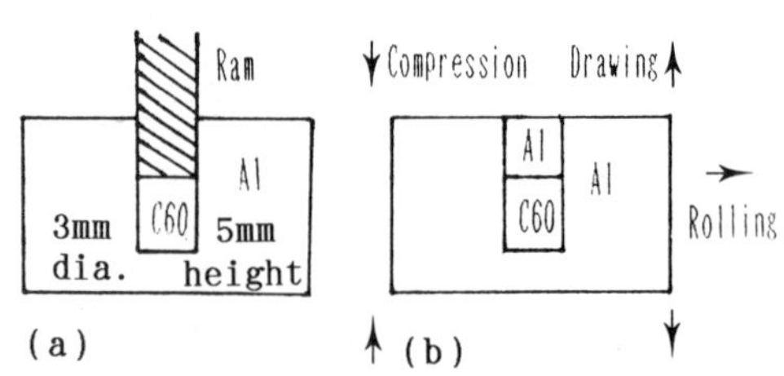

Fig.1 Schematic presentation of compaction (a) and deformation (b) procedure.

The specimens were cemented to glass plate, polished, etched by benzene, and observed with optical microscope. Debye rings from the small specimens cemented to glass needles were recorded on imaging plate by use of nickel filtered copper K alpha ray running horizontally as the incident beam. In the case of compressed specimens, the compression axis was at first held vertically and then the specimens were tilted about the axis which was perpendicular both to the incident beam and to the vertical line.

Intensities along the Debye ring were corrected with respect to absorption and background. The pole density distribution as a function of angle between compression axis and diffracting plane normal was calculated under the assumption that it might be symmetric with respect to the compression axis. The inverse pole figure was computed through a vector method from the pole density distributions of (111), (220), and (311) diffractions. A mean distribution of several Debye ring measurements obtained by changing the tilting angle was used in practice of the computation. Only qualitative pole figures were drawn in the case of rolled specimens.

3. RESULTS AND DISCUSSION

Shape and size of crystals as grown are shown in figure 2. In the section of compressed specimen(figure 3), some lineage

Fig.2 Section of as grown crystals embedded in half transparent cementing material.

Fig.3 Optical micrograph of a compressed specimen. Compression axis is parallel to the vertical line.

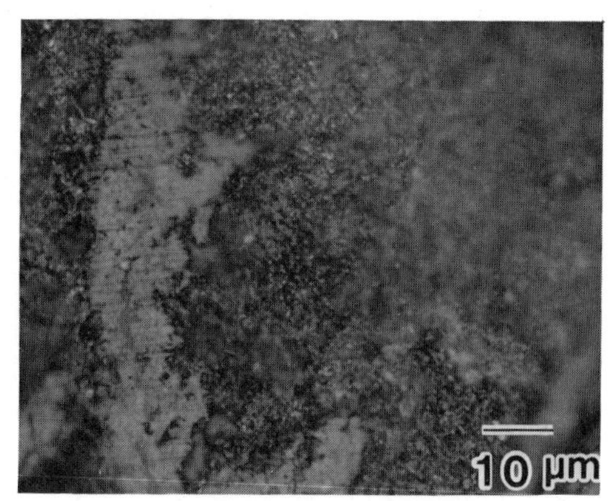

Fig.4 Optical micrographs of a compressed and annealed specimen. The section is parallel to the plane of compression.

structures are visible. They are thought to be interfaces between original crystals. The annealing yields network-like regions of a transformed structure(figure 4). They are at moment interpreted as "recrystallizing grains" developing from the interface between original grains(cf. figure 3). These features are observed just in the same manner in the specimen subjected to rolling.

An example of the diffraction pattern obtained from a drawn and annealed specimen is shown in figure 5, where Laue spots originate from the aluminium sheath annealed together with the specimen. Though intensities are not uniform along the rings, the distribution patterns are independent of the diffraction indices and the non-uniformity is certainly due to the absorption effect of the sheath. Any preferred orientation has been found neither in the drawn nor in the drawn and annealed specimen.

Figure 6 shows an example of the diffraction pattern obtained from a compressed specimen. Computed inverse pole figures of

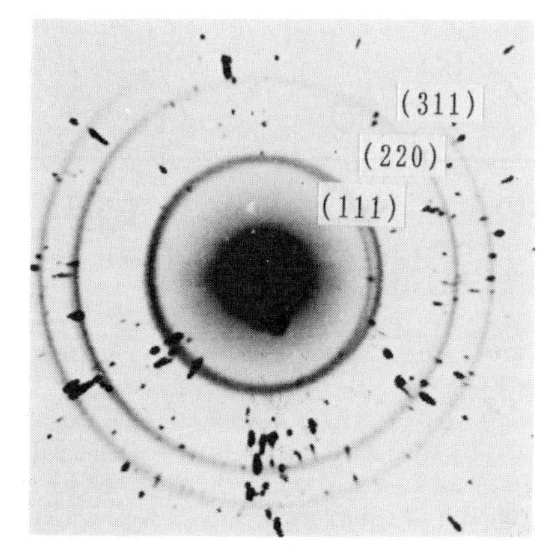

Fig.5 Diffraction pattern from a drawn and annealed specimen with aluminium sheath.

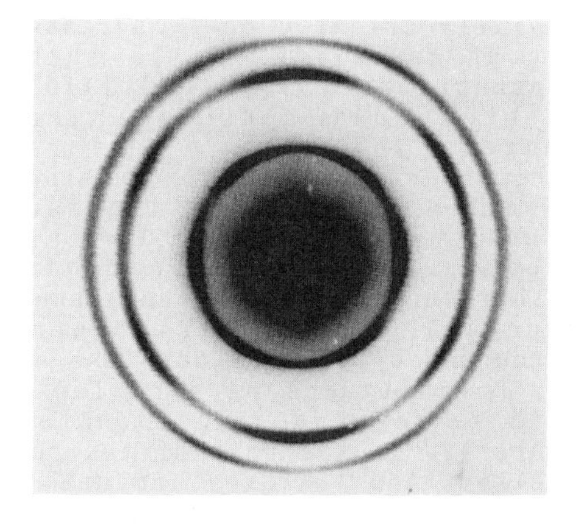

Fig.6 Diffraction pattern from a compressed specimen. The compression axis is parallel to the vertical line.

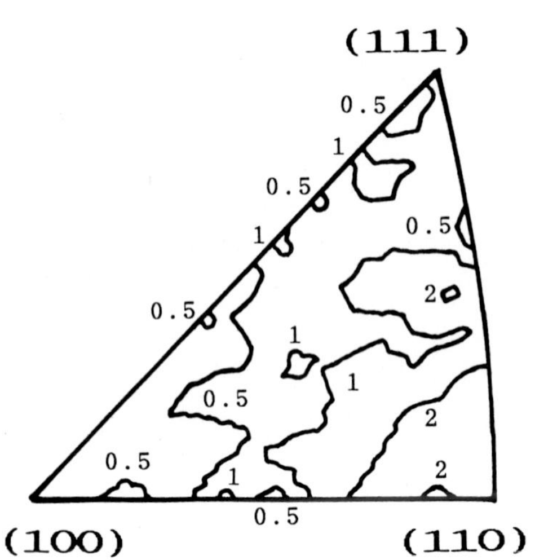

Fig.7 Inverse pole figure of a compressed specimen. The mean intensity is unity.

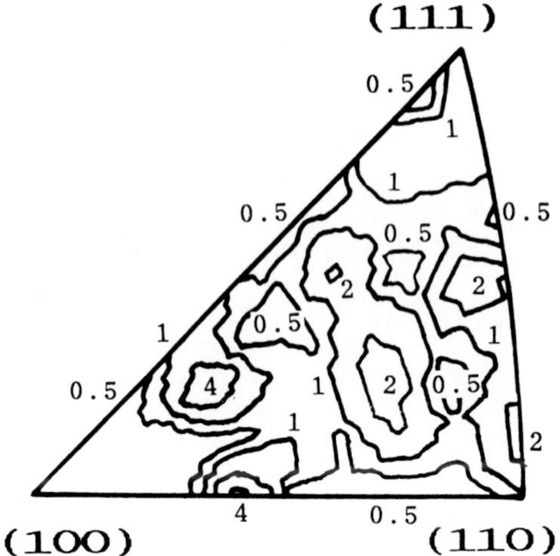

Fig.8 Inverse pole figure of the compression axis of a compressed and annealed specimen.

compressed as well as compressed and annealed specimen are shown by figure 7 and figure 8, respectively. Corresponding ideal orientations are shown in figure 9. The compression texture, where near [110] orientations prevail, is similar to the texture of compression rolled fcc metals, especially to that of a low stacking fault energy alloys[2]. Some components like [932], [310], and [111] oriented ones increase through annealing, while such components as [320] and [110] oriented ones decrease. The change in the texture corresponds to formation of the network-like transformed structure(figure 4). It is, however, at moment not clear whether any orientational and structural change has occurred also in the interior region surrounded by the network-like structure.

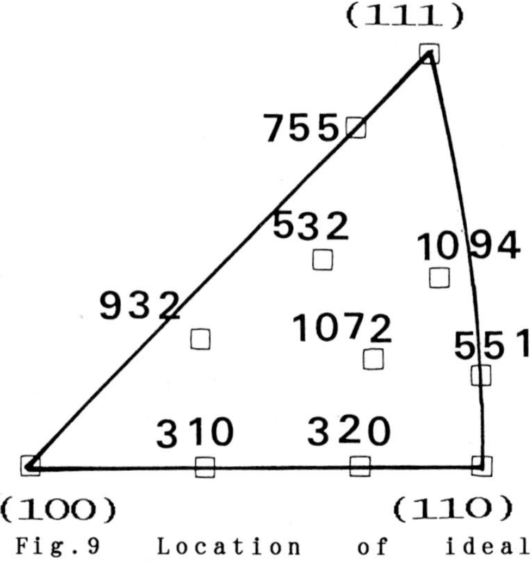

Fig.9 Location of ideal orientations.

Pole figures of a rolled specimen are shown by figure 10, while those of a rolled and annealed specimen are presented by figure 11. In the case of the rolled as well as rolled and annealed specimen, the textures are weak and are in contrast to the case of compression rather different from those of fcc metals and alloys. The observed asymmetry may be due to the method of sample preparation of unidirectional rolling. The preferred orientation of the rolling texture is located near to the (111)[$\bar{1}1\bar{2}$] and (111)[$10\bar{1}$] orientation, while that of the annealing texture appear

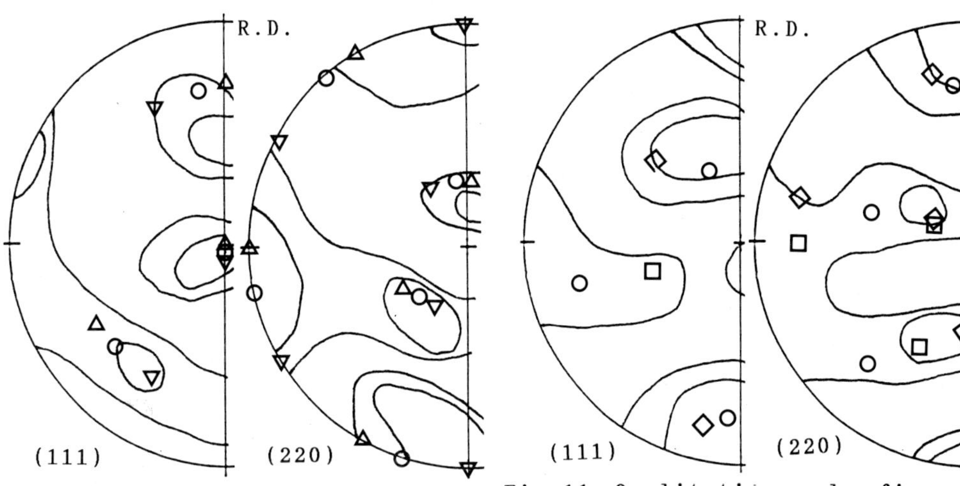

Fig.10 Qualitative pole figures of a rolled specimen. The right hand side was undetermined.
△(111)[11$\bar{2}$] ○(111)[21$\bar{3}$]
▽(111)[10$\bar{1}$]

Fig.11 Qualitative pole figures of a rolled and annealed specimen.
◇(11 20 18)[$\bar{2}$2$\bar{1}$] ○(012)[$\bar{2}$2$\bar{1}$]
□($\bar{4}$ 1 10)[$\bar{2}$2$\bar{1}$]

near to the (012)[$\bar{2}$2$\bar{1}$] orientation. The change in the preferred orientations by annealing seems to have occurred in both cases of compression and rolling by preferential growth of structural components in certain orientations in the deformed structure.

The reason why no preferred orientations are formed by drawing deformation is not clarified yet. It might be related to the difference in deformation behaviour of the material. It deforms at room temperature by tensile and compression loading under several MPa and several tens MPa, respectively.

4. SUMMARY

The drawing deformation of aggregate of carbon sixty crystals in the fcc structure embedded in a hole in aluminium block has formed no preferred orientations in the deformed state, nor after subsequent annealing.

In the case of rolling deformation, weak preferred orientations which are different from those of metals in the same crystal structure are found. They are located near to the (111)[21$\bar{3}$] and to the (012)[$\bar{2}$2$\bar{1}$] orientation in the as deformed and in the annealed state, respectively.

The compression deformation has developed relatively strong near [110] preferred orientations similar to the fcc metals. Some components like [320] oriented one decrease on annealing while other components like [932] oriented one increase.

The microstructure has been correspondingly observed by optical microscope to change by annealing.

REFERENCES

1) P.A.Heiney et.al.:Phys.Rev.Lett.,1991,66,2911.
2) C.S.Barret and T.B.Massalski:Structure of Metals, third edition, 1966,547.

Materials Science Forum Vol. 157-162 (1994) pp. 1423-1434

INHOMOGENEOUS TEXTURES IN TANTALUM SHEETS

H. Klein [1], C. Heubeck [2] and H.J. Bunge [1]

[1] Institut für Metallkunde und Metallphysik, TU Clausthal, Grosser Bruch 23, D-38678 Clausthal-Zellerfeld, Germany

[2] Diehl GmbH & Co., Nürnberg, Germany

Keywords: Inhomogeneous Textures, Local Texture Measurement, Global Texture Measurement, Coordinate Transformation, Cross-Rolling, Tantalum

Abstract

Tantalum sheets produced by pseudo cross–rolling /1/ with intermediate annealing treatments develop extraordinarily strong texture inhomogeneities in through–thickness direction as well as in the sheet plane. These inhomogeneities were studied by x–ray and neutron diffraction texture analysing using ODF.
The "inplane" inhomogeneities must be attributed to inhomogeneous deformation of the sheets when passing the rolls in two mutually perpendicular directions.
The through–thickness inhomongeneities can be attributed to local different recrystallization in a particular layer of the sheet. This in turn must be due to through–thickness inhomogeneities of rolling deformation.
The inhomogeneities have strong influence on the mechanical properties of the sheet which were measured and calculated from the texture.

Introduction

Tantalum sheets were commercially produced by pseudo–cross–rolling with intermediate heat treatments. This procedure leads to a complicated texture inhomogeneity in the sheet plane as well as in the sheet normal direction. These inhomogeneities play an important role when physical properties e.g. elastic or plastic are to be considered. It was the purpose of the present work to study these texture inhomogeneities in more detail and to estimate their influence on the materials properties.

Experimental Procedure

The material was commercial produced tantalum sheet of 4 – 5 mm thickness. It had been cross–rolled with several intermediate annealing treatments. The rolled slabs had dimensions approximately of 1.0m by 0.70m. The samples were taken at different places as shown in

Fig.1.
Texture studies were carried out in three different ways:

1) x–ray reflection measurements parallel to rolling plane at different depths s beneath the surface.

2) x–ray reflection measurements with composite samples cut perpendicular to one of the sheet rolling direction.

3) Neutron measurements using composite samples.

X-ray measurements were done with a computer operated texture goniometer using $Cu_{K\alpha}$ radiation. Reflection samples of 30mm Ø were measured in steps of $\Delta\alpha = 5°, \Delta\beta = 3.6°$ up to $\alpha_{max} = 70°$. Three pole figures i.e. (110), (200), (211) were taken.
Neutron measurements were carried out with a texture diffractometer at National Institute of Standards and Technology, Gaithersburg, USA. Pole figure measurements were made in steps of $\Delta\alpha = 5°, \Delta\beta = 5°$.
From pole figures, ODF were calculated using the series expansion method and assuming orthorhombic sample symmetry. Series expansion was used up to L = 22 including odd–order terms obtained with the iterative positivity method /2/. Measurements taken according to method 2), i.e. in a plane perpendicular to the rolling plane were transformed into the rolling plane. This was done in terms of the texture coefficients $C_\lambda^{\mu\nu}$ rather than directly in the pole figures. If the measuring coordinate system K_M is rotated with respect to the reference system K_A (Fig.2) (i.e. the rolling–plane system) by the rotation g^M then the coefficients C_l^{mn} in K_M can be transformed into the ones C_l^{mn} in K_A by the relation ship

$$^{KA}C_l^{mn} = \sum_{s=-l}^{+l} {^{KM}C_l^{ms}} \cdot T_l^{sn}(g^M) \tag{1}$$

In order to study the expected strong through–thickness inhomogeneity surface layers $s = \frac{d}{d_0}$ were removed from the surface by machining grinding and subsequent etching (Fig.3).

Results

In Fig.4 (110) pole figures are shown obtained in sample A at different depth s. These pole figures show a strong through–thickness variation but they also show that orthorhombic sample symmetry is not fullfilled.
ODF sections calculated from the pole figures are shown in Fig.5. They exhibit several texture components which can be related to deformation and recrystallization textures of bcc materials e.g. a cube component (001)[100] the rotated cube component (001)[110] with extension into the α–fibre (hkl)[110] and particularly a strong γ–fibre (111)[uvw].
The through–thickness inhomogeneity is drastically illustrated by the orientation density in the orientation $\{\varphi_1 = 0°, \Phi = 54.7°, \varphi_2 = 45°\}$ corresponding to {111}[110] which is shown in Fig.6.
The average texture over all depths s is obtained from the composite sample measured at the cross–section plane (and transformed into the rolling–plane system). It is shown in Fig.7 for the sample position A and in Fig.8 for the sample position B. These textures can be understood as averages of some kind of the textures shown in Fig.5. It is however also seen that the textures of two different places in the rolling plane are different with the differences

being significantly greater than the experimental error.
Finally, the most accurate measurement of the through–thickness average texture is obtained by neutron diffraction as shown in Fig.9 (sample position C). This texture shows clearly the strong γ–fibre component seen in the local texture at $s = 0.45$ in Fig.5 superposed to the other texture components dominating in the other s–sections.
The influence of texture inhomogeneities on mechanical properties is drastically illustrated in Fig.10 which shows the Lankford parameter r as a function of the angle toward the rolling direction. Mechanically measured values are compared with values calculated from texture coefficients $C_{\lambda}^{\mu\nu}$ using the full constrained Taylor model and $\{hkl\}\langle 111\rangle$ glide. All three curves of this figure were obtained with samples from different places in the sheet.

Conclusions

The investigated tantalum sheets show inhomogeneities in the rolling plane as well as in through–thickness direction which are high in the rolling plane and even larger in the through–thickness direction.
In rolled sheets, texture inhomogeneities along the rolling direction are normally found at the beginning and at the end of the sheet. Here, the lenght of the sheet is nearly the same as its width. Hence, the start and the end inhomogeneities follow each other immediately. Also in width direction inhomogeneities are conventionally found near the rim of the band. Here, these inhomogeneities superpose with the lenght inhomogeneities when the sheet is turned for cross rolling. As a result, a complicated texture *field* develops in the sheet of which we have shown only some feature in this paper. A more comprehensive investigation of this texture *field* will be published elsewhere /3/.
The extraordinary form of the through–thickness inhomogeneity illustrated in Fig.5 is due to local different recrystallization in a limited layer from $s \approx 0.25$ to $s \approx 0.5$. The strong γ–fibre found in this layer is a typical recrystallization texture component of bcc metals. This is also corroborated by optical micrographs showing a bigger grain size in this region. The textures of the other parts of through–thickness direction are within the variation range of rolling textures of bcc metals.
A through–thickness profile of the type of Fig.6 poses a serious problem in the calculation of average mechanical properties such as the r–value. Strictly speaken, the sample is a composite sample in which each layer has another r(α) curve.
Besides this, the recrystallization layer also has a lower absolute value of yield stress. This must be taken into consideration in the calculation of the resulting r(α)–curve which is then no longer the linear average over the textures of the layers. Hence, in this particular case the neutron texture measurement is not necessarily the "correct" one for the macroscopic r(α)–curve. This effect will be discussed later /3/.
The question why part of the cross section of the material is recrystallized whereas other parts are not, cannot be answered without further investigations. It may however be assumed that in the recrystallized part the local deformation degree was higher than in the non–recrystallized ones.

Acknowledgements

The authors would like to thank the Cabot Corp., Boyertown Pa. U.S.A. for the neutron diffraction pole figures of tantalum.

References

/1/ A. Böcker, H. Klein, H.J. Bunge: Development of Cross–Rolling Textures in Armco–Iron, in: Textures and Microstructures, (1990), Vol.12, pp.103–123.

/2/ H.J. Bunge, Calculation and Representation of Complete ODF, Proc. of ICOTOM–8, Ed.: J.S. Kallend and G. Gottstein, The Metallurgical Society of AIME (1988) 69–78

/3/ H. Klein, C. Heubeck, H.J. Bunge, in: Textures and Microstructures, in preparation

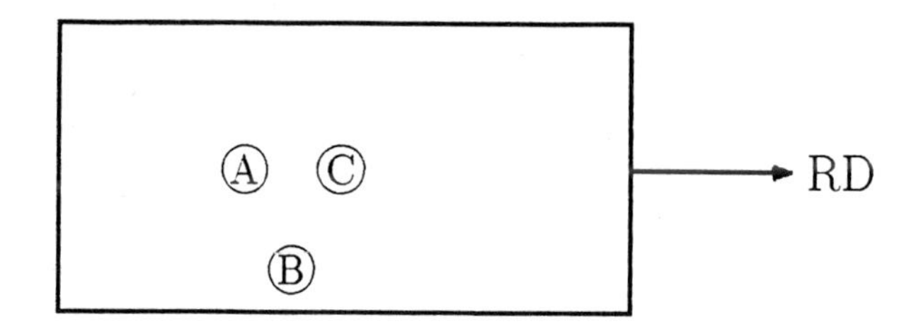

Fig.1 Positions of samples in the sheet.

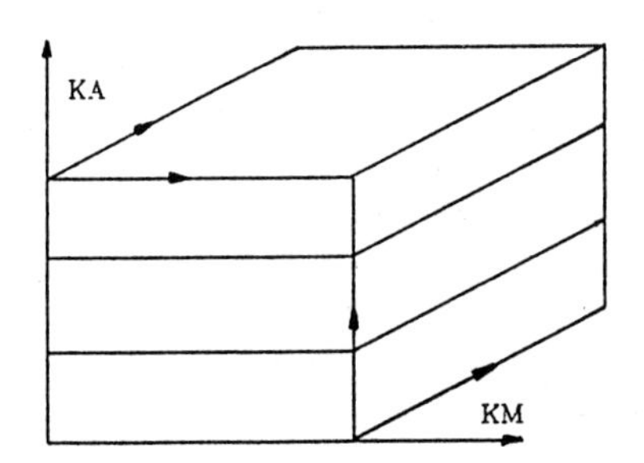

Fig.2 Relationship of measuring and rolling–plane coordinate systems with composite samples.

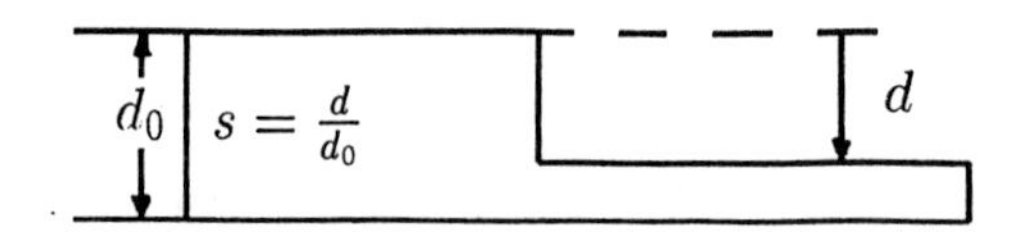

Fig.3 Definition of the depths s beneath the surface.

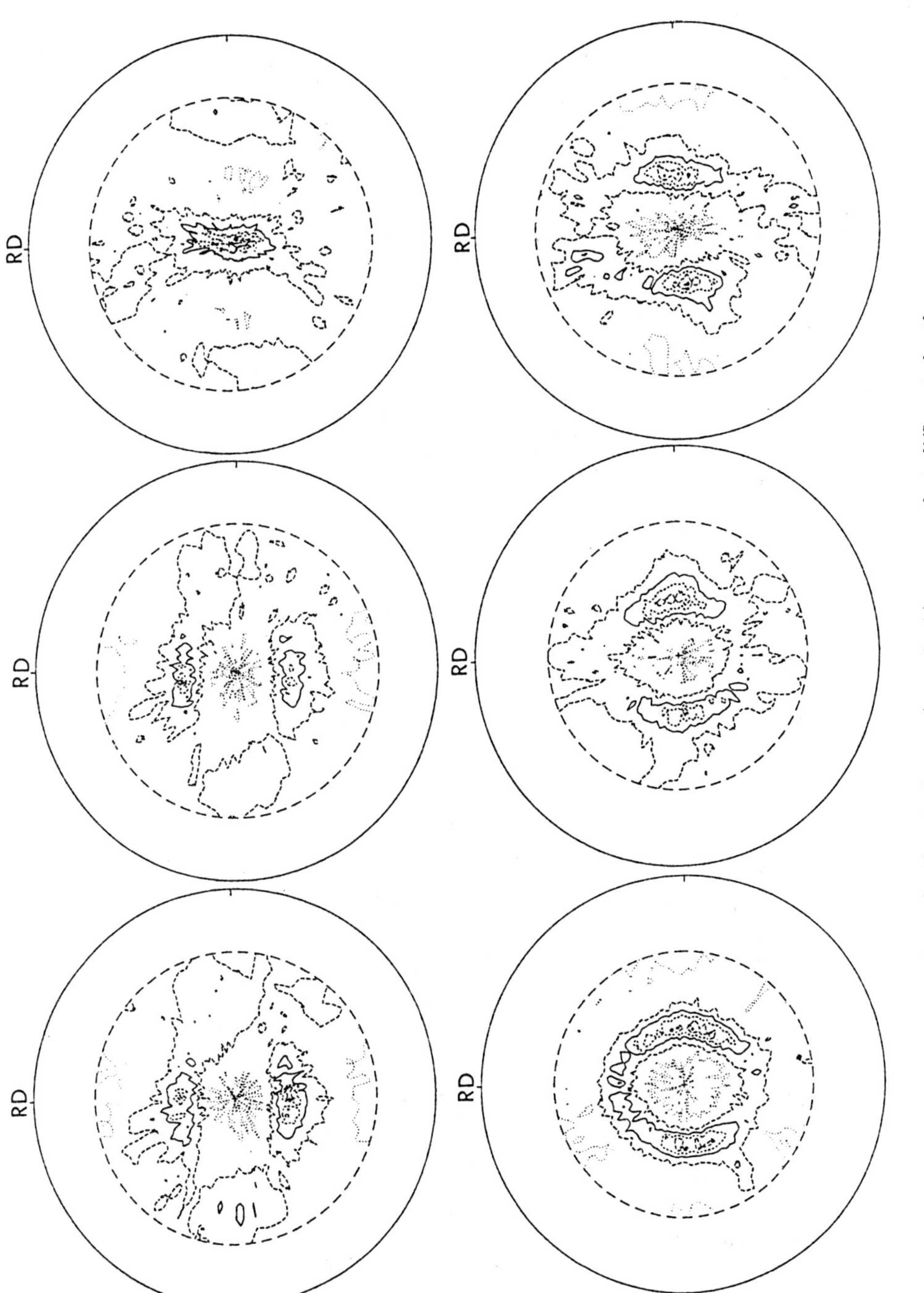

Fig.4 (110) pole figures (sample A) measured at different depths s:
a) 0.025, b) 0.075, c) 0.275, d) 0.45 , e) 0.52 , f) 0.85

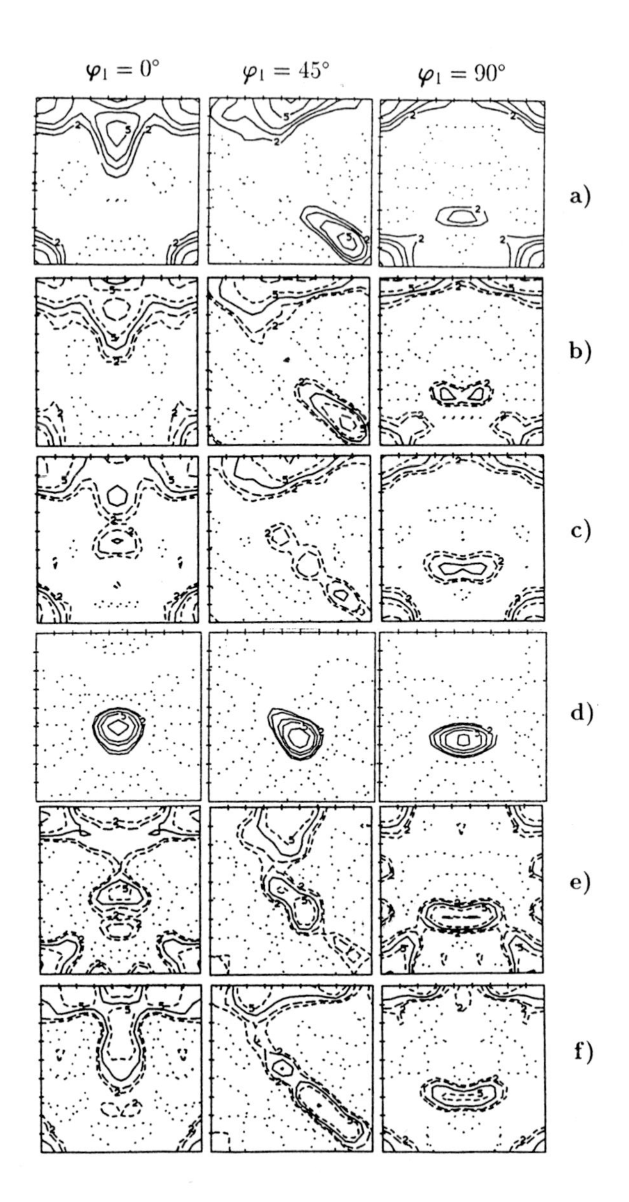

Fig.5 Orientation Distribution Functions of sample A in $\varphi_1 = 0°, 45°$ and $90°$ sections. The measurement was carried out in layers at different depths s.
a) 0.025, b) 0.075, c) 0.275, d) 0.45 , e) 0.52 , f) 0.85

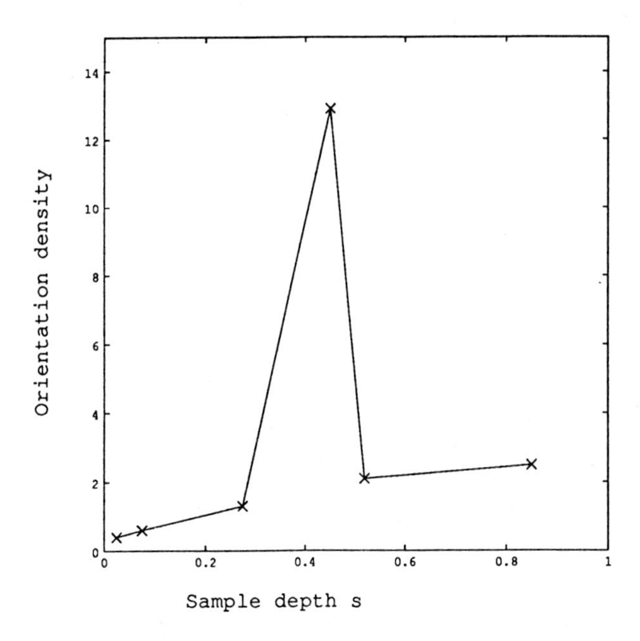

Fig.6 Orientation density of the {111}[110] component as a function of the depth s in the sheet.

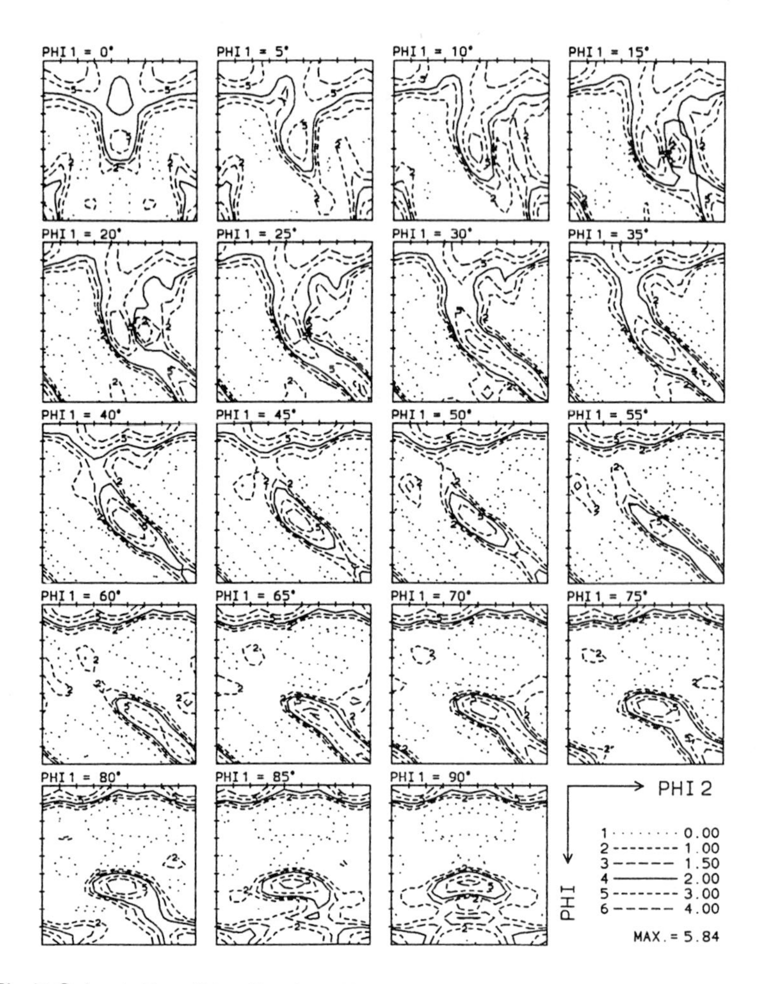

Fig.7 Orientation Distribution Function of the composite transverse sample A (x-ray measurement) in φ_1–sections.
Coordinate transformation by texture coefficients.

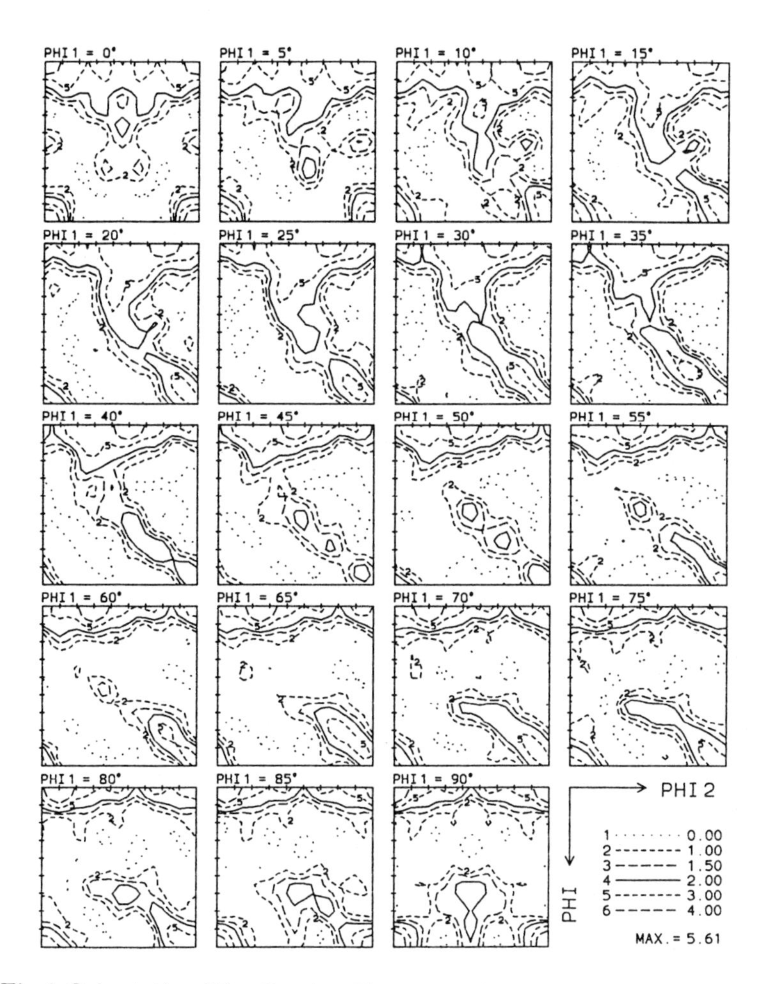

Fig.8 Orientation Distribution Function of the composite transverse sample B in φ_1–sections (x-rax diffraction).
Coordinate transformation by texture coefficients.

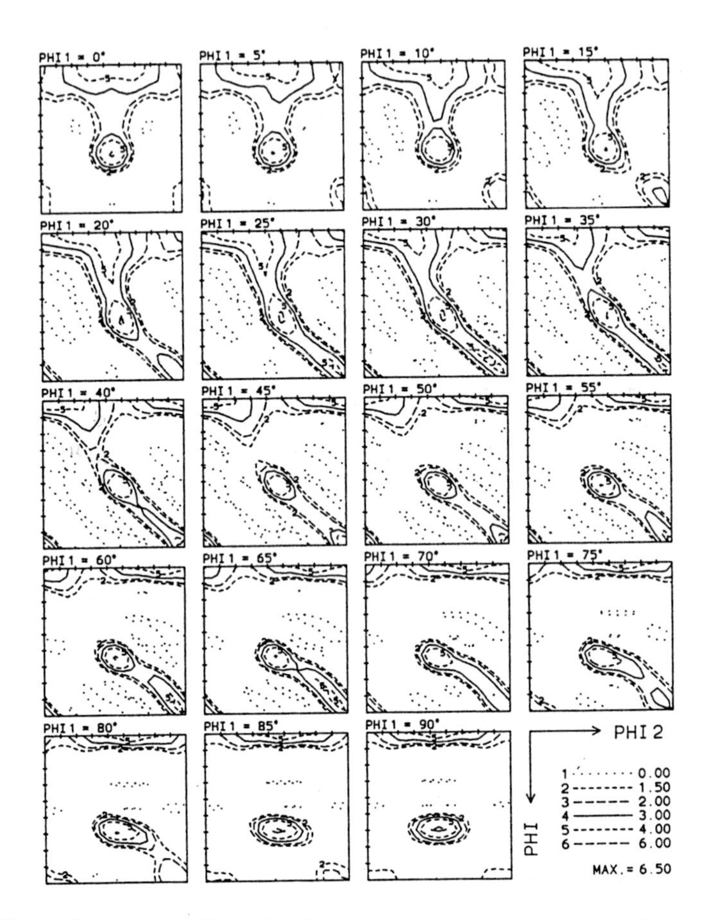

Fig.9 Orientation Distribution Function in φ_1–sections of the sample C (Neutron diffraction)

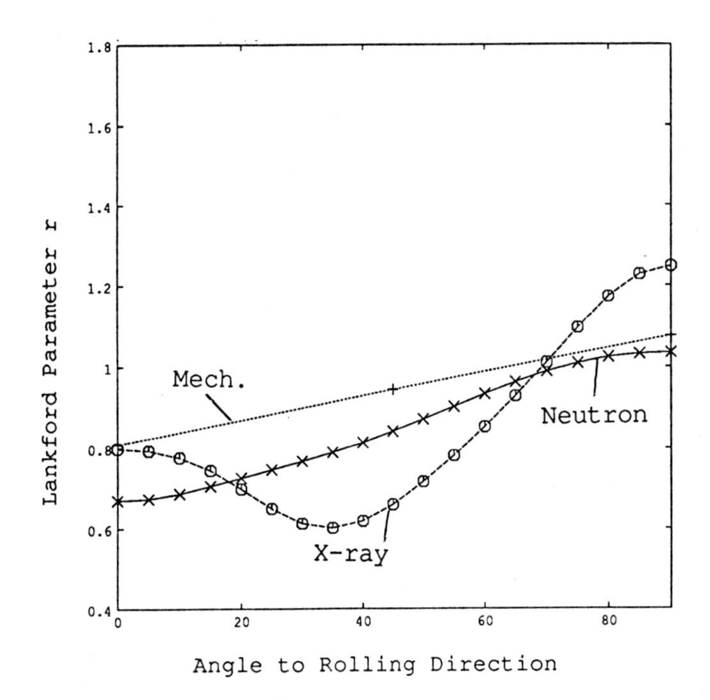

Fig.10 Lankford parameter as a function of the angle α towards the rolling direction. Calculated from global x–ray measurement, Neutron diffraction and mechanical measurement.

Materials Science Forum Vol. 157-162 (1994) pp. 1435-1442

TEXTURE AND GRAIN STRUCTURE EFFECTS ON THE RELIABILITY OF MICROELECTRONIC INTERCONNECTS

D.B. Knorr [1] and K.P. Rodbell [2]

[1] Materials Engineering Department and Center for Integrated Electronics, Rensselaer Polytechnic Institute, Troy, NY 12180-3590, USA

[2] IBM T.J. Watson Research Center, PO Box 218, Yorktown Heights, NY 10598, USA

Keywords: Interconnects, Reliability, Electromigration, Stress Voiding, Hillocks, Fiber Texture, Aluminum Alloys

ABSTRACT

Fine lines of aluminum-based alloys are fabricated on single crystal silicon wafers to interconnect devices. Progressively finer lines and multiple layers of interconnect wiring facilitate higher levels of function in successive generations of electronic devices. As the line widths decrease to 0.5 μm in the latest generation of devices, manufacturability and reliability in-service are greater concerns. Four failure modes are discussed: 1) thermal hillocks, 2) grain collapse, 3) stress voiding, and 4) electromigration. Texture and grain structure are shown to exert a critical influence on all four processes. The grain size/line width ratio and the texture control the electromigration and stress voiding failures where transport along grain boundaries implies that local texture and grain misorientations are important. Anisotropy of deformation within the grains on a local scale controls the extent of thermal hillocking and grain collapse so local grain orientations control these processes.

INTRODUCTION

Aluminum-based thin film metallizations, typified by Al-(0.5 to 4)Cu, Al-(0.5 to 1)Si, and Al-(0.5 to 2)Cu-(0.5 to 1)Si, are most widely used to interconnect devices in silicon single crystal wafers. The subtractive fabrication process consists of several steps: a blanket film is deposited, usually by sputtering, lines are defined in a photoresist, and excess metal is removed by gaseous reaction (Reactive Ion Etch, RIE). The line structures are coated by an insulating dielectric, apertures are opened as vias between metal levels or as contacts to the silicon devices, and the holes are filled with a conductive metal (tungsten in the present technology) to provide electrical continuity between all levels of the structure. This processing sequence is repeated to interconnect all the devices with as many as four levels of wiring and to establish large bond pads on the chip periphery. Often, the conductor consists of an underlayer/Al alloy/overlayer where the underlayer and/or overlayer function as a diffusion barrier or as a redundant conductor to enhance reliability. Common thin film barriers are Ti, TiW alloy, Ti/TiN, TiN, or W. Current leading edge technology utilizes a minimum line width of <0.5 μm, but future integrated circuits with line widths <0.2 μm are envisioned.

The microstructure of the Al-based metallization contributes to high manufacturing yields and to the reliability of devices in-service. It is recognized that several microstructural attributes are important: 1) precipitate type, volume fraction, size, and distribution, 2) solute distribution, 3)

multilayer conductor configuration, if present, 4) grain size and grain size distribution, and 5) texture. The first two depend on alloy selection while the last two are related to the polycrystalline grain structure. All factors are heavily influenced by processing (deposition and annealing) which distributes the alloying additions, establishes the grain size, and develops the texture. Subsequent discussion will consider how texture and, to a lesser extent, grain structure influence several failure mechanisms in thin film metallization. An ultra smooth surface on the blanket film is preferred during processing. Hillocks and grain collapse, which are caused by thermally-induced stress during processing, degrade surface quality. Stress voiding [1] and electromigration [2] occur in line structures from exposure to stress and high current density, respectively. All four phenomena are known or suspected to depend on texture [3]. The next section will describe each one and summarize the state of understanding on the texture dependence. The texture in aluminum metallization is usually composed of random + (111) fiber components [3,4]. A stronger texture results from a decrease in the random volume fraction and/or a decrease in the spread of the (111) fiber distribution.

THIN FILM DAMAGE MECHANISMS

Stress develops in thin aluminum-based films because of thermal expansion mismatch between the silicon substrate (α_{Si} = 3 ppm/°C) and aluminum (α_{Al} = 23 ppm/°C). The stress is balanced biaxial for a blanket film on a silicon substrate, and tensile triaxial in narrow lines that are covered by an uniform layer of SiO_2 passivation [5]. When subjected to a thermal cycle during processing, a blanket Al-based film develops a characteristic stress-temperature profile illustrated in figure 1 [6,7]. The compressive stress during heating can result in the thermal hillocks while the tensile stress on cooling drives the grain collapse and stress voiding phenomena. A description of each damage mechanism is given below with additional discussion on the textural aspects of failure. Electromigration receives the most extensive consideration because more work has been done to characterize this phenomenon than the sum of the work on the other three.

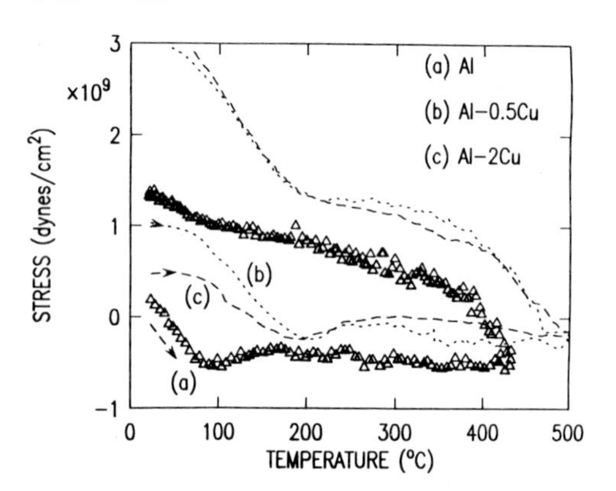

Fig. 1. Stress vs. temperature for 1 μm thick sputtered aluminum-based films on silicon.

Thermal Hillocks

The formation of thermal hillocks results from compressive stresses during deposition or during the compressive portion of a thermal cycle as shown in figure 1. Hillocks appear as surface protrusions or side protrusions in lines [8] due to a stress relief mechanism where grain boundary diffusion and plastic flow are implicated as the deformation processes. Several studies report conflicting results on orientation effects where a stronger texture (not accurately measured) either increases [9] or decreases [10] the number of hillocks. Modeling [11] indicates that only misoriented grains are responsible for hillock formation because inelastic deformation by slip is easier in grains misaligned from the (111) orientation. Early measurements on the orientation of individual hillocked grains [12] in tetragonal tin showed essentially all hillocks were non-random and oriented well away from the c-axis. The volume and height of hillocks increase with the breadth of the (111) distribution in an Al-Si-Ti alloy [10]. Definitive results on the origin of hillocks [13,14] proves that most hillocks are associated with grains oriented well away from the (111) distribution. Hillocks on grains near the (111) orientation are due to grain growth that has consumed the original misoriented grain [14,15].

Grain Collapse

A study [16,17] of this phenomenon reveals that local deformation and transport on the scale of an individual grain causes that grain to be depressed relative to the surrounding grains. In extreme cases, the grain disappears entirely [18,19]. The same phenomenon is usually observed, to some extent, in any film that is thermally cycled such that the stress becomes tensile on cooling. The only experimental evidence of orientation shows that well oriented (111) grains are involved [19]. The mechanism is unresolved but diffusion and slip are possibilities [18]. Modeling work suggests that texture is responsible because grains with a (111) orientation are stronger than grains with other orientations [11,14,20]. Two scenarios are possible to justify a texture effect. A non-(111) grain surrounded by better oriented (111) grains would experience initial yielding while (111) grains remain primarily elastic because the (111) grains are stronger. The accommodation of the inelastic deformation accounts for the collapse. This argument is advanced by Smith et al. [18] to explain their data and observations. Another scenario is that an (111) grain surrounded by non-(111) or misoriented (111) grains will be under higher elastic stress. Diffusive flow of mass from the (111) grain will cause it to sink [19]. Further experimental work is required to identify the dominant mechanism; it is conceivable that both might be observed, but under different conditions.

Stress Voiding

A critical reliability concern in interconnect structures with lines < 1 μm wide is stress voiding. A tensile triaxial stress state develops in lines encapsulated with insulating dielectric such as SiO_2 or Si_3N_4. The stress is relieved during high temperature exposure by diffusive flow that opens slit and wedge-type voids at grain boundaries. Because of surface energy minimization, {111} planes are found along the faces of the voids [21] so specific boundaries are prone to damage where void volume is minimized [22], so it is speculated that control of local orientation would be an effective method to minimize stress voiding. No definitive experimental data or modeling studies exist to quantify the extent of a texture effect.

Electromigration

Electromigration is the mass transport of a conductive material in an electric field. Both the electric field and an "electron wind" are responsible for the mass transport. The reliability of fine interconnect lines depends on the mass transport. If sufficient material is transported, an open line results; if the local deposition of material extrudes a hillock or whisker, a short with a nearby line occurs. Electromigration hillocks are distinct from the thermal hillocks discussed previously. Mass transport usually occurs along grain boundaries and, under some circumstances, along interfaces. Damage by electromigration is associated with regions of "flux divergence" in a line. Damage at grain boundaries, often triple points, occurs because of a mass flux imbalance: more material entering than leaving a location creates an hillock while more material leaving than entering results in a void. Both grain topology and anisotropy of boundary transport are contributing factors. Line thinning is caused by variations in interfacial diffusion. In very narrow lines that are one grain wide, failure can occur by transport of an entire segment of line.

Electromigration testing involves the fabrication of narrow line structures as single lines or as parallel line arrays [23]. Since time to failure varies widely between samples, a statistically meaningful sample of 10 to 30 lines is tested. Data are correlated on a probability plot where the logarithm of time to failure of individual samples is plotted as a cumulative probability. A straight line implies a lognormal distribution so median time to failure and the standard deviation of the lognormal distribution (slope of the plot) are obtained. An improvement in reliability is realized by an increase in the mean time to failure and a decrease in the standard deviation, which means that the most dangerous early fails are delayed. Failure distributions that are not monomodal can be especially damaging because a small population of early failures might determine the lifetime of an integrated circuit.

Three microstructural attributes influence the electromigration lifetime: 1) grain size, 2) grain size distribution, and 3) texture. Texture effects must be considered in the context of the grain structure where four regimes are defined by the relationship between line width, w, and grain size, d [24,25]. These regimes are:

1) *Polycrystalline*: w/d > 1 such that the line is several grains across.
2) *Near bamboo*: the line is composed of a mixture of segments that alternate, in a random fashion, between a single grain across the width and polycrystalline segments.
3) *Bimodal*: a mixture of large grains with w/d < 1 and small grains with w/d >1 [26].
4) *Bamboo*: w/d << 1 such that a line is a chain of single grains [27].

The distinction between near bamboo and bimodal involves both the grain size distribution and the relative line width, w/d. Near bamboo is usually associated with a monomodal grain size distribution due to normal grain growth where line width effects are found in narrow lines. Bimodal refers to a grain size distribution of both small and large grain size distributions where the large grains traverse the relatively wide line and the small grains do not. Texture effects on electromigration should be interpreted cautiously by considering the grain structure regime. The failure statistics are sensitive to the grain structure where polycrystalline and bamboo grain size distributions usually have monomodal failure distributions while bimodal and near-bamboo grain size distributions often have bimodal or multimodal failure distributions. In the latter cases flux divergences, where small grains meet large grains across the line, are implicated in the early failures. Bimodal and multimodal failure distributions are attributed to vias between levels of metallization [24], to a near bamboo structure in the line [24,28], and to texture [29-31].

Most studies that examine texture effects presume or confirm a polycrystalline grain structure in the interconnect line. Attardo and Rosenburg [26] were the first to report a substantially improved median time to failure (MTF or t_{50}, which is the time required for 50% of the test population too fail) for strongly (111) textured films compared to weakly textured films, but an unambiguous correlation was not possible because the grain size was very different between their two conditions. Similarly, a study by Nagasawa et al. [32] did not separate the effects of grain size, grain size distribution, and texture. Li et al. [33] showed a substantial increase in electromigration lifetime as (111) texture became stronger; however, the study lacks statistical significance since only one line was tested in each condition. The most complete study of microstructural effects in electromigration is by Vaidya and Sinha [34]. Their work produced a correlation for MTF with grain size, grain size distribution, and texture:

$$\mathrm{MTF} \propto S/\sigma^2 \log[I_{(111)}/I_{(200)}]^3 \qquad (1)$$

where S is the median grain size, σ is the standard deviation of the log-normal grain size distribution, and $I_{(111)}$ and $I_{(200)}$ are the X-ray intensities of the (111) and (200) peaks, respectively. An important limitation is that this correlation was developed *only for polycrystalline lines*. Furthermore, texture correlations are done with Bragg-Brentano intensities which is a *very* incomplete measures of texture [3]. Thus, the important role of texture in electromigration is well established, but additional rigorous texture analyses are required both to develop correlations with the texture archetypes [3,4] and to understand the local misorientation effects on microscopic transport processes [35].

Recent studies use pole figure analysis to isolate the texture effect on electromigration behavior in pure aluminum [29,30,35], Al-Cu metallurgies [36], Ti/Al-Cu structures [36], Al-Si [25], and Al-Mg [37]. The results on pure aluminum best exemplify the texture effect since grain structure is constant as polycrystalline lines for three materials with different textures. Fiber texture plots [3,4] and electromigration results at 225°C are shown in figures 2 and 3, respectively. Substantial increases in failure times are observed as the texture becomes stronger. The slopes on the failure probability

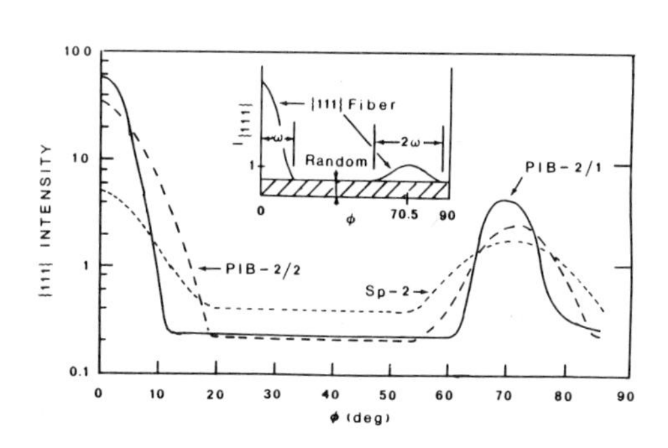

Fig. 2. Fiber texture plot for three pure aluminum conditions, 1μm in thickness. Texture parameters are: random = 0.23, ω = 11° for PIB-2/1; random = 0.22, ω = 17° for PIB-2/2; random = 0.42, ω = 17° for Sp-2.

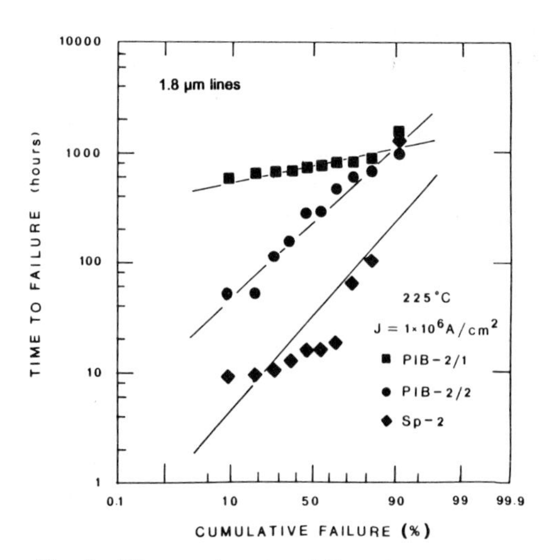

Fig. 3. Electromigration failure data for three pure aluminum conditions with 1.8 μm wide lines. Solid lines are regression fits assuming a monomodal lognormal failure distribution.

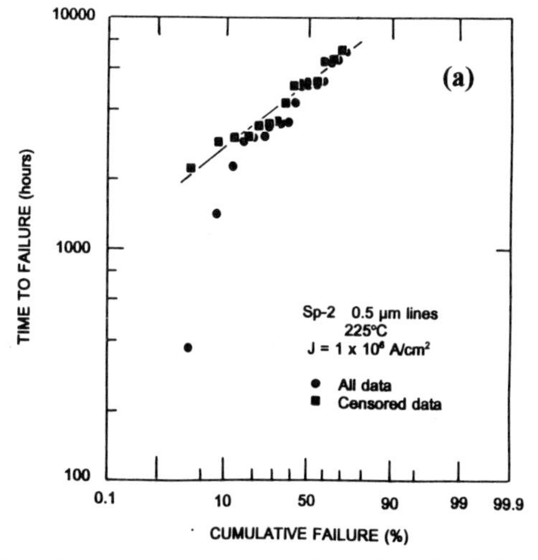

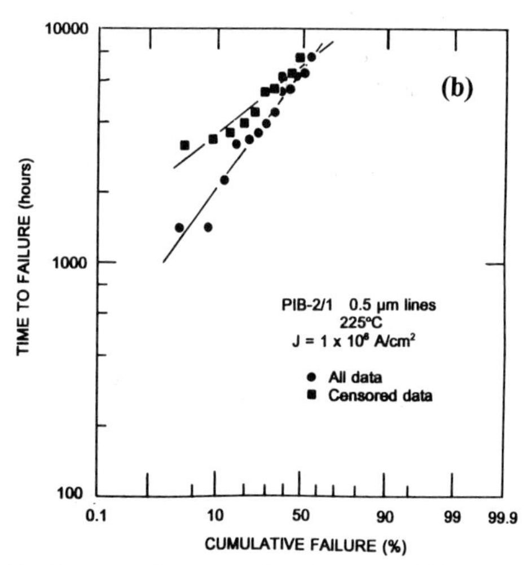

Fig. 4. Electromigration failure data for 0.5 μm wide lines on the pure aluminum conditions: a) weakly textured Sp-2 where the solid regression line is fit to data where the two earliest fails are removed from the distribution; b) strongly textured PIB-2/1 the upper line is a regression fit on the failure population where the first three failures are removed from the data.

plot are very different depending on the texture, where a bimodal distribution is observed in the weakest texture. The texture effect extends over the temperature range of 150°C to 250°C [35]. The relatively weak textures seem to be more susceptible to a bimodal failure distribution [25,35]. A study of Al-Si [25] shows that the texture effect is important in near bamboo grain structures and for conditions where texture differences are relatively modest. Narrow line (0.5 μm) electromigration studies [38], where the grain structure is almost bamboo, were done on two pure aluminum conditions analogous to the wide line study in figure 3. Electromigration failure data in figure 4 show that the texture effect is still evident but is not as pronounced. Bimodal failure distributions with early fails are also apparent. The most important conclusion of these texture studies is that *both the width of the (111) distribution and the random fraction are important* where stronger texture correlates with improved lifetimes. Studies to date indicate that a higher random component is more damaging than a broader (111) fiber distribution.

DISCUSSION

Both the orientation of individual grains and the grain misorientation (grain boundary texture) are important to the understanding of the four damage mechanisms. In both thermal hillocks and grain collapse, grain orientation determines whether inelastic deformation occurs on a local scale to drive these processes. In stress voiding and electromigration, transport and voids occur along grain boundaries where specific boundaries are selected to accumulate damage. Our understanding of these phenomena are too incomplete at this time to correlate damage with boundary structure. It is plausible that special boundaries are involved because their transport properties are different from general boundaries [39]. Anisotropy of transport properties would contribute to the flux divergence. Kirchheim [40] has noted that transport occurs perpendicular to (111) tilt boundaries in (111) oriented grains implying that boundary transport is relatively slow and relatively uniform in a tight texture, which would account for the superior electromigration resistance of these textures. Thorough evaluation of damage sites will be required to elucidate these local orientation effects.

An optimum texture that would suppress all damage mechanisms in interconnects would be highly desirable. The current state of knowledge is incomplete so only some trends can be identified. A large fraction of grains misoriented from the (111) orientation is clearly detrimental. A tight (111) distribution is very advantageous for electromigration and, probably, thermal hillock formation [41]. The case is less clear for grain collapse and stress voiding. Too little is understood about the texture effects in these latter damage mechanisms.

SUMMARY

The effects of texture on damage to aluminum-based thin film metallurgies in microelectronic interconnect applications have been considered. Grain collapse and thermal hillocks depend on the local grain orientations while stress voiding and electromigration depend on the local misorientation of grains for the accumulation of damage. Other important factors, in addition to texture, are grain structure (size and distribution in both blanket regions and narrow lines) and stress level, which drives many of these processes. The most critical need in this field is for data on local orientation measurements to identify the grain orientations and configurations that are most susceptible to damage.

REFERENCES

1. **Stress-Induced Phenomena in Metallization**, C.-Y. Li, P. Totta, and P. Ho, eds., (Am. Instit. Phys. Conf. Proc. 263, New York, 1992).
2. Kwok, T. and Ho, P.S.: in **Diffusion Phenomena in Thin Films**, 1988 (Noyes, Park Ridge, NJ), 369.
3. Knorr, D.B.: in **Materials Reliability Issues in Microelectronics III**, 1993, K.P. Rodbell, W.F. Filter, H.J. Frost, and P.S. Ho, eds., Vol. 309 (MRS, Pittsburgh, PA), 75.
4. Knorr, D.B. and Tracy, D.P., this proceedings.
5. Greenebaum, B., Sauter, A.I., Flinn, P.A., and Nix, W.D.: Appl. Pgys. Lett., 1991, 58, 1845.
6. Flinn, P.A., Gardiner, D.S., and Nix, W.D.: IEEE Trans. on Electron Devices, 1987, ED-34, 689.
7. Gardiner, D.S. and Flinn, P.A.: J. Appl. Phys., 1990, 67, 1831.

8. Pico, C.A. and Boniface, T.D.: J. Mater. Res., 1993, 8, 1010.
9. Minkiewicz, V.J., Moore, J.O. , and Eldridge, J.M.: J. Electrochem. Soc., 1992, 139, 271.
10. Bacconnier, B., Lormand, G., Papapietro, M., Achard, M., and Papon, A.-M.: J. Appl. Phys., 1988, 64, 6483.
11. Sanchez, Jr., J.E. and Arzt, E.: Scripta Metall. et Mater., 1992, 27, 285.
12. Chaudhari, P.: J. Appl. Phys., 1974, 45, 4339.
13. Gerth, D. and Schwarzer, R.A.: Materials Science Forum, 1992, 113-115, 625.
14. Schwarzer, R.A. and Gerth, D.: J. Elect. Mater., 1993, 22, 607.
15. Ericson, F., Kristensen, N., Schweitz, J.-A.: J. Vac. Sci. Tech. B, 1991, 9, 58.
16. Kristensen, N., Ericson, F., Schweitz, J.-A., and Smith, U.: J. Appl. Phys., 1991, 69, 2097.
17. Kristensen, N., Ericson, F., Schweitz, J.-A., and Smith, U.: Thin Solid Films, 1991, 197, 67.
18. Smith, U., Kristensen, N., Ericson, F., Schweitz, J.-A.: J. Vac. Sci. Tech. A, 1991, 9, 2527.
19. Shin, H.: Proc. VMIC Conf., 1991, (IEEE, New York), 292.
20. Thompson, C.V.: J. Mater. Res., 1993, 22, in press.
21. Kaneko, H., Hasunuma, M., Sawabe, A., Kawanoue, T., Kohanawa, Y., Komatsu, S., and Miyauchi, M.: in Proc. of the 28th Reliability Physics Symposium, 1990, (IEEE, New York), 194.
22. Ogawa, S. and Inoue, M.: in **Stress-Induced Phenomena in Metallization**, C.-Y. Li, P. Totta, and P. Ho, eds., Am. Instit. Phys. Conf. Proc. 263, (New York, 1992), 21.
23. Cho, J. and Thompson, C.V.: Appl. Phys. Lett., 1989, 54, 2577.
24. Thompson, C.V. and Kahn, H.: J. Elect. Mater., 1993, 22, 581.
25. Campbell, A.N., Mikawa, R.E., and Knorr, D.B.: J. Elect. Mater., 1993, 22, 589.
26. Attardo, M.J. and Rosenberg, R.: J Appl. Phys., 1970, 41, 2381.
27. Pierce,J.M. and Thomas, M.E.: Appl. Phys. Lett., 1981, 39, 165.
28. Cho, J. and Thompson, C.V.: J. Elect. Mater., 1990, 19, 1207.
29. Knorr, D.B. and Rodbell, K.P.: in **Submicron Metallization: The Challenges, Opportunities, and Limitations**, T. Kwok, T. Kikkawa, and K. Sendai, eds., 1993, Proc. SPIE 1805, (Bellingham, WA), 210.
30. Knorr, D.B., Tracy, D.P., and Rodbell, K.P.: Appl. Phys. Lett., 1989, 59, 3241.
31. Knorr, D.B., Rodbell, K.P., and Tracy, D.P.: in **Materials Reliability Issues in Microelectronics**, 1991, J.R. Lloyd, F.G. Yost, and P.S. Ho, eds., Vol. 225 (MRS, Pittsburgh, PA), 21.
32. Nagasawa, E. and Okabayashi, H.: in Proc. of the 17th Reliability Physics Symposium, 1979, (IEEE, New York), 64.
33. Li, P., Yapsir, A.S., Rajan, K., and Lu, T.-M.: Appl. Phys. Lett., 1989, 54, 2443.
34. Vaidya, S. and Sinha, A.K.: Thin Solid Films, 1981, 75, 253.
35. Knorr, D.B. and Rodbell, K.P.: in **Materials Reliability Issues in Microelectronics II**, 1992, C.V. Thompson and J.R. Lloyd, eds., Vol. 265 (MRS, Pittsburgh, PA), 113.
36. Rodbell, K.P., Knorr, D.B., and Tracy, D.P.: in **Materials Reliability Issues in Microelectronics II**, 1992, C.V. Thompson and J.R. Lloyd, eds., Vol. 265 (MRS, Pittsburgh, PA), 107.
37. Licata, T.J., Sullivan, T.D., Bass, R.S., Ryan, J.G., and Knorr, D.B.: in **Materials Reliability Issues in Microelectronics III**, 1993, K.P. Rodbell, W.F. Filter, H.J. Frost, and P.S. Ho, eds., Vol. 309 (MRS, Pittsburgh, PA), 87.
38. Knorr, D.B. and Rodbell, K.P.: in **Materials Reliabiliy Issues in Microelectronics III**, 1993, K.P. Rodbell, W.F. Filter, H.J. Frost, and P.S. Ho, eds., Vol. 309 (MRS, Pittsburgh, PA), 345.
39. Kaur, I., Gust, W., and Kozma, L.: in **Handbook of Grain and Interphase Boundary Diffusion Data**, 1989, Ziegler, Stuttgart.
40. Kirchheim, R.: in **Materials Reliabiliy Issues in Microelectronics III**, 1993, K.P. Rodbell, W.F. Filter, H.J. Frost, and P.S. Ho, eds., Vol. 309 (MRS, Pittsburgh, PA), 101.
41. Yamada, Y., Inokawa, H., and Takagi, T.: J. Appl. Phys., 1984, 56, 2746.

Materials Science Forum Vol. 157-162 (1994) pp. 1443-1448

TEXTURES IN ALUMINUM AND COPPER THIN FILMS

D.B. Knorr and D.P. Tracy

Materials Engineering Department and Center for Integrated Electronics,
Rensselaer Polytechnic Institute, Troy, NY 12180-3590, USA

Keywords: Fiber Texture, Aluminum Alloy, Copper, Underlayer

ABSTRACT

Texture is an important microstructural attribute of metals deposited by physical vapor deposition. Aluminum alloys and copper films are deposited, primarily by sputtering, on either amorphous SiO_2 or a polycrystalline underlayer of Ti, Ti/TiN, or TiN that is < 30 nm in thickness. Typical textures are categorized for films on an amorphous substrate. Most Al-alloy films form a two component texture of random + (111)-based fiber. Copper films have a wider variety of textures where a three component texture of random + (111) fiber + (200) fiber is most commonly observed. In both systems the underlayer texture is "inherited" by the metal film subsequently deposited. Differences in deposition conditions result in a wide variation in the details of the texture distributions without altering the typically observed components. Annealing tends to sharpen the texture established by deposition conditions but does not alter the basic character of the distribution.

INTRODUCTION

Textures are very commonly observed in thin films deposited by almost any technique [1]. Thin ($\leq$ 2 μm) films deposited by physical vapor deposition (PVD) have received considerable attention because of potential used in thin coatings, optical applications, magnetic applications, and microelectronic interconnects. The work presented in subsequent discussion will concentrate on the last application. Aluminum-based alloys are almost universally used for on-chip interconnections. Copper is under development as the interconnect metallization system of the future. The texture in both systems will be considered.

The concept of symmetry is important in films, where the texture of most PVD samples on a non-oriented substrate is fiber. An high degree of sample symmetry exists so it is inconvenient and redundant to plot the entire pole figure since no variation in normalized intensity exists by an azimuthal rotation. Texture data are best displayed as *fiber texture plots* [2,3] or as inverse pole figures calculated from the Orientation Distribution Function (ODF). A fiber texture plot takes a "cut" through the pole figure to represent the texture as normalized intensity (or log I) versus tilt angle ϕ measured from the fiber axis. Additional quantification is provided in figure 1 by defining two metrics for a simple texture of random + (111) fiber components. The fraction of randomly oriented grains is one measure, while the width ω of the (111), or any (hkl), fiber component is the second.

A less ambiguous definition of the component width involves an equal area integration of the fiber component to calculate the width for 50%, 63%, 90%, or 95% of the fiber component volume fraction as ω_{50}, ω_{63}, ω_{90}, or ω_{95}, respectively. The total width of the distribution is $2\omega_i$. An alternate metric, which is not used here, is the width of the fiber distribution at half maximum.

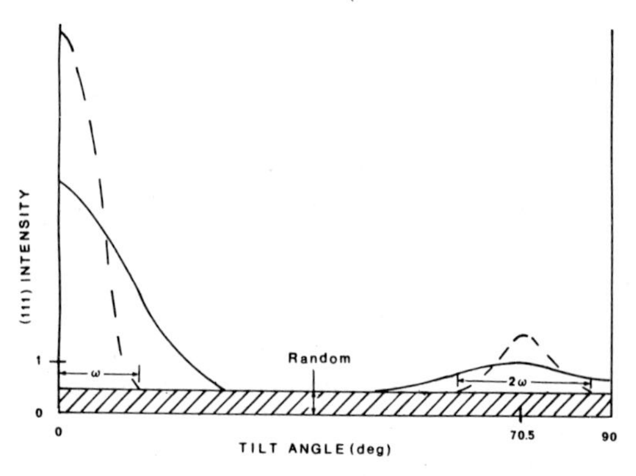

Fig. 1. Schematic representation of typical (111) fiber textures. The two components that define this texture are the half width of the (111) fiber component, ω, and the volume fraction of randomly orientated grains.

TEXTURES IN ALUMINUM-BASED THIN FILMS

The textures in aluminum-based metallurgies are relatively uncomplicated because they are composed of only two components. All experimental observations can be characterized into four texture *archetypes* based on the pioneering work of Wassermann and Grewen [4] who carried out extensive analyses on the fiber textures of drawn and annealed wires. The four designations are shown schematically as both (111) pole figures and fiber texture plots in figure 2.

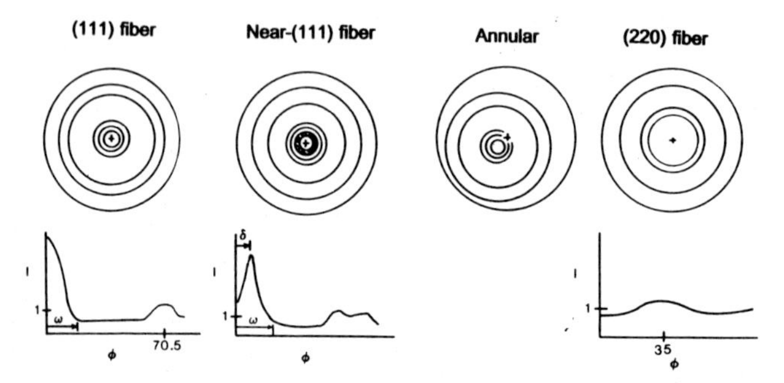

Fig. 2. Texture archetypes in aluminum-based metallurgies. The (111) pole figure and (111) fiber plot are shown under each texture type.

The list of archetypes is in an order representing their frequency of occurrence in our database:

1) **(111) fiber** - the fiber axis corresponds to the sample normal direction; the spread of (111) orientations and the volume fraction of (111) vary dramatically depending on deposition and annealing conditions.

2) **Near-(111) fiber** - a fiber texture where the peak (111) intensity is not in the sample normal direction but is symmetrically distributed with a maximum value at an angle δ from the sample normal meaning that an irrational plane is parallel to the sample surface; the sample normal direction is the fiber axis.

3) **Annular** - this (111) texture has the fiber axis located in a direction that is displaced by several

degrees from the sample normal direction; a correlation between the direction of displacement of the fiber axis and the sample normal requires a detailed knowledge of the deposition geometry.

4) **(220) fiber** - this texture is observed in only a few instances and is weak compared to most (111)-type textures; it demonstrates that a non-(111) texture is feasible in vapor deposited films.

The film thickness varies from 0.4 to 1.3 μm, but 1 μm is most typical. For textures with any type of (111) component, the random fraction is as large as 0.5, but more typical values are in the range 0.2 to <0.04. It is not known whether the random component truly goes to zero because 0.04 is our limit of experimental sensitivity by X-ray diffraction. In (111) fiber textures the (111) fiber component width ω_{95} varies between 3° in very strongly textured samples to ~30° for diffuse textures. A low ω_{95} is likely to have a low random fraction while a high ω_{95} probably has a higher random component. In near-(111) textured films the angle of displacement for peak (111) intensity δ is as great as 10° with values <5° most commonly observed. Likewise, the displacement of the fiber axis from the sample normal direction in the annular textures is 5° or less, although angles as high as 12° are observed. The (220) fiber texture is a special case with random fractions between 0.5 to 0.8 and broad distributions, i.e., an $\omega_{95}^{(220)}$ is typically 20-30°. This implies that the (220) fiber textured samples, as a class, do not deviate substantially from being randomly textured.

Annealing sharpens the texture by reducing the magnitude of the random component, reducing the width of the (111) distribution, or both [5,6,7]. However, no components are introduced or eliminated by this process. Normal grain growth occurs during these anneals where grain size increases from an as-deposited value in the range 0.15 - 0.25 μm to a value of 1 - 1.5 μm for an 1 μm thick film. The grain size stagnates (i.e., growth rate decreases substantially) because the grain boundaries are pinned by surface grooves [8]. No substantial change in the texture is observed if secondary grain growth is initiated in stagnated primary grains [11]. If the primary grains are pinned by particles rather than by surface grooves, secondary grain growth can completely change the texture from (111) to (220) [10].

Several typical (111)-type textures are shown in figure 3. Their definitions and texture parameters are: A) Sharp (111), ω_{95} = 5.1°, random fraction $\leq$ 0.08; B) Strong (111) fiber, ω_{95} = 16.6°, random fraction = 0.10; C) Moderate (111), ω_{95} = 20.1°, random fraction = 0.20; D) near-(111), δ = 8°, ω_{95} = 18.9°, random fraction $\leq$ 0.08. All samples were deposited by sputtering, but both deposition conditions and sputtering geometry vary widely between the specimens.

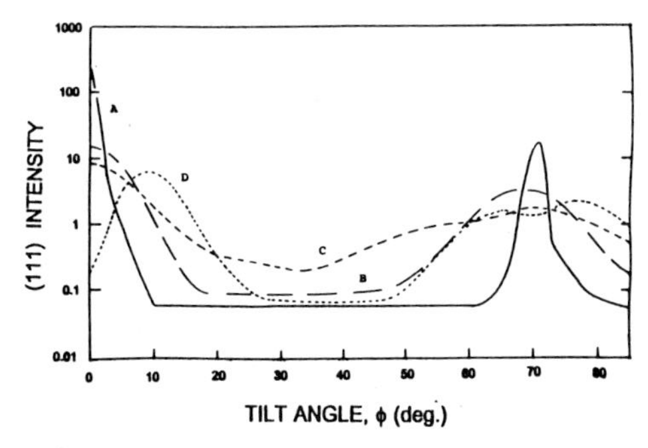

Fig. 3. (111) fiber texture plots for four Al samples. Note how the texture components vary among the samples.

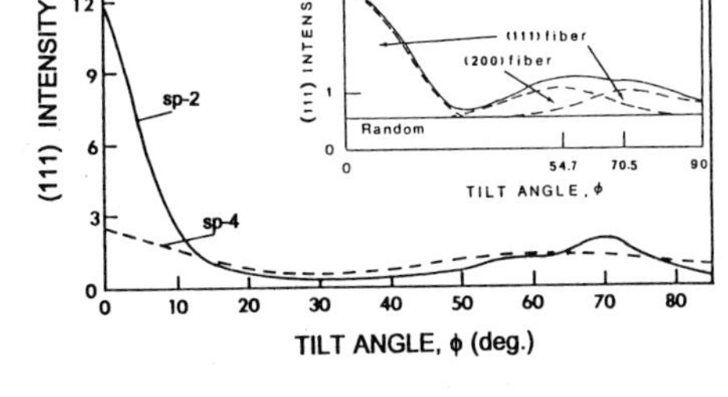

Fig 4. (111) fiber plot for two Cu samples. The schematic defines the Cu texture components.

TEXTURES IN COPPER THIN FILMS

Texture components in copper films differ from those observed in aluminum-based metallurgies where a typical copper film contains a (200) fiber component along with the (111) fiber and random components. These and other components are observed for copper films deposited by several techniques [11]: sputtering, ion assisted, chemical vapor deposition (CVD), evaporation, and electroplating. The film thicknesses range from 0.3 μm to 2 μm. The volume fractions of the three components vary from sample to sample: 0.17 to 0.82 for the (111) and 0.08 to 0.67 for the random. The width of the (111) texture component, as defined by ω_{63}, ranges from ~5° to 15°, but values are typically about 10°. The width of the (200) component is usually broader than the (111) component. An (111) fiber plot for two sputtered conditions is shown in figure 4; Sp-2 is deposited on oxidized silicon, and Sp-4 is deposited on a tantalum nitride underlayer. A schematic illustrating the location of the texture components on the fiber texture plot is included. The figure exemplifies the variation in the textural attributes for copper: Sp-2 (111) fraction = 0.65, random fraction = 0.27, $\omega_{63}^{(111)}$ = 9.4°; Sp-4 (111) fraction = 0.38, random fraction = 0.52, $\omega_{63}^{(111)}$ = 14.1°. Compared with aluminum films, an higher random component and a wider distribution of the (111) fiber texture component are typically observed in copper films.

Besides the (111) fiber, (200) fiber, and random components, other texture components and types are observed in copper: (511) fiber and (220) fiber components and annular textures. Twinning in a strongly textured film generates the (511) component that is distinctly observed as peaks in the fiber plot. Multiple twinning results in additional peaks. A (220) fiber component is occasionally observed in vapor deposits, but more commonly in copper electroplated films and is accompanied by a considerable random component. Annular textured films are relatively less common than in aluminum-based films but have the typical (111) fiber + (200) fiber + and random components with the fiber axis displaced by ~5°.

UNDERLAYER EFFECTS

Underlayers are important in contemporary interconnect systems where they are used as diffusion barriers and as redundant conductor structures. The barrier/adhesion layers will be required in copper interconnect systems since copper diffuses into silicon, SiO_2, and some polymers such as the polyimides [12]. Furthermore, no exposed surfaces are permitted because copper corrodes when exposed to the ambient. The textural development in both aluminum alloy and copper metallurgies is influenced by the underlayers. In aluminum alloy films with 25 nm titanium underlayers, a much stronger, sharper texture is attributed to the strong texture in the titanium [5]. In the case of an AlPd metallurgy, a broad (220) fiber + substantial random texture becomes a strong (111) fiber texture when it is deposited on titanium [13].

		Al-0.5Cu			Cu		
		Vol. Fract.			Vol. Fract.		
Underlayer	Underlayer ω_{90}	Random	ω_{90}	Type	(111)	Random	ω_{90}
Oxide	NA	0.25	16.3 °	near(111) δ=5°	0.55	0.27	18.7°
Ti	10.2°	<0.12	11.4°	(111)	0.74	0.14	12.2°
TiN	~random	0.4	18.8°	near(111) δ=3°	0.26	0.32	19.7°
Ti/TiN	9.2°	≤0.17	11.8°	(111)	0.57	0.15	12.9°

Table 1. Texture components and distributions for Al-0.5Cu and Cu films on various underlayers.

The results of a well controlled deposition experiment are given in table 1. Al-0.5Cu (0.52 μm) and pure copper (0.5 μm) were deposited on oxide (baseline condition), 25 nm Ti, 32 nm TiN, and Ti/TiN. The same Ti deposition conditions were used for Ti only and the Ti in the Ti/TiN. Likewise, the same TiN deposition condition were used. The presence of the textured Ti underlayer results in a highly textured, rather than random TiN. For AlCu or for Cu deposition, the underlayer texture is "inherited" by the metal film.

DISCUSSION

The reasons for the substantial variation in the textures of aluminum-based thin films are obscure because vapor deposition is carried out far from equilibrium and is influenced by a large number of controlled **and** uncontrolled variables. A list of process variables and adjustable parameters includes, but is not be limited to, background vacuum and system cleanliness, working gas pressure, substrate/target geometry, power, deposition rate, and substrate temperature. Despite the complexity of the problem, an extensive study of specimens deposited in a variety of commercial deposition systems reveals some trends (specific exceptions can be found but are infrequent):

- evaporated films generally have both a higher random volume fraction and a larger ω sputtered films [5].
- stronger textures result from a higher substrate temperature and higher power when all other deposition parameters are held constant.
- no systematic trend between film composition and texture is found among pure Al, Al-(0.5 to 4)wt.%Cu, Al-1wt.%Si, and Al-1wt.%Si-(0.5 to 1)wt.%Cu metallurgies because sharp textures are possible in all systems.
- Al- (0 to 1)wt.%Si-(0.15 to 0.3)wt.%Pd differ substantially from the above metallurgies by often having a weak (220) or a weak (111) texture [13].

The copper texture components are independent of deposition process. The typical (111) fiber, (200) fiber, and random components have been observed for sputtered, evaporated, CVD, ion assisted, and electroplated film deposition. To date, no systematic study of processing effects on copper film texture has been carried out. One noticeable effect on the copper film texture is the texture of the underlayer (table 1). Compared with the SiO_2 and TiN underlayers, copper deposited on textured titanium has a stronger, sharper (111) texture component. Depositing copper on Ti/TiN also strengthens and sharpens the (111) component compared to copper deposited on the randomly textured TiN underlayer alone. The relative fraction of (111) fiber and (200) fiber change depending on the textured underlayer. The (111) is most prominent on Ti while (200) is greatest on Ti/TiN. Fiber texture components within these ranges are normally observed so insufficient data are available to ascribe the effect exclusively to the underlayer.

Experience with titanium thin films of 25 to 30 nm thickness [5,13] shows that the titanium is highly (0002) textured. This characteristic is passed to the aluminum-based overlayer which is, in effect, seeded by the underlayer. Comparisons are direct since the same deposition system and conditions are used so only the substrate underlayer is different. Such behavior is easily explained since (0002) titanium and (111) aluminum are very closely lattice matched resulting in a "granular epitaxy" during the initial stages of aluminum film formation. Titanium nitride deposited on titanium has a strong (111) texture component as seen in table 1. The strengthening of both the aluminum and copper deposited on Ti/TiN is due to a change from a random to an (111) fiber texture for TiN. Textured TiN formed by gaseous reaction of ammonia with titanium is also known to produce textured TiN which passes its strong orientation to an aluminum alloy film deposited subsequently [14]. An interesting comparison is the metal texture on amorphous SiO_2 and on the TiN only underlayer. The texture of both AlCu and Cu is weaker on a random, crystalline underlayer than on an amorphous substrate. This implies that the seeding effect is effective from very early in the deposition process because the texture is so efficiently carried through during film growth. If the substrate did not influence nucleation and coalescence of the film, very similar textures would likely be observed for both amorphous SiO_2 and TiN underlayers [1].

REFERENCES

1. Knorr, D.B.: this proceedings.
2. Knorr, D.B. and Rodbell, K.P.: in **Submicron Metallization: The Challenges, Opportunities, and Limitations**, T. Kwok, T. Kikkawa, and K. Sendai, eds., 1993, Proc. SPIE 1805, (Bellingham, WA), 210.
3. Knorr, D.B.: in **Materials Reliability Issues in Microelectronics III**, 1993, K.P. Rodbell, W.F. Filter, H.J. Frost, and P.S. Ho, eds., Vol. 309 (MRS, Pittsburgh, PA), 75.
4. Wassermann, G. and Grewen, J.: in **Texturen metallisher Wertstoffe**, 1962 (Springer-Verlag, Berlin), 5.
5. Rodbell, K.P., Knorr, D.B., and Tracy, D.P.: in **Materials Reliability Issues in Microelectronics II**, 1992, C.V. Thompson and J.R. Lloyd, eds., Vol. 265 (MRS, Pittsburgh, PA), 107.
6. Knorr, D.B., Tracy, D.P., and Lu, T.-M.: Textures and Microstructures, 1991, 14-18, 543.
7. Knorr, D.B., Tracy, D.P., and Lu, T.-M.: in **Evolution of Thin Film and Surface Microstructures**, 1991, C.V. Thompson, J.Y. Tsao, and D.J. Srolovitz, eds., Vol. 202 (MRS, Pittsburgh, PA), 199.
8. Mullins, W.W.: Acta Metall., 1958, 6, 414.
9. Knorr, D.B. and Rodbell, K.P.: unpublished research, 1993.
10. Longworth, H.P. and Thompson, C.V.: J. Appl. Phys., 1991, 69, 3929.
11. Tracy, D.P. and Knorr, D.B.: J. Electronic Materials, 1993, 22, 611.
12. Ho, P.S.: in **Principles of Electronic Packaging**, 1989, D.S. Seraphim, R. Lasky, and C.-Y. Li, eds., (McGraw-Hill, New York), 809.
13. Rodbell, K.P., Knorr, D.B., and Mis, J.D.: J. Electronic Mater., 1993, 22, 597.
14. Fu, K.-Y., Kawasaki, H., Olowolafe, J.O., and Pyle, R.E.: in **Submicron Metallization: The Challenges, Opportunities, and Limitations**, T. Kwok, T. Kikkawa, and K. Sendai, eds., 1993, Proc. SPIE 1805, (Bellingham, WA), 263.

Materials Science Forum Vol. 157-162 (1994) pp. 1449-1454

CONTROLLED TEXTURE DEVELOPMENT IN HOT-FORGED LITHIUM FLUORIDE

K.L. Kruger and K.J. Bowman

School of Materials Engineering, Purdue University, 1289 Mat. Electr. Eng. Bldg., West Lafayette, IN 47907-1289, USA

Keywords: Ionic Ceramics, Doping, Hot-Forging, Dynamic Recrystallization

ABSTRACT

The evaluation of self-consistent models for predicting textures in ceramics requires experimental data for deformation textures in materials with known slip behavior. Lithium fluoride provides a unique opportunity for studies in this area. Axisymmetric forgings of LiF at 400°C (0.6 T_m) have shown substantial two-component fiber textures, 110 and 100. By doping with a divalent cation, 800 ppm Mg^{2+}, slip activity can be altered, effectively "controlling" slip-based texture development. For three levels of purity, an obvious reversal of the dominant texture component is observed in axisymmetric forging of LiF. Additional contributions to texture arising from dynamic recrystallization are discussed.

INTRODUCTION

Slip behavior of lithium fluoride has been extensively studied by a number of researchers [e.g. 1-6], including investigations of dislocation motion, temperature dependencies of dislocation glide and climb, and effects of purity on yield and work hardening behavior. Most of these studies have focused on the single crystal behavior of LiF of varying purities, with many considering only primary slip systems, {110}<110>. An example of the changes in primary slip behavior caused by small additions of divalent ion impurities is shown in Figure 1. Mitchell and Heuer have shown calculations estimating that the tetragonal distortion of the lattice around the divalent cation-cation vacancy should cause the observed hardening from dopant additions [6]. Obviously, the distortion will affect dislocations on dissimilar planes to varying degrees, thus leading to preferential hardening for the different families of slip systems. Some studies concentrating on secondary slip, i.e. {100}<110>, have reported strong slip anisotropy [e.g. 2] or have investigated hardening behavior [e.g. 4,5], but have failed to correlate the doping effects on the overall plastic behavior in a manner that would aid in predicting the polycrystal response, particularly with regard to deformation textures. The combined changes in slip anisotropy, strain rate sensitivity, and preferential hardening of the different slip systems should have a significant influence on polycrystalline deformation behavior, and will ultimately affect texture evolution.

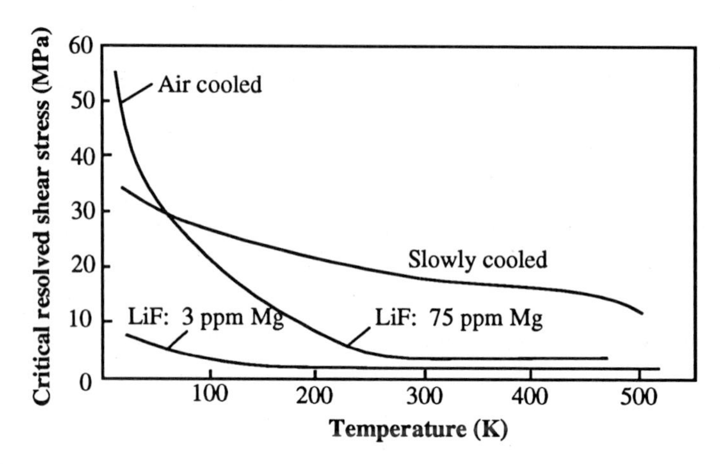

Figure 1. Variation of the critical resolved shear stresses of the primary slip system as a function of temperature at two different doping levels, 3 and 75 ppm [after 7]. The slowly cooled sample was doped at 75 ppm.

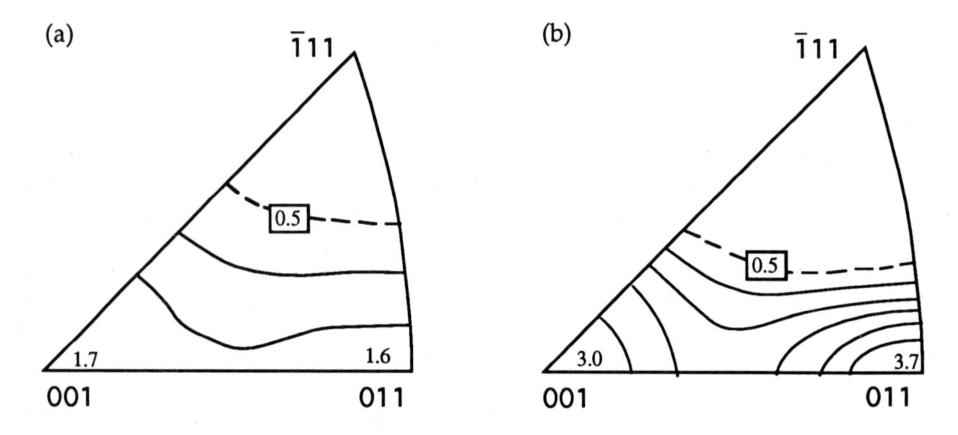

Figure 2. Inverse pole figures of commercially pure LiF axisymmetrically forged to true strains of (a) -0.5 and (b) -1.0 at 400°C which show typical two component textures, 100 and 110 [8].

The inverse pole figures in Figure 2 are representative of commercially pure LiF samples axisymmetrically forged to true strains of (a) -0.5 and (b) -1.0. This previous work reveals substantial two-component textures, 110 and 100, apparently produced through a combination of slip and dynamic recrystallization [8]. However, the precise strengthening effect of the impurities in this material are not known. By capitalizing on the changes in slip anisotropy through doping with divalent impurities, changes in texture development should reveal the relative behavior of the two families of slip systems.

EXPERIMENTAL PROCEDURE

Three LiF powders, commercially pure (99.5%), pure (99.99%) and doped (pure with 800 ppm $MgCl_2$ added), were procured from Alfa Johnson Matthey. The processing of the commercially pure powder is described elsewhere [8]. The pure and doped powders were cold-pressed, sintered and hot-pressed at 600°C into 11 mm diameter cylindrical billets. End faces of the hot-pressed samples were polished metallographically and examined by x-ray diffraction and optical microscopy to determine initial texture and initial grain size and morphology. Uniaxial forging of the billets was performed at strain rates of 1×10^{-4}/sec to various strain levels at 400°C. Specimens were sectioned through the center to measure deformation textures and perform microstructural analysis. Incomplete (0-75°) pole figures measured by x-ray diffraction employed Cu Kα radiation with a four-circle goniometer.

RESULTS AND DISCUSSION

In Figure 3, the 220 and 200 pole figures for pure (a and b), commercially pure (c and d), and doped (e and f) LiF samples reveal trends in texture development with increasing impurity content. Each of the samples experienced nearly identical deformation conditions (true strain of -0.8, strain rates of 1×10^{-4}/sec at temperatures of 400°C). The high purity LiF shows a strong 100 component with a weaker 110 component. The opposite is true of the doped sample. The commercial purity LiF shows nearly equal contribution of both components. These trends in texture maxima are summarized more concisely in Figure 4, which shows a histogram of the experimental peak intensities from the centers of each of the pole figures.

Clearly, impurity content plays a key role in the texture evolution for these materials. Although the commercially pure LiF is seemingly less pure than the doped LiF, the impurities are apparently not the divalent type which significantly alter slip behavior. In fact, all alkaline metals tested by inductively-coupled plasma (ICP) spectroscopy were nonexistent or present only in trace amounts. As discussed above, a divalent impurity is necessary to create the tetragonal distortion produced from association with the charge-balancing vacancy. The 110 texture component becomes increasingly dominant with increasing divalent impurities, and slip on the primary {110} system is presumably more strongly affected by dopants than the {100} system.

Since single crystal experiments pose difficulties in obtaining orientations that allow only secondary slip systems (i.e. {100} slip planes) to be active, it is hoped that texture measurements like those discussed here can be coupled with self-consistent (SC) modelling to determine changes in shear stress ratios with varying purity levels and deformation conditions. As previously observed [9], the primary difficulty in using SC models is that they assume a single stress exponent for all families of slip systems. The above results plainly show a disparity in the response of individual slip systems to dopant additions. Further, it seems reasonable that the thermal activation required to overcome obstacles by different dislocations, particularly in a material which has strong elastic anisotropy like LiF,

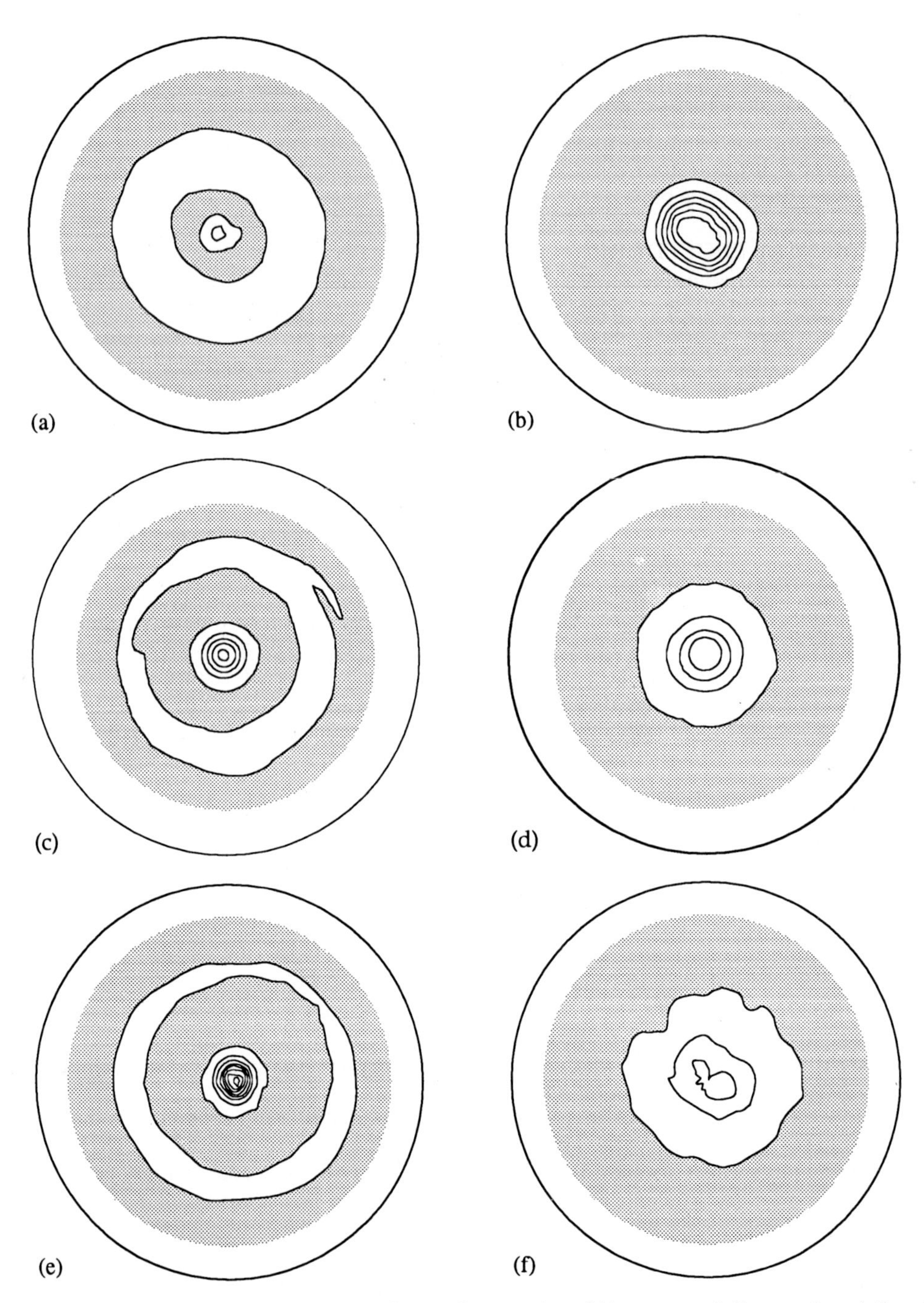

Figure 3. 220 (left) and 200 (right) pole figures for pure (a and b), commercially pure (c and d) and doped (e and f) LiF deformed in axisymmetric forging at 400°C. Contour interval is 1.0; shaded areas are below random (i.e. 1.0).

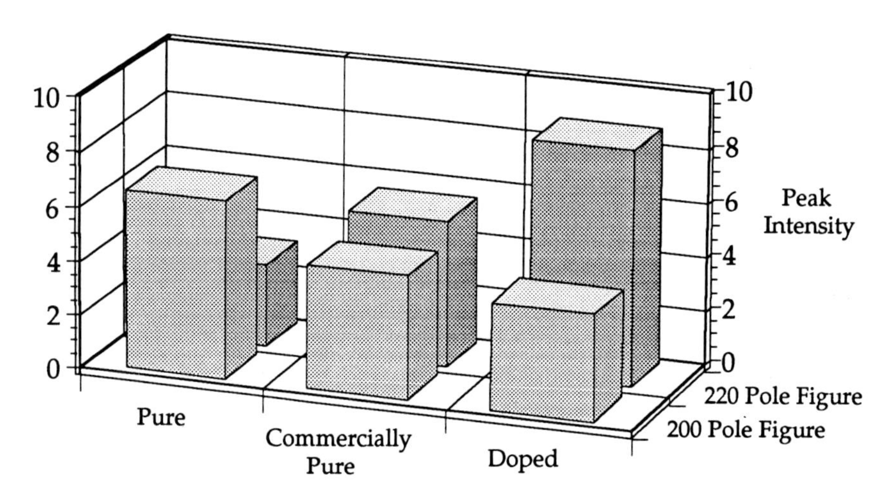

Figure 4. Histogram revealing measured peak intensity data from the center ($\chi = 0°$) of each pole figure shown in Figure 3.

would vary for each slip system. Texture results may be a more viable alternative to discern some of the complex effects impurities have on different dislocation interactions. Clearly, further experiments and modelling are necessary to understand these issues.

ACKNOWLEDGMENTS

K.L. Kruger is supported in part by a Purdue Research Foundation Fellowship. This research is supported by the U.S. National Science Foundation under DMR 91-21948.

REFERENCES

1. J.J. Gilman and W.G. Johnston, *J. Appl. Phys.*, 1956, **27**, 1018.
2. Z.G. Liu and W. Skrotzki, "Slip on {100} Planes in LiF," *Physica Status Solidi*, **A70**, 433, (1982).
3. R.B. Day and W.A. Johnston, *J. Amer. Cer. Soc.*, 1969, **52**, 595.
4. H.L. Fotedar and T.G. Stoebe, *Phil. Mag.*, 1971, **23**, 859.
5. W. Skrotzki, R. Steinbrech, and P. Haasen, *Mat. Sci. and Eng.*, 1978, **32**, 55.
6. T.E. Mitchell and A.H. Heuer, *Mat. Sci. and Eng.*, 1977, **28**, 81.
7. R. W. Davidge, Mechanical Behavior of Ceramics, Cambridge Univ. Press, Cambridge, England, (1979).
8. K.L. Kruger and Keith J. Bowman, *Scripta met. mat..*, 1992, **26**, 1227.
9. Yuechu Ma and K.J. Bowman, *Proc. of Special NATO Workshop on Polyphase Polycrystal Plasticity*, Palm Springs, CA, to be published in a special volume of *Mat. Sci. and Eng A.*.

Materials Science Forum Vol. 157-162 (1994) pp. 1455-1462

TEXTURE ANALYSIS IN ZIRCALOY CLADDING TUBE MATERIAL FOR NUCLEAR FUEL

S.D. Le Roux [1] and D.J. Van der Merwe [2]

[1] Atomic Energy Corporation of South Africa Ltd.,
PO Box 582, Pretoria 0001, Republic of South Africa

[2] Business Management Services CC,
PO Box 73601, Lynnwood Ridge 0040, Republic of South Africa

Keywords: Zircaloy, Preferred Orientation, Nuclear Fuel Cladding, Sample Geometry

ABSTRACT - Zirconium-based alloys are used commonly world-wide in the manufacturing of cladding tubes for fuel elements in light-water nuclear reactors. The pilger mill plant at the AEC in South Africa has been producing Zircaloy-4 cladding tubes for the Koeberg Pressurized Water Reactors (PWR's) near Cape Town since the mid-eighties. The specific thermomechanical history of a cladding tube determines the final texture which makes it possible to tailor the texture according to service requirements. The range of pole figures obtained from different production stages at the AEC plant is presented. Tube sample preparation was reduced to a single surface etch. Prism plane pole figures yielded information about grain behaviour during thermomechanical processes. The pole figure data is used to calculate several different texture parameters which serve to characterize the texture quantitatively and for comparison with other manufacturers.

1. INTRODUCTION

Low alloyed zirconium (Zircaloy) have been used since the 1950's in various types of fission reactors, both light and heavy water types for different applications, the main ones being cladding tubes for nuclear fuel, fuel assembly grids, channels in boiling water reactors (BWRS) and pressure tubes in pressurized heavy water reactors (PHWRS) [1,2].
The inherent anisotropy of the hcp Zircaloy crystals together with the limited number of slip systems and the strict crystallographic orientation relationships, for slip and twinning provides for the manifestation of a high degree of preferred orientation (texture) in these materials [2,3]. Zircaloys have a low c/a axes ratio but strong textures occur throughout the family of hcp metals with varying c/a ratios, for example Ti, Zn, Cd, Co, Mg, etc. The anisotropic behaviour of textured materials can be explored by detailed measurements of the texture [4-7]. Production of complete Zircaloy-4 cladding tube assemblies for the two 900 MW PWRS at Koeberg just north of South Africa's picturesque city of Cape Town started in the mid 1980's at the Pelindaba plant near Pretoria, some 1000 km north-east of Cape Town. The pilger mill plant manufacture route and related texture measurements on the various tubes produced by this plant forms the subject of this paper. Special emphasis is placed on the final cladding tube products.

2. CLADDING TUBE MANUFACTURE ROUTE

The cladding tube manufacturing process consists of several cold pilger mill tube reductions as is shown in the flow chart, figure 1. The original feedstock is a thick wall tube shell with OD = 80 mm and wall thickness 14.5 mm. The cladding tube final product, designated P5, has an OD of 9.5 mm and a wall of 0.57 mm. A similar manufacturing route is used for the guide thimble tubes, designated P8 (see right-hand branch of flow chart). Each pilger step is followed by annealing at 650°C in vacuum or inert atmosphere, except for the final pilger step which is concluded with stress relief heat treatment at a lower temperature of about 475°C. The main purpose of the latter treatment is to conform to the mechanical property specifications of the client, Electricity Supply Commission (ESCOM), for the Koeberg Nuclear Reactors.

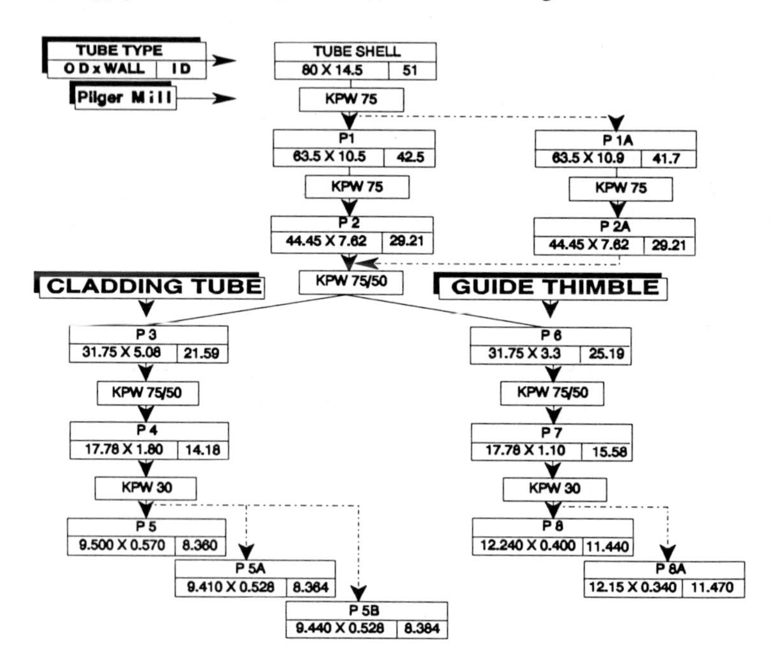

Figure 1: Flow Chart for Cold Pilger Mill Steps in Cladding Tube Manufacture

3. EXPERIMENTAL PROCEDURE

3.1 X-RAY DIFFRACTION

Direct pole figure data in the back reflection mode were obtained on a modified Siemens Pole Figure Diffractometer. Typically a Zircaloy-4 tubing specimen was mounted on a movable carriage and tilted in steps of 5° by a vertical circle (ψ angle), which covered the range $0 \leq \psi \leq 85°$. For each ψ position the sample was rotated one complete revolution in its own plane (ϕ angle). The four circle goniometer stepping motors and associated x-ray scintillation counter were linked to a personal computer. The hardware protocol was similar to that used by Hofer at KWU [4]. Extensive software development by the authors provided for plotting of the pole figures in colour on the video screen, as well as for calculation of associated texture parameters (see section 5). Typical x-ray generator operating conditions were 40 kV and 30 mA for the Cu K_α tube, with a Ni-filter in front of the counter. Circular collimation of 0,4 mm diameter was used on the x-ray tube side and a ¼° slit at the counter. The software could also facilitate recording and indexing of ordinary powder diffraction patterns.

3.2 SAMPLE PREPARATION

Cladding tubes (P5), intermediate tube shell samples (P1-P4) and guide thimbles (P8), preferably of 2,5 cm length, were analyzed. If necessary, samples up to 5cm length could be accommodated. A short etch in 10% HF, 40% HNO_3 and 50% H_2O solution was needed to remove worked metal from the surface. The etch step is essential for the recording of representative pole figures in the case of x-rays. With neutron diffraction no such sample preparation is required. It should be mentioned that some controversy exists as to the nature and effect of a thin surface layer of worked material on the corrosion properties of Zircaloy. P5 tube samples could be analyzed without axial sectioning. The tube shell samples had to be sectioned to fit into the sample holder. Composite samples with flat surfaces, each of which consisted of several precut segments, were prepared in order to compare texture results with those of curved-surface samples, as seen in figure 2. Machining followed by selective etching [8] may also be used to prepare specimens in order to probe through the wall texture gradients.

4. TEXTURELESS STANDARD

A textureless specimen is used as a reference material for defocusing and absorption corrections. The following steps were followed to prepare flat and P5 cylindrical reference samples: about 5 g of ZrH powder was sintered for 3 hours at 850°C in vacuum.

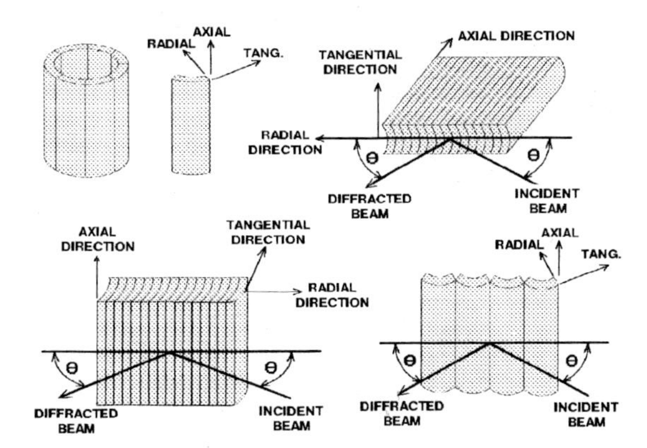

Figure 2: Sample Sectioning Procedure and Corresponding Diffraction Geometries

5. TEXTURE PARAMETERS

Several texture parameters were quantified from the direct pole figures. These parameters, which are described in detail in Reference 8, are the f-parameter (the Kearns factor [9]), the location of the maximum basal pole intensity (α), the texture factors F_T and F_A, standard deviation $(SD)_T$ and the integrated intensity ratio S.

6. RESULTS

6.1 POLE FIGURES

Representative basal and prism plane pole figures are given in figure 3, starting from thicker wall P2 tubing, down to the P5 final cladding and P8 guide thimble.

The following observations were made:

(i) The P8 tube has the thinnest wall and shows basal poles close to the radial direction. It is true in general that the thermomechanical deformation parameter, which mainly determines the texture of tubing, is the ratio of reduction in wall thickness to diameter R_W/R_D. For $R_W/R_D > 1$ the basal poles are oriented preferentially nearer to the radial direction, as in the case of P8. For $R_W/R_D \leq 1$ the basal poles are distributed in the R-T plane.

(ii) Independent of the pilger production process for the semi-finished products, a [100] direction will be parallel to the deformation direction (tube axis), as is evident from the prism plane pole figures for the pre-heat-treated cases.
(iii) Annealing above 500°C causes recrystallization [11] with the nett effect being that prism planes become parallel to the tube axis rather than perpendicular, as in the cold-worked case. Annealing leaves the basal plane pole figures relatively unchanged.

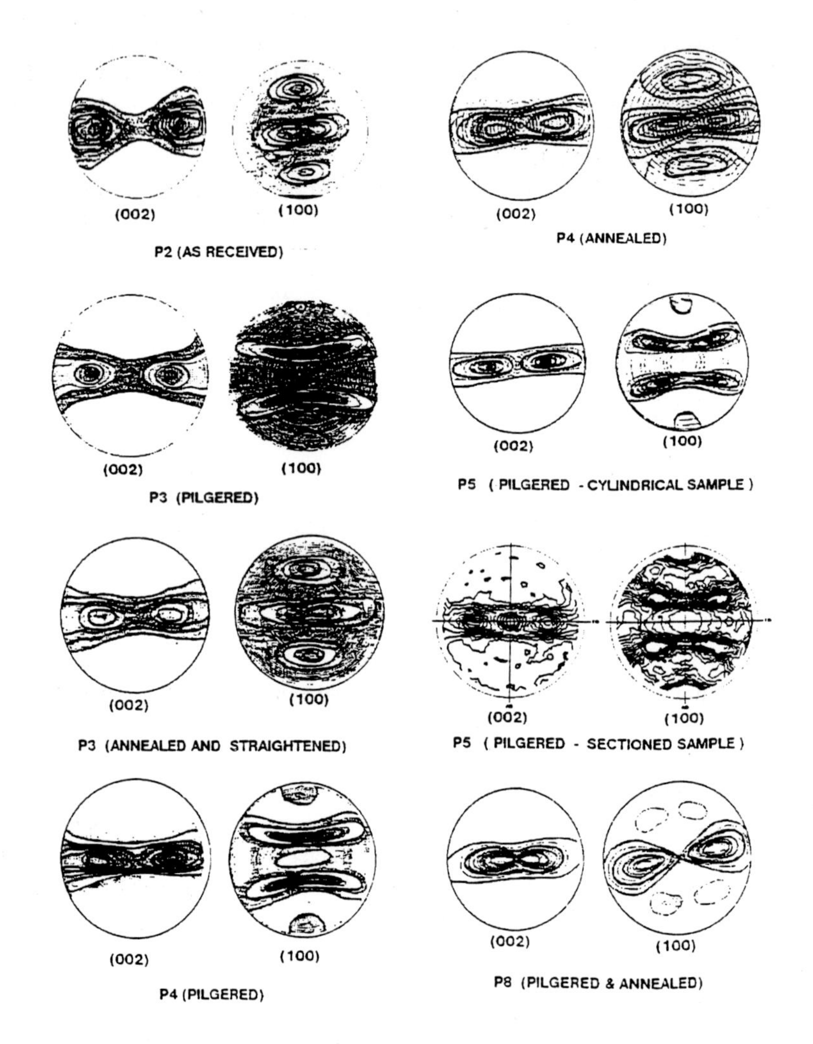

Figure 3: Pole figures for various stages in the cladding tube production line.

6.2 REPRODUCIBILITY

To confirm the reproducibility of the measuring process, samples were taken from various tubes and five measurements of texture parameters were independently taken on each sample. Ten measurements were taken on unflattened P5 tubing samples. Table 1 contains the results. The standard deviation of a parameter is given in parentheses.

TABLE 1: TEXTURE PARAMETERS FOR DIFFERENT TUBING SIZES

SAMPLE	†n	$f_r(002)$	$f_r(100)$	α in degrees	F_A	F_T	$(SD)_T$	S
P3 Pilgered	5	0,393(1)	0,407(2)	51,756(24)	0,692(4)	-0,035(1)	0,272(1)	0,175(2)
Annealed	5	0,441(1)	0,389(1)	48,417(77)	0,725(3)	0,084(2)	0,294(1)	0,259(1)
P4 Pilgered	5	0,461(1)	0,373(1)	48,933(65)	0,765(1)	0,055(2)	0,272(1)	0,194(1)
Annealed	5	0,522(1)	0,378(8)	44,287(61)	0,805(1)	0,210(2)	0,291(1)	0,274(2)
P5 Pilgered	10	0,486(4)	0,306(7)	46,773(565)	0,691(19)	0,111(17)	0,266(1)	0,231(3)
Stress relieved	10	0,511(5)	0,305(11)	45,522(214)	0,740(4)	0,154(7)	0,274(1)	0,299(10)
Straightened	10	0,513(10)	0,302(8)	46,299(767)	0,732(12)	0,137(24)	0,271(3)	0,237(14)
P6 Pilgered	5	0,471(7)	0,368(1)	48,738(691)	0,744(12)	0,061(26)	0,272(7)	0,261(15)
Annealed	5	0,498(10)	0,349(12)	45,823(638)	0,759(8)	0,163(17)	0.292(1)	0,312(6)
P7 Pilgered	5	0,485(2)	0,350(2)	47,982(117)	0,794(2)	0.082(4)	0.270(1)	0,221(2)
Annealed	5	0,599(1)	0,335(2)	42,064(63)	0,830(2)	0,274(3)	0.288(1)	0,271(6)
P8 Pilgered	5	0,567(11)	0,278(13)	42,560(523)	0,787(5)	0,238(16)	0.258(1)	0.229(5)

†n≡number of measurements on each sample

The statistics of the f-parameter for cladding tubes (P5) were also compared with statistics from an overseas round robin study [10]. The results are shown in Table 2.

TABLE 2: COMPARISON OF f-PARAMETERS FOR AEC AND ROUND ROBIN STUDY [10].

PLANE	SD(%)a	SD(%)b	SD(%)c
(002)	1,95	1,6	4,8
(100)	3,4	3,2	14,5

The standard deviations are defined as follows:
SD(%)=(SD/x)100 where x is the average value of the f parameter in a particular study.
SD(%)a: Present study. From Table 1 data for unflattened P5 samples.
SD(%)b: Repeated over period of years by single laboratory in the round robin study (composite flats).
SD(%)c: Mean of independent laboratories in the round robin study (elastically flattened samples).

6.3 ACCURACY

To verify the accuracy of the measuring process, sections were taken in the principal directions (T,L,R) of a P5 tube and prepared into compound samples. Theoretically, the sum of the three f_i-parameters must be unity. The sum of the basal plane f-parameter and twice the prism plane f-parameter must also be unity. Each of the 3 samples was analyzed 5 times for the two planes (002) and (100). Table 3 shows the results with the standard deviation of the 5 analyses shown in parentheses.

Accuracy checks were also attempted by means of an interlaboratory comparison: Measurements on similar type P5 tubing by CEZUS in France are shown in Table 4 and compared with local results:

TABLE 3: ACCURACY CHECK BY MEANS OF f_r-PARAMETER SUMS.

SAMPLE DIRECTION	$f_r(002)$	$f_r(100)$	$f_r(002)+2f_r(100)$
T	0,375(4)	0,345(3)	1,065
L	0,082(8)	0,475(11)	1,032
R	0,499(2)	0,272(8)	1,043
Σf	0,956	1,092	

TABLE 4: COMPARISON OF AEC AND CEZUS f_r RESULTS

SAMPLE DIRECTION	$f_r(002)$	CEZUS $f_r(002)$
T	0,372	0,401
L	0,095	0,057
R	0,498	0,543
Σf	0,965	1,001

6.4 TUBULAR vs FLAT SAMPLES

To show that the method of analyzing the P5 tubular sample in as-received condition, agrees with the conventional method of analyzing the flattened sample, ten samples were taken from a P5 tube. Five were analysed as is, while the other five were first flattened. The results are shown in Table 5. The W-test was done on each set of values to determine whether the data has a normal Gaussian distribution. The paired t-test was used to determine whether two sets of data were statistically significantly different.

TABLE 5: COMPARISON BETWEEN FLATTENED AND CYLINDRICAL SAMPLES.

	CYLINDRICAL			FLATTENED		
n	$f_r(002)$	α	$f_r(100)$	$f_r(002)$	α	$f_r(100)$
1	0,511	44,835	0,290	0,506	44,970	0,297
2	0,521	44,653	0,297	0,510	45,321	0,300
3	0,504	45,439	0.300	0,526	43,985	0,287
4	0,521	43,993	0,300	0,497	45,945	0,283
5	0,528	42,634	0,300	0,476	46,133	0,302
x	0,517	44,311	0,297	0,503	45,271	0,294
SD	0,009	1,070	0,004	0,018	0,858	0,008
W	0,942	0,934	0,722	0,978	0,939	0,891
t	1,157	-1,146	0,755			

6.5 CORRELATION BETWEEN PILGERED, STRESS RELIEVED AND STRAIGHTENED TUBING TEXTURE

The paired t-test was evaluated to determine the correlation between pilgered (P) and stress-relieved (SRI and SR2) and straightened tubing (ST) data sets. The t-values for the various parameters are given in Table 6.

TABLE 6: t-VALUES OF PILGERED, STRESS RELIEVED AND STRAIGHTENED TUBING.

	$f_r(002)$	$f_r(100)$	F_A	F_T	α	$(SD)_T$	S
t(P/SR1)	2,630	-2,328	11,878	2,530	-2,258	2,248	1,231
t(ST/SR2)	-2,169	-1,769	-4,262	-1,812	1,761	-2,391	-1,820

7. CONCLUSIONS

7.1 Reproducibility of measurements, as indicated by $(SD)_T$ in Table 1, shows maximum variances of 0.013 for the prism plane f-parameter and 0.011 for the basal plane f-parameter. These variances are acceptable for all the different types of tubing measured.

7.2 The sum of the three f-parameters in the mutual orthogonal directions are within 9% of unity according to the data in Table 3. The sums of $f_r(002) + 2f_r(100)$ are unity within only 7% in the t-direction.

7.3 As indicated by t in Table 5 the difference in the results for tubular and flat samples are not statistical significant ($t < 2,78$ with 95% risk for $n-1 = 4$).The tubular sample analysis are therefore pursued with confidence.

7.4 Table 6 shows that there is no statistical difference between data sets for pilgered, stress relieved and straightened tubing, except for the F_A parameter. This indicates that the stress relieving and straightening processes have no major effect on the texture.

8. ACKNOWLEDGEMENTS

The authors are grateful to the Atomic Energy Corporation of South Africa who made this investigation possible. The support of Dr G.A. Eloff in preparing this manuscript is appreciated.

9. REFERENCES

[1] Knödler, D.; Reschke, S.; Weidinger, H.G.: *Kerntechnik*, 1987, 50(4), 255.
[2] Murty, K.L.: *Zirconium in the Nuclear Industry*: 8th Int. Symposium, ASTM STP 1023, 1989, 570.
[3] Tenckhoff, E.: *Zirconium in the Nuclear Industry*: 5th Conference, ASTM STP 754, 1982, 5.
[4] Hofer, G.: *Experimental Techniques of Texture Analysis* (H.J. Bunge, Ed.), Informationsgesellschaft Verlag, 1986, 331.
[5] Bunge, H.J.: *Directional Properties of Materials*, Informationsgesellschaft Verlag, 1988.
[6] Bunge, H.J.: *Texture Analysis in Materials Science*, Mathematical Methods, Butterworths, London, 1982.
[7] Bunge, H.J.; Esling, C.: *Quantitative Texture Analysis*, Deutsche Gesellschaft für Metallkunde, 1982.
[8] Knorr, D.B.; Pelloux, R.M.: *J. Nucl. Mat.*, 1977, 71, 1.
[9] Kearns, J.J.: WAPD-TM-472, TID-4500, Nov. 1965.
[10] Lewis, J.E.; Schoenberger, G.; Adamson, R.B.: *Zirconium in the Nuclear Industry*: 5th Conference, ASTM STP 754, 1982, 39.
[11] Eloff, G.A.: *PhD Thesis*, University of the Orange Free State, Republic of South Africa, March 1993.

Materials Science Forum Vol. 157-162 (1994) pp. 1463-1468

NATURAL FIBRE TEXTURES IN A NATURALLY TEXTURED MATERIAL

A. Oscarsson [1] and U. Sahlberg [2]

[1] Institutet för Metallforskning,
Drottning Kristinas väg 48, S-114 28 Stockholm, Sweden

[2] STFI (Swedish Pulp and Paper Research Institute),
Box 5604, S-114 86 Stockholm, Sweden

Keywords: Wood, Microfibril Angle, Chain Texture, Cellulose

ABSTRACT

Properties of wood are strongly influenced by the microfibril angle defined as the angle between the longitudinal direction of the wood fibre and the (0, b, 0) direction of the monoclinic cellulose. Chains of this cellulose have a strong influence on the mechanical properties of wood. Samples were cut out from different positions in a tree and pole figures were measured by X-ray reflection to determine the texture for each one. The textures are frequently of fibre character but with different microfibril angles depending on the position in the tree. A quick and simplified method was developed to measure the average microfibril angle. This method uses a direct measurement of the (0, 4, 0) peak in reflection, instead of the more commonly used (0, 0, 2) peak in transmission which only gives an indirect measure of the microfibril angle. Results from X-ray measurements in reflection are compared with results obtained by optical microscopy.

INTRODUCTION

Wood can briefly be described as consisting of a regular pattern of small cells where each one is constructed as a long rectangular tube with walls consisting of different layers. These layers mainly contain cellulose, hemicellulose and lignin. The cellulose, which is made up of a monoclinic crystal structure, is present as fibrous chains rotated as a helix around the long direction of the wall. The angle between the longitudinal direction of wood fibre and the (0,b,0) direction of the cellulose has conventionally been defined as the microfibril angle. This angle is of course distributed around different directions in space. The chain texture of the cellulose can therefore be described by the Euler angles φ_1 and Φ with respect to a sample coordinate system.

The microfibril angle is one of the basic ultrastructural characteristics of woody cell walls. It influences wood properties such as modulus of elasticity, creep and dimensional stability. Each growth ring of the tree has its individual characteristic texture or distribution of microfibril angles in space depending on position in the tree and growth conditions. Furthermore, the crystallinity as well as the density at different positions also varies considerably.

The presence of textures in wood and other organic fibres were first reported in 1913 (1) but has seldom if ever previously been reported at ICOTOM. There is no standard and generally accepted method today to evaluate the average microfibril angle.

Previously reported X-ray diffraction, XRD, techniques (2 - 5) determine the average microfibril angle of cellulose crystallites using XRD in transmission from intensity distribution of paratropic planes such as (002), $(10\bar{1})$ and (101). These techniques have fundamental drawbacks because they are an indirect measurement of microfibril angle. A calibration procedure has to be done for conversion to microfibril angles and problems arise when subtracting background intensities.

A direct X-ray technique for measuring microfibril angle has been developed by Lofty et al (6) using the (040) diffraction pattern in transmission. It compensates for the weak contribution to the net intensity caused by other crystallographic planes with d-values nearby. However, this technique only considers distributions of microfibril angle rotated around the radial axes. A study (7) has also been made comparing different types of technique which points out the importance of proper correction of measured average microfibril angles. A technique (8) has also been established for determining orientation distributions of cellulose crystalites. It uses several complete pole figures measured in both reflection and transmission. However, this is a rather time consuming method if many positions in a tree are to be examined. It is most desirable to have a quick and reliable method for determining the mean microfibril angle in a number of growth rings. The purpose is to establish such a technique by using X-ray diffraction in reflection viewed from the longitude direction of the tree.

EXPERIMENTAL

Wood specimens of Scandinavian spruce (Picea abeis) were investigated. X-rays of Co Kα from a point focus were used and the geometry for the goniometer set up is shown in fig. 1. The size of the beam and loading directions of the samples could be changed when necessary. Pole figures were measured from different viewpoints using the Schulz reflection method. Plotted pole figures are corrected for background intensity and normalised. Chi scans were done while measuring the intensity for the (040) reflection continuously from -60° to +60°. In this way different positions in each growth ring could be examined considering the microfibril angle distribution rotated around the tangential direction of the tree.

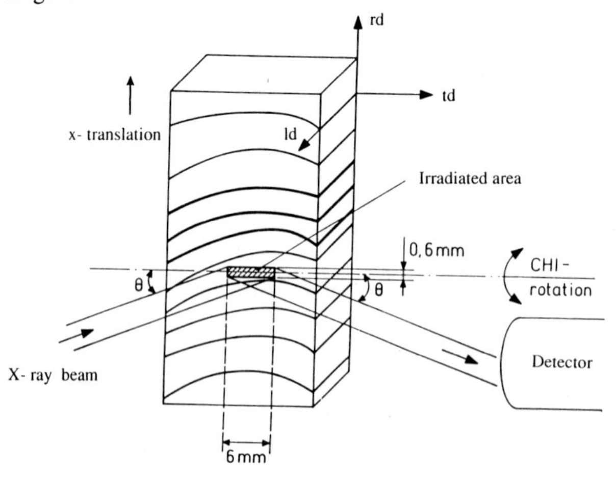

Figure 1
Principal geometry of wood samples and texture goniometer. For pole figure measurements of individual annual rings, the irradiated length was changed to ~2 mm.

The microfibril angle was also measured in optical microscopy by the method of (9). A number of slices were then cut from each growth ring and measurements were carried out on tangential walls. It is then possible to measure the variation of microfibril angle with position in the ring. Usually five measurements were made on each slice and about 40 slices were taken from each growth ring. Measurements close to pores were avoided.

RESULTS AND DISCUSSION

Typical pole figures for normal wood and compression wood are shown in figs. 2 and 3 respectively. Normal wood has rectangular and rather thin cell walls while compression wood has thicker and more rounded cell walls. Compression wood is developed when the tree is brought out of its equilibrium position in space by natural forces such as wind or snow. Both samples show a well developed chain texture with (0,b,0) rotated around the longitudinal direction of the tree. This chain texture is sharper in normal wood than in compression wood. The (040) pole figure for compression wood clearly shows a larger scatter for rotations around the tangential direction than around the radial direction. The same effect, but less pronounced, exists also in the normal wood.

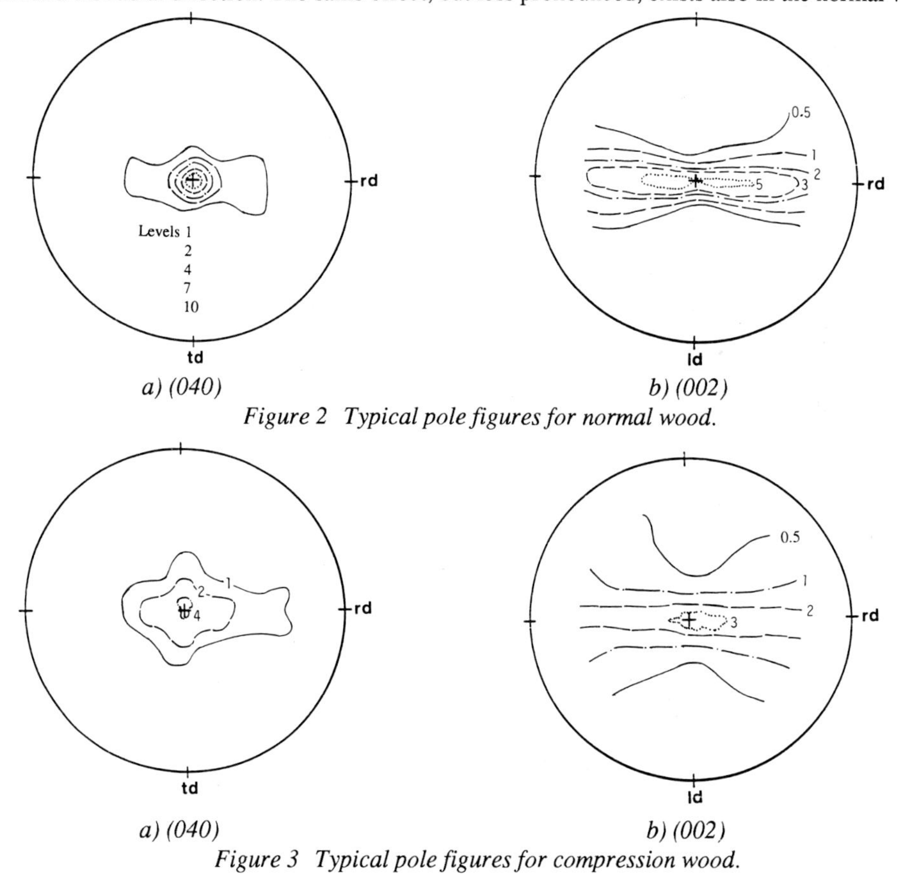

a) (040) *b) (002)*

Figure 2 Typical pole figures for normal wood.

a) (040) *b) (002)*

Figure 3 Typical pole figures for compression wood.

This is in a reasonable qualitative agreement with previously reported results (3, 7, 10) from different species of wood. It has been proposed that the microfibrils have mainly a scatter of rotations around the tangential axis in radial walls and radial axis in tangential walls due to the geometry of the walls. The larger rotation around the tangential axis has previously been explained by more bordered pits on the radial walls and that the microfibrils tend to form whorls around these. These results have clearly shown that a complete description of orientation distribution of microfibril angles need two angles.

It is often very difficult to measure more than two pole figures properly in wood. This in combination with the low symmetry of the monoclinic crystal makes calculation of a complete Orientation distribution function hazardous. However, in many cases, it seems sufficient to describe the chain texture only with the microfibril with one angle, Φ, between the longitudinal direction of the tree and the (0,b,0) direction of the monoclinic cellulose. Fig. 4 shows the difference in intensity distribution of (040) and (002) reflections when rotated around the tangential or radial direction of the sample. The curves are compensated for background intensity and compared to correct curves. These correct curves were achieved by using the average intensity when rotating the sample around its normal direction while tilting around the chi direction. No single rotation alone gives a proper correlation with the correct curve for the microfibril angle distribution.

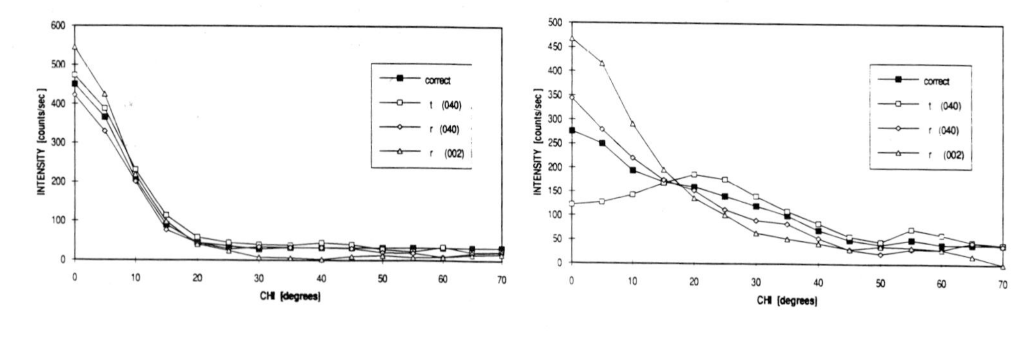

a) normal wood *b) compression wood*

Figure 4 Intensity of (040) and (002) planes vs tilting angle (chi) for rotation around radial (r) and tangential (t) direction.

The geometry of annual rings in wood makes it convenient to measure the distribution of microfibril angles as rotated around the tangential direction. This is acheived when the sample is loaded as shown in fig. 1. In this way a large number of measurements from different positions in the tree can be made. The method used for estimating the average microfibril angle is shown in fig. 5. This method to determine the mean microfibril angle was originally proposed by Cave (3) and has been widely used for measurements done in transmission of the (002) peak.

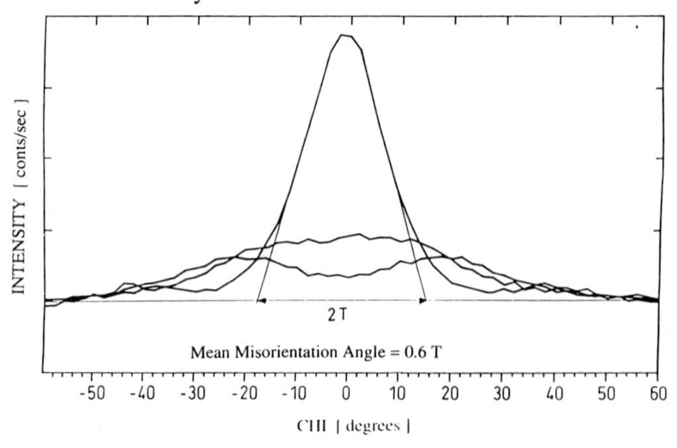

Figure 5 Schematic diagram showing how the mean microfibril angle was obtained from diffraction data.

Fig. 6 shows the variation of microfibril angle through an annual ring of compression wood where the results have been achieved by XRD and by the microscopy method. Both methods show a transition at the same position from high microfibril angle in early wood (Spring) to low microfibril in late wood (Summer). The obtained angles are somewhat higher for the XRD method. This can be explained by a) XRD measures the microfibril angle rotated around the tangential direction preferably in radial walls, while the microscopy method was done in tangential walls where the microfibril angle is shown to be smaller. b) all angles are projected on a tangential plane in the microscopy method which will lead to equal or lower estimated values than the real ones. Furthermore, very large microfibril angles caused by pores etc. were avoided in the microscopy measurements.

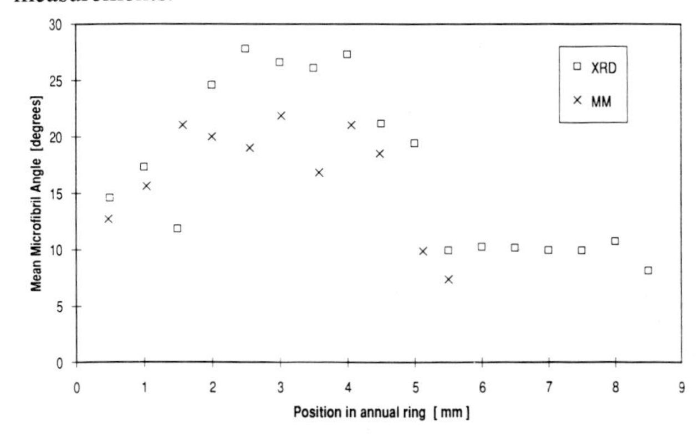

Figure 6
Variation of mean microfibril angle through an 9 mm thick annual ring of compression wood. Measurements were carried out by X-ray diffraction (XRD) and the microscopy method (MM).

The intensity of the (040) peak when no tilting is applied to the wood sample can often be used to separate different types of wood in the tree. An example of that is shown in fig. 7 for normal and compression wood when the sample was simply moved in its radial direction. Normal wood is very consistent with higher intensity in late wood than in early wood. For compression wood low intensity is recorded when the microfibril angle is high and it is then impossible to separate early wood from late wood. However, these kind of diagrams are very useful for deciding at which positions in the sample the microfibril angle is going to be measured. The microfibril angle was then measured for a great number of positions in a tree with 1 - 2 mm wide annual rings. The results plotted in fig. 8 show higher microfibril and a number of annual rings (45 - 60) with typical compression wood. A few annual rings were also analysed by the microscopy method. Comparison with XRD results showed again somewhat lower values for the microscopy method for the same reasons as mentioned before.

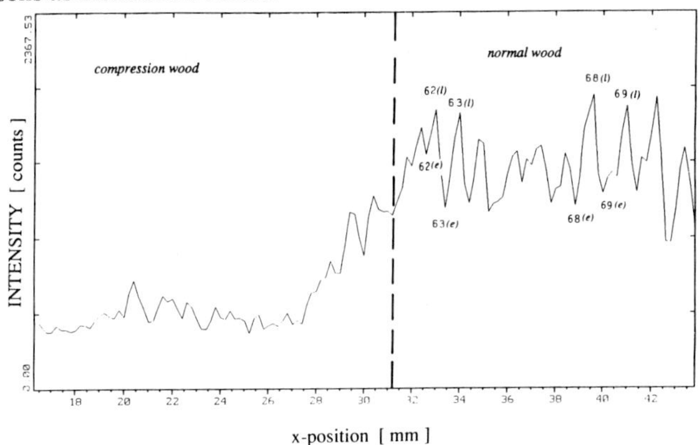

Figure 7
Intensity of (040) planes parallel to the longitudinal direction of the tree at different positions in the tree. Given figures indicate annual ring and whether it is early wood (e) or late wood (l).

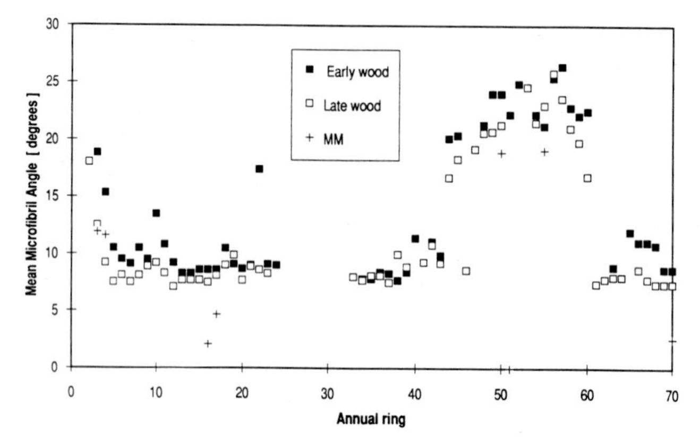

Figure 8
Mean microfibril angle for different positions in the tree measured by XRD. The microfibril angle was also determined by the microscopy method (MM) for a few annual rings as indicated.

CONCLUSIONS

This preliminary study of microfibril angle in wood by X-ray diffraction has shown:

- The (0,b,0) direction of the monoclinic cellulose is arranged in a chain texture which often needs two angles be properly described.
- The chain texture varies considerably with position in the tree.
- A quick and simplified method has been established for measuring the microfibril angle at different positions in a tree. It uses directly the (040) peak in reflection instead of the commonly used (002) peak in transmission.

ACKNOWLEDGEMENTS

The authors are grateful to Ms. U. Jonsson for carrying out microscopy measurements. We also wish to thank Dr. L. Salmén, Swedish Pulp and Paper Research Institute, and Prof. W.B. Hutchinson, Swedish Institute for Metals Research, for useful discussions.

REFERENCES

1. Nishikawa, S. and Ono, S., Proc. Mathematical and Physical Society (Tokyo), 1913, 7, 131.
2. Cave, I.D., Forest Products Journal, 1966, 16, 37.
3. Meylan, B.A., Forest Products Journal, 1967, 17, 51.
4. Boyd, J.D., Wood Sci. Technol., 1977, 11, 93.
5. Paakari, T. and Serimaa, R., Wood Sci. Technol., 1984, 18, 79.
6. Lotfy, M., El-Osta, M., Kellogg, R.M., Foschi, R.O. and Butters, R.G., Wood and Fiber, 1973, 2, 118.
7. Sobae, N., Hirai, N., Asano, I., J. Japan Wood Res. Soc., 1971, 2, 44.
8. Tanaka, F., Takaki, T., Okamura, I. and Koshijima, T., Wood Research, 1980, 66, 17.
9. Crosby, C.M., Mark, R.E., Svensk papperstid., 1974, 17, 636.
10. Tanaka, F. and Koshijima, T., Wood Sci. Technol., 1984, 18, 177.

Materials Science Forum Vol. 157-162 (1994) pp. 1469-1474

MACROTEXTURES OF STAINLESS Fe-Cr STEELS

D. Raabe and K. Lücke

Institut für Metallkunde und Metallphysik,
RWTH Aachen, Kopernikusstr. 14, D-52056 Aachen, Germany

Keywords: Ferritic Stainless Steel, ODF, Inhomogeneity, Rolling, Annealing

ABSTRACT

The texture development of ferritic stainless steels was inspected during hot rolling, cold rolling and annealing. During hot rolling a strong, recovered "cold rolling" texture (α-fibre) is formed in the sheet center, whereas in the subsurface layer shear components as Goss and {4 4 11}<11 11 8> are generated. During cold rolling the inherited texture sharpens in the center layer and the subsurface components decrease by rotating towards the α-fibre and towards {111}<112>. The final annealing treatment leads to a γ-fibre and to a weak {001}<110> component which is retained by recovery.

INTRODUCTION

Ferritic high grade steels with 11%-17%Cr (FeCr) establish an important group of construction materials, due to their good mechanical properties and corrosion resistance. Several insufficiencies in the industrial processing of these steels are dependant on the texture.
The phenomenon of ridging seems to result from topological aspects of the texture. The inhomogeneity through the thickness of the cold rolled and recrystallized steel is due to the inherited texture profile of the hot rolled band. During recrystallization (RX) of microalloyed FeCr steels, deviations from the γ-fibre lead to a decrease of the r-value. These technological shortcomings give rise to a survey on the texture development in FeCr steels. For this purpose quantitative texture analysis and metallography has been applied to investigate various industrial FeCr alloys with respect to hot rolling, cold rolling and RX.

EXPERIMENTAL

Five steels with a Cr content between 11% and 17% have been analyzed. Four steels are microalloyed with Nb or Ti, respectively (Table 1). All steels were industrially hot rolled and subsequently annealed at 1000°C. Cold deformation was carried out ε=40%, 50%, 60%, 70%, 80% and 90%. After cold rolling the samples were annealed for 5 minutes at 850°C or 950°C in a salt bath. All textures were examined by computing the orientation distribution function (ODF) by

Table 1: Investigated alloys in weight %

Sample	Cr	C	N	Ti	Nb
Cr11Ti	11.3	0.01	0.01	0.21	-
Cr11Nb	10.5	0.01	0.02	0.11	0.32
Cr17	16.5	0.06	0.03	0.01	0.01
Cr17Ti	16.5	0.02	0.01	0.47	0.01
Cr17Nb	16.6	0.02	0.02	0.01	0.64

means of the series expansion method from four pole figures [1]. Bcc metals tend to develop strong fibre textures, so that for many cases it is convenient to present the textures by isointensity diagrams in φ_1 sections or by fibre diagrams [2,3] (Table 2):

Table 2: Some characteristic texture fibres of ferritic stainless steels

Fibre	Fibre Axis	Orientations on the Fibre
α-fibre	<110> ∥ RD	{001}<110>-{112}<110> and {111}<110>
γ-fibre	<111> ∥ ND	{111}<110>-{111}<112>
ζ-fibre	<110> ∥ ND	Goss, {110}<112> and {110}<110>
ε-fibre	<110> ∥ TD	{001}<110>, {112}<111>, {4 4 11}<11 11 8>,{111}<112>, Goss

Since the textures of FeCr are inhomogeneous [2,3], all samples were measured in several layers. Each layer is described by the s parameter, which indicates the distance between layer and sample center devided by the half thickness (s=0 center, s=1 surface).

RESULTS AND DISCUSSION

Hot Rolling

In figure 1 the hot band texture is presented in Δs=0.1 through thickness steps for CR17NB. Between s=0.7-1.0 the texture is characterized by Goss, {110}<112> and {4 4 11}<11 11 8>,

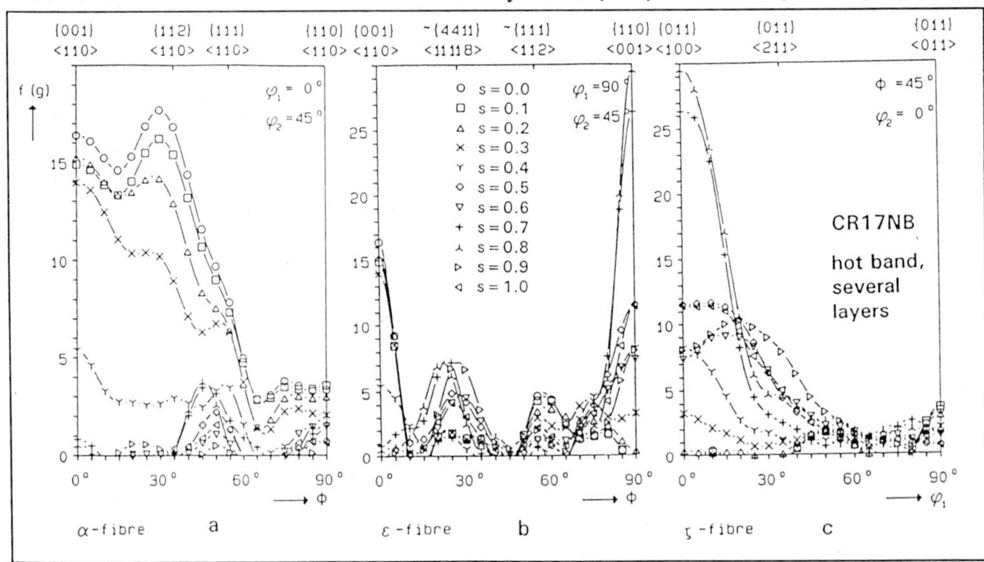

Figure 1: Hot band texture, Δs=0.1 steps through thickness, sample CR17NB.

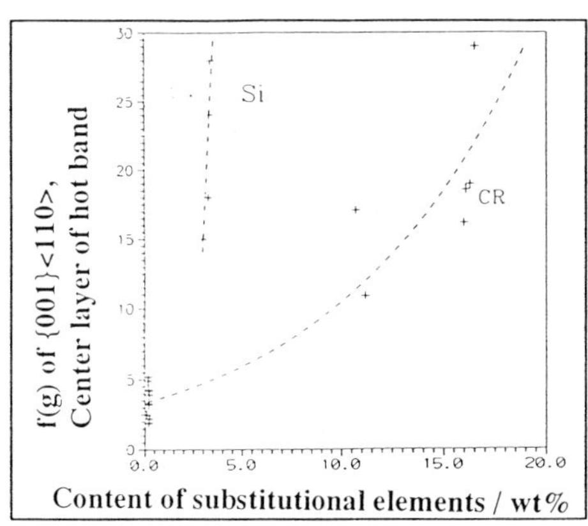

Figure 2: Influence of alloy content.

which arise from the there occuring strong shear deformation (figures 1b,c). At s=0 the sheet is deformed by plane strain [4] and a sharp α-fibre is formed (figure 1a). The water spray in front of each rolling gap lowers the surface temperature of the sheet. This leads to a higher stored dislocation energy and thus to RX at s=0.7-1. At s=0 dynamic recovery, which prevents RX, so that a "cold rolling" texture is formed. Due to the high Cr and low C content 17% Cr alloyed steels are not undergoing transformation during hot rolling. The 11% Cr alloyed steels reveal phase transformation up to 20vol% during the first hot rolling steps [5]. With increasing amount of transformation the texture is randomized, i.e. weakened (figure 2).

Cold Rolling

The cold rolling textures of CR11NB are shown in figure 3. In the center a strong α-fibre with maxima at {001}<110> and {112}<110> is developed (figure 3a), whereas at s=0.8 the α-fibre

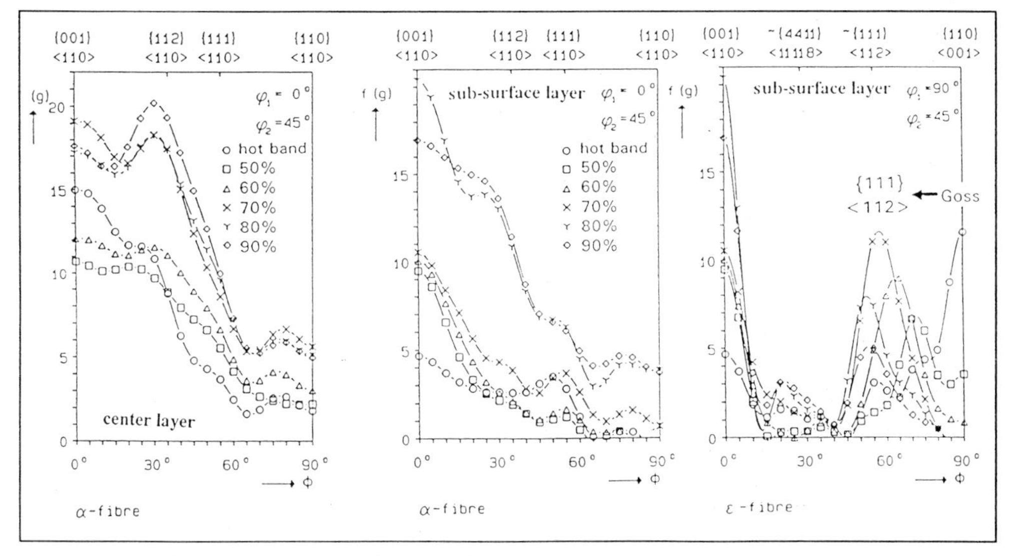

Figure 3: Cold rolling texture, sample CR11NB.

develops much slower, due to the initial shear texture (figure 3b) [2,3]. On the ε-fibre the rotation of Goss towards {111}<112> and {001}<110> is presented (figure 3c), leading to a strong maximum on the γ-fibre after 70%. The variation of the Nb and Ti content did not lead to distinct changes in the rolling textures. Concerning the Cr content, differences can be seen for the location of {112}<110>, which is shifted from Φ=35° (17%Cr) to Φ=30° (11%Cr). The surface textures of the 11%Cr alloys reveal a slower sharpening of the α-fibre, than the 17% Cr alloys, so that after 90% rolling still a strong gradient through the thickness occurs [3].

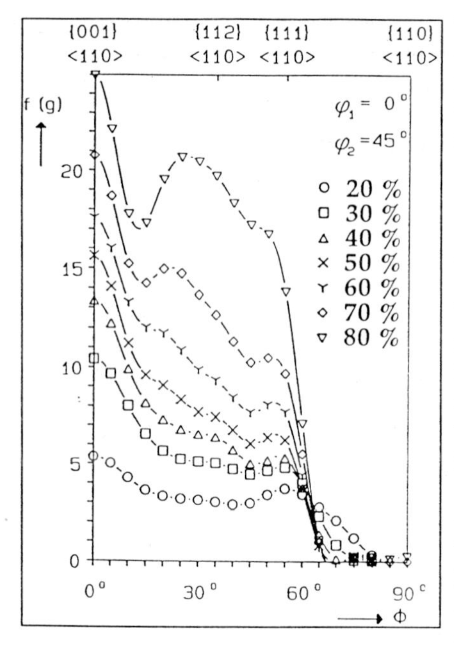

Figure 4: *Taylor Simulations.*

The cold rolling textures can be simulated by means of the relaxed constraints Taylor theory [6]. The simulations presented in figure 4 were carried out with relaxed longitudinal and transverse shear and consideration of dislocation movement on {110}, {112} and {123} slip planes.

Annealing

The annealing textures depend on the initial cold rolling texture and microstructure. In figure 5 the RX textures are shown for sample CR11TI. The maximum at {111}<112> after primary RX which is well known for bcc steels [2,3], can be interpreted by the ≈32°<110> rotation relationship between {111}<112> and the strong rolling component {112}<110>. This relationship is close to the ideal 27 °<110> rotation, characterized by a Σ19a coincidence. Due to the high mobility of this special boundary the {111}<112> nuclei can selectively grow into the {112}<110> matrix [7].

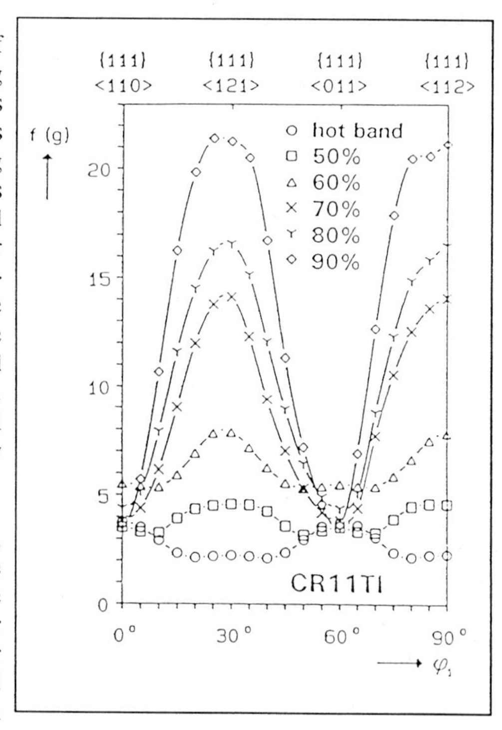

Figure 5: *Recrystallization texture.*

After rolling of 50% to 75% also the increase of the Goss component can be seen in the annealing texture (figure 6). The maximum of the Goss after 70% deformation and annealing correlates to the maximum of {111}<112> in the rolling texture (figure 7) at s=0.8. In this layer the Goss component was inherited from the hot rolled material and rotated during cold rolling under plain strain deformation towards {111}<112> and {001}<110>, respectively (figure 3). These divergent rotations supposingly led to the maintenance of transition bands with preserved Goss orientation between the rotated fragments. Due to the fine cellstructure and high local misorientations, it is thus expected that Goss may nucleate in the transition bands [3].

In former works especially on Fe3%Si textures, the correlation between {111}<112> in the rolling texture and Goss in the annealing texture was explained by nucleation of Goss in shear bands [8]. In FeCr however only very weak shear banding occurs, due to the small grain size and the interstitial free matrix, so that nucleation in transition zones is considered to represent the main mechanism of Goss formation.

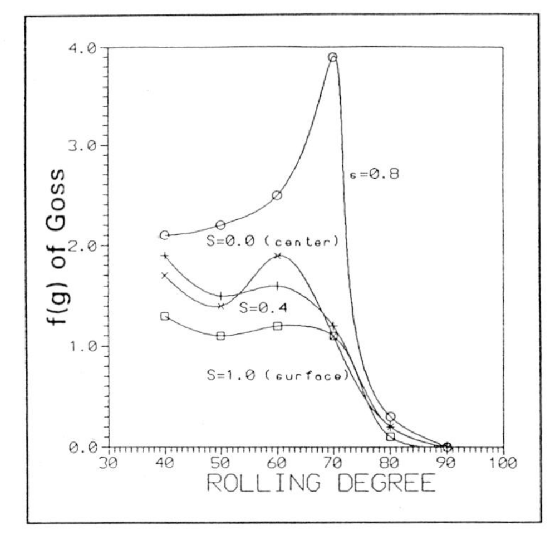

Figure 6: Intensity of Goss in CR11NB after RX for various layers.

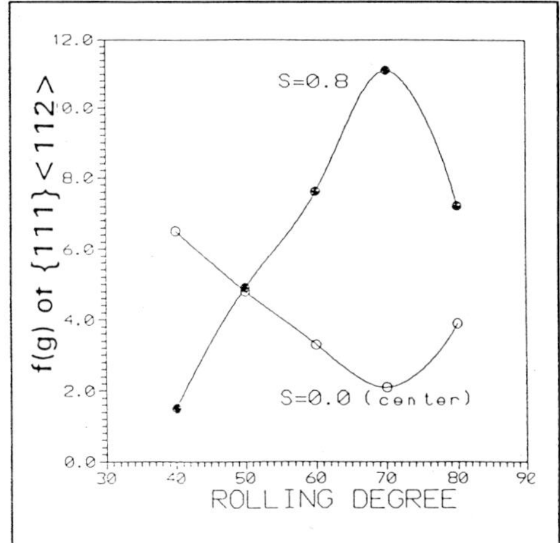

Figure 7: Intensity of {111}<112> in CR11NB after cold rolling.

Besides the formation of {111}<112> and {011}<100> in FeCr alloys, the RX of samples with stable particles can lead to a new component ≈{557}<583> instead of the classical {111}<112> maximum [9].

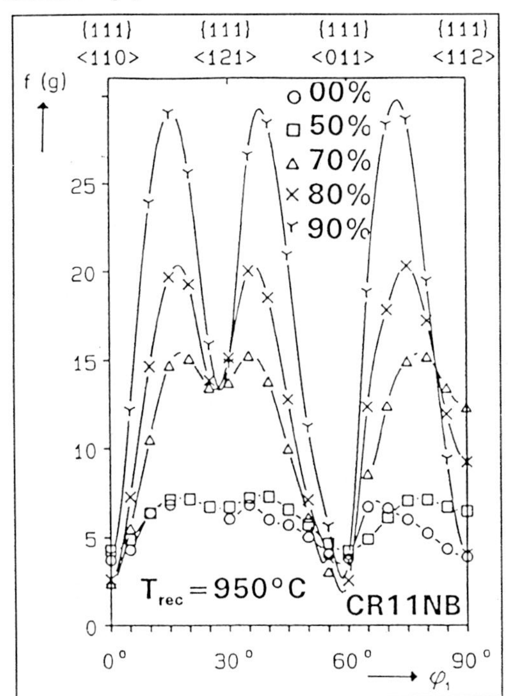

Figure 8: RX texture of sample CR11NB, 5min at 950°C.

In figure 9 the increase of {557}<583> is presented for RX of CR11NB at 950°C. The maximum is shifted about 10° from {111}<112> and increases with the rolling degree. The measurements present a good correlation between the intensity of {112}<110> in the rolling texture and {557}<583> in the annealing texture. The exact positions of both orientations reveal an ideal 27°<110> rotation relationship. The RX texture of CR11TI however shows a classical maximum at {111}<112> (figure 5).

The development of this component can be explained by particle induced growth selection [9]. The Ti and Nb carbonitrides are precipitated before hot rolling and hence randomly distributed in the microstructure without necessarily being connected to subboundaries or large angle grain boundaries. In this case the formula for particle drag can be written as $F_p = (6 \cdot \gamma \cdot f) / (\pi \cdot R)$, where F_p is the particle drag force, γ is the grain boundary energy, f the precipitated volume fraction of particles and R the average particle radius.
The driving forces for primary recrystallization

of FeCr steels are of the magnitude $F_r \approx 10^5$ Pa, because the precipitations extend the incubation period, so that strong recovery can take place. It is supposed that the back driving force which is imposed by the precipitations can reach the same magnitude as the driving force for primary RX. It has however to be considered that the back driving force is proportional to the grain boundary energy [9]. Therefore non-special grain boundaries which have a high energy are much more hindered by particles than coincidence boundaries which have a low energy. In case of CR11TI the particle drag forces have supposingly been too weak to gain influence on growth selection.
In case of fine dispersion, the retarding force of the particles thus is responsible for a strong growth selection during RX, which leads to the growth of {557}<583> nuclei into the {112}<110> deformation matrix because of their 27°<110> coincidence relationship.

CONCLUSIONS

1. FeCr hot bands reveal shear textures at the surface and "cold rolling" textures with α-fibre in the center layer.
2. Cold rolling leads to a strong α-fibre texture in all layers with a maximum in the center.
3. Increase of Cr content shifts {112}<110> to higher Φ angles on the α-fibre.
4. Cold rolling textures can be simulated by means of relaxed constraints Taylor models.
5. Recrystallization of 50%-70% cold rolled samples with strong {111}<112> leads to Goss especially in the sub-surface layer.
6. The development of Goss during annealing is dependant on the existence of transition bands.
7. Recrystallization of cold rolled samples without stable particles or with low dispersion rates leads to formation of {111}<112> by growth selection.
8. The presence of finely dispersed stable microcarbonitrides can influence the recrystallization texture by selective particle drag.

REFERENCES

1) Bunge, H.J.: Z. Metallkunde, 1965, 56, 872
2) Hölscher, M., Raabe, D., Lücke, K.: Steel Research 1991, 12, 567
3) Raabe, D., Lücke, K.: Mat. Science and Techn., 1993, 9, 302
4) Beynon, J.H., Ponter, A.R.S, Sellars, C.: Proc. Mod. of Met. Form., Kluwer Publ., 1988, 321 Steels, Chiba, Japan, ed. ISIJ, (1991), 877-884.
5) Houdremont, E.: Handbuch der Sonderstahlkunde, 3.ed., Springer Verlag, 1956
6) Honneff, H., Mecking, H.: Proc. ICOTOM 5, Aachen, 1978, 457, 233
7) Ibe, G., Lücke, K.: Arch. Eisenhüttenwesen, 1969, 39, 33
8) Matsuo, M.: ISIJ Int., 1989, 29, 809
9) Raabe, D., Lücke, K.: Steel Res., 1992, 10, 457

Materials Science Forum Vol. 157-162 (1994) pp. 1475-1480

NEUTRON DIFFRACTION MEASUREMENTS OF TEXTURE IN SILICON CARBIDE WHISKER-REINFORCED ALUMINUM COMPOSITES

J.H. Root [1] and H.J. Rack [2]

[1] AECL Research, Chalk River Laboratories, Chalk River, Ontario KOJ 1JO, Canada

[2] Department of Mechanical Engineering, Clemson University, Clemson, SC, USA

Keywords: Neutron Diffraction, Metal-Matrix Composites, Silicon Carbide, Whiskers

ABSTRACT

Neutron diffraction analysis provides a rapid method to analyze the whisker orientation distribution function in SiC whisker-reinforced aluminum composites. The ambiguity in relating cubic crystallographic and whisker directions is removed because the (111) and ($\bar{1}$11) peak shapes are easily distinguished. The SiC pole figures satisfy neither cubic nor hexagonal symmetries, but to quantify the whisker orientation distribution function, the whiskers can be treated as hcp objects, with the basal plane-normal parallel to the axis. A CODF is obtained from the two available complete pole figures by the series expansion method, with a maximum l of 22. Three additional pole figures, which are needed for this calculation, are determined by iteratively fitting the available data. The degree of whisker alignment is strongly coupled to the extrusion ratio but weakly coupled to the volume fraction of whiskers in the composite material.

INTRODUCTION

Aluminum that contains silicon carbide whisker reinforcements is a light-weight, strong metal-matrix composite. Preferred alignments of the whiskers are created when processing the material. The resulting anisotropy in the mechanical properties of the composite may be optimized through insights obtained from a quantitative analysis of the whisker orientation distribution. However, such analyses are normally performed by a painstaking process of examining micrographs of carefully prepared surfaces, counting whiskers and constructing histograms of volume fraction in coarsely divided bins of whisker-axis direction. Typically, one to three days are needed to obtain an analysis of the volume fraction of whiskers versus direction by microscopy. A question of statistical accuracy also arises because sectioning and etching specimens tends to remove broken or short whiskers from the population that can be viewed in the microscope.

Since neutrons easily penetrate through aluminum, the preferred crystallographic orientations of the silicon carbide whiskers can be determined in bulk material with no effect on the whisker population due to specimen preparation [1]. A scan of the volume fraction of crystal plane-normals versus direction about one axis of a specimen requires between five and fifteen minutes. SiC whiskers are prepared as growth-faulted single crystals with a predominant cubic-diamond structure [2]. The [111] direction is parallel to the whisker axis and a [2$\bar{2}$0] direction is perpendicular to the axis. The [$\bar{1}$11] direction is 70.53° from the whisker axis. An alternative crystallographic phase of SiC is hexagonal-close-packed (hcp), in which the [0002] direction is parallel to the whisker axis and a [10$\bar{1}$0] direction is perpendicular to the axis. However, a neutron powder diffraction pattern of the whiskers produced by ACMC [2] revealed no significant volume fraction of the hcp crystallographic phase. In a well-ordered three-dimensional crystal lattice, the (111) and ($\bar{1}$11) diffraction peaks are indistinguishable. For this reason, Juul Jensen *et al* pointed out that a unique relationship between crystallographic orientations and whisker orientations could not be made [1]. However, the "single-crystal" whiskers contain a high frequency of stacking faults along the whisker axis. Like a roll of coins, the whisker is a stack of thin lamellae, each of thickness less than 20 nm [3]. This lamellar structure breaks the long-range order of the ($\bar{1}$11) planes but preserves the long-range order of the (111) planes. The resulting diffraction peaks exhibit distinctive lineshapes, so the ambiguity in relating crystal and whisker orientations can be removed.

EXPERIMENT

Metal matrix composites of various compositions were obtained from ACMC [2]. Extruded-plate materials contained 5, 10 and 20 v/o of whiskers in an Al-2124 alloy matrix. The 5 v/o material was made with grade F-8 whiskers, which contain less than 10% of the particulate morphology, and its extrusion ratio was 20:1. The 10 v/o and 20 v/o materials were made with grade F-9 whiskers, containing less than 20% particulates, and were reduced by only 11.5:1 during extrusion. From each plate, a cylindrical specimen was machined, with a diameter of 10 mm, a length of 40 mm and the axis parallel to the original transverse direction.

On the L3 neutron diffractometer, measurements were made of the peak intensity versus direction, (χ,η), where χ was the angle of tilt from the cylinder axis, and η was the azimuthal angle from the extrusion direction. The L3 diffractometer, which is attached to the NRU reactor at AECL Research in Chalk River, was configured to permit measurements of texture as well as whisker-matrix interaction stresses. Soller-slit collimators restricted the angular divergences of the incident and diffracted beams to be 0.4°, but the beams had cross sections of 50 mm x 50 mm, to fully illuminate the specimen volume at every orientation. The specimen orientations were set by an Eulerian cradle, and peak intensities were collected with a single detector of cross sectional area 50 mm x 50 mm. Each pole figure required a data collection time of 3 hours.

Detailed measurements of the diffraction peak shape were later made on the DUALSPEC neutron powder diffractometer, also at Chalk River Laboratories [4]. For the composite specimen with 5 v/o of whiskers, the region of the SiC (111) diffraction peak was examined in directions near 0°, 70° and 90° from the extrusion direction, which correspond, respectively, to crystallographic directions of the highly-aligned whiskers: [111] (or [0002]), [$\bar{1}$11] and [10$\bar{1}$0].

RESULTS

Detailed neutron diffraction peak shapes are presented in figure 1. The (111) diffraction peak near the extrusion direction is very intense and narrow, as expected. Extrusion aligns many whisker axes parallel to the extrusion direction and the (111) crystal planes have long range order, despite the lamellar structure of the whiskers. The diffraction peak measured 70° from the extrusion direction corresponds to the ($\bar{1}$11) crystal planes, for which long range order has been broken by the lamellar structure of the whiskers. The resulting ($\bar{1}$11) diffraction peak is clearly different in shape from the (111) peak. This distinctive peak shape can be analyzed to determine the volume-averaged thickness of the lamellae in the ensemble of whiskers. The diffraction peak measured 90° from the extrusion direction is also shown. No peak is expected in this direction if the whiskers are purely cubic and highly aligned in the extrusion direction. The observed (10$\bar{1}$0) peak, indicates that between 5 and 10 v/o of the silicon carbide in the composite material can be characterized as hexagonal close-packed. It cannot be determined from these data whether the hcp volume fraction arises from some of the lamellae in some whiskers, or arises from some number of purely hcp whiskers or is associated with the fraction of the silicon carbide that is in the particulate morphology. However, such a determination is not essential to the focus of this paper, which addresses the orientation distributions of whiskers rather than their crystallographic character.

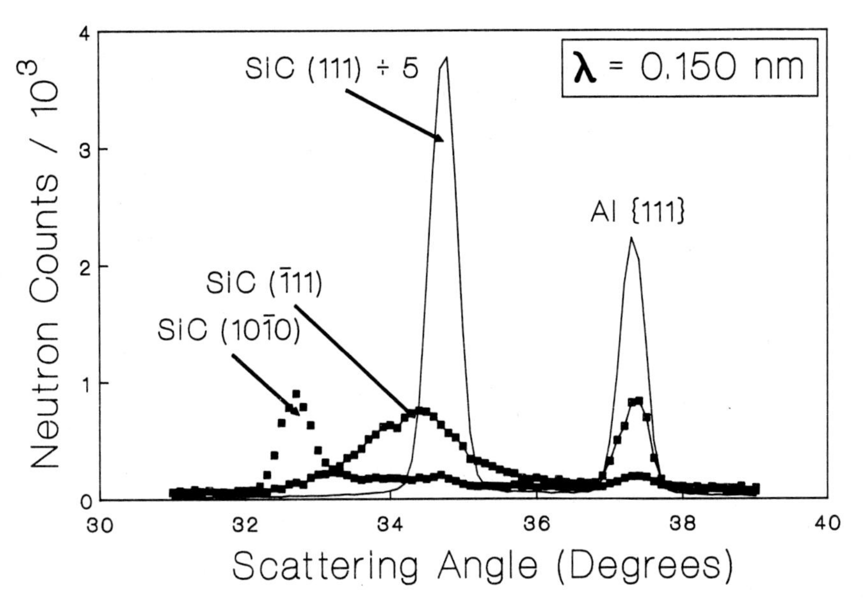

Figure 1. Diffraction peaks from 5 v/o SiC whisker-reinforced aluminum. The (111), ($\bar{1}$11) and (10$\bar{1}$0) peaks were measured 0°, 70° and 90°, respectively, from the extrusion direction.

The crystallographic texture of the SiC phase in the 5 v/o composite material is presented in figures 2 (a) and 2 (b), which show stereographic (111) and (220) pole figures, respectively. Contour lines join points of equal intensity, with the dashed line indicating a level of 0.5 x random, the heavy line indicating the level of 1 x random and successive continuous lines indicating 2, 4, 8 and 16 x random. The high (111) intensity near the extrusion direction (E) and the strong band of (220) intensity in the normal-transverse (N-T) plane correspond to a whisker orientation distribution where the axes are concentrated near E and preferred orientations

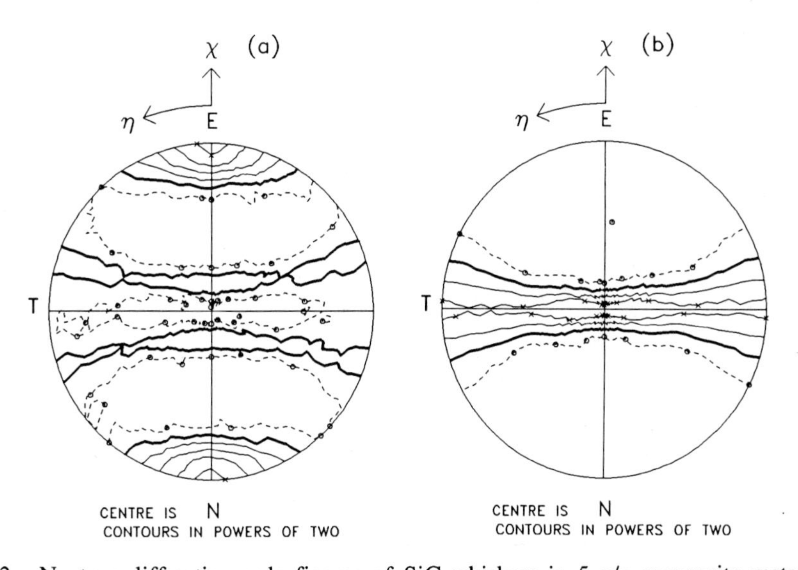

Figure 2. Neutron diffraction pole figures of SiC whiskers in 5 v/o composite material.

about the axes are minimal. A striking feature of these pole figures is the absence of a definite crystallographic symmetry. If the silicon carbide whiskers exhibited ideal cubic symmetry, the bands of intensity 70° from E in the (111) pole figure would be about three times as intense as observed. Similarly, there would be a second band of intensity 30° from E in the (220) pole figure, as intense as the one observed in the N-T plane. If the silicon carbide whiskers exhibited ideal hcp symmetry, there would be no intensity bands other than the major features observed in the pole figures. The crystallographic symmetries of the pole figures fall between these two ideals.

Despite the lack of crystallographic symmetry in the pole figures, a meaningful analysis of the whisker orientations can still be made. The (111) intensity near E is uniquely associated with whisker axes and the measured neutron intensity is directly proportional to the volume fraction of whiskers whose axes lie in a given direction. Histograms of the volume fraction in 10°-wide bins of whisker direction in the E-T and E-N planes are readily constructed and are shown in figures 3 (a) and 3 (b), respectively. A comparison is made between the results of neutron diffraction and microscopy for specimens of 5, 10 and 20 v/o composites. The agreement between the two techniques is excellent, except that the histograms obtained by microscopy on the 5 v/o materials are slightly more peaked in the extrusion direction than those obtained by neutron diffraction. In both figures, it is clear that the whisker axis distributions for the 10 and 20 v/o materials are similar, but the 5 v/o material is about twice as intense in the extrusion direction. Evidently, the sharpness of the whisker axis distribution is only weakly coupled to the volume fraction of whiskers in the composite, but strongly coupled to the extrusion ratio, which was 20:1 for the 5 v/o material and 11.5:1 for the 10 and 20 v/o materials. To acquire sufficient statistical precision, the collection of the microscopy data was an extremely labour-intensive, and lengthy exercise. In contrast, the neutron technique lends itself well to automation, no special preparation of the specimen is required and the whisker axis distribution is determined a hundred times faster than by microscopy.

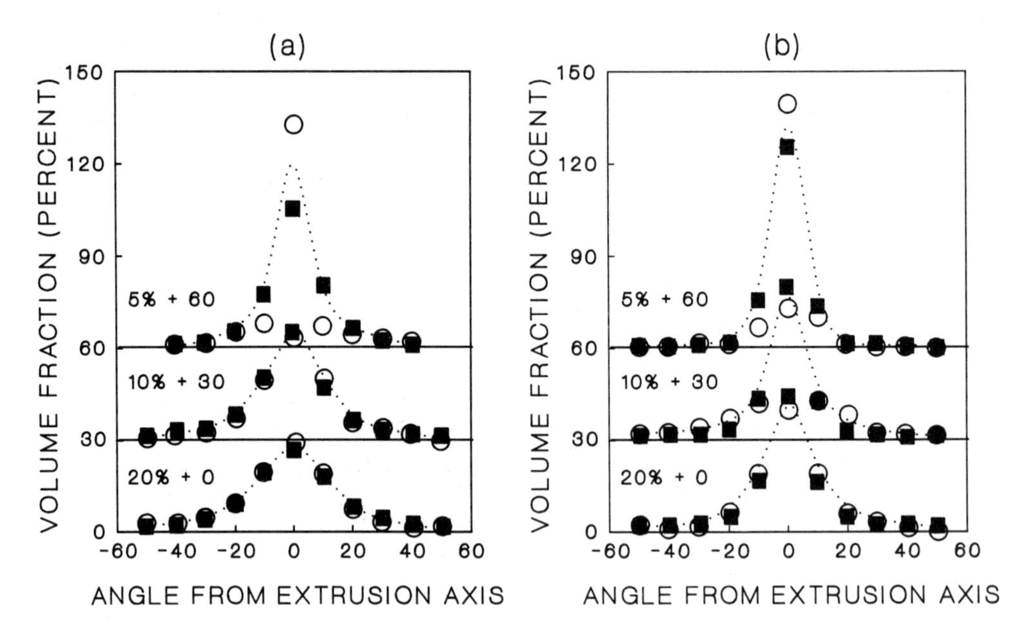

Figure 3. Whisker axis distributions determined by neutron diffraction (filled squares) and by microscopy (open circles). Histograms for 5, 10 and 20 v/o materials are separated by vertical shifts. Dotted lines are guides to the eye.

DISCUSSION

A complete description of whisker orientations requires a whisker orientation distribution function, w(ψ,θ,ϕ), analogous to a CODF. Calculation of w(ψ,θ,ϕ) from the measured crystallographic pole figures is complicated by two factors. First, the weak bands of intensity 70° from E in the (111) pole pole figure arise from the defective crystal structure of the whiskers and not from the whisker axes. Similarly, intensities in regions other than the (N-T) plane of the (220) pole figure do not relate to directions perpendicular to the whisker axes. To obtain pole figures that relate only to directions parallel and perpendicular to the whisker axes, intensities found more than 40° away from the maximum value are reset to zero. The axis-parallel pole figure then contains only one, broad peak of intensity near E. The axis-perpendicular pole figure contains only one broad band of intensity near the (N-T) plane.

The statistical symmetries of the axis-parallel and axis-perpendicular pole figures are, respectively, the same as (0002) and (11$\bar{2}$0) pole figures that might be obtained from a hexagonal close-packed crystallographic material. Therefore, despite the fact that the whiskers are crystallographically cubic, the (non-crystallographic) whisker orientation distribution function, w(ψ,θ,ϕ), can be calculated with a standard CODF calculation that is adapted for hcp symmetry [5,6]. However, the second complication arises at this point. High-order expansion coefficients, W_{lmn}, are needed to characterize the sharp texture of the whiskers by a series of harmonic basis functions. To obtain a least-squares solution for the set, $\{W_{lmn}\}$, where the maximum l is 22, for example, requires at least five pole figures [7], whereas only two complete pole figures are available. The missing pole figures are constructed by iteration. Three, random-textured "phantom" pole figures are added to the two existing "real" pole figures. The phantoms are designated as (10$\bar{1}$0), (10$\bar{1}$1) and (10$\bar{1}$2). The total of five pole figures are submitted to the series-expansion part of the CODF program to obtain $\{W_{lmn}\}$ with a maximum l of 22. The set, $\{W_{lmn}\}$, is then applied to recalculate the phantoms, which now acquire some of the structure that is consistent with the real pole figures. The new phantoms and the original real pole figures are re-submitted to the series-expansion to obtain an improved $\{W_{lmn}\}$. With each iteration of this process, the phantoms converge towards pole figures that are symmetrically consistent with the real ones. Typically, the set, $\{W_{lmn}\}$, converges within ten iterations. The procedure was tested by extracting a set of expansion coefficients from five real pole figures, (10$\bar{1}$0), (0002), (10$\bar{1}$1), (10$\bar{1}$2) and (11$\bar{2}$0) from a rolled plate of (hcp) Zircaloy-2. Expansion coefficients were also obtained from only two of the pole figures, (0002) and (11$\bar{2}$0) through the iteration procedure. Some of the results are compared in Table I, where excellent agreement, within the uncertainties of least-squares fitting, is obvious.

Table I

Comparison of W_{lmn} Values in Rolled Zircaloy-2 Plate

Numbers in brackets indicate the uncertainties derived from least-squares fits of the pole figure data. Coefficients $W^{(2)}_{lmn}$ and $W^{(5)}_{lmn}$ are obtained from 2 and 5 pole figures, respectively.

l	m	n	$W^{(2)}_{lmn}$	$W^{(5)}_{lmn}$
2	0	0	0.0127(5)	0.0129(6)
2	2	0	-0.0081(1)	-0.0081(3)
4	0	0	0.0010(4)	0.0008(6)
4	2	0	-0.0080(3)	-0.0082(6)
4	4	0	0.0042(1)	0.0042(1)
6	0	0	-0.0011(2)	-0.0011(3)
6	0	6	-0.0015(1)	-0.0018(10)
⋮			⋮	⋮
6	6	6	0.0038(1)	0.0037(4)
⋮			⋮	⋮

Not surprisingly, the whisker orientation distribution function, $w(\psi,\theta,\phi)$, derived from the axis-parallel and axis-perpendicular pole figures through the interation process exhibits little dependence on ϕ, which is the orientation about the whisker axis. Slices of $w(\psi,\theta,\phi)$, with $\phi = 30^\circ$, for 5, 10 and 20 v/o materials are presented in figure 4. As observed in the axis distributions (figure 3), the maximum value of $w(\psi,\theta,\phi)$ is similar in the 10 and 20 v/o materials, 22 x random and 18 x random, respectively, but attains a much higher value, 42 x random, in the 5 v/o material.

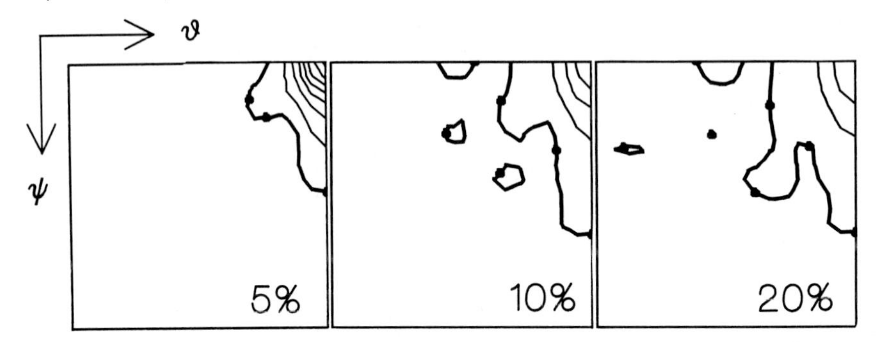

Figure 4. Slices of the whisker orientation distribution function at $\phi = 30^\circ$ for 5, 10 and 20 v/o materials. The contours are in intervals of five times random.

CONCLUSIONS

Neutron diffraction texture analysis is an effective means to characterize the whisker orientation distribution function of silicon carbide whisker-reinforced aluminum composites. The degree of preferred orientation of whiskers is strongly coupled with the extrusion ratio of the composite material, but only weakly coupled to the volume fraction of whiskers in the matrix.

ACKNOWLEDGEMENTS

We gratefully acknowledge the contributions of Dr. Z. Tun of AECL Research and Dr. M. Dahms of GKSS Forschungszentrum, whose stimulating discussions helped greatly to shape this paper.

REFERENCES

1) Juul Jensen, D., Lilholt, H., Withers, P.J., in "Mechanical and Physical Behaviour of Metallic and Ceramic Composites", S.I. Andersen, H. Lilholt and O.B. Pedersen eds., Risø National Laboratory, Roskilde, Denmark, 1988.

2) Advanced Composite Materials Corporation, Greer, SC.

3) Nutt, S.R.: J. Amer. Ceram. Soc., 1984, 67, 428

4) Konyer, N.B. and Root, J.H., "DUALSPEC Powder Diffractometer User's Guide", Report RC-922, AECL Research, Chalk River, Ontario, Canada, K0J-1J0, (1992).

5) Roe, R.J.: J. Appl. Physics, 1965, 36, 2024

6) Root, J.H. and Holden T.M., "Programs to Determine the Crystallite Orientation Distribution Function from Diffraction Intensity Pole Figures", Report ANDI-29, AECL Research, Chalk River, Ontario, Canada, K0J-1J0, (1990).

7) Bunge, H.J., "Texture Analysis in Materials Science", Translated by P.R. Morris, Butterworths & Co. (Publishers) 1982.

Materials Science Forum Vol. 157-162 (1994) pp. 1481-1486

FORMATION OF OBLIQUE SHAPE AND LATTICE PREFERRED ORIENTATION IN A QUARTZ BAND OF A GNEISSIC MYLONITE

W. Skrotzki [1], J. Dornbusch [2], F. Heinicke [3] and K. Ullemeyer [4]

[1] Institut für Kristallographie und Festkörperphysik,
Technische Universität Dresden, Mommsenstr. 13, D-01062 Dresden, Germany

[2] Institut für Geologie und Dynamik der Lithosphäre,
Universität Göttingen, Goldschmidtstr. 3, D-37077 Göttingen, Germany

[3] GKSS Forschungszentrum Geesthacht,
Max-Planck-Str. 1, D-21494 Geesthacht, Germany

[4] Joint Institute for Nuclear Research, Laboratory for Neutron Physics,
141980 Dubna, Moscow District, Russia

Keywords: Quartz Mylonite, Texture Components, Domainal Texture

ABSTRACT

Microstructural and textural investigations on a dynamically recrystallized quartz band in a gneissic mylonite reveal an oblique shape and lattice preferred orientation. The bulk texture is characterized by a slightly kinked c-axis oblique girdle and an asymmetric a-axis maximum normal to it. It consists of texture components prevailing in certain domains. The grain shape foliation is inclined at about 16° with respect to the overall foliation. Both asymmetric microstructural and textural elements indicate a large component of simple shear with dextral shear sense. The results are compared with predictions of "Sachs-" and "Taylor-type" models on polycrystalline plasticity combined with dynamic recrystallization.

INTRODUCTION

In dynamically recrystallized rocks oblique shape and lattice preferred orientation with respect to the overall foliation and lineation is a common feature. It has been described from shear zones in many different rock types including quartzites [e.g. 1 to 7], carbonates [8, 9] and peridotites [10], as well as from experimentally deformed analoque materials such as ice [11] and octachloropropane [12, 13]. To further study the mechanism of its formation microstructural and textural investigations have been carried out on a quartz band in a gneissic mylonite.

GEOLOGICAL FRAME

The mylonitic rock investigated originates from the Portalegre Shear Zone (NE-Brazil) which characterizes an important tectono-metamorphic event under upper greenschist facies conditions (350 - 450°C, 0.2 - 0.5 GPa) following the emplacement of the Panafrican/Brasilian granites [14]. The rock is of granitic composition with coarse k-feldspar and oligoclase porhyroclasts embedded in a fine-grained matrix of quartz, feldspar and biotite. It contains a ≈3 mm thick quartz band parallel to the overall rock foliation S_A defined by the alternation of bands with different degree of recrystallization.

EXPERIMENTAL

The grain structure of the quartz band has been analyzed quantitatively with an image analyzing program developed by Duyster [15]. The parameters calculated from the digitized grain boundaries are the ratio R_f between the long and short grain axis, the grain size as well as the orientation Φ of the grain long axis with respect to the stretching lineation L_A.

The microstructure of the quartz band has been studied by conventional transmission electron microscopy (TEM) on a Siemens Elmiskop 101 operating at 100 kV acceleration voltage. The TEM-foils used have been prepared by Ar-beam ion-thinning and coating with a thin amorphous carbon layer.

Pole figures have been measured locally (grain by grain) with a universal stage (c-axis, [0001]) and globally in four sections parallel to S_A on a X-ray pole figure goniometer (m, {10-10}; a, {11-20}; r+z, {10-11}+{01-11}; {10-12}; {10-14}; {20-21}). The X-ray pole figures have been analyzed quantitatively using the component method developed by Helming and Eschner [16].

RESULTS

The microstructure of the quartz band is typical for dislocation creep of a dynamically recrystallyzed material consisting of a network of grain and subgrain boundaries and a high density of free dislocations with predominantly <a>-Burgers vectors (figures 1 and 2). The recrystallized grain size measured optically (diameter of spherical grains of same area) is about 45 μm. TEM measurements yield about 5 μm and $5 \cdot 10^8/cm^2$ for the subgrain size and the dislocation density, respectively.

The grains and subgrains have a fairly oblate shape with the flattening plane being inclined to S_A by about 16° and defining a grain shape foliation S_B (figures 1 and 2). The maximum aspect ratio R_f is about 5 and 3 in thin sections parallel L_A/normal to S_A and parallel to S_A, respectively (figure 3). However, there is no preferred orientation of the grain long axes in the section parallel to S_A.

Figure 1: Optical micrograph (crossed polarizers) of a thin section cut parallel to the lineation L_A and normal to the foliation S_A showing the grain structure of the quartz band investigated. The grain shape foliation S_B is indicated. Light and dark areas represent domains of similar c-axis orientations.

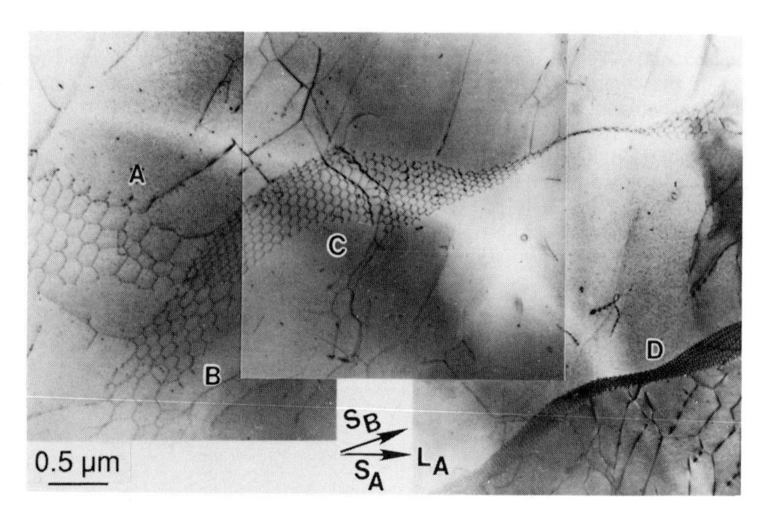

Figure 2: TEM micrograph of the microstructure with subgrain boundaries indicated A to D. Boundaries A to C are low-angle twist boundaries close to the basal plane characterized by a hexagonal network of screw dislocations with <a>-Burgers vectors. (Bright field image taken with g = 11-20 reflection)

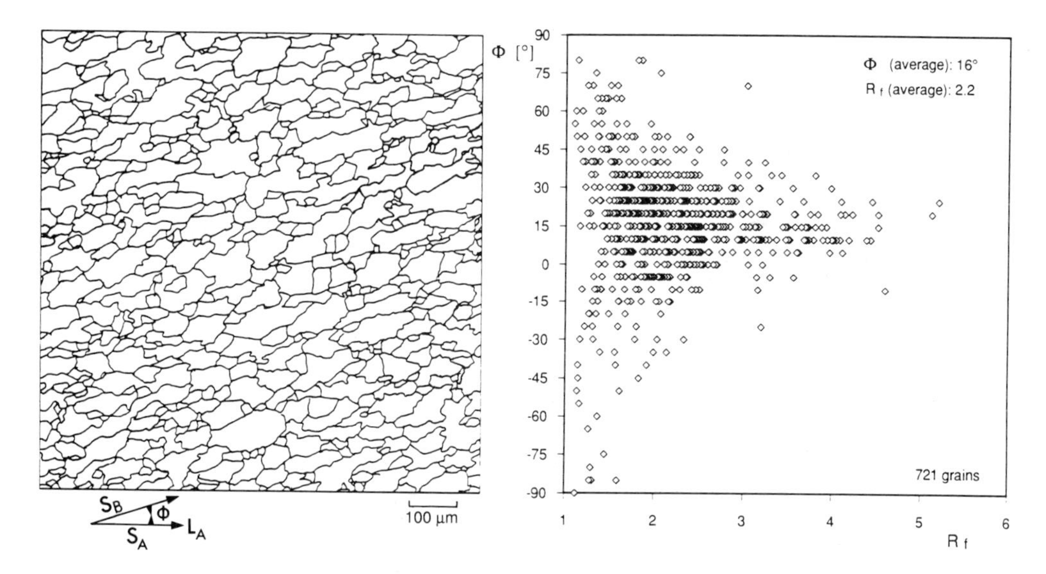

Figure 3: Diagram showing the relation between the aspect ratio R_f and the orientation Φ of the grain long axis with respect to L_A derived from the grain structure of the given section parallel to L_A and normal to S_A.

The bulk texture has been derived by superposition of all texture components of the four sections parallel to S_A with volume fractions greater than 1%. It can be well described by six components. The c- and a-axis pole figures recalculated with these components are shown in figure 4. The bulk texture consists of c-axis maxima on a slightly kinked oblique girdle with the pole inclined to L_A by about 20°. Consistently the a-axis pole figure yields inclined girdles with a strong asymmetric maximum normal to the c-axis girdle. The main components characterizing this texture are given in figure 4 and table 1. They are of different half width (sharpness) and volume fraction in the bulk. The main c-axis maximum is composed of two texture components separated by 30° rotation about the common c-axis. To get some information about texture variation in the four sections the volume fraction of the components has been calculated at fixed orientation and constant bulk half width by fitting the components to the experimental data. Table 1 shows that the volume fraction of the components varies from one section to the next. There is no systematic variation across the quartz band, i.e. from section A to D. Differences in volume fraction in the four sections indicate a domain structure of related orientations. This is confirmed by similar extinction in large elongate areas of the quartz band parallel to L_A (figure 1). Optical measurements of the c-axes in these domains yield different maxima on an oblique girdle (figure 5); dark domain: components 1, 2, 4; light domain: 3 and/or 5, 6. The rotations about the c-axis as observed for components 3 and 5 cannot be distinguished optically.

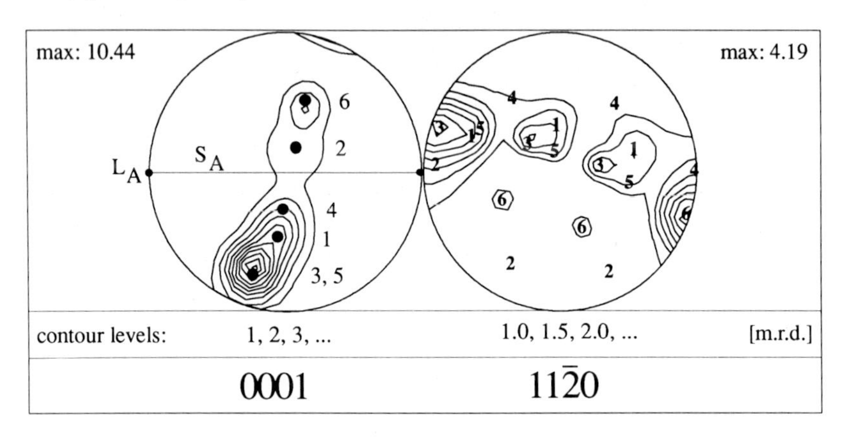

Figure 4: c- and a-axis pole figures representing the bulk texture recalculated with the six components indicated. The diagrams are the superposition of the pole figures of four sections parallel to S_A. (Equal area projection, lower hemisphere)

Table 1: Half width (in degrees) and volume fraction (in percent) of the main components composing the texture of the bulk and of four sections parallel to S_A (see figure 4). Sections A to D represent a traverse across the quartz band. Euler angles are given in the Roe-convention.

component	orientation	half	volume fraction				
[nr]	{ α, β, γ }	width	bulk	A	B	C	D
1	{261.11, 51.38, 69.37}	28	21	10	25	25	15
2	{ 67.13, 19.23, 49.83}	41	17	19	27	20	14
3	{249.98, 72.44, 92.65}	21	14	21	6	4	26
4	{270.40, 31.88, 30.40}	32	11	17	24	18	12
5	{250.91, 73.71, 69.75}	23	10	7	13	16	16
6	{ 72.63, 52.58, 86.33}	22	7	15	5	12	7

Both asymmetric microstructural and textural elements indicate a large component of simple shear with dextral shear sense.

DISCUSSION

Schmid and Casey [17] have suggested that by changing from coaxial to non-coaxial straining or by increasing the strain in simple shear the c-axis pole figure changes from a symmetrical cross-girdle to an asymmetrical single girdle with the shear direction given by the main a-axis maximum (figure 6). The texture observed may represent a transitional stage. As revealed by the component method the components 2 and 4 have one a-axis maximum parallel to L_A which is inclined to the assumed shear direction. The corresponding a-axis maximum of components 1 and 5 deviate from

L_A and the shear direction by about 30°, respectively. These deviations may reflect the higher stability of some components during the transition from a symmetrical cross-girdle to an asymmetrical single girdle.

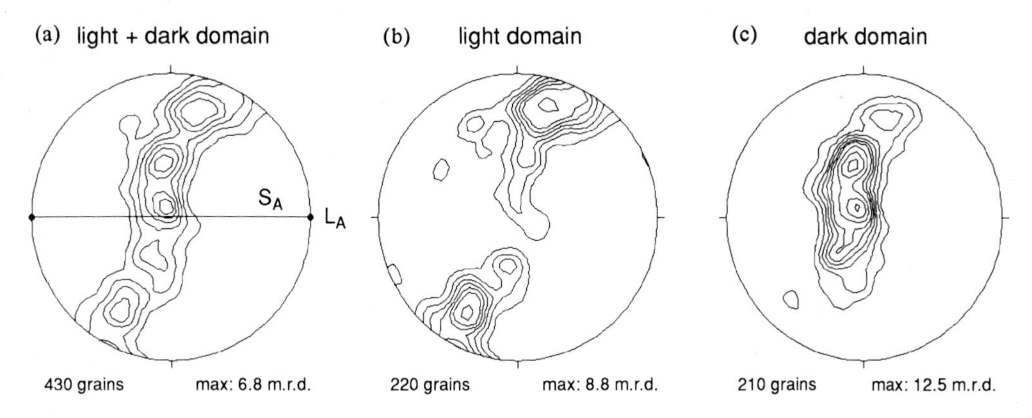

Figure 5: Optically measured c-axis pole figures. (b) and (c) differentiate between light and dark domains of figure 1.

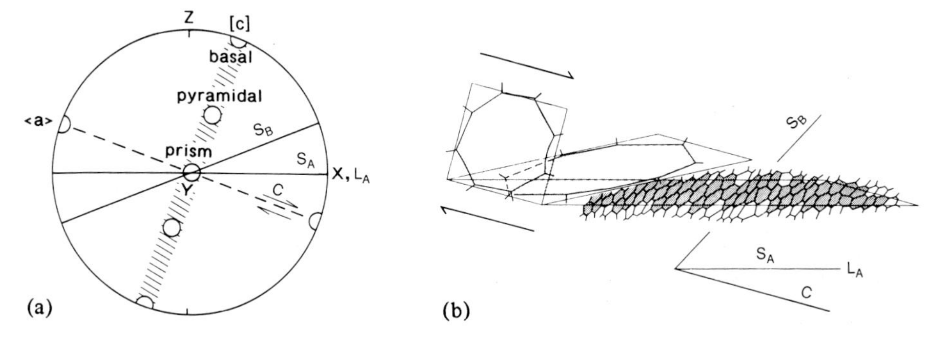

Figure 6: (a) Schematic sketch summarizing the microstructural and textural results. The plane C and the direction given by the a-axis maximum are assumed to be the shear plane and shear direction, respectively. Single slip end positions are indicated. (b) Model suggested by Lister and Snoke [2] for the development of a grain shape foliation. Dynamic recrystallization periodically resets the "finite-strain clock".

Two different models on polycrystalline plasticity may be considered to explain the development of the domainal texture observed.

(i) "Sachs-type" deformation requiring stress equilibrium (Etchecopar and Vasseur [18])

In this case an originally coarse-grained aggregate, with the grains representing later recrystallized domains each with similar orientation pattern (figure 6b), has been deformed by single slip on the slip systems with the highest Schmid factor. Depending on the orientation of the grains the predominant slip systems activated have been either basal-<a>, prism-<a> or pyramidal-<a>. Preferred alignment of these slip systems during shearing with the slip plane parallel to the shear plane and the slip direction parallel to the shear direction leads to a domainal "single slip texture" with the end positions indicated in figure 6a. The unconstrained "Sachs-type" deformation is due to dynamic recrystallization by subgrain rotation and minor grain boundary migration.

(ii) "Taylor-type" deformation requiring strain homogeneity combined with dynamic (migration) recrystallization (Jessell and Lister [19])

In this case an originally fine-grained aggregate has been deformed by multiple slip on five independent slip systems. Depending on the plastic anisotropy and intensity of migration recrystallization with increasing strain in simple shear the constrained deformation leads from a c-axis cross-girdle (symmetrical to the maximum finite elongation) to a single girdle normal to the shear direction with certain maxima at single slip positions. The maxima result from the preferred growth of grains with less stored dislocation energy (soft orientations). At large strains the microstructure becomes domainal and exhibits a strong grain shape foliation.

There is a better agreement with the predictions of the hybrid model by Jessell and Lister [19], in particular, because this model is able to explain texture components deviating from the ideal single slip positions. However, with increasing temperature climb and cross slip as well as other processes may relax the constraints on polycrystalline deformation. As a consequence, there is a change from "Taylor-" towards "Sachs-type" deformation [20]. This fact might be better simulated by a self-consistent model (for a review see Wenk and Christie [21]) in combination with dynamic recrystallization.

The interplay of rotation and migration recrystallization as foliation-weakening processes [11] leads to a uniform grain shape foliation S_B with respect to the shear plane which may be an oscillating or a steady state foliation [2, 3, 4, 13]. In contrast, with increasing strain the overall foliation S_A rotates towards the shear plane (figure 6b). This effect has been observed by Law et al. [3].

ACKNOWLEGDMENTS

We would like to thank Dipl. Geol. F. Heidelbach and Dr. G. Zarske for universal stage c-axis measurements and Dr. K. Helming (TU Clausthal) for helpful discussions about his component method. The rock sample has been kindly provided by Prof. Dr. P.C. Hackspacher (Rio Claro, Brasil). The work has been supported by the Deutsche Forschungsgemeinschaft through a Heisenberg-fellowship (W. S.).

REFERENCES

1) Simpson, C. and Schmid, S.M.: Geol. Soc. Am. Bull., 1983, 94, 1281
2) Lister, G.S. and Snoke, A.W.: J. Struct. Geol., 1984, 6, 617
3) Law, R., Knipe, R.J. and Dayan, H.: J. Struct. Geol., 1984, 6, 477
4) Burg, J.P.: J. Struct. Geol., 1986, 8, 123
5) Knipe, R.J. and Law, R.D.: Tectonophysics, 1987, 135, 155
6) Law, R.D., Schmid, S.M. and Wheeler, J.: J. Struct. Geol., 1990, 12, 29
7) Dell'Angelo, L.N. and Tullis, J.: Tectonophysics, 1989, 169, 1
8) Schmid, S.M., Panozzo, R. and Bauer, S.: J. Struct. Geol., 1987, 9, 747
9) De Bresser, J.H.P.: Geologie Mijnb., 1989, 68, 367
10) van der Waal, D., Vissers, R.L.M., Drury, M.R. and Hoogerduijn Strating, E.H.: J. Struct. Geol., 1992, 14, 839
11) Burg, J.-P., Wilson, C.J.L. and Mitchell, J.C.: J. Struct. Geol., 1986, 8, 857
12) Jessell, M.W.: J. Struct. Geol., 1986, 8, 527
13) Ree, J.-H.: J. Struct. Geol., 1991, 13, 1001
14) Hackspacher, P.C. and Legrand, J.M.: Rev. Bras. Geoci., 1989, 19, 63
15) Duyster, J.: Ph.D. thesis, Univ. Göttingen, 1991, 141 pp
16) Helming, K. and Eschner, Th.: Cryst. Res. Technol., 1990, 25, K 203
17) Schmid, S.M. and Casey, M.: AGU, Geophys. Monogr., 1986, 36, 263
18) Etchecopar, A. and Vasseur, G.: J. Struct. Geol., 1987, 9, 705
19) Jessell, M.W. and Lister, G.S.: Geol. Soc. Spec. Pub., 1990, 54, 353
20) Skrotzki, W.: In: Textures in Geological Materials. DGM Press, 1994, in press
21) Wenk, H.-R. and Christie, J.M.: J. Struct. Geol., 1991, 13, 1091

Materials Science Forum Vol. 157-162 (1994) pp. 1487-1494

TEXTURE OF ALUMINIUM AND IRON THIN FILMS AND TEXTURE OF BILAYERS Al/Fe DEPOSITED ON SILICON

A. Tizliouine [1], J. Bessières [1], J.J. Heizmann [1] and J.F. Bobo [2]

[1] LMPC/ISGMP, Université de Metz, Ile du Saulcy, F-57045 Metz Cédex 1, France

[2] Laboratoire CNRS Saint Gobain, Centre de Recherche de Pont à Mousson, F-54704 Pont à Mousson, France

Keywords: Low Incidence, Grazing Incidence, Bragg-Brentano Geometry, Thin Film, Bilayer, Aluminium and Iron Texture, Pole Figures, Cathodic Pulverization, Bias

ABSTRACT:

Thin metallic layers and bilayers are used in electronic industry as well as protecting layers in metallurgy and glass industries.

We present a study on aluminium and iron thin films deposited, by cathodic deposition in vacuum, on silicon wafer or iron sheet and bilayers Al/Fe also deposited on silicon.

We show how the elaboration conditions, and mainly the Bias effect [1], influence the texture of thin films. Structural relationships between the substratum and the different layers are given.

I) INTRODUCTION

The production techniques of thin layers have recently been much improved.

In this paper, we present results about a system constituted by two metals, iron and aluminium.

Several X-Rays diffractometric methods as $\theta/2\theta$, rocking curve, grazing incidence and texture goniometry using, low and Bragg-Brentano incidences, are used to determine the organization of the layers.

Depending on each experimental technique and on the path of X-rays, diffracted intensities are corrected in order to obtain the same diffracting volume during the experiment.

II) TECHNIQUES

Metallic deposition was carried out with an Alcatel SCM 650 sputtering set-up [2]. The targets are made of pure metal. A computer driven system leads the sample holder in front of each target in turn. Fully automatised deposit cycles may be thus effectued. Iron is set by a 600w power radio frequency sputtering, Aluminium is set by direct current sputtering (400 w power D.C.

magnetron). Several types of samples have been elaborated their characteristics are listed on the following table.

SAMPLE	LAYER	THICKNESS	SUBSTRATUM	TECHNIQUE
Monolayer N°1	Aluminium	4000 Å	Iron	Sputtering
Monolayer N°2	Aluminium	4000 Å	Iron	Sputtering with Bias
Monolayer N°3	Iron	1500 Å	Silicon	
Bilayer N°4	Aluminium / Iron	1000 Å / 1500 Å	Silicon	

III) STUDY OF ORIENTATION OF ALUMINIUM THIN LAYERS

For the 0.4µm aluminium thin layer studied here, the intensity diffracted according to Bragg - Brentano's conditions is given by the relation:

$$I_{\theta,t} = \frac{i.So}{2\mu}\left[1 - \exp\left(\frac{2\mu t}{\sin\theta}\right)\right]$$

i = *intensity diffracted by an unit volume located on the surface,* So = *section of the incident beam,* µ = *linear absorption coefficient,* t = *thickness of the layer,* θ = *Bragg is angle of the diffracting plane .*

For a sample of the same type but of an infinite size, the experimental conditions (Bragg-Brentano) being the same, the diffracted intensity is:

$$I_{\theta,\infty} = \frac{i.So}{2\mu} = i.\,V_{\theta,\infty}$$

$V_{\theta,\infty}$ being the diffracting volume of an "infinite" thickness sample observed in Bragg-Brentano conditions.

When the sample is observed with an incidence angle α the intensity diffracted by the layer becomes [3]:

$$I_{\alpha.t} = I_{\theta,\infty}\left[1 - \frac{tg\omega}{tg\theta}\right] * \left[1 - \exp\left\{-\mu t\left(\frac{1}{\sin(\theta+\omega)} + \frac{1}{\sin(\theta-\omega)}\right)\right\}\right] = I_{\theta,\infty} * f(\theta,t,\omega) = i * V_{\alpha.t}$$

where $\omega = \alpha - \theta$ stands for the angular difference between this geometry and the Bragg-Brentano's one. This angle will be positive if the incidence angle is superior to Bragg's angle and negative if the angle is inferior to it .

1) θ/2θ and grazing incidence diffractogrammes

On The two aluminium samples studied, we have recorded the diffraction θ/2θ spectrum and the spectrum collected by grazing incidence . The first one (N°1) was prepared by cathodic pulverization (fig.1a), the second was deposited by same technique, but a low negative voltage (Bias effect) was applied on substratum (fig.1b).

- Sample N°1

In the case of grazing incidence, all the iron and / or aluminium diffracting planes are present. However, in Bragg's position, the aluminium line (311) is missing. This observation would mean that plane (311)Al are not parallel to the sample surface. On the contrary, the grazing incidence observation of line (311) would show that the planes (311) make approximately an angle ($\theta-\alpha$) equal to 52° with the surface.

- <u>Sample N°2</u>

In Bragg's position, the diffracting planes are the same as those previously observed. But the value of the intensity diffracted by the planes (111)Al is much higher than the value detected for sample N°1. We could suppose a preponderant orientation of (111)Al planes being parallel to the sample surface.

In the case of grazing incidence, aluminium lines (111) and (311) are missing. Aluminium lines (200), (220) and (222) being superposed on iron lines, we can not know if they are present or not.

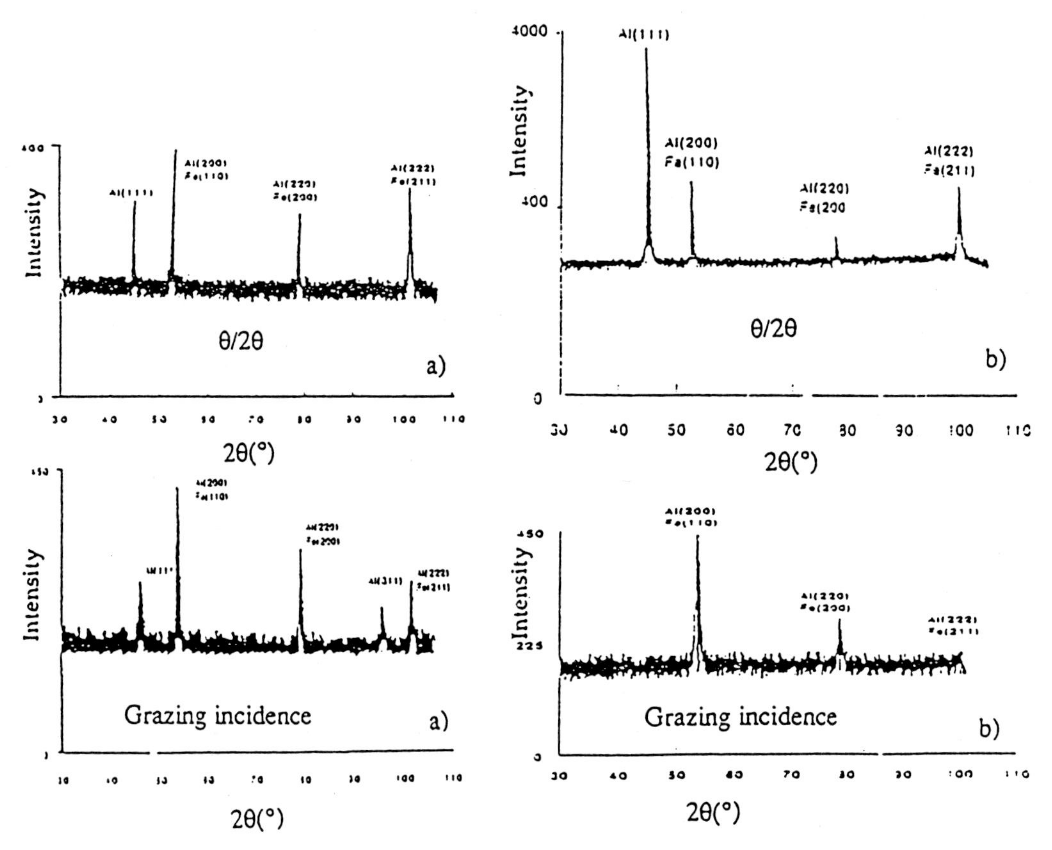

<u>Figure 1:</u> *Diffraction patterns of Aluminium monolayer (0,4μm) deposited on Iron substratum by cathodic pulverization, a) Without Bias effect, b) With Bias effect.*

<u>2) Rocking curves.</u>

In order to optimise the characterisation of the (111)Al plane orientation which is supposed to be parallel to the surface of the samples, rocking curves of these samples have been recorded. The peak width thus obtained characterises the average desorientation of (111) planes with respect to the sample surface. Sample N° 1 and N°2 rocking curves (obtained with a cobalt anti-cathode and for which Bragg angle θ of aluminium planes (111) is equal to 22°51) are shown on fig.2. It is important to note that the measured rocking curve is deformed because of the geometrical diffraction conditions changes the diffracted intensity (fig.3). These curves show the Bias effect on the average desorientation. It is very low, ranging between 2°7 and 3° for sample N°2, whereas it is more important for sample N°1 ranging from 13°5 to 14°.

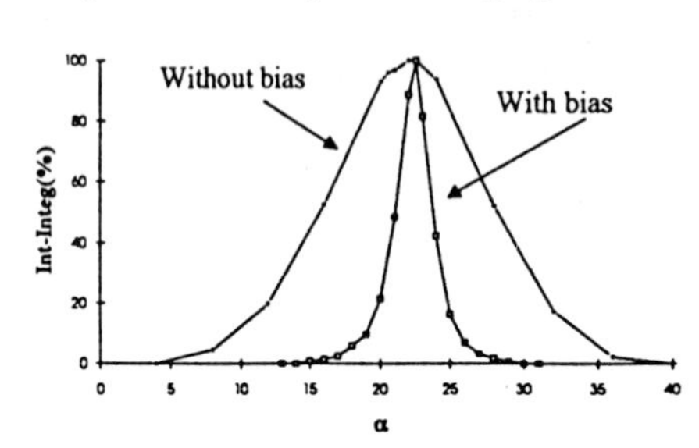

Figure2 :*Rocking curves of {111}Al planes*

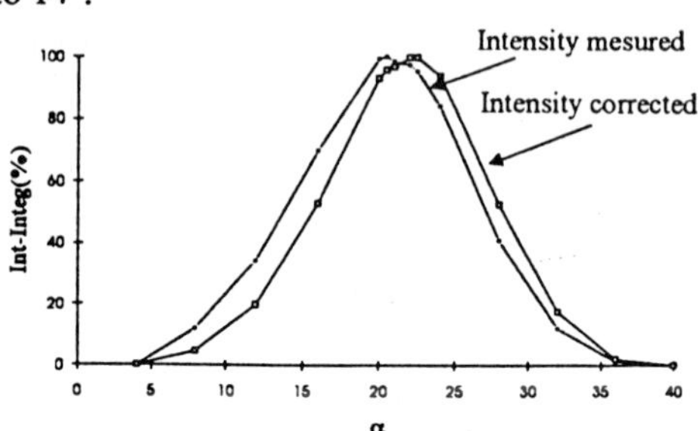

Figure3: *Rocking curves of {111}Al without bias befor and after intensity corrections*

3) Poles figures

The use of multipole figures technique [4] allows to get simultaneously several pole figures corresponding to several lattice planes. Location corrections of the information and corrections of intensities are necessary for all lattice planes.

After doing all these corrections, a pole figure makes under any incidence angle, correspond to the pole figure which would have been effectuated on an infinite thickness film in Bragg-Brentano conditions.

IV) RESULTS

-Sample N°1

The corrected and normalised (111) aluminium pole figure, prepared without bias, corresponds to a fiber texture, the [111] fiber axis is perpendicular to the sample surface (fig.5). The ring, located at 70° from the center, caracteristic of this fiber, represents the other (111) planes. There is no other aluminium orientation.

-Sample N°2

The figures 6a and 6b correspond to the (111) aluminium poles distribution, before and after intensity corrections. We can notice the necessity to correct the intensities. The same fiber texture is observed, nevertheless it is stronger. The Bias effect reinforces the [111] fiber texture.

1) Fe/Si monolayer.

An Iron layer 0.15μm thick is deposited on the (111) lattice plane of a silicon single crystal (wafer). The pole figure shows that iron planes (110) are parallel to the (111) silicon plane. This orientation is confirmed by the presence of three rings (30°, 55° and 78°) on the (211) pole figure not presented here. These pole figures are caracteristic of a fiber texture which axis [110] is normal to the (111) plane of the substratum.

2) Al/ Fe bilayer

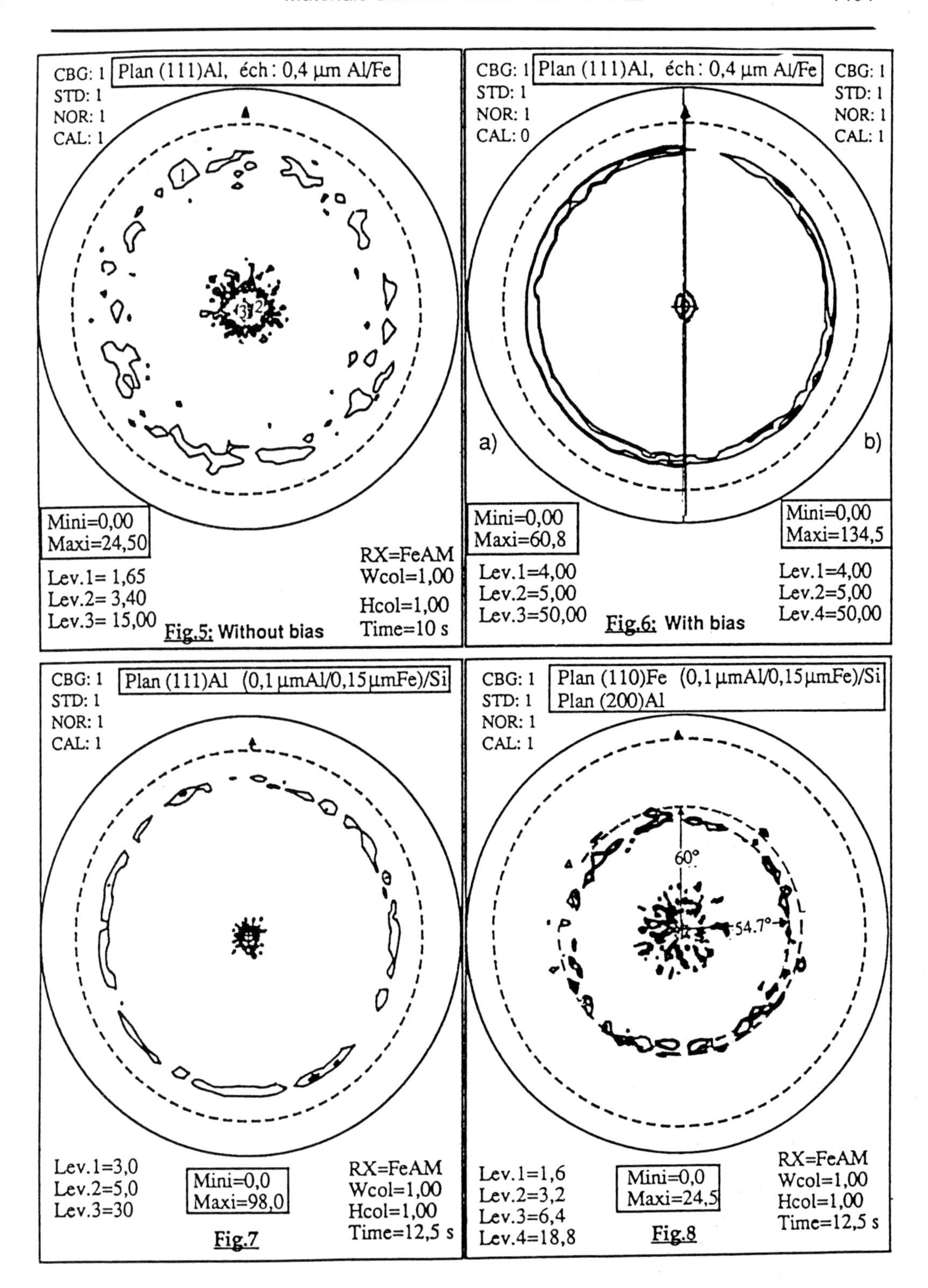

Fig.5: Without bias

Fig.6: With bias

Fig.7

Fig.8

-Intensity of iron

Let us consider a bilayer of two materials A and B with respective thicknesses X and Y. These kinds of bilayers often happens for instance as oxides layers on thin films, bilayer deposits of metal (copper + zinc) to make a thin layer of brass.

The A material can be treated as a thin film. For the B material we must take in account the variation of the diffracting volume of the film B and we must also take in account the filter effect of the film A which apparent thickness evolves during the movements of the goniometer.

Then the intensity diffracted by iron layer and collected by the detector, if we consider that A and B are, respectively, Aluminium and Iron, is:

$$I_{\alpha,\psi,y(Fe)} = I_{\theta_{Fe},\infty}\left(1-\frac{\tan(\alpha-\theta_{Fe})}{\tan\theta_{Fe}}\right).\exp\left(-\frac{\mu_{Al}.x.K_{Fe}}{\cos\psi}\right)\left[1-\exp\left(-\frac{\mu_{Fe}.y.K_{Fe}}{\cos\psi}\right)\right]$$

$$\text{with } K_{Fe} = \frac{1}{\sin\alpha} + \frac{1}{\sin(2\theta_{Fe} - \alpha)}$$

$I_{\theta_{Fe},\infty}$ *is the intensity diffracted by a volume of an infinite thickness of Iron sample observed in Bragg-Brentano conditions.*

Figure (4 a,b) show a bilayer observed either in Bragg-Brentano geometry or in low incidence (5°) geometry. We can see for example that for great tilt angle ψ, X Rays cannot reach the layer to be observed when the incidence angle is too small. So because the filter effect low incidence is not necessary the best geometry. On these figures $D\alpha$ is the ratio between $I_{\alpha\psi}$ and $I_{\theta,\infty}$.

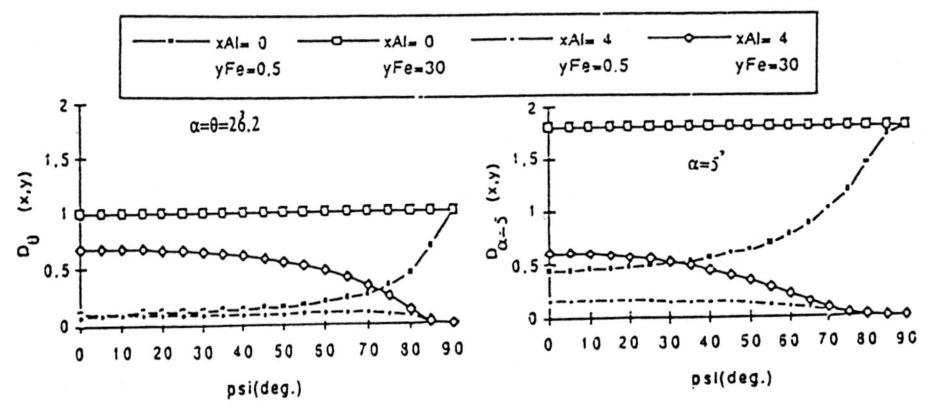

Figure 4: $D\alpha$ *evolution for an Iron layer located ander Aluminium layer, a) Bragg-Brentano incidence, b) Low incidence*

In the Fe/Al bilayers, the aluminium planes (111) and iron planes (110) are parallel to the substratum and there is an epitaxial relationship (111)Al // (110)Fe and [011]Al // [001]Fe, for which there is an accommodation of two crystal lattices [5].

V) CONCLUSION

The θ/2θ, grazing incidence spectrums and rocking curves show that Bias effect increases the crystallographic organisation of the deposited layers. Texture goniometry either in Bragg-Brentano geometry or low incidence geometry needs intensity corrections, because the diffracting volume evolves during the measurement. When bilayer are studied, the low incidence goniometry is not necessary the best geometry because the first layer is a filter which masks the second layer more than in the Bragg-Brentano geometry. The studied Aluminium / Iron bilayers shows an epitaxial relationship between the two layers.

REFERENCES

1) DUPUIS, V., RAVET, M. F., TETE, C., PIECUCH, M.: J. Appl. Phys., 1990, **7** ,3348.
2) ESAKI, L., "Synthetic Modulated Structures", CHANG LEROX, GIESEN, B.: Academic Press, 1985.
3) HEIZMANN, J.J.,VADON, A., SCHLATTER, D., BESSIERES, J.: Advance in X ray analysis, BARRET, C. , 1989, **32**, 285.
4) HEIZMANN, J.J., LARUELLE, C.: J. Appl. Cryst., 1986, **19**, 467.
5) VAN DER MERWE, J. H., BRAUN, N. W. H.: Appl. Surf. Sci., 1985, **223**, 545.

Materials Science Forum Vol. 157-162 (1994) pp. 1495-1500

X-RAY DIFFRACTION STUDY OF TEXTURE EVOLUTION IN ELECTRODEPOSITED ZINC LAYERS

I. Tomov

Institute of Physical Chemistry, Bulgarian Academy of Sciences,
Acad. G. Bontchev bl. 11, BG-1040 Sofia, Bulgaria

Keywords: X-Ray Diffraction Method, Texture Evolution, Zinc Layers, Oriented Nucleation, Oriented Growth, Growth Selection

ABSTRACT

An X-Ray diffraction method for the monitoring of texture evolution in electrodeposited and vapor-deposited layers was developed. It was applied to the study of zinc deposits. An information was obtained for the effect of amorphous substrate on the texture formation in samples deposited under equal conditions.

INTRODUCTION

It is well known that the texture arising in the layers is related to continuous microstructural changes in normal direction to the substrate. Van der Drift [1] distinguished two contradictory ways of how the texture sharpness changes with the layer thickness. The texture sharpness change depends on different mechanisms of texture formation. In the case of oriented nucleation, no matter whether the initial preferred orientation has arisen on a single crystal or a polycrystalline or an amorphous substrate, the texture sharpness decreases during the vertical growth. In the case of oriented growth because of the growth selection process the texture sharpness increases during the vertical growth. Consequently the mechanism of texture formation could be revealed if experiments were carried out in which the microstructural changes were monitored during the layer growth (or even just in some stages of its growth). In this connection an X-ray diffraction method for controlling the texture evolution in layers was proposed [2]. Thus the aim of the present study is its application for obtaining information on the mechanism of texture formation in electrodeposited zinc layers grown on amorphous substrate.

EXPRESSING THE POLE DENSITY OF A THIN SURFACE SUBLAYER

The texture evolution can be traced by means of X-ray diffraction control of pole density in each of the ideal crystal directions. We assume that we know the increase in the layer thickness Δt_m (m =

1,2,3...n) in each stage of its growth. Each Δt_m defines the thickness of the respective m-sublayer. To solve the problem it is necessary to know the individual diffraction contribution of each sublayer to the measured integrated intensity. It is possible by means of X-ray diffraction to determine only the mean value of the pole density of a given sublayer. Thus the proposed model contains the approximation that the pole density inside the sublayer does not change with its thickness, i.e. the structural changes come about with a jump at the border between two adjacent sublayers.

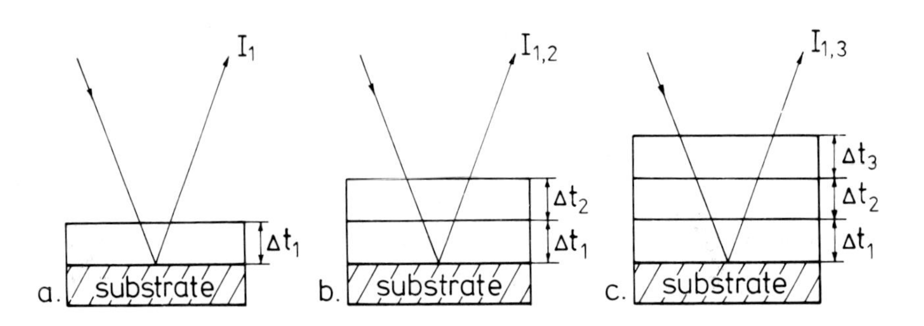

Figure 1. A schematic presentation of the diffraction by thin sublayers. The integrated intensities I, diffracted by: a) one sublayer (I_1); b) two sublayers ($I_{1,2}$); c) three sublayers ($I_{1,3}$) with thicknesses of Δt_1, Δt_2 and Δt_3, respectively.

Let us denote by I_1 the integrated intensity diffracted from the first sublayer Δt_1, located on the substrate (figure 1a). If a layer with Δt_2 is grown over it (figure 1b), then the total intensity $I_{1,2}$, diffracted from the two sublayers will be expressed by the equation

$$I_{1,2} = I_2 + I_1 \exp(-\frac{2\mu}{\sin\theta} \Delta t_2) \qquad (1)$$

where I_2 is the intensity diffracted by the second sublayer with a thickness of Δt_2, while the next expression is the contribution to the $I_{1,2}$ intensity, due to the first sublayer which is covered by the sublayer Δt_2. The exponential factor accounts for the attenuating of X-rays intensity as a result of their absorption in the second sublayer.

If a third sublayer is grown with Δt_3 (figure 1c), the total intensity measured $I_{1,3}$ is expressed by the equation

$$I_{1,3} = I_3 + I_2 \exp(-\frac{2\mu}{\sin\theta}\ \Delta t_3) + $$

$$+ I_1 \exp[-\frac{2\mu}{\sin\theta}\ (\Delta t_2 + \Delta t_3)] \qquad (2)$$

In the general case of a layer which has been obtained by the growth of n sublayers, the total intensity $I_{1,n}$ is a superposition of the diffraction contribution of all sublayers i.e.

$$I_{1,n} = I_n + \sum_{m=1}^{n-1} I_m \exp[-\frac{2\mu}{\sin\theta}\ (\sum_{k=m+1}^{n} \Delta t_k)] \qquad (3)$$

where the right-hand side of equation 3 shows the contributions of all m-sublayers [1 ≤ m ≤ n] to the $I_{1,n}$ intensity. Experimentally measurable are I_1 and all $I_{1,n}$ intensities.

The intensity I_n, diffracted only by the surface sublayer could be expressed by the equation 4 as shown

$$I_n = I_{1,n} - \sum_{m=1}^{n-1} I_m \exp[-\frac{2\mu}{\sin\theta}\ (\sum_{k=m+1}^{n} \Delta t_k)] \qquad (4)$$

I_n depends both on the sublayer thickness and the pole density in the respective crystal direction. Then the integrated intensity I_n^{∞} of an "infinitely" thick layer, having the same structure as the n-th sublayer is expressed by the equation

$$I_n^{\infty} = \frac{I_n}{1 - \exp\ (-\frac{2\mu\ \Delta t_n}{\sin\theta})} \qquad (5)$$

Its value does not depend on the sublayer thickness. The knowledge of I_n^{∞} allows us to express the pole density P_n in the respective crystal direction by the equation

$$I_n^{\infty} = P_n\ N \qquad (6)$$

where N is the normalization factor of the same pole figure.

An average value will be obtained for the pole density of each sublayer. This value does not reveal the pole density variation with the increase of the sublayer thickness. It is evident that the thinner the sublayers are the less "smeared out" the results for the respective pole densities will be. An absolutely correct information for the texture evolution would be obtained in the case of infinitely thin sublayers. In practice this could be realized if the X-ray diffraction measurements would be carried out during the course of electrodeposition.

EXPERIMENTAL AND DISCUSSION

The models of our study were bright electrodeposited zinc layers. The conditions of deposition are given in a previous paper [2]. The preparation was carried out on electrodeposited amorphous nickel-iron-phosphorus substrate. In this sense the current study of texture evolution rests on the presumption that there is no substrate influence on the preferential nucleation and growth.

At the unchangeable deposition conditions the samples were prepared seriatim with stepwise thickness increasing. Then each m-sublayer thickness Δt is defined as difference between thicknesses of two neighboring samples, i.e.

$$\Delta t_m = t_m - t_{m-1} \tag{7}$$

The layer thicknesses of each samples were determined by using an X-ray diffraction method. The measurements of the zinc diffraction lines were performed with an X-ray diffractometer Philips with CuK-alpha radiation (graphite focusing monochromator).

In the most general case the texture arising in a layer may be divided into three stages: the initial, transition and final texture [3]. With respect to their origin, there are two different standpoints considering possible causes of texture formation in the electrodeposited and vapor deposited layers: namely, oriented nucleation [4] and oriented crystal growth [1,3,5,6]. The role of each of them may be checked with our method which enables the establishing of the pole density distribution in normal direction to the layer surface. The method gives a possibility for an objective monitoring of pole density evolution in many low indexes crystal directions.

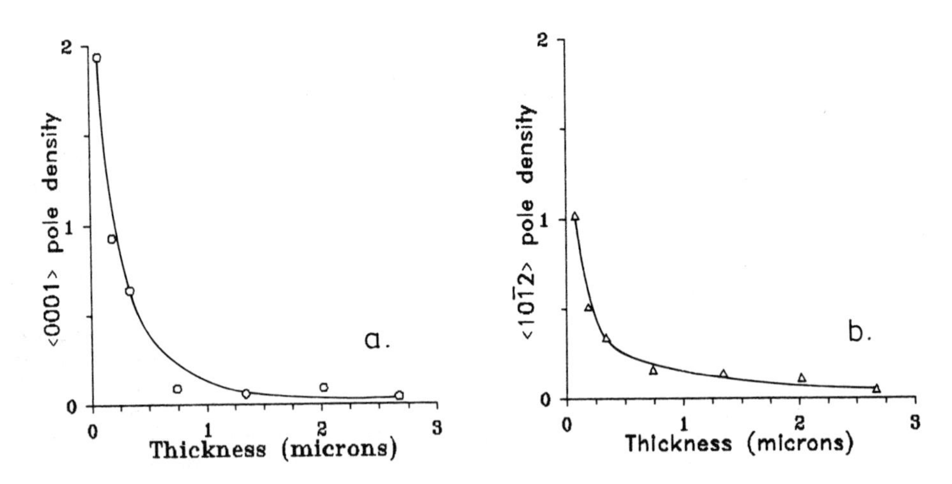

Figure 2. Initial stage of <0001> and <10$\bar{1}$2> pole density evolution vs. layer thickness t in electrodeposited zinc grown on amorphous nickel-iron-phosphorus substrate.

It was only monitoring the pole density evolution in these crystal directions in which its values were measurable. Each of the figures represents a pole density evolution vs. layer thickness. Figure 1a shows the $\langle 0001\rangle$ pole density evolution. It is obviously that it pertains to the $\langle 0001\rangle$ nucleation orientation which rapidly decreases with the layer thickness. Probably this is the first observed nucleation orientation of crystallites with h.c.p. lattice [7]. Now in addition to it we are reporting an $\langle 10\bar{1}2\rangle$ nucleation orientation (figure 2b).

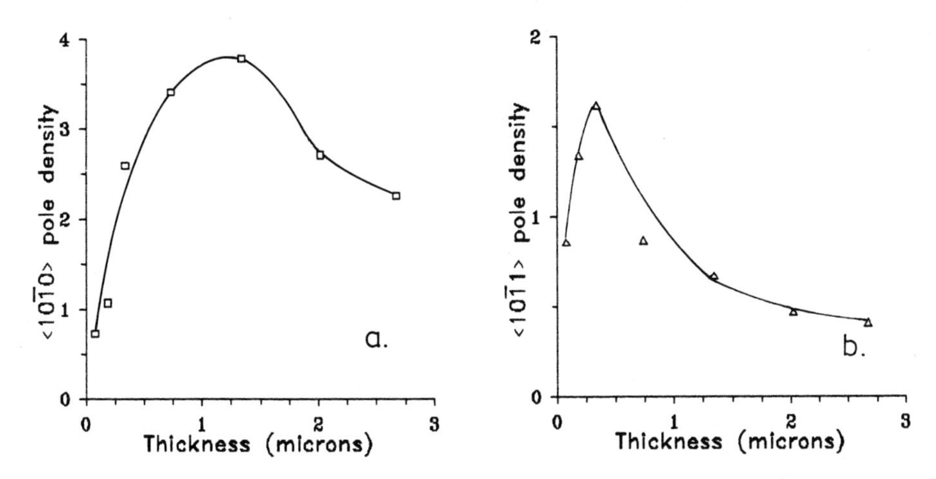

Figure 3. Initial stage of $\langle 10\bar{1}0\rangle$ and $\langle 10\bar{1}1\rangle$ pole density evolution vs. layer thickness t in electrodeposited zinc grown on amorphous nickel-iron-phosphorus substrate.

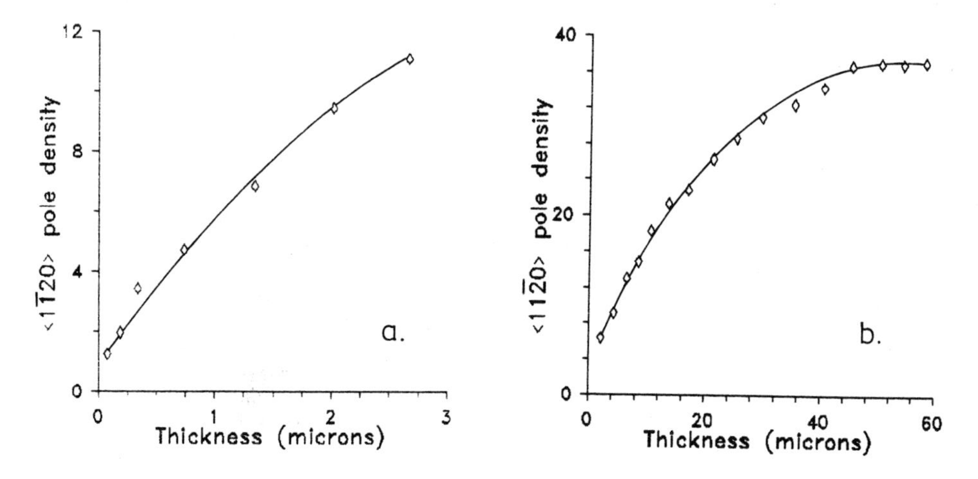

Figure 4. $\langle 11\bar{2}0\rangle$ pole density evolution vs. layer thickness t in electrodeposited zinc grown on amorphous nickel-iron-phosphorus substrate (a) initial stage; (b) transition and final stages.

Besides of the nucleation orientations it has been found two other ideal orientations whose pole densities increase to certain level and than decrease which is due to the change in the growth rates of these directions (figure 3). The coarse of the curves is probably due to the changes in the selective adsorption of brightener agents during the initial stage of electrodeposition. In principle the texture evolution could be attributed to the existence or the formation of different chemical species in the cathode-electrolyte interface during the cathodic process which are selectively absorbed on the continually renewed metal surface and impose on the oriented growth of the crystallites [8].

The $\langle 11\bar{2}0 \rangle$ orientation develops as growth orientation and overgrows as final orientation of the texture. Figure 4 shows all the stages of texture evolution in $\langle 11\bar{2}0 \rangle$-direction. Such a final orientation is the most frequent for h.c.p. metals. It was first reported by Volmer [9].

CONCLUSION

In the initial stage of texture formation it mainly arises randomly oriented crystallites and some low indexes preferred orientations which predetermine a multicomponent texture. But the final texture is usually single component. Consequently in each particular case the type of the observed texture depends on the layer thickness.

The texture of the layer in all stages of its formation depends on the competition between the oriented nucleation and the oriented growth. In this sense the plateau of the curves should be considered as a steady-state region for the aforesaid processes.The results reveal the dominant role of the growth selection in the texture formation.

In general each of the two different standpoints considering the possible reasons of texture origin in electrodeposited layers is in an agreement with some of the experimental results. That is, each of the ideal orientations in a layer could reasonably be explained by one of them.

REFERENCES

1) van der Drift,A.: Philips Res.Repts., 1967, 22, 267
2) Tomov,I. Banova,R. and Surnev,S.: Textures and Microstructures, 1992, 19, 189
3) Bauer,E.: Z.Kristallogr.Kristallgeom., 1956, 107, 72
4) Pangarov, N.A.: J.Electroanal.Chem., 1965, 9, 70
5) Gorbunova,K.: Tr.Vtoroi Konf.Corroz.Metal.,Moskow, 1943, 2, 142
6) Laemmlein,G.: Dokl.Akad.Nauk.USSR, (Physics), 1945, 48, 177
7) Evans, D.M. and Wilman,H.: Acta Cryst., 1952, 5, 325
8) Kollia,C. Loizos,Z. and Spyrellis,N.: Surface Coat.Technology, 1991, 45, 155
9) Volmer, M.: Z.Phys., 1921, 5, 31

Materials Science Forum Vol. 157-162 (1994) pp. 1501-1506

THE EFFECT OF PRECIPITATES ON TEXTURE DEVELOPMENT

H.E. Vatne [1], O. Engler [2] and E. Nes [1]

[1] Metallurgisk Institutt, NTH, Alfred Getz vei 2b, N-7034 Trondheim, Norway

[2] Institut für Metallkunde und Metallphysik,
RWTH Aachen, Kopernikusstr. 14, D-52056 Aachen, Germany

Keywords: Aluminium, Recrystallisation Textures, Precipitation, Cube/PSN Competition

ABSTRACT

The effect of precipitation on the evolution of recrystallisation textures in commercial purity aluminium and an Al-1wt%Mn alloy has been studied by means of ODF-analysis. Through different preannealing treatments various conditions characterised by different particle distributions and solid solution levels were obtained. After rolling to various strains the specimen were annealed at different temperatures in order to vary the amount of precipitates. The development of recrystallisation textures is interpreted as being due to a competition between randomly oriented PSN-nuclei and cube-oriented nuclei from cube bands. At lower temperatures where a heavy precipitation reaction takes place during the recrystallisation, strong effects on texture and grain size are observed. The precipitates seem to have a stronger influence on the PSN-nuclei, which leads to a sharpened cube texture. The role played by the precipitates in the competition between cube- and PSN-sites has been modelled.

INTRODUCTION

Annealing of a supersaturated, heavily deformed metal, may result in a concurrent precipitation reaction that interferes with the recovery and recrystallisation reactions. It is well established that such a precipitation has a profound influence on recrystallisation kinetics and final grain size. The effect on texture, however, has not been given similar attention, although several investigations indicate that precipitation affects nucleation and growth of different texture components to different extents [1-4]. Accordingly, precipitation may alter the evolution of the recrystallisation texture compared to conditions where no precipitation occurs. In the present paper, the effect of a concurrent precipitation reaction (i.e. precipitation taking place during recrystallisation) on the recrystallisation process of the two alloys AA1145 (commercial purity Al) and AA3103 (AlMn) has been studied. The texture evolution has been followed by means of ODF-analysis, and also optical microscopy including grain size measurements has been used in order to improve the understanding of the transformation process. The main objective of the work has been to investigate whether the precipitates have a stronger influence on certain recrystallisation texture components than others.

EXPERIMENTAL

Two commercial grade alloys with the following compositions (wt%) have been investigated: AA1145 (0.43Fe, 0.09Si) and AA3103 (1.1Mn, 0.56Fe, 0.18Si). Both alloys were DC-cast, homogenised and hot rolled at Alcoa Technical Center (Pittsburg), and were received as hot rolled sheets 25mm thick. For the purpose of this investigation the two alloys were given the following preannealing treatments:

> Condition A: annealing for 24 hours at T=620°C followed by slow cooling to 500°C, further 24 hours annealed at this temperature and slowly cooled to room temperature. With this heat treatment an equilibrium structure with coarse particles is achieved.
>
> Condition B: annealing for 24 hours at T=620°C and slowly cooled to 500°C, held at this temperature for another 24 hours and quenched to room temperature. Through the quenching, a state with a higher level of solid solution than condition A is reached.

Following the thermal pretreatment, both alloys were cold rolled to the following reductions: $\varepsilon=1$, 2, 3, and 4. Annealing was performed in a salt bath furnace, except for alloy 1145B which also was given a batch annealing treatment (labelled 1145B-ba) with a heating rate of 100°C/h.

Hardness measurements were done in order to decide the appropriate annealing times for 100% recrystallisation (for details, see [5]). Optical microscopy was used to investigate grain-shape and size. Conductivity measurements were carried out to control the amount of precipitation. All textures were measured in the center section of the sheet. The textures of the samples were analysed by pole figures and 3-dimensional ODFs, f(g). Here g denotes the orientation, given in the form of the three Euler angles $(\varphi_1,\Phi,\varphi_2)$. The ODFs were produced from four incomplete pole figures ({111}, {200}, {220}, {113}), which were measured by means of an automatic X-ray texture goniometer [6] and corrected with respect to background intensity and defocusing errors. They were calculated by the series expansion method [7] and ghost corrected according to the method of Lücke and co-workers [8]. Pole figures and ghost corrected ODFs are presented in the form of iso-density lines normalised to random intensity in stereographic projection or in constant φ_2 sections in 5° steps in Euler space.

RESULTS AND DISCUSSION

Figure 1a) shows the grain sizes (measured in the rolling direction in the longitudinal-transverse section) for the three 1145-alloy-conditions as a function of deformation. Grain sizes decrease with increasing strain, and the largest grains are obtained for the batch annealed 1145B-ba. The small bulge of the grain size curve of 1145B at intermediate strains should be noticed. Fig. 2 shows the typical recrystallisation texture of 1145, where the cube together with a retained rolling texture is observed. The retained rolling texture consists of the R-component and a shifted C-component. The sharpness seems to increase a little with increasing degree of deformation and annealing temperature. The intensity of the cube component as a function of deformation is shown in Fig. 1b). For 1145A and B the cube is relatively constant, while it makes a significant maximum for intermediate strains for the batch annealed condition (1145B-ba). As expected, conductivity measurements show that precipitation has been strongest for the batch annealed specimen ($\sigma=34.5\mathrm{m}/\Omega\mathrm{mm}^2$ and $\sigma=33\mathrm{m}/\Omega\mathrm{mm}^2$, respectively).

Figure 3a) shows the grain sizes for the 3103-states at different temperatures (300°C, 425°C) and strains. The grain sizes decrease with increasing deformation, with one exception for the supersaturated 3103B at the lower annealing temperature, where the grain sizes increase for intermediate strains. Conductivity measurements confirm a strong precipitations for 3103B at this low temperature (conductivity increases from $\sigma=21\mathrm{m}/\Omega\mathrm{mm}^2$ to $\sigma=29\mathrm{m}/\Omega\mathrm{mm}^2$ during the recrystallisation process). The micrographs in Fig. 4 show elongated grains for 3103B, which is often observed in the case of precipitation, while 3103A has a more equiaxed shape at the same temperature and strain.

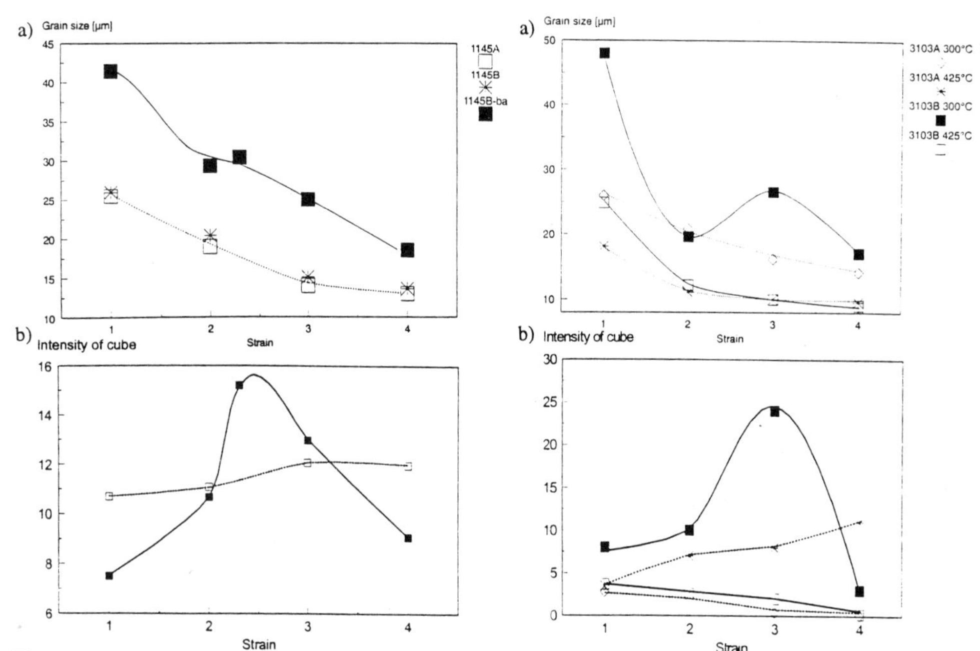

Fig. 1: 1145: a) Grain size b) cube intensity vs. strain

Fig. 3: 3103: a) Grain size b) cube intensity vs. strain

Typical ODFs are also shown in Fig. 4, where 3103A and B are compared at the low temperature. The only texture component observed is cube in addition to the random texture (and a small tendency of P and ND-rotated cube at high temperatures and strains). The textures are summarised in Fig. 3b) where the cube intensity is plotted as a function of temperature and strain. The most interesting result is the strong cube at the low temperature for 3103B. It reaches a maximum at intermediate strains which coincides with a large grain size. Like in the case of 1145, this indicates a connection between heavy precipitation and preference of cube texture.

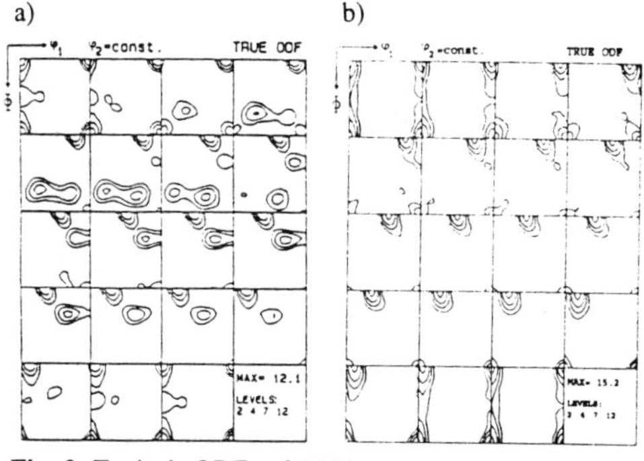

Fig. 2: Typical ODFs of 1145, ε=3, T=350°C: a)1145A b)1145B-ba

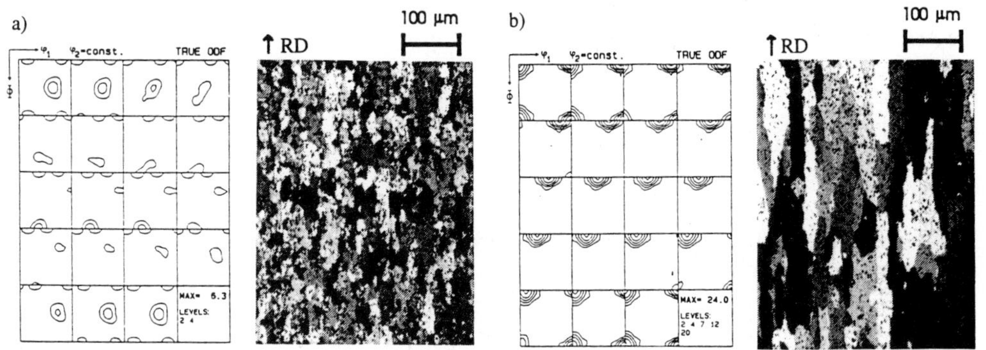

Fig. 4: ODFs and micrographs of 3103, ε=3, T=300°C: a) 3103A b) 3103B

All recrystallisation textures found in this work belong to the well known cube, retained rolling and random textures. In aluminium, the cube is known to be an important recrystallisation texture component, which will be discussed later. However, if the nucleation and the growth of the cube-orientation is retarded, e.g. through alloying of small amounts of iron to purity Al (like 1145), the grain boundary mobility might dramatically decrease (e.g. [9]). In such cases a retained rolling texture with a sharp R is often observed (R is very close to the S-texture of the rolling texture) [10]. It was already pointed out by Beck [11] that the R-orientation can develop from the rolling texture either through extended recovery in which no orientation changes occur, so that a more or less retained rolling texture results, or through discontinuous recrystallisation through nucleation from nuclei already existing in the rolling structure. The ODF-analysis of alloy 1145 shows that the position of the R-component does not overlap with that of S, so that the latter assumption of Beck is probably the case here. Thus, the retained rolling texture of 1145 develops directly from the deformation texture, probably as a consequence of a nucleation process at preexisting high angle boundaries.

For the AlMn alloy the cube texture dominates. The cube grains are assumed to be nucleated from cube-oriented transition bands [12]. There is a competition between grains nucleated from cube bands and randomly oriented grains presumably nucleated in the deformation zones around coarse particles (PSN). PSN is known to be a strong and important nucleation mechanism [13]. The particles' ability to act as nucleation sites is dependent on their size, the driving pressure due to the stored energy, P_D, and the Zener-drag, P_Z, the critical diameter (of the deformation zone) for nucleation becoming

$$\lambda_C = \frac{2\gamma}{P_D - P_Z} \qquad (1)$$

where γ is the grain boundary energy. Due to the coarse particles leading to randomly oriented PSN-grains in both alloys, rather weak textures are observed.

The only nucleation site that seems to be able to compete with PSN, are the cube bands. The cube texture sharpens remarkably for the supersaturated 3103B at the low annealing temperature, and the results of both 1145 and 3103 clearly show that a strengthening of the cube is associated with a heavy precipitation reaction taking place during recrystallisation. Interestingly, this seems to be so only for the intermediate strains ($2<\varepsilon<3$). The reason for this is unclear, but may be found in the effect of deformation on the recrystallisation kinetics. At the largest strains the high driving pressure will speed up the nucleation and growth of new grains. The precipitation reaction at 300°C is quite slow, thus allowing nucleation to take place more unaffected by precipitation (due to the higher driving pressure). This speculation is to some extent confirmed by the fact that also the grain size reaches a maximum at intermediate strains. Based on the grain size curves and textures in the cases of a concurrent precipitation reaction, it seems like the nucleation of PSN-grains is heavily retarded at intermediate strains and speeded up at higher. The reasoning above is based on the hypothesis that cube grains are less affected by precipitates than PSN-grains. It has been shown [14, 15] that the subgrain structure within cube bands is much coarser than that of other texture components of the rolling texture in aluminium. In the following it will be shown that this size advantage might be determining for the competition between cube and PSN nucleation sites in the case of a concurrent precipitation reaction. Thus, the rest of the paper will be attributed to a first-approach-model for the PSN/cube competition, when a precipitation reaction takes place during recrystallisation. A similar approach has been given by Daaland [14].

The critical diameter λ_C for a subgrain in a cube band to bow out and start growing into the surrounding matrix is given by

$$\lambda_C = \frac{4\gamma_{GB}}{P_D^M} = \frac{4\gamma_{GB}}{(\kappa/\delta_M)\gamma_{SB}} \qquad (2)$$

where γ_{GB} is the grain boundary energy (of the nucleus) and $P_D^M=(\kappa/\delta_M)\gamma_{SB}$ is the driving pressure due to the stored energy in the matrix substructure. κ is a geometric constant, δ_M the matrix subgrain size and γ_{SB} is the energy of a subboundary. The latter can be estimated from the Read-Schockley relation, giving

$$\lambda_C=4[\frac{\kappa}{\delta_M}\frac{\theta}{\theta_C}\ln(\frac{e\theta}{\theta_C})]^{-1} \quad (3)$$

where θ is the sub-boundary misorientation and θ_C the maximum value of this. For a heavily cold rolled aluminium alloy ($\varepsilon>1$), typical values are $\kappa=3$, $\theta=3°$ and $\theta_C=15°$ which gives $\lambda_C=2.2\delta_M$. Daaland [14] found cube subgrains of size $3\text{-}4\delta_M$ already at the as deformed stage. Thus, viable cube nuclei may already at the onset of annealing be present in the material, so that the nucleation from cube bands is unaffected by the precipitation reaction.

For PSN to succeed, a new grain will first have to consume the deformation zone around the particle after which it must be capable of further growth into the matrix. Humphreys [13] has shown that the critical stage of PSN is not the growth within the zone, but the growth out of it. However, the growth of the new grain within the zone will require time. During this period (τ_{PSN}) precipitation will take place. Hence, the grain will be exposed to a Zener-drag when it starts growing out of the deformation zone and the critical particle radius for PSN is then given by Eq. (1). The situation is illustrated schematically in Fig. 5.

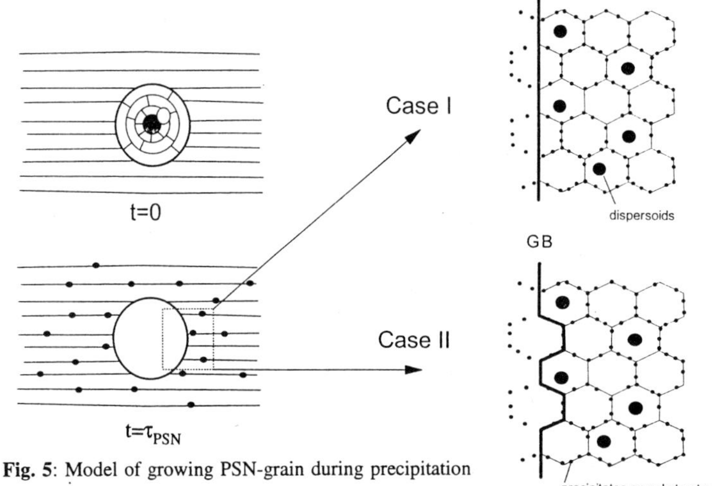

Fig. 5: Model of growing PSN-grain during precipitation

There are two contributions to the Zener-drag:

i) from dispersoids, i.e. particles precipitated during rolling, randomly distributed
ii) from small particles precipitated on the deformation substructure during recrystallisation

The Zener-drag from the dispersoids is given by P_Z(dispesoids)$=(3/2)\gamma_{GB}f/r$, where f is the volume fraction and r the radius of the dispersoids. Accoring to Strid [16], typical values are f=0.001 and r=0.05μm which gives a retarding pressure of the order P_Z(dispersoids)$\sim 10^4\gamma_{GB}$. For the small particles, precipitated during recrystallisation, one has to estimate the fraction of particles being in contact with the moving grain boundary. The cases I and II in Fig. 5 represent the two limiting cases: I) a straight boundary which shape is unaffected by the the precipitates and II) a flexible boundary adjusting itself to the precipitates on the substructure. The two cases lead to Zener-drags of the form

$$P_{Z,I}=2\pi n\gamma r^2\frac{\kappa}{\delta_M} \quad \text{and} \quad P_{Z,II}=\pi n\gamma r \qquad (4)$$

where r is the radius of the precipitates and n their area density. Typical values are $r=10^{-8}$m and $n=10^{14}m^{-2}$ which lead to drags of the order $P_{Z,I}=P_Z^{min}(prec.)\sim2\cdot10^5\gamma_{GB}$ and $P_{Z,II}=P_Z^{max}(prec.)\sim3\cdot10^6\gamma_{GB}$. The true value will lay somewhere between these limits, probably closest to $P_{Z,II}$ due to the flexibility of the boundary. This means that the Zener-drag due to the concurrent precipitation reaction is of the order $P_Z(prec.)\sim10^6\gamma_{GB}$. This is considerably higher than that caused by the dispersoids, thus the total retarding pressure from precipitates is of the order $P_Z\sim10^6\gamma_{GB}$. The driving pressure outside the deformation zone is the same as that for growth of cube grains and has in Eq. (2-3) been estimated to $P_D\sim10^6\gamma_{GB}$ which is of the same order as P_Z. This means that the PSN-grains will have great difficulties growing out of the deformation zone because of the Zener-drag from precipitates. On the other hand, cube grains may already in the as deformed condition be of the size of the critical Gibbs-Thompson radius or larger, so that only the growth and not the nucleation is influenced by the precipitates. This explains the sharp cube texture observed in the cases of a concurrent precipitation reaction.

CONCLUSION

1. The evolution of recrystalliation textures in 1145 and 3103 is determined by the competition between randomly oriented PSN nuclei and cube nuclei from transition bands.

2. Precipitates seem to have a higher retarding effect on the nucleation in deformation zones around particles than on nucleation from transition bands. Hence, cube grains are favoured in cases where precipitation takes place during recrystallisation.

Acknowledgement: The authors thank the Norwegian Council for Research, Deminex and NTH/SINTEF through the "Strong Point Center for Light Metals" for financial support and Alcoa for providing material.

REFERENCES

1. E. Nes, Proc. Symp. on Microstructure Control during Processing of Al-alloys, AIME, New York, (1985)
2. W.B. Hutchinson, A. Oscarsson, Å. Karlsson, Mat. Sc. & Tech., **5**, 1118, (1989)
3. O. Daaland, P.-E. Drønen, H.E. Vatne, S.E. Næss, E. Nes, Proc. "Recrystallization '92", (ed. M. Fuentes, J. Gil Sevillano), Trans Tech Publications, San Sebastian, (1992)
4. O. Engler, K. Lücke, Scripta metal., **27**, 1527, (1992)
5. H.E. Vatne, Diploma-Thesis, RWTH Aachen, (1992)
6. J. Hirsch, G. Burmeister, L. Hoenen and K. Lücke: in "Experimental techniques of texture analysis", (ed. H.J.Bunge), 63, Oberursel, DGM, (1986)
7. H.J. Bunge, "Mathematische Methoden der Texturanalyse", Berlin, Akademie, (1969)
8. K. Lücke, J. Pospiech, K.-H. Vernich and J. Jura: Acta metall., **29**, 167, (1981)
9. N. Hansen, D. Juul Jensen: Met. Trans. A, **17A**, 253, (1986)
10. Z. Ito, K. Lücke, R. Rixen: Z. Metallkde. **67**, 338, (1967)
11. P.A. Beck, H. Hu: "Recrystallization, Grain Growth and Textures", (ed. H. Margolin), American Society of Metals, A.S.M. Metals Park Ohio, (1966)
12. J. Hjelen, R. Ørsund, E. Nes, Acta metal., **39**, 1377, (1991)
13. F.J. Humphreys, Acta metall., **25**, 1323, (1977)
14. O. Daaland, PhD-Thesis, The Norwegian Institute of Technology, Trondheim, (1993)
15. H.E. Vatne, O. Daaland, E. Nes, This conference
16. J. Strid, unpublished work

Materials Science Forum Vol. 157-162 (1994) pp. 1507-1512

ON THE INFLUENCE OF A WEAK PREFERRED ORIENTATION ON THE STRENGTH OF ALUMINIUM OXIDE CERAMIC

W. Winter [1], H.-G. Brokmeier [2] and H. Siemes [1]

[1] Institut für Mineralogie und Lagerstättenlehre,
RWTH Aachen, Wüllnerstr. 2, D-52056 Aachen, Germany

[2] Institut für Metallkunde und Metallphysik der TU Clausthal,
GKSS Forschungszentrum, Max-Planck-Str. 1, D-21502 Geesthacht, Germany

Keywords: Aluminum Oxide, Ceramic, Strength, Tension, Diametral Compression Test, Weibull Diagram, Crystallographic Preferred Orientation, Texture

ABSTRACT

The crystallographic preferred orientation (texture) of a slip casted aluminium oxide quarter-disk, fired at 1600°C, was measured on 7 specimens by means of neutron diffraction. The strength of three series of different oriented specimens of this quarter-disk was determined by means of the diametral compression test. The tensile strength for two orientations one parallel to the c-axis maximum and one perpendicular to the c-axis maximum, characterised by their Weibull parameters, differs about 25 %.

INTRODUCTION

The strength of several brittle materials had been evaluated by means of the diametral compression test. Object of the study was a detailed and systematic investigation of the diametral compression test for the determination of the tensile strength of ceramic materials. The effect of several parameters on the tensile strength was determined: i) the quality of the specimens surface, ii) the nature of the specimen mounting especially in view of the application of the load, iii) the loading rate, iv) the specimen diameter and the ratio of thickness to diameter, v) the location of the fracture initiation using conductor tracks. The tests were carried out by means of an electromechanical Instron Universal Testing Equipment Type 8562. The selected materials were glass, stoneware ceramic, aluminium oxide and silicon carbide. In contrast to the other materials the reproducibility of the strength parameters for the aluminium oxide was less clear. The reason seemed to be an anisotropy of the material (Winter, 1992).

CHARACTERISATION OF THE TESTED MATERIAL

The tested material had been produced by slip casting of aluminium oxide into a quarter-disk mould and fired at a temperature of 1600°C. Mayor phase of the material is α-Al_2O_3, minor phase β-$NaAl_{11}O_{17}$ with a porosity of $\pm 10\%$. Because of the presumed anisotropy of the material cylindrical samples (length 28 mm and diameter 20 mm) in two orientations, one perpendicular and the other parallel to the disk surface (Fig. 1), were prepared for neutron texture analysis (Brokmeier, 1989).

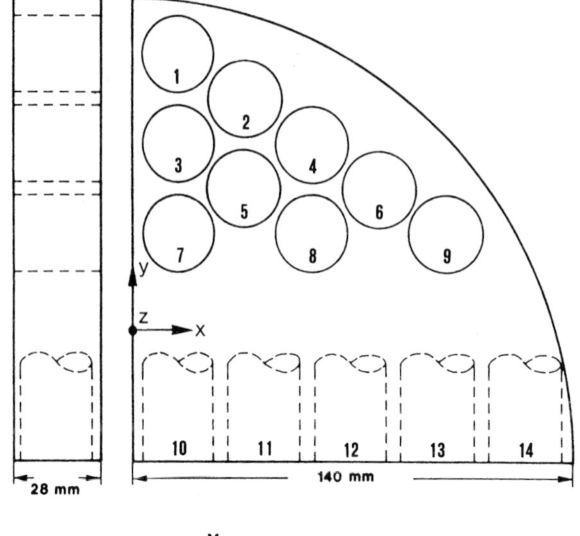

Fig. 1:
Tested Al_2O_3 quarter-disk. Indicated is the orientation of the specimens for neutron texture analysis and diametral compression test.

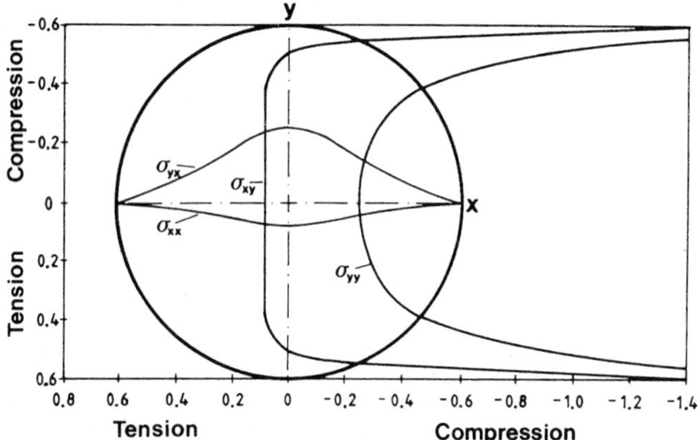

Fig. 2:
Stress distribution in a solid disk (after Fairhurst, 1964): σ_{xx} - tensile stress along the x-axis, σ_{yx} - compressive stress along the x-axis, σ_{xy} - tensile stress along the y-axis, σ_{yy} - compressive stress along the y-axis.

CRYSTALLOGRAPHIC PREFERRED ORIENTATION (TEXTURE) OF Al_2O_3

The central part of the ceramic quarter-disk (specimen 8) reveals a fiber texture with quite perfect axial symmetric pole figures of the (0006)-reflection (Fig. 3c) and the $(30\bar{3}0)$-reflection (Fig. 4c). Close to the borders and corners of the disk the (0006)-maxima become weaker and they are elliptical shaped (Figs. 3a,3b,3d) or the (0006)-pole figures show more elongated maxima with extension to great circle distributions (Figs. 3e,3f,3g). In the first instance the basal planes are preferentially aligned with the main surface of the disk. The intensity and shape of the pole figures of the specimens 1,7,9 are influenced by one or two side walls of the mould during the casting process. The specimens 10,12,14 are placed at the topside of the mould wall where the mould had been charged. All pole figures were smoothed and for comparison the pole figures of the differently oriented specimens were rotated parallel to the disk surface (Traas et al. 1993).

DIAMETRAL COMPRESSION TEST

The diametral compression test is based upon the state of stress developed when a cylindrical specimen is compressed between two diametrically opposed generators across the diameter. This ideal loading produces a stress distribution within the specimen (Fig. 2) which is assumed to be

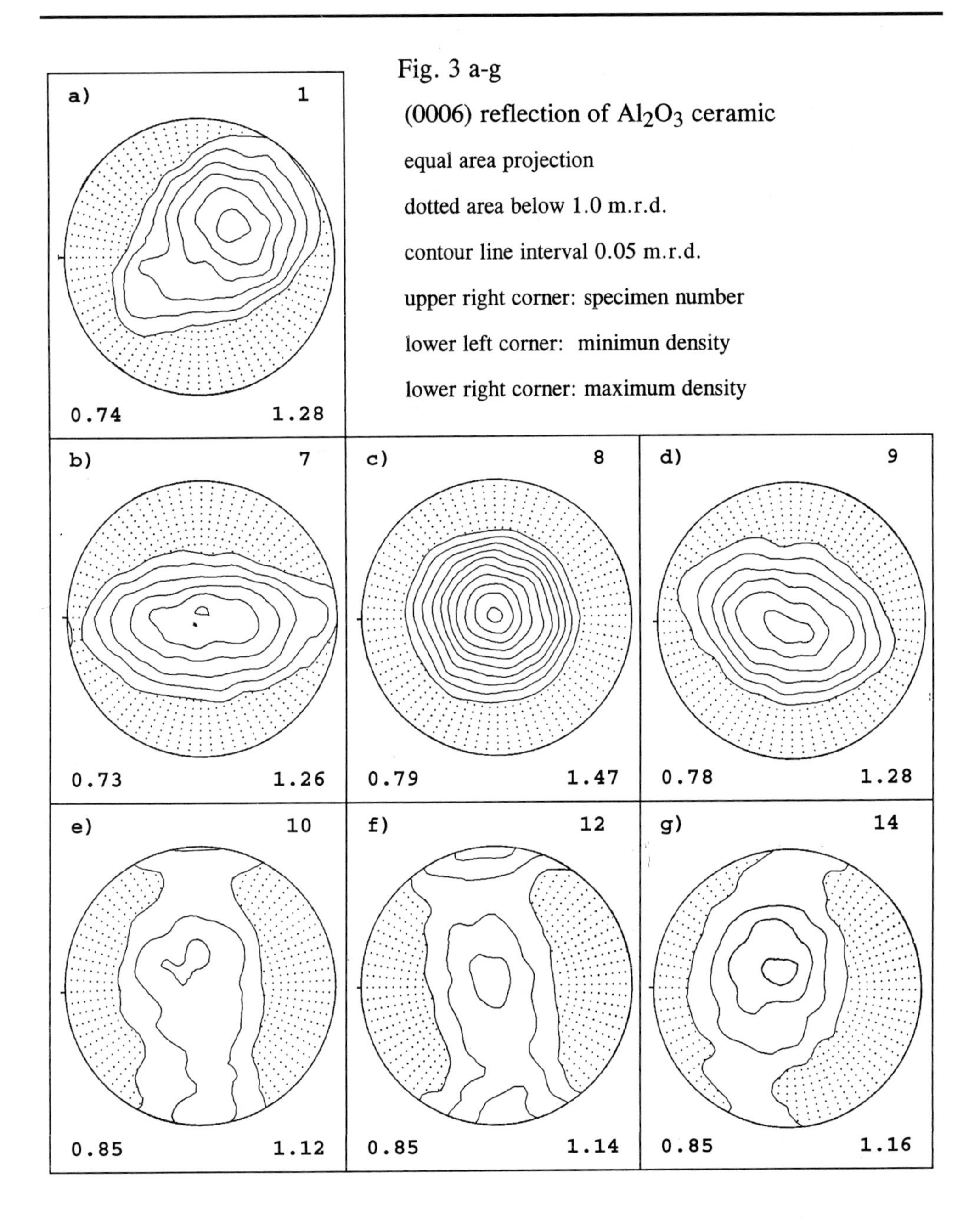

Fig. 3 a-g

(0006) reflection of Al_2O_3 ceramic

equal area projection

dotted area below 1.0 m.r.d.

contour line interval 0.05 m.r.d.

upper right corner: specimen number

lower left corner: minimun density

lower right corner: maximum density

biaxial (compression and tension). The tensile stress usually initiates the fracture in the central part of the loaded diameter as measured by the electronic device of the conductor tracks. For details and a review of the literature see Winter, 1992.

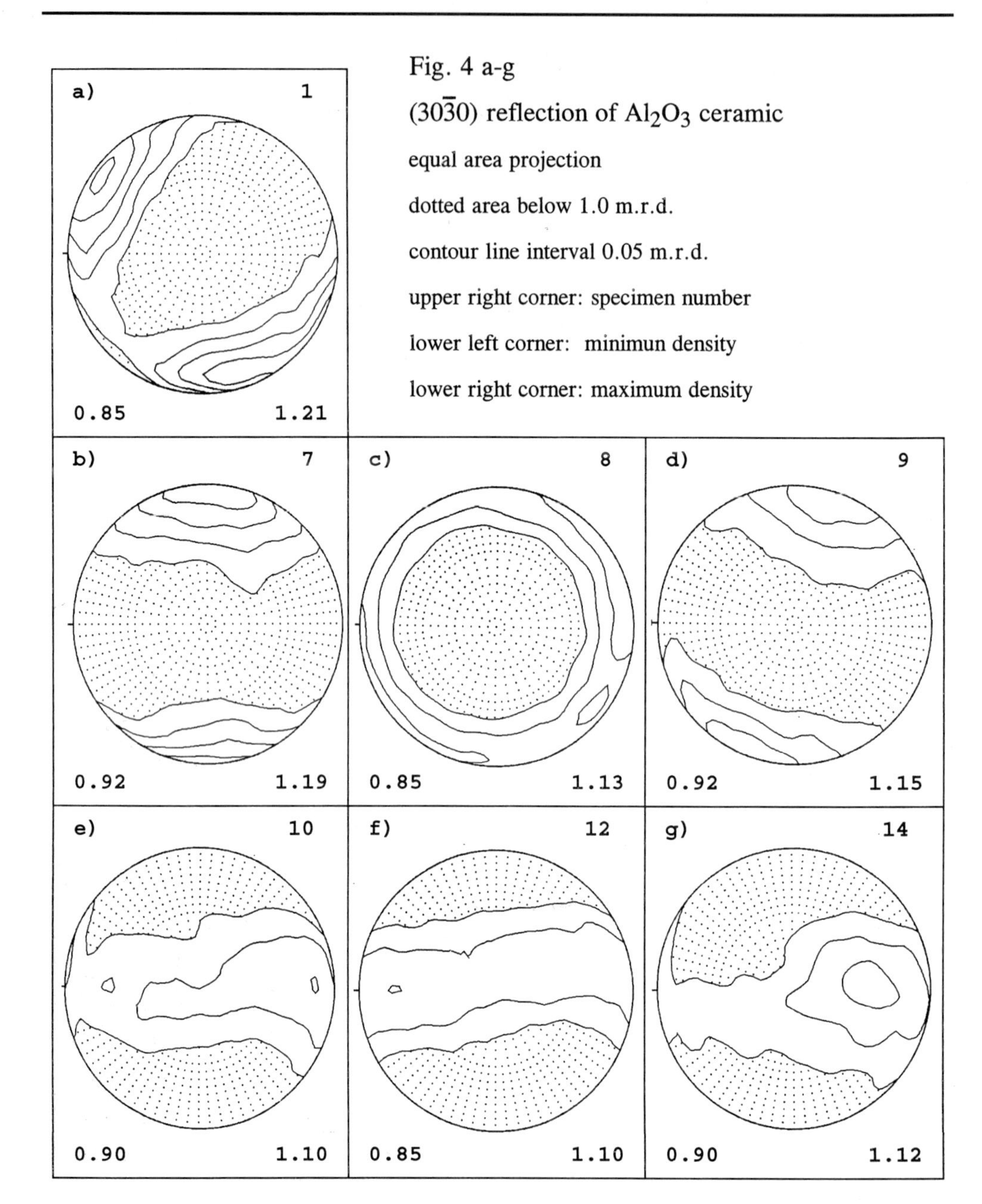

Fig. 4 a-g

$(30\bar{3}0)$ reflection of Al_2O_3 ceramic

equal area projection

dotted area below 1.0 m.r.d.

contour line interval 0.05 m.r.d.

upper right corner: specimen number

lower left corner: minimun density

lower right corner: maximum density

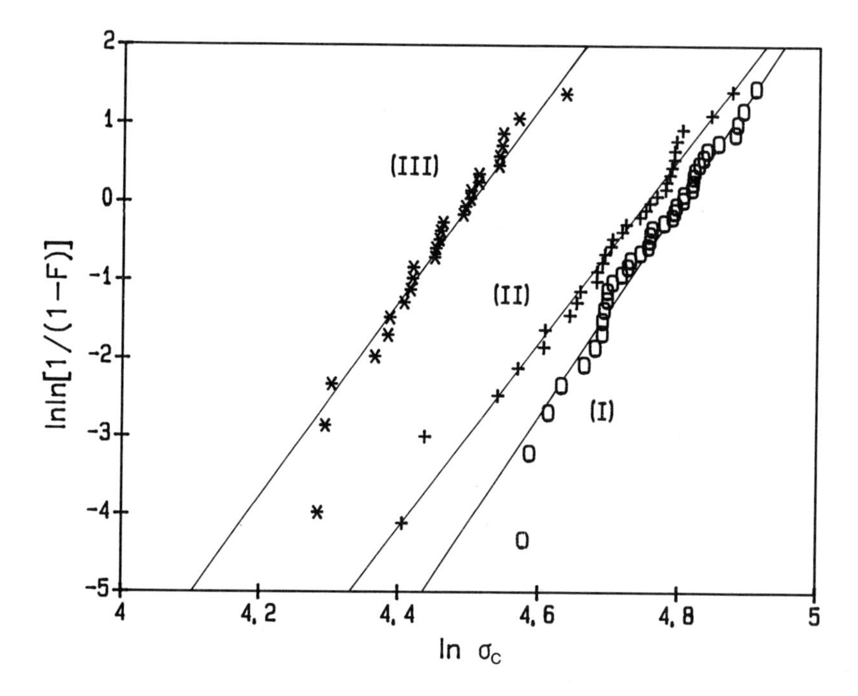

	Orientation of the disks	n	m	σ_0 [MPa]	$\hat{\sigma}$ [MPa]
I	compression ‖ x tension ⊥ xz	38	13.27	121.28	117.98
II	compression ‖ z tension ⊥ yz	31	11.81	115.53	112.00
III	compression ‖ x tension ⊥ xy	27	12.23	89.80	87.15

Fig. 5:
Weibull diagram and estimated parameters of the diametral compression tests on Al_2O_3 specimens: n - number of observations, m and σ_0 - Weibull parameters, $\hat{\sigma}$ - median of the Weibull distribution.

DATA ANALYSIS

Three series each consisting of more than 20 specimens (diameter 20 mm, thickness 7.5 mm) were tested under the same experimental conditions. A Weibull distribution was fitted to the strength data σ_c and the data were plotted in the related diagram. The Weibull parameters m and σ_0 of the distribution were estimated by means of the maximum likelihood method (Munz & Fett, 1989).

One series of oriented cylindrical disks was tested (I) with the diametral compression axis parallel to x (tension axis perpendicular to xz in sections parallel to the disk surface), one series (II) with the diametral compression axis parallel to z (tension axis perpendicular to yz across the disk) and one series (III) with the diametral compression axis parallel to x (tension axis perpendicular to xy). The Weibull diagram (Fig. 5) and the values of the Weibull estimates indicate that the strength in case I is approximately 5% greater than in case II. In both cases the tension axis is oriented parallel to the maximum of the basal planes i.e. perpendicular to the prism planes and the difference of 5 % may be due to the different oriented areas of the stressed planes in the quarter-disk. In case III where the tension axis is perpendicular to the preferentially oriented basal planes the strength is approximately 25 % lower than in case I and II.

DISCUSSION

Axial symmetric preferred orientations with a strong preferential alignment of the basal planes of Al_2O_3 parallel to the surface of substrates (approximately 0.6 mm thick) are described by DiMarcello et al. (1972) and Böcker et al. (1991). In contrast to Böcker et al. (1991) who measured several reflections by X-rays and calculated the pole figure of the basal plane DiMarcello et al. (1972) directly measured the (0006) reflection of fired and unfired material. They propose as explanation for the origin of the texture in the unfired substrate that the preferred orientation is related to the platelike crystallites with the basal plane parallel to the platelet surfaces. The final texture seems to be enhanced by the firing process. This was confirmed by Böcker et al. (1993). The explanation of the development of preferred orientation is consistent with the measured textures of the quarter-disks. The statement of DiMarcello et al. (1972) that 'Al_2O_3 has a highly anisotropic fracture behaviour, with fracture occurring along prism planes but never along basal planes' is in contradiction with the present investigation. The orientation dependent differences in the tensile strengths may be not only caused by the crystallographic preferred orientation. It is not unlikely that a preferred alignment of pores parallel to the surface of the quarter-disk i.e. parallel to the basal planes contributes to the low strength in this orientation.

ACKNOWLEDGEMENTS

The testing equipment was funded by the German Federal Ministry of Research and Technology (BMFT) under the contract number 03 M 2035 7. The aluminium oxide ceramic was provided by the Schunk Ingenieurkeramik GmbH, Willich-Münchheide, Germany.

REFERENCES

Böcker, A., Brokmeier, H.-G., Bunge H.J. (1991) Determination of Preferred Orientation Textures in Al_2O_3 Ceramics. *Jour. Europ. Ceramic Soc.* **8**, 187-194.
Böcker, A., Bunge H.J., Ernst, J., Krahn, W. (1993) Texture Evolution in Al_2O_3 Sheets by Grain Growth During Sintering. *Mat. Sci. Forum* **113-115**, 575-578.
Brokmeier, H.-G. (1989) Neutron diffraction texture analysis of multi-phase systems. *Textures and Microstructures* **10**, 325-346.
DiMarcello, F.V., Key, P.L., Williams, J.C. (1972) Preferred Orientation in Al_2O_3 Substrates. *Jour. Amer. Ceramic Soc.* **55**, 509-514.
Fairhurst, C. (1964) On the Validity of the 'Brasilian' Test for Brittle Materials. *Int. Jour. Rock Mech. Min. Sci.* **1**, 535-546.
Munz D., Fett, T. (1989) Mechanisches Verhalten keramischer Werkstoffe. Werkstoff-Forschung und -Technik **8**, Springer Verlag Berlin, pp. 244.
Traas, C., Siemes, H., Schaeben, H. (1993) Smoothing pole figures using tensor products of trigonometric and polynomial splines. This volume.
Winter, W. (1992) Experimentelle Bestimmung der Zugfestigkeit spröder Werkstoffe (Glas, Keramik) im Scheiben-Druck-Versuch. Dissertation, RWTH Aachen, pp. 245.

Materials Science Forum Vol. 157-162 (1994) pp. 1513-1520

INFLUENCE OF SEGREGATION OF PHOSPHORUS ON TEXTURE DEVELOPMENT IN COLD-ROLLED Fe-P ALLOYS DURING ANNEALING

L. Zhang [1], L. Xiong [1], H. Ning [2], D. Ye [1] and B.J. Duggan [2]

[1] Institute of Corrosion and Protection of Metals, Academia Sinica, Shenyang, China

[2] Department of Mechanical Engineering, University of Hong Kong, Hong Kong

Keywords: Fe-P Alloy, Lattice Image, Segregation of Phosphorus, Texture

ABSTRACT : The relations between the non-equilibrium segregation process of the texture in Fe-P alloys have been studied by analytical electron microscope and orientation distribution function. It was shown that P segregated preferentialy on the { 1 1 0 } slip planes, the P segregation structures with repeating cycle a = 1.582nm form at 450 °C. < 0 0 1 > || ND direction abated. <111> || ND direction heightened. And { 1 1 1 } < 1 1 0 > has a tendancy to transform into { 1 1 1 } < 1 4 3 > texture in recovering process.{ 1 1 1 }< 1 4 3 > direction transforms into { 1 1 1 } < 1 1 2 > direction after recrystalizaing. A model to describe the effects of non-equilibrium segregation structure of P on orientation change was proposed and employed to interpret the experiment results.

The state of P existing in steels has been intensively studied. It has been pointed out by many authors that P segregates at grain boundaries [1]. However, it is difficult to explain the positive function of P in steels, for example, it has been confirmed by many authors that P promotes the formation of advantageous textures and improves the deep drawing properties of the steel sheets through forming { 1 1 1 } < 1 1 2 > textures and increasing the volume of < 1 1 1 > fiber textures[2]. However,a definite conclusion has not been made about how P plays a role in it although many viewpoints have been put forward and a lot of work has been done. Recently, it was found [3,4] under the given treatment condition, that anomalous metallograghs with P segregation within grains were found, and a regular microstructure were observed from where P segregated in dislocation lines, for this reason a kind of segregation process must exist, which affects the formation of texture. In this paper, these aspects were studied by analytical electron microscope and X-ray diffraction orientation distribution function (ODF) , for clarifying the segregation state and its relation with the orientation change during annealing process.

1. EXPERIMENTAL

10 kg ingots with different P were prepared in vacuum induction furnace, the chemical composition (wt-%) of steels is shown in table 1.

Table 1 Chemical composition of samples, wt --%

No.	C	P	Mn	S	Si	Al	N
1	0.010	0.007	0.07	0.016	<0.05	<0.03	0.007
2	0.0094	0.09	0.03	0.016	<0.05	<0.03	0.010

The samples after 70% reduction by cold rolling were annealed at 200, 450, 500 and 700°C respectively, in a common vacuum tubular furnace. The foils used for TEM analyses were firstly ground to 30--50 μm and then thinned by double jet electro-polishing. The microstructures were studied by Philips EM420 analytical electron microscope. The composition was analyzed by EDAX, and the structure was observed by selected area diffraction and high resolution electron images. The samples used for texture measurement were firstly ground off a quarter of thickness and polished chemically to eliminate the effect of surface layer textures.The textures are identified automatically using three incomplete pole figures, (1 1 0), (2 0 0) and (2 1 1) by the reflection method, then ODFs were calculated using "two step method".

2. RESULTS

2.1 NON-EQUILIBRIUM SEGREGATION STRUCTURE OF P

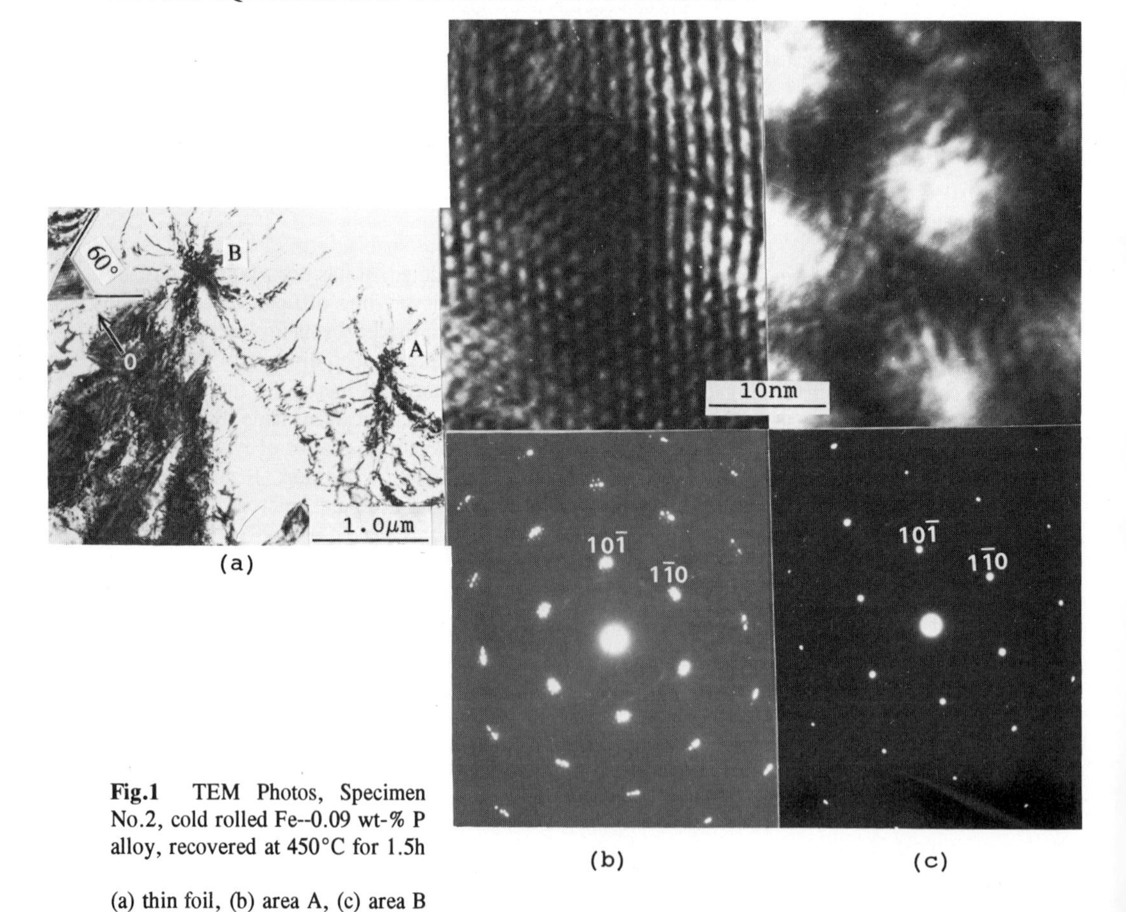

Fig.1 TEM Photos, Specimen No.2, cold rolled Fe--0.09 wt-% P alloy, recovered at 450°C for 1.5h

(a) thin foil, (b) area A, (c) area B

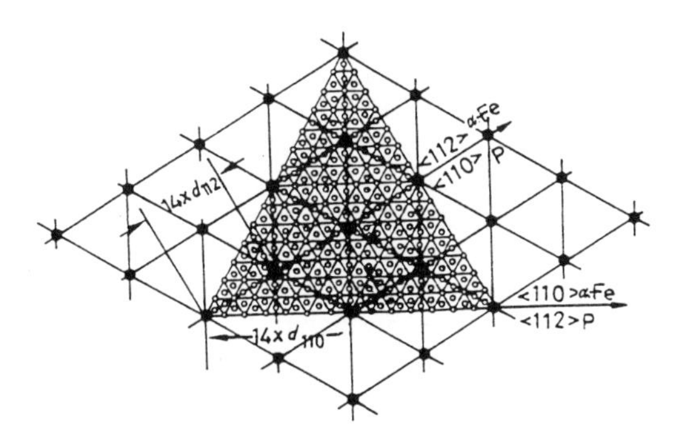

Fig.2 Segregation structure of P at {110} slip planes

TEM images of the foil sample were shown in Fig.1, in which there are two obvious different regions A and B, with only bend interference fringes caused by stress in A, but besides the fringes there are obvious net in B. The shape of the nets, with the same angle of 60° between the fringes, were similar to the image of the ends of the dislocation lines. According to the orientation analysis on the diffraction pattern, these fringes are on the { 1 1 0 } planes and elongated in < 1 1 2 > direction.

The regions A and B were measured and analyzed carefully by EDAX, the counts rates of P in B were 2-3 times higher than that in A, for example, the mean value for three measurments was 0.309 in A, but it is 0.706 in the B, this is similar to the result obtained by electron probe analysis in Ref.[4]. This indicates that the characteristics of P segregation within the scope of metallography also is within the scope of TEM, which further proves that P in the center of dislocations segregates to the denser section of dislocations together with the movement of them which finally results in the formation of the non-equilibrium segregation structure.

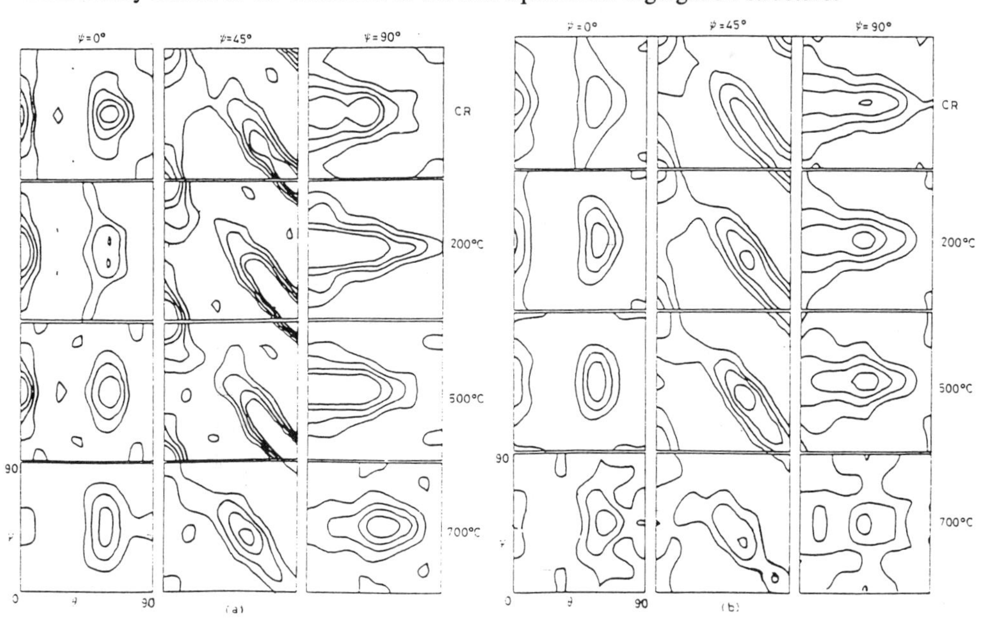

Fig.3 ODF sections, specimens cold rolled or annealed at various temperatures for 1.5 h

(a) pure Fe, (b) Fe—0.09 wt—%P alloy

The results of regions A and B observed and analysed by high resolution electron microscope (HREM) are shown in Figs.1b and 1c, in which the diffraction pattern of A and B are both α-Fe < 1 1 1 > reciprocal plane. The HREM shows typical lattice image in B, but without special image in A. In addition, it can be carefully seen in the diffraction pattern of B that there are some regular extra spots around every main spot. The repeating cycles and directions of the extra spots in reciprocal lattice are quite similer to those in lattice image of real space, with a=1.582nm in reciprocal space and a=1.591nm in real space respectively. Fig.2 shows the relationship in real space lattice between the main spots and extra spots. Small empty dots reflected by main diffraction spots is the position of Fe atoms on { 1 1 1 } planes, but big black dots represent the lattice image which were reflected by extra diffraction spots. The distances of lattice spot are 1.638nm and 2.804nm respectively, and it is similar to the values 1.591nm and 2.817nm calculated from real space image. The orientation of the lattice image on the { 1 1 1 } plane of α-Fe is exactly 30° different from α-Fe. According to the situation of higher P content in B region, it can be considered that the formation of lattice image is caused by P segregation on { 1 1 0 } slip planes.

The structure of the segregation is the product formed during the process of dislocation movement. The repeating cycle is changeable, the magnitude of which obviously relates to the different residual stresses and original dislocation distribution. Of course, it is likely that P segregates on other slip planes such as { 1 1 2 } and { 3 2 1 }, but what have been seen at present is the segregation on { 1 1 0 } slip planes.

2.2 ORIENTATION VARIATION IN RECOVERING AND RECRYSTALIZING PROCESS

ODF figures of samples 1 and 2 under different treatment conditions with certain ψ=0°, 45° and 90° are shown in Figs.3a and 3b from which it can be seen that the cold rolled sheet of sample 2 has changed during recovery. From figure ψ=0°, the orientation density of cold rolling texture (0 0 1) [1 $\bar{1}$ 0] is higher than that of ($\bar{1}$ 1 1) [1 $\bar{1}$ 2], but it changes after annealing at 200°C with a little decrease of (0 0 1) [1 $\bar{1}$ 0] and a slight increase of ($\bar{1}$ 1 1) [1 $\bar{1}$ 2].

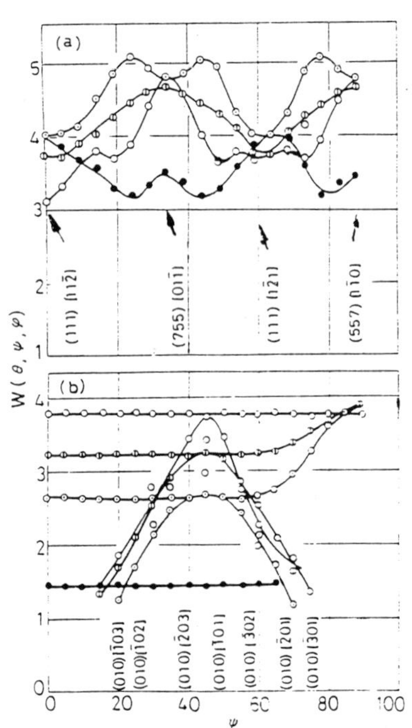

Fig.4 Deviation of skeleton line for fiber texture in Fe—0.09 wt—%P alloy (a) <111> // ND. (b) <010> // ND

However, the orientation density of ($\bar{1}$ 1 1) [1 $\bar{1}$ 2] after sample 2 annealed at 500°C is inversely higher than that of (0 0 1) [1 $\bar{1}$ 0]. The orientation after recrystallization has changed completely, and (0 0 1) [1 $\bar{1}$ 0] almost disappears but the orientation density of ($\bar{1}$ 1 1) [1 $\bar{1}$ 2] increases obviously. The results of sample 1 are different, no above change happens during the recovering.

To show the effects of P, the trends of the skeleton line along direction of ODF of sample 2 are given in Fig.4 (describing the change of the maxium orientation density). The skeleton line of < 1 1 1 > ‖ ND fiber textures in Fig.4a are usually called A texture composition, from which it can be seen that at cold rolling { 1 1 1 } < 1 1 0 > are the strongest and { 1 1 1 } < 1 1 2 > are the weakest in A composition; after recovering at 200°C, the orentation desities in all A composition are all increased; after annealing at 500°C, the maxium orientation density of ψ=35° splits into two peak values with 10° deviation from it, which corresponds to the two counterpart positions among textures { 1 1 1 } < 1 4 3 > and besides it the orientation density of A increases further; after annealing at 700°C, the recrystallization have happened and orientation changes completely at ψ=0° and ψ=60° the orientation density changes from the minimum to maxium before and after recrystallization, which shows that the < 1 1 1 > fiber texture consisting mainly of { 3 3 2 }< 1 1 3 > forms.

The skeleton lines of the partial fiber texture of < 0 1 0 > ‖ ND are shown in Fig.4b, (a group of reserved V curves), in which it can be seen that they are opposite to the change of < 1 1 1 > fiber textures, the volume of < 0 1 0 > textures reduces with the annealing temperature up to near disappearance after recrystallization,especially to the textures { 0 1 0 } < 1 1 0 >. Another group of skeletion lines of < 1 1 0 > ‖ RD partial fiber texture are shown in Fig.4b. Its variance is similar to the < 0 1 0 > ‖ ND fiber texture, as the temperature rises the value of orientation density decreases.

It can be seen from the above results that P plays an important role in formation of texture for cold rolling, recovering and recrystallization.

3. DISCUSSION

3.1 FORMATION OF NON-EQUILIBRIUM SEGREGATION STRUCTURE OF P

The P-rich and P-poor regions with high and low density of dislocations began to form after alloy No.2 was annealed above 300°C, which has been pointed out in Ref.[4]. It is possible that P enrichment is due to the directional movement of dislocations carrying P atoms and they may accumulate in some places.

Because different speed of deformation in different layers perpendicular to the surface of sheet during cold rolling, residual stresses paralled to rolling plane are caused. In this case, the maximum shear stress direction should be in the region of cone angle 45° with axis normal to plane. In this work two usual crystal direction < 1 1 1 > and < 0 0 1 >, normal to the rolling plane, in which 24 slip planes and 4 slip directions are found within 30° -- 60° region; however, only 9 slip planes are found in the crystal with < 1 1 1 > normal to the rolling plane, and all the slip directions are out 60°. It can be thought that the dislocations in the crystal with < 0 0 1 > normal to the rolling plane move more easily than that of < 1 1 1 >. There are three main groups of fiber texture in cold rolled sheets of α-Fe, which connected each other. It is the continuity the variance of orientation in crystal that makes the dislocation line carried P in crystals with { 0 0 1 } < 1 1 0 > texture finally transfer into the crystals with { 1 1 1 } < 1 1 0 > texture under the action of residual stress, by means of the orientation crystal of the < 1 1 0 > ‖ RD fibric texture, then, the P-rich region of < 1 1 1 > ‖ ND orientation crystal and the P-poor region of < 0 0 1 > ‖ ND orientation crystal form as shown in Fig.5.

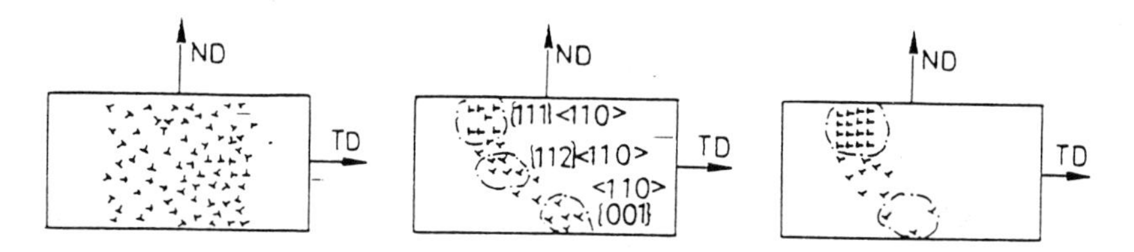

Fig.5 Non−equilibrium segregation process of P

It is obvious that the segregation of P in slip plane is the result of directional movement of dislocation lines carring P atoms. Considering the decrease of the system energy, it is much easier for P in dislocation lines to segregate at the joining dot of { 1 1 0 } slip plane (the position of big black dot in Fig.2.). In this way, the triangle prism could be taken as a segregation structure unit, supposed the apex of prism were occupied by the segregation atoms of P, the segregation amount of P were estimated to be 0.23wt-%, which is similar to the result of practical measurement.

3.2 EFFECT OF NON-EQUILIBRIUM SEGREGATION OF P ON THE ORIENTATION VARIATION

It can be seen from the above analyses that the non-eqilibrium segregatiion process, in principle, is the result of directional movement of dislocations towards edge dislocations, this is actually the transformation of atom planes. Therefore, the directional slip of dislocations is accompanied by the transformation of materials , in other words , some atoms in crystal of < 1 1 0 > ∥ RD or < 0 0 1 > ∥ ND transform into crystal of < 1 1 1 > ∥ ND. Which can explain the present experimental results that why the fiber textup373X⊗f1 1 0 > ∥ RD decreases and the fiber texture of < 1 1 1 > ∥ ND increases during recovering in cold rolled Fe-P alloy.

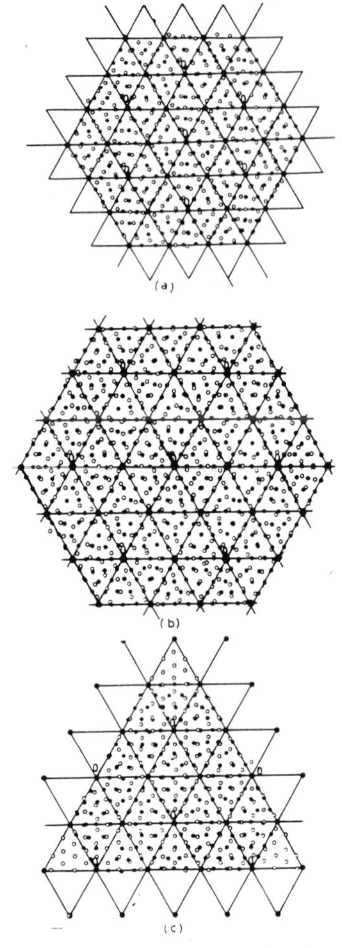

Fig.6 Segregation structure (a) {321} plane. introduced by orientation variation. (b) {321}→{110} plane. (c)-{110} plane

3.3 EFFECT OF NON-EQUILIBRIUM SEGREGATION OF P ON THE TEXTURE VARIATION

The same characteristics of non-equilibrium segregation structures are in the form of triangle frame with P as joint points. A definite angle are formed between the P triangle and Fe atom triangle on the { 1 1 1 } planes of α-Fe, the angle is 0° between two triangles segregated in { 1 1 2 } slip planes, the angle is 10.89° between two triangles segregated in { 3 2 1 } slip planes, the angle is 30° between two triangles segregated in { 1 1 0 } slip planes. If we consider that the P triangle frame is relatively stable, the Fe atoms within the triangles should be in the unstable position; but these Fe atoms can move to the regular arrangement positions. The state of segregation { 3 2 1 } plane is shown in Fig.6a. All the o points in the figure are equivalent points with the same environment, so if Fe atoms turn relatively about o points, as a center axis, near to about 10°, the structure can be rearranged completely. The course of atomic rearrangement will result in the variation of crystal orientation, the characteristics of which are that the orientation turns around the < 1 1 1 > fiber texture axis, i.e., the orientation parallel to the rolling direction changes while the orientation parallel to the rolling plane keeps unchanged.

The orientation of the sample after recovering at 500°C, changes from { 1 1 1 } < 1 1 0> into { 1 1 1 } < 1 4 3 > with an angle of 10.89° between < 1 4 3 > and < 1 1 0 >, which agrees with two < 1 4 3 > orientation peaks symmetrical about < 1 1 0 > orientation caused by the orientation of original crystal into { 3 2 1 } slip segregation structure. The orientation changes greatly after annealing at 700°C. It can be seen by careful analyses that the peaks of original orientation transfers into right and left after annealing, near to 20°, turning the peak into valley with orientation variation from < 1 5 7 > into < 1 1 3 >, actually from < 1 4 3 > into < 1 1 2 > taking < 1 1 1 > as the turning axis. It can be explained that the crystal with { 3 2 1 } segregation structure formed during recovering transfers into P triangle frame arrangement with { 1 1 0 } plane segregation structure. Obviously, it is possible that the orientation position of segregation structure of the { 1 1 2 } plane turns into { 1 1 0 } slip plane directly as shown in Fig.6c, but the atomic displacement is bigger than former, it cannot occur easily.

4.CONCLUSIONS

(a) During recovering of cold rolled Fe-P alloy, P lies in the center of dislocation lines preferentially, and accompanies the moving dislocation lines giving P segregation on slip planes.

The structure of P segregation with repeating cycle a=1.591nm forms at 450°C, a stable skeleton structure of P forms in the deformed matrix of α-Fe.

(b) P plays an important role in the variation of orientation of cold rolled Fe-P alloy during recovery and recrystallization. There are two types of orientation variation during recovery: first < 1 1 1 > ∥ ND and < 0 0 1 > ∥ ND fiber texture increase and decrease respectively with the increase of recovering temperature; second { 1 1 1 } < 1 1 0 > texture splits into two { 1 1 1 } < 1 4 3 > texture. After recrystallization, the change of { 1 1 1 } < 1 4 3 > into { 1 1 1 } < 1 1 2 > direction happens.

(c) In Fe-P alloy, the course and the structure of P non-equilibrium segregation are important factors controlling the variation of orientation during recovering and recrystallization.

REFERENCES

1) Grabke H J.: Steel Res., 1986, 57, 178
2) Hu H.: Metal trans. 1977, 8A, 1567
3) Zhang Lixin, Qu Zhe, Li Liguang et al.: Acta Metall Sin., 1989, 2A, 119 (English Edition)
4) Ning Hua, Zhang Lixin.: Acta Metall Sin., 1991, 4, A108 (English Edition)
5) Li Zhengxiu, Liu Xingpu, et al.: Control and Decision, 1986, 3, 16 (in Chinese)
6) Zhang Lixin, Li Liguang, Wei Wenduo.:Sheet Steels and Forming Technology for Automobile, Chinese Deep Drawing Research Group, 1990, 300 (in Chinese)
7) Liang Zhide, Xu Jiazhen, Wang Fu.: Proc 6th Conf on Textures of Materials, ISIJ, Tokyo, 1981, 2, 1259 (in Chinese)

4. Texture and Directional Properties of Materials

Materials Science Forum Vol. 157-162 (1994) pp. 1523-1536

ANISOTROPIC PROPERTIES OF MINERALS AND ROCKS

H. Kern

Mineralogisch-Petrographisches Institut, Olshausenstr. 40, D-24098 Kiel, Germany

Keywords: Elastic Anisotropy of Rock-Forming Minerals, Crack-Related and Texture-Related Velocity Anisotropy at Pressure and Temperature, Shear Wave Splitting, 3-D Velocity Calculations, Geophysical Significance of Seismic Anisotropy

ABSTRACT: This paper focuses largely on the elastic anisotropy of rock-forming minerals and of crustal and mantle rocks, because an understanding of the directional dependence of elastic wave velocities is most important in the earth sciences. The role of crack-related and texture-related velocity anisotropy in polyphase crustal and mantle rocks at high and low pressure and temperature conditions will be discussed and particular emphasis will be placed on shear wave propagation and shear wave splitting which is a direct indicator for fabric-related anisotropy. Two different approaches can be used to determine the elastic rock properties: (1). Experimental determination of compressional (Vp) and shear wave velocities (V_{S1},V_{S2}), and (2) 3D modelling based on the lattice preferred orientation (LPO) of major minerals, the single crystal elastic constants and the modal composition of the rock. Geophysical implications of data derived from both approaches will be discussed.

INTRODUCTION

Means of measuring physical properties across large sections of the in-situ earth's crust and upper mantle are provided through a variety of geophysical techniques. Most important is seismic refraction and reflection profiling, and both techniques have been successfully used (independently or in combination with data from electrical, gravity and magnetic investigations) to probe the deep crust and upper mantle. Because seismic (elastic) waves are the most powerful tool that geophysicists have at their disposal to investigate the in-situ physical properties of the earth's deep interior, this paper focuses largely on elastic properties of rocks and rock-forming minerals.

Rocks, in contrast to metals, have a great variety in composition and structure, generally containing several phases with different grain size and different physical and chemical properties. In addition, rocks contain numerous microcracks and pores. An important property of most of the rocks constituting the earth's crust and upper mantle is anisotropy, that means that physical properties are different in different directions. Anisotropy of physical properties in rocks depends basically on the anisotropic properties of the rock-forming minerals, their lattice preferred orientation (LPO) (texture) and the distribution and size of microfractures. The effectiveness of microcracks is largely controlled by pressure and temperature and the presence or absence of fluids.

ELASTIC ANISOTROPY OF ROCK-FORMING MINERALS

The elastic properties of minerals depend on both, the chemical composition and the crystal structure. They are defined by the stiffness tensor C_{ijkl} and the compliance tensor S_{ijkl} relating stress σ_{ij} and strain ε_{kl},

$$\sigma_{ij} = C_{ijkl}\,\varepsilon_{kl}, \qquad \varepsilon_{ij} = S_{ijkl}\,\sigma_{kl} \qquad (1)$$

For practical purposes, the fourth-rank tensors are generally represented by equivalent symmetrical 6x6 matrices C_{mn} and S_{mn} with 21 independent components for crystals exhibiting triclinic symmetry [1,2]. The number of independent coefficients is reduced to 3 for cubic symmetry.

Most data on elastic wave velocities of single crystals cited in the literature were determined experimentally by dynamic methods using high-frequency vibrations in crystals. Widely used is the pulse-transmission technique [3,4], and more recently, the Brillouin-scattering technique. For relevant rock-forming minerals, both sets of elastic constants along with the mineral densities are available [e.g.5]. Using the Christoffel equation, these data allow the calculation of elastic wave velocities for arbitrary directions of wave propagation, and the data can be presented in stereograhic projections. (Fig. 1).

The maximum, minimum and average velocities of compressional and shear waves along with the coefficients of anisotropy (A) [$A = 200\,(V_{Pmax} - V_{Pmin})/V_{Pmean}$) (%)] for the most important rock-forming minerals are listed in Table 1. [For further data see 6]. The corresponding directions of wave propagation are given by Miller indices, and the minerals are arranged from ortho- and ring-silicates via chain-silicates and framework silicates to non-silicates. The elastic anisotropy is generally closely linked with the crystal structure and crystal symmetry. The directional dependence of elasticity in a crystal is highly dependent on the strength of the interatomic bond in the corresponding direction of the crystal structure [7]. In a crystal having a marked structural anisotropy, the elastic moduli will be higher in the direction in which the structure has the strongest bonds [8]. This is clearly documented by the sheet silicates which exhibit the highest velocity anisotropies of the relevant rock-forming minerals (Table 1). In the structure of this mineral group, layers of SiO_4-tetrahedrons are bonded by relatively weak interlayer forces so that the P-wave velocities are $V_{P[100]} \approx V_{P[010]} \ll V_{P[001]}$.

An important characteristic of anisotropic crystals (and rocks) is shear wave splitting (acoustic birefringence). Upon entering an anisotropic medium, a single shear wave will be vectorially split in two orthogonally polarised components travelling along the same propagation path with different velocities [9,10]. This phenomenon is analogous to optical birefringence in the anisotropic minerals. It is important to note that the polarisation of the split shear wave is closely related to distinct planes of the crystal structure (Fig.1).

Elastic-wave velocities (V_p, V_s) and coefficients of elastic anisotropy (A) in rock-forming minerals

Mineral	Essential feature of structure	Density	Compressional waves					Shear waves					Source of data
		[g/cm^3]	V_p max [km/s]	Propagation direction	V_p min [km/s]	Propagation direction	A [%]	V_s max [km/s]	Propagation direction	V_s min [km/s]	Propagation direction	A [%]	
Ortho- and Ring Silicates													
Olivine (Fo_{92})	isolated SiO_4-Tetrahedra	3,31	9,89	[100]	7,72	[010]	24,6	5,53	$\vartheta=45°$ $\varphi=0°$	4,42	[010]	22,3	[40]
Chain Silicates													
Pyroxenes													
Bronzite	single chains	3,34	8,30	[100]	7,04	[010]	16,4	4,90	[010]	4,27	$\vartheta=90°$ $\varphi=45°$	15,6	[41]
Diopside		3,31	8,60	[001]	6,94	[101]	21,4	4,83	[011]	3,94	[110]	20,3	[42, 43]
Hornblende	double chains	3,15	8,13	[001]	6,18	[101]	27,2	4,60	[011]	3,37	[001]	30,8	[44]
Sheet Silicates													
Muscovite	single layers	2,79	8,06	[110]	4,44	[001]	57,9	5,01	[010]	2,03	[001]	84,7	[7]
Biotite		2,89	7,80	[010]	4,21	[001]	59,7	5,06	[010]	1,34	[010]	116,3	[45]
Framework Silicates													
Feldspars													
Microcline	tetrahedral framework	2,56	8,15	[010]	5,10	[100]	46,0	4,96	[011]	2,14	[010]	79,4	[46]
Orthoclase	tetrahedral framework	2,54	7,64	[010]	4,76	[101]	46,4	4,45	[011]	2,33	[001]	62,5	[47]
Albite (An_9)	tetrahedral framework	2,61	7,26	[010]	5,31	[101]	31,0	4,63	[011]	2,59	[001]	56,5	[48]
Anorthite	tetrahedral framework	2,76	8,61	[010]	6,01	[101]	35,6	4,96	[011]	2,91	[010]	52,0	[49]
Quartz	tetrahedral framework	2,66	7,00	$\vartheta=130°$ $\varphi=90°$	5,36	$\vartheta=72°$ $\varphi=90°$	26,5	5,06	[100]	3,35	[100]	40,1	[50]
Non-Silicates													
Calcite	isolated CO_3-triangles	2,72	7,55	$\vartheta=70°$ $\varphi=90°$	5,43	[001]	32,7	4,77	$\vartheta=126°$ $\varphi=90°$	2,66	$\vartheta=36°$ $\varphi=90°$	56,7	[51]

ϑ and φ determine directions which cannot be defined by simple indices (for explanation, see Fig. 3.1. [6])

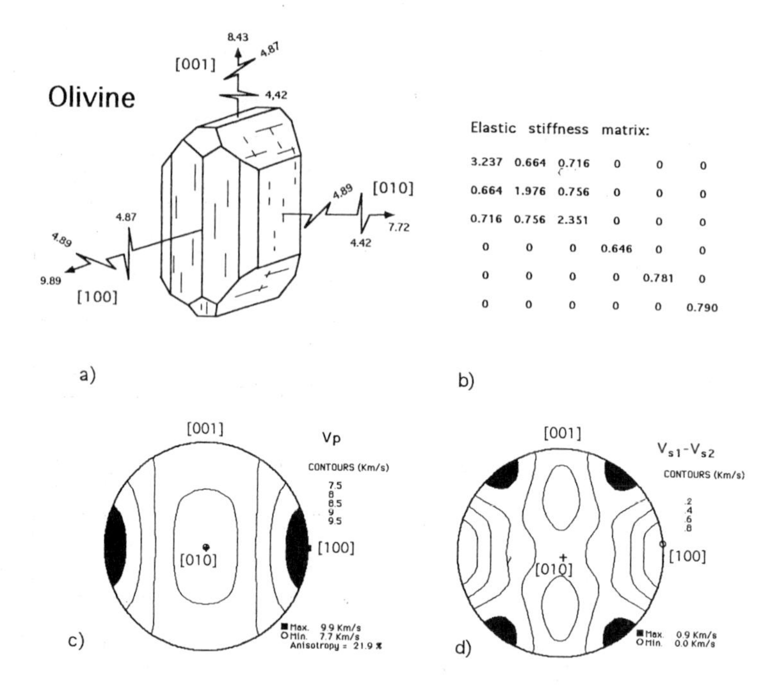

3.237	0.664	0.716	0	0	0
0.664	1.976	0.756	0	0	0
0.716	0.756	2.351	0	0	0
0	0	0	0.646	0	0
0	0	0	0	0.781	0
0	0	0	0	0	0.790

Fig. 1. Compressional wave velocities and shear wave splitting in a monocrystal of olivine
a) P-wave velocities, shear wave polarisation and shear wave splitting along basic crystallographic directions,
b) Elastic stiffness matrix (Mbar),
c,d) Calculated 3-D variations of P-wave velocities and shear wave splitting (equal area projection).

VELOCITY ANISOTROPY IN ROCKS AND THEIR PRESSURE AND TEMPERATURE DEPENDENCE

Determination of compressional and shear wave velocities and their directional dependence as a function of both pressure and temperature is important in seismic modelling of the crust and upper mantle.

Experimental determination of elastic wave velocities

Up to now most velocity determinations at elevated pressure and temperature have been carried out on dry samples and were done exclusively by the method of ultrasonic pulse transmission [e.g. 3,11]. Pressure may be applied to the samples either by gaseous, fluid or solid pressure transmission. The use of triaxial anvil presses [e.g.11,12,13]) offers the advantage of simultaneous velocity measurements in the three mutually orthogonal directions and direct control of length changes of the sample with variation of pressure and temperature. In general, measurements of P- and S-wave velocities along the three principle axes are insufficient to describe the elastic properties completely, but permit the estimate of maximum and minimum velocities and thus the coefficient of anisotropy. This, particularly, holds for the case that the sample reference system has been related to inherent fabric elements such as foliation and lineation (see inset of Figs. 2 and 3).

Crack-related and texture-related velocity anisotropy at pressure and temperature

Pressure and temperature are competing parameters and, therefore, they affect elastic wave velocities oppositely. This especially holds with respect to microcracks, because elastic wave propagation through dry natural rocks is very sensitive to the state of microfracturing. There is a close relationship between velocity increase and closure of microcracks respectively velocity decrease and opening of microcracks due to changes in pressure and temperature [e.g. 13,14,]. Figure 2a shows the compressional wave velocities as a function of pressure for the three perpendicular propagation directions in a sample of biotite-sillimanite gneiss recovered from the KTB pilot hole, Germany (Deutsche Kontinentale Tiefbohrung). The lower velocity curve was measured normal to the foliation plane, whereas the two higher velocity curves are velocities within the foliation plane. The velocity/pressure relations display the well-known initial steep increase of velocity with increasing confining pressure resulting from progressive closure of microcracks. The effect of crack closure is clearly documented by the strong correlation of the slopes of the velocity vs. pressure curves and the linear and volumetric strain curves vs. pressure curves (not shown here; see [15]). The change of velocity anisotropy of P-waves for a number of KTB rocks (gneisses, metagabbros and amphibolites) as a function of pressure at room temperature is shown in Fig. 2b. In most cases, anisotropy decreases markedly with increasing confining pressure until a stable value is reached. From the pressure dependence we infer that oriented microcracks are a major contribution to velocity anisotropy at low pressure. The residual anisotropies (intrinsic) observed at elevated confining pressure originate primarily from preferred mineral orientation (texture) [e.g.9,16,17], because most of oriented cracks are closed above about 200 MPa. It is important to note that the texture-induced velocity anisotropy is not significantly affected by temperature [17,18].

Figure 3 presents shear wave splitting (a) and shear wave splitting anisotropy (b) as a function of pressure (room temperature) for a KTB gneiss sample. A marked shear wave splitting occurs for the shear wave propagating parallel to the foliation, and, importantly, the fast and slow shear wave are polarised parallel and normal to the foliation plane, respectively. For the shear wave propagating normal to the foliation plane, shear wave splitting is low.

The fabric-related (intrinsic) anisotropy is mostly a result of ductile deformation. This is particularly apparent from the mylonites of the Santa Rosa Mylonite Zone, California [17]. During tectonism, these rocks were progressively deformed from a granodiorite protolith (quartz, plagioclase, microcline, biotite) to mylonite and ultimately phyllonite without a change of mineral composition. With increasing strain, the fabric changed, specifically the grain size and grain shape (Fig.4a), and a strong preferred orientation of biotite developed (Fig. 4b). Increase of biotite preferred orientation (indicated by (001) pole densities) (Fig.4b) is strongly correlated with an increase of intrinsic P-wave velocity and shear wave splitting (Fig. 4c), whereas average P-wave velocities are more or

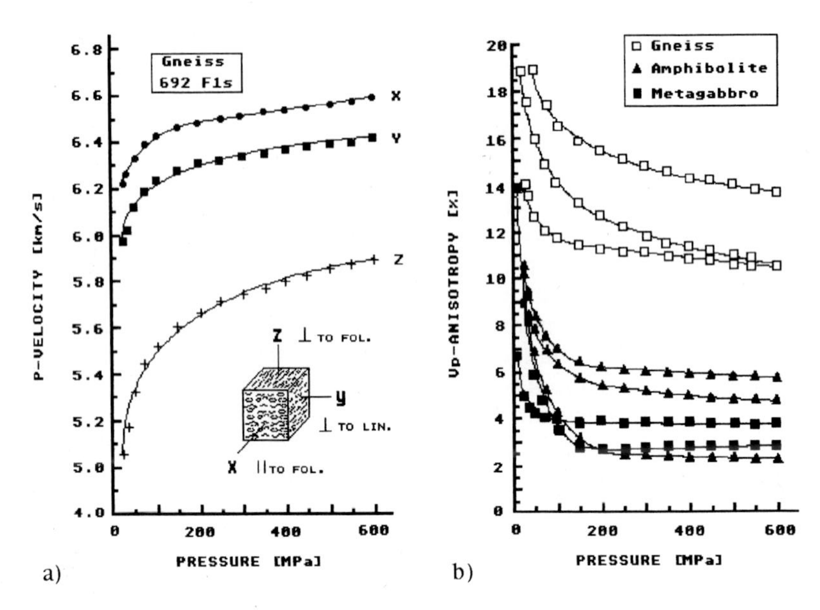

Fig. 2. Directional dependence of P-wave velocities in a gneiss (a) and velocity anisotropy in a number of gneisses, metagabbros and amphibolites (b) recovered from the KTB-pilot hole as a function of pressure at room temperature [15].

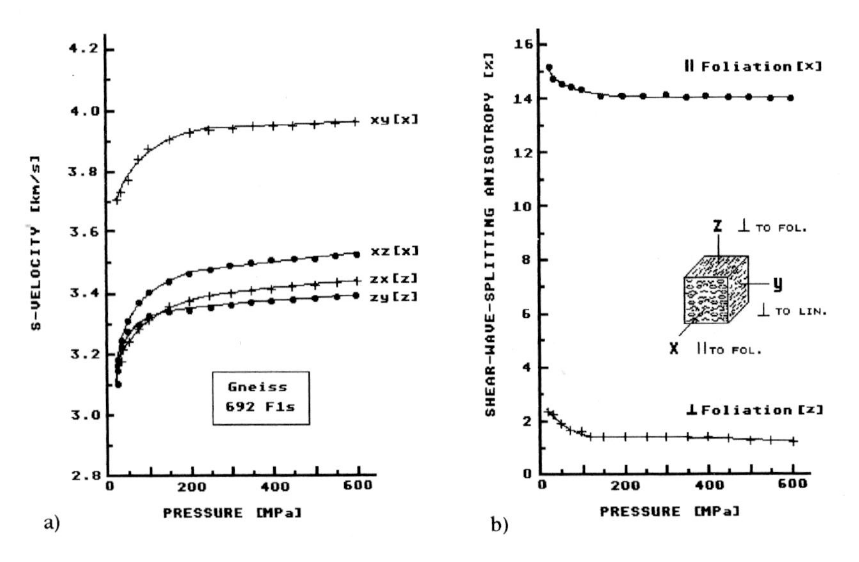

Fig. 3. Measured shear wave splitting (a) and shear wave splitting anisotropy (b) in the KTB gneiss 692F1s. Note the pronounced shear wave splitting parallel to the foliation (parallel to [X]) and the strong relationship of shear wave polarization to the foliation plane) [17].

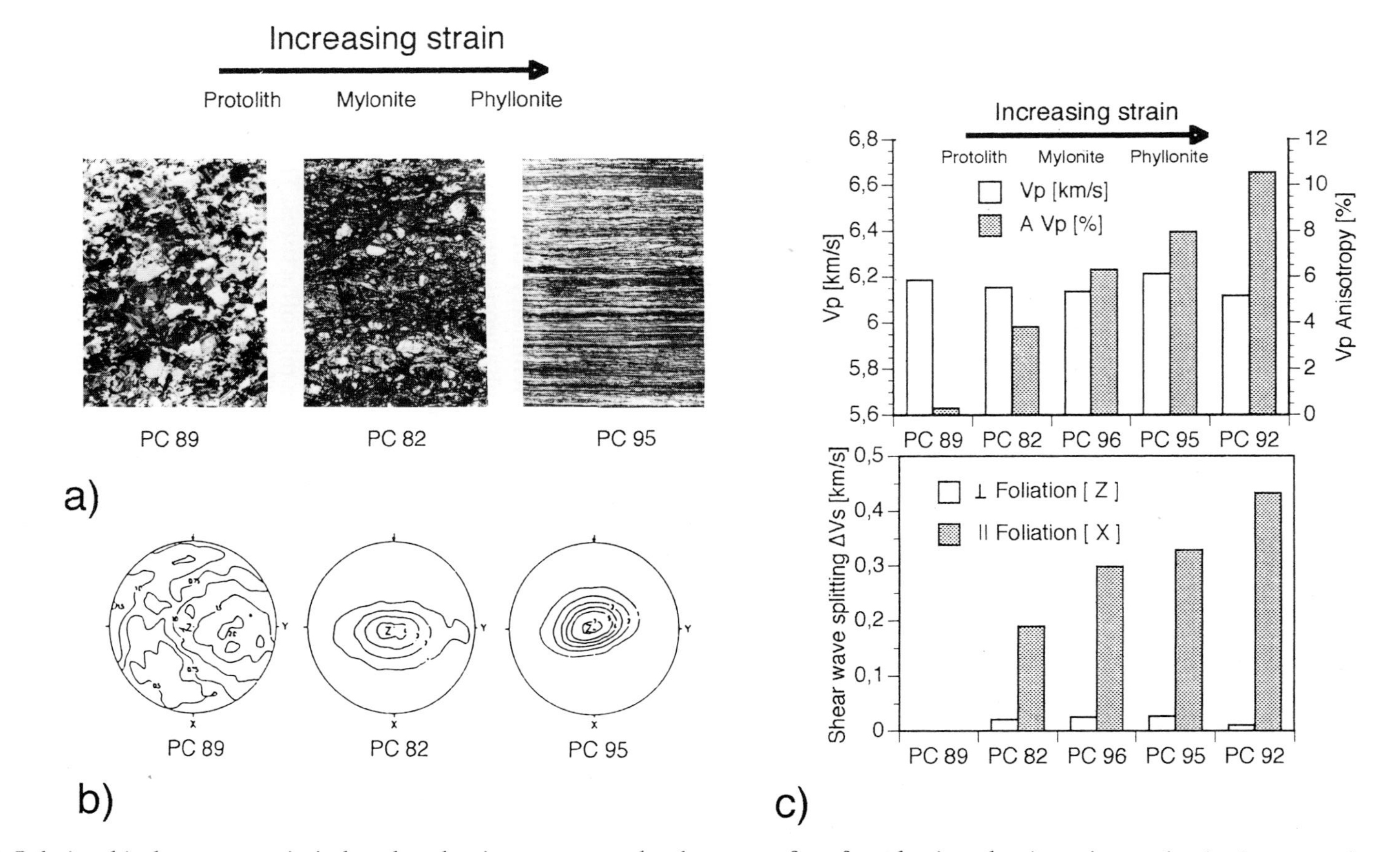

Fig. 4. Relationship between strain-induced rock microstructure, development of preferred mineral orientation and seismic properties of a granodiorite which has been progressively deformed from an undeformed protolith to mylonite and finally to phyllonite. (a) Change of microstructure during progressive deformation with reduction in grain size and development of foliation. (b) Development of preferred orientation of biotite during progressive strain. Biotite (001) pole figures projected on to the foliation plane (equal area).(c) Comparison of average velocities (open) and velocity anisotropies (shaded) in ductily sheared granodiorite (top) and shear wave splitting (below) ordered with increasing deformation [17].

less the same in all samples (Fig. 4c, top). Again, shear wave splitting is most pronounced for the shear waves propagating parallel to the foliation plane with the fast shear wave being polarised parallel to the foliation. Normal to the foliation plane there is practically no shear wave splitting observed.

In rocks constituting the earth's crust, the fabric-related (intrinsic) anisotropy is highest in foliated rocks and is mainly due to the alignment of phyllosilicates and hornblende. Typical values are 10.7% for quartz-mica shists, 7.3% for felsic amphibolite gneisses, 5.5% for granulite-facies metapelites, and 5.4 for amphibolites [19]. The variations of P- wave velocities in crustal rocks due to changes in composition are substantially greater than velocity variations due to texture-related anisotropy [6, 20]. Importantly, the reverse is true in the olivine-dominated upper mantle rocks (dunites, lherzolite, peridotite) where the ranges of P-wave velocities due to preferred orientation of olivine are greater than the ranges of P-wave velocities due to variability in composition. P-wave anisotropy is almost between 3 and 6%, but can reach 10% [21].

CALCULATION OF 3-D VELOCITIES FOR ANISOTROPIC AGGREGATES

For seismic purposes and in order to directly compare fabric anisotropy (pole figures) with velocity anisotropy, it is useful to present P- and S-wave velocities in the form of contour stereograms. The 3-D seismic properties can either be calculated from direct measurements of seismic velocities in many directions or from lattice preferred orientation (LPO) data of the constituent minerals.

Estimation of elastic constants from laboratory measurements

Laboratory measurements of P-and S-wave velocity in critical symmetry directions provides the means to directly calculate the elastic constants of natural rocks for particular pressure and temperature conditions [5, 22]. Many natural rocks exhibit more or less hexagonal or rhombic aggregate symmetry. Thus, their elastic properties are uniquely defined by 5 and 9 independent constants, respectively. For the case of hexagonal aggregate symmetry, the stiffness matrix can be obtained from one experiment by the simultaneous measurement of V_P, V_{S1} and V_{S2} in the three orthogonal directions of the sample cube, with the simplifying assumption that the P-wave velocity in a direction 45° inclined to the axis of symmetry [Z] corresponds to the average of the maximum and minimum velocity [23]. For rhombic aggregate symmetry, two experiments with different sample orientations are needed for the determination of the full elastic tensor. This kind of data describes the bulk rock properties resulting from the properties of the matrix minerals <u>and</u> the effective microfractures. The data are thus of importance for simulating conditions of shallow depth (low pressure). Figure 5 shows the 3-D velocity calculations of a KTB biotite-gneiss (hexagonal symmetry) for particular pressure conditions corresponding to a depth of 412m and 6044 m and 600 MPa, respectively [24]. Comparison with the respective 600 MPa velocity and anisotropy data which are considered to represent intrinsic (matrix) properties shows that the in-situ seismic properties are significantly affected by oriented microcracks, in particular at conditions of shallow depth (low pressure). Under low pressure conditions oriented cracks are more or less kept open, and the anisotropic properties of the oriented (effective)

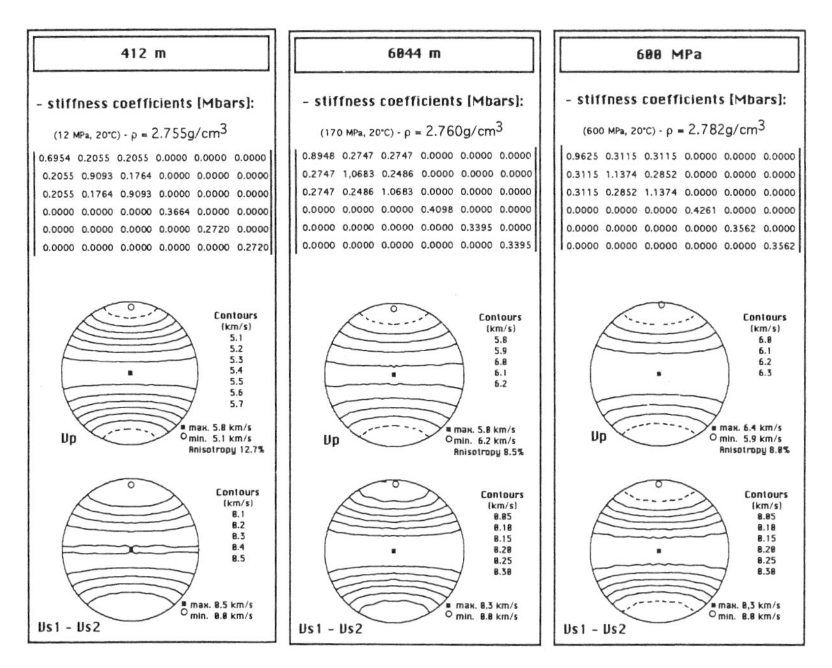

Fig. 5. Elastic stiffness matrices (Mbars) derived from the experimental velocity measurements and calculated 3-D velocity distribution of Vp and shear wave splitting (ΔV_S) in a KTB biotite gneiss at pressures corresponding to depth of 412 m and 6044m, respectively, and at 600 MPa representing intrinsic properties [24].

microcracks and the oriented matrix minerals (texture) can be additive and enhance anisotropy. In mica-gneisses oriented cracks are closely related with the biotite (muscovite) fabric, occurring mainly parallel to the morphologic sheet plane (001).

Calculation of the seismic rock properties from the lattice preferred orientation data

A second approach to calculate elastic rock properties is based on the LPO of major minerals, the volume percentage of major minerals and the single crystal stiffness coefficients, and the Voigt average has been widely used for the velocity calculations of the oriented polyphase rocks [e.g. 25, 26, 27, 28]. The LPO of crystals in the rocks described by an Orientation Distribution Function (ODF) [29] can be measured by volume diffraction techniques (X-ray diffraction, neutron diffraction) or by individual orientation measurements (optical microscope, electron channelling, electron backscattered patterns) [e.g.30, 31, 32]. The LPO-based calculations represent the intrinsic elastic properties, i.e. the seismic properties of a crack free material, probably occurring at greater depth levels.

Figure 6 shows the seismic properties of an amphibolite based on the hornblende LPO and the plagioclase LPO [33]. It is clear from the diagram that the hornblende minerals constituting the main part of the rock (60 Vol.-%), control the P-wave distribution. The direction of maximum P-wave velocity (7.2 km/s) corresponds to the hornblende [001]-maximum (parallel to lineation) and the

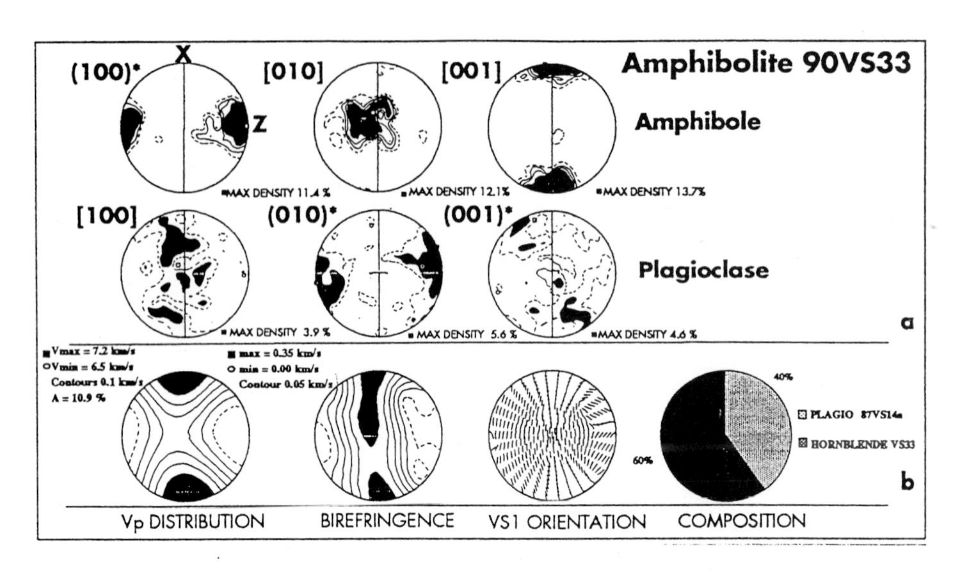

Fig. 6. Pole figures of hornblende and plagioclase [100]-, [010]-, and [001]-axes for an amphibolite of the Ivrea Verbano Zone presented on equal area projection (a) and calculated 3-D variations of the elastic properties (b.) P-wave velocities (left), variations of shear wave splitting (middle), and orientations of the polarization plane of the fast split shear wave (right); each short arc segment is a part of a great circle corresponding to the orientation of the polarization plane [33].

minimum (6.5 km/s) correspond to the pole of the foliation. The significant seismic anisotropy is, therefore, largely due to the hornblende fabric. When summed together, the hornblende and plagioclase fabric have some destructive effect on the resulting anisotropy, because the direction of maximum P-velocity in hornblende corresponds to the direction of minimum velocity in plagioclase (compare respective single crystal data in Table 1). Shear wave propagation in the rock is also significantly controlled by the hornblende LPO. Shear wave splitting is highest for an S-wave propagating parallel to the maximum of hornblende [001]-axes (parallel to lineation and foliation), and the fast split shear wave (V_{S1}) is polarised parallel to the foliation plane.

GEOPHYSICAL SIGNIFICANCE OF SEISMIC ANISOTROPY

The close relationship between seismic anisotropy (in particular, shear wave splitting and shear wave polarisation) and strain-induced preferred mineral orientation (and foliation) may provide a powerful method of studying internal deformations of the lithosphere, as well as motions in the mantle associated with plate tectonics [34]. Figure 7 illustrates the effect of the olivine texture and the symmetry of the olivine fabric on P- and S-wave propagation in a peridotitic upper mantle as may be inferred from experimental and calculated data. In a lithospheric plate exhibiting horizontal orientation of the foliation plane and an orthorhombic olivine fabric (Fig. 7, left), significant seismic anisotropy should be shown by shear waves propagating vertically (e.g. teleseismic S-waves such as SKS), i. e. parallel to [010]. In this case, the fast split shear wave

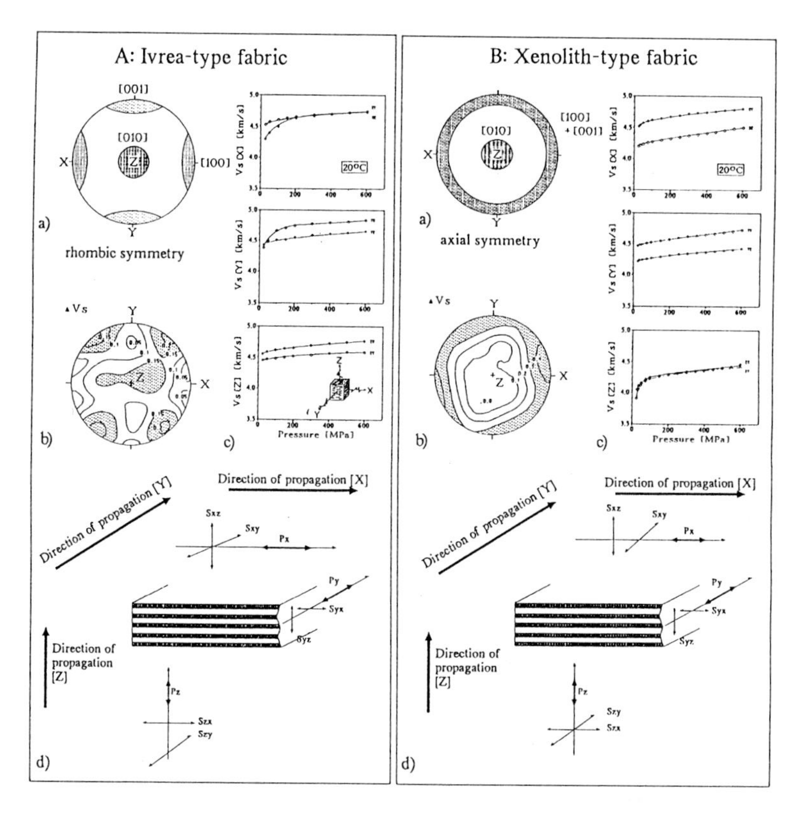

Fig. 7. Relationship between olivine preferred orientation, fabric symmetry and shear wave splitting in mantle derived rocks and conceptual models illustrating P- and S-wave propagation in an olivine-dominated upper mantle [18,35].

a) Olivine preferred orientation (schematic drawings); for measured pole figures see the original papers),
b) Calculated 3-D shear wave splitting orientation based on the olivine LPO,
c) Experimental data,
d) Model for texture-related P- and S-wave propagation in a peridotitic upper mantle exhibiting orthorhombic (Ivrea-type) fabric symmetry (left) and hexagonal (xenolith-type) fabric symmetry (right).

will be polarised normal to the foliation and parallel to the lineation, i. e. parallel to the probable fossil flow direction. A marked shear wave splitting can also be expected for shear waves propagating horizontally and normal to the lineation with the fast shear wave being polarised parallel to the foliation. By contrast, there is practically no shear wave splitting to be expected for shear waves propagating parallel to the flow direction (parallel to the fast direction of P-waves). The Ivrea-type fabric thus mimics single crystal behaviour (see Fig. 1). In a lithospheric plate exhibiting axial symmetry of the olivine fabric with the symmetry axis normal to the (horizontal) foliation plane (Fig.7, right), shear wave splitting will not be shown by vertically propagating S-waves, i.e. parallel to the olivine [010]-axes (parallel to the slow direction of P-waves). However, for S-waves propagating horizontally (parallel to the foliation), both shear wave components of motion have significantly different velocities and the leading shear wave is polarised parallel to the foliation plane.

From seismic experiments the delay time (difference in arrival time of the split shear waves), which is a function of the thickness and anisotropy of the medium, and the orientation of the polarisation planes can be derived [34,36,37]. The seismic experiment ISO 89 II [38, 39] at the KTB drill site has revealed significant seismic anisotropy and a marked shear wave splitting. The fast shear wave was found to be polarised parallel to a steep dipping (SW) plane which follows the overall azimuth and dip of the foliation. This strong relationship between polarization of the fast split shear wave and texture-related foliation is in excellent agreement with the experimental and theoretical findings. The close association between S-wave polarization and the dominant geological fabric can thus provide information on deformation and deformed structures at depth associated with the last tectonic event [36].

ACKNOWLEDGEMENTS

I am grateful to T. Popp for compiling the reference list and for draughting a number of figures on the computer. T. Goldbach compiled Table 1.

REFERENCES

1. Voigt, W.: Lehrbuch der Kristallphysik, 1928, B.G. Tubner, Leipzig
2. Nye, J. F.: Physical properties of crystals. Their representation by tensors and matrices, 1972, Oxford University Press, Oxford
3. Birch, F.: J. Geophys. Res., 1960, 65, 1083-1102
4. Anderson, O. L. &. L., R.C.: Sound velocities in rocks and minerals. 1966.
5. Gebrande, H.: Elastic wave velocities and constans of elasticity of rocks and rock-forming minerals. Landolt-Bornstein, Numerical data and functional relationships in science and technology. 1, Phys. properties of rocks: 1-99, 1982.
6. Babuska, V. and M. Cara: Seismic Anisotropy in the Earth, 1991, Kluwer Academic Publishers, Dordrecht
7. Aleksandrov, K. S. and T. V. Ryzhova: Izv. Acad. Sci. USSR, Geophys.Phys. Solid Earth, 1961, 1165-1168
8. Wooster, W.: Rep. Progr. Physics, 1953, 16, 62-68
9. Christensen, N. I.: J. Geoph. Res., 1966, 71 (4), 3549-3556
10. Crampin, S.: Geophysics, 1985, 50(1), 142-152
11. Kern, H. and M. Fakhimi: Tectonophysics, 1975, 28, 227-244
12. Fielitz, K.: Geophys., 1971, 37, 943-956
13. Kern, H., in: High Pressure Researches in Geoscience, W. Schreyer (ed.), Schweizerbart´sche, Stuttgart, 15-45
14. Birch, F.: J. Geophys. Res., 1961, 66, 2199-2224
15. Kern, H., R. Schmidt and T. Popp: Scientific Drilling, 1991, 2, 130-145
16. Christensen, N. I.: J. Geophys. Res., 1965, 70(24), 6147-6164
17. Kern, H. and H. R. Wenk: J. Geophys. Res., 1990, 95 (B7), 11213-11223
18. Kern, H.: Phys. Earth Planet. Int., 1993, 78, 245-256
19. Holbrook, W. S., W. D. Mooney and N. I. Christensen, In: Continental Lower Crust, (Fountain, Arculus and Kay, Eds.), 1992, Elsevier Science Publishers, 1-43
20. Babuska, V.: Geophys. J. R. astr. Soc., 1984, 76, 113-11
21. Christensen, N. I.: Geophys. J. R. astr. Soc., 1984, 76, 89-111
22. Seront, B., D. Mainprice and N. I. Christensen: J. Geophys. Res., 1993 (in press)
23. Christensen, N. I. and R. S. Crosson: Tectonophysics, 1968, 6 (2), 93-107
24. Kern, H., T. Popp and R. Schmidt: Surv. Geophys., (subm.)

25. Crosson, R. S. and J. W. Lin: J. Geophys. Res., 1971, 76 (2), 570-578
26. Baker, D. W. and N. L. Carter, in: Flow and fracture of rocks., Heard, Borg, Carter and Raleigh., Eds.), 1972, Geophys. Monogr. am. Geophys. Un., 157-166
27. Peselnick, L., A. Nicolas and P. R. Stevenson: J. Geophys. Res., 1974, 79, 1175-1182
28. Siegesmund, S., T. Takeshita and H. Kern: Tectonophysics, 1989, 157, 25-38
29. Bunge, H. J.: Texture analysis in material science, 1982, Butterworths, London
30. Kern, H.: Die Geowissenschaften, 1988, 9, 257-264
31. Mainprice, D., M. Casey and S. Schmid: Mémoires de la Société Géologique de France, 1990, 156, 85-95
32. Lloyd, G. E., C. C. Ferguson and R. D. Law: Tectonophysics, 1987, 135, 243-249
33. Barruol, G. and H. Kern: Geophys. J. Int., (subm.)
34. Silver, P. G. and W. Chan: Nature, 1988, 335, 34-39
35. Kern, H.: Phys. Earth Planet. Int., 1993, 79, 113-136
36. Silver, P. G. and W. W. Chan: J. Geophys. Res., 1991, 96(B10), 16429-16454
37. Shih, X. R., R. P. Meyer and J. F. Schneider: Geology, 1991, 19, 807-810
38. Gebrande, H., Bopp, M., Meichelböck, M. and Neurieder, P.: KTB-Report, 1990, 90-4, 531
39. Lüschen, E., Söllner, W., Hohrath, A. and Rabbel, W.: KTB-Report, 90-6b/Dekorp Reports, 85-134
40. Kumazawa, M. and O. L. Anderson: J. Geophys. Res., 1969, 74(25), 5961-5980
41. Kumazawa, M.: J. Geophys. Res., 1969, 74(25), 5973-5980
42. Aleksandrov, K. S., T. V. Ryzhova and B. P. Belikov: Sov. Phys. Crystallogr., 1963, 8, 738-741
43. Volarovich, M. P., Bayuk, E.I. and Effimova, G.A.: Elastic properties of minerals at high pressure (in Russian), 1975, Nauka, Moscow
44. Aleksandrov, K. S. and T. V. Ryzhova: Izv. Acad. Sci. USSR, Geophys.Phys. Solid Earth, 1961, 1339-1344
45. Belikov, B. P., Aleksandrov, K.S. and Ryzhova, T.V.: Elastic properties of rockforming minerals and rocks (in Russian), 1970, Nauka, Moscow
46. Aleksandrov, K. S. and T. V. Ryzhova: Acad. Sci. USSR, Geophys. Ser, 1962, N°2, 186-189.
47. Ryzhova, T. V. and. Aleksandrov., K.S.: Izv. Acad. Sci, USSR, Geophys. Ser., 1965, 12, 1799-1801
48. Ryzhova, T. V.: Akad. Nauk. USSR, Bull. Geophys. Ser., 1964, 7, 633-635
49. Aleksandrov, K. S., U. V. Alchikov, B. P. Belikov, B. I. Zaslavskii and A. I. Krupnyi: Izv. Acad. Sci. USSR, Geol. Ser., 1974, 10, 15-24
50. Huntington, H. B., in: Solid State Physics, (Seitz, Ed.), 1958, Academic Press, New York, 213-285
51. Hearmon, R. F. S.: Adv. Phys., 1956, 5, 523-382

Materials Science Forum Vol. 157-162 (1994) pp. 1537-1542

THE INFLUENCE OF MICROSTRUCTURE AND TEXTURE ON FRACTURE TOUGHNESS IN TITANIUM ALLOY CORONA-5

S. Benhaddad, C. Quesne and R. Penelle

Laboratoire de Métallurgie Structurale, URA CNRS 1107,
Bât. 413, Université Paris-Sud, F-91405 Orsay Cédex, France

Keywords: Corona 5, Fracture Toughness, Microstructure, Texture

ABSTRACT:

The different routes studied for the present alloy lead to various microstructures and textures. They consist of a forging followed by an homogeneization and a tempering. All the studied states present microstructures which can be classified in accordance with the morphology of the α phase: equiaxed, lamellar or mixed and a texture with two components evolving with forging conditions and thermal treatments. Comparaison of evolution of crystallographic and morphological anisotropy with the value of toughness J1c has respectively shown effects of these two parameters. The two favorable factors to high values of toughness are a lamellar structure and a texture near the orientation {1010}<1120>, obtained after forging in the β field. The equiaxed structure more ductile with a texture {1010}<0001> obtained by forging in ($\alpha+\beta$) field, lead to low values of J1c.

INTRODUCTION:

The influence of heat and thermomechanical treatments on mechanical properties in α/β titanium alloys was carried out for several alloys. As the titanium alloys present structural heredity phenomena, these treatments govern the final microstructure and the crystallographic texture. Therefore, mechanical characteristics of titanium alloys vary with features of α and β phases.
For each transformation route, the aim of the study was on one hand to characterize the resulting α crystallographic texture and on the other hand to relate these textures to the corresponding fracture toughness values. In the same way the effect of the microstructure on fracture toughness evolution has been performed.

EXPERIMENTAL PROCEDURE:

On the same material, whose nominal composition is Ti-4.5Al-5Mo-1.5Cr, six transformation routes were carried out. These processes involve one forging at different temperatures and one annealing at two different temperatures followed by an ageing [1].

Table 1: Transformations studied

Forging	Field	Low αβ	High αβ	High αβ	β	β	High β
Annealing	Field	Low αβ	High αβ	Low αβ	High αβ	Low αβ	High αβ
				AGEING			
Studied process		A	B	C	D	E	F

The fracture toughness J1c was measured after each process with partial unloadings on CT-15 samples [2]. After the last unloading, each specimen was maintained for 10 minutes at 500°C to oxidize the crack propagation zone. Then all CT specimen were broken in liquid nitrogen.
The texture was determined from four poles figures (p.f), obtained by X.ray diffraction or neutron diffraction, orientations distribution function (O.D.F) was calculated using Roe notation.
For both microscopic observations of structures and p.f measurements, the plane considered was always the one parallel to crack plane of the CT specimens.

RESULTS AND DISCUSSION

1. Microscopic observations
Microscopic observations of different microstructures reveal effects of two parameters:
-forging temperature
-annealing temperature
* Three fields of forging temperature give three different microstructures (Figures 1a,b & c):
-the process A results in primary α- phase nodular structure within (α+β) acicular structure. The primary α phase becomes nodular with the sub transus forging.
-the processes D, E and F induce an acicular structure inside prior β whose grain boundaries are marked by primary α phase.
-the bimodal structure (process B and C).is obtained by forging at high temperature in (α+β) field.

*Comparatively to the structure of process C (low annealing), the structure of process B (high annealing) contains less of residual β phase. During cooling, the β phase transforms in acicular (α+β) because the low concentration of beta elements.

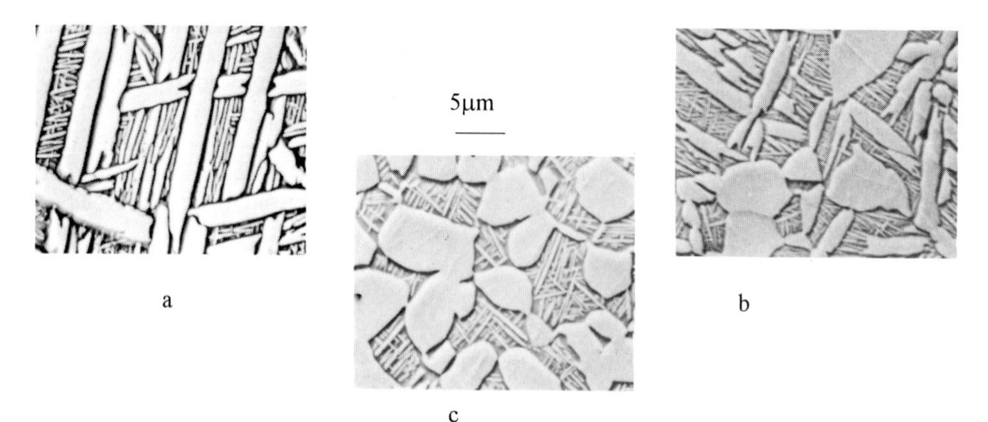

Figure 1: Microstructures resulted after: a) β forging, b)high (α+β) forging, and c)low (α+β) forging

2.Textures

All p.f (0002) of studied specimens have a similar cross arrangement of (0002) reinforcements (Figure2). The textures are represented by the principal orientation {hkil}<uvtw> where:

{hkil} is the crystallographic plane parallel to crack plane (f,t) and <uvtw> is the crystallographic direction parallel to crack propagation direction df.

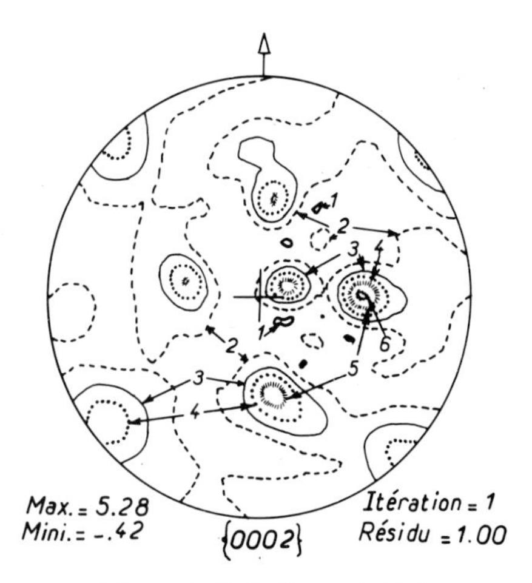

Figure 2: Pole figure (0002)

These textures have a mixed character with two components Cdt and Cdf.. The Cdt component corresponds to basal planes whose <0001> axis are in the (n,t) plane. The Cdf component corresponds to <0001> axis in the (n,f) plane.

The intensity of these two components varies with both solution treatment and forging temperatures. The angle γ between the axis <0001> of every component and the normal axis n vary also with these two temperatures.

Table 2:Texture components.

Process	Cdf	γ(°)	F(g)max	Cdt	γ(°)	F(g)max
A	{2021}<1013>	78	3.00	---------------	-----	-------
B	{3032}<1013>	68	3.30	{1012}<1210>	42	2.20
C	{2023}<1011>	50	7.10	{2023}<1210>	50	8.50
D	{1012}<1011>	42	8.90	{1012}<1210>	42	10.60
E	{0001}<1120>	0	6.00	{1011}<1210>	60	8.10
F	{1103}<2201>	32	3.50	{1101}<1120>	60	3.20

3.Toughness test:

The J1c value of each process is the average of three tests. These values are given with an error about $\pm 15 kJ/m^2$..

Table 3:Fracture toughness

Process	A	B	C	D	E	F
J1c(kJ/m^2)	113	125	144	145	152	166

Fracture toughness evolution is given in figure 3 according to forging temperature of each process. Two graphs correspond respectively to one of these two groups:

-high annealing
-low annealing

These two graphs have the same evolution: higher the forging temperature, better is the fracture toughness. Acicular structure is obtained after forging at high temperature in the β field. This tangled lamellar structure of the primary α phase reduces crack propagation and so leads to high values of J1c (3). Conversely, a nodular shape of α phase (process A) does not deviate very much crack propagation. Finally, bimodal structures (B and C) present intermediate values of fracture toughness.

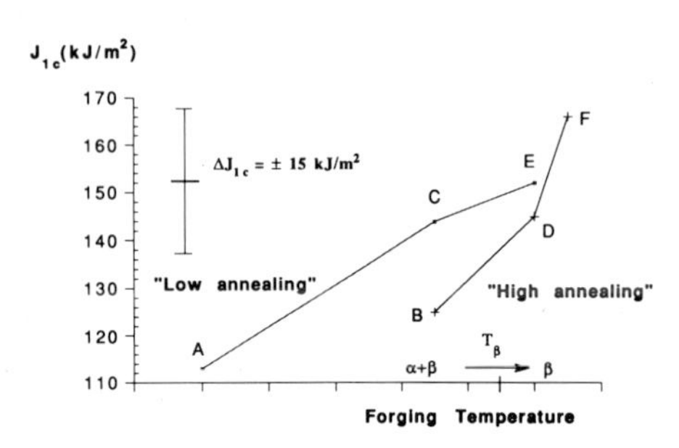

Figure 3: Fracture toughness variation with forging temperature.

*Usually, (α+β) titanium alloys present a ductile fracture at room temperature. This fracture proceeds in three steps: nucleation, growth and finally coalescence of microvoids which leads to the final cracking.

It is very important to consider any element which affects fracture toughness for each of these three steps. Outside the microstructure, two other parameters have to be considered:

-A difference of stress triaxiality amount between an α nodule and an α lamella can have an influence on microvoid growth.

-the crystallographic texture of the α phase.

In a recent study in titanium alloy Ti-6246 (Ti-6%Al-2%Sn-4%Zr-6%Mo), A.S.Béranger [4] has not found any difference between stress triaxility rates of α nodules and α lamellae. So, in this study we consider only crystallographic texture and its influence on fracture toughness.

So the fracture mechanism of the present material seems depend strongly on preferential orientation of crystals. Figure 4 shows fracture toughness J1c evolution as a function of textural components Cdt and Cdf for each process. The processes are separated in "high annealing" and "low annealing", to show influence of annealing treatments on fracture toughness.

A similar evolution of these two graphs can be noticed: nearer the Cdt component to the transverse texture {1010}<1210>, higher is the fracture toughness.

The crystallographic plane of equivalent single crystal parallel to crack plane and crystallographic direction parallel to crack direction, allow one indixes to determine the active slip systems at the crack tip.

The Cdt and Cdf components evolve inversely with forging temperature to the respective orientations {1010}<1210> and {1010}<0001>. The first orientation is most favorable to a deformation with prismatic slip systems, which are preponderant in titanium. The second orientation favours pyramidal slip. These slip systems have a high resolved shear stress, so they involve a less important plastification and a lower fracture energy, so the fracture toughness is lower than in the first case.

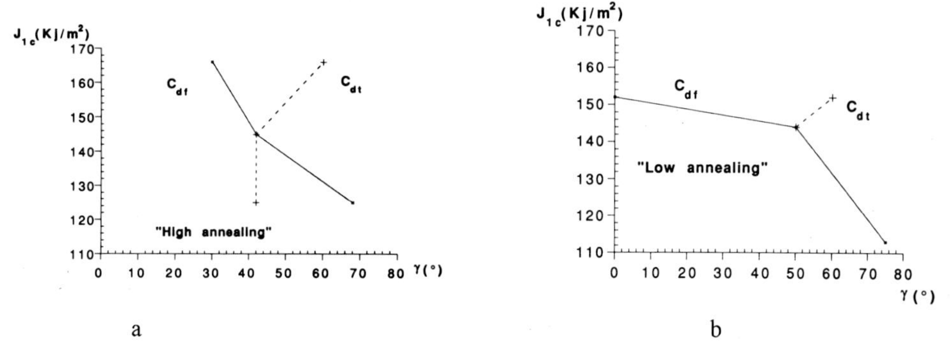

a b

Figure 4: Fracture toughness variation with different components of texture.

Several authors, particularly Tchorzewski et al [5] have observed a similar evolution in some other alloys. Bowen [6,7] gave another explanation for the low toughness values of the {1010}<0001> orientation, he calculated a critical resolved shear stress for this orientation for prismatic and pyramidal glide systems. He found that the <1120> slip is always the most prefered slip mode. In the present work it can be noticed that:

-process F and E lead to highest fracture toughness values, with a Cdt component {1011}<1210> near to the transverse orientation.

-process D leads with respect to process E to a Cdt component far from the transverse orientation. The same results are obtained for all "high annealing" with respect to corresponding "low annealing". This difference from transverse orientation reduces fracture toughness of "high annealing".

-process A and B have close textures with principal component Cdf and same crack direction <1013>. These orientations do not enable a strain accommodation on the specimen width and then involve a low fracture toughness.

-process C and D present similar fracture toughness values because they have a silmilar Cdt component.

CONCLUSION

Both structures and textures evolve in the same way with different process. They give jointly their contribution to fracture toughness variation. A distinction between effects of each feature was difficult.

1.Acicular structure leads to the highest fracture toughness. A tangled laths structure improves fracture toughness but a nodular structure decreases it. The α nodules present a low resistance to crack propagation.

2.The acicular structure presents a texture near to transverse orientation. This orientation favours prismatic glide which contributes to high toughness.

3.A forging in low (α+β) field leads to a nodular structure and to a texture component favourable to deformation on pyramidal slip system. Both features explain the low fracture toughness due to this process.

4.Forgings in high (α+β) field lead to a bimodal structure and two texture components. The structures differ in the α and β phase proportion and in textures. So they present different toughness.

5. After a same forging, a "low annealing" in ($\alpha+\beta$) field gives a higher proportion of acicular primary α phase than a "high annealing" and a texture similar to a β forging. Thus, the "low annealing" gives rise to an higher fracture toughness than a "high annealing".

REFERENCES:

1.S.Benhaddad Thèse Doctorat Paris XI (1992)
2.G.A.Clarke, W,R.Andrews, P.C.Paris and D.W.Shmidt A.S.T.M STP 590, 27 (1976)
3.F.H.Froes, J.C.Chesnutt, C.G.Rhodes et J.C.Williams A.S.T.M STP 651 (1978) 115
4.A.S.Béranger Thèse Doctorat U.T.C (1991)
5.R.M.Tchorzewski and W.B.Hutchinson Metall.Trans 9A, 08 (1978) 113
6.A.W.Bowen Acta Met 26 (1978) 1423
7.A.W.Bowen 6th World Conference on Titanium, Ed.P.Lacombe, R.Tricot et G.Béranger Cannes (France) 1988, 93

Materials Science Forum Vol. 157-162 (1994) pp. 1543-1550

ORIENTATION DEPENDENCE OF THE PERMEABILITY IN TEXTURED SOFT MAGNETIC MATERIALS

M. Birsan and J.A. Szpunar

Department of Metallurgical Engineering, McGill University,
3450 University Street, Montreal, Quebec H3A 2A7, Canada

Keywords: Soft Magnetic Material, Permeability, Anisotropy, Domain Structure

Abstract: The permeability at H = 10 Oe is calculated for a polycrystalline soft magnetic material at any direction in the sheet plane. The theoretical development is based on the tendency of individual crystallites to take a domain structure to reduce the magnetostatic energy. The averaging procedure uses the ODF as a weight function. The predicted results describe well the particularities of two different textured samples.

INTRODUCTION

The crystallographic texture in the most important commercial soft magnetic materials determines their magnetic properties. The advantages of these textured materials led to intensive studies to improve their permeability and reduce their power losses.

It is generally agreed that the permeability or induction measured at H = 10 Oe (hereafter B_{10}) is linked to the angular distribution of grains. Extensive work on this subject has concluded that B_{10} is generally insensitive to impurity content, internal stress and grain size, and is only dependent on the statistical distribution of the grain orientation [1]. These results made it possible to calculate B_{10} as function of texture parameters.

Investigations on the relationship existing between B_{10} and the average deviation of grain orientation from (110)[001] texture in 3% Si-Fe [1] has found that B_{10} decreases non-linearly with increasing angular spread. B_{10} was measured only in rolling direction (RD) which is the symmetry axis for the spread of angular distribution of grains. Other papers [2],[3] give an analytical expression for the magnetization of the specimen at the knee of the magnetization curve and we will use these results to analyze the permeability at 10 Oe in a polycrystalline, textured material. Up to this date, no investigation which takes into account the realistic orientation distribution of grains has been undertaken to study the anisotropy of B_{10} in polycrystalline 3% Si-Fe.

DISCUSSION

Crystallographic texture is defined by the orientation distribution function (ODF) of the crystallites in the material. Using this function as a parameter characterizing the material (weight function) the anisotropy of physical properties of polycrystals may be predicted from the corresponding properties of the individual crystallites. In some cases, e.g.

magnetocrystalline energy, the average value of the single crystal properties is analytically obtained. In this paper it is assumed that polycrystals contribute to the property in direction β proportional to the volume fraction V_i of crystals oriented in this direction and the averaging is numerically done.

To calculate the magnetization in a cubic single crystal it is necessary to know the distribution of domains among the six easy directions of magnetization for each value of the applied field. Considering a strip of oriented iron placed in a magnetic field (H = 10 Oe) lying in the plane of the sheet, one introduces a constraint on the domain structure such that only the three easy directions closest to the applied field can be occupied. This field cannot rotate the magnetization vector out of the easy direction because it is much smaller than the anisotropy field (H_{an} = 560 Oe for silicon iron). In order to avoid high demagnetizing energy in each crystallite, the magnetization distributes itself between the easy directions so that the component of magnetization perpendicular to the specimen surface is zero .

In this case the volume fractions magnetized along each of the easy directions can be uniquely determined. The net component of magnetization B_p of the specimen parallel to the applied field is only a function of crystal orientation:

$$\frac{B_p}{B_s} = \frac{1}{(\alpha_1 + \alpha_2 + \alpha_3)} \tag{1}$$

where the applied field has direction cosines α_1, α_2 and α_3 with the three closest easy directions, and $B_s = 4\pi I_s$, I_s being the saturation magnetization.

One can generalize the above relation for a known distribution of grain orientations. If one grain orientation is given by the Euler angles (ϕ_1,Φ,ϕ_2) and the internal induction vector **B** is parallel to the applied field direction **H**, the rotation matrix [g] defined in [4] can be used to determine α_1, α_2 and α_3 :

$$\begin{aligned}
\alpha_1(\beta) &= |(\cos\varphi_1 \cos\varphi_2 - \sin\varphi_1 \sin\varphi_2 \cos\phi) \cos\beta + \\
&\quad +(\sin\varphi_1 \cos\varphi_2 + \cos\varphi_1 \sin\varphi_2 \cos\phi) \sin\beta \,| \\
\alpha_2(\beta) &= |(-\cos\varphi_1 \sin\varphi_2 - \sin\varphi_1 \cos\varphi_2 \cos\phi) \cos\beta + \\
&\quad +(-\sin\varphi_1 \sin\varphi_2 + \cos\varphi_1 \cos\varphi_2 \cos\phi) \sin\beta)| \\
\alpha_3(\beta) &= |(\sin\varphi_1 \sin\phi \cos\beta - \cos\varphi_1 \sin\phi \sin\beta)|
\end{aligned} \tag{2}$$

where β is the angle that **H** makes with the sheet rolling direction.

With these direction cosines (1) gives the magnetization of grain "i" having orientation $g_i = (\phi_1, \Phi, \phi_2)_i$ and the volume fraction V_i :

$$B_i(\beta) = V_i \frac{B_s}{[\alpha_1(\beta) + \alpha_2(\beta) + \alpha_3(\beta)]} \tag{3}$$

$$\textit{with:} \quad \sum V_i = 1$$

The sum of the contribution from all the grains weighted by their volumes gives the overall magnetization of the polycrystal in the direction of the magnetic field :

$$B_{10}(\beta) = B_s \sum_i \frac{V_i}{[\alpha_1(\beta) + \alpha_2(\beta) + \alpha_3(\beta)]_i} \quad (4)$$

The volume fraction V_i is related to the ODF intensity $f(g_i)$ by the following equation:

$$V_i = \frac{1}{8\pi^2} f(g_i) \sin\phi_i \, \Delta\varphi_1 \, \Delta\phi \, \Delta\varphi_2 \quad (5)$$

Because of the ODF normalization, one can write the final relation:

$$B_{10}(\beta) = B_s \frac{\sum_i \frac{f(g_i) \sin\phi_i}{[\alpha_1(\beta) + \alpha_2(\beta) + \alpha_3(\beta)]_i}}{\sum_i f(g_i) \sin\phi_i} \quad (6)$$

It should be interesting to compare the results given by the above equation with the fourth-order texture parameter $F_4(\beta)$ [5]:

$$F_4(\beta) = \frac{1}{\sqrt{2}} C_4^{11} \overline{P_4^0}(90^\circ) + C_4^{12} \overline{P_4^2}(90^\circ) \cos 2\beta + C_4^{13} \overline{P_4^4}(90^\circ) \cos 4\beta \quad (7)$$

where $P_4^i(90°)$ are the normalized associated Legendre functions and C_4^{11} , C_4^{12} and C_4^{13} are the ODF coefficients calculated from standard X-ray pole figure measurements. Since a linear relationship (in the first approximation) exists between B_{10} and F_4, the variation in permeability (ΔB_{10}) is proportional to the variation in $F_4(\beta)$, (ΔF_4). If two test values of permeability are available, the following equation can be used to calculate the magnetization in β_j direction:

$$B_{10}(\beta_j) = B_{10}(RD) + [F_4(\beta_j) - F_4(RD)] \frac{\Delta B_{10}}{\Delta F_4} \quad (8)$$

EXPERIMENT

Two different qualities of Goss oriented steels were investigated. Magnetization measurements for the grain oriented 3% Si-Fe samples were provided by Dofasco Research Laboratory. The ODFs were calculated from the (110), (200) and (211) X-ray pole figures (Schultz technique). The $f(g_i)$ results are in the form of a three dimensional matrix with equal steps $\Delta\phi_1 = \Delta\Phi = \Delta\phi_2 = 5$ deg. These values of $f(g_i)$ were used to calculate $B_{10}(\beta)$ according to equation (6).

The two samples A (good quality) and B (poor quality) have only one texture component, i.e. (110)[001] orientation. The main differences between the two samples are illustrated in Table 1 using the normalized ODF intensities close to the ideal (110)[001] = (0°,45°,0°) orientation.

Table 1.:

ϕ_1	Φ	ϕ_2	A	B
0	45	0	76	379
5	45	0	69	346
10	45	0	26	213
0	40	0	59	44
0	50	0	52	65

The calculated and experimental permeabilities are plotted in fig.1. and 2. for samples A and B respectively. It should be stressed that the values experimentally obtained are in good agreement with the ones describing the theoretical magnetization. Some differences may be caused by statistical reliability of measurements: the experimental magnetic permeability is the average value of 10 different Epstein strips while the sample for X-ray ODF determination is about 4 cm^2 large. For both samples the theoretical and experimental data have the same features: in sample B the maximum permeability is not in the RD as experimentally tested, while in sample A, which has less sharp texture, the permeability is higher with the maximum in the RD. The major contributing factor to these differences is the different amount of material near the ideal (110)[001] orientation for these two samples.

For example, one can define the relative amount of material at orientation g = (10°,45°,0°) as the ratio between the ODF intensity at (10°,45°,0°) and the intensity at (0°,45°,0°). This ratio has the value of 0.34 for sample A and 0.56 for sample B and it determines the value of permeability near the rolling direction.

In comparison, the values of permeability calculated according to equation (8) and the experimental data, are plotted in figs.3 and 4 for sample A and B respectively. It was assumed that B_{10}(RD) and B_{10}(TD) are the two values to be used for appropriate calibration. As one can see, $F_4(\beta)$ describes with reasonable accuracy the experimental data (if two or more values of test permeability are known). However, the $F_4(\beta)$ calculation does not predict the shift of maximum from RD.

CONCLUSION

A formula for grain oriented sheets has been developed to calculate the magnetization at 10 Oe as function of the direction of the applied field (β) using the ODF as a weight function. This formula is derived directly from the magnetization of a crystal having a certain orientation in the sample reference frame. Relation (6) gives the absolute value of permeability, for different directions. It is suitable to describe magnetic materials having any degree of texture. The formula may be applied in the study of soft magnetic materials where the dependence of magnetization in the direction of applied field is important.

It has been shown that the amount of material near the ideal (110)[001] orientation influences both the value and the position of permeability maximum.

In the first approximation the fourth-order texture parameter characterizes well the variation of permeability with angle β. Since the formula of $F_4(\beta)$ includes only three ODF expansion coefficients it cannot describe the details of the $B_{10}(\beta)$ curve.

REFERENCES:

1. M.F.Littmann, J.Appl.Phys. 38, 1104 (1967).
2. J.W.Shilling and G.L.Houze Jr., IEEE Trans.Mag., M10, 195 (1974).
3. D.J.Craik and D.A.McIntyre, IEEE Trans.Mag., M5, 378 (1969).
4. H.J.Bunge, Texture Analysis in Materials Science, Butterworths, London 1982.
5. P.R.Morris and J.W.Flowers, Texture of Crystalline Solids, vol.4, 129 (1981).

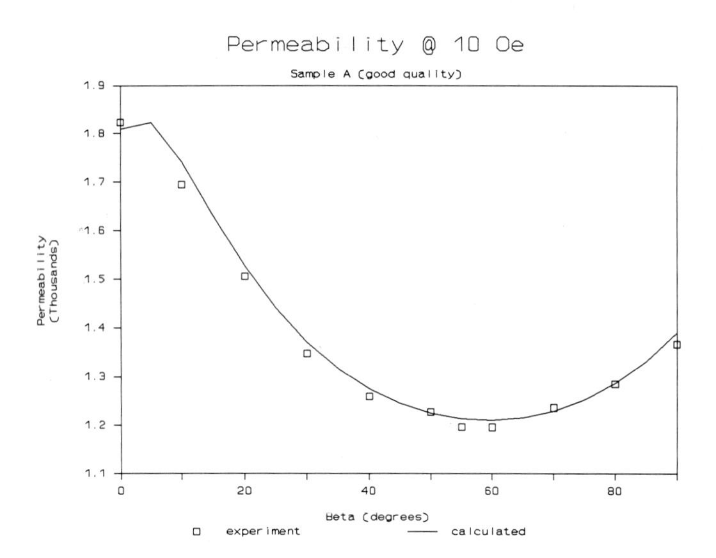

Fig.1. Data for sample A (good quality) derived from eq.(6).

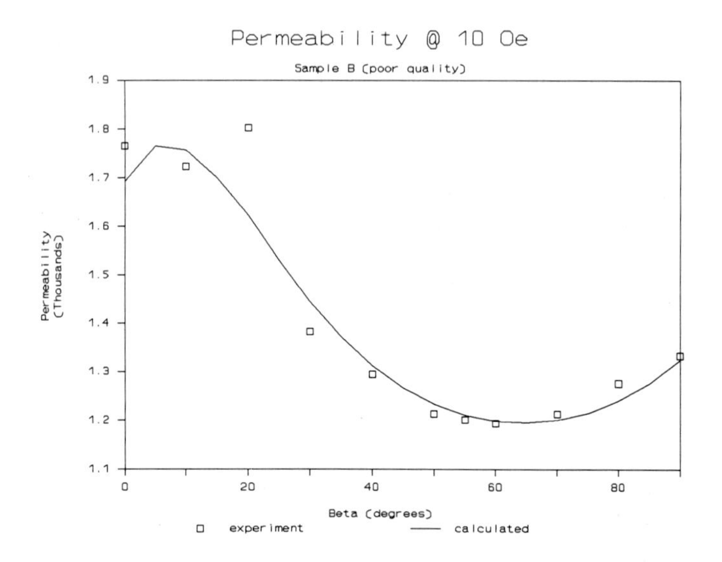

Fig.2. Data for sample B (poor quality) derived from eq.(6).

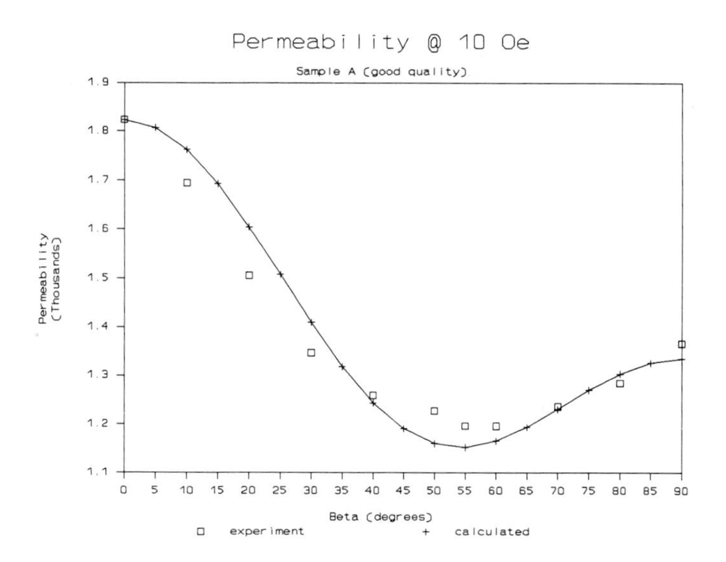

Fig.3. Data for sample A compared to the fourth-order texture parameter eq.(8).

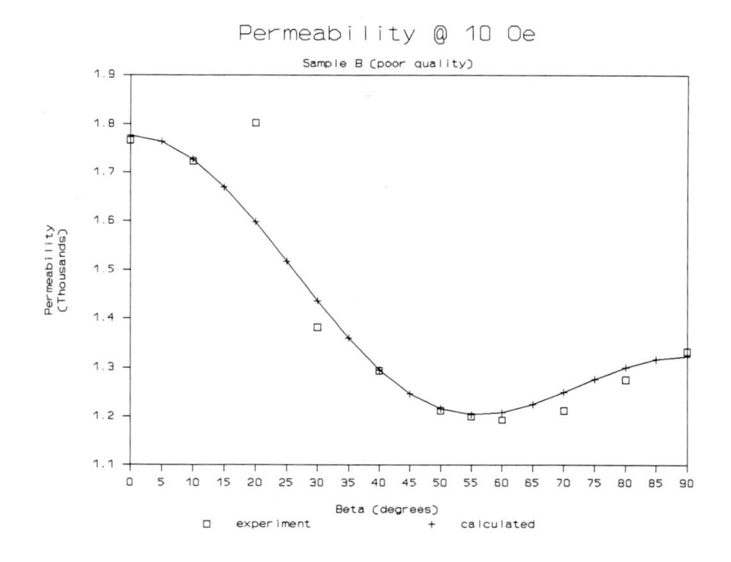

Fig.4. Data for sample B compared to the fourth-order texture parameter eq.(8).

Materials Science Forum Vol. 157-162 (1994) pp. 1551-1558

ANOMALIES OF YOUNG'S MODULUS IN Fe-Cu COMPOSITES AFTER HIGH DEGREES OF DEFORMATION

W. Böcker, H.J. Bunge and T. Reinert

Institut für Metallkunde und Metallphysik der TU,
Grosser Bruch 23, D-38678 Clausthal-Zellerfeld, Germany

Keywords: Composites, Plastic Deformation, Recovery Annealing, Residual Stresses, Young's Modulus, Nonlinear Elasticity Theory, Texture Changes

ABSTRACT

The development of Young´s modulus after strong cold plastic deformation in Composites of iron and copper is studied. A strong increase of Young's modulus with maximum values up to 30 % above the expected value according to the rule of mixture is observed. After annealing at 200° C this anomaly disappears nearly completely. The effect cannot be attributed to texture changes during annealing which was checked by neutron diffraction texture measurements. The effect can, however, be understood in terms of the stress-dependence of Young's modulus combined with internal stresses of opposite sign in copper and iron i.e. in terms of non-linear elasticity theory.

INTRODUCTION

Several metallic two-phase composites are showing an anomalous increase of Young´s modulus after strong cold plastic deformation. This effect was observed at the first time in Fe-Ag sheets after cold rolling [1,2] . It was estimated that texture sharpenig during deformation is not able to cause changes of Young´s modulus up to 40% .
Since this effect disappears after recovery annealing, it thus may be understood in terms of internal stresses of opposite sign in the order of 1% of Young´s modulus, taking the stress-dependence of the elastic properties into account [3]. In order to study this effect, Fe-Cu composites were produced by extrusion and deformed by wire drawing [4,5].

EXPERIMENTAL PROCEDURE

Iron-copper composites and the pure materials were prepared from powders with particle sizes of ~ 13 μm (Fe: BASF-CF) and ~ 32 μm (Cu: NA-FS) [4,5]. The powders were compacted and hot extruded at 850°C with a deformation degree of η = 93.35%. After that, the rods were cold drawn with true strains up to φ = 5 (corresponding to η = 99.3%).

Finally, a part of the samples was given recovery annealings at 200° C, 220°C and 250° C, respectively.
Young´s moduli were measured using the "Elastomat" (Förster) with longitudinal as well as transversal vibration modes. The accuracy of the measurements was estimated to about 0.1%.
Texture measurements were carried out by neutron diffraction using the texture diffractometer TEX-2 at the research reactor FRG-1 at GKSS Geesthacht [6]. Pole densities of the reflexes (111) (200) (220) of copper and (110) (200) (211) of iron were measured in steps of $\Delta\alpha = 5^0$. Orientation distribution functions (ODF) were calculated, using the series expansion method with an expansion degree of L=22. Because of the axial symmetry, these textures are completely represented by inverse pole figures, calculated with the same method.

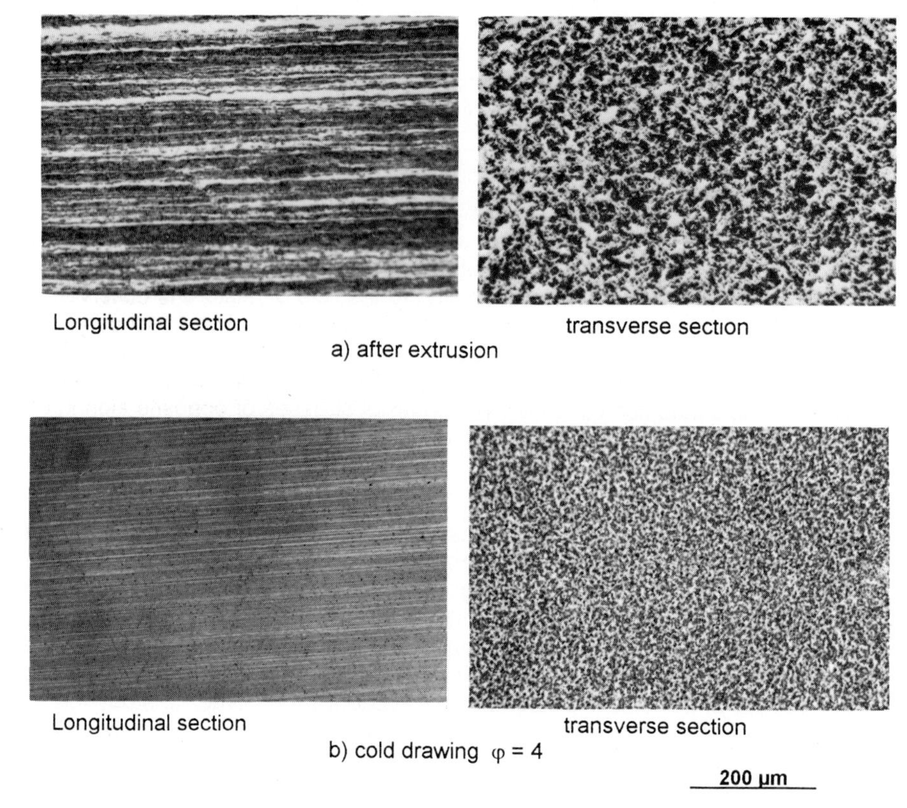

Fig. 1: Microstructure of the composite Fe50Cu50

RESULTS

The structure of the samples was studied metallographically. Fig.1 shows the longitudinal and the transverse sections of an extruded rod and of a cold drawn wire of the composition Fe50Cu50. The diameters of the particles are in the order of 1 to 10 µm, whereas the fibre lengths are in the mm-range.

Young´s modulus as a function of the deformation degree is shown in Fig. 2 for iron, copper and the composition Fe50Cu50. In all cases a strong increase at higher deformation degrees occurs. Fig.2 also contains the values of Young´s modulus measured subsequently after annealing at 200°C, i.e. without recrystallization. It is seen that the heat treatment eliminates the mayor part of the anomalies.

Young´s modulus of the undeformed, the deformed, and the annealed state is plotted in Fig. 3 as a function of composition. This figure shows more clearly that the mayor part of the anomalies is eliminated by annealing at 200° C, although a smaller part still remains. The relative changes of Young´s modulus E/E_0 with respect to the value E_0 of the undeformed state are shown in Fig. 4. The maximum of ~ 1.3 is reached at a composition of Fe75Cu25. Annealing at 200° C eliminates most of the effect but even after 250° C small differences still prevail.

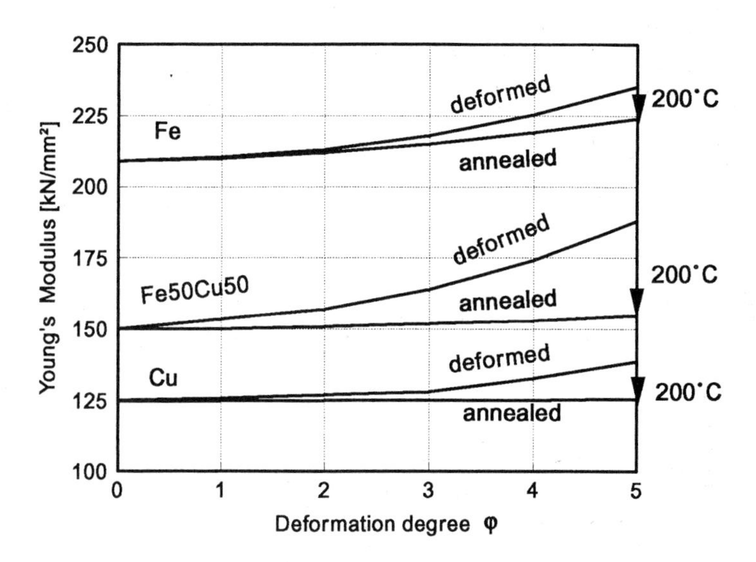

Fig. 2: Young´s Modulus as a function of true strain as well as after subsequent annealing 1 h at 200°C

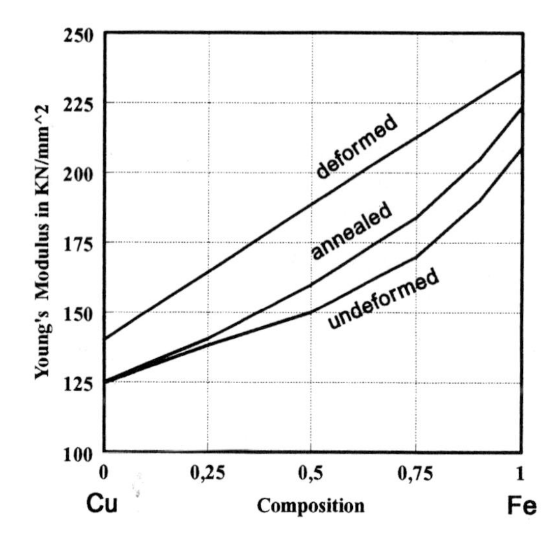

Fig. 3: Young´s modulus of the undeformed state, the deformed state (φ = 5) and after annealing for 1 h at 200°C as a function of composition

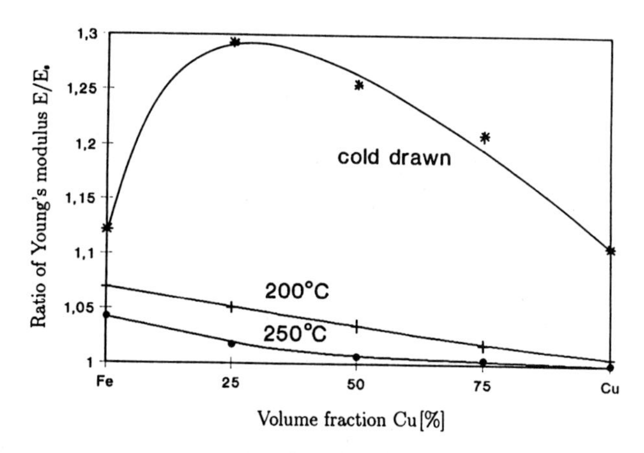

Fig. 4: Young´s modulus ratio E/E_0 of the deformed state as well as after annealing at 200°C and 250°C as a function of composition. E_0 is the value before cold deformation.

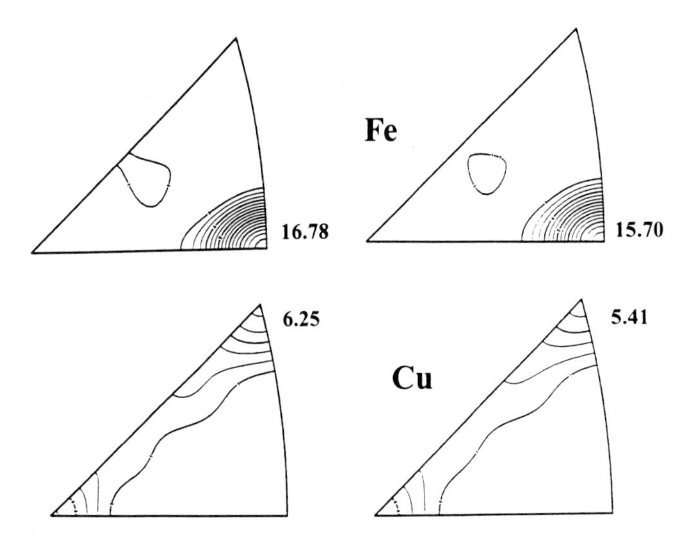

Fig. 5: Inverse pole figures of the copper and iron phase of a composite Fe50Cu50 after deformation φ = 4 and after subsequent annealing at 220°C

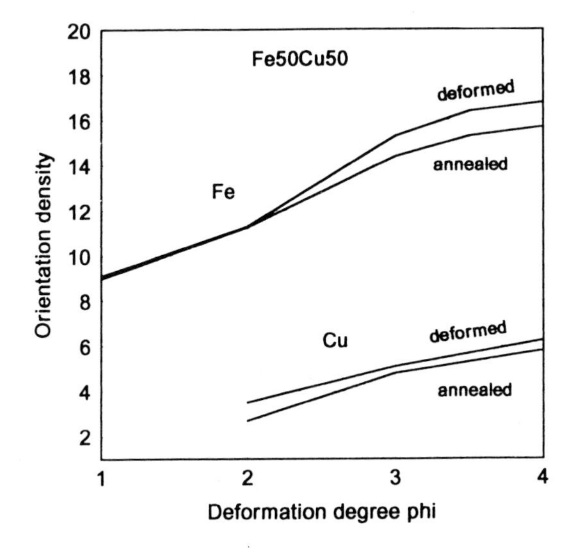

Fig. 6: Maximum pole densities of the copper and iron texture of a composite Fe50Cu50 after deformation and after subsequent annealing at 220°C

The textures of both phases are shown in Fig. 5 in terms of inverse pole figures. The iron phase exhibits the well known {110} texture whereas the copper phase shows the double fibre texture {100} + {111}.
These texture types are found in all cold drawn composites as well as in the pure materials. It is seen that they remain virtually unchanged after annealing at 220° C.
Texture development during deformation and subsequent annealing is shown in Fig.6 for the composite Fe50Cu50 as an example. Similar results were also found in the other composites. The textures of both phases remain essentially constant within the limits of experimental error.

DISCUSSION OF THE RESULTS

hardness tests it can be concluded that annealing at 200° C does not lead to recrystallization. Hence, the texture developed by deformation remains essentially constant as it is proved by Figs. 5,6. Nevertheless, the greatest part of the anomalies of Young´s modulus introduced by plastic deformation is eliminated. This definitely excludes texture as the reason of the anomaly and strongly corroborates residual stresses which are relaxed by annealing.
In a single phase polycrystalline material and even more so in a two-phase material residual stresses as a function of the position in the material will be developed after plastic deformation. In the absence of external stresses positive and negative values must balance each other such that the integral vanishes.
The elastic constants are known to be stress-dependent. In the presence of an internal stress σ_r the actual value $E(\sigma_r)$ can be expressed in the form

$$E(\sigma_r) = E_0 + \frac{\delta E}{\delta\sigma} \cdot \sigma_r \tag{1}$$

Hence, the (Voigt) average E_r of the elastic constants in the presence of internal stresses $\sigma_r(\vec{r})$ is given by

$$E_r = \frac{1}{V} \int E_0(\vec{r})\, dV + \frac{1}{V} \int \frac{\delta E(\vec{r})}{\delta\sigma} \cdot \sigma_r(\vec{r})\, dV \tag{2}$$

The <u>first term</u> in eq (2) is the conventional volume average (Voigt approximation) of the local elastic constants. Thereby E_0 depends on the position $\vec{r}$ according to the local phase (Fe or Cu) as well as on the local crystallographic orientation (i. e. on the texture). This term is assumed to describe the undeformed state and the mayor part of the annealed state in Figs. 3, 4, 5. The difference between undeformed and annealed state may be attributed to texture changes during deformation.

The <u>second term</u> in eq (2) contains the local variations of the elastic constants due to residual stresses. Although the stresses themselves must integrate up to zero this is not required for the product of the stresses with the stress-sensitivity factor. In fact, the stress-sensitivity factor is different in Fe and Cu, and because of crystal anisotropy it also depends on the local crystal orientation within each of the phases. Hence, the second term in eq (2) is, in general, not zero.
This is illustrated for a simple case (consisting of equal amounts of the phases α and β and constant residual stresses within them) in Fig. 7.

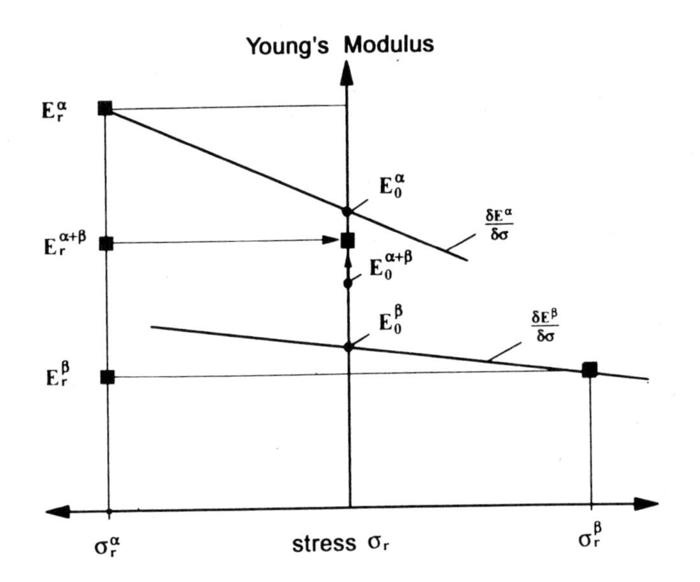

Fig. 7: Young´s modulus of the phases α and β as well as of a composite $\alpha 50\ \beta 50$ with and without internal stresses

This second term is thus assumed to be responsible for the difference in Young´s modulus between deformed and annealed state. (It must be admitted that annealing at 200° C may not have eliminated the residual stresses σ_r completely. Hence, the difference between undeformed and annealed state may still contain a contribution from residual stresses.)

According to Fig. 4, this second term contains a large contribution from the two-phase nature of the material. The residual stresses assumed to be responsible for this part are thus stresses of first kind (which are constant within the polycrystalline particles of the Cu- and Fe-phase, respectively).

As it is, however, also seen in Fig. 4, the single phase samples Fe and Cu, too, show an effect attributed to the second term of eq (2). In this case, residual stresses of second kind varying with crystal orientation are assumed to be responsible.

In principle, also stresses of third kind varying within the crystallites themselves must be taken into consideration. The present measurements, however, do not allow to distinguish between these two latter kinds of stresses.

Taking measured values of the third order elastic constants into account it can be estimated that residual stresses of the order of one or two percent of Young´s modulus are needed to explain the observed anomaly effects. Residual stresses of this order are well within the experimentally established range.

CONCLUSIONS

The present results show clearly, that texture changes are not responsible for the observed changes of Young´s modulus in two-phase composites after strong plastic deformation and particularly the subsequent elimination of these changes by annealing at recovery temperatures.
On the other hand, the virtually complete elimination of the incremental term of Young´s modulus after recovery is in favour of a model based on residual stresses and on the stress-dependence of elastic properties in terms of non-linear elasticity theory.

ACKNOWLEDGEMENT

This work has been supported by the Bundesminister für Forschung und Technologie (BMFT) under the contract number 03-BU3CLA A.1-K18.

REFERENCES

[1] P. I. Welch, L. Ratke, G. Wassermann
The Effect of Deformation on Young´s Modulus in Iron-Silver Sheet Composites
In: Proc. 7-th Intern. Conf. on Textures of Materials, Netherlands Soc. Mat.
Eds. C. M. Brakman, T. Yongenburger, E. Mittemeyer (1984) 675 - 680

[2] L. Ratke
Young´s Modulus of Iron-Silver Sheet Composites
Textures and Microstructures (1989), 10, 227 - 242

[3] H. J. Bunge
Model Calculations of Elastic Moduli in Highly Deformed Composites
Textures and Microstructures (1989), 10 , 153 -164

[4] R. Beusse, W. Böcker, H. J. Bunge
Anomalies of Young´s Modulus in Highly Deformed Iron-Copper Composites
Scripta Met. et. Mat. (1992), 27, 767 - 770

[5] W. Böcker, R. Beusse, H.G. Brokmeier, H. J. Bunge
Influence of Residual Stresses on Young´s Modulus in Fe-Cu Composites
In: Proc. Europ.Conf. on Residual Stresses
DGM Informationsgesellschaft, Oberursel (1993)

[6] Opoku,M.
Diploma Work, TU Clausthal (1992)

Materials Science Forum Vol. 157-162 (1994) pp. 1559-1564

STRESS BIREFRINGENCE IN TEXTURED SILVER CHLORIDE

P. Dietz and H. Gieleßen

Institut für Maschinenwesen, TU Clausthal,
Robert-Koch-Str. 32, D-38678 Clausthal-Zellerfeld, Germany

Keywords: AgCl, Silver Chloride, Photoplasticity, Photoelasticity, Crystal Photoelasticity, Transparent Metal, Texture Modelling

ABSTRACT

Experimental and theoretical studies have been carried out in order to visualize elastic anisotropy by means of photoelasticity. A simple method has been found to grow doubly refracting single-crystals of silver chloride, which then have been cut in various orientations. These were submitted to tensile stress and relative retardation and extinction angles observed in polarized monochromatic light to show conformity to theories which quantitatively relate the state of stress and optical phenomena. Single crystals then have been processed further to obtain quasi-isotropic fine-grained specimens as a well defined starting point for texture investigations. Textures of cold-rolled as well as of recrystallized silver chloride specimens were determined using an X-ray goniometer. Texture determining parameters as degree of rolling and recrystallisation time have been varied.

INTRODUCTION

For about a hundred years photoelastic studies of all kinds of machine parts have been carried out in order to evaluate stress distributions. It was and still often is not possible to measure stresses in regions below the surface of the part directly. Most components have a far too complicated shape to allow for analytical calculation and even numerical computation is restricted to simplified and discrete models. For this reason analogies have been developed which permit use of transparent models which, when loaded by external or internal forces, allow investigation of stresses in inner parts of the structure. All of the commonly used model materials have an amorphous structure so that they are only suitable for studying problems of elasticity theory in an homogeneous and isotropic body as they are homogeneous and isotropic in their structure, mechancial and optical properties themselves. The majority of materials used in engineering however are metals and thus of crystalline structure. Anisotropic properties have normally been induced by various types of thermal and mechanical processing these parts have usually been submitted to. Another drawback of photoelastic materials, which are mostly plastics of different kinds, is that their mechanical behaviour above the yielding point is totally different from yield of metals. Plastic deformation of polymers takes place as breaking, forming and rearranging of the polymeric chains. There are no slip-planes, twinning or work-hardening. Furthermore deformation is dependant of load times and strain rates. For this reasons there was the need of a transparent material exhibiting textural effects, slip-planes, work-hardening, recrystallisation, workability and sufficient stress-optical sensivity. Materials to meet all these requirements had to be crystalline. In [1] silver and thallium halide salts were described for the first time to be suited for photoelastic studies. Stepanov [2], who was the first one to give a description of the preparation of silver chloride specimens for photoelastic purposes formed the term "transparent metals" for these substances. As they are of cubic structure they are not naturally doubly refracting and their birefringence gives a direct picture of magnitude and direction of internal stresses. These transparent metals could be rolled, forged, cut, annealed and recrystallized just like metals. Also texture formation seemed to take place in the same way. However the early researchers were interested in textural effects only because they wanted to avoid them. A method was suggested how to obtain nearly texture-free polycrystalline specimens [3] which was developed further by Weber [4]. In the 1960s the interest in silver chloride decreased because of its difficult handling. The material is sensitive to visible light and therefore used in photography. Besides it is strongly corrosive to most metals. The Institut für Maschinenwesen started its research on silver chloride in the late 70s when there was the need for a stress-optical material with high plasticity to simulate metal forming processes. At present time our work concentrates on the creation of mechanical anisotropy in silver chloride that is similar to anisotropy in rolled aluminium, to relate this to certain textures and to formulate the fundamentals of photoelasticity with textured material.

PREPARATION OF SILVER CHLORIDE SPECIMENS

For the stress-optical investigations to be carried out silver-chloride specimens were needed both in monocrystalline and small-grained polycrystalline form. While the procedure to obtain monocrystalline material is quite standard a way had to be sought to produce fine-grained polycrystaline bars with reproducable structure. From earlier investigations [5] it was apparent that casting silver chloride does only produce material with a well defined structure and of good optical quality when the casting was done in chlorine. So casting was not considered further and a certain sequence of mechanical and thermal processing was chosen to transform single crystals into polycrystals.

The raw material was bought as white granules which were then compacted into cylindrical pellets until sintering occured. This considerably reduced shrinking during the melting process and also the formation of silver oxides. A number of pellets with a total weight of approx. 350 grams were filled into crucibles of gastight Al_2O_3 and of quartz glass with an internal diameter of 30 and 34 mm respectively. Crucibles of Al_2O_3 were annealed at 900°C for 12 hrs to remove impurities and quartz crucibles were cleaned with a solution of 1% HF and 15% HNO_3. At first melting was performed in normal atmosphere. Later we switched to argon which considerably improved the quality of the ingots. The melting furnace was of Bridgman-Stockbarger type with a maximum temperature of 525°C. The rates of lowering the container with the melt were 7.5 and 3.7 mm/hr. Impurities of the raw material were collected as a layer of dross at the upper end of the ingot. The size of this layer which was initially 5-7 mm could be reduced to a thin film during further development of our technique. Analysis showed that the film contained small amounts of Al, Mg, Fe, Si and possibly K and As in case of an ingot grown in an Al_2O_3 crucible. Independent of the material of the crucibles each of them had to be destroyed to detach the ingot. Figure 1 shows a monocry-

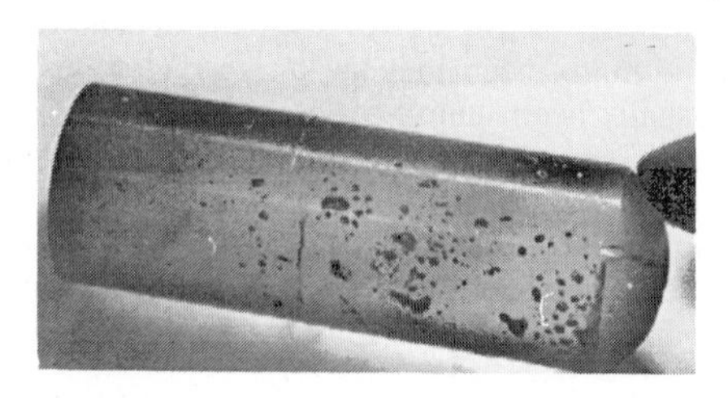

Fig.1: Silver chloride ingot after turning off on the lathe

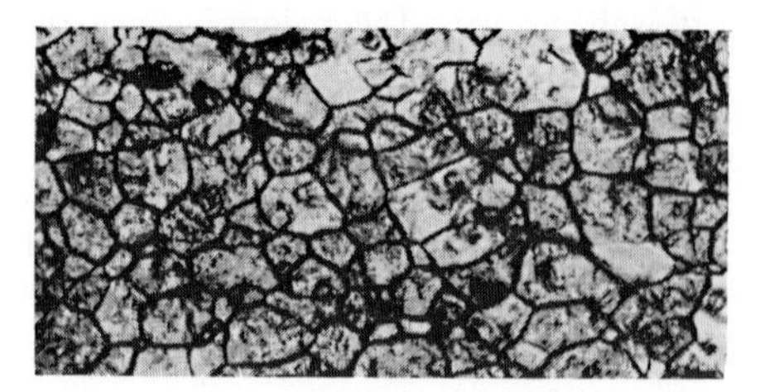

Fig.2: Polycrystalline silver chloride (mag. 200)

stalline ingot of 3 cm diameter and 10 cm length. In general ingots grown in quartz containers showed less impurities but were more difficult to free from the crucible. Due to the adherence of the ingots and the high thermal expansion coefficient of silver chloride there was a considerable amount of residual stress found after cooling which, if not removed by annealing, caused the ingots to crack within a few days. Transparency and homogeneity of the crystals was good, there was only a slight reddish or yellow colouring. These single-crystals were then cut in different crystallographic orientations. Grinding and polishing was performed on a round table polishing machine using SiC abrasive paper and diamond paste. Alkohol was used as lubricant. Because of the softness of silver chloride extreme care had to be taken not to deform the surface during the preparation process.

There seems to be no easier way to obtain single crystals of sufficient size. Attempts to crystallize dissolved silver chloride failed as the size of the crystallites did not exceed 1mm. Elecrolytical precipitation is not possible. Polycrystalline material for texture studies was produced by extrusion of monocrystalline material and alternating cold rolling and recrystallisation. By extrusion in a 5.6-ton hydraulic press bars of a cross section of 12 x 30 mm were obtained. These were recrystallized at a temperature of 105°C. Multiple passes of colled rolling and subsequent recrystallisation produced polycrystalline material with grain sizes of less than 0.1 mm as is shown in figure 2.

STRESS-OPTICAL EFFECT IN SILVER CHLORIDE

To be able to relate photoelastic effects and texture it is necessary to quantify the relationship of the stress state and the photoelastic effects i.e. relative retardation and extinction angle. For this reason single-crystal studies have been carried out. In general, the directions of the principal stress, the principal axes of the indicatrix or of the Fresnel-ellipsoid, the crystal axes, the directions of the incident wave normal and the orientation of the sample do not coincide. It is no restriction to choose the sample, stress and the wave normal directions to be the same and also to measure crystal orientations with reference to these which makes coordinate transformations simpler. In this case one only has to know the orientation of the Fresnel-ellipsoid and the crystal orientation in relation to the sample coordinate system. The extinction angle is measured by an ordinary polariscope and the crystal orientation by an x-ray goniometer. The angle between the Fresnel-ellipsoid and the sample depends on the stress level. Thus it is necessary to determine the relative phase retardation. This is achieved by the compensation method of Sénarmont which is an in-built feature of most polariscopes.

A plane-parallel plate of silver chloride in a plane state of stress is to be considered. The plate is cut in an arbitrary way, so there are three coordinate systems to take into account. These are denoted as XYZ - crystal coordinates, X'Y'Z' - system of principal stresses with Z' perpendicular to the plate surface and X''Y''Z'' - coordinate system of the optical ellipsoid. The Fresnel ellipsoid expressed in the X'Y'Z'-system has the general form

$$B'_{11}x'^2 + B'_{22}y'^2 + B'_{33}z'^2 + 2B'_{23}y'z' + 2B'_{13}x'z' + 2B'_{12}x'y' = 1 \qquad (1)$$

For the plane case with Z' = 0 this simplifies to an ellipse of the form

$$B'_{11}x'^2 + B'_{22}y'^2 + 2B'_{12}x'y' = 1 \qquad (2)$$

The B'_{ii} describe the intersection of the ellipse and the coordinate system whereas B'_{12} is a measure for the angle α which in this case is the optical isocline or the extinction angle.

$$\tan 2\alpha = \frac{2B'_{12}}{B'_{11} - B'_{22}} \qquad (3)$$

The Fresnel-ellipsoid and the stress state, which is an ellipsoid as well, have to be related to each other. This requires a system of 36 stress-optical coefficients c_{ij} in the general case. Because of cubic symmetry only three coefficients remain. With the assumption that the changes of the coefficients B'_{ij} of the ellipsoid are linear functions of the stresses [6] these coefficients can be expressed as

$$\begin{aligned} B'_{11} - v_0^2 &= c'_{11}\sigma_1 + c'_{12}\sigma_2 \\ B'_{22} - v_0^2 &= c'_{12}\sigma_1 + c'_{22}\sigma_2 \\ 2B'_{12} &= c'_{16}\sigma_1 + c'_{26}\sigma_2 \end{aligned} \tag{4}$$

Herein the σ_i are the pricipal stresses, v_0 is the speed of light in unstressed AgCl and the c_{ij}' are functions of the three photoelastic coefficients c_{11}, c_{12} and c_{44} in the crystal coordinate system. If these are substituted into (4) and then into (3) α is expressed as a function of the photoelastic coefficients, the principal stresses and the direction cosini of the angles between the crystal and the sample coordinate system all of which are known.

The angle α of the optical isocline gives the planes of polarization of the two waves emerging from the crystal. The relative path difference of these two waves can be measured by a compensation method by adding or subtracting an artificial path difference until extinction occurs for a particular wavelength. When this is done in monochromatic light an accuracy of better than 0.02 fringe orders can be achieved. The path differences as well as the optical isocline depend of the state of stress as was stated earlier. Therefore the relation between the path differences and the state of stress has to be formulated. If we write (4) in the optical coordinate system X''Y''Z'' we obtain

$$\begin{aligned} B''_{11} - v_0^2 &= c''_{11}\sigma_{x''} + c''_{12}\sigma_{y''} + c''_{16}\tau_{x''y''} \\ B''_{22} - v_0^2 &= c''_{12}\sigma_{x''} + c''_{22}\sigma_{y''} + c''_{26}\tau_{x''y''} \\ 2B''_{12} &= 0 = c''_{16}\sigma_{x''} + c''_{26}\sigma_{y''} + 2c''_{26}\tau_{x''y''} \end{aligned} \tag{5}$$

The B''_{ii} represent the speeds of the polarized light waves emerging from the crystal. According to [5] they can be expressed as

$$\begin{aligned} B''_{11} &= v_0^2(1 + \frac{p}{v_0^2}\sigma_{x''} + \frac{q}{v_0^2}\sigma_{x''}) \\ B''_{22} &= v_0^2(1 + \frac{q}{v_0^2}\sigma_{x''} + \frac{r}{v_0^2}\sigma_{x''}) \end{aligned} \tag{6}$$

where p, q and r are functions of the photoelastic coefficients and the orientation of the optical coordinate system in relation to the crystallographic system. The speeds of the light waves then have to be expressed in terms of refraction indices. Transforming stresses back into principal stresses leads to the conditional equation (7) for the relative optical path difference δ.

$$\delta = d\frac{V}{2v_0^3}(((p-q)\cos^2\alpha - (r-q)\sin^2\alpha)\sigma_1 + ((p-q)\sin^2\alpha - (r-q)\cos^2\alpha)\sigma_2) \tag{7}$$

V is the speed of light in vacuum and d is the thickness of the sample. With these relations the plane state of stress can be calculated from photoelastic measurements or on the other hand photoelastic constants of the material can be calibrated.

EXPERIMENTAL STUDY OF SINGLE-CRYSTALS UNDER TENSILE STRESS

The silver chloride single-crystals that were used were 30 mm long, 10 mm wide and 3 mm thick. Small clamps to hold the specimens were constructed. They had the special feature that the contact pressure was proportional to the tensile load in order not to deform the fixing points more than was necessary. These were mounted into a loading frame together with a mechanical load-sensing device. The load was increased in steps of 50 N to a limit of 15 N/mm^2. One of the samples cracked when reaching this limit. Plastic flow started at 5-6 N/mm^2. Figure 3 shows two systems of slip bands when the specimen was loaded just above the yield point. The angle between the two systems is 64°. The (001)-direction was tilted about 13° from the sample normal direction. At higher loads birefringent bands appeared which were parallel to one of the two slip systems. The intersections of the slip planes with the specimen surface form a dense pattern of surface lines if the slip planes are not parallel to the surface.

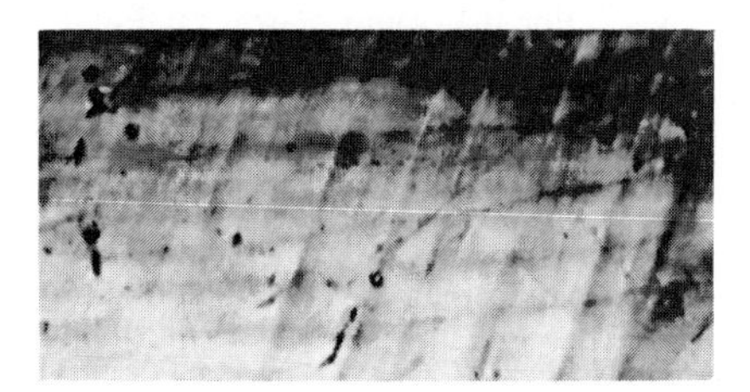

Fig.3: Onset of plastic deformation

Fig 4: Glide lines and two birefringent bands

Figure 4 shows surface lines and two birefringent bands in a specimen that was loaded nearly until fracture. The angle between the two slip systems (one of them is hardly visible due to photographic reproduction) is 58°. This is quite close to the theoretical value of 60°. Load direction was horizontal in this case. The fact that these birefringent bands can be seen only in certain parts of the sample with different densities and contrast is an indication of different glide planes being active. Those birefringent bands are interpreted as remaining stresses between successive glide zones. Thus their contrast and visibility differs with their inclination to the sample normal as they might overlap and their phase retardations average. From similar experiments Nye [7] states that glide direction in silver chloride is always ‹110› but that glide planes do not exactly coincide with low-indexed crystallographic planes but are in most cases near {111}.

The loading frame with the specimens was placed in a magnifying projection polariscope and extinction angles and relative retardation measured. While the load was increased in steps the strains in both tranverse direction were measured. Shearing of the sample occured in all directions. Therefore the deformation of the specimen's cross section was measured as well and the true stresses calculated. Figure 5 shows the curves of stress versus relative retardation for three samples. They show a good linearity even in the plastic range which begins at a stress rate of 5 N/mm^2. As strain increases superproportional with the load this indicates that the photoelastic effect depends on stress rather than on strain which is contrary to most photoelastic polymers.

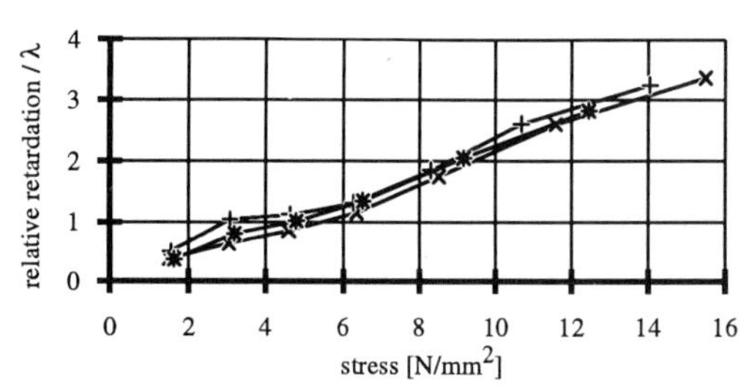

Fig 5: Relative retardation at λ=546nm versus stress of specimens K1, K12 and K18

TEXTURE DEVELOPMENT BY ROLLING AND RECRYSTALLIZATION

Silverchloride single-crystals have been processed with the method described above in order to produce polycrystalline samples. By alternated rolling in different directions with small degrees of reduction the single crystals were cracked into material without significant initial texture. This was checked optically by submitting the specimens to a constant load and observing the isochromatics pattern in a polariscope while changing the angle of the planes of the polarized light. Equal patterns for all different angles indicated that the samples were of optical quasi-isotropy. The so prepared specimens were then colled rolled with degrees of rolling of up to 92% and part of them recrystallized at the recrystallization starting temperature of 105° from 3 to 96 hrs.

Figure 6 shows a (200) pole figure of a specimen that was rolled in 15 subsequent steps from 12 to 4.6 mm thickness. Further rolling increases the sharpness of the texture to a certain degree. At even higher degrees of rolling the texture weakens again. As can be seen from the center of the pole figure the material is still too coarse-grained for exact analysis. For that same reason ODF-calculation does not show clear results at the moment as peaks do not exceed two times random distribution. There seems to be a certain resemblance to the ODF of cold rolled and annealed Fe30Ni sheets as given in [8] but this has to be investigated further. The formation of recrystallization textures has been investigated by annealing specimens of different degrees of rolling for different times at a temperature of 105°C. Up to a degree of rolling of 50% no cube texture occurred, independent of annealing time. At 75% there was a clear cube texture which was even stronger at 92%. Figure 7 shows the cube texture after 3 hours of recrystallisation. In general the sharpness of the texture increased with the annealing time. This is in good accordance to [9] et.al. where the same is stated for different fcc metals. The spread of the cube texture around RD was found to be about 20° and less than 15° around TD for the specimen with the highest degree of rolling and the longest annealing time. This is somewhat higher than would be expected in the case of metals but the ratio of the spreads is similar.

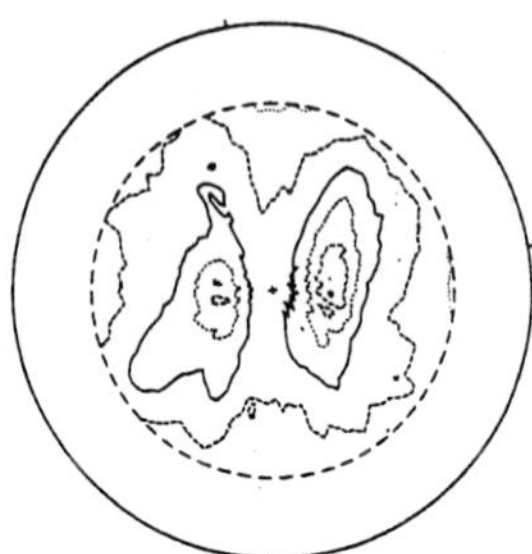

Fig 6: (200) pole figure of cold rolled AgCl

Fig 7: AgCl recrystallization texture

CONCLUSIONS

Progresses in the application of the technique of photoelasticity have been presented in this paper. For the first time metal working processes have been used to create mechanical and optical anisotropy in a doubly refracting material in order to simulate mechanical anisotropy in metal parts. Though research is still at its beginning it is apparent that texture formation in silver chloride takes place in the same way as it does in technical metals. Further development will concentrate on the description of stress-birefringence in dependence of the texture and the relation of mechanical and optical anisotropy.

REFERENCES

1) Tamman, G.: Naturwissenschaften, 1932, 20, 958
2) Stepanov, A.V., Zhitnikov, R.A.: Soviet Physics - Technical Physics, 1956, 754
3) Stepanov, A.V., Zhitnikov, R.A.: Soviet Physics - Technical Physics, 1956, 769
4) Weber, H.J.: Eigenschaften des spannungsoptisch aktiven Modellwerkstoffes Silberchlorid und seine Anwendung auf die Untersuchung von Spannungszuständen mit teilweise plastischer Verformung, Dissertation, Clausthal, 1986
5) Moeller, R.D., Schonfeld, F.W., Tipton, C.R., Waber, J.T.: Transactions of the ASME, 1951, 43, 39
6) Krasnov, V.M., Stepanov, A.V.: J. Exp. Theor. Phys. (USSR), 1952, 23, 199 (in russian)
7) Nye, J.F.: Proc. Royal Soc. London, 1949, A198, 190
8) Bunge, H.J., Esling, C.: Quantitative Texture Analysis, Oberursel, 1982, 456
9) Wassermann, G., Grewen, J.: Texturen metallischer Werkstoffe, Berlin, 1962, 307

Materials Science Forum Vol. 157-162 (1994) pp. 1565-1570

LIMITED FIBRE COMPONENTS IN TEXTURE ANALYSIS: TEXTURE CONTRIBUTION IN ANISOTROPY OF PHYSICAL PROPERTIES

V.N. Dnieprenko and S.V. Divinskii

Institute of Metal Physics, Ukrainian Academy of Sciences,
36 Vernadsky str., Kiev 142, 252 680 Ukraine

Keywords: Simulation, Limited Fibre Components, ODF, Properties Anisotropy

ABSTRACT: The present approach is based on a mathematical treatment of the Wassermann's hypothesis about texture description by a superposition of limited fibre components. With this aim in view, an own local coordinate system, which is rigorously determined by the component texture axis, is introduced for each texture component. In this case, a grain orientation $g_B=g_B(\varphi_1,\Phi,\varphi_2)$ in *k*-th component is determined by three independent sets of Eulerian-type angles $G = g_0^{(k)}\cdot g_\gamma\cdot g_1^{(k)}$, where the $g_0^{(k)}$ and $g_1^{(k)}$ are some constants. Spread of the *k*-th limited fibre component is in fact formed by rotations on $g_\gamma=g_\gamma(\gamma_1,\gamma_2,\gamma_3)$. Basing on this, the calculations of anisotropy of elastic properties for copper are carried out, with contributions of different components in total anisotropy being determined also for each separately.

Anisotropy of properties of polycrystalline aggregates is determined by a texture, or by a distribution of structure components on orientations. Thus, the general analytic presentation of the texture is to be known to predict the properties. Up to now, faithful averaging can be carried out only in two limiting cases, namely for single crystals and for random distributions of grains on orientations.

The only numerical method, which is based on series expansions of experimental pole figures, has been derived for calculation of the anisotropy of polycrystalline aggregates with texture [1]. Nevertheless, this technique can not be used for prediction purposes, since in such an approach one can not handle a quantitative composition of textures.

With this aim in view an analytic presentation of the texture should be used. Despite of the great number of suggested up to now models the textures, *e.g.* the copper texture, are not faithfully simulated. This concerns especially the regions with "abnormally" high pole density on the {111} pole figures of the copper-type textures. At these points the pole density reaches 10-15 in terms of random distribution. Moreover, different authors point out different number of texture components.

This ambiguity may be eliminate by taking into account of a correlation of the texture and microstructure. In the case of the fibre textures of aluminum Vandermeer [2] carried out electron microscopic investigations of the dislocation structures and established that the <100> and <111> components are characterized by different microstructures. In much the same way for copper rolling, a careful investigation of the dislocation structure with the simultaneous determination of orientations of regions having the characteristic microstructure has been done [3]. Investigations of the sections normal to plane rolling reveal that four different types of the dislocation structures are present. The type of dislocation structure was characterized mainly by the cell dimensions and their orientations with respect to the external directions of a sample. However, two of the four which correspond to the orientations close to {110}<1$\bar{1}$2> and {112}<11$\bar{1}$> are characteristic for the most part of the grains (up to 90%); therefore, it is quite natural to first concentrate our attention only on these components and to describe the copper texture based mainly on them. This is possible, in principle, with the Wassermann approach [4], which uses the hypothesis of partial axiality of rolling textures. In this case the regions mentioned above having an abnormally high pole density on the {111} pole figures of copper rolling are due to the positions of the axes of the partial axial components (PACs). Nevertheless, up to now this approach has been developed only schematically.

THEORY

Specific directions, namely the axes of PACs, exist when textures are treated as the superposition of the partial axial components. Moreover, the particular crystallographic position of the given axis corresponds to each texture component. Hence, along with the sample's coordinate system, K_S, we introduce the "local" coordinate system, K_T, which is connected with the texture axis of the component selected. We then assume that the $0z$ axis of K_T coincides with the direction of the texture axis and, based on that assumption, the grain orientation will be characterized by the following set of angles of the Eulerian type:

- $g_o^{(k)} = (\Psi_o^{(k)}, \Theta_o^{(k)}, \varphi_o^{(k)})$, which specifies transition from K_S to $K_T^{(k)}$ for the k-th component;
- $g_\gamma = (\gamma_1, \gamma_2, \gamma_3)$, which specifies the grain spread relatively to the texture component axis;
- $g_1^{(k)} = (\Psi_1^{(k)}, \Theta_1^{(k)}, \varphi_1^{(k)})$, which specifies the position of the k-th PAC relatively to the coordinate directions of the sample. The set of angles $G^{(k)} = (\Psi_o^{(k)}, \Theta_o^{(k)}, \varphi_o^{(k)}, \gamma_1, \gamma_2, \gamma_3, \Psi_1^{(k)}, \Theta_1^{(k)}, \varphi_1^{(k)})$ introduced by us and the set of the Eulerian angles g_B usually applied in texture analysis and which have been determined in the sample's coordinate system K_S are connected by the relation

$$g_B = g_o^{(k)} \cdot g_\gamma \cdot g_1^{(k)} . \tag{1}$$

Assuming that the spread of partial axial textures depends separately on γ_2 and $\gamma_1 + \gamma_3$, we may consider that the first angle, γ_2, characterizes the radial spread on the angle which has been counted from the texture axis; the second combination, or $\gamma_1 + \gamma_3$, characterizes the spread on the angle of rotation around the texture axis. Taking the functions of spreads as Gaussian ones, and allowing for the possibility of the existence of regions with a constant orientation density (the orientation tubes, in other words), we may write the orientation distribution function (ODF) $f^{(k)}$ of the k-th PAC as:

$$f^{(k)} \sim \exp\left\{-\frac{1}{2}\left[\frac{\gamma_2}{o_1^{(k)}}\right]^2\right\} \cdot \exp\left\{-\frac{1}{2}\left[\frac{\big||\gamma_1+\gamma_3|-o_3^{(k)}\big| + |\gamma_1+\gamma_3|-o_3^{(k)}}{2\,o_2^{(k)}}\right]^2\right\}, \tag{2}$$

where $o_1^{(k)}$, $o_2^{(k)}$, and $o_3^{(k)}$ are the spread parameters, and, as can be seen from (2), $o_3^{(k)}$ characterizes the extent of the region with a constant ODF.

Taking into account the presence of N texture components, we find the pole density distribution $P_{\{\bar{h}\}}(\bar{y})$ on the $\{\bar{h}\}$-type pole figure from

$$P_{\{\bar{h}\}}(\bar{y}) = \sum_{i=1}^{M} \sum_{k=1}^{N} \frac{p_k}{2\pi M} \int_0^{2\pi} f^{M}\left(\gamma_1^{(ik)}, \gamma_2^{(ik)}, \gamma_3^{(ik)}\right) d\delta , \qquad (3)$$

where $g_\gamma^{(ik)}$ is obtained from the relation:

$$g_\gamma^{(ik)} = \left[g_o^{(k)}\right]^{-1} \cdot g_B^{(i)} \cdot \Omega(\delta) \cdot \left[g_1^{(k)}\right]^{-1} . \qquad (4)$$

Here $g_B^{(i)}$ - is the matrix of rotations for that crystallite orientation where $\bar{y}=\bar{h}_i \cdot g_B^{(i)}$ or, in other words, the direction of $\bar{h}_i$ coincides with $\bar{y}$ after rotations on $g_B^{(i)}=(\varphi_1^{(i)}, \Phi^{(i)}, \varphi_2^{(i)})$, and $\Omega(\delta)$ is the matrix of rotations on angle δ, which specifies the rotation of the crystallite around an axis coinciding with $\bar{h}_i$ when $\bar{h}_i \| \bar{y}$; $\bar{h}_i$, $i=1,...M$ are equivalent planes from the $\{\bar{h}\}$ family, and p_k are the relative amounts of the texture components.

The ODF, f_B, in a space of routine Eulerian angles, g_B, is determined by the following expression:

$$f_B(\varphi_1, \Phi, \varphi_2) = \sum_{k=1}^{N} \frac{p_k}{M_k} \sum_{l=1}^{M_k} f^{M}(\gamma_1^{(kl)}, \gamma_2^{(kl)}, \gamma_3^{(kl)}) , \qquad (5)$$

where $l=1,...,M_k$ is the index consistently numbering the equivalent orientations in the texture component, and M_k is the number of these equivalent orientations. Values of angles $g_\gamma^{(kl)}=(\gamma_1^{(kl)}, \gamma_2^{(kl)}, \gamma_3^{(kl)})$ can be found from (1) in a similar way as in (4) (where the matrix $\Omega(\delta)$ may be simply omitted).

The (2)-(5) expressions can be considered as the mathematical techniques for description of textures within the PAC framework.

A full body of mathematics is stated in [5], where special problems are discussed:

- approximation of distributions on pole figures;
- form of separate maxima on the pole figures;
- choice of the texture axis of a component;
- correspondence with others models in an extreme case of spheric distributions (or $o_1^{(k)} = o_2^{(k)}$, and $o_3^{(k)}=0$ in equation (1));

and others.

RESULTS

In the simulation of the copper-type rolling texture, the iteration approach, constructed on the basis of (2)-(5), shows that the typical copper rolling texture is described mainly by a sum of the $\{110\}\langle 1\bar{1}2\rangle|1\bar{1}1|$ and $\{112\}\langle 11\bar{1}\rangle|111|$ PACs (about 85%). Here the designation $|u'v'w'|$ is introduced to denote the direction of texture axis of the components. Model {111} and {100} pole figures and model ODF are presented in Fig.1 and 2, respectively. Precise qualitative composition of the texture is presented in [6]. At the same place the correlation of the even part of series-expansion of the model ODF with the ODF reconstructed by the Bunge method is discussed.

Now, ODF being known in analytical form, texture contribution in anisotropy of physical properties can be found.

E.g. the texture-average tensor $\langle S_{ijkl}\rangle$ of compliancies will be

$$\langle S_{ijkl}\rangle = \frac{1}{8\pi^2}\int f_{\mathbf{B}}(g_{\mathbf{B}})\cdot\alpha_{im}(g_{\mathbf{B}})\cdot\alpha_{jn}(g_{\mathbf{B}})\cdot\alpha_{kp}(g_{\mathbf{B}})\cdot\alpha_{mq}(g_{\mathbf{B}})\cdot S^{\circ}_{mnpq}\,dg_{\mathbf{B}}, \qquad (6)$$

where $\alpha_{im}(g_{\mathbf{B}})$ are the directional cosines, S°_{mnpq} are the compliancies in the crystallographic coordinate system (single crystal values).

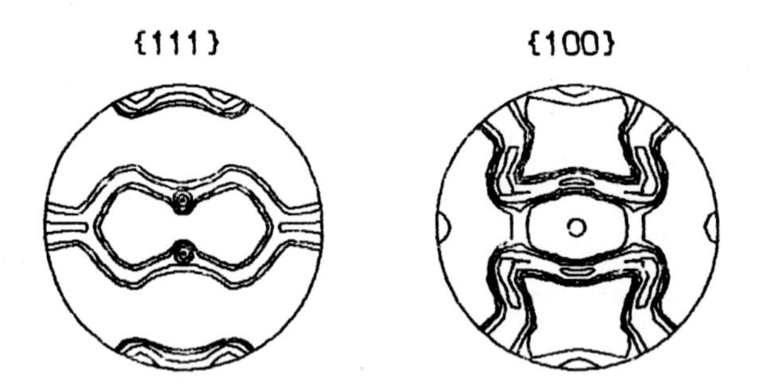

Fig.1. Model pole figures for copper rolling texture

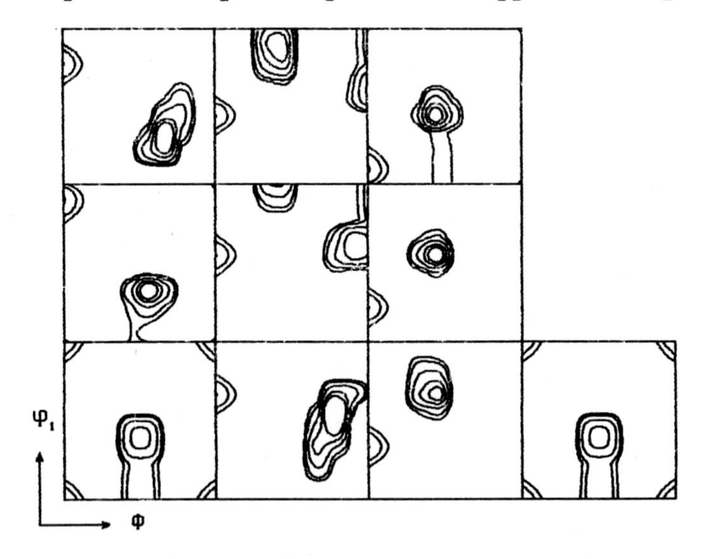

Fig.2. Model ODF for copper rolling texture

The tensor $\langle S_{ijkl}\rangle$ being averaged, the anisotropy of the Young modulus E can be determined. It should be emphasized that the E modulus both for each separate texture component and for the whole of the textured sample can be calculated. Thus, the anisotropy of E for the the $\{110\}\langle 1\bar{1}2\rangle|1\bar{1}1|$ and $\{112\}\langle 11\bar{1}\rangle|111|$ PACs is presented in Fig.3a. Corresponding three-dimensional images of the Young modulus are shown in Fig.3b.

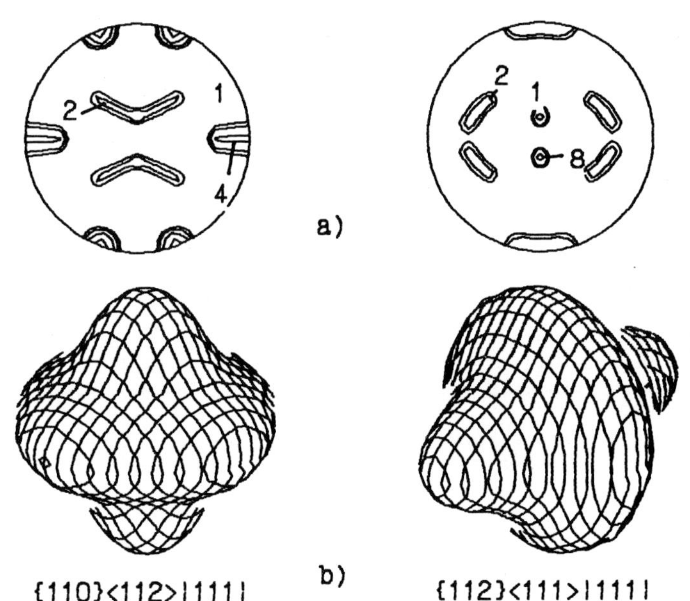

Fig.3. Pole figures (a) and Young modulus (b) for components

The correspondence of the texture components with the structure composition allows to evaluate the fine effects of influence of temperature dependence of defects in different texture components on the elastic properties. In particular, this allowed to explain the behaviour of the Young modulus of rolled copper under low-temperature annealing. Evolution of point defects, which give rise to a dislocation fixing, turns out to be substantially different in two main copper components, namely in $\{110\}\langle 1\bar{1}2\rangle|1\bar{1}1|$ and $\{112\}\langle 11\bar{1}\rangle|111|$. Defect of the Young modulus observed experimentally and calculated theoretically for different components and the whole of the sample are presented in Fig.5a,b, respectively. In these figures the Young modulus defects, which are normalized on their maximum values, are shown. To compute a compliancies defect, which is caused by existence of a dislocation with the Burgers vector b, the following relation was used [7]:

$$\Delta S_{iklm}=\frac{D\cdot l}{4\varkappa}\left(\varepsilon_{ipq}\, b_k+\varepsilon_{kpq}\, b_i\right)\left(\varepsilon_{lrt}\, b_m+\varepsilon_{mrt}\, b_l\right)n_p n_t \tau_q \tau_r, \tag{7}$$

where D is the dislocation density, l is the average length of a dislocation segment, $\varkappa$ is a quasi-elastic factor, ε_{ipq} is the unit pseudo vector, τ is the unit tangent vector.

Computer simulation of the copper rolling texture development shows that in polycrystal copper the formation of the $\{112\}\langle 11\bar{1}\rangle|111|$ component is connected with action of the $\{111\}\langle 110\rangle$ and $\{100\}\langle 011\rangle$ slip system types, and $\{110\}\langle 1\bar{1}2\rangle|1\bar{1}1|$

component is connected with slipping on planes of the only {111} type [8,9]. Thus, it can be supposed that different dislocation assembles exist in the different texture components. Accounting for these facts (for details see [8,9]), it can be shown that anisotropy of the Young modulus defect in the {110}<1$\bar{1}\bar{2}$>|1$\bar{1}$1| and {112}<11$\bar{1}$>|111| components differs. Analysis of the experimental and theoretical data in Fig.4 suggests that annealing of point defects starts at different temperatures in the components.

By the same means, it may be supposed that energy revealing under annealing starts at different time in these components. This is schematically shown in Fig.5.

Therefore, the simulation by the limited fibre components allows not only to model more faithfully the grain distribution on orientations, but also to explain fine effects of evolution of properties of textured materials under external actions.

REFERENCES

1) Bunge,H.J.: *Krist. und Techn.* 1968,3,431.
2) Vandermeer,R.A. and McHarque,C.J.: Trans. Met. Soc. AIME,1964,230,667.
3) Dnieprenko,V.N., Larikov,L.N. and Stoyanova,E.N.:Metallofizika,1982,4,5,58 (in Russian).
4) Wassermann,G. and Grewen,J.: Texturen Metellischer Werkstoffe, Springer-Verlag, Berlin 1962.
5) Divinski,S.V. and Dnieprenko,V.N.:Texture and Microstructure. In press.
6) Dnieprenko,V.N. and Divinski,S.V.: Scripta Met.,1992,27,1617.
7) Dnieprenko,V.N., Divinski,S.V., Usov,V.V. and Brukhanov,A.A.:Fizika Tverdogo tela, 1992, 34,1872.
8) Divinski,S.V. and Dnieprenko,V.N.:Texture and Microstructure. In press.
9) Divinski,S.V. and Dnieprenko,V.N.:Prec. of ICOTOM-10, 1993.

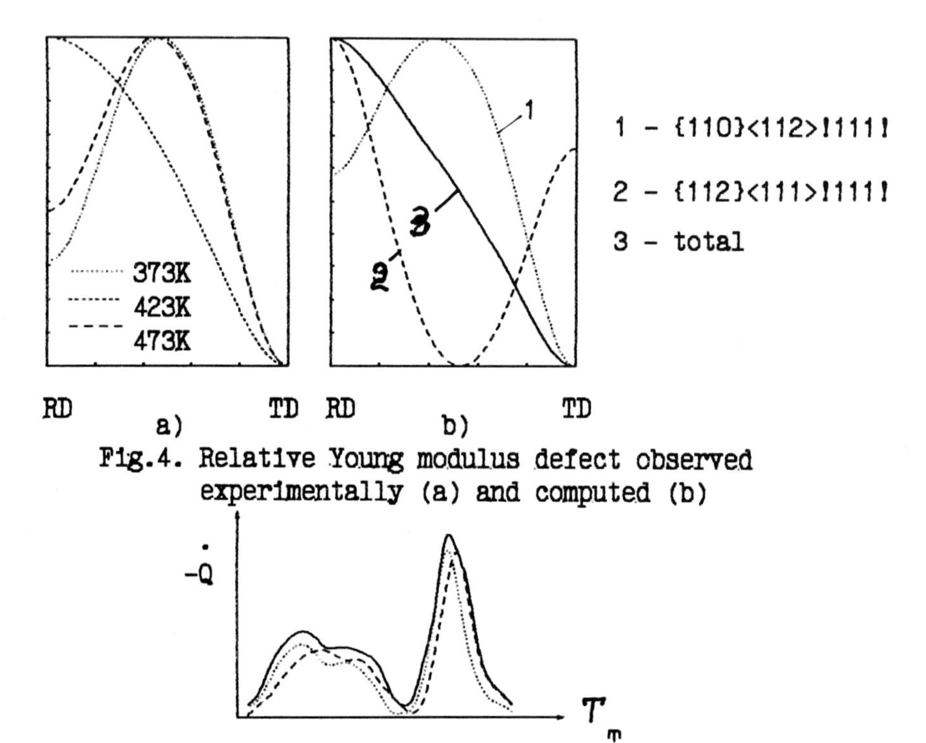

Fig.4. Relative Young modulus defect observed experimentally (a) and computed (b)

Fig.5. Scheme of heat revealing in copper under annealing

Materials Science Forum Vol. 157-162 (1994) pp. 1571-1576

LOCAL MECHANICAL PROPERTIES IN THIN ALUMINIUM LAYERS ON SILICON SUBSTRATES CALCULATED FROM MEASURED GRAIN ORIENTATIONS

D. Gerth and R.A. Schwarzer

Institut für Metallkunde und Metallphysik der TU,
Grosser Bruch 23, D-38678 Clausthal-Zellerfeld, Germany

Keywords: Grain Growth, Thin Films, Stress, Strain, Local Texture, Al-1%Si, Electron Diffraction, Kikuchi Patterns, TEM

ABSTRACT

During rapid temperature changes mechanical stress is induced in thin aluminium layers on silicon substrates due to the mismatch in the thermal expansion coefficients. Although thermal expansion is isotropic in cubic crystals, thermomechanical stress is distributed inhomogeneously in the layer, as a result of the spatial arrangement and specific lattice orientations of grains in the microstructure. Lattice orientations of single grains in heated 800 nm Al-1%Si layers on oxidised silicon substrates were measured in the TEM with the Kikuchi pattern method. They were used for the calculation of elastic constants and elastic stored energy of these grains by using the biaxial strain model and the biaxial stress model, respectively. The results are visualised by grey-step images of the grain structure. The dependence of resolved glide stress on grain orientation results in a preferential growth of grains deviating from the <111> fibre texture of the layer.

I. INTRODUCTION

During temperature changes a mechanical stress is induced in thin aluminium layers on silicon substrates by the mismatch between the thermal expansion coefficients of layer and substrate. The mechanisms of stress relaxation have been described by dislocation glide, dislocation climb and diffusional creep [1]. Only the development of global stress in the layer has yet been investigated, disregarding microstructure and local inhomogeneities of stress and strain and its effect on microstructure. In earlier investigations [2] some evidence was found for an effect of thermal stress on grain growth in thin aluminium-alloy films. This effect should be a local one due to inhomogeneities of stress relaxation and, as a consequence, an inhomogeneous change of the microstructure. An instantaneous mechanism of grain growth (known as grain coalescence) was found, which is not a diffusion process [2, 3]. Transmission electron microscopy (TEM) is an appropriate means to study the morphology and crystallographic orientation on a grain-specific scale. From TEM investigations statistical values and local differences of elastic constants can be computed in dependence of the grain orientation during the heating process.

II. EXPERIMENTAL

Investigations were carried out with 800 nm thick Al-1%Si layers on thermally oxidised silicon substrates (SiO_2 thickness 100 nm). The layers were sputter deposited at room temperature. The samples were heated under vacuum at various temperatures and for different periods. Two heating regimes were used:

- direct heating to a fixed temperature with a further annealing for 30 minutes (named isothermal annealing),
- stepwise heating up to 550°C in steps of 50°C and 6 minutes.

The samples for transmission electron microscopy were prepared by chemical polishing. The morphology and the crystallographic orientation of the grains in the layers were investigated by a TEM at an accelerating voltage of 300 kV (PHILIPS EM 430). The crystal orientations were measured on-line using Kikuchi patterns [4]. The micrographs of the areas with the measured grains were digitised in order to obtain the grain boundary coordinates. Statistical values and local differences of elastic constants were calculated on the basis of the measured grain orientations.

III. CALCULATION OF MECHANICAL CONSTANTS

A. Strain-stress relation

The calculation of stress and strain is carried out in a linear theory of elasticity. Hook's law is given by:

$$\varepsilon_{ij} = S_{ijkl}\,\sigma_{ij} \qquad (i,j,k,l = 1,2,3) \quad , \tag{1}$$

where the ε_{ij} and σ_{ij} are the components of the second rank tensor of strain and stress, and S_{mnop} are the components of the fourth rank tensor of elastic compliance. Here and in the following repeated indices are summed. The conversion of the components from the tensor to matrix coefficients with regard to a Cartesian coordinate system is given by:

$$S_{ijkl} = \frac{s_{mn}}{[(2-\delta_{ij})(2-\delta_{kl})]} \qquad (i,j,k,l = 1,2,3;\ m,n = 1,...,6) \quad , \tag{2}$$

with $\delta_{uv} = 1$ for $u = v$ and $\delta_{uv} = 0$ for $u \neq v$.

In the case of cubic crystals it is possible to reduce the independent components of $\boldsymbol{s}$ to s_{11}, s_{12}, s_{44}. The values of s_{11}, s_{12} and s_{44} refer to a (001)[100] oriented crystal. For other crystallographic orientations it is necessary to transform the tensor by a transformation matrix $\boldsymbol{g}$ which describes the rotation of the (001)[100] coordinate system into the orientation of the crystal in a sample reference frame:

$$S((hkl)[uvw])_{ijkl} = g_{im}\,g_{jn}\,g_{ko}\,g_{qp}\,S((001)[100])_{mnop} \qquad (i,j,k,l,m,n,o,p = 1,2,3) \quad . \tag{3}$$

The transformation matrix $\boldsymbol{g}$ may be described for example by Miller indices $(hkl)[uvw]$. The (hkl) plane of the crystal is parallel with the surface of the substrate surface of the sample, and the $[uvw]$ direction of the crystal is parallel with a reference direction in the sample surface. The transformation matrix $\boldsymbol{g}$ is then given by [5]:

$$\mathbf{g}((hkl)[uvw]) = \begin{pmatrix} \frac{u}{n} & \frac{v}{n} & \frac{w}{n} \\ \frac{kw-lv}{nm} & \frac{lu-hw}{nm} & \frac{hv-kw}{nm} \\ \frac{h}{m} & \frac{k}{m} & \frac{l}{m} \end{pmatrix} \qquad m = \sqrt{h^2+k^2+l^2}\,, \qquad n = \sqrt{u^2+v^2+w^2}\,, \tag{4}$$

B. Biaxial strain model

During heat treatments of a thin layer on a substrate a thermal strain is generated as a consequence of the difference between the thermal expansion coefficients of layer and substrate:

$$\varepsilon = \int_{T_1}^{T_2} (\alpha_{sub} - \alpha_{film})\, dT \quad , \tag{5}$$

where α_{sub} and α_{film} are the thermal expansion coefficients of substrate and thin layer. They are independent of crystal orientation for cubic crystals. T_1 and T_2 are the temperatures before and during thermal annealing. As a consequence of the thermal strain a stress is induced. The relation between thermal strain and the components of stress and strain is described by the elastic constants. In general equation 1 consists of six equations and 12 unknown continuous parameters. Therefore six boundary conditions are required in order to solve equation 1. Boundary conditions are deduced from the origins of strain and/or stress acting on the layers. In the following the boundary conditions of the biaxial strain model [6] are described:

- Thermal strain is biaxial due to the isotropic character of the thermal expansion coefficient: $\varepsilon_1 = \varepsilon_2 = \varepsilon$
- The strain acting on the layer is applied at the interface between layer and substrate, and consequently all stress components standing normal to the interface are zero: $\sigma_3 = \sigma_4 = \sigma_5 = \sigma$
- Since the layer is assumed to adhere rigidly on the substrate, it follows: $\varepsilon_1 = 0$.

In the calculation of the relation between the components of stress or strain and the thermally induced biaxial lateral strain the following assumptions are made as a first approximation:

1. A thin layer with uniform thickness is tightly bonded to a rigid substrate. The thickness of the substrate is large as compared to the layer.
2. The slight bending of the substrate by thermal stress is neglected.
3. There is no plastic stress relaxation (only the elastic deformation region is considered).
4. The interaction between neighbouring crystals is neglected.
5. The edge regions of the layer are not considered.

The calculation of the components of stress and strain as a function of biaxial strain is possible by solving Hook´s law. In the equations the elastic compliance for known crystallographic orientations has to be used. The resulting values are the elastic constants σ_1/ε, σ_2/ε, σ_6/ε, $\varepsilon_3/\varepsilon$, $\varepsilon_4/\varepsilon$ and $\varepsilon_5/\varepsilon$. The values of σ_4/ε, σ_5/ε and σ_6/ε are nearly zero. The elastic energy per unit volume u is given by: $u = 1/2 \cdot \varepsilon_i \cdot \sigma_i$.

If the resolved glide stress is exceeded, plastic stress relaxation will be initiated by dislocation glide. Some grains are expected to deform preferably by dislocation glide, depending on their crystal orientation. The residual shear stress on a given glide plane in a given glide direction can be calculated by a twofold rotation of the calculated stress tensor. For fcc cubic crystals the glide systems are the {111} planes and the <110> directions. The first rotation yields the coincidence of the σ_{11} component with the [100] direction of the crystal, and of the σ_{33} component with the [001] direction. The transformation matrix is the inverse matrix of equation 4. The second rotation aligns σ_{22} with the normal on the {111} glide plane and of σ_{11} with the <110> glide direction. The transformation matrix $\boldsymbol{g}$ is given by:

$$g((hkl)[uvw]) = \begin{pmatrix} \frac{u}{n} & \frac{v}{n} & \frac{w}{n} \\ \frac{r}{o} & \frac{s}{o} & \frac{t}{o} \\ \frac{vt-ws}{no} & \frac{wr-ut}{no} & \frac{us-vr}{no} \end{pmatrix} \qquad o = \sqrt{r^2+s^2+t^2}\,, \qquad n = \sqrt{u^2+v^2+w^2}\,, \qquad (6)$$

where r, s, t stand for the Miller indices of the glide plane and u, v, w for those of the glide direction. Now the shear stress on the glide plane in the glide direction of the dislocations is given by σ_{12}. Since the components of the calculated stress tensor are related to the biaxial stress, the glide stress is likewise related to it: σ_{12}/ε. The calculation is carried out for the four {111} glide planes and three <110> glide directions. Because each glide direction exists in two glide planes, 6 different values have to be considered. Among all possible glide planes and glide directions, the one will predominate the glide process which exerts the greatest resolved shear stress. Therefore the maximum value of these 6 values is selected as the maximum glide stress acting on a {111} glide plane in a <110> glide direction. It is an elastic constant, named glide stress, σ_{gl}/ε, which describes the relation between strain and shear stress on a glide system.

In figure 1 the distributions of elastic energy per unit volume per squared biaxial strain, vertical strain per strain, lateral stress per strain and resolved glide stress per strain are depicted for aluminium in the standard triangle of the stereographic projection. The values of the elastic compliance of aluminium are $s_{11} = 1.592 \cdot 10^{11}$ Pa, $s_{12} = -0.577 \cdot 10^{11}$ Pa, $s_{44} = 3.546 \cdot 10^{11}$ Pa [7]. The maximum and minimum values of the elastic constants are given in table 1.

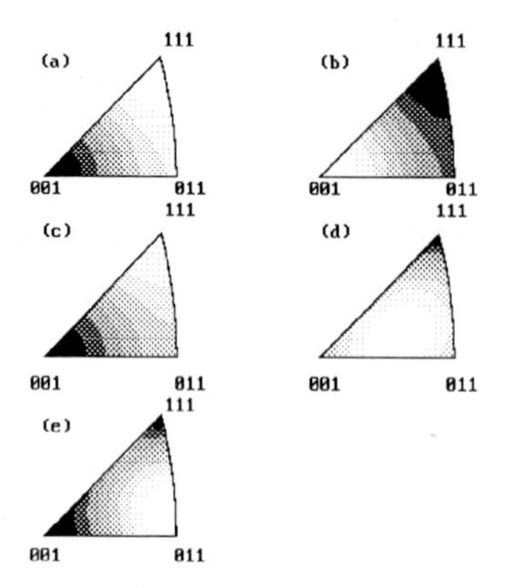

Fig. 1 Distribution of elastic constants for Al calculated by biaxial strain model.
a) elastic energy per unit volume and squared strain, u/ε^2,
b) vertical strain per strain, $\varepsilon_3/\varepsilon$,
c) lateral stress per strain, σ_l/ε,
d) maximum resolved glide stress per strain, σ_{gl}/ε,
e) range of values of lateral stress per strain.

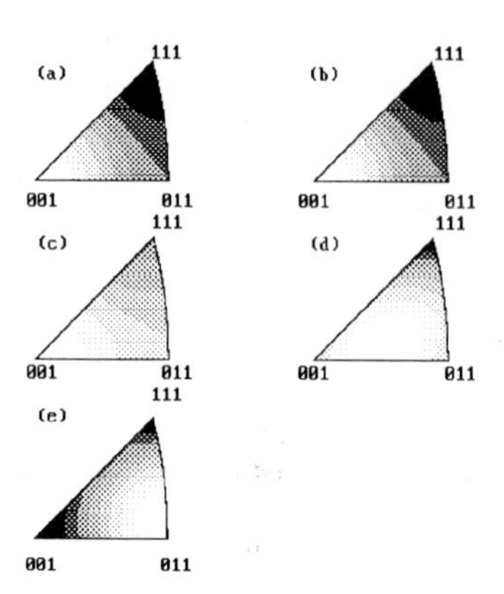

Fig. 2 Distribution of elastic constants for Al calculated by biaxial stress model.
a) elastic energy per unit volume and squared stress, u/σ^2,
b) vertical strain per stress, ε_3/σ,
c) lateral stress per stress, σ_l/σ,
d) maximum resolved glide stress per stress, σ_{gl}/σ,
e) range of values of lateral stress per stress.

An important result is that the maximum value of resolved glide stress per strain occurs when the grain interface with the substrate is close to {135}, rather than on an edge of the standard triangle. In contrast to the nearly isotropic behaviour of the other elastic constants, a larger difference between the maximum and the minimum value of resolved shear stress was found, due to the specific orientation of glide systems to the acting force in dependence of the crystal orientation.

C. Biaxial stress model

A uniform biaxial strain and a discontinuous stress is possible in a layer only on the interface with the substrate. The elastic stress relaxation is reduced due to the rigid adhesion to the substrate. The larger the distance from the interface the higher is the possibility for an elastic relaxation of stress. A complete relaxation of elastic stress leads to a uniform biaxial stress on the surface (and probably also in the layer above a critical distance from the interface). The value of the biaxial stress is equal to the average value of the stresses in the grains induced by biaxial strain on the interface. Then strain gradually decreases from the top to the bottom of the layer. Strain differences between the value at the surface and that at the interface have been measured by x-ray diffraction [8].

Table 1 Maximum and minimum values of elastic constants and supporting surfaces of aluminium layers. The calculations were performed by using the biaxial strain model and the biaxial stress model, respectively.

	Biaxial strain model				Biaxial stress model			
	u/ε^2 [10^9 Pa]	$\varepsilon_3/\varepsilon$	σ_l/ε [10^9 Pa]	σ_{gl}/ε [10^9 Pa]	u/σ^2 [10^{-12} Pa^{-1}]	ε_3/σ [10^{-12} Pa^{-1}]	ε_1/σ [10^{-12} Pa^{-1}]	σ_{gl}/σ
minimum value	98.52	-1.008	98.52	30.86	8.83	8.90	8.17	0.2725
supporting surface	{001}	{111}	{001}	{111}	{111}	{111}	{111}	{111}
maximum value	113.30	-1.137	114.99	51.52	10.15	11.54	10.15	0.500
supporting surface	{111}	{001}	{111}	~{135}	{001}	{001}	{001}	~{135}

As a complement to the biaxial strain model a biaxial stress model [8] has been developed to calculate local stress and strain. In order to solve Hook's law, boundary conditions have to be considered in a critical distance from the interface:

- The stress is biaxial: $\sigma_1 = \sigma_2 = \sigma$.
- All other stress components are zero: $\sigma_3 = \sigma_4 = \sigma_5 = \sigma_6 = \sigma$.

The same assumptions are made here as in the case of the biaxial strain model. The values resulting from the calculation are the elastic constants ε_1/σ, ε_2/σ, ε_3/σ, ε_4/σ, ε_5/σ and ε_6/σ. The constants ε_4/σ, ε_5/σ and ε_6/σ are nearly zero. The elastic energy per unit volume, u, is related to the biaxial stress, u/σ^2. In the same way as in the biaxial strain model described above, the resolved shear stress can be calculated. It is related to the biaxial stress, σ_{gl}/σ.

In figure 2 the distributions of elastic energy per unit volume and squared biaxial stress, vertical strain per stress, lateral strain per stress and resolved glide stress per stress are given in the standard triangle of the stereographic projection. The maximum and minimum values are given in table 1. There is a maximum value of elastic energy per unit volume and stress squared for a {100} interface of the crystal and a minimum value for a {111} interface (table 1). This finding is contrasting to the biaxial strain model. At biaxial strain the largest stress is induced in "hard" grains. Hence the largest elastic energy in these grains is accumulated. The elastic stress relaxation, generated in a critical distance from the interface, leads to a uniform biaxial stress. As a result the relaxation is the largest in "hard" grains leading to the smallest strain as compared to all other grains. This means that elastic energy is also smallest in "hard" grains, since stress is assumed equal in all grains. As a consequence the difference between the elastic energy at the interface and the surface is highest in grains with a {111} supporting surface.

The widest range of values is for resolved glide stress per stress. The other elastic constants play a marginal role, due to the low elastic anisotropy of aluminium. This is the same finding as in the biaxial strain model.

IV. DEVELOPMENT OF ELASTIC CONSTANTS DURING GRAIN GROWTH

Elastic constants of the heated aluminium layers were calculated under the approximations of the described models above. Figure 3 shows the development of the lateral stress per strain, lateral strain per stress, the resolved glide stress per strain and the glide strain per stress in terms of the two models. There is only a small change of the lateral stress per strain and lateral strain per stress. The resolved glide stress per strain increases with temperature. Hence glide stress and stress relaxation by plastic deformation are reached at decreasingly smaller temperature intervals, i.e. at lower strain, with rising temperature. This finding is independent of the model of calculation and therefore of the location of glide processes in the grains.

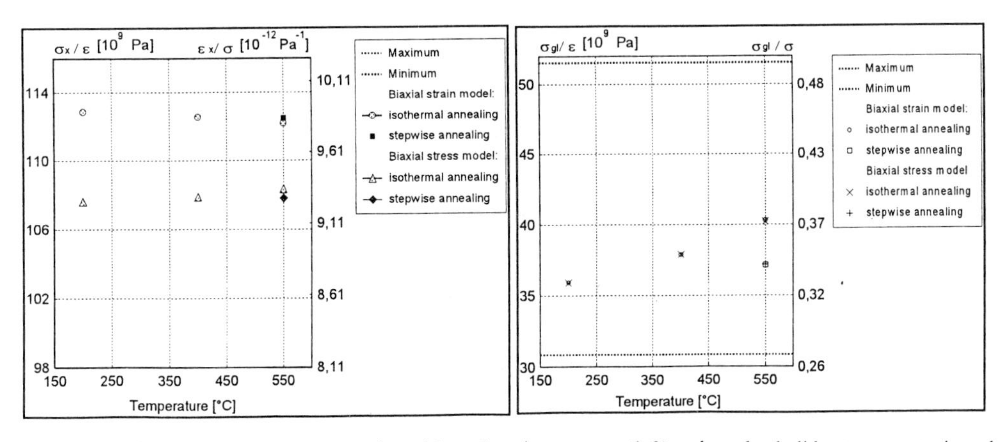

Fig. 3 Dependence of lateral stress per strain and lateral strain per stress (left) and resolved glide stress per strain and per stress (right) on temperature in 800 nm Al-1%Si layers on oxidised silicon substrates.

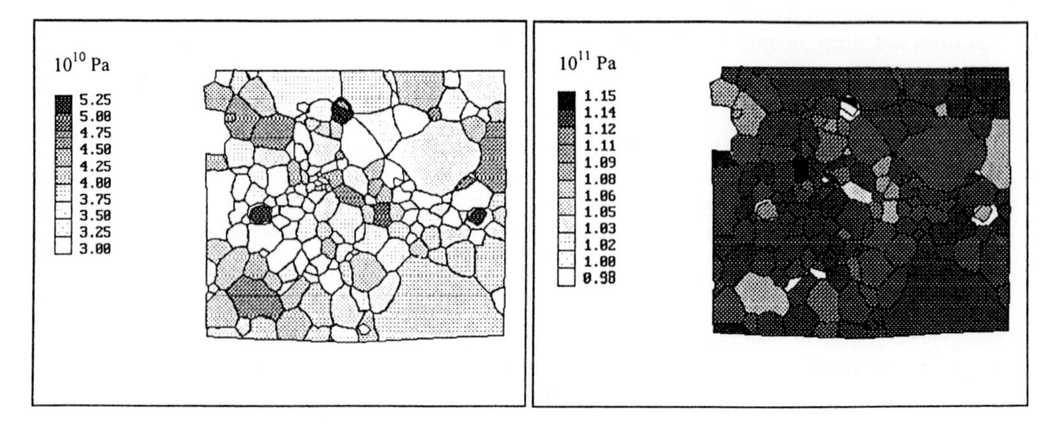

Fig. 4 Local distribution of resolved glide stress per strain, σ_{gl}/ε, (left) and lateral stress per strain, σ_l/ε, (right) of an 800 nm Al-1%Si layer on an oxidised silicon substrate, stepwise heated to 550 °C. The values were calculated using the biaxial strain model.

The local inhomogeneities of elastic constants can be illustrated by maps of the digitised grain structure. The values of the elastic constants in each grain are indicated by a specific grey shade. As an example the lateral stress per strain and the glide stress per strain of the aluminium layer, heated stepwise to 550 °C, are given in figure 4. The sharp <111> fibre texture results in only small differences in lateral stress per strain between neighbouring grains (<5 % of the maximum value). Higher inhomogeneities, however, were found for the glide stress. Hence plastic stress relaxation will start at different temperatures and different sites during the heating process, depending on the individual crystal orientation. Stress inhomogeneities and differences in the dislocation density are produced. These differences may lead to different deformation energies as the dominant driving force for grain growth. Not the absolute value but the local differences of glide stress between neighbouring grains is decisive here. At each temperature step the grains with the highest values of glide stress (i.e. grains oriented with the <135> direction standing normal on the plane) start with plastic stress relaxation and have, as a consequence, a lower dislocation density and deformation energy. The result is a weaker <111> fibre texture and a higher mean value of glide stress.

The deformation mechanism maps of thin aluminium films [9] indicate that at a fixed temperature a larger fraction of stress relaxation is accomplished by dislocation glide, if stress is high. This condition is met by annealing in one step, i.e. for a high heating rate. A smaller stress, however, is induced at each temperature interval during stepwise heating. Hence the texture is weaker after isothermal annealing than after stepwise heating.

ACKNOWLEDGMENT

The authors are indepted to the Deutsche Forschungsgemeinschaft for financial support (DFG project Bu 374/22-1).

REFERENCES

[1] Murakami, M.: Thin Solid Films, 1978, 55, 101
[2] Gerth, D., Katzer, D., and Krohn, M.: Thin Solid Films, 1992, 208, 67
[3] Thon, A., and Brokman, A.: J. Appl. Physics, 1988, 63, 5331
[4] Schwarzer, R. A.: Textures and Microstructures, 1990, 13, 15
[5] Bunge, H.J.: Texture Analysis in Materials Science. Butterworths London, 1982
[6] Vook, R. W., and Witt, F.: J. Appl. Physics, 1965, 36, 2169
[7] Boas, W., and Mackenzie, J. K.: Anisotropy in Metals. in Chalmers, B.: Progress in Metal Physics. Butterworth, London, 1950
[8] Murakami, M.: Acta Metallurgica, 1978, 26, 175
[9] Koleshko, V. M., Belitsky, V. F., and Kiryushin, I. V.: Thin Solid Films, 1986, 142, 199

Materials Science Forum Vol. 157-162 (1994) pp. 1577-1584

TOUGHNESS ANISOTROPY IN BRITTLE MATERIALS

M.D. Grah and K.J. Bowman

School of Materials Engineering, Purdue University,
1289 Mat. Electr. Eng. Bldg., West Lafayette, IN 47907, USA

Keywords: Anisotropy, Brittle Fracture, Toughness, Biaxial

ABSTRACT

Anisotropic fracture behavior is prevalent in textured materials. In textured brittle materials either intergranular or transgranular fracture can cause anisotropic toughness. Two toughening mechanisms discussed widely in the ceramics literature, bridging and crack deflection, appear to be particularly sensitive to oriented crack propagation. In this paper, fracture processes in textured aluminum sheets that have been embrittled by liquid gallium are discussed. Biaxial fracture experiments on thin, locally embrittled specimens resulted in preferential crack orientations.

I. INTRODUCTION

Brittle fracture in textured materials should result in some degree of fracture anisotropy for either intergranular or transgranular fracture propagation. Recently, several examples of toughness enhancements and toughness anisotropies have been shown for textured of ceramic and ceramic composite materials [1-4]. Correlating texture to intergranular fracture provides several possibilities for anisotropy, with the resulting anisotropy dependent on the mechanisms of toughening. Mechanisms for enhancing toughness in ceramic materials by microstructural control frequently involve introduction of anisometric grains or composite reinforcements that can provide toughening via bridging or crack deflection.

In toughening of brittle materials from bridging, the crack passes some grains that remain across the fracture plane as shown in Fig. 1(a). These grains are clamped in place due to thermal expansion mismatch or by frictional forces. The energy required to completely separate these grains across the fracture surface add to the work of fracture.

The schematic in Fig. 1(b) shows the mechanism of crack deflection. In this mechanism, the toughening enhancement arises from injecting Mode II and Mode III stress intensities in addition to Mode I crack propagation. This change in stress intensity occurs from a combination of tilt and twist of the crack front [e.g. 2, 5].

In addition to these examples, anisotropic fracture should also occur in textured materials without anisometric grains. Discussed below is the conceptual basis by which anisotropic fracture should occur with isometric or equiaxed grains along with some results of model experiments intended to demonstrate brittle fracture in textured materials with equiaxed grains.

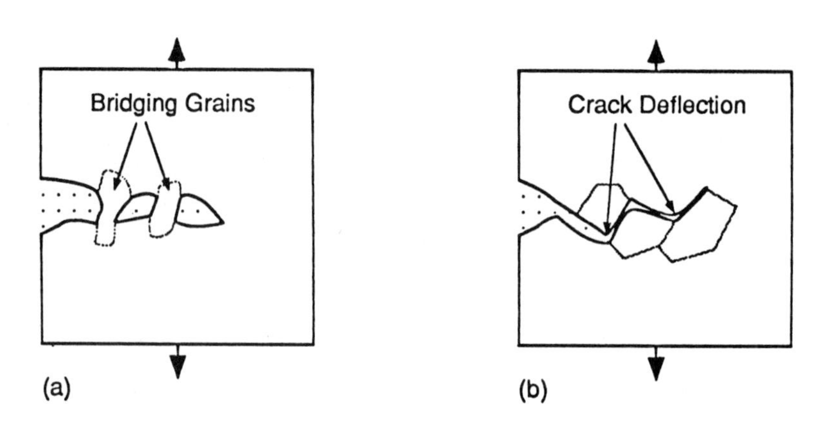

Figure 1 The two toughening mechanisms (a) crack bridging and (b) crack deflection are illustrated by the diagrams showing overall Mode I fracture.

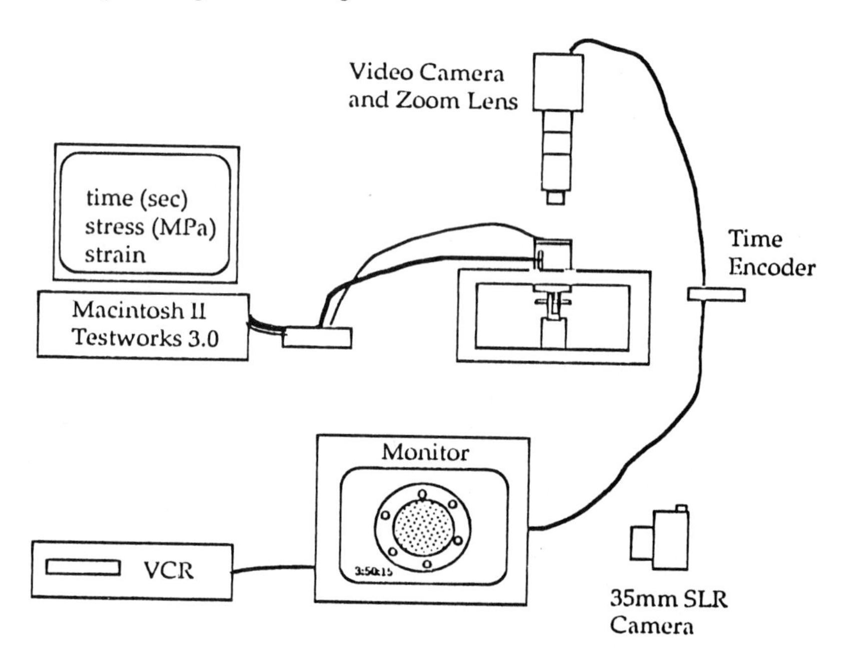

Figure 2 Schematic representation of the loading device, measuring apparatus and recording apparatus.

II. BACKGROUND

The requirements for crack bridging in a single phase material are only that grains must bridge the crack behind the point of furthest progress of a mode I crack. In materials with anisometric grains or reinforcements the crystallographic texture is often directly related. Ceramic materials with isometric grains and crystal structures with non-cubic symmetry can possess some toughening contribution from anisotropic thermal expansion [e.g. 6]. All such materials can have two independent thermal expansion coefficients. If textured materials have crystal symmetry lower than hexagonal, the number of independent thermal expansion coefficients can become as high as three. Under these circumstances textured polycrystals might readily demonstrate an anisotropy in the thermal expansion mismatch contribution to bridging. A simple example demonstrating this might be a basal fiber texture in a tetragonal material. If we consider a simple orientation difference between grains, for example a (110)||(100) grain boundary, which is normal to the basal plane, the maximum and minimum thermal expansions for directions normal to [001] would coincide. Clamping of the bridging grain could occur if only one orientation of the grain fails to match the stress-free displacements of its neighbors.

Another possible route for bridging is anisotropic grain boundary energies at crystal interfaces. This anisotropic interfacial energy can arise at the boundaries between grains with direct contact or in materials with a grain boundary phase. Under these conditions variability in the fracture resistance of particular grain boundaries could lead to grains that remain in contact across the crack face.

In order for a brittle cubic material with isometric grains to demonstrate fracture anisotropy when fracture takes place by intergranular fracture, anisotropic interfacial or grain boundary energy appears to be the only possibility. Under these circumstances the occurrence of special grain boundaries that coincide with the texture might enhance the likelihood of fracture anisotropy. In fact, particular crystallographic textures apparently enhance the occurrence of special grain boundaries [e.g. 7]. Similarly, the preferred orientation of special grain boundaries would be anticipated to influence crack paths, resulting in crack deflection. Also, such alignment of fracture resistances should result in fracture initiation that coincides with particular orientations within a sample. Of course, in real materials the same processes that introduce the texture may also align defects that lead to failure.

III. MODEL EXPERIMENTS IN BIAXIAL FRACTURE

Although anisotropic fracture could be investigated by performing a number of uniaxial fracture experiment at a variety of orientations, a more effective way to find the weakest orientations is in biaxial fracture of thin sheet specimens [8]. Unfortunately, specimen preparation and equibiaxial loading are extremely difficult for intrinsically brittle materials. One approach that appears to be successful for carrying out such experiments is localized embrittlement of otherwise ductile, textured materials. We have prepared thin sheet specimens having several types of microstructures from both silicon-iron and commercial purity, 1100 series aluminum alloys [8,9]. In both of these materials grain boundary fractures can be enhanced by embrittlement. For the experiments discussed here, gallium embrittlement of 1100 series aluminum was employed.

Specimen Preparation

The material used in this investigation for all sheet specimens was 1100-O series aluminum foil from AESAR/Johnson Matthey (Ward Hill, MA). The expected chemical impurity content is listed in Table I. Large sheets were procured with nominal thicknesses of 254 μm and 1000 μm. Aluminum sheet manufacture was described as successive cold rolling and recrystallization anneals ending with a full recrystallization anneal. As-received average grain size was measured by the linear intercept method to be 20 μm. Sheet specimens were cut into 75 mm x 75 mm strips so that they could be used in the experimental set-up shown in Figure 2 which includes the biaxial loading device illustrated in Figure 3. The as-received stock was recovery annealed at 225°C for 16 hr to remove minor cold work introduced during shipment, cutting and handling. Stereological measurements of grain size after the recovery anneal revealed an average 20 μm grain size, identical to that of the as-received sheets.

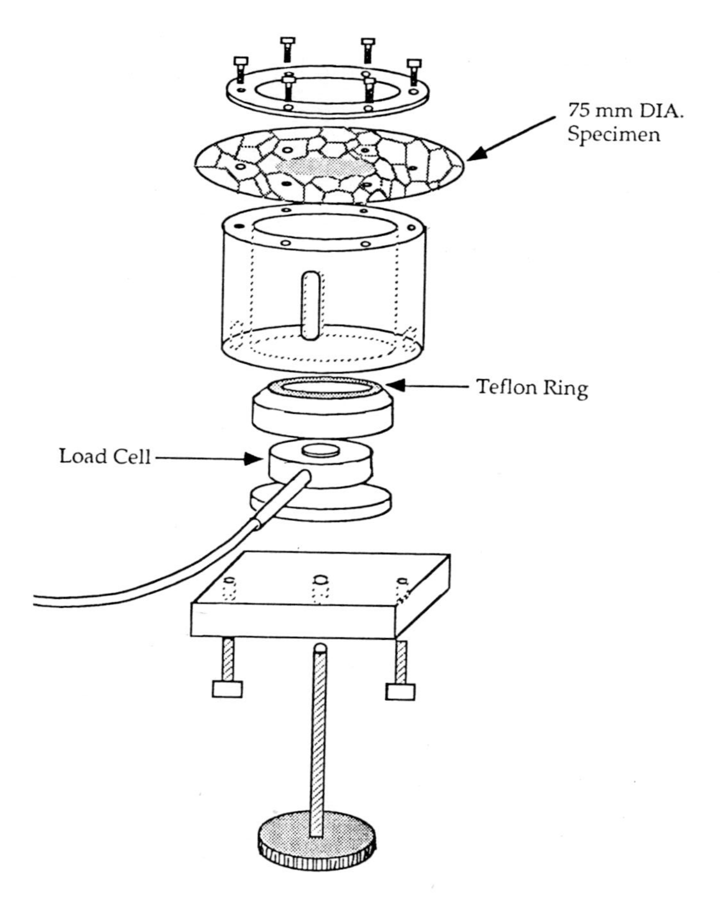

Figure 3 Schematic diagram of the biaxial loading device.

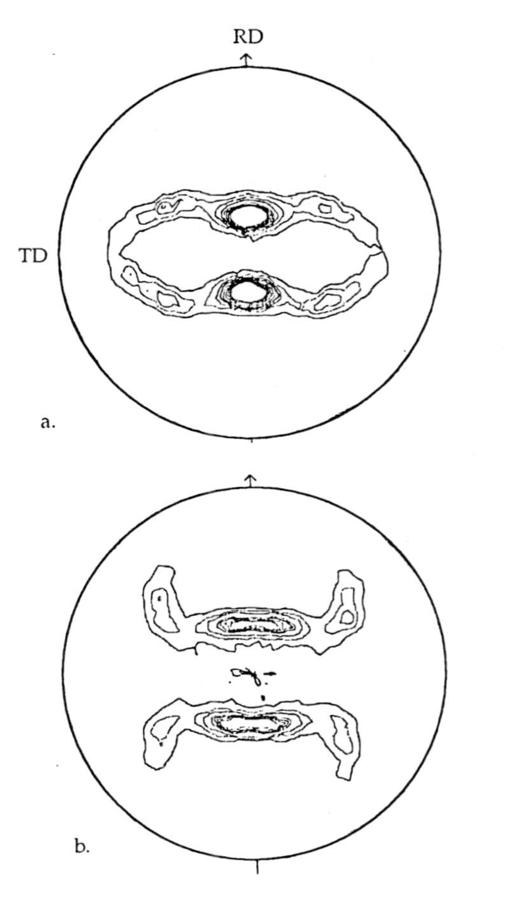

Figure 4 (a) 111 and (b) 002 experimental pole figures for the as-received and recovery annealed aluminum material. In both pole figures the calculated maximum intensity was nine times random.

Table I. Nominal Chemical Impurity Content of 1100-O Series Aluminum.

Element	Nominal Composition (Wt.%)
Si+Fe	.95
Cu	.05-.20
Mn	<.05
Zn	<.10

Specimens were embrittled by wetting them with 99.999% pure gallium (AESAR/Johnson Matthey). Gallium was applied in a circular patch 20±2 mm in diameter centered on one side of each sheet specimen. Wetting was facilitated by lightly scratching the thin oxide on the surface through the liquid gallium so that it was in direct contact with the metal. Once in contact with the aluminum, gallium proceeds rapidly along grain boundaries. Excess gallium was allowed to remain on the wetted surface at 50°C for 20 minutes.

After embrittlement, most excess gallium on the surface was wiped off with a paper towel. Gallium remaining on the surface was removed by immersion of the sheet in a beaker of distilled water for 20 sec at 23°C. Specimens were annealed for two minutes at 150°C to partially de-embrittle them enough to permit handling.

Biaxial Testing

The biaxial testing device was designed so that crack propagation could be observed and videotaped in the embrittled region during biaxial loading. The high resolution camera shown schematically in Figure 2 sends images to a high resolution 14" video monitor. Micrographs of fracture surfaces were recorded by photographing images on the monitor with a 35 mm single lens reflex camera with a 50 mm lens. Additional experimental details are in Reference 8.

Experimental Results

A significant crystallographic texture was expected in the as-received sheets. This texture was confirmed through X-ray diffraction and measurement of incomplete pole figures. Incomplete 111 and 002 pole figures of the recovery annealed aluminum sheet are displayed in Figure 4. The rolling direction and transverse directions of the sheet are marked. A strong recrystallization texture is evident in both pole figures. This texture is consistent with deformation textures normally introduced by extensive rolling reductions in most high stacking fault energy metals. The primary ideal textural component is the C {112}<111> component. The {123}<364> R component is also evident, but is much less intense. The highest intensity contours occur at 9 times random centered over the poles representing the C component.

The localized embrittlement of aluminum strips and the design of the loading rig allowed application of stresses where the sheet is ductile and transfer of those stresses directly to the embrittled region. The time of fracture events was videotaped simultaneously with biaxial stress and time files recorded on the computer. Later, the use of the data files in conjunction with the fracture videotape allowed graphical reconstruction of the stresses where specific fracture events occurred.

A group of thirteen specimens was fractured. They all exhibited well behaved and consistent fracture behavior. Average crack initiation stresses fell between 5.0 and 8.9 MPa with an average of 6.7 MPa. The initial or primary crack grew very rapidly by the apparent coalescence of many small cracks that initiated co-linearly and almost simultaneously along the length of the embrittled region. At approximately 1.4 times the initial fracture stress, transverse cracks initiated and grew at approximately right angles to the original radial crack. The number of secondary cracks that initiated was usually two; one on each side of the primary crack. The primary and secondary crack orientations are shown in Fig. 5. The cracks have a strong tendency to grow almost parallel or perpendicular to the rolling direction of the as-received sheets. The large data points represent the angle of primary crack formation from the rolling direction whereas

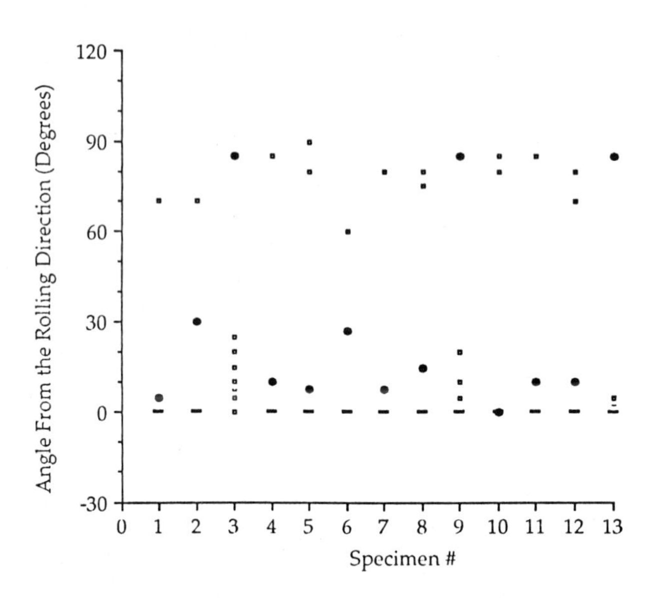

Figure 5 The angular deviation of primary and secondary cracks from the rolling direction for all specimens. The dotted line represents the rolling direction, the large dots represent primary cracks and the small dots represent secondary cracks.

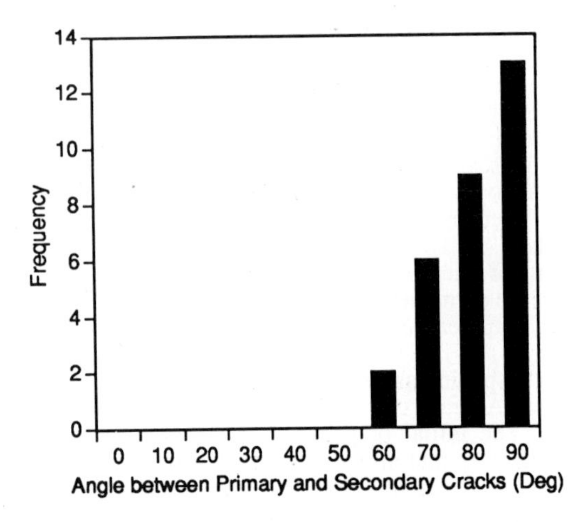

Figure 6 Frequency distribution of the angles between the primary and secondary cracks.

the small data points represent the angular deviation of the secondary cracks from the rolling direction. The average angle between primary crack and the rolling direction was 11° in 69% of the specimens. In the remaining 31% of specimens, the average angle between primary crack and rolling direction was 85°.

Secondary cracks generally formed nearly perpendicular to the direction of the primary radial crack. The extent of this orthogonality is illustrated by the frequency distribution of the difference between primary and secondary crack angles shown in Figure 6. The average angle between primary and secondary cracks is 83°.

IV. DISCUSSION

The biaxial loading conditions applied in the above experiments show strong directionality in the fracture orientation despite equiaxed grains in a cubic material. Why nearly three fourths of the initial cracks propagated almost along the rolling direction and one fourth propagated near the transverse direction is clearly related to the deformation and thermal history of the aluminum sheets and an interaction with the embrittlement process. Although cracks propagating along the rolling directions might be attributed to aligned flaws, the lack of any microstructural causes and that the remaining fractures initiate nearly across the rolling direction do not support such a conclusion.

The interaction of the liquid gallium with the grain boundaries will depend on the character of specific grain boundaries. In related experiments on similar materials with large, through-thickness grains special boundaries were embrittled less than random grain boundaries [10]. Also, these larger grain materials showed a weaker orientation dependence of fracture due to fracture initiated along large grain facets. Since grain boundary character was not determined in the smaller grain specimens of this investigation, the supposition that more random grain boundaries may lie along the rolling and transverse directions cannot be demonstrated.

V. ACKNOWLEDGEMENTS

This research was supported by the U.S. Air Force Office of Scientific Research Grant No. 89-0423, National Science Foundation Grant No. 91-2121948 and a Department of Defense Fellowship supporting M. D. Grah.

VI. REFERENCES

1. T. Hansson, R. Warren and J. Wasen, "Fracture Toughness Anisotropy and Toughening Mechanisms of a Hot-Pressed Alumina Reinforced with Silicon Carbide Whiskers," *J. Am. Ceram. Soc.*, 76 [4] 841-48 (1993).
2. F. J. Lee, M. S. Sandlin and K. J. Bowman, "Toughness Anisotropy in Textured Ceramic Composites," *J. Am. Ceram. Soc.*, 76[7], 1793-1800 (1993).
3. Y-S. Chou and D. J. Green, "Silicon Carbide Platelet/Alumina Composites," *J. Am. Ceram. Soc.*, 76[6] 1452-58 (1993).
4. Y. Goto and A. Tsuge, "Mechanical Properties of Unidirectionally Oriented SiC-Whisker-Reinforced Si_3N_4 Fabricated by Extrusion and Hot-Pressing," *J. Am. Ceram. Soc.* 76[6] 1420-24 (1993).
5. K. T. Faber and A. G. Evans, "Crack Deflection Processes-I, Theory," *Acta Metall.*, 31[4] 565-76 (1983).
6. J. Hay and K. White, "Grain Bridging Mechanisms in Monolithic Alumina and Spinel," *J. Am. Ceram. Soc.* 76[7] 1849-54 (1993).
7. T. Watanabe, et al., "Texture and Grain Boundary Character Distribution in Rapidly Solidified and Annealed Fe-6.5 w/o Si Ribbons," *Phil. Mag. Letters*, 59[2] 47-52 (1989).
8. M. D. Grah, Ph.D. Thesis, Purdue University, December, 1993.
9. M. D. Grah, K. J. Bowman and M. Ostoja-Starzewski, "Fabrication of Two-Dimensional Microstructures in Fe-3.25 Si Sheet," *Scripta Met,* 26[2] 429-34 (1992).
10. M. D. Grah, M. D. Vaudin and K. J. Bowman, Unpublished Results.

Materials Science Forum Vol. 157-162 (1994) pp. 1585-1592

NUMERICAL ANALYSIS OF STIFFNESS IN SHEET STEEL PRODUCTS BASED ON CRYSTAL ANISOTROPY

S. Hiwatashi [1], T. Hatakeyama [2], K. Ushioda [3] and M. Usuda [1]

[1] Steel Research Laboratories, Nippon Steel Corporation,
20-1 Shintomi, Futtsu, Chiba 299-12, Japan

[2] Yawata Works, Nippon Steel Corporation,
1-1 Tobihata-cho, Tobata-ku, Kitakyushu, Fukuoka 804, Japan

[3] Kimitsu R&D Laboratory, Nippon Steel Corporation,
1 Kimitsu, Kimitsu, Chiba 299-11, Japan

Keywords: Elastic Anisotropy, Sheet Steel Product, Finite Element Method, Stiffness

ABSTRACT

Some steel sheets with strong anisotropy produced by cold–rolling and continuous annealing can be applied to some structural components, such as automotive body parts, to improve their stiffness. This paper concerns a numerical analysis method that can evaluate the stiffness of some components made of steel sheets by considering the elastic anisotropy. In this method, an elastic stress–strain equation for a polycrystalline material is defined in advance according to its orientation distribution. Then the constitutive equation is taken into the conventional finite element program, MARC, and the three-dimensional analysis is carried out using shell elements. The anisotropy of calculated Young's moduli was similar to that of the experimental results. The panel stiffness of a cylindrical shell and the flexibility of a pipe were analyzed. The stiffness depends on the direction of the sheet applied to the component, and this system makes it easy to decide the most effective way to apply the anisotropic steel sheets.

1. INTRODUCTION

Elastic moduli of metal sheets have planar anisotropy determined by crystalline texture [1]. As a single crystal of iron has remarkably strong anisotropy, texture control is effective in raising Young's modulus of steel sheets in some directions on the plane [2]. Such steel sheets can be used for structural components, such as automotive body parts, to improve their stiffness. This paper concerns a numerical analysis that can evaluate the stiffness of some components made of steel sheets by considering the elastic anisotropy and that is useful in designing some industrial products.

Various approximations for calculating the elasticity of a random–oriented polycristalline material from elastic coefficients of a single crystal have been proposed. Voigt assumed constant strain in all grains of the polycrystal [3], while Reuss assumed constant stress [4]. Hill showed that these two approximations give upper and lower bound, respectively, for the elastic moduli experimentally observed [5]. Hashin and Shtrikman applied variational principles to improve the Voigt and Reuss bounds [6]. Hershey [7] and Kröner [8] obtained better approximations by the 'self–consistent' method, which takes the boundary conditions of stress and strain in each grain into account [7,8]. The elastic anisotropy of a polycrystal with texture was calculated by Bunge who generalized the Voigt–Reuss–Hill approximations [9,10]. The elastic anisotropy of the steel sheets with texture, however, has been applied to few industrial products to improve their stiffness because the structural design using anisotropic materials is difficult.

In this paper, the stiffness of some components made of anisotropic steel sheets is analyzed. First, the constitutive equation for elasticity is calculated on the Voigt or Reuss assumption where the elastic compliance or stiffness matrix of a single crystal is transformed and weight–averaged according to the orientation distribution of the crystals. Then the obtained constitutive equation is taken into the finite element program, MARC, and the three–dimensional analysis on stiffness of some components is carried out. As examples, this system was applied to the stiffness analysis of a cylindrical shell and a pipe.

2. CONSTITUTIVE EQUATION

2.1. Elastic anisotropy of a single crystal

In the generalized Hook's law, constitutive equations are expressed as follows:

$$\sigma_{ij} = C_{ijkl}\, \varepsilon_{kl} \tag{1}$$

where σ_{ij} , ε_{kl} , and C_{ijkl} are the tensors of stress, strain, and elastic coefficients, respectively. In a local coordinate system (x_1' , x_2' , x_3') of a cubic single crystal, equation 1 is conventionally expressed as follows:

$$\begin{Bmatrix} \sigma_{11}' \\ \sigma_{22}' \\ \sigma_{33}' \\ \sigma_{12}' \\ \sigma_{23}' \\ \sigma_{31}' \end{Bmatrix} = \begin{bmatrix} C_{11} & C_{12} & C_{12} & 0 & 0 & 0 \\ C_{12} & C_{11} & C_{12} & 0 & 0 & 0 \\ C_{12} & C_{12} & C_{11} & 0 & 0 & 0 \\ 0 & 0 & 0 & C_{44} & 0 & 0 \\ 0 & 0 & 0 & 0 & C_{44} & 0 \\ 0 & 0 & 0 & 0 & 0 & C_{44} \end{bmatrix} \begin{Bmatrix} \varepsilon_{11}' \\ \varepsilon_{22}' \\ \varepsilon_{33}' \\ \varepsilon_{12}' \\ \varepsilon_{23}' \\ \varepsilon_{31}' \end{Bmatrix} \tag{2}$$

where C_{ij} is a component of the stiffness matrix $[C']$ of a cubic crystal.

The relation between the global coordinate system (x_1 , x_2 , x_3) and a local coordinate system can be expressed by equation 3.

$$x_p' = l_{pi}\, x_i \tag{3}$$

The transformation of stress and that of strain are obtained as equations 4 and 5, respectively.

$$\sigma_{ij} = l_{ip}\, \sigma_{pq}'\, l_{qj} \tag{4}$$

$$\varepsilon_{pq}' = l_{pi}\, \varepsilon_{ij}\, l_{jq} \tag{5}$$

Equations 4 and 5 are conventionally expressed by equations 6 and 7.

$$\{\sigma\} = [\Phi]\, \{\sigma'\} \tag{6}$$

$$\{\varepsilon'\} = [\Psi]\, \{\varepsilon\} \tag{7}$$

Here, $[\Phi]$ has the relationship with $[\Psi]$ as described in equation 8.

$$[\Phi] = [\Psi]^T \tag{8}$$

The constitutive equation of a single crystal in the global coordinate system is conventionally rewritten by equation 9.

$$\begin{aligned} \{\sigma\} &= [\Phi]\{\sigma'\} \\ &= [\Psi]^T [C']\{\varepsilon'\} \\ &= [\Psi]^T [C'][\Psi]\{\varepsilon\} \\ &= [C]\{\varepsilon\} \end{aligned} \tag{9}$$

The compliance matrix is obtained by inverse matrix of $[C]$ as equation 10.

$$[S] = [C]^{-1} \tag{10}$$

2.2. Stress and strain equation of a polycrystal

There are a few methods to analyze the texture of polycrystalline metals, and the vector method is used here to analyze the crystalline orientation distribution from measurements using X-ray diffraction. In the vector method, the orientation space is divided into 36 x 36 equivalent elements, and the volume fraction of each orientation element is expressed in terms of the ratio to random distribution of orientation. The ratios are called components of texture vector [11].

From the orientation of each element, the stiffness of the crystal is defined by equation 9. According to the Voigt (constant strain) model [3], the constitutive equation of a polycrystal is expressed as follows:

$$\begin{aligned} \{\sigma\} &= \sum \frac{y_i}{\Sigma y_i}\{\sigma\}_i \\ &= \sum \frac{y_i}{\Sigma y_i}[C]_i\{\varepsilon\}_i \\ &= \sum \frac{y_i}{\Sigma y_i}[C]_i\{\varepsilon\} \\ &= [C]\{\varepsilon\} \end{aligned} \tag{11}$$

where y_i is the i-th component of the texture vector. Similarly, the compliance matrix in the Reuss (constant stress) model [4] is obtained as equation 12.

$$[S] = \sum \frac{y_i}{\Sigma y_i}[S]_i \tag{12}$$

3. APPLICATION TO STIFFNESS ANALYSIS

3.1. Analytical method

The panel stiffness of a cylindrical shell part and the flexibility of a pipe were analyzed. Figures 1 and 2 show the schematic illustrations of these simulated models. In the panel stiffness analysis, the displacements of sides A–B and C–D were fixed, and 50 N of the point load was given at the center of the panel. The third component of the displacement vector at the center was used to evaluate the panel stiffness. In the analysis of pipe flexibility, nodes E and F were fixed and 50 N of point load was given at node G. The third component of the displacement at the node G was used in flexibility evaluation.

In finite element analyses on MARC, a four–node thick shell element (element 75) was used. The panel was divided into 240 square elements and the pipe was divided into 380 rectangular elements.

Elastic anisotropy is expressed in user–subroutine HOOKLW, where the compliance matrix is given. This is calculated in advance according to equation 12 or the inverse of $[C]$ in equation 11. The elastic constants C_{ij} of a single crystal are assumed to be the same as those of pure iron as shown in table 1 [12].

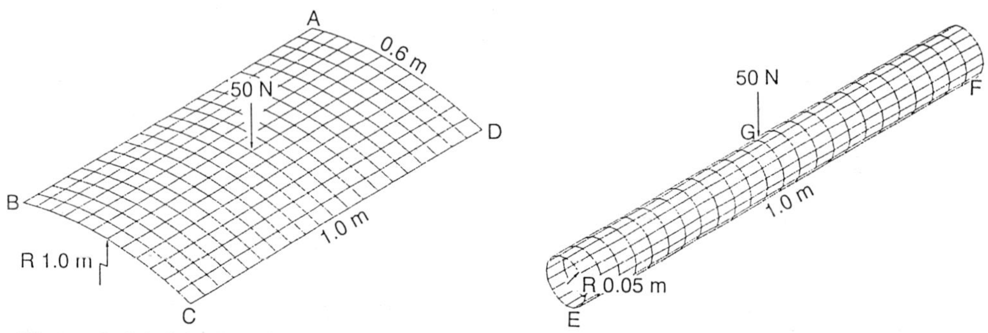

Figure 1. Model of analyzed panel part

Figure 2. Model of analyzed pipe

Table 1. Elastic constants C_{ij} of pure iron [12]

C_{11}	C_{12}	C_{44}
237 GPa	141 GPa	116 GPa

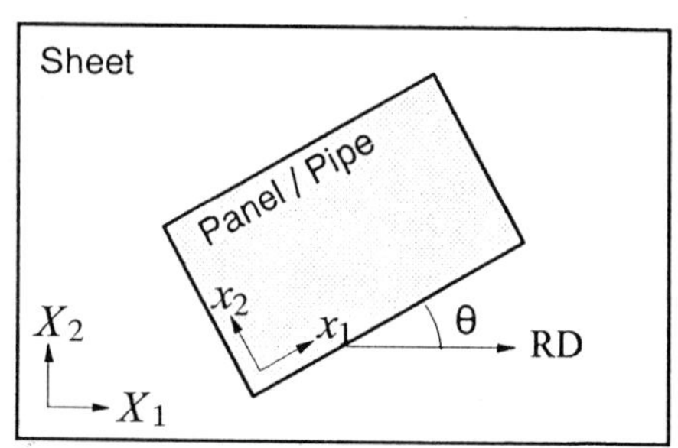

Figure 3. Relation between the coordinate system (X_1, X_2, X_3) on sheet and the global coordinate system (x_1, x_2, x_3) on panel or pipe

Let us make some comments on how to obtain the transformation matrix of equation 3. The orientation of each texture vector element can be expressed in Miller indices, $(h\ k\ l)[u\ v\ w]$. A coordinate system (X_1, X_2, X_3) on the sheet is defined, where the rolling direction (RD), the transverse direction (TD), and the normal direction (ND) are parallel to X_1, X_2, and X_3 directions, respectively. Then the matrix [Π] that transforms the sheet coordinates to local coordinates is expressed as equation 13, where (u', v', w'), (p, q, r), and (h', k', l') are the unit vectors parallel to X_1, X_2, and X_3 directions, respectively.

$$[\Pi] = \begin{bmatrix} u' & p & h' \\ v' & q & k' \\ w' & r & l' \end{bmatrix} \tag{13}$$

To cut the panel out of this sheet, θ is defined as the angle between A–B direction and RD (X_1 direction) as shown in figure 3. In the pipe simulation, θ is the angle between the longitudinal direction E–F and RD. The global coordinate on the component is rotated to the sheet coordinate by the matrix [Θ].

$$[\Theta] = \begin{bmatrix} \cos\theta & -\sin\theta & 0 \\ \sin\theta & \cos\theta & 0 \\ 0 & 0 & 1 \end{bmatrix} \tag{14}$$

Therefore, the transformation matrix of equation 3 is obtained as equation 15.

$$\{x'\} = [\Pi]\ [\Theta]\ \{x\} \tag{15}$$

3.2. Analytical results

Two steel sheets, A and B, with different textures were prepared for the numerical analysis. Their chemical compositions and heat treatments are shown in table 2 and figure 4, respectively. They were cold-rolled to 80% reduction before the heat treatments. The (200) pole figures measured by the X-ray diffraction are shown in figure 5.

Table 2. Chemical compositions (mass%) of steels used

Steel	C	Si	Mn	P	S	Al	N	Ti	Nb
A	0.0080	0.046	1.52	0.014	0.0039	0.024	0.0024	0.071	0.027
B	0.022	0.014	0.11	0.007	0.006	0.075	0.0017	-	-

Furthermore, those Young's moduli are measured in seven directions at 15° intervals to RD using the transverse vibration method. The results are shown in figure 6 with the calculated Young's moduli. The effects of the oriented-grain arrangement, the grain boundary, and the alloy elements have to be considered for more accurate analyses [2]. But figure 6 shows that the anisotropy of calculated Young's moduli is similar to that of the experimental results and the calculated values by the Reuss model agree with the measured values very well. Then the panel stiffness was analyzed based on the Reuss model.

The analytical results of panel stiffness and pipe flexibility at various angles are shown in figures 7 and 8. The displacements of anisotropic materials depend on the shape and boundary condition of the component and the angle θ. For reference, the panel stiffness and the pipe flexibility of a completely random-oriented steel sheet is also shown in these figures. Since all of the components of its texture vector become unity, the elasticity should be isotropic.

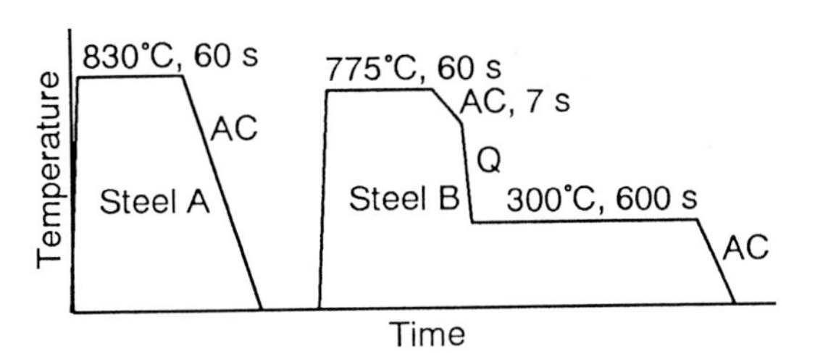

Figure 4. Heat treatment scheme

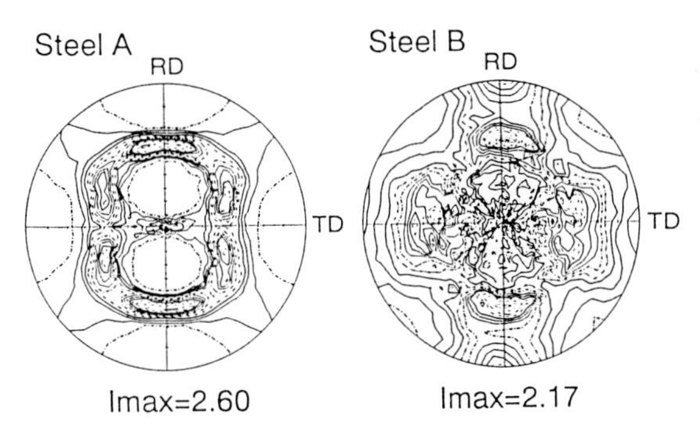

Figure 5. (200) pole figures of steels A and B

The third component of displacement in the panel is reduced more than 12% by applying the anisotropic materials at some angles as shown in figure 7. The most effective angles, 90° in steel A and a little under 60° in steel B, agree with the angle of the highest Young's modulus. The angles show that, in these cases, panel stiffness is most effectively improved when the direction of the highest Young's modulus coincides with the least rigid direction (x_1 direction) of the panel.

Figure 8 shows that the anisotropic sheets at some angles improve the pipe flexibility more than 10% in comparison with isotropic sheets. The displacement is affected by the rigidity of cross-section perpendicular to the longitudinal direction and the flexibility changes with the Young's modulus at the angle along the circumference on the cross-section (parallel to x_2 direction).

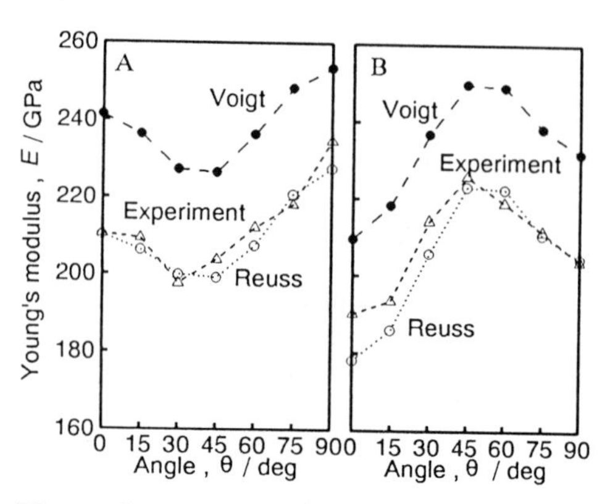

Figure 6. Young's moduli of steels A and B

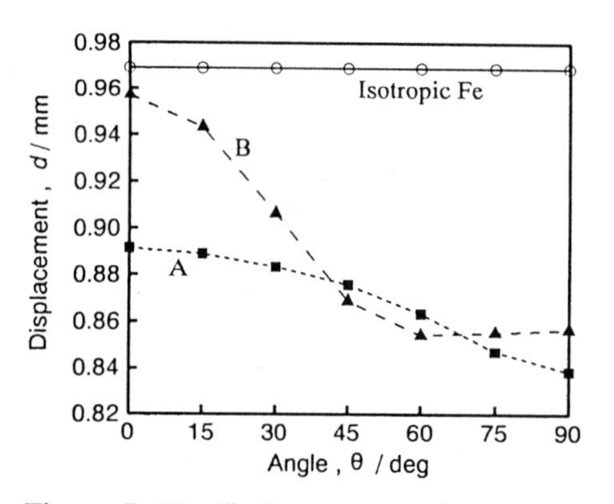

Figure 7. The displacements at the center of panels with three kinds of orientation distribution analyzed at various angles

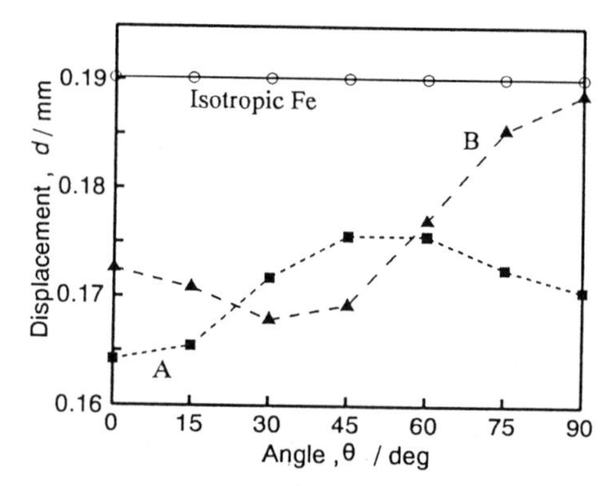

Figure 8. The displacements at the node G of pipes with three kinds of orientation distribution analyzed at various angles

4. CONCLUSION

A computer simulation method has been developed with which the stiffness of the anisotropic polycrystalline metal sheet components can be analyzed for instance from the orientation distribution determined by X-ray diffraction. The stiffness depends on the shape of the component and the angle of the anisotropic sheet applied to the component. The method makes it easy to use a strongly anisotropic steel sheet on structural components, such as automotive parts, for the purpose of weight reduction.

ACKNOWLEDGMENT

The authors are grateful to Professor Shin–ichi Nagashima for his advice about the calculation of elastic anisotropy.

REFERENCES

1) Bunge, H.J. and Roberts, W.T.: *J. Appl. Cryst.*, 1969, **2**, 116
2) Nagashima, S., Shiratori, M., Matsukawa, K., and Nakagawa, R.: Proc. JIMS–4, 1986, 833
3) Voigt, W.: Lehrbuch der Kristallphysik, 1928, 962
4) Reuss, A.: *Z. angew. Math. Mech.*, 1929, **9**, 55
5) Hill, R.: Proc. Phys. Soc. Lond. 1952, **A65**, 349
6) Hashin, Z. and Shtrikman, S.: *J. Mech. Solids*, 1962, **10**, 343
7) Hershey, A.V.: *J. Appl. Mech.*, 1954, **21**, 236
8) Kröner, E.: *Z. Physik*, 1958, **151**, 504
9) Bunge, H.J.: *Kristall und Technik*, 1968, **3**, 431
10) Spalthoff, P., Wunnike, W., Nauer–Gerhard, C., Bunge, H.J., and Schneider, E.: *Textures and Microstructures*, 1993, **21**, 3
11) Ruer, D. and Baro, R.: *J. Appl. Cryst.*, 1977, **10**, 458
12) Tegart, W.J.M.: Elements of Mechanical Metallurgy, 1966

Materials Science Forum Vol. 157-162 (1994) pp. 1593-1598

THE DEPENDENCE OF THE ANISOTROPY AND TEXTURE PROPERTIES OF INJECTION MOULDED LIQUID CRYSTAL POLYMER PARTS ON MOULDING PARAMETERS

W. Hufenbach and M. Lepper

Institut für Technische Mechanik, Technische Universität Clausthal,
Graupenstr. 3, D-38678 Clausthal-Zellerfeld, Germany

Keywords: Liquid Crystal Polymers, LCP, Injection Moulding, Mechanical Anisotropy

ABSTRACT

Thermotropic liquid crystal polymers are supposed to be one of the most interesting materials developed in the last decade. Their potential fields of use reach from electronics and apparatus construction to automotive industry as well as aero- and astronautics. Liquid crystal polymers show a strong anisotropy of their mechanical properties. The direction of the preferential material orientation in injection moulded parts is basically influenced by the flow history of the melt. Thus, the properties of parts made of LCP are determined not only by the material used but also by the mould design and the moulding parameters.

STRUCTURE OF THERMOTROPIC LCP

The distinctive feature of liquid crystal polymers are their rigid rod–like chain molecules, which are partially ordered in the melt, in contrast to the mostly disarranged molecules of conventional thermoplastic polymers. Between the solid and the liquid phase, LCP show an intermediate, liquid crystalline phase.

The rigid chain molecules in a liquid crystalline melt organize themselves into entropically stable domains with homogeneous orientation, thus producing a polydomain structure. During processing these domains break up and the molecules are aligned so as to form a monodomain structure (fig. 1). While cooling down, this structure freezes to the specific morphology of solid LCP with marked fibers and fibrils, similar to the morphology of wood. As fibers develop directly from the matrix material, LCP are so–called self–reinforcing polymers.

ORIENTATION INDUCING EFFECTS

Injection moulding is an efficient and economical process for producing complex and technically demanding parts. During this process the melt is subjected to both shear and elongational forces which cause an alignment of the polymer molecules. This oriented state of LCP is energetically advantageous. Therefore the orientation relaxation time is much longer than that of conventional

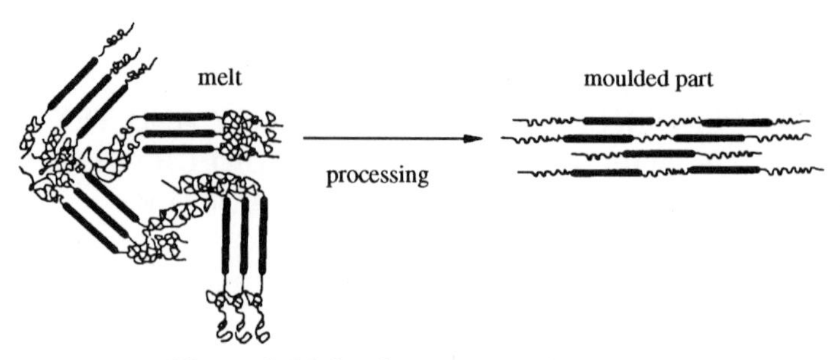

Figure 1: Molecular structure of LCP [1]

polymers. For this reason the orientation of LCP molecules is perfectly frozen in the solidifying melt and preserved in the finished part.

The fluid mechanics for injection moulding of LCP are rather complex, due to the unsteady two–dimensional flow, heat transfer and the existance of a moving free surface at the melt front. The flow in the mould cavity can be described as a superposition of shear flow and clongational flow. In the case of pure two–dimensional shear flow, also known as Poiseuille flow, the velocity profile $v_1 = f(x_2)$ (fig. 2) is characterized by the shear rate $\dot{\gamma}$ and the rate of deformation tensor $\mathbf{D}$ is of the following shape:

$$\mathbf{D} = \frac{1}{2}\left(\frac{\partial v_i}{\partial x_j} + \frac{\partial v_j}{\partial x_i}\right) = \frac{1}{2}\begin{bmatrix} 0 & \dot{\gamma} & 0 \\ \dot{\gamma} & 0 & 0 \\ 0 & 0 & 0 \end{bmatrix} \quad (1)$$

Since this flow is not free of rotation (rot $v \neq 0$), the rigid rod–like LCP molecules are rotating at a rotating speed dependent on their orientation φ [2]:

$$\dot{\varphi} = \frac{\dot{\gamma}}{2}(1 - \lambda \cdot \cos 2\varphi) \quad (2)$$

Here λ is a shape factor of the molecules assuming a value between λ=0 for ideal spheres and λ=1 for slim rods. Hence a preferential orientation of the molecules in the flow direction (φ=0°) follows.

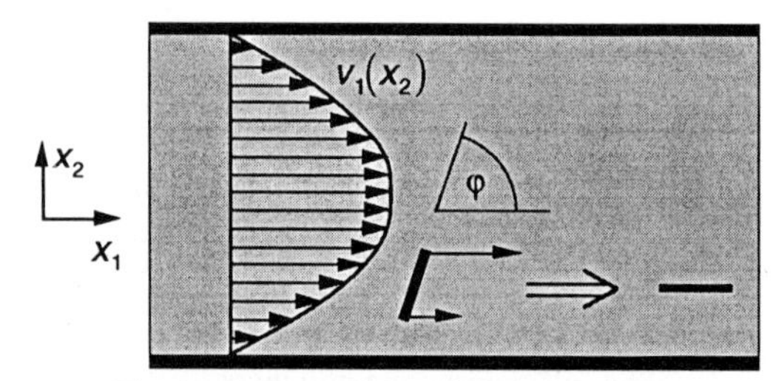

Figure 2: Two–dimensional shear flow

To some extent homogeneous elongational flow makes up the counterpart to shear flow. It is distinguished by the fact that the fluid elements don't rotate (rot v = 0) but are elongated or compressed in three perpendicular directions. Hence the rate of deformation tensor has diagonal shape:

$$\mathbf{D} = \frac{1}{2}\begin{bmatrix} \dot{\varepsilon}_1 & 0 & 0 \\ 0 & \dot{\varepsilon}_2 & 0 \\ 0 & 0 & \dot{\varepsilon}_3 \end{bmatrix} \quad (3)$$

By the elongation the LCP molecules are stretched and, for this reason, exhibit almost perfect orientation in the extension direction [3]. In a converging mould, the molecules are oriented in the direction of flow; in a diverging mould, they are stretched perpendicular to the flow direction. If, however, the mould is neither converging nor diverging, the previously coined orientation of the molecules remains unchanged (fig. 3). Thus, the flow history plays an important role for the elongation induced orientation [4].

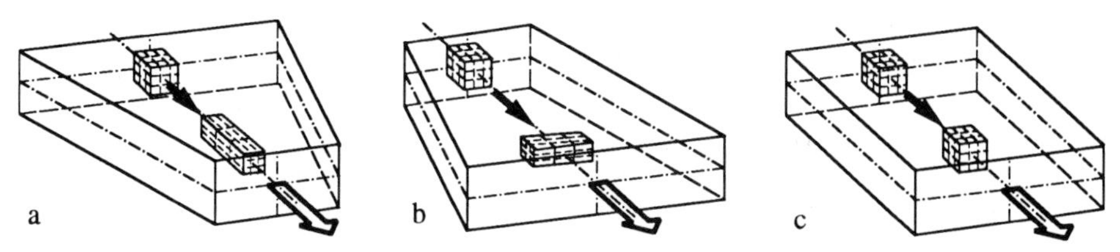

Figure 3: Elongation induced orientation (a: converging, b: diverging, c: parallel mould)

At the advancing melt front, a complicated flow pattern, known as "fountain flow", develops: The fluid elements from the core of the moulding reach the surface of the moving front and are deposited on the walls of the mould cavity (fig. 4). During this type of flow the polymer molecules in the fluid are elongated while flowing along the streamlines towards the walls. Since the temperature at the cavity surface is low, the molecules are frozen in this region with an orientation in the direction of flow.

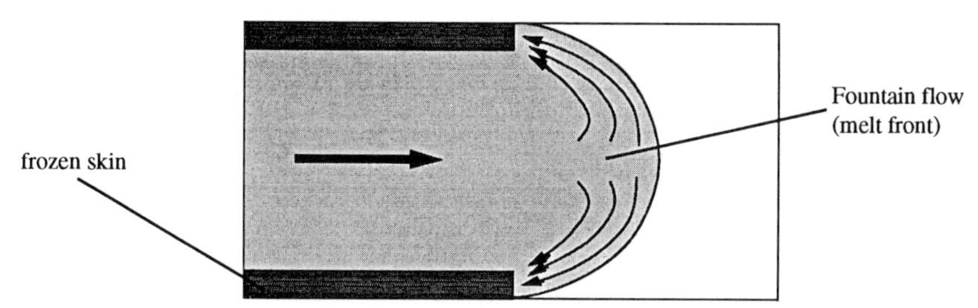

Figure 4: "Fountain flow" effect at the melt front

LAYER STRUCTURE IN INJECTION MOULDED LCP

In the case of non–isothermal flow, as in the filling of a cold mould, it has been shown [5] that the velocity profile has an inflection point due to the lower temperatures and the higher viscosity in the region close to the walls (fig. 5). As the shear rate attains a maximum at the location of the inflection, orientation of molecules due to shear effects takes place mainly in the region around the inflection point. The core region is dominated by elongation induced orientation.

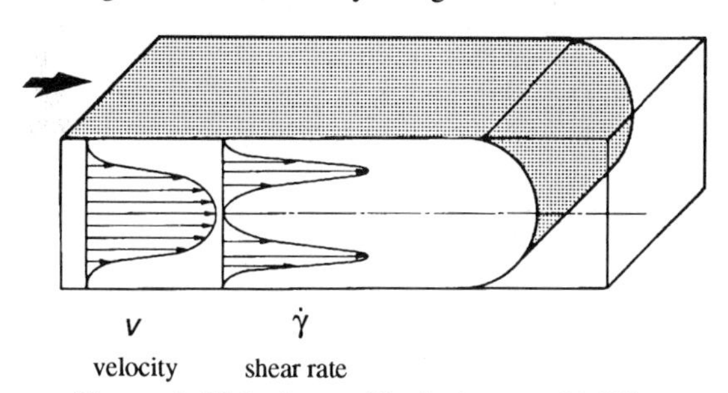

Figure 5: Velocity profile during mould filling

Due to the different flow regimes three distinct areas with different texture can be observed in moulded LCP samples (fig. 6). The development of this layer structure and the anisotropy in a moulded part fundamentally depend upon the moulding process. Even the mould design influences the anisotropy, because the type of the runner and of the gate determines the progress of the melt in the mould cavity (fig. 7) and thus the preferential orientation of the LCP molecules in the different layers. The expanding flow after a point gate, for example, produces a radial pattern of orientation, that flattens only at a certain distance from the gate, whereas a film gate at a rectangular plate leads to a parallel melt front quite from the start. In the same way obstacles in the melt flow affect the anisotropy and the development of fibers in injection moulded LCP. In the dead water areas behind the insert, for example, turbulences strongly disturb the texture (fig. 8).

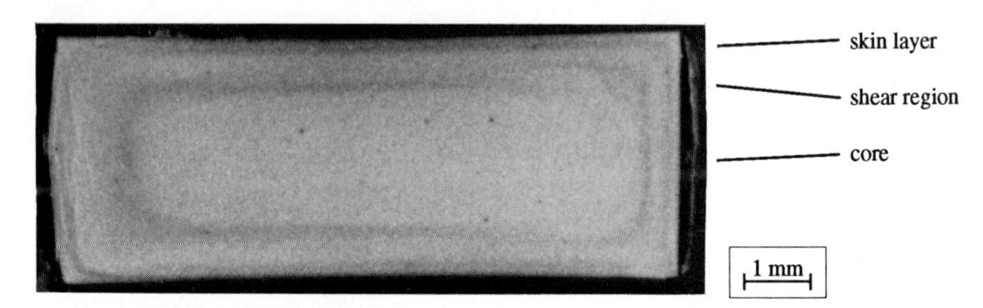

Figure 6: Layer structure of injection moulded LCP

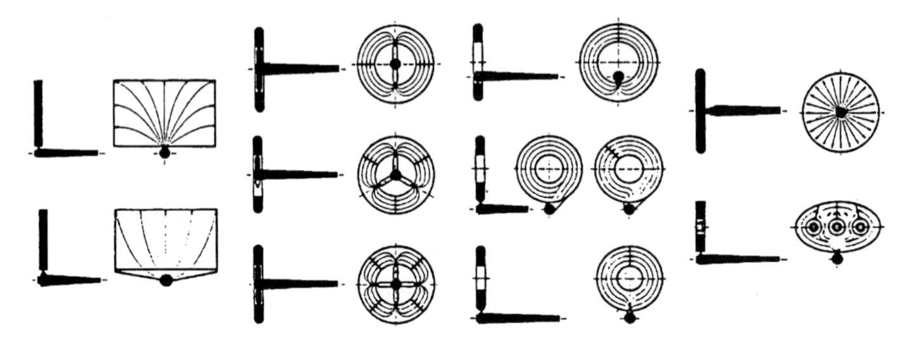

Figure 7: Flow pattern for different types of gates

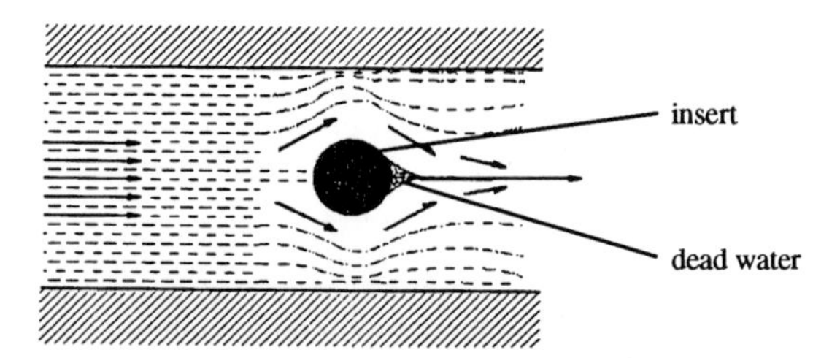

Figure 8: Dead water area behind an insert in the flow

The thickness and the texture in the different layers are substantially controlled by the moulding parameters melt temperature, mould temperature and volume flow rate.

As the "fountain flow" is an elongational type of flow, the orientation in the skin layer is always highly pronounced [3], while the thickness of the skin basically depends on the heat transfer from melt to mould [6]. Hence, the lower the melt and mould temperatures are and the slower the melt advances into the mould the thicker the skin should be.

In the shear flow region the degree of anisotropy depends on the shear rate and the time of exposure to shear effects, as can be seen from eq. (2). For this reason, when volume flow is faster, stronger texture in the shear layer could be found. In the same way, the texture in the shear layer increases with the distance from the gate.

Both the skin and the shear layer always exhibit a preferential orientation in the direction of flow, whereas the orientation of the core region is determined by the flow history (cf. fig. 3). Because of the long orientation relaxation time of LCP, the strong texture of the core layer induced by elongational flow shows only a small dependence on temperatures.

EXPERIMENTAL RESULTS

For our present studies on the layered properties of injection moulded LCP bending test bars according to DIN 53452 were moulded from ®Vectra A 950, an unfilled thermoplastic LCP supplied by Hoechst AG. The process parameters melt temperature, mould temperature and fill rate were varied within the range suggested by the material supplier.

These samples were characterized in a three point bending test. Due to the stress distribution induced during this test, the flexural modulus is well correlated with the thickness of the outermost region [7]. This is especially pronounced, because the fibers and fibrils in the fountain flow skin are strongly oriented in the flow direction, which coincides with the main stress direction for the samples under consideration. Therefore the flexural modulus exhibits the same behaviour as the skin thickness: With increasing melt temperature and increasing mould temperature the flexural modulus shows an evident declination (fig. 9 and 10). The same holds for an acceleration of the fill rate (fig. 11).

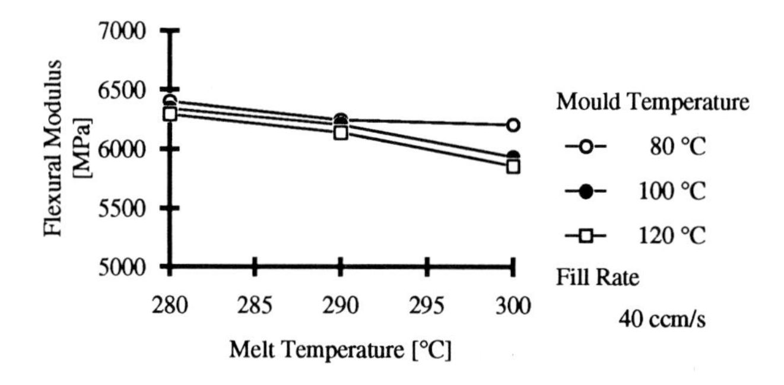

Figure 9: Dependence of flexural modulus on melt temperature

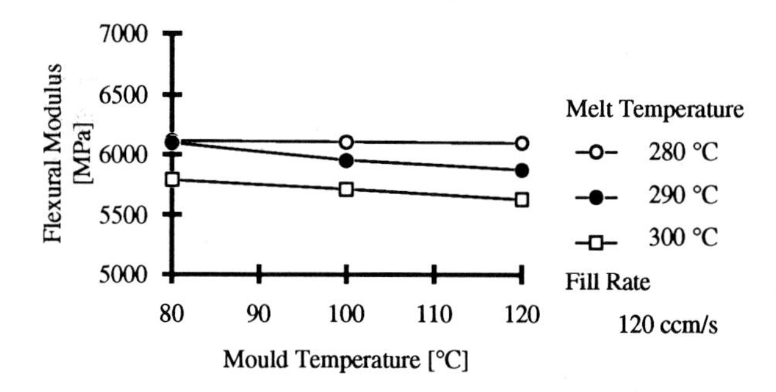

Figure 10: Dependence of flexural modulus on mould temperature

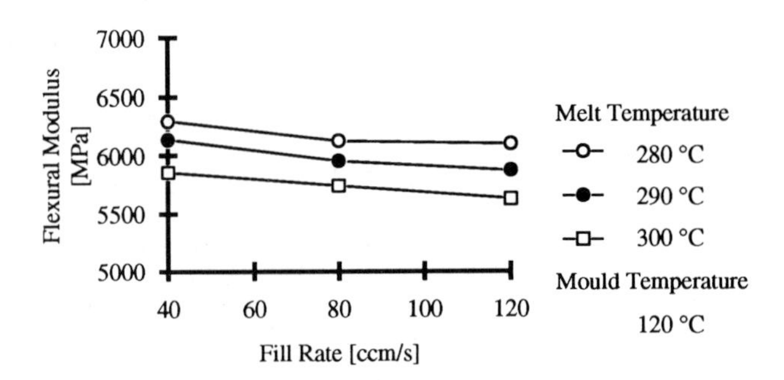

Figure 11: Dependence of flexural modulus on fill rate

DESIGN OF LCP PARTS

Our experiments have shown, that the layer thickness and the texture of injection moulded LCP can be controlled by a specific adjustment of the moulding process. Determining the local textures of the individual layers in a texture goniometer enables conclusions about the mechanical properties of the layers to be drawn. This can be used as a basis for the design of complex parts made of LCP.

For the assessment of the melt flow in the mould and the resulting preferential molecule directions, a numerical simulation can be a valuable tool. Modern software packages already allow a comfortable calculation of the injection moulding process on the basis of an adapted finite element method (fig. 12)[8].

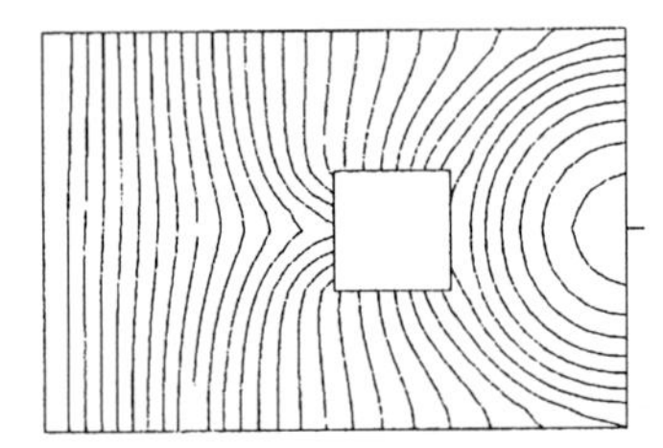

Figure 12: Numerical simulation of a melt front pattern (isochrones)

Knowing about the anisotropic layer properties and the layer structure and orientation, these results can be transferred to the global structure of complete parts by means of the homogenization technique and the laminate theory. Thus valuable proposals for their construction and load–adapted lay out can be given.

Financial support from the Deutsche Forschungsgemeinschaft (DFG) is gratefully acknowledged.

REFERENCES

1) Bangert, H.: Kunststoffe, 1989, 79, 1327
2) Jeffery, G. B.: Proc. Roy. Soc. A, 1922, 102, 161
3) Ide, Y.; Ophir, Z.: Polym. Eng. and Sci., 1983, 23, 261
4) Plummer, C. J. G.; Zülle, B.; Demarmels, A.; Kausch, H.–H.: J. Appl. Polym. Sci., 1993, 48, 751
5) Kamal, M. R.; Kenig, S.: Polym. Eng. and Sci., 1972, 12, 302
6) Kenig, S.; Trattner, B.; Anderman, H.: Polym. Comp. 1988, 9, 20
7) Garg, S. K.; Kenig, S.: in "High Modulus Polymers" (eds. Zachariades, A. E.; Porter, R. S.), Dekker, New York and Basel, 1988
8) Hufenbach, W.; Lepper, M.: Maschinenmarkt, 1993, 99, No. 16, 44

Materials Science Forum Vol. 157-162 (1994) pp. 1599-1602

USE OF IRREDUCIBLE SPHERICAL TENSORS IN THE CALCULATION OF THE MEAN ELASTIC PROPERTIES OF POLYCRYSTALS

M. Humbert [1], L. Zuo [2], J. Muller [1] and C. Esling [1]

[1] LM2P/ISGMP, Université de Metz, Ile du Saulcy, F-57045 Metz Cédex 1, France

[2] On leave from Department of Materials Sciences and Engineering, Northeastern University of Technology, Shenyang 110006, PR China

Keywords: Mean Values, Elastic Properties, Irreducible Spherical Basis

ABSTRACT

We have expressed the elastic tensors in an irreducible spherical basis. This allowed us to calculate the mean values of these tensors from the texture very easily, whatever the crystal symmetries. From the theoretical point of view, we have demonstrated symmetry related properties of the mean values in the case of cubic crystal symmetry.

INTRODUCTION

Generally, the mean values of elastic properties of a polcrystalline material, according to Voigt's or Reuss's assumption, are obtained from the texture by averaging the elastic tensors (compliance and stiffness) expressed in a Cartesian basis. These calculations can be performed in a similar way whatever the basis. Nevertheless, the practical calculations are made easier when the basis is well chosen. It is obvious that an irreducible spherical basis is well suited for rotation operations. Therefore, we propose to perform the averaging of the elastic tensors in an irreducible spherical basis. Moreover, the use of irreducible spherical tensors offers the possibility to elegantly deduce the number of independent components of the mean tensors in the case of different symmetries.

TENSORS OF RANK n IN AN IRREDUCIBLE SPHERICAL BASIS

A tensor B of rank n is a mathematical quantity with 3^n components, expressed in a Cartesian basis ($\mathbf{e_1}$, $\mathbf{e_2}$, $\mathbf{e_3}$) by:

$$B = B_{i_1 i_2 .. i_n} e_{i_1} e_{i_2} .. e_{i_n} \qquad (1)$$

$$\left(i_1, i_2, .., i_n = 1,2,3 \right)$$

When dealing with rotations the use of an irreducible spherical basis is more convenient

The expression of the tensor in an irreducible spherical basis reads [2]:

$$B = \sum_{\substack{\alpha\beta..\gamma \\ l m}} B_l^m(\alpha\beta..\gamma) E_l^m(\alpha\beta..\gamma) \qquad (2)$$

The $E_l^m(\alpha\beta..\gamma)$ are the new basis vectors and the notation $(\alpha\ \beta\ ..\ \gamma)$ indicates the sequence of intermediate irreducible representations used in the construction of the final basis.

TRANSFORMATION RULE

As already mentioned, the irreducible spherical basis is well adapted for rotation operations. The components of the tensor in a given reference system K^A are easily expressed from their components in the crystal coordinate system K^B [2] :

$$(-1)^n B_l^n(l_1 l_2) = \sum_{m=-l}^{l} (-1)^m B_l^m(l_1 l_2) T_l^{mn}(g) \qquad (3)$$

In this relation, $T_l^{mn}(g)$ are the generalized spherical harmonics associated with the rotation g which puts k^A in coincidence with K^B [3].

TENSORS OF ELASTICITY

We have built the irreducible spherical basis adapted to the tensors of rank 4, being subjected to the additional symmetries of the elastic properties. According to these symmetries the number of components is reduced to 21. Therefore, an elastic tensor (compliance or stiffness) can be expressed by [4]:

$$\begin{aligned} B = {} & B_0^0(0\,0)\, E_0^0(0\,0) + B_0^0(2\,2)\, E_0^0(2\,2) \\ & + \sum_{m=-2}^{2} B_2^m(2\,0) E_2^m(2\,0) + \sum_{m=-2}^{2} B_2^m(2\,2) E_2^m(2\,2) \\ & + \sum_{m=-4}^{4} B_4^m(2\,2) E_4^m(2\,2) \end{aligned} \qquad (4)$$

Only even ranks l up to 4 are involved in this expression.
For further calculations, it is necessary to express the spherical components of the tensors from the traditional Cartesian components and conversely. The relation between both components can be expressed through a unitary matrix U which was detailed elsewhere [4].

CALCULATION OF MEAN VALUES

Generally, the macroscopical properties of polycrystals can be well approximated by mean values calculated from the Orientation Density Function (O.D.F. denoted as f(g)). The mean value of a given quantity symbolically reads [3] :

$$\langle B\rangle = \oint B(g)\, f(g)\, dg \tag{5}$$

with

$$f(g) = \sum_{l=0}^{l_{max}} \sum_{\mu=1}^{M(l)} \sum_{n=-l}^{l} C_l^{\mu n} \dot{\dot{T}}_l^{\mu n}(g) \tag{6}$$

Transformation relation (3) along with the orthogonality properties of the generalized spherical harmonics $\dot{\dot{T}}_l^{\mu n}(g)$ (adapted for a given crystal symmetry) allows us to easily calculate the components of the average tensor [4]:

$$\langle B(l_1 l_2)\rangle_l^n = \frac{(-1)^n}{2l+1} \sum_{\mu=1}^{M(l)} \sum_{m=-l}^{l} (-1)^m \dot{A}_l^{\mu m} B_l^m(l_1 l_2)\, C_l^{\mu n *}(g) \tag{7}$$

where quantities $\dot{A}_l^{\mu m}$ are the crystal symmetry coefficients [3].

The corresponding Cartesian components of the mean value can be found by using matrix U.

EXAMPLE OF APPLICATION

The use of a spherical basis allows us to show some properties very efficiently. This section details one example. The elasticity of cubic single crystals is completely described by 3 independent Cartesian components: B_{1111}, B_{1122}, B_{1212}. By using matrix U, it is easy to find the corresponding spherical components :

$$B_0^0(0\,0) = B_{1111} + 2\,B_{1122}$$

$$B_0^0(22) = \frac{2}{\sqrt{5}}\left(B_{1111} - B_{1122} - 3B_{1212}\right)$$

$$B_4^4(22) = B_{1111} - B_{1122} - 2B_{1212}$$

$$B_4^0(22) = \frac{7}{\sqrt{70}} B_4^4(22)$$

According to relation (7), only rank l = 0 and 4 are involved in the case of cubic symmetry.

This is due to the fact that the symmetry coefficients $\dot{A}_l^{\mu m}$ differ from 0 for l = 0 and 4 . Moreover, m are multiples of 4, otherwise these quantities are null. Relations (4) and (7) directly show that the mean value of the elastic tensor of a cubic polycrystal without sample symmetry has only 11 independent components (rank $l = 2$ is not involved). The equivalent analysis, performed in a Cartesian basis, is quite tedious [5-6].

CONCLUSION

The use of a unitary matrix U allows us to relate the Cartesian and irreducible spherical components of elastic tensors. The calculation of rotated tensors as well as that of texture averaged tensors is made straightforward by the use of a spherical basis whatever the crystal symmetry of the investigated materials. From a theoretical point of view the use of irreducible spherical bases offers the possibility to demonstrate a number of symmetry-related properties in a very convenient way.

REFERENCES

1) Brink, D.M. & Satchler, G.R. : *Angular Momentum*, 1968, Oxford Library of Physical Sciences.
2) Jones, H.N. : *Spherical Harmonics and Tensors For Classical Field Theory*, Butterworths, 1985, London.
3) Bunge, H.J. : *Texture Analysis In Materials Science*, Butterworths, 1982, London.
4) Humbert, M. , Zuo, L., Muller, J. & Esling, C. : Phil. Mag. A, 1993, 68, 575.
5) Zuo, L., Xu, J. & Liang, Z. : J. Appl. Phys., 1989, 66, 2338.
6) Zuo, L., Humbert, M. & Esling, C. : J. Appl. Phys., 1993, 26, 422.

Materials Science Forum Vol. 157-162 (1994) pp. 1603-1608

DERIVATION OF YIELD CRITERIA OF CUBIC METALS FROM SCHMID'S LAW

H.-T. Jeong, D.N. Lee and K.H. Oh

Department of Metallurgical Engineering and Center for Advanced Materials Research, Seoul National University, Shinrim Dong, 151-742 Kwanak-Ku, Seoul, Korea

Keywords: Yield Locus, Schmid's Law, Von Mises Yield Surface

ABSTRACT

Yield loci of isotropic fcc and bcc polycrystals have been calculated by averaging yield loci of randomly oriented fcc or bcc single crystals, each of which is calculated using the generalized Schmid's law. Two different averaging procedures were used. One is to average the yield loci of single crystals along constant stress paths and another is to average along strain paths normal to the yield surfaces of single crystals. As the orientation distribution approaches the randomness, the calculated yield surface approaches the Von Mises yield criterion when averaging is performed along the stress paths and is similar in shape to that obtained from the Taylor-Bishop-Hill method when averaging is made along the strain paths.

INTRODUCTION

The Von Mises and Tresca yield criteria have been widely used for plastic behavior of isotropic materials. The criteria are basically empirical and are lack of physical picture. The Taylor-Bishop-Hill theory for active slip systems[1,2] has been used to calculate yield criteria for isotropic fcc and bcc metals[2,3]. The theory assumes homogeneous deformation of each grain in a polycrystal. Therefore the result is an upper bound solution. A lower bound approach has been attempted by Piehler[4]. But he calculated only loci for fcc and bcc planar isotropic materials.

The purpose of this work is to calculate lower bound solutions of yield loci for isotropic fcc and bcc polycrystals starting from Schmid's law.

CALCULATION

The orientation of each grain in a fcc or bcc polycrystal may be given by the Euler angles (ψ_1, ϕ, ψ_2) in the range of $0 \le \psi_1, \psi_2 \le \pi/2$, $0 \le \phi \le \pi/2$ [5].

All the stress states can be defined by the following three principal stresses[6].

$$\begin{aligned} \sigma_I &= k\cos(\pi/6+\theta)+\sigma_m \\ \sigma_{II} &= k\sin\theta+\sigma_m \\ \sigma_{III} &= -k\cos(\pi/6-\theta)+\sigma_m \end{aligned} \tag{1}$$

where θ is angle, σ_m hydrostatic stress and $k = (4J_2/3)^{1/2}$ with J_2 being the quadratic stress invariant. A given θ value determines σ_I, σ_{II} and σ_{III}. The principal stresses defined in the above satisfy the relation of $\sigma_I + \sigma_{II} + \sigma_{III} = 3\sigma_m$. The stress states defined in the case of $\sigma_m = 0$ are on the π plane.

The {111}<110> slip systems for fcc metals and the {110}<111>, {112}<111> and {123}<111> slip systems for bcc metals have been considered in the calculation.

For a single crystal, the shear stress acting on the h'th slip system can be calculated using the following equation.

$$\tau_h = \left| (\overline{a_1}\cdot\overline{p_h})(\overline{a_1}\cdot\overline{d_h})\sigma_I + (\overline{a_2}\cdot\overline{p_h})(\overline{a_2}\cdot\overline{d_h})\sigma_{II} + (\overline{a_3}\cdot\overline{p_h})(\overline{a_3}\cdot\overline{d_h})\sigma_{III} \right| \tag{2}$$

where $\overline{p_h}$ and $\overline{d_h}$ are the unit vector normal to the slip plane and the unit vector along the slip direction, and $\overline{a_1}$, $\overline{a_2}$ and $\overline{a_3}$ are the unit vectors along σ_I, σ_{II} and σ_{III}, respectively. Equation 2 shows that the shear stresses acting on the slip systems are naturally independent of the hydrostatic stresses. Therefore, the calculation of the yield locus on the π plane (σ_m=0) is good enough.

The yield locus calculation procedures are as follows : The values of σ_I, σ_{II} and σ_{III} are first determined by a given θ value. Then the shear stresses on all the slip systems are calculated using equation 2. The acting slip systems are those which give rise to the maximum shear stress. The yield condition is obtained by equating the maximum τ_h to the critical resolved shear stress, τ_{CR}, then k is expressed in terms of τ_{CR}. The k value is substituted into equation 1 to give the stresses at yielding. The same procedure is applied at every θ value to give a yield locus of the single crystal with a specific orientation.

Specifically, the principal stress axes of a single crystal are expressed relative to its orientation represented by the Miller indices (figure 1). The Euler angles of the crystal are transformed to the Miller indices. The stresses obtained for the single crystal at various orientations are averaged to give the yield condition of the polycrystal.

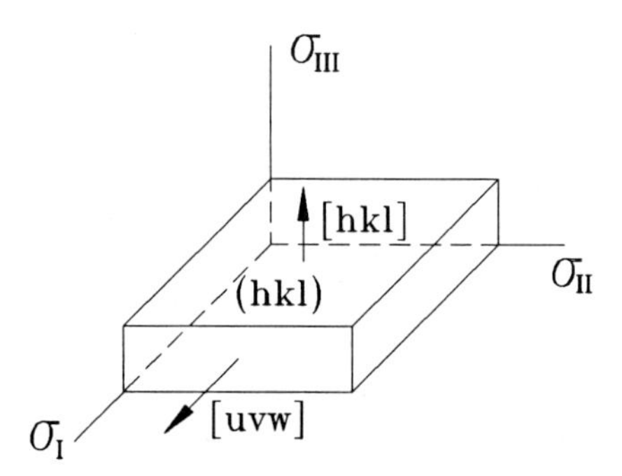

Figure 1 Relation between principal stress axes and orientation represented by the Miller indices.

Two averaging procedures have been used to calculate the yield loci of the polycrystal. One (method A) is to average the yield loci along constant stress paths, and another (method B) is to average along strain paths normal to the yield surface (the strain vector normality).

An example of the averaging procedure along the stress path (method A) is explained in figure 2. Figure 3 shows the averaging procedure depending on the strain vector normality (method B).

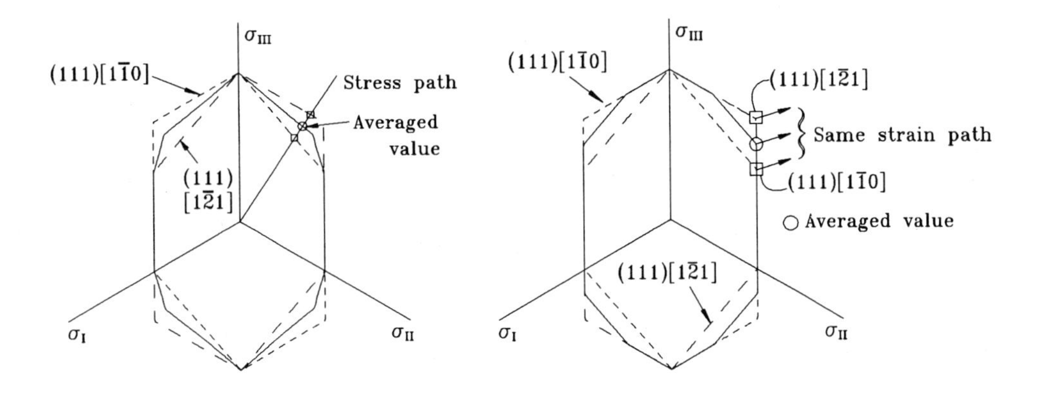

Figure 2 Averaging procedure of yield loci of single crystals having the (111)[1$\bar{1}$0] and (111)[1$\bar{2}$1] along a stress path (Method A).

Figure 3 Averaging procedure of yield loci of single crystals having the (111)[1$\bar{1}$0] and (111)[1$\bar{2}$1] along a strain path (Method B).

The yield surface of an isotropic polycrystal may be obtained by averaging yield loci of the randomly distributed single crystals. The yield loci are averaged using the following equation.

$$\bar{\sigma} = \int \sigma_s \, dg \tag{3}$$

where $\bar{\sigma}$ is averaged stresses at yielding, σ_s the stresses at yielding of single crystal s and g the orientation of the crystal s. If we express the orientation using the Euler angles, equation 3 can be expressed as

$$\sigma(\theta) = \frac{1}{8\pi^2} \int\int\int \sigma(\theta, \psi_1, \phi, \psi_2) \sin\phi \, d\psi_1 \, d\phi \, d\psi_2 \tag{4}$$

where θ defines the stress path for method A and defines the strain path satisfying the strain vector normality for method B.

RESULTS AND DISCUSSION

Figure 4 shows the yield loci of fcc single crystal calculated for the (001)[1$\bar{1}$0], (110) [$\bar{1}$10] and (111)[1$\bar{1}$0] orientations. Figure 5 shows the yield loci calculated for fcc and bcc polycrystals composed of randomly oriented single crystals whose orientation differences are 90°, 45° and 15° in the Euler space. The randomness of the polycrystal can be improved by decreasing intervals of the Euler angles. When the intervals of the Euler angles were less than 10°, the calculated results were almost identical. Figures 6 and 7 show the yield loci of fcc and bcc polycrystals calculated at intervals of 5°.

The yield loci calculated using method A approached the Von Mises yield criterion whereas method B yielded the yield loci similar in shape to those obtained by the Taylor- Bishop-Hill method (figure 8).

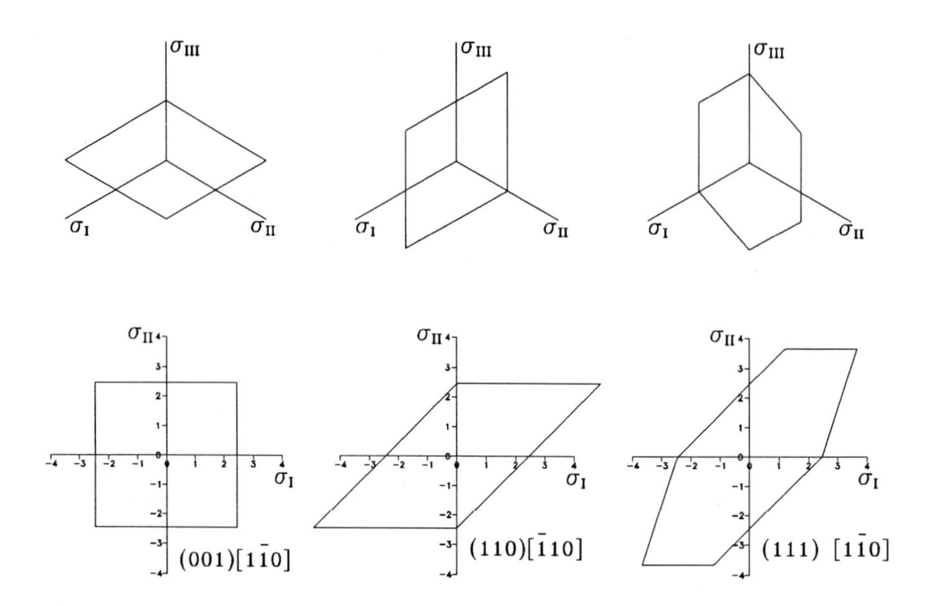

Figure 4 Calculated yield loci of fcc single crystals.

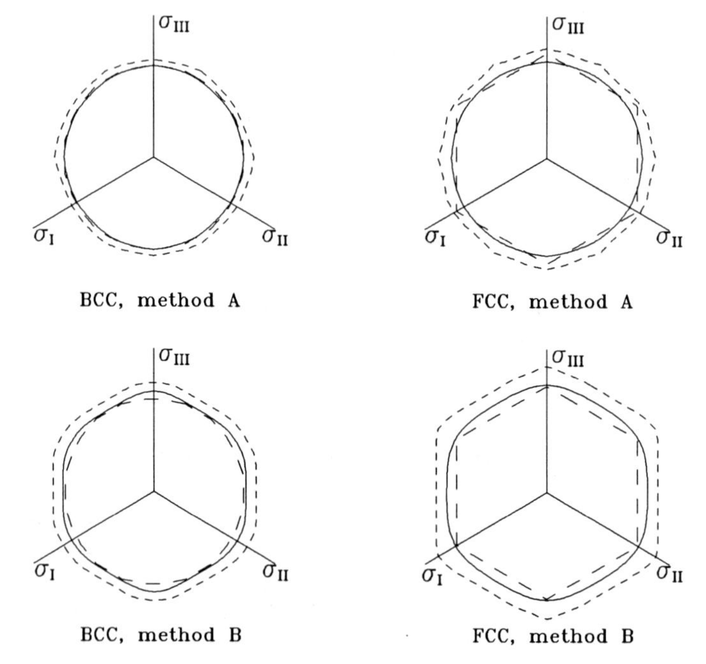

Figure 5 Yield loci calculated for fcc and bcc polycrystals composed of randomly oriented single crystals. The intervals of the Euler angles are (- - -) 90°, (-----) 45° and (——) 15° .

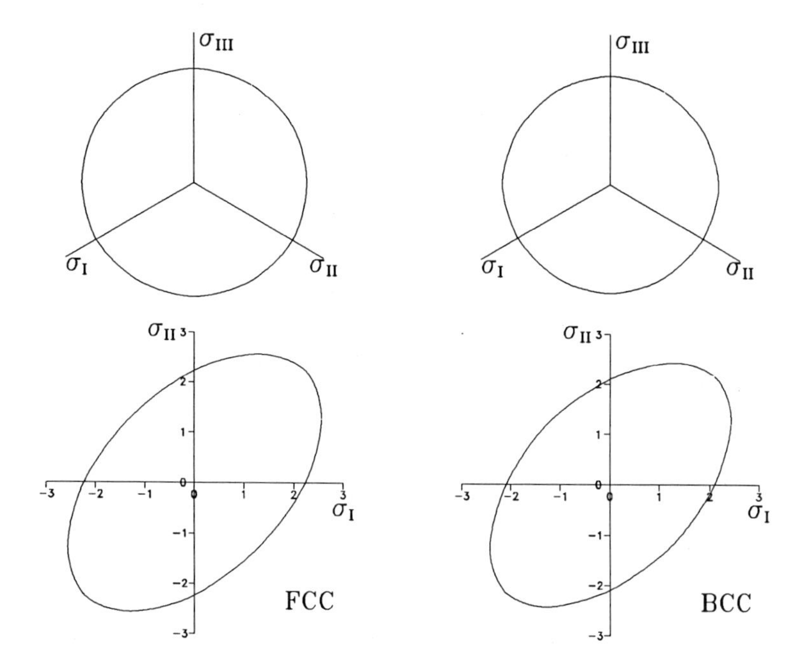

Figure 6 Yield loci of isotropic fcc and bcc polycrystals calculated using method A.

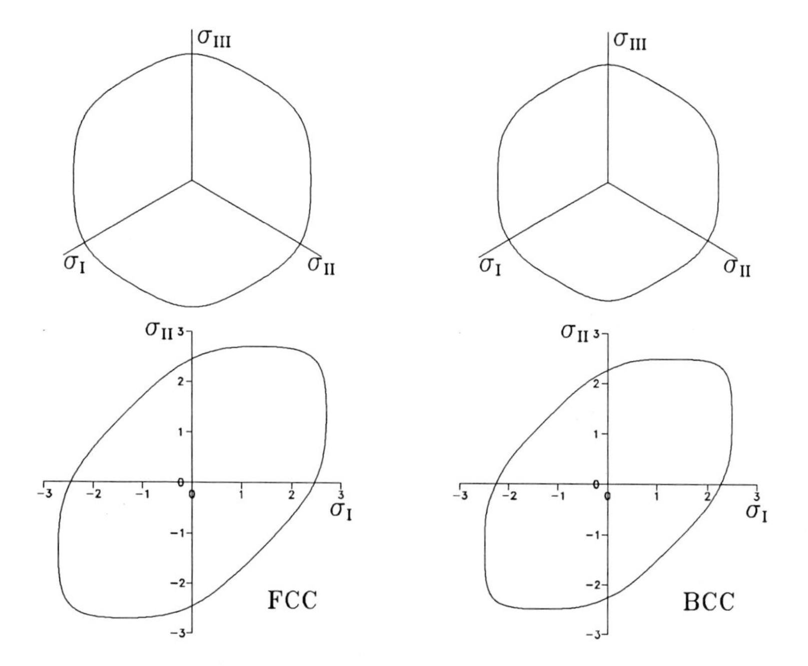

Figure 7 Yield loci of isotropic fcc and bcc polycrystals calculated using method B.

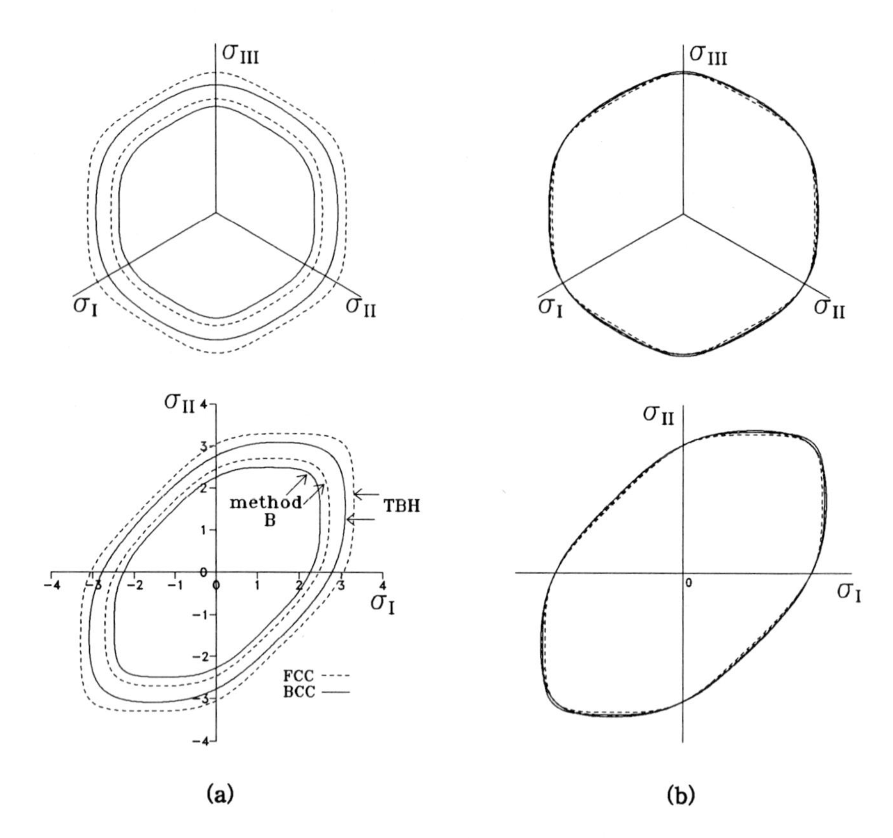

Figure 8 Yield loci of isotropic fcc and bcc polycrystals calculated using the Taylor-Bishop-Hill method and method B. Figure (a) show th loci in the same scale and (b) comparison between shapes of the different loci.

ACKNOWLEDGEMENTS

This work was supported by Korea Science and Engineering Foundation through Research Center for Thin Film Fabrication and Crystal Growing of Advanced Materials, Seoul National University.

REFERENCES

1) Taylor, G. I. : J. Inst. Met., 1938, 62, 307.
2) Bishop, J. W. F. and Hill, R. : Phil. Mag., 1951, 42, 414 and 1298.
3) Hutchinson, J. W. : J. Mech. Phys. Sol., 1964 12, 25.
4) Piehler, H. R. : Sc. D. Thesis, Dept. of Metallurgy, M. I. T. Cambridge, Mass., 1967.
5) Bunge, H. J. : "Texture analysis in Material Science", Butterworths, London, 1982.
6) Chakrabarty, J. : "Theory of plasticity", McGraw-Hill, New York, 1987, 58.

Materials Science Forum Vol. 157-162 (1994) pp. 1609-1622

CALCULATION OF AVERAGE ELASTIC MODULI OF POLYCRYSTALLINE MATERIALS INCLUDING $BaTiO_3$ AND HIGH-T_C SUPERCONDUCTORS

H. Kiewel and L. Fritsche

Institut für Theoretische Physik B, TU Clausthal,
Leibnizstr. 10, D-38678 Clausthal-Zellerfeld, Germany

Keywords: Effective Elastic Moduli, Ceramics

ABSTRACT

We report a new method that allows the calculation of effective elastic constants of polycrystalline materials - which may in general exhibit texture - within an iterative self-consistent scheme. The conceptual idea of our method consists in simulating the average elastic properties of the material by embedding a small single crystal of the polycrystalline structure in a suitably chosen fictional homogeneous medium and by averaging over a sufficiently large number of crystal orientations. We report results on the displacement field of an embedded single crystal and the resulting mean elastic moduli of a set of well defined polycrystalline materials. The calculations were then extended to some ceramics of practical importance to provide data on the ideal polycrystalline properties of these materials.

An important advantage of our approach over presently available methods pursuing similar goals is that our scheme can be generalized to embedded clusters of small crystals so that the effect of true grain boundaries can be accounted for.

INTRODUCTION

In many practical applications the mechanical properties of polycrystalline materials are described by well defined elastic moduli which represent averages of the elastic constants of the individual crystals that constitute the material. First attempts to gain estimates of these averages were made by Voigt [1] and Reuss [2]. In performing the average of the elastic constants over the various grain orientations these authors neglect any interaction between the crystals as it actually occurs at the grain boundaries. Hill [3] was able to show that these estimates by Reuss and Voigt are lower and upper bounds, respectively, of the true mean value of the elastic constants. The interaction between the grains has first been taken into consideration by Kröner [4] in a summary fashion which amounted to embedding a spherically shaped sample grain in a homogeneous elastic medium whose coupling to the grain is controlled by the familiar boundary conditions at the sphere. The elastic constants of the embedding medium were determined by requiring the "polarizability" of the sphere to vanish after averaging over all grain orientations. This model had already been discussed in a preliminary form by Hershey [5] four years earlier but not as consistently condensed into a practicable scheme as Kröner's theory which was used by Kneer [6], [7] to calculate average elastic moduli for a variety of materials with impressive accuracy.

However, the quantitative success of this method hinges crucially on the validity of the assumption that the grains of the material are well described by spheres on the average. If there is a sizeable departure from sphericity of the grain shape, one can only calculate lower and upper bounds for the mean elastic moduli provided that two averaged tensors, that are connected to the polarizabilities and their inverse, are simultaneously positive or negative definite. The latter cannot be ensured by general criteria.

The present paper was motivated by recognizing the limitations of the Kröner-Kneer approach in the context of strongly textured materials. Our work pursues the objective to primarily gain access to the strain field of a cluster of grains embedded in a homogeneous elastic medium. The shape of the grains and the relative orientation of their crystallographic axes are allowed to vary within a sizeable range to simulate realistic materials. We still use the Hershey/Kröner idea of embedding an anisotropic body in a homogeneous matrix and subjecting this composite system to asymptotically homogeneous strains as a method of eventually defining macroscopic averages for the true polycrystalline material. In order to obtain the mean elastic moduli of a textured sample one has to perform angular averages over the orientation space of the cluster by invoking orientation distribution functions from the pertinent experiments. The work of Kumar [8] is also aimed at calculating macroscopical averages of elastic moduli for polycrystalline materials. The theoretical concept of this author differs, however, considerably from ours in that he is using a representative (and hence sufficiently large) section of the material whose response to certain stresses is treated within a finite element method.

The mathematical procedure employed in our method amounts to expanding the displacement inside each grain of the embedded object in terms of basis functions that individually satisfy the pertinent differential equation governing the static equilibrium in an elastic medium. These functions are required to remain finite at the origin which coincides with the center of the pertinent grain.

In case one is dealing with an elastically isotropic embedding medium, the displacement in this area can also be expanded in terms of basis functions which individually satisfy the respective differential equation but vanish asymptotically as the distance from the embedded object tends to infinity. If the embedding medium is anisotropic a linear combination of these functions (which apply to isotropic media) is no longer suitable to represent the displacement in this area. Instead of finding new basis functions that solve the new differential equation we use plane waves which, apart from analytical simplicity, offer the additional advantage of presenting a so-called "unbiased basis set". Since these functions are not solutions to the differential equation of the embedding medium, the associated expansion coefficients can only be determined by solving the equivalent variation problem, i. e. by minimizing the excess deformation energy with respect to a reference energy that would occur in the system without the embedded object. This minimization is subject to the constraint that the force and displacement mismatch at the grain boundaries attains a minimum as well. This mismatch occurs as a result of the error in the displacement introduced by the truncation of the expansions which the practical calculations naturally require. In carrying out the minimization one arrives at a general eigenvalue problem the solution of which yields the set of expansion coefficients for the sought-for displacement. (Actually one obtains a set of solutions out of which one has to select the one of physical significance by searching for the smallest deformation energy.) The procedure bears close resemblance to the so-called Linear Rigorous Cellular (LRC-) method [9] used in electronic structure calculations for solids.

In Section 2 we develop the formal scheme of our method in full generality but confine ourselves to reporting only results on polycrystalline materials which conform sufficiently well to the assumption of a spherical grain shape. This is to assess the functioning of our method within a proven testing ground. To demonstrate further capabilities of the new scheme we also present results on the spatial dependence of the trace of the strain tensor for a cubical crystal embedded in a homogeneous matrix. The results are altogether discussed in Section 3.

THEORY

The fundamental equations governing the behavior of elastic media in a state of static equilibrium are given by

$$\operatorname{div} \underline{\underline{\sigma}}(\underline{r}) = 0 \tag{1}$$

and by Hooke's law

$$\sigma_{ij}(\underline{r}) = \sum_{k,l} c_{ijkl}\, \epsilon_{kl}(\underline{r}) \tag{2}$$

as long as the strains described by the tensor $\underline{\underline{\epsilon}}$ are small compared to unity. The stress tensor is denoted by $\underline{\underline{\sigma}}$, and the quantities c_{ijkl} stand for the elastic constants. Inserting equation 2 into equation 1 one obtains a differential equation for the components s_k of the displacement $\underline{s}(\underline{r})$

$$\sum_{j,k,l} c_{ijkl} \frac{\partial^2 s_k}{\partial x_j\, \partial x_l} = 0 \text{ for } i = 1,2,3\,. \tag{3}$$

This equation may be viewed as the Euler equation of the variation principle that the deformation energy

$$W = \int u(\underline{r})\, dV \tag{4}$$

where

$$u(\underline{r}) = \frac{1}{2} \sum_{i,j} \sigma_{ij}(\underline{r})\, \epsilon_{ij}(\underline{r}) \tag{5}$$

attains an extremum under the constraint of certain boundary conditions which render the system uniquely defined at static equilibrium.

To gain access to the most general situation we first consider the case of isotropic crystal symmetry. Equation 3 may then be cast as

$$- c_{11} \operatorname{grad} \operatorname{div} \underline{s} + c_{44} \operatorname{curl} \operatorname{curl} \underline{s} = 0 \tag{6}$$

where we have used Voigt's notation for the elastic constants. One can break up $\underline{s}$ into a curl-free component and a remainder with vanishing divergence

$$\underline{s} = - \operatorname{grad}\phi + \operatorname{curl} \underline{A} . \tag{7}$$

This decomposition is unique once $\underline{A}$ has been gauged. For the present purpose it is convenient to set

$$\operatorname{div} \underline{A} = 0 . \tag{8}$$

On substituting expression 7 into equation 6 and using equation 8 we obtain

$$- \operatorname{grad}\hat{\phi} + \operatorname{curl} \hat{\underline{A}} = 0 \tag{9}$$

where $\hat{\phi}$ and $\hat{\underline{A}}$ may be viewed as inhomogeneities of Poisson-type equations for ϕ and $\underline{A}$:

$$\begin{aligned} c_{11} \triangle \phi &= -\hat{\phi} \\ c_{44} \triangle \underline{A} &= -\hat{\underline{A}} . \end{aligned}$$

As one can see on forming, respectively, the divergence and the curl of equation 9, $\hat{\phi}$ and $\hat{\underline{A}}$ are solutions to the Laplace equation. For that reason the scalar elastic potential $\phi(\underline{r})$ can be expanded

$$\begin{aligned} \phi(\underline{r}) = & \sum_{l=0}^{l_{max}} \sum_{m=-l}^{l} c_{lm}^{(1)} r^l Y_{lm}(\vartheta,\varphi) + \sum_{l=0}^{l_{max}} \sum_{m=-l}^{l} d_{lm}^{(1)} r^{l+2} Y_{lm}(\vartheta,\varphi) \\ & + \sum_{l=0}^{l_{max}} \sum_{m=-l}^{l} c_{lm}^{(2)} r^{-(l+1)} Y_{lm}(\vartheta,\varphi) + \sum_{l=0}^{l_{max}} \sum_{m=-l}^{l} d_{lm}^{(2)} r^{-(l-1)} Y_{lm}(\vartheta,\varphi) \end{aligned} \tag{10}$$

where we have introduced spherical coordinates r, ϑ, φ, and $Y_{lm}(\vartheta,\varphi)$ denotes spherical harmonics. Obviously each term under the first and third sum satisfies the Laplace equation. If one expands $\hat{\phi}(\underline{r})$ in analogy to these sums and forms the Poisson integral, one obtains the second and fourth sum in equation 10. The vector potential $\underline{A}(\underline{r})$ may be expanded similarly, the only difference being that one has to observe equation 8. It is elementary to show that $r^l Y_{lm}(\vartheta,\varphi)$ may be reexpressed as a linear combination of products $x^i y^j z^k$ where $i+j+k=l$ and x, y, z denote Cartesian coordinates. If one takes advantage of this property and inserts $\phi(\underline{r})$ and $\underline{A}(\underline{r})$ into equation 7, the result may be written

$$\underline{S}^{(\alpha)}(\underline{r}) = \sum_{l,m} a_{lm}^{(\alpha)} \underline{s}_{lm}^{(\alpha)}(\underline{r}) \tag{11}$$

where $\alpha = 0$ refers to the embedding region and $\alpha = 1, 2, \ldots, N$ numbers the embedded grains. The new expansion coefficients $a_{lm}^{(\alpha)}$ are linear combinations of the previously introduced coefficients, and the new basis functions are defined:

$$\underline{s}_{lm}^{(\alpha)}(\underline{r}) = \sum_{i=0}^{l}\sum_{j=0}^{l-i} b_{lmij}^{(\alpha)}\, x^i\, y^j\, z^{l-i-j} \quad \text{for } \alpha = 1,2,\ldots,N \tag{12}$$

$$\text{and} \quad \underline{s}_{lm}^{(0)}(\underline{r}) = r^{-2l-3}\sum_{i=0}^{l+2}\sum_{j=0}^{l+2-i} \underline{b}_{lmij}^{(0)}\, x^i\, y^j\, z^{l+2-i-j}\,. \tag{13}$$

Although expansion 11 has been obtained from considering the case of isotropic crystal symmetry only, it may be used to find basis functions in terms of which the displacement inside the grains may be expanded in the more general case of elastic anisotropy. To this end, we insert expression 12 into equation 3. One obtains a sum of products $x^i y^j z^k$ with certain prefactors that are c_{ijkl}-dependent linear combinations of the coefficients $b_{lmijk}^{(\alpha)}$. In equating these linear combinations to zero one arrives at a set of $l(l+1)/2$ equations which can be solved for $b_{lmijk}^{(\alpha)}$. The number of the latter for fixed k, l and m is $(l+2)(l+1)/2$. Since

$$(l+2)(l+1)/2 \; - \; l(l-1)/2 \; = \; 2l+1\,,$$

we can construct $2l+1$ different sets of these coefficients. Each set can be labelled by the index m which runs, e. g. from $-l$ to $+l$. Inserting these coefficients into equation 12 we obtain for each grain altogether $3(l_{max}+1)^2$ basis functions $\underline{s}_{lm}^{\alpha}(\underline{r})$ whose simple algebraic form considerably alleviates the calculation of expressions that involve these functions.

If the embedding material is isotropic, the difference between the number of coefficients $b_{lmijk}^{(0)}$ and the number of independent equations is again $2l+1$. In determining these coefficients we may therefore proceed as before.

At the boundaries between neighboring grains and between the embedding material and adjacent grains one has to require the displacements $\underline{S}$ and the normal stresses $\underline{\sigma}_n$ to be continous, that is

$$\underline{S}^{(\alpha')}\Big|_{A_{\alpha',\alpha}} = \underline{S}^{(\alpha)}\Big|_{A_{\alpha',\alpha}} \tag{14}$$

$$\text{and} \quad \underline{\sigma}_n^{(\alpha')}\Big|_{A_{\alpha',\alpha}} = \underline{\sigma}_n^{(\alpha)}\Big|_{A_{\alpha',\alpha}} \tag{15}$$

where $A_{\alpha',\alpha}$ denotes the associated boundaries. Since the expansions for $\underline{S}$ in the pertinent regions have in general to be truncated at some l_{max}, the conditions 14 and 15 cannot rigorously be fulfilled. One has therefore to resort to the weaker condition that the boundary mismatch defined by

$$E^2 := \sum_{\alpha',\alpha}\int_{A_{\alpha',\alpha}} \{|\underline{S}^{(\alpha')}(\underline{r}) - \underline{S}^{(\alpha)}(\underline{r})|^2 + \mu|\underline{\sigma}_n^{(\alpha')}(\underline{r}) - \underline{\sigma}_n^{(\alpha)}(\underline{r})|^2\} d^2r \tag{16}$$

attains a minimum. This requirement will be used throughout this article as a substitute for equations 14, 15. The real-valued quantity μ may be viewed as a Lagrange parameter.

We first consider the simplest embedding case defined by a spherical grain in an isotropic environment. This model is also underlying Kröner's treatment [4] which, however, differs from ours in the use of the mathematical means. One envisages the system to be subject to well defined stresses which lead to homogeneous strains ϵ_{0ij} at sufficiently large distances from the sphere. If the elastic moduli of the isotropic embedding medium (e. g. the bulk modulus B and the shear modulus G) have the appropriate average values of the polycrystalline material, the mean value $\bar{\epsilon}_{ij}$ of the strain averaged over the sphere volume and the orientation space of the grain must by definition be identical to ϵ_{0ij}. This is to say that the quantity

$$\triangle\epsilon_{ij} = \frac{1}{N_\beta} \sum_{\beta=1}^{N_\beta} \bar{\epsilon}_{ij}^{(\beta)} - \epsilon_{0ij} \tag{17}$$

where the sum runs over a sufficiently large number of uniformly distributed crystal orientations, vanishes when B and G attain their proper values.

We expand the displacement outside and inside the sphere ($\alpha = 0, 1$) according to equation 11 except that we explicitly separate off an asymptotic displacement $\underline{S}_0$ for $\alpha = 0$ so that

$$\underline{S}^{(0)} = \underline{S}_0 + \sum_{l,m} a_{lm}^{(0)} \underline{s}_{lm}^{(0)} . \tag{18}$$

The displacement $\underline{S}_0$ is assumed to be a linearly increasing function that gives rise to a homogeneous strain field. Clearly, the overall displacement $\underline{S}(\underline{r})$ has in this case odd parity if the sphere center coincides with the origin of the coordinate system. Consequently, all coefficients for even l in expansion 11 must vanish. As has been shown by Eshelby [10], the strain field inside the sphere is also homogeneous. One may therefore break off the expansion for $\underline{S}^{(1)}(\underline{r})$ at $l = 1$, and the average quantities $\bar{\epsilon}_{ij}^{(\beta)}$ in equation 17 may be identified with the true values of $\epsilon_{ij}^{(\beta)}$ inside the sphere. It furthermore turns out that expansion 18 breaks off at $l = 3$ and that the boundary conditions 14, 15 can be fulfilled exactly as a consequence of this. The non-zero coefficients are obtained from solving the associated linear set of inhomogeneous equations. Due to the properties of $\underline{S}^{(0)}(\underline{r})$, the expansion of the elastic potentials ϕ and $\underline{A}$ for the outside region contain only components for $l = 0$ and $l = 2$ which may be viewed as describing a superposition of a monopole and a quadrupole field. This is nicely made evident by some plots discussed in Section 3. Once the strains $\epsilon_{ij}^{(\beta)}$ inside the sphere have been determined for a set of 3^3 to 21^3 crystal orientations β, we form $\triangle\epsilon_{ij}(B, G)$ according to equation 17 and search for the common zero of these quantities.

As mentioned above, the expansion coefficients $\underline{b}_{lmij}^{(0)}$ in equation 13 can only be determined if the embedding material is isotropic. In the more general case when one is dealing with a textured material, the basis functions $\underline{s}_{lm}^{(0)}(\underline{r})$ defined by equation 13 can no longer be used in expanding the displacement in the embedding region. Instead we partition $\underline{S}^{(0)}(\underline{r})$ in the form

$$\underline{S}^{(0)} = \underline{S}_0 + \underline{S}_{PW} , \tag{19}$$

where $\underline{S}_0$ has the same meaning as in equation 18 and $\underline{S}_{PW}$ is a plane wave expansion

$$\underline{S}_{PW} = \sum_{k=1}^{3} \sum_{\underline{K}} \left(a_{\underline{K}+}^{(0)k} \cos\underline{K}r + a_{\underline{K}-}^{(0)k} \sin\underline{K}r \right) \underline{e}_k \tag{20}$$

which describes the response field caused by the embedded object. In order to make this expansion discrete, we have introduced a sufficiently large orthorhombic unit cell as a periodicity region such that $\underline{S}_{PW}$ has (on the scale of interest) dropped to zero at the cell boundaries. The $\underline{K}$-vectors in expansion 20 denote vectors of the reciprocal lattice associated with the primitive orthorhombic real-space lattice. Unit vectors along the Cartesian coordinate axes $k = 1, 2, 3$ are denoted by $\underline{e}_k$.

As opposed to the previous basis functions $\underline{s}_{lm}^{(0)}(\underline{r})$ the plane waves in the expansion 20 do not individually satisfy the differential equation 3. To determine the expansion coefficients we therefore employ the variation principle defined in the context of equations 4 and 5. If one only minimizes the extra deformation energy

$$W_{PW} = \int_{V^{(0)}} u_{PW}(\underline{r})\, dV \tag{21}$$

under the constraint that E^2 - defined by equation 16 - attains simultaneously a minimum, one arrives at a general eigenvalue problem of the form

$$\underline{\underline{M}}_{PW}\, \underline{a} + \lambda \left(\underline{\underline{M}}_S\, \underline{a} + \mu\, \underline{\underline{M}}_\sigma\, \underline{a} \right) = 0\,. \tag{22}$$

Here $\underline{a}$ denotes a vector that contains the coefficients a_0, $a_{lm}^{(\alpha)}$, $a_{\underline{K}^+}^{(0)k}$, $a_{\underline{K}^-}^{(0)k}$ where a_0 represents a factor in front of the linear function describing the real-space dependence of $\underline{S}_0$. The symmetrical matrices $\underline{\underline{M}}_{PW}$, $\underline{\underline{M}}_S$ and $\underline{\underline{M}}_\sigma$ result from inserting the expansions 20 and 11 into equation 21 and equation 16, respectively, and performing the integrals over the real-space dependent portions. The real-valued quantity λ occurs as an additional Lagrange parameter which attains certain values in solving the eigenvalue problem given by equation 22. As for the previously introduced parameter μ, one has first to find an estimate for the ratio of the contributions to E^2 associated with the displacement mismatch and the normal stress mismatch, respectively. The value of μ is then chosen such that the first term in the curly brackets of equation 16 is of the same magnitude as the μ-weighted second term. The integral in equation 21 extends only over the embedding region whose volume is denoted by $V^{(0)}$.

Since we only require W_{PW} to attain a minimum one may expect the resulting displacement $\underline{S}_{PW}(\underline{r})$ to drop automatically to zero as one moves towards the cell boundaries.

RESULTS

To make contact to the results of Kneer we have first treated the case of a spherical grain embedded in an isotropic medium. If one wants to calculate the average moduli B and G by searching for the zero of $\triangle\epsilon_{ij}$ according to equation 17, it is sufficient to consider strains associated with an asymptotic uniform compression

$$\underline{\underline{\epsilon}}_0^{(I)} = \begin{pmatrix} \epsilon & 0 & 0 \\ 0 & \epsilon & 0 \\ 0 & 0 & \epsilon \end{pmatrix}$$

and an asymptotic pure shear, e. g.

$$\underline{\underline{\epsilon}}_0^{(II)} = \begin{pmatrix} 0 & \epsilon & 0 \\ \epsilon & 0 & 0 \\ 0 & 0 & 0 \end{pmatrix},$$

where $\epsilon \neq 0$. The results are listed in Table 1 together with those obtained by Kneer [6], [7]. The experimental data are due to Bradfield and quoted in Kneer's thesis as "private communication". The agreement between the theoretical results is almost perfect. But their departure from the experimental data is also less than 1% which lends strong support to the reliability of our method. We have extended our calculations to some ceramics of practical importance. As regards these materials, the calculation of average elastic moduli appears to

Table 1. Theoretical and experimental results for the bulk (B) and shear modulus (G) of polycrystalline metals in GPa.

	Bulk modulus B			Shear modulus G		
	Theory		Expt.	Theory		Expt.
	This work	Kneer	Bradfield	This work	Kneer	Bradfield
Cu	137.64	137.64		48.27	48.26	48.3
Ti	104.49	104.49		43.84	43.84	44.2
Zn	69.400	69.405	69.4	41.853	41.868	41.8

Table 2a. Elastic constants (in GPa) of some compound single crystals measured at room temperature.

Material	c_{11}	c_{22}	c_{33}	c_{23}	c_{13}	c_{12}
$BaTiO_3$	211	211	160	114	114	107
La_2CuO_4	171.9	171.2	200.0	73.1	72.7	90.4
$YBa_2Cu_3O_7$	223	244	138	93	89	37
$Bi_2Sr_2CaCu_2O_8$	125.2	125.2	75.8	56.0	56.0	78.9

Table 2b.

Material	c_{44}	c_{55}	c_{66}	Source
$BaTiO_3$	56.2	56.2	127	[11]
La_2CuO_4	65.6	65.8	96.8	[12]
$YBa_2Cu_3O_7$	61	47	97	[13]
$Bi_2Sr_2CaCu_2O_8$	15.8	15.8	50.4	[14],[15]

be particularly desirable since the preparation of ideal polycrystalline samples constitutes a serious practical problem. In Table 2a,b we have compiled experimental results on the elastic constants of the pertinent single crystals. (Note that Table 2b is just meant to continue the listing of Table 2a.)

The above data deserve some comments. As $BaTiO_3$ is a ferroelectric material, certain elastic constants depend sizeably on whether the measurements are performed at zero electric field or at zero dielectric displacement. Our data refer to zero electric field. The experimental error ranges from 3% to 7%. As for the (1,2,3) and (2,2,1,2) high-T_C materials the error is even larger and in particular for the (2,2,1,2)-compound greater than 10%. In the work of Boekholt et al. [14] the value of c_{66} is not explicitly listed. The authors infer the validity of the condition of isotropy which interconnects c_{11} and c_{12} with c_{66}. As follows from the experimental work of Wu et al. [15], this assumption is incorrect. It would only apply to hexagonal crystal symmetry whereas one is actually dealing with tetragonal symmetry in this case. We have therefore used the value of c_{66} measured at T=250K [15] which seems to be admissible because of its weak temperature dependence (less than 1% from 80K to 250K). The elastic constants for La_2CuO_4 are relatively well known within an error of less than 1%. All the data listed in Table 2a,b refer to room temperature.

In Table 3a,b we compare our results with the respective experimental data. Using the results of Table 2a,b, we have also calculated Hill values for the lower and upper bounds of the elastic moduli which are listed in the columns "Reuss" and "Voigt". As has to be expected

Table 3a. Hill bounds and calculated mean elastic moduli (in GPa) of ideal polycrystalline materials composed of crystals which Table 2a,b refers to. The experimental data refer to the respective ceramics.

	Bulk modulus B			
	Theory			Expt.
	Voigt	Reuss	This work	
$BaTiO_3$	139.1	136.7	138.4	106 [16]
La_2CuO_4	112.83	112.76	112.8	122.0 [17] 87 [18]
$YBa_2Cu_3O_7$	115.9	113.7	115.1	120 [19] 68.5 [20]
$Bi_2Sr_2CaCu_2O_8$	78.7	69.8	74.7	30.8 [21]

Table 3b.

Shear modulus G				Young's modulus E			
Theory			Expt.	Theory			Expt.
Voigt	Reuss	This work		Voigt	Reuss	This work	
64.3	52.0	57.5	44 [16]	167.3	138.4	151.5	115 [16]
66.1	61.3	63.8	62.92 [17] 63 [18]	165.9	155.6	161.0	161.1 [17] 152 [18]
66.7	54.5	60.9	58.5 [19] 57.4 [20]	168.0	141.0	155.2	151 [19] 135 [20]
25.4	20.8	22.5	29.5 [21]	68.8	56.6	61.4	67.1 [21]

from Hill's work [3], our calculated mean elastic constants lie, in fact, between the Reuss- and the Voigt-value. A comparison with the experimental results is impeded by the fact that the ceramics in question are more or less porous which gives rise to a sizeable scatter of the data after the actually measured values have been corrected for zero porosity. The inaccuracy of the relatively simple extrapolation used in obtaining these corrections affects particularly the bulk modulus. Sometimes, as for the (1,2,3)-compounds, there is an additional uncertainty of the experimental results due to the possible occurrence of more than one phase [19]. One arrives at considerably more reliable values for B and G in that case if one calculates the so-called bond compression modulus B_{bc} (see e. g. [19]) and assumes that the ratio B_{bc}/B is the same as for $BaTiO_3$ in the perovskite structure, i. e. $B_{bc}/B = 2.4$. Furthermore, one may safely assume that Poisson's ratio is also the same for the two compounds. The values for B, G and E thus determined are listed in Table 3a,b. The experimental data for $BaTiO_3$ have been obtained from inverting the compliances s_{ij} which are due to Berlincourt and Jaffe [16]. Since these data are not porosity corrected they are much smaller than our theoretical mean values.

The feasibility and (in terms of computer coding) practical simplicity of the Kröner-Kneer approach resides in the assumption of a spherical or ellipsoidal grain so that the homogeneous strains inside and far outside the grain are interconnected by a tensor whose 3^4 elements are real-space independent. In that particular form the scheme does not yield any information about the strains outside the grain. If one wants to calculate average elastic moduli for more general grain shapes or clusters, one looses the advantage of dealing with a homogeneous strain field inside the embedded object even when the asymptotic strain is homogeneous. In that case there is no way of avoiding a detailed calculation of the entire strain field. As has been elaborated in Section 2, our approach provides the means of treating this more general problem. To demonstrate the functioning of our scheme we have calculated the displacement field for an anisotropic spherical grain of cubic symmetry which is embedded in an isotropic environment. The system is subject to a uniaxial strain

$$\underline{\underline{\epsilon}}_0 = \begin{pmatrix} \epsilon & 0 & 0 \\ 0 & 0 & 0 \\ 0 & 0 & 0 \end{pmatrix} .$$

The symmetry axes of the grain are chosen to coincide with those of the coordinate system. Figure 1 shows, for the plane z=0, a black and white coded distribution of the relative changes of the volume

$$\frac{\delta V}{V} = \mathrm{Tr}\,(\underline{\underline{\epsilon}} - \underline{\underline{\epsilon}}_0) .$$

One clearly recognizes the quadrupole-type deformation outside the sphere and the homo-

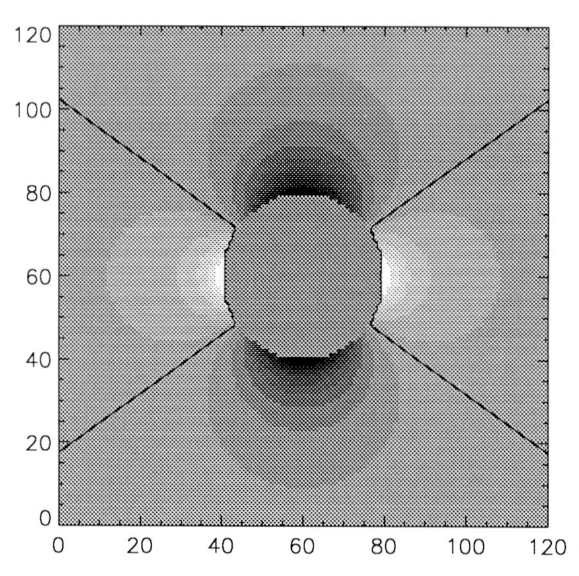

Figure 1: An anisotropic sphere in an isotropic environment subject to a uniaxial strain. The black and white shading refers to the relative changes of the volume along the plane z=0.

geneous strain (i. e. constant $\frac{\delta V}{V}$) inside. The displacement field in the embedding medium has in the present case been expanded in terms of the basis functions defined by equation 13.

In order to assess the plane wave approach envisaged to deal with the more general problem of a non-spherical grain we have first used this approach to reproduce the results for the embedded spherical grain. This is what figure 2 refers to, which, again, shows the relative changes of the volume in the plane z=0. Obviously the near field features agree quite satisfactorily with those in figure 1. Farther away from the sphere, one observes, however, a serious departure from the true asymptotic behavior. This is due to the fact that the asymptotic

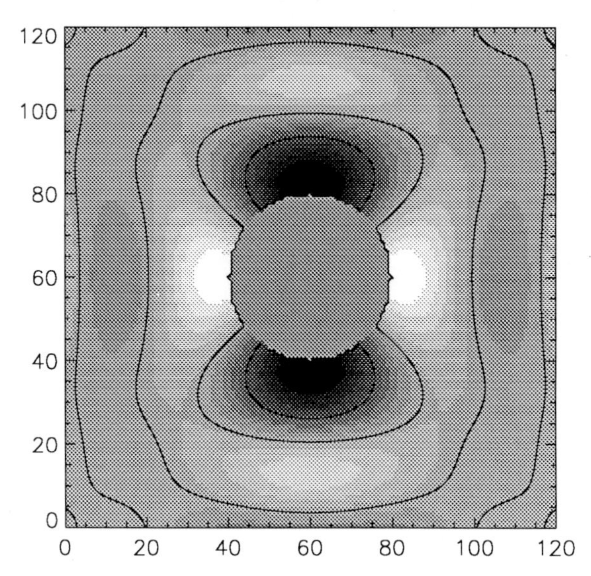

Figure 2: An anisotropic sphere in an isotropic environment subject to a uniaxial strain as in figure 1 (relative changes of the volume along the plane z=0). The associated difference displacement for the embedding region has in this case been expanded in terms of plane waves. This difference is referenced to the one that follows from an asymptotic homogeneous strain field.

dependence of $\underline{S}^{(0)}(\underline{r})$ on r (distance from the sphere center) scales as r^{-2} which requires a size of the unit cell considerably larger than the one we have actually chosen. In order to remove this deficiency it appears to be indispensable that one introduces a second peripheral embedding that consists of the same material and in which the displacement is expanded in terms of the isotropic basis functions 13 so that the correct asymptotic behavior can be ensured as in figure 1. In that case one has, of course, to fulfil another boundary condition at the additional surface. This can be incorporated into the minimization of $W_{PW} + \lambda E^2$ by adding the respective E^2-mismatch. Moreover, there is now an extra deformation energy W_{SW} associated with the spherical wave type functions 13 which has to be added as well so that the minimization of the resulting new expression leads automatically to the optimum set of expansion coefficients for the peripheral region, thereby ensuring that the expansion becomes very close to the exact solution of the differential equation.

Figure 3 shows the behavior of a cubical grain subject to the same uniaxial strain as in figure 2. Apart from similar deficiencies of the plane wave expansion in the region farther away from the embedded object, the structure of the near field is obviously well described, in particular certain features that are associated with the occurrence of corners. The strain field

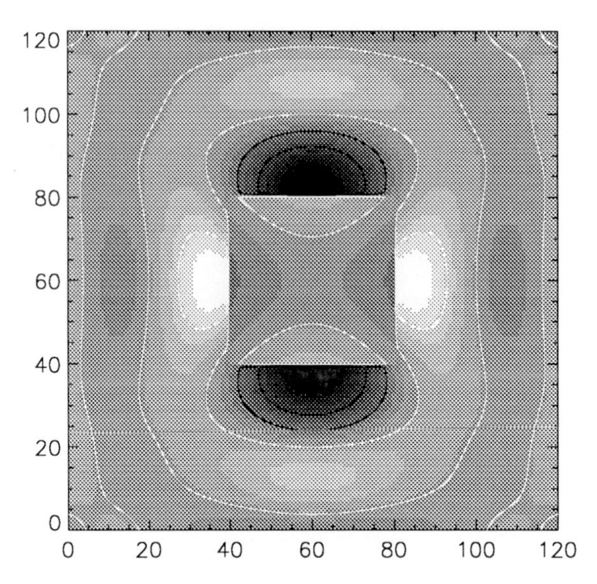

Figure 3: An anisotropic cubical grain in an isotropic environment subject to a uniaxial strain as in the preceding figures. The difference displacement in the embedding region is expanded in terms of plane waves as in figure 2.

inside the grain is, of course, no longer homogeneous which demonstrates the capability of our method to handle grain shapes that are less artificial.

REFERENCES

1) Voigt, W.: Lehrbuch der Kristallphysik (Teubner, Berlin), 1910
2) Reuss, A.: Z. angew. Math. Mech., 1929, 9, 49
3) Hill, R.: Proc. Phys. Soc. A, 1952, 65, 351
4) Kröner, E.: Z. Phys., 1958, 151, 504
5) Hershey, A. V. and Dahlgren, V.: J. appl. Mech., 1954, 21, 236
6) Kneer, G.: Thesis (Technische Universität Clausthal), 1964
7) Kneer, G.: phys. stat. sol., 1965, 9, 825
8) Kumar, S.: Thesis (The Pennsylvania State University), 1992
9) Fritsche, L., Rafat-Mehr, M., Glocker, R. and Noffke, J.: Z. Phys. B, 1979, 33
10) Eshelby, J. D.: Proc. Roy. Soc. A, 1957, 241, 376
11) Li, Z.,Chan, S. K., Grimsditch, M. H. and Zouboulis, E. S.: J. Appl. Phys., 1991, 70, 7327
12) Migliori, A., Visscher, W. M., Brown, S. E., Fisk, Z., Cheong, S. W., Alten, B., Ahrens, E. T., Kubat-Martin, K. A., Maynard, J. D., Huang, Y., Kirk, D. R., Gillis, K. A., Kim, H. K. and Chan, M. H. W.: Phys. Rev. B, 1990, 41, 2098
13) Ledbetter, H. and Ming, L.: J. Mater. Res., 1991, 6, 2253
14) Boekholt, M., Harzer, J. V., Hillebrands, B. and Guentherodt, G.: Physica C, 1991, 179, 101

15) Jin,W. , Yening, W., Pingsheng, G., Huimin, S., Yifeng, Y. and Zhongxian, Z.: Phys. Rev. B, 1993, 47, 2806
16) Berlincourt, D. and Jaffe, H.: Phys. Rev., 1958, 111, 143
17) Ledbetter, H., Kim, S. A., Violet, C. E. and Thompson, J. D.: Physica C, 1989, 162-164, 460
18) Fanggao, C., Cankurtaran, M., Saunders, G. A., Al-Kheffaji, A., Almond, D. P., Ford, P. J. and Ladds, D. A.: Phys. Rev. B, 1991, 43, 5526
19) Bridge, B.: J. Mater. Sci. Lett., 1989, 8, 695
20) Cankurtaran, M. and Saunders, G. A.: Phys. Rev. B, 1989, 39, 2872
21) Ledbetter, H. and Kim, S.: Physica C, 1991, 185-189, 935

Materials Science Forum Vol. 157-162 (1994) pp. 1623-1628

TEXTURES AND PLASTIC STRAIN RATIOS OF PLANAR ISOTROPIC SHEET METALS

I.S. Kim

Department of Materials Science and Engineering, Kum Oh National University of Technology, 188 Sinpyung-dong, 730-701 Kumi, Kyung Buk, Korea

Keywords: Yield Locus, Plastic Strain Ratio, Planar Isotropy, Texture

ABSTRACT

The plastic strain ratios of planar isotropic sheet specimens were studied using unidirectionally solidified commercial purity Al, Sn and Al-Cu alloy sheets and electrodeposited Cu sheets under various electolysis conditions.
There was a substantial discrepancy between the measured plastic strain ratios of the unidirectionally solidified commercial purity Al and Al-Cu alloy sheets and the electrodeposited Cu sheets and the calculated plastic strain ratios in planar isotropic cubic metals.
The measured plastic strain ratios of [100] planar isotropic sheets using unidirectionally solidified Al are about 0.45 and [110] planar isotropic sheets using unidirectionally solidified Sn are about 2.4.

1. INTRODUCTION

The directional properties of materials are divided into two kinds, anisotropy and isotropy.
Anisotropy is reflected in property values that change, isotropy is that do not change with the direction of measurement.
Theoretical and experimental works of anisotropy on the plastic strain ratios (or R-values), the ratio of the true strain of width and thickness direction, and yield stresses of sheet metals have been undertaken by Hill[1] based on the continuum mechanics, by Sachs[2], Bishop and Hill[3,4], Taylor[5], Vieth and Whiteley[6], Hosford and Backofen[7], Dillamore[8,9], Piehler and Backofen[10], Bunge[11] and Lee[12,13] based on the crystallographic orientation.
These studies had successful results in theoretical and experimental aspects.
On the other side, the theoretical work on the planar isotropy has been done by Piehler[14], he calculated the yield loci for planar isotropy of cubic metals under the combined stresses.
Based on the Piehler′ s yield loci, the equivalent R-values for planar isotropy around [100], [110] and [111] rotation axes were calculated by Backofen[15].
The calculated plastic strain ratios for planar isotropic sheets did not have been proven by the mearured values, because the planar isotopic materials are difficult to get.
The purpose of this paper is to compare the plastic strain ratios calculated by Beckofen with the measured plastic strain ratios using the planar isotropic sheet metals.

2. CALCULATION METHOD

Piehler [14] represented the two anisotropic yield loci for textured cubic metals deforming by restricted-glide.
One is called the planar-stress yield loci and the other is the planar-strain yield loci.
Planar-stress yield loci for restricted-glide can be found directly from the generalized Schmid′ s law. Planar-strain yield loci for restricted-glide can be expressed from the maximum work of Bishop and Hill[16].
Where the yield loci determined under conditions of planar-stress and planar-strain are lower and upper bounds of the yield locus which principal axes of stress and strain are assumed to coincide.

Piehler[14] calculated the restricted-glide loci and the rotationally symmetric loci for the three common textures, (100)[001], (110)[001] and (111)[112], which have been described as the cube-on-face, cube-on-edge and cube-on-corner textures, respectively.
Backofen [15] calculated the strain ratios of planar isotropic materials using the the rotationally symmetric loci which are calculated by Piehler[14] and equation(1).

$$\frac{\sigma_{y(nn)}}{\sigma_{y(rr,tt)}} = \left(\frac{1 + R}{2} \right)^{1/2} \quad \text{- - - - - - - -} \quad (1)$$

where $\sigma_{y(nn)}/\sigma_{y(rr,tt)}$ is directly calculated from the rotaionally symmetric loci.
The strain ratio for planar isotropy around [100], [110] and [111] rotation axis is shown in Table 1.

3. EXPERIMENTAL PROCEDURES

The planar isotropic sheet metals can be obtained by the unidirectionally solidification and electrodeposited layer. The unidirectionally solidification apparatus is shown in Fig.1.
It is composed of the melting furnace, chill and adiabatic furnace.
The planar isotropic sheet metals around [100] rotation axis were gotten by the unidirectionally solidified commercial purity Al and Al-Cu alloy and around [110] rotation axis were gotten by commercial purity Sn(β, BCT).
The experimental electrolysis apparatus is shown in Fig.2. It is composed of the D.C.power supply, anode (electrolytic copper), cathode (AISI 304 stainless steel), electrolytic cell (electrolyte : $Cu(BF_4)_2$ solution), magnetic pump, heating cell, etc.
I have obtained the planar isotropic sheets around [100], [110] and [111] rotation axis from the experimental electrolysis apparatus in Fig.2.
The variations of electrodeposition are current density, electrolyte temperature and electrolyte concentration which are shown in table 2.
Diffraction diagrams of all planar isotropic sheet specimens were measured by X-ray diffractometer and texture coefficient(T.C.) of each plane calculated by the equation (2) [17].

$$\text{T.C.}(hk\ell) = \frac{\dfrac{I\,(hk\ell)}{I_o(hk\ell)}}{\dfrac{1}{n}\sum \dfrac{I\,(hk\ell)}{I_o(hk\ell)}} \quad \text{- - - - - - - - - -} \quad (2)$$

where I : measured integral intensity of plane $\{hk\ell\}$ in specimen , I_o : measured integral intensity of plane $\{hk\ell\}$ in standard powder.
The calculated texture coefficient of each planar isotropic specimen was shown in table 3.
For microstructural characterization by optical microscopy, the planar isotropic specimens were cut off the cross section.
The ASTM standard tensile specimens for planar isotropic sheets were prepared by mechanical machining and replaced by photo-etched 2mm square on the surface of tensile specimens.
All tensile tests were carried out in an Instron machine at cross-head speed of 1mm/sec.
After the tensile test, the strain in width and length direction was measured by trevelling optical microscopy and got the plastic strain ratio.

4. RESULTS AND DISCUSSION

X-ray diffraction diagrams of unidirectionally solidified commercial purity Al, Al-Cu alloy and commercial purity Sn specimens are shown in Figs 3 a), b) and c), respectively.
X-ray diffraction diagrams exhibit that (200)plane developed well in Figs 3 a) and b), (110) plane developed well in Fig. 3 c)
X-ray diffraction diagrams of electrodeposited copper specimens are shown in Fig. 4.
Each diagram exhibits that (200), (220) and (111) plane developed well, respectively.
To confirm the planar isotropy, I measured the pole figures of all kinds of specimens and these are shown in Figs 5 and 6.
The pole figures of unidirectionally solidified commercial purity Al, Al-Cu alloy and Sn specimens exhibit good planar isotropic sheets around [100] and [110] axis in Fig.5.
The pole figures of electrodeposited Cu specimens exhibit good planar isotropic sheets around [100], [110] and [111] rotation axes in Fig.6, respectively.

Metallographic cross sections of all planar isotropic sheets are shown in Fig.7.
Metallography of unidirectionally solidified Al-Cu alloy exhibits dendrite growth in Fig.7 b), commercial purity Al and commercial purity Sn exhibit the same as polycrystalline microstructure in Figs 7 a) and c).
Metallography of electrodeposited Cu specimens exhibit to have the different grain size and microstructure in Figs 7 d), e) and f)
The grain of planar isotropic sheet around [100] rotation axis exhibits the biggest size in Fig.7 d), e), f). The grain of planar isotropic sheet around [110] rotation axis is shown needle-like shape in Fig. 7 e). The (111) plane layer of planar isotropic sheet around [111] rotation axis exhibits the parallel to the sheet plane in Fig. 7 f).
I had the tensile tests and measured the plastic strain ratios which were shown in table 3.
There was a substantial discrpancy of plastic strain ratios between mearsured and calculated, between unidirectionally solidified and electrodeposited specimens, between unidirectionally solidified commercial purity Al and Al-Cu alloy having the same isotropic plane in table 1 and 3
The plastic strain ratios of electrodeposited Cu sheets are analogized with calculated values having the same plane, but they have higher values than that of unidirectionally solidified specimens having the same planar isotropic plane around [100] rotation axis.
The discrepancy is due to the effect of the pin holes of electrodeposited Cu specimens and the mechanical fibering of the dendrite growth in unidirectionally solidified Al-Cu alloy specimens.
Pin holes grow to thickness direction and electrodeposited sheet will be become like honeycomb.
Deformation of honeycomb will be offered the effects increasing the strain in width direction and decreasing the strain in thickness direction and then the plastic strain ratio will be increased.
But, unidirectionally solidified commercial purity Al and Sn specimens do not have pin hole and dendrite and these are shown like polycrystalline microsturucture of metal in Figs 7 a) and c).
Therefore, unidirectionally solidified commercial purity Al and Sn sheets are good specimens that can be used to measure the planar isotropic plastic strain ratios.
The mesured plastic strain ratios of unidirectionally solidified commercial purity Al sheets having planar isotropy around [100] rotation axis are about 0.45 and of Sn (β type, BCT) sheets having planar isotropy around [110] rotation axis are about 2.4.

5. CONCLUSIONS.

1. The measured plastic strain ratios of unidirectionally solidified purity Al sheets having planar isotropy around [100] rotation axis are about 0.45.
2. The measured plastic strain ratios of unidirectionally solidified purity Sn (BCT) sheets having planar isotropy around [110] rotation axis are about 2.4.

ACKNOWLEDGEMENTS

The support by Korea Research Foundation(1991) is greatfully acknowledged.

6. REFERENCES

1. R.Hill : Plasticity, Pergamon Press, London, (1950)
2. G.Sachs : Z.V.D.I., 72 (1928) 734
3. J.F. Bioshop and R. Hill : Phil. Mag., 42 (1951) 42
4. J.F. Bioshop and R. Hill : Phil. Mag., 42 (1951) 1298
5. G.I. Taylor : J. Inst. Metals, 62 (1963) 307
6. R.M. Vieth and R.L. Whiteley : IDDRG Colloquim (1964)
7. W.F. Hosford and W.A. Backofen : Fundamental of Deformation Processing, Proceedings of the 9th Sagamore Conference, Syracuse Univ. press, Syracuse Univ. N.Y., (1964)
8. I.L. Dillamore : Met. Trans., 1 (1970) 2463
9. I.L. Dillamore, E. Butler and D. Green : Met. Sci. J., 2 (1968) 161
10. H.R. Pichler and W.A. Backofen : Met. Trans., 2 (1971) 249
11. H.J. Bunge : Texture Analysis in Materials Science : Mathematical Methods, Butterworths, London (1982)
12. D.N. Lee : Korean Isnt. of Metals and Materials 20, 9 (1982) 586
13. D.N. Lee, I. Kim and K.H. Oh : J. of Mater. Sci. 23 (1988) 4013
14. H.R. Piehler : ScD. Thisis, M.I.T., (1967)
15. W.A. Backofen : Deformation Processing, Addisson-Wesley Series, (1972)
16. J.F. Bishop and R.Hill : Phil. Mag., 42 (1951) 414
17. C.S. Barrett and T.S. Massalski : Structure of Metals, 3rd ed. McGraw Hill book Co., (1966)

Table 1. Equivalent R-values for planar isotropy around directions given : from Eq.(1)

Axis of Rotation	Lower Bound $\sigma_Y(nn)/\sigma_Y(rr,tt)$	R	Upper Bound $\sigma_Y(nn)/\sigma_Y(rr,tt)$	R	R
[100]	1.04	1.3	1.02	1.08	1.2
[110]	1.00	1.0	1.18	1.8	1.4
[111]	1.67	4.6	1.27	2.2	3.4

Table 2. The composition of electrolyte

Solution No.	S.G	$Cu(BF_4)_2$(gr/l)	Cu^{++}	HBF	H_3BO_3
1	1.30	373	100	25	25
2	1.24	298	80	20	20
3	1.18	224	60	15	15
4	1.09	112	30	7.5	7.5

Table 3. Texture coefficient of reflection planes and strain ratios

a) Unidirectionally solidified commercial purity Al, Al-Cu alloy and commercial purity Sn

No.	materials	(111)	(200)	(220)	(311)	measured strain ratio
1	Al	0.25	3.68	0.00	0.07	0.52
2	Al	0.22	3.70	0.00	0.08	0.42
3	Al+1.82Wt%Cu	0.00	4.00	0.00	0.00	0.21
4	Al+1.76Wt%Cu	0.04	3.90	0.00	0.06	0.17

No.	materials	200	110	211	321	measured strain ratio
5	Sn	0	2.24	1.06	0.70	2.27
6	Sn	0	2.21	1.20	0.59	2.62

b) Electrodeposited copper

No.	TEMP (℃)	C.D (A/dm²)	Sol. NO.	(111)	(200)	(220)	(311)	measured strain ratio
7	35	7.41	4	0.12	2.82	0.41	0.64	1.53
8	35	7.41	4	0.14	3.07	0.34	0.44	1.83
9	40	30	1	0.00	0.01	3.98	0.01	1.38
10	30	20	2	0.01	0.08	3.55	0.36	2.05
11	25	10	3	0.05	0.00	3.95	0.04	1.50
12	10	7.16	4	3.45	0.11	0.18	0.26	2.85
13	10	6.67	4	3.29	0.13	0.26	0.33	2.61

STAINLESS STEEL
HEATER
INSULATION
out
out
WATER
COLD SINK
in

Fig 1. Schematic diagram of unidirectionally solidification apparatus.

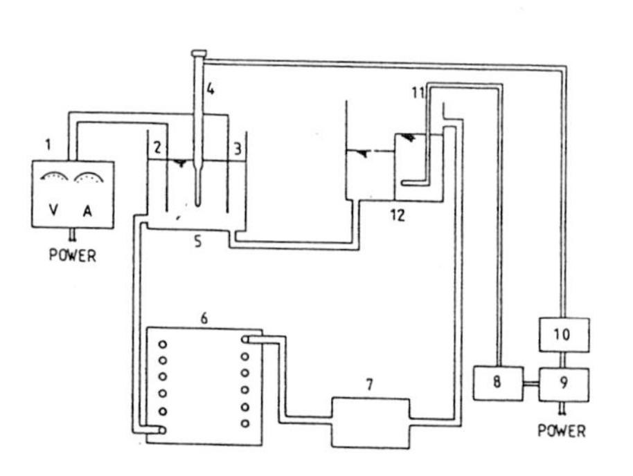

Fig 2. Schematic diagram of electrolytic experimental apparatus.
1. D.C. Power Supply 2. Anode
3. Cathode 4. Thermoregulator
5. Electrolytic Cell 6. Filtering Column
7. Magnetic Pump 8. Transformer
9. Magnetic Switch 10. Mini Relay
11. Immersion Heater 12. Heating Cell

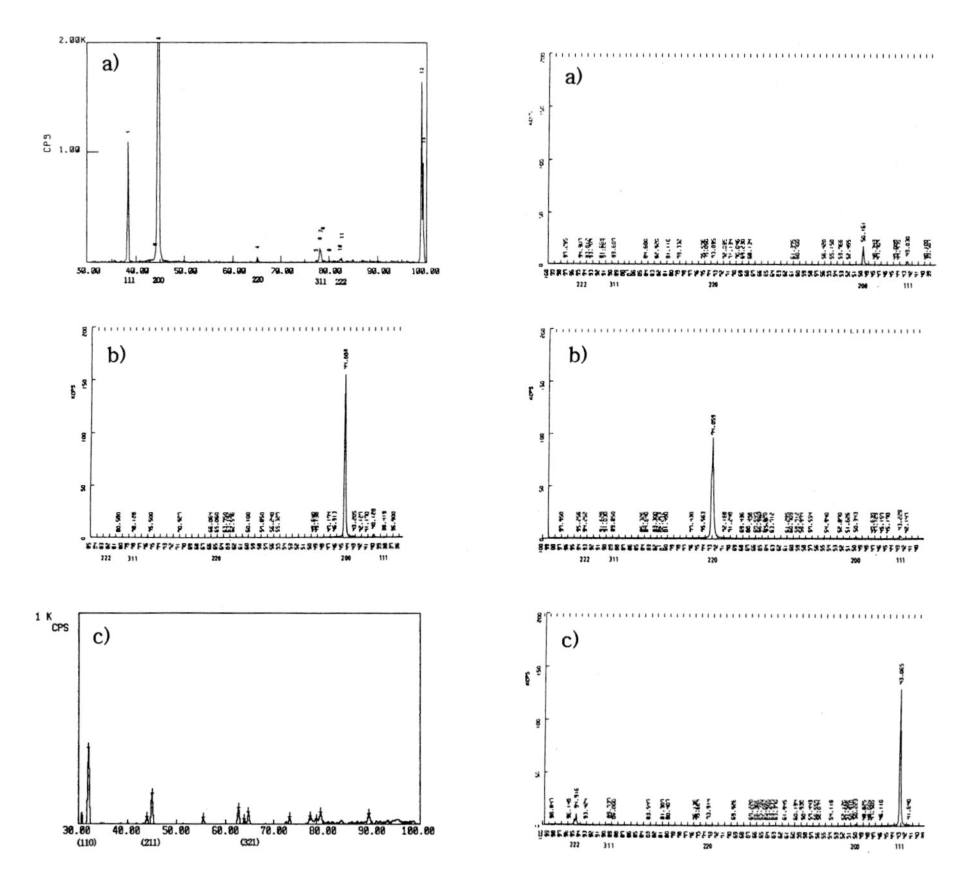

Fig 3. X-Ray diffraction diagrams of unidirectionally solidified specimens. a) commercial purity Al (specimen No.1) b) Al-Cu alloy (specimen No.3) diagrams and c) commercial purity Sn (specimen No.5) in table 3.

Fig 4. X-Ray diffraction diagrams of electrodeposited coppers. a) specimen No.8, b) specimen No.9 and c) specimen No.12 in table 3.

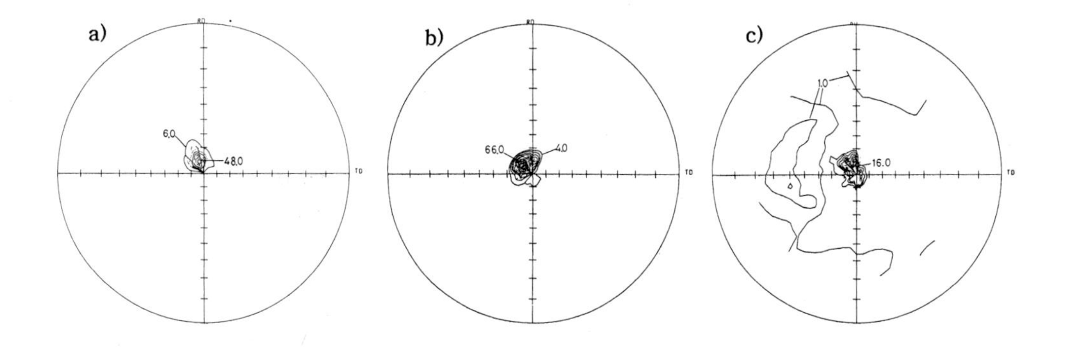

Fig 5. Measured pole figures of unidirectionally solidified specimens. a) (200) pole figure of Al, b) (200) pole figure of Al-Cu alloy, c) (110) pole figure of Sn

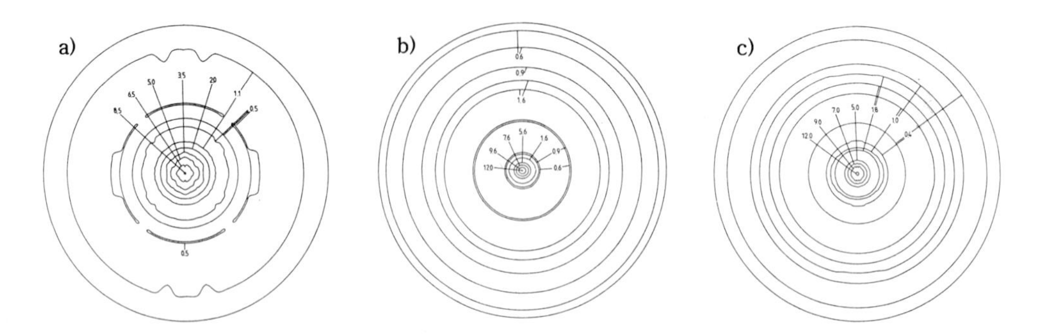

Fig 6. Measured pole figures of electrodeposited coppers. a) (200) pole figure of specimen No.8, b) (220) pole figure of specimen No.9, c) (111) pole figure of specimen No.13 in table 3

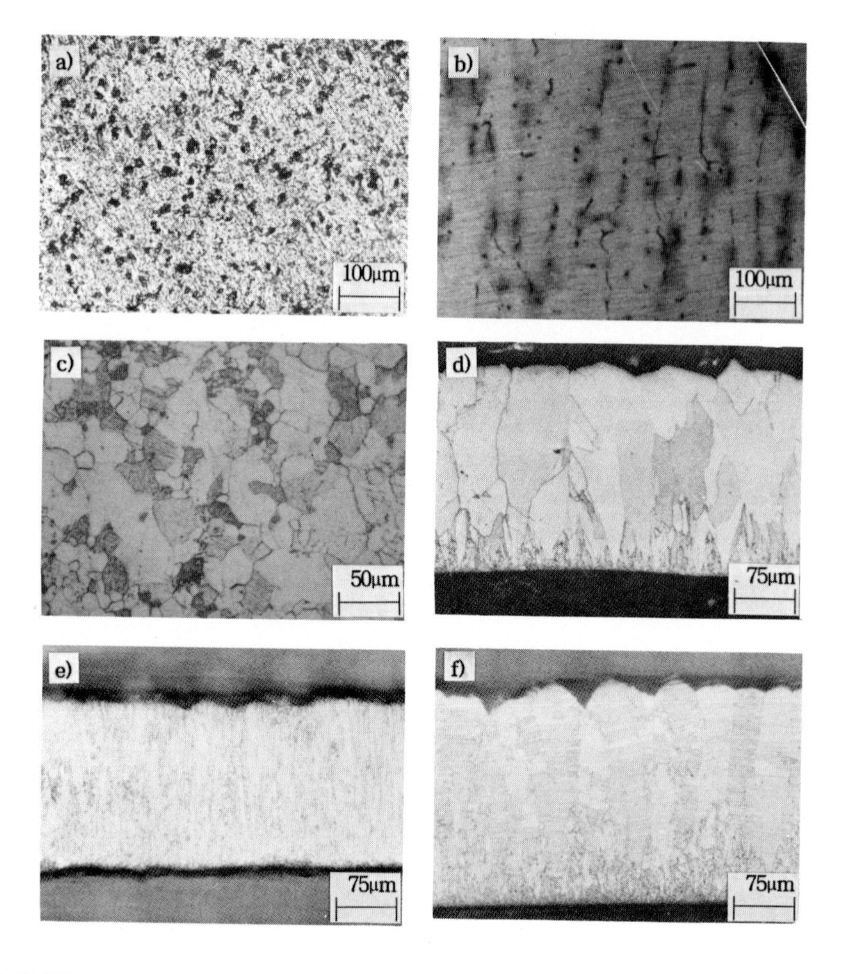

Fig 7. Microstructure of a) unidirectionally solidified (100) planar isotropic Al, b) unidirectionally solidified (100) planar isotropic Al-Cu alloy, c) unidirectionally solidified (110) planar isotropic Sn, d) electrodeposited (100) planar isotropic copper, e) electrodeposited (110) planar isotropic copper, f) electrodeposited (111) planar isotropic copper.

Materials Science Forum Vol. 157-162 (1994) pp. 1629-1634

TEXTURE AND FORMABILITY OF ALUMINUM SHEETS

S. Kohara

Department of Materials Science and Technology,
The Science University of Tokyo, Noda, Chiba 278, Japan

Keywords: Aluminum Sheet, Anisotropy, Formability, Forming Limit Curve, Mechanical Properties, Texture

ABSTRACT

The mechanical properties, the forming limit curves and the textures of A1050 aluminum sheets, as rolled and annealed, have been determined. The mechanical properties change remarkably when recrystallisation occurs after annealing at 300°C. Correspondingly, the limit strains increase greatly at 300°C. The rolling texture of the original sheet retains up to 250°C and the cube texture evolves after annealing at 300°C. The anisotropy of the sheet increases as the cube texture becomes stronger. However, the change in r-value, accordingly the texture, has little effect on the forming limit curve.

INTRODUCTION

The assessment of forming properties is important for the wide industrial applications of sheet materials. The forming limit curve (FLC) /1//2/ is a useful tool to predict forming properties of metal sheets. It is considered that the forming limit curve is affected by the plastic characteristics of the sheets. The effects of plastic properties, such as r-value and n-value, on the forming limit curve of sheet materials have been studied /3/. In gneral, metal sheets have the anisotropy in mechanical properties caused by the evolution of textures. The effect of the texture on the biaxial deformation of sheet metals have been reported /4//5/. However, the correlation between the texture and the forming limit curve in sheet metals is not clear.

The purposes of the present study are to determine the anisotropy in mechanical properties and to reveal the correlation of the anisotropy with the forming limit curve in aluminum sheets. To realize the wide range of mechanical properties and forming limit curves, aluminum sheet specimens, as cold-rolled and annealed at various temperatures, were used in this study.

EXPERIMENTAL

In the present study, A1050-H18(hardened state) aluminum sheets of 1 mm thickness were used. Test specimens were cut from the sheets and annealed at 150, 200, 250, 300, 350 and 400 ^{o}C for 1 hour. The mechanical properties of aluminum sheets were determined by tension test using the specimens cut at angles 0°, 45°, 90° to the rolling direction of the sheets. The n-values and the r-values of the specimens were calculated from the results of tension test. The forming limit curves were determined with hydraulic bulge test, stretch-forming test and tension test using a circular grid printed on the specimens. The details on the determination method of the formoing limit curve have been described elsewhere /6/.

RESULTS

The changes in tensile strength and elongation with annealing temperature are shown in Fig. 1. The tensile strength of hardened aluminum decreases with annealing temperature, but the elongation increases with temperature gradually up to 250°C and abruptly at 300°C and again gradually above 300°C.

The change in n-value with annealing temperature is shown in Fig. 2. The n-value increases with annealing temperature a little up to 250°C and greatly at 300°C and about the same at temperatures above 300°C. The change in n-value well corresponds to that in elongation.

The change in r-value with annealing temperature is shown in Fig. 3. The r-value in the directions 0°, 45° and 90° to the rolling direction of the sheets and the planar mean value, $\bar{r}$ and the planar anisotropy, Δr are also shown in Fig. 3. The r-values in the three directions are almost the same in the original sheet, but r_{90} becomes highest by annealing up to 250°C and r_0 becomes highest after the annealing at 300°C and higher temperatures. Correspondingly, the planar anisotropy Δr is nearly zero up to 250°C and becomes positive value at 300°C and higher temperatures. This means that the cold-rolled sheets are less anisotropic than the full annealed sheets.

The pole figures of the aluminum sheets as received and annealed are shown in Fig. 4. The texture of the sheets as received is a typical rolling texture of aluminum and it retains up to 250°C. The cube texture appears at 300°C and becomes stronger at higher temperatures.

The change in the forming limit curve with annealing temperature is shown in Fig. 5. The curve shifts to the upper position as softened by annealing. The limit strains increase a little at 250°C and greatly at 300°C, but do not change so much at temperatures higher than 300°C. The increase in the limit strains by annealing at 300°C is remarkable compared to the change in the other temperature range.

DISCUSSION

The formability of sheet materials can be assessed by the forming limit curve. The limit strains are correlated with the mechanical properties of the sheet. The effects of n-value and r-value on the forming limit curve have been studied, and it has been reported that the n-value affected the position and the

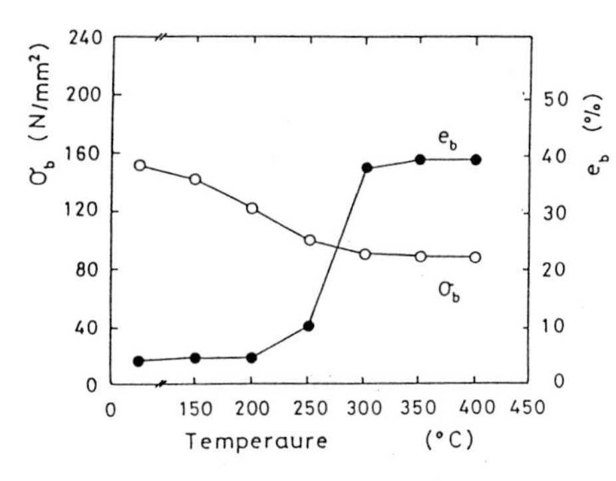

Fig.1 Changes in tensile strength and elongation with annealing temperature.

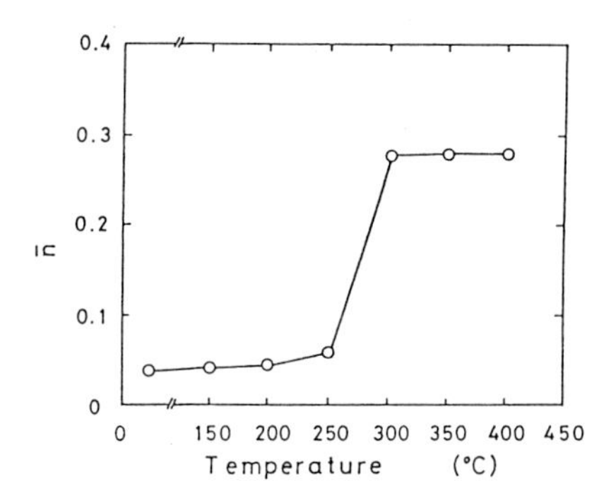

Fig.2 Change in n-value with annealing temperature.

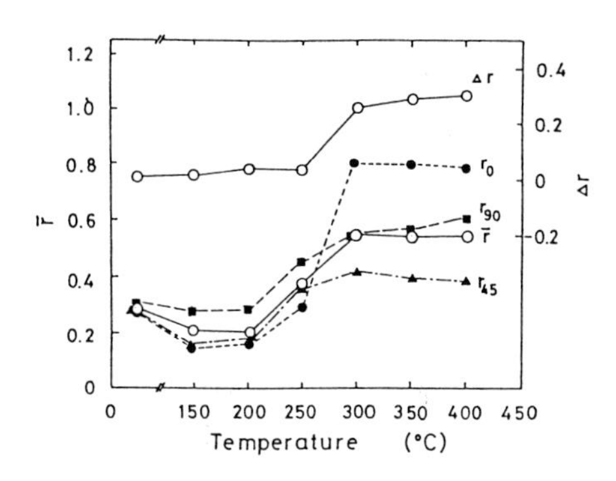

Fig.3 Change in r-value with annealing temperature.

r-value affected the shape of the forming limit curve /3/.

All the measured mechanical properties show the large change at 300°C. The limit strains are also largely increased by annealing at 300°C. These changes may be attributed to the recrystallization occured in the sheets by annealing at 300°C. The pole figure for the sheet annealed at 300°C shows the evolution of cube texture which is an evidence of the recrystalization. Then, it is clear that the limit strains are increased very much when recrystallization occurs in the sheet.

Among the mechanical properties measured in the present study, the change in n-value best corresponds to the change in forming limit curve. It can be said that the level of forming limit curve depends on the n-value of the sheet.

The r-value varies with the anisotropy of the sheets caused by the texture evolution. The planar anisotropy Δr is nearly zero up to 250°C, but changes to a positive value at 300°C and gradually increases above 300°C. This means that the sheets annealed at lower temperatures are less anisotropic than the

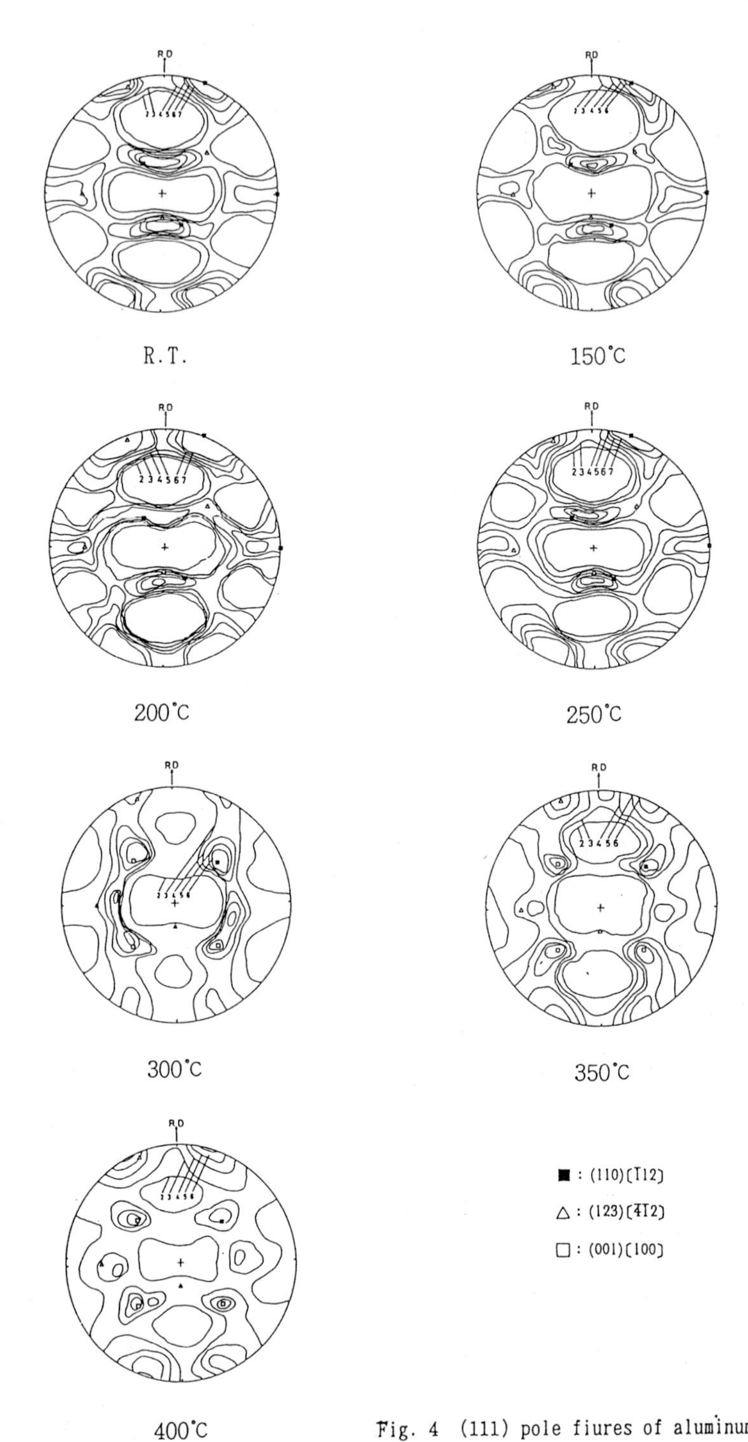

Fig. 4 (111) pole fiures of aluminum sheets.

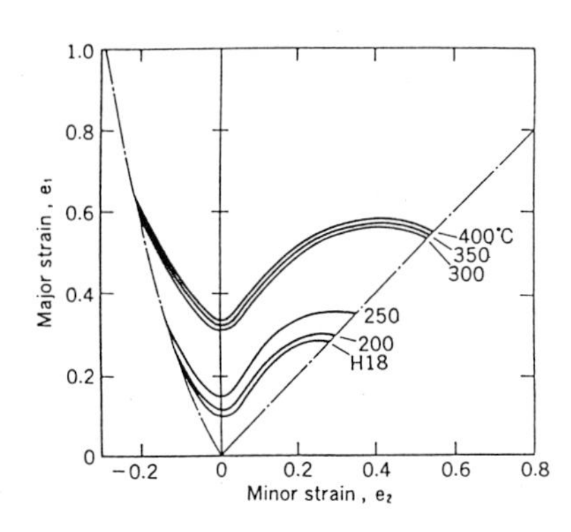

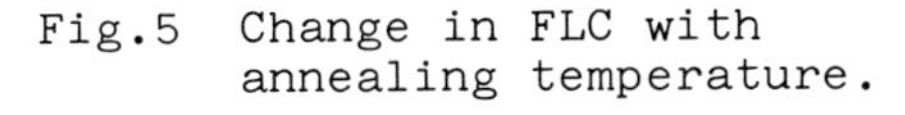
Fig.5 Change in FLC with annealing temperature.

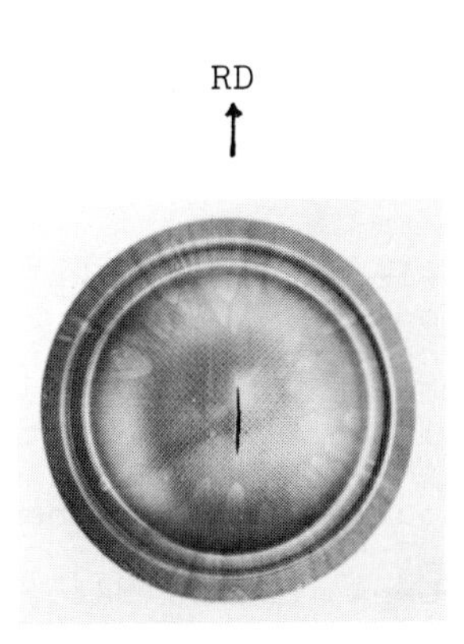

Fig.6 Fracture of a bl in biaxial tensi deformation.

recrystallized sheets and the anisotropy gradually increases for the sheets annealed at temperatures higher than 300°C. However, these differences in anisotropy appear not to be reflected on the forming limit curves.

In the biaxial tensile deformation, such as hydraulic bulging, the fracture of blanks of aluminum sheets mostly occurs parallel to the rolling direction irrespective of the anisotropy in the mechanical properties as shown in Fig. 6. Accordingly, the right hand side (the equi-biaxial tensile deformation side) of the forming limit curve would not be affected by the anisotropy of the sheets. The fracture behaviour is an important factor controlling the forming limit /7/. The limit strains in the plastic forming processes of sheet materials concern with the conditons for fracture rather than for deformation. Accordingly, the anisotropy in mechanical properties, and the texture as well, has little effect on the forming limit curve in aluminum sheets.

CONCLUSIONS

The mechanical properties, the pole figures and the forming limit curves of A1050 aluminum sheets as cold-rolled and annealed were determined and the correlations between those properties were investigated in the present study. The following conclusions were obtained.

1) The forming limit curve of hardened aluminum sheet shifts to the upper position when the sheet is softened by annealing and the limit strains are remarkably increased when recrystallization occurs in the sheet.

2) The change in the forming limit curve well corresponds to the change in the n-vaue among the plastic properties of aluminum sheets. The n-value affects the level of the forming limit curve.

3) In most cases, the fracture of aluminum sheets in the biaxial tensile deformation occurs parallel to the rolling direction of the sheets irrespective of the anistropy of the sheets. Therefore, the equi-biaxial deformation side of the forming limit curve is not affected by the r-value.

4) From these results, the anisotropy of plastic properties, correspondingly the texture, has little effect on the forming limit curve of aluminum sheets.

REFERENCES

1) Keeler, S.P., Backofen, W.A.:Trans. ASM, 1963, 56, 25.
2) Goodwin, G.M.: SAE Paper 680093, 1968.
3) Woodthorpe, J., Pearce, R.: Sheet Metal Ind., 1969, 46, 1061.
4) Wilson, D.V., Roberts, W.T., Rodrigues, P.M.B.: Metall. Trans., 1981, 12A, 1603.
5) Stout, M.G., Staudhammer, K.P.: Metall. Trans., 1984, 15A, 1607.
6) Kohara, S.: J. Mater. Proces. Tech., 1993, 38, 723.
7) Pearce, R.: Sheet Metal Ind., 1971, 48, 1061.

Materials Science Forum Vol. 157-162 (1994) pp. 1635-1640

STRAIN LOCALIZATION AND FRACTURE IN ANISOTROPIC FCC METAL SHEETS: SHEAR BANDS

J. Kusnierz

Instytut Podstaw Metalurgii, PAN, ul. Reymonta 25, PL-30-059 Kraków, Poland

Keywords: Rolling Texture, Plastic Anisotropy, Localized Neck, Fractography

ABSTRACT

The subject of the study is the effect of crystallographic texture on the anisotropy of mechanical and plastic properties and also on the localization of plastic strain during tension in a plane sample. The strength properties $R_{0.2}$, R_m, σ_m and r-value for aluminium, copper and their alloys are compared with the values calculated utilizing the orientation distribution function. The relations based on the theory of plastic deformation of an anisotropic body to calculate the inclination angle of localized neck are proposed.

INTRODUCTION

The present study describes an attempt at an experimental verification of the anisotropic theory of the plasticity of a continuous medium [1] taking into consideration the crystal concept of the deformation [2-4] with reference to the measurements of the strength properties as well as plastic quantities, in particular the coefficient of plastic anisotropy. The global values occuring in the theory of plastic deformation of an anisotropic continuous medium have been obtained by averaging the values from the crystallite taking advantage of the orientation distribution function ODF of crystallites and of the adopted model of deformation. This method of relating the anisotropy parameters has been used to calculate the inclination angle of the necking localized in flat sample undergoing tension, as well as to analyse the inclination angle of plastic instability during process of cold rolling of sheets (for rolling and experimental details compare [5]).

In the experiments there were taken into account the changes in the stacking fault energy SFE and also in the technological parameters. The measurements were carried out for aluminium, copper and CuZn37 alloy.

CRYSTALLOGRAPHIC TEXTURE AND PLASTIC ANISOTROPY

In the analysis of the sheets, the ODF [6,7] is expanded in a series of spherical functions T(g) invariant with respect to the symmetry of the sample (one dot) and with respect to the cubic symmetry of crystallite (two dots) [6]

$$f(\varphi_1,\phi,\varphi_2) = \sum_{L=0}^{L} \sum_{\mu=1}^{M(l)} \sum_{\nu=1}^{N(l)} C_l^{\mu\nu} \, \dot{\dot{T}}{}_l^{\dot{\mu}\nu} (\varphi_1,\phi,\varphi_2) \qquad (1)$$

To discuss the plastic deformation of the sheet the deformation model as proposed by Taylor [2] and developed by Bishop and Hill [3,4] (T-B-H theory) was considered. According to this proposal, plastic strain is realized by a set of five slip systems selected in agreement with the principle of minimal internal work. Bishop and Hill provided a theoretical justification of the adopted assumptions by applying the equivalent principle of maximal external work. They have demonstrated that this theory is consistent with the yield criterion of plastic flow. The shear stresses attain critical values only in the anticipated active systems. This statement is essential since in other deformation models the selected systems do not satisfy the yield criterion or do not conform with the preset deformation.

According to Bunge's approach [8] the Taylor factor M resulting from T-B-H theory is expanded into a series of spherical functions as in case of ODF, which gives a form convenient for calculations. For a sample cut out from a sheet with the orientation g_o the mean value of Taylor factor is expanded into the series analogous to (1) with the functions $\ddot{T}_l^{\mu\delta}(g_o)$ taking into account the orthorhombic symmetry in both systems and q- the contraction ratio.

Pulling in tension of a sample cut out from a metal sheet does not dictate the magnitude of the contraction ratio ($q=\varepsilon_2/\varepsilon_1$ where ε_1 determines elongation and ε_2 contraction in width direction). It is assumed that the contraction ratio is equal to such a quantity q_{min} for which the mean value of the Taylor factor assumes the minimal value and the coefficient of plastic anisotropy r is equal

$$r = q_{min} / (1-q_{min}) \quad (2)$$

EXPERIMENTAL STUDY OF TEXTURE AND PLASTIC ANISOTROPY

In metals with f.c.c lattice the development of texture during cold rolling is independent of SFE up to the reduction $z\cong40\%$ ($\varepsilon\cong0.5$). Further increase of deformation reveals individual features of the orientation distribution of crystallites. These individual features of the investigated textures can be found in figure 1, which shows the sections $\varphi_2=45^\circ$ ($1^\circ = \pi/180$ rad) of the aluminium, copper, CuZn37 brass after rolling including a scheme for $\varphi_2=45^\circ$ of the components distribution.

The dependence of the plastic anisotropy coefficient r on the degree of reduction by rolling for the investigated metals is shown in figure 3-left. Good egreement between the values calculated and determined experimentally is obtained already at the expansion up to L=4 ; hence the expansion of the plastic anisotropy coefficient r up to L=10 is regarded as satisfactory [9]. The growing reduction in area during cold rolling increases the anisotropy in the plane of the sheet (plane anisotropy).

The Taylor factor M, the value of which is used to calculate the coefficient of plastic anisotropy, is a measure of plastic resistance and it may be employed in a direct way to compare the anisotropy of the strength properties measured in a tensile test. The calculated values of the factor M, shown in figure 2, represent well the changes in the measured values of $R_{0.2}$ for aluminium, copper and brass (especially in a recrystallized state).

PLASTIC ANISOTROPY OF CONTINUOUS MEDIUM - HILL'S THEORY

Anisotropic theory of plastic flow, according to the concept of Hill [1], is based on the introduction of the constants of anisotropy H,F,G,N,M and L into the plasticity criterion of Huber-v.Mises-Hencky. In the deformation of sheets, the considerations can be restricted to the analysis of a plane state of stress in the plane X,Y /in the rolling direction RD and the transverse direction TD of the sheet, respectively/ with the stress in the direction of normal ND $\sigma_z=0$. The plasticity criterion takes the form

$$(G+H)\sigma_x^2-2H\sigma_x^2\sigma_y^2+(H+F)\sigma_y^2 +2N\tau_{xy}^2 = 1 \quad (3)$$

Application of a tensile yield stress σ in the direction of tension, lying in the sheet plane and inclined at an angle α to RD, is equivalent to the action of stress in the directions X and Y

$$\sigma_x=\sigma\cos^2\alpha,\quad \tau_{xy}=\sigma\sin\alpha\cos\alpha,\quad \sigma_y=\sigma\sin^2\alpha \qquad (4)$$

As in the Levy-v.Mises equations of flow for an isotropic medium one obtains the dependences relating the strain increments with stresses:

$$d\varepsilon_x=\sigma d\lambda[(G+H)\cos^2\alpha-H\sin^2\alpha],\quad d\varepsilon_z=-\sigma d\lambda[F\sin^2\alpha+G\cos^2\alpha]$$
$$d\varepsilon_y=\sigma d\lambda[(F+H)\sin^2\alpha-H\cos^2\alpha],\quad d\gamma_{xy}=\sigma d\lambda(N\sin\alpha\cos\alpha) \qquad (5)$$

where $d\varepsilon_x$, $d\varepsilon_y$, $d\varepsilon_z$ and $d\gamma_{xy}$ are the components of the strain tensor.

The considerations of the anisotropic flow of a sheet in the directions inclined to the rolling direction RD at an angle α equal to: 0 rad /r_1/, $\pi/4$ rad /r_2/ and $\pi/2$ rad /r_3/ allowed the author to determine the plastic anisotropy constants (for numerical values see [5]) from the equations (5):

$$H=r_1/(1+r_1)/\sigma_o^2,\quad G=H/r_1,\quad F=H/r_3,\quad N=H(r_2+0.5)(1/r_1+1/r_3) \qquad (6)$$

where σ_o denotes the yield strength in the direction RD.

LOCALIZATION OF STRAIN AND FRACTURE OF A TENSILED PLANE SAMPLE

In a flat sample with the width/thickness ratio greater then 5 [1] theoretical analysis leads to the conclusion that failure should occur along a plane perpendicular to the sample surface, lying along the characteristics of the plastic flow (equations 5), which in this case form the straits lines inclined to the tension axis at a certain angle θ. With the anisotropic constants (equations 6), the value of the angle θ for a sample cut out from a metal sheet in the direction α can be obtained from the relation

$$tg\theta = -b/a \pm \sqrt{(b/a)^2 + c/a} \qquad (7)$$

where the constants a,b and c, dependent on α, and r_1, r_2, r_3, define the state of plastic anisotropy at the limit of uniform elongation [5].

Figure 3-middle illustrates by way of example the values of θ obtained experimentally and calculated using formula (7). The measurements of the necking inclination angle θ depending on the angle α of the sample cut-out for strain hardened aluminium and brass show that the measured values of θ are as a rule a little higher than the calculated mean values of θ_{sr}. The reason for this may be the measurement of the angle θ on the fractured test piece the failure of which proceeded along a shear band. The proof of shearing was the distinct shifting of fragments of the fractured test piece and also the alongated shape of voids in the shearing direction. The difference between the particular metals was manifested in the size and the number of voids; in aluminium the diameter of the voids can be estimated to be equal to 2-5μm, in copper and brass - to be equal to 1-3μm [5].

A different behaviour is observed in case of recrystallized metals. As far as in the CuZn37 brass the tendency did not change after annealing, in case of aluminium and copper the changes are essential. In annealed aluminium Al2W the fracture proceeded along an irregular surface the inclination of which might be characterized by at least 2 values of the angles θ (fig. 3-middle).

Although in an annealed metal one can expect differences between the experimental measurements and the theoretically predicted values on account of the formation of a tensile texture, yet the irregularity in the fracture surface may result from inhomogeneous deformation of a particular grain which, especially in case of coarse grained flat sample, is noted as wavy surface.

The observed different course of changes in the measured values of θ was confirmed in the image of the fracture surface. The rupture of an annealed aluminium Al2W pulled in tension occurred in such a way that the failure surface took the shape of a "knife-edge", although a very irregular one. Similar behaviour is observed in case of copper C5W.

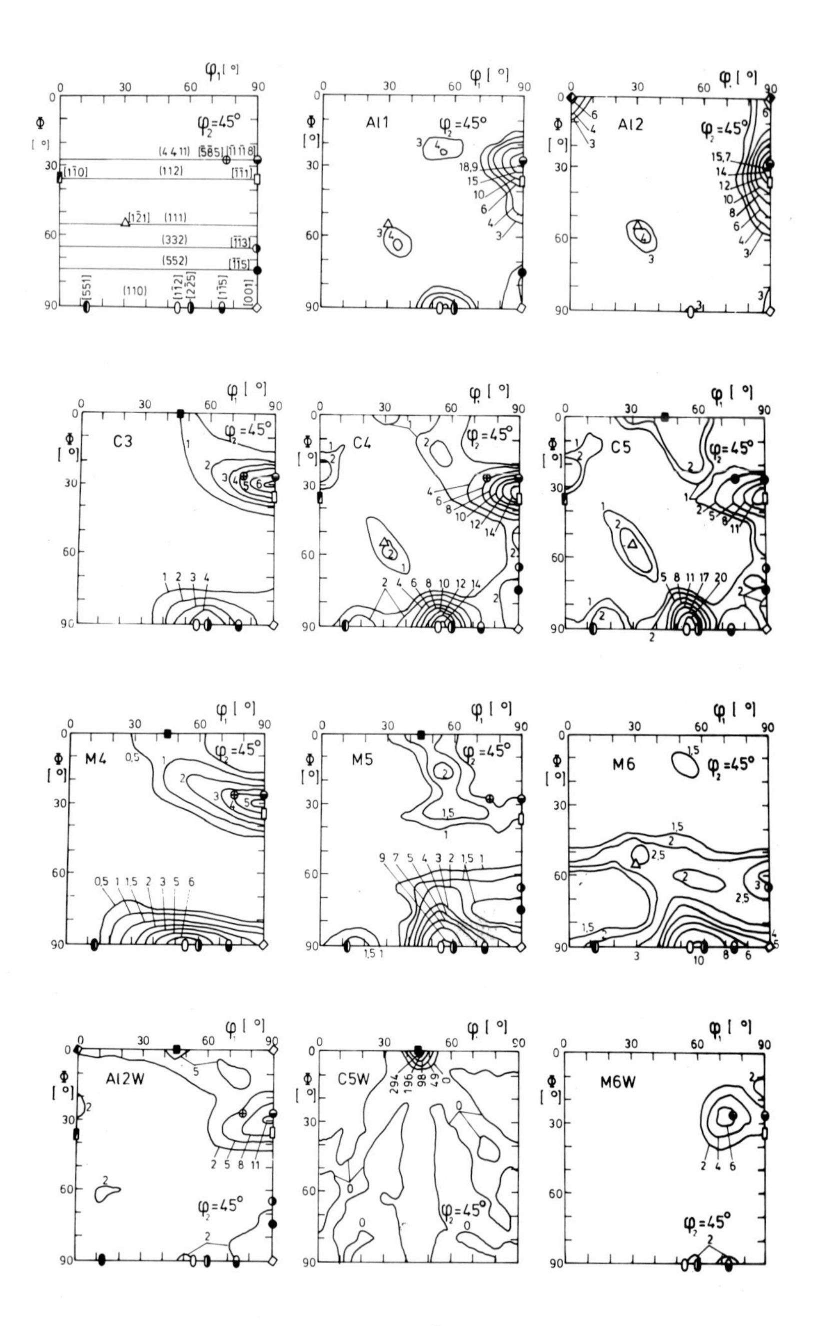

Figure 1. Sections of ODF for φ2=45°. Aluminium: Al0-hard, unstressed, Al1(z=75%), Al2(z=95%), Al2W-recr. at 623K; Copper: C3-hard, unstressed, C4(z=75%), C5(z=93%), C5W-recr. at 773K; Brass CuZn37: M4-hard, unstressed, M5(z=75%), M6(z=92%), M6W-recr.at 673K.

CONCLUSIONS

The anisotropy of the strength properties resulting from the crystallographic texture was distinctly observed in all examined cases of the rolling textures, the maximal differences in the calculated values of the coefficient M not exceeding 20% of the mean value. This defines the limits of plane anisotropy of the measured strength properties of rolled aluminium, copper and brass CuZn37 for industrial practice.

Besides its purely scientific aspect this research study aims to present a method of predicting the anisotropy of the strength and plastic properties as well as the localization of fracture in a tension test on the basis of the knowledge of the orientation distribution of crystallites.

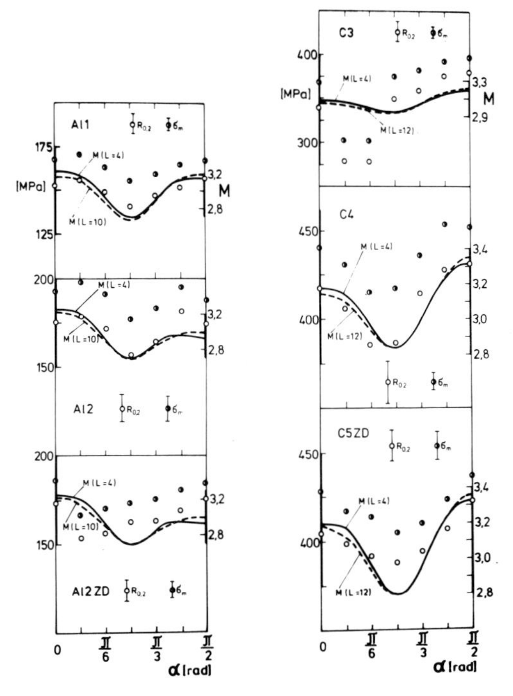

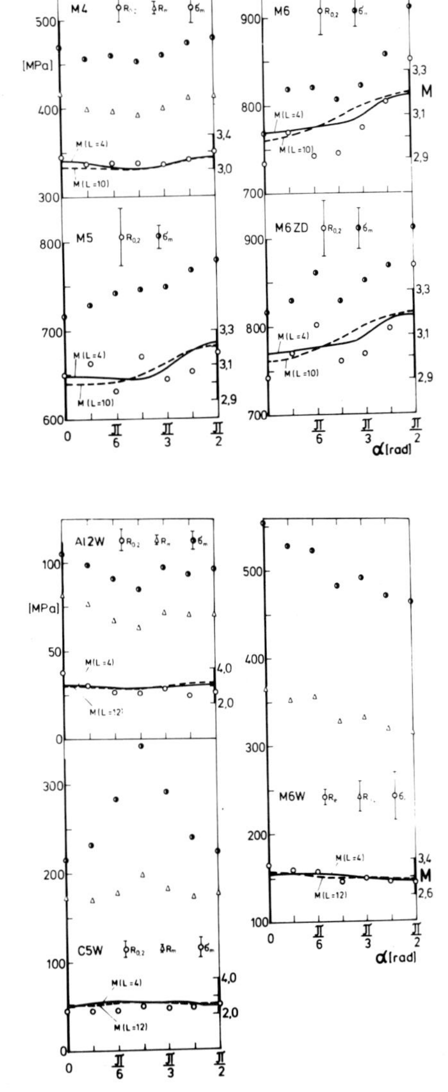

Figure 2. Dependence of the strength properties and the Taylor factor M on the angle α for rolled sheets. Materials- see figure 1.

REFERENCES

1. Hill R.: Mathematical Theory of Plasticity, Clarendon Press, Oxford, 1950
2. Taylor G.I.: J. Inst. Metals, 1938, 61, 307
3. Bishop J.F.W., Hill R.: Phil. Mag., 1951, 42, 414 and 1298
4. Bishop J.F.W.: Phil. Mag., 1953, 44, 51
5. Kusnierz J.: Archives of Metall., 1992, 37, 203
6. Bunge H.J.: Z. Metallkunde, 1965, 56, 872
7. Roe R.J.: J. Appl. Phys., 1965, 36, 2024
8. Bunge H.J.: Kristall u. Technik, 1970, 5, 145
9. Bunge H.J.: Directional Properties of Mat., DGM, Germany, 1988

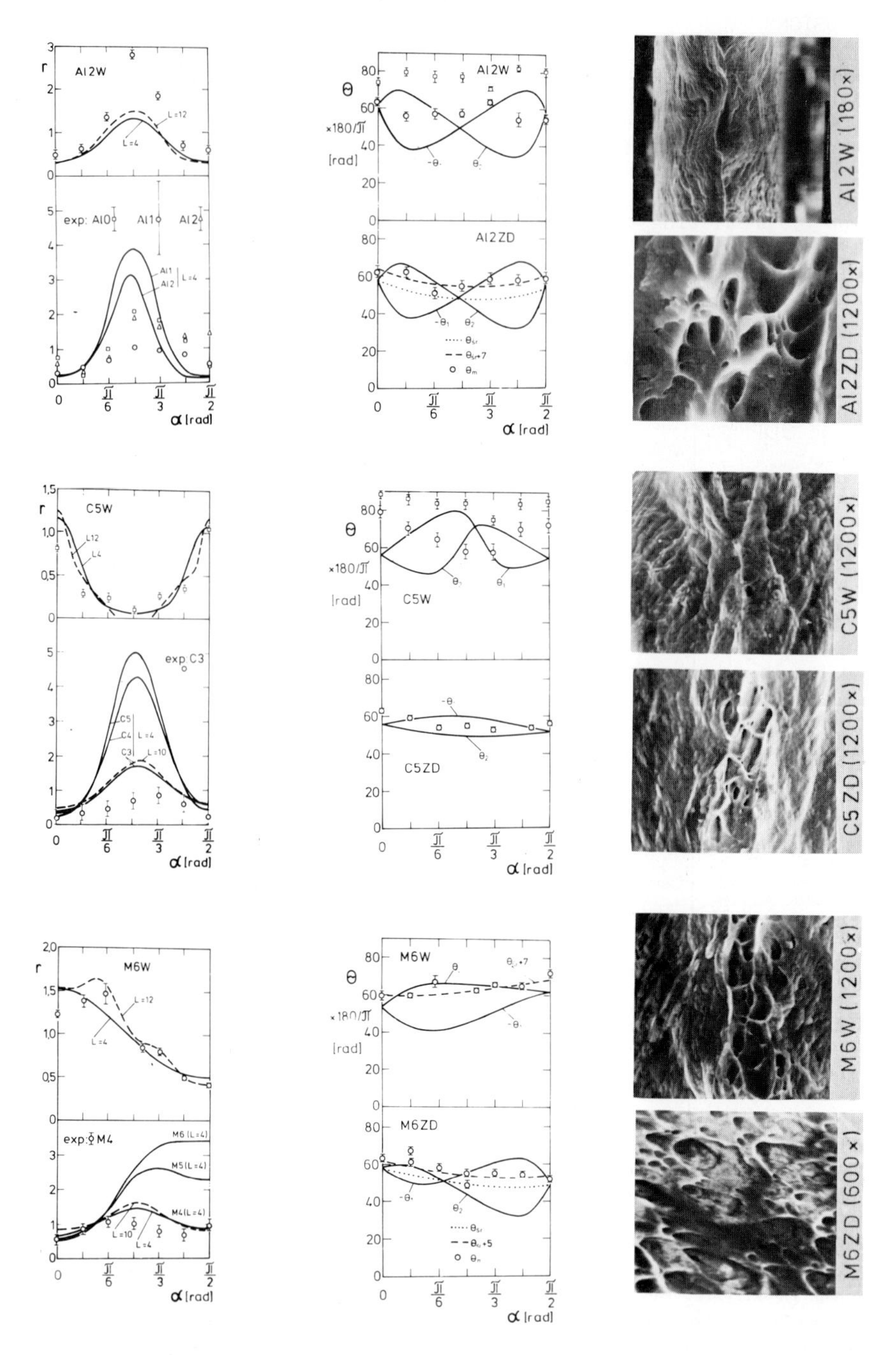

Figure 3. Left- Dependence of plastic anisotropy coefficient r on angle α relating to RD. Middle- The inclination angle θ of the localized neck in the tensiled sample. Right- Microphotograph of the fractured samples obtained in a scanning microscope. Materials- see fig. 1 (Al2ZD, C5ZD, M6ZD-recov. at 423K).

Materials Science Forum Vol. 157-162 (1994) pp. 1641-1646
© 1994 Trans Tech Publications, Switzerland

ON SOME METHODICAL DEVELOPMENTS CONCERNING CALCULATIONS PERFORMED DIRECTLY IN THE ORIENTATION SPACE

S. Matthies [1,2,3] and G.W. Vinel [4]

[1] Université de Metz, L2MP, F-57045 Metz Cédex 1, France

[2] UC, Berkeley, Department of Geology and Geophysics, Berkeley, CA 94720, USA

[3] Permanent address: Liebigstr. 24, D-01187 Dresden, Germany

[4] FZR Rossendorf, Postfach 51 01 19, D-01314 Dresden, Germany

Keywords: ODF Calculation, Tensorial Properties, Smoothing, Filtering, Single Orientation, Correlations

ABSTRACT : As a consequence of the new level of productivity of computer techniques, available even for small laboratories, there are no more restrictions to use the original expressions of quantitative texture analysis for concrete calculations. They are formulated directly in the orientation (G) space and don't need any translations and retranslations by harmonic expressions with their known restrictions and effects concerning numerical accuracy. The use of "the whole G-space" in calculations leads to drastical simplifications and speed of the direct algorithms. The qualitatively new possibilities are demonstrated for typical problems of QTA like ODF reproduction, tensorial properties, pole figure moments, smoothing (filtering) of practically unlimited numbers of single orientation and correlation data.

1. INTRODUCTION

As is well known, new sophisticated tools, which become available, may open possibilities for the consideration of problems or for the practical realization of solutions or ideas having been formulated already years earlier.

The results published resently by the so-called "orientation imaging microscopy" [1] are an impressive example. The new capacity of single orientation measurements improves the number and density of measuring points in such a manner that a statistically well founded consideration of correlation functions of various kind or phenomena like nature of grain boundaries became real now.

The availability of computer controlled goniometers with stepper motors opened the possibility to realize any grid of measuring points on the pole sphere. E.g. regular sets with equal area cells (instead of equal angular nets) may save measuring time near the "north pole" without loss of resolution. The systematic net of hexagonal cells [2] adds the moments of continuously adjustable resolution and the possible use of (continuous up to second derivatives) cubic splines for interpolations in the equal area projection plane in order to fit INPUT FORMAT's of any software.

Below some other analogous effects in the frame of common quantitative texture analysis (QTA) will be discussed. Except for some new ideas, all are in principle more or less connected with the new level of speed and volume of operative memory for PC's typical even on a "truly personal" level.

All ODF's shown are given in σ-sections and equal area projection.

2. FOURIER SPACE OR DIRECT SPACE CALCULATIONS ?

Comparing "methods" we keep in mind the basic principle that "two different, but mathematically clean and unique approaches must in the end lead to the same results of a well defined problem ". But one apparatus may be more adapted to a given problem and given computing conditions than the other. It's not new knowledge that the harmonic method [3] (excellent in theoretical analysis and characterizations in connection with symmetries, especially for linear problems) has some difficulties to guarantee correct (or simply positive) numerical values for relative sharp textures or nonlinear problems.

Due to limited speed, memory and accuracy of a given computer there exists always an upper $l = L$ with the known series truncation effects. But more serious are the effects for great l in connection with the drastically increasing number of positive and negative terms of very different order of magnitude having to be added in the numerous sums typical for the method [4]. They are difficult to estimate.

The harmonic method lives by the (original complex) spherical functions. These can be constructed by sums of products of e.g. only 144 $\sin x_i$ and $\cos x_i$ values ($x_i = 0°, 5°, 10°, ..., 355°$) for a 5°-angular grid. But the critical quantities of interest are the necessary coefficients at the product terms. The coefficients themselves are connected with spherical functions of special argument a.s.o. I.e. libraries cannot be avoided and their volume rises with increasing l.

The direct algorithms (e.g for the calculation of a pole figure from a given ODF by simple integrals over a projection path g_{pp} [5]

$$P_{\boldsymbol{h}_i}(\boldsymbol{y}) = \int_0^{2\pi} f(g_{pp}(\widetilde{\varphi}))\, d\widetilde{\varphi}/2\pi \;; \qquad g_{pp}(\widetilde{\varphi}) = \{\boldsymbol{h}_i, \widetilde{\varphi}\}^{-1}\{\boldsymbol{y}, 0\} \tag{1}$$

or for ODF reproduction from pole figures using e.g. WIMV [6]) contain as a rule positive, real numbers only.

A great advantage is the apparent connection between cell dimensions in the G- or Y-space and resolution. But in past the direct algorithms also needed rather large and symmetry specific libraries of complex structure. One of the reason for the libraries was the exaggerated overestimation of the role of the symmetry related elementary G-space regions for computing purposes. Also it was related to the limited operative memory of computers. E.g. a tetragonal/orthorhombic ODF was considered in a $19 \times 19 \times 19$ array for the common 5°-steps of EULERian angles.

This leads to a complicated form e.g. for a projection path (1) as it is schematically shown in the left upper corner of figure 1. All addresses of the G-space cells belonging to this path had to be stored in a library, and there is a large number of paths... Therefore for a long time direct methods were commonly considered as too complicated and nonelegant.

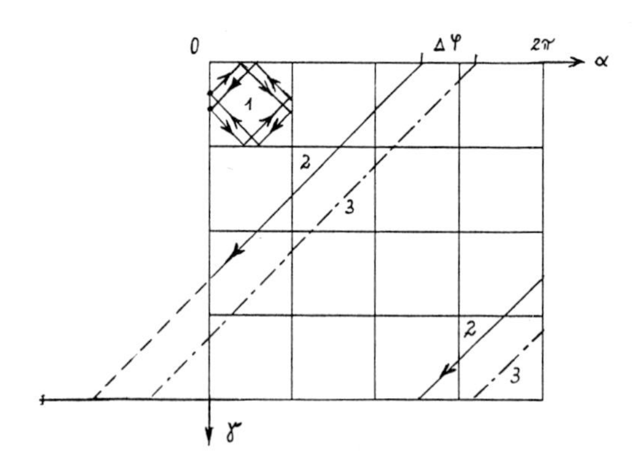

Figure 1. Schema of projection paths in the G-space. 1 - path reflected into an elementary region; 2 - original path in the whole G-space; 3 - path for another pole with the same polar angle ϑ.

3. "WORKING IN THE WHOLE ORIENTATION SPACE"

The situation drastically changed (but it was scarcely noticed by the "texture community") when it became possible to use operatively the whole G-space (e.g. $73 \times 37 \times 73 = 197173$ 5°-cells), i.e. filling it redundantly $4 \times 2 \times 4$ times with the above information for the tetragonal/orthorhombic example. What are the consequences ?

For instance it is obvious that the original projection path (cf. "2" in figure 1) does not know anything about "tetragonal/orthorhombic" symmetries. This information is completely included in the ODF values. Moreover, from the explicit expression of $g_{pp}(\tilde{\varphi})$ (1) follows that the G-space cell addresses for paths with the same polar angle ϑ ($\boldsymbol{y} = (\vartheta, \varphi)$) are connected in a trivial manner, as is also indicated in figure 1. This can be used now and means e.g. that a "path library" for a whole pole figure $P_{\boldsymbol{h}}(\boldsymbol{y})$ would contain (assuming 2°-steps in $\tilde{\varphi}$) a maximum of $181 \times 37 = 6697$ cell numbers only.
It turned out that with a PC-486/33MHz the calculation of a pole figure from an ODF with any crystal and sample symmetry takes only about 2 seconds without any path library. Stored data lead to savings less than 20%. Analogous effects appear also for other typical "library burdened" problems of QTA as will be described below.
As a result of the new concept the extremely short and explicitely symmetry independent "inner iteration cycle" leads to a simplified and universal formulation of the WIMV algorithm. PC adapted 5°-WIMV programs for any crystal and sample symmetry up to 100 pole figures (complete, incomplete, with or without overlaps) exist now and are available.
The typical starting point of any calculations needing an ODF is now an ODF in its symmetry dependent "WIMV STANDARD" elementary region [7] that is immediately converted into the whole G-space.

4. AVERAGING OF PHYSICALLY TENSORIAL PROPERTIES

A great number of rather sophisticated analysis has been made in the past in order to express the arithmetic mean of tensors of various rank and kind according to the ODF by its symmetrized FOURIER coefficients $C_l^{\mu\nu}$ (cf. e.g. [8, 9, 10]). But now concrete numerical calculations can be done directly and in a less complicated way using the simple and quite transparent starting expression, with E° - the single crystal material constants :

$$\overline{E}^a_{i_1 i_2 \dots i_n} = \sum_{i'_1 \dots i'_n = 1}^{3} [\cdots]\, E^\circ_{i'_1 i'_2 \dots i'_n}\,; \quad [\cdots] = \int_G g_{i'_1 i_1}\, g_{i'_2 i_2} \cdots g_{i'_n i_n}\, f(g)\, dg\,. \tag{2}$$

The object of interest $[\cdots]$ leads to about $(n+1)^3$ integrals of the simple type

$$\int_0^{2\pi} d\alpha \int_0^{\pi} d\beta \int_0^{2\pi} d\gamma\, \sin^q \alpha\, \cos^r \alpha\, \sin^{s+1} \beta\, \cos^t \beta\, \sin^u \gamma\, \cos^v \gamma\, f(g)\,; \quad 0 \le q, r, \dots, v \le n \tag{3}$$

Using a cell structure of the G-space the integrals can be replaced with sufficient accuracy by simple sums. For a tensor of fourth rank and a 5°-grid the calculation takes a few minutes. Again the crystal and sample related symmetries of importance are included in the numerical values of the ODF or of the E° components. It may be of some interest that we succeeded to solve for symmetric tensors (and to implement corresponding universal routines with direct algorithms) [11] the old open problem [12, 13] (in the frame of (2)) to realize an averaging procedure "—" that identically obeys the physical condition

$$\left(\overline{E^{-1}}\right)^{-1} = \overline{E}\,. \tag{4}$$

Up to now this has only been achieved on a macroscopic level for arbitrary orientation distributions with complicated self-consistent calculations [14]. For more details see the separate paper [15].

5. INTERNAL STRESS - POLE FIGURE MOMENTS

By $\theta|2\theta$ variations in a diffraction experiment with the sample direction $\boldsymbol{y}$ for the scattering vector $\boldsymbol{N}$ and the projection direction $\boldsymbol{h}_i$ (commonly characterizing a pole figure) information about the strain tensor component $\varepsilon'_{zz}(\boldsymbol{h}_i, \boldsymbol{y})$ in the scattering coordinate system (with $Z' \| \boldsymbol{N}$) can be obtained. Combining this with a REUSS approximation of the compliance tensor for the crystals just in reflecting position information about the stress tensor in the scattering volume can be extracted [16]. The ODF-average enters the problem in form of integrals of the type

$$g_{3j}\, g_{3k} \int_0^{2\pi} f(\tilde{g}^{-1} \cdot \{\boldsymbol{y}, 0\})\, \tilde{g}_{tl}\, \tilde{g}_{pm}\, d\tilde{\varphi}\,, \quad \tilde{g} = \{\boldsymbol{h}_i, \tilde{\varphi}\}\,. \tag{5}$$

Describing this by symmetrized spherical functions very complicated expressions appear (cf. e.g. [17]). From the sight of the direct apparatus the argument of the ODF is simply g_{pp} - the projection path well known from (1). Considering additionally the $\widetilde{\varphi}$-dependence of the $\widetilde{g}$-matrices it immediately follows that all the "stress specific, ODF related" information is included in only 4 numbers, "pole figure moments", or "momentum pole figures"

$$\begin{array}{lll} M_s(\boldsymbol{h}_i, \boldsymbol{y}) & \sim & \int_0^{2\pi} f(g_{pp}(\widetilde{\varphi})) \quad \sin\widetilde{\varphi} \quad d\widetilde{\varphi} \\ M_c & & \cos\widetilde{\varphi} \\ M_{s2} & & \sin 2\widetilde{\varphi} \\ M_{c2} & & \cos 2\widetilde{\varphi} \end{array} \qquad (6)$$

The calculation of the momentum pole figures needs only seconds for any ODF using the corresponding universal routine.

6. SMOOTHING AND FILTERING IN THE G-SPACE

The l-series truncation in the harmonic method leads automatically to some kind of ODF smoothing. After extraction of the F-coefficients [3] or equivalent information up to $l = L$ specific frequencies, all connections to the "noisy" experimental data are, as a rule, completely lost.
The direct methods of ODF reproduction always rest upon the original pole figure data and possess usually a higher resolution. This often results in noisy distributions in the G-space as is shown in figure 2 for a recrystallized Al sample with a relatively sharp cube component.

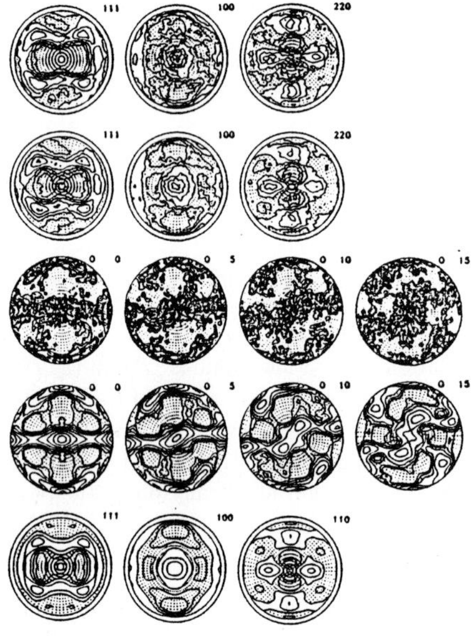

Figure 2. Recrystallized Al. Rows: 1 - noisy experimental PF's, max. 3.25, min. 0.26; 2 - WIMV recalc. PF's, max. 4.63, min. 0.27, RP/1 = 17/14%, 3 - noisy WIMV-ODF, max. 35.5, min. 0.18; 4 - smooth $l \leq 22$ harmonic, l-even ODF, max 6.79, min 0.01; 5 - harmonic recalc. PF's, max. 2.82, min. 0.26, RP/1 = 24/17%.

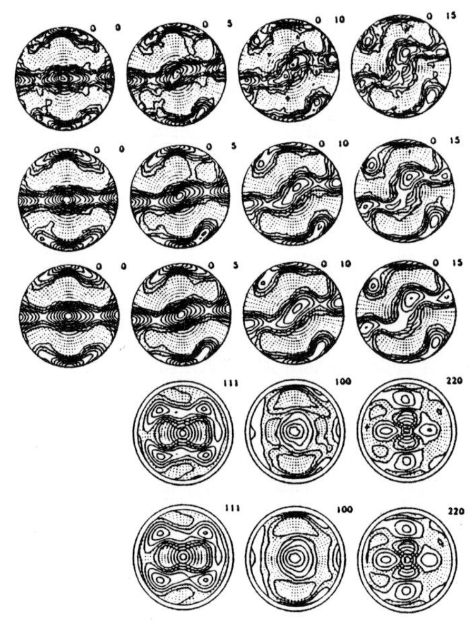

Figure 3. Demonstration of filtering. Rows: 1 – 7.5°-filtering of figure 2.3, max. 17.89, min. 0.19; 2 - twice filtered ODF, max. 15.04, min. 0.18; 3 - triple filtered ODF, max. 14.5, min. 0.18; 4 - recalc. PF's from ODF-1, max. 3.66, min. 0.27, RP/1 = 18/18.5%; 5 - recalc. PF's from ODF-3, max. 3.82, min. 0.26, RP/1 = 19, 21%.

Obviously some sort of smoothing would be desirable for visual interpretations. In a well defined manner this can be done by folding with an orientation distance ω-depending bell curve B of halfwidth b :

$$f^{sm}(g) = \int_G f(g_0)\, B(b,\omega)\, dg_0 ; \quad B(b,\omega) \geq 0 ; \quad \omega = \omega(g, g_0) . \tag{7}$$

The new concept opens the possibility in a maximum way to use the inner symmetries of the $\omega(g, g_0)$ relation and to realize (8) with a reasonable computing effort now.
Any smoothing is connected with a broadening of the texture components. The starting width of the main components can be preserved by a second "sharpening step" :

$$f^{sh}(g) = f_{min} + (1 - f_{min})\, (f^{sm}(g) - f_{min})^{x_0} \,/\, [\int_G (f^{sm}(g') - f_{min})^{x_0} dg'] , \tag{8}$$

with $x_0 > 1$ determined e.g. by the condition of equal texture index for $f(g)$ and $f^{sh}(g)$. The combination of both steps is equivalent to a true noise filter, only slightly deforming the main components. The filter can be applied several times enhancing the main features of a given distribution as is shown in figure 3.

7. SINGLE ORIENTATIONS AND CORRELATIONS

Putting any number of single orientations (possibly weighted) into the corresponding cells of the symmetry specific elementary G-space region, after normalization smoothing operations can be done using the algorithm (8). The orientations can also be multiplied in order to represent exactly e.g. an assumed sample symmetry. Using this tool at last systematic numerical considerations are possible of the nontrivial questions of type

- How many statistical single orientations N are necessary in order (after smoothing, using b; $b = b(N)$?) to represent satisfactorily an ODF of a certain type ?
- How many are necessary in order to establish if there is (or is not) a random orientation ? a.s.o.

The corresponding treatment of correlation data does not differ greatly from that of single orientations. When the corresponding data have to be "multiplied" for the right-sided "sample" symmetry simply the crystal symmetry is used plus the correlation specific $F(g) = F(g^{-1})$ symmetry.
There exist three universal routines for the corresponding class of problems. Fig. 4 - 6 demonstrate some aspects of their use.

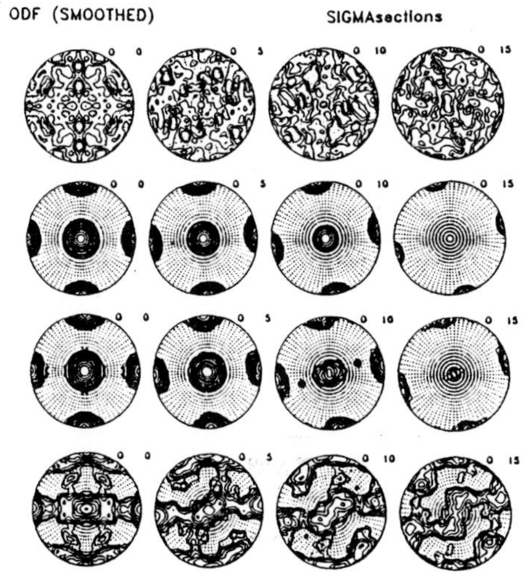

Figure 4. Smoothing of single orientations (400 grains, $b = 10°$, cubic/orthorh. symmetry). Rows: 1 - simulation of a random distribution, max. 2.45, min.0.14; 2 - standard cube position, $b_{st} = 20°$, $phon = 0.1$; 3 - simulation of the previous ODF; 4 - simulation of figure 3.3.

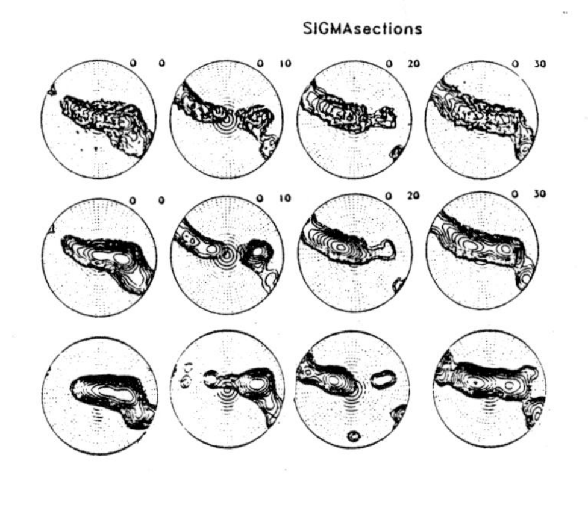

Figure 5. ODF of a quartz sample. Rows: 1 - WIMV-ODF from 7 PF's derived by neutron scattering, $2 - 7.5°$-filtered ODF, $3 - 15°$ -smoothed 40,000 single orientation measurements of the same sample (cf.[18]).

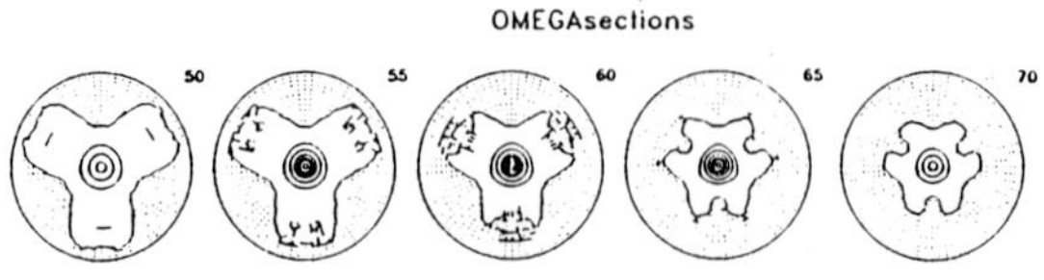

Figure 6. Misorientation distribution. Dauphiné twins ($\omega = 60°$, $\boldsymbol{n} = 001$) in the sample of figure 5. 30,000 correlations of neighbouring grains, 10°-smoothing.

8. TRANSFER OF VARIABLES

All algorithms described use the $\{\alpha, \beta, \gamma\}$ representation of orientations. It is obvious that using the whole cyclic G-space all transfer operations into other variables (e.g. $\varphi_1, \phi, \varphi_2; \sigma, \delta, \kappa; \omega, \vartheta, \varphi$) are simplified in an optimum manner.

9. WHY "100" POLE FIGURES ?

The characteristics of the new or revised direct algorithms permit to approach more complicated problems in near future.

d-spacing spectra (e.g. from neutron TOF measurements) contain as a rule a volume of information used only very poorly up to now. The problem is to extract the great number of overlapped pole figure signals that may additionally belong to several phases. But knowing (or iteratively refining) all necessary structural and instrumental parameters the powerful high-developed algorithms of modern powder diffraction can then be used to extract pole figure weighted texture information from complicated samples. The first steps of this concept suggested 1991 by R.Wenk, Berkeley and realized in cooperation with L.Lutterotti, Trento and the authors look promising. New active and creative partners for this challenging project are welcome.

ACKNOLEDGEMENT: The analysis and reformulation of the direct universal algorithms in a relatively short period was decisively supported by the creative atmosphere in the host laboratories of S.M. in Metz (C.Esling) and Berkeley (H.R. Wenk).

References

[1] Adams, B.L., Wright, S.I. and Kunze, K.: Metall. Trans., 1993, **24A**, 819

[2] Matthies, S. and Wenk, H.R.: phys. stat. sol.(a), 1992, **133**, 253

[3] Bunge, H.J.: Texture Analysis in Materials Science (1982), Butterworths, London

[4] Matthies, S.: Textures and Microstructures, 1988, **8&9**, 115

[5] Matthies, S.: phys. stat. sol.(b), 1979, **92** , K135

[6] Matthies, S. and Vinel, G.W.: phys. stat. sol.(b), 1982, **112**, K111

[7] Matthies, S., Vinel, G.W. and Helming, K.: Standard Distributions in Texture Analysis (1987), Akademieverlag, Berlin

[8] Brandmueller, I. and Winter, F.X.: Z. f. Kristallographie, 1985, **172**, 191

[9] Ganster, J. and Geiss, D.: phys. stat. sol.(b), 1985, **132**, 395; 1988, **147**, 191

[10] Zuo, L., Humbert, M. and Esling, C.: J. Appl. Cryst., 1992, **25**, 751

[11] Matthies, S. and Humbert, M.: phys. stat sol(b), 1993, **177**, K47

[12] Aleksandrov, K.S. and Aizenberg, L.A.: Dokl. Akad. Nauk. SSSR, 1967, **167**, 1028

[13] Morawiec, A.: phys. stat. sol.(b), 1989, **154**, 535

[14] Hirsekorn, S.: Textures and Microstructures, 1990, **12**, 1

[15] Matthies, S. and Humbert, M.: This conference, paper P15

[16] Van Houtte, P. and De Buyser, L.: Acta metall. mater., 1993, **41**, 323

[17] Brakman, C.M.: J. Appl. Cryst., 1983, **16**, 325

[18] Kunze, K., Adams, B.L., Heidelbach, F. and Wenk, H.R.: This conference, paper L12

Materials Science Forum Vol. 157-162 (1994) pp. 1647-1652

ON THE GEOMETRIC MEAN OF PHYSICAL TENSORS USING ORIENTATION DISTRIBUTIONS

S. Matthies [1,2] and M. Humbert [1]

[1] Laboratoire de Métallurgie des Matériaux Polycristallins, ISGMP, Université de Metz, F-57045 Metz Cédex 1, France

[2] Permanent address: Liebigstr. 24, D-01187 Dresden, Germany

Keywords: Properties, Elastic Constants, Arithmetic Mean, Geometric Mean

ABSTRACT: A practicable, simple averaging procedure of even rank symmetric tensors is described, realizing an idea of Aleksandrov and Aizenberg (1967). It possesses the properties of a geometric mean, identically obeying the physical condition $\overline{E} = [\overline{E^{-1}}]^{-1}$. The orientation distribution function $f(g)$ enters the calculations in form of the well known arithmetic mean. The general case is completed by the consideration of the specific (twice symmetric) fourth rank elastic tensors. Calculations with real and modelled orientation distributions lead to results close to those of much more sophisticated self consistent schemas.

1. INTRODUCTION

The arithmetic mean is the simplest variant to estimate the polycrystalline properties of a sample (characterized by an orientation distribution $f(g)$) using the single crystalline data of the corresponding tensor E° of rank m $(g \mathrel{\widehat{=}} K_A \longrightarrow K_B)$:

$$\overline{E}^{a} = \int_G E^\circ(g)\, f(g)\, dg\,, \qquad \int_G dg = 1\,, \tag{1}$$

with

$$E^\circ_{t_1 t_2 \ldots t_m}(g) = \sum_{t'_1 \ldots t'_m = 1}^{3} g_{t'_1 t_1} \cdots g_{t'_m t_m}\, E^\circ_{t'_1 t'_2 \ldots t'_m}\,. \tag{2}$$

$E^\circ(g)$ represents the single crystalline property described in the sample coordinate system K_A. The techniques to determine (1) are well developed in the frame of the harmonic apparatus [1],[2]. The integral for $\overline{E}^{a}$ can also directly be calculated for an ODF $f(g)$ given in the whole orientation space G without any problems [3]. Therefore below this step in the corresponding expressions is not of primary interest.

Using $\overline{E}^{a}$ in order to characterize a property of a polycrystal the contributions of all crystallites are considered as absolutely independent. This does not hold as a rule in reality. An obvious demonstration of this fact is connected with the mean values of "inverse properties" (E and $H \equiv E^{-1}$) like electrical resistivity and conductivity or elastic stiffness (C) and compliance (S).

As is well known [1] $\overline{S}^a$ (Reuss approximation) may considerably differ from the Voigt approximation $[\overline{C}^a]^{-1}$, whereas the experimental data obey the relation

$$\overline{E} = [\overline{E^{-1}}]^{-1} = [\overline{H}]^{-1} \tag{3}$$

on the macroscopic level.
The last condition is one of the constructive components of any iterative self-consistent schema. Such an approach contains the exact description of the interaction of a single grain with a continuous matrix possessing effective polycrystalline properties [4],[5]. The last ones are tuned step by step using the response of the matrix to a grain of given orientation and frequency.
Two moments are of interest in this connection. In order to be able to describe the microscopic level, model assumptions are necessary e.g. about the grain sizes or forms. On the other hand it is known that averaging mechanisms of large physical ensembles may be quite insensitive to such details if there are not strong interactive effects between the constituents of the sample.
I.e. in a system consisting for instance of almost spherical grains (form isotropy relative to orientation characteristics) it may appear that averaging procedures using ODF-dependent steps of type (1), but identically obeying (3) will lead to results close to the reality much simpler than relatively complicated self-consistent schemas.
The first trivial attempt in this direction was Hill's approximation

$$\overline{E}^H = [\overline{E}^a + [\overline{H}^a]^{-1}]/2\,, \tag{4}$$

but it does not exactly obey (3).
A procedure that should exactly obey (3), possessing the character of a geometric mean, was schematically proposed 1967 by Aleksandrov and Aizenberg [6] for even rank ($m = 2n$) symmetric tensors, but has not been noticed by texture specialists at that time. Moraviec [7] returned to this idea and derived some valuable results concerning the case of random orientation distribution. However, his consideration of the general case did not lead to a practicable schema.
A straightforward solution of the problem has shortly been described by the authors in [8]. The implemented universal routines for any orientation distributions and symmetries are free available. Below the most important connections will be given, explicitly considering the specific case of the elastic, twice symmetric tensors also.

2. THE GENERAL SCHEMA

The components of even rank symmetric tensors $E_{i_1 i_2 \ldots i_n; j_1 j_2 \ldots j_n}$ $(i_k, j_k = 1,2,3)$ can formally be considered as elements of a symmetric quadratic matrix of order $N = 3^n$:

$$\| E \| \,;\; E_{IJ} = E_{JI} \,;\; I,J = 1,2,\ldots,N. \tag{5}$$

The connection between I, J and i_k, j_k is given by code tables

$$l_k(I) \longrightarrow i_k \,;\quad l_k(J) \longrightarrow j_k \,;\quad k = 1,2,\ldots,n\,. \tag{6}$$

The single crystal data matrix $\| E^\circ \|$ can be diagonalized by an orthogonal matrix P° :

$$\left[P^{\circ -1}\, E^\circ\, P^\circ \right]_{IJ} = \lambda_I^\circ \, \delta_{IJ}\,. \tag{7}$$

For typically physical tensors the eigenvalues λ_I° are real numbers, greater than zero. The matrix of the inverse property $H^\circ = E^{\circ -1}$ is also diagonalized by P° and its eigenvalues are simply equal to $1/\lambda_I^\circ$.
The arithmetic mean (1) is given by the new symbols as

$$\overline{E}_{IJ}^a = \sum_{I',J'=1}^{N} \overline{WW}^a_{IJ,I'J'}\, E^\circ_{I'J'} \,;\; \overline{E}^a = \overline{WW}^a : E^\circ\,, \tag{8}$$

$$\overline{WW}^{a}_{IJ,I'J'} = \int_G W(g)_{II'}\, W(g)_{JJ'}\, f(g)\, dg\,, \tag{9}$$

$$W(g)_{II'} = g_{\imath_1(I')\imath_1(I)}\, g_{\imath_2(I')\imath_2(I)} \cdots g_{\imath_n(I')\imath_n(I)}\,. \tag{10}$$

It can be shown that the matrix $W(g)$ is orthogonal. For $E^\circ(g)$ it follows

$$E^\circ(g) = W(g)\, E^\circ\, W(g)^{-1} = [\,W(g)\, P^\circ\,] \begin{bmatrix} \lambda_1^\circ & & \\ & \ddots & \\ & & \lambda_N^\circ \end{bmatrix} [\,W(g)\, P^\circ\,]^{-1}\,. \tag{11}$$

Introducing the "logarithmic matrix"

$$Ln\,[E^\circ(g)] \equiv [\,W(g)\, P^\circ\,] \begin{bmatrix} \ln\lambda_1^\circ & & \\ & \ddots & \\ & & \ln\lambda_N^\circ \end{bmatrix} [\,W(g)\, P^\circ\,]^{-1} \equiv W(g)\, Ln\, E^\circ\, W(g)^{-1} \tag{12}$$

we get for its arithmetic mean in analogy to (8)

$$\overline{Ln\,E}^{a} = \overline{WW}^{a} : Ln\, E^\circ\,. \tag{13}$$

The matrix $\overline{Ln\,E}^{a}$ is diagonalized by the orthogonal matrix $\widetilde{P}$ and possesses the eigenvalues $\widetilde{\lambda}_I$. The final expression for the geometric mean "< >" reads

$$< E >_{IJ} = \sum_{K=1}^{N} \widetilde{P}_{IK}\, e^{\widetilde{\lambda}_K}\, \widetilde{P}_{KJ}\,. \tag{14}$$

Using $E^{-1} = H$ for E, beginning with (7), it is simple to show that

$$< E^{-1} >_{IJ} = \sum_{K=1}^{N} \widetilde{P}_{IK}\, e^{-\widetilde{\lambda}_K}\, \widetilde{P}_{KJ}\,; \quad < E > \cdot < E^{-1} > = I\,, \tag{15}$$

i.e. the condition (3) is identically obeyed. The eigenvalues of $< E >$ (cf. (14)) can be represented in the form

$$< \lambda_I^E > = e^{\widetilde{\lambda}_I} = \prod_{K=1}^{N} (\lambda_K^\circ)^{\widetilde{W}(I,K)} \;;\quad \widetilde{W}(I,K) \geq 0\,;\ \sum_{K=1}^{N} \widetilde{W}(I,K) = 1\,, \tag{16}$$

verifying the term "geometric mean".

3. TWICE SYMMETRIC TENSORS

The numerous possibilities of symmetry relations for twice symmetric tensors can hardly be described in a general way. Therefore the case of the elastic tensors was chosen to demonstrate the principle how to manage equivalent cases.
The fourth rank compliance tensor S $(n = 2\,;\, N = 9)$ connects the strain and stress tensors of second rank

$$\epsilon_{i_1 i_2} = \sum_{j_1 j_2 = 1}^{3} S_{i_1 i_2\,;\,j_1 j_2}\, \sigma_{j_1 j_2}\,. \tag{17}$$

Additionally to the $I \leftrightarrow J$ symmetry (4) the symmetry properties of ϵ and σ induce a symmetry within the i and j sets of indexes :

$$S_{i_1 i_2\,;\,j_1 j_2} = S_{i_2 i_1\,;\,j_1 j_2} = S_{i_2 i_1\,;\,j_2 j_1} = S_{i_1 i_2\,;\,j_2 j_1}\,. \tag{18}$$

As a consequence the determinant (i.e. some eigenvalues also) of the 9×9 S matrix will be zero. Therefore the general schema just derived can not directly be applied.
It is not a new knowledge that elastic problems can be considered in a 6×6 frame of independent variables. But the point is to transfer to new variables using an orthogonal transformation and not the traditional Voigt or Wooster schemas [9]. So we introduce renormalized ("r") effective strain and stress vectors and correspondingly elastic matrices S^r and C^r using the T matrix given in figure 1 ($i_1 i_2 \leftrightarrow I$ code according to Voigt [9]) :

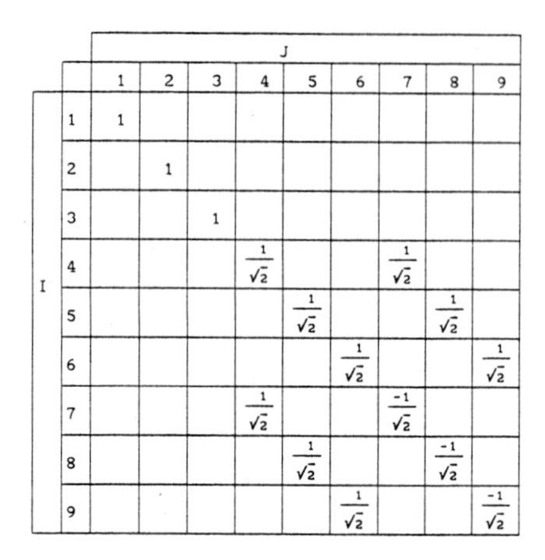

I \ J	1	2	3	4	5	6	7	8	9
1	1								
2		1							
3			1						
4				$\frac{1}{\sqrt{2}}$			$\frac{1}{\sqrt{2}}$		
5					$\frac{1}{\sqrt{2}}$			$\frac{1}{\sqrt{2}}$	
6						$\frac{1}{\sqrt{2}}$			$\frac{1}{\sqrt{2}}$
7				$\frac{1}{\sqrt{2}}$			$\frac{-1}{\sqrt{2}}$		
8					$\frac{1}{\sqrt{2}}$			$\frac{-1}{\sqrt{2}}$	
9						$\frac{1}{\sqrt{2}}$			$\frac{-1}{\sqrt{2}}$

Figure 1. The T-matrix for transformations to renormalized quantities

$$\sigma^r = T\sigma; \quad \epsilon^r = T\epsilon; \quad S^r = TST^{-1}; \quad C^r = TCT^{-1}. \tag{19}$$

As it turns out all components with $I, J > 6$ are identical zero in this representation, i.e. the schema of the previous paragraph can completely be applied. We have only to change $N = 9 \rightarrow 6$, $I, J \rightarrow \mu\nu = 1, ...6$ and $X^r \rightarrow X'$. Especially valuable is the property $C' = S'^{-1}$ in the direct sense of an inverse 6×6 matrix. The transformation $W(g) \rightarrow W'(g)$ is to be done with some care in order not to lose index relations with a geometric meaning to which the ODF and g matrix refer.
The free energy F (in Hook's approximation) is given by the new variables as the quadratic form

$$F = \frac{1}{2}\sum_{\mu,\nu=1}^{6} S'_{\mu,\nu}\,\sigma'_\mu\,\sigma'_\nu = \frac{1}{2}\sum_{\mu,\nu=1}^{6} C'_{\mu,\nu}\,\epsilon'_\mu\,\epsilon'_\nu\,. \tag{20}$$

From its positivity the necessary positivity of the corresponding eigenvalues of $S^{\circ\prime}$ or $C^{\circ\prime}$ follows. After calculating e.g. $< S' >$ (6×6) the return to $< S^r >$ (9×9) is trivial. The transfer to the common variables uses the T-matrix again (c.f. (19)). Some more mathematical details will be given in [10].
Figure 2 shows the Young modulus according to the variation of texture sharpness for Biotite known by its remarkable anisotropic single crystal data. In order to demonstrate the "whole ODF spectrum" a GAUSSian component was chosen at $g = \{0°, 0°, 0°\}$ possessing the halfwidth b. For $b = 0°$ all averaging procedures must lead to the single crystal data. $b \gg 1$ corresponds to the other limit - the random distribution $f(g) = 1$.
As was expected, the simply to calculate geometric mean values are very close to the corresponding self consistent results, calculated by the schema described in [11].
Figure 3 shows the dependence of the Young modulus on the direction in the rolling plane for Zn. Again the geometric mean is practically identical with the self consistent results. The experimental data concerning the sample have also been published in [11].
Obviously the schema of geometric averaging can be applied to other weighted problems too, like e.g. to multiphase samples.

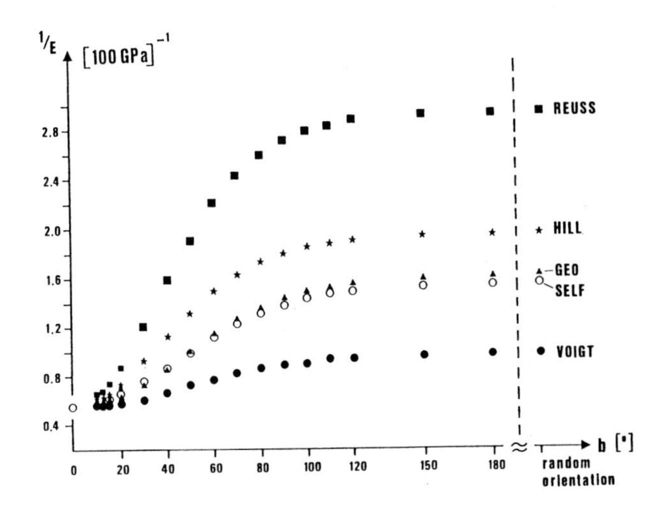

Figure 2. Mean values of Young's modulus for Biotite depending on a model parameter determining the sharpness of orientation distributions; cf. text

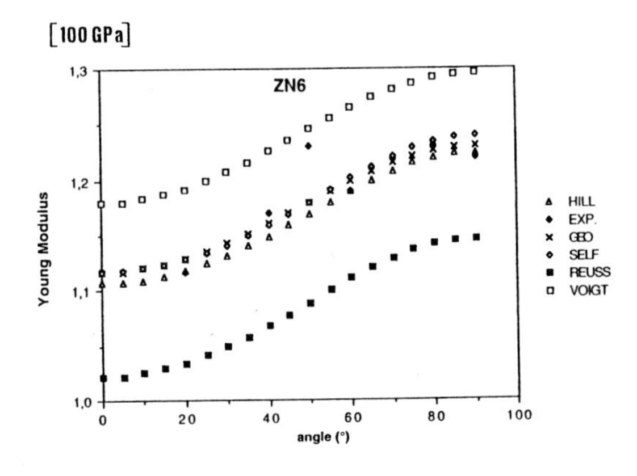

Figure 3: Mean values of Young's modulus for a rolled Zn sheet in dependence on the angle between the rolling and experimental directions

References

[1] Bunge, H.J.: Texture Analysis in Materials Science (1982), Butterworths, London

[2] Ganster, J. and Geiss, D.: phys. stat. sol.(b), 1985, 132, 395; 1988, 147, 191

[3] Matthies, S. and Vinel, G.W.: This conference

[4] Kröner, E.: J. Phys., 1958. 151, 504

[5] Hirsekorn, S.: Textures and Microstructures, 1990, 12, 1

[6] Aleksandrov, K.S. and Aizenberg, L.A.: Dokl. Akad. Nauk. SSSR, 1967, 167, 1028

[7] Morawiec, A.: phys. stat. sol.(b), 1989,154, 535

[8] Matthies, S. and Humbert, M.: phys. stat sol(b), 1993, 177, K47

[9] Nye, J.F.: Physical Properties of Crystals(1959), Clarendon Press, Oxford

[10] Matthies, S. and Humbert, M.: 1993, in preparation

[11] Humbert, M. and Diz, J.: J. Appl. Cryst., 1991, 24, 978

Materials Science Forum Vol. 157-162 (1994) pp. 1653-1662

THE EFFECTS OF TEXTURE AND STRAIN ON THE R-VALUE OF HEAVY GAUGE TANTALUM PLATE

Ch. Michaluk [1], J. Bingert [2] and C.S. Choi [3]

[1] Cabot Performance Materials, County Line Road, Boyertown, PA 19512, USA

[2] Los Alamos National Laboratory, PO Box 1663, MS G770, Los Alamos, NM 87545, USA

[3] U.S. Army, ARDEC, c/o NIST, Bldg. 235, Rm. E-151, Gaithersburg, MD 20899, USA

Keywords: Tantalum, TA-2.5W, Texture, R-Value

ABSTRACT

Previous work by other researchers suggests that the r-value measured from thick-gauge tantalum plate do not correspond to the predicted r-values calculated from ODF coefficients. To understand this behavior, bulk texture analysis using neutron diffraction techniques was conducted on annealed samples of Ta and Ta-2.5W plate: the pure tantalum exhibited a primary {111} type texture whereas the alloy contained a cube texture. For the pure tantalum, the r-values calculated from the texture of as-annealed and the deformed specimens were similar and correlated well with measured data. The r-value of the Ta-2.5W exhibited a greater amount of strain-sensitivity, such that the calculated r-values based on the initial texture did not represent those calculated or measured after tensile deformation. The strain sensitivity of r-values in Ta and Ta-2.5W plates is shown to relate to the generation of a <$\bar{1}$10> fiber texture during deformation.

INTRODUCTION

Tantalum is a high-density bcc refractory metal which is used in a variety of specialized applications[1] which range from high-reliability capacitors for electronic components to corrosion resistent conduits and linings for the chemical processing industry. Tantalum's combination of low flow stress, high ductility, and favorable texture lends itself for use in deep-drawing: thin-gauge tantalum strip (>0.40 mm) is deep drawn to form wet capacitor cans; thin-gauge plate (0.40-4.0 mm) is conventionally drawn into laboratory crucibles; and moderate-gauge plate (4.0mm-6.6mm) is explosively formed into armor-piercing long-rod penetrators.

There has been recent interest to use heavy-gauge tantalum and Ta-W alloy plate (<4.0mm) for deep-drawing. However, researchers found that the r-value data calculated from X-ray diffraction pole figures did not reflect those measured instantaneously from tensile samples.

BACKGROUND

The formability of a metal sheet, specifically its ability to be deep-drawn into a cup or conical shape, is critically dependent on the isotropy of the material. Lankford *et al* [2] are recognized as the first to correlate the drawability of steel to a strain ratio, termed an r-value:

$$r = \frac{\varepsilon_W}{\varepsilon_T} = \frac{\ln(1+e_W)}{\ln(1+e_T)} \tag{1}$$

Typically, the r-value (also known as the Lankford parameter) will vary within the plane of the sheet with respect to the rolling direction. For this reason, an "average" r-value, termed r_m, is often cited as a measure of the normal anisotropy in a plate:

$$r_m = \frac{r_0 + 2r_{45} + r_{90}}{4} \tag{2}$$

A material exhibiting a high r_m will preferentially deform in the plane of the sheet by resisting through-thickness thinning. Hence, increased r_m-values have been related to improved drawability[2] as well as to greater dent resistance[3] and reduced edge splitting[4] in steel sheets.

The effect of a variation in r-value with directions parallel to the plane of the sheet results in the formation of ears on the flange during deep-drawing. The tendency of earing is generally represented by Δr:

$$\Delta r = \frac{r_0 - 2r_{45} + r_{90}}{2} \tag{3}$$

The development of mechanical anisotropy and r-values in sheet steels is related to the crystallographic texture of the annealed product. Numerous researchers have shown that an increased r-value is obtained in steels which have a large volume fraction of (111) planes oriented in the plane of the sheet, while a significant (100) texture component normal to the surface is considered detrimental to deep drawability[5,6,7,8,9,10]. Held[11] has demonstrated that the r-value can be correlated to a ratio of the intensities of the (111) and (100) texture components measured by X-ray diffraction. Hence, much of the research in alloy development on processing of sheet steel focuses on increasing the concentration of the (111) component and/or decreasing the level of (100).

With the advent of the Orientation Distribution Function (ODF), a method of calculating theoretical r-values directly from pole figure data was derived using a mean Taylor factor and ODF coefficients[12]:

$$r(\beta) = \overline{r^0} + (r^1 \cdot C_4^{11}) + (r^2 \cdot C_4^{12} sin2\beta) + (r^3 \cdot C_4^{13} sin4\beta) \qquad (4)$$

where $r(\beta)$ = r-value along the β direction in the plate
r^i = constants which can be determined from Taylor Theory
$C_\lambda^{\mu\nu}$ = ODF Coefficients

For bcc materials, such as low-carbon steel, which slip on (hkl)<111> systems (pencil glide), Equation 4 has been successful in approximating the variation in r-values measured along different β directions[13,14], but has been found to overpredict r-values when compared to measured data[15]. Therefore, the objective of this effort was to gain insight into the mechanism responsible for the r-value variation resulting from uniaxial strain.

EXPERIMENTAL PROCEDURE

The tantalum used for this effort was produced from a triple electron beam melted (3EB) electrode which was vacuum arc remelted (VAR) to yield a 305 mm (12") diameter ingot. The ingot was press forged and machined into slabs. The mults were annealed and cold rolled perpendicular to the ingot axis to a total reduction of 92%. The Ta-2.5 material was fabricated from a 255 mm (10") diameter 3EB melted ingot which was side forged, cut into slabs, machined, then annealed. The Ta-2.5W slab was transverse rolled to a cold reduction of 87%. The rolling direction for the Ta-W alloy was defined as the rolling direction of the last pass through the mill. The Ta and Ta-W alloy were annealed at temperatures which yielded a common grain size of 50-55 μm, as measured by the linear intercept method. Chemical analyses of the pure and alloy plate are provided in Table I.

Texture analysis was conducted on samples removed from both the center and edge of each plate and mounted on a four-circle neutron diffractometer at the National Institute for Standards and Technology (NIST). Using 0.208 nm neutron radiation, the (200), (110), and (211) pole figures were measured over the entire orientation hemisphere at 5° intervals. The raw diffraction data was corrected for background radiation and absorption, then normalized to provide pole figures in

Table I: Chemistry of Ta and Ta-2.5W Plate (ppm by weight except where noted).

	O	C	N	H	W	Nb	Mo	Fe	Cu
Ta	<10	<10	<10	<5	35	<25	<5	<5	<5
Ta-2.5W	<10	<10	<10	<5	2.9%	1160	<5	10	<5

units of random distribution densities. ODF calculations were performed using the *popLA* program developed by Los Alamos National Laboratory[16]. The theoretical r-values were calculated by harmonic analysis.

Tensile specimens were machined with their axes oriented in the longitudinal (0°), transverse (90°), and 45° directions with respect to the rolling directions. Three tensile bar designs having a 25.4mm (1.0") gauge length but differed in cross-section geometry were employed for this study: design 1 had a 7.62 +/- 0.13mm square cross-section; design 2 had a 6.35 +/- 0.13mm diameter cross section; and design 3 had 5.59 +/- 0.13mm square cross-section. Samples were pulled on a screw-driven Instron tensile machine at a crosshead speed of 1.27mm/min. (0.05"/min.). Widths and thickness strains were measured at the center of the gauge section using a pin micrometer (strain gauges could not be used due to the high ductility of the materials). Measurements were taken at 1 minute increments until ultimate strain was reached. The $r_m^{0.3}$-values reported herein represent the average of the measurements at 0.26, 0.30, and 0.34 true strain.

To evaluate the effect of strain on the texture of each material, a longitudinal tensile specimen (design 1) of Ta and Ta-2.5W was uniaxially strained to $\varepsilon=0.30$ then analyzed using neutron diffraction. R-values were recalculated using the texture data measured after straining.

RESULTS

The pure tantalum contained strong $\{111\}\langle\bar{1}\bar{1}2\rangle$, $\{334\}\langle\bar{8}43\rangle$, and $\{111\}\langle uvw\rangle$ textures and moderate $\{001\}\langle uvw\rangle$ and $\{114\}\langle\bar{1}10\rangle$ components. The tensile specimen deformed to a true strain of 0.30 contained equally strong $\{111\}\langle\bar{1}10\rangle$, $\{223\}\langle\bar{1}10\rangle$, $\{111\}\langle uvw\rangle$, $\{001\}\langle\bar{1}10\rangle$, and $\{114\}\langle 110\rangle$ components. The as-annealed Ta-2.5W consisted of a primary $\{001\}\langle\bar{1}00\rangle$ cube texture and a weak $\{116\}\langle\bar{6}01\rangle$ component, whereas the deformed sample contained a primary $\{001\}\langle\bar{1}10\rangle$ texture and the weak $\{116\}\langle\bar{1}10\rangle$ component. Little difference was evident in the texture measured at the center and edge locations of the pure Ta or Ta-2.5W plates.

The r-values predicted from the texture data obtained from each of the starting plates and from longitudinal tensile specimens deformed in tension to a true strain of 0.30 is presented in Table II. Based on texture of the as-annealed and the deformed tantalum, little variation is predicted in r_0 with strain. However, the r_m and Δr values are predicted to vary with strain. Caution must be taken with regard to these predicted r_m and Δr values since the r_{45} and r_{90} values used in the calculations are based on the texture of one sample deformed in the rolling direction. A better assessment on the effects of strain on the calculated r_m and Δr values should be made using texture data from samples deformed in transverse and 45° directions.

For the Ta-2.5W alloy, due to the differences between the undeformed and deformed materials, a decrease in r_0 is expected during deformation. The variation in r_0 with strain for Ta and Ta-2.5W is presented in Figures 1 and 2, respectively. The r_0 values for each material calculated from the texture of the as-annealed (ε=0) and the deformed samples (ε=0.3) is represented as horizontal lines in Figures

Table II: R-Values of Ta and Ta-2.5W Plate. Calculated Values Determined By Harmonic Analysis of Diffraction Data Assuming Pencil Glide; Measured Values Determined by Tensile Testing.

CALCULATED					
n^0 SAMPLE	r_0	r_{45}	r_{90}	r_m	Δr
Ta (Center)	1.1	0.7	1.7	1.0	0.6
Ta (Edge)	1.2	0.8	1.8	1.1	0.7
Ta (ε=0.30)	1.2	1.0	0.5	0.9	-0.1
Ta-2.5W (Center)	0.9	0.4	1.3	0.8	0.7
Ta-2.5W (Edge)	1.2	0.4	1.3	0.8	0.8
Ta-2.5W (ε=0.30)	0.8	0.5	0.8	0.6	0.3
MEASURED					
TENSILE SAMPLE	$r_0^{.30}$	$r_{45}^{.30}$	$r_{90}^{.30}$	$r_m^{.30}$	$\Delta r^{.30}$
Ta (Design 1)	1.2	0.7	1.3	1.0	0.6
Ta (Design 2)	1.4	0.6	1.3	1.0	0.8
Ta (Design 3)	1.2	1.1	1.3	1.2	0.2
Ta-2.5W (Design 1)	0.8	0.5	1.0	0.7	0.4
Ta-2.5W (Design 2)	0.6	0.5	0.9	0.6	0.3
Ta-2.5W (Design 3)	0.7	0.6	1.1	0.7	0.4

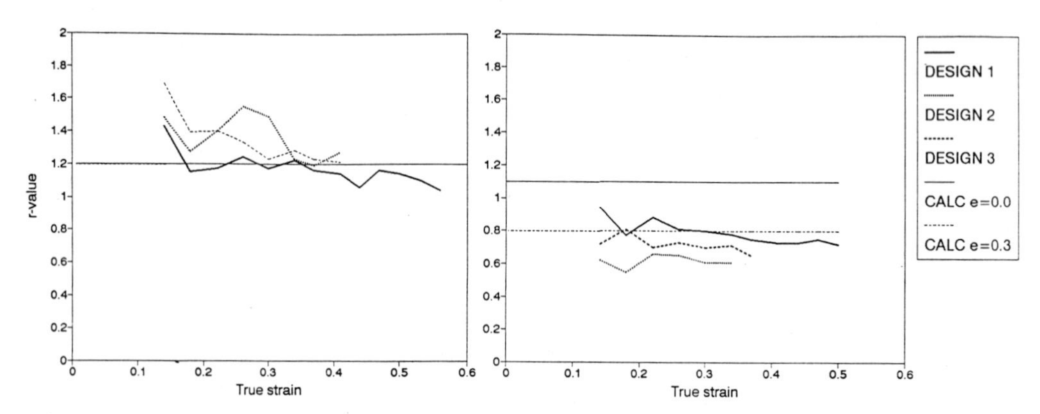

Figure 1: Variation in r_0 with Strain for Ta Plate.

Figure 2: Variation in r_0 with Strain for Ta-2.5W Plate.

NOTE: Heavy Solid and Dashed Lines Represent r_0-Values Predicted from Texture Analysis (Assuming Pencil Glide) of As-Annealed and Strained Specimens, Respectively.

1 and 2. It can be generalized that the measured r_0 decreases with strain; the trend is most apparent in the (111) textured Ta. The r_0-values predicted from the texture of the as-annealed Ta-2.5W alloy is an overestimate of the r_0-value measured during plastic deformation. However, the r_0 calculated from the texture measured from a longitudinal tensile specimen deformed to a true strain of 0.30 closely resembles those measured at a similar strain level.

DISCUSSION

It has been observed that the r-value of many materials changes with increasing strain. Comparisons of the strain dependance of r-values in aluminum-killed and titanium bearing steels lead researchers to conclude that a variation in r-value with strain (beyond the Luders range) is associated with the interstitial content of the metal[17]. Yet, it is unlikely that interstitials are a major factor in the strain sensitivity of r-values in Ta and Ta-2.5W because of the low concentrations measured in the materials.

Hu[18] demonstrated that the change in r-value with strain was dependent on the initial r-value of the steel being tested: r-values decreased with strain for strongly (111) textured steel and increased with strain for steels having an initial r-value less than unity. This behavior was rationalized by Hu based on the difference in through-thickness and in-plane work-hardening exponents of the deep-drawing steel. (111) textured material exhibited a higher flow stress and lower work-hardening exponent (which Hu correlated to a greater work-hardening rate) in the through-thickness direction than in the plane of the sheet; the opposite was true for the (001) textured, low r-value sample[19]. However, the work-hardening rate, determined by differentiating the Ludwig equation, is dependent on both n and K, and n alone is insufficient to describe the work-hardening rate (θ).

Polycrystalline (111) tantalum and Ta-2.5W have been shown to exhibit a greater flow stress in the normal direction than in the in-plane directions; the work hardenling rates in the normal direction for Ta and Ta-2.5W were similar to those found along the longitudinal direction of each alloy[20]. It unclear why the (100)<001> textured Ta-W alloys exhibit the same trends in orientation-dependent mechanical behavior as gamma-fiber textured tantalum. However, these results indicate that differential work-hardening behavior alone does not predict the strain-sensitivity behavior of Ta and Ta-2.5W.

Truszkowski[21] speculated that the change in r-value with strain in fcc aluminum and nickel is due to the formation of a fiber texture during tensile deformation. From the comparison of the texture results on as-annealed and deformed Ta and Ta-2.5W, it is shown that tensile strain acts to orient all slip directions to $<\bar{1}10>$. Rolling direction inverse pole figures of as-annealed and axially strained tantalum and Ta-2.5, provided in Figures 3 and 4, reveal the reorientation of the tensile axes toward $<\bar{1}10>$.

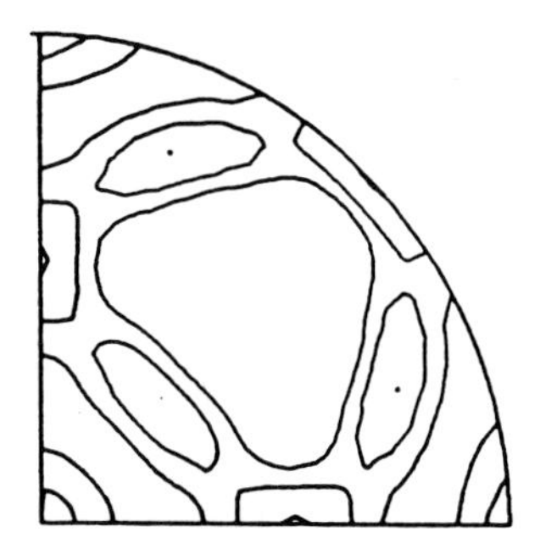

Figure 3a: Rolling Direction Inverse Pole Figure of As-Annealed Ta.

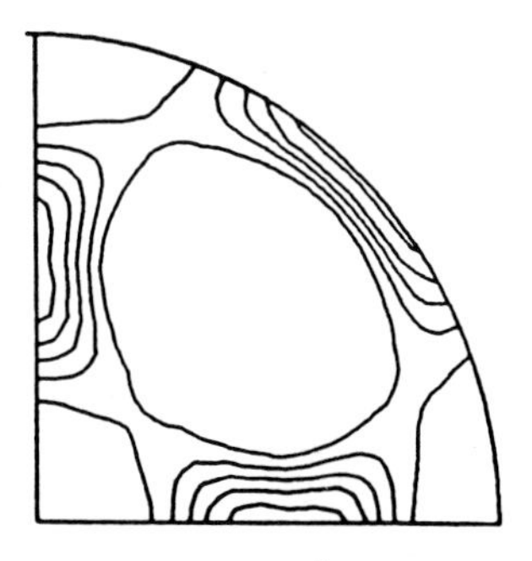

Figure 3b: Rolling (Tension) Direction Inverse Pole Figure of Ta Deformed ε=0.30 in Tension.

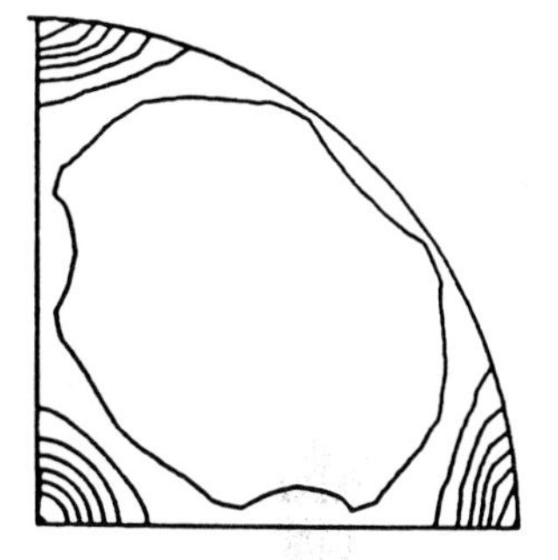

Figure 4a: Rolling Direction Inverse Pole Figure of As-Annealed Ta-2.5W.

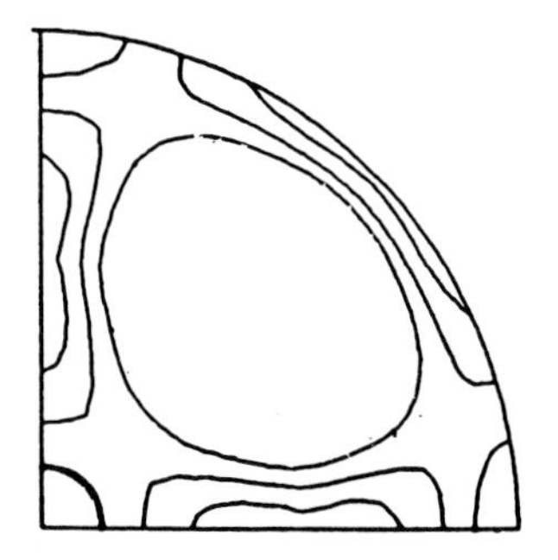

Figure 4b: Rolling (Tension) Direction Inverse Pole Figure of Ta-2.5W Deformed ε=0.30 in Tension.

For the pure material, the rotations of the primary components $\{111\}\langle\bar{1}\bar{1}2\rangle$ and $\{334\}\langle\bar{8}43\rangle$ to $\{111\}\langle\bar{1}10\rangle$ and $\{223\}\langle\bar{1}10\rangle$ likely do not have an overbearing influence on the strain sensitivity of the r-value. However, the $\{001\}\langle\bar{1}10\rangle$ orientation has been predicted to have relatively low r_0 value[22]. The formation of the $\{001\}\langle\bar{1}10\rangle$ from a $\{001\}\langle uvw\rangle$ fiber may account for the slight decrease in r_0 with strain as measured in Ta. Since there is a relatively low concentration of $\{001\}\langle uvw\rangle$ component in the as-annealed Ta, it is expected that the fiber component would be quickly consumed at the initial stages of deformation, and the r-value would then soon be a constant value with respect to subsequent strain. The finding herein that the calculated r-values of the as-annealed and deformed Ta material are similar, while the measured r_0-value (for full thickness tensile specimen) is nearly constant beyond a true stain of 0.20. support this claim.

In the Ta-2.5W plate, the final $\{001\}\langle\bar{1}10\rangle$ texture is expected to have lower r_0-value than expected from the initial $\{001\}\langle\bar{1}00\rangle$ texture[22]. Therefore, the reduction in r_0 with strain for Ta-2.5W is due to the rotation of the slip directions towards $\langle\bar{1}10\rangle$. Since $\{001\}\langle\bar{1}00\rangle$ represents a relatively large volume fraction of the annealed Ta-2.5W alloy, a significant decrease in r_0-value is expected during deformation. Therefore, the strain-sensitivity of the r-value can be understood by comparing the texture components of the as-annealed and strained alloys.

CONCLUSIONS

The r_0-value predicted using texture information from the as-annealed Ta is similar to both that predicted from the texture of an axially deformed tensile specimen and that measured (beyond a true strain of 0.20). The lack of strain-sensitivity of the Ta has been associated with similarities in r-values behavior based on the initial texture components and the $\langle\bar{1}10\rangle$ fiber components identified after deformation.

The strain-sensitivity of the Ta-2.5W alloy is due the rotation of the as-annealed $\{001\}\langle 100\rangle$ cube texture towards $\{001\}\langle\bar{1}10\rangle$. The r_0-value of the $\{001\}\langle\bar{1}10\rangle$ orientation is lower than that for the $\{001\}\langle 100\rangle$, and r-value decreases with strain while rotation occurs.

REFERENCES

1. S.M. Cardonne, P. Kumar, C.A. Michaluk, and H.D. Schwartz, *Adv. Mat. Proc.*, **142**, No.3 (Sept. 1992) 16.

2. W.F. Lankford, S.C. Snyder, and J.A. Bauscher, *Trans. ASM*, **42** (1950) 1179.

3. D.V. Wilson, *J. Inst. Met.*, **94** (1966) 84.

4. A.J. Klein and E.W. Hitchler, *Met. Eng. Q.*, **13** (1973) 25.

7. H. Era, T. Tomokiyo, H.C. Chen, and M. Shimizu, *Scr. Metall.*, **23** (1989) 173.

6. K. Suzuki, *J. Mech. Work. Technol.*, **15** (Sept. 1987) 131.

7. D.B. Lewis and F.B. Pickering, *Met. Technol.*, **10**, No.7 (July 1983) 264.

8. M. Betzl, *Acta Crystall. A*, **34 (S4)** (Aug. 1987) S385.

9. K. Matsukura and K. Sato, *Trans. ISIJ.*, **21** (1981) 783.

10. W.B. Hutchinson, *Int. Met. Rev.*, **29** (1984) 25.

11. J.F. Held, *Mechanical Working and Steel Processing IV*, Paper 3, The Metallurgical Society of AIME, New York (1963).

12. H.J. Bunge, M. Shulze, and D. Grzesik, *Calculation of the Yield Locus of Polycrystalline Materials According to the Taylor Theory*, Peine + Salgitter Beriche, Sonderheft (1980).

13. H.J. Bunge and W.T. Roberts, *J. Appl. Crystall.*, **2** (1969) 116.

14. H.J. Bunge, *Textures in Non-Ferrous Metals and Alloys*, The Metallurgical Society of AIME, Warrendale, PA. USA, (1985) 145.

15. D.N. Lee, I. Kim, and K.H. Oh, *J. Mat. Sci.*, **23** (1988) 4013.

16. J.S. Kallend, U.F. Kocks, A.D. Rollett, and H.R. Wenk, *Mat. Sci. Eng.*, **A132** (1991) 1.

17. R.P. Arthey and W.B. Hutchinson, *Met. Trans. A*, **12A** (1981) 1817.

18. H. Hu, *Met. Trans. A*, **6A** (1975) 2307.

19. H. Hu, *Met. Trans. A*, **6A** (1975) 945.

20. C.A. Michaluk, Master's Thesis, Drexel University, 1993.

21. W. Truszkowski, *Met. Trans. A*, **7A** (1976) 327.

22. P. Lequeu, F. Montheillet, and J.J. Jonas, *Textures in Non-Ferrous Metals and Alloys*, The Metallurgical Society of AIME, Warrendale, PA. USA, (1985) 189.

Materials Science Forum Vol. 157-162 (1994) pp. 1663-1670

DETERMINATION OF YOUNG'S MODULUS IN TEXTURED CuZnAl SHAPE MEMORY ALLOYS

N.J. Park and H.J. Bunge

Institut für Metallkunde und Metallphysik, TU Clausthal,
Grosser Bruch 23, D-38678 Clausthal-Zellerfeld, Germany

Keywords: Young's Modulus, Reuss-Voigt-Hill Approximation, Shape Memory Alloy, Cu-Zn-Al

ABSTRACT

CuZnAl shape memory alloys are elastically highly anisotropic (e.g. anisotropy parameter A $\sim$ 15) [1]. This effect plays an important role for the size of the transformation hysteresis loop [2,3,4]. The elastic anisotropy of a textured polycrystalline material can be calculated from the single crystal anisotropy and the texture using, for instance, the Voigt - Reuss - Hill approximation [6]. Hence, the measurement of the texture can yield a good insight into the shape memory processes themselves and it may help to produce a starting texture in the material which gives rise to the maximum possible shape change for a particular application.
The Young's - modulus was measured in several directions in the rolling plane using the "Elastomat" and longitudinal vibration mode. The results are in reasonable agreement with calculated values obtained from the texture in the Voigt - Reuss - Hill approximation.

FUNDAMENTALS

The elastic modulus E of a cubic single crystal in a crystallographic direction $[hkl]$ can be calculated [5] :

$$\frac{1}{E_{[hkl]}} = \frac{C_{11}+C_{12}}{(C_{11}-C_{12})(C_{11}+2C_{12})} - 2\left(\frac{1}{C_{11}-C_{12}} - \frac{1}{2C_{44}}\right)(h^2k^2+k^2l^2+l^2h^2) \qquad (1)$$

In a polycrystalline materials the properties depend, in general, on the properties of the crystallites and on the distribution of the crystallites. The average properties in a sample direction $\mathbf{y}$ of a polycrystalline material can be determined by the axis distribution function $A(hkl, \mathbf{y})$ respectively by the inverse pole figure $R_{\mathbf{y}}(hkl)$ [6]:

$$\bar{E}(\mathbf{y}) = \frac{1}{4\pi} \oint E(hkl) \cdot A(hkl, \mathbf{y}) d(hkl) = \frac{1}{4\pi} \oint E(hkl) \cdot R_{\mathbf{y}}(hkl) d(hkl) \qquad (2)$$

The function $A(hkl, \mathbf{y})$ or $R_{\mathbf{y}}(hkl)$ is needed as the weight function. In order to calculate the $\bar{E}(\mathbf{y})$ the functions $E(hkl)$, $A(hkl, \mathbf{y})$ and $R_{\mathbf{y}}(hkl)$ are developed into a series of sperical functions [6],

$$E(hkl) = \sum_{l=0}^{p} \sum_{\mu=1}^{M(l)} e_l^{\mu} \cdot k_l^{\mu}(hkl) \qquad (3)$$

$$A(hkl, \mathbf{y}) = R_{\mathbf{y}}(hkl) = \sum_{l=0}^{\infty} \sum_{\mu=1}^{M(l)} \sum_{\nu=1}^{N(l)} \frac{4\pi}{2l+1} C_l^{\mu\nu} \dot{k}_l^{*\mu}(hkl) \dot{k}_l^{\nu}(\mathbf{y}). \qquad (4)$$

One obtains from equations (3), (4) and (2)

$$\bar{E}(\mathbf{y}) = \sum_{l=0}^{p} \sum_{\nu=1}^{N(l)} \bar{e}_l^{\nu} \cdot \dot{k}_l^{\nu}(\mathbf{y}) \qquad (5)$$

with the averaged property coefficients

$$\bar{e}_l^{\nu} = \frac{1}{2l+1} \sum_{\mu=1}^{M(l)} C_l^{\mu\nu} \cdot e_l^{\mu} \qquad (6)$$

In the case of the tensor properties, p is equal to the rank of the tensor. The elastic properties are described by a fourth - rank tensor, thus $p = 4$.

The true average value may depends on the orientation of the considered crystallite but also on shape, size and other structural parameters. An approximation to the true average can be obtained for the elastic properties in the following way: Hook's law can be formulated in either of two forms

$$\varepsilon_{ij} = S_{ijkl} \cdot \sigma_{kl} \qquad (7)$$

$$\sigma_{ij} = C_{ijkl} \cdot \varepsilon_{kl} \qquad (8)$$

Correspondingly, for the polycrystalline materials

$$\bar{\varepsilon}_{ij} = \tilde{S}_{ijkl} \cdot \bar{\sigma}_{kl} \qquad (9)$$

$$\bar{\sigma}_{ij} = \tilde{C}_{ijkl} \cdot \bar{\varepsilon}_{kl} \qquad (10)$$

The polycrystal quantities $\tilde{S}_{ijkl}$ and $\tilde{C}_{ijkl}$ in equations (9) and (10) differ, in general, from the simple average values $\bar{C}_{ijkl}$ and $\bar{S}_{ijkl}$. We set approximately

$$\tilde{C}_{ijkl} \approx \bar{C}_{ijkl} = \bar{C}_{ijkl}^{V} \quad ; \quad \bar{S}_{ijkl}^{V} = (\bar{C}_{ijkl}^{V})^{-1} \qquad \text{Voigt} \qquad (11)$$

$$\tilde{S}_{ijkl} \approx \bar{S}_{ijkl} = \bar{S}_{ijkl}^{R} \quad ; \quad \bar{C}_{ijkl}^{R} = (\bar{S}_{ijkl}^{R})^{-1} \qquad \text{Reuss} \qquad (12)$$

this corresponds to the approximation used by Voigt [7] and Reuss [8].

The equation (11) is correct if the strain ε is the same in all crystallites. Equation (12) is correct in the case of equal stress σ in all crystallites. Normally neither of these cases is realistic. The averages of these two extremum values can be used

$$\tilde{S}_{ijkl} \approx \frac{1}{2}[\bar{S}^{R}_{ijkl} + \bar{S}^{V}_{ijkl}] = \bar{S}^{H}_{ijkl} \qquad \text{Hill(1)} \tag{13}$$

$$\tilde{C}_{ijkl} \approx \frac{1}{2}[\bar{C}^{R}_{ijkl} + \bar{C}^{V}_{ijkl}] = \bar{C}^{H}_{ijkl} \qquad \text{Hill(2)} \tag{14}$$

These Hill [9] averages are usually very close to the true polycrystal values.

In the case of the cubic crystal and orthorhombic sample symmetry one obtains with the coefficients of the texture $C_4^{1\nu}$ from the ODF - analysis

$$\bar{S}_{ijkl} = S^{0}_{ijkl} + S_a[\bar{a}^{11}_{0}(ijkl) - t^{0}_{ijkl} + \sum_{\nu=1}^{3} \bar{a}^{1\nu}_{4}(ijkl)C^{1\nu}_{4}] \tag{15}$$

$$\bar{C}_{ijkl} = C^{0}_{ijkl} + C_a[\bar{a}^{11}_{0}(ijkl) - t^{0}_{ijkl} + \sum_{\nu=1}^{3} \bar{a}^{1\nu}_{4}(ijkl)C^{1\nu}_{4}] \tag{16}$$

Finally, one obtains the Young's modulus $\tilde{E}(\mathbf{y})$ in an arbitrary sample direction $\mathbf{y} = (y_1, y_2, y_3)$ from these average values [5,6]

$$\begin{aligned}\frac{1}{\tilde{E}(\mathbf{y})} &= y_1^4\tilde{S}_{1111} + y_2^4\tilde{S}_{2222} + y_3^4\tilde{S}_{3333} + 2y_1^2y_2^2(\tilde{S}_{1122} + 2\tilde{S}_{1212}) \\ &+2y_1^2y_3^2(\tilde{S}_{1133} + 2\tilde{S}_{1313}) + 2y_2^2y_3^2(\tilde{S}_{2233} + 2\tilde{S}_{2323})\end{aligned} \tag{17}$$

EXPERIMENTALS

An alloy with the composition Cu - 26.75 wt.% Zn - 4.0 wt.% Al - 0.5 wt.% Ti was melted in a medium frequency induction furnace. The Ti - content was used for grain refinement. The melt was cast into a steel mold of 78 mm diameter. It was homogenized at 850°C 4 hours and extruded to a bar of 35mm $\times$ 6.8mm equivalent to a deformation of 95%. Then this bar was hot - rolled at 800°C down to a final thickness of 0.65 mm equivalent to a rolling degree of 90%. Thereby several intermediate annealing treatments were necessary. The sheet was then cooled down to room temperature whereby it remains as a cubic austenite with a lattice parameter $a_o = 5.866$Å

The Young's - moduli were calculated in angle steps of 2.5° to the rolling direction in the rolling plane using the Voigt - Reuss - Hill approximation. The elastic constants, $C^0_{1111} = 130kN/mm^2$, $C^0_{1122} = 118.4kN/mm^2$, $C^0_{1212} = 86kN/mm^2$ [1], were taken into account. Texture coefficients $C_4^{1\nu}$ were taken from texture analysis with three pole figures.
In order to compare with the experimental values the Young's - moduli were measured in several directions (0°, 22.5°, 45°, 67.5° and 90°) in the rolling plane using the "Förster - Elastomat 1.042" and longitudinal vibration mode.

TEXTURE ANALYSIS

Samples for texture analysis were prepared from the plane perpendicular to the rolling direction. This section is often advantageous to other sections for the measurement of the whole texture by the sheet. Three incomplete pole figures, i.e. (2 2 0), (4 0 0) and (4 2 2) were measured in steps $\Delta\alpha = 5°$, $\Delta\beta = 3.6°$ up to a tilt angle $\alpha_{\max} = 70°$ using Co K_α - radiation. The measured pole figures are shown in figure 1. The complete ODF was then calculated using an iterative positivity method with a maximum series expansion degree $L_{max} = 23$ (figure 2). The texture resembles strongly to the rolling texture found in steel which can be described by two continuous lines of preferred orientation, i.e. by a $\langle 111 \rangle$ and a $\langle 110 \rangle$ fibre texture. The line of maximum orientation density (skeleton line) represented by the orientations of the rolling and normal directions relative to the crystal coordinate system is shown in figure 3. The orientation density along the skeleton lines is given in figure 4.
The stronger fibre $\langle 110 \rangle$ with a maximum value of 13.51 is shown in cuts $\varphi_1 = 0°$. The fibre-axis lies parallel to the rolling direction. It extends from the orientation $\{001\}\langle 110 \rangle$ to $\{112\}\langle 110 \rangle$. The second fibre $\langle 111 \rangle$ whose axis lies parallel to the normal direction, is seen through all the φ_1 cuts. It runs from $\{112\}\langle 110 \rangle$ to $\{111\}\langle 121 \rangle$ and $\{211\}\langle 011 \rangle$ with different orientation density. The maximum density ($f(g) = 7.4$) was developed in this tube at the orientation $\{211\}\langle 011 \rangle$, where it merges into the $\langle 110 \rangle$ fibre.

RESULTS AND DISCUSSION

The calculated Young's moduli with the Reuss-, Voigt- and Hill-approximation as a function of the angle to the rolling direction are represented in figure 5. The measured Youngs's moduli using the "Elastomat" and logitudinal vibration mode are shown also in this figure (indicated with star). The values are given in table 1.

Angle to R.D.	0.0°	22.5°	45.0°	67.5°	90.0°	N.D.
Experimental	71.6	56.0	51.9	59.9	72.6	–
Approximation by	with texture coefficients					
Reuss	38.1	33.9	31.4	37.7	45.5	32.3
Voigt	138.3	127.9	121.2	137.3	154.5	124.0
Hill (S)	59.7	53.6	49.8	59.2	70.3	51.2
Hill (C)	90.7	83.1	78.2	90.0	103.0	80.2
Hill (M)	75.2	68.3	64.0	74.6	86.7	65.7

Table 1: Estimated Young's moduli in several directions $[kN/mm^2]$

Because of the big difference between both Hill-approximations, Hill(S) and Hill(C), the mean value of these was also calculated and given in table 1, Hill(M). Is is seen that the Young's moduli using the Hill-approximation with the constants S, Hill(S), are in good agreement with the measured values.

A minimum value of the measured Young's modulus 51.9 kN/mm^2 was found at 45° to the rolling direction in the 95% extruded and 90% hot-rolled CuZnAlTi shape memory alloy and a maximum value 72.6 kN/mm^2 at 90°. Hereby the technical anisotropic constant, $A_{tech} = \frac{E_{max}}{E_{min}}$, is only 1.4. The low value of A_{tech} may be caused by the weaker texture which can be clearly recognized by inverse pole figures (figure 6). At 45° to the rolling direction which shows the lower E-modulus the crystallites clustered with 3.24 $\times$ random around the ⟨001⟩-orientation (figure 6(c)). At 90° to the rolling direction which shows the high E-modulus the crystallites clustered with 1.82 $\times$ random around the ⟨111⟩- and ⟨110⟩-orientations (figure 6(e)).

The Young's moduli of the single crystal were calculated according to equation (1) with the elastic constants and presented in figure 7. One can see that the values show very big differences between the orientations. The minimum value of 17.1 kN/mm^2 was determined at ⟨001⟩-orientation, the maximum 209.0 kN/mm^2 at ⟨111⟩-orientation and thus A_{tech} is 12.2. These values can be measured in a polycrystalline material with an extremely strong texture. If the crystallites lie totally random, the Young's modulus with the Hill-approximation is 60.0 kN/mm^2 and the technical anisotropic constant A_{tech} is 1.

For the technological application of the materials such as shape-memory-motor, the 45° to the rolling direction can be used where the minimum value of Young's modulus was found. Similar calculations also allow to estimate the optimum texture - in this sense - for a particular application.

References

[1] G. Guénin, M. Morin, P.F. Gobin, W. Dejonghe, L. Delaey, Scripta Met., 11 (1977), 1071–1075

[2] I. Müller, Continuum Mech. Thermodyn., 1, (1989), 125–142

[3] I. Müller, H. Xu, Acta Met., 39, (1991), 263–271

[4] T.W. Duerig, K.N. Melton, D. Stöckel, C.M. Wayman (Ed.), Engineering Aspects of Shape–Memory–Alloys, 1990, Butterworth-Heinemann, London

[5] J.F. Nye, Physical Properties of Crystals, 1957, Clarendon Press, London

[6] H.J. Bunge, Texture Analysis in Material Science, 1982, Butterworth, London

[7] W. Voigt, Lehrbuch der Kristallphysik, 1928, B.G. Teubner Verlag, Leipzig

[8] A. Reuss, Z. Angew. Math. Mech. 9, (1929), 49–58

[9] R. Hill, Proc. Phys. Soc. A65, (1952), 349–354

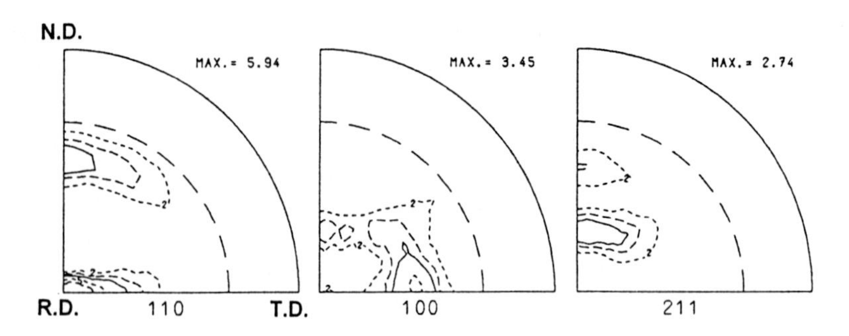

Figure 1: Measured pole figures of Cu-26.75 wt.% Zn-4.0 wt.% Al-0.5 wt.% Ti

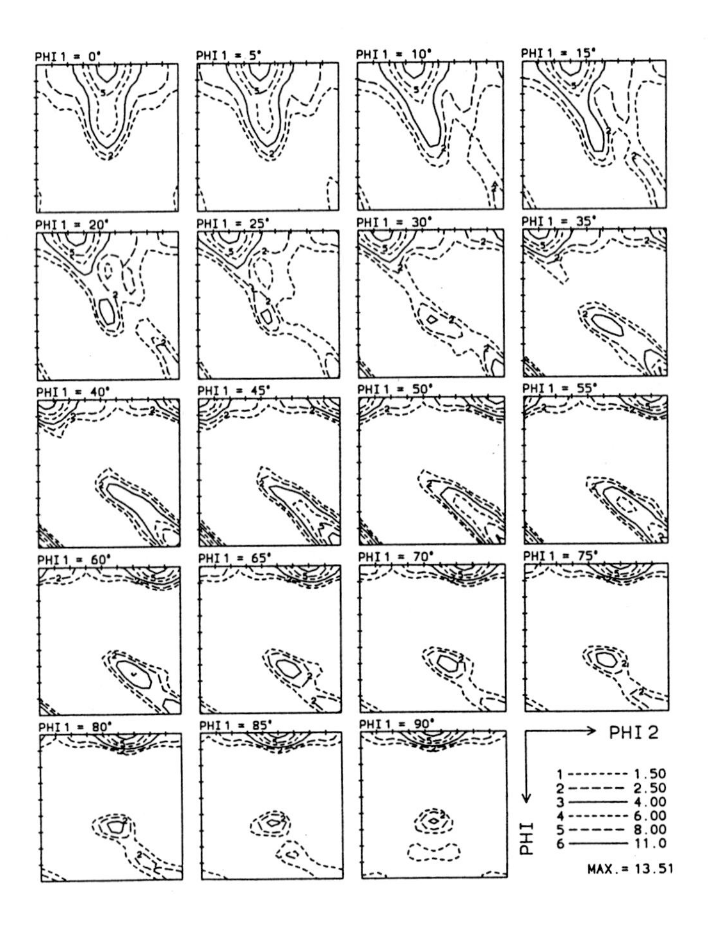

Figure 2: ODF of Cu-26.75 wt.% Zn-4.0 wt.% Al-0.5 wt.% Ti

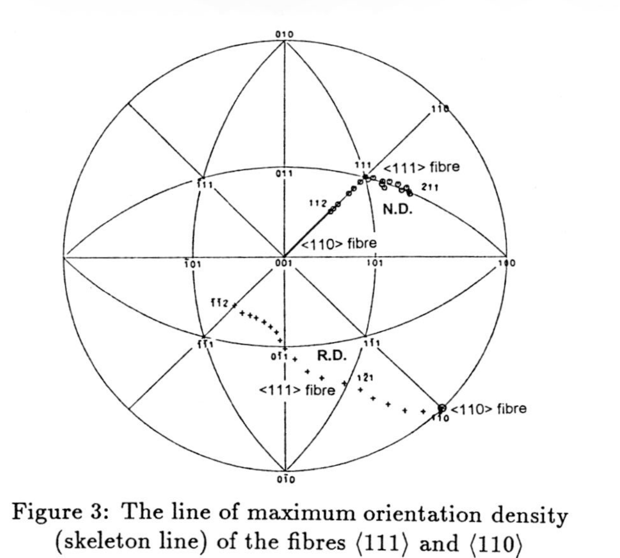

Figure 3: The line of maximum orientation density (skeleton line) of the fibres ⟨111⟩ and ⟨110⟩

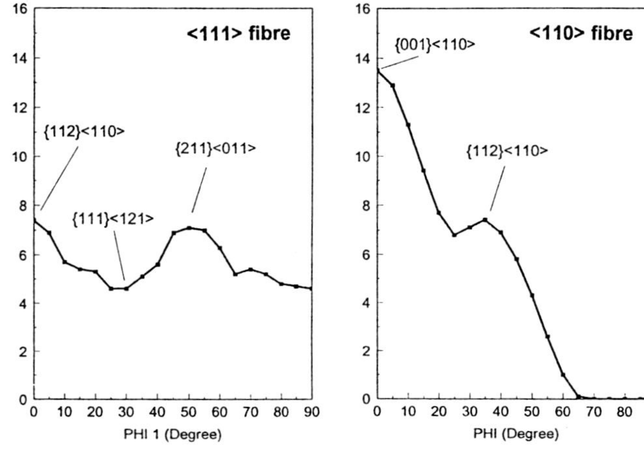

Figure 4: The orientation density along the skeleton lines

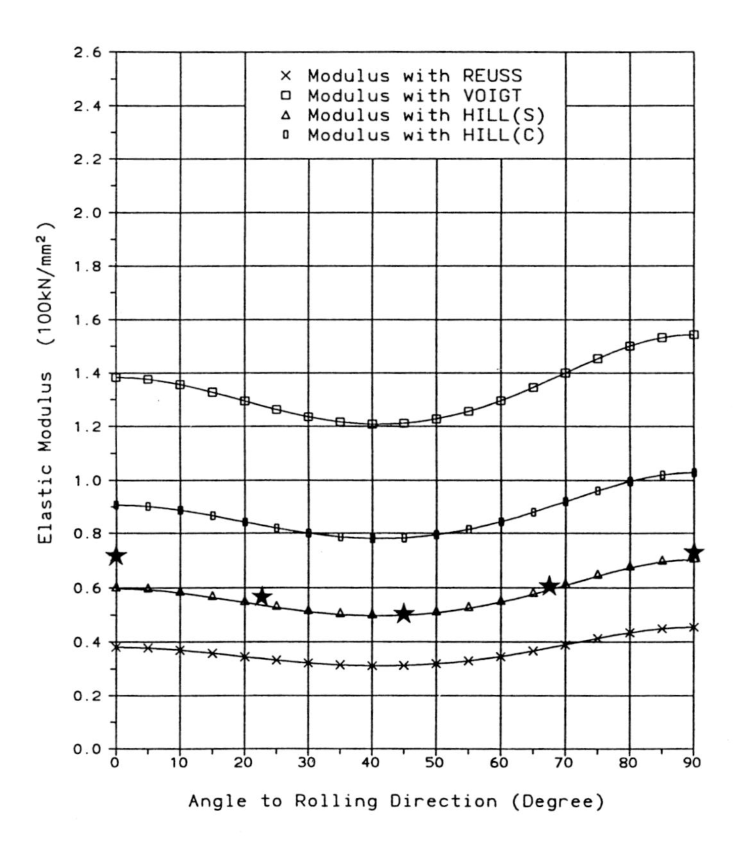

Figure 5: Young's modulus of a extruded and hot rolled CuZnAlTi alloy as a function of angle to the rolling direction

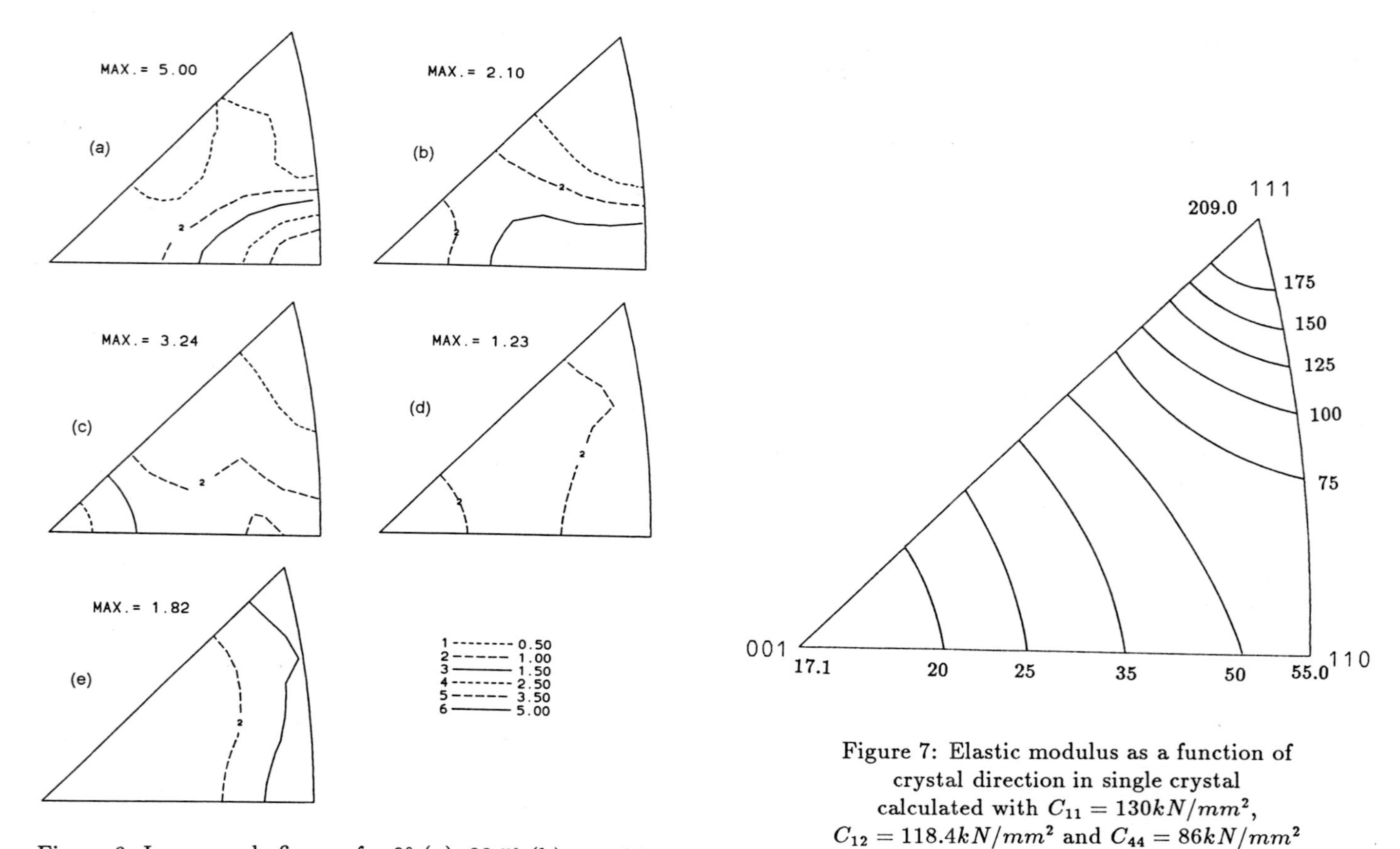

Figure 6: Inverse pole figures for 0° (a), 22.5° (b), 45° (c), 67.5° (d) und 90° (e) to the rolling direction

Figure 7: Elastic modulus as a function of crystal direction in single crystal calculated with $C_{11} = 130kN/mm^2$, $C_{12} = 118.4kN/mm^2$ and $C_{44} = 86kN/mm^2$

Materials Science Forum Vol. 157-162 (1994) pp. 1671-1680

EFFECT OF TEXTURE AND MICROSTRUCTURE ON THE MECHANICAL PROPERTIES OF Zn ALLOYS

M.J. Philippe [1], I. Beaujean [1], E. Bouzy [1], M. Diot [1], J. Wegria [2] and C. Esling [1]

[1] LM2P/ISGMP, Université de Metz, Ile du Saulcy, F-57045 Metz Cédex 1, France

[2] Union Minière, B.P. 1, F-59950 Auby les Douai, France

Keywords: Deformation, Recrystallization, Zinc, Bendability, Process Optimization

ABSTRACT

The zinc-copper-titanium alloys present usually a poor bendability at temperatures below 7 °C, since the ductile-brittle transition temperature of these alloys is too high. We have studied the effect of some alloying elements on the texture and microstructure of these alloys. In addition, we have explained the effect of the texture on the low temperature cracking. Optimized texture and microstructure permit to lower the ductile-brittle transition temperature down to -15 °C.

INTRODUCTION

The zinc-copper-titanium alloys present usually a poor bendability at temperatures below 7 °C, since the ductile-brittle transition temperature of these alloys is too high. In view of decreasing this transition temperature and thus improving the low temperature bendability, earlier studies (1, 2) have shown the importance of the grain size and also the importance of the crystallographic textures. The aim of this study is to underscore parameters having an effect on the bendability and to define the most suitable textures and microstructures in order to improve the bendability at low temperature and to reduce the brittle-ductile transition temperature of these alloys.

THE STUDIED MATERIALS AND THEIR THERMOMECHANICAL TREATMENTS

The study was carried out on a series of ZnCuTi alloys containing in weight 0.16% Cu, 0.076% Ti, and containing a weight of grain refiner which increases according to the following references: Zn1, Zn2, ... Zn6. These alloys have all undergone the same thermomechanical treatments yielding 1 mm sheets. In particular, the final rolling passes were performed at about 100 °C.

TEXTURES AND MICROSTRUCTURES OF THE VARIOUS 1 MM SHEETS

The texture of the sheet Zn1 (Zn Cu Ti alloy refiner free) shows two components with the c-axes tilted at 20° from Normal Direction (ND) to the Rolling Direction (RD). The peaks on the recalculated Pole Figures (PFs) are relatively low (6.5) and spread out [Fig. 1]. By increasing refiner alloying, the peaks become sharper and narrower.

Thus, for the Zn6 alloy, the same peaks on the recalculated PFs show a higher intensity of 10.9 [Fig. 2]. However, when the amount of refiner is high, a new component appears, with c-axes lying in RD.

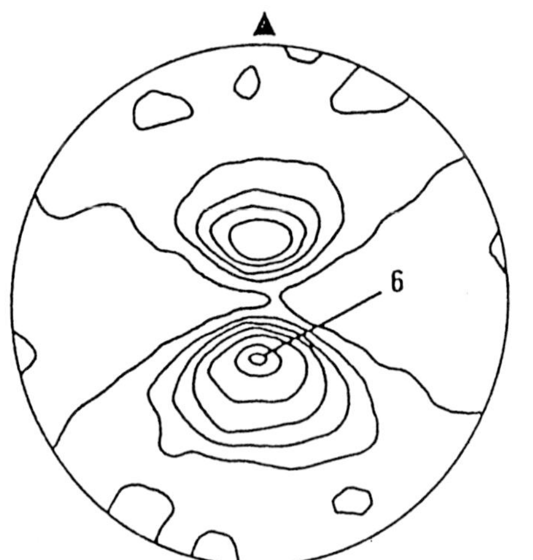

Fig. 1 00.2 Recalculated Pole Figure for Zn1.

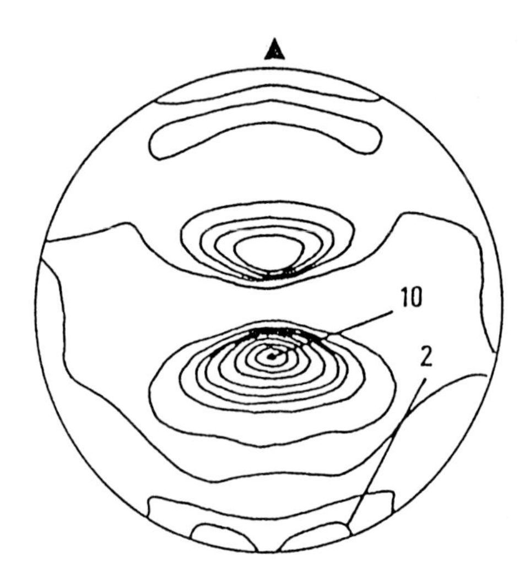

Fig. 2 00.2 Recalculated Pole Figure for Zn6.

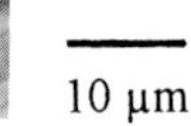

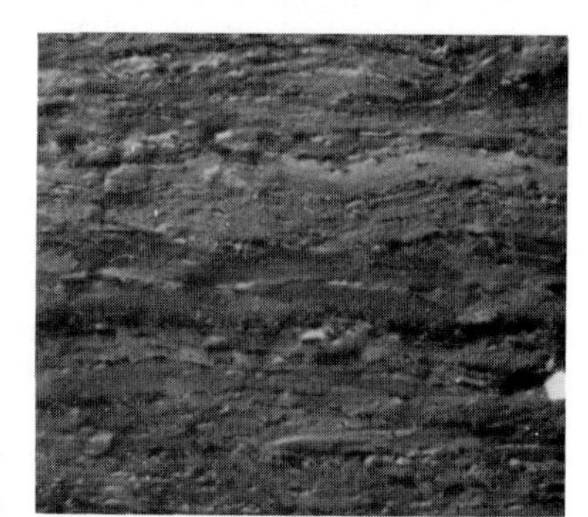

Fig. 3 a, b Microstructures of Zn1 and Zn6.

The intensity of this component also increases with the amount of refiner. The microstructural study shows an equiaxed microstructure with a mean grain size of about 10 µm for the Zn1 alloys. By adding the refiner, the grain size decreases down to 2 - 3 mm reached in Zn6 [Fig. 3 a, b]. Then the grains are not equiaxed, but rather keep their elongated shape coming from the rolling.

THE EFFECT OF THE GRAIN REFINER ON THE RECRYSTALLIZATION DEGREE AND THE TEXTURE

A previous study (3, 4) showed that the texture evolution during cold rolling of Zn alloys could be modelized. Thus, it could be shown that during cold rolling the components with c-axes tilted 20° from ND to RD sharpened and that the component with c-axes in RD increasingly appeared. This last component which is due to the activation of pyramidal glide, is always present during the cold rolling and increases with the deformation degree. In the first stage of recrystallization, the most important effect on the texture consists in lowering the intensity of the components with c-axes in rolling direction. After that, when either time or temperature is increased, the components with c-axes in ND-RD at 20° to ND decrease slowly. In fact, the grain refiner also increased the recrystallization temperature of the alloys from 73 °C for Zn1 to 130 °C for Zn6. Whereas the Zn1 alloy was fully recrystallized, the Zn6 alloy was only partially recrystallized, showing therefore a texture close to the cold rolling one.

MECHANICAL PROPERTIES

Table 1 summarizes the bendability tests performed on the alloys at different temperatures. The bendability is characterized by the total crack length measured on a given sample length. Thus, the lower the crack percentage is, the better the bendability is. The refiner free sample has a poor bendability below 7 °C. The Zn4 sample shows a good bendability even up to -15 °C. When the amount of refiner is too high (Zn6), the bendability is debased again, although the mean grain size of this sample is the smallest one.

Table 1 Bendability (% of cracks) at different temperatures for these alloys.

Zinc alloy	Zn1	Zn4	Zn6
temperature			
7 °C (% cracks)	4.85	1.00	20.00
4 °C (% cracks)	43.50	1.75	18.50
0 °C (% cracks)	100.00	2.00	22.50
-15 °C (% cracks)	100.00	2.36	2.00

INDIVIDUAL GRAIN ORIENTATIONS AND GRAIN-BOUNDARY MISORIENTATIONS

A previous study (2) showed also that brittle fractures were initiated either by intergranular crackings at high-angle boundaries, or by cleavage on basal (00.2) planes in grains whose c-axes were tilted to a high angle (more than 60°) from the ND of the sheet.

So, in order to compare these results to those previously obtained for the ZnCu and ZnCuTi alloys, we have performed individual orientation measurements to characterize the misorientation between grains. The individual orientation measurements have been performed by means of a SIT camera installed on an SEM to visualize and analyze the Electron Back Scattering Patterns (EBSP). Thus these measurements could confirm the occurrence of grains with high tilted c-axes in the Zn6 sample. Indeed, the Misorientation Density Function (MODF) of the Zn1 sample shows its maximum for $\Delta g = 90°$ (Fig. 4), whereas that of the Zn6 sample shows its maximum for $\Delta g < 30°$ (Fig. 5).

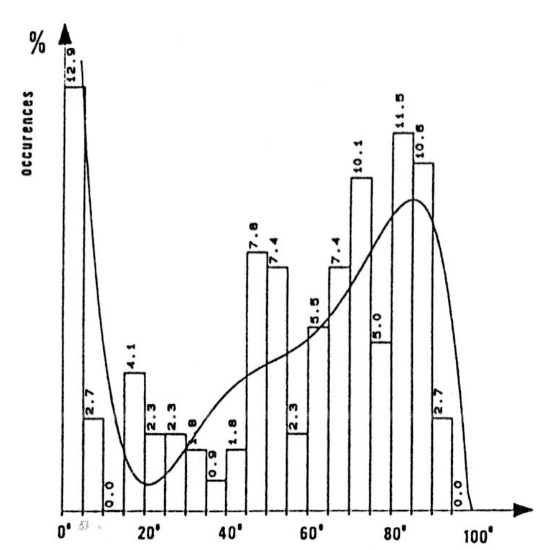

Fig. 4 Statistical Distribution of Misorientations Δg Between the Grains in Zn1.

Fig. 5 Statistical Distribution of Misorientations Δg Between the Grains in Zn6.

CONCLUSION

The grain refiner free samples present large grains, weak textures and relatively high-angle grain boundaries. The intergranular crackings can then be initiated at temperatures even higher than 0 °C.

For the Zn6 sample with high refiner rate, even if the grains are small and the texture sharp, this texture is close to a cold rolling texture. Especially the grains with high-tilted c-axes easily undergo fractures by cleavage.

Finally, the Zn4 sample with intermediary refiner rate presents an intermediary grain size, a relatively marked texture, but no texture component with high-tilted c-axes. Moreover, the misorientations between neighbouring grains are low. As a result, this sample shows a good low temperature bendability up to -15 °C.

REFERENCES

1) WEGRIA, J.: "Thesis", UST Lille (France), 1985.
2) PHILIPPE, M.-J., FUNDENBERGER, J.J., GALLEDOU, Y., HUMBERT, M., WEGRIA, J., and ESLING, C.: "Proc. ICOTOM 9", Textures and Microstructures, 1991, 14-18, 471.
3) PHILIPPE, M.J., MELLAB, F.E., WAGNER, F., WEGRIA, J., and ESLING, C.: Proceedings of the 13th RISØ International Symposium, "Modelling of Plastic Deformation and its Engineering Applications", Risø (Denmark) 1992, pp. 385.
4) PHILIPPE, M.J., WAGNER, F., MELLAB, F.E., ESLING, C., and WEGRIA, J.: Acta Met. and Mater., 1994, 42, 1, 239.

Materials Science Forum Vol. 157-162 (1994) pp. 1675-1680

DIAMAGNETIC ANISOTROPY OF PRECAMBRIAN QUARTZITES (MOEDA FORMATION, TAQUARAL VALLEY, MINAS GERAIS, BRAZIL)

H. Quade, T. Reinert and D. Schmidt

Institut für Geologie und Paläontologie, TU Clausthal,
Leibnizstr. 10, D-38678 Clausthal-Zellerfeld, Germany

Keywords: Quartz and Mica Textures, Grain-Shape, Diamagnetic AMS, Itacolomite

ABSTRACT

In the Taquaral valley in the Iron Quadrangle, Brazil, Lower Proterozoic quartzites are exposed over a large area. The orthoquartzites consist of almost pure quartz layers separated by thin interlayers of muscovite. A special facies of this quartzite ("itacolomite") is characterized by high mechanical flexibility, resulting from the strong shearing of the quartzite. The overall geometry of the quartz- and mica-c-axis as well as the orientation of the grain-shape ellipsoids is controlled by the mineral-lineation. The finite deformation intensities are generally low, resulting from the intense recrystallization of quartz. The anisotropy of the magnetic susceptibility of the quartzite is negative, with values varying between -3 and $-12*10^{-6}$ SI units/unit volume. This implies that the bulk AMS is dominantly defined by the diamagnetic quartz.

Geological setting

In the Vale do Taquaral (i.e. Taquaral valley), located north of Ouro Preto in the Quadrilátero Ferrífero (i.e. Iron Quadrangle) of Minas Gerais, Brazil, Lower Proterozoic quartzites (Moeda Formation, Fig. 1) are exposed in a southeast dipping plate overthrust from the southeast by Archean metavolcanic schists [1]. The orthoquartzites are well layered and exhibit smooth surfaces which represent the penetrative foliation corresponding to the plane of strong simple shear. A strong stretching (mineral) lineation is omnipresent and indicates an up-dip relative movement towards north to northwest.

The Taquaral quartzites consist of almost pure quartz layers separated from each other by thin interlayers of muscovite and disperse cyanite; the bulk content of these components varies considerably (mica > 5 %, cyanite < 15 %), whereas the amount of iron minerals is extremely low. The mica is hydro-muscovite and appears in two generations: porphyroclasts (> 400 μm) aligned parallel to the mineral lineation and flakes (100-500 μm) adjusted to the boundaries of quartz grains. A special facies of this quartzite is known as "itacolomite", a variety that splits in thin plates of high mechanical flexibility. This peculiar feature results from sequences of interfingering pure quartz alternating with continuous layers of mica of the second generation. The Taquaral quartzite is strongly sheared and has a hypidiomorphic mosaic texture, with muscovite flakes orientated always parallel to the penetrative foliation. Oriented samples were taken from quartzites at different sites in the Taquaral valley as well as from the Moeda Formation superjacent with normal contact upon the overthrusted schists.

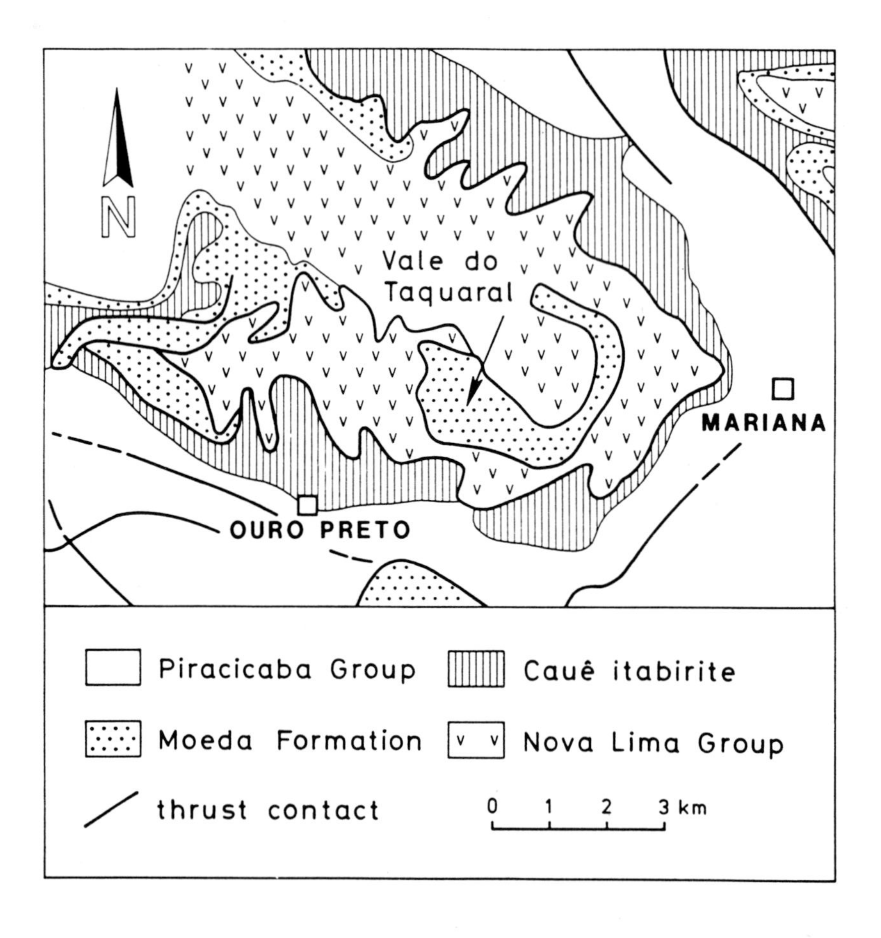

FIG. 1: Geological map of the Vale do Taquaral, Minas Gerais, Brazil.

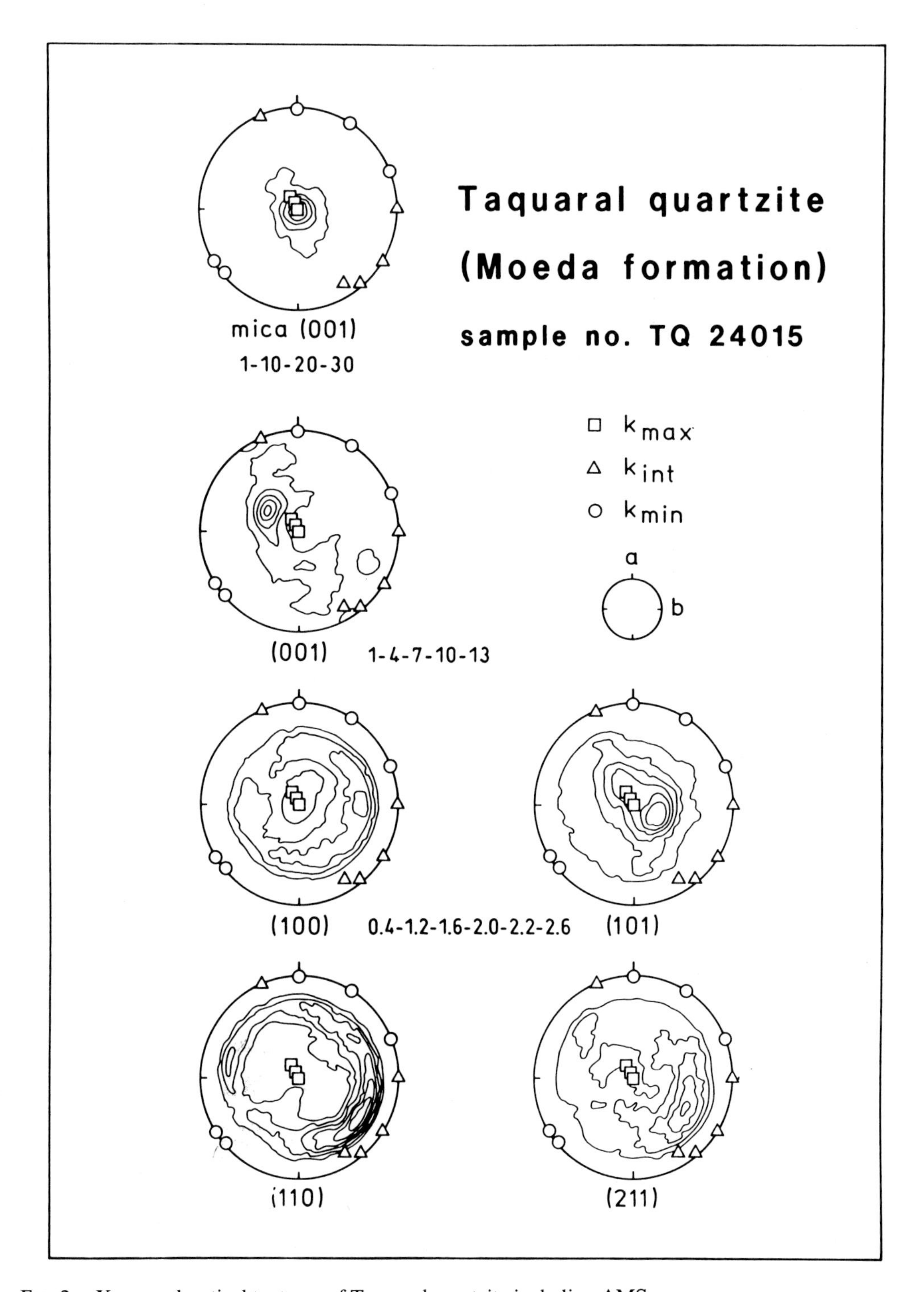

FIG. 2: X-ray and optical textures of Taquaral quartzite including AMS-axes.

Mica and quartz textures

Mica pole diagrams were determined by optical methods; they are characterized by elongated point clusters around the pole to the penetrative foliation which is the dominant shear plane. The maxima are stretched in the direction of the mineral lineation. The degree of preferred orientation [8] of mica is about 90 %.

Orientation of quartz c-axes was determined using the universal stage; complementary texture data were obtained by X-ray measurements (Fig. 2). The resulting c-axes diagrams show either girdles sub-perpendicular to the foliation plane (when the degree of preferred orientation is below ca. 30 %) or with more or less irregular clusters resulting from continuous recrystallization. In specimens with a degree of preferred orientation above 45 % small circles around the foliation pole prevail. The more homogeneous the distribution of mica in the sample the lower is the degree of preferred orientation of quartz c-axes. However, a correlation between the mica content and the degree of preferred orientation does not exist. The overall geometry of the arrangement of axes is controlled by the omnipresent southeast plunging stretching (mineral) lineation.

Grain shape anisotropy

Grain shape analysis was executed applying Surfor and Inverse Surfor methods [5], [6]. The Surfor method uses the length of the polygonal grains projected onto a base line and calculates the maximum grain diameter as a strain indicator in the corresponding plane. The Inverse Surfor method starts at the evaluation of the number of intersections of grain boundaries with traversal lines on the thin sections.

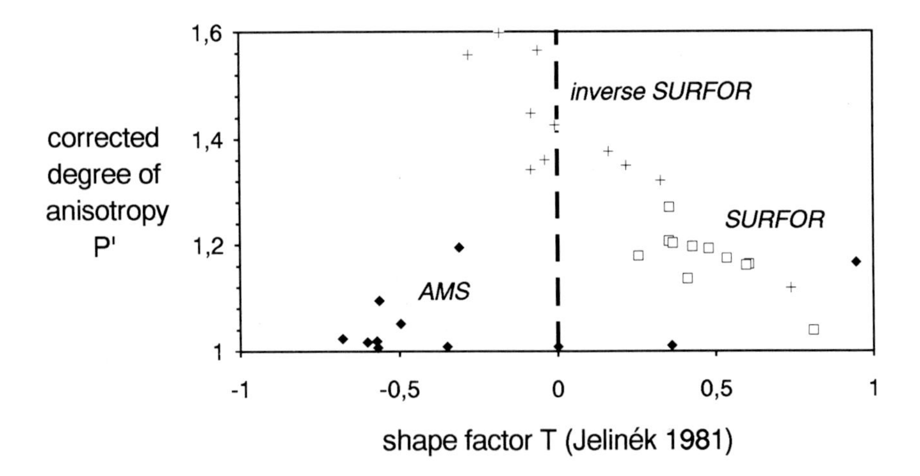

FIG. 3: Corrected degree of anisotropy [3] vs shape factor [3] of the grain shape and the AMS-ellipsoids.

The strain ellipsoids obtained by the Surfor method are all oblate and indicate that the quartz grains are generally flattened, whereas ellipsoids from Inverse Surfor data suggest more or less triaxial shapes (Fig 3). In samples of itacolomite the XY plane of the ellipsoid is the foliation, whereas in "normal" Taquaral quartzites the XY plane is inclined by 10-30° to the foliation plane which contains X. The strain data concerning the grain shape of quartz imply that the quartz texture greatly results from recrystallization and the c-axes of the quartz individuals are oriented perpendicular to the foliation which is the plane of mica alignment. In general, there is still a marked geometrical relationship between the quartz texture and the pre-existing shear fabric as X, the maximum strain axis, runs parallel to the mineral lineation. This fact coincides with the above observation that the ellipoid forms obtained by the Surfor method still depict the original shear-induced flattening of grains. The finite deformation intensities are generally low and indicates an intense coaxial recrystallization of quartz.

Magnetic anisotropy

The anisotropy of the magnetic susceptibility was determined using the Kappabridge KLY-2.02 and sample cubes of 8 ccm. The bulk susceptibility of all specimens is negative, with values ranging from $-3{*}10^{-6}$ SI (units per unit volume) to $-12{*}10^{-6}$ SI (quartz single crystals: $-13.5{*}10^{-6}$ SI). This implies that the AMS is dominantly defined by the diamagnetic quartz and that the influence of the paramagnetic (hydro)mica [2], [3], [7] and cyanite is generally low. Obviously, in the presence of high amounts of quartz, the volume susceptibility of muscovite and cyanite is significantly lower than the susceptibility of monominerals. The AMS ellipsoids are dominantly prolate, corresponding to the strong orientation of the quartz c-axes perpendicular to the foliation plane; only two specimens yielded oblate shapes, and in one case a triaxial shape was obtained (Fig 3). The anisotropy configuration [4] out of all is of type IIIa (only k_{max} tightly clustered) and type Ia (only k_{min} tightly clustered).

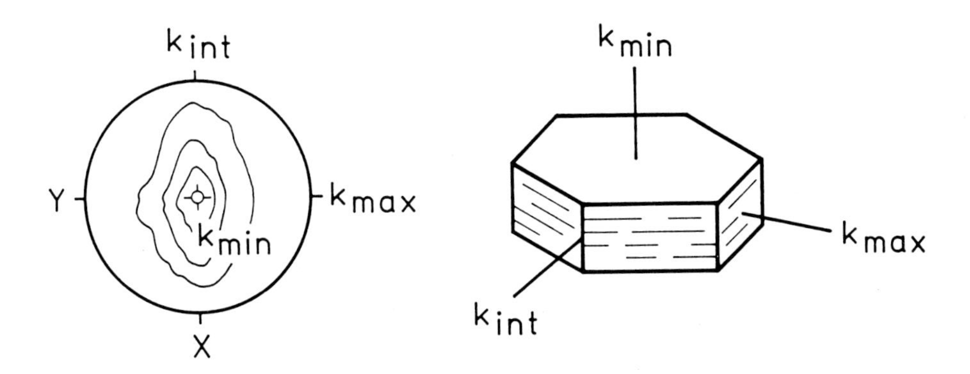

FIG 4: Relationship between the mica-texture and the orientations of the AMS-axes.

Normally, the relationship between the orientations of mica-containing quartzites and the AMS-ellipsoid is governed by relations as shown in Fig. 4. The Taquaral quartzite is one of the rare natural rock examples which exhibits a total anisotropy affected substantially by quartz as a carrier of the susceptibility. The influence of the associated mica is restricted to lowering the bulk susceptibility; the effect of the cyanite, which is present in some specimens is still unknown because reference data are not available. The combination of strain and AMS analysis, however, provided evidence of a complex deformation and recrystallization history of the Taquaral quartzite depicted by the mica and quartz fabrics, the former indicating simple shear and the latter subsequent coaxial recrystallization.

Acknowledgement

Financial support for this project was provided by the Deutsche Forschungsgemeinschaft (Forschergruppe "Textur und Anisotropie kristalliner Stoffe"). H. Quade expresses his sincere thanks to Wilson José Guerra of the Universidade Federal de Ouro Preto, Brazil, for assistance during field work.

References

1) Dorr, J.V.N. (1969): Physiographic, stratigraphic and structural development of the Quadrilátero Ferrífero, Minas Gerais, Brazil. - USGS Prof. Pap., 641-A: 110 pp.

2) Dunlop, D.J. (1981): The rock magnetism of fine particles. - Phys. Earth Planet. Int., 26: 1-26.

3) Jelinék, V. (1981): Characterization of the magnetic fabric of rocks. - Tectonophysics, 79: T63-T67.

4) Kligfield, R., Owens, W.H. & Lowrie, W. (1977): Magnetic susceptibility anisotropy as a strain indicator in the Sudbury basin, Ontario. - Tectonophysics, 40: 287-308.

5) Panozzo, R. (1984): Two-dimensional strain from the orientation of lines in a plane. - J. Struct. Geol., 6: 215-221.

6) Panozzo, R. (1987): Two-dimensional strain determination by the inverse SURFOR wheel. - J. Struct. Geol., 9: 115-119.

7) Uyeda, S., Fuller, M.D., Belshé, J.C. & Girdler, R.W. (1963): Anisotropy of magnetic susceptibility of rocks and minerals. - J. Geophys. Res., 68: 279-291.

8) Wallbrecher, E. (1985): Tektonische und gefügeanalytische Arbeitsweisen - Graphische, rechnerische und statistische Verfahren. - 244 pp, Stuttgart (Enke).

Materials Science Forum Vol. 157-162 (1994) pp. 1681-1688

MAGNETIC ANISOTROPY AND TEXTURE OF BANDED HEMATITE ORES

H. Quade and T. Reinert

Institut für Geologie und Paläontologie, TU Clausthal,
Leibnizstr. 10, D-38678 Clausthal-Zellerfeld, Germany

Keywords: Banded Hematite Ores, Iron Quadrangle, AMS, Hematite Textures, Demagnetization

ABSTRACT

Banded hematite ores covering the spectrum of types encountered in Precambrian areas world-wide, have been studied using measurements of the anisotropy of low field magnetic susceptibility (AMS). While k_{max}/k_{min} ratios of single crystals vary between 20:1 and 1500:1, samples of polycrystalline protores and of massive ores yield k_{max}/k_{min} ratios of 1.0-1.2:1.0-0.8. The geometry of AMS-ellipsoids is well correlated with that of texture patterns: k_{min} lies parallel to the *c*-axis maximum. The mean susceptibility of protores and massive hematite ores permits a differentiation of ore types according to ore composition. Protores and rich ores of the quartzose association yield a mean susceptibility of 3,000 - 13,000 SI units/unit volume, the mean susceptibility of carbonate ores varies widely between 20,000 - 250,000 SI units/unit volume. Partial demagnetization at 600-1000 Oe resulted in a slight modification of the susceptibility and of the axis-orientation.

Introduction

Banded to laminated hematite ores compose thick sequences in Precambrian metavolcanic and metasedimentary successions and consist of layers rich in iron minerals alternating with beds rich in quartz, carbonate and/or phyllosilicates. Such "protores" suffered a certain diagenetic enrichment by small-scale mobilization of iron and impurities [10] and may contain irregular to concordant masses of internally laminated high-grade hematite ores which result from tectonometamorphic differential enrichment of iron. Both protores and high-grade ores display a pronounced planar anisotropy of physical properties due to compositional layering, differences in grain shape, orientation of crystallites, and pore distribution.

The primary phase of iron minerals in banded hematite ores is magnetite. Magnetite alters in a very complex way, either to martite-hematite by addition of oxigen or to metal-deficient spinel phases which retain the structure of the original magnetite. The martite-forming process consists in diffusion of ferrous iron to internal crystal surfaces giving rise to the formation, by reaction with oxigen, of hematite on or between the octahedral parting planes. This "air oxidation" theoretically leads to a 2% volume increase [10]. Under pronounced anoxic conditions, oxidation proceeds by removal of Fe^{2+} from the crystal structures to circulating fluids and yields metal-deficient spinels. In the presence of adsorbed or hydrated water and at low temperatures magnetite is oxidized to maghemite when small grain sizes dominate, whereas larger grains are converted to martite and hematite, particularly when temperatures increase to more than 400° C [4], [9]. Ore types that suffered low-grade metamorphism and dominant flattening by pure shear are characterized by fine-grained hematite ensembles and textures of more or less randomly oriented crystals and often still contain considerable amounts of maghemite and relics of magnetite. The intermediate phase between magnetite and maghemite is kenomagnetite which in nearly all BIF-derived iron ores occurs instead of magnetite. The transition to kenomagnetite involves minimal volume change, but induces a density change as a result of the removal of up to one iron atom from each nine in the crystal [10]. (In reflected light, magnetite is grey while maghemite is greyish blue and kenomagnetite pinkish-brown). Increased temperatures and progressive simple shear trigger general hematitization, so that itabirites and derived high-grade ores largely consist of platy (specularitic) hematite aligned with (001) preferably parallel to surfaces of pronounced shear which correspond to penetrative foliation, thrust, fault or cleavage planes. The lattice-preferred orientation pattern of hematite *c*-axes [001] is controlled by degree and style of deformation, pure shear causing subcircular maxima (see Fig. 2) and simple shear to maxima elongated in the shear direction; both patterns are of approximately orthorhombic symmetry [5], [13], [14].

Magnetic susceptibility of hematite

Hematite is basically antiferromagnetic with weak spin-canted ferromagnetism superimposed [8], [16]. The iron atoms are arranged in almost plane ferromagnetic sheets perpendicular to the trigonal axis which are coupled antiferromagnetically. Each layer is magnetized spontaneously to saturation in a definite direction, and successive layers are magnetized alternately in opposite directions.

Hematite has an enormous intrinsic crystallographic magnetic anisotropy with the maximum and intermediary susceptibilities parallel to its basal plane, very low susceptibility in the direction of the trigonal axis and a ratio of maximum to minimum of >100. Within the basal plane a certain anisotropy exists, however without any systematic pattern of anisotropy [16]. Hematite yields a magnetic fabric to the rock that is influenced primarily by the crystallographic alignment of the magnetically anisotropic grains. In hematite ores, the minimum susceptibility direction lies in the maxima of the *c*-axis concentrations [7]. The mean susceptibility values of single hematite crystals vary considerably between about $2*10^{-3}$ SI and $8*10^{-2}$ SI (SI units per unit volume). The cause of this variability is not yet clear, but there is a striking difference of values obtained from specularitic (platy) single crystals and those obtained from more isometric ones [16]. Obviously hematite possesses a certain shape anisotropy [1], [6] which, however, shows substantial variations between specimens and can originate from the trigonal axis symmetry in the twinned volume of the crystal [11], [12].

The hard isotropic stable magnetization found in natural hematite material decreases with grain size [2]. The remanent magnetization of fine-grained hematite particles is characterized by high coercive forces [12], whereas large crystals have low coercive forces [15]. Magnetic interaction in single-domain particles yields an increase of initial susceptibility and an decrease of coercive force but remains constant with increasing concentration or packing [3]. The parasitic ferromagnetism consists of an anisotropic moment due to spin-canting and an isotropic defect moment which can be altered by stress and annealing (heating). Investigations on the domain structure of hematite single crystals yield wall spacing of 100-500 μm; the critical single-domain size probably is in the range of 10-100 μm. Wall motions controls the coercivity in large, multi-domain grains, whereas magnetization reverses by rotation in the single-domain particles [2]. It appears likely that the partial demagnetization behavior of pure hematite ores (up to 1 kOe, see Fig. 3) can be explained by the effects of grain-size.

The fabric of hematite in iron ores can be investigated by the magnetic anisotropy method applying low alternating fields or the anisotropy of the magnetization energy to saturation in large steady fields. The distribution of axes of maximum and minimum susceptibility or hard and easy magnetization, respectively, defines the magnetic fabric. The determination of the anisotropy ellipsoid by the low field method yields an average orientation of a large number of crystals. Because of the crystallographic anisotropy of hematite, even weakly aligned hematite crystals cause a significant magnetic anisotropy. As hematite in iron ores frequently appears in platy forms parallel to the basal plane, the minimum susceptibility should be perpendicular to the bedding or shear plane.

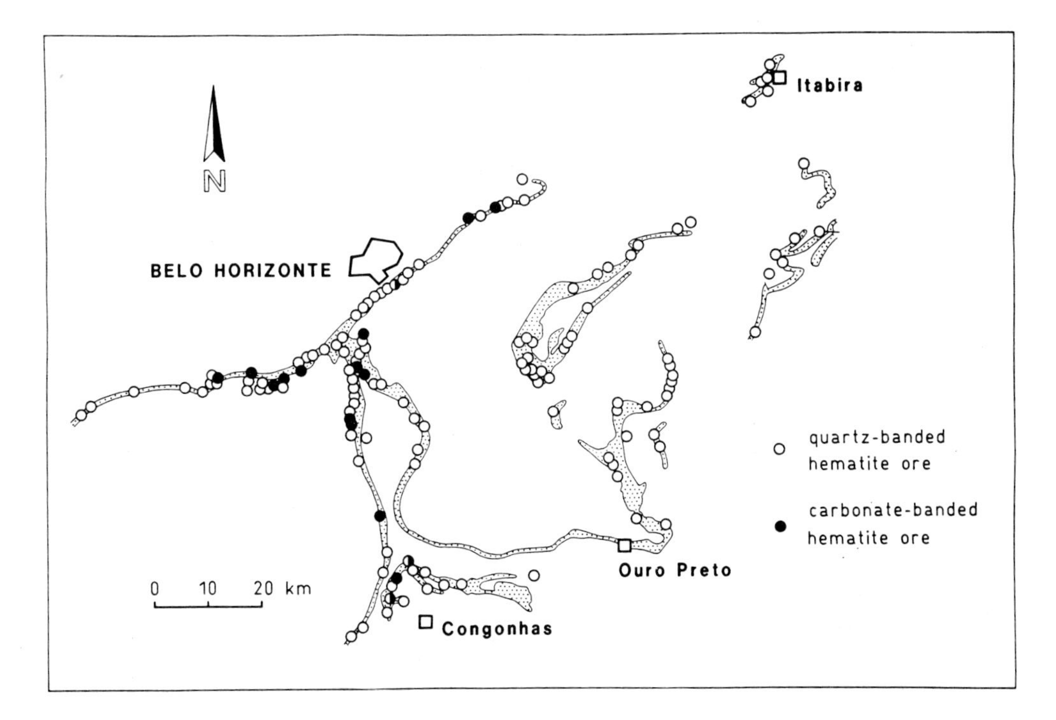

FIG. 1: General map of the Quadrilátero Ferrífero (Brazil) with outcrops of hematite ore and locations of iron ore mines.

Banded hematite ores from diverse deposits, covering more or less the spectrum of types encountered in Precambrian areas worldwide, have been studied using measurements of the anisotropy of low field magnetic susceptibility (Kappabridge KLY-2.02, Sapphire SI-2). In single hematite crystals, the directions of easy (k_{max}) and hard (k_{min}) magnetization lie in the basal plane and along the c-axis, respectively. While k_{max}/k_{min} ratios of single crystals vary between 20:1 and 1500:1, samples of polycrystalline protores as well as of massive ores yield k_{max}/k_{min} ratios in the range of 1.0-1.2 to 1.0-0.8, in extreme cases up to 1.6:0.6. The geometry of ellipsoids of magnetic anisotropy (AMS) is well correlated with that of texture patterns: k_{min} lies in the center of the c-axis maximum, whereas k_{int} indicates the elongation and k_{max} the narrowing of the maximum. The ellipsoid is oblate (pure shear), prolate (simple shear) or triaxial (transitional between pure and simple shear or owing to recrystallization); uniaxial oblate or prolate ellipsoids have never been encountered.

Susceptibility field of the Quadrilátero Ferrífero (Brazil)

The Quadrilátero Ferrífero (i.e. Iron Quadrangle) in Minas Gerais, central Brazil (Fig. 1), is an enclave of Precambrian supracrustal sequences surrounded by an older basement composed of varied gneisses. Overlying an Archean metavolcanic and metasedimentary suite, the ore-bearing succession commences with quartzites (Moeda formation) interfingering with carbonate beds towards the west and passing to a guide horizon of graphitic phyllites. The itabiritic iron ores compose a sequence (Cauê formation), some hundreds of meters thick, which for about 200 years has been one of the principal sources of marketable iron ores in the world. In the Quadrilátero Ferrífero itself they are overlain by carbonates (Gandarela formation), while in nearby regions, for example in the Itabira area, phyllitic meta-arenites are superjacent. The Quadrilátero Ferrífero was subjected to a low- to medium-grade metamorphism and suffered a complex deformation consisting of a SE to NW directed detachment, followed by uplifting of basement blocks around and within the enclave and finally by E to W directed high-level thrusting which transported the easterly adjacent basement upon the supracrustal succession.

The iron ores of the Cauê Formation are represented by low-grade itabirites with thin quartz layers, by well bedded hematite ores alternating with carbonate bands (particularly in deposits towards the western-northwestern edge of the Quadrilátero Ferrífero) and by compact high-grade ores bound to lenticular domains of heterogeneous shear deformation which may be fault zones or hinges and crests of vergent folds. While the average iron content is about 35-37% in bulk itabirites and carbonate protores, it can attain 69% in high-grade ores. In all ore types hematite is the dominant iron mineral; it results from the primary magnetite by oxidation, with maghemite as an intermediary mineral phase. In strongly sheared ores (itabirites) hematite appears as coarse-grained platy specularite (> 100 μm), whereas in carbonate ores the original intergrowth of fine-grained hematite crystals and the primary porosity is mostly preserved. Most of the ores show a relatively high degree of preferred orientation (Fig. 2 shows a sample from Conceição mine).

The mean or bulk susceptibility of protores and massive hematite ores encountered in the Quadrilátero Ferrífero varies dependent on the protore composition. This diagnostic aspect is of special metallurgical interest since quartz-bearing ores and derived massive ores differ considerably in their reduction behavior from carbonate ore varieties. Protores and rich ores of the quartzose association yield a mean susceptibility in the order of $2000\text{-}13{,}000 * 10^{-6}$ SI, whereas the mean susceptibility of carbonate ores varies between $20{,}000 * 10^{-6}$ SI and $450{,}000 * 10^{-6}$ SI. Obviously, hematite, which is of nearly single-domain grain size, is not the only carrier of susceptibility in carbonate ores; kenomagnetite, the metal-deficient spinel, is omnipresent and contributes a considerable remanent

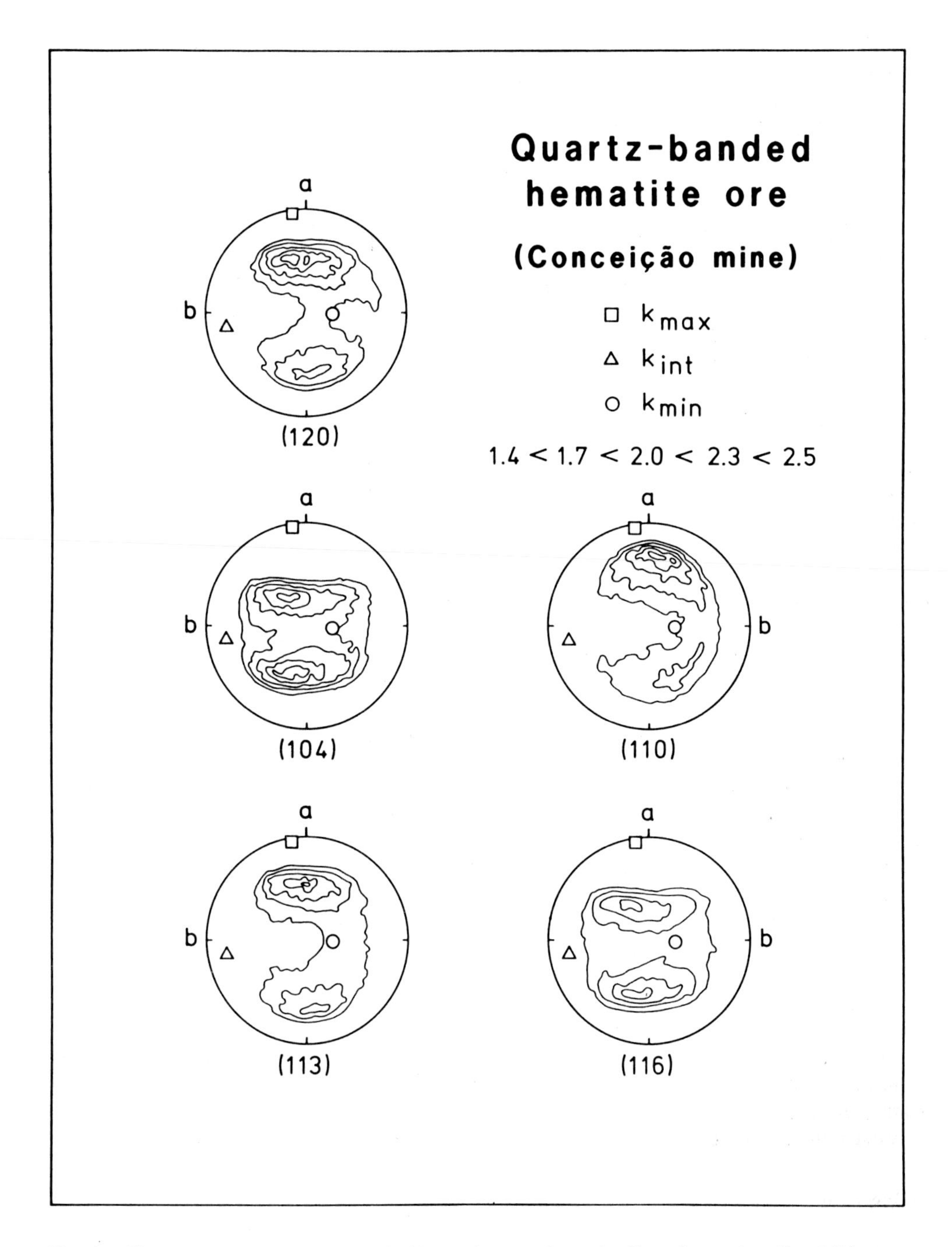

FIG. 2: X-ray textures of quartz banded hematite ores from the Conceição mine (Quadrilátero Ferrífero) including AMS-axes.

magnetic component. In some deposits at the northwestern border of the Quadrilátero Ferrífero, the susceptibility may even attain values above 1 SI, indicating the presence of significant amounts of magnetite relics. In the more intensely sheared quartz-itabiritic ores and derived high-grade ores, on the other hand, strongly aligned specularite prevails and causes a low bulk susceptibility, possibly due to antiparallel ordering of crystal stacks.

Prolate ellipsoids are found where the local influence of folding is significant, whereas oblate ellipsoids are either due to shearing (itabirites) or to compositional layering in carbonate-hematite ores in which stress was largely accommodated in carbonate bands and hematite bands are considerably less deformed. Ores stamming from deposits near the eastern border and from enclaves in the easterly adjacent basement area, where the influence of thrusting is dominant, exhibit a significant higher planarity of the magnetic fabric than ores from occurrences on the western and northern side of the Quadrilátero Ferrífero.

Susceptibility field of the Casa de Pedra mine

The Casa de Pedra mine is located at the southwestern edge of the Quadrilátero Ferrífero and belongs to the occurrences that suffered little deformation and low-grade metamorphism. The AMS of the different high-grade ore types investigated is highly variable and exhibits prolate and oblate ellipsoid shapes but differences in the anisotropy degree are generally small. The bulk susceptibilities indicate that low-susceptibility iron ores of the itabirite types are present as well as high-susceptibility ores of carbonate provenance, with maximum susceptibilities of about $250,000 * 10^{-6}$ SI.

In order to ascertain the magnetic properties of the iron ores specimens, as well as the susceptibilities the magnetic moments were determined. They vary considerably between about 500 nT and more than 12,000 nT. Demagnetization tests using fields of 1 kOe aimed at lowering the remanent magnetization of the high-grade ores and studying its effect on the susceptibilities (Fig. 3). In all cases only partial demagnetization was obtained since the applied field was not high enough for total saturation of the specimens. As magnetite is only preserved in relics and is mostly altered to kenomagnetite, the principal carrier of the high remanent magnetization should be kenomagnetite, with a slight contribution of maghemite. The bulk susceptibilities after demagnetization are nearly the same as before but the specific susceptibilities in given directions are always modified, from which result certain but not significant modifications of the anisotropy parameters. A general decrease in the magnetic moment, however, is to be registered, more pronounced in derivates of carbonate-bearing ores than in those of quartz-bearing ones. The orientation of the AMS ellipoids changed slightly by the demagnetization but remains dependent on the respective structural position of the specimen in the ore body, indicating that the remanent magnetic moment was induced before deformation and block rotation occurred. During the deformation the platy hematite component obviously underwent only translation without significant rotation of the grains. Thus the degree of the magnetic anisotropy and the shape of AMS ellipsoids are solely controlled by the alignment of hematite, whereas the magnetude of magnetic properties of the Casa de Pedra high-grade ores are largely determined by the content of the isotropic magnetite-derived spinels.

Acknowledgement

The study was supported by the Deutsche Forschungsgemeinschaft (Forschergruppe "Textur und Anisotropie kristalliner Stoffe"). We thank Carlos Alberto Rosière and João Henrique Grossi Sad of the Universidade Federal de Minas Gerais, Belo Horizonte, Brazil, for providing ore samples from the Quadrilátero Ferrífero.

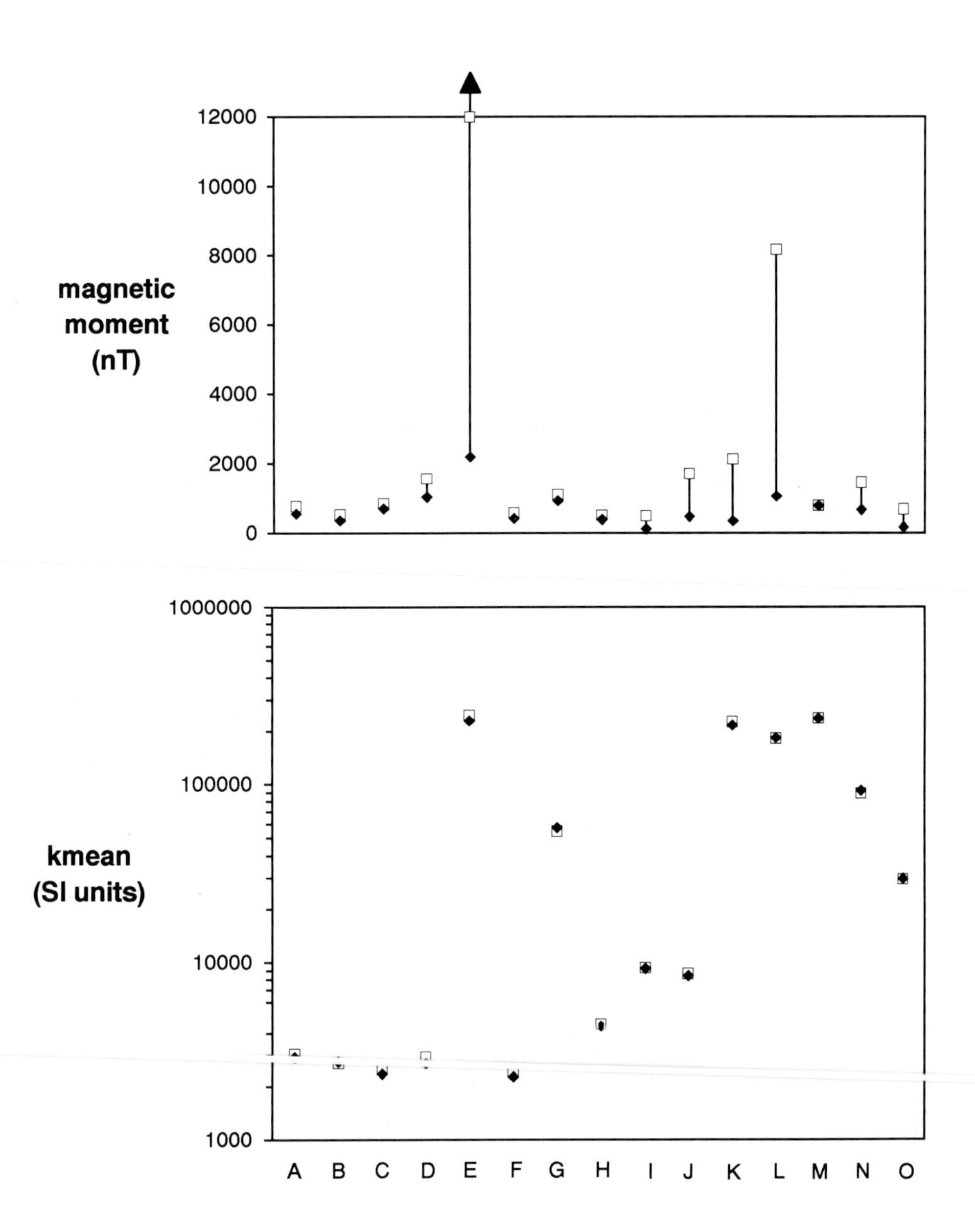

Fig. 3: Samples of hematite ores from the Casa de Pedra mine;
above: magnetic moments (nT); below: bulk susceptibilities (SI units).
Square: value before demagnetization; rhombus: value after demagnetization at 1 kOe.

References

1) Banerjee, S.K. (1963): An attempt to observe the basal plane anisotropy of hematite. - Phil. Mag., 8: 2119-2120.

2) Dunlop, D.J. (1971): Magnetic properties of fine-grained hematite. - Ann. Géophys., 27: 269-293.

3) Dunlop, D.J. (1981): The rock magnetism of fine particles. - Phys. Earth Planet. Int., 26: 1-26.

4) Elder, T. (1965): Particle-size effect in oxidation of natural magnetite. - J. Appl. Phys., 36: 1012-1013.

5) Esling, C., Quade, H., Wagner, F. & Walde, R. (1981): Pesquisa textural de minério hematítico da jazida de Águas Claras (MG) pelo método de difração de neútrons. - Rev. Bras. Geociências, 11: 84-90.

6) Flanders, P.J. & Schuele, W.J. (1964): Anisotropy in the basal plane of hematite single crystals. - Phil. Mag., 9: 485-490.

7) Hrouda, F., Siemes, H., Herres, N. & Hennig-Michaeli, C. (1985): The relationship between the magnetic anisotropy and the c-axis fabric in a massive hematite ore. - J. Geophys., 56: 174-182.

8) Jelinék, V. (1981): Characterization of the magnetic fabric of rocks. - Tectonophysics, 79: T63-T67.

9) Johnson, H.P. & Merrill, R.T. (1972): Magnetic und mineralogical changes associated with low-temperature oxidation of magnetite. - J. Geophys. Res., 77: 334-341.

10) Morris, R.C. (1985): Genesis of iron ore in banded iron-formation by supergene and supergenemetamorphic processes. A conceptual model. - In: Wolff, K.H. (Ed.), Handbook of stratabound and stratiform ore deposits. Part IV. Volume 13. Regional studies and specific deposits: 72-260; Amsterdam (Elsevier).

11) Porath, H. & Raleigh, C.S. (1967): An origin of the triaxial basal planes anisotropy in hematite crystals. - J. Appl. Geophys., 38: 2401.

12) Porath, H. (1968): The magnetic anisotropy of the Yampi Sound hematite ores. - Pure Appl. Geophys., 69: 168-178.

13) Quade, H. (1988): Natural and simulated (10.4) pole figures of polycrystalline hematite. - Textures and Microstructures, 8/9: 719-736.

14) Quade, H. & Taugs, R. (1988): Textural anisotropy of banded hematite ores and its influence on reduction behavior. - In: Bunge, H.J., Directional properties of materials: [illegible]; Oberursel (DGM Informationsgesellschaft).

15) Sunagawa, I. & Flanders, P.J. (1965): Structural and magnetic studies in hematite single crystals. - Phil. Mag., 11: 747-761.

16) Uyeda, S., Fuller, M.D., Belshé, J.C. & Girdler, R.W. (1963): Anisotropy of magnetic susceptibility of rocks and minerals. - J. Geophys. Res., 68: 279-291.

Materials Science Forum Vol. 157-162 (1994) pp. 1689-1694

THE WORK HARDENING OF PEARLITE DURING WIRE DRAWING

P. Watté [1], P. Van Houtte [1], E. Aernoudt [1], J. Gil Sevillano [2], W. Van Raemdonck [3] and I. Lefever [3]

[1] Department of Metallurgy and Materials Engineering, KU Leuven, de Croylaan 2, B-3001 Leuven, Belgium

[2] C.E.I.T. and Universidad de Navarra, San Sebastian, Spain

[3] N.V. Bekaert S.A., Zwevegem, Belgium

Keywords: Eutectoid Pearlite, Wire Drawing, Work Hardening, Taylor Modelling, Patenting Structure

ABSTRACT.

An advanced Taylor type polycrystalline model has been developed for the simulation of the evolution of the flow stress and the deformation texture of pearlitic ferrite in forming processes. Due to its generality, it lends itself well for the study of the work hardening of pearlite. The influence of microstructural parameters such as the crystallographic texture, the initial ferrite spacing and the distribution of obstacles to dislocation glide in the ferrite can be assessed. The macroscopic stress required for the plastic deformation of an aggregate of pearlite colonies has been predicted for wire drawing deformation. A distinction has been made between the surface layers and the central layers of the wire because of different constraint conditions. The predictive capabilities of the model have been evaluated by comparing the numerical simulation results with the work hardening properties revealed by experimental studies.

INTRODUCTION.

The mechanical requirements of cold drawn in situ composite wires become more demanding. Particularly in the processing of high strength steel cord, the performance characteristics of the wire filaments should be enhanced. The mastering of the design of these filaments is of considerable technical importance in the wire industry. The very high strengths these materials exhibit after cold work renders them very attractive for reinforcement applications.

The understanding of the work hardening of pearlite would enable to control and to predict their strength in forming operations. All experimental studies point out that the work hardening of in situ composites is substantially different from the work hardening of their individual phases [1-4]. This is one of the major reasons why their apparent strength may deviate from the one predicted by an isostrain rule of mixtures. Such rule indeed makes a volumetric weighted average of the strength of their individual components but the results obtained are rather inadequate since this approach does not account for the distinct work hardening of the separate phases within an in situ composite.

Eutectoid pearlite has been subject to extensive experimental investigations of the microstructural evolution under different deformation modes (wire and strip drawing, rolling, compression)[1-2]. The values of the critical resolved shear stress (CRSS) of pearlitic ferrite are expected to be non-uniform because of the orientation dependent slip spacing on different slip planes. Because of the small mean free path of the dislocation movement, the activated slip systems in pearlitic ferrite are in general different from those in free ferrite of the same orientation. In order to simulate the deformation in pearlitic ferrite one can make use of models which are based on a flow stress theory which links the flow stress evolution with the crystallographic properties of the material. A quantification of the flow stress evolution of pearlitic ferrite has been attempted by Gil Sevillano using a Taylor model [5]. The validity of his predictions was restricted to randomly oriented {112} pearlite, i.e. to as-transformed pearlites where the {112} ferrite habit plates predominate. A Hall-Petch hardening law was proposed for the CRSS, in analogy with the experimentally measured flow stress properties of pearlite. The Taylor hypothesis was adopted to yield upper bound approximations for the calculated stress in the ferrite.

The model presented here is a refinement of the former approach. It can operate under different constraint hypotheses. Moreover, the calculation of the distribution of lamellar spacing is performed on a more general basis. It assumes that in a given pearlite colony, the deformation is equal in the ferrite and the cementite phase. Hence, compatibility between adjacent ferrite and cementite lamellae is assured. It provides information about the texture evolution, the values of the CRSS on the activated slip systems, but only for the ferrite phase. The texture and flow stress of ferrite have been calculated incrementally. After each strain increment, the crystallographic rotation of the ferrite and the rotation of the cementite plates were envisaged. In pearlite, the dislocation accumulation process in the ferrite is the most dominant during forming. It is generally accepted that the cementite interfaces act as barriers to slip propagation in the ferrite. As a consequence, the strain induced dislocation accumulation in the ferrite will be affected by the geometrical distribution of these barriers. Therefore, a simulation of the work hardening of pearlite should focus on the role of this distribution in the strengthening process. In the present model this was done taking the redistribution of the lamellar planes in each crystallite into account.

THE PEARLITE MODEL.

The basic structure of the Taylor model which has been used is described in detail in [7]. The change in the Taylor factor M concomitant with strain is directly related with the texture evolution of the material during deformation. The macroscopic flow stress can be assessed from the deviatoric stress which is acquired by multiplying the CRSS with this orientation factor M. As referred above, the material flow in in situ composites, and in particular in pearlite, is controlled by the distribution of obstacles to dislocation glide[3-6]. It is widely accepted that in pearlite the ferrite-cementite interfaces act as such obstacles. In account of work hardening this distribution of obstacles must be envisaged and updated in the course of the simulation. This concept has been incorporated into our model. It requires the implementation of slip system hardening. Thereto, the Taylor code has been extended with a physical hardening law. Gil Sevillano argued that this can be done by using an appropriate CRSS law. A simulation of the work hardening of pearlitic ferrite is only meaningful when the CRSS, τ^c, is a function of the orientation of the cementite planes and the phase spacing. He proposed a CRSS law for the dislocation movement in ferrite of the form

$$\tau^c - \tau^0 = \frac{A.G.b}{2.\pi.S.v_f}.\ln(\frac{S.v_f}{b}) \qquad (1)$$

in which A=1.21 is a constant for a line dislocation, G is the ferrite shear modulus equal to 6.4 10^4 MPa, b=0.248 nm is the Burgers vector and v_f is the volume fraction of ferrite equal to 0.87 in eutectoid pearlitic steel. τ^0 is a friction stress equal to 60 MPa. S=d/sin λ , where λ is the angle between the vectors normal to the lamellae and normal to the ferrite slip plane, d being the true minimum spacing of the pearlite colony. As such, S depends both on the colony structure orientation (through d) and on the ferrite slip plane orientation (through λ).

The introduction of this equation in the Taylor model is rather straightforward. In order to enable the quantitative prediction of work hardening, the model used consists of two parts: a flow stress theory for a given structure and a theory for the evolution of the material structure with strain. The calculation of the phase space changes and the alignment of the lamellae has been done with the aid of continuum mechanics.

The model which is presented will calculate the redistribution of lamellar spacing. It makes use of two discrete textures: a crystallographic one as common in texture modelling and a morphological pearlite texture. To each crystallite only one pearlite colony is associated in which the lamellar spacing is assumed to be homogenous. All lamellae in each colony are aligned in a direction which is allowed to vary during deformation. For the starting textures in the simulation this direction was chosen at random.

The calculation of the reorientations of the lamellae is inspired by the March model [9]. It can be performed as follows. Let a finite strain be described by a second order tensor **F** such that $\mathbf{y} = \mathbf{F}\cdot\mathbf{x}$. **x** and **y** are vectors within the lamellar plane, respectively before and after deformation. The tensor **F** is related to the strain tensor and the rigid body tensor through:

$$\mathbf{dK_{ij}} = \mathbf{d\varepsilon_{ij}} + \mathbf{d\omega_{ij}} \text{ and } (\mathbf{F} + \mathbf{dF}) = (\mathbf{1} + \mathbf{dK})\cdot\mathbf{F} \qquad (2)$$

Let $\mathbf{x}^N$ and $\mathbf{y}^N$ be vectors respectively normal to x and y. From the normality between **x** and **y** it follows that

$$\mathbf{y}^N \cdot \mathbf{F} \cdot \mathbf{x} = 0 \qquad (3)$$

Equation 4 must be valid for an arbitrary vector x in the lamellar plane. This is only fulfilled when:

$$\mathbf{y}^N \cdot \mathbf{F} = \lambda.\mathbf{x}^N \qquad (4)$$

with λ an arbitrary scalar. In matrix notation equation 4 can be rewritten as

$$\left[y_i^N\right]^T . \left[F_{ij}\right] = \lambda . \left[x_j^N\right]^T \qquad (5)$$

If λ is set equal to 1 and by defining $\left[G_{ij}\right] = \left[F_{ij}\right]^{-1}$ equation 5 becomes

$$\mathbf{y}^N = \mathbf{x}^N \cdot \mathbf{G} \qquad (6)$$

Equation 6 allows to find the direction of a normal to a particular lamella as the result of a finite deformation. A unit vector $\hat{\mathbf{y}}^N$ obtained after normalizing y is further used to calculate the ferrite phase spacing. Let $\mathbf{x}^d$ be a vector normal to the lamellae before deformation and going from one lamellae to another. If d is the lamellae spacing then one can write

$$\mathbf{x}^d = d.\hat{\mathbf{x}}^N \qquad (7)$$

in which $\hat{\mathbf{x}}^N$ is a unit vector normal to the lamellae before deformation. After deformation, the lamellae spacing has become d'

$$\mathbf{y}^d = d'.\hat{\mathbf{y}}^N = \mathbf{F}\cdot\mathbf{x}^d \qquad (8)$$

The new lamellae spacing d' is obtained by projecting the vector $\mathbf{y}^d$ on the unit vector normal to the lamellae after deformation. From equation 7 and 8 it follows that

$$\frac{d}{d'} = \mathbf{F}\cdot\hat{\mathbf{x}}^N \cdot \hat{\mathbf{y}}^N \qquad (9)$$

IMPLEMENTATION OF THE PEARLITE MODEL INTO A TAYLOR SIMULATION.

In the simulation, the same convention for the choice of the sample reference system has been adopted as in [10], namely x_1= wire axis, x_2=tangential direction and x_3=radial direction, The calculation starts with a set of 400 random crystallographic orientations. This initial crystallographic texture is complemented by a set of lamellae orientations and lamellar spacings. Each orientation from the first set is associated to a single pearlite colony of which the lamellar orientation is reflected by the vector $\mathbf{x}^N$. These vectors have also been chosen at random. The initial interlamellar spacing d_0 was taken homogeneous in each colony, which is representative for as-patented pearlite.

The velocity gradient tensors that were fed into the Taylor program were provided by a simple flow line model [10]. It enables to calculate the strain rate tensors which describe the material flow of a particle during its trajectory in the deformation zone of the drawing dies. When knowing the drawing schedule one can translate the strain history along flow lines at the surface and at the centre of the wire in terms of these strain tensors. The model allows for the calculation of the flow stress at the surface and in the centre of the

wire. The flow line pattern at the surface of the wire is dictated by the geometry of the dies in the drawing line. In the centre of the wire one can assume that the flow lines are straight.
Care was taken that $\Delta\varepsilon$ was smaller than 0.02 for all incremental steps. Assuming full constraint conditions, the change in interlamellar spacing can be calculated separately from the texture calculation using the current formalism. The formulas for the finite deformations are then applied to find the new lamellar directions and new interlamellar spacings. This data allows to calculate the new τ^c values on the {112}[111] and {110}[111] slip systems. The described procedure is then repeated for each increment. The constraint hypotheses to impose to the crystallites are interrelated with the flow line pattern. At the surface the material flow deviates from pure axisymmetric deformation. In order to model it one can assume full constraint conditions. In the central layers of the wire, the deformation is pure axisymmetric. There, the development of a [110] fibre texture will induce a curly microstructure. Hosford [11] has shown that pearlite crystals with former orientation have a strong energetic advantage to deform in plane strain, rather than by axisymmetric deformation. In order to achieve geometrical compatibility, the colonies will bend over each other, and this is known as the curling effect. Curling can be accounted for by introducing relaxed constraint conditions. These abandon partly that the crystallites undergo the same uniform deformation or that the macroscopic finite strain tensor F_{ij} not necessarily equals f_{ij}, the strain tensor to prescribe to the crystallites. The validity of the RC model only holds when grains have become flat and elongated. It can be shown that curling can be simulated by dropping two constraints, namely $f_{13}=0$ and $f_{33}-f_{22}=0$ [12]. A relaxed constraint Taylor model will find values for it by using a minimum internal power dissipation criterion. In the relaxed constraint case, the calculation of the redistribution of lamellar spacings is somewhat more complicated. During each increment step the local deformation of the crystallites is sought. It is then used to find the new lamellae orientations and the interlamellar spacings.

RESULTS AND DISCUSSION.

A series of patenting structures has been chosen for which the true initial ferrite spacing varied between 0.07 and 0.11 μm. The drawing schedule which has been used was the same for all simulations. It consisted of 22 drawing passes with varying reductions per pass. The reductions were such that their maximum did not exceed 20% per pass, the semi die angle in each draw was about 4°.
The flow stress of the ferrite phase of pearlite yields from the average deviatoric stress provided by the simulation. Thereto, we averaged the stress states in the crystallites which were found from the plasticity analysis pertaining to the Taylor model. An equivalent ferrite flow stress is introduced given by:

$$\sigma_{flow}^{ferr} = s_{11} - \frac{1}{2}.(s_{22} + s_{33}) \tag{10}$$

in which s_{ij} are the deviatoric stress components which are averaged over all crystallites which constitute the deformation texture. Equation 10 is based on an equal plastic power dissipation concept. The equivalent stress, σ_{eq}, can be found from

$$\sigma_{ij} \cdot \varepsilon_{ij} = \sigma_{ij} \cdot \begin{bmatrix} 1 & 0 & 0 \\ 0 & -0.5 & 0 \\ 0 & 0 & -0.5 \end{bmatrix} . \varepsilon_{eq} = \sigma_{eq} \cdot \varepsilon_{eq} \tag{11}$$

The flow stress of drawn pearlitic wire shows a Hall-Petch type relationship with the mean transverse ferrite spacing. According to Gil Sevillano, this Hall-Petch behaviour can be reproduced by the equation:

$$\sigma_{flow}^{pearl} = \sigma_{flow}^{ferr}.(1 - v^{cem}) + \sigma_{flow}^{cem}.v^{cem} \tag{12}$$

with v^{cem} the volume fraction of cementite. Former equation expresses that the flow stress of pearlite is a superposition of the ferrite flow stress σ_{flow}^{ferr}, controlled by the need of multiplying dislocations in the interlamellar space, and a constant contribution of the cementite phase, which was taken equal to 700 MPa. The results of the calculation are depicted in figure 1. The pearlite flow stresses were fitted to an exponential hardening law of the following type

$$\sigma - \sigma_0 = c_1 . \exp(c_2 . \varepsilon) \tag{13}$$

in which c_1 and c_2 are fitting constants. These were estimated by means of linear regression. The FC

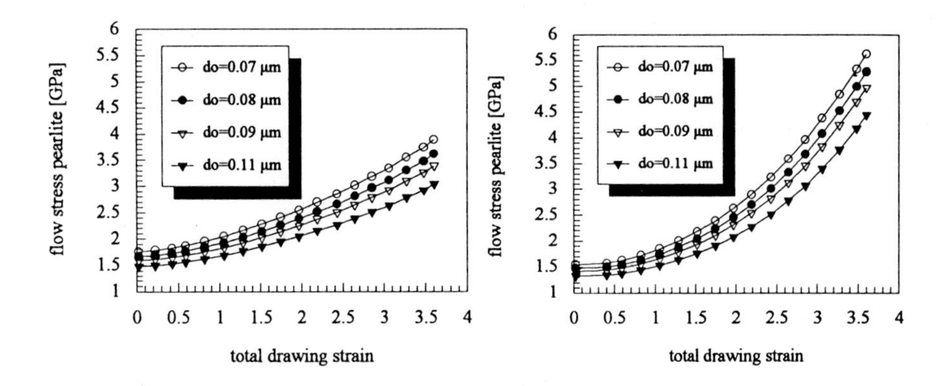

fig. 1: evolution of the pearlite flow stress with drawing strain. Right: flow stress evolution in the central layer of the wire yielding from a relaxed constraint simulation taking the curling of the pearlite colonies into account. Left: flow stress in the surface layer of the wire yielding from a full constraint simulation. d_0 is the mean interlamellar spacing of the patenting structure which was taken into consideration

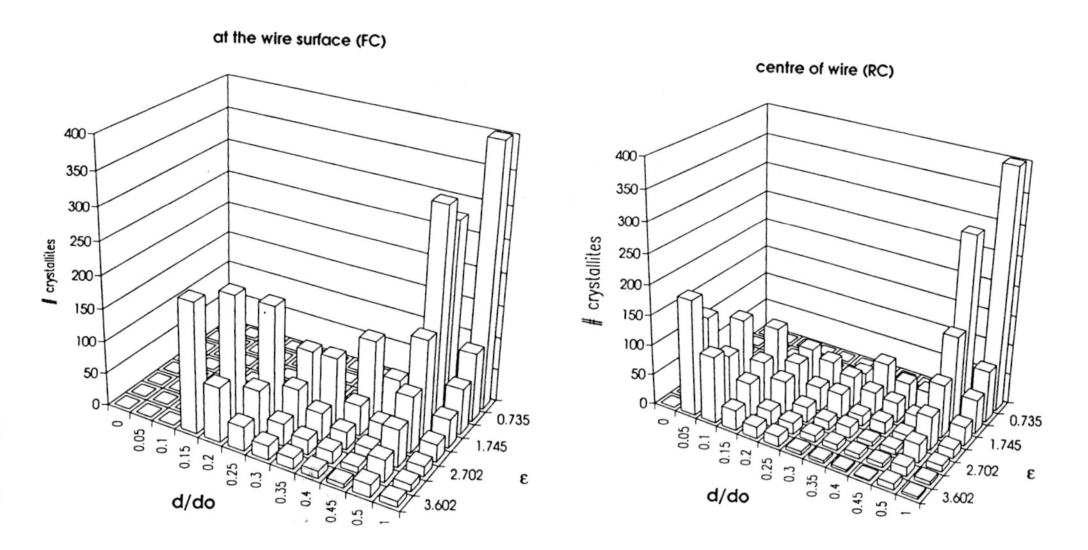

fig 2: distribution of interlamellar spacing over the 400 crystallites taken up in the simulation. Left plot of the distribution of the ferrite interlamellar spacing. over the crystallites at the surface of the wire. Right plot id. for the central layer of the wire.

simulations give values of c_2 close to 1/4, in agreement with the experimental results of work hardening studies of pearlite, whereas a value of c_2 close to 1/3 has been found for the RC simulations. It is apparent that the simulation model predicts increases of the flow stress to the same extent for both constraint hypotheses FC and RC. Changing the mean ferrite spacing as-patented from 0.11 μm to 0.7 μm is predicted to result in a flow stress increase of about 30%. In addition, the redistribution of interlamellar spacings over the crystallites was calculated for the intermediate drawing strains. The evolution of these spacings is modelled assuming the same drawing schedule as the flow stress simulations. Figure 3 depicts the evolution in the interlamellar spacing by means of a plot of the relative ferrite spacing d/d_0 versus the drawing strain. The introduction of curling is found to give rise to the smallest interlamellar spacing.

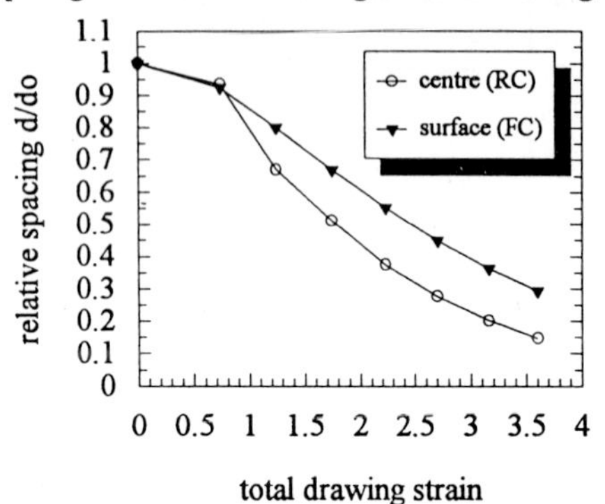

fig 3: evolution of the relative interlamellar spacing d/do with drawing strain. Full constraint conditions are assumed for the surface layer whereas relaxed conditions were assumed for the centre of the wire

According to Langford [2], for fine pearlite, the transverse spacings fit well to $\frac{d}{d_0} \propto \exp(-K.\varepsilon)$ with K=0.5. Fitting the FC results gives a value of K equal to 0.413 whereas K equals 0.611 for the RC modelled spacing change. The histograms in figure 2 label the distribution of spacings at intermediate strains. These are very similar to the distribution of lamellar spacing obtained from Langford's extensive T.E.M. study of the pearlite structure measured at cross sections of wires [2]. In this study, the experimentally determined ferrite spacings are initially widely distributed. The width of this distribution enlarges to become skewed towards finer spacings. This tendency is also found in our predictions, although the skewness of the distribution appears to be more accentuated.

The result that the FC model predicts smaller flow stresses than the RC model is quite surprising. One should expect smaller flow stresses predicted by the RC model, which is not the case here. The observed phenomenon can be explained in the light of the substructure model. The evolution of the lamellar spacing in figure 3 points out that the constraint conditions have a substantial influence on the change in ferrite phase spacing. Curling turns out to promote higher interlamellar spacing decreases. Although the Taylor factors are lower for the RC predictions, the multiplication with a higher CRSS explains the excess in flow stress to the FC model.

REFERENCES.

1) Embury, J.D. and Fisher R.M., Acta metall., 1966, 14, 147
2) Langford, G., Metall. Trans., 1977, 8., 861
3) Funkenbush, P.D. and Courtney, T.H., Acta metall.,1985, 33, 913
4) Spitzig, W.A., Pelton, A.R. and Laabs, F.C., Acta metall., 1987, 35, 2477
5) Gil Sevillano, J.et al, ICOTOM 5, Vol II, 1978, Gotstein, G. and Lücke, K, eds., Springer Verlag, 495
6) Gil Sevillano, J., J. de Physique III, 1991, 1, ,967
7) Van Houtte, P., Textures and Microstructures, 1988, 8 & 9, 313.
8) Gil Sevillano, J. et al, Progr. Mat. Sc., 1980, 25, 69
9) March, A., Z. für Kristallografie, S1 Bd, 1938,19
10) Van Houtte, P. et al, this conference
11) Hosford, W.F., Trans. Met. Soc. AIME, 1963, 230, 12.
12) Van Houtte, P., ICOTOM 7, 1984, Vol 1, Brakman et al, eds.,7

Materials Science Forum Vol. 157-162 (1994) pp. 1695-1700

TEXTURE GRADIENT EFFECTS IN TANTALUM

S.I. Wright [1], A.J. Beaudoin [2] and G.T. Gray III [1]

[1] Materials Technology: Metallurgy, Los Alamos National Laboratory, Mail Stop K762, Los Alamos, NM 87545, USA

[2] Reynolds Metals Company, Richmond, VA 23261, USA

Keywords: Inhomogeneous Textures, Microdiffraction, Finite-Element Modeling

ABSTRACT

The effects of inhomogeneities in texture on the mechanical response of a tantalum plate were investigated. Through-thickness compression samples exhibited an hourglass shape after deformation. The compression tests were modeled using a finite-element approach. A polycrystal plasticity model is embedded at each element of the mesh enabling the simulation to account for the spatial distribution of texture and its evolution. The texture was characterized using spatially specific individual lattice orientation measurements. The orientations were measured using automatic analysis of electron backscatter Kikuchi diffraction patterns. The influence of the spatially nonuniform distribution of texture on the experimental and simulated results is discussed. This work demonstrates the capability of currently available tools for characterizing inhomogeneous microstructures and modeling the effects of variations in texture on mechanical behavior.

INTRODUCTION

In a previous paper[1], the characterization of the texture gradient in a tantalum plate was described. The spatial distribution of texture was characterized using individual lattice orientation measurements obtained through automatic analysis of backscattered electron Kikuchi diffraction patterns[2, 3, 4].. The plate exhibited a {100}<001> (cube) texture at the surface and a {111}<110> texture at the centerline. This paper described compression tests of through-thickness cylindrical specimens. After compression, these samples possessed an hourglass shape. The hourglass shape was ascribed to the through-thickness texture gradient present in the plate. The shape was effectively modeled using a Taylor polycrystal plasticity model where strain is assumed to be uniform throughout the polycrystal. To model the hourglass shape, two hypothetical samples were constructed. One sample possessing the texture at the surface of the plate and the second the texture at the centerline. The same amount of work in compression was numerically applied to both samples. The hypothetical sample with the plate surface texture strained more than the sample with the centerline texture. The ratio of the area at the surface of the sample to the area at the midplane was predicted to within a few percent of the measured value. However, the paper suggested that a more complete approach to modeling the effect of the variation in texture on the mechanical response could be achieved using the finite-element approach developed by Beaudoin and coworkers[5, 6].

The present work reviews some of the experimental details of the compression testing and the microtextural characterization presented in more detail in the previous paper. The finite-element modeling technique employed is also reviewed. This study details the mapping of the individual orientation measurements into the finite element model. Results of the simulation are presented and discussed.

EXPERIMENTAL DETAILS

The material for this study was cross-rolled and annealed arc-remelted tantalum plate. The nominal chemistry is reported in table I.

Table I – Chemical analysis of the test material (weight percent in PPM).

O	N	C	H	Fe	Ni	Cr	Ca	Cu	Si	Ti	Mo	W	Nb	Ta
38	24	5	<1	27	39	18	<2	<1	<8	<5	<5	<40	<10	Bal.

Through-thickness cylindrical test specimens (diameter = 5.0mm, length = 6.2mm) were electrodischarge machined from the plate. The specimens were compressed at strain rates of $0.001s^{-1}$ at room temperature. After straining, the samples were slightly hourglass shaped rather than remaining cylindrical or barreling slightly as commonly observed in compression samples.

A section plane passing through the center axis of a compression sample prior to deformation was prepared for the orientation measurements. Approximately 11,000 orientation measurements were made using an automatic technique based on computer analysis of electron backscatter diffraction patterns[2, 3, 4]. The measurements were made on a 1mm × 4mm regular hexagonal grid with 20μm spacing between measurement points as shown in figure 1.

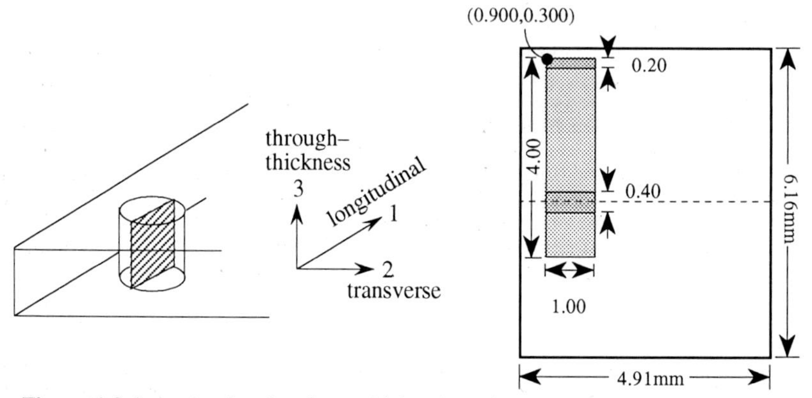

Figure 1 Schematic of region from which orientation measurements were obtained.

Numerical simulation of the compression deformation was carried out using a finite element code which derives material properties from polycrystal plasticity theory. In this approach, element constitutive properties are derived from an idealized polycrystal aggregate situated in the element. The polycrystal aggregate is typically represented by hundreds of orientations. The code utilizes massive parallel computations to carry out the numerous computations inherent in such a formulation.

An octant of the test specimen was discretized using a mesh of 5184 brick elements. Symmetry was assumed along specimen mid-planes with normals in the through-thickness, rolling, and transverse plate directions. The mesh consisted of twelve layers, each of which was comprised of 432 elements. Correspondingly, the experimental orientation measurements were partitioned into twelve groups of roughly 900 measurements. Symmetry operations were applied to each measurement so as to achieve an orthorhombic sample description in agreement with the symmetries and boundary conditions prescribed in the mesh definition. This resulted in a set of approximately 3600 orientations assigned to each layer of the finite element mesh. The assignment, providing an initialization of material state, was carried out by taking a random selection of 256 orientations for each finite element from the appropriate layer measurement set.

A polycrystal viscoplasticity approach was used where the single crystal slip system constitutive behavior was assumed to follow a power law relationship as follows:

$$\dot{\gamma} = a|\tau / \hat{\tau}|^{1/m} |\tau| / \tau \tag{1}$$

where $\dot{\gamma}$ is the slip system shear rate, τ is the resolved shear stress and $\hat{\tau}$is a scaling parameter that defines the hardness of the slip systems. The rate sensitivity m was taken as 0.06 based on compression test data[1],. The hardening was described using a Voce law formulation[7] where the strain-hardening rate for a crystal is given as a function of the resolved stress τ as follows:

$$d\hat{\tau} / dt = \theta_0 [1 - (\hat{\tau} - \tau_0) / \tau_V] |\dot{\gamma}^*| \tag{2}$$

This definition requires three material constants to parameterize the hardening behavior: the reference stress τ_0 at yield, the rate of decrease of $d\hat{\tau} / dt$ with $\hat{\tau}$, characterized by the Voce stress τ_V and an adjustable parameter θ_0 that sets the rate of hardening in the evolution equation. $\dot{\gamma}^*$ is the sum of shear rates on all slip systems in the crystal. The parameters used in the simulation were 66 MPa for τ_0, 170.8 MPa for $\theta_0$and 90 MPa for τ_V.

The simulation was conducted using uniform time steps of 10s. As detailed above, each crystal was associated with the state variables describing orientation (three Euler angles) and hardness $\hat{\tau}$. The evolution of these state variables was computed for over 1.3 million crystals. The simulation was carried out to a strain of 0.26. The simulation was conducted on the Thinking Machines CM-5 computer at the Advanced Computing Laboratory at Los Alamos National Laboratory. The calculations were completed in four hours using one-eighth of the CM-5 processing power.

RESULTS

Figure 2 shows the distribution of {100} and {111} type grains in the sampling region. Orientation measurements having {111} poles within 15° of the plate normal are highlighted in black in the left hand figure and measurements having {100} poles within 15° of the plate normal are highlighted in the right figure. Clearly, the centerline of the plate is primarily composed of grains having {111} poles normal to the surface of the plate and the {100}type grains are primarily located at the surface of the sample. This is the cause of the hourglass shape after deformation of the compression samples. The {100} type grains are in a "softer" orientation than the {111} grains. Thus, more of the strain is accommodated by the grains at the surface of the sample than the grains at the center. The boundaries in figure 2 were formed by drawing a line segment between neighboring measurement points when the misorientation between the measurement points exceeded 15°. The measurements are too coarse to adequately reconstruct the grain boundary map as has been done from automatic orientation measurements in previous works[3, 4]. However, a nonuniform grain size distribution is evident in the figure and confirmed by metallographic observations.

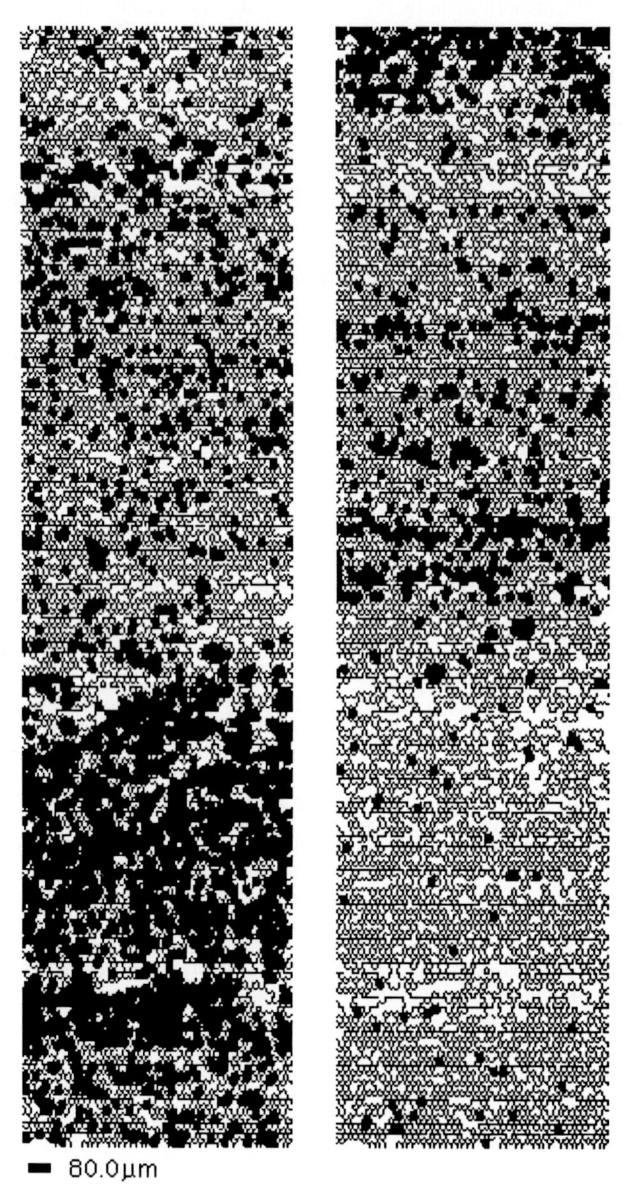

Figure 2 Grain boundary maps reconstructed from the orientation measurements. Measurements are highlighted having (111) (left) and (100) (right) poles normal to the plate surface.

An interesting feature in the modeling effort was the correlation between the distribution of strain (and strain rate) and the distribution of orientation in the sample. Figure 3 shows the distribution of strain rate at a strain of 0.13 and later at a strain of 0.26.

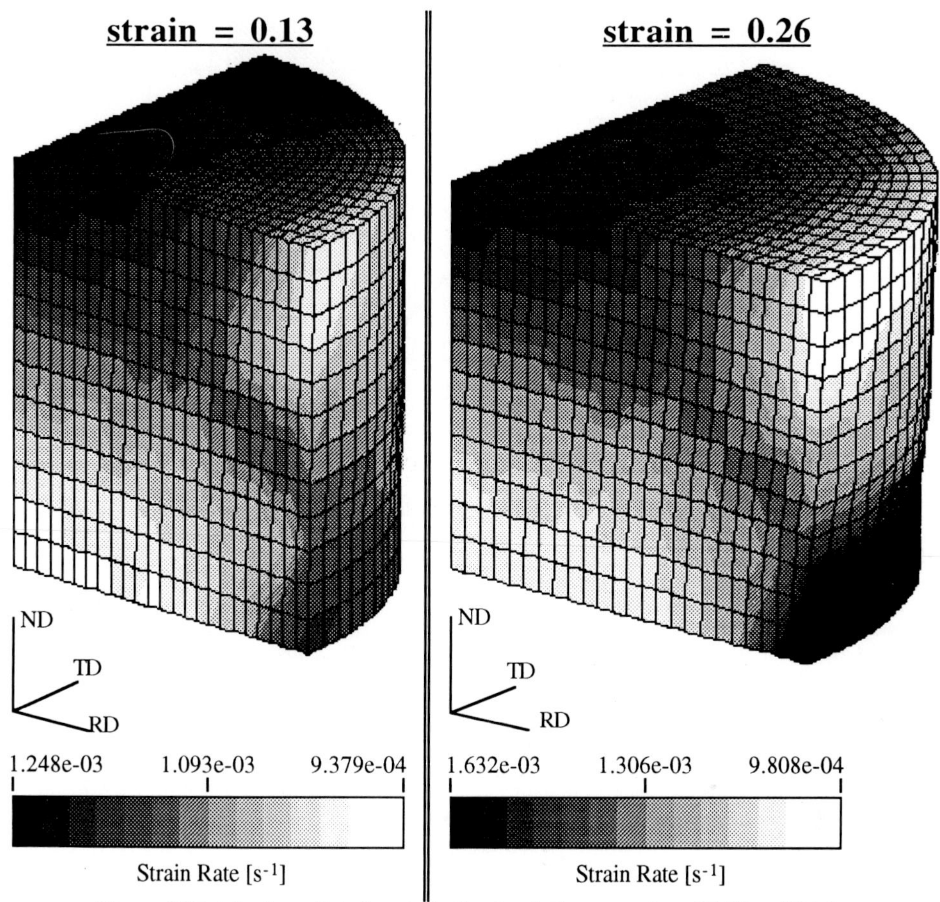

Figure 3 Distribution of strain rate in the simulation at strains of 0.13 and 0.26.

At earlier stages of the deformation, elements lying along the transverse direction at the top surface of the specimen exhibit the highest deformation rate. The maximum deformation rate is located at the extreme position along the transverse axis – this will be the position of the most extreme hourglass effect. At a strain of 0.26, the position of maximum deformation rate has moved to the specimen through-thickness mid-plane at the extreme position of the rolling direction axis. The inhomogeneity of the deformation field is a result of texture gradients along the specimen axis; the strain rate would be uniformly distributed in a sample without texture variations.

Figure 4 shows a profile of a section plane in the compression sample. The particular plane shown is normal to the rolling direction of the plate. The symmetrized-experimental curve was formed by averaging the experimental curve with its mirror image (through the vertical). The experimental curve is not symmetric because either the texture gradient is not symmetric about the midplane of the plate or the midplane of the sample is not coincident with the midplane of the plate.

In addition to the hourglass shape of the compression specimen, there was also measurable ovaling. The ovaling measured on the experimental specimen at the mid-section was 98.0%. The ovaling predicted by the simulation was 97.9%, in close agreement with the experimental result. The ovaling measured at the surface was 98.3% and predicted by the simulation was 98.9%.

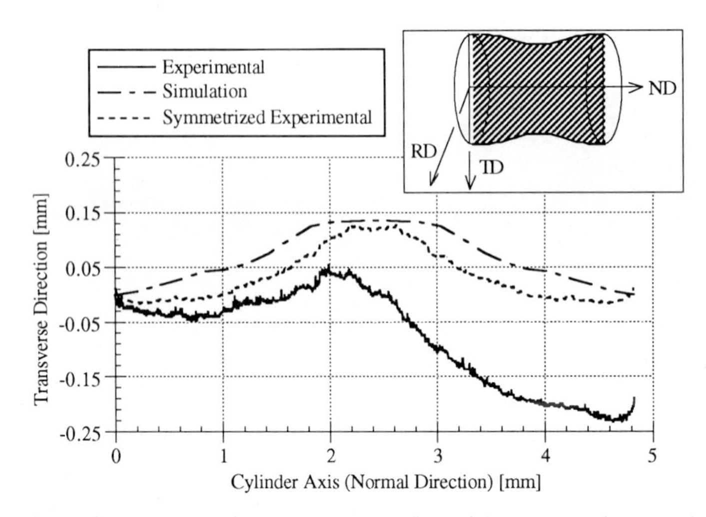

Figure 4 Experimental and predicted profiles of the compression sample.

DISCUSSION

The simulation captures the sense of the hourglass effect. However, the simulation results do not precisely match the experimental results. One cause is that the texture not only varies through the thickness of the plate but in the plane of the plate as well. Thus, each sample will differ in mechanical behavior simply due to in-plane variations in texture. Another cause for discrepancies is the description of the work hardening behavior used in the simulation. Polycrystalline plasticity models of this type generally require good single crystal work hardening data to achieve optimum results. In this case, the work hardening behavior was obtained from polycrystalline measurements and the single crystal data inferred using the polycrystal texture.

Nonetheless, the observed trends are well predicted by the simulation. The coupling of spatially specific orientation measurements with polycrystalline plasticity modeling within the finite-element regime provides an excellent way of using the orientational aspects of microstructure to quantitatively model the mechanical response of polycrystalline materials.

ACKNOWLEDGMENTS

The authors gratefully acknowledge B. L. Adams and K. Kunze for providing access to the automatic single orientation measurement facility at Brigham Young University. Helpful discussions with U. F. Kocks and A. D. Rollett of Los Alamos National Laboratory are acknowledged. A. J. Beaudoin is supported by Reynolds Metals Company. This work has been performed under the auspices of the United States Department of Energy.

REFERENCES

1) Wright, S. I., Gray III, G. T. and Rollett, A. D.: Metall. Trans. A, 1993, (submitted).
2) Kunze, K., Wright, S. I., Adams, B. L. and Dingley, D. J.: Textures Microstruct., 1993, 20, 41.
3) Adams, B. L., Wright, S. I. and Kunze, K.: Metall. Trans., 1993, 24A, 819.
4) Wright, S. I.: J. Computer Assisted Microscopy, 1993, 5, 207.
5) Beaudoin, A. J., Mathur, K. K., Dawson, P. R. and Johnson, G. C.: Intl. J. Plasticity, 1993, (in press).
6) Beaudoin, A. J., Dawson, P. R., Mathur, K. K., Kocks, U. F. and Korzekwa, D. A.: Comp. Meth. Appl. Mech. Eng., 1993, (submitted).
7) Kocks, U. F., Stout, M. G. and Rollett, A. D., in *Strength of Metals and Alloys (ICSMA 8)* p. 25, P. O. Kettunen, T. K. Lepistö, M. E. Lehtonen, Eds. Pergamon Press, New York, (1988).

5. Mathematical Modelling of Texture Formation and Materials Properties

Materials Science Forum Vol. 157-162 (1994) pp. 1703-1712

FINITE ELEMENT MODELING OF POLYCRYSTALLINE SOLIDS

P.R. Dawson [1], A.J. Beaudoin [2] and K.K. Mathur [3]

[1] Sibley School of Mechanics & Aerospace Engineering, Cornell University, Upson Hall, Ithaka, NY 14853-7501, USA

[2] Reynolds Metals Company, Richmond, VA 23261, USA

[3] Thinking Machines Corporation

Keywords: Polycrystal Plasticity, Finite Element Method, Metal Forming, Parallel Computing

ABSTRACT

The finite element method provides a powerful complement to polycrystal plasticity theory for analyzing the deformations of polycrystalline solids. Anisotropy of the flow can be computed based on the slip characteristics of individual crystals and included in finite element formulations as the constitutive description of the material. A variety of approaches exist for merging finite element formulations and polycrystal plasticity, and depending of the intended application the two may have different relationships to each other. In this review, we summarize two regimes that we refer to as large and small scale applications. We discuss some important issues associated with each and briefly outline contributions to each reported in the literature.

INTRODUCTION

The finite element method is a powerful complement to polycrystal plasticity theory for modeling the non-uniform deformation of crystalline solids. Polycrystal plasticity provides a micromechanical model for slip dominated plastic flow and serves as a constitutive theory for engineering analyses. The finite element method provides a numerical means to solve partial differential equations, such as the field equations of elasticity or plasticity. The two can be combined in different ways depending on the goals of a modeling effort.

One possibility is to embed polycrystal theory within a finite element formulation as a constitutive theory in much the same way as is currently done for continuum elastoplasticity models. For this purpose we refer to a crystal ensemble as an aggregate. An aggregate underlies a continuum point and is interrogated at appropriate times in a simulation to evaluate flow properties. In turn, hardening of crystal slip systems and crystal rotations follow from the continuum fields of stress and deformation. Polycrystal plasticity can be viewed in this context as a state variable constitutive theory in which the orientation distribution of crystal lattices, along with the crystal hardnesses, comprise the description of state.

A second combination of finite elements and polycrystal plasticity is associated with more detailed study of the crystal ensembles themselves. In this case, finite elements discretize the crystals and balance laws are applied to form an idealization of the microstructure. For this case, a crystal ensemble is called a polycrystal. Within an element the single crystal relations hold, with the finite element solution providing insight into the distribution of deformation and details of load sharing in the polycrystal.

We will call these large and small scale applications, respectively. The two have many issues in common, including the effective methods for parallel computer architectures to exploit the concurrencies inherent with such models. However, there also are issues that are either distinctly related to either large or small scales applications, or are much more important in one than the other. For small scale applications, the smoothness of properties cannot be assumed, as crystal interfaces exist and properties are not continuous across them. Further, the initialization of orientations and polycrystal topology involve their true spatial variations. For the large scale, the linking of micro and macro length scales arises through assumptions made in partitioning the deformation among crystals of an aggregate. The relationship between the dimensions of the aggregate and those of the body is a concern, but is not explicitly defined as in the small scale.

In this paper we review recent progress in these areas. We organize our paper according to the small versus large scales. Examples of each are discussed, although specific results are shown only for our own efforts. There are other combinations of polycrystal plasticity and the finite element method that we do not touch on here. One is the finite element discretization of the domain that defines the material properties (state). Some work has been done along these lines: the evolution of texture has been simulated by the integration of the orientation distribution using appropriate piecewise basis functions. Such work is only in its formative stages, however. Another area deals with composite materials, in which domains of the individual phases may be large enough to be treated accurately as isotropic continua. The focus of such efforts may be, and has been, the impact of local nonuniformity of the deformation caused by the interaction of dissimilar phases on the texture evolution. However, polycrystal plasticity is not explicitly implemented in the finite element formulation and consequently we will not review that work here.

IMPLEMENTATIONS

Like models for polycrystal plasticity, finite element formulations come in many different forms. Because of this it is not advisable, nor even feasible, to attempt to establish the best wedding of microstructural theory with numerical treatment of a boundary value problem. For example, finite element codes are available based on explicit dynamic formulations as well as implicit static methodologies. Both displacements and velocities have been chosen as the primitive kinematic variables. Some formulations neglect elasticity while others are constructed around its existence. Introduction of a polycrystal plasticity model into a finite element formulation introduces further variants; polycrystal plasticity models have been developed that incorporate an extended Taylor hypothesis while others utilize self consistent premises to allow for richer variety of grain interactions. It it not surprising that various combinations of these have been investigated, nor is it surprising that no single combination has demonstrated clearly its overall superiority.

One special concern arising with formulations that include elasticity is the effective integration of the stress. This involves the crystal kinematics, especially the decomposition of motion in term of elastic and plastic components. Both explicit and implicit procedures have been devised. Peirce, Asaro and Needleman [1] developed an explicit procedure. Implicit procedures have been reported by Kalidindi, Bronkhurst, and Anand [2] and by Maniatty, Dawson and Lee [3].

Finite element formulations that employ polycrystal plasticity model are computationally demanding. If a large scale application is pursued on a single processor computer architecture the calculations related to the crystal equations can take up to 95% of the execution time. For these models the materials related computations dominate. This is a switch in the primary computational demand in finite element model from most of the resources being consumed satisfying the balance laws to most of the resources being used for materials related computations. However, parallel computer architectures are ideal for these models as they contain a high degree of concurrency. Examples of the application of parallel computation will be given in the following sections.

LARGE-SCALE APPLICATIONS

Large-scale applications are those in which the body, herein called the workpiece, is very much larger than individual crystals. Examples include the rolling of plates, sheet forming, and forging. A variety of technical issues arise as the focus of these applications, such as unsymmetric deformations (earing), texture gradients within the workpiece (and consequential effects on mechanical properties), macroscopic shear banding, and residual stresses.

Finite element formulations require the material response as part of the integration of elemental stiffness and force matrices arising from the residual on the balance of linear momentum. Polycrystal plasticity provides a microstructurally motivated constitutive model for the flow which defines the yield condition and flow rule for a material element comprised of an aggregate of crystals. The distribution of lattice orientations, along with the hardness of the crystals, defines the material state. The balance laws are not applied to the crystal of the aggregate *per se*, but rather the aggregate is interrogated to define the stress and flow properties (as needed to evaluate the stiffness). Constitutive assumptions, such as a Taylor hypothesis or self consistent theories, are made in deciding how to project the macroscopic deformation onto the crystals of an aggregate and subsequently to average the resulting responses (stress) of those crystals to obtain a macroscopic value.

As a state variable representation for plastic flow, polycrystal plasticity theory has several distinct advantages. Implicit with the use of state variable models is the ability to initialize the state. Orientation distributions are directly accessible via x-ray measurement and well-established methodologies for interpreting those measurements. The crystal hardness distributions are not as direct, but for the restricted assumption of common hardening among all crystals of an aggregate the hardness may be initialized from simple compression testing. Polycrystal theory also provides a direct means for updating the material state via integration of the evolution equations for the crystal lattice orientation and the hardness.

In large-scale applications, the aggregate is not a body, but rather only an abstract representation of the microstructural state. Care must be exercised in assuring consistency between the macroscopic material element size and the aggregate dimensions. It is possible to think of macroscopic deformation gradients within the workpiece that are large, but yet only vary slowly across the dimension of an aggregate. As such we may consider the aggregate to be subjected to a uniform deformation locally. Locally here refers to a point on macroscopic scale, so that we permit only single (tensor) values of stress and velocity gradient. Thus the dimension of a crystal must be small compared to the dimension over which the macroscopic velocity gradient changes appreciably. In turn, the aggregate of crystals must contain a sufficient number and inherent appropriate symmetry relations such that the above arguments of homogenization are justified.

In most finite element implementations, isoparametric elements are employed in which gradients of velocity (or displacement) may exist across an element. Further, elemental integrations are performed using quadrature in which the integrand is evaluated at specific points within an element. One can consider the material in the vicinity of a quadrature point as a region on which the homogenization is performed on the crystal responses of an aggregate. Sufficient crystals must reside in this region for the homogenization to be valid, as discussed in the previous paragraph. Variations in the velocity gradient within an element imply, however, that all of the quadrature points are not experiencing the same strain rate. The material's rate sensitivity, texture, and hardening characteristics become important then because of the influence they exert on the variations in stress in the presence of differences in the strain rate over the domain of the element. For convergence of the finite element solution for the motion, the stress should approach a smooth distribution within an element and across its boundaries to neighboring elements. Smoothness of stress will not be achieved without some smoothness of the properties. This requires that the number of crystals within an aggregate be sufficiently large to represent an orientation distribution well, and that the distributions within an element not vary greatly. In fact, the use of a single aggregate within each element, placed at its centroid, provides excellent performance in finite element simulations because it restricts the possible variations of stress. For consistency then, as elements become sufficiently small that smooth stresses are achieved, they each must still remain sufficiently large to physically contain an aggregate that is representative of the complete distribution and over which a legitimate homogenization may be performed.

The rolling of plate and drawing of wire using a steady state Eulerian formulation were reported by Mathur, Dawson, and Kocks [4,5,6] using both Taylor and relaxed constraints assumptions for aggregates comprised of FCC crystals. Here the material was assumed to have an initially uniform orientation distribution and texture gradients in the products were predicted. van Bael, van Houtte, Aernoudt, Hall, Pillinger, Hartley, and Sturgess [7] have implemented a procedure for computing the anisotropic stiffness from an approximation to the average Taylor factor written using a series expansion of terms involving the strain rate. The expansion parameters may be computed from the texture and thus updated with continued deformation. They have implemented the formulation in a Lagrangian elastoplastic formulation and simulated rolling of bcc metals. Kalidindi, Bronkhurst and Anand [2] reported on the implementation of a polycrystal model based on Taylor assumption into a commercial elastoplastic displacement based code. They compared the textures computed to those measured for laboratory microforgings of copper. Chastel, Dawson, and Kocks [8,9] modeled the rolling of initially textured silicon steels with an Eulerian formulation using a model for pencil glide slip modes. Detailed comparisons of through thickness texture variations were made with experiment. Chastel, Dawson, Kristin and Wenk [10] used the same Eulerian viscoplastic formulation to study texturing of rocks in the earth's mantle. Here, because of low crystal symmetry an equilibrium based model was employed to partition the deformation among crystals of an aggregate. Comparisons were made between computed and measured seismic velocities through the mantle. Maniatty [11] extended this Eulerian viscoplastic formulation to include elasticity and examined the influence of initial texture on the residual stress distribution in the product. Smelser and Becker [12] have studied drawing of cups with a commerical code that calls the polycrystal theory through a user defined module, both using single crystal and polycrystal representations.

Recent progress in parallel computing has been exploited for large scale applications. Parallel computing hardware, together with rewriting of algorithms for this architecture, have made it possible to simulate 3-d deformations involving several thousand elements each with hundreds of crystals defining every aggregate. For example, Beaudoin, Mathur, Dawson and Johnson [13] examined the evolution of texture in compression test specimens; hydroforming was studied by Dawson, Beaudoin, Mathur, Kocks and Korzekwa [14,15] respect to the formation of ears. The hydroforming simulation demonstrated the ability to predict both the location and strength of ears in the initially textured aluminum sheet.

Limited dome height tests also have been investigated simulated with respect to localization of strain. Shown in Figure 1 is a deformed mesh from the LDH application, depicting strain contours and showing the zones of greater thinning. This simulation used a mesh comprised of 3600 brick elements, each with an aggregate of 256 crystals. The full operation was simulated on a CM-5 using

40 time steps. Given in Figure 2 are near surface strains along the long dimension of the LDH specimen. Part failure is observed experimentally near the location of greatest strain magnitude shown in this plot. These strains are clearly affected by the initial orientation of the testpiece. With the rolling direction oriented along the long dimension of the testpiece, shear strain in the rolling-normal plane is larger as compared to the case of the transverse direction oriented lengthwise. The opposite trend is demonstrated by the normal strains, though not to the same extent.

SMALL SCALE APPLICATIONS

Small scale applications are ones in which the crystal dimensions are comparable to the entire body. Finite elements typically have volumes similar to that of a single crystal, and in many cases are one-to-one. In defining the mesh, one also is prescribing the topology of the polycrystal. Gradients of the deformation are permitted over an element, and thus over individual crystals as well. In contrast to large scale applications, the balance laws are applied directly to a polycrystal via the weighted residuals of the finite element formulation. The solution to the field equations renders the partitioning of deformation among the crystals.

The single crystal anisotropy is the source both of interesting behaviors (and thus the focus of many studies) and of numerical difficulties in the solution of the model equations. For example, the nonuniformity of straining is of particular importance. The tendency of deformations to localize into shear bands as the preferred mode of deformation has received considerable attention. Also, an understanding of crystal to crystal variations in the deformation on texture development is necessary for improvements to models which employ simpler mean field assumptions such as the Taylor or Sachs hypotheses.

The discontinuity of properties across the elemental boundaries may be a source of difficulty for simulation. Conventional kinematically based finite element formulations rely on increased resolution of the discretization to reduce traction discontinuities at elemental interfaces, and thereby approach convergence. However, as property discontinuities are an inherent feature of interfaces within polycrystal, special attention to the element boundaries is necessary to assure converged solutions.

Just as with the large scale applications, the initialization of the textures is a central issue. However, this is compounded for small scale applications by the feature in polycrystals that the spatial relation of crystals must be defined. This may be done by randomly assigning orientations to elements, or with recently developed experimental capability the elements of a finite element mesh can be associated with specific orientations measured within a material sample.

Small scale simulations were reported as early as 1978 by Gotoh [16], in which 125 fcc crystals were modeled using a planer arrangement of elements. Gotoh studied the initial yielding of this assembly assuming that crystals uniformly covered orientation space and were randomly positioned in the mesh. Needleman, Asaro, Lemond and Peirce [17] implemented the rate dependent formulation in a large strain elastoplastic framework and studied the localization of thin strips comprised of crystals with planer slip modes. Harren and Asaro [18] studied the validity of the Taylor assumption using a similar formulation to Needleman. McHugh, Asaro, and Shih [19] applied this formulation to the study of metal matrix composites, showing the effect of strain localization around harder phases on the overall mechanical response of the composite. Becker [20] and Kalladindi, Bronkhurst and Anand [21] have performed similar studies, but applied full fcc deformation modes to the crystal kinematics, although the meshes were constrained to two-dimensional deformation modes (as intended to replicate plane strain compression). In both studies the polycrystal models were implemented as user defined material subroutines in a general purpose commercial finite element code. Becker [22] examined a similar problem with the same formulation in his analyses of the strain distributions in the deformation zone of a thin sheet subjected to bending. Slip system activity has been investigated using finite elements by Havlicek, Tokuda, Hino, and Kratochvil [23] and by Yao and Wagoner [24].

Beaudoin, Mathur, Dawson and Kocks [25] have proposed an alternative to conventional kinematically based finite element formulation utilizing hybrid methodologies. In this work the equilibrium condition is enforced through a weighted residual formed on the tractions across element interfaces. Trial functions are used both for the velocity and the stress (hydrostatic and deviatoric), with the stress interpolation satisfying equilibrium within elements *a priori*. The crystal constitutive equations also enter the formulation in weak form, that is, as a weighted residual over each elemental volume. The stress trial functions are not required to be continuous across element boundaries, to that the equations for the constitutive response may be inverted element by element to give the stress in terms of the motion. The stress, in turn, is used to write the weak form of equilibrium entirely in terms of the velocity, which is required to exhibit interelement continuity. The formulations is well-suited for a massively parallel computer architecture, as the crystal computations through to the solution of the stresses has a high degree of concurrency.

The hybrid formulation has been applied to the study of texture evolution of fcc polycrystals subjected to plane strain compression. Each element is assigned a randomly chosen orientation from a uniform distribution. The polycrystal constructed consists of a 10 × 10 × 10 array of 3-D brick elements centered in a 16 × 16 × 16 element mesh (Figure 3). The three layers of elements represent emulate a surrounding medium isolating the polycrystal from the external boundary conditions. The applied boundary conditions were chosen to approximate a channel die compression. Shear deformations in the inner polycrystal are not explicitly prescribed; each may vary freely to be determined by the governing field equations. This will in turn impact upon the subsequent development of crystallographic texture.

It is well known that the realistic development of certain rolling texture components in FCC metals requires shear deformations in crystals. The development of the brass, copper, and S texture components result from shears induced by the varying constitutive response that exists from grain-to-grain. The development of the brass, copper, and S texture depends on the presence of shears in the transverse-rolling, normal-rolling, and normal-transverse planes, respectively [26,27,28,29]. Shown in Figure 4 are pole figures for plane strain compression developed using the finite element polycrystal model and the relaxed constraints model. The relaxed constraints idealization relaxes the shears in the normal-rolling and normal-transverse planes leading to development of copper and S shears respectively. Development of the brass texture, not predicted by the relaxed constraints model, is clearly evident in the finite element polycrystal result. Incorporation of additional kinematic degrees of freedom through inclusion of the out-of-plane dimension in the 3-D model leads to a development of the brass texture.

SUMMARY

Finite element formulations that utilize polycrystal plasticity as the constitutive description have progressed significantly over the past decade. Gains has been reported both in the use of polycrystal descriptions to better understand fundamental aspects of the deformation behavior of crystalline solids and in the investigation of commerical forming operations for these materials. While considerable progress has been made, a variety of issues remain open as to the best way to merge the two. The relative size of crystals and elements plays an important role in this respect. In all cases, the computational demands are heavy, but the opportunity for concurrency is great. Formulations can exploit this concurrency with implementations designed for massively parallel computer architectures.

ACKNOWLEDGEMENTS

This work was supported through the Office of Naval Research under contract NOOO14-90-J-1810. Access to computing resources was provided by the Advanced Computing Laboratory at Los Alamos Laboratories and the National Center for Supercomputing Applications at the University of Illinois.

REFERENCES

1. Peirce, D., R.J. Asaro, and A. Needleman, Acta Met., 1983, **31**, 1951.
2. Kallidindi, S.R., C.A. Bronkhurst, and L. Anand, J. Mech. Phys. Solids, 1992, **40**, 537.
3. Maniatty, A.M., P.R. Dawson, and Y. S. Lee., Int.J.Num.Meth.Engr., 1992, **35**, 1565.
4. Mathur, K.K. and P.R. Dawson, Int. J. Plast., 1989, **5**, 67.
5. Mathur, K.K. and P.R. Dawson, J.Engr.Mat.Tech., 1990, **112**, 292.
6. Mathur, K.K., P.R. Dawson and U.F. Kocks, Mech.Mat., 1990, **10**, 183.
7. van Bael, A., P. van Houtte, E Aernoudt, F.R. Hall, I. Pillinger, P. Hartley, and C.E.N. Sturgess, Textures and Microstructures, 1991, **14**, 1007.
8. Chastel, Y.B. and P. R. Dawson, Modeling Def. Cry. Solids (TMS), 1991, 225.
9. Chastel,Y.B., P.R. Dawson and G.B. Sarma, Proc. Int. Conf. Rolling Metals, London, Sept. 1993, in press.
10. Chastel, Y.B. P.R. Dawson, K. Kristin and H.-R. Wenk, J.Geophy.Res., in press.
11. Maniatty, A.M., Ph.D. Dissertation, Cornell Univ., 1991
12. Smelser R.E. and R. Becker, Proc. ABACUS Usr's Group, Oxford, Sept11-13, 1991, 457.
13. Beaudoin, A.J., K.K. Mathur, P.R. Dawson, and G.C. Johnson, Int.J.Plast., in press.
14. Dawson, P.R., A.J. Beaudoin, and K.K. Mathur, Proc. NUMIFORM '92, Sept, Valbonne, 1992, 25.
15. Dawson, P.R., A.J. Beaudoin, K.K. Mathur, U.F. Kocks and D.A. Korzekwa, Num.Meth.Sim.Ind.MetalFormingProc., 1992, ASME-AMD **156**, 1.
16. Gototh,M., Int.J.Num.Meth.Engr., 1978, **12**, 101.
17. Needleman, A. R.J. Asaro, J. Lemond and D. Peirce, Comp.Meth.AppliedMech.Engr., 1985, **52**, 689.
18. Harren, S.V. and Asaro, R.J., J. Mech. Phys. Solids, 1989,**37**,191.
19. McHugh, P.E., R.J. Asaro and C.F. Shih, 1993, Acta Metall.Mater.,1993, **41(5)**, 1465.
20. Becker, R., Acta Metall. Mater., 1991, **39**, 1211.
21. Kallidindi, S.R. and L. Anand, Int. J. Mech. Sci., 1992, **34**,309.
22. Becker, R., Modeling Def.Crys.Solids (TMS), 1991, 249..
23. Havlicek, F., M. Tokuda, S. Hino, and J. Kratochvil, Int.J.Plast., 1992, **8**, 477.
24. Yao, Z. and R.H. Wagoner, Acta Metall.Mater., 1993, **41(2)**, 51.
25. Beaudoin, A. J., K.K. Mathur, P.R. Dawson and U.F. Kocks, Int.J.Plasticity, submitted.
26. Fortunier, R. and J.H. Driver, Acta Metall., 1987, **35**, 1355.
27. Leffers, T., R.J. Asaro, J.H. Driver, U.F. Kocks, H. Mecking, C.N. Tome, and P. van Houtte, Proc. 8th ICOTOM (TMS), 1988, 265.
28. Hirsch, J. and K. Lucke, Acta Metall., 1988, **36**, 2883.
29. Becker, R., Textures and Microstructures, 1991, **14-18**, 145.

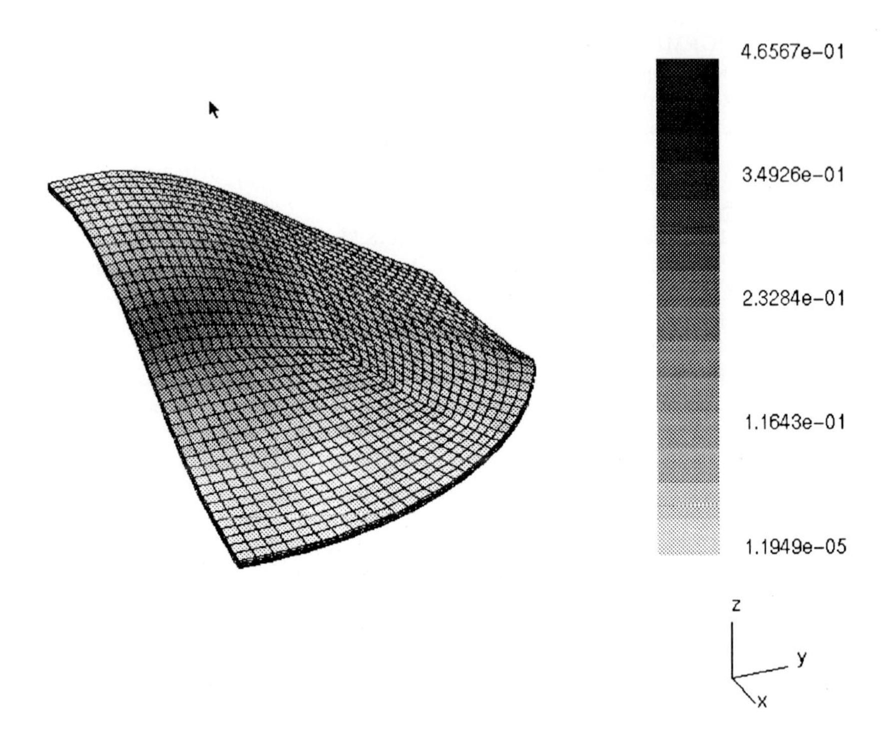

Figure 1: Deformed mesh of the limiting dome height specimen. Shading shows strain contours corresponding to the deformed mesh.

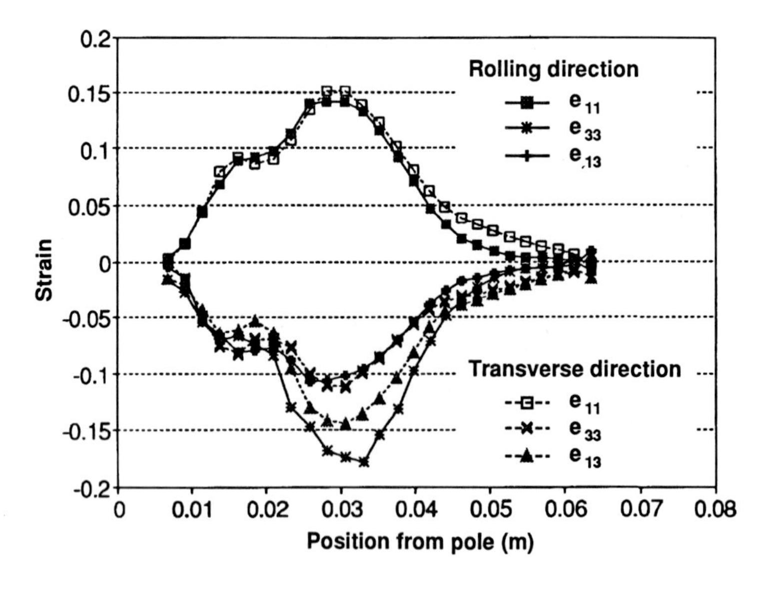

Figure 2: Strain profiles along the strip centerline after deformation.

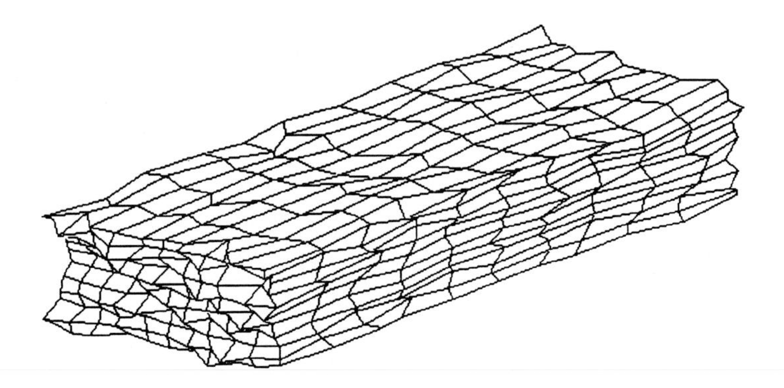

Figure 3: Polycrystal finite element mesh after compression deformation.

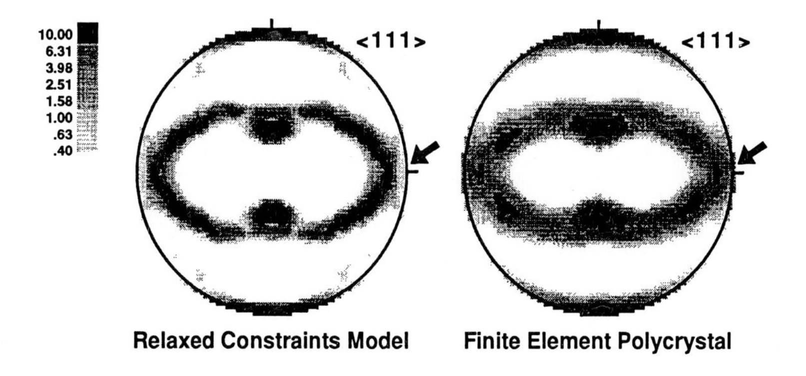

Figure 4: $< 111 >$ pole figure corresponding to the finite element polycrystal and a relaxed constraints model prediction.

Materials Science Forum Vol. 157-162 (1994) pp. 1713-1730

MODELLING THE EFFECTS OF STATIC AND DYNAMIC RECRYSTALLIZATION ON TEXTURE DEVELOPMENT

J.J. Jonas [1], L.S. Tóth [2] and T. Urabe [3]

[1] Department of Metallurgical Engineering, McGill University, 3450 University Street, Montreal, Quebec H3A 2A7, Canada

[2] Laboratoire de Physique et Mécanique des Matériaux ISGMP, Université de Metz, F-57045 Metz, France

[3] Materials and Processing Research Center, NKK Corporation, 1-1 Minamiwatarida-cho, Kawasaki-ku, Kawasaki, 210, Japan

Keywords: Dynamic Recrystallization, Static Recrystallization, Texture Modelling, Variant Selection, Recrystallization Textures, Selective Growth

ABSTRACT

Samples of OFHC Cu were twisted in the temperature range 200 to 300°C leading to the initiation of dynamic recrystallization. Analysis of the ODF's determined on samples strained up to γ =11 indicates that both oriented nucleation and selective growth take place. The former involves growth of the minimum Taylor factor ($A/\bar{A}$) component, the latter that of nuclei characterized by ± 40° rotations about particular <111> axes. Of the four possible <111> axes in each deformed fcc grain, the one perpendicular to the most highly stressed slip system is the one that operates. Cold rolled IF steels were annealed at 720°C and the ODF's corresponding to successive stages of static recrystallization were determined. Nuclei characterized by ± 22° rotations around particular <110> axes grew rapidly. Of the six possible <110> axes in each deformed bcc grain, those which are normal to the most active slip systems play the largest roles. This type of variant selection suggests that accelerated impurity diffusion along the dislocation sheets that are perpendicular to high angle tilt boundaries are responsible for their high mobility.

INTRODUCTION

Modelling the orientation changes produced by recrystallization is considerably more difficult than predicting the lattice rotations attributable to dislocation glide because of the complex effects of composition, grain size and precipitate morphology. Nevertheless, there is much to be gained by advancing the understanding of these mechanisms because of the commercial importance of annealing operations. The availability of reliable recrystallization models for fcc and bcc materials now makes it possible to design overall models for texture development during all the stages of steel processing. These must incorporate a total of *five* sub-models, those describing the effects of: i) austenite rolling, ii) austenite recrystallization, iii) the γ-to-α transformation, iv) ferrite rolling, and ferrite recrystallization. While texture models that predict the effects of rolling and transformation are now common, there has been less progress with regard to the modelling of recrystallization. It is therefore the purpose of the present review to describe some models that have been developed recently which permit the grain rotations taking place during austenite (fcc) and ferrite (bcc) recrystallization to be simulated and predicted.

ORIENTED NUCLEATION

Although oriented nucleation and selective growth are sometimes seen as competing processes, they can in fact both operate concurrently and are more usefully viewed as complementary mechanisms. In the present context, oriented nucleation can be considered to involve particular orientations that recover more easily than others, and are thus able to get a head start on the growth process. The recovery events of importance in this case are the annihilation of dislocations within the potential nucleus and the growth of this nucleus to a significantly larger size than that of the potential nuclei of other orientations. For the simulations described below, the view was taken that regions with the lowest possible Taylor factor M have this property because, for a given von Mises equivalent strain increment $\Delta\varepsilon$, the lowest M grains have experienced the least total microscopic slip [1]:

$$\Sigma \, \Delta \, \Gamma = M \Delta\varepsilon \qquad (1)$$

They are therefore likely to have undergone the least work hardening or increase in dislocation density, and so to be the most highly recovered at any particular strain.

Taylor factor maps for the rolling of fcc and bcc metals are presented in Fig. 1. These were calculated by the full constraint method, but the results differ only in detail when other glide models, such as those of the relaxed constraint method are employed. In the fcc case, it can be seen that the lowest Taylor factor (2.39) is observed at the {001}<110> and {110}<001> orientations. In the bcc case, the lowest Taylor factor (2.12) is again associated with the rotated cube {001}<110> and Goss {110}<001> orientations. In steels, the Goss orientation is often nucleated at deformation bands, so that oriented nucleation occurs *intragranularly* in this case [2]. For the torsion of fcc materials, as shown in more detail in ref. [1], the lowest possible value of M (1.73) corresponds to the {111}<110> or A/$\bar{A}$ component and M = 2.0, 2.45, and 3.0 for the {111}<112> (A_1^*/A_2^*), {112}<110> (B/$\bar{B}$), and {001}<110> (C) components, respectively. In regions removed from the deformation fibre, M can attain values as high as $\sqrt{18} = 4.24$ for the 'cube' component.

MODELLING SELECTIVE GROWTH

Selective growth during annealing involves the rapid increase in size of nuclei characterized by particular misorientations with respect to the matrix. In fcc metals, the orientations of these nuclei generally differ by ± 40° rotations around appropriate <111> axes [3]. Other rotation angles (e.g. ± 23°) and rotation axes (e.g. <100>) are also possible. In bcc metals, the nuclei generally differ by ± 27° rotations around the <110> axes.

VARIANT SELECTION

As shown in Fig. 2, there are four geometrically equivalent <111> axes in fcc materials and six equivalent <110> axes in bcc materials. Given that positive and negative rotations are equally likely about each of these axes, there are eight possible recrystallization variants in the former metals and twelve in the latter. It can be readily shown that when all of these variants are equally possible, the intensities in even the sharpest deformation textures are significantly reduced by recrystallization, leading to annealing textures which are nearly random. By contrast, the textures of recrystallized samples are often more intense than those of the deformed specimens from which they are derived. It is thus clear that some sort of variant selection is taking place during selective growth.

In the experiments described below, it was apparent that one particular *pair* of variants, involving both positive and negative rotations about a *single* axis, always played the major role during the reorientations associated with selective growth. This generalization applies to both the dynamic recrystallization of an fcc metal (Cu) as well as to the static recrystallization of a bcc metal (Fe). In each case, the reorientation axis was the one that was perpendicular to the most active slip plane (or to the active slip plane most nearly parallel to the macroscopic plane subjected to the highest critical resolved shear stress). In some of the calculations, one or two additional axes were employed, that is, those that were perpendicular to the second most active, and even the third most active planes. In these cases, the probability of the occurrence of the "secondary" and "tertiary" variants was taken to be substantially less than those of the "primary" variants.

The occurrence of such variant selection naturally leads to the question of the physical basis for this phenomenon. The present type of nucleus is surrounded by both tilt as well as twist (and mixed) boundaries. However, the view is taken here that the tilt boundaries are considerably more mobile than the twist boundaries [4,5] because of the relative rapidity of g.b. diffusion within these interfaces. They also contain lower concentrations of impurities when the rotation angle is close to a coincidence relationship. The high boundary mobility of the successful variants is associated in turn with the ease of impurity diffusion *perpendicular to* these boundaries and within the deformed microstructure. The latter arises because of the ease of pipe diffusion along the active slip planes, which is able to reduce the extent of solute drag in this way. These features of the present model are depicted schematically in Fig. 3.

It should be noted that the "active slip planes" in the figure are those which are closest to the planes of maximum shear stress and along which there is appreciable slip activity. An alternative interpretation of the figure involves the presence of dense "dislocation sheets" [6,7], "microbands," and dense dislocation walls (DDW's) [8] approximately parallel to the slip planes or to the planes of maximum shear stress. Thus the rotation axis of interest is also approximately perpendicular to the dislocation sheets, so that the latter could be the physical feature that is actually responsible for the ease of impurity pipe diffusion, and therefore for the high mobility of the boundaries against which they abut.

According to the above model, nuclei which have the appropriate tilt relationships with respect to the deformed matrix but which are not bounded by tilt boundaries that are intersected by dislocation sheets or bands are unable to grow. In other words, it is the absence of dislocation sheets perpendicular to the boundaries of the inactive nuclei that is responsible for their low rates of growth.

EXPERIMENTAL OBSERVATIONS OF DYNAMIC RECRYSTALLIZATION DURING THE HOT TORSION OF COPPER

In the experiments described below, copper bars of 99.95% purity were twisted to various strains on the free end testing machine at North Carolina State University and then rapidly quenched. One of the points of interest in these tests concerned the length changes that take place when cubic metals are deformed in torsion. At ambient temperatures, they generally lengthen, although lead is an exception, in that it shortens after a brief initial lengthening period [9]. However, room temperature is equivalent to a homologous temperature of 0.5 for Pb, and other cubic metals, such as Cu, also shorten when deformed at homologous temperatures in excess of 0.3 or 0.4. The lengthening behaviour of these metals is readily explained in terms of the conventional glide models of crystal plasticity [10]. An explanation of the shortening behaviour has been more elusive, although it has recently been ascribed to the occurrence of dynamic recrystallization [11].

Prior to testing, the samples were annealed at 550°C in vacuo for 30 min., which led to a mean grain size of about 30μm. The torsion tests were performed at room temperature, as well as at 125, 200 and 300°C. The textures produced were determined by measuring pole figures in the Siemens texture goniometer at McGill University. ODF's were then calculated by means of the spherical harmonic method.

The shear textures present at large strains ($\gamma = 11$) are presented in Fig. 4 in the form of (111) pole figures. To conserve space, only the room temperature and 300°C results are reproduced here, and the reader is referred to Ref. [11] for the 125°C and 200°C pole figures. The room temperature texture [Fig. 4a] shows a strong B fibre and the A^*_1 fibre is somewhat stronger than the A^*_2 fibre, although the situation is opposite at low strains [11]. Even though the texture at 300°C [Fig. 4b] resembles those obtained at lower temperatures, distinct differences can be observed. In particular, there is a sharp decrease in the intensity of the C component (central part of the B fibre); this is accompanied by a decrease in the strength of the A^*_2 component, and a considerable intensification of the $A/\bar{A}$ component (see below).

The ODF for twisting at room temperature to a shear strain of $\gamma = 11$ is presented in Fig. 5a. Here it is evident that the sharpest orientations are the A^*_1, $A/\bar{A}$, $B/\bar{B}$ and C. The only ideal orientation that is relatively weak is the A^*_2, but this was strong at lower strains [11], and was gradually rotated into the C and A^*_1 orientations, as called for by the operation of the glide model. The ODF for twisting to $\gamma = 11$ at 300°C is displayed in Fig. 5b. Here it can be seen that the fibre texture described above and attributable to deformation glide is still evident. However, the C component has all but disappeared, and several new components have appeared which are not part of the deformation fibre. These differences indicate that some additional process (not only plastic slip) is contributing to the orientation changes which lead to the disappearance of the C component. This new mechanism is dynamic recrystallization, which is unable to operate when straining is carried out much below $0.4T_m$.

MODELLING THE ORIENTATION CHANGES PRODUCED BY DYNAMIC RECRYSTALLIZATION

In order to model oriented nucleation, it was assumed that certain subgrains located within the lowest Taylor factor or $A/\bar{A}$ grains become supercritical and begin to grow [12]. Their rates of growth can be described in terms of the volume fraction increase occurring within the material during each period of simulated recrystallization; e.g. 120% (increase in volume fraction)/25%(shear strain). The spread about the exact $A/\bar{A}$ orientation over which such growth can proceed must also be specified. In the present case, a gaussian of 15° was used, with the growth rate falling off exponentially with angular distance from the ideal orientation.

Selective growth was modeled by assuming that ± 40° rotations took place around the <111> axis which is perpendicular to the most highly stressed slip system [1,12]. The computer model was based on three subcomponents, one each for glide, oriented nucleation, and selective growth. The glide model operates on its own until the critical strain for the initiation of DRX is reached ($\gamma = 1.5$ in the present case). From this point on, intervals of shear deformation ($\gamma = 0.25$) are alternated with the corresponding elapsed time intervals ($\Delta t = 0.25/\dot{\gamma}$), one being allocated to each of the two recrystallization mechanisms in turn. The kinetics of recrystallization processes are then adjusted to correspond to realistic "recrystallization strains".

The deformation component was simulated using the rate sensitive glide model of Tóth et al. [10] and a rate sensitivity $m = 0.05$. The incremental shear strain of $\gamma = 0.25$ was applied in five discrete strain steps of $\Delta\gamma = 0.05$. Various kinetics were employed to represent the oriented nucleation step; for the results presented below, the volume fraction of near $A/\bar{A}$ ($M = 1.73$) orientations was permitted approximately to double during each time interval corresponding to $\gamma = 0.25$. The possibility of oriented nucleation was largely limited to grains within a gaussian spread of misorientations of 15°, which corresponds to an increase in M up to approximately 2.6 [1]. Over this interval, the permissible increase in volume fraction was progressively reduced from 120% per 0.25 shear strain to 37% in an exponential fashion.

An M-dependence was also introduced into the selective growth model, where the highest growth rates correspond to grain boundary migration into the regions of highest dislocation density (i.e. of highest $\Sigma\ \Delta\Gamma = M\Delta\varepsilon$). Various rates of recrystallization were used; the results presented below were obtained with a maximum (highest M) recrystallization rate of 20% during the time interval corresponding to the shear strain increment of $\gamma = 0.25$. Grains of lower M were consumed at lower rates, which varied linearly according to the difference between the highest ($M=\sqrt{18}$) and lowest ($M=\sqrt{3}$) possible values of M. For $M=\sqrt{3}$, the rate was 0.4 times that for $M=\sqrt{18}$.

The calculations described here were carried out on a population of 800 individual grains, whose initial orientations were calculated from the initial texture by means of a suitable discretization method. This population was changed and updated every $\Delta\gamma = 0.25$, as required by the glide simulation. Once the two recrystallization models began to operate, each of these contributed further changes to the distribution. In the selective growth model, the total number of grains increased because the partial replacement of an orientation being consumed led to the introduction of two new orientations, that is to those rotated by + 40° and by - 40° about the <111> axis in question. For this reason, the ODF was recalculated and a rediscretization was carried out after each combination of five glide steps + oriented nucleation + selective growth, and the three sets of simulations were always performed on 800 grains.

Finally, it should be pointed out that the oriented nucleation process, when modeled as described above, leads to the reduction in volume fraction of all the non-favourably oriented grains; i.e. it consumes the matrix indiscriminately. Conversely, the selective growth mechanism by its nature only removes from the distribution those grains into which the nucleus is 'growing', as defined by the 40° <111> rotation principle.

COMPARISON OF PREDICTIONS AND EXPERIMENTAL OBSERVATIONS

Comparison of the experimental (111) pole figures corresponding to samples twisted to various strains at 300°C and the simulated pole figures showed that the model reproduces all the essential features of the experimental diagrams quite faithfully. The experimental and simulated ODF's are also compared in Fig. 6. Here, the close correspondence between observations and calculations is particularly striking. These similarities support the validity of the model described above. This is because simulated ODF's are highly sensitive to the parameters selected to represent a given mechanism, so that distinctly different ODF's are obtained when the model assumptions are changed significantly.

The grain rotations produced by dynamic recrystallization involve ± 40° <111> rotations away from high M factor orientations, as described above. This is accompanied by the intensification of the $A/\bar{A}$ orientations. These reorientations extract grains which are more heavily concentrated in the lengthening regions of Euler space and deposit them more frequently in shortening than in lengthening locations. It is the rigid body and glide spins that place the grains predominantly in the lengthening regions; conversely, it is the grain rotations attributable to dynamic recrystallization that move these grains out of the lengthening and into the shortening regions.

Similar considerations and explanations apply to the *inverse* Swift effect, in which previously twisted wires undergo clockwise or counterclockwise twisting when stretched under "free-end" tensile testing conditions [13]. Copper samples previously twisted at room temperature undergo continued forward twisting under free extension; conversely, samples twisted at elevated temperatures undergo "untwisting" during free end extension. Such behaviour is readily understood in terms of the grain rotations produced by dynamic recrystallization; these move individual grains from "twisting" to "untwisting" regions of the twist map and in this way change the inclination of the yield locus normal at the tensile axis from positive to negative shear.

The relative importance of oriented nucleation and selective growth in producing the results presented above was estimated by summing the new volume fractions introduced by each mechanism when these were permitted to operate at the selected shear strain intervals of 0.25 [1]. The kinetic parameters specified above led to a ratio of about 2:1 over the whole strain interval, signifying that oriented nucleation is the more important mechanism in copper twisted at 300°C. In terms of the "necklacing" phenomenon of DRX [14], oriented nucleation corresponds to the growth into their neighbors of near $A/\bar{A}$ oriented grains, whereas selective growth corresponds to the volume increase of grains with the required 40° <111> misorientation. Experiments are now under way employing the recently developed microdiffraction and EBSP techniques to investigate these two mechanisms in more detail, and to elucidate the effects of strain rate and temperature on the early stages of these two fundamental processes.

TEXTURE CHANGE DURING THE STATIC RECRYSTALLIZATION OF A COLD ROLLED IF STEEL

The material used in this study [15] was a commercially processed hot band of a Ti-Nb IF steel with the following chemical composition, all in wt pct: 0.0018%C, 0.02%Si, 0.14%Mn, 0.004%P, 0.003%S, 0.042%sol.Al, 0.0023%N, 0.079%Ti, 0.010%Nb. Samples of this material were subjected to cold rolling reductions of 70% and 85% and then annealed at 720°C in a salt bath for times ranging from 100 to 700 s. Texture measurements were carried out on mid-thickness specimens on a Siemens D-500 texture goniometer using a molybdenum tube. ODF's were calculated from three incomplete pole figures, {110}, {200} and {112}, which had been measured up to an azimuth of 85°.

Ghost corrected φ_2 = 45° sections of the ODF's of the 70% and 85% cold-rolled and annealed specimens are presented in Fig. 7. In the deformation textures, higher cold reductions strengthen both the RD//<110> and ND//<111> fibers. In the recrystallization textures, the RD//<110> fiber is weakened and the ND//<111> fiber is appreciably strengthened when the cold reduction is increased.

The present selective growth model involves the rapid increase in size of nuclei characterized by particular misorientations with respect to the deformed matrix. When recrystallization takes place, each discretized v_{Di} in the experimental deformation texture is converted into new orientations with specific rotation relationships around selected <110> axes. The volume fraction $v_{R_{ij}}$ of each these recrystallization variants can be expressed in terms of the volume fraction v_D of the orientation in the deformation texture that is being replaced using the relation:

$$v_{R_{ij}} = p_{ij} \cdot v_D \quad (\mathrm{i} = 1,.., 6, \mathrm{j} = 1, 2) \tag{2}$$

Here p_{ij} is the probability factor for growth of the component rotated about the i-th <110> axis and in the j-th rotation direction; the sum of the p_{ij}'s is equal to one: $\sum_{i=1}^{6}\sum_{j=1}^{2} p_{ij} = 1$. After recrystallization according to this discrete model, the C-coefficients for the simulated recrystallization texture are calculated from the orientations of the new grains and from their volume fractions, v_{Rij}.

Instead of employing a variant selection model, the probability factors p_{ij} can be assumed to be equal for all the <110> axes and for both positive and negative rotations ($p_{ij} = p_0$). Using the experimental C-coefficients of the 85% cold-rolled material, the recrystallization texture was simulated by means of ± 27° rotations, leading to the results presented in Fig. 8a. It can be seen that, in the absence of variant selection, the recrystallization texture is very nearly randomized. Thus, if a selective growth model is to be employed, specific <110> axes and rotation directions must be chosen so as to increase the ODF intensities along the ND//<111> fiber. In the present model, the probability factors p_{ij} were considered to be related to the slip rates in the six {110} slip planes which are perpendicular to the six possible <110> rotation axes. In order to calculate these probabilities, a crystal plasticity analysis was first carried out to derive the slip rates in the different slip systems.

SLIP SYSTEM ACTIVITY

The stress state for rolling was assumed to be parallel in all the crystals (Sachs approach). The resolved shear stress in slip system s can then be obtained from:

$$\tau^s = \left((\sigma)\,\mathbf{n}^s\right)\mathbf{b}^s = \sigma\left(b_1^s n_1^s - b_3^s n_3^s\right) \tag{3}$$

where $\mathbf{n}^s$ and $\mathbf{b}^s$ are the slip plane normals and slip directions, respectively. The quantities $b_1^s n_1^s$ and $b_3^s n_3^s$ in eq. (3) represent the orientations of the slip systems of interest with respect to the two perpendicular planes of maximum shear stress in the sample. This factor $\left(b_1^s n_1^s - b_3^s n_3^s\right)$ is therefore a geometric one and is essentially the Schmid factor for rolling: $m^s = b_1^s n_1^s - b_3^s n_3^s$. Equation (3) can then be rewritten as:

$$\tau^s = \sigma m^s \tag{4}$$

The rate sensitive power law is used here to relate the resolved shear stress and the slip rate in a given slip system:

$$\dot{\gamma}^s = \dot{\gamma}_0 \left(\frac{\tau^s}{\tau_0}\right)^h \tag{5}$$

In this law, h is the inverse of the rate sensitivity, τ_0 is a reference shear stress, and $\dot{\gamma}_0$ is a reference shear rate. These are material constants and they are assumed to be equal in all the crystals. The combination of eq. (4) and eq. (5) leads to:

$$\dot{\gamma}^s = \sigma^h \dot{\gamma}_0 \tau_0^{1/h} \left(m^s\right)^h \tag{6}$$

By introducing $C_\sigma = \sigma^h \dot{\gamma}_0 \tau_0^{1/h}$, the slip rates can be expressed in more concise form as

$$\dot{\gamma}_s = C_\sigma \left(m^s\right)^h \tag{7}$$

VARIANT SELECTION

In the present variant selection model, the probability of occurrence of each variant is taken to be proportional to the relative activity on the slip plane which is normal to the rotation axis of interest. Denoting the shear rate associated with a given {110} plane by $\left|\dot{\gamma}_i^1\right| + \left|\dot{\gamma}_i^2\right|$ (where *1* and *2* are the slip directions and *i* is the index of the plane, $i = 1,..., 6$), the probability factor associated with each <110> rotation axis, F_i can be written as:

$$F_i = \frac{\left|\dot{\gamma}_i^1\right| + \left|\dot{\gamma}_i^2\right|}{\sum_{k=1}^{6}\left(\left|\dot{\gamma}_k^1\right| + \left|\dot{\gamma}_k^2\right|\right)} \tag{8}$$

When positive and negative rotations are assumed to be equally likely and the probability factors for the twelve variants, p_{ij} are given solely by F_i, the recrystallization texture produced by ± 27° rotations is shown in Fig. 8b On the φ_2 = 45° section, strong intensities are observed along the ND//<111> fiber (Φ = 55°), with the peak intensity located at the {111}<112> component. Although the tendencies of the ND//<111> and RD//<110> fibers are similar to those displayed in the experimental recrystallization textures, Fig. 7, the calculated ND//<111> fiber intensities are much weaker than the experimental ones.

THE AVAILABILITY OF RECRYSTALLIZATION NUCLEI

In the present recrystallization model, there is a further factor that governs the progress of recrystallization, and that is the availability in the deformed material of nuclei of the appropriate new orientation, N_{ij}. The latter factor is simply the ODF intensity of the deformation texture at the rotated position of interest, g_{ij}, normalized by the sum of the intensities of the twelve possible rotated positions:

$$\mathbf{N}_{ij} = \frac{\mathbf{f}\,(\mathbf{g}_{ij})}{\sum_{k=1}^{6}\sum_{l=1}^{2}\mathbf{f}\,(\mathbf{g}_{kl})} \tag{9}$$

When the probability factor p_{ij} is simply assumed to be equal to N_{ij} without any variant selection, the calculated recrystallization texture depends strongly on the starting texture, as shown in Fig. 8c. This type of model leads to RD//<110> fibers that are stronger than the ND//<111> fibers, a result which is inconsistent with the experimental observations.

PROBABILITY FACTOR

Because of the inadequacies of models based solely on variant selection or solely on the availability of nuclei, as described above, a combined model was adopted in the present analysis. In this case, the probability factor that governs the likelihood of finding the i-th variant rotated in the j-th direction, p_{ij}, is given by the normalized product of F_i and N_{ij}:

$$p_{ij} = \frac{F_i \cdot N_{ij}}{\sum_{k=1}^{6}\sum_{l=1}^{2} F_k \cdot N_{kl}}$$

(10)

EFFECT OF ROTATION ANGLE

In the detailed account of this work [15], results are presented for a range of rotation angles, beginning with ± 15° and continuing on up to ± 35°. It is of interest that the Σ (19a) CSL boundary relationship of ± 27° did not give the best fit to the experimental data. Instead, the most satisfactory overall results were obtained with ± 22°, which led to the ODF, see Fig. 9, which most closely resembled the recrystallization texture. This $\phi_2 = 45°$ section was calculated from the discretized texture of the 85% cold rolled material. It is evident from this diagram that when the "most active slip system" law is combined with a factor representing the volume density of available nuclei, the main features of the experimental ODF's are satisfactorily reproduced.

CONCLUSIONS

1. When twisted Cu undergoes dynamic recrystallization, both oriented nucleation and selective growth appear to occur. Oriented nucleation involves the growth of the minimum Taylor factor ($A/\bar{A}$) shear component; this mechanism plays a larger role than selective growth.

2. Selective growth involves the rapid increase in size of nuclei characterized by ± 40° rotations around particular <111> axes. Of the four possible <111> axes available in each deformed grain, the one perpendicular to the most highly stressed slip system is the one that operates.

3. When fine grained, cold rolled IF steels are annealed, nuclei characterized by ± 22° rotations around particular <110> axes appear to grow rapidly. Of the six possible <110> axes in each deformed grain, those which are normal to the most active slip systems play the largest roles.

4. In addition to such variant selection, the bcc recrystallization model includes a probability factor proportional to the availability of potential nuclei. This leads to different overall probabilities for positive and negative rotation about a particular axis, and to predictions which are in reasonable agreement with the experimental textures.

5. The existence of the variant selection law suggests that accelerated impurity diffusion along the dislocation sheets which are perpendicular to moving tilt boundaries are responsible for the high mobility of these boundaries.

ACKNOWLEDGEMENTS

The authors are grateful to the Natural Sciences and Engineering Research Council of Canada, the Canadian Steel Industry Research Association, and the NKK Corporation for financial support.

REFERENCES

1. Jonas, J.J. and Tóth, L.S.: *Scripta Metall. et Mater.*, 1992, 27, 1575.
2. Ibe, G. and Lücke, K.: *Arch. Eisenhuttenwesen*, 1968, 29, 693.
3. Beck, P.A.: *Acta Metall.*, 1953, 1, 230.
4. Aust, K.T. and Rutter, J.W.: in "Recovery and Recrystallization of Metals", ed. L. Himmel, J. Wiley and Sons, N.Y., 1963, p. 131.

5. Aust, K.T.: in Conf. on "Interfaces", Melbourne, Australia, Australasian Institute of Mining and Metallurgy, 1969.
6. Fernandes, J.V. and Schmitt, J.H.: *Phil. Mag. A*, 1983, 48, 841.
7. Rauch, E.F. and Schmitt, J.H.: *Mater. Sci. Eng.*, 1989, A113, 441.
8. Rosen, G.I., Jensen, D.J. and Hansen, N.: *Mater. Sci. Forum*, 1993, 113-115, 201.
9. Swift, H.W.: *Engineering*, 1947, 163, 253.
10. Tóth, L.S., Jonas, J.J., Gilormini, P. and Bacroix, B.: *Int. J. Plasticity*, 1990, 6, 83.
11. Tóth, L.S., Jonas, J.J., Daniel, D. and Bailey, J.A.: *Textures and Microstructures*, 1992, 19, 245.
12. Tóth, L.S. and Jonas, J.J.: *Scripta Metall. et Mater.*, 1992, 27, 359.
13. Shrivastava, S., Jonas, J.J. and Tóth, L.S.: in "Modelling the Deformation of Crystalline Solids", ed. Terry C. Lowe, Anthony D. Rollett, Paul S. Follansbee and Glenn S. Daehn, The Minerals, Metals & Materials Society, of AIME (American Institute of Metallurgical Engineers), Warrendale, PA, 1991, p. 205.
14. Jonas, J.J. and Sakai, T.: in "Deformation, Processing and Structure", ed. G. Krauss, American Society for Metals, Metals Park, Ohio, 1984, p. 185.
15. Urabe, T. and Jonas, J.J.: *ISIJ Int.*, 1994, 34, in press.

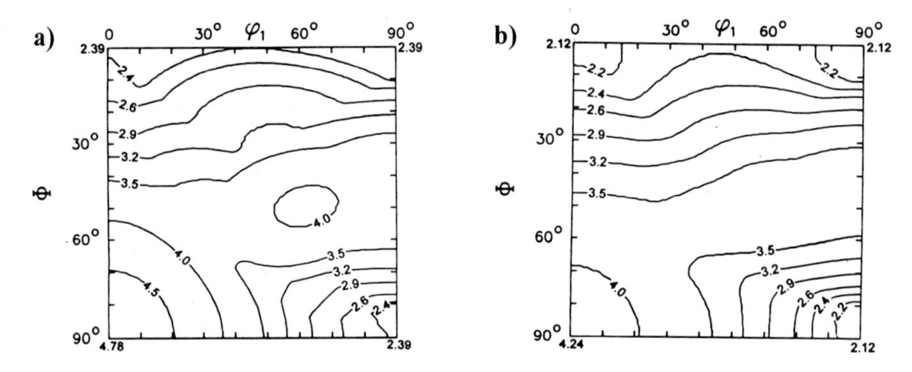

Figure 1 $\phi_2 = 45^\circ$ sections of Euler space showing the Taylor factors pertaining to full constraint deformation; calculated using a rate sensitive model with $m = 0.05$. (a) fcc metals (b) bcc metals.

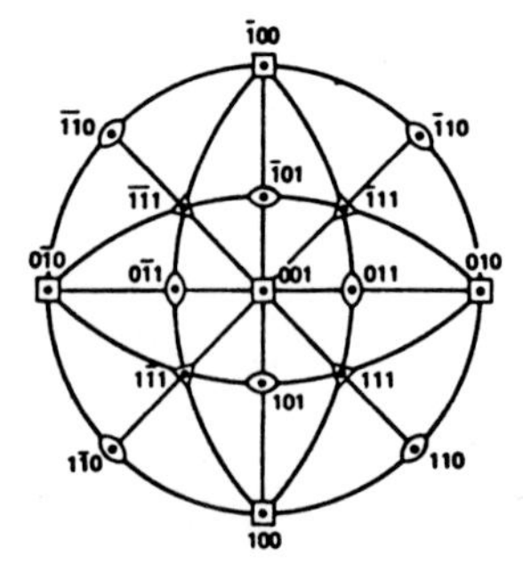

Figure 2 Standard stereographic projection of a (001) cubic crystal on which the four <111> and six <110> type axes are displayed.

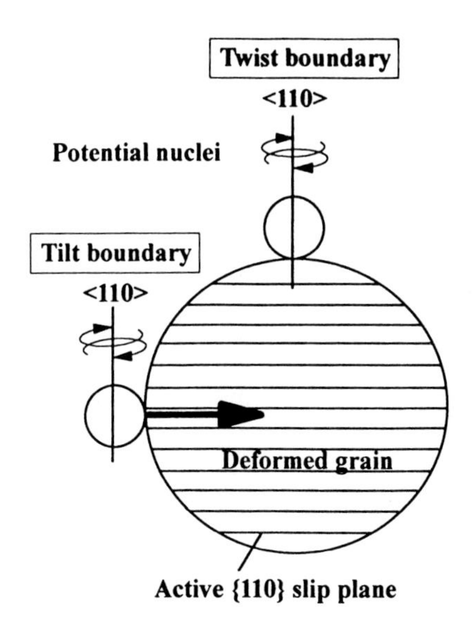

Figure 3 Schematic diagram of deformed bcc grain containing one set of active {110} slip planes; two potential recrystallization nuclei are shown. Growth of the upper one involves the motion of a <110> twist boundary; that of the lateral one, the motion of a <110> tilt boundary. The orientation of the active slip plane in the deformed grain determines the particular rotation axis about which recrystallization takes place.

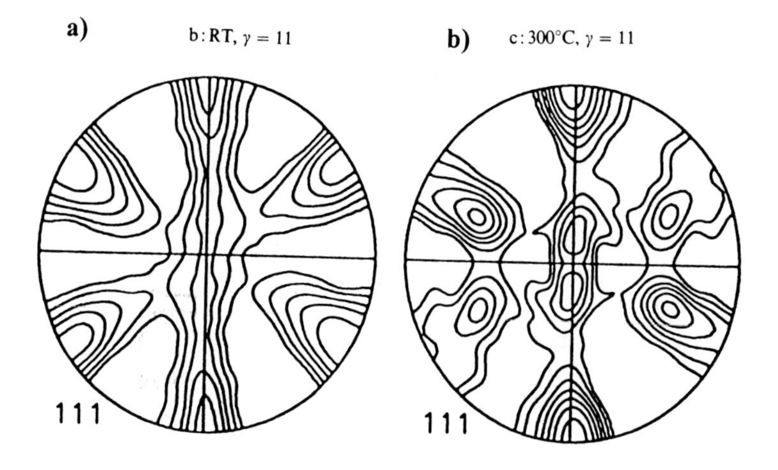

Figure 4 Measured (111) pole figures determined on OFHC Cu twisted to a shear strain of $\gamma = 11$. Isovalues: 0.8, 1.0, 1.3, 1.6, 2.0, 2.5, 3.2.

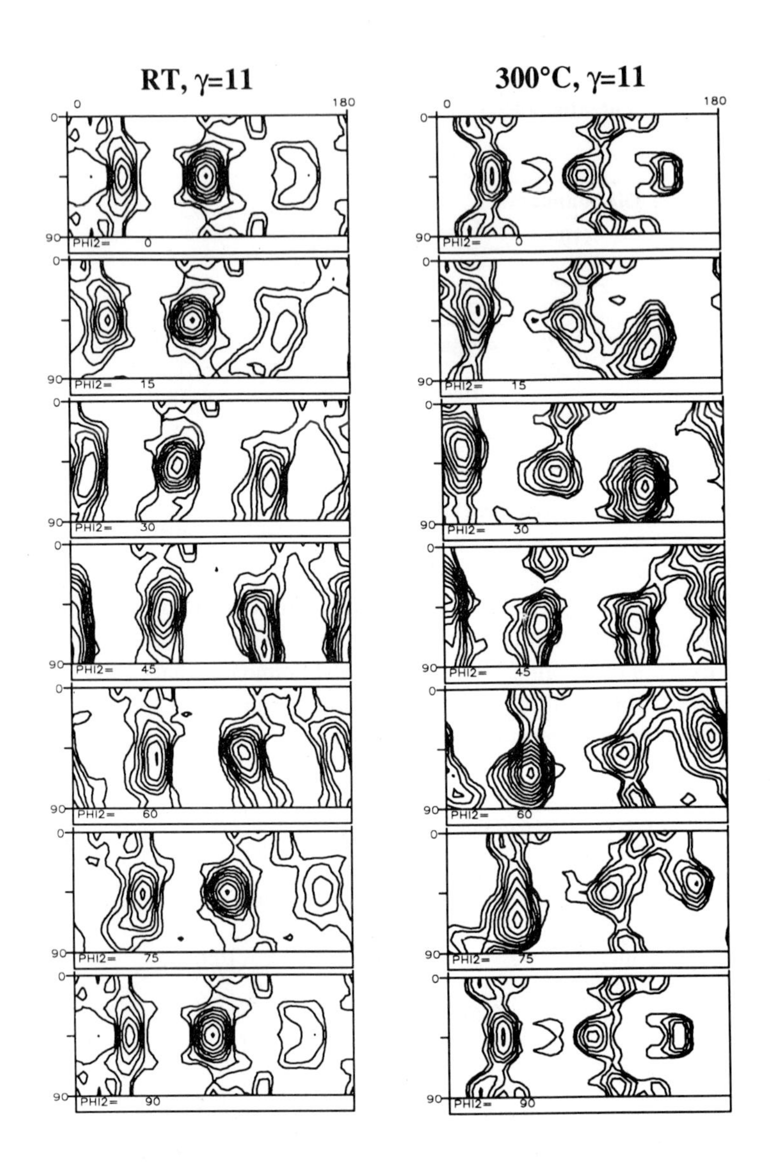

Figure 5 ODF sections obtained from (a) a room temperature test performed to $\gamma = 11$; (b) an elevated temperature test carried out at 300°C, $\gamma = 11$. Isovalues: 0.8, 1.0, 1.3, 1.6, 2.0, 2.5, 3.2, 4.0, 5.0, 6.4.

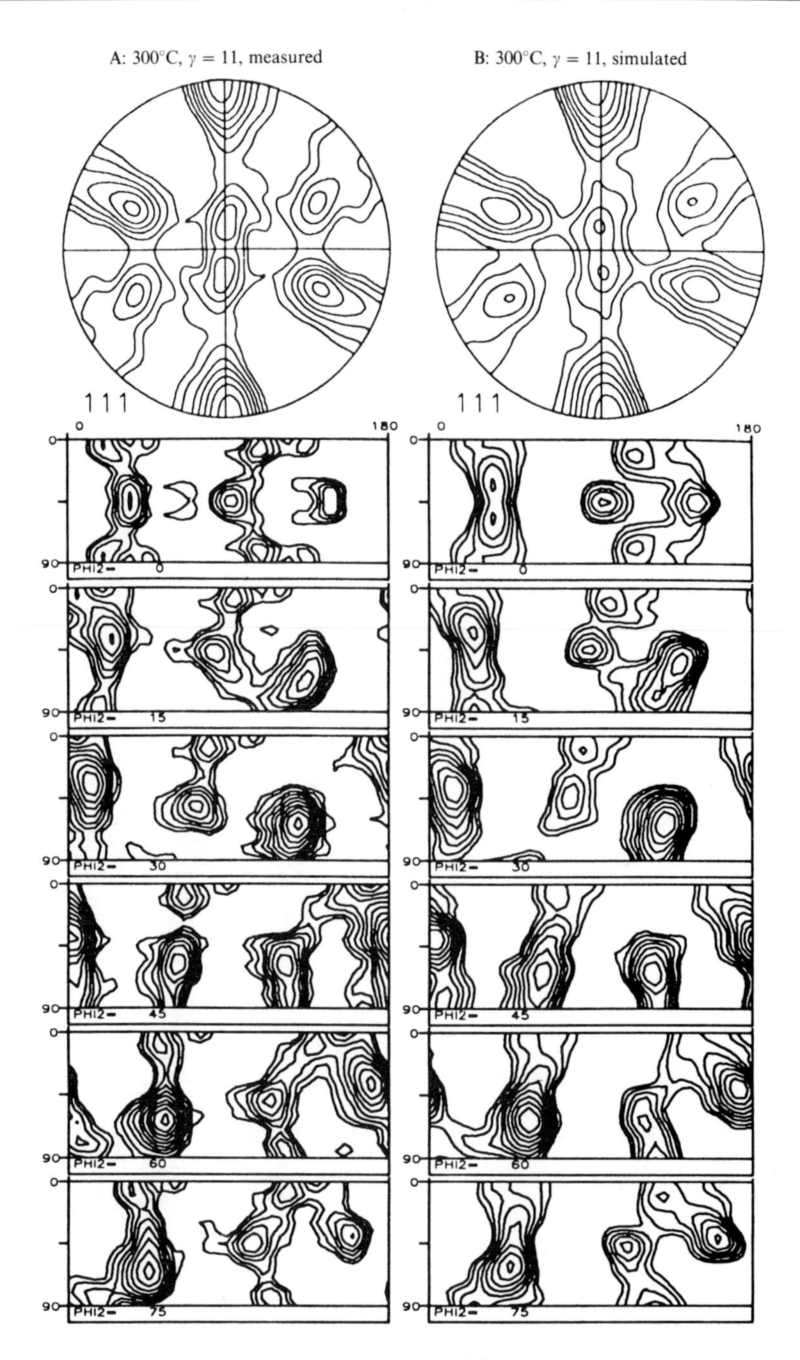

Figure 6 A: Measured (111) pole figure and ODF of the sample twisted at 300° C to $\gamma = 11$. B: Simulated texture starting with the experimental initial texture using crystallographic slip and dynamic recrystallization. Isovalues on all diagrams: 0.8, 1.0, 1.3, 1.6, 2.0, 2.5, 3.2, 4.0, 5.0, 6.4.

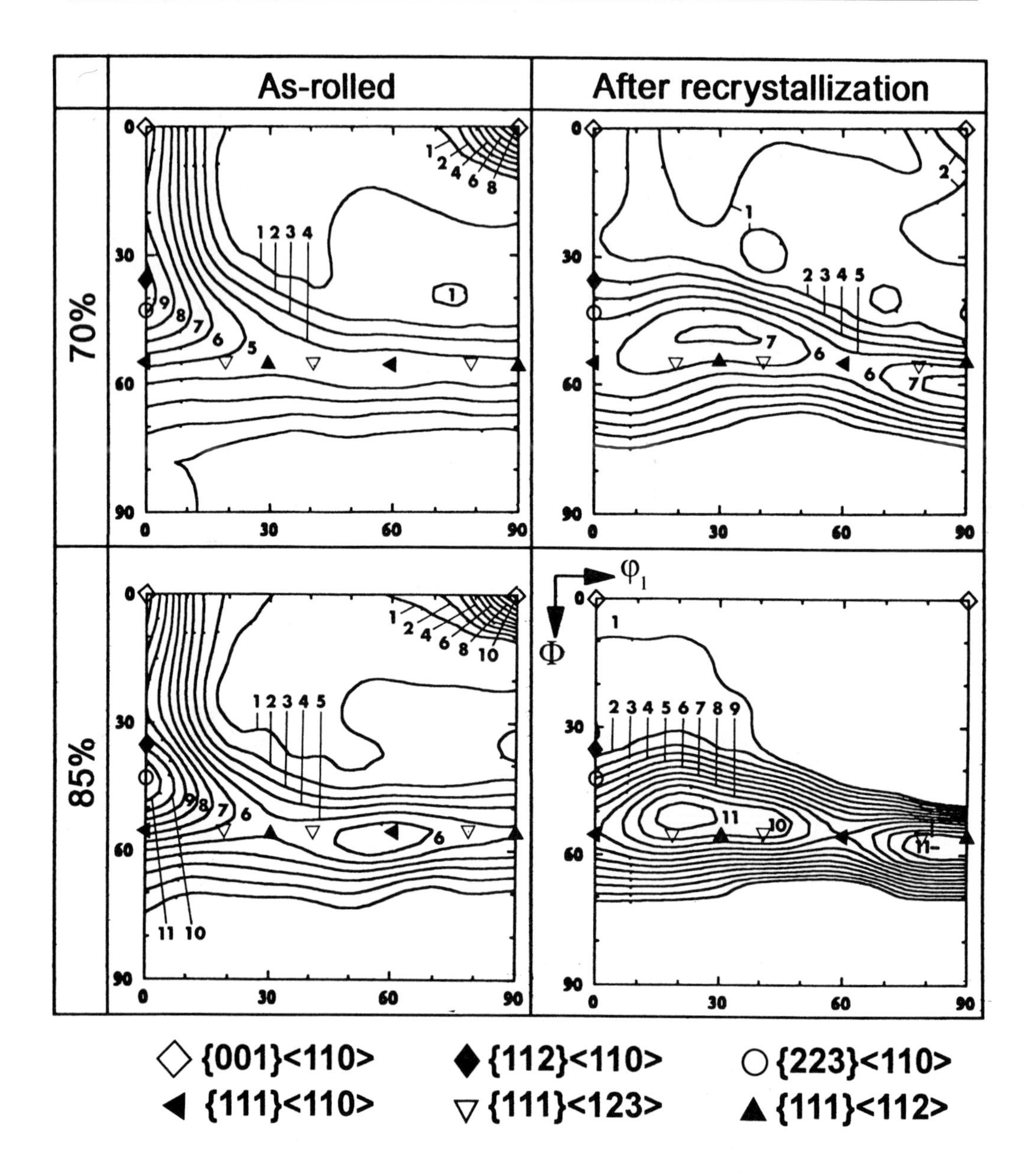

Figure 7 $\phi_2 = 45^{\circ}$ sections of the ODF's of the cold rolled (left hand column) and annealed (right hand column) specimens. The upper and lower rows correspond to 70% and 85% reduction, respectively.

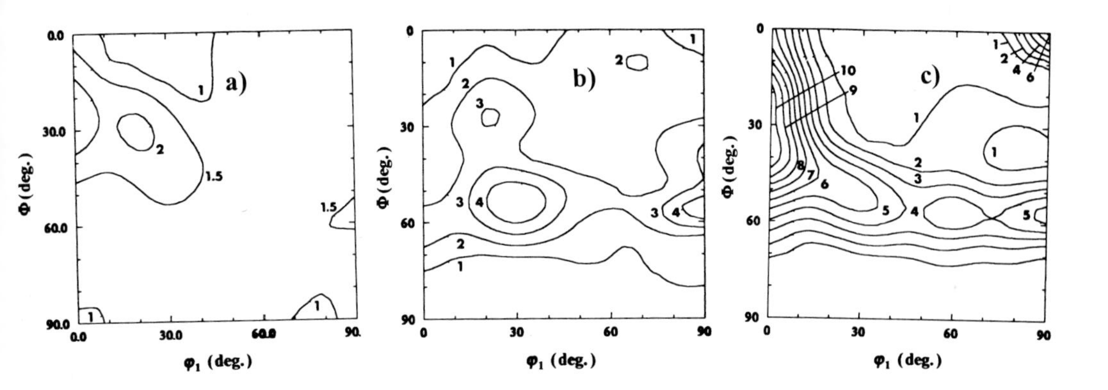

Figure 8 $\phi_2 = 45^o$ sections of the recrystallization ODF (rotation angle =27°) calculated from the experimental texture of the 85% cold rolled material. In (a), the same probability factor, $p_{ij} = p_0$, was applied to all the twelve possible <110> rotated variants. In (b), the variant selection rule based solely on slip activity, $p_{ij} = F_{ij} / \sum_{k=1}^{6}\sum_{l=1}^{2} F_k$ was used. In (c), the probability of growth depended only on the availability of nuclei, $p_{ij} = N_{ij}$, and no variant selection was used.

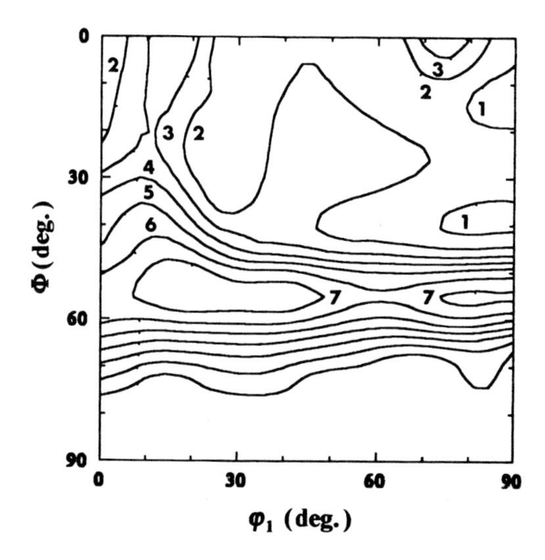

Figure 9 $\phi_2 = 45^o$ section of the recrystallization ODF calculated from the experimental texture of the 85% cold rolled material. Eq. (10) was employed to model both variant selection and the availability of nuclei.

Materials Science Forum Vol. 157-162 (1994) pp. 1731-1738

THEORY OF GRAIN BOUNDARY STRUCTURE EFFECTS ON MECHANICAL BEHAVIOUR

B.L. Adams [1], T.A. Mason [2], T. Olson [1] and D.D. Sam [2]

[1] Department of Manufacturing Engineering and Engineering Technology, Brigham Young University, 435 CTB, Provo, UT 84602, USA

[2] Department of Mechanical Engineering, Yale University, New Haven, CT 06520, USA

Keywords: Mechanical Behaviour, Elasticity, Yielding, Texture, Microstructure

ABSTRACT

Described in this paper are elements of emerging theories for the mechanical behavior of polycrystals which influence elastic and plastic behavior through details of the spatial placement of lattice orientation. The approach taken here is through the two-point statistics of placement of lattice orientation. Although explicit connections between statistical and exact representations of microstructure are not achieved, it is argued that grain boundary structure plays a dominant role, in theories which extend beyond the well-known first-order theories which incorporate only a volume-fraction representation of microstructure. As an example the effects which are predicted by the theory, the spatial placement of the well-known copper and brass components of rolling texture in fcc metals is considered. It is demonstrated that spatial placement of lattice orientation can have strong effects upon components of the effective elastic stiffness tensor, yield strength, the local curvature of the yield surface, and the scatter of these properties.

INTRODUCTION

It is known that most properties of polycrystalline materials are related to the internal microstructure. A central aim of the discipline of materials science is to discover this linkage between microstructure and properties. Further, it is then desirable to use such knowledge to guide processing in order to produce microstructures which are tailored for specific applications of the material.

Polycrystalline microstructures are particularly challenging in that their microstructures are inherently statistical in character. The sizes and shapes of grains, their connectivities, spatial placement and the structure of intergranular boundaries, the distribution of lattice orientation and phase, among other features, are all known to have stochastic character. Emerging microscopies are capable of measuring 10^7 distinct lattice orientations, 10^{15} distinct structures of grain boundaries, and a truly astronomical number of details of the microstructure. It is evident that some elements of this vast sea of microstructural detail will be more important than others in affecting properties of interest. Robust theory, connecting microstructure and properties, is required to classify and study polycrystalline microstructures; the absence of such theory leaves the investigator awash in details, many of which may have little importance to the behavior of interest.

In this paper we consider some connections which have been developed between certain representations of polycrystalline microstructure and mechanical properties. For ease of presentation the focus will be upon single-phase polycrystals, although there is no inherent limitation of the theory proscribing a broader approach. No attempt is made to derive or otherwise motivate the primary results upon which the theory is founded; rather, the approach is to explore ramifications of these results which can guide the materials engineer in the production of microstructures which could achieve better mechanical performance. We shall limit our remarks to the important mechanical properties of elastic stiffness and yield strength.

BOUNDS ON ELASTIC PROPERTIES

Recently, Olson [1] derived a convenient, explicit formulation of the generalized Hashin-Shtrikman bounding theory incorporating only knowledge of the one- and two-point statistical representation of microstructure. The resulting bounds are expressed as

$$\begin{aligned} C^* &\lesseqgtr C^r + A_1 (A_1 + A_2)^{-1} A_1 \\ A_1 &= \langle C(\mathbf{x}) \rangle - C^r \\ A_2 &= \left\langle C'(\mathbf{x})\, \Gamma(\mathbf{x} - \mathbf{x}')\, C'(\mathbf{x}') \right\rangle \end{aligned} \quad (1)$$

where C^r denotes a constant isotropic reference stiffness tensor based on the bulk modulus of the material and either the larger or smaller shear modulus depending upon which bound is sought. The notation <( )> denotes volume or ensemble averaging. A_1 and A_2 are both fourth-order tensors, with A_1 depending only upon the one-point statistics of lattice orientation (i.e., the odf) and A_2 depending upon the two-point statistics, or the spatial correlation of lattice orientation. The upper-bound requires that A_1 be negative semi-definite, or that the larger shear modulus be used in the calculation. Substituting the smaller shear modulus would yield a positive semi-definite A_1 and the reversal of the inequality.

Mason [2] has shown, from careful calculations using the theory expressed by relations (1) (incorporating extensive measurements of the one- and two-point statistics) and independent measurements of six independent components of the elastic stiffness tensor, that a factor of five narrower bound is achieved in the higher-order theory as compared with ordinary Voigt and Reuss averages.

A_2 can be written as two parts. One part involves a kernel with Dirac quality requiring only one-point information while the other requires an integration of a Green's function with the full two-point statistics over the volume of the body (see Kröner [3]). We shall refer to the first term as the E-term and the second as the Φ-term. This may be written as

$$\begin{aligned} A_2 = E\, \delta(x - x')\Big(\left\langle C(x)\, C(x') \right\rangle - \left\langle C(x) \right\rangle \left\langle C(x') \right\rangle \Big) \\ + \int_V \Phi\big(x - x' \big) \left\langle C'(x)\, C'(x') \right\rangle dV\,. \end{aligned} \quad (2)$$

Thus, for a fixed orientation distribution, the E-term is constant but the Φ-term can vary depending upon the placement of lattice orientation in the microstructure.

We note that the kernel of the Green's function integration expressed in the Φ-term is isotropic, and varies as $|\mathbf{x}-\mathbf{x}'|^{-3}$. It follows that short-range correlation of lattice orientation is much more important to the strength of the Φ-term than is long-range correlation in the two-point statistics. Short-range correlation effects are strongly linked to grain boundary structure. *It follows that grain-boundary structure distribution is the dominant aspect of microstructure affecting bounds on the elastic stiffness beyond the orientation distribution.* It is this observation that motivates our approach to the study of grain-boundary structure effects on elastic properties.

It is also important to note that the Φ-term is found to be precisely zero if the two-point statistics of lattice orientation correlation are isotropic. More precisely, if $<C'(\mathbf{x})C'(\mathbf{x}')>$ is a function of $|\mathbf{x}-\mathbf{x}'|$ only, and not of $\mathbf{x}-\mathbf{x}'$, then the Φ-term is zero. Stated another way, within the context of a two-point Hashin-Shtrikman bounding theory, it is only anisotropy of the grain-boundary structure distribution which gives rise to any appreciable alteration in the bounds obtained from the orientation distribution.

If the nature of the two-point statistics for a given single-phase material is explored, insight into the possible variation in elastic bounds which can be caused by a preferential occurrence of certain grain boundaries can be obtained. To illustrate this, pairs of orientations drawn from the well-known orthotropic variants of the fcc rolling texture have been considered. The idealized copper and brass orientations were placed at the tip and tail of vectors oriented in various directions relative to the sample reference frame. We shall identify the (100), (010) and (001) directions as the rolling, transverse and normal directions in the sample. The strength of contributions from these various combinations of orientations and directions were noted and compared for the <100>, <110> and <111> directions in the sample reference frame. Those orientations used are expressed below in terms of the Euler angles of Bunge [4].

Cu^{I} : (90° , 35.26°, 45°) Cu^{II} : (39.23°, 65.91°, 26.56°)
Bs^{I} : (35.26°, 45°, 0°) Bs^{II} : (54.74° , 90° , 45°)

We emphasize that contributions to the Φ-term in the bounding relationship for a particular <uvw> direction in the point-pair statistics will always be dominated by the occurrence of grain boundaries lying perpendicular to the particular <uvw> direction of interest. We shall consider the effects of specific point-pairs of the various copper and brass components on the bounds. (To a first-order approximation such effects can be blended with a full 2-point representation of the microstructure, but here only the isolated effects are considered.) It must be remembered that within the context of the present theory, short-range correlation induced by altering grain shape or by adding certain grain boundaries is linked to such alterations of the statistical mix characteristic of the microstructure. The specific required alterations are not, however, fully explored.

The effect of the E and the Φ-term on the C_{1111} and C_{1212} components of the elastic stiffness tensor will be reported. Two primary influences are of interest: either that the bounds would be shifted up and/or narrowed or that they are driven down and/or broadened by the specific combinations of orientations considered. All reported values were calculated using copper single crystal stiffnesses and are in units of GPa. When the effect of the added point-pair statistics to both the upper- and lower-bounds is positive then the set of bounds shifts upward. This can be a very beneficial effect if higher stiffness is desirable in the application of the material. If both bounds are reduced the bounds shift downward. If they are of different sign, the bounds may shift up or down while either narrowing or widening depending on which contribution is larger. *We believe that narrowing the bounds is also of considerable significance since it implies that the natural scatter in measured properties (from samples exhibiting the same one- and two-point statistics) may be commensurately narrowed.*

The general result of the study of Φ-term contributions is that any correlation of preferred grain boundaries comprised of combinations of the four mentioned orientations in the three major sample directions; (100), (010) or (001), has a very detrimental effect on the magnitude and width of C_{1111} bounds and only a moderate shift upwards and narrowing of the bounds on C_{1212}.

When point-pairs parallel to one of the <110> or <111> directions were considered all of the orthotropically equivalent directions were also evaluated and the result averaged. It was found that many combinations of orientations in the (101) direction (averaged with its three equivalents in the 1-3 plane, where 1-3 refers to the plane with normal parallel to (010)) strongly shifted the lower bound on C_{1111} up while only slightly lowering its upper bound. C_{1212} saw a moderate increase in the lower bound with a more modest increase of the upper bound. The other <110> type directions in the 1-2 and 2-3 planes proved to drastically widen the bounds on C_{1212} and C_{1111}, respectively. The results of considering the <111> directions was that all combinations of the four orientations greatly narrowed the bounds on C_{1111} due to the upper bound being slightly decreased in most cases while the lower bound greatly increased in all cases. On the other hand, C_{1212} saw only moderate narrowing when the orientation pairs were Cu^I/Cu^I and Bs^{II}/Bs^{II}. All other combinations featured no change or a general shift downward of the bounds. Thus, grain elongation or grain boundary placement in non-principal directions can result in beneficial increases in elastic stiffness and narrowing of bounds in some instances; but it can also lead to harmful effects in other instances.

Some of the most significant of these trends are summarized in Table 1. It is emphasized that elongated grains in the direction of rolling, or the other two main axes of the sample, do not theoretically increase the stiffness of the material in that direction. On the contrary, it is the spatial correlations of lattice orientations in non-principal directions in the material which increase the stiffness.

Table 1. Contributions of Φ terms to Bounds on C_{1111} and C_{1212} (GPa).

Direction	Combination	Upper Bound C_{1111}	Lower Bound C_{1111}	Upper Bound C_{1212}	Lower Bound C_{1212}
(100)	Bs^I - Bs^I	-1.15	-26.7	1.5	9.5
	Bs^{II}- Cu^{II}	-0.2	-55.7	1.90	3.41
(010)	Bs^I - Bs^I	1.01	-0.6	2.2	4.2
	Bs^{II}- Cu^{II}	0.11	-11.5	0.94	-0.22
(001)	Bs^I - Bs^I	1.01	-0.6	1.3	7.3
	Bs^{II}- Cu^{II}	0.27	-10.5	-0.18	-2.90
(110)	Bs^I - Bs^I	0.83	9.60	-1.08	-8.17
	Bs^{II}- Cu^I	0.4	22.5	-5.53	-0.25
(101)	Bs^I - Bs^I	0.83	9.60	-1.08	-8.17
	Bs^{II}- Cu^I	-0.21	9.48	1.42	-0.87
(011)	Bs^I - Bs^I	-1.87	-12.2	1.13	5.44
	Bs^{II}- Cu^I	0.11	-11.15	1.40	-2.08
<111>	Bs^{II} - Bs^{II}	-0.05	17.25	-1.05	1.55
	Cu^I- Cu^I	-0.05	17.25	-1.05	1.55
	Bs^{II} - Cu^I	0.27	18.54	-2.14	-2.84
	Cu^I- Cu^{II}	-0.23	16.5	-0.94	-1.49

We must also inquire as to the effects of adding various copper-brass point-pairs on the E-term. In this instance nothing about spatial placement alters the effects since only one-point statistics are important. Table 2 summarizes the effects. It can be seen that all of these six combinations cause the bounds to be shifted upwards and narrowed. Also, examination of the average orientation, <C>, for the these orientation-pairs indicates that the effects are higher than that for most texture components (Mason [2]). It can be seen that the addition of any of these types of grain boundaries into a material will alter the one point statistics in such a way as to shift the volume averaged stiffness and compliance (i.e., Voigt and Reuss averages) upward.

Table 2. Contributions of E term to Bounds on C_{1111} and C_{1212} (GPa).

Combination	Upper Bound C_{1111}	Lower Bound C_{1111}	Upper Bound C_{1212}	Lower Bound C_{1212}
Bs^I - Bs^{II}	3.15	9.13	2.94	8.51
Bs^I - Cu^I	1.44	4.18	1.44	4.18
Bs^I - Cu^{II}	1.87	5.41	1.98	5.72
Bs^{II}- Cu^I	1.07	3.09	2.78	8.04
Bs^{II}- Cu^{II}	1.23	3.56	2.51	7.27
Cu^I - Cu^{II}	0.59	1.70	2.94	8.51

Although the sensitivity study we have performed is very far from complete, it is evident that spatial placement of lattice orientation might be exploitable to obtain improved elastic performance in polycrystalline materials. A more complete study is possible, and it would be helpful to make comparisons of these theoretical predictions with experimental observations of actual placement of lattice orientation in typical polycrystals. For this the emergence of Orientation Imaging Microscopy [5] can play a key role.

BOUNDS ON PLASTIC YIELDING

Recently, Olson [6,7] introduced a new variational principle for calculating upper- bounds on the effective yielding behavior of heterogeneous rigid-plastic materials. The variational principle takes the form of a bound on the dot product of the macroscopic stress Σ^o with the macroscopic strain rate D^o (i.e., the rate of plastic working), and depends on a suitable choice of a linear elastic comparison medium. The bound can be written as follows:

$$\left(D^0 \cdot \Sigma^0\right)^2 \leq \left[D^0 \cdot (S^{-1})^* D^0\right]\left[\frac{1}{vol\,(\Omega)}\int_\Omega \max_{\sigma(\mathbf{x})\in P(\mathbf{x})}\left(\sigma \cdot S(\mathbf{x})\sigma\right) d\mathbf{x}\right] \tag{3}$$

where $S(\mathbf{x})$ represents the local elastic compliances of the comparison material and $(S^{-1})^*$ is the effective stiffness of this material. Here Ω represents the region occupied by the material.

To evaluate this bound we must choose a reference medium and some method for bounding it's effective properties. Here this is accomplished by implementing the linear bound involving 2-point statistics given in equation (1). Thus we derive 2-point bounds on the yielding behavior of a given polycrystal. Furthermore, we generate an explicit formula for this bound which is calculable for an arbitrary geometry. It is instructive to choose a reference medium which will guarantee that our bound will be at least as good as that of Taylor [8]. Two conditions must be satisfied to retrieve Taylor's result from relation (3). First, we require a reference medium which satisfies a linear constitutive law given by $D^o = S(\mathbf{x})\sigma^c(\mathbf{x})$, where $\sigma^c(\mathbf{x})$ is the *corner stress state* of the yield function at $\mathbf{x}$ derived from D^o which satisfies the maximum external work criteria of Bishop & Hill [9,10]. Second, we require $(S^{-1})^*$ to be bounded using the arithmetic-mean of $S(\mathbf{x})$. A tighter bound on $(S^{-1})^*$ results in a bound which is tighter than the Taylor bound. In the work presented here, we have incorporated two-point statistics of the field $S(\mathbf{x})$ to narrow the bound according to relations (1). Computational details for this bound in the case of fcc polycrystals are explained by Sam, et al. [11].

To illustrate the behavior of this 2-point bound, we have generated hypothetical composite materials consisting of two alternating layers of fcc crystal with the brass-1 and brass-2 orientations described in the previous section; the volume fraction of each component is identical. A yield surface was computed using the method of Taylor illustrating the classical upper-bound. Taylor's model depends only upon volume fractions, and not upon spatial placement of the components. A second calculation, making the assumption that the components are equiaxed and randomly mixed, obtains a tighter-bound for the yield surface by application of the method explained by Walpole [12,13] for bounding $(S^{-1})^*$. Such an assumption leads to the absence of contribution from the Φ-term in the linear bounding calculation. Finally, we incorporated the full 2-point statistics for a composite structure consisting of alternating, planar layers of the two components. This results in a new bound. These models are illustrated in Figure 1a,b,c.

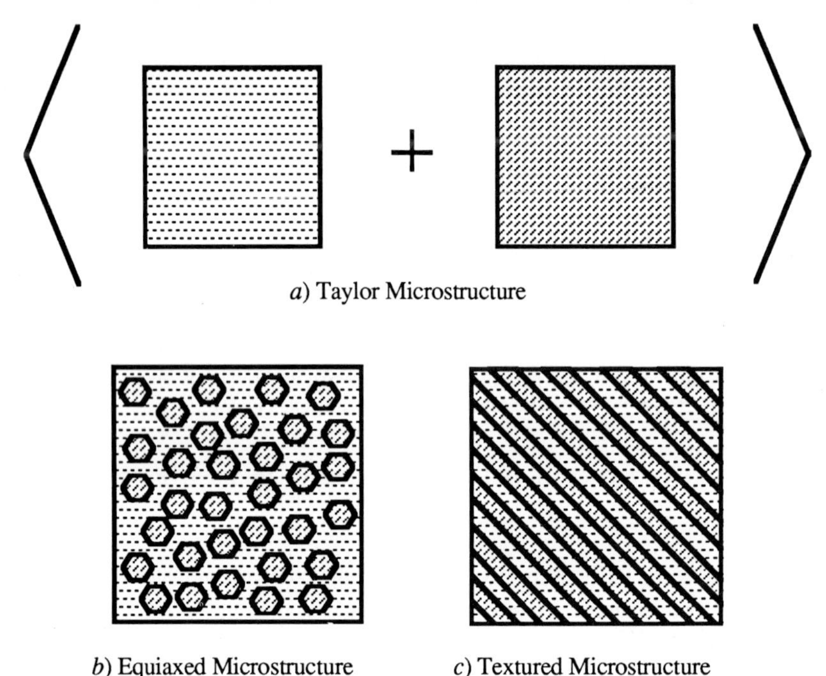

a) Taylor Microstructure

b) Equiaxed Microstructure *c*) Textured Microstructure

Figure 1. Three microstructural models considered in bounding calculations.

We have chosen to show 2-dimensional sections of the full yield surface in the σ_{11}-σ_{22} plane. Stresses are normalized to the critical-resolved-shear-stress, τ_c, according to the usual convention. The yield surfaces corresponding to each of the model calculations are shown in Figure 2a,b,c. In order to visualize the comparison between these three calculations, curves were drawn from the inner envelope of tangent lines for each calculation; these are shown in Figure 3, where the inner surface is the full 2-point statistical model, the intermediate curve is the equiaxed model, and the outer surface is the Taylor model.

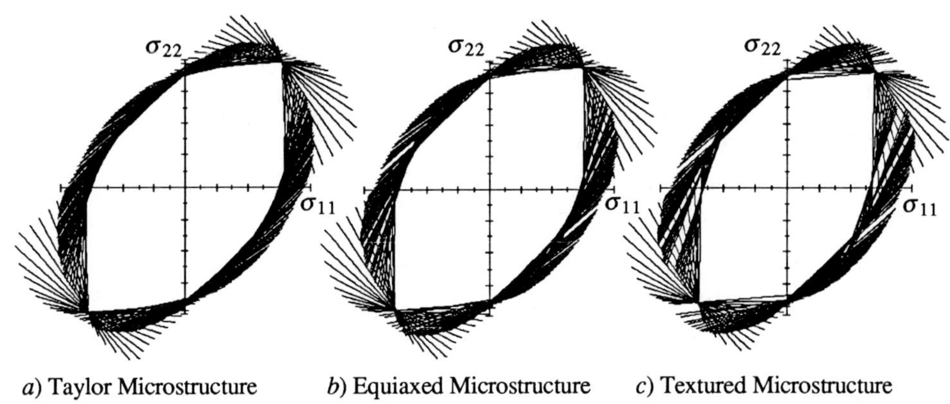

a) Taylor Microstructure *b*) Equiaxed Microstructure *c*) Textured Microstructure

Figure 2. Tangent yield surfaces computed for three microstructural models.

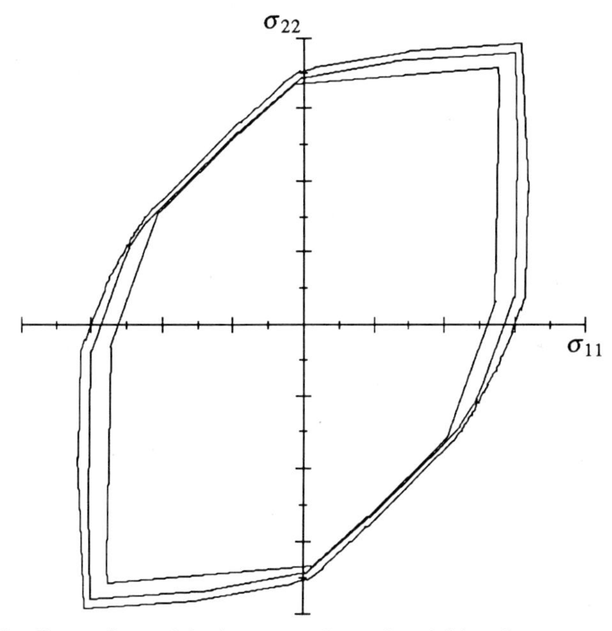

Figure 3. Comparison of the inner-envelopes for yield surfaces computed for three microstructural models.

The essential point to be emphasized in comparing these three models is the role of spatial organization of orientation components. In the Taylor model this aspect of microstructure is ignored entirely. It is evident that as we begin to incorporate information about spatial placement, the bound moves inward. For the case studied, a reduction of ~20% is apparent in the equi-biaxial stress state direction. Of equal, and perhaps even greater interest, however, is the sharpening of corners, and the appearance of new corners in the yield surface. Such corners are of great interest in questions of plastic stability, shear band formation, and bifurcation. It is evident that sharp corners are predicted by the classical Taylor model, but additional corners are predicted as we proceed from an equiaxed assumption about the microstructure, to an extreme case of coherence represented by the laminar structure. Clearly the spatial arrangement of lattice orientation appears to be of fundamental importance to plastic yielding.

ACKNOWLEDGEMENT

This work was sponsored by a Materials Research Groups award from the National Science Foundation. The guidance of E. Turan Onat and M. Beran is gratefully acknowledged.

REFERENCES

1) Olson, T.: *Acta metallurgica et materialia*, 1993, submitted.
2) Mason, T.: Dissertation, Yale University, 1993, in preparation.
3) Kröner, E.: *Modelling Small Deformations of Polycrystals*, ed. J. Gittus and J. Zarka, Elsevier, London, 1986, 229.
4) Bunge, H. J.: *Texture Analysis in Materials Science*, Butterworths, London, 1982, 5.
5) Adams, B. L., S. I. Wright and K. Kunze: *Metallurgical Transactions*, 1993, 24A, 819.
6) Olson, T.: *Journal of the Mechanics and Physics of Solids*, 1993, in press.
7) Olson, T.: *Materials Science and Engineering*, 1993, in press.
8) Taylor, G. I.: *Journal of the Institute of Metals*, 1938, 62, 307.
9) Bishop, J. F. W. and R. Hill: *Philosophical Magazine*, 1951, 42, 414.
10) Bishop, J. F. W. and R. Hill: *Philosophical Magazine*, 1951, 42, 1298.
11) Sam, D. D., T. Olson and B. L. Adams: *Proc. Fourth International Symposium on Plasitcity and Its Current Applications*, ed. A. Khan, Baltimore, MD, 1993.
12) Walpole, L. J.: *Journal of the Mechanics and Physics of Solids*, 1966, 14, 151.
13) Walpole, L. J.: *Journal of the Mechanics and Physics of Solids*, 1966, 14, 289.

Materials Science Forum Vol. 157-162 (1994) pp. 1739-1746

SIMULATION OF ROLLING AND DEEP DRAWING TEXTURES IN FERRITIC STEELS: APPLICATION TO EAR PROFILE CALCULATION IN DEEP DRAWING

D. Ceccaldi [1], F. Yala [1], T. Baudin [1], R. Penelle [1] and F. Royer [2]

[1] Laboratoire de Métallurgie Structurale, Université Paris-Sud, URA CNRS 1107, Bât. 413, F-91405 Orsay Cédex, France

[2] IPEM/CMSR, 1 Bvd Arago, F-57078 Metz Cédex 3, France

Keywords: Rolling, Deep Drawing, Textures, Ear Profile, Steels

ABSTRACT

Two models are used to calculate low deformation textures (deep drawing) and high deformation textures (rolling) of ferritic steels. The first model is the TAYLOR model with mixed glide conditions {110}<111>+{112}<111> , the second one is a non-homogeneous model (with the same mixed glide conditions). An original method is used to compute the deformation texture F'(g) (the initial texture being F(g)) with a high accuracy because of absence of truncation errors related to the harmonic method. An ear profile calculation is also presented in deep drawing taking into account texture evolution.

I. INTRODUCTION

In a previous work, CECCALDI et al.(1) have simulated deformation texture in uniaxial tension and calculated the LANKFORD coefficient of a low carbon steel in the framework of TAYLOR assumptions with mixed {110}<111> and {112}<111> glide conditions. A direct calculation of the orientation distribution function O.D.F., after deformation from an initial texture F(g) at the undeformed state, was carried out without using series expansion method. The so calculated O.D.F., whatever the symmetry of the texture,is accurate because of absence of truncation errors. So it becomes possible to calculate ear profile occuring during deep drawing taking into account texture evolution at each point of a net under the blank holder.GRUMBACH et al.(2) performed a similar calculation in low carbon steels but without introducing texture evolution and consequently that of yield loci; more recently BARLAT et al.(3) presented a simplified calculation of ear profile in the case of aluminium alloys.

The present paper deals with simulation of deformation texture for high strain amount, $\varepsilon > 1$, by rolling using a non-homogeneous model and simulation of both deformation texture and ear profile for medium strain amount $\varepsilon < 1$ by deep drawing using the homogeneous TAYLOR model.

II. THE HOMOGENEOUS TAYLOR MODEL AND THE NON-HOMOGENEOUS MODEL.

In the present work, glide is assumed to occur only on the {110}<111> and {112}<111> slip systems. The combination of active slip systems is deduced from the analysis of the mixed yield polyhedron calculated by ORLANS-JOLIET et al.(4); this polyhedron is similar to that of

BISHOP and HILL (5,6,7) in the case of {111}<110> glide in F.C.C. materials. The ratio ξ of the critical shear stresses :

$$\xi = \tau_C\{112\} / \tau_C\{110\} \quad \text{with} \quad \sqrt{3}/2 < \xi < 2/\sqrt{3}$$

is an adjustable parameter which is taken equal to 0.93.

For each step of deformation $\{\Delta\varepsilon_n\}$, the variation $\Delta g_n = \{\Delta\psi_n, \Delta\vartheta_n, \Delta\varphi_n\}$ of the orientation g of any grain in the EULER space G (where ψ, ϑ, φ are the EULER angles) is calculated and, $g' = g + \Sigma\,\Delta g_n$ can be computed. So the trajectory of each point g of a grid (5°x5°x5°) in the EULER space can be determined. After having defined a final grid identical to the initial one, the final orientations g' are redistributed in the new grid and weighted by the initial O.D.F., F(g).Using a three dimensional interpolation the final texture F'(g) is then calculated.

In the case of the non-homogeneous model, the procedure is similar but in this case the average strain rate ε^c in the crystallite c is allowed to differ from the macroscopic strain rate ε. In the present version a neighbourhood C of the deformation tensor $\Delta\varepsilon$ is then defined and for each orientation g the non-homogeneous deformation $\Delta\varepsilon_g$ is calculated by minimizing the deformation work with respect to two parameters Δq and Δg which determine $\Delta\varepsilon_g$ within C of $\Delta\varepsilon$. This model is similar to the ARMINJON model (8,9).

III.DEEP DRAWING EAR PROFILE.

Assuming that the whole deformation occurs under the blank holder, the equation of stress equilibrium of the flange is:

$$\sigma_{rr} + \sigma_{\alpha\alpha} + r\frac{\partial \sigma_{rr}}{\partial r} = 0 \qquad (1)$$

with the condition $\sigma_{rr} = 0$ for r = RE(t) where RE is the radius of the flange; the axis r is along a radius, the axis α is perpendicular to the former in the sheet plane (figure 1). A friction stress can also be taken into account, RAULT(10),that is:

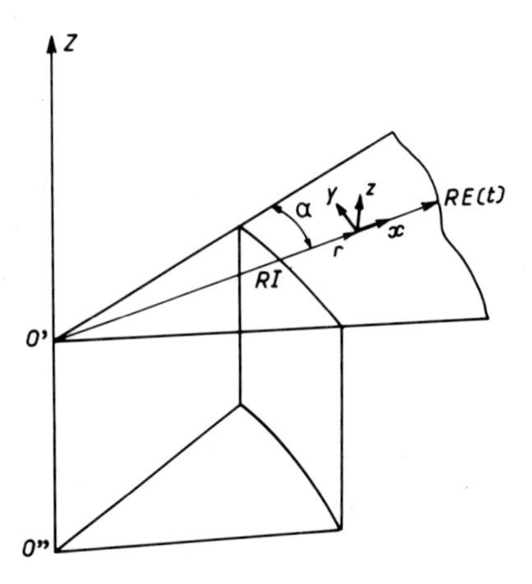

Figure 1 : Theoretical geometry of a deep drawing test

$$\sigma_{fr} = \frac{\mu F}{\Pi e\, RE(t)} = \sigma_o \frac{RE(o)}{RE(t)} \qquad (2)$$

μ is the friction coefficient between the sheet and the blank holder; F is the applied force on the blank holder; e is the sheet thickness. As μ and F are unknown σ_o is taken as an adjustable parameter, that is $\sigma_o = 0.1\ \tau_c\{110\}$.

Knowing the yield locus, $q(r, \alpha) = -\dfrac{d\varepsilon_{\alpha\alpha}}{d\varepsilon_{rr}}$ can be calculated.

After GRUMBACH et al.(2),

$$\dot{\varepsilon}_{rr} = -\frac{VI}{r}\exp\left(\int_{RI}^{r} -\frac{dr}{qr}\right) \tag{3}$$

Where VI is the punch velocity.Then,

$$\dot{\varepsilon}_{\alpha\alpha} = -q(r,\alpha)\,\dot{\varepsilon}_{rr}(r,\alpha) \tag{4}$$

and $$\dot{\varepsilon}_{zz}(r,\alpha) = -\left(\dot{\varepsilon}_{\alpha\alpha} + \dot{\varepsilon}_{rr}\right) \tag{5}$$

The equation (5) gives the thickness variation ε_{zz} (r, α).

In order to calculate the yield locus at each point under the blank holder, the evolution of the texture F(g,α,ε_{rr},t) is taken into account by computing in a first programm the evolution of two textures, $F_1(g,\alpha_n,\varepsilon_n)$ and $F_2(g,\alpha_n,\varepsilon_n)$ with

$0 \le \varepsilon_n \le 0.7$ and $0 \le \alpha_n \le 90°$

For that, the homogeneous TAYLOR model is used with two extreme deformation matrix :

$$\Delta\varepsilon_1 = \Delta\varepsilon_{rr}\begin{bmatrix}1 & 0 & 0\\ 0 & -1 & 0\\ 0 & 0 & 0\end{bmatrix} \quad (6) \qquad \Delta\varepsilon_2 = \Delta\varepsilon_{rr}\begin{bmatrix}1 & 0 & 0\\ 0 & -2 & 0\\ 0 & 0 & 1\end{bmatrix} \quad (7)$$

and for each of these textures (and each values of α_n and ε_n), the yield loci CLEP $_{1,2}$ (α_n,ε_n) are calculated.

For each value of r,α and ε_{rr}, q_1 (r,α,ε_{rr}) and q_2 (r,α,ε_{rr}) are calculated (with two small extrapolations for $\varepsilon_n \le \varepsilon_{rr} \le \varepsilon_{n+1}$ and $\alpha_n \le \alpha \le \alpha_{n+1}$).

We consider $x = \dfrac{\varepsilon_{\alpha\alpha}}{\varepsilon_{rr}} + 2$ (8)

With $0 \le x \le 1$ (x is 0 or 1 if x is out of this condition).

So $q(r, \alpha) = q_1(r,\alpha)\, x + q_2(r, \alpha)(1-x)$ (9)

IV.RESULTS.

IV.1.DEFORMATION TEXTURES

In the case of rolling corresponding to high deformation $\varepsilon = 1.38$, the non-homogeneous model gives better results than the classical TAYLOR model; figure 2a corresponding to the section at $\phi = 55°$ shows that the texture presents a sharp maximum in the case of the homogeneous TAYLOR model whereas the <111> fiber appears clearly in the case of the non-homogeneous model

Figure 2b ; however it has been shown that the classical TAYLOR model is sufficient for low deformations. It must be mentionned that these calculated textures differ from textures calculated with the harmonic method because in the present work there is no truncation errors, let us moreover mention that the function F'(g) is a total function.

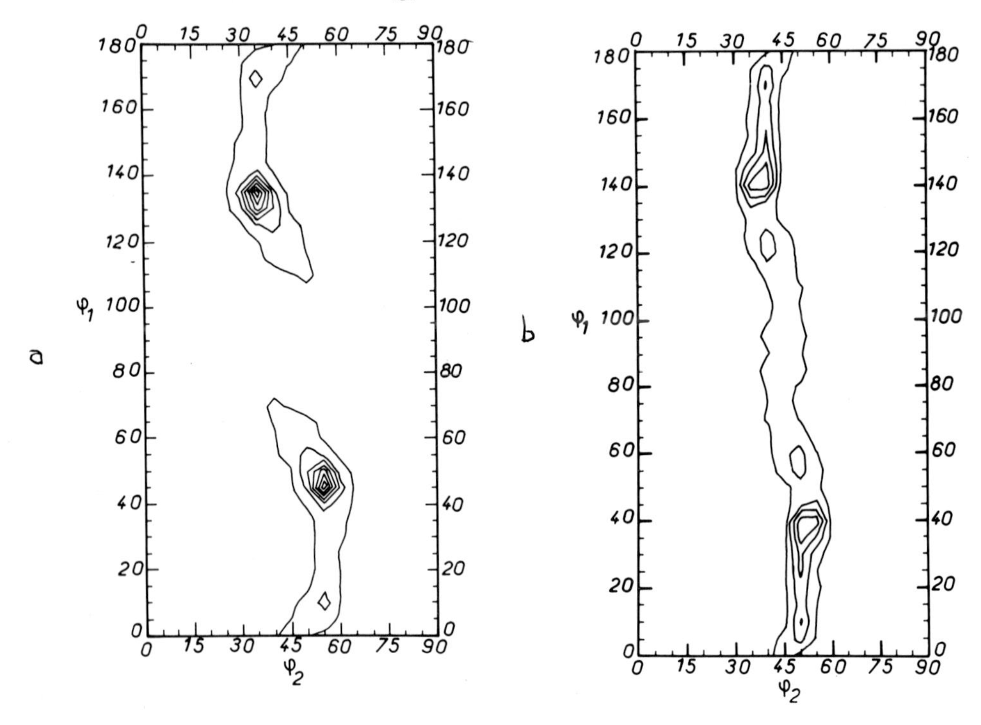

Figure 2 : Rolling texture a) Taylor model, b) non homogeneous model

In the case of deep drawing the homogeneous TAYLOR model is used, the initial texture is given figure 3 (section at $\vartheta = 45°$) and the deep drawing texture at $\alpha = 45°$ is given figure 4a. The simulated texture corresponding to a deformation amount of 30% with the matrix 1 is, as it can be

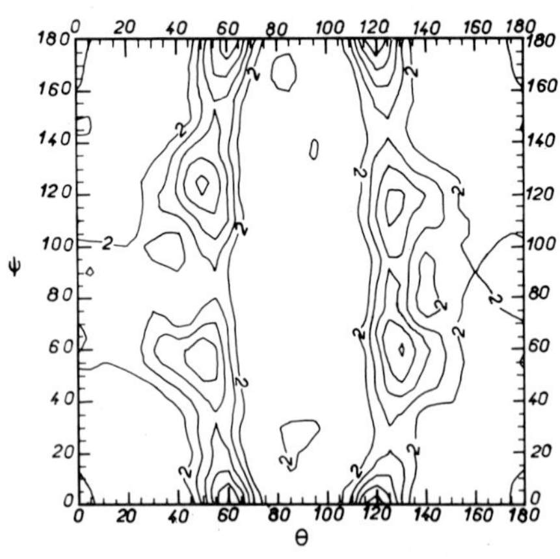

Figure 3 : Initial texture before deep drawing

seen figure 4b in agreement with the experimental one. However this experimental texture is a mean texture and the ear profile computation should need to use the two matrix 1 and 2.

IV.2.EAR PROFILE.

The yield locus is defined as the envelop of straight lines :

$$\frac{\sigma_{rr}}{\tau_c} - \frac{\sigma_{\alpha\alpha}}{\tau_c} = \overline{M}(q, a) \qquad (10)$$

where $\overline{M}(q,\alpha)$ is the mean TAYLOR factor. An example of yield locus is given figure 5, ear profile for the studied steel is shown in figure 6. The radii of the punch and the flange are respectively RI = 25 mm and RE = 50 mm. The profile is expressed as a function of $H(\alpha)$ (height from the bottom of the cup to the top at α°of the rolling direction).The calculated profile is given by :

$$\frac{H(\alpha) - H_M}{\overline{H}} = \frac{\Delta H(\alpha)}{\overline{H}} \qquad (11)$$

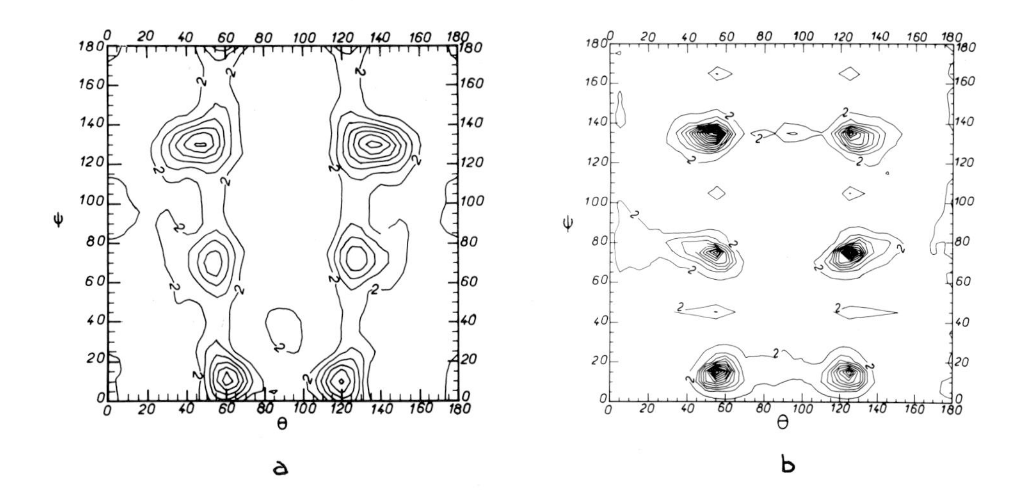

Figure 4 : Deep drawing texture (α = 45°)
a) experimental b) theoretical

In conclusion the shape of the ear profile is correct figure 7, the magnitude of the calculated $\Delta H(\alpha)$ variations is smaller than the experimental one. If texture evolution is not taken into account, the magnitude is similar to the previous one but the agreement of the calculated profile is a little less good figure8.

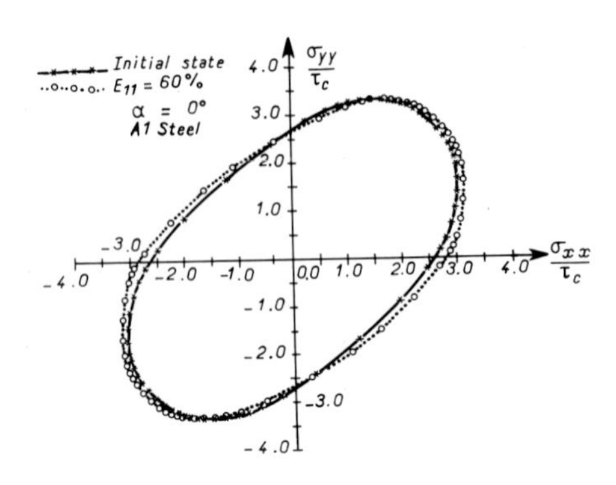

Figure 5 : Computed yield locus,

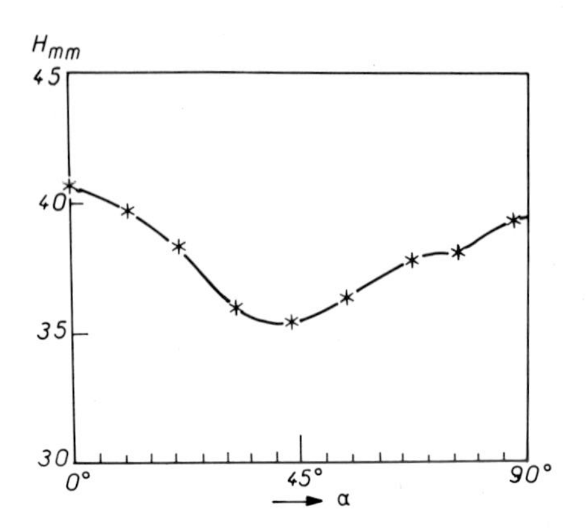

Figure 6 : Experimental ear profile for the studied steel

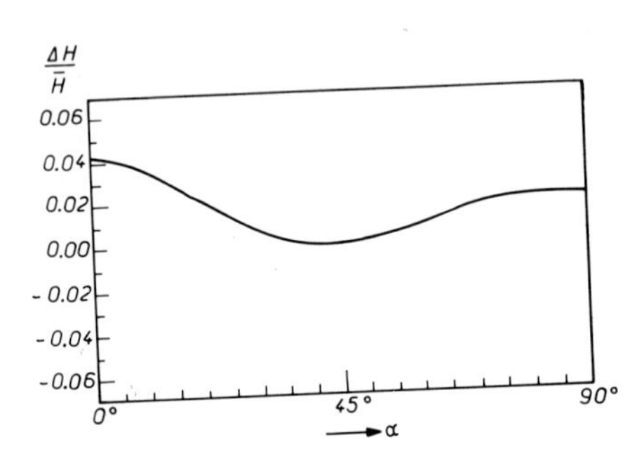

Figure 7 : Computed ear profile with texture evolution

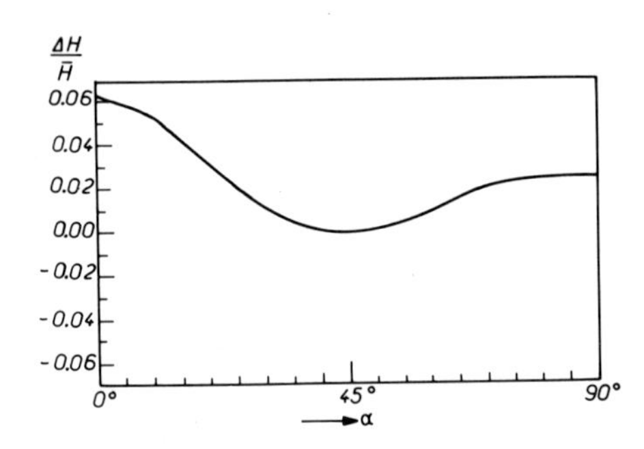

Figure 8 : Computed ear profile without texture evolution

BIBLIOGRAPHY.

(1) CECCALDI,D.,YALA,F.,BAUDIN,T.,PENELLE,R.,and ROYER,F.: Acta Met.Mater.,1992,40, n°6,1177.

(2) GRUMBACH,M., PARNIERE,P., ROESCH,L., and SAUZAY,C.: Mém. Sci. Rev. Mét.,1975, 72,241.

(3) BARLAT,F., PANCHANADEESWARAN,S., and RICHMOND, O. Metall. Trans.A,1991,22A,1525.

(4) ORLANS-JOLIET,B., BACROIX,B., MONTHEILLET,F., DRIVER,J.H., and JONAS,J.J.: Acta Met.,1988, 36, 1365

(5) TAYLOR,G.I.: J. Inst. Met.,1938,62,307.

(6) BISHOP,J.F.W., and HILL,R.: Phil.Mag.,1951, 42, 414.

(7) BISHOP,J.F.W.: Phil.Mag, 1953, 44, 51.

(8) ARMINJON,M.:Acta Metall.,1987,35,615.

(9) ARMINJON,M.,and DONADILLE,C.:Mém.Sci.Rev.Métall.,1990, 87, n°6, 359.

(10) RAULT,D.:"Mise en forme des métaux et alliages" Ed.CNRS,1976, 297.

Materials Science Forum Vol. 157-162 (1994) pp. 1747-1752

AN EQUILIBRIUM-BASED MODEL FOR ANISOTROPIC DEFORMATIONS OF POLYCRYSTALLINE MATERIALS

Y.B. Chastel and P.R. Dawson

Sibley School of Mechanical and Aerospace Engineering,
Cornell University, Ithaca, NY 14850, USA

Keywords: Polycrystal Plasticity, Equilibrium-Based Model, HCP Materials, Minerals

ABSTRACT

An equilibrium-based model for the plastic flow of rate-dependent polycrystalline materials is presented. The model satisfies equilibrium identically within an aggregate by requiring a uniform state of stress among all crystals. The average deformation rate of the crystals is required to equal the macroscopic imposed deformation rate. The model provides a fixed state lower bound of the stress for any polycrystal subjected to an applied rate of deformation. Its principal applications, however, are for single phase aggregates comprised of low symmetry crystals and multiphase aggregates in which the crystals have appreciably different strengths and/or low symmetry .

INTRODUCTION

Several approaches have recently been investigated for handling kinematically constrained crystalline structures. An alternative model based on strict enforcement of equilibrium within a polycrystal has been developed. The values of the stress components are determined so that the same stress exists in each crystal and the differing deformation rates from crystal to crystal average to the desired macroscopic value [1]. This latter constraint forms the basis of a nonlinear system of equations from which the five stress components can be determined [2].

The model is mathematically complete for all crystallographic structures provided there is at least one slip system in each crystal and at least five independent orientations in an aggregate. Materials with low symmetry crystals and multiphase aggregates often display appreciable crystal to crystal variations in the deformation and are not modeled well using the Taylor assumption. Using the equilibrium-based model, the response of hexagonal close-packed crystalline aggregates with only basal and prismatic glide permitted is shown as an example of a single phase rank deficient system. The behavior of olivine and enstatite minerals is presented to illustrate the predicted response of a two phase system in which each phase is rank deficient.

MODEL EQUATIONS

An equilibrium state can be achieved in a trivial manner by requiring that the stress in a body be uniform. For a polycrystal, we enforce this condition simply by requiring that the stress in every crystal be the same. Because the macroscopic stress is the average of the crystal stresses, we obtain:

$$\boldsymbol{\sigma}' = \boldsymbol{\sigma}'^{c}, \tag{1}$$

where $\boldsymbol{\sigma}'$ is the deviatoric Cauchy stress and the superscript c refers to a crystal quantity. This relationship is straightforward to apply if it is the stress is that always is known, but more complicated if the macroscopic deformation rate is the known quantity. For the average of the crystal deformation rates to match the macroscopic value, we must also require:

$$\boldsymbol{d} = \frac{1}{N}\sum_{c} \boldsymbol{d}^{c}, \tag{2}$$

where $\boldsymbol{d}$ is the deformation rate and N is the number of crystals in the aggregate. [1] As we seek a description of the material for arbitrary conditions of stress or deformation rate, we will enforce both equations at all times. Equations 1 and 2 provide a link between the crystal aggregate and the macroscopic scale in which the overall motion is defined. Within an aggregate, equilibrium is achieved identically, but compatibility may be violated as no continuity constraint is placed on the motions at the interfaces of crystals.

Equations 1 and 2 are not sufficient by themselves to determine the stresses and deformation rates within the aggregate, however. To complete the system of equations three single crystal equations are introduced. First, the crystal deformation rate is the linear combination of shearing rates, $\dot{\gamma}$, of the active slip systems:

$$\boldsymbol{d}^{c} = \sum_{\alpha} \boldsymbol{P}^{(\alpha)} \dot{\gamma}^{(\alpha)}. \tag{3}$$

Second, the resolved shear stress, τ, on a given slip system may be computed by projecting the crystal stress onto the system:

$$\tau^{(\alpha)} = \boldsymbol{P}^{(\alpha)} : \boldsymbol{\sigma}'^{c}. \tag{4}$$

Third, the resolved stress on each system is related to the corresponding shearing rate by a nonlinear kinetic expression in which the resolved shear stress is scaled by a measure of the crystal's strength, $\hat{\tau}$:

$$\dot{\gamma}^{(\alpha)} = \dot{a}\left(\frac{\tau^{(\alpha)}}{\hat{\tau}}\right)\left|\frac{\tau^{(\alpha)}}{\hat{\tau}}\right|^{\frac{1}{m}-1} = f(\tau^{(\alpha)}, \hat{\tau})\tau^{(\alpha)}. \tag{5}$$

In Equations 3 and 4 the symmetric part of the Schmid tensor appears, which is given by:

$$\boldsymbol{P}^{(\alpha)} = \mathrm{sym}\left(\boldsymbol{b}^{(\alpha)} \otimes \boldsymbol{n}^{(\alpha)}\right), \tag{6}$$

where $\boldsymbol{n}^{(\alpha)}$ is the slip plane normal and $\boldsymbol{b}^{(\alpha)}$ is the slip direction. $\dot{a}$ and m are constitutive parameters and slip systems are denoted by the superscript α.

Equations 1 to 5 form a system whose solution renders the macroscopic stress in terms of the macroscopic deformation rate and the material state. To write the system in a form convenient for obtaining such a solution, we substitute Equation 3 into Equation 2, and then eliminate the slip system shearing rates using Equations 4 and 5:

$$\boldsymbol{d} = \frac{1}{N}\sum_{c}\sum_{\alpha} \boldsymbol{P}^{(\alpha)}\dot{\gamma}^{(\alpha)} = \frac{1}{N}\sum_{c}\sum_{\alpha} f(\tau^{(\alpha)}, \hat{\tau})\boldsymbol{P}^{(\alpha)}\boldsymbol{P}^{(\alpha)} : \boldsymbol{\sigma}'^{c}. \tag{7}$$

At this point the equal stress premise (Equation 1) is introduced to replace the separate crystal stresses with the macroscopic stress. As the macroscopic stress is identical for all crystals of an

[1] The relative importance of each crystal could be independently weighted if desired without changing the overall structure of the model.

aggregate, it may be factored out of the summations over both the slip systems of a crystal and the crystals of an aggregate. Thus, we obtain:

$$\boldsymbol{d} = \frac{1}{N}\left(\sum_{c} \mathcal{M}^c\right) : \boldsymbol{\sigma}', \tag{8}$$

where for each crystal:

$$\mathcal{M}^c = \sum_{\alpha} f(\tau^{(\alpha)}, \hat{\tau}) \boldsymbol{P}^{(\alpha)} \boldsymbol{P}^{(\alpha)}. \tag{9}$$

This may be written in terms of a (tensor-valued) residual as:

$$\mathcal{F}(\boldsymbol{\sigma}') = \boldsymbol{d} - \frac{1}{N}\left(\sum_{c} \mathcal{M}^c\right) : \boldsymbol{\sigma}' = 0. \tag{10}$$

$\mathcal{M}^c$ is implicitly dependent on the stress through τ, making the equation nonlinear. We obtain solutions using a Newton iteration, which has demonstrated robust convergence for a variety of crystal systems.

The mechanical states of crystals in an aggregate are defined by the lattice orientations and crystal hardnesses. These are updated by integrating appropriate evolution equations. For the lattice orientations the evolution equation is derived from the single crystal kinematics and from the requirement that the Euler spins are equal among all crystals [3]. Increases in the crystal hardnesses are described by a Voce hardening relationship [4].

APPLICATION TO HCP CRYSTALS

In many HCP materials slip along basal and prismatic systems occurs at resolved shear stresses considerably lower than that needed to induce slip on the pyramidal systems. Basal and prismatic slip systems alone contain only four independent slip systems and allow no extension in the c-axis direction of the crystal. In such cases the Taylor model fails since the flow stress would be unresolved for some orientations of the crystals.

Simulations were performed using an aggregate of 1000 initially randomly oriented grains for loadings corresponding to plane strain compression. Only basal and prismatic slip systems were allowed for the HCP crystals. The critical resolved shear stress was chosen to be the same for all slip systems and for all grains at all times, therefore allowing only for geometric hardening due to the reorientations of the grains. The texture evolution for two values of the rate-sensitivity parameters (m=0.06 and m=0.12) is shown in Figure 1 for a total plane strain compression of 1.0. The $< 0001 >$ pole figures show similar trends. As reported by Prantil et al. [5] the texture evolutions predicted by the model vary with rate sensitivity. When the rate-sensitivity is higher, a greater number of those crystals that remain non-aligned with the compression axis share the deformation for a fixed orientation distribution.

The deformation up to a strain of 1.0 is accommodated by only some of the grains. This appears clearly on the pole figures : some grains migrate along the rolling direction and approach fibers at 15 to 20 degrees from the normal direction while the others do not reorient. Grains with c-axes initially perpendicular to either the rolling or normal directions are the last ones to partake in accommodating the motion. By construction the model allows for strain heterogeneities within the aggregate. Over an increment of deformation only few grains deform plastically and rotate. These reorientations will often induce higher Taylor factors. As a result, subsequent deformations might be accommodated by other grains with more favorable orientations. Thus the texture of the whole aggregate will evolve over large strain deformations.

As an application we computed the texture evolution for a cross-rolled HCP aggregate. The first pass of plane strain compression was followed by a second pass in a direction rotated 90 degrees about the normal direction. Both passes were performed for a total strain of 1.0. The evolutions are shown as $< 0001 >$ pole figures in Figure 2. Interestingly, the trends for monotonic plane strain compression are repeated so that final textures exhibit symmetries about the rolling and the transverse directions.

APPLICATION TO MINERALS

One most promising application for the equilibrium-based model is for materials with low lattice symmetry, few available slip systems, and high rate-sensitivity. These are the properties of many geological minerals such as the two phase system consisting of olivine and enstatite. Olivine and enstatite slip systems are described the same way as in a previous study on plastic deformation of peridotites [6]. Both minerals possess kinematic rank deficiencies; that is, they have insufficient sets of slip systems to permit any arbitrary deformation.

In our simulations we assume that a polycrystalline aggregate is comprised of 200 crystals representing a sample of orientations. A value of 2/7 was chosen for m based on experiments [7]. We neglect work hardening since extensive recovery and climb are common in geophysical situations. Thus the slip system hardness remains constant during the deformations and all hardening of an aggregate can be attributed to geometric effects. We assume an initial random texture and simulate several monotonic deformation paths (pure shear and simple shear).

Strong preferred orientations develop as can be observed on the (100), (010) and (001) pole figures after 40% equivalent strain (Figures 3 and 4). As in aggregates deformed experimentally by pure shear, the slip plane normals of the soft slip systems ((010) for olivine and (100) for enstatite) align with the compression direction. Also, the slip directions of these soft slip systems ([100] for olivine and [001] for enstatite) orient quickly in the plane perpendicular to the compression axis. In simple shear the slip plane normal and the slip direction rotate in the sense of the macroscopic deformation. For simple shear deformations we predict asymmetric texture maxima for (010) in the case of olivine and (100) in the case of enstatite. These maxima are displaced against the sense of shear. Their positions do not change appreciably with increasing strain, just as described earlier. The predicted texture patterns are very similar to the results of viscoplastic self-consistent models [6], but develop more quickly.

SUMMARY

An equilibrium based model for the mechanical response of a polycrystalline aggregate is outlined. The model can serve as a lower bound on the stress for an applied deformation at fixed material state. It also serves as a physically reasonable representation for low symmetry crystal types having kinematically rank deficient slip geometries and high slip system rate sensitivities. Two examples are presented to illustrate the textures computed by the model under monotonic loading conditions.

REFERENCES

1. Sachs, G., VDIZ, 1928, **12**, 134.
2. Chastel, Y.B., Ph.D. Dissertation, Cornell Univ., 1993
3. Mathur, K.K., P.R. Dawson and U.F. Kocks, Mech.Mat., 1990, **10**, 183.
4. Chastel, Y.B. P.R. Dawson, K. Kristin and H.-R. Wenk, J.Geophy.Res., in press.
5. Prantil, V.C., Y.B. Chastel and P.R. Dawson, submitted to Acta Met.
6. Wenk, H.-R., K. Kristin, G.R. Canova and A. Molinari, J.Geophy.Res., 1991, **96**, 8337.
7. Bai, Q., S.J. Mackwell and D.L. Kohlstedt, J.Geophy.Res., 1991, **96**, 2441.

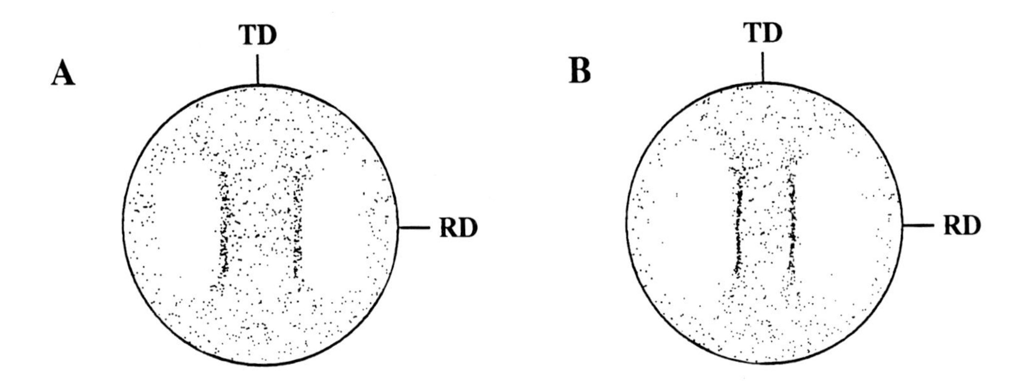

Figure 1: HCP polycrystal. < 0001 > pole figure after a total strain of 1.0 in plane strain compression. Rate-sensitivity parameter m = 0.06 (**A**) and m=0.12 (**B**).

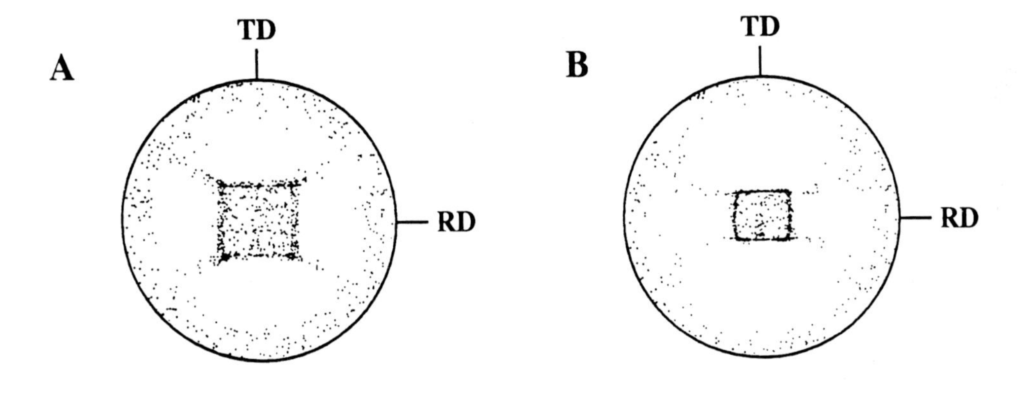

Figure 2: HCP polycrystal. < 0001 > pole figure after cross-rolling (strain of 1.0 each pass). Rate-sensitivity parameter m = 0.06 (**A**) and m=0.12 (**B**).
RD indicates the rolling direction for the first pass.

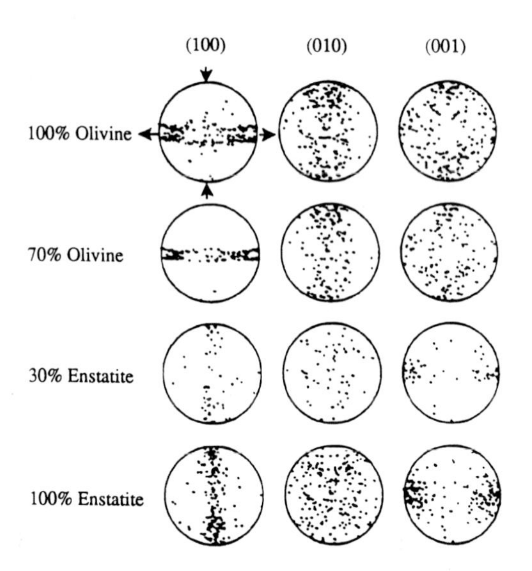

Figure 3: < 100 >, < 010 > and < 001 > pole figures after 40% equivalent strain in pure shear for pure olivine, 70%olivine-30%enstatite, and pure enstatite.

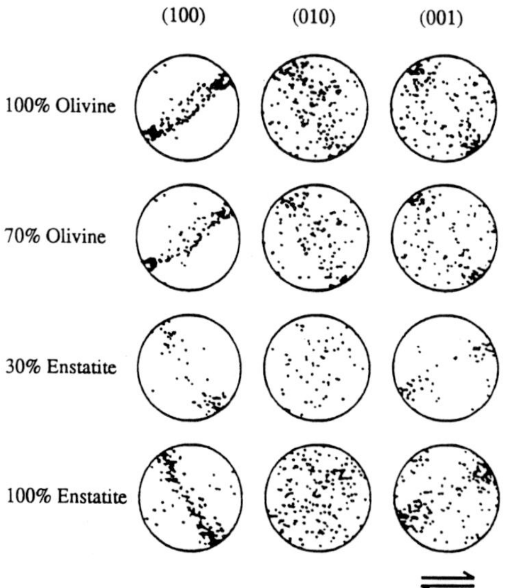

Figure 4: < 100 >, < 010 > and < 001 > pole figures after 40% equivalent strain in simple shear for pure olivine, 70%olivine-30%enstatite, and pure enstatite.

Materials Science Forum Vol. 157-162 (1994) pp. 1753-1758

INTERCONNECTION OF TEXTURE DEVELOPMENT AND ALTERNATIVE SLIPPING OF DIFFERENT TYPES OF SLIP SYSTEMS: COMPUTER SIMULATION

S.V. Divinskii and V.N. Dnieprenko

Institute of Metal Physics, Ukrainian Academy of Sciences,
36 Vernadsky str., Kiev-142, 252680, Ukraine

Keywords: Simulation, Limited Fibre Components, ODF, Properties Anisotropy

ABSTRACT: The model being intermediate between Sachs and Taylor approaches is developed in the present investigation. It was revealed that if to monitor on crystallographic indices of acting slip system in each elementary act of plastic deformation within the Sachs model, it may detect, simulating a texture development under rolling, that beginning with some (typically 20-30%) deformation stage an alternative slipping is developed. The parts of development of the main texture components in FCC and BCC polycrystals under rolling is obtained. BCC single crystals of different orientations are studied. Correlation of these results and experimental data is discussed.

Computer simulation of texture formation under plastic deformation both FCC and BCC metals was the aim of numerous studies. Nevertheless, the texture is usually considered as a whole, without accounting for consequences of different parts of formation of different texture components. These differences may result in different microstructure development.

Addition of the well known Sachs model by some new presentations acquired from experimental data is considered in the present work. It has been determined by the electron microscopy in a number of papers, *e.g.* in [1], that the correlation of structure and texture development is observed during the plastic deformation. Virtually each texture component can be characterized by its own type of the dislocation structure. It was faund [2] that two main types of dislocation structures, which are clearly distinguished in the plane normal to the rolling direction and can be attributed to the $\{110\}\langle\bar{1}12\rangle$ and $\{112\}\langle 11\bar{1}\rangle$ preferred orientations, respectively, may be unambiguously separated in the rolled copper. If we adopt the hypothesis of Wassermann [3] about the partial axiality of rolling textures we may restrict our analysis to consideration, mainly, of these two orientations. In [2] it was established that the $\{110\}\langle 1\bar{1}2\rangle$ texture component (called as "brass" one), as it might be expected, was developed through rotations induced by dislocation slips in the {111}-type planes. In turn, the development of second main texture component, namely, $\{112\}\langle 11\bar{1}\rangle$ or copper-type one, can be explained with the allowance for extra slips in the {100} planes. The direct electron-microscopical evidences of existence of such slip type were presented, *e.g.*, in [4]. More earlier, the assumption of possible copper-type component development as a result of the only {100}-type slips was suggested by Haessner [5] with hypothetical consideration of the crystallographic aspects of slips. Nevertheless, in the Taylor-type computer simulation of the texture development carried out in [6] the allowance for the {100} slips did not result in any substantial changes of the final crystal distribution in comparison with the use of only {111} slips.

In the present work the calculations were carried out based on the Sachs technique. This allows to reveal new features of development of different texture components both FCC and BCC metals.

In our computer model 12 octahedral slip systems $\{111\}\langle 110\rangle$ and 6 cubic slip systems $\{100\}\langle 011\rangle$ were taken into account for FCC metals and 12 systems $\{110\}\langle 111\rangle$ and 12 systems $\{112\}\langle 111\rangle$ did for BCC metals. Accounting for the "pencil"-like slip or others slip systems in the BCC metals was established to not result in significant changes of obtained rules.

The slip system with maximum value of $\Phi^{(s)}\cdot\nu^{(t)}$, where $\nu^{(t)}$ is the efficiency of activation of the *t*-th type slip system (*i.e.* $\nu^{(cub)}$ for cubic $\{100\}\langle 011\rangle$ systems and $\nu^{(oct)}$ for octahedral $\{111\}\langle 1\bar{1}0\rangle$ ones) and $\Phi^{(s)} = \sum n_i^{(s)}\cdot\sigma_{ij}\cdot b_j^{(s)}$ is the Schmid factor for *s*th slip system ($\bar{n}^{(s)}$ is the unit vector normal to the slip plane, and $\bar{b}^{(s)}$ is the unit vector coinciding with the slip direction; σ is the stress tensor), is considered as active one in this work. The choice of the $\nu^{(cub)}$ and $\nu^{(oct)}$ values was based on comparison of experimental and calculating model textures.

After the slip system had been selected, a certain amount of slip was performed (in the present calculations the amount of shear per slip event was varied in the range 0.01-0.05). The rotation accompanying each slip event was computed as in [7]. The calculations in a stepwise manner (slip system choice - rotation - slip system choice - rotation ...) had been carried out until the desired reduction of the thickness of the crystal was obtained. In a majority of the calculations the number of the crystals in question was 300.

The results presented below were obtained with using of the routine biaxial stress tensor, where only $\sigma_{xx}=-\sigma_{zz}\neq 0$ (here *0x* coincides with the RD and *0z* coincides with the normal to the rolling plane). The simulation showed that the small additions of various shear components cause no significant changes of the results obtained.

FCC POLYCRYSTALS

In the case of action of only $\{111\}\langle 1\bar{1}0\rangle$-type slip systems the texture being described by the $\{110\}\langle 1\bar{1}2\rangle$ and weak $\{110\}\langle 001\rangle$ components is developed, see Fig.1a.

Addition of the $\{100\}\langle 011\rangle$ slip systems as a possible mechanism of the plastic deformation results in the development of extra component in addition to mentioned above which is close to the $\{112\}\langle 11\bar{1}\rangle$ one, and, therefore, the copper-type texture is formed, see Fig.1b. Coinciding of the model component composition with experimental one for rolled copper is reached at $\nu^{(cub)}/\nu^{(oct)}=0.87$.

If we take into account slip systems of the only cubic type the texture reminding the one presented in Fig.1a and typical for the $\{111\}\langle 1\bar{1}0\rangle$-type slip will be developed. Thus, the copper component in Fig.1b is developed by the way which differs from simple slips in the cubic planes as it might be expected. For establishing the nature of the development of this component the more detailed analysis is to be carried out.

To establish the parts of the main texture components development we monitored the indices of the slip system with maximum loading (which will be therefore active at this moment) during every elementary act of the plastic deformation and for every grain. Plastic deformation by sequential slips in a single slip system was faund to be replaced by the alternative slips in a few slip systems after reaching the average strain of ~30%.

It was faund that sets of acting slip systems under formation of the $\{110\}\langle 1\bar{1}2\rangle$ and $\{112\}\langle 11\bar{1}\rangle$ is principally different. The $\{110\}\langle 1\bar{1}2\rangle$ component is developed as a result of the alternative actions of at least two octahedral slip systems, *e.g.* $(1\bar{1}1)[110]$ and $(11\bar{1})[101]$, whereas the copper component is developed by the alternative slips in the $\{111\}\langle\bar{1}10\rangle$ and $\{100\}\langle 011\rangle$ slip system types. The regularities of the slip system changes was shown by special symbols in Fig.1c and they divide strongly grains into components. Is is obvious that the Fig.1c is more informative than Fig.1b, although they represent the same texture.

The division was made as follows. The last 25 elementary deformation acts (before reaching the given reduction) were recorded. If a regularity in change of

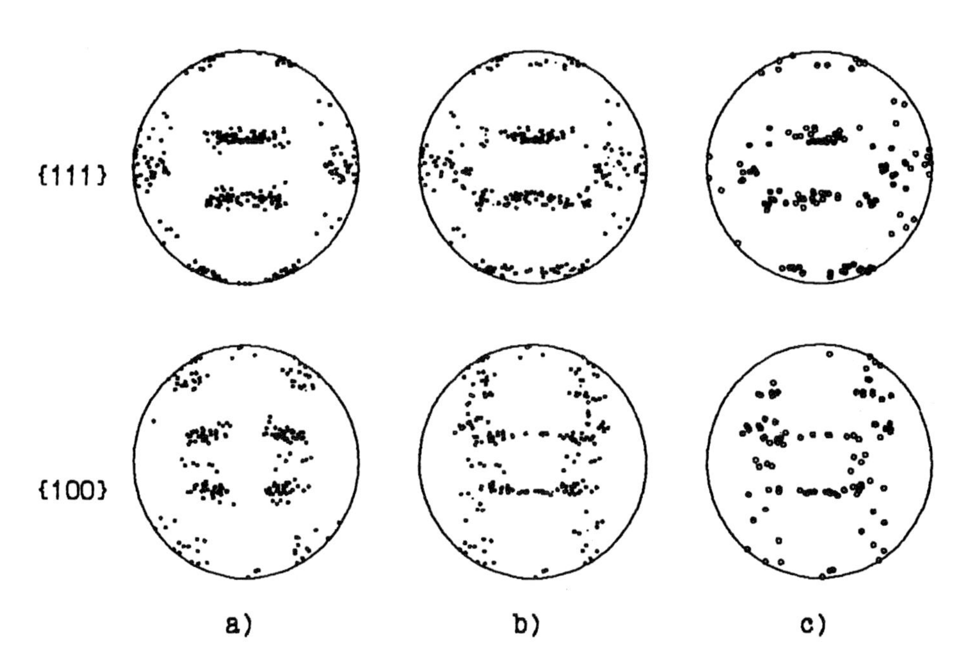

Fig.1 Model textures for different acted slip systems : {111}<110> (a); {111}<110> and {100}<011> (b). c) presents the same as (b) with marking of acted alternatively slip systems: {111}<110>+{100}<011> (.) and {111}<110>+{111<110> (o)

slip systems of different types (octahedral on cubic and reverse) is observed, this fact is marked as {111}<1$\bar{1}$0>+{100}<011>. Then slip systems of the same type is altered the symbol {100}<011>+{100}<011> or {111}<1$\bar{1}$0>+{111}<1$\bar{1}$0> is used. If no regularity is observed the symbol depends on the last acted system ({111}<1$\bar{1}$0> or {100}<011>).

The detail analysis of the number of the quasi-simultaneously (changing by turn one another) active slip systems revealed that both two and three (sometimes greater) independent slip systems may change regularly one another during last deformation stages.

The main components of the copper-type rolling texture would develop near the boundaries of regions of action of the different slip systems as the analysis carried out shows. These regions are pointed in Fig.2, the system with the maximum value of $\Phi^{(s)} \cdot \nu^{(t)}$ being considered as the active one. The systems with the {111} and {100} types slip planes were taken into account. The positions of the preferred orientations are in agreement with the regions of the grain orientation stability to such slip [8]. These regions are also marked in Fig.2.

Attentive analysis of grain distribution in Fig.1c and corresponding ODF shows that the spread of the grains is anisotropic. More rigorously, the spread of the grain orientations in the direction being perpendicular to the boundary of the regions of the action of different slip systems is clearly smaller than the spread along this boundary. This is in agreement with the concept of limited fibre components [9].

Thus, textures may be quite strictly resolved into separate components even with the allowance for their spreading up to overloading. The specific deformation mechanism of each grain plays the key role for this.

BCC POLYCRYSTALS

Dislocation slips mainly in two types of planes : {110} and {112} are considered in the analysis of acting slip systems in BCC metals. Slipping in other plane types also being considered, for example in the {123} planes, can be presented as some combinations of slips in the above mentioned main systems, as this was verified by a preliminary simulation.

Let us will mark the grain orientations by special symbols in dependence of slip system types acted on last deformation stages : ∘ – {110}<111>+{112}<111>; • – {112}<111>+{112}<111>, · – others.

If we assume that equal possibilities of activation of the {110}<111> and {112}<111> slip systems is in a case of equal threshold shear stresses, then a texture presented in Fig.3a will be taken for ε = 70%. Comparison of the pole figure with multiple experimental data reviewed in [3] proves that its only qualitatively corresponds pole figures typically observed in the BCC metals under rolling. The difference is in unusually greater part of grains oriented around of $\{112\}\langle\bar{1}10\rangle$. Only one case is known for us then an experimental pole figure corresponds enough to the Fig.3a. This is the case of cubic deformation martencite in a *Cr-Ni*-steel under rolling [11].

To achieve an agreement with experimental data [8] the following values of effectivities should be used : $\nu_{\langle 110\rangle}$ = 0.9, $\nu_{\langle 112\rangle}$ = 1.0. Model texture is shown in Fig.3b. It is seen, that in this case the pole density is in a good agreement with an experiment for majority of rolling textures in various BCC metals. Note, that slip systems of the mainly {112}<111> types are active here. And the {100}<011> texture component is connected with the {112}<111>+{112}<111} slip, and the {111}<112> one does with the {112}<111>+{110}<111> slip. The {112}<110> texture component is practically absence.

TEXTURE FORMATION UNDER BCC CRYSTALS ROLLING

It is most convenient to consider an interconnection of grains orientation, acting slip systems and microstructure being developed on the case of single crystals deformation. From our point of view the dislocation structure of rolled b.c.c. single crystals with the (001)[110] orientation is one of the most interesting. In this case it is formed no cellular dislocation structure. Let us consider dislocation slip and texture formation in BCC single crystals of some orientations from point of view of the approach suggested in the present work.

The single crystal was simulated as a polycrystal with a small spread around main orientation. The approach of [9] was used to ODF modeling.

(001)[110]. Simulation of the rolling texture shows that deformation is carried out by alternative slip of dislocations in two systems : $(112)[11\bar{1}]$ and $(11\bar{2})[111]$ and produce practically no dispersion of the initial orientation, see Fig.4a. It may be noted that the Burgers vectors of the dislocations and normals to the slip planes lie in the same plane formed by the rolling and normal directions. This is a reason of the fact that a dislocation structure in rolling planes consists randomly distributed dislocation segments and it forms no regular pile-ups of dislocations (since line of transaction of the slip planes is perpendicular to RD). It may be noted that in this case the picture of dislocation distribution in the plane perpendicular to transversal direction should be at least more regularly.

(001)[100]. Results of simulation of texture formation (Fig.4b) are in agreement with experimental data of non-stability of the orientation under rolling. The microstructure may be characterized by existence of two structural compounds in accordance with well known experimental data. As it is seen from Fig.4b, there are generally two deformation modes : (1) $\{112\}\langle 11\bar{1}\rangle+\{11\bar{2}\}\langle 111\rangle$ and (2) $\{\bar{1}12\}\langle\bar{1}11\rangle$+–$\{10\bar{1}\}\langle 111\rangle$. It should be noted that approximately equal parts of the modes (1) and (2) are realized at the value of ratio of effective activation probabilities $\nu_{\langle 110\rangle}/\nu_{\langle 112\rangle}$ = 0.9 adopted above. And practically one mode (2) is realized at

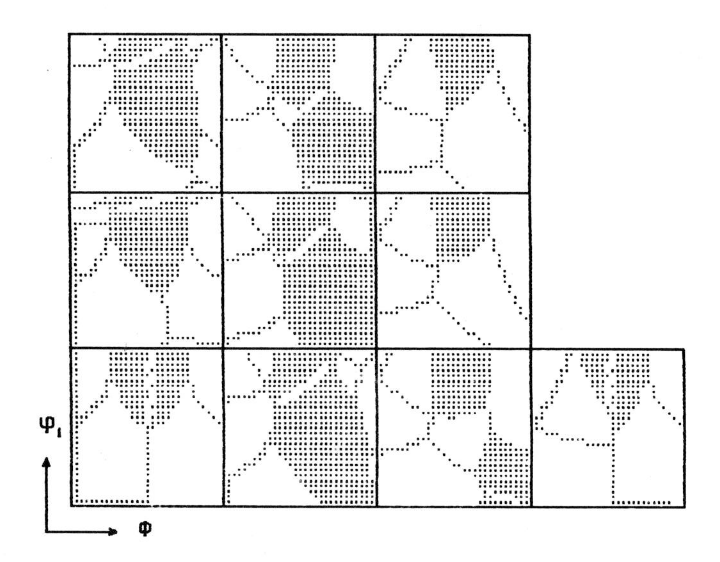

Fig.2. Regions of action of different slip system types:
dark - {100}<011>
ligth - {111}<110>

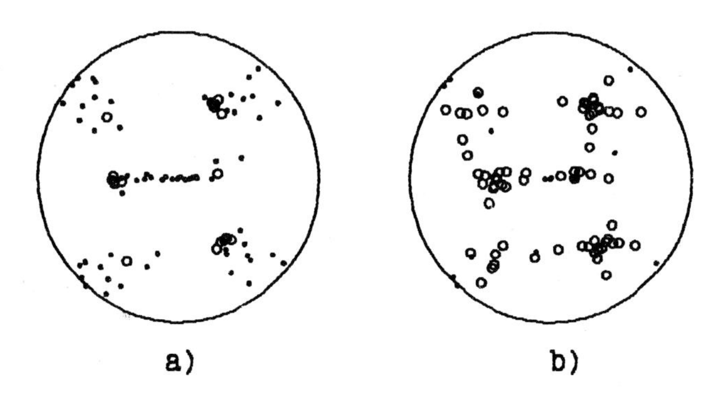

Fig.3. Model pole figures {100} of rolled BCC polycrystals.
a) presents the case of $\nu_{<110>} = 0.9$, $\nu_{<112>} = 1.0$ and
b) does $\nu_{<110>} = \nu_{<112>} = 1.0$. Slip systems acted alternatively :
∘ - {110}<111>+{112}<111>; • - {112}<111>+{112}<111>, · - others.

$\nu_{\{110\}}/\nu_{\{112\}} = 1.0.$

Analysis shows that the alternative slipping in the {112} type planes may give rise to random dislocation structure, and the alternative slipping in the {112} and {1$\bar{1}$0} type planes may form a substructure with subboundaries.

Thus, the developed approach to consideration of acted deformation mechanisms permits to derive separate components from a total texture and, in some cases, to predict a dislocation structures morphology.

REFERENCES

1)Vandermeer,R.A. and McHarque,C.J.: *Trans.Met.Soc.AIME*,1964,230,667.
2)Dnieprenko,V.N., Larikov,L.N. and Stoyanova,E.N.: *Metallofizika*,1982,4,58 (in Russian).
3)Wassermann,G. and Grewen,J.: *Texturen-Metallischer Werkatoffe*, Springer-Verlag, Berlin, 1962.
4)Wierzbanowski,K., Hihi,A., Berveiller,M. and Clement,A.: *Proc. 7th ICOTOM*, 1984,179.
5)Karthaller,H.P.: *Phil.Mag.*,1972,38,367.
6)Haeßner,F.: *Z.Metallk.*,1963,54,98.
7)Leffers,T.: *Phys.Stat.Sol.*,1962,25,337.
8)Divinski,S.V. and Dnieprenko,V.N.:Texture and Microstructure,1993. In press.
9)Dnieprenko,V.N. and Divinski,S.V.: Scripta Met.,1992,27,1617.
10)Dnieprenko,V.N., Karpovich,V.V. and Larikov,L.N.:Metallofizika,1987,9,1,15 (in Russian).

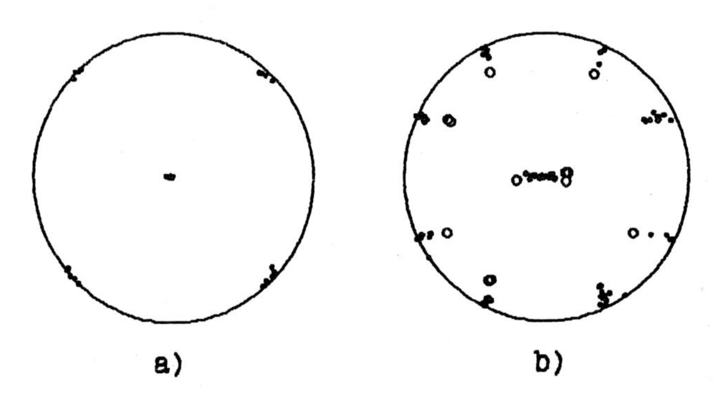

Fig.4. Model textures for rolling of (001)[110} (a) and (001)[100] (b) BCC single crystals. Marks the same as in Fig.3

Materials Science Forum Vol. 157-162 (1994) pp. 1759-1764

A NEW THEORY OF THE FCC ROLLING TEXTURE TRANSITION

B.J. Duggan [1], C.S. Lee [2] and R.E. Smallman [2]

[1] Department of Mechanical Engineering,
University of Hong Kong, Pokfulam Road, Hong Kong, Hong Kong

[2] School of Metallurgy and Materials,
University of Birmingham, Edgbaston, Birmingham B15 2TT, UK

Keywords: Texture Transition, Deformation Bands, Stability, Texture Components, SFE

ABSTRACT

Deformation banding is observed to be an important deformation mode and its inclusion into texture modelling in a Taylor framework produces textures in which S, C and B coexist. This result is based on octahedral slip and does not require any inhomogeneous deformation mechanism other than deformation banding. Extensive SEM and TEM microstructural investigation of α brass showed that S, B and G are present in significant untwinned quantities in the rolling range 60-85%. It is concluded that these are formed by normal slip plus deformation banding, in an identical manner to that in copper. The major difference in textures between high and low SFE metals arise, therefore, from the behaviour of C.

INTRODUCTION

Wassermann made the critical contribution to explaining the texture transition by insisting that normal slip in both copper and brass produces similar textures centred on C,S and B at 40%-50% reductions, but mechanical twinning in C crystals reorients them to {552}<115> which then rotates by octahedral slip to B via G. This picture remained substantially unchanged until interest focused in the 1970s on shear bands which form in brass after ~50% rolling and increase in number and volume as strain increases. Detailed work by Duggan et.al. (2) and Hutchinson et.al. (3) on shear banding has lead to a modified version of the Wassermann theory. The five stages identified in the development of the brass texture in their study are: (i) 0-50% reduction - deformation occurs by {111} slip in both copper and brass, producing similar copper-like textures; (ii) 50-60% reduction - extensive twinning occurs in brass in orientations close to {112}<111>; twins are fine, 0.02-0.2 μm thick and represent ~25% of the volume of twinned regions; (iii) 50-80% reduction - in twinned volumes slip is restricted by the twin boundaries to planes parallel to those boundaries, leading to overshooting such that orientations of the type {111}<uvw> are formed by coupled rotation; this component is detected in pole figures; (iv) 60-95% reduction - shear bands divide heavily twinned volumes, destroying {111} components. Coupled rotation and shear banding compete to produce a maximum 111 intensity in the range 80-90%, after which it declines. Shear bands are composed of crystallites of 0.02-0.3 μm in diameter in sheets ~1-2 μm wide, and there is some evidence of {110}<001> in the bands; (v) above 85% reduction - homogeneous slip processes in the increasing volume of shear band material is believed to lead to the stable {110}<112> texture. This model for the development of texture in α-stress is deficient in a number of ways, the most important concerning the behaviour of the orientations B and S at intermediate (i.e. 50%-80%) rolling reductions in low SFE materials.

EXPERIMENTAL OBSERVATIONS

(a) Copper

Figure 1 shows a SEM micrograph from the longitudinal section of a coarse grained copper specimen after etching by the Köhlhoff technique in which {111} planes of copper alloys are most slowly attacked and hence revealed. Layers 1 to 50 μm thickness are observed, the orientations of which are typical of rolled copper, i.e. S, B, G and C. Boundaries between the layers are often crystallographically sharp. The initial grain size was 3000 μm and the average grain thickness after 85% reduction should be ~450 μm but the average layer thickness was measured at four locations on two different specimens and found to be ~17 μm. This means that each grain has broken up to give ~25 layers of distinct orientation. Examination of the rolling plane section revealed the three dimensional nature of deformation banding (DB) and supports the idea that each grain has subdivided by DB into a large number of elements of different orientation. Orientation relationships allow twinning to be eliminated as a formation mechanism. Copper with a grain size of 50 μm was also prepared, rolled to the same strain and etched. On average each grain was split 2.5 times in the LS, so there is some evidence of a grain size dependence on deformation banding. A theory for this is developed in (4). Textures showed a higher concentration of {112}<111> in the coarse grained material.

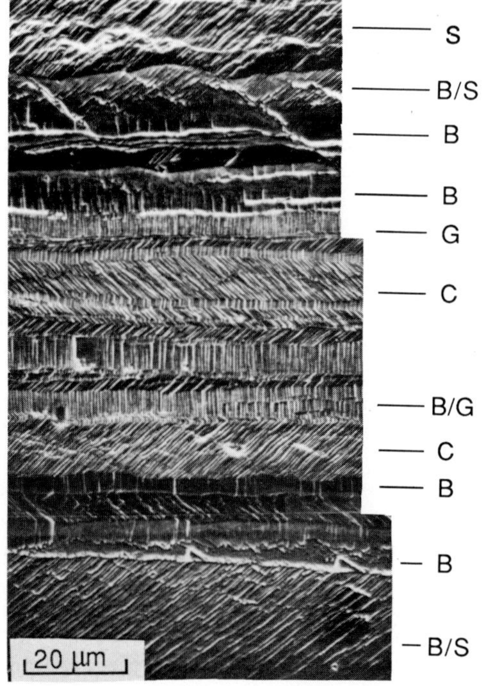

Figure 1 Coarse grained copper CR 85% showing deformation banding in the LS section.

(b) Brass

The components SG and B are readily found in brass of random starting texture when cold rolled in the range 50-85%, using both Köhlhoff etching and standard TEM methods. Figure 2 shows a TEM of fine grained brass rolled 85% with both twinned and non twinned S. B exists at this stage as homogenous blocks. A brass specimen with a strong B starting texture was prepared and rolled. B was found to be present throughout the rolling scheme, Fig. 3.

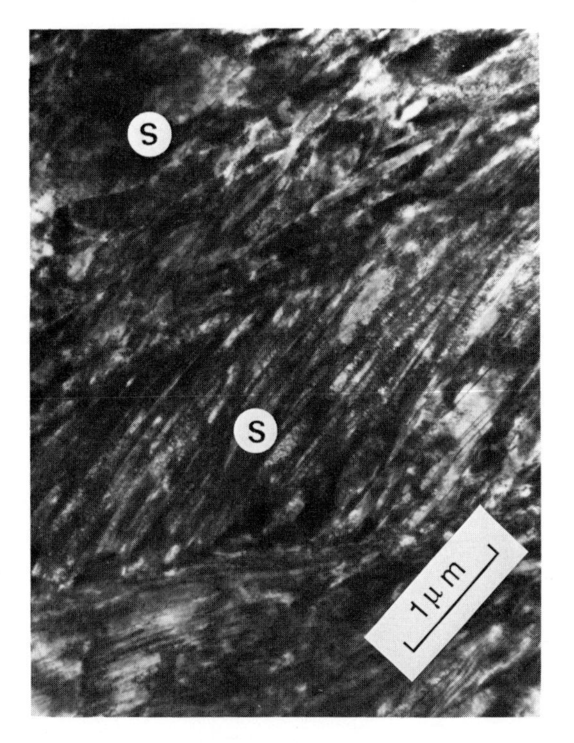

Figure 2 Fine grained brass cold rolled 85% showing twinned and non-twinned S.

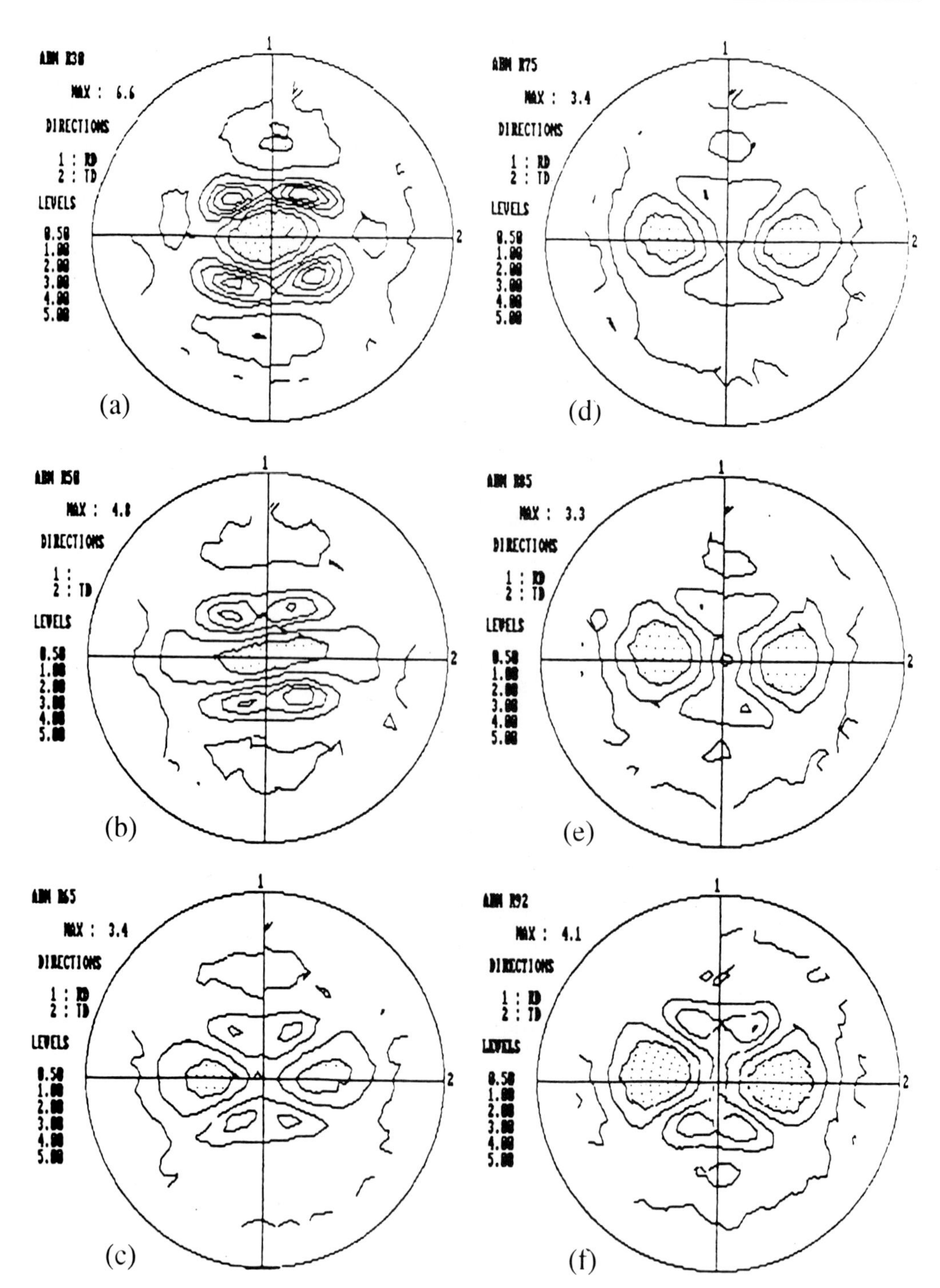

Fig. 3 111 pole figures showing persistence of B in 250 μm grain sized α brass from an initially strong B texture. (a) 30% (b) 50% (c) 65% (d) 75% (e) 85% (f) 92%

RESULTS AND SUMMARISING DISCUSSION

Models of texture formation based on Taylor theory have been reviewed by Hirsch and Lücke (5) who showed that in order to produce model textures similar to experimental ones, RC Taylor theory is essential. Different relaxations predict the major components C, S and B, but not their coexistence under the same conditions. By including DB into Taylor theory in a rather crude way, i.e. DB is allowed to relax e_{32} and RC to relax e_{21} and e_{31}, the coexistence of B, S and C naturally arises (6).

Stability of B arises because at intermediate strains it is evident from Köhlhoff etching that B forms complementary volumes, B and B' by DB. These complementary oriented volumes are common in copper in this strain range Fig. 4. Of course, there is nothing in the theory to suggest that low SFE material should behave differently, and experimentally B is found in the strain range 50-85% in abundance. If B does not form complementary volumes, it must be instable, but Fig. 3 shows that B is stable. This means that in principle, twinning and shear banding are not required for its formation. Hence its presence at low strains in brass can be explained by deformation banding. However, homogeneous slip in the fine crystallites was suggested as the mechanism considered to be responsible for the tremendous strengthening at high strain. This idea is developed further by noting that Chung et.al. (7) found extensive shearing to be occurring in this material, even though etched shear bands did not appear. It is therefore suggested here that homogeneous shear deformation of the fine crystallites in addition to homogeneous plane strain, produces {110}<112> at very high rolling reduction.

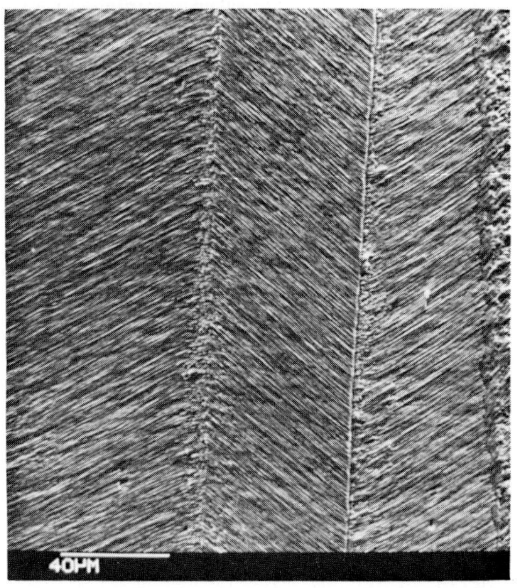

Fig. 4 Coarse grained copper cold rolled 70%, Rolling Plane section, TD parallel to micron marker. Complementary B oriented deformation bands showing sharp boundaries.

S in the brass texture

TEM studies show that there is a considerable volume of S in brass after 85% (Lee 1981) even though this component is not considered to be part of the brass texture. It can be shown that S twins to an orientation close to itself, and so is not destroyed by this process.

Deformation banding likewise produces an orientation close to S, and only after high strains (~90%) does it rotate away towards B. There is some evidence that S twins less frequently and at higher strain that C and these twinned grains behave similarly to {112}<111> in that co-rotation occurs. It is thus clear that S is rather stable in brass. The question and why it is rarely considered as part of the brass texture is considered by the authors to be due to the low resolution of 111 pole figures for this particular orientation, and the fact that, because of this lack of distinction it is included in the spread of orientations from C and B. Figure illustrates how a significant volume fraction of S and near S appears in a 111 pole figure.

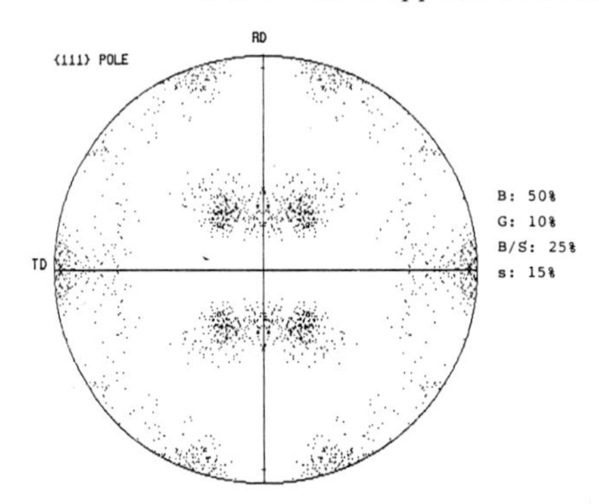

Fig. 5 Computer generated pole figure from four ideal orientations B, G, B/S and S with volume fractions of 50, 10, 25 and 15% respectively. The figure is plotted with 100 orientations generated by a standard deviation of 5° rotations on the ideal components about the ND, RD and TD axis.

The Texture Transition

B, C and S form progressively as deformation proceeds. B and S strengthen in the strain range 50-85% in both high and low SFE materials. B is stabilised by DB to form complementary volumes and S is relatively stable in all FCC materials. Several orientation produce G. In low SFE materials C twins, and it is the removal of C by this and the process of co-rotation which is responsible for the significant differences between low and high SFE materials known as the texture transition.

ACKNOWLEDGEMENT

C.S. Lee gladly acknowledges provision of EM facilities and support in terms of a studentship and Fellowship.

REFERENCES

1. G. Wassermann, Z. Metall., 1963, 54, 61.
2. B.J. Duggan, M. Hatherly, W.B. Hutchinson and P. Wakefield, Met. Sci., 1978, 12, 343.
3. W.B. Hutchinson, B.J. Duggan and M. Hatherly, Met. Technal., 1979, 6, 398.
4. C.S. Lee, B.J. Duggan and R.E. Smallman, Acta Metall et Mater., 1993, 41, 2265.
5. J. Hirsch and K. Lücke, Acta Metall., 1988, 36, 2883.
6. C.S. Lee and B.J. Duggan, Acta Metall., accepted for publication, 1993.
7. C.Y. Chung, B.J. Duggan, M.S. Bingley and W.B. Hutchinson in Proc. 8th Int. Conf. on Strength of Metals and Alloys, 1988, Vol. 5, 319.

Materials Science Forum Vol. 157-162 (1994) pp. 1765-1770

EFFECT OF CUBE NUCLEUS DISTRIBUTION ON CUBE TEXTURE

B.J. Duggan [1] and C.Y. Chung [2]

[1] Department of Mechanical Engineering,
University of Hong Kong, Pokfulam Road, Hong Kong, Hong Kong

[2] Department of Physics and Materials Science,
City Polytechnic of Hong Kong, Hong Kong

Keywords: Cube Texture, Nucleation, Grain Size, Strain, Growth Range

ABSTRACT

It is shown that cube oriented layers appear once in 10 to 20 layers, a range which is not strongly affected by original grain size nor rolling reduction in the range 90-98%, but the layer thickness for all orientations depends on the original grain size, being thicker in coarse grained copper. Growth range of viable cube nuclei is an important factor in determining the strength of the recrystallization cube texture. A model correlating initial grain size, deformation strain, the measured growth range and volume fraction of cube grains in annealing is proposed.

INTRODUCTION

Many factors influence the development of cube texture in FCC materials, such as rolling temperature, purity, rolling degree and original grain size etc. The development of a crystallographic etch for copper in which 111 planes are revealed has allowed new kinds of information to be gained, and it is the purpose of this paper to describe the effect of cube deformed layer distribution with increasing strain, on cube texture development.

EXPERIMENTAL

A plate of 10 mm thick phosphorus-deoxidised Direct Chill cast copper with 1mm x 1 mm columnar grains through the thickness was prepared which had a weak <100> fibre texture parallel to ND, **(Batch A)**. Part of the material was processed to give 20 μm equi-axed grains and weak textures **(Batch B)** and high purity copper with initial grain size 50 μm and random texture comprised **Batch C.** These materials were then cold rolled to more than 90% reduction and annealed at 100°C or 300°C in an air circulation furnace. Edge sections were examined in the SEM using an etch developed by Köhlhoff (1) in which 111 planes are exposed, thus allowing orientation to be determined at a resolution of 2°. Resultant grain sizes were measured from an image analyser.

RESULTS AND SUMMARISING DISCUSSION

1. DEFORMED MICROSTRUCTURE

All the rolled materials show a typical layer structure which is strain as well as initial grain size dependent because of deformation banding.

A comparison between the measured layer thickness and the expected grain thickness estimated from homogeneous deformation is shown in Table 1.

	Phosphorus-deoxidized DC-copper				High Purity Copper	
	1 x 1 x 10 mm³ **A** columnar grain		20 μm equiaxed grain **B** fine grains		50 μm equiaxed grain **C** medium grains	
	Calculated	Measured	Calculated	Measured	Calculated	Measured
90% CR	1000 μm	100 μm	2 μm	1.9 μm	5 μm	2 μm
92% CR	--	--	--	--	4 μm	1.7 μm
95% CR	500 μm	55 μm	1 μm	1 μm	--	--
96.6% CR	--	--	--	--	1.7 μm	1.1 μm
97% CR	300 μm	28 μm	0.6 μm	< 1 μm	1.5 μm	< 1 μm

Table 1 Comparison between the estimated grain thickness and the measured average layer thickness after heavy cold rolling reduction.

For Batch **B** materials, the measured layer thickness is close to the expected grain thickness. This implies that the deformation of individual grains conforms to the homogeneous deformation assumption. However, the estimated grain thicknesses of **A** and **C** materials are larger than the measured layer thicknesses. Grain subdivision by deformation banding is therefore extensive and is also completed at rolling strains below 90% in **A** and **C**. It is clear that deformation banding occurs more frequently in large grain sized copper. The ratio of layers expected/no. of layer measured, are shown in Table 2, which for comparison purposes includes results obtained by Lee et al. (2).

Initial grain size	20 μm, B	50 μm, C	Columnar grain, A	30 μm	500 μm	3000 μm
Deformation banding per grain	1	2	10	1.7*	4.5*	25*

Table 2 Relationship between initial grain size and frequency of deformation banding.
* *Result from Duggan et al. ( )*

Cube oriented material was clearly visible in a section parallel to LS and at 45° to ND as unambiguous stripes (3) and it was possible to determine area fraction of cube material. In addition, the average distance between cube bands was measured and is shown in Table 3. Cube bands are rare in the original columnar material and so the average distance between cube layers could not be determined. Instead the shortest distance between cube bands was recorded for reference, and the thickness of these cube bands was 5 and 3.5 μm after 90 and 95% cold rolling reduction respectively.

	Columnar grains A		20 μm equiaxed grains B		50 μm equiaxed grains C			
% Cold rolling	90%	95%	90%	95%	90%	92%	96.6%	97%
Average distance between cube layers (μm)	500*	115*	20	12**	69	62	35	16
Average number of layer between cube layers	--	--	10	12	34	36	32	--

Table 3 The amount of cube layer found in the cold rolled polycrystalline copper.
* *Cube oriented bands are rare in columnar materials. The figure quoted is the only data recorded.*
** *Cube bands in rolled fine-grained copper are too thin to be determined accurately.*

Cube bands were found frequently in the rolled 20μm and 50μm polycrystalline copper. The length of the cube bands was between 100 and 1000 μm at 95% rolling strain for the 50μm equiaxed grains. It was found that cube oriented layers appear once in about 30 layers (average = 15 grains, standard deviation = 5 grains), and the lineal fraction of cube oriented layers is about 3.3% and these cube layers were distributed randomly. This result is in agreement with Sindel et. al. (4) who used X ray methods who also reported that the volume fraction of cube bands remained unchanged when the rolling strain was increased to 98%. Thus cube oriented layers are produced (or retained) by cold rolling from an initially random texture.

Orientation	Volume Fraction (%)	Orientation	Volume Fraction (%)
B-{1 1 0}<1 1 2>	30.0	G-{1 1 0}<0 0 1>	14.9
S-{1 2 3}<6 3 4>	11.1	C={1 1 2}<1 1 1>	9.4
S/C	9.4	C/S	4.1
Cube-{0 0 1}<1 0 0>	3.3	B/G	1.6
G/B	0.3	Others	15.9
		Total:	**100.0**

Table 4 Volume fraction of various orientations after 92% cold rolling reduction as determined from the longitudinal section and 45°-longitudinal section (initial grain size 50 μm).

The volume fraction of other orientations for 92% cold rolled copper with initial grain size 50 μm was also determined and were classified into 10 orientations ($\pm$15° misorientation) as shown in Table 4. Grains with orientations between C and S were classified as C/S when they were closer to C etc. Some elongated grains were too thin to allow determination, and other etch patterns were not clear; these ambiguous orientations were classified as 'others'. These direct measurements should be compared with the indirect X-ray measurements (5).

2. GENERAL ANNEALING BEHAVIOUR

For batch A, cube annealing texture did not develop. Köhlhoff etching of the 45°-LS section revealed that recrystallized cube grains were rare and contributed to not more than 2% by volume. However, after 99.1% cold rolling, cube volume increased to about 5%. The size of cube grains was similar to grains with other orientations in the materials. For batch B, cube annealing texture was not strong even after 95% rolling. Cube grains occupied about 12% of the volume. The recrystallized grains were all equiaxed with a nominal size of 9.4 μm, compared with the average grain size of 5.5 μm. C gave the strongest cube annealing texture. The cube volume fraction after annealing at 300°C for 1 hour against rolling strain is plotted in Fig. 1. Cube grains are large and usually elongated along RD, while the grains with other orientations are equiaxed which suggests that the cube grains are derived from the elongated flat layers in the deformed material.

The growth of cube grains was followed using a lower annealing temperature of 100° which allowed better measurements but did not change the final annealing texture cube volume fraction. After 92% rolling, annealing at 100°C for 10 hours gave 27 cube nuclei in an area covering 1200 grains. The number of cube nuclei was increased to 50 after 20 hours. It was further found that the number of cube nuclei was always about 50 in 1200 grains for the strain range from 92% to 96%. This implies that approximately 1 out of 25 grains give cube nuclei in C.

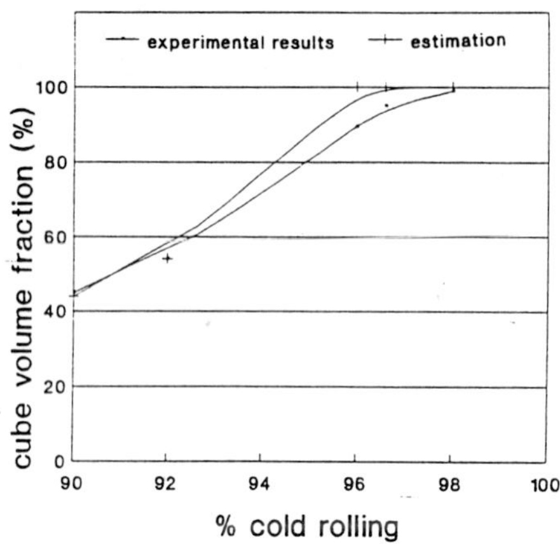

Fig. 1 *Experimental area fraction of cube oriented materials vs rolling strain. Also shown is an estimated volume fraction based on growth range theory. (D_L=25 μm, initial grain size = 25 μm, 4% of the grains give cube nuclei)*

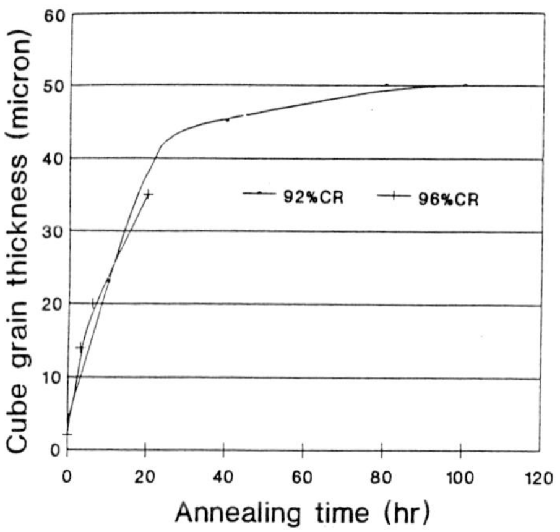

Fig. 2 *Average cube grain thickness vs annealing time after different rolling strains.*

The above results show that the number of cube layers is about 1.6 times higher than the cube nuclei formed in the partially annealed state. A selection process is obviously operating. Similar results were reported by Duggan et al., (3) recently. They found that cube layers are contained between a wide variety of neighbouring orientations and successful cube nuclei are cube layers sharing a common <111> with the adjacent misoriented materials, a process they designated as Micro Growth Selection.

3. EFFECTS OF STRAIN ON CUBE

Average thickness of cube grains increased during annealing was measured and it was found that the boundary migration rates in ND do not differ much between the rolling strain 92% and 96%, Fig. 2. Cube grains in the 96% cold rolled copper impinged after 20 hours of annealing. Beyond this point, cube grains joined together into much larger blocks which made the determination of average cube grain thickness very difficult. However, sometimes, slight misorientation existed between two adjacent recrystallized cube grains and this allowed the average cube grain thickness to be roughly determined. An interesting result from this is that the average cube grain thickness when fully recrystallized was always about 40 to 50 μm for all rolling strains above 90%, Table 5. This means that the grain size of cube grains is independent of rolling strain.

Rolling strain (%)	90	92	96	97	98
Average cube grain thickness, RD (μm)	43	50	50*	40*	45*

Table 5 *Average cube grain thickness with respect to rolling strain.*
* *Results determined from the cube grains with slight misorientations.*

Assuming most cube grains grow parallel to ND, the maximum distance that the cube grain boundaries can migrate is about 20-25 μm. This suggests that, there exists *limiting growth distances* D_L for cube grains. We now use the idea of a *limiting growth distance*, D_L, for cube grains. Clearly, D_L is the impingement distance when cube layers are relatively close, i.e. at very high strains. At lower strains D_L is determined by solute drag, reduction in driving forces due to recovery, and the successful nucleation of crystals of other orientations. But in the strains investigated here, growth rates are identical, hence recovery etc. is no more important after 92% than after 96% rolling.

Taking the reported density of cube nuclei giving grains (approximately 1 out of 25 grains give cube nuclei), the average distance between cube nuclei L_{cube} can be estimated as a function of rolling reduction.

$$L_{cube} = (1 - \frac{n}{100})\, Gd \tag{1}$$

where G is the initial grain size, n is the percentage cold reduction and 1 out of d grains (25 in this case) will give cube nuclei. Assuming that D_L is constant, the volume fraction of cube grains V_{cube} can be approximated by only considering boundary migration along the direction ND, and this gives

$$V_{cube} = \frac{2 D_l}{L_{cube}} \tag{2}$$

A comparison between the estimated cube volume fraction and experimental results after annealing with respect to rolling strain is plotted in Fig. 1 using 25 μm for D_L. It is clear that the simple formula provides a good prediction of cube volume fraction in the heavily rolled copper. From Fig. 2 D_L could be calculated and, D_L was fairly constant at strain lower stains, Fig. 3. At higher strains, cube impingement occurs and D_L drops accordingly. Impingement distance, I.D., will be shorter when every cube layer nucleates successfully, I.D. (100%), while it will be longer when less cube layers nucleate. For example, the I.D. (50%) line in Fig. 3 represents the impingement distance when 50% of the cube layers nucleate and grow successfully.

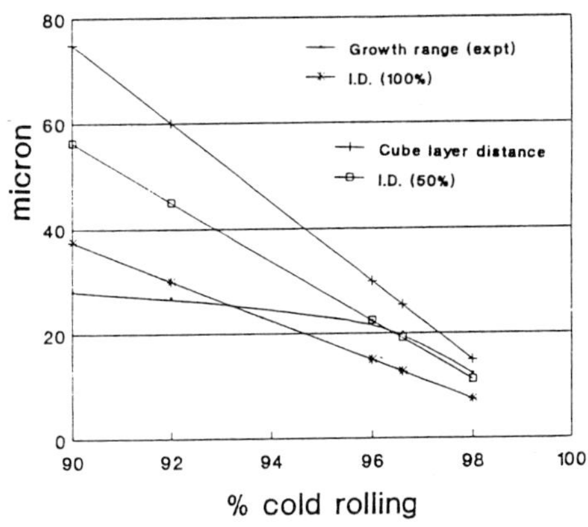

Fig. 3 *A plot of growth range against % cold rolling. Also shown is the average distance between cube deformed layers and impingement distance, I.D. (100%) and I.D. (50%). (Initial grain size = 50 μm, 1 out of 15 grains will produce cube layer after heavy deformation)*

4. THE ROLE OF GRAIN SIZE IN CUBE TEXTURE FORMATION

Consider a volume of polycrystal containing some grains with special properties but the average size of the special grains is the same as others. It can be shown that the volume fraction of the special grains is equal to the corresponding areal and lineal fractions. Cube bands were found in every 10-20 layers after heavy rolling and only about 60% of them can become viable nuclei. Therefore, the average distance between two cube layers is 17 grain thicknesses. After heavy rolling reduction, all grains become pancake shape. The average distance between cube nuclei giving grains is shortened along ND, lengthened along RD, and remains unchanged along TD. The average distance between cube nuclei along ND is only 43 μm if the initial grain size is 50 μm. As the cube nucleation sites are pancake shape, the growing front in the direction ND is most significant because the growth in direction gives most contribution to the increase in the recrystallized volume. If we further imposed a *limiting growth distance* to all strains and grains size, the final cube volume fraction can be predicted accordingly. The result is shown in Fig. 4. Hence the initial grain size will determine the density of cube nuclei and thus the cube volume fraction after rolling and annealing. The fine grained copper will always give strong cube recrystallization texture. The strength of cube in the normal and medium grain size copper depends very much on the rolling strain, but the cube strength of coarse grained copper is always low. When the impurity effect is high (equivalent to a small D_L), cube volume fraction will be significantly lowered. The effect is shown in Fig. 4. Cube volume fraction will be significant for a limiting growth distance of 5 μm only if the rolling strain is very high or the initial grain size is very fine, Fig. 4.

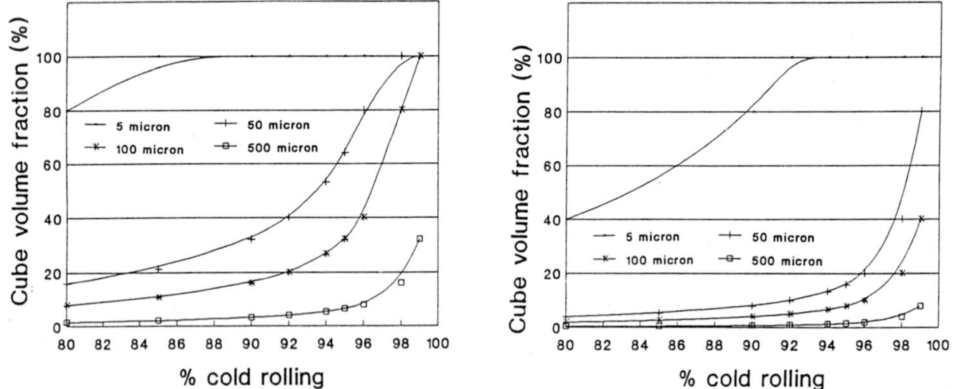

Fig. 4 The esitmated variation of cube volume fraction with respect to rolling for materials with different initial grain size and limiting growth distance D_L for cube grains. About 4% of the grains are assume to give cube nuclei. (a) D_L = 20 μm, (b) D_L = 5 μm.

REFERENCES

1) G.D. Köhlhoff, X. Sun and K. Lücke, ICOTOM 8, 1988, 183.
2) C.S. Lee and B.J. Duggan, Acta Met et Mater., 1993, 41, Sept.
3) B.J. Duggan, K. Lücke, G.D. Köhlhoff and C.S. Lee, Act Met et Mater., 1993, 41, 1921.
4) M. Sindel, G.D. Kohlhoff, K. Lücke and B.J. Duggan, Textures and Micros., 1990, 12, 37.
5) J. Hirsh and K. Lücke, Acta Metall., 1988, 36, 2863.

Materials Science Forum Vol. 157-162 (1994) pp. 1771-1776

TEXTURE DEVELOPMENT AND SIMULATION OF INHOMOGENEOUS DEFORMATION OF FeCr DURING HOT ROLLING

A.I. Fedosseev, D. Raabe and G. Gottstein

Institut für Metallkunde und Metallphysik,
RWTH Aachen, Kopernikusstr. 14, D-52056 Aachen, Germany

Keywords: Superposition of Harmonic Currents, Simulation, Hot Rolling, Stainless Steel

ABSTRACT

In present paper the through thickness inhomogeneity of deformation and texture development of a hot rolled FeCr steel was experimentally inspected and simulated with respect to the geometry of the rolling gap and the friction between band and rolls.

INTRODUCTION

The texture can be determined from pole figure measurements and subsequent computation of the ODF [1]. Textures of FeCr steels are very inhomogeneous through the thickness. Close to the surface a strong Goss component and in the center layer a strong α-fibre texture, i.e. a "cold rolling" texture is developed. For a more reliable assessment of this inhomogeneity a systematic analysis of the involved parameters and its correlation with the inhomogeneity of texture is required.

MODELLING OF THE ROLLING PROCESS

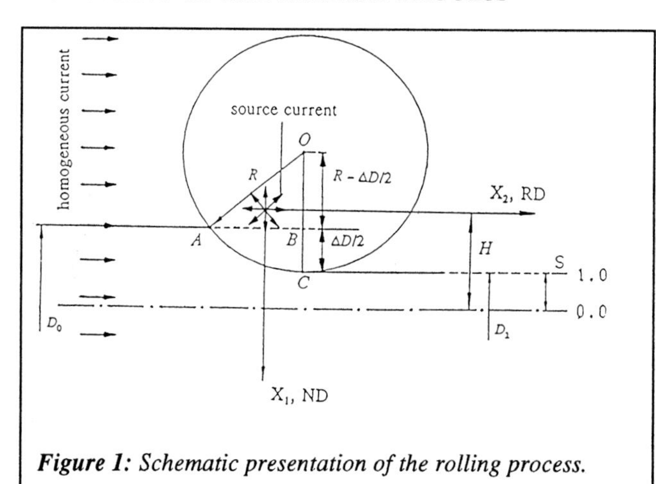

Figure 1: Schematic presentation of the rolling process.

The rolling process was modelled by a new computational method which is referred to as the method of **S**uperposition of **H**armonic **C**urrents (SHC) [2]. In comparison to FEM, the SHC-method is less CPU time consuming. SHC is based on concepts of fluid dynamics [3]. In the current study rolling without broadening was assumed so that a 2-dimensional simulation was adequate. This geometry can be treated in the complex plane. The rolling process was simulated by superposition of a homogeneous current along the positive direction of the imaginary axis (rolling direction: $RD \equiv X_2$) with an infinite number of source currents located equidistantly (pints z_n) on the real axis (normal direction: $ND \equiv X_1$) $z_n = \pm 2nH$ with $n \in N$ and H - scale factor (figure 1). The flux equation allows an analytical solution, obtainable by functions for complex variables [4]. The complex potential

$w=\varphi(X_1,X_2)+i\Psi(X_1,X_2)$ of material flow can be described as:

$$w = \frac{I}{2\pi}\, ln\; sin\, \frac{\pi Z}{2H} - iV_\infty Z, \tag{1}$$

$$with \quad I = 2\,H\,V_0\,C_1\,; \quad C_1 = \frac{D_0 - D_1}{2\,D_1}\;; \quad Z = X_1 + iX_2\,;$$

$$V_\infty = V_0\,C_2\,; \quad C_2 = \frac{D_0 + D_1}{2\,D_1}\,,$$

V_0 - velocity of rolls, X_i - complex coordinates, D_0 and D_1 - thickness before and after rolling.

The scale factor H relates the flow parameters (the distance between the sources, the position of the coordinate center) to the rolling parameters (R, V_0, D_0, D_1), where R is the radius of the rolls. For determination of H the transcendental equation R=f(H) was numerically solved by means of the Newton method such that the curvature of the surface flow line fits the shape of the rolls :

$$R = \frac{H\,C_1\,K_r^{3/2}\, sin^2 \dfrac{\pi\, r}{2\,H}}{\pi\; C_2}\,, \tag{2}$$

$$with \quad K_r = 1 + \frac{C_2^2}{C_1^2}\, ctg^2\, \frac{\pi\, r}{2\,H}\,, \quad r = \frac{D_0 D_1}{D_0 + D_1}\,.$$

The imaginary part of the current potential $\Psi(X_1,X_2)$ is the scalar flow function. The flow velocities in ND, V_1, and in RD, V_2, are calculated by the Cauchy-Riemann conditions. The strain rate tensor along RD can be computed by differentiation of V_1 and V_2 assuming steady state flow:

$$\dot{\varepsilon}_{11} = -\dot{\varepsilon}_{22} = -V_0\, \frac{\pi\, C_1}{H}\, \frac{cos\dfrac{\pi\, X_1}{H}\; ch\dfrac{\pi X_2}{H} - 1}{\left(ch\dfrac{\pi\, X_2}{H} - cos\dfrac{\pi\, X_1}{H}\right)^2}\,; \tag{3}$$

$$\dot{\varepsilon}_{12} = \dot{\varepsilon}_{21} = V_0\, \frac{\pi\, C_1}{H}\, \frac{sin\dfrac{\pi\, X_1}{H}\; sh\dfrac{\pi\, X_2}{H}}{\left(ch\dfrac{\pi\, X_2}{H} - cos\dfrac{\pi\, X_1}{H}\right)^2}\,.$$

The coefficient $\Delta\Lambda_\Sigma$ represents the "inhomogeneity of deformation history" and describes the arc length of the deformation path in the different layers through the thickness.

In the current work it is studied, whether $\Delta\Lambda_\Sigma$ is a suitable parameter for assessing the texture inhomogeneity of a hot rolled Fe16%Cr band. The parameter Λ describes the deformation history between the start t_s and the end t_e of deformation (eq. 4).

$$\Lambda = \int_{t_s}^{t_e} \eta(\tau)\, d\tau, \qquad (4)$$

with η - shear strain rate [5] as function of the second invariant of the deviatoric strain rate (eq.3). As a degree for deformation inhomogeneity we selected the coefficient of strain inhomogeneity $\Delta\Lambda$:

$$\Delta\Lambda = \frac{\Lambda_{max} - \Lambda_{min}}{\Lambda_{max}} 100\%, \qquad (5)$$

where Λ_{max} and Λ_{min} are the maximum and minimum values of the parameter Λ (eq.4) through the sheet thickness. The limiting case ($\Delta\Lambda=100\%$) corresponds to maximum inhomogeneous deformation. The coefficient Λ_{min} is equal to zero in the central layer of the sheet (eq.5), i.e. the deformation (shear and without shear) does not proceed to the central layer. The other limiting case $\Delta\Lambda=0\%$ corresponds to homogeneous deformation, i.e. deformation is constant through the sheet thickness $\Lambda_{max}=\Lambda_{min}$ (eq.5). In the most general case of rolling the parameter $\Delta\Lambda$ depends on the rolling schedule. The coefficient of accumulated strain inhomogeneity $\Delta\Lambda_{\Sigma}$ according to the SHC method after n passes then reads:

$$\Delta\Lambda_{\Sigma} = \frac{ln\left[\prod_{i=1}^{n}\left(tg\frac{\pi D_{0i}}{4H_i}\, ctg\frac{\pi D_{1i}}{4H_i}\right) / (D_0/D_1)\right]}{\Lambda_{\Sigma max}} 100\%. \qquad (6)$$

It is emphasized that eq.(6) depends only on the rolling parameters and can be used only, if the geometry of the rolling gap is the sole cause of inhomogeneous deformation.

EXPERIMENTAL PROCEDURE, RESULTS AND DISCUSSION

The Fe16%Cr steel (D_0=24 mm) was industrially hot rolled to a thickness 2.3 mm with different reductions per pass (sequence 1, in Table 1) (roll diameters 750 mm for the first three passes and 650 mm for the other four passes). The temperature during rolling was decreased from 1143°C to 960°C. The different rolling sequences 2 and 3 (Table 1) were choosen, since the inhomogeneity is strongly influenced by the choice of thickness reductions. The parameter $\Delta\Lambda_{\Sigma}$ (eq.6) predicts a smaller inhomogeneity of deformation ($\Delta\Lambda_{\Sigma}$=4.96%) for sequence 1 (decreasing reductions per pass), a larger inhomogeneity of deformation ($\Delta\Lambda_{\Sigma}$=8.66%) for 3 (increasing reductions) and a

Table 1: The reduction regimes for the three rolling sequences

Nº	Thickness after i^th pass, reduction for i^th pass D_i /mm; $(D_{i-1}-D_i)/D_{i-1}$x100 /%						
	1st pass	2nd pass	3rd pass	4th pass	5th pass	6th pass	7th pass
1	14.6; 39.2	9.2; 37.0	6.1; 33.7	4.3; 29.5	3.2; 25.6	2.5; 21.9	2.3; 8.0
2	17.2; 28.3	12.3; 28.5	8.8; 28.5	6.3; 28.4	4.5; 28.6	3.2; 28.9	2.3; 28.1
3	20.0; 16.6	16.0; 20.0	12.0; 25.0	8.5; 29.2	5.7; 32.9	3.8; 33.3	2.3; 39.5

mean value ($\Delta\Lambda_{\Sigma}$=6.19%) for sequence 2 (equal reductions). Figure 2 shows the through thickness distributions of Λ for different rolling sequences from pass to pass. The texture in the case of the rolling sequence 1 was determined with an automated X-ray texture goniometer by means of pole figure ({110}, {200}, {112} and {103}) measurements and subsequent ODF computation and was analysed in Δs=0.1 steps to show the successive change of the texture of the hot rolled band (figure 3). The maximum shear texture, i.e. the Goss orientation {011}<100> is located in the sheet layers close to the surface (s≈0.78). Approaching the center layer the intensity

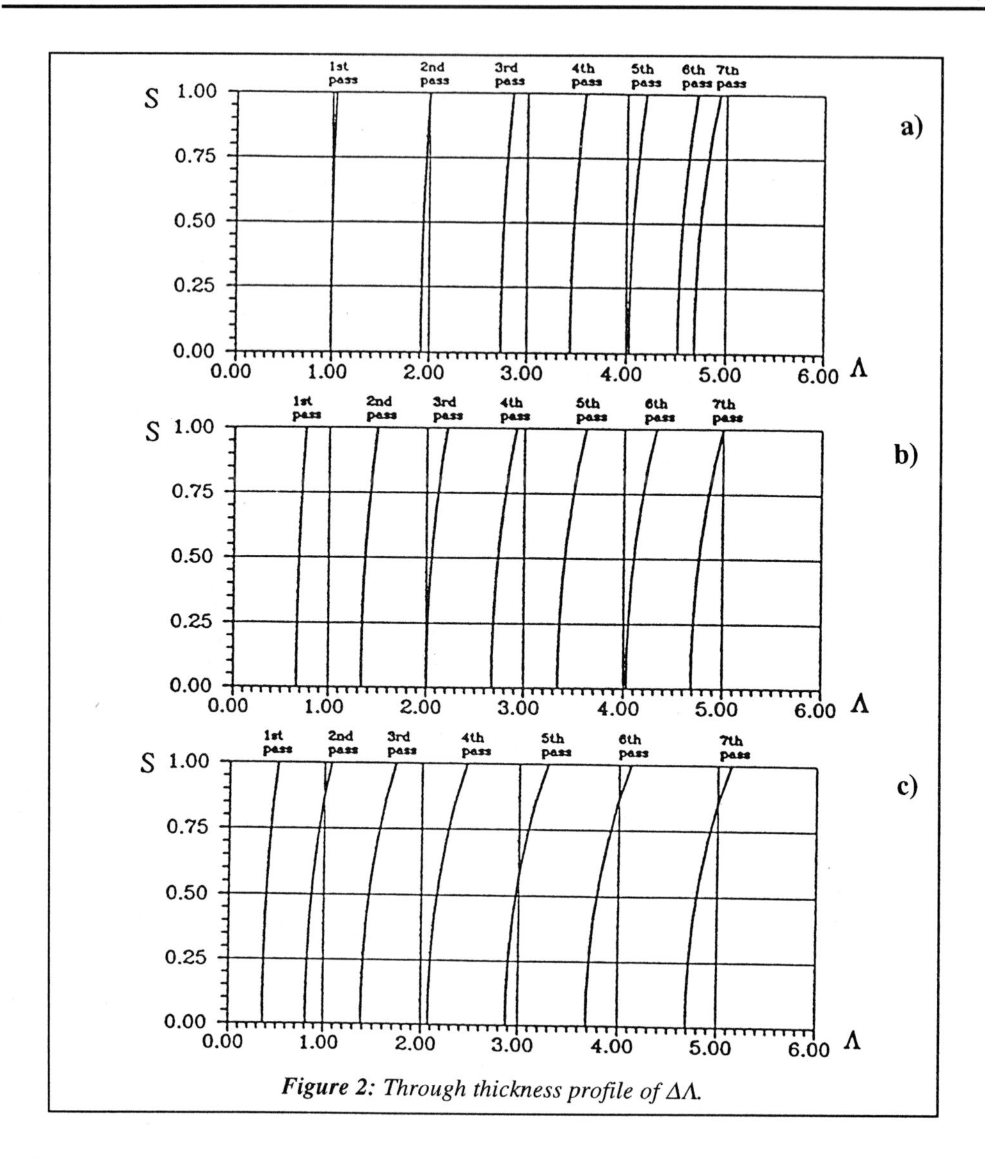

***Figure 2:** Through thickness profile of ΔΛ.*

of Goss is decreased to zero. The rolling components {001}<110> and {112}<110> are increased towards the centre layer of the sheet. The contour lines of the coefficient ΔΛ as defined by eq.(6) (in the case $n=1$) in the field of rolling geometry, which is described in terms of two independent parameters, namely the ratio of contact length to sample thickness l_d/D_0 and the degree of rolling $\Delta D/D_0$, as defined in the previous section are plotted in figure 4. A set of these two parameters identifies the geometry of the rolling gap (diameter of the rolls, thickness before and after rolling). An simultaneous increase of l_d/D_0 and $\Delta D/D_0$ leads to the decrease of ΔΛ and therefore to a more homogeneous deformation.

A shortcomming of the here presented SHC results is the confinement to the rolling geometry, the absence of friction between the roll and the surface of the sample and the neglected influence of the inhomogeneous temperature distribution in the rolled material. The friction changes proportionally to the normal pressure (Coulomb's law) which is proportional to $\Delta D/D_0$. The presence of a zone of adhesion by hot rolling of steel sheet changes not the friction. Therefore, the

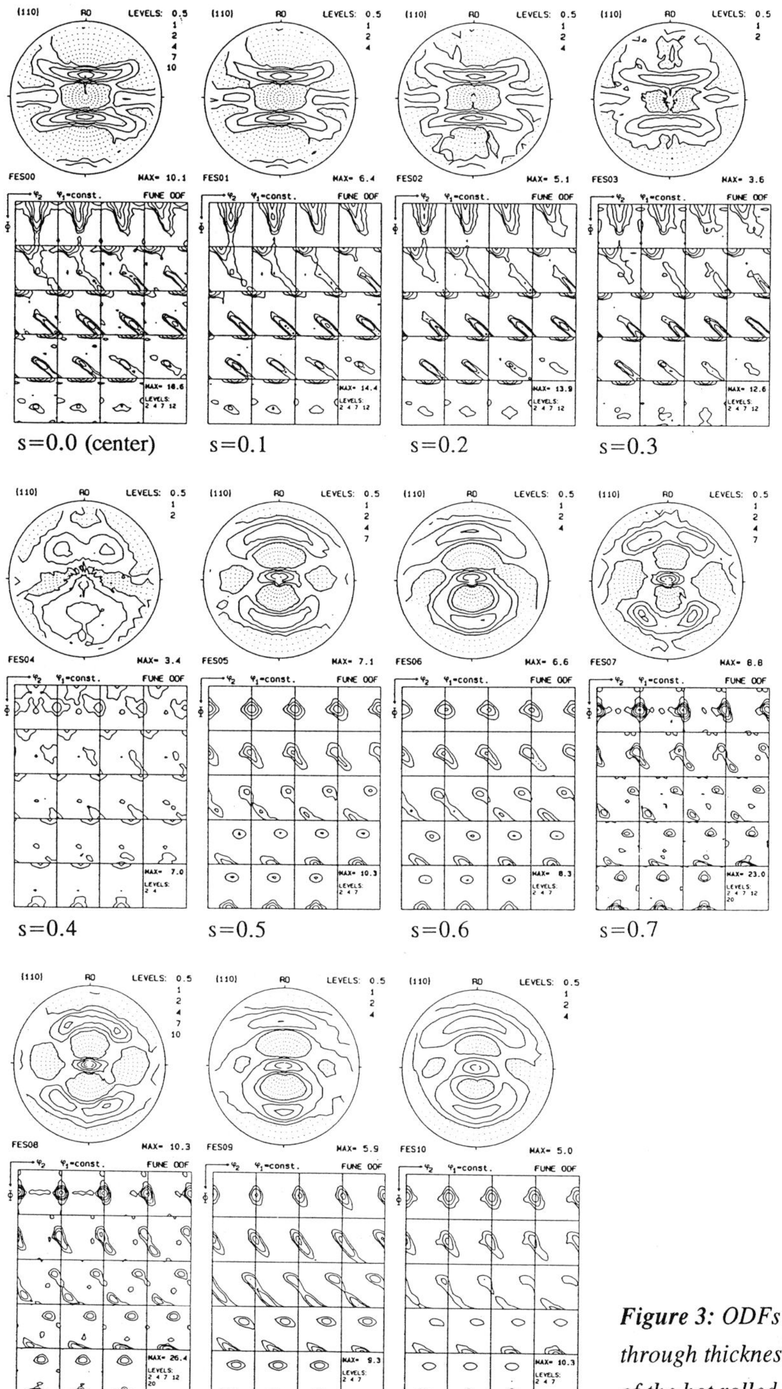

***Figure 3:** ODFs of the through thickness texture of the hot rolled FeCr.*

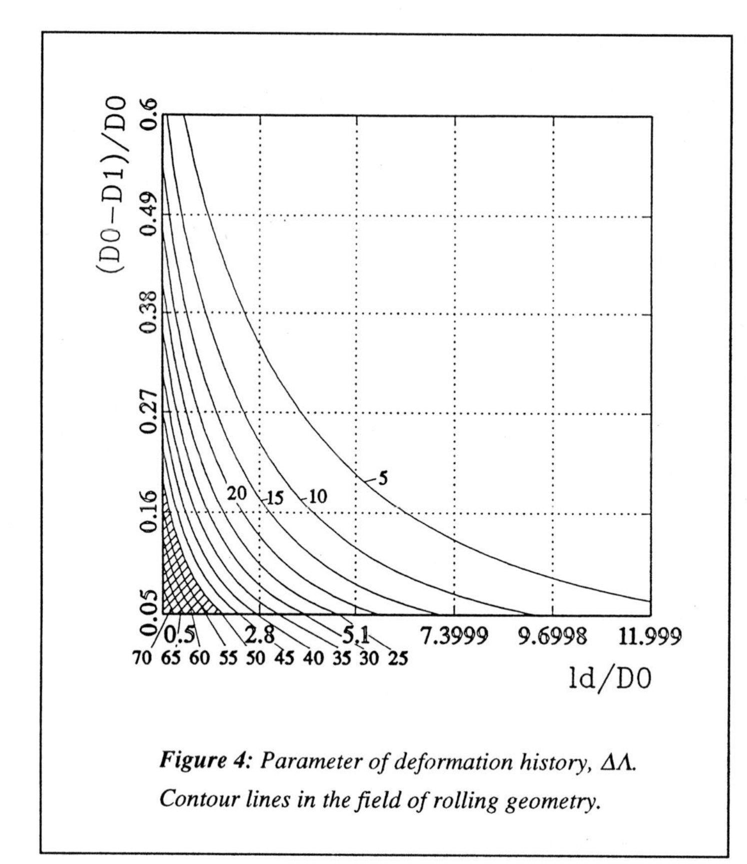

Figure 4: *Parameter of deformation history, $\Delta\Lambda$. Contour lines in the field of rolling geometry.*

increase of $\Delta D/D_0$ leads to the amplification of influence of friction on the texture inhomogeneity. The superposition of the influence of friction and of the rolling geometry results in the formation of a transition zone.

Although the through sheet thickness profile of the temperature was not taken into account, it is suggested from microstructure investigations [5], that recrystallization occurs at the surface which is indicated by a nearly random texture.

CONCLUSIONS

The hot rolling process of ferritic stainless steels was simulated by means of a recently introduced analytical method making use of the superposition of harmonic currents (SHC). This method is less CPU time consuming when compared to the FEM techniques. The results show a strong shear deformation close to the sheet surface and a plain strain state in the center of the sheet. This through thickness profile of deformation gives rise to a strong texture inhomogeneity of hot bands of ferritic stainless steels. The present assessment of the texture inhomogeneity acoording to eq.(6) can be used only on condition, that the influence of the inhomogeneity of temperature distribution in the rolling gap and the friction between band and rolls are small. For an improvement of prediction of through thickness texture distribution by deformation models, a more quantitative analysis of inhomogeneous development of dynamic recristallization in the rolling gap is necessary.

REFERENCES

1) Bunge, H.J.: "Mathematische Methoden der Texturanalyse", 1969, Akad.-Verlag, Berlin
2) Fedosseev,A.I., Wagner,P., Gottstein, G.: in 'Mitteilungen aus dem Institut für Mechanik, 78, ed. Bruhns,O.T., Inst. für Mechanik, Ruhr-Univers. Bochum, 1991, 102
3) Koutcheriaev, B.V., Paptschenko, V.I., Fedosseev, A.I.: Technology legkih splavov (in Russian), 3, 1985, 5
4) Lawrentyev, M.A., Shabat, B.V.: "Theory of complex function" (in Russian), Nauka, Moscow, 4th ed., 1973, 736
5) Raabe, D., Lücke, K.: Mat. Science and Techn, 1993, 9, 302

Materials Science Forum Vol. 157-162 (1994) pp. 1777-1782

ON THE EFFECT OF GRAIN ORIENTATION ON DEFORMATION TEXTURE

J. Hirsch

VAW Aluminium AG Bonn, R&D,
Georg-von-Boeselager-Str. 25, D-53117 Bonn 1, Germany

Keywords: Deformation Texture, Rolling Texture Simulation, Taylor-Model

ABSTRACT

The texture evolution during rolling (plane strain) deformation of FCC material was simulated with <110> (111) slip systems using an improved relaxed constraints model by introducing orientation dependent boundary conditions. The magnitude of shear relaxation of the individual grain orientation is varied between full constraints ("FC") and fully relaxed constraints ("RC") conditions, controlled by the difference in Taylor factor. A significant improvement of texture prediction is achieved by simultating a two component (C and B) type rolling texture which agrees with the experimentally observed splitting of the rolling textures into separate components.The exact Euler angle Θ of the C component between the stable FC ($\Theta = 27°$) and the RC ($\Theta = 35°$) orientation is used to adjust the model to experimental observations ($\Theta = 32°$).

INTRODUCTION

In high stacking fault energy (FCC) materials, like aluminium, the texture after high rolling deformation consists of a few stable texture components ({124}<211>"S", {112}<111>"C" and {011}<211>"B") which are aligned along the stable 'β-fibre [1].Their individual strengths are strongly affected by the initial texture. Taylor type models using relaxed constraints (RC) boundary conditions have been applied successfully to predict the major features of this texture formation, but still some discrepancies exist [1, 2]. To explain the necessity to compensate the occurring shear strains often grain shape effects are discussed [3]. However, the (local) shearing behaviour of orientations involved is as important, as shown by the single orientation measurements of Driver et.al. [4] and by the FEM calculations of Becker [5]. They demonstrate that during deformation in a polycrystal aggregate the grains are influencing each other due to their strain geometry, i.e. their shear behaviour.

Therefore, in order to improve also bulk texture simulations it is necessary to consider the individual tendency for shear deformation (called "shear capacity") of each grain which is a function of its orientation.This parameter often controls the effective shear strain of each grain as shown by the comparison between local orientation changes and theoretical (FC and RC) predictions [4].

Results of rolling texture simulations using conventional FC and RC-models have been discussed in detail in previous papers [1, 2]. The lack of the B component is the major problem in conventional FCC rolling texture simulations caused by the fact that under the initial full constraints conditions only few grains rotate towards B. This orientation requires the relaxation of the shear ε_{TR} ("ε_B") [1] which is most improbable when only grain shape effects are considered.

THE SHEAR CAPACITY

The tendency of a grain for shear deformation (its "shear capacity") can be assessed by comparing the stress required to deform a grain under full constraints conditions with that obtained when relaxing a specific shear strain component. It can be quantified by the ratio r_i of the orientation (Taylor) factor for FC condition (M^{FC}) and that for the specific RC conditions

$$r_\varepsilon = \frac{M^{FC}}{M_\varepsilon^{RC}}$$

with ε being the various types of shear ε_S, ε_C, ε_B (ε_{NT}, ε_{NR}, ε_{TR}) [1] and

N = normal direction, R = rolling direction, T = transverse direction.

The shearing capacity r_ε indicates the reaction stress necessary to suppress the specific shear. It is plotted in Figure 1 for the three principal relaxed constraints conditions in the 3-D Euler space. It is strongly orientation dependent and can be quite significant (up to ~30 %). Only for very few orientations (those close to the dotted lines in Figure 1), no reaction stresses occur.

For the S shear (Figure 1a) high values are found for orientations along the <110> = RD fibre (along Θ at $\varphi_1 = 0°$, in the $\varphi_2 = 45°$ section) and some other orientations, but all of them are far from the typical rolling texture components. The same is valid for the C shear capacity (Figure 1b) except for some high values in the vicinity of the C orientation (the maximum is somewhat TD rotated). The third shear ε_B, (Figure 1c) has its most pronounced capacity value exactly at the B orientation (with a certain scattering).

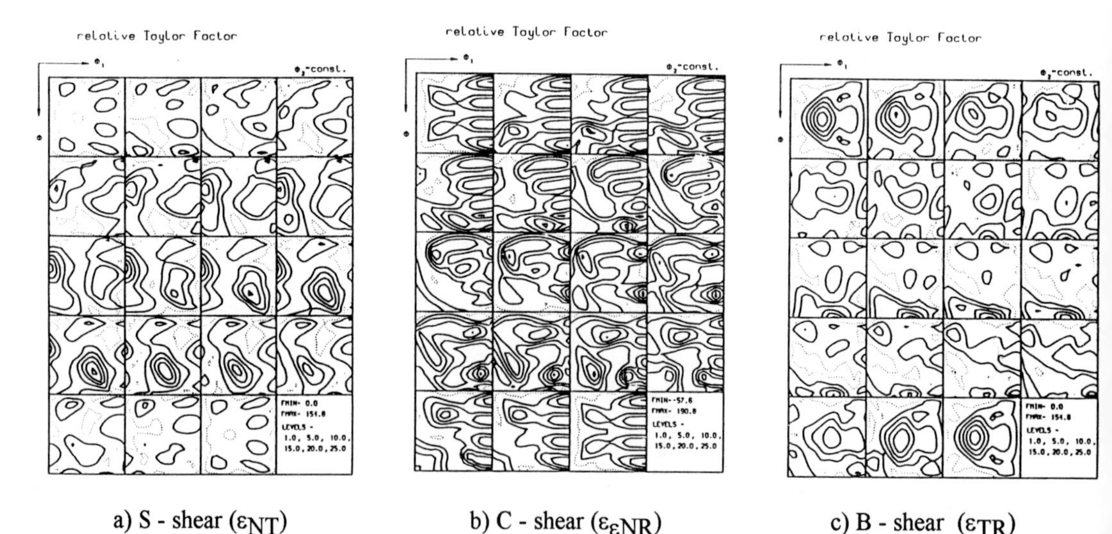

a) S - shear (ε_{NT}) b) C - shear ($\varepsilon_{\varepsilon NR}$) c) B - shear ($\varepsilon_{TR}$)

Fig. 1 Shear capacity levels $r_i = M^{FC}/M_i M^{RC}$ plottet in the 3-D Euler space (φ_2-sections)

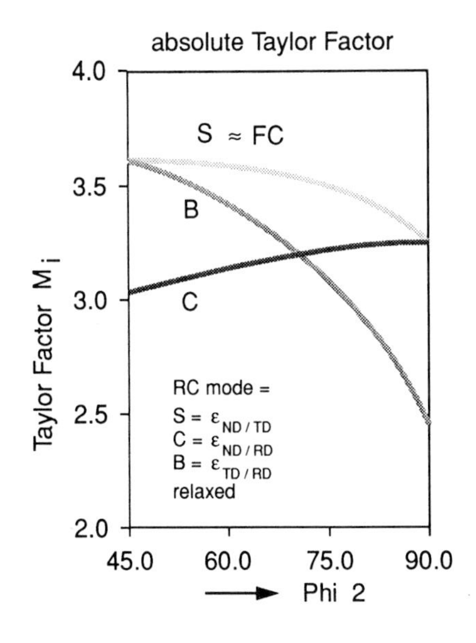

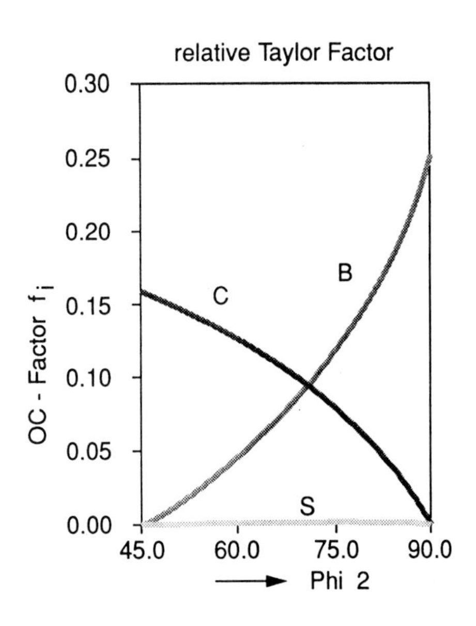

a) absolute Taylor-Factor M_i b) relative Taylor-Factor = $(M^{FC}-M^{RC})/M^{FC}$

Fig. 2 Taylor factors for rolling texture orientations (along the β-fibre)

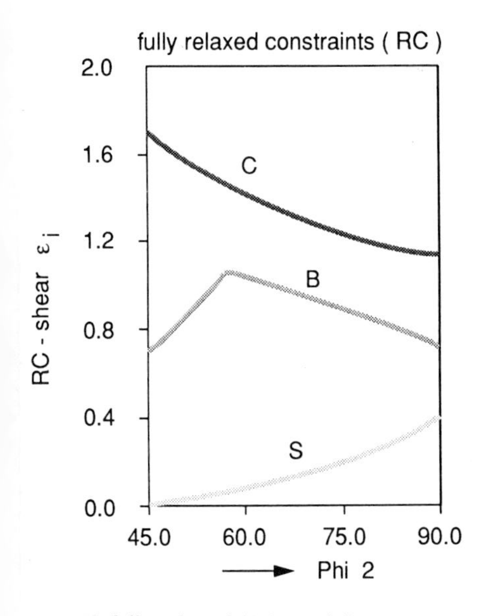

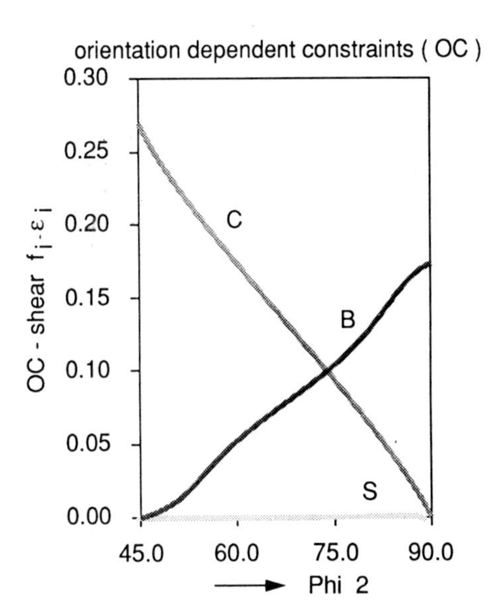

a) fully relaxed (RC-model) b) partly relaxed (OC-model)

Fig. 3 Amount of shear (per unit strain) for rolling texture orientations (β-fibre)

THE OC-MODEL

For the "OC-model" the shear strain components ε are assumed to be fractions f of the fully relaxed shear ε^{RC}: $\varepsilon^{OC} = f_i^{OC} \cdot \varepsilon_i^{RC}$. For each orientation i this fraction is controlled by the shear capacity:

$$f_i^{OC} = K \cdot \left(\frac{M^{FC} - M_i^{RC}}{M^{FC}} \right) = K \left(1 - \frac{1}{r_\varepsilon(i)} \right) \leq 1$$

The variation of the factor K controls its amount, which can be different depending on the spatial grain arrangement [2].

For K < 1, shear deformation will be more constrained while, for K > 1, it will be more relaxed. In the intermediate range, the type and amount of relaxation is orientation dependent. As a first approach a fixed K-value, constant for all orientations has been applied here which was found by comparison the resultant C orientation with experimental orientation data (see below). (At a later stage, K might also be adjusted individually for different orientations or for different volumes of grains to also account for different local boundary conditions.)

Figure 2 illustrates the Taylor factors M_i^{RC}, OC-factors f_i^{OC} and Figure 3 the resulting RC- and OC-shears for the main rolling texture (β-fibre) orientations along φ_2 [1]. It can be noted in Fig. 2b that high reaction stresses are necessary to hinder the B orientation ($\varphi_2 = 90°$) from shearing in ε_B and the C orientation ($\varphi_2 = 45°$)from shearing in ε_C. This explains their relative strong amount calculated by the OC model (Fig. 3b). (The ε_S shear plays no important role in stable FCC rolling textures.) Along the β-fibre a continuous change from dominant ε_C to ε_B shear takes place.

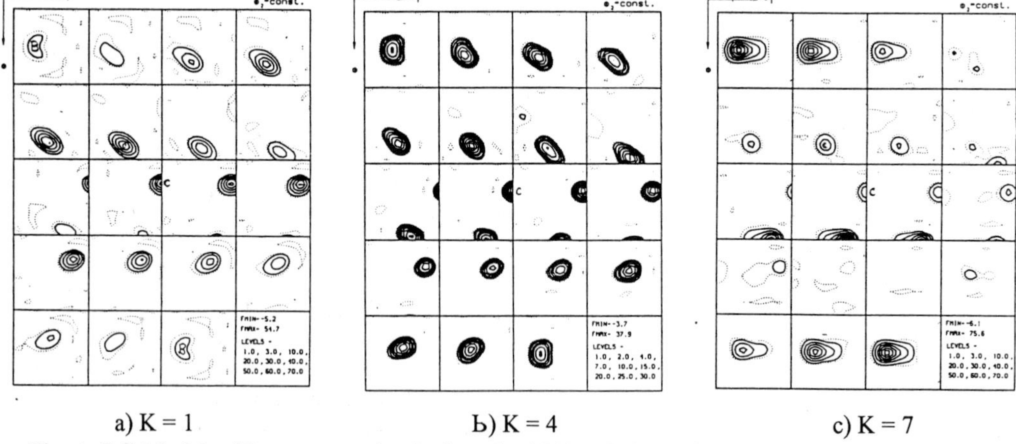

a) K = 1 b) K = 4 c) K = 7

Fig. 4 OC-Model rolling texture simulations for 95% red. ($\varepsilon = 3$)

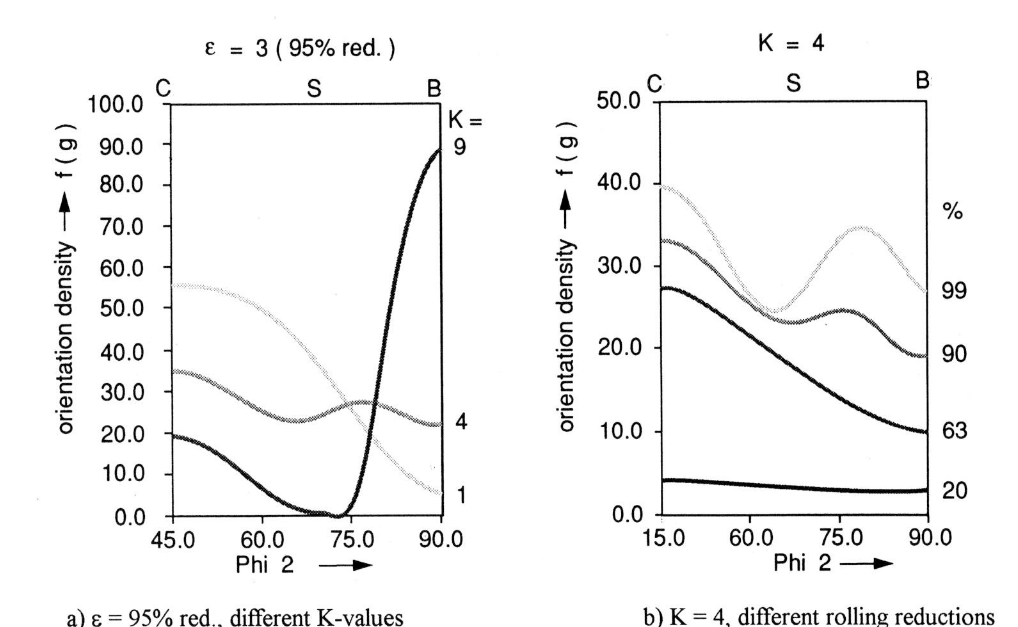

a) ε = 95% red., different K-values b) K = 4, different rolling reductions

Fig. 5 Results of OC-model calculations (resulting orientation density along the β-fibre)

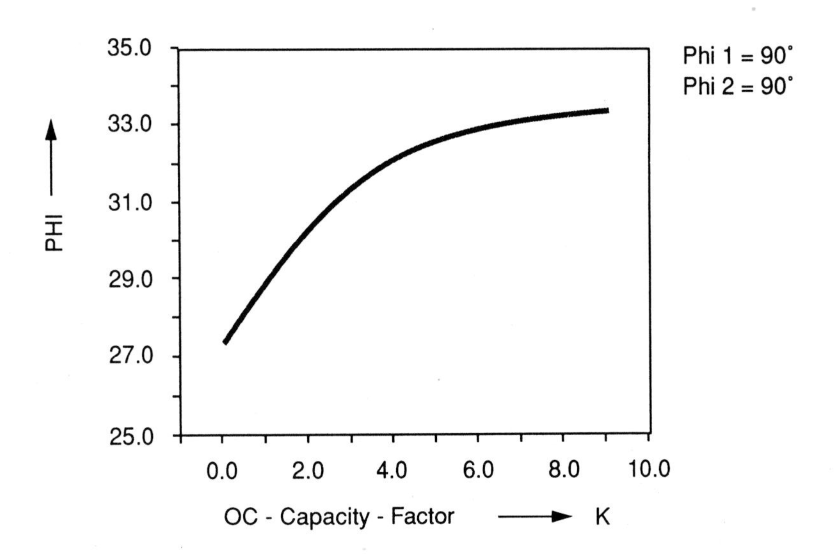

Fig. 6 C-Orientation stability (Θ) as function of OC-factor K

BULK TEXTURE SIMULATION

Figure 4 shows some results of the OC-model for rolling texture simulations after 95 % reduction ($\varepsilon = 3$) using 936 initial orientations a random initial texture [1] and different values for K. The resulting ODF shows the typical β-fibre with two distinct maxima at the C and near the B orientation and a systematic shift of the β-fibre position in the orientation space (Fig. 5). From the comparison with experimental observations of the exact C orientation of $30° < \Theta < 33°$ [1] a valid range for the OC shear capacity factor of $2 < K < 7$ can be derived as appropriate to describe the actual deformation conditions of C oriented grains (Fig. 6). As K increases and more shear is permitted (for $K > 4.3$ grains with high shear capacities begin to achieve their full amount of RC-shear) slip system selection obviously prefers grain rotation towards the B orientation (Fig. 5a) and the number of orientations contributing to C decreases.

Figure 5b shows the rolling texture evolution with strain for an OC factor of $K = 4$, i.e. the transition range where the OC conditions are most effective. The predicted stable C orientation at $\Theta = 32°$ (for 95 % reduction) is in good agreement with experimental observations [1]. In this case, the model predicts first the preferred formation of a C component (at $\varepsilon < 80$ % reduction) and later the formation of an additional peak near B and S orientation.

CONCLUSIONS

The integration of orientation dependent differences in reaction stresses to shear deformation is shown to be an essential feature to predict slip system selction and details of texture evolution during rolling in bulk texture models. The first results using a generalized form already demonstrate the validity of this concept by reproducing the experimental observation of seperate texture components and their exact orientation. An intermediate state between full and relaxed constraints conditions is shown to be the best form to describe an average shearing behaviour and to predict bulk texture effects. The method can further be developed to include initial texture effects and corresponding variations in local boundary conditions without the need for FEM calculations.

REFERENCES

1. J. R. Hirsch, Overview No. 76 Part II Acta Metall. Vol 36, No. 11 (1988), p. 2883.
2. J. R. Hirsch, Material Science and Technology Vol. 6 (1990), p. 1048.
3. U. F. Kocks and G.R. Canova, Acta Metall. (1982), Vol. 30, p. 665.
4. J. H. Driver, A. Skalli, M. Winterberger, Philosophical Magazine A1984, Vol. 49,4, p. 505
5. R. C. Becker, Textures and Microstructures Vol. 14 (proc. ICOTOM 9) 1991, p. 145
6. J. R. Hirsch and K. Lücke, in "Al-Alloys, Their Physical and Mechanical Properties", ed. by. E.A. Starke and T. H. Sander, Chameleon Press, London (1986), p. 1725.

ACKNOWLEDGEMENT

Special thanks to B. Kayser and S. Dietrich for careful preparation of the manuscript. Fruitful discusssions with R. Fortunier and R. Becker are gratefully acknowledged.

Materials Science Forum Vol. 157-162 (1994) pp. 1783-1790

EXPERIMENTAL AND THEORETICAL STUDY OF THE RECRYSTALLIZATION TEXTURE OF A LOW CARBON STEEL SHEET

L. Kestens [1], U. Köhler [2], P. Van Houtte [1], E. Aernoudt [1]
and H.J. Bunge [2]

[1] Department of Metallurgy and Materials Engineering,
Katholieke Universiteit Leuven, de Croylaan 2, B-3001 Leuven, Belgium

[2] Institut für Metallkunde und Metallphysik, TU Clausthal,
Grosser Bruch 23, D-38678 Clausthal-Zellerfeld, Germany

Keywords: Steel Sheet, Rolling Texture, Recrystallization Texture, Recrystallization Model

ABSTRACT.
A low carbon steel sheet was industrially cast and hot rolled. Several samples have been cold rolled on a laboratory mill. The cold rolling direction was varied with respect to the hot rolling direction (longitudinal, transverse , cross and two-stage rolling). After rolling, the sheets were annealed in a salt bath furnace at 720°C during 3 min.
The variation of cold deformation modes gave rise to different textures prior to primary recrystallization. An attempt was made to model the texture selection during recrystallization by means of a model which allows as well for oriented nucleation as for oriented growth. In the case of the longitudinally rolled material a good qualitative correspondence between the experimental and the modelled texture could be obtained by application of the model under conditions of random nucleation and oriented growth. The oriented growth law was modelled according a preferred <110>27° orientation relationship in combination with a <110>84° orientation relationship. For the mixed and transverse rolled material a different recrystallization texture has been observed, corresponding with a different deformation texture. In this case the observed <110>27° rotation from deformed to recrystallized orientations was submitted to a pronounced variant selection favouring the <110>-axis closest to the transverse direction of rolling. The experiment clearly demonstrated the influence of the macroscopic strain path on the nature of the activated recrystallization mechanism.

INTRODUCTION.

For various applications texture is a very important physical parameter determining the technical quality of the material. More specifically for low carbon steel sheet, texture control is a major concern of the manufacturer. The recrystallization during an annealing treatment on a cold rolled sheet is perhaps the most decisive event governing the final texture of the fully processed material. The orientation selection mechanisms which determine the transformation of the initial deformation texture to the final annealing texture are however not yet fully understood. Different physical metallurgical theories have been proposed in literature, ranging from oriented nucleation to oriented growth [1]. Over the last decade several recrystallization models have been developed [2], some of them also taking into account texture phenomena [3,4,5]. None of the models however, has ever been applied for a simulation of the recrystallization texture of a real material. In the present paper an attempt is made to explain the results of a recrystallization experiment by applying an established recrystallization model on the experimental rolling textures.

EXPERIMENTAL PROCEDURE.

The investigated material was a low carbon, non-oriented electrical steel sheet, commonly used as magnetic flux carrier in small electrical motors. In as-received condition the material was industrially cast (C: 0.036; Mn: 0.153; Si: 0.575; P: 0.010; S: 0.004; N: 0.004; Al_{met}: 0.292 mass%) and hot rolled to a final thickness of 2.50mm. To eliminate texture gradient effects the surface of the sheet was mechanically removed (25% at both sides) by grinding in small successive steps of 10µm.
In order to vary the initial texture prior to recrystallization four different deformation modes have been applied by varying the rolling direction with respect to the hot rolling direction : longitudinal rolling (// hot rolling direction); transverse rolling (⊥ hot rolling direction), two stage rolling (longitudinal rolling followed by transverse rolling) and cross rolling (subsequent longitudinal and transverse rolling passes). The rolling experiments have been carried out on a laboratory mill, accurately lubricated to avoid shear deformation components. For each rolling mode the total reduction was fixed at 70% (ε=1.2). After rolling the samples were annealed in a salt bath furnace at approximately 720°C for 3 min. The detailed experimental set up is shown in table 1.

Sample	Deformation Mode	Rolling Direction	Rolling Reduction (%)	Heating Time (s)	Annealing Time (s)	Annealing Temperature (°C)
K1	Rolling	Longitudinal	70.5	7	180	707±2
K2	Rolling	Longitudinal + Transverse	70.5 48+42	10	180	734±1
K3	Rolling	Transverse	70.2	5	180	730±2
K4	Rolling	Cross	72.7	10	180	709±2

Table 1 : Experimental procedure for applied rolling and annealing treatments. The reference direction on the sheet was the hot rolling direction.

After each processing step (as-received, cold rolled, annealed) the texture of the samples was examined by measuring four incomplete pole figures on an X-ray texture goniometer. From those pole figures the ODF was calculated [6], using an appropriate ghost correction procedure [7].

TEXTURE RESULTS.

BCC rolling and recrystallization textures are commonly represented in φ_2=45° sections of Euler space, which contain the most important BCC texture components. The hot rolling texture is characterised by a very strong partial α-fibre with an intense maximum of 33x random on the {001}<110> component. The cold rolling textures are displayed in figure 1. After longitudinal rolling (figure 1a) the intensity along the partial α-fibre has been levelled, causing a decrease of the intensity maximum on the {001}<110> component. In each of the other rolling experiments a very sharp and unique {001}<110> component emerges. For the transverse rolled case the maximum even raises above 150x random (figure 1c).

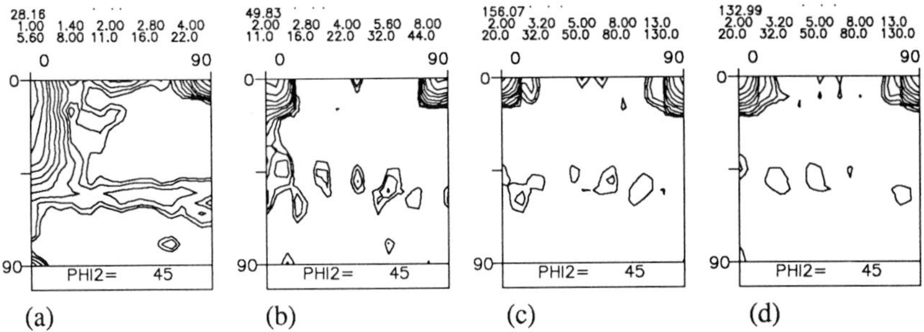

(a) (b) (c) (d)

Figure 1 : Rolling Texture : (a) = Longitudinally rolled (x_1//RD); (b) = Two-stage rolled (x_1//RD transverse stage); (c) = Transverse rolled (x_1//RD); (d) = Cross rolled (x_1//longitudinal RD).

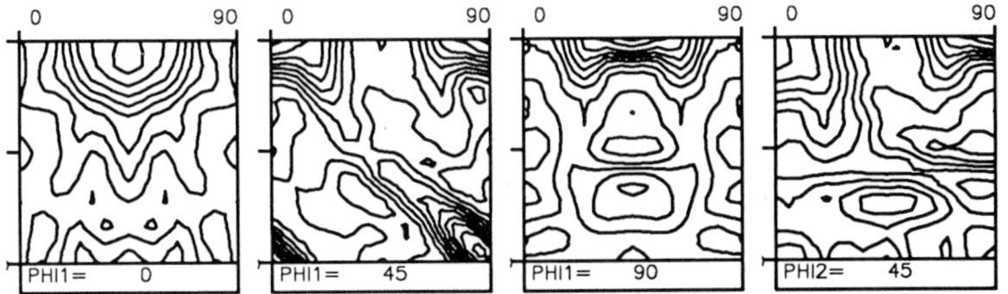

Figure 2 : Measured ODF of the longitudinally rolled sheet, after recrystallization. (isolevels : 0.3-0.6-0.9-1.2-1.5-1.8-2.1-2.4-2.7-3.0)

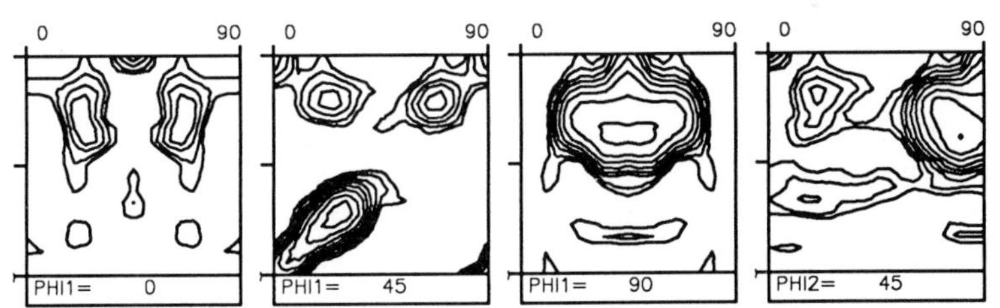

Figure 3 : Measured ODF of the two stage rolled sheet after recrystallization (x_1= RD transverse stage) (isolevels : 0.8-1.0-1.3-1.6-2.0-2.5-3.2-4.0-5.0-6.4)

As a common feature of the recrystallization textures they are drastically weakened in comparison with the deformation textures, with maxima decreasing from >100x random to <5x random. For the longitudinally rolled material a rather classic BCC texture (figure 2) is observed consisting of a (partial α + γ) fibre. For the alternative rolling modes a quite different type of recrystallization texture is measured, which is however quite similar for all non-exclusively longitudinally rolled sheets. Figure 3 shows the ODF for the two-stage rolled material, as a representative example for the annealing textures of this type. Already by visual inspection it is seen that a near <110>30° orientation relation exists between the deformation component and the recrystallization component. At the same time a remarkable difference can be noticed in the intensities on the symmetrical equivalent orientations which are obtained by rotating the {001}<110> rolling component over all crystallographic equivalent <110>30° variants. The <110> axis parallel to the transverse direction of rolling is preferentially selected among all other variants, while the <110> axis parallel to the rolling direction is preferentially suppressed.

THEORETICAL BASIS OF THE SIMULATION.

In order to obtain a more detailed understanding of the experimental results several hypotheses, concerning the underlying physical mechanisms, were tested by application of a recrystallization model. The model has already extensively been described elsewhere [3,4,5]. Assuming a certain nucleation texture as well as a local growth law, the recrystallization texture is calculated from the deformation texture. The nucleation texture is represented by the ODF of the recrystallization nuclei, while the local growth law fits the dependence of the nucleus growth rate on the orientation relation between the nucleus and the surrounding deformed matrix grains.

The simulations have been executed on the experimental rolling textures shown in figure 1. Concerning the nucleation texture and the local growth law, the assumptions of oriented growth and oriented nucleation were considered. The oriented growth hypothesis was implemented by assuming a random nucleation texture and a local growth law with gaussian peaks on the preferred orientation relations. Two different growth laws have been applied : a preferential <110>27° growth and a combined preferential <110>27° + <110>84° growth [8]. The oriented nucleation hypotheses was modelled by assuming a preferred texture for the recrystallization nuclei. Three different alternatives have been considered : deformation texture nucleation, subgrain coalescence (SGC) and strain induced boundary migration (SIBM). The first case simply assumes that the nuclei have the same texture as the deformed matrix in which they are embedded. SGC and SIBM nucleation textures have been modelled by assuming that the Taylor factor can be considered as a measure for the stored plastic energy in a certain orientation, for a given deformation mode. As a first approximation a simple linear relation is assumed between the Taylor factor and the nucleation texture. Since the SGC mechanisms favours nucleation of hard orientations [9] the following relation was applied :

$$f^N(g) \propto \beta\left(\frac{M(g)-M_{\min}}{M_{\max}-M_{\min}}\right)f^D(g)+1 \qquad (1)$$

with $f^N(g)$: the nucleation ODF, $f^D(g)$: the deformation ODF, $M(g)$: the Taylor factor, M_{min} and M_{max}: respectively the minimal and maximal Taylor factors of the orientation domain. The parameter β allows for the variation of the weight of the oriented nucleation effect. To model the SIBM mechanism, equation 1 was reversed, thus favouring the nucleation of soft orientations :

$$f^N(g) \propto \beta\left(\frac{M_{\max}-M(g)}{M_{\max}-M_{\min}}\right)f^D(g)+1 \qquad (2)$$

The nucleation texture was not combined with an entirely flat local growth law, but with a growth law reflecting the decreased mobility of low angle grain boundaries by imposing a Gaussian cusp on the $\Sigma 1$ orientation relationship.

SIMULATION RESULTS.

The graph of figure 4 compares the oriented growth with the oriented nucleation hypothesis by means of volume fractions of the dominant components, for the longitudinally rolled material. The volume fractions are calculated by convoluting the modelled ODFs with the ODFs of the ideal components of which the volume fractions are to be determined. It is seen that none of the simulations predicts correct quantitative values for the volume fractions. On the other hand, figure 4 shows that the model calculations based on oriented growth approximately simulate the correct mutual ratios of the volume fractions, what is obviously not the case for the simulations based on oriented nucleation. A comparison by visual inspection, of the simulated ODFs with the experimental ODFs turns out in favour of the oriented growth hypothesis as well. The best fit of the experimental results could be obtained by a simulation in the condition of random nucleation and oriented growth according a bimodal growth law with gaussian peaks on the <110>27° and the <110>84° positions, with respective weights of 80% and 20%. The ODF sections of figure 5 show that the major features of the experimental texture are correctly reproduced by the model, although quantitatively, the agreement is poor. Comparison of the simulation on base of a monomodal <110>27° growth law with a bimodal <110>27°(80%) + <110>84°(20%) growth law showed the necessity to add the minority component.

Also for the non-exclusively longitudinally rolled materials the best correspondence with the measured recrystallization textures could be obtained on base of the oriented growth hypothesis. In contrast with the longitudinally rolled sheets, the influence of the minor <110>84° component in the local growth law is negligible in this case. As can be seen on the simulated ODF-sections of figure 6, the major problem with the modelled texture is that the intensity is equally distributed among all symmetric equivalent <110>27° related orientations with the dominant {001}<110> component of the rolling texture. In the experimental texture (cfr. figure 3) however, a clear favouring of the <110>//TD direction was observed. The actual implemented growth laws do not allow for this type of variant selection, because none of them is related to the deformation mode.

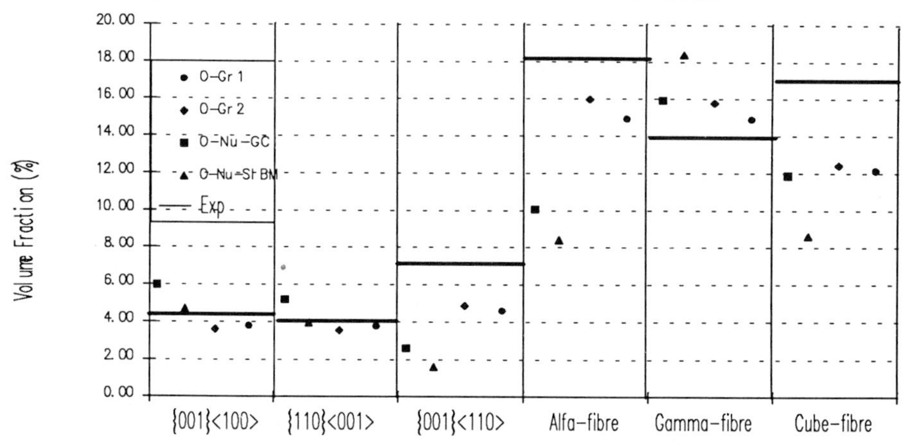

Figure 4 : Volume fraction of the dominant BCC rolling components for the simulated textures, in comparison with the experimental values (O-Gr1: <110>27° oriented growth; O-Gr2: (80%)<110>27°+(20%)<110>84° oriented growth; O-Nu1 : SIBM oriented nucleation; O-Nu2: SGC oriented nucleation)

DISCUSSION.

The experimental rolling textures all can be understood in terms of the Taylor theory of plasticity. This was checked by applying a relaxed constraints Taylor mode [10] on the initial hot rolling texture. For each of the deformation modes a satisfying correspondence with the experimental rolling textures was obtained. Although four different rolling modes were applied, basically only two types of rolling textures were developed : the longitudinal and the non-longitudinal type. Such classification also imposes itself after recrystallization. This gives evidence to the fact that the

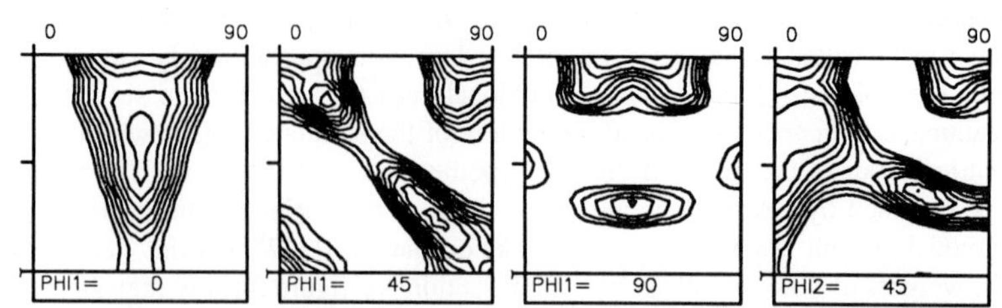

Figure 5 : Simulated recrystallization texture for the longitudinally rolled material, under assumption of random nucleation and combined oriented growth: 80%<110>27° + 20%<110>84° (isolevels : 1.00-1.05-1.10-1.15-120-1.25-1.30-1.35-1.40-1.45)

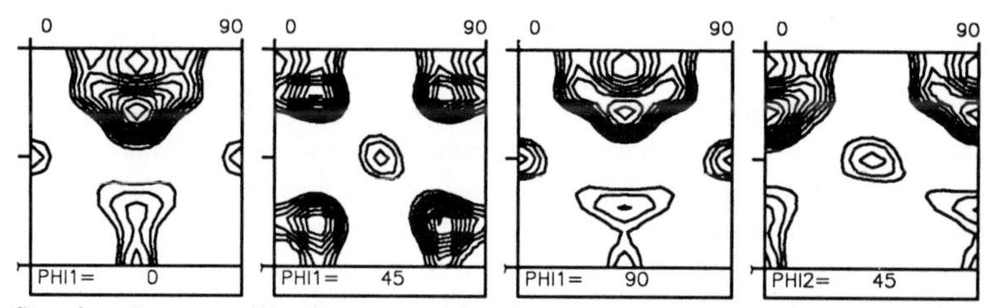

Figure 6 : Simulated recrystallization texture for the two-stage rolled material (x_1//transverse RD), under assumption of random nucleation and combined oriented growth : (80%)<110>27°+(20%)<110>84° (isolevels : 1.05-1.10-1.15-1.20-125-1.30-1.35-1.40-1.45-1.50)

deformation texture, independent of the rolling mode, determines the recrystallization texture to a large extent. If it is assumed that the variation of rolling mode from two stage rolling, over transverse rolling, to cross rolling affects the substructure of the material, it has to be concluded that none of this influence is reflected in the orientation selection at the moment of recrystallization.

None of the simulated textures have shown a satisfying numerical correspondence with their experimental counterparts. At best a qualitative fit could be obtained for the recrystallization texture of the longitudinally rolled material. This best fit is modelled under the assumption of a random nucleation and a peaked oriented growth law, indicating that the alternative of oriented nucleation has not been active. It has to be noted however that the test of the hypothesis of oriented nucleation is heavily dependent on the implementation of this hypothesis in the model, more precisely on the exact numerical form of equations 1 and 2. If the pure nucleation textures were solely considered, without application of the growth law, a better correspondence with the experimental recrystallization texture was revealed. An entirely flat growth law however makes no sense, because it is generally accepted that low angle grain boundaries have a decreased mobility. In this way the result also indicates the sharp impact of the applied growth law on the simulated recrystallization texture, which was additionally confirmed by the fact that a variation of the free parameter β in the range 0.1 to 2 had nearly no effect on the simulated recrystallization texture.

If the hypothesis of oriented growth is accepted the problem remains that the physical origin of the oriented growth law is not yet understood. The Σ19-CSL nature of a <110>27° orientation relation has been mentioned [8] as a sufficient ground for increased grain boundary mobility. It is however not yet understood why only this CSL seems to be active, while there is a multitude of CSL relationships available in the assumption of a random nucleation texture. Also in the case of the transverse and cross rolled sheets this <110>27° orientation relation reappears, but with the important modification that only the <110>//TD-axis is favoured. This was already observed before, on recrystallization of Fe-3%Si single crystals [11]. Whatever the origin of the observed relationship might be, the experiments have given evidence of a firm influence of the deformation mode, which cannot be explained in terms of the CSL theory. Therefore, an additional or alternative mechanism must be active, which is however impossible to trace by texture measurements only, without a thorough investigation of the microstructure.

CONCLUSIONS.

A rolling and annealing experiment was carried out on a low carbon non oriented electrical steel sheet. Corresponding the initial rolling texture different types of recrystallization textures were observed, revealing a clear distinction between the longitudinal rolled material and the material which was partially or entirely transverse rolled. In this way it was proved that the annealing texture was mainly determined by the deformation texture, irrespective of eventual submicrostructural differences between the rolled sheets.

An established recrystallization model was applied to test the classical hypotheses of oriented growth and oriented nucleation. None of the model simulations could reproduce numerically correct recrystallization ODFs. Only for the longitudinally rolled material a qualitatively satisfying correspondence could be obtained, under the assumption of random nucleation and oriented growth according a bimodal growth law with preferential growth for <110>27° and <110>84° relationships. None of the recrystallization textures could be simulated on the basis of an oriented nucleation model.

If the same conditions of random nucleation and oriented growth were applied in the case of non-exclusively longitudinally rolled material, the model calculations failed to simulate the experimentally observed variant selection. This highly discredits the CSL theory on the physical origin of a <110>27° preferential growth.

ACKNOWLEDGEMENTS.

N.V. Sidmar is gratefully acknowledged for the ready disposal of the material.

REFERENCES.

1. Vanderschueren D.: PhD Thesis, 1991, 9-42.
2. Humphreys F.J.: Mat. Science & Techn., 1992, 8, 135.
3. Bunge H.J., Plege B.: Theoretical Methods of Texture Analysis, 1987, ed. H.J. Bunge, DGM, Informationsgesellschaft, Oberursel, 289.
4. Bunge H.J., Plege B.: Annealing Processes, Recovery, Recrystallization and Grain Growth, ed. N. Hansen et.al., RisØ, 261.
5. Köhler U., Dahlem-Klein E., Klein H., Bunge H.J.: Text. & Micr., 1992, 125.
6. Van Houtte P.: Text. & Micr., 1984, 6, 137.
7. Van Houtte P.: Text. & Micr., 1991, 13, 199.
8. Ibe G., Lücke K.: Archiv. Eisenhüttenwesen, 1968, 39, 693.
9. Hutchinson W.B.: Int. Met. Rev., 1984, 29, 25
10. Van Houtte P.: Text. & Micr., 1988, 8&9, 313.
11. Dunn C.G., Koh P.K.: Trans. AIME, J. of Metals, August 1956, 1017.

Materials Science Forum Vol. 157-162 (1994) pp. 1791-1796

MODELLING CYCLIC DEFORMATION TEXTURES WITH ORIENTATION FLOW FIELDS

H. Klein and H.J. Bunge

Institut für Metallkunde und Metallphysik, TU Clausthal,
Grosser Bruch 23, D-38678 Clausthal-Zellerfeld, Germany

Keywords: Orientation Flow Field, Taylor Theory, Cyclic Deformation, Fatigue

ABSTRACT

Plastic deformation by glide leads to continuous orientation changes of the crystallites of a polycrystalline material. These orientation changes can be expressed by an orientation flow field in the orientation space. Assuming a small deformation step $\Delta\eta \cdot \varepsilon_{ij}$ the flow field shifts the orientation g into the nearby orientation g'. To model cyclic deformation textures an alternatingly reverse deformation step $-\Delta\eta \cdot \varepsilon_{ij}$ must be carry out. Non-unique flow fields lead to stronger effect as model calculations with unique ones. Since the magnitude of non-unicity in flow fields is quite different in different orientations, increasing and decreasing orientation densities in different orientations occur. Particularly, an originally random orientation distribution thus develops into a distinct cyclic deformation texture.

INTRODUCTION

Orientation changes of the crystallites of a polycrystalline material by plastic deformation can be described by an orientation flow field in the orientation space, i.e. a flow vector $\vec{v}(g)$ /1/. This flow field depends on the applied strain. Assuming a small deformation step $\Delta\eta \cdot \varepsilon_{ij}$ the flow field vector $\vec{v}$ shifts one orientation g into the nearby orientation g'/2/. Because of the orientation dependence of $\vec{v}(g)$ the reverse deformation $-\Delta\eta \cdot \varepsilon_{ij}$ does not exactly shift g' back into g. It is $\vec{v}(g, \Delta\eta \cdot \varepsilon_{ij}) \neq -\vec{v}(g', -\Delta\eta \cdot \varepsilon_{ij})$.
This effect may lead to texture formation by cyclic deformation. The effect is, however, small and decreases to zero with $\Delta\eta \longrightarrow 0$ (assuming $\sum \mid \Delta\eta \mid= const.$).
A second, much stronger effect occurs in non-unique flow fields as, for instance, in the Taylor theory. Fig.1 shows schematically a multivalued flow field in the orientation space. Here, the same strain $\Delta\eta \cdot \varepsilon_{ij}$ allows a whole bunch of flow vectors $\vec{v}(g, \Delta\eta \cdot \varepsilon_{ij})$ in the same orientation g. Hence, the starting orientation g is shifted and at the same time spread out into a whole orientation distribution g'_x. During the reverse strain $-\Delta\eta \cdot \varepsilon_{ij}$ each of these orientations is again spreads out into an orientation distribution, and so on in each deformation cycle. This effect gives rise to a much stronger texture change during cyclic

deformation and it does not vanish with $\Delta\eta \longrightarrow 0$ (assuming$\sum \mid \Delta\eta \mid = const.$). This is demonstrated schematically in Fig.2.
Since the magnitude of non-unicity of the flow field is quite different in different orientations g, this effect leads to increasing and decreasing orientation densities in different orientations. Particularly, an originally random orientation distribution thus develops into a distinct cyclic deformation texture.
In the present model calculations the deformation rate tensor was chosen to be plane strain

$$\varepsilon_{ij} = d\eta \cdot \begin{vmatrix} 1 & 0 & 0 \\ 0 & 0 & 0 \\ 0 & 0 & -1 \end{vmatrix}$$

Using this deformation rate tensor and$\{111\}\langle 110\rangle$ glide the flow field was calculated under the assumption of the classical, full constrained Taylor model /3/.

NUMERICAL RESULTS

An example of a cyclic deformation texture is shown in Fig.3. Here deformation steps were used with $d\eta = \pm 5\%$ alternatingly. Glide systems were $\{111\}\langle 110\rangle$ and the starting texture was random and the number of cycles was 100. The spread of the flow field was approximated by a Gaussian distribution. It is seen that the random texture does not remain random. The maximum orientation density of 2.8 is rather low compared with unidirectional deformation. Nevertheless, stable and unstable orientations are clearly found. The flow towards and away from certain orientations during cyclic deformation is illustrated in Fig.4. Here the relative change of the maximum orientation density is given as a function of number of cycles. The starting orientations were some prominent ideal orientations with Gaussian spread of $\omega = 10°$. Growth or shrinkage of the maximum density is clearly demonstrated.
If the spread of the starting texture is changed then part of the distribution may fall into a growth area others into a shrinkage area as may be estimated from Fig.3.
Hence, growth or shrinkage of the maximum density depends strongly on the spread width of the original Gauss distribution around the ideal orientations. This is illustrated in Fig.5.
The texture change of an experimental texture during cyclic deformation was also modelled, Fig.6 shows the orientation density along the skeleton line of an Al–rolling texture and its change after an increasing number of cycles.

CONCLUSION

- Using a sequence of deformation steps $(+d\eta, -d\eta, ...)$ the important case of a fatique texture can be modelled by means of orientation flow fields
- After cycling deformation stable and unstable orientations can clearly be found.
- The consideration of a non–unique flow field gives a stronger texture change during cyclic deformation.

REFERENCES

/1/ H. Klein, C. Esling and H.J. Bunge in: Textures and Microstructures 14–18, 1079 (1991)

/2/ H. Klein and H.J. Bunge in: Steel Research 62, 548 (1991)

/3/ G. Sevillano, P. van Houtte, E. Aernoudt in: Progress in Material Science 25 (1980)

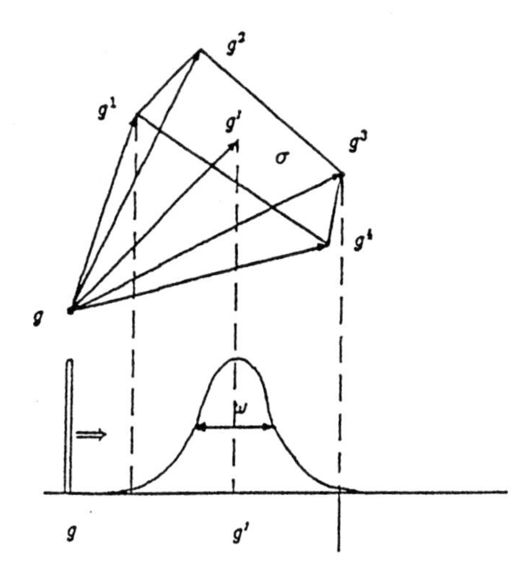

Fig.1 Flow field in the orientation space (schematic representation).

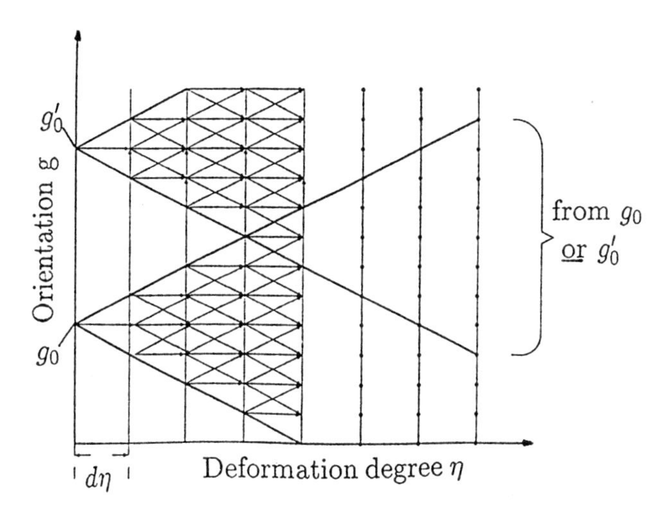

Fig.2 Schematic representation of texture development with increasing number of deformation steps. Using a multi-valued flow field.

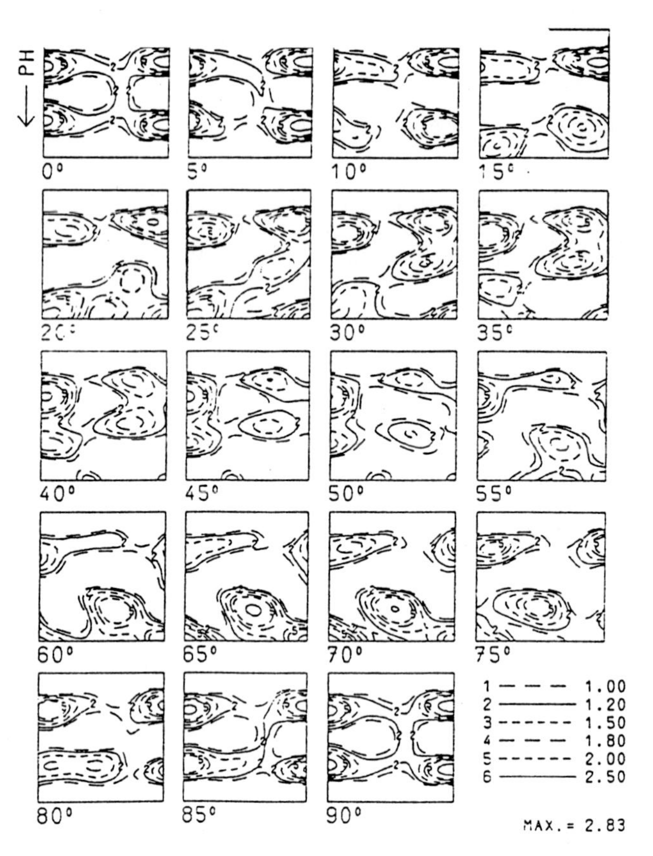

Fig.3 Orientation distribution function in φ_2–sections after cyclic deformation using $\{111\}\langle110\rangle$ glide, $d\eta = \pm5\%$, starting texture: random.

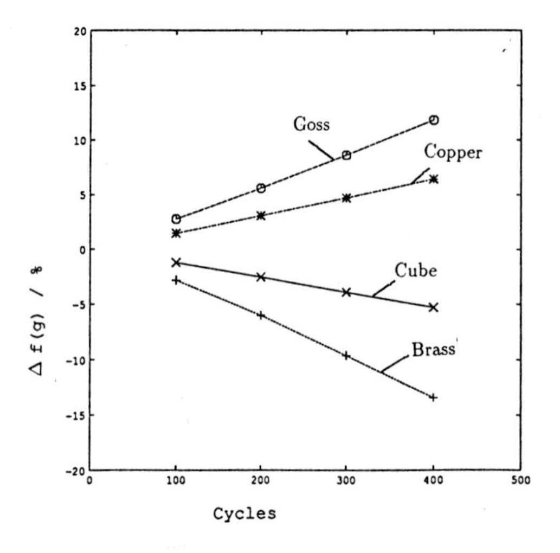

Fig.4 Texture change of the orientation density of ideal orientations ($\omega = 10°$) as a function of the number of deformation cycles.

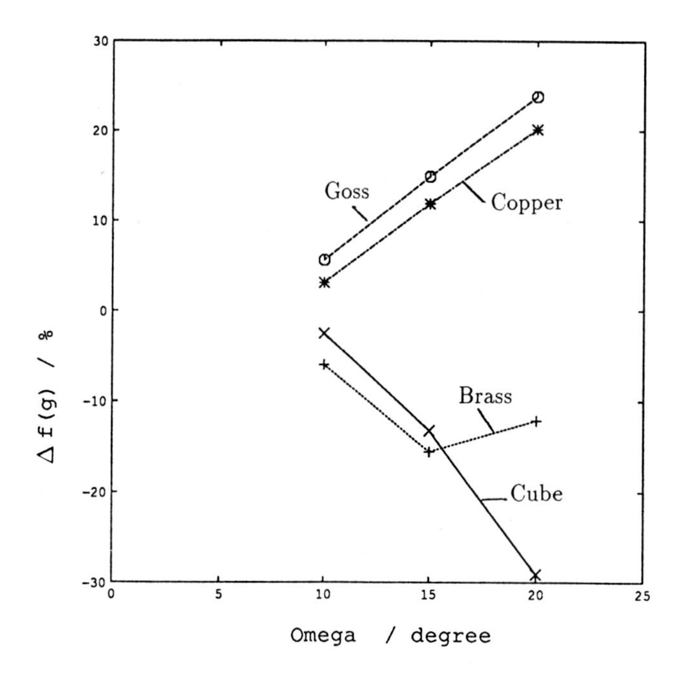

Fig.5 Texture change of ideal orientations as a function of the Gaussian spread width. Number of cycles = 100.

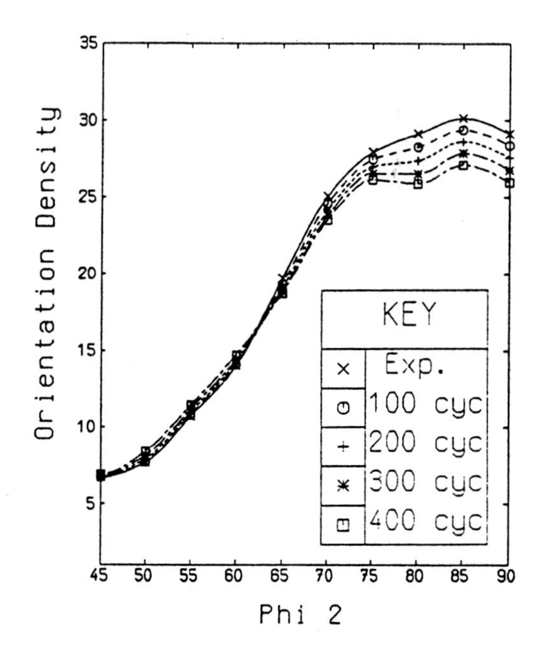

Fig.6 Orientation densities along the skeleton line according to a rolling texture of Al after different number of deformation steps ($\Delta\eta = \pm 5\%$).

Materials Science Forum Vol. 157-162 (1994) pp. 1797-1802

FAILURES TO MODEL THE DEVELOPMENT OF A CUBE TEXTURE DURING THE HIGH TEMPERATURE COMPRESSION OF Al-Mg ALLOYS †

U.F. Kocks [1], S.R. Chen [1] and P.R. Dawson [2]

[1] Center for Materials Science, Los Alamos National Laboratory, Mail Stop K765, Los Alamos, NM 87545, USA

[2] Engineering Theory Center, Cornell University, Ithaca, NY 14853, USA

Keywords: Cube Texture, Al-Mg, Non-Octahedral Slip, Rate Sensitivity, Climb, Dynamic Recrystallization

ABSTRACT

A strong {100} fiber texture has been observed under some circumstances after the compression and tension of Al-Mg alloys. It is not due to static recrystallization after the test, but probably connected with some form of dynamic recrystallization. Possibilities that this unusual texture might arise from some unconventional deformation mechanisms were searched for but not found.

INTRODUCTION

At temperatures and strain rates where viscous deformation with a stress exponent n=3 is observed, one expects the slip system distribution, and therefore the grain rotations and the development of deformation texture, to be different than when the flow stress is rate insensitive. This effect had been observed during tension of Al-3%Mg single crystals at 360 and 410°C, and qualitatively explained in this way.[1] In addition, glide on non-conventional slip planes and the presence of climb should influence rotations and texture development.[2]

---

† Work supported by the U.S. Department of Energy

	n>>4	n≈4	n≈3
Al-2%Mg	T=25°C, $\dot{\varepsilon}=10^{-2}s^{-1}$ T=282°C, $\dot{\varepsilon}=10^{-1}s^{-1}$	T=282°C, $\dot{\varepsilon}=10^{-5}s^{-1}$ T=380°C, $\dot{\varepsilon}=10^{-3}s^{-1}$	T=360°C, $\dot{\varepsilon}=5\cdot10^{-5}s^{-1}$ T=440°C, $\dot{\varepsilon}=10^{-4}s^{-1}$
Al-5%Mg	T=25°C, $\dot{\varepsilon}=10^{-2}s^{-1}$ T=300°C, $\dot{\varepsilon}=5\cdot10^{-2}s^{-1}$	T=473°C, $\dot{\varepsilon}=1s^{-1}$ T=380°C, $\dot{\varepsilon}=10^{-3}s^{-1}$	T=473°C, $\dot{\varepsilon}=10^{-2}s^{-1}$ T=473°C, $\dot{\varepsilon}=10^{-4}s^{-1}$ **(Tension <u>or</u> Compression)**

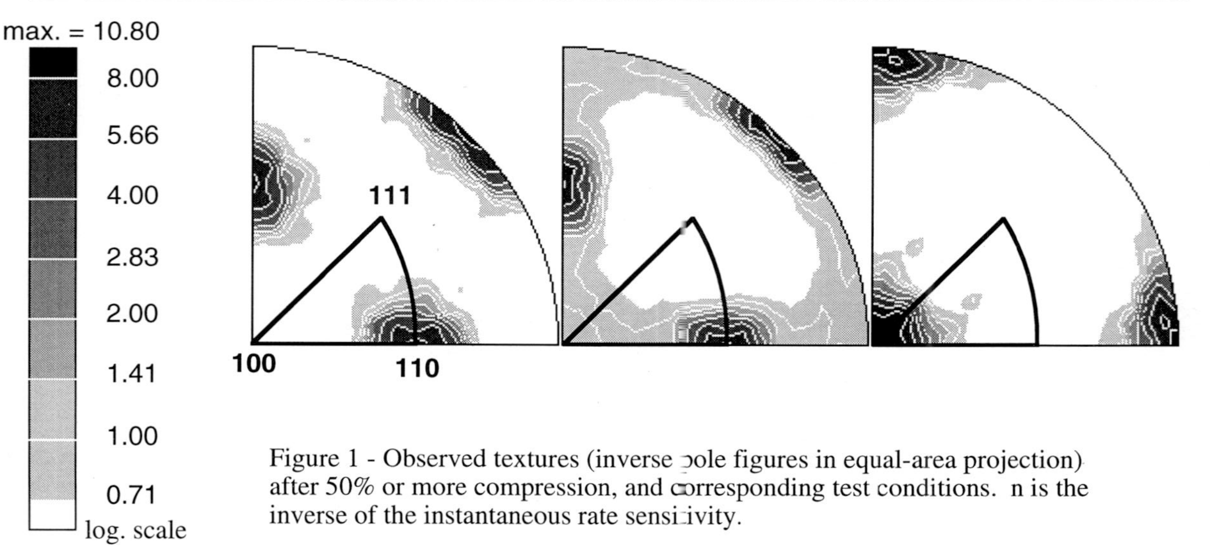

Figure 1 - Observed textures (inverse pole figures in equal-area projection) after 50% or more compression, and corresponding test conditions. n is the inverse of the instantaneous rate sensitivity.

OBSERVATIONS

We undertook a series of compression tests on Al-2%Mg [3] and Al-5%Mg polycrystals in a range of temperatures and strain rates, both where n=3 and where it was higher. The observed textures are displayed in figure 1. At low temperatures, the standard compression texture, essentially a {110} fiber, is seen. As the stress exponent approaches 3, a small peak at the cube fiber position develops, along with some general dispersion of the texture, but without losing the {110} fiber. Eventually, at 440°C and $10^{-4}s^{-1}$, the texture is a pure cube fiber - broad, but with no intensity left anywhere else.

We now believe, for a number of reasons, and after various supplementary experiments, that this texture is due to some form of dynamic recrystallization. One reason for this conviction is that the *same* texture develops in tension as in compression. Nevertheless, we attempted to model this texture on the basis of deformation only.

SIMULATIONS

We prescribed uniaxial *strain*, a stress exponent of 3, and a <110> slip direction. The Los Alamos polycrystal plasticity (LApp) simulation code [4] was used, except in case 2. The results are shown in figure 2. Both the rotation paths of a few characteristic crystals during the first 10% strain (arrows) and polycrystal textures after a strain of 1.0 (symbols) were investigated.

1. Slip on {111} plane only - full constraints (upper bound). This Taylor-type simulation was carried out for the rate-insensitive case (figure 2a) and for a stress exponent of 3 (figure 2b). The rotations were very similar in both cases, only slower in the latter; thus, the final texture was similar, only slightly more diffuse. This result had also been obtained by Bacroix et al.[2].

2. Slip on {111} only, lower bound. This case was calculated with a new model that keeps all stress components the same everywhere and finds their magnitude such as to satisfy the imposed average strain rate [5]. The conditions under which this lower bound may be a reasonable approximation include high rate sensitivity. The resulting texture (figure 2c) was similar to the upper-bound texture in that the strongest component was {110}; the tail toward {511} was, however, moved to {311}, i.e. *away* from {100}.

3. Slip on {111}, plus climb. This was modeled by progressing from single to multiple slip and stopping at the third slip system; the remaining prescribed strain components can then be satisfied by climb, which does not contribute to the rotations [6]. The direction of all rotations is similar, only the rate is faster (figure 2d). Such behavior should be expected when fewer slip systems operate.

4. Slip on {111} and {100}. There are only minor differences to slip on {111} only, and the rotations are somewhat slower (figure 2e).

5. Slip on {100} and {110} planes. In this case (figure 2f), the {100} corner becomes preferred; the maximum is actually near {511}, with a tail toward {510}; and a significant remnant at {110}. This result could be marginally in agreement with the experiments; but we cannot fathom why {111} slip should cease entirely.

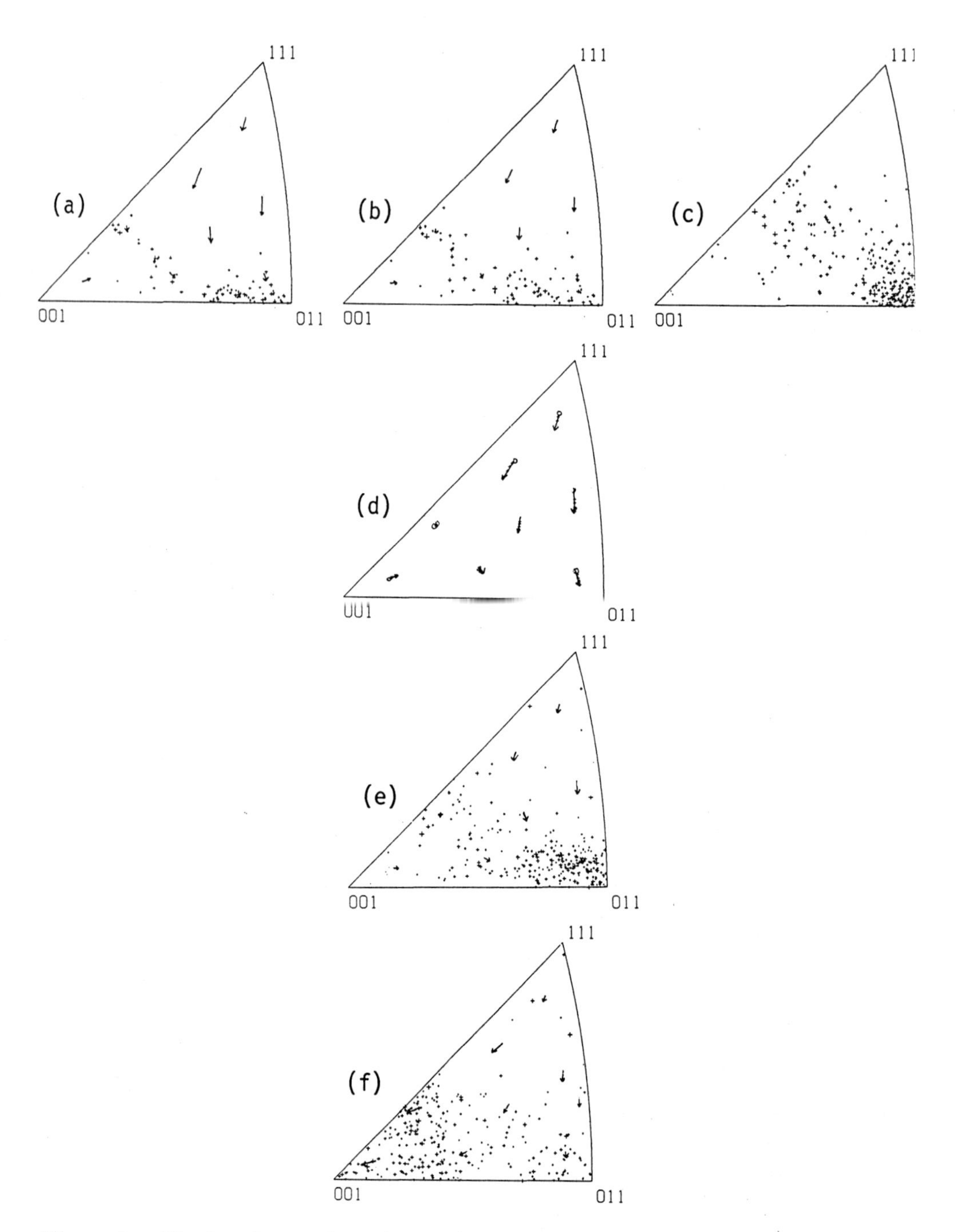

Figure 2 – Simulated rotations (arrows) and end textures (symbols) under various assumptions. Taylor model with stress exponent n=33 (a) and n=3 (b) [4]. Lower bound, n=3 (c) [5]. Allowing for climb (d) or for different slip planes (e and f), all with n=3.

SUMMARY

In summary, on the basis of deformation simulation only, the appearance of a strong {100} component in compression can be predicted only under unreasonable conditions; the disappearance of the {110} cannot be explained in any way we know.

REFERENCES

1. M. Otsuka and R. Horiuchi: J. Japan Inst. Metals **36**, 809 (1972)
2. B Bacroix, P. Franciosi, and A. Mecif: Risø Int'l. Symposium on Modelling of Plastic Deformation, S.I. Andersen et al., eds. (1992) p. 205
3. S.R. Chen and U.F. Kocks: in *Hot Deformation of Aluminum Alloys*, T.G. Langdon et al.eds., The Minerals, Metals and Materials Society, Warrendale, Pa. (1991) p.89
4. U.F. Kocks, G.R. Canova, C.N. Tomé, and A.D. Rollett: Los Alamos Polycrystal Plasticity simulation code (Los Alamos National Laboratory LA-CC-88-6)
5. Yvan B. Chastel and P.R. Dawson: this volume.
6. G.W. Groves and A. Kelly: Phil. Mag. **19**, 977 (1969)

Materials Science Forum Vol. 157-162 (1994) pp. 1803-1808

CALCULATION OF THE RECRYSTALLIZATION TEXTURES OF CUBIC METALS

U. Köhler and H.J. Bunge

Institut für Metallkunde und Metallphysik, TU Clausthal,
Grosser Bruch 23, D-38678 Clausthal-Zellerfeld, Germany

Keywords: Recrystallization, Modelling, Oriented Growth, Oriented Nucleation, Compromise Texture, Cube Texture

ABSTRACT

Primary recrystallization textures of cubic metals were calculated according to a model based on oriented nucleation and oriented growth. Model calculations of the texture formation in fcc metals showed, that it was possible to reproduce the main features of the recrystallization texture formation with the assumption of random nucleation and oriented growth. Calculations of the primary recrystallization texture in bcc metals, based on the mechanism of oriented growth alone, showed some strong deviations between experimental and simulated results. In this case the mechanism of oriented nucleation had to be introduced into our model calculations.

INTRODUCTION

The development of primary recrystallization textures depends on a lot of parameters, for example the composition of the material, its microstructure, the deformation process and the annealing treatment. All these parameters have a certain influence on the mechanisms taking place during recrystallization, so that it does not seem possible to create a generally applicable model for this process. On the other hand, experimental results have shown, that some characteristic texture developments are in a certain range independent of variations of these parameters.

The primary recrystallization textures $f^R(g)$ of cubic metals were calculated according to a model based on oriented nucleation and oriented growth [1]:

$$f^R(g) = const. \cdot f^N(g) \cdot \overline{W}(g)^3 \qquad (1)$$

Here $f^N(g)$ is the orientation distribution function (ODF) of nuclei. We have different possibilities to take oriented nucleation into account. The ODF of nuclei can be assumed to be orientation independent ($f^N(g) = 1$) or we can introduce a certain nuclei-distribution. This may be either a measured nuclei-distribution or a simulated one (taking different nucleation mechanisms, e. g. twinning into account).

The nuclei start to grow at the same time with the mean linear growth rate $\overline{W}(g)$. They stop growing when all the deformed grains are consumed. It is assumed that the distance between neighbouring nuclei is large compared with the grain size, so that each nucleus has to grow through several grains with different orientations. In this way the "compromise" texture is formed. The average growth rate $\overline{W}(g)$ can be calculated from the orientation dependent driving force $P(g^D)$ and the local grain boundary mobility $m(\Delta g)$, which depends on the orientation difference between the growing grain g^R and the matrix g^D:

$$\frac{1}{\overline{W}(g^R)} = \oint \frac{f^D(g^D)}{P(g^D) \cdot m(\Delta g)} dg^D \quad ; \quad g^R = \Delta g \cdot g^D \tag{2}$$

That means that we have a rather complex system of input parameters for our calculations: The deformation texture $f^D(g)$, the ODF of nuclei $f^N(g)$, the driving force $P(g^D)$ and the local mobility $m(\Delta g)$. For first calculations we assumed the driving force $P(g^D)$ to be constant. The deformation texture and the local mobility were simulated by model functions:

$$f^D(g) = \sum \nu_i \cdot f_i(g) \quad ; \quad f_i(g) = f^{Gauss}(g_o, \omega_D) \tag{3}$$

$$m(\Delta g) = m_o + \sum \nu_i \cdot m_i(\Delta g) \quad ; \quad m_i(\Delta g) = m^{Gauss}(\Delta g_o, \omega_G) \tag{4}$$

The ODF of the nuclei was either assumed to be random or to correspond with the deformation texture. The maximum of the local growth rate law was chosen according to growth selection experiments, e. g. the well-known $40^o\langle 111\rangle$-orientation relationship for fcc metals [2] or a $27^o\langle 110\rangle$-orientation relationship for bcc metals [3].

RESULTS

For fcc metals starting deformation textures were simulated by model functions according to equation 3. With the assumption of random nucleation the recrystallization texture is represented by the mean linear growth rate: $f^R(g) = const. \cdot \overline{W}(g)^3$.

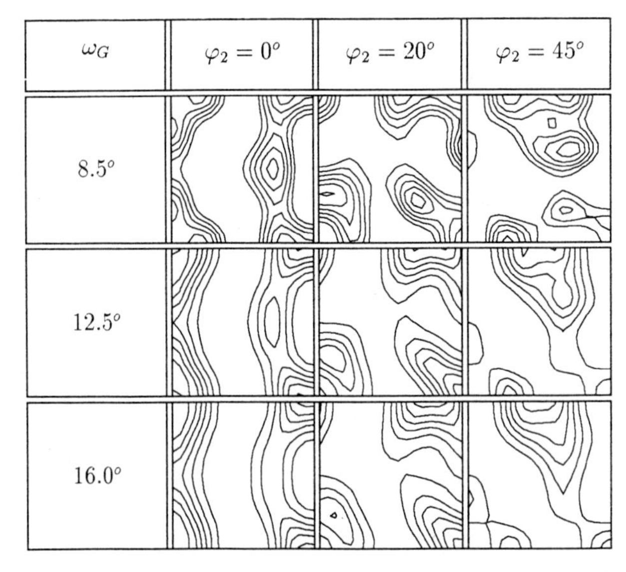

Figure 1
Sections $\varphi_2 = 0^o, 20^o$ and 45° of the recrystallization texture, for a deformation texture consisting of equal amounts of the three components Copper, S and Brass, calculated with a maximum of the local growth rate law at a $40^o\langle 111\rangle$ orientation relationship with a varying spread ω_G

Rolling textures of fcc metals with a high stacking fault energy show an orientation tube, consisting of the ideal orientations Brass $(\varphi_1, \Phi, \varphi_2) = (35^o, 45^o, 0^o)$, S $(59^o, 37^o, 63^o)$ and Copper $(90^o, 35^o, 45^o)$. At first the recrystallization texture was calculated for a starting deformation texture with equal amounts of all three deformation texture components. The maximum of the local grain boundary mobility Δg_o was assumed as a $40^o\langle 111\rangle$-orientation relationship, while the spread ω_G was varied. The results in figure 1 show a strong influence of this parameter. While a low spread ω_G causes a high growth rate for rotated cube components besides some other orientations, an increase of ω_G favours the cube orientation, at the cost of the other components.

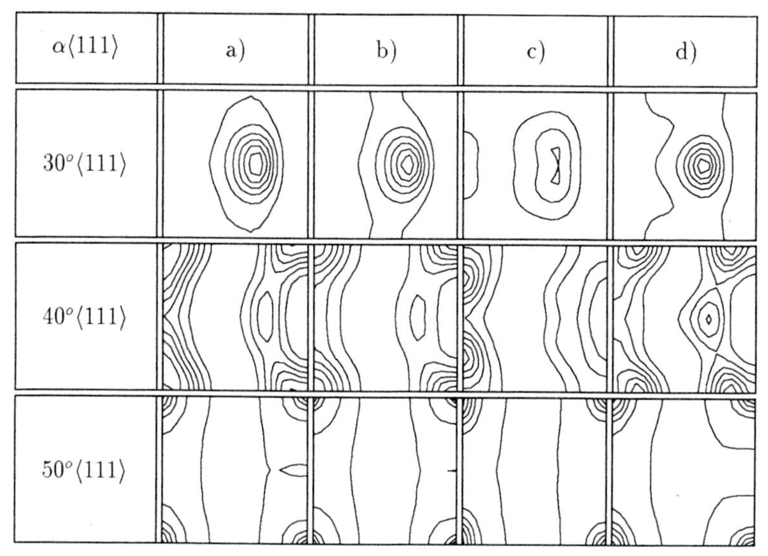

Figure 2: Sections $\varphi_2 = 0^o$ of the recrystallization textures calculated for different mixtures of the three components Copper, S and Brass as deformation texture and varied orientation relationships Δg_o for the local growth law. The composition of the deformation textures is:
a) 33 % S, 33 % Br, 33 % Cu, b) 50 % S, 25 % Br, 25 % Cu,
c) 25 % S, 50 % Br, 25 % Cu, d) 25 % S, 25 % Br, 50 % Cu

Experimental observations showed, that the composition of the rolling texture has a strong influence on the development of the recrystallization texture [4]. The influence of a change of the deformation texture composition on the formation of the cube texture is shown in figure 2. Besides the deformation texture composition the rotation angle of the maximum in the local growth law was varied. Results gained with a $30^o\langle 111\rangle$ orientation relationship have no similarities with experimental results at all. The development of the cube texture during recrystallization can not be explained with this growth law. An increase of the rotation angle leads to a higher growth rate of cube or rotated cube components. With a rotation angle of 40° we obtain a strong cube texture with the S orientation as major deformation texture component, whereas a high amount of the other components leads to the development of rotated cube components. A further increase of the rotation angle causes a shift of these components towards the ideal cube component. With a $50^o\langle 111\rangle$ orientation relationship we get strong cube textures, independent of the deformation texture composition.

Similar calculations were made for fcc metals with a low stacking fault energy. It was shown that it was possible to predict the formation of the Brass recrystallization texture with a starting deformation texture consisting of an amount of about 70 % Br and 30 % Goss orientation and a local growth law with a $40^o\langle 111\rangle$ orientation relationship as maximum [5].

The rolling textures of bcc metals usually are described by two fibre components, an incomplete α-fibre with the direction $\langle 110\rangle$ parallel to the rolling direction and the complete γ-fibre with the $\langle 111\rangle$–direction parallel to the normal direction. The recrystallization process leads in general to the disappearance of the α-fibre, wheras the γ-fibre is strengthened, especially the orientation $\{111\}\langle 112\rangle$ [6].

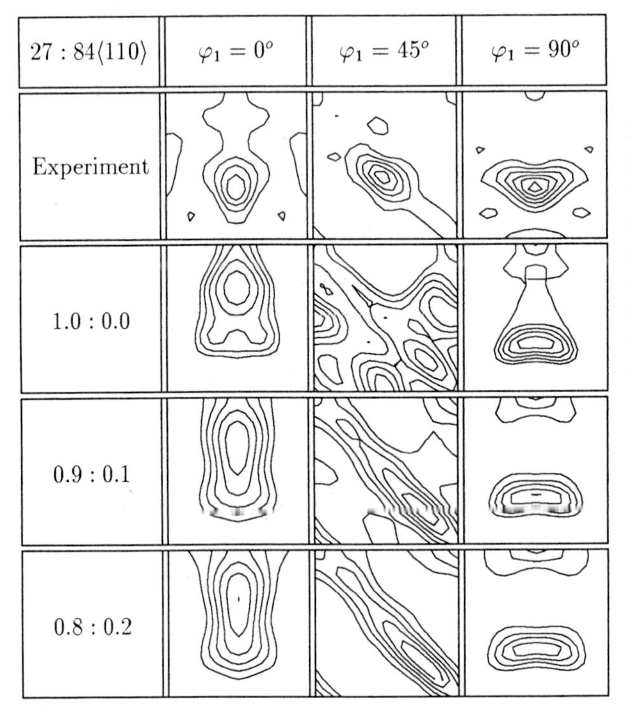

Figure 3
Sections $\varphi_1 = 0^o, 45^o$ and 90° of the experimental and some calculated recrystallization textures for ARMCO-Iron. The nucleation was assumed to be random, and the local growth law was combined of the two orientation relationships $27^o\langle 110\rangle$ and $84^o\langle 110\rangle$.

Because of the complexity of the deformation texture we introduced experimental data of ARMCO-Iron, 90 % cold rolled as starting deformation texture. For first calculations we assumed the orientation distribution of nuclei to be random. The maximum of the local mobility was chosen according to experimental results by Ibe and Lücke [3]. They found out, that about 90 % of fast growing grains had a $27^o\langle 110\rangle$ orientation relationship to the surrounding matrix, while the other 10 % showed a 84° rotation about the same axis. In first calculations the influence of the second, minor orientation relationship on the formation of the recrystallization texture was investigated. Figure 3 shows the simulated results as function of the composition of the growth rate law, compared with the experimental recrystallization texture. We see, that with a pure $27^o\langle 110\rangle$ orientation relationship we get a high growth rate in the orientation $\{111\}\langle 112\rangle$, as it is observed in the experimental texture, but in the section $\varphi_1 = 45^o$ some other components become visible, which have no experimental proof. An addition of the second growth relationship leads to a suppression of these unexpexted components, but simultaneously the growth rate of the orientation $\{111\}\langle 112\rangle$ decreases. Independent of the composition of the local growth law the growth rate of the orientations of the α-fibre is too high.
Another possibility to suppress the components in the section $\varphi_1 = 45^o$ is to consider oriented nucleation in matrix-orientation. This mechanism is very often refered to as a reason for the lack of new texture components after recrystallization [7]. But this mechanism alone is not suitable for the explanation of the texture development, because usually some characteristic

changes of the texture during recrystallization can be observed. When we superimpose a high probability of nucleation in the range of rolling texture components with the oriented growth mechanism we get the results in figure 4.

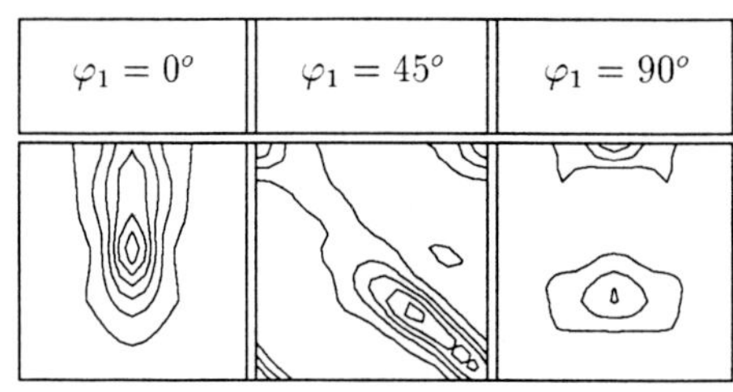

Figure 4
Sections $\varphi_1 = 0^o, 45^o$ and 90° of the recrystallization texture of ARMCO-Iron,calculated with the assumption of preferred nucleation in matrix orientation and a combination of 90 % $27^o\langle 110\rangle$ and 10 % $84^o\langle 110\rangle$ for the local growth law.

Although new orientations do not appear, we even get a stronger deviation from experimental results as with oriented growth alone, because the orientations of the α-fibre are strengthened, whereas the intensity of the orientation $\{111\}\langle 112\rangle$ is weakened considerably.
Another attempt to improve the results was made by varying the basic model we used for our calculations. Experiments of Kern showed, that ARMCO-Iron in the deformed state showed three types of microstructure with different etching behaviour and different crystallographic orientations [8]: "smooth" regions containing the orientations of the α-fibre, $\{100\}\langle 110\rangle$ to $\{112\}\langle 110\rangle$, and some grains with orientations ranging from $\{100\}\langle 110\rangle$ to $\{100\}\langle 100\rangle$, "grooved" regions with a higher degree of distortion, consisting of grains with the orientations $\{111\}\langle uvw\rangle$, $\{111\}\langle 112\rangle$ predominantly, and "splintered" regions with a very high degree of distortion, where small volume fractions of different orientations were found. The recrystallization process was rather inhomogeneous [9]: It started with a recovery process in the "smooth" regions, followed by classical recrystallization in the "splintered" and a little bit later in the "grooved" regions. At last the "smooth" regions were consumed by neighbouring grains. These observations were introduced in our calculations by the way of splitting the deformation texture into the three different partial textures. Then the "compromize" texture was calculated for the two regions where classical recrystallization has been observed. The results of this procedure were introduced as nuclei distribution into the calculation of the recrystallization texture of the third part. The final recrystallization texture was composed of the results of the three parts, taking their volume fractions into account. With the assumption of random-nucleation and oriented growth we got the results in figure 5a. Compared with the results from our "classical" model we do not get better results, on the contrary the deviation from the experimental data is even stronger. The next step was to introduce oriented nucleation in matrix orientation, but only in the two regions which showed classical recrystallization (again the calculated recrystallization texture for these regions was introduced as nuclei distribution for the third region). Now the results showed a better correspondence to experimental results (figure 5b), although the maximum again is not in the orientation $\{111\}\langle 112\rangle$. Nevertheless it is difficult to determine, whether oriented growth or oriented nucleation is the dominant process. Although the calculations with oriented growth alone can explain some important features of the texture development, it is still the question whether this mechanism is of importance. When we consider oriented nucleation in matrix orientation, but only in the two regions where classical recrystallization was observed, we obtain a nuclei distribution as in figure 5c, which shows already a rather good correspondence with the experimental recrystallization texture.

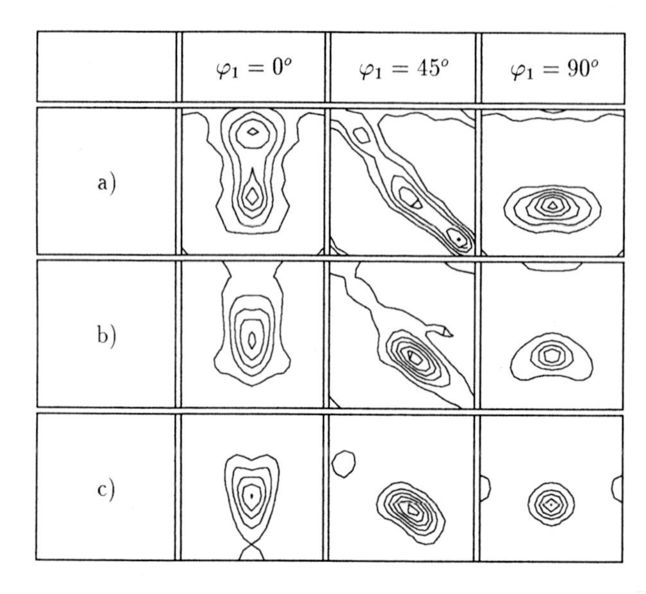

Figure 5: Sections $\varphi_1 = 0^o, 45^o$ and 90° of some recrystallization textures for ARMCO-Iron, calculated with the varied model, taking the inhomogeneous microstructure into account. The maximum for the local growth law was chosen at a combination of 90 % $27^o\langle 110\rangle$ and 10 % $84^o\langle 110\rangle$, respectively.
a) oriented growth
b) oriented nucleation and oriented growth
c) oriented nucleation

CONCLUSION

The results of our calculations showed, that the formation of the final recrystallization texture in fcc metals can be explained by the model of "compromize" texture formation. As well the formation of the cube texture as the formation of the Brass recrystallization texture can be explained by oriented growth and random nucleation. These results must be confirmed by calculations with experimental data, because small variations of the deformation texture composition have already a strong influence on the results of our calculations.
In contrast, the texture development in bcc metals was not so easy to reproduce. In this case oriented nucleation seems to be the dominant mechanism. To which extent oriented growth plays a part, can not be determined yet.

REFERENCES

[1] H. J. Bunge, B. Plege, in: Theoretical Methods of Texture Analysis, Ed.: H. J. Bunge, pp. 289

[2] B. Liebmann, K. Lücke, G. Masing, Z. Metallkunde 47 (1956), pp. 57

[3] G. Ibe, K. Lücke, in: Recrystallization, Grain Growth and Textures Ed: H. Margolin, New York 1966, pp. 434

[4] J. Hirsch, E. Nes, K. Lücke, Acta Metallurgica 35 (1987), pp. 427

[5] U. Köhler, E. Dahlem-Klein, H. Klein, H. J. Bunge, Textures and Microstructures 19 (1992), pp. 125

[6] F. Emren, U. v. Schlippenbach, K. Lücke, Acta Metallurgica 34 (1986), pp. 2105

[7] I. L. Dillamore, C. J. E. Smith, T. W. Watson, Metals Science Journal 1 (1967), pp. 49

[8] R. Kern, H. J. Bunge, in: The 7th International Conference on Textures of Material, Noordwijkerhout 1984, pp. 89

[9] R. Kern, J. Grewen, H. J. Bunge, in: The 7th International Conference on Textures of Material (ICOTOM), Noordwijkerhout 1984, pp. 257

Materials Science Forum Vol. 157-162 (1994) pp. 1809-1814

COMPARISON OF A SELF-CONSISTENT APPROACH AND A PURE KINEMATICAL MODEL FOR PLASTIC DEFORMATION AND TEXTURE DEVELOPMENT

R.A. Lebensohn and R.E. Bolmaro

Instituto de Física Rosario, National Council of Research of Argentina,
Rosario National University, Bv. 27 de febrero 210 bis, 2000 Rosario, Argentina

Keywords: Texture, Finite Deformations, Self-Consistent Models, Polycrystalline Plasticity

ABSTRACT

The present work is concerned with the comparison of a visco-plastic self-consistent (VPSC) approach and a new purely kinematical scheme called Grain Axes Coincidence Model (GACM). Both formulations are conceived for modeling plastic deformation and texture development of polycrystals and lead to spin equations for each grain that make evident the interaction between grain strain and a homogeneous effective medium strain. The interaction term is linear in the difference between both strains with a coefficient that is, for the VPSC model, an integral function of the anisotropic visco-plastic compliances and some form factors related with the current degree of deformation. Meanwhile, the coefficients of the GACM are, as expected, only function of the principal axes of the ellipsoid representative of the grain shape. The comparison of both kinds of interaction terms will be presented for rolling and torsion tests in cubic and hexagonal materials.

INTRODUCTION

In the field of texture development the influence of the grain shape has been recognized and modeled by many authors [1-3]. Each grain must keep compatibility with the matrix guaranteeing coincidence of shape and size with the hole lodging the grain. It means that the principal axes of the ellipsoid representative of the grain must be collinear with the principal axes of the ellipsoid representative of the hole. The models based in the Taylor hypothesis, assuming equal deformation for each grain and equal to the macroscopic strain, do not follow the shape and orientation of such ellipsoids. The Relaxed Constraints (RC) models approach the problem allowing some of the strain components to be different from the average when the grain is far from the initial spherical shape. The self-consistent models approach the problem right from the beginning and they will be briefly discussed in the second section. In the present work we compare a VPSC model with a pure kinematical model named Grain Axes Coincidence Model (GACM) for modeling both the plastic deformation and texture development in polycrystals. The GACM model has the Taylor model as a limit for spherical grains and it is related with the RC model in ways that will be discussed.

THE GRAIN AXES COINCIDENCE MODEL

In this model the grain is considered as part of a continuum media subject to finite deformations [4]. The deformation gradient **F** can be decomposed in two components: a) Proper plastic deformation reached by activation of crystal slip systems $\mathbf{F^P}$. b) Elastic deformation and "any other cause" for extra reorientations $\mathbf{F^*}$. They can be decomposed, by the polar decomposition, in the product of a pure rotation matrix **R** and a pure stretching **V**.

$$\mathbf{F} = \mathbf{F^* F^P} = \mathbf{R^* V^* R^P V^P} \tag{1}$$

The elastic deformation is regarded negligible compared with the plastic one and it does not appreciably reorient crystal or grain axes. **R** can be differentiated by the chain rule and transposed:

$$\dot{\mathbf{R}} = \dot{\mathbf{R}}^* \cdot \mathbf{R^P} + \mathbf{R^*} \cdot \dot{\mathbf{R}}^{\mathbf{P}} \qquad \mathbf{R^T} = \mathbf{R^{PT}} \cdot \mathbf{R^{*T}} \tag{2}$$

that leads to

$$\mathbf{\Omega} = \dot{\mathbf{R}} \cdot \mathbf{R^T} = \dot{\mathbf{R}}^* \cdot \mathbf{R^{*T}} + \mathbf{R^*} \cdot \dot{\mathbf{R}}^{\mathbf{P}} \cdot \mathbf{R^{PT}} \cdot \mathbf{R^{*T}} = \mathbf{\Omega^*} + \mathbf{R^*} \cdot \mathbf{\Omega^P} \cdot \mathbf{R^{*T}} \tag{3}$$

where $\mathbf{\Omega}$ is the time derivative of the orientation called "spin". The magnitudes representing $\mathbf{\Omega}$-spins can be written in terms of the antisymmetrical component of its velocity gradient and certain functions α_{ij} of the eigenvalues λ_i , representative of the ellipsoid axes, multiplying the component of the strain rate tensor. Rearranging we can obtain an expression slightly different from the usual in Taylor models. The extra term depends on the grain deformation and the difference between the average strain rate and the grain strain rate.

$$\Omega^*_{ij} = \Omega_{ij} - \Omega^P_{ij} = w_{ij} - w^P_{ij} - \frac{\lambda_i - \lambda_j}{\lambda_i + \lambda_j}(D_{ij} - d^P_{ij}) = w_{ij} - w^P_{ij} - \alpha_{ij}(D_{ij} - d^P_{ij}) \tag{4}$$

The α_{ij} coefficients are identically null for spherical grains, no matter how large the differences between the strain rates are. They grow asymptotically from zero to one for large deformations. This is the expected behavior in RC models for some components. The anisotropy of the matrix is not taken in account but only the grain anisotropy, through the slip systems activated. A similar equation is obtained for the SC models.

THE VISCO-PLASTIC SELF-CONSISTENT MODEL

The viscoplastic equation relating the plastic velocity gradient with the deviatoric stress applied to each crystal (grain) can be written:

$$d^P_i = \dot{\gamma}^0 \sum^{s} m^s_i (m^s_j \sigma'_j / \tau^s_c)^n \tag{5}$$

summing over all the slip systems. m_i^s is the Schmid tensor, σ_k' is the deviatoric stress in each grain, τ_c is the critical stress, $\dot{\gamma}^0$ is a normalization factor and n is the inverse of the strain rate sensitivity of the material (n>>1). The single index magnitudes d_i^P , σ_j' and m_j^s are the vectorized form of d_{ij}^P, σ_{ij}' and m_{ij}^s. Equation (5) can be rewritten as:

$$d_i = [\dot{\gamma}^0 \sum^{s} (m_i^s m_j^s / \tau_c^s)(m_j^s \sigma_j' / \tau_c^s)^{n-1}] \sigma_j' = M_{ij}^c(\sigma')\sigma_j' \quad (6)$$

where M^c is the grain viscoplastic compliance modulus. An analogous pseudo linear relation can be written at the polycrystal level:

$$D_i = M_{ij}(\Sigma')\Sigma_j' \quad (7)$$

where Σ'_i and D_i are the deviatoric stress and the velocity gradient in the polycrystal in the vectorized form and M_{ij} is the polycrystal viscoplastic compliance modulus. The equations (6) and (7) are the constitutive equations of an inclusion and a viscoplastic matrix. Applying the Eshelby formalism for the resolution of the inclusion problem we obtain the interaction equation:

$$D_i - d_i^P = \tilde{d}_i = -\tilde{M}_{ij}\tilde{\sigma}_j = -\tilde{M}_{ij}(\Sigma_i - \sigma_i^P) \quad (8)$$

where $\tilde{\sigma}$ and $\tilde{d}$ are the deviations in stress and velocity gradient of the grain with respect to the macroscopic magnitudes and $\tilde{M}$ is the interaction tensor defined as $n(I-S)^{-1}SM$. The viscoplastic Eshelby tensor S is function of the macroscopic modulus M and of the grain shape. The self-consistent equation is:

$$M = \langle M^c(M^c + \tilde{M})^{-1}(M + \tilde{M}) \rangle \quad (9)$$

The SC formulation can be used for texture development modeling [5]. An incremental step of deformation is imposed to a polycrystal composed of a discrete number of ellipsoidal grains, fixing a velocity gradient D in a time interval Δt. A first try guess is proposed for the stresses in each grain and iteratively the solution of equation (9) can be obtained. Each grain can be reoriented by:

$$\Omega_{ij}^* = w_{ij} - w_{ij}^P - \Pi_{ijkl} S_{klmn}^{-1}(D_{mn} - d_{mn}^P) \quad (10)$$

where we have gone back to the tensorial notation and w_{ij} is the antisymmetrical part of the macroscopic velocity gradient, w_{ij}^P is the antisymmetrical part of the distortion due to plastic slip and the third term is the local spin. The tensor Π is the antisymmetrical complementary component of the symmetric Eshelby tensor **S**. Its product by $\mathbf{S^{-1}}$ is a tensor that linearly combines the differences in strain rates. They are a set of interaction coefficients evaluated numerically in each deformation step.

RESULTS

Fig. 1 shows the evolution of the α_{ij} parameters proposed by the GACM for rolling up to a Von Mises equivalent deformation of 2.00. The rolling, normal and transversal directions are assumed to be along the axis 3, 1 and 2 respectively. This model, being purely kinematical, cannot be used to calculate the grain reorientation without an additional constitutive equation connecting the spin with the stresses or, alternatively, allowing the calculation of the strain rate differences. The asymptotic evolution of the coefficients becomes evident as the accumulated deformation increases.

The SC simulations were performed starting from a polycrystal with equiaxed grains and random distribution of orientations. Rolling deformation was simulated for copper polycrystal up to a Von Mises equivalent strain of 2.00. Equation (10) shows that the coefficients potentially significant are more than one for each spin component. Nevertheless, it is numerically observed that the coefficients presented in Fig. 2, which correspond one to one to the kinematically calculated coefficients, are at least two orders of magnitude higher than the others. Their behavior is also similar to the behavior of the α_{ij} coefficients except because they approach faster to the asymptotic value. Fig. 3 shows the evolution of the standard deviations (SD) of the respective strain rates. They represent the average differences between the strain rates of the grains and the polycrystal strain rates. The component SD(d_{13}) grows with the deformation stage and it is the most important one.

The kinematical coefficients for torsion are shown in Fig. 4. They evolve slower than in rolling because the shape does not change at the same pace. A similar VPSC simulation was performed for torsion in copper and the coefficients are shown in Fig. 5. They evolve similarly to the rolling coefficients but closer to the kinematical ones.

A VPSC simulation was performed for rolled zirconium (hexagonal symmetry) at high temperature (no twinning has been considered) and the coefficients are shown in Fig. 6. They behave similarly to the kinematical ones keeping the other coefficients close to zero.

The VPSC model also takes into account the anisotropy of the matrix. The coefficients include the influence of such anisotropy simultaneously with the shape influence. The shape evolution can be turned off in the code like if the grains were kept spherical during the deformation process and allowing just for crystal reorientation and consequent texture evolution. The anisotropy influence can be extracted and it should be complementary of the kinematical shape influence. Fig. 7 shows the calculated coefficients for rolled copper without grain shape evolution. At low deformation they show the right evolution to be complementary of the kinematical coefficients. The disagreements at higher deformations with the simple subtraction of Fig. 1 and 2 can be attributed to the fact that the grains, deprived of shape evolution, reorient differently from the shape evolving case and the consequent anisotropy is different. The evolution of anisotropy can also explain the differences between copper in rolling and torsion. The anisotropy evolves not only different but also slowly in torsion tests than in rolling because the texture reaches lower strength for the same deformation. For hexagonal crystal symmetry the coefficients are also strongly influenced by the anisotropy.

CONCLUSIONS

It has been shown that a pure kinematical model can give a better insight about the reorientation mechanisms in texture development. It does not provide the kinetics of crystal reorientation and an extra constitutive relationship has to be used in order to calculate the grain strain deviation. Self-consistent models are suitable for that purpose and simultaneously they provide the values of the interaction coefficients, which also include the anisotropy influence. The machine time consumption could be reduced by using the kinematical coefficients but further tests are necessary to check its influence in the final texture. For RC approaches the GACM model provides the value of the coefficients, at any deformation step, with no divergence problems for spherical grains.

REFERENCES

1) Tiem S., Berveiller M. and Canova G.R.. Acta metall.,1986, **34**, 2139.
2) Kocks U.F. and Chandra H., Acta metal., 1982, **30**, 695-709.
3) Kocks U.F. and Dawson P.R., Aristotle Conference on Mechanics, Physics and Structure of Materials in Thessaloniki, Greece, August 1990.
4) Bolmaro R.E.. To be published.
5) Lebensohn R.A. and Tome C.N., Acta metall. mater.. In press.

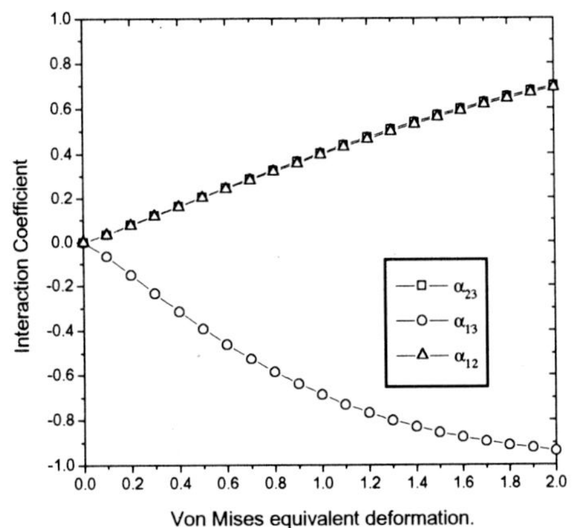

Fig. 1: Interaction coefficients calculated by the GACM for rolling.

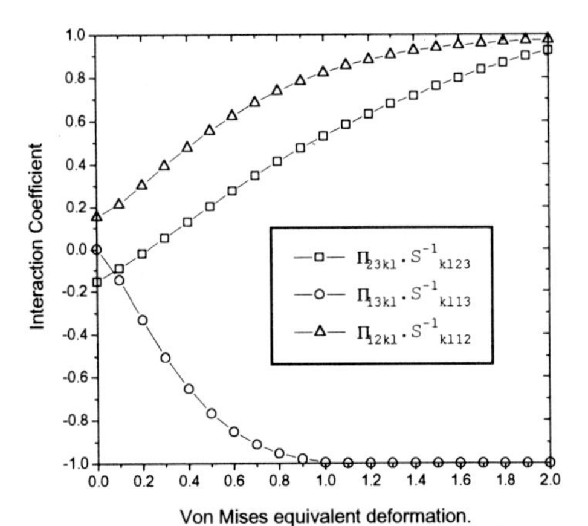

Fig. 2: Interaction Coefficients calculated by the VPSC model for copper. Rolling with shape and anisotropy evolution. Starting from randomly distributted spherical grains.

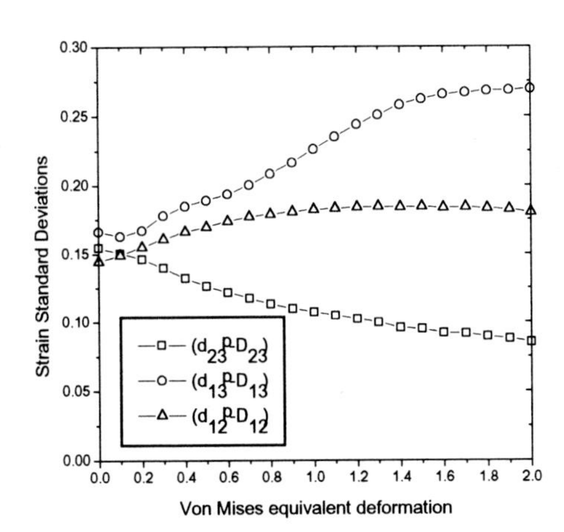

Fig. 3: Standard deviations for the three components significantly different from the average. Rolling in copper by VPSC simulation.

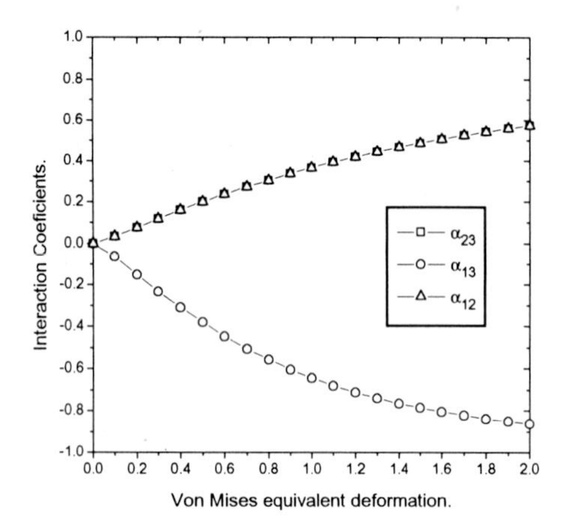

Fig. 4: Interaction coefficients calculated by GACM for torsion.

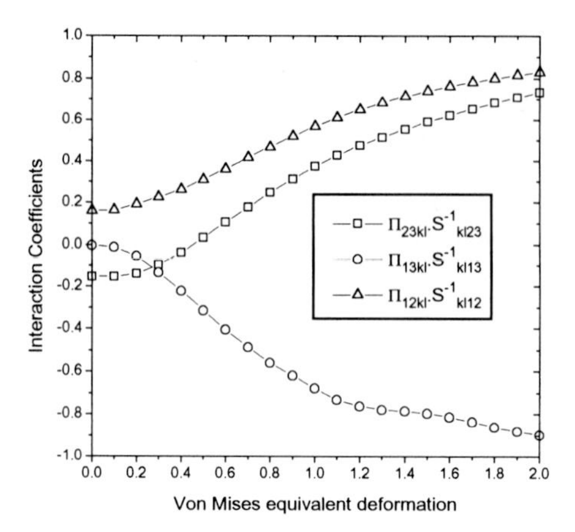

Fig. 5: Interaction coefficients calculated by VPSC model. Torsion in copper with initially spherical grains.

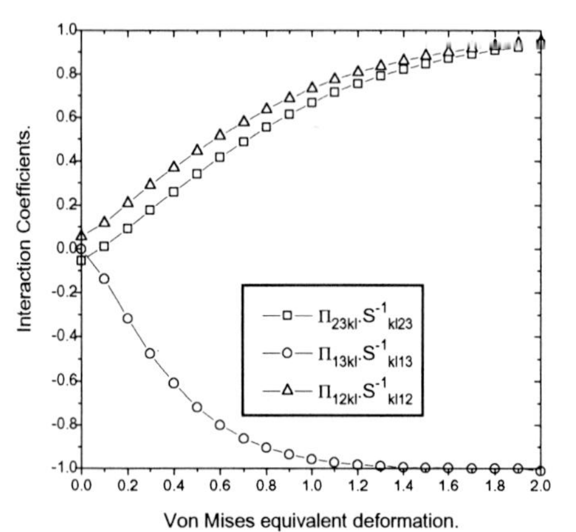

Fig. 6: Interaction Coefficients for initially spherical grains. VPSC calculation for zirconium rolled at high temperature.

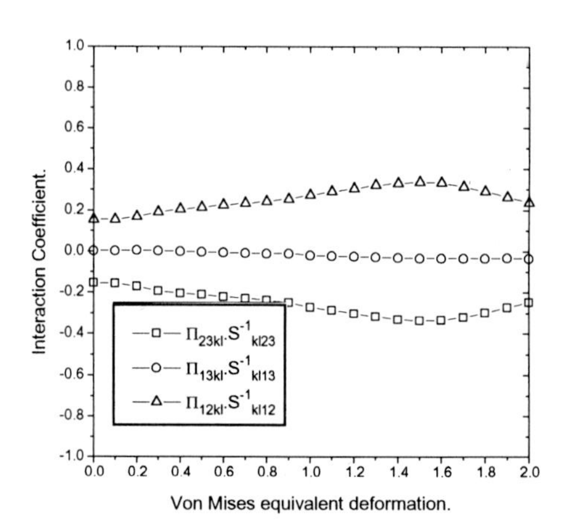

Fig. 7: Interaction Coefficients for rolling without shape updating. VPSC model for spherical copper grains.

Materials Science Forum Vol. 157-162 (1994) pp. 1815-1820

LATTICE ROTATIONS DURING PLASTIC DEFORMATION WITH GRAIN SUBDIVISION

T. Leffers

Materials Department, Risø National Laboratory,
PO Box 49, DK-4000 Roskilde, Denmark

Keywords: Lattice Rotation, Deformation Texture, Grain Subdivision, Plastic Deformation, Full Constraint, Relaxed Constraint

ABSTRACT. In a new quantitative model for plastic deformation with grain subdivision the traditional rules for lattice rotation during plastic deformation do not apply directly. The paper describes the necessary modifications of the rules.

INTRODUCTION

In normal texture calculations/simulations the individual "crystallites" do not interact with each other. They interact more or less with a surrounding continuum matrix. In the Taylor model [1] the crystallites interact very strongly with the matrix. In the relaxed-constraint Taylor model [2-4] there is a somewhat weaker but still strong interaction. In the Sachs model [5] and the modified Sachs model [6] there is only a weak interaction, and in Hosford's "extreme lower bound model" [7] the interaction is very weak or absent.

In the full-constraint Taylor model the strong interaction with the continuum matrix leads to a unique lattice rotation in the individual crystallites (which are the grains)*. For the relaxed-constraint models with the relaxation ascribed to the grain shape and a direct relation between grain shape and the directions of the macroscopic strain, e.g. [3, 4], there is only one logical solution for the lattice rotation (even though it is strictly speaking not a unique solution). In the models with weaker interaction with the matrix there is no unique solution for the lattice rotation in the crystallites/grains, cf. [8]. In neither of the cases described above one has to consider the specific interaction between the individual crystallites/grains.

Recently, the present author has suggested a quantitative model for rolling deformation with subdivision of the individual grains into "cell blocks" or bands [9, 10]. In this model the individual bands, and not the individual grains, act as crystallites, and the model considers the specific interaction between the bands. Therefore, one cannot consider the lattice rotation in the individual bands independently of the surrounding bands. For instance, there may be large variations in strain between the bands, which means that there would be large variations in lattice rotation between the bands if they were considered individually. This, in turn, would mean that the bands would not fit together after the lattice rotation (as to be described in details later).

* It should be noticed that we are only dealing with the lattice rotation corresponding to a specific slip pattern; the ambiguity in the selection of the slip pattern is not considered.

It is the aim of the present work to describe the special set of rules for the lattice rotation which has been formulated in order to cope with the special situation of grain subdivision.

THE MODEL

The model refers to experimental observations on rolled aluminium [11-13] (see also [14] for related observations on nickel). The individual grains are subdivided into "cell blocks" with the shape of flat bands approximately parallel to the transverse direction making angles in the range 32°-49° with the rolling plane. There are different combinations of active slip systems in the different cell blocks as reflected in different lattice orientations. There is a clear trend for at +/- relation for the orientation differences between the cell blocks: if there is a certain positive orientation difference between two cell blocks in the stack of cell blocks, there will be approximately the same absolute orientation change to the next cell block but now negative. It is suggested that the number of active slip systems in the individual cell block is less than the five required in the Taylor model[1], i.e. that the individual cell block does not follow the macroscopic strain. It is also suggested that the individual grains approximately follow the macroscopic strain via a scheme of collaboration between the cell blocks. The net effect of the grain subdivision is suggested to be a reduced work hardening rate because of the reduced number of slip systems operating at any specific point in the material.

In some cases the situation is more complex than that described above: there may be two systems of intersecting band-shaped cell blocks. However, in the present work we are only going to deal with the simple situation with one set of parallel cell blocks.

As sketched in figure 1 the subdivision into band-shaped cell blocks is given a very simple interpretation in the present model: there are two families of bands, band family 1 and band family 2, each with one specific combination of slip systems and specific shears on the different slip systems. The model maintains strict strain continuity between the two band families, and it imposes the Taylor condition of identical strain in the individual grains for the *average* strain in each grain. The bands are approximately (within 15°) parallel to the transverse direction and their traces on the longitudinal plane (parallel to the rolling and the normal directions) make an angle of 38° with the rolling direction. The equations for strain continuity are expressed in the two coordinate systems in figure 1.

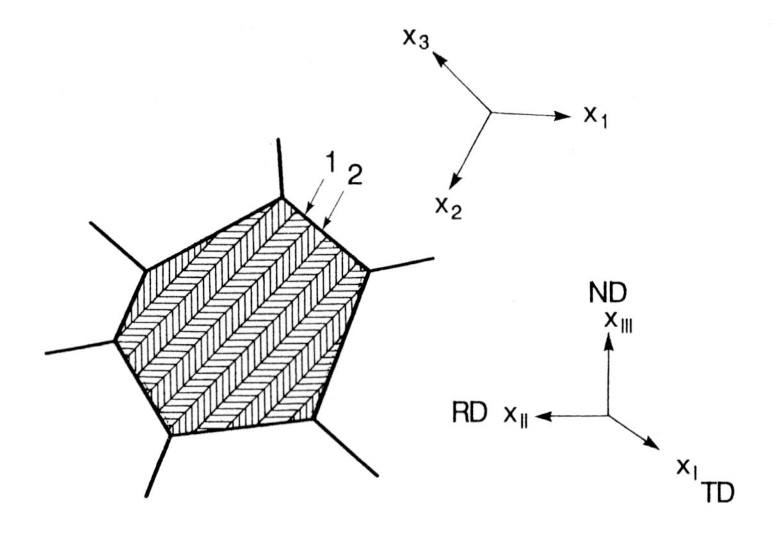

Figure 1. The subdivision of a grain into band families 1 and 2 with the band and the sample coordinate systems indicated.

One system, characterized by Arabic indices (the band system), has the x_3 axis perpendicular to the plane of the bands and the x_2 axis perpendicular to the transverse direction. The other system, characterized by Roman indices (the sample system), has the axes parallel to the transverse, the rolling and the normal directions of the rolled plate. The continuity equations are:

$$d\varepsilon_{11}(1) = d\varepsilon_{11}(2) \quad (1)$$
$$d\varepsilon_{22}(1) = d\varepsilon_{22}(2) \quad (2)$$
$$d\varepsilon_{12}(1) = d\varepsilon_{12}(2) \quad (3)$$
$$d\varepsilon_{I\,I}(1) + d\varepsilon_{I\,I}(2) = 0 \quad (4)$$
$$d\varepsilon_{II\,II}(1) + d\varepsilon_{II\,II}(2) = 2\,dE \quad (5)$$
$$d\varepsilon_{I\,II}(1) + d\varepsilon_{I\,II}(2) = 0 \quad (6)$$
$$d\varepsilon_{I\,III}(1) + d\varepsilon_{I\,III}(2) = 0 \quad (7)$$
$$d\varepsilon_{II\,III}(1) + d\varepsilon_{II\,III}(2) = 0 \quad (8)$$

dE being the macroscopic normal strain in the rolling direction. Equations 1-3 express the continuity between the band families 1 and 2. Equations 4-8 express the continuity between the grains. The equations are a combination of relaxed constraints and full constraint: relaxed constraints for the bands in accordance with their flat shape (equations 1-3) and full constraint for the grains in accordance with their equiaxed shape (equations 4-8). This implies that the model refers to moderate reductions for which the grains remain approximately equiaxed. At high strains the microstructure changes; it is dominated by a laminar structure parallel to the rolling plane, e.g. [15].

As already mentioned, the aim of the present work is to describe the set of rules which has been formulated for the lattice rotations in the special situation defined by the present model. Therefore, the general implications of the model are not to be considered. For details about these implications the reader is referred to [10] - and to subsequent publications to appear. It should be mentioned, however, that there are two types of solutions to equations 1-8: 4+4 solutions with four active slip systems in each band family and 5+3 solutions with five active slip systems in one band family and three in the other. Of these only 4+4 solutions are considered in the present work.

THE ROTATION RULES

Figure 2 demonstrates that the application of the standard rules for lattice rotation on the individual bands in the situation of the present model (with substantial differences in the "Arabic strains" ε_{13} and ε_{23} between the bands in family 1 and family 2, e.g. figure 1) would lead to an impossible situation with partial overlap of the bands and partial formation of empty space between them. This can only be avoided by the implementation of a new set of rules for the lattice rotations. The new

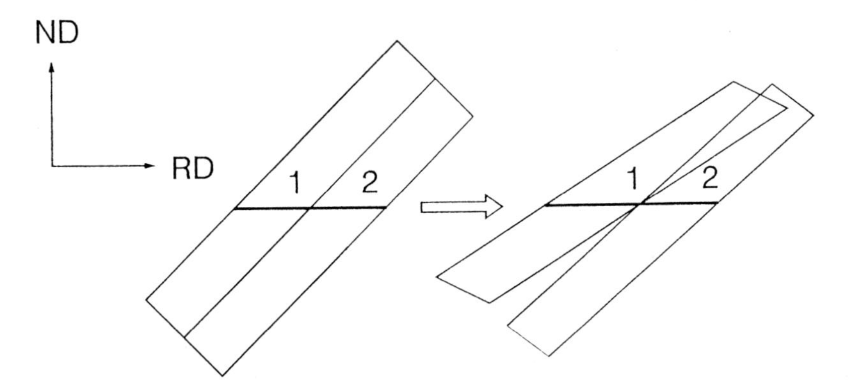

Figure 2. Rotations of bands belonging to family 1 and family 2 according to the standard rules when there is a great difference between the two band families ($d\varepsilon_{32}(1) > d\varepsilon_{32}(2)$).

rules have two parts: (i) special rules which make the bands in the two band families rotate relative to one another so that their major faces remain parallel without any change in the centre of gravity of the two orientations and (ii) the normal rules for the lattice rotation in the full-constraint Taylor case applied collectively on the two band families (with change of the centre of gravity of the two orientations).

Part (ii) is pure routine which does not require any further explanation (it should be added, though, that the lattice rotation in any of the two band families is governed by the slip patterns in both families). Part (i) refers to the difference between the slip patterns in band family 1 and band family 2. If there is no difference, there is no part (i). If there is a difference (as there is normally), the two band families have part (i) lattice rotations in opposite directions. In the actual computer program the lattice rotations are expressed via Apq which is the q coordinate in the "new" coordinate system of the p axis in the "old" coordinate system ("old" means before and "new" means after lattice rotation). The expression for the crucial Apq's for part (i) are given in equations 9-12 in terms of the Arabic coordinates of slip plane normal (n1,n2,n3) and slip direction (d1,d2,d3) and the shears $\delta\gamma$, i and j referring to family 1 and family 2, respectively:

Family 1

$$A12 = \frac{1}{4}\sum_{i=1}^{4}(n2_i\, d1_i - n1_i\, d2_i)\delta\gamma_i - \frac{1}{4}\sum_{j=1}^{4}(n2_j\, d1_j - n1_j\, d2_j)\delta\gamma_j \tag{9}$$

$$A13 = \frac{1}{2}\sum_{j=1}^{4} n1_j\, d3_j\, \delta\gamma_j - \frac{1}{2}\sum_{i=1}^{4} n1_i\, d3_i\, \delta\gamma_i \tag{10}$$

$$A21 = -A12 \tag{11}$$

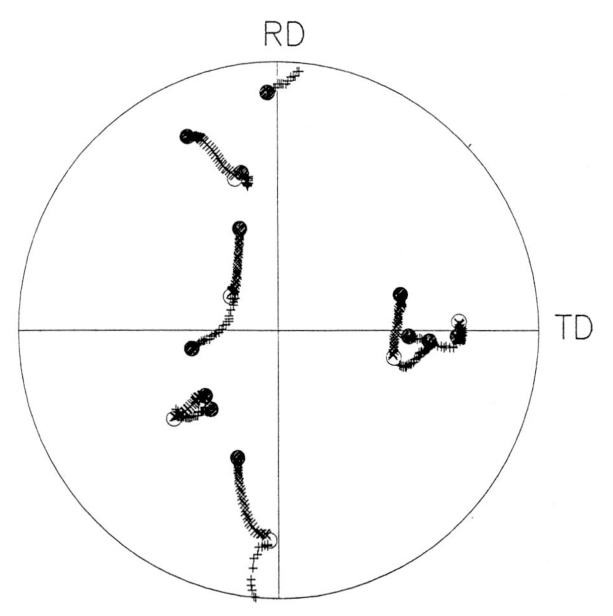

Figure 3. Two grains with initially random orientations split up each into two band families during rolling deformation to 50% reduction as illustrated by the {200} poles. Open circles represent the initial orientations.

$$A23 = \frac{1}{2}\sum_{j=i}^{4} n2_j \; d3_j \; \delta\gamma_j - \frac{1}{2}\sum_{i=1}^{4} n2_i \; d3_i \; \delta\gamma_i \qquad (12)$$

Family 2

All signs change

The actual lattice rotations in two grains with initial orientations selected at random are shown in figure 3. One very clearly sees the splitting of the grains into two band families.

DISCUSSION

It has been demonstrated that one can formulate a rational set of rules for the lattice rotations when the individual grains split up into bands with different deformation patterns as sketched in figure 1.

The reason why a new set of rules is necessary is that the underlying model considers the specific interaction between specific crystallites (the bands): the model is not a 1-point model but rather a 2-point model [16]. In principle the problem will appear for any n-point model (n > 1), but it may be very difficult to formulate specific rules for more complex n-point models. The advantage of the present model is that the interaction between the two band families is so well defined. It should be mentioned that Lee, Duggan and Smallman [17, 18] are presenting a model which addresses a problem similar to that in the present model, viz. the subdivision of the grains into "deformation bands". Lee and Duggan are less specific about the interaction between the bands into which the grains are subdivided than the present author, which means that their model may be seen as a 1-point model. This is consistent with the fact that Lee and Duggan do not specifically refer to any problems with the calculation of the lattice rotations.

The present work addresses the specific problems about the rules for the lattice rotation. The actual simulation of deformation (and hence texture formation) with grain subdivision involves a number of problems as described in[10]. The present work does not aim at solving these problems. As an illustration of the present state the final figure (figure 4) shows the simulated {111} pole figure for

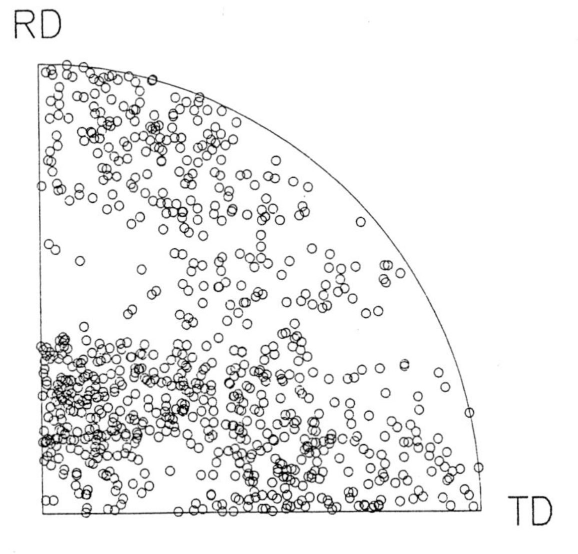

Figure 4. {111} pole figure for 100 grains of initially random orientation after simulated rolling reduction of 50% with grain subdivision.

100 grains of initially random orientation deformed to 50% reduction with grain subdivision into two band families. Compared with standard simulated pole figures according to the full-constraint or the relaxed-constraint Taylor models, the pole figure in figure 4 (simulated for deformation with grain subdivision) reflects a decrease in texture sharpness which is in agreement with experimental observations.

ACKNOWLEDGEMENTS

Thanks are due to J. Mortensen for typing the manuscript. The work is supported by the Danish Materials Technology Development Programme.

REFERENCES

1. Taylor, G.I.: J. Inst. Met., 1938, 62, 307.

2. Leffers, T.: Risø Report No. 184, 1968.

3. Mecking, H.: Strength of Metals and Alloys (Pergamon Press, Oxford, 1980) p. 1573.

4. Kocks, U.F. and Canova, G.R.: Deformation of Polycrystals (Risø National Laboratory, Roskilde, 1981) p. 35.

5. Sachs, G.: Z. verein. deut. Ing., 1928, 72, 734.

6. Pedersen, O.B. and Leffers, T.: Constitutive Relations and Their Physical Basis (Risø National Laboratory, Roskilde, 1987) p. 147.

7. Hosford, W. and Galdos, A.: Textures and Microstructures, 1990, 12, 89.

8. Hosford, W.F.: Texture Cryst. Sol., 1977, 2, 175.

9. Leffers, T.: Strength of Metals and Alloys (Freund Publishing House, London, 1991) p. 615.

10. Leffers, T.: Proceedings of Plasticity '93 (Baltimore, July 1993) in press.

11. Bay, B., Hansen, N. and Kuhlmann-Wilsdorf, D.: Mater. Sci. Eng., 1989, A113, 385.

12. Hansen, N.: Mater. Sci. Techn., 1990, 6, 1039.

13. Bay, B., Hansen, N., Hughes, D.A. and Kuhlmann-Wilsdorf, D.: Acta Metall. Mater., 1992, 40, 205.

14. Hansen, N., Juul Jensen, D. and Hughes, D.A.: In These Proceedings.

15. Bay, B., Hansen, N. and Kuhlmann-Wilsdorf, D.: Mater. Sci. Eng., 1992, A158, 139.

16. Molinari, A., Canova, G.R. and Ahzi, S: Acta Metall., 1987, 35, 2983.

17. Lee, C.S. and Duggan, B.J.: Acta Metall. Mater., 1993, 41, 2691.

18. Duggan, B.J., Lee, C.S. and Smallman, R.E.: In these Proceedings.

Materials Science Forum Vol. 157-162 (1994) pp. 1821-1826

TEXTURE OF MICROSTRUCTURES IN BCC METALS FOR VARIOUS LOADING PATHS

X. Lemoine[2], M. Berveiller[1] and D. Muller[1]

[1] Laboratoire de Physique et Mécanique des Matériaux, URA CNRS 1215, ENIM, Ile du Saulcy, F-57045 Metz Cédex 1, France

[2] Laboratoire d'Etude et de Développement des Produits Plats, SOLLAC, Avenue des Tilleuls, F-57191 Florange, France

Keywords: Dislocation Structure, Self-Consistent Scheme, Texture of Microstructure, Slip System, Micromechanical Model

ABSTRACT : We study the orientation distribution of the most active slip planes during several loads simulated by an elastoplastic self-consistent model for an initially non-textured BCC polycrystal. This orientation distribution is compared to the orientation of dislocation microstructures. A simplified micromechanical analysis linking the stress rates to the deformation of a dislocation network is applied to the case of a cell structure.

I. INTRODUCTION

The anisotropy evolution of the elastoplastic behaviour of polycrystalline metals during forming processes is usually related to crystallographic and morphological textures, latent hardening and second order residual stresses [1].

While the effects of crystallographic and morphological textures appear for large prestrains, whatever the offset of plastic strain used to define the flow stress, residual stresses induced by low or high plastic strain are responsible for the anisotropy inside the microplasticity range (low and medium offsets 0-1%) [2]. However, the intracrystalline strain-induced dislocation microstructure constitued of low and high dislocation density regions (the cells and walls) contributes also significantly to the overall plastic anisotropy [3,4].

At first, the morphological and topological characters of the cell structure are more related to the overall strain path than to the polycrystalline aspects (crystallographic texture ...)[5,6]. Next, dense dislocation walls and regions of low dislocation density may be considered as a two-phase material or composite with a high non-local behaviour since plastic strain due to dislocation motion within the cell interiors contributes to hardening through the storage of dislocations inside the cell walls [7]. Thus, the heterogeneity of the flow stress between soft (cell interior) and hard (cell walls) parts of the metal, becomes higher than the one between crystallographic misorientated grains.

The modelisation of this non-local material needs the knowledge of the non-local hardening relations, the evolution of the third order internal stresses and the orientation of the induced microstructure.

This last one follows, at least partly, from a general principle of minimization of free (elastic) energy [8,9]. Using such kind of hypothesis, it is easy to show that dislocation walls are parallel to the planes of the most actives slip systems, at least when single slip is predominant and perfect topological accomodation is allowed. This evaluation is confirmed by several experimental observations [10,11,12] despite some discrepancies for a few number of grains.

In this work, an elastic self-consistent model [2,13] is used in order to analyse the orientation distribution of the planes of the most active slip systems of an initially non-textured BCC polycrystal. During several loading paths, the slip activity of each system, in the whole polycrystal, is calculated, which allows to define for each grain the orientation of the dislocation walls. The distribution of these orientations with respect to the sample axis, is called texture of the dislocation microstructure and may be represented in form of pole figures.

In the last part, an simplified micromechanical analysis linking the stress rates to the deformation of a dislocation network is proposed and applied to the case of a cell structure.

II. TEXTURE OF THE MICROSTRUCTURE FOR BCC METALS.

The calculations are performed for a 1000 grains BCC polycrystal without initial textures. Isotropic elasticity (μ = 80000 MPa, ν = 0,3) and slip sytems of type {110}<111> and {112}<111> are assumed. The initial critical shear stress is supposed identical for all the systems and the linear hardening matrix takes into account latent as well as self hardening [14]. During the loading, the rotations of the lattices are calculated which allow to account for the rotation of the walls related to the crystallographic texture formation. Various monotoneous loading paths (uniaxial tension (UT), equibiaxial expansion (EEB), shear (ST) and rolling test (RT)) are simulated by the following applied stress rates:

$$\text{UT}\quad \dot{\Sigma}\begin{pmatrix}1&0&0\\0&0&0\\0&0&0\end{pmatrix}\qquad \text{EEB}\quad \dot{\Sigma}\begin{pmatrix}1&0&0\\0&1&0\\0&0&0\end{pmatrix}\qquad \text{ST}\quad \dot{\Sigma}\begin{pmatrix}0&1&0\\1&0&0\\0&0&0\end{pmatrix}\qquad \text{RT}\quad \dot{\Sigma}\begin{pmatrix}1&0&0\\0&0&0\\0&0&-1\end{pmatrix}$$

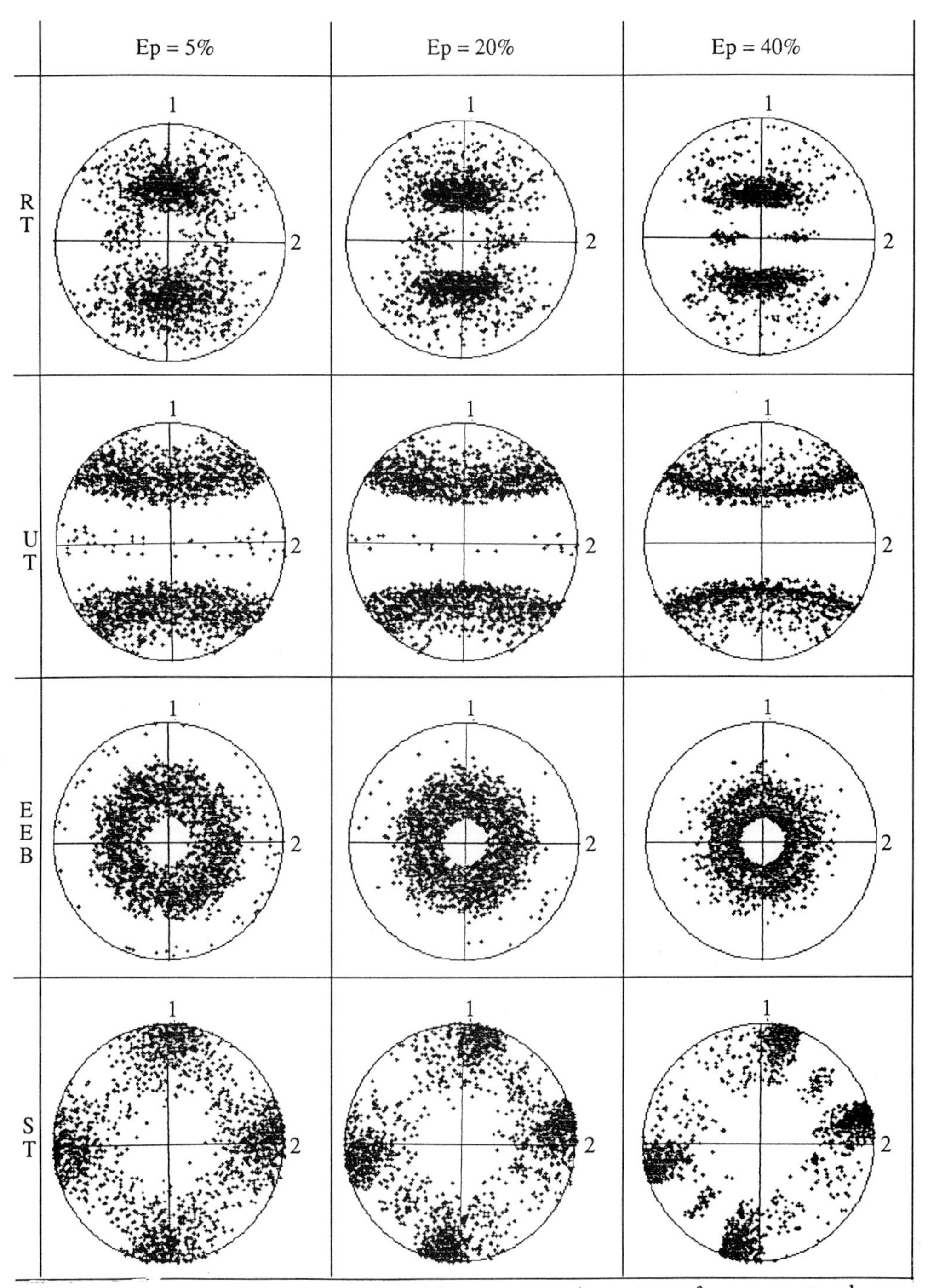

Figure 1: Pole figures of the orientation of the two most active systems, for a non-textured BCC polycrystal, for several loading paths and at different plastic strains.

The orientation of the planes corresponding to the two most actived systems inside each grain is represented by their poles in the sheet plane (1,2), for different levels of plastic strain (5%, 20%, 40%).

The results are presented in figure 1. For each loading, the distribution of the microstructure appears to be concentrated in some clouds or zones immediatly after plastic flowing (E^p = 5%) despite the isotropy of the crystallographic texture. The shape and orientation of these zones or clouds depend strongly on the loading paths and follow approximatively the Schmid law. When the plastic strain increases (E^p = 20% to 40%), these zones or clouds become sharper and some "ideal" orientations or "fiber", different from the initial orientation at low plastic strain E^p (5%) are observed.

III. STRESSES RESULTING FROM DEFORMATION OF DISLOCATION NETWORKS.

This polarized dislocation structure induces an heterogeneous lattice distortion, i.e. lattice deformation and rotation leading to texture and internal stresses.

If a metal containing such kind of microstructure is plastically deformed, this dislocation network does not contribute to plastic flow but is convected, at least partially, by mobile dislocations. The deformation of dislocation networks induces changes in stress and orientation of the lattice.

In the following, a simple micromechanical model is proposed in order to evaluate such effects.

A classical plastic inclusion model [15,16] is used in order to describe the stress field, and dislocation density for a cell structure. From an initial state corresponding to a nearly uniform single crystal, a plastic distortion field $\beta^p(r)$ assumed piece wise constant equal to β^{p_c} inside the cells and $\beta^{p_w} \approx 0$ inside the walls with volume fraction f is imposed on the crystal. The topology of the cell structure is simply described by an inclusion-matrix problem so that the total (elastic + plastic) strain ε^c inside the inclusion (representing the cell interior) is given by :

$$\varepsilon^c = E^T - S^1\left[(1-f)\varepsilon^{p_c} - \varepsilon^{p_c}\right] \qquad (1)$$

where $E^T = E^e + E^p$ is the overall total strain including elastic loading E^e and $E^p = (1-f)\varepsilon^{p_c}$ is the mean value of the $\varepsilon^p(r)$ field. S^1 is the usual Eshelby tensor. The dislocation density $\alpha(r)$ related to the incompatible field $\beta^p(r)$ is defined by Nye [17] or Kröner [18] like :

$$\alpha_{ij} = -\epsilon_{imk}\beta^p_{kj,m} \qquad (2)$$

For the inclusion-matrix problem, the $\beta^p(r)$ field is formulated by :

$$\beta^p_{kj}(r) = \beta^{p_0}_{kj} + \left(\beta^{p_c}_{kj} - \beta^{p_0}_{kj}\right)\theta(r) \qquad (3)$$

$$\beta^{p}_{kj}(r) = \beta^{p_0}_{kj} + \Delta\beta^{p_c}_{kj}\theta(r) \quad (4)$$

where $\beta^{p_0} = (1-f)\beta^{p_c}$ and $\theta(r)$ is the indicatrix function for the inclusion of volum V_I ($\theta = 0$ if $r \notin V_I$, $\theta = 1$ if $r \in V_I$). From equation 2 and equation 4, the dislocation density becomes :

$$\alpha_{ij} = -\epsilon_{imk}\Delta\beta^{p_c}_{kj}\delta(S)n_m \quad (5)$$

which is a superficial dislocation density located inside the boundary of V_I, i.e. the dislocation walls. From equation 1, the stress σ^c inside the inclusion is related to the applied stress $\Sigma = CE^e$ and the heterogeneous plastic strain field by :

$$\sigma^c = \Sigma - Cf(I - S^1)\varepsilon^{p_c} \quad (6)$$

If a small change $\delta\varepsilon^{p_c}$ is superimposed on the predeformation introduced by the preloading path, the change $\delta\sigma^c$ resulting, from the change $\delta\Sigma$ and the deformation of the dislocation network located on the inclusion surface is given by :

$$\delta\sigma^c = \delta\Sigma - Cf(I - S^1)\delta\varepsilon^{p_c} + Cf\,\delta S^1\varepsilon^{p_c} \quad (7)$$

where elastic isotropy is assumed so that $\delta C = 0$ and the change δf is neglected. Now, the change of the inclusion shape is only due to the superimposed plastic strain $\delta\varepsilon^{p_c}$ so that :

$$\delta S^1 = \frac{\partial S^1}{\partial \varepsilon^{p_c}}\delta\varepsilon^{p_c} \quad (8)$$

or

$$\delta S^1 = S^2\delta\varepsilon^{p_c} \quad (9)$$

where S^2 is a new Eshelby type tensor which can be calculated by usual micromechanical methods [16]. From the Green function G, S^2 is obtained [19] by :

$$S^2_{nmklpq} = -\int_{V_I} C_{ijkl}G_{jn,imp}(R)x'_q dV' \quad (10)$$

Finally, the stress change inside the cells becomes :

$$\delta\sigma^c = \delta\Sigma - Cf(I - S^1)\delta\varepsilon^{p_c} + Cf\,S^2\delta\varepsilon^{p_c}\varepsilon^{p_c} \quad (11)$$

The new term $Cf\,S^2\,\delta\varepsilon^{p_c}\varepsilon^{p_c}$ appearing in equation 11, corresponds to the stress change due to the deformation of dislocation networks.
Such a simple formulation applied to a single inclusion problem has clearly shown strong induced internal stresses [19]. Shear banding are probably related to this effects.

References :

1) Hansen, N., Leffers, T.: Rev. Phys. Appl., 1988, 23, 519.
2) Lipinski, P., Krier, J., Berveiller, M.: Rev. Phys. Appl., 1990, 25, 361.
3) Haasen, P.: Phil. Mag., 1958, 3, 384.
4) Gil Sevillano, J., Van Houtte, P., Aernoudt, E.: Prog. Mater. Sci., 1980, 25, 69.
5) Fernandes, J.V., Schmitt, J.H.: Phil. Mag. A., 1983, 48, 841.
6) Mughrabi, H.: Rev. Phys. Appl., 1988, 23, 367.
7) Berveiller, M., Muller, D., Kratochvil, J.: Int. J. Plasticity, 1993, 9, 633.
8) Hansen, N., Kuhlmann-Wilsdorf, D.: Mat. Scie. Eng., 1986, 81, 141.
9) Berveiller, M., Bouaouine, H.: 32e col. de Metallurgie, Ed. Rev. de Metallurgie, 1989, 89.
10) Inagaki, H.: Z.Metallkde, 1990, 81, 474.
11) Thuillier, S.: Thesis, 1992, INPG Grenoble.
12) Schmitt, J.H.: Thesis, 1986, INPG Grenoble.
13) Lipinski, P., Berveiller, M.: Int. J. Plasticity, 1989, 5, 149.
14) Francioisi, P.: Rev. Phys. Appl., 1988, 23, 383.
15) Kröner, E.: Acta Met., 1961, 9, 155.
16) Eshelby, J.D.: Prog. in Sol. Mech. North Holland, Amsterdam t2, 1961, 89.
17) Nye, J.F.: Acta Met., 1953, 1, 153.
18) Kröner, E.: Summer School Hrazanai, Ed. Academia Prague, 1966.
19) Muller, D., Lemoine, X., Berveiller, M.: Mecamat 93, Int. Seminar on Micromech. of Mat., Moret-sur-Loing (France), Ed. Eyrolles, EDF Collection, 1993, 84, 128.

Materials Science Forum Vol. 157-162 (1994) pp. 1827-1838

MODELLING OF THE TEXTURE FORMATION IN ELECTRODEPOSITION PROCESS

D.Y. Li and J.A. Szpunar

Department of Mining and Metallurgical Engineering, McGill University,
3450 University Street, Montreal, Quebec H3A 2A7, Canada

Keywords: Texture, Electrodeposition, Hydrogen Co-Deposition

ABSTRACT

Electrodeposition process has been simulated using a Monte Carlo technique. In the simulation, a two dimensional hexagonal lattice is used to map the microstructure at a cross section of the deposit. Criterion for grain growth is based on the energy state of the lattice site, which is set by counting the interactions between the site and its nearest neighbours. The texture formation is modelled by taking into account the surface-energy anisotropy. Since the surface energy of metals can be affected by hydrogen adsorption, texture of a deposit may vary if co-deposition of hydrogen takes place during the metal deposition. This effect has also been simulated. In addition, the present model can provide information about the influence of deposition rate on the microstructure and porosity of the deposit as well as its texture.

1. INTRODUCTION

Since Anderson et al [1-6] suggested a Monte Carlo statistical model to simulate grain growth in 1984, the Monte Carlo technique has attracted great interest and found extensive application in materials science [5,7-11]. Using the Monte Carlo technique, Srolovitz et al [5] modelled the recrystallization process and also the thin film formation in deposition (VD) [11]. This technique was applied later by Spittle and Brown [7-9] to model the solidification process. Recently, Tavernier and Szpunar applied the Monte Carlo technique to model texture development in the recrystallization processes [10].

In this paper, the Monte Carlo technique has been applied to electrodeposition. Although electrodeposition is a process in which mass transfer is accompanied by a charge transfer [12-14], formation of a deposit is also a nucleation and grain growth process [14,15]. It is therefore possible to apply the Monte Carlo technique to simulate this process.

In the electrodeposition, current density is the main parameter controlling the deposit formation [12,14]. The deposition rate is proportional to the current density. This relation is well described by Faraday's law [12]:

$$\frac{dh}{dt} = \frac{h}{t} = \frac{i_c \omega E_{el}}{60\zeta} \quad (1)$$

where h is the thickness of deposit, and E_{el} is the electrochemical equivalent of the deposited metal. t, ζ and i_c are the deposition time, density of the deposit and the current density, respectively. ω is the current efficiency and its value is dependent on overvoltage, pH, temperature and the type of cathode, as well as other factors. Although electrodeposition is affected by many factors, the current density nevertheless plays the most important role in the deposit formation. Therefore, in this simulation, the current density is considered as the controlling parameter.

Modelling of the texture formation is the major task of this simulation. Although several theories are available to explain the deposit texture, e.g., the geometrical selection theory and the two-dimensional nucleation theory [16-18], none of them can provide a fully satisfactory explanation as to the mechanism of texture formation in electrodeposition. The present simulation presents a different approach. The texture formation is modelled based on the minimum energy principle. A two-dimensional hexagonal lattice is used to map the microstructure at a cross section of the deposit. Criterion for grain growth through the occupation of lattice sites is based on the state-energy of the lattice site. A lattice site can be occupied, if by doing so, the state-energy of the site is decreased. To simulate the texture formation, the surface-energy anisotropy is taken into account.

In the texture formation, hydrogen co-deposition plays an important role because cathodic metal deposition is often accompanied by simultaneous hydrogen deposition [12,13] and also because hydrogen adsorption can change metals' surface energy [19]. When the co-deposition of hydrogen takes place, deposit's texture may change.

Hydrogen adsorption on a metal's surface can lower its surface energy [19]. As well, the degree of this surface-energy lowering varies with different crystal planes because of their different adsorption abilities. It has already been demonstrated [20] that the greater the distance between adjacent metal atoms, the less the activation energy for hydrogen adsorption. This is to say that the lower the reticular density of a lattice plane, the greater is the content of the adsorbed hydrogen [17]. For a BCC lattice, the reticular density increases in the sequence of $D^{111}<D^{211}<D^{100}<D^{110}$ [21], the content of adsorbed hydrogen, C_H^{hkl} would therefore have a reverse sequence: $C_H^{111}>C_H^{211}>C_H^{100}>C_H^{110}$. When the metal surface energy is lowered by hydrogen adsorption, planes having different reticular densities would have different rates for the lowering of surface energy. As a result, the surface-energy anisotropy might change, leading to a variation in the deposit's texture type.

The hydrogen co-deposition and in turn the deposit's texture are affected by a number of factors such as applied voltage, pH, temperature and hydrogen overvoltage. It was observed that changes in these factors can result in different textures [22]. Pangarov used a two-dimensional nucleation model to explain this change in texture type. This model, however, is controversial and does not agree with some other experiments [23-25]. Reddy [17] applied the geometrical selection theory to explain the variation of texture type. The prediction of texture from Reddy's theory is dependent on the growth mode. However, this theory can not answer what determines the mode of crystal growth.

In this paper, the effect of hydrogen adsorption on the surface-energy anisotropy of iron and in turn on iron deposit's texture have also been simulated. Influences of applied voltage, pH, temperature and hydrogen overvoltage on the hydrogen co-deposition and consequent influences on the texture formation have been analyzed and discussed.

2. THE MONTE CARLO SIMULATION MODEL

2.1 The deposition without hydrogen co-deposition

In this initial step, the current efficiency ω is assumed to be 100%. A two dimensional hexagonal lattice containing 12000 lattice sites is used to map the microstructure at a cross section of the deposit. Each site is assigned a number P, corresponding to the orientation of the grain in which the site is embedded. At the beginning, the P value of each site is assigned zero, indicating that the site is unoccupied. During the deposition process, the lattice sites become occupied by particles which are deposited layer by layer. The occupation of a site is indicated by a change of its P value from zero to a positive number.

Initially, nuclei are generated at the layer closest to the substrate. The lattice is then scanned layer by layer from bottom to top. In the scanning of a layer, a site is randomly selected and its P value is tested. If $P=0$, the site will be occupied and its P value will change to a positive value. If the site has $P>0$, the site is then skipped and no changes in P value are made.

Grain growth occurs through the occupation of lattice sites, and the decision which P value is selected is based on the initial-state energy of the site, H_0, and its final-state energy, H_n. The energy of a site is set by considering the interaction between the site and its neighbours using the following formula:

$$H=-J\sum_{nn}(\delta_{P_iP_j}-1)-J\sum_{mm}{}'\Delta(P_i\neq P_j) \tag{2}$$

where J is one half of the bond energy between two neighbouring sites. P_i represents the P value of site i. δ_{PiPj} is the Kronecker delta. If $P_i=P_j$, $\delta_{PiPj}=1$ and if $P_i\neq P_j$, $\delta_{PiPj}=0$. $\Delta(P_i\neq P_j)$ is a parameter representing the surface energy anisotropy. For each pair of sites (i,j), Δ is non-zero only when $P_i\neq P_j$ and one of the two sites is unoccupied (i.e., the interface between these two sites represents a surface). We call such a bond connecting an occupied site and an unoccupied site a "surface bond". The value of $\Delta(P_i\neq P_j)$ depends on P value of the occupied site.

The first term in equation (2) gives the state-energy which depends only on the number of bonds between the testing site and its neighbours without taking the surface energy anisotropy into account. The surface energy anisotropy is considered in the second term. The sum Σ is taken over all of the six nearest neighbour sites, while Σ' is taken over only those sites which are connected to the testing site by "surface bonds". The bond energy between a pair of sites is 2J if the two sites have unlike P values, and zero if have the same P. If the pair is connected by a "surface bond", the bond energy will be $2J(1-\Delta(P_i\neq P_j))$ where the modification term,

$\Delta(P_i \neq P_j)$, depends on the P value of the occupied site.

It should be noted that the value of the bond energy 2J is different for the "surface bond" and the bond connecting two occupied sites having different P values. The "surface bond" relates to the surface energy γ_s, while the other is related to the grain boundary energy γ_{gb}. In the present simulation, iron deposition is chosen as an example. For iron at 0°C, γ_s=2360 ergs/cm^2 and γ_{gb}=2106 ergs/cm^2 (γ_{gb}=756 ergs/cm^2 at 1350°C, $d\gamma_{gb}/dT$=-1.0 [26]), and in this case, there is no significant difference between γ_s and γ_{gb}. To simplify the calculation, we ignore the difference between γ_s and γ_{gb}, and assume that the energy of a bond connecting any two sites having unlike P values is 2J.

The initial-state energy, H_0, of the selected site is calculated using equation (2) (in fact, H_0 is calculated using equation (4), an explanation will be given later). Then, P value of the site is changed to the P value of its nearest neighbours and a trial final-state energy, H_n, is calculated using equation (2) again. P values of all nearest neighbours are used in the same way to obtain a number of trial H_n values, and the lowest H_n is chosen as the final-state energy of the selected site. The energy associated with the change from the initial state to the final state is thus given by

$$\Delta H = H_n - H_0 \tag{3}$$

If $\Delta H<0$, the new P value is used for the selected site. If $\Delta H \geq 0$, this change in P value will not be accepted. $\Delta H \geq 0$ usually occurs at sites on grain boundaries or at sites which neighbour with pores, where the nucleation rate is always high. Therefore, if $\Delta H \geq 0$, we generate a fresh nucleus with a random orientation.

It should be noted that the deposition can be considered as a phase transformation process in which independent ions arrive at the cathode to form a solid deposit. The "vapour" phase consisting of independent ions should have a higher free energy than the solid phase. Therefore, an additional energy H_L should be added to H_0 when the testing site is unoccupied, ensuring that the free energy of the "vapour" phase is higher than that of the solid phase. So, when calculating H_0, we actually use equation (4) instead of equation (2),

$$H_0 = -J\sum_{nn}(\delta_{P_iP_j}-1) - J\sum_{mm}{}'\Delta(P_i \neq P_j) + H_L \tag{4}$$

It has been turned out [27] that value of H_L is equal to 3J if we assume that a critical nucleus is composed of four lattice sites. An unoccupied site, therefore, has an intrinsic energy of 3J.

In the simulation, scanning of one layer is defined as one Monte Carlo step. In each Monte Carlo step, sites in the layer are randomly selected to test. Each Monte Carlo step contains 200 iterations. The deposition rate is modelled using different number of Monte Carlo steps involved in scanning each layer. The deposition rate is lower if the number of Monte Carlo steps is higher. In a realistic electrodeposition precess, the main parameter controlling the deposition rate is the current density. Since the deposition rate is proportional to the current density (see equation (1)) and it decreases with an increase in the number of Monte Carlo steps involved in scanning each layer, it is therefore possible to apply the model to a realistic deposition process, assuming that the current density, i_c, is inversely proportional to n, the number of Monte Carlo steps involved in scanning one layer. That is to say

$$i_c = \alpha \cdot (1/n) \tag{5}$$

where, α is a conversion constant, which may be determined for a specific case. For instance, if the current efficiency, deposition time, thickness and the density of deposit are determined, one may estimate α by combining equation (5) and (1) in the following way,

$$\alpha = n \cdot i_c = n\left(\frac{h}{t}\right)\frac{60\zeta}{\omega E_{el}} \tag{6}$$

2.2 Results and discussion

Nine different surface energies for iron have been chosen to model the growth of nine differently oriented grains. These orientations and corresponding parameters Δs, are listed in table I. Three deposition rates (corresponding to $n=(\alpha/i_c)=4$, 6 and 8) are used in the modelling. Final microstructure at a cross section of the deposit is illustrated in figure 1. At low deposition rate (i.e, at low current density), the deposit is coarse-grained, while at high deposition rate (i.e., at high current density), the deposit is fine-grained. It can be seen that the deposition rate, and thus the current density, is an important parameter controlling the microstructure of the deposit. In fact, at low current densities the process of discharge of ions at the cathode occurs slowly, which provides time for the metal nuclei to grow and there is little scope for the creation of new nuclei. Therefore, the deposits are coarse-grained. Whereas at high current densities, the rate at which ions are discharged is raised and new nuclei formation is therefore enhanced. As a result, the deposits are fine-grained.

The Monte Carlo model can provide information about the porosity introduced during the deposition as a function of the deposition rate. Figure 1 illustrates that the density of porosity increases with increasing the deposition rate. Volume fractions of porosity introduced at different deposition rates are presented in figure 2. At higher deposition rates, ions are discharged and pile up quickly, more pores can thus be introduced. Experimentally, it has been observed indeed that the porosity increases with an increase in the current density [28]. It should be pointed out that in this simulation the current efficiency is assumed to be 100%, therefore the effects of hydrogen co-deposition and other factors on the porosity formation are not taken into consideration.

Explaining texture formation is the major task of our simulation. As discussed elsewhere [29], the anisotropy of surface energy plays an important role in the texture formation of a deposit. Grains having higher surface energies tend to decrease their surface areas while those having low surface energies increase their surface areas so that the overall surface energy of the deposit reaches a minimum. As a result, grains having a high surface energy are hindered from growing by the growth of those having a low surface energy.

Nine orientations are used to describe the texture formation in the deposition. Volume fractions of differently oriented grains are calculated to represent the intensity of each texture component. The simulation demonstrates that grains having lower surface energies have correspondingly larger volume fractions. The fraction of {011} component, which corresponds to the lowest surface energy, is the strongest. This agrees with experimental observations [29] which demonstrates that the {011} fibre texture exists in electrodeposited iron foils. It can also be seen from figure 2 that the {011} component becomes stronger when the deposition rate decreases. This is because at low deposition rates, the deposition is close to the equilibrium process, and therefore selective grain growth based on the minimum surface energy requirement occurs more easily in contrast with grain growth in those processes at higher deposition rates more removed from equilibrium.

3. THE EFFECT OF HYDROGEN CO-DEPOSITION ON TEXTURE FORMATION OF IRON DEPOSIT

3.1 Texture variation with the content of adsorbed hydrogen

The Monte Carlo simulation demonstrates that the anisotropy of a metal's surface energy plays an important role in the texture formation of the metal deposit. When hydrogen co-deposition takes place, the surface energy and its anisotropy will change. As mentioned earlier, surface energies of lattice planes having lower reticular densities decrease more rapidly with an increase in hydrogen content than those having higher reticular densities. To simulate the effect of hydrogen co-deposition on texture formation, it is necessary to know the relation between the reticular density and the change in surface energy due to the hydrogen adsorption. At present, accurate description of this relation is not available. However, a reasonable qualitative prediction can be made on the assumption that the change in surface energy of a (hkl) plane is inversely proportional to its reticular density. Therefore, the anisotropy parameter, $\Delta(P_i \neq P_j)$, in equation (2) and (4) is replaced by $\Delta^*(P_i \neq P_j)$ which counts the hydrogen adsorption:

$$\Delta^*(P_i \neq P_j) = \Delta(P_i \neq P_j) + \frac{\partial\gamma_s(C_H)}{\gamma_s} \cdot \frac{1}{D(P_i \neq P_j)} \tag{7}$$

where $\Delta(P_i \neq P_j)$ is the anisotropy parameter without the hydrogen adsorption. $D(P_i \neq P_j)$ represents the reticular density whose value depends on the orientation of the occupied site. $\partial\gamma_s(C_H)$ represents the decrease in the surface energy which is a function of the adsorbed hydrogen content. Six orientations have been chosen to simulate the growth of corresponding grains. Values of the corresponding Δs and reticular densities Ds are listed in Table II. Based on Table II, the effect of hydrogen adsorption on the distribution of these six texture components in an iron deposit has then been simulated using the new anisotropy factor Δ^* expressed in (11). In the simulation, $\partial\gamma_s(C_H)$ is taken from Petch's results [19]. Results of the simulation demonstrate that the iron deposit changes its texture from {110}, through {112}, to {111} type with an increase in the content of the adsorbed hydrogen (see figure 3).

3.2 Influences of applied voltage, pH, temperature and hydrogen overvoltage on the deposit's texture

Hydrogen co-deposition is dependent upon the deposition condition. Applied voltage (E_{appl}), hydrogen overvoltage (η_H), pH and temperature of the bath (T) all can affect the hydrogen co-deposition. Therefore, all these factors may affect the deposit texture. In fact, it was already observed that in an iron electrodeposition, texture of the iron deposit changed from {110}, through {112}, to {111} type with decreases in pH and T as well as with an increase in the deposition overvoltage [22]. It is known that the equilibrium voltage for hydrogen deposition is a function of pH, T and η_H [13]:

$$E^H_{eq}=-\frac{RT}{F}pH-\eta_H \tag{8}$$

where F and R are the Faraday's constant and gas constant. When E_{appl} is greater (i.e., more negative) than E^H_{eq}, hydrogen co-deposition takes place. The hydrogen coverage, θ, which is defined as the fraction of the surface covered with a monolayer of hydrogen [30] at the cathode surface increases with an increase in the difference between E_{appl} and E^H_{eq}, i.e., with an increase in

$$\eta' = E_{appl} - E^H_{eq} \tag{9}$$

This relation may be represented as [30-32]:

$$\theta \propto \exp(-\lambda\eta') \qquad (\eta'<0) \tag{10}$$

where λ is a constant. This expression is suitable for moderate η', but at higher η', θ tends to approach a constant value [32].

From the above equations, it is seen that θ increases with an increase in E_{appl} (i.e., E_{appl} becomes more negative) and with decreases in pH and T. According to [33], the adsorbed hydrogen content C_H at the cathode surface is proportional to the hydrogen coverage θ, i.e., $C_H=k''\theta$ (for iron, k" has an order of 10^{-6} mol/cm^3) . One can therefore calculate the content of adsorbed hydrogen at the cathode surface as a function of (E_{appl}, T, pH, η_H). According to previous experimental data and theoretical work on the hydrogen co-deposition in iron deposition [33-35], we have estimated the hydrogen content (C_H) with different (E_{appl}, T, pH), and then compared these results with the data of texture ~ C_H obtained from the Monte Carlo simulation (see figure 3) to get information about the effects of E_{appl}, T and pH on the texture of iron deposit. The results have been summarized in Table III.

In the above calculation, η_H is assumed a constant and have a value of -0.5 V. In fact, η_H can also be affected by the current density [12]. According to Pangarov and Vitkova's measurement [22], the current density increased from about 1 A/cm^2 to 25 A/cm^2 when iron deposit's texture changed from {110} to {111} type. This variation of current density may cause a change in η_H about 0.1~0.2 V [12]. Therefore, the calculation of $C_H=C_H(\eta')$ given here is approximate. More accurate calculation is possible if the quantitative relation between η_H and i_c and that between i_c and E_{appl} are known.

3.3 Discussion

The simulation demonstrates that an iron deposit changes its texture from {110}, through {112}, to {111} type with an increase in the applied voltage E_{appl} or decreases in T and pH, due to the hydrogen co-deposition. This result is consistent with experimental observations made by Pangarov and Vitkova [22]. They observed that the texture of an iron deposit changed from {110}, through {112}, to {111} type when the pH and/or T decreased or when the deposition overvoltage increased with increasing E_{appl}. In their experiments, some {310} component appeared in intermediate stages together with the {112} component. The reasons behind the existence of the {310} component is unclear. It is possible that the {310} component was induced by other chemisorbable species (like OH^-, metallic anions, additives, etc) which might also influence the anisotropy of metal's surface energy. In any case, the basic agreement between the Monte Carlo simulation and the experimental observation leads us to believe that hydrogen co-deposition in electrodeposition plays a major role in the texture type variation.

4. SUMMARY

Texture formation in electrodeposition has been simulated using a Monte Carlo model. The simulation demonstrates that the surface-energy anisotropy plays an important role in the texture formation, and that hydrogen co-deposition may induce changes in the texture type. In addition, the simulation provides the information about microstructure and porosity of the deposit and also their changes with the deposition rate.

REFERENCES

1. M.P. Anderson, D.J. Srolovitz., G.S. Grest and P.S. Sahni, Acta Metall., 1984, 32, 783
2. D.J. Srolovitz, M.P. Anderson, P.S. Sahni and G.S. Grest, Acta. Metall., 1984, 32, 793
3. D.J. Srolovitz, M.P. Anderson, G.S. Grest and P.S. Sahni, Acta Metall., 1984, 32, 1429
4. G.S. Grest, D.J. Srolovitz and M.P. Anderson, Acta Metall., 1985, 33, 509
5. D.J. Sorolovitz, G.S. Grest and M.P. Anderson, Acta Metall., 1986, 34, 1833
6. D.J. Srolovitz, G.S. Grest, M.P. Anderson and A.D. Rollett, Acta Metall., 1988, 36, 2115
7. S.G.R. Brown and J.A. Spittle, Mater. Sci. Technol., 1989, 5, 362
8. J.A. Spittle and S.G.R. Brown, J. Mater. Sci., 1989, 23, 1777
9. J.A. Spittle and S.G.R. Brown, Acta Metall., 1989, 37, 1803
10. Ph. Tavernier and J.A. Szpunar, Acta Metall., 1991, 39, 549, 557
11. D.J. Srolovitz, J. Vac. Sci. Technol., 1986, A4(6), 2925
12. E. Raub and K. Müller, Fundamentals of Metal Deposition, Elsevier Publishing Co., Amsterdam, 1967
13. N.V. Parthasaradhy, Practical Electroplating Handbook, Prentice-Hall, Inc., Englewood Cliffs, New Jersey, 1989
14. Frank C Walsh and Maura E Herron, J. Phys. D: Appl. Phys., 1991, 24, 217
15. M.Y. Abyaneh, M.Fleischmann and M. Labram, Proc. of the Symposium on Electrocrystallization, edited by R. Weil and R.G. Barradas, The Electrochemical Society, Inc., Pennington, NJ, P.1, 1981
16. H. Wilman, Trans. Inst. Metal. Finishing, 1955, 32, 281
17. A.K.N. Reddy, J. Electroanal. Chem., 1963, 6, 141
18. N.A. Pangarov, J. Electroanal. Chem., 1965, 9, 70
19. N.J. Petch, Phil. Mag., 1956, 1, 331
20. G. Okamoto, J. Horiuti and K. Hirota, Sci. Papers Inst. Phys. Chem. Res., (Tokyo), 1936, 29, 223
21. Gösta Wranglén, Acta Chemica Scandinavica, 1955, 9, 661
22. N.A. Pangarov and S.D. Vitkova, Electrochem. Acta, 1966, 11, 1719
23. I.Epelboin, M. Froment and G. Maurin, Plating, 1969, 56, 1356
24. I.A. Menzies and C.X. NG, Trans. Inst. of Met. Fin., 1969, 47, 156
25. J.R. Park and D.N. Lee, J. Korean Inst. Metals, 1976, 14, 359

26. Lawrence E. Murr, Interfacial Phenomena in Metals and Alloys, Addison-Wesley Publishing Company, Inc., Advanced Book Program, Reading, Massachusetts, USA, 1975
27. D.Y. Li and J.A. Szpunar, to be published
28. A.T. Vagramyan and Z.A. Solov'eva, Technology of Electrodeposition, 1959, translated by A. Behr, Robert Draper Ltd., Teddington, 1961, P.280
29. D.Y. Li and J.A. Szpunar, J. Electr. Mater., 1993, 22, 645
30. C.D. Kim and B.E. Wilde, J. Electrochem. Soc., 1971, 118, 202
31. M. Zamanzadeh, A. Allam, H.W. Pickering and G.K. Hubler, J. Electrochem. Soc., 1980, 127, 1688
32. M.A.V. Devanathan and Z. Stachurski, J. Electrochem. Soc., 1964, 111, 619
33. Rajan N. Iyer, Howard W. Pickering and Mehrooz Zamanzadeh, J. Electrochem. Soc., 1989, 136, 2463
34. Rajan N. Iyer, Howard W. Pickering and Mehrooz Zamanzadeh, Scripta. Metall., 1988, 22, 911
35. C. Kato, H.J. Grabbe, B. Egert and G. Panzner, Corros. Sci., 1984, 24, 591
36. J.K. Mackenzie, A.J.W. Moore and J.F. Nicholas, J. Phys. Chem. Solids, 1962, 23, 185

Table I. Δ Factors of Different Crystallographical Planes Based on the Broken-Bond Theory (For BCC Metals) [36]

Crystallographical planes	Δ
(100)	0.00
(113)	0.10
(111)	0.13
(233)	0.15
(133)	0.20
(168)	0.24
(177)	0.25
(056)	0.26
(011)	0.29

Table II. Δ and D Factors of Different Crystallographical Planes (For BCC Metals)

Crystallographical Planes	Δ	D
(110)	0.29	1.41
(100)	0.00	1.00
(211)	0.19	0.82
(310)	0.06	0.63
(111)	0.13	0.58
(210)	0.11	0.89

Table III. Effects of E_{appl}, pH and T on the Texture of Iron Electrodeposit

V_{appl} / T	−0.8 (V)			−0.9 (V)			−1.0 (V)		
	pH=2	pH=3	pH=4	pH=2	pH=3	pH=4	pH=2	pH=3	pH=4
0 °C	C_H=0.174 {112}	C_H=0.162 {112}	C_H=0.058 {110} {112}	C_H=0.436 {111}	C_H=0.270 {110}	C_H=0.167 {112}	C_H=1.14 {111}	C_H=0.706 {111}	C_H=0.405 {111}
20 °C	C_H=0.167 {112}	C_H=0.094 {110}	C_H=0.047 {110}	C_H=0.406 {111}	C_H=0.244 {110} {112}	C_H=0.144 {112}	C_H=1.102 {111}	C_H=0.612 {111}	C_H=0.352 {110}
100 °C	C_H=0.124 {112} {110}	C_H=0.054 {110}	C_H=0.017 {110}	C_H=0.307 {110}	C_H=0.159 {112}	C_H=0.119 {110}	C_H=0.887 {111}	C_H=0.386 {111} {110}	C_H=0.201 {112}
150 °C	C_H=0.101 {110} {112}	C_H=0.035 {110}	C_H=0.006 {110}	C_H=0.258 {110}	C_H=0.119 {110}	C_H=0.046 {110}	C_H=0.661 {111}	C_H=0.297 {110}	C_H=0.140 {112} {110}

C_H -- Hydrogen concentration (c.c./100g)

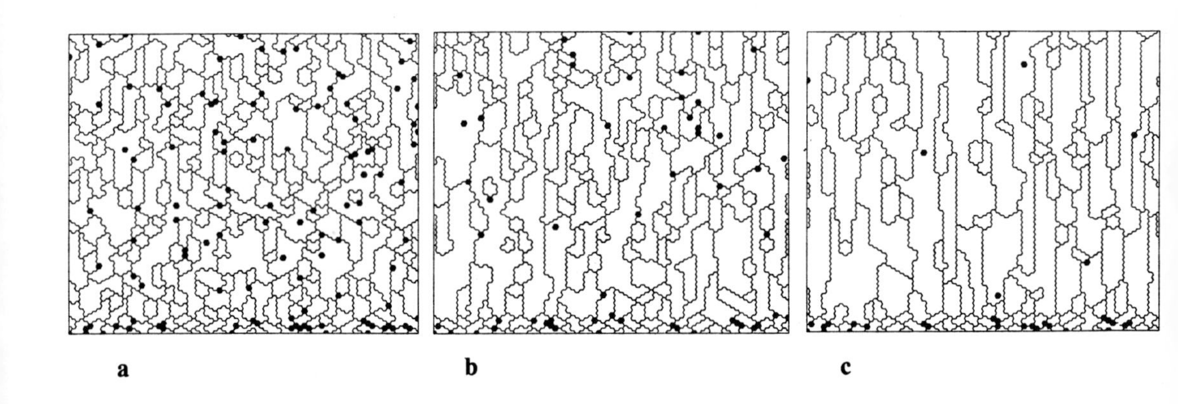

Fig.1 Simulated microstructure at a cross section of deposit obtained at three deposition rates. Numbers of Monte-Carlo steps (for scanning of one layer) corresponding to the three deposition rates are respectively (a) n=4; (b) n=6; (c) n=8. Dark spots represent the porosity.

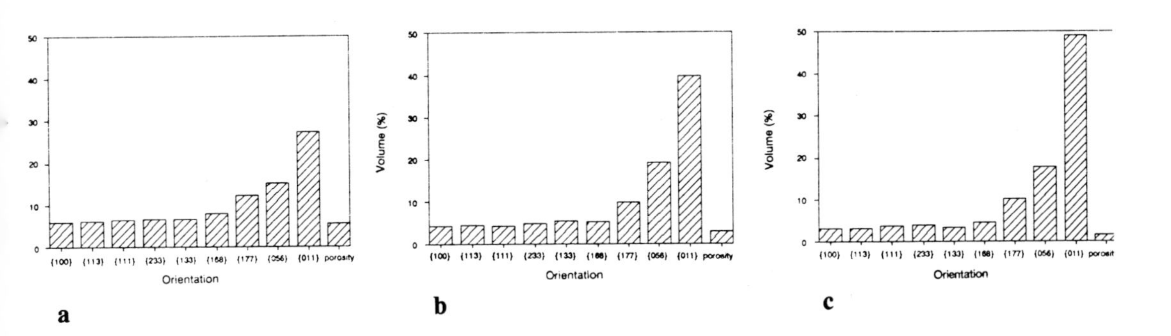

Fig.2 Volume fractions of differently oriented grains and volume fractions of porosity simulated at different deposition rates. (a) n=4; (b) n=6; (c) n=8.

Effect of the Hydrogen Adsorption on Texture Components

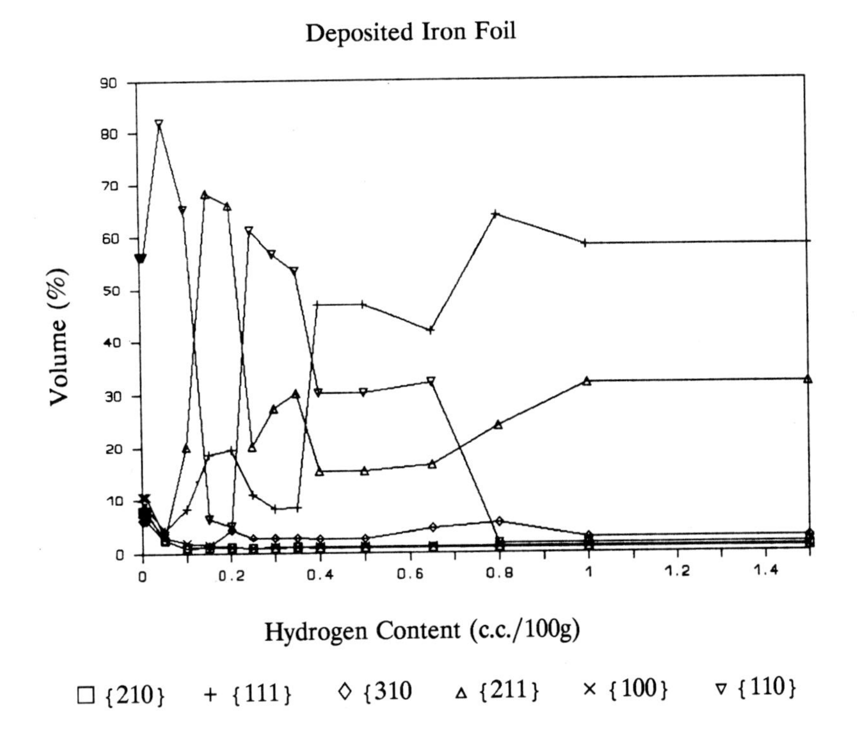

Fig.3 The texture of iron deposit changes from {110}, through {112}, to {111} type with an increase in the content of adsorbed hydrogen.

Materials Science Forum Vol. 157-162 (1994) pp. 1839-1846

SIMULATION OF RECRYSTALLIZATION TEXTURE IN COPPER

Y.S. Liu, F. Wang, J.Z. Xu and Z.D. Liang

[1] Department of Materials Science and Engineering,
Northeastern University of Technology, Shenyang, Liaoning 110006, PR China

Keywords: Recrystallization, Texture Simulation, Annealing Texture, Microstructure Simulation

ABSTRACT

Cold rolling textures in copper were quantitatively introduced in the simulated microstructure. The stored energy of grains with different orientations was derived by the Taylor model. Orientation transitions during the recrystallization were determined by the stored energy. The simulated results are displayed by ODF sections and compared with the experimental results. Qualitative agreements are obtained.

1. INTRODUCTION

Recrystallization is an important process in controlling the properties of materials. The understanding of recrystallization can be improved by modelling the physical phenomenon. Monte-Carlo computer simulation technique proposed by Anderson and collaborators is one of the most promising method to analyse the process in detail[1]. For thin plate product, recrystallization texture is a predominant problem. Because recrystallization texture is effected by many factors such as initial cold rolling texture and stored energy, it is very difficult to simulate the process with incorporating every detail by Monte Carlo Method. However if a model can capture certain essentials with neglecting other unimportant features, it will be a reasonable approach[2]. Tavernier and Szpunar has proposed a method for simulating texture development during recrystallization in low carbon steel by considering several main texture components[3].

In the present paper we propose a method which quantitatively introduces the texture information into the microstructure. The method is not rigorous but it provides an effective method to couple the texture with microstructure.

2. CONSTRUCTING MICROSTRUCTURE WITH TEXTURE

The microstructure was plotted by the method described by Srolovitz[1]. It is mapped onto a discrete lattice points and the lattice point is assigned a number corresponding to the orientation of the grain. If one wants the microstructure to represent a certain texture distribution, the appearing frequency of the lattice number must proportion to the volume fraction of the texture component . In order to couple the number on the lattice with ODF, the orientation space is divided into N equal volume cells[4], and the volume fraction in i cell is written as

$$P_i = \int_{\Delta V_i} \omega(\theta, \psi, \varphi) \sin\theta \, d\theta \, d\psi \, d\varphi \qquad (1)$$

where $\omega(\theta, \psi, \varphi)$ is the true ODF, ΔV_i is the volume of i cell. If the number i represents the grain orientation in i cell, Pi will represent the appearing probability of number i in the microstructure , and equation 2

$$\sum_{i=1}^{N} P_i = 1 \qquad (2)$$

is always right for true ODFs.

Usually a microstructure is constructed with thousands of lattice points. If a big number K is multipled with Pi and the result is integrated(decimal part is removed), that is

$$f_i = \mathrm{INT}(K\,P_i)\,, \qquad (3)$$

fi will represent the appearing time of i number on the microstructure. Let fo=0, a sample space, {Fk} can be set up with its elements

$$F_k = i \qquad k = k_1, \ldots k_2 \qquad (4a)$$

where

$$K_1 = \sum_{g=0}^{i-1} f_g \qquad (4b)$$

$$K_2 = \sum_{g=0}^{i} f_g \qquad (4c)$$

represent that in the sample space from element F_{k_1} to F_{k_2} there are fi elements with i orientation. Write out the all elements

$$\{Fk\}: \{1,1,1,\ldots1,2\ldots2,\ldots,\ldots,i,i,i\ldots i,\ldots,\ldots,N,N,N\ldots N\} \qquad (5)$$

Now if misorientation between grains is assumed to be random, a microstructure with MxMo grains can be constructed by randomly taking the elements, the numbers in sample space,and then putting

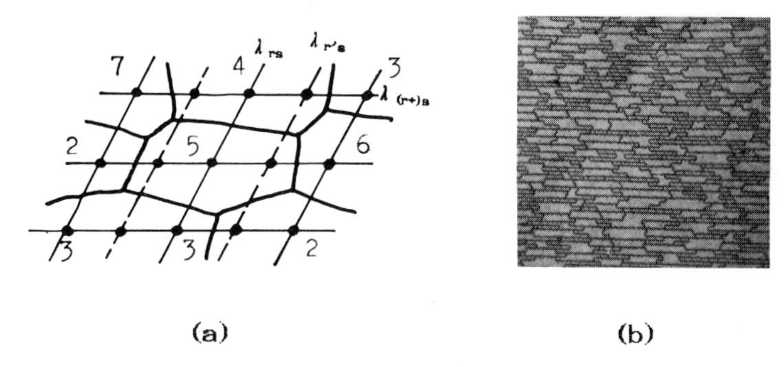

(a) (b)

Figure 1 (a) Random interpolation of lattice points,
(b) Cold rolled microstructure of simulation.

them onto the lattice, that is

$$\lambda_{rs}=F_R \qquad (R=RND(1,2,3,...K)) \qquad (6)$$

The subscripts indicate the order of lattice point along x- and y-direction respectively. The above process produces a microstructure of one grain with one lattice point(figure 1a).

In fact it is not suitable to represent a grain with one lattice point. One grain in a cold rolled specimen contains great number of subgrains. In order to represent the phenomenon , random interpolation of lattice points is necessary. Let λ_{rs} ,$\lambda_{(r+1).s}$ be the two numbers of two neighbour lattice points, and the newly developed point between them may randomly take one of them, that is

$$\lambda_{r's}=RND(\lambda_{rs},\lambda_{(r+1).s}) \qquad (7)$$

where RND means randomly taking the two numbers in the parenthesis.By recycling this step along x- and y- direction the lattice points in a grain can quickly increase to the number of

$$d=2^{u+v} \qquad (8)$$

with u,v as interpolation times(figure 1a).

Due to the interpolation, the local number of lattice points with certain orientation is changed. In the view of statistics the appearing frequency of the orientation in the entire specimen is not changed because a microstructure is make up by thousands of lattice points or grains. The grain shape can be easily controlled by u and v.The newly developed lattice may be thought as subgrains. By the above method the microstructure of cold rolled copper with 90% reduction was constructed with 100x400 lattice points as shown in figure 1b. The true ODF of the simulated microstructure is illustrated in figure 2a and is very similar to that of experiment determined by the maximum entropy method (figure 2b).

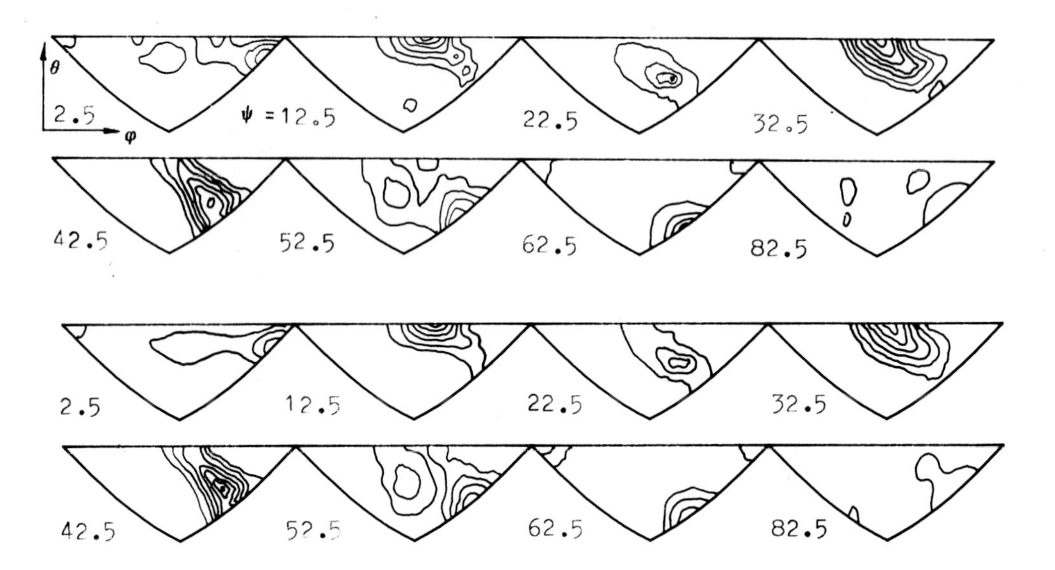

Figure 2 True ODF sections of the simulated microstructure (a) and of the cold rolled specimen with 90% reduction (b). Levels:1,3,5,7...

3. SIMULATION

The simulation is proceeded directly on the generated microstructure(figure 1b). The energy of every lattice point is very important in the simulation. It is assumed that the energy is just concerned with grain boundary energy and stored energy of cold working. As indicated by Anderson [5] the grain boundary energy is calculated by

$$E_p = J \sum_{nn} (1-\delta_{S_p S_q}) \qquad (9)$$

where S_p is one of the N orientations on lattice point $p(1<S_p<N)$ and δ_{ab} is the Kronecker delta. J is a constant proportional to the grain boundary energy per unit length. The sum is taken over nearest neighbor points(nn). Equation 9 is not only suitable for calculating the grain boundary energy of cold rolled specimens but also for those of recrystallization. For the two states, J has different value. The stored energy is indicated by

$$G_p = H_p\, f(S_p) \qquad (10)$$

H_p is a constant of stored energy per unit area. $f(S_p)$ is a function of stored energy related with orientation or the number S_p of the lattice point. Dillamore has pointed out that stored energy is a function of texture and is related with the Taylor factor[6]. Some experimental data of stored energy show that the stored energy is proportional to the Taylor factor[7]. Here f(Sp) is assumed to be the Taylor factor and Hi is used as a converting constant. The Taylor factor of every orientation can be calculated by the model proposed by Van Houtte [8].

In the simulation, the lattice point is the smallest size, the nucleus dimension is also the same even it is rather big compared with experimental recrystallization. The simulation of recrystallization is the process that the number $n=1\sim N$ which represents orientation of cold rolling state turns into the number $z=1\sim Q$ which indicates the recrystallization texture. For oriented nucleation $z=1\sim Q$ is the fixed orientation observed in experiment and for selective growth $z=1\sim Q$ represents every possible preferred orientation. The transition probability W is given by

$$W = \exp(-\Delta E/K_b T) \qquad \Delta E>0 \qquad (11a)$$
$$W = 1 \qquad \Delta E<0 \qquad (11b)$$

where ΔE is the energy changing before and after transition. Nucleation and growth use the same transition ruler of equation 11.

For oriented nucleation mechanism, here are several assumptions:

1) nucleation is a continuous process and nuclei are produced during a sequence of time.

2) only those orientations observed in the experiment result of recrystallization are assigned to $z=1\sim Q$.

For the selective growth mechanism, reorientation of recrystallized lattice points is allowed. This needs experiment data of anisotropic grain boundary energy and it will be our future work. The present paper only introduced oriented nucleation mechanism.

The true ODF of simulated recrystallization is calculated by

$$\omega(\theta_I, \psi_I, \varphi_I) = \sum_{r=1}^{X_m} \sum_{s=1}^{Y_m} \delta_{\lambda_{rs} i} / (X_m Y_m \Delta V_I) \qquad (12)$$

By this way the texture and microstructure can be combined together and the simulated results are displayed in figure 3~5.

The fully developed recrystallization texture of copper is almost a pure cube component, so $z=1\sim Q$ is simply assigned to the cube orientation. Let Q=64 in order to avoid curl effect.

4. RESULTS AND DISCUSSION

For cold rolling texture, the ODF analysis shows that the texture components are mainly {112}<111>, {123}<634> and {110}<112>. The rolling plane and direction scatter in strap shape around the above ideal orientations. The level of the maximum intensity point is about 15, and the half intensity scattering width is about 15°. The cube component of cold rolled copper shows an intensity only about 2. The simulated original microstructure and the cold rolled specimen have the similar texture distribution as shown in figure 2.

As the simulating annealling progressing, the intensities of cold rolling texture components gradually reduce and the strength of cube component increases quickly (figure 3). The calculation of the volume fraction of cube component shows that at the beginning and the final stage of recrystallization the transition is very slowly and in the middle stage the transition is rather fast. The transition curve is displayed in figure 4. The transition obeys Avrami equation. The microstructure shows that the cold rolled

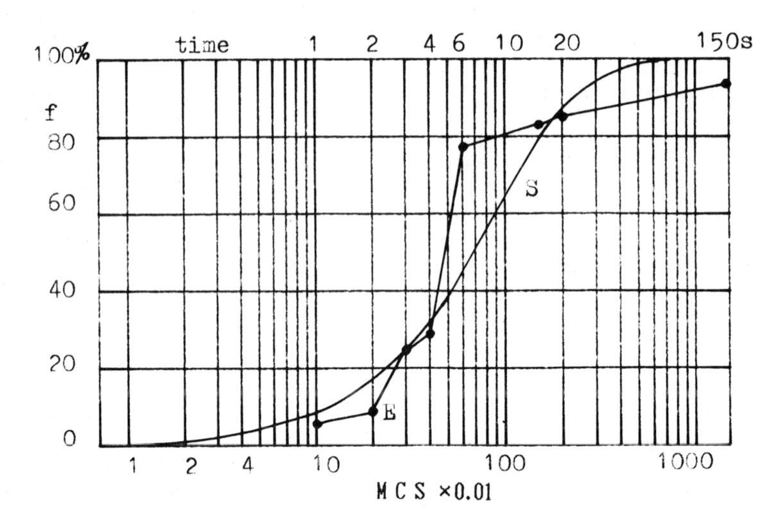

Figure 4 The change of volume fraction during recrystallization.
S-Simulation and E-experiment

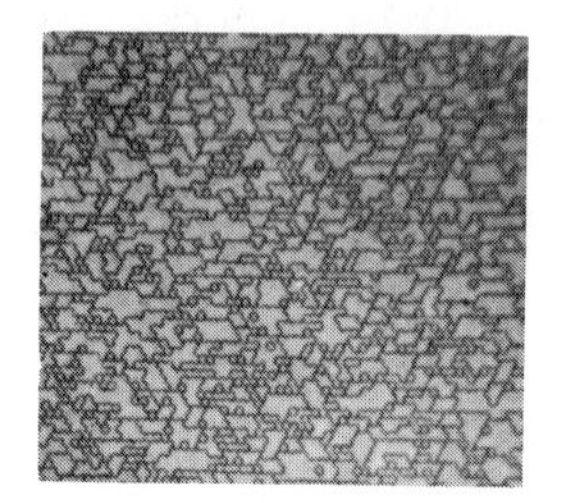

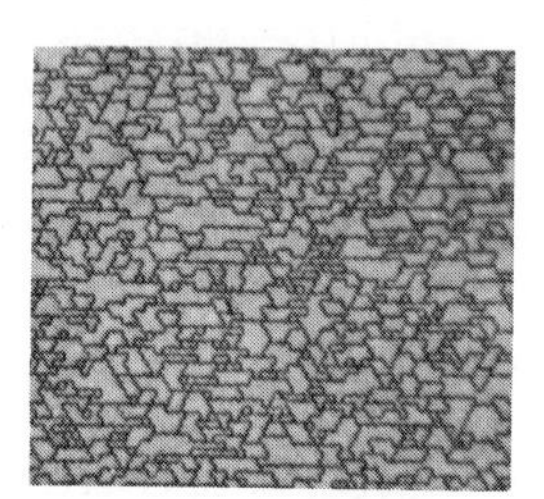

Figure 5 The evolution of microstructure during recrystallization
(a) simulated 1 unit time (b) simulated 3 unit time
(c) experiment specimen 2s/250°C.

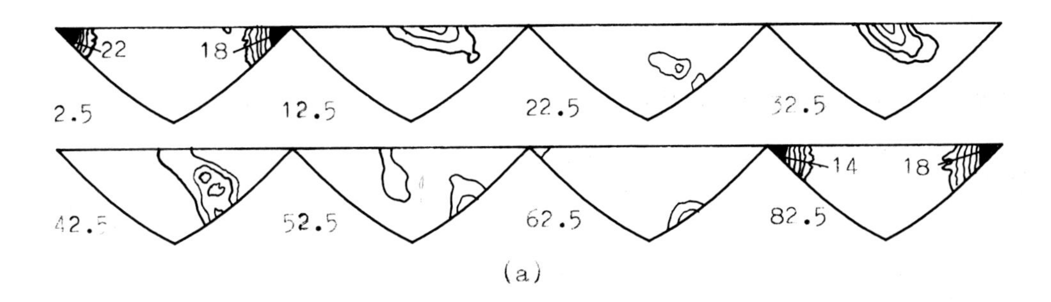

(a)

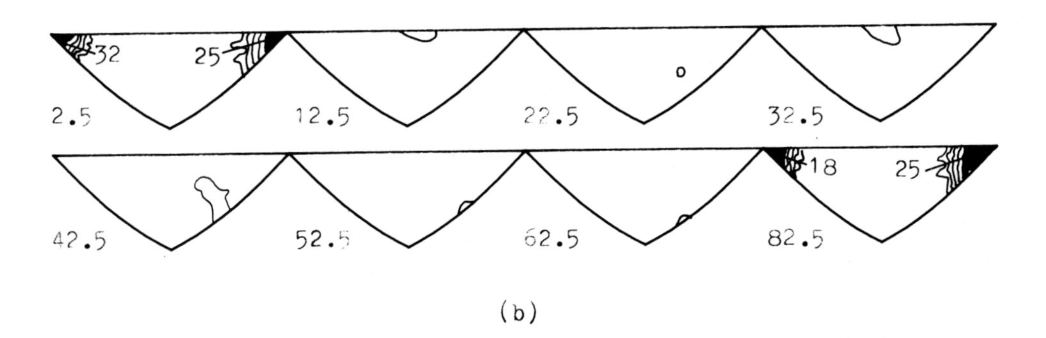

(b)

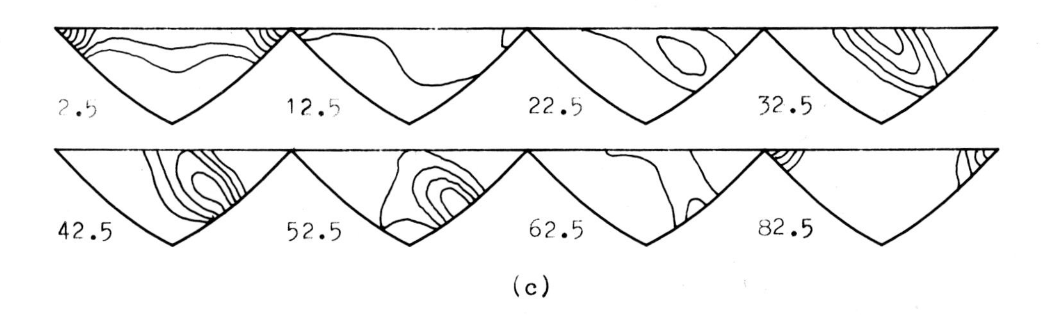

(c)

Figure 3 ODF constant sections of recrystallization textures (a) simulated 1 unit time , (b) simulated 2 unit time; (c) 3s/250C recrystallized copper specimens. Levels: 1, 3, 5, ...

enlonged grains are gradually eaten out by the newly recrystallized lattice points. The full recrystallization results in a single cube component with high intensity and high volume fraction.

In order to verify the result of simulation, the simulated results are compared with the experiment results of commercial copper. The experiment results show that the cube component quickly reaches a high intensity with the consumption of cold rolling components. The microstructure evolution also shows that deformed matrix is gradually occupied by the recrystallized structure. The typical enlonged feature of cold working disappears and a new microstructure develops . The difference is that the final texture of experiment appears a weak twin relation texture component and this is not considered in the simulation.

This model is very simple and is rather phenomenistic. The reason of this is that for copper only one orientation, the cube which characterizes recrystallization is introduced in the simulation. At least,from this work it may be concluded that the model enables to simulate both texture transition and microstructure evolution.

More improvement is needed on the basis of experiment information. The assumptions used in the simulation may cause some deviation from experiment. For example , the factor of stored energy J is assumed to be isotropic, but in experiment it is anisotropic. In the simulation the initial microstructure only has the information of ODF. If misorientation is introduced, a more reasonable approach probably will be obtained.

5. CONCLUSION

(1) The interpolation method is suitable for two-dimensional microstructure simulation.
(2) The copper recrystallization transition simulated with the simplified oriented nucleation theory is qualitatively agreeing with the experiment.

RFERENCES

1) Srolovitz D.J, Anderson M.P.,Grest G.S. and Sahni P.S.:Scripta Metall.,1983,17,241
2) Ashby M.F.:Mater. Sci. and Tech.,1992,8,102
3) Tavernier Ph. and Szpunar J.A.:Acta Metall. Mater.,1991,39,549
4) Liu Y.S.,Wang F.,Xu J.Z.,Liang Z.D.: J.Appl.Cryst.,1993,26,268
5) Srolovitz D.J.,Grest G.S.,Anderson M.P.,Rollett A.D.:Acta Metall. Mater.,1988,36,2115
6) Dillamore I.L.,Katoh H.:Metall Sci.,1974,8,21
7) Tavernier Ph.,Szpunar J.A.:Acta Metall. Mater.,1991,39,557
8) Van Houtte P.,Aernoudt E.:Z.Metallkd.,1975,66,202
9) Beck P.,Hu H.:ASM Seminar on Recovery,Recrystallization and Grain Growth,Met.Park,Ohio,1966
10) Engler O.,Lucke K.:Scripta Metall. Mater.,1992,27,1527
11) Doherty R.D.,Samajdar I,Kunze K.,Scripta Metall. Mater,1992,27, 1459
12) Adams B.L.,Wright S.I.,Kunze K.:Metall. Trans. A,1993,24A,819

Materials Science Forum Vol. 157-162 (1994) pp. 1847-1854

GRAIN GROWTH SIMULATION BY MONTE CARLO METHOD IN A HiB Fe 3% Si ALLOY

P. Paillard [1,2], R. Penelle [2] and T. Baudin [2]

[1] UGINE SA Centre de Recherche d'Isbergues, F-62330 Isbergues, France

[2] Laboratoire de Métallurgie Structurale, Université Paris-Sud, Bât. 413, URA CNRS 1107, F-91405 Orsay Cédex, France

Keywords: Simulation, Monte Carlo, Normal and Abnormal Grain Growth, FeSi Alloy

ABSTRACT :

As many physical properties of materials depend of texture and microstructure, it is important to control texture and microstructure evolution. Monte Carlo simulation is applied to the problem of normal and abnormal grain growth in a HiB Fe Si 3% alloy, on a two dimensional lattice. Energy and boundary mobility anisotropy between two grains with different crystallographic orientations were introduced in the simulation. So it is possible to simulate growth laws and to compare them with the experimental ones.

INTRODUCTION :

The Fe Si electrical steels present a final {110}<001> texture, which is obtained by abnormal grain growth during the final annealing under hydrogen atmosphere. The present paper deals with simulation of grain growth [1,2,3] of such materials starting from the primary recrystallization experimental texture.

MICROSTRUCTURE MODELLING :

In order to incorporate the complexity of grain boundary topology, the microstructure is mapped onto a discrete lattice (figure 1).In a two dimension lattice, a grain has six neighbors [4].

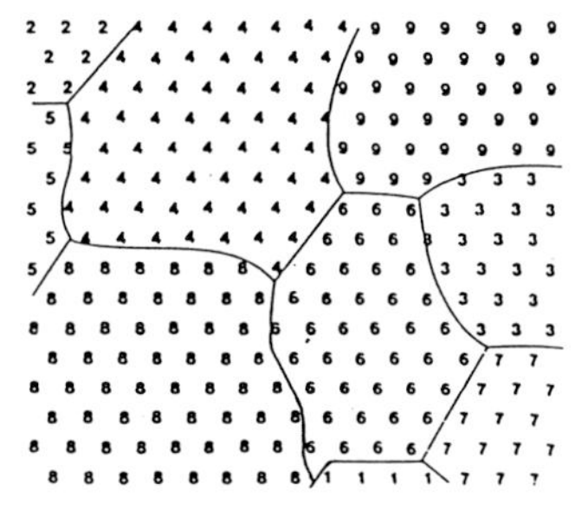

Figure 1 : Sample microstructure mapped onto a triangular lattice.The integers denote orientations and the lines represent grain boundaries.

The chosen lattice is triangular of 200 per 200 sites, so one site has six nearest neighbors. Each lattice site is assigned a number between 1 and Q corresponding to the orientation of the grain in which it is embedded. Q is chosen large enough so that grains like orientation impinge infrequently.

For other simulations, Q value is 48 or 64. Though a Q increase increases the computer time, a value of 1000 is used. This high value allows one to correctly simulate grain growth of material with 4 texture components.The starting texture is the primary recrystallization texture given in table 1. The simulation is performed for a alloy model assuming that its texture is that of an HiB without the presence of second phase particles (AlN, MnS).

	{554}<225>	{100}<012>	Random	{110}<001>
Component % [5]	45.0	15.0	39.9	0.1
Q values	1 to 450	451 to 601	602 to 999	1000

Table 1 : Recrystallization texture and Q corresponding values.

A grain boundary segment is defined to lie between two sites of unlike Q orientation value.The grain boundary energy is specified by defining an interaction between lattice sites as nearest neighbor. E_{ij} is the energy of site (i,j) (i,j is the position of the site on the lattice).

$$E_{i,j} = -J \sum_{1^{st}\ neighbors} (\delta_{Q_{i,j},Q'} - 1) \qquad (1)$$

$Q_{i,j}$ is the orientation of i,j site, Q' is the first neighbor site orientation, J is a specific grain boundary constant and $\delta_{a,b}$ is Kronecker function. The principle of the Monte Carlo simulation used in the present work is : a site is selected at random on the lattice and reoriented to a randomly chosen orientation between 1 and Q_{max}. The energy of the site is calculated with the equation 1 before and after reorientation. If the change in energy associated with the reorientation, ΔE, is less than or equal to zero the reorientation is accepted. However, if the energy variation is higher than zero, the reorientation is accepted with the probability :

$$\exp(-\frac{\Delta E}{k_B\ T}) \qquad (2)$$

where k_BT is the thermal energy. The time, in this simulation, is proportional to the number of reorientation attemps. N (here 200x200 sites) reorientation attemps are used as the unit of time and are referred to as 1 Monte Carlo Step (MCS).When a site at a grain boundary is reoriented with success, then it defines a boundary migration.

NORMAL GRAIN GROWTH :

In the first step, normal grain growth (in the artificial Fe Si alloy, without second phase particles) was simulated, using equations 1 and 2. In this case, the simulation is performed without taking into account the influence of the anisotropy of grain boundary energy due to the texture. The figure 2 shows microstructural evolution as a function of the computer time. The corresponding grain size distributions show an homogeneous behaviour, the mean grain diameter variation as a function of the computer time (figure 3a) presents a plateau for about 10^5 times Monte Carlo Steps (MCS), corresponding to a growth steady state. Plotting the same data in logarihmic scale (figure 3b), the points are lined up (figure 3b). The growth law is standard model :

$$D = D_0.t^n$$

where D is mean diameter of grains, D_0 is a constant and t the computer time.

The n value is equal to 0.43 for this simulation at 950° K. This value is near that found by other authors [1,4,6,7]. Figures 4a and 4b show that there is no growth of one of texture components to the expense of the others.The Goss grains in figure 2 are not indicated because of the randomly selection of the very small volume fraction of these grains (0.1%)

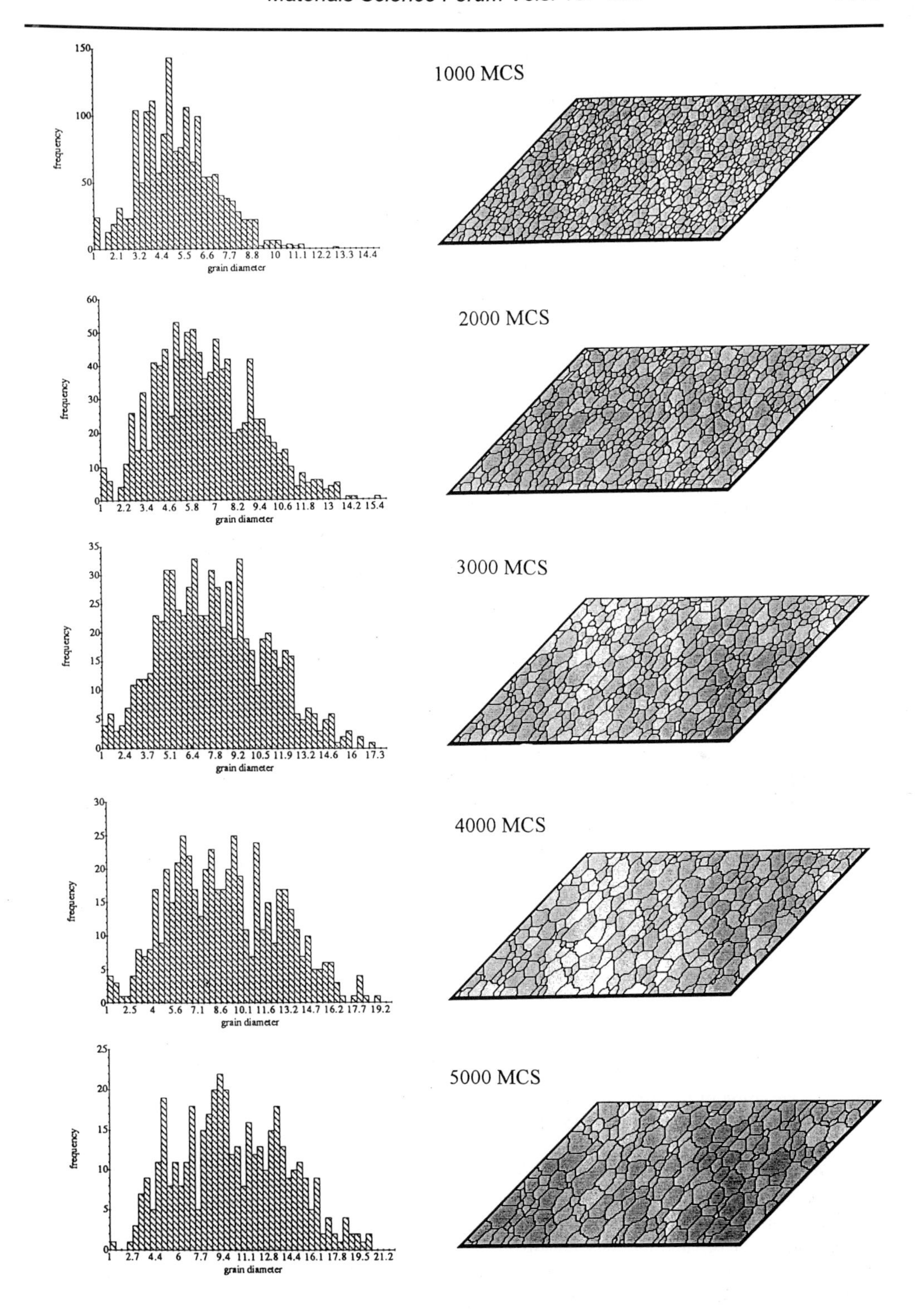

Figure 2 : Microstructural evolution and corresponding grain size (in site number) distribution as a function of the time. Normal grain growth.

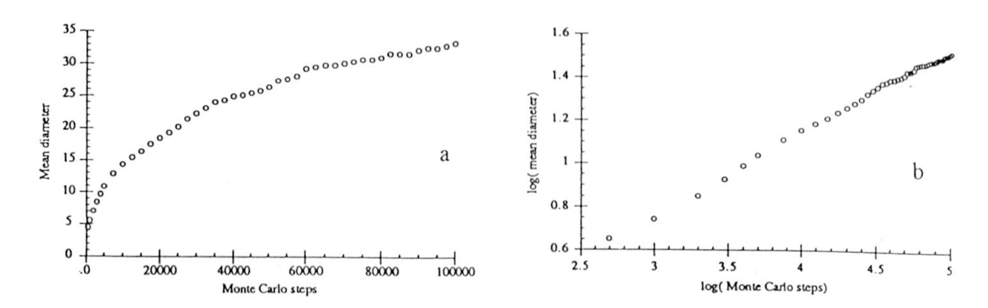

Figure 3 a) Variation of the mean diameter (in site number) as a function of Monte Carlo Steps.
b) Variation of the mean diameter as a function of Monte Carlo Steps, in logarithmic scale.

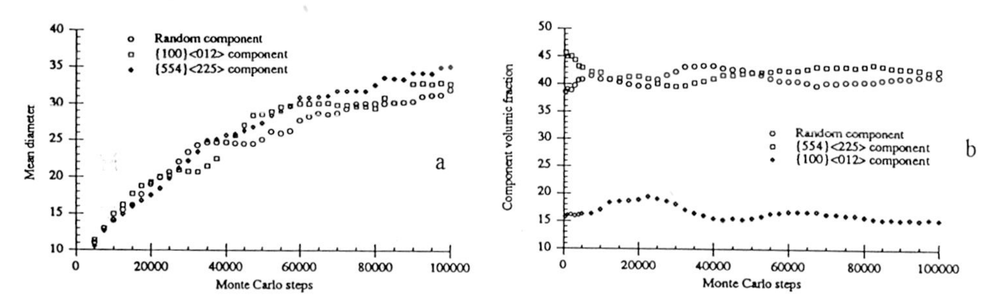

Figure 4 : a) Variation of the mean diameter (in site number) as a function of Monte Carlo Steps for the different texture components.
b) Variation of texture component volume fraction as a function of Monte Carlo Steps.

Using equation 2, it is possible to simulate growth as a function of temperature (figure 5). As for the previous case, a normal grain growth behavior is found, but in this case with n values within 0.35 and 0.45.

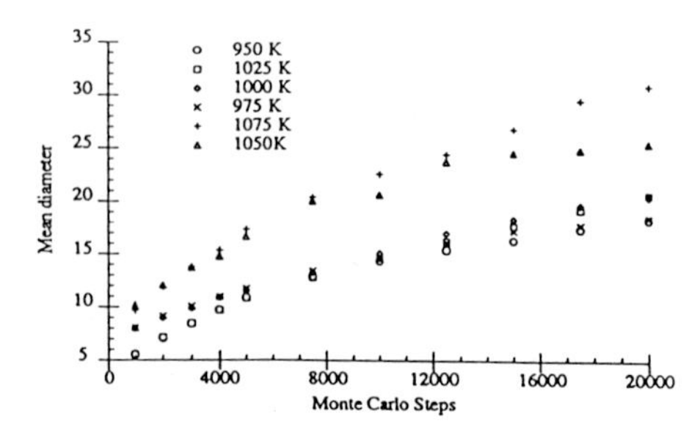

Figure 5 : Variation of the mean diameter (in site number) as a function of Monte Carlo Steps for the different simulation temperatures.

ABNORMAL GRAIN GROWTH :

The simulation procedure is the same as for normal grain growth. The only difference is for the site energy calculation and for the reorientation probability. The presence of a texture involves a grain boundary energy and mobility anisotropy. As for instance, the grain boundary energy between two grains {554}<225> and {110}<001>is different from that between two grains having the orientation {554}<225>, it is necessary to introduce in the simulation some different energy constants J according to the boundary type.

Then the chosen site energy is calculated with equation 3 :

$$E_{i,j} = - \sum_{1^{st}\ neighbors} (J_1, J_2 \ldots J_{10})(\delta_{Q_{i,j},Q'} - 1) \tag{3}$$

The grain boundary mobility takes place in acceptation probability of reorientation. Now if the change in energy associated with the reorientation, ΔE, is less than or equal to zero the reorientation is accepted with the probability :

$$\frac{1}{\mu^*} \tag{4}$$

μ* is mobility coefficient assigned to i,j site, surrounded by 6 nearest neighbors.If the energy variation higher than zero, the reorientation is accepted with the probability :

$$\frac{1}{\mu^*} \exp(-\frac{\Delta E}{k_B\ T}) \tag{5}$$

Energy and mobility coefficient values are given in table 2 [8].

Boundary type	{110}<001> {554}<225>	{110}<001> {100}<012>	{110}<001> Random	{554}<225> {100}<012>	{554}<225> Random	{100}<012> Random
Energy **J** arbitrary unit	330	260	300	360	380	350
Mobility **μ** arbitrary unit	2.5	0.1	2.5	2.3	2.5	2.5
Boundary type	{110}<001> {110}<001>	{554}<225> {554}<225>	{100}<012> {100}<012>	Random Random		
Energy **J** arbitrary unit	400	450	600	650		
Mobility **μ** arbitrary unit	0.1	0.2	0.1	0.4		

Table 2 : Energy and mobility coefficients for an iron silicon steel.

In the case of abnormal grain growth, the mean grain size variation as a fonction of computer time is different from the variation of normal grain growth (figure 6a).The curve does not present the same behaviour than that of normal grain growth. There is no growth steady state.

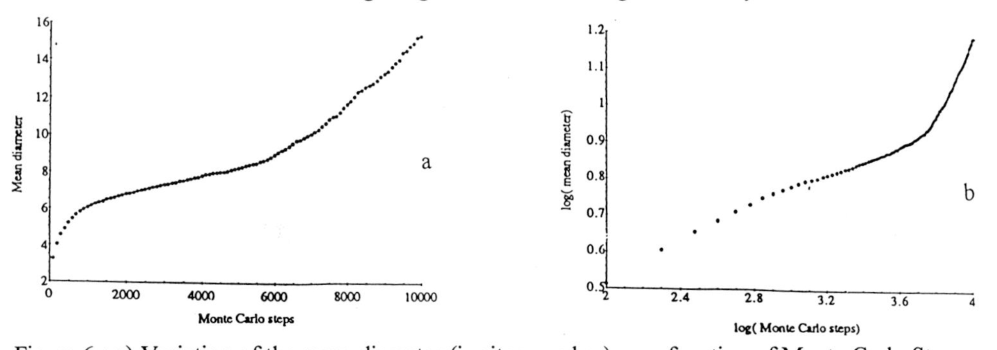

Figure 6 : a) Variation of the mean diameter (in sites number) as a function of Monte Carlo Steps.
b) Variation of the mean diameter as a function of Monte Carlo Steps, in logarithmic scale.

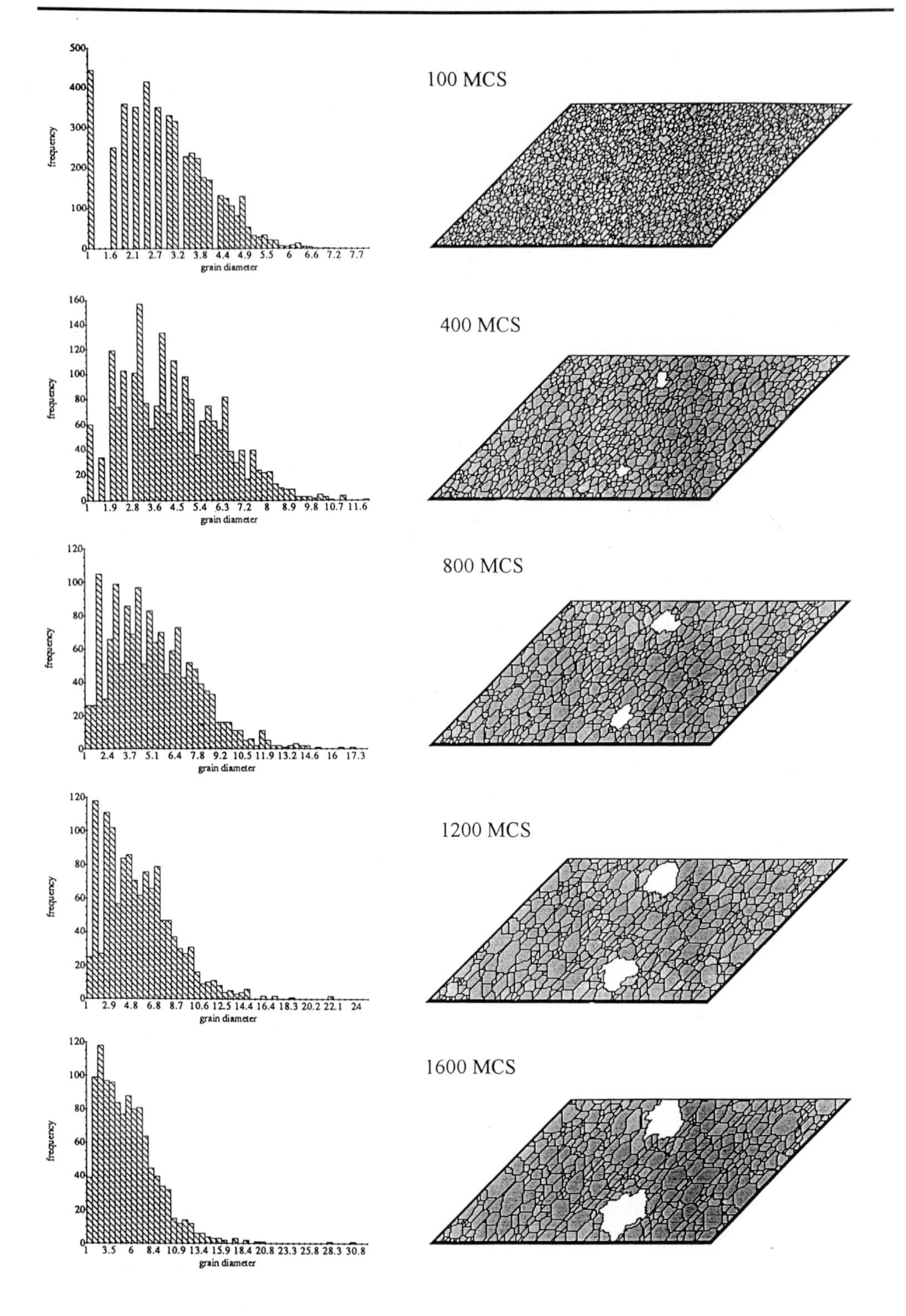

Figure 7 : Microstructural evolution and corresponding grain size distribution as a function of the time. Abnormal grain growth.

This variation in logarithmic scale (figure 6b) presents two growth rates. The first part of the curve has a constant n less than 0.25. The coefficient of the second part is higher than 1.3. This high growth rate is due to the exaggered growth of Goss grains. The microstructures and corresponding grain size distributions for abnormal grain growth are shown figure 7. The Goss grains are in white. As in the experiment, the Goss grains grow faster than the other grains of the matrix.

As for the normal grain growth, it is possible to study the mean grains size as a function of Monte Carlo steps, for each texture component (figure 8a). Except for the {110}<001> orientation grains, there is grain growth stability. The discontinuity perceptible on the Goss grains curve is due to the disappearance of some Goss grains but there is no discontinuity of the volume fraction (figure 8b).

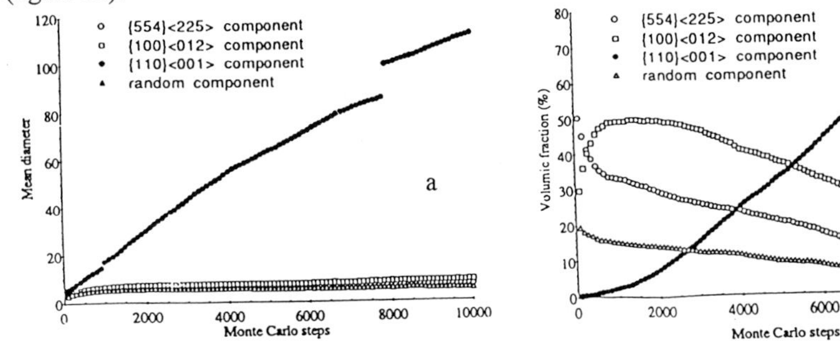

Figure 8 : a) Variation of the mean diameter (in sites number) as a function of Monte Carlo Steps for the different texture components.
b) Variation of volume fraction of the different components as a function of Monte Carlo Steps.

It has been experimentally found that a {554}<225> neighbourhood is favourable to the growth of Goss grains [9]. As the volume fraction of {554}<225> is important (45%) after primary recrystallization, the Goss grains grow faster than the other orientations. However the {100}<012> grains, which are also stable, are the last to be consumed by the Goss grains. This experimental observation is simulated as well.

Simulation as well as the experiment firstly shows a strong decrease of the {554}<225> component and a increase of both {110}<001>, {100}<012> components, then the Goss grains consumed the {100}<012> grains. Simulation also points out a decrease of the random component and finaly the Goss grains consume the initial matrix.

CONCLUSION :

The simulation allowed one to predict the evolution of grain size and texture for primary recrystallized samples with several texture components. For normal grain growth, law as $D = D_0.t^{0.4}$ is found. For abnormal grain growth, the growth rate and disappearance of the different texture components are in good agreement with the experimental observations.

This study is the first step of the growth simulation possibilities. The next step of the simulation will be to combine both texture influence (grain boundary energy and mobility anisotropy) and dragging force of second phase particles (Zener force). These simulations can be obviously adapted to other FeSi alloys [10].

REFERENCES

1. M.P. ANDERSON, D.J. SROLOVITZ, G.S. GREST and P.S. SAHNI.
Acta Metallurgica, Vol 32, n° 5, p783, 1984.

2. D.J. SROLOVITZ, M.P. ANDERSON, P.S. SAHNI and G.S. GREST.
Acta Metallurgica, Vol 32, n° 5, p793, 1984.

3. G.S. GREST, D.J. SROLOVITZ and M.P. ANDERSON.
Acta Metallurgica, Vol 33, n° 3, p509, 1985.

4. M. HILLERT.
Acta Metallurgica, Vol 13, p227, 1965.

5. N. ROUAG.
Thèse de Docteur en Sciences, Orsay, France, 1988.

6. G.S. GREST, M.P. ANDERSON and D.J. SROLOVITZ.
Computer Simulation of Microstructural Evolution. October 13-17 1985, Toronto (Canada).

7. Y. ENEMOTO and R. KATO.
Acta Metallurgica and Materiala, Vol 38, n° 5, p765, 1990.

8. G. ABBRUZZESE, A. COMPOPIANO and S. FORTUNATI.
Textures and Microstructures, Vol 11, p775, 1991.

9. N. ROUAG and R. PENELLE.
Textures and Microstructures, Vol 14-18, p203, 1989.

10. P. PAILLARD, T. BAUDIN and R. PENELLE.
Proceeding of ICOTOM 10, 20-24 October 1993, Clausthal (Germany), in press.

Materials Science Forum Vol. 157-162 (1994) pp. 1855-1860

EFFECT OF DEEP DRAWING ON TEXTURE DEVELOPMENT IN EXTRA LOW CARBON STEEL SHEETS

J. Savoie [1], D. Daniel [2] and J.J. Jonas [1]

[1] Department of Metallurgical Engineering, McGill University, 3450 University Street, Montreal, Quebec H3A 2A7, Canada

[2] Pechiney - Centre de Recherches de Voreppe, BP 27, F-38340 Voreppe, France

Keywords: Texture Formation, Deep Drawing, IF Steel, Crystal Plasticity

ABSTRACT

Although the development of rolling textures in steels has been studied by several authors [1-4], the evolution of texture during deep drawing has been much less investigated [5,6]. The results of an experimental investigation of these effects were presented in a recent publication [6]. In the present work, computer simulations were carried out using a rate dependent theory for mixed {112}- and {110}<111> slip. Several grain interaction models were employed, based on the full constraint (FC), relaxed constraint (RC) and least work (LW) theories. It is shown that the experimentally observed grain reorientations are reproduced reasonably well using the full constraint and least work models. The orientation dependences of the preferred deformation modes are described, together with the relative stabilities of the expected end textures.

DEFORMATION SIMULATIONS

In the present calculations, the first step involved discretizing the initial sheet ODF so that the texture could be represented by ~500 different grains. $\phi_2=45°$ section showing the main orientations present in IF steel sheets is illustrated in figure 1a. A similar ODF section representing the texture of the starting material is presented in figure 1b. Simulations of deformation along the RD and TD directions of the rolled sheet were carried out using the discretized initial texture (with an initial scatter width of 4°). For this purpose, a rate sensitive crystal plasticity code (with m=.05) was employed. Mixed {110}-{112}<111> slip was assumed to take place with a CRSS ratio $\tau_{112}/_{110}$ of 0.95. Several grain interaction models were used, but only the results obtained with the FC, pancake and LW models will be discussed in the next section. The least work model produces textures that combine features of those obtained from the FC and pancake models [4]. Deformed ODF's were calculated from the simulation data using final scatter widths of 7°-8°.

SIMULATION RESULTS AND DISCUSSION

Flange textures

The simulated $\phi_2=45°$ ODF sections corresponding to drawing ($\varepsilon=0.29$) along the RD and TD directions of the sheet obtained using the FC model are illustrated in figures 2a and 2b. A thickness increase of 20% in the flange was assumed in order to produce good agreement with the experimental results (figures 2c and 2d). The FC model reproduced the experimental texture along RD exactly, in both position and strength: here the most intense component is located at $\Phi=46°$ ({334}<110>). Along TD, however, the main calculated texture component ($\Phi=55°$) lay 3° away from the experimental one (Φ =52° {10 10 11}<110>) (Tables I and II).

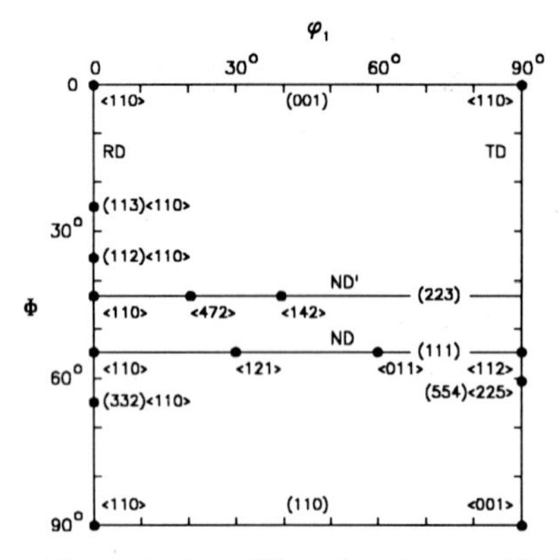

Figure 1a: ϕ_2=45° section showing ideal initial orientations in bcc materials.

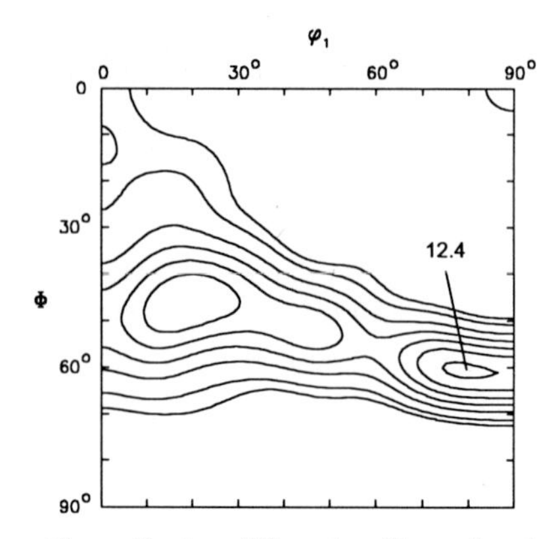

Figure 1b: ϕ_2=45° section illustrating the texture of the IF steel. Levels:1,2,4,6,8,10,12.

In the experiments, grain reorientations took place faster along TD than along RD, leading to TD textures that were more intense ($f(g)_{RD}=24$ vs $f(g)_{TD}=34$) [6]. In the simulations, when these were carried out to equal strains, the calculated maximum intensity difference between RD and TD was not as large ($f(g)_{RD}=24$ vs $f(g)_{TD}=26$). The probable cause of this discrepancy is that the strains along RD and along TD are not equal in the flange. This effect originates in the plastic anisotropy of the material. The present IF steel displayed 6 ear behavior at 0°, 60°. This corresponds to an ear at 0°, a hollow at 30° and an ear again at 60°. As the ears are associated with higher Taylor factors than the hollows [6], such "hard" RD regions undergo less plastic deformation than the "softer" TD regions. For this reason, simulations along TD were carried out to 20% larger strains ($\varepsilon=0.35$) than were employed in the RD calculations (see Table II). In this way, the intensities of the calculated TD textures reached values which were close to the experimental ones.

The least work criterion [4] was also employed for modelling texture development in the flange. The difference in the plastic work rates required to switch completely from the FC to the pancake deformation mode was set to 20%. This level was chosen because it led to peak positions in the simulated textures which were in good agreement with experiment. The LW predictions are similar to those obtained from the FC model (Tables I and II).

Table I. Main texture components obtained in the flange after drawing along RD

	ϕ_1	Φ	ϕ_2	f(g)
experimental RD	**0**	**46**	**45**	**24**
FC	**0**	**46**	**45**	**24**
LW Δ=20%	**0**	**45**	**45**	**22**

Table II. Main texture components obtained in the flange after drawing along TD. Same strain path as for RD, but deformed to $\varepsilon=0.35$

	ϕ_1	Φ	ϕ_2	f(g)
experimental TD	**0**	**52**	**45**	**34**
FC	**0**	**55**	**45**	**30.5**
LW Δ=20%	**0**	**55**	**45**	**29**

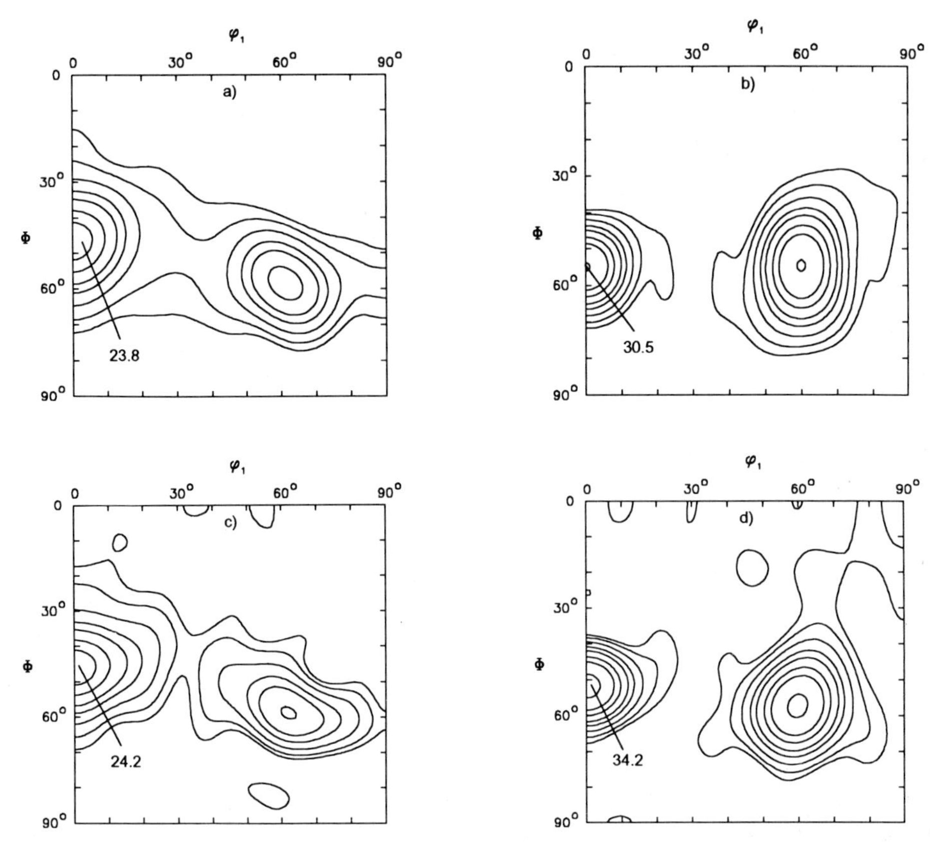

Figure 2: $\phi_2=45°$ ODF sections of the simulated textures in a sample initially containing the measured sheet textures (20% thickness increase). The calculations were carried out using the FC model (a) RD ($\varepsilon=0.29$) and (b) TD ($\varepsilon=0.35$). $\phi_2=45°$ ODF sections of the IF steel measured near the rim of the flange after drawing ($\varepsilon=0.29$) (c) RD and (d) TD. Levels: 1, 2, 4, 7, 10, 15, 20, 30.

Wall textures

In this case, the FC model reproduced the experimental textures relatively well, but to a lesser extent than in the case of flange deformation. Along RD, the position of the peak coincided exactly with the experimental one: $\Phi=45°$ ({557}<110>). Along TD, the calculated texture component ($\Phi=50°$) lay 2° away from the experimental one ($\Phi=48°$ {445}<110>). However, the intensities are higher and the scatter around the peaks is less pronounced than in the experimental ODF's (see Tables III and IV).

The least work criterion was also employed to model texture development in the wall (figures 3a and 3b). In this case, the difference in the plastic work rate required to switch completely from FC to pancake deformation was set to 15%. Under these conditions, the positions and strengths of the simulated textures were in better agreement with the experimental results (figures 3c and 3d). The predictions of this model differ significantly from those obtained with the FC program: the sharpness of the texture is reduced considerably and the scatter around the main orientations is increased. This indicates that the release of the shear rate components at sufficiently high deformations in deep drawing ($\varepsilon=0.60$ in the wall) is more pronounced because of the greater flatness of the grains.

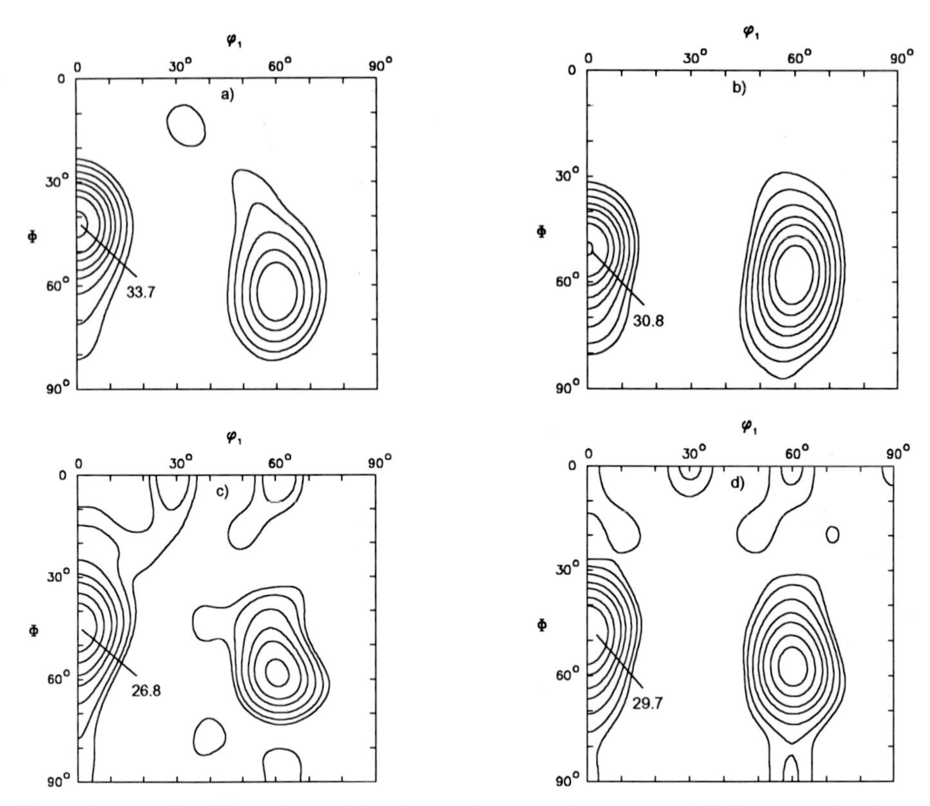

Figure 3: $\phi_2=45°$ ODF sections of the simulated textures in a sample initially containing the measured sheet textures and deformed to $\varepsilon=0.60$ (20% thickness increase). The calculations were carried out using the LW model ($\Delta=15\%$): (a) RD and (b) TD. $\phi_2=45°$ ODF sections of the IF steel measured in the wall after drawing ($\varepsilon=0.60$) (c) RD and (d) TD. Levels: 1, 2, 4, 7, 10, 15, 20, 30.

Table III. Main texture components obtained in the wall after drawing along RD

	ϕ_1	Φ	ϕ_2	f(g)
experimental RD	**0**	**45**	**45**	**27**
FC	**0**	**45**	**45**	**43**
LW $\Delta=15\%$	**0**	**42**	**45**	**34**

Table IV. Main texture components obtained in the wall after drawing along TD

	ϕ_1	Φ	ϕ_2	f(g)
experimental TD	**0**	**48**	**45**	**30**
FC	**0**	**50**	**45**	**36**
LW $\Delta=15\%$	**0**	**50**	**45**	**31**

ORIENTATION STABILITY IN DEEP DRAWING

The texture simulations described above indicate the orientations towards which the texture evolves. However, they do not provide information regarding the stability of these orientations and the expected end textures. For this reason, rotation field maps were calculated; some examples are presented in figure 4. They show the directions and rates of lattice rotation of individual orientations. A point of stability is reached when all orientations converge into that point. For the FC model (figure 4a), two stable components are expected: {223}<110> (0°,43.8°,45°) and {110}<110> (0°,90°,45°). In the case of the pancake model (figure 4b), two orientations are again stable: {112}<110> (0°,33.7°,45°) and {110}<110> (0°,90°,45°). For both the RD and TD initial textures, orientations with $\Phi>75°$, which are expected to rotate towards {110}<110> in the FC model, are absent. This is not the case for the pancake model, where orientations with $\Phi>60°$-65°, which are present in both the RD and TD initial textures, can rotate towards {110}<110> (figure 4b).

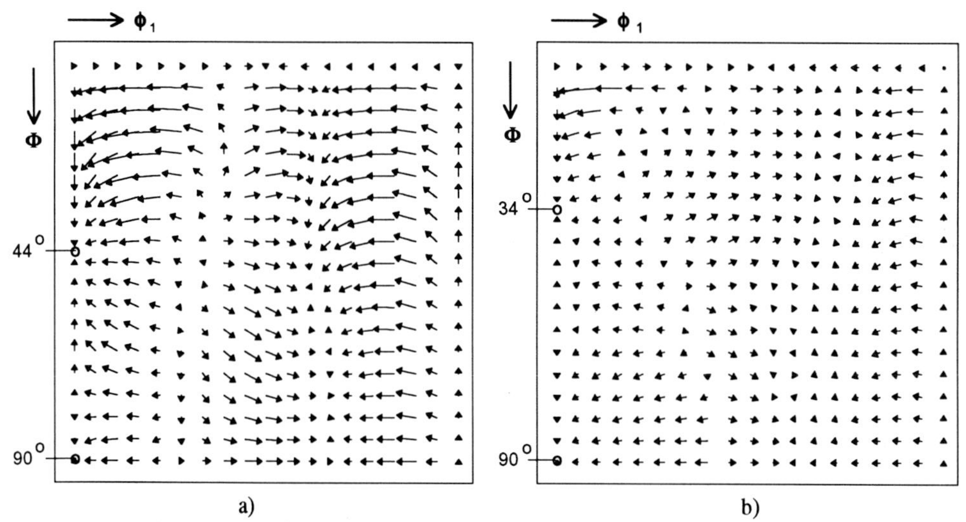

Figure 4: $\phi_2=45°$ sections showing the rotation rate maps for flange deformation with a 20% thickness increase (m=.05, CRSS ratio=.95) for the (a) FC and (b) pancake models.

Thus, in the FC case, the grains rotate towards a single orientation near $\Phi=45°$. This is the behaviour that was observed in the measured textures. Along RD, at the beginning of deformation, the peak was located at $\Phi=45°$, and it remained at this location during straining. Along TD, it began at $\Phi=60°$ in the initial texture, attained $\Phi=52°$ in the flange, and reached $\Phi=48°$ in the wall. (If no thickness increase were to occur in the flange, such low Φ values would not be reached; the unique stable orientation for the FC model in this case is {111}<110> (0°,54.7°,45°)). When the shears are relaxed (LW model), grains in the Φ range 65° to 75° move in the direction of increasing Φ, with the result that the texture is broadened along the RD fibre towards the {110}<110> component.

To examine the stability of the initial texture components in more detail, the rotation rates, their gradients and the divergences [7] were calculated for the main orientations found in the initial textures: i.e. {554}<225> and {223}<472> in the RD samples, and {554}<110> and {223}<274> in the TD samples. For these calculations, the FC model with a 20% thickness increase was employed (see Table V). For comparison purposes, the two stable orientations called for by the FC model are included in the table: {223}<110> and {110}<110>.

From this table, it can be seen that the two stable orientations satisfy the first stability condition of zero rotation rate. However, the second condition, i.e. three dimensional convergence, is properly satisfied only for the {110}<110> component. Although stable, the {223}<110> orientation only displays two dimensional convergence (i.e. $\delta\dot{\phi}_2/\delta\phi_2 > 0$). In this last case, an arbitrary orientation with a ϕ_2 that differs slightly from 45° (but with $\phi_1=0°$ and $\Phi=43.8°$) will first retreat from the stable position, and then return via a combination of converging ϕ_1, Φ and ϕ_2 displacements. Although the two TD orientations are formally unstable, they display negative div **g**. values, which implies that they are less unstable than the two RD components, which display positive div **g**. values. These characteristics are in good agreement with the experimental observations.

Table V Rates of change $\dot{\phi}_i$, gradients ($\delta\dot{\phi}_i/\delta\phi_i$), and divergences (div $\dot{\mathbf{g}}$) predicted by the FC model with a 20% thickness increase.

	{554} <225>	{223} <472>	{554} <110>	{223} <274>	{223} <110>	{110} <110>
$\dot{\varphi}_1$	0.0	-0.6	0.0	-1.0	0.0	0.0
$\dot{\Phi}$	-0.2	-0.1	-0.4	0.2	0.0	0.0
$\dot{\varphi}_2$	0.0	-0.2	0.0	0.6	0.0	0.0
$\delta\dot{\varphi}_1 / \delta\varphi_1$	5.3	3.7	-5.1	-3.7	-4.9	-5.2
$\delta\dot{\Phi} / \delta\Phi$	0.4	-0.6	-0.8	-1.2	-1.7	-3.0
$\delta\dot{\varphi}_2 / \delta\varphi_2$	0.8	-0.8	0.2	0.8	0.6	-0.3
div $\dot{\mathbf{g}}$	6.4	2.3	-5.8	-4.1	-6.0	-8.6

CONCLUSIONS

When deep drawing operations are simulated using either the full constraint or the least work mode, good agreement with the experimental flange textures requires the thickness in the flange to increase by 20%. Under these conditions, the initial orientations rotate towards {223}<110>. Because of the larger strains associated with the walls of cups, however, the texture changes occurring at this location during deformation can no longer be modelled using the full constraint mode. Instead, the least work model with significant shear relaxations accurately reproduces the experimental textures. At these larger strains, the initial orientations rotate towards *two* stable orientations, the {112}<110> and {110}<110>, which is why the texture broadens and its intensity does not increase significantly as drawing proceeds.

REFERENCES

1) Hutchinson, J.W.: *Proc. R. Soc. London A*, 1976, 348, 101
2) Inagaki, H.: *Trans. Iron Steel Inst. Jpn.*, 1984, 24, 266
3) Von Schlippenbach, U., Emren, F. and Lücke, K.: *Acta Metall.*, 1986, 34, 1289
4) Toth, L.S., Jonas, J.J., Daniel, D. and Ray, R.K.: *Metall. Trans. A*, 1990, 21A, 2985
5) Dabrowski, W., Karp, J. and Bunge, H.J.: *Arch. Eisenhüttenwes.*, 1982, 53, 361
6) Daniel, D., Savoie, J. and Jonas, J.J.: *Acta Metall. Mater.*, 1993, 41, 1907
7) Zhou, Y., Tóth, L.S. and Neale, K.W.: *Acta Metall. Mat.*, 1992, 39, 3179

Materials Science Forum Vol. 157-162 (1994) pp. 1861-1868

CONTRIBUTION OF EBSP TO THE DETERMINATION OF THE ROTATION FLOW FIELD

M. Serghat, M.J. Philippe, C. Esling and B. Bouzy

[1] LM2P/ISGMP, Université de Metz, Ile du Saulcy, F-57045 Metz Cédex 1, France

Keywords: EBSP, Rotation Flow Field, Titane, Twinning, Rolling

ABSTRACT

Over the 10 last years, different authors have contributed to determine and explain the texture evolution in hexagonal materials using models of polycristalline plasticity. Another approach to the study of global texture evolution consists in determining the orientation of individual grains, which allows to have acces to:

1- the individual rotation of a grain as a function of its initial orientation,

2- the comparison between the rotation of 2 grains with the same initial orientation but different orientations of neighbouring grains,

3- the type of twinned part in a grain, its orientation and orientation evolution with respect to the matrix.

INTRODUCTION

A crystalline grain undergoing plastic deformation changes its lattice orientation. This rotation leads to the development of the deformation texture. The grain lattice rotation and thus the texture evolution are governed by crystallographic slip and twinning. In the case of slip, the continuous rotation of the crystallites can be described by means of a continuous rotation flow field (1 to 6). Over the ten last years, different authors have contributed to the determination and interpretation of the texture evolution, mainly in cubic but also in hexagonal materials using models of polycrystalline plasticity. These models have been used to describe the texture evolution during cold rolling. However, in the specific case of hexagonal materials, some hypotheses can be called into question. Thus to have a more precise information about the individual rotations associated with an orientation (rotation field) and to justify the hypotheses of the model, we have measured the local rotation of individual grains during deformation.

EXPERIMENTAL PROCEDURE

The investigated material is a T40 alloy with a low oxygen contents (1000 ppm). The various observations were made on samples prepared from a 1.5 mm thick sheet of T40 alloy manufactured by the company CEZUS. The chemical composition of this sheet is indicated in table 1.

Table 1: Chemical composition of the T40 alloy.

Element	C	Fe	H	N	O	Si
Composition (ppm)	52	237	3	41	1062	<100

In this work we used two recrystallized samples. In order to obtain a large grain size (150 mm), the recrystallization treatment was performed in the α phase at a temperature of 850°C for 3 hours. This treatment in the α phase was followed by a standard cooling (350°C per hour). Rolling tests were performed in two directions : the initial rolling direction and the transverse direction (see Fig. 2). The samples were electropolished before deformation. We have made cold rolling on these samples up to 9% and 18% deformation successively. The grain orientations have been measured before deformation and we mapped the position and orientation of each grain.
The grain orientation was also measured after the two steps of deformation.
The scanning electron microscopy was performed with a JEOL electron microscope using an accelerating voltage of 30 kV. A very low-light TV camera is mounted in the SEM on the rear port to detect EBSPs from a 50 mm diameter phosphor screen. The screen is mounted vertically in the SEM chamber (see Fig. 1).

RESULTS AND COMMENTS

In order to determine the grain rotation during rolling of titanium alloys, we measured the orientation of about two hundred grains per sample with the EBSP technique. Measuring the initial grain orientation gives access to the individual rotation of a grain during deformation. The measurements carried out also permit the comparison between the rotation of two grains with the same initial orientation but having neighbouring grains with different orientations. In parallel the orientation of twinned parts in the grain can be determined. These results make it also possible to follow the evolution of the twinned part in individual grains.

The initial (00.2) PF of the samples after annealing at 850°C is presented in Fig 2. We have also determined the texture evolution during rolling of the two samples at each stage of deformation for the two types of deformation (i.e. 9% and 18% deformation). After 18% reduction, we observe two stable texture components for the longitudinal rolling and for the transverse rolling. We notice that rolling in the initial transverse direction increases more the sharpness of the texture than in the longitudinal direction. These stable components correspond to the current component after rolling (7 to 10).

We can also represent the rotation of the c axes during deformation (Fig 3). As the c axes of the grains in a stable position practically do not move, only the rotations of the grains which c axes are far from a stable position will be drawn. We can afterwards compare these results to the rotation fields calculated with different models and for different CRSS ratios.

With these measurements we can also determine the type of mechanical twinning activated during deformation. We only have to determine the angle between the c axes of the twin and matrix and to compare them with the theoretical values (see Table 2). The two samples rolled in the former longitudinal, respectively transverse direction both show twins. The samples rolled in TD present a great number of $(11\bar{2}1)$, $(11\bar{2}2)$, and $(10\bar{1}2)$ twins. However the sample rolled in RD shows no $\{11\bar{2}1\}$ twin. When observing the orientations of the twinned grains on the (0002) stereographic projection, we can determine the grain orientations in which the different types of twins were activated (Fig. 4). Thus when the c axis is tilted to a high angle, we obtain $(10\bar{1}2)$ tension twins and when the c axis is central, we obtain compression twins. Fig. 3 presents the reorientations of the c axes of a large number of grains. The rotation of the twinned part is different from that of the matrix and brings the matrix and the twin orientation back towards the stable components of the texture.

Table 2: The different twinning modes in titanium and the corresponding rotation angles of the c-axes (11).

Twin plan	$\{10\bar{1}2\}$	$\{11\bar{2}2\}$	$\{11\bar{2}1\}$
Rotation axes	$<\bar{1}2\bar{1}0>$	$<\bar{1}100>$	$<\bar{1}100>$
Rotation angle	94° 52'	63° 58'	34° 54'

We also study the rotation of grains having very close initial orientations but different neighbouring orientations (Fig. 5). For instance, the orientation of the neighbouring grains can be close to the considered grain or rather different (Fig. 6). Such a difference can induce quite different lattice rotations, especially at the early stage of deformation. This is due to the surroundings of the grains and a detailed study of the possible couplings between the grains is to be carried out.

Different domains undergoing different lattice rotations may appear even within one and the same grain. So Fig. 7 presents a grain which two parts deformed differently. One domain was surrounded by grains which c axes were little misoriented, the other by grains which c axes were heavily misoriented. We can see here the respective rotations of the domains 1, 2 and 3. For this example too, we can measure the influence of the close neighbours.

CONCLUSION

This work has already allowed to point out the deformation inhomogeneities which appear at the early stages of deformation. However, it is also to be stressed that these inhomogeneities decrease with the deformation.

The first results of this work, which is so far essentially a qualitative approach concerning a limited number of grains, will be complemented by :

- a statistical approach carried out on a large number of grains e.g. in the order of one thousand grains
- a study of the deformation inhomogeneities within the grain and of the rotation differences observed for two grains having the same orientation. For the latter grains the rotation differences will be correlated to the various deformation systems activated in the grains and to the neighbouring grains.

The expected results will contribute to the first experimental determinations of the rotation flow field which was previously only predicted in the frame of a model of polycrystalline plasticity.

REFERENCES:

1) BUNGE H.J., Kristall Technic, 5, 145.
2) CLEMENT A., Mat. Sci. Eng., 1982, 55, 203.
3) WIERBANOWSKI K. and CLEMENT A., Cryst. Res. Technol., 19, 201.
4) WIERBANOWSKI K. and CLEMENT A., Phil. Mag. A, 1985, 51, 145.
5) KLEIN H., DAHLEM E., ESLING C. and BUNGE H.J., Theo. Meth. of Text. Anal., BUNGE H.J. (Edt.), 1987, 259.
6) BACZMANSKI A., WIERBANOWSKI K., JURA A., HAISE W.G., HELMHOLDT R.B., and MANIAWSKI F., Phil. Mag. A., 1993, 67 N° 1, 155.
7) MELLAB F.E., Thèse de 3ème cycle, ISGMP, Uni. Metz, 1992.
8) TOME C.N., LEBENSOHN R.A. and KOCKS U.F., Acta. Met. Mater., 1991, 39, 11, 2667.
9) SERGHAT M., PHILIPPE M.J. and ESLING C., Proc. of Conf. on Adv. Mater. and Proc. Techn., August 1993. (to be published)
10) PHILIPPE M.J., *Texture formation in hexagonal materials*. Invited conference ICOTOM 10, this book.
11) Partridge P.G., Int. Met. Reviews, 1967, 12, 169.
12) HJELEN J., Proc. Inst. of Techn., Trondheim, Norway, 1992, 2, 408.

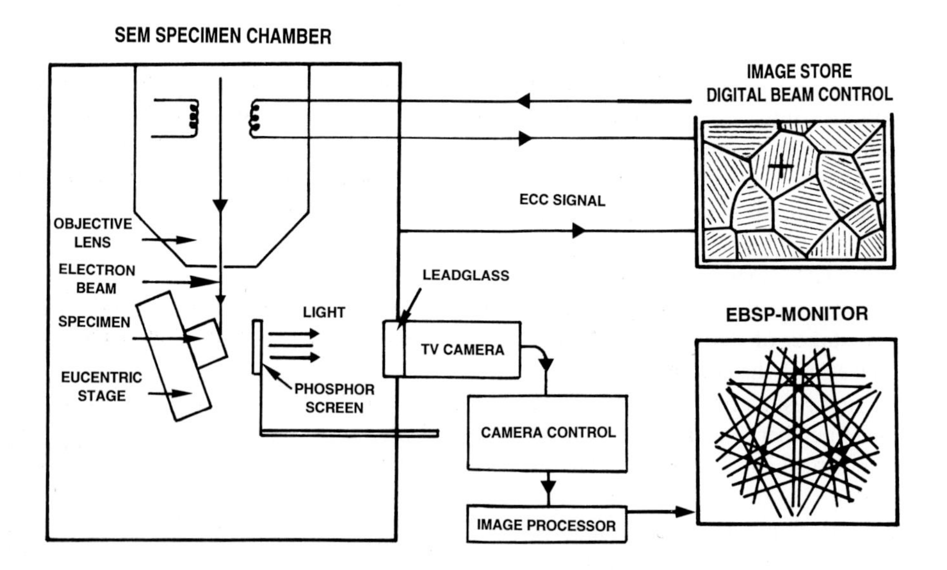

Figure 1: Sketch of a specimen chamber attached with a side entry EBSP, image store, camera, and an image processor (12).

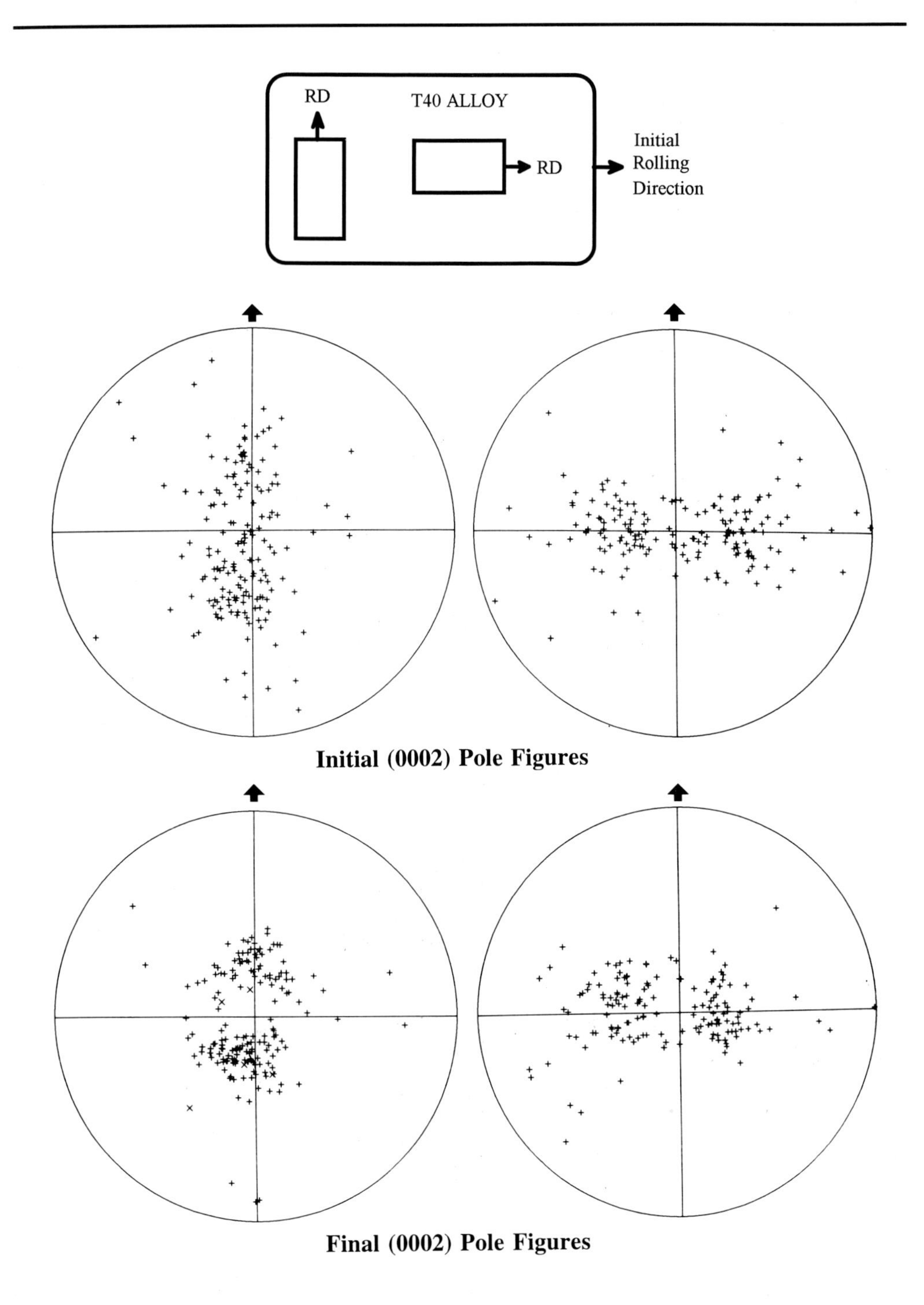

Figure 2: Positions of the samples in the sheet, and initial and final (0002) Poles Figures.

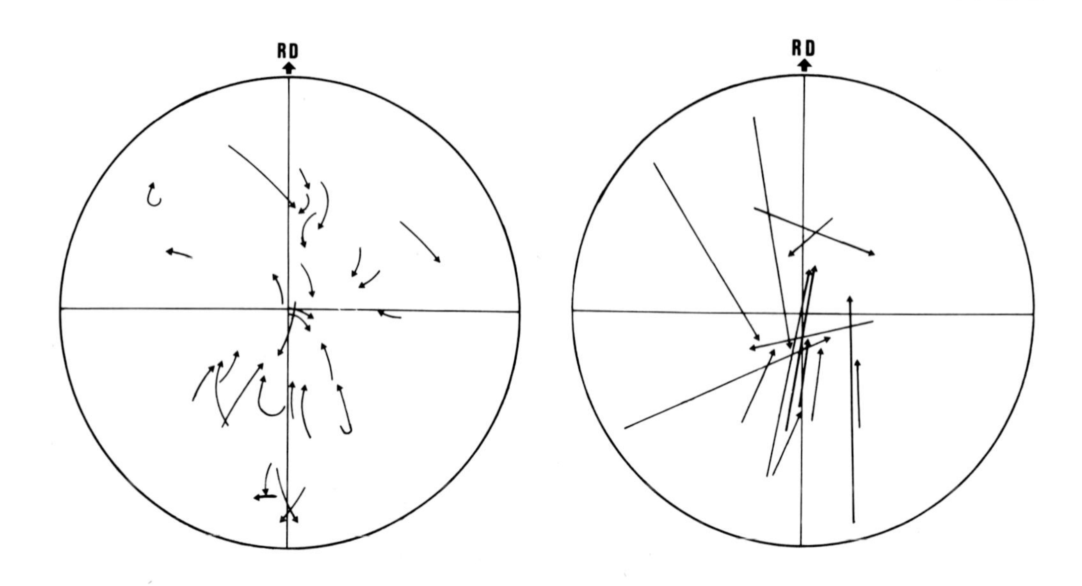

Figure 3: a) Rotation of grains due to slip in (0002) Pole Figure.
b) Rotation of the twinned part after the first deformation.

Figure 4: Types of twins as a function of the c-axes positions of the grains in the (0002) Pole Figure.

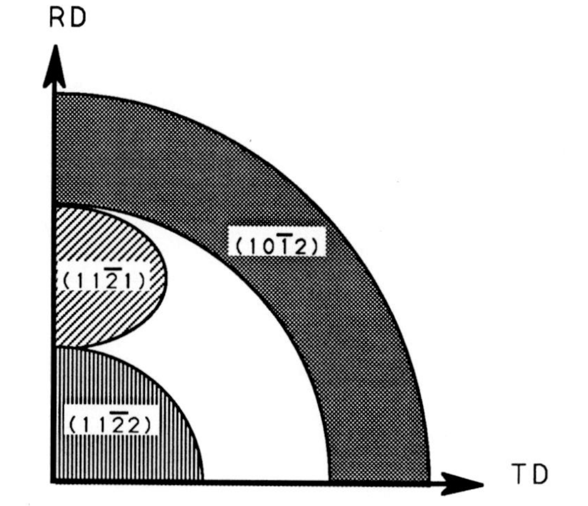

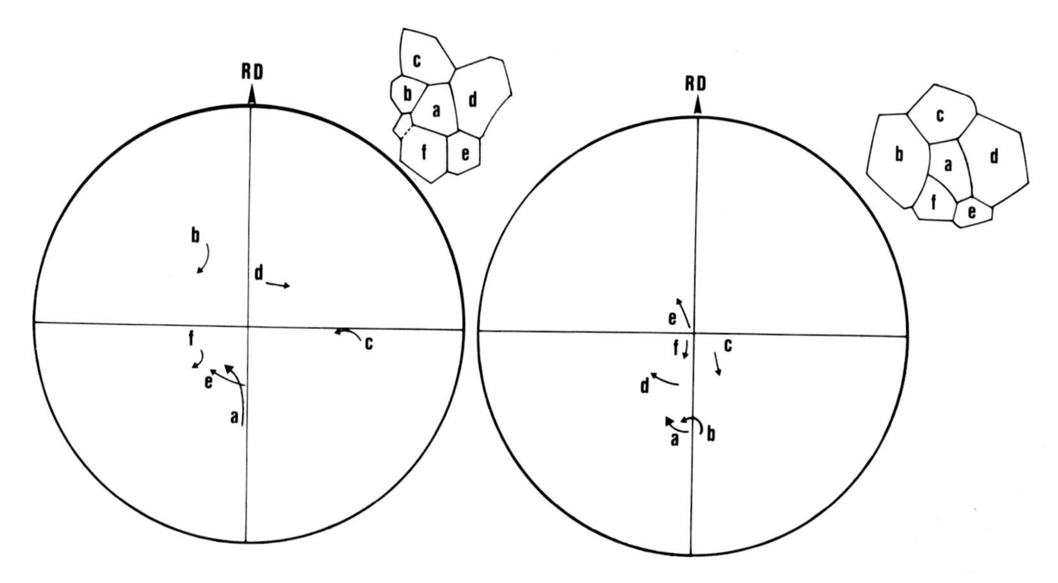

Figure 5: Influence of the environment on the rotation of two grains with the same initial orientation.

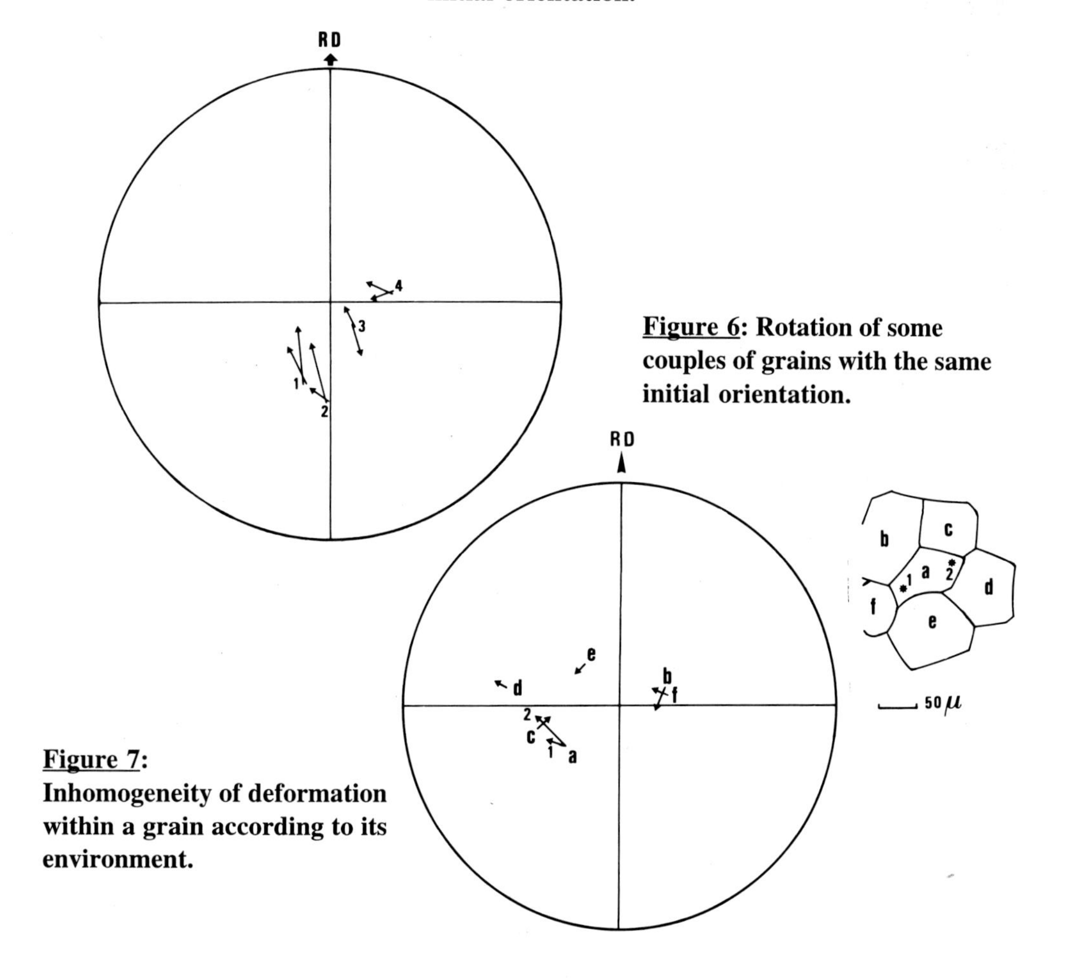

Figure 6: Rotation of some couples of grains with the same initial orientation.

Figure 7: Inhomogeneity of deformation within a grain according to its environment.

Materials Science Forum Vol. 157-162 (1994) pp. 1869-1874

A MODIFIED SELF CONSISTENT VISCOPLASTIC MODEL BASED ON FINITE ELEMENT RESULTS

L.S. Tóth and A. Molinari

Laboratoire de Physique et Mécanique des Matériaux, Université de Metz,
Ile de Saulcy, F-57045 Metz Cédex 1, France

Keywords: Self Consistent Modelling, Tangent Approach, Secant Approach, Polycrystal Deformation, Texture Development, Rate Sensitive Slip, Torsion Textures, Grain Shape Effects

ABSTRACT

A tuning of the self consistent model of Molinari et. al (Acta Metall., 1987, 35, 2983) has been carried out using the finite element results of Gilormini & Germain (Int. J. Solids Structures, 1987, 23, 413). The modification implies the introduction of a new scalar parameter in the interaction law of the self consistent approach. A simple relation between the new parameter and the value of the strain rate sensitivity has been found. The new model has been applied for the large strain simulation of torsion texture development and compared with experimental results. This approach leads to results which are much nearer to the experimental ones than those obtained by the Taylor model or by the pure tangent formulation of the self consistent model. An important effect of the grain shape changes on the evolution of anisotropy at large strains is also confirmed.

INTRODUCTION

In recent years, the modelling of plastic behavior of polycrystalline materials involves the application of more and more finite element technique [1,2]. Although detailed local information can be obtained by these numerical calculations, their use in the engineering plasticity is not yet common. The reason is in their complexity when solving 3 dimensional problems. An alternative method is to derive only macroscopic average behavior, for which purpose different self-consistent models have been developed [3-7]. These models employ a homogenization scheme in which the grain interactions with the matrix are taken into account. The basic element of these approaches therefore is the inclusion problem in an infinite homogeneous matrix. To solve this problem, an assumption has to be made concerning the strength of the interaction between the matrix and the inclusion in case of viscoplastic deformation. Two basic approaches are used for this purpose, the so called secant [3] as well as the tangent [4-7] formulations.

The above approaches and the classical Taylor and static cases are related to each other. A tuning of the interaction law is carried out using the finite element results of Gilormini and Germain [8]. The tuning involves one scalar parameter only which depends on the value of the strain rate sensitivity. It has been shown that a good correspondence can be obtained in this way between the predictions of the present modified self consistent approach and the results of the finite element simulations.

Detailed description of the theoretical part can be found in [9]. The new model is then applied to the texture development in copper bars subjected to large strain torsion. Grain shape effects and microscopic hardening are also incorporated. For the interaction between the matrix and a given grain, the isotropic version of the numerical code was used, for the reason that the tuning with the finite element results was also done for the isotropic case. The results show that the new model, which uses the interaction law tuned to finite element results, can well describe the experimental behavior of polycrystalline copper. Comparisons with the results of the Taylor predictions as well as with the pure tangent moduli approach are also made. An important effect of the grain shape on texture development has also been found at large strains. More details about the texture modelling part of the present work can be found in [10].

THE SELF CONSISTENT MODEL

A polycrystalline material is considered with the ususal power law slip for the slip systems. Elastic deformations are neglected. The interaction law introduced by Molinari et. al [4] between a given grain and the equivalent homogeneous medium can be written in the following form:

$$\mathbf{s}^g - \mathbf{S} = \mathbf{s}^g - \mathbf{A}^s{:}\mathbf{D} = \alpha(\Gamma^{sgg^{-1}} + \mathbf{A}^s)(\mathbf{d}^g - \mathbf{D}) \tag{1}$$

where $\mathbf{s}^g$ and $\mathbf{d}^g$ are the local stress and strain rates in the grain, similarly, $\mathbf{S}$ and $\mathbf{D}$ are the corresponding macroscopic quantities on the sample, respectively. $\mathbf{A}^s$ is the macrocopic reference (secant) modulus of the material defined by $\mathbf{S}=\mathbf{A}^s{:}\mathbf{D}$, and $\Gamma^{sgg^{-1}}$ is the grain shape dependent interaction tensor [4,10]. α in equation (1) is the new scalar parameter. It can be shown that by changing the value of α, different modellings follow: $\alpha=0$; static model, $\alpha=m$; tangent model, $\alpha=1$; secant model, and $\alpha=\infty$; Taylor model [4].

TUNING PROCEDURE

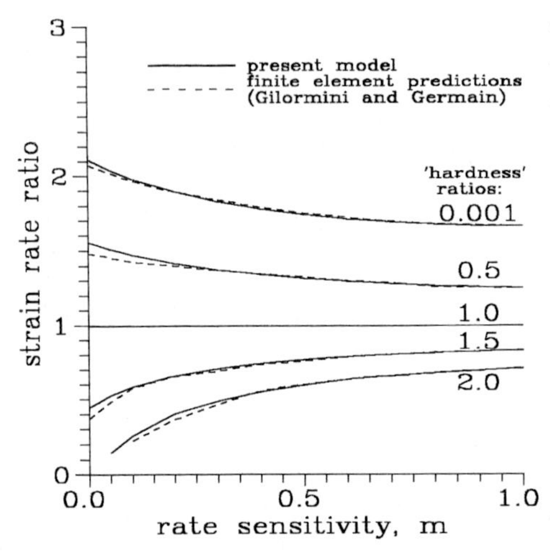

Figure 1
Average results from the finite element simulations [8] and predictions from the present tuned self consistent model.

The tuning of the self consistent model involves the finding of the appropriate value of α in equation (1) using the results of some finite element modelling. For this purpose, the simulations of Gilormini and Germain [1] are the most suitable which were carried out for the case of spherical inclusions embedded into an infinite matrix when both the inclusion and the matrix are viscoplastic. Some of their results are reproduced in figure 1 (represented by the broken lines). They have calculated the average equivalent strain rates in the inclusion (i) and in the matrix (M) and plotted their ratio $\chi = D_i^{eq} / D_M^{eq}$ as a function of the strain rate sensitivity exponent m, at constant "hardness" ratios η. This latter quantity is defined as the ratio of the average equivalent stress in the inclusion with respect to the average equivalent stress in the matrix, both corresponding to the same strain rate: $\eta = S_i^{eq}(D_M^{eq}) / S_M^{eq}(D_M^{eq})$. The results obtained by the finite element method (figure 1) *do not depend* on the specific form of the strain rate tensor $\mathbf{D}_M$. Rewriting then equation (1) for the case of isotropy and for von Mises behavior with equivalent quantities, the α parameter can be expressed as [9]

$$\alpha = 2(\eta\chi^m - 1)/3(1-\chi) \qquad (2)$$

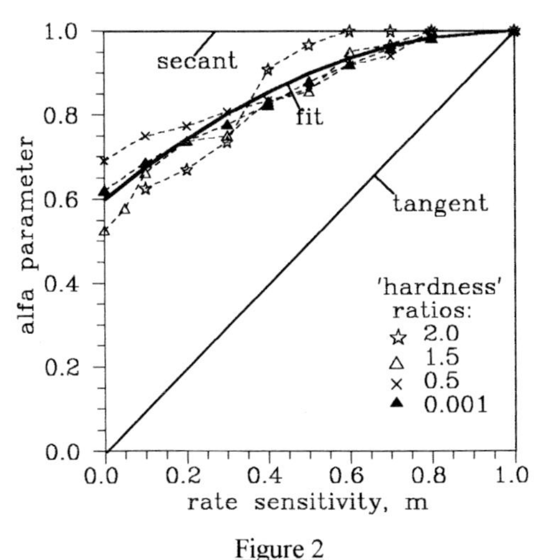

Figure 2
The α parameter obtained from the finite element results. The fit represents equation (3).

The α parameter obtained in this way is plotted at constant stress ratios (η) in figure 2 as a function of the strain rate sensitivity m. It can be seen in this figure that although α depends on η, figure 2 shows no systematic dependence. Therefore this dependence will be neglected for our purposes and the α parameter is approximated with the following simple function:

$$\alpha = 1 - 0.4(m-1)^2 \qquad (3)$$

It can be seen in figure 2 that this function describes the variations in α with acceptable precision at all levels of η. The recalculated behavior, using the present self consistent approach in conjunction with (3), is also displayed in figure 1 (continuous lines). We can conclude that the agreement between the two kinds of curves is satisfactory.

APPLICATION TO TORSION TEXTURE DEVELOPMENT

For the purpose of testing the present self consistent approach, simulations were carried out for the case of large strain torsion of OFHC copper. For this investigation, detailed texture measurements are available [11] which were obtained during the free end torsion of copper bars. The textures were measured near the surface before deformation and at subsequent large strains. In the present application, the textures corresponding to shear strain of γ=2 and γ=5.5 were selected for the modelling and comparisons with experiments. Figure 3 shows the experimental textures in ODF form (only the ϕ_2=45° sections are shown for all the textures here as they contain all the ideal components of shear textures).

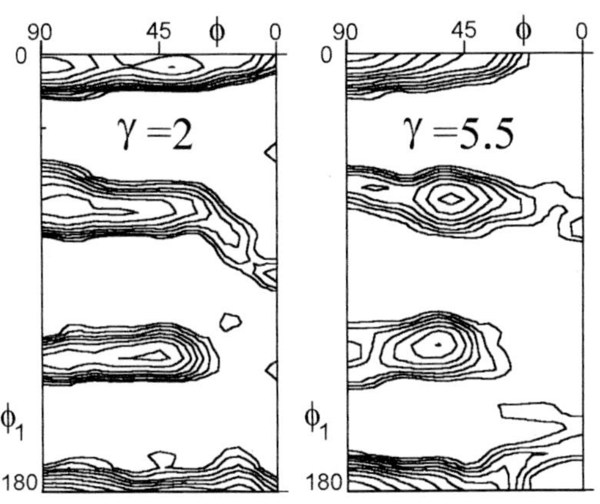

Figure 3
Measured shear textures in ODF form at shears of 2 and 5.5 (ϕ_2=45° sections). Isovalues: 0.8, 1, 1.3, 1.6, 2, 3.2, 4, 5, 6.4.

As can be seen in figure 3, all texture components are well developed at γ=2, while at γ=5.5 the so called C component becomes the most important (located at (ϕ_1,ϕ,ϕ_2)=(0,90°,45°)). The B/$\overline{B}$ components stay also relatively strong (located at (ϕ_1,ϕ,ϕ_2)=(60°,55°,45°) and at (ϕ_1,ϕ,ϕ_2)=(120°,55°,45°), respectively).

For the purpose of the simulations, the initial experimental texture (not shown here) was discretized to 400 grains and then the texture development was simulated incrementally. The value of the rate sensitivity was taken to m=0.143, so the corresponding α parameter was 0.706 (see equation 3). This relatively large value was used because this value led to a rate of texture development in agreement with the experiments. Lower values (e.g. m=0.02) led to similar textures but with significantly lower rate for the evolution of the texture.

Self and latent hardening of the slip systems were considered in the same way as in [2]. This latter approach uses the following hardening law:

$$\dot{\tau}_o^\alpha = \sum_\beta H^{\alpha\beta}\left|\dot{\gamma}^\beta\right|, \quad \alpha,\beta = 1,...,12, \qquad \text{where} \quad H^{\alpha\beta} = q^{\alpha\beta}\, h_0 \left\{1-(\tau_0{}^\alpha / \tau_{sat})\right\}^a$$

Here $q^{\alpha\beta}$ is a 12x12 matrix. Introducing α_0 and β_0 so that they can take only the values $\alpha_0=\beta_0$=2,5,8,11, $q^{\alpha\beta}$=is defined so that $q^{\alpha\beta}$=1 if $\alpha=\beta=\alpha_0=\beta_0$ and also for $(\alpha,\beta)=(\alpha_0\pm1, \beta_0\pm1)$. For all the other elements $q^{\alpha\beta}$=q, where q is a scalar. $\tau_0{}^\alpha$ is the reference stress in the rate sensitive constitutive law for crystallographic slip:

$$\dot{\gamma}^\alpha = \dot{\gamma}_0{}^\alpha\,(\text{sign}\,\tau^\alpha)\left|\,\tau^\alpha / \tau_0{}^\alpha\,\right|^{1/m}$$

Here τ^α is the resolved shear stress in slip system α and $\dot{\gamma}_0{}^\alpha$ is a constant reference slip rate. The initial values of the $\tau_0{}^\alpha$ were all set equal to each other at zero strain: $\tau_0{}^\alpha=\tau_0$. There are five parameters in the above hardening law; τ_0, h_0, τ_{sat}, a and q, which were obtained by several large strain simulations and comparisons with the experimental stress-strain curve. The following parameters have led to a satisfactory agreement: τ_0=16 MPa, h_0=220 MPa, τ_{sat}=200 MPa, a=2 and q=1.4. The experimental and the predicted hardening curves are displayed in figure 4. It can be seen, that the self consistent approach reproduces the hardening fairly well up to shears of 3.5. Above this strain the stress level increases. This is a consequence of the formation of a strong texture component (i.e. the C, see below) which corresponds to a high Taylor factor.

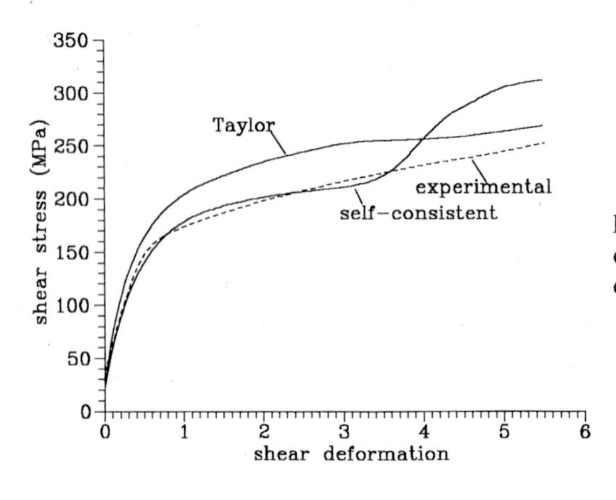

Figure 4

Experimental [11] and predicted hardening curves for large strain shear of OFHC copper.

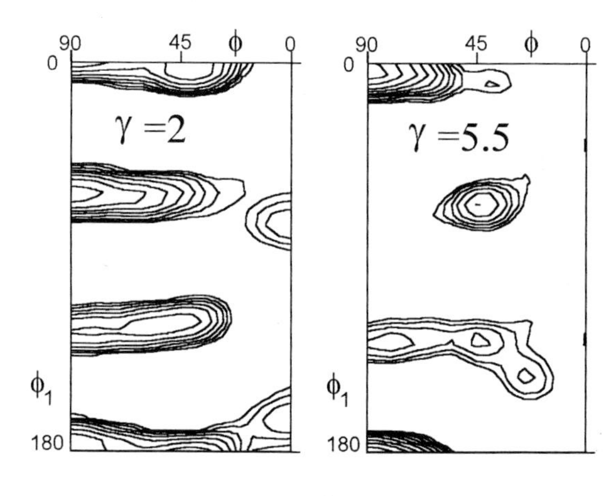

Figure 5
Predicted textures using the finite-element-tuned self consistent model. Isovalues: 0.7, 1, 1.4, 2, 2.8, 4, 5.6, 8, 11.

The predicted textures are displayed in figure 5. As can be seen, all texture components are present at shear strain of 2 and have good relative intensities compared to the measurement, see figure 3. At γ=5.5, the C orientation is the most important component of the texture, together with the less intense B/$\overline{B}$ components, again in agreement with the experiments. It can be seen, however, that the C component is too strong at this large strain. For the case of the Taylor deformation mode (see detailed results in [10]), together with the C orientation, a strong A2* component (located at $(\phi_1,\phi,\phi_2)=(35°,90°,45°)$) is predicted at shear strain of 5.5, which is very weak in the experiments, so the Taylor deformation mode can be exluded at large strains in torsion.

It has to be noted here, that the α parameter was tuned for the case of spherical grains, while during large strain, the form changes. Although the form-changes were accounted for by the term Γ^{sgg} in (1), a possible dependence on α was neglected. In fact, when the grain shape changes are not considered, then the predicted texture gives a weak C component in disagreement with the experiments (see the predictions in [10]). It indicates that the shape effects are very important at large shear strain and a dependence of the α parameter on grain shape should be taken into account. Such an investigation needs more finite element modelling to explore the variations in the α parameter. Another approximation was the assumption of isotropy for the macroscopic behavior. Nevertheless, in spite of all these assumptions, the texture development could be relatively well modelled with the use of the finite-element-tuned self consistent model.

SUMMARY

In the present paper a new self consistent approach has been presented which is based on finite element results. The new approach has been tested on the development of the texture of OFHC copper in torsion up to very large strains (γ=5.5). It has been found that the use of the tuned interaction parameter in the self consistent approach leads to satisfactory texture development. The polycrystal model simulations reproduced all important features of the experimental textures. An important effect of the grain shape on the texture development has been found which explains the existence of a strong C component at large strains in shear textures.

ACKNOWLEDGEMENT

The authors are grateful to Professor P. Van Houtte for providing his ODF software which has been used for the presentation of the simulated textures.

REFERENCES

1. K.K. Mathur and P.R. Dawson, Int. J. Plasticity, 1989, 5, 67.
2. S.R. Kalidinki, C.A. Bronkhorst and L. Anand, J. Mech. Phys. Solids, 1992, 40, 537.
3. M. Berveiller and A. Zaoui, Res. Mech. Let., 1981, 1, 119.
4. A. Molinari, G.R. Canova and S. Ahzi, Acta Metall., 1987, 35, 2983.
5. S. Ahzi, A. Molinari and G.R. Canova, Yielding, Damage, and Failure of Anisotropic Solids, EGF5 (Ed. J.P. Boehler), Mechanical Engineering Publications, London (1990) 425.
6. H.R. Wenk, G. Canova, A. Molinari and H. Mecking, Acta Metall., 1989, 37, 2017.
7. R.A. Lebensohn and C.N. Tomé, Acta Metall. et Mater. in press.
8. P. Gilormini and Y. Germain, Int. J. Solids Structures, 1987, 23, 413.
9. A. Molinari and L.S. Tóth, submitted to Acta Metall. et Mat., 1993.
10. L.S. Tóth and A. Molinari, submitted to Acta Metall. et Mat., 1993.
11. L.S. Tóth, J.J. Jonas, D. Daniel and J.A. Bailey, Textures and Microstructures, 1992, 19, 245.

Materials Science Forum Vol. 157-162 (1994) pp. 1875-1880

PREDICTION OF FORMING LIMITS OF TITANIUM SHEETS USING THE PERTURBATION ANALYSIS WITH TEXTURE DEVELOPMENT

L.S. Tóth, D. Dudzinski and A. Molinari

Laboratoire de Physique et Mécanique des Matériaux, Université de Metz,
Ile de Saulcy, F-57045 Metz Cédex 1, France

Keywords: Perturbation Analysis, Forming Limits, Diffuse Necking, Texture Development, Yield Potential of Polycrystals, Titanium, Hexagonal Crystal Deformation, Taylor Modelling, Rate Sensitive Slip, Twinning in Hexagonal Metals

ABSTRACT

The perturbation analysis of plastic deformation developed by Dudzinski and Molinari (Int. J. Solids Structures, 1991, 27, 601) has been applied to predict forming limits of a titanium alloy (T40). Deformation by slip and twinning was modelled using a rate sensitive Taylor model and the yield potentials were computed for proportional strain paths. These polycrystal stress potentials were fitted locally with Hill-ellipsoids at the loading points and the perturbation analysis was applied as deformation proceeded. Using the perturbation technique only the beginning of diffuse necking can be modelled. The predicted diffuse neck-limits agree well with the measured ones.

INTRODUCTION

Dudzinski and Molinari [1] have developed a new approach to calculate the forming limits of sheet metals using the perturbation analysis of deformation. The model employs the Hill approximation of the yield locus [2]. In the present paper, the perturbation analysis is developed for the case of *polycrystalline hexagonal materials with a known texture*. For this purpose, first a polycrystal yield locus routine is developed in which the yield locus is approximated by a Hill-one but only at the vicinity of the loading point. In this way it is not only possible to carry out simulations for arbitrary yield locuses but even the *development* of anisotropy with the deformation can be taken into account. It should be noted here that such a simulation is difficult in the approach of the initial defect theory (Marciniak-Kuczinsky analysis, see a review on forming limits by Ferron and Molinari [3]). The reason is that in the M-K approach the deformation path in the imperfection differs from the prescribed strain path on the specimen so the texture development is also different. There is, however, only bulk-behavior in the perturbation analysis, so assuming proportional strain path, the texture development can be incorporated. Further advantage, that employing the local fitting of Hill coefficients, a large part of the analysis can be carried out analytically, and the numerical part requires negligible calculation time.

It should be emphasized here that the perturbation method, unlike the Marciniak-Kuczinsky initial defect theory, does not allow to follow the evolution of deformation within the band. It is restricted to the prediction of the *appearance* of a necking process, that is, it predicts the strain levels which correspond to the initiation of *diffuse necking*. As it is known, the M-K theory predicts the final limit strains, which correspond to the end of the *localized* necking process but is not able to predict the initiation of diffuse necking. The two methods therefore are not competing ones, rather complementary.

In the following first the main elements of the perturbation analysis is shown, then the description of the new polycrystal code is given. Finally, results of the simulations and comparisons with the predictions of the M-K approach are presented.

MAIN ELEMENTS OF THE PERTURBATION ANALYSIS [1]

A biaxial deformation mode, described by the strain rates D_{xx}° and D_{yy}°, is imposed at the remote boundaries of a sheet (figure 1). The straining path is linear and defined by the ratio

$$\frac{D^0_{yy}}{D^0_{xx}} = \rho = \text{const.} \tag{1}$$

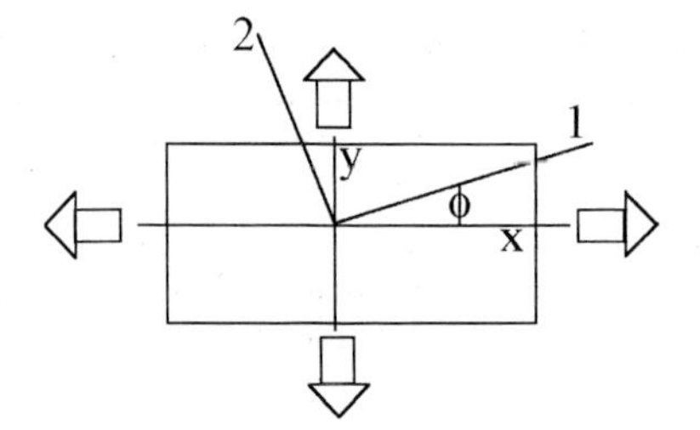

Figure 1.
Geometry of the sheet under deformation. The directions of the applied strain rates are x and y.

The initial thickness is uniform $h=h^{\circ}(t=0)$. Knowing the hardening law and the yield surface, a homogeneous solution can be obtained in the rotated reference system (1-2, see figure 1) at any moment of the deformation. This solution is denoted by the subscript 0:

$$S^0 = (D^0_{11}, D^0_{22}, \sigma^0_{11}, \sigma^0_{22}, \sigma^0_{12}, \bar{\sigma}^0, \bar{\varepsilon}^0, h^0) \tag{2}$$

The stability of this solution is then analysed at a time t_0 by the superposition of a small perturbation $\delta S = (\delta D_{11}, \delta D_{22}, \ldots, \delta h)$ where the perturbation is assumed to have the following form:

$$\delta S_i = \delta S^0_i \exp\left[\eta(t - t_0)\right] \exp(i\xi x_1) \tag{3}$$

where $\delta S^0 = (\delta D^0_{11}, \delta D^0_{22}, \ldots, \delta h^0)$. The spatial modulation is periodic and defined by the wave number ξ. The factor η caracterizes the rate of growth of the perturbation. The perturbed solution reads:

$$\mathbf{S} = \mathbf{S}^0 + \delta\mathbf{S} \tag{4}$$

The equations of the above perturbation problem can be linearized and written in the following form:

$$\mathbf{L}\,\delta\mathbf{S}^0 = 0 \tag{5}$$

A non-zero solution exists if the determinant of the linear equation system (equation 4) is null. In this way, a cubic equation can be obtained for the parameter

$$\eta' = \eta / \bar{\varepsilon}^{0} \tag{6}$$

The present linearized theory predicts an effective instability when one of the roots of the cubic equation has a real and large enough value:

$$\mathrm{Re}(\eta') > e \tag{7}$$

Here e is the intensity of the instability. When the above condition is attained, it means that a necking process starts in a band inclined at the given angle ϕ (see figure 1). The critical angle, to which the minimal deformation corresponds, is obtained by varying the orientation of the band (ϕ).

THE POLYCRYSTAL YIELD LOCUS CODE

One of the input of this code is the measured crystalline orientation distribution function of the given sample, in form of discrete orientations (about 700). The main characteristic of the texture in this material is a preference of the basal poles near the normal direction with some spread along the transverse direction, see figure 2.

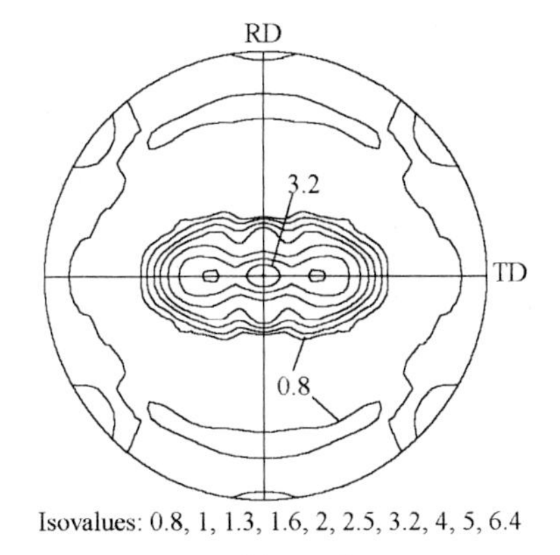

Figure 2
(0002) pole figure showing the distribution of the basal poles in the titanium sheet before deformation (T40, 1mm thickness).

The other input is the slip and twinning systems together with their relative critical strengths (crss values). These crss values are themselves can be obtained from other experiments and subsequent simulations [4,5].

For the given loading direction, which is specified by the imposed velocity gradient on the sample, a polycrystal simulation is carried out to obtain the corresponding stress components. For this purpose, the rate sensitive polycrystal code of Tóth [6] has been employed with m=0.05.

One peculiarity of the time dependent deformation theory is that there is no yield locus in its classical sense. This is because there can be plastic deformation at any applied non-zero stress level. There exist, however, a yield potential function, which, for $m \to 0$, tends to the one known in the rate independent case. This yield potential function is defined with a constant rate of plastic work along its surface. This constant value (W_0) has been chosen to be the rate of plastic work corresponding to equibiaxial stretching of the given sample and for the following strain rate: $D^0_{11} = D^0_{22} = 1 s^{-1}$. For any prescribed strain rate $\mathbf{D'}$ which is different from this one, the stress level σ^{sp} corresponding to the chosen rate of plastic work can be obtained as follows. Hutchinson has shown [7] that the stress level changes with the magnitude of the strain rate according to the relation:

$$\sigma(\lambda \mathbf{D}) = \lambda^{m} \sigma(\mathbf{D}) \tag{8}$$

It can readily be shown using this relation that the stress components on the chosen yield potential can be obtained from:

$$\sigma^{sp} = \left(\frac{W_0}{W(\mathbf{D}')}\right)^{\frac{m}{m+1}} \sigma(\mathbf{D}') \tag{9}$$

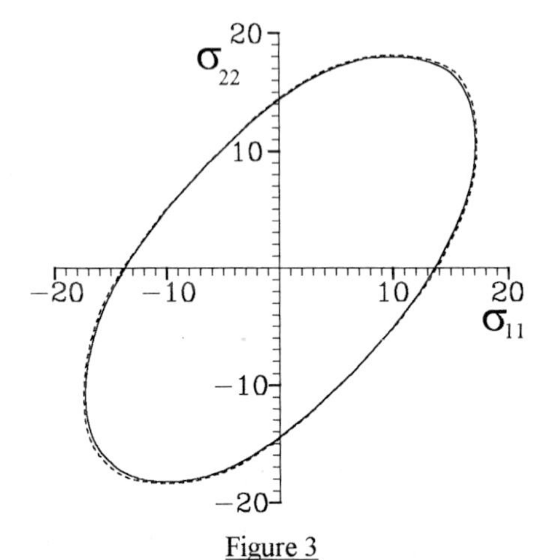

Figure 3
The yield potential function of the T40 material before stretching (continuous line) for m=0.05. Broken line: the rate insensitive yield locus.

The polycrystal simulations have been performed with the rate sensitive polycrystal code of Tóth [6] for proportional strain paths. For the slip systems and the relative critical stresses the following systems and relative values have been adopted [8]: prismatic: 1, pyramidal $\langle a \rangle$: 4.5, pyramidal $\langle c+a \rangle$: 9, $10\bar{1}2$ twins: 8.5, $11\bar{2}2$ twins: 15. The measured texture in the T40 1 mm sheet was used as the input distribution which was represented by 700 grain orientations. Hardening was neglected. The mechanism of twinning was modelled by the Monte-Carlo technique of Van Houtte [9]. For any strain path, however, the contribution of twinning to the plastic deformation was less than 5%. The yield potential function before deformation is displayed in figure 3. As can be seen that it does not differ much from the rate insensitive yield locus.

In order to obtain the parameters of a Hill ellipsoid tangent to the polycrystal yield potential function, 4 points of it have been calculated in the vicinity of the given loading point. One point taken exactly at the loading point, and 3 points correspond to strain rate directions which are 5° away from the loading strain rate direction. These 4 points include also one shear component: D_{12}. For each of these points the stress coordinates have to statisfy the Hill-equation [2]:

$$F(\sigma_{22}-\sigma_{33})^2+G(\sigma_{33}-\sigma_{11})^2+H(\sigma_{11}-\sigma_{22})^2+2N\sigma_{12}^2 = 1 \tag{10}$$

The four stress states therefore completely define the F, G, H and N parameters of the Hill-ellipsoid through a linear equation system.

EVOLUTION OF THE YIELD POTENTIAL DURING LARGE PLASTIC DEFORMATION

In the forming operation the strain path is biaxial deformation, which is defined by the parameter ρ with the convention that $D_{xx}=1s^{-1}$. The interval of interest is $-1<\rho<1$ for the purpose of forming limit predictions, within which the simulations were repeated for 21 selected strain paths. After each incremental step, the parameters of the local tangent Hill-ellipsoid were also computed which were facilitated in the forming limit predictions.

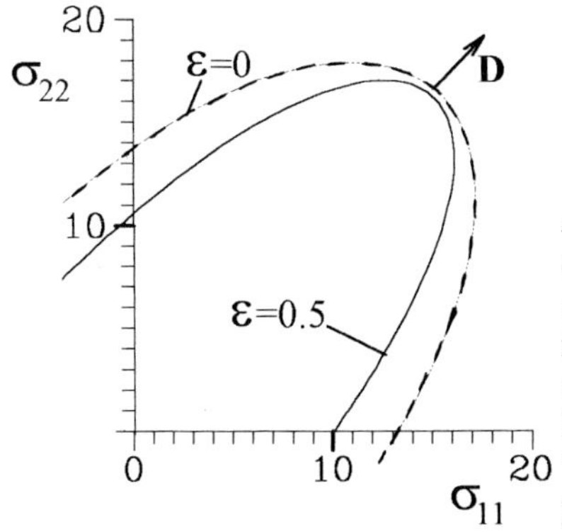

Figure 4
Tangent Hill-ellipsoids before deformation (broken line) and after 0.5 strain (continuous line) for equibiaxial stretching.

As a representative example, we show here the evolution of the tangent Hill-ellipsoid for $\rho=1$ in figure 4. A significant change in the curvature of the yield potential function can be observed. Our simulations therefore support the phenomenon suggested earlier in the literature, that during plastic deformation a so called "vertex" can develop on the yield surface (or on the yield potential, in the present case).

FORMING LIMIT PREDICTIONS

The perturbation analysis was peformed for the same strain paths as the polycrystal simulations, so the evolution of the tangent Hill ellipsoids were continuously taken into account. Figure 5 shows the forming limits for an e parameter of 2.5 and for n=0.14 and m=0.05. Here n is the strain hardening exponent, measured by a tensile test. The broken line represents the forming limits when the evolution of the yield potential is neglected, that is, when at any instant only the initial potential function is used (i.e. the one in figure 2). The solid line in figure 5 corresponds to the general case when the evolution of the texture has been taken into account (variable F,G,H,N values). The experimental limit strain curve and the curve corresponding to the beginning of diffuse necking are also displayed in figure 5. The predicted strains in the perturbation analysis correspond to the latter one. It can be seen that the agreement between the measured and the simulated diffuse-necking curve is rather satisfactory.

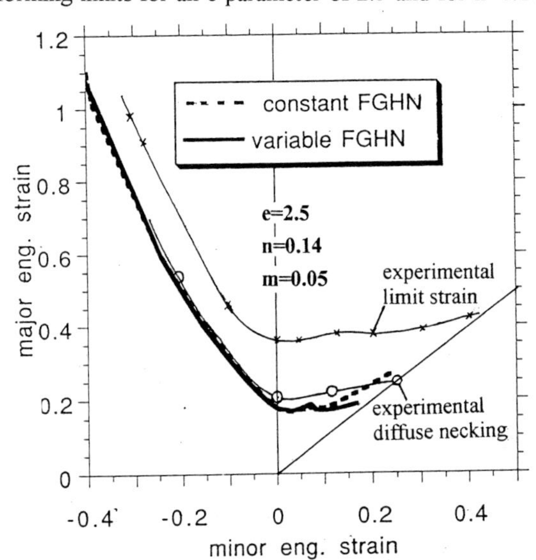

Figure 5
Predicted diffuse necking strains by the perturbation analysis. The experimental measurements for the limit strains and for the appearence of necking are also displayed [10,11].

According to figure 5, the effect of the evolution of the texture is very important in the stretching range, in the vicinity of equibiaxial loading. In the contraction range, no significant changes take place. The reason of the decrease in the forming limit for equibiaxial stretching is due to the increase in the curvature of the yield potential, see figure 4. Such a relation between the local curvature and forming limit has already been suggested in several studies, but here it is clearly related to the development of the texture during plastic deformation.

For comparison purposes, we have also derived the forming limit curves using the Marciniak-Kuczinsky type polycrystal code developed Van Houtte and Tóth [10]. For three values of the rate sensitivity, i.e. m=0, m=0.01 and m=0.05, and using the same crss values, the obtained results are displayed in figure 6 (for an initial defect parameter of 0.996). We can see that the limit strain depends very much on m. As the perturbation method-results were computed for m=0.05, the curve corrresponding to m=0.05 in figure 6 has to be compared with the experimental **limit strain** curve in this case. The correspondence is very good in the contraction range and for plane strain, however, near the biaxial region, the level of the predicted limit strains is somewhat overestimated. This may originate from the fact that the development of the texture was neglected in the M-K approach.

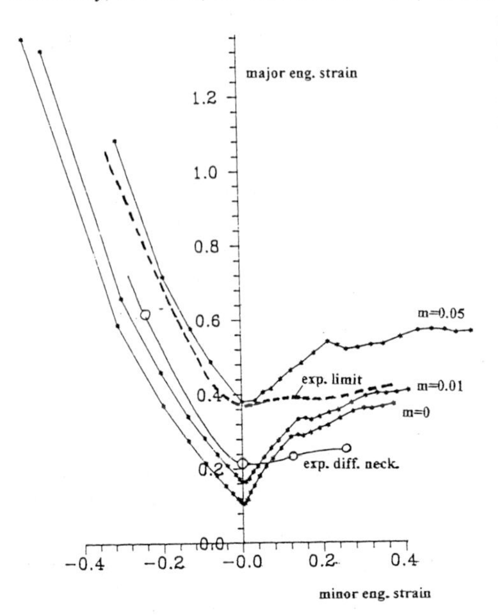

Figure 6
Predicted limit strains using the Marciniak-Kuczinsky polycrystal forming limit code.

CONCLUSION

The appearance of diffuse necking can be well modelled by the perturbation analysis in hexagonal titanium taking into account the texture development. The Marciniak-Kuczinsky initial defect theory in turn is applicable to predict the final limit strains during the stretching of sheet metals.

ACKNOWLEDGEMENT

This work has been supported by the European Community under the contract Brite Euram N°011C.

REFERENCES

1. D. Dudzinski and A. Molinari, Int. J. Solids Structures, 1991, 27, 601.
2. R. Hill, "The mathematical theory of plasticity", Oxford University Press 1950.
3. G. Ferron and A. Molinari, Forming Limit Diagrams: Concepts, Methods and Applications, Eds. R.H. Wagoner, K.S. Chan and S.P. Keeler, The Minerals, Metals & Materials Society, 1989.
4. A. Molinari, L.S. Tóth and D. Dudzinski, Brite Euram Project on Titanium (N° 011C), in Final Report: "Prediction of forming limits and texture development in titanium alloys", 1993.
5. Driver and N. Cheneau-Spath, "Mécanismes de déformation des métaux hexagonaux", ibid.
6. L.S. Tóth, 'LSTAYLOR' Rate Sensitive Polycrystal Code, Université de Metz, 1993.
7. J.W. Hutchinson, Proc. R. Soc. Lond. A, 1976, 348, 101.
8. M.J. Philippe, personal communication.
9. P. Van Houtte, Acta Metall., 1978, 26, 591.
10. P. Van Houtte and L.S. Tóth, "Generalization of the Marciniak-Kuczynski defect model for predicting forming limit diagrams", Proc. Asia-Pacific Symposium on Advances in Engineering Plasticity and its Application (W.B. Lee Editor), 15-17 Dec., 1992, Hong Kong, 1013.
11. L.S. Tóth and Van Houtte, Forming limit experiments on titanium sheets, partly published in [10], Katholieke Universiteit Leuven, 1992.

Materials Science Forum Vol. 157-162 (1994) pp. 1881-1886

TAYLOR SIMULATION OF CYCLIC TEXTURES AT THE SURFACE OF DRAWN WIRES USING A SIMPLE FLOW FIELD MODEL

P. Van Houtte [1], P. Watté [1], E. Aernoudt [1], J. Gil Sevillano [2], I. Lefever [3] and W. Van Raemdonck [3]

[1] Department MTM, Katholieke Universiteit Leuven, de Croylaan 2, B-3001 Leuven, Belgium

[2] CEIT, Universidad de Navarra, San Sebastian, Spain

[3] N.V. Bekaert, Zwevegem, Belgium

Keywords: Taylor Theory, Deformation Textures, Surface Textures, Cyclic Textures, Wire, Pearlite, Carbon Steel, Flow Field, Visioplasticity, Finite Strain

ABSTRACT

The deformation textures which develop during wire drawing at the surface of pearlitic carbon steel wires were simulated by means of the Taylor theory. These surface textures are so-called "cyclic textures", which do not have a full rotational symmetry such as the fibre textures found in the centre of drawn wires. This effect was believed to be caused by the more complex material flow which exists near the surface in comparison with the centre. The total plastic deformation undergone by a volume element of the material during passage through the plastic zone near the surface was split up into a large number of small deformation increments. Each of these was described by a different displacement gradient tensor. These tensors were fed to the Taylor program. In order to obtain this strain history, it was necessary to use a simplified model for the flow field inside the wire drawing die. This model was analysed by means of techniques taken from the visioplasticity theory. An example of the cyclic textures in the ferrite phase predicted by this method is given. The example is specific for pearlitic high carbon steel; a special variant of the Taylor model was used to that purpose. The result shown agrees qualitatively with a cyclic texture observed by other authors near the surface (but not *at* the surface) of a drawn wire.

INTRODUCTION

During wire drawing, the deformation is purely axisymmetric at the core of the wire. This is not true in other locations in the wire, especially not in regions close to the surface. An elementary volume of the material which is close to the surface of the wire undergoes various types of deformation as it travels through the die. As a result, the deformation texture that develops during successive drawing passes in the centre of the wire is different from that at the surface. Those textures also have a different "sample symmetry" At the centre, where the deformation is axisymmetric, the texture is also axisymmetric: it is a so-called *fibre texture*. Pole density distributions measured at the surface often are not axisymmetric, but have either *one* symmetry plane (the plane containing the radial direction and the wire axis) or *three* (the above with, in addition, the plane tangent to the wire surface and the transverse plane). Hence the sample symmetry of the ODF is either *monoclinic* or *orthorombic*. In the past such textures have been called *cyclic textures*. The present authors wanted to investigate the surface textures since it was expected that the behaviour of these wires during torsion testing was correlated to the surface texture [1]. Unfortunately, the wires were too thin for experimental determination of the cyclic textures at the surface. A method for the measurement of textures located at the surface of thin wires was available [2] but was of not much help, since it assumed the textures to be fibre textures. Therefore it was attempted to predict these cyclic surface textures by means of a Taylor simulation.

Taylor's assumption of homogeneous strain had to be partly abandoned to that purpose. The development of the deformation texture in an infinitesimal volume element of the material which is drawn through the die had to be simulated. It was not assumed that all volume elements would undergo the same deformation; nor was it assumed that the velocity gradient which describes the rate of deformation of a particular volume element would be constant in time. But it was still assumed that such volume element would consist of thousands of crystallites, which together have the "local" deformation texture, and that at a given moment in time, the crystallites of a particular volume element would undergo the same strain. That strain is a function of time; so the Taylor theory can only be applied if the strain history of a particular volume element is known during its passage through the die. An analysis will be presented below, which allows for an approximative calculation of the strain history for volume elements which are close to the surface of the wire.

When describing surface textures, a reference system will be used which depends on the point at the surface which is considered. x_3 will be the radial direction, x_2 the circumferential direction and x_1 is parallel to the wire axis, its positive sense in principle coinciding with the wire drawing direction. The reference system defined in this way corresponds to the one shown by figure 1.

SIMPLIFIED GEOMETRICAL DESCRIPTION OF FINITE DEFORMATION

Consider a infinitesimal volume element of wire material which travels along a flow line through the plastic zone of a drawing pass. Figure 1 shows a (hypothetical) example of such flow line. Let us for the sake of the present analysis define as "undeformed state" the state of the material before it enters the plastic zone (time t_0). Let **dx** be a vector connecting two points of the material within the volume element. At a time t, the volume element may be well inside or even beyond the plastic zone. The two points are now connected by the vector **dy**. The change of length and orientation of **dx** as it is transformed into **dy** can be described by:

$$\mathbf{dy} = \mathbf{F} \bullet \mathbf{dx} \tag{1}$$

F is a second order tensor. For a perfect axisymmetric deformation, it would have the following expression in the reference system $x_1x_2x_3$ defined above:

$$\left[F_{ij}\right] = \begin{bmatrix} e^{\varepsilon} & 0 & 0 \\ 0 & e^{-\varepsilon/2} & 0 \\ 0 & 0 & e^{-\varepsilon/2} \end{bmatrix} \tag{2}$$

in which ε is the "true strain" achieved in longitudinal direction. A tensor such as **F** describes a finite deformation. Equation 2 is valid for a flow line which is at the centre of the wire, but not for a flow line close to the surface. In the latter case, **F** is a function of the time t . **F** is related to the strain tensor and the rigid body rotation. Suppose that $\mathrm{d}\varepsilon_{ij}$ and $\mathrm{d}\omega_{ij}$ are the tensors describing the strain and the rigid body rotation that take place between t and t+dt, using the the configuration at time t as reference (Lagrangian tensors for small strains). The components of the corresponding displacement gradient tensor **dK** are given by:

$$\mathrm{d}K_{ij} = \mathrm{d}\varepsilon_{ij} + \mathrm{d}\omega_{ij} \tag{3}$$

According to the strain theory for small strains, the operator (**I+dK**) achieves the transformation of a material vector between t and t+dt. **I** is the unit tensor. Let **F+dF** describe the finite strain at time t+dt. Then:

$$\mathbf{F+dF=(I+dK)\bullet F} \tag{4}$$

which can be used to obtain the evolution of **F** from a series of increments **dK**.

APPROXIMATIVE CALCULATION OF THE STRAIN HISTORY OF A VOLUME ELEMENT

The strain history of a volume element that passes through the die along a flow line could perhaps be calculated accurately by a finite element analysis of the wire drawing process if a fine mesh is used. This is not an easy task and the present authors wished to avoid it. Instead, they assumed a very simple strain history for flow lines close to the die surface (figure 1). The flow line was assumed to consist of three straight segments:

a) (at the left of figure 1) An incoming segment, parallel to the wire axis. No deformation takes place along this segment.

b) (between AA' and BB', figure 1) A segment which makes an angle δ with the wire drawing axis. At the surface, it is parallel to the (conical) surface of the die.The material undergoes a strain as it moves along this segment (see below).

c) (at the right of figure 1) An outcoming segment, parallel to the wire axis. No deformation takes place along this segment.

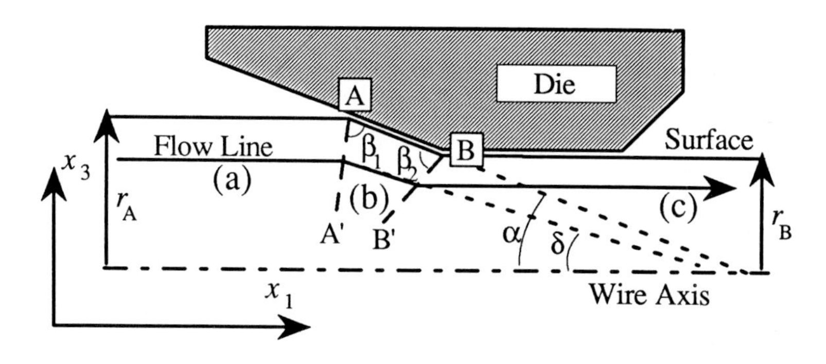

Fig. 1 Model for the material flow in a wire drawing pass. AA' and BB' are velocity discontinuity lines.

A reference system $x'_1x'_2x'_3$ is introduced at each point along a flow line. x'_1 is parallel to the flow line and x'_3 is normal to it. x'_2 is the circumferential direction. It is assumed that in a limited region underneath the surface of the wire, all the flow lines are as described above. The segments (c) are not parallel to each other, the angle δ gradually decreasing for flow lines closer to the centre. Such set of flow lines can be analyzed by means of the visioplasticity theory [3], which is essentially based on the incompressibility of the material. This analysis shows that in the region of the segments (c), the strain components ε'_{11}, ε'_{22}, ε'_{33} and ε'_{13} can be different from zero. In the present analysis, it is assumed that ε'_{13} is zero, which means that the effect of the friction along the die on the flow pattern is neglected. For segment (c) of the flow line at the surface, visioplasticity theory also shows that the total "true strain" in circumferential direction as a volume element moves from A to B, is equal to $-\ln(r_A/r_B)$. r_A and r_B are the distances of A and B from the wire axis (figure 1). Note that a volume element undergoes a large strain at the moment that the flow line changes direction on passage through the lines AA' and BB'. These lines are *velocity discontinuity lines (VDL)*; the normal component of the velocity is the same on both sides of such line, but the tangential component is not. The material is sheared off on passage through such VDL. The shear planes are not necessarily normal nor parallel to the wire axis. This means that the shear strain *can* contribute to the elongation in wire drawing direction.

A flow model like this cannot be valid from surface to centre throughout the wire; it can at most be an approximative description for flow lines close to the surface. It has been inspired by the flow pattern ('hodograph') predicted by a slip line field analysis for sheet drawing [4], and by flow patterns of extruded bars observed by means of a visioplasticity analysis [5]. The main advantage of the proposed geometrical model is, that it respects the geometrical boundary conditions as well as the incompressibility of the material.

The flow model presented here can only be useful if it leads to simple expressions for the strain history. Therefore it is essential to be able to describe the deformation that takes place when a volume element flows through a VDL. This deformation is essentially a simple shear. The shear plane is the plane that contains the VDL (as shown in figure 1: AA' or BB') and the circumferential direction (x_2) (figure 1). The shear direction is parallel to the VDL. The amount of shear can be calculated: it must be such that a material vector which was parallel with the flow line before passage through the VDL becomes parallel to the new direction of the flow line after. For the present application, it is most convenient to describe the tensor **F** which characterizes this deformation in a reference system which moves with the material and is co-rotational with the flow line. $x'_1x'_2x'_3$ as described above is such reference system since by definition, x'_1 is always tangential to the flow line. In such description, $[x'_i]$ would be the matrix representation of a material vector **x** (within an volume element) just before passage through the VDL, and expressed with respect to the frame $x'_1x'_2x'_3$ which is valid at that moment. After passage, the shear has transformed **x** into **y**. The matrix representation of the latter is $[y'_i]$, however with respect to a new reference system $x'_1x'_2x'_3$ which has rotated together with the flow line. $[y'_i]$ is obtained by multiplying $[x'_i]$ at the left side with $[F'_{ij}]$, which is an unconventional representation of **F**. It can easily be understood that $[F'_{ij}]$ must be of the following form:

$$\left[F'_{ij}\right]=\begin{bmatrix} e^{\varepsilon'} & 0 & \gamma' \\ 0 & 1 & 0 \\ 0 & 0 & e^{-\varepsilon'} \end{bmatrix} \tag{5}$$

Indeed, if F'_{21} or F'_{31} would be non-zero, then the operator $[F'_{ij}]$ would transform a vector $[x'_i]$ which is be parallel to the incoming flow line into a vector $[y'_i]$ which is not parallel to the outcoming flow line. Any vector which is parallel to $x_2=x'_2$ should be not deformed at all, which explains the scond column of $[F'_{ij}]$. Finally the expressions for F'_{11} and F'_{33} have been chosen in such way that the volume is kept constant. Note that ε' is equal to the "true strain" along the flow line. ε' and γ' depend on β_1, δ and α (for AA', figure 1) or on β_2, δ and α (for BB'). Expressions for ε' and γ' as a function of these parameters have been developed. It was found that the von Mises equivalent strain associated to this deformation was minimum for the combinations of the parameters which made ε' zero. This situation corresponds to the case where the VDL is the bisector of the angle formed by the two directions of the flow line (case $\eta_1=\eta_2$ in figure 2). It was decided to choose the parameters always in this way, since it seemed a rudimentary form of minimization of plastic work, as would be done in an upper bound analysis. If in addition to this, δ is set equal to α, which means that the flow line at the surface of the wire is considered, the following values for β_1 and β_2 are obtained:

$$\beta_1=\frac{\pi}{2}-\frac{\alpha}{2} \tag{6}$$

$$\beta_2=\frac{\pi}{2}+\frac{\alpha}{2} \tag{7}$$

This then results in the following values for γ' (ε' being zero):

$$\text{At AA':}\quad \gamma'=-2\tan\left(\frac{\alpha}{2}\right) \tag{8}$$

$$\text{At BB':}\quad \gamma'=2\tan\left(\frac{\alpha}{2}\right) \tag{9}$$

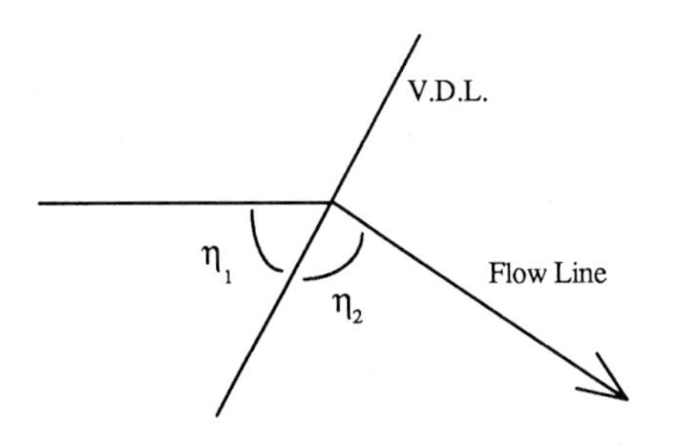

Fig. 2 A flow line passing through a velocity discontinuity line (VDL)

These values (together with $\varepsilon'=0$) can now be used in equation 5 to find the $[F'_{ij}]$ matrix. In order to simulate the texture change during the passage through a VDL, this finite strain has to be split up into a series of small deformation increments, each described by a displacement gradient tensor, since the available Taylor code [6] uses this kind of data. Suppose that one wants to achieve the finite strain in *n* steps. Then each step is described by the following displacement gradient:

$$\left[K'_{ij}\right]=\begin{bmatrix}0 & 0 & \gamma'/n\\ 0 & 0 & 0\\ 0 & 0 & 0\end{bmatrix} \qquad (10)$$

Note that, since the angles β_1 and β_2 have been chosen in such way that $\varepsilon'=0$, the two passages through the VDL do not contribute to the elongation nor to the strain in the direction normal to the flow lines. This means that all strains of this type have to be achieved in segment (b) of the flow lines. Hence the operator $[F'_{ij}]$ which describes the total strain in that segment must be equal to $[F'_{ij}]$ as given by equation 2, in which $\varepsilon=2\ln(r_A/r_B)$. The only "contribution" of the two VDL's to the strain is that the material is sheared backwards as it flows into the plastic zone, to be sheared again in forward direction when it emerges from the plastic zone.

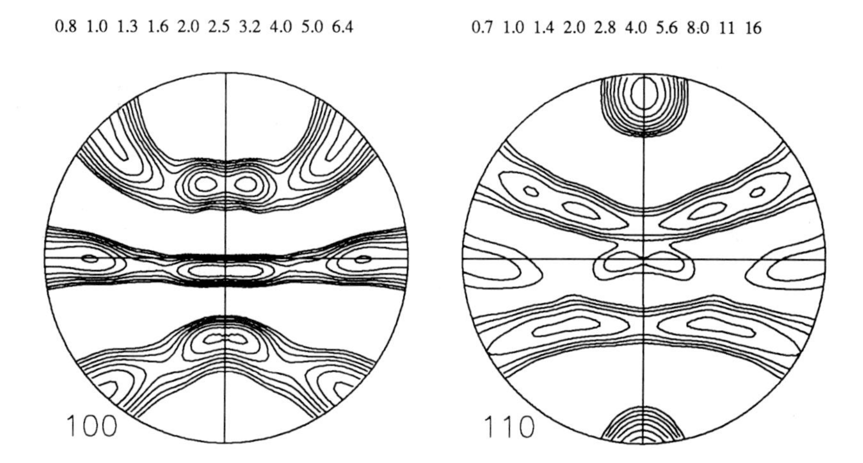

Fig. 3 Pole figures of a simulated cyclic texture of the ferrite phase of a pearlitic wire. The "Pearlite Model" (a variant of the Taylor model) was used. The radial direction (x_3) is at the centre of the pole figures, and the wire drawing direction (x_1) is pointing downwards. 26 drawing passes were simulated. In each pass, the true strain was 0.1525 and the half-die angle α was 8°.

TAYLOR SIMULATION OF THE SURFACE TEXTURE

The Taylor program [6] was used for the simulation of one wire drawing pass by first simulating the passage through the VDL AA' (figure 1), using the displacement gradients given by equation 10, then simulating an axisymmetric deformation in the ordinary way (segment (b) between A and B, figure 1) and finally simulating the passage through the VDL BB'. The parameters that have to be chosen are:

- the half-die angle α;
- the diameter reduction r_A/r_B.

This scheme was then repeated several times in order to simulate a real drawing operation consisting of several passes. Figure 3 shows the results of such simulation for pearlitic steel. The texture development of the ferrite phase was simulated. The {110}[111] and the {112}[111] slip system families were both considered. A special variant of the Taylor model was used: the "pearlite model", which also simulates the morphology changes of the cementite lamellae, and uses that information in order to find a corrective factor on the CRSS of the individual slip systems [7].

The simulated surface texture has a monoclinic sample symmetry. The dyad axis is x_2, and the x_1x_3-plane is a symmetry plane of the pole density distributions. However, the x_2x_3-plane is almost a symmetry plane as well. The pole figures show a strong [110]-fibre texture component, with some individual components on top of it: a (001)[110] and a $(1\bar{1}0)[110]$ component. These are the "cyclic components". The latter has been reported by Kanetsuki and Ogawa [8] in an intermediate layer close to the surface, but not at the surface, of 2 comparatively thick pearlitic carbon steel wires which were cold drawn to a total drawing strain (true strain) of 1.63 and 2.46, respectively.

Other simulations have also been carried out. It was possible to see the influence of wire drawing angle and reduction per pass on the resulting deformation textures. For certain combinations of these parameters, a $(1\bar{1}2)[110]$ cyclic texture component was found instead of the $(1\bar{1}0)[110]$ component.

CONCLUSIONS

It is possible to explain the presence of cyclic textures at the surface of drawn pearlitic carbon steel wires by taking the material flow into account while doing a simulation with the Taylor theory. This can already be done by a simple method, which is based on the ideas of visioplasticity and for which no sophisticated software is required. The predicted textures have in principle a monoclinic sample symmetry, but the deviation from orthorombic sample symmetry is not large. This corresponds to experimental observations. Several cyclic texture components were predicted, of which one at least has been observed experimentally.

In principle, the method would make a parametric study (influence of die angle and wire drawing reduction per pass) possible. There is however a need for a better validation of the model using experimental results. Since in the present study, only very thin wires have been studied for which it is extremely difficult to measure the surface texture, no such validation exists as yet.

REFERENCES

1) Van Houtte, P., Watté, P., Aernoudt, E., Gil Sevillano, J., Martin Meisozo, A., Van Raemdonck, W., Lefever, I.: in Modelling of Plastic Deformation and Its Engineering Applications (Proc. 13th Risø International Symposium on Material Science), S.I. Andersen et al., Eds., Risø National Laboratory, Roskilde, Denmark (1992), pp. 485-490.

2) Langouche, F., Aernoudt, E., Van Houtte, P.: J. Appl. Cryst., 1989, 22, 533.

3) Medrano, R.E., Gillis, P.P.: J. of Strain Analysis., 1972, 7, 170.

4) Johnson, W., Sowerby, R., Venter, R.D.: Plane-Strain Slip Line Fields for Metal-Deformation Processes, Pergamon Press, Oxford 1982.

5) Van Houtte, P.: Metall. , 1978, 32, 144.

6) Van Houtte, P.: Textures and Microstructures, 1988, 8-9, 313.

7) Watté, P., Van Houtte, P., Aernoudt, E., Gil Sevillano, J., Lefever, I., Van Raemdonck, W.: "The Work Hardening of Pearlite During Wire Drawing", this conference, 1993.

8) Kanetsuki, Y., Ogawa, R.: Proc. Sixth Intntl. Conf. on Textures of Materials (ICOTOM 6), vol. 2, S. Nagashima, ed., The Iron and Steel Institute of Japan, Tokyo (1981), pp. 1127-1136.

Materials Science Forum Vol. 157-162 (1994) pp. 1887-1894

MODELING MICROSTRUCTURAL EVOLUTION OF MULTIPLE TEXTURE COMPONENTS DURING RECRYSTALLIZATION

R.A. Vandermeer [1] and D. Juul Jensen [2]

[1] Physical Metallurgy Branch Code 6320.1, Naval Research Laboratory, 4555 Overlook Ave., SW, Washington DC 20375-5343, USA

[2] Materials Department, Risø National Laboratory, DK-4000 Roskilde, Denmark

Keywords: Recrystallization Textures, Modeling, Microstructural Evolution, Aluminum

ABSTRACT -- The microstructural evolution of multiple texture components during recrystallization of heavily cold-rolled commercial aluminum was studied experimentally using stereological point and lineal measurements of microstructural properties in combination with EBSP analysis for orientation determinations. Geometrical models were formulated based on a nucleation and growth, microstructural path methodology (MPM) in an effort to characterize recrystallization of each texture component as well as the whole. A uniform impingement MPM model failed to characterize recrystallization in this material. A successful linear / uniform, two stage impingement MPM model was developed in which early grain impingement in one dimension is followed much later by general impingement in the other two grain dimensions as recrystallization goes to completion; the grains essentially grew two dimensionally after the first stage impingement.

INTRODUCTION -- Many metals develop multi-component crystallographic textures during recrystallization annealing. This is typical of heavily deformed, polycrystalline metals. On recrystallization, new grains belonging to different texture components may nucleate and grow according to different mechanisms. In that case model development based on a single component premise inevitably will give the wrong result.

Various methods are used to model recrystallization. Statistically-based, geometric models, whose genesis goes back to the work of Kolmogorov [1], Johnson and Mehl [2] and Avrami [3] (KJMA), formulate analytical expressions for the changes in microstructural properties. The latest versions of these are the microstructural path models (MPM), e.g.[4,5]. Other types of models are based on computer simulations which utilize either discrete lattice Monte Carlo techniques, e.g.[6], a cellular automata approach, e.g.[7] or a geometric scheme akin to the analytical method, e.g.[8]. Each of these modeling approaches has strengths and weaknesses.

The geometrical models seek to account for impingement by employing statistical mathematics. The KJMA models are formulated based on the assumption that the recrystallized grains are nucleated at random, thus, impingement occurs ***uniformly*** throughout the volume. The method allows analytical expressions for V_V, the volume fraction recrystallized, as a function of annealing time to be derived subject to additional assumptions about nucleation and growth behavior.

The MPM improves and extends the KJMA approach. This is made possible by employing additional microstructural properties in the analysis and, if need be, relaxing the uniform impingement constraint. Thus, in addition to V_V, the primary framework of MPM models incorporate the microstructural property, S_V, the grain boundary area per unit volume separating a recrystallized grain from the deformed matrix [9, 10]. Newer MPM models [11] focus on treating cases of more complex grain arrangements (linear and planar colonies of grains). The MPM models are designed to be combined and integrated with experimental measurements of microstructural properties, thus, allowing calculation of the nucleation and growth characteristics of recrystallization.

Only a few efforts at modeling multiple component recrystallization have been attempted [12, 13, 14]. In this paper we explore the possibility of applying MPM to construct a model for multi-component recrystallization in commercial aluminum. Two important objectives must merge to allow modeling with MPM to be successful. First, meaningful experimental measurements of the microstructural properties V_v (i) and S_v (i) must be obtained for each defined recrystallization texture component. (Here and throughout the paper, i, refers to a specific texture component.) Second, realistic geometrical models based on nucleation and growth premises, which take into account the mutuality of impingement between components and which can be expressed mathematically, must be developed. In the case of the first objective, the recently developed back-scattered electron pattern (EBSP) analysis [15] employed in conjunction with stereological measurements provides the means to determine the necessary microstructural properties using a scanning electron microscope (SEM) [16]. The EBSP technique allows the rapid orientation determination of recrystallized regions $\approx$ 1 μm and larger. The location of the grain boundaries of recrystallized grains which is needed for the stereological measurements is established by virtue of the fact that EBSP's are diffuse with meandering poles in deformed regions but sharp and uniform in recrystallized grains.

It is the purpose of the present paper to demonstrate by way of summary that the second objective noted above can also be realized. Recrystallization data obtained from commercially pure, heavily-rolled aluminum and isothermally annealed will be analyzed with the MPM until a reasonable model, consistent with as many microstructural observations as possible, is developed.

EXPERIMENTAL DETAILS

Material -- Commercially pure aluminum (99.6 wt %) containing 0.33 wt % iron and 0.09 wt % silicon as principal impurities was used in this study. The aluminum contained about 0.005 volume fraction of large (0.2-7 μm) $FeAl_3$ precipitate particles. The final deformation was by cold rolling to a 90% reduction in thickness. Recrystallization anneals were carried out at 253 °C for periods of time ranging from 0.9 ks to 14.4 ks. Additional details regarding specimen preparation and heat treatment are given in Reference [17].

New grain orientation measurements and quantitative microscopy -- The partially recrystallized specimens were mechanically, chemically and electrolytically polished and mounted in a JEOL 840 SEM operated in the electron back scatter mode with a spatial resolution of $\approx$ 1 μm. Each specimen was examined while linearly traversing the electron beam across the specimen. Recrystallized grain orientations and microstructural property measurements were determined simultaneously by combining the EBSP analysis and point and lineal counting techniques. Deformed regions were identified by diffuse EBSP while the recrystallized grains produced sharp patterns. For recrystallized grains encountered during each traverse, the grain orientations were cataloged into one of three orientation classes. ***Cube*** grains were those whose observed orientations fell within 15 degrees of the cube orientation, i.e. $\{100\}\langle 001\rangle$, while ***rolling*** grains were those with orientations within 15 degrees of the rolling components, i.e. $\{110\}\langle 112\rangle$, $\{123\}\langle 634\rangle$ or $\{112\}\langle 111\rangle$. All other recrystallized grains comprised the ***random*** component. The individual intercept-free grain lengths (chord lengths), λ_i were measured for each recrystallized grain encountered as well as for the deformed regions between them by noting the vernier positions of the stage where the EBSP changed during each traverse. Also recorded was N_i, the number of grains of the ith orientation class intersected during traverse and the number of times, n_i, a transition between a deformed region and a recrystallized grain of the ith class occurred and conversely. If L_λ is the total line length of the traverse, the following microstructural properties could be calculated from this data: $V_v(i) = \Sigma \lambda_i / L_\lambda$, the ith component volume fraction recrystallized; $S_v(i) = 2 n_i / L_\lambda$, the ith component grain boundary area per unit volume (deformed / recrystallized); $N_\lambda(i) = N_i / L_\lambda$, the number of ith component grains per unit length of test line; and $\langle\lambda\rangle_i = \Sigma \lambda_i / N_i$, the mean intercept-free grain length of ith component grains.

MODEL DEVELOPMENT

Uniform impingement model -- The first attempt at developing an MPM recrystallization model was to invoke the uniform impingement criteria. This was done by adapting and extending the shape preserved nucleation and growth transformation model of Vandermeer, Masumura and Rath [10] to the case of multi-texture-component recrystallization.

In the model the recrystallized grains are considered to impinge uniformly on one another irregardless of which texture component is encountered (mutuality of impingement). Thus, for the ith -texture component, the impingement equations are defined as

$$\frac{d\,V_v(i)}{d\,V_{vex}(i)} = 1 - V_v \quad (1) \qquad \text{and} \qquad \frac{S_v(i)}{S_{vex}(i)} = 1 - V_v \quad . \ (2)$$

The quantities with the subscript vex refer to the corresponding volume fractions and grain boundary areas as imagined in the abstraction of ***extended*** space as discussed by KJMA, i.e. grains are allowed to grow through one another without regard to impingement etc. The V_v with no i-index refers to the total amount of recrystallization. The recrystallized grains are represented as spheroids that do not change eccentricity during growth. Thus, for the ith component at any annealing time, t, the semi-major axis , $\mathbf{a}_i$, of a grain nucleated at time, τ, and the nucleation rate are given by

$$\mathbf{a}_i = \mathbf{G}_i \cdot (t - \tau)^{\alpha_i} \quad (3) \qquad \text{and} \qquad \dot{\mathbf{N}}_i = \mathbf{N}_i^* \cdot \tau^{\delta_i - 1} \quad (4)$$

respectively where the $\mathbf{N}_i^*$,$\mathbf{G}_i$, α_i and δ_i are experimentally determined model constants. The working equations for $V_v(i)$ and $S_v(i)$ which are to be fitted to the experimental data are derived using the methods described in Reference [19] or [20] . The results for the ith texture component are

$$\ln \frac{1}{1-V_v(i)} = \int_0^t \left(\frac{1-V_v}{1-V_v(i)} \right) \cdot k_i \cdot B_i \cdot t^{k_i - 1} \cdot dt \quad (5) \qquad \text{and} \qquad \frac{S_v(i)}{(1 - V_v)} = K_i \cdot t^{m_i} \quad . \ (6)$$

where B_i, K_i, k_i and m_i are a new set of constants related to $\mathbf{G}_i$, $\mathbf{N}_i^*$, α_i and δ_i . There will be one set of equations like Eq. 3 to 6 for each recrystallization texture component considered. The model parameters α_i and δ_i are related to k_i and m_i according to the formulas [4,5,10]

$$\alpha_i = k_i - m_i \qquad \text{and} \qquad \delta_i = 3 \cdot m_i - 2 \cdot k_i$$

A general curve fitting procedure was applied to the experimental data allowing the constants B_i , K_i , k_i and m_i for each of the texture components to be determined. (To evaluate the integral in Eq. 5 , the time dependence of the function between the parentheses was determined experimentally by fitting the values to a polynomial.) Some of the calculated model parameters and their standard deviations are tabulated in Table 1. The rather small standard deviations are indicative that the data apparently fit the equations rather well. However, when these numbers are scrutinized carefully,

Table 1.
Experimentally Determined Model Constants:
The Uniform Impingement Multi-component Model

i-component	α_i	δ_i	$q_i\ (= m_i / k_i)$
Random (1)	0.55 ± 0.17	-0.61 ± 0.45	0.47 ± 0.13
Rolling (2)	0.74 ± 0.07	-0.48 ± 0.17	0.57 ± 0.03
Cube (3)	0.88 ± 0.06	-1.01 ± 0.14	0.46 ± 0.03

it must be concluded that, on the contrary, the model is inappropriate. The simple reason is that the δ_i ***are forbidden to have values less than zero*** [4,5,10]. This point is brought home in another way by examining the parameter q_i ($= m_i/ k_i$). According to the work by Vandermeer, Masumura and Rath [10], uniform impingement models require $q_i \geq 0.67$. From Table 1, it may be seen that the q_i values are nearer to 0.5 than to 0.67. They remain outside the limits of such a model even when the experimental errors in the parameters are considered. Relaxing the shape preservation re-striction and using the uniform impingement equations of Vandermeer and Rath [18] for shape changing spheroids, does not provide a resolution of the difficulty and inconsistencies still prevail. It is clear from such behavior that the experimentally measured microstructural properties do not support a uniform impingement recrystallization model. Thus, these parameters are at best, merely an empirical representation of the data that do not represent a relevant model.

Linear / uniform impingement model -- How a more realistic MPM model for recrystallization in heavily deformed commercial aluminum may be devised is provided by considering the q_i parameters in the light of the predictions of a two stage impingement model having linear / uniform impingement characteristics similar to a grain edge nucleation model developed by Vandermeer and Masumura [11]. They showed that when, ***early in the recrystallization process,*** the new grains formed in linear arrays and impinged in one dimension, but not in the other two, a q_i near one-half should be expected. The partially recrystallized microstructure for such a case can be envisioned geometrically as many "string-of-bead" (each bead is a recrystallized grain) segments or colonies embedded randomly in a directional sense in the volume of deformed material. Impingement with other recrystallized grains ("beads") is, therefore, a two stage process. Early impingement in one dimension (linearly along the "string") formint the colony is followed much later by general impingement in the other two dimensions between the colonies as recrystallization goes to completion. The grains essentially grow two dimensionally after the first stage impingement. The "string" directions are considered to be randomly distributed, i.e. evidence of colonies will be visible on any sample section examined microstructurally. The latter impingement (between colonies), therefore, may be treated as uniform impingement when developing the mathematical model.

Just such a model adapted for the commercial aluminum can explain many of the observed microstructural features. One microstructural ramification of such a model is that the recrystallized grains should show a tendency to be clustered in colonies even at early stages of the process. Sketches of the recrystallized grain distributions on the plane of polish revealed that at volume fractions below 0.10, such a tendency was present. The distribution of the $FeAl_3$ precipitate particles around which many recrystallized grains were nucleated, contributed to the clustering of recrystallized grains.

The characteristics of the linear / uniform impingement microstructural model which was decided upon for commercial aluminum is based on the following premises: 1) All the recrystallization "nuclei" preexist and begin growing at time, $\tau = 0$ (This is suggested by the uniform impingement model calculations , i.e. low k_i and m_i values and the observation that $\alpha_i \approx m_i \approx k_i / 2 \approx - \delta_i$). 2) The number of recrystallization "nuclei" of each component, $\mathbf{N_i}$, are distributed randomly in approximately linear arrays ("strings of beads" analogy) where the total array length per unit volume of material is, L. The array length is a somewhat ficticious quantity as it has not been possible to associate it with a microstructural feature of the deformed material as was the case of grain edge nucleation of recrystallization of moderately deformed high purity aluminum [19]. 3) The component grains will be assumed to be shape-preserved, prolate spheroids of minor semi-axis, $\mathbf{b}_i$, which grow with annealing time, t , according to Eq. 3 with $\tau = 0$ and $\mathbf{b_i}$ substituted for $\mathbf{a_i}$. The major semi-axis of each prolate spheroid, $\mathbf{a_i}$, is assumed to be parallel to L and the eccentricity, ε_i, is equal to $[1 - (\mathbf{b}_i / \mathbf{a}_i)^2]^{1/2}$. 4) The migration rate of an i-component grain in the $\mathbf{b_i}$ direction is given as a function of time by

$$v_i = \alpha_i \cdot \mathbf{G_i} \cdot t^{\alpha_i - 1}$$

The mathematical procedure for estimating microstructural properties for the model, i.e. the $V_v(i)$ and $S_v(i)$, is a modification of the method described by Vandermeer and Masumura [11]. Details of the derivations will be presented elsewhere.

Application of the linear / uniform, double impingement model to experiment -- Trial and error methods were employed until a reasonable fit of the data to a model was achieved. To simplify the procedure, it was assumed throughout that $\alpha_{\mathbf{cube}} = \alpha_{\mathbf{rolling}} = \alpha_{\mathbf{random}} = \alpha$. This is not unreasonable in view of an earlier analysis of this data [12] in which the Cahn-Hagel growth rates [20] of the three texture components were all estimated to be proportional to $t^{-0.3}$, i. e. $\alpha = 0.7$. In the present case, $\alpha = 0.6$ gave a better fit than an α of 0.7. Initially, it was also assumed that all the eccentricities (ε_i) were zero, i.e. the grains were spherical. In the end it was only necessary to require the cube-component grains to be spheroids with an eccentricity of 0.57, i.e. an aspect ratio of about 5/6. The model that was selected out of many that were tried, and the data fit that was achieved, is illustrated in Figs. 1 and 2. In these plots are shown the time dependence of the

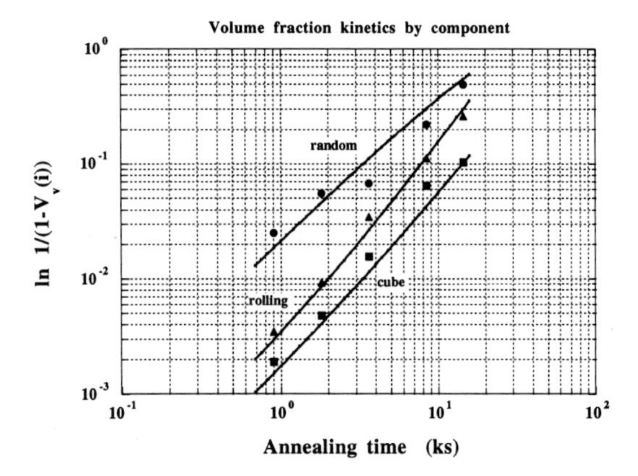

Fig. 1 Fit of component geometrical path model with experiment -- the volume fraction kinetic functions.

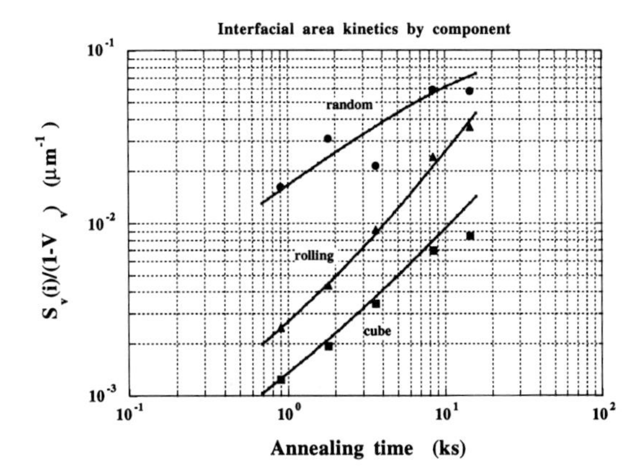

Fig. 2 Fit of component geometrical path model with experiment -- the inerfacial area kinetic functions.

volume fraction recrystallized (Fig. 1) and the grain boundary area (Fig. 2) . Table 2 lists the model constants that were necessary to generate this match.

Discussion of the Linear / uniform impingement model -- Several noteworthy observations should be made relative to the nucleation and growth characteristics of recrystallization in heavily worked, commercial aluminum as deduced from the model: 1) There is a preference for the re-crystallized grains to have random orientations - the random grains are 8 times more prevalent than

Table 2.

Linear / Uniform Impingement Model Constants			
Texture Component	Random	Rolling	Cube
α_i	0.6	0.6	0.6
G_i ($\mu m \cdot ks^{-0.6}$)	3.42	3.53	3.50 (b-direction) 4.24 (a-direction)
N_i (μm^{-1})	0.328	0.041	0.021
ε_i	0	0	0.57
L (μm^{-2})		0.0010	

rolling grains and 16 times more so than cube grains; 2) Rolling grains and cube grains have a growth rate advantage over random grains but only by about three percent; 3) This growth rate advantage where random grains tend to be "squeezed out", is the reason why the cube and rolling volume fractions tend to "catch up" with the random volume fraction towards the end of recrystallization even though there are many fewer of these grains at the beginning and the number of grains does not change as time goes on; 3) All the growth rates decrease with time; 5) Initially, at least, the recrystallized grains have shapes which, on average, resemble cylinders where the early stage impingement is in the cylinder length direction; 6) The number of recrystallized grains per unit volume for the ith-texture component is given by the product, $L \cdot \mathbf{N_i}$; and 7) Statistically speaking the recrystallized grain "nuclei" on average are located along the "string" direction in each colony every $(\Sigma \mathbf{N_i})^{-1}$ length per grain or 2.56 μm.

The suitability of the linear / uniform, double impingement model can be scrutinized even further. The modeling outcome that the recrystallized grains are approximately cylindrical in shape allows the possibility of calculating from the various model constants, $N_\lambda(i)$, the number of i-component grains intersecting a unit length of test line. A direct comparison with the experimentally measured values of these quantities is, therefore, possible. Fullman [21] derived a number of relationships between average dimensions measured on a polished cross section and the spatial dimensions of grains dispersed randomly as uniform cylinders. In terms of the present microstructural model, the equation relating N_λ, the number of cylinders intersected per unit length of test line, and N_v the number of cylinders embedded in a unit volume, when written in terms of the model constants, becomes for the ith-texture component approximately

$$N_\lambda(i) = \frac{\pi}{2} \cdot L \cdot \mathbf{N_i} \cdot \left(\mathbf{G_i^2} \cdot t^{2\alpha_i} + \langle h(i) \rangle \cdot \mathbf{G_i} \cdot t^{\alpha_i} \right) \quad (7)$$

where <h(i)> is the mean height of the ith-texture component grains and $N_v(i) = L \cdot \mathbf{N_i}$. This equation is valid only for low volume fractions of recrystallization - before significant impingement of the cylinders in the radial directions. The quantity <h(i)> which because of early impingement should be approximately constant, may be evaluated from the model constants. Using Eq. (7) the ratios, $N_\lambda(i) / \Sigma N_\lambda(i)$, were formed and calculated. The model calculations are compared with experimental estimates for the two earliest annealing times in Table 3. The agreement is excellent.

One last comparison of the model with experiment pertains to the grain size of each recrystallized texture component. Experimentally the grain size may be taken as $\langle \lambda \rangle_i$, the mean intercept free grain length (chord length). Except for the cube grains at the largest grain sizes, there was fair agreement between the model and the experiment. The model in its present form is unable to account for the very large sizes of the cube grains as recrystallization approaches completion.

Table 3.

$N_\lambda(i) / \Sigma N_\lambda(i)$ - Comparison of Model with Experiment			
TEXTURE COMPONENT	Vv = 0.03	Vv = 0.068	MODEL
Random	0.82	0.82	0.83
Rolling	0.12	0.13	0.11
Cube	0.06	0.05	0.06

It should be noted that in MPM modeling, there is no absolute assurance that the model constants arrived at by matching the model equations with experimental microstructural properties are a unique set. The model presented above represents only one of several that were deduced; each yielded plots very similar to Figs. 1 and 2. However, the others when inspected carefully could not explain the totality of experimental observations as well as this one.

Acknowledgements --- The authors wish to express their grateful appreciation to Drs. N. Hansen and R. A. Masumura for helpful discussions. A portion of this work was performed at the Naval Research Laboratory under the sponsorship of the Office of Naval Research of the US Department of Navy whose support is gratefully acknowledged.

REFERENCES

1. Kolmogorov, A.N.: *Izv.Akad.Nauk.USSR-Ser.Matemat.*, (1937), **1**(3), pp 355-59.
2. Johnson, W.A. and R.F.Mehl: *Trans.AIME*, (1939), **135**, pp 416-30.
3. Avrami, M.: *J.Chem.Phys.*, (1939), **7**, pp 1103-09; *ibid.*, (1940), **8**, pp 212-24; *ibid.*, , (1941), **9**, pp 177-84.
4. Vandermeer, R. A. and B. B. Rath:, *Met Trans A,* (1989), **20A**, p. 391.
5. Vandermeer, R. A. and B.B. Rath: in *Materials Architecture*, edited by J. B. Bilde-Sørensen, N. Hansen, D. Juul Jensen, T. Leffers, H. Lilholt and O. B. Pedersen, Risø National Laboratory, Roskilde, Denmark, (1989), p.589.
6. Srolovitz, D. J. , G. S. Grest and M. P. Anderson: *Acta Metall. et Mater.*, (1986), **34**, p. 1833.
7. Hesselbarth, H. W. and I. R. Gobel: *Acta Metall. et Mater.*, (1991) **39**, p. 2135.
8. Mahin, K. W. , K. Hanson, and J. W. Morris: *Acta Metall. et Mater.*, (1986), **34,** p. 981.
9. Vandermeer, R.A. : *Scripta Metall. et Mater.*, (1992), **27**, pp 1563-1568.
10. Vandermeer, R. A. , R. A. Masumura and B.B. Rath: *Acta Metall. et Mater.*, (1991), **39,** p. 383.
11. Vandermeer, R. A. and R. A. Masumura: *Acta Metall. et Mater.*, (1992), **40,** p. 877.
12. Juul Jensen, D. : *Scripta Metall. et Mater.,* (1992), **27,** p. 1551.
13. Tavernier, Ph. and J. A. Szpunar: *Acta Metall. et Mater.*, (1991), **39,** p. 549; *ibid*, (1991), **39,** p. 557.
14. Juul Jensen, D. and N. Hansen: in *Recrystallization '90,* edited by T. Chandra, TMS, Warrendale, PA, (1990), p.661.
15. Schmidt, N. H. , J. B. Bilde-Sørensen and D. Juul Jensen: *Scanning Microscopy* , (1991), **5,** p. 637.
16 Juul Jensen, D. : *Scripta Metall. et Mater.,* (1992), **27,** p. 533.
17. Juul Jensen, D. , N. Hansen and F. J. Humphreys: *Acta Metall. et Mater.*, (1985), **33,** p. 2155.
18. Vandermeer, R. A. and B. B. Rath: in *Recrystallization '90,* edited by T. Chandra, TMS, Warrendale, PA, (1990), p. 49.
19. Vandermeer, R. A. and P. Gordon: *Trans. TMS-AIME,* (1959), **215**, p. 577.
20. Cahn, J. W. and W. C. Hagel: in *Decomposition of Austenite by Diffusional Processes,* edited by Z.D.Zackey and H.I.Aaronson, Interscience Publ., NY, (1960), p.131.
21. Fullman, R. L. : Trans AIME , (1953), **197**, p.1267.

Materials Science Forum Vol. 157-162 (1994) pp. 1895-1900

ON THE THEORY OF COMPROMISE TEXTURE

H.E. Vatne [1], T.O. Sætre [2] and E. Nes [1]

[1] Metallurgisk Institutt, NTH, Alfred Getz vei 2b, N-7034 Trondheim, Norway

[2] Hydro Aluminium a.s., R&D Centre, Karmøy, N-4265 Håvik, Norway

Keywords: Compromise Texture, Oriented Growth, Local Mobility Variations, Computer Simulations

ABSTRACT

The influence of local mobility variations on a growing recrystallisation front has been simulated. The growing grain is assumed to be exposed to a variety of deformation texture components. On the assumption that texture has an effect on grain boundary mobility, the growth rate will vary along the transformation front. The shape and average mobility of the front have been calculated as a function of local mobilities and volume fractions of the surrounding deformation texture components. Both the local growth behaviour and the shape of the recrystallised grain were found to be significantly influenced; the extent depending on the mobility variations around the growing grain.

INTRODUCTION

Over the years, the concepts of oriented growth and compromise texture have been extensively used in order to explain the development of recrystallisation textures in metals. The basic idea of oriented growth is that out of a broad spectrum of nuclei formed, those with the best growth conditions with respect to the surrounding deformed matrix will determine the recrystallisation texture. In aluminium single crystals, for example, grains with a 40°<111> orientation relationship to the neighbouring grains were found to possess an especially high growth rate [1,2]. Applying this to polycrystalline metals one has to consider that the grains have to grow not into one orientation only, but into several orientations corresponding to the different components of the deformation texture (cf. Fig. 1). Therefore, those grains which have the maximum growth orientation with respect to one deformation component do not necessarily consume the largest volume. Those with the largest volume fraction will be the ones with an approximate maximum growth orientation to several components of the rolling texture. The resulting recrystallisation texture from such a process is denoted as the compromise texture.

Although the ideas of oriented growth were already introduced by Beck in 1949 [3] and compromise texture by Beck and Hu in 1952 [4,5] and further developed by Bunge [6] and Lücke et al. [7], these concepts have up till now been applied on a qualitative level only.

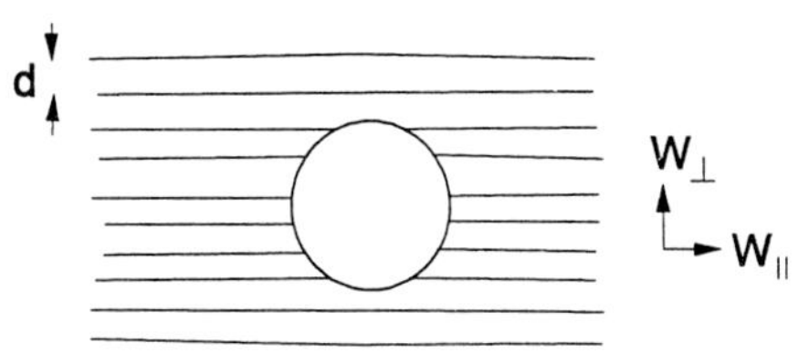

Fig. 1: Growing grain in heavily deformed structure, schematically

A first attempt to present an analytically based quantitative theory has recently been published by Bunge et al. [8]. Although there are several weaknesses concerning their description, it is a welcomed attempt to rise the understanding of oriented growth and compromise texture to a higher scientific and quantitative level. Their work is based on the idea that a growing grain which is exposed to a matrix of a multicomponent texture can be assigned an isotropic and average growth rate. No justification of this idealised assumption was given.

The objective of the present paper is to analyse from first principles the growth behaviour of a grain characterised by a boundary with a mobility which varies from component to component around its periphery. This should be regarded as a step towards a quantitative formulation of the oriented growth/compromise texture phenomena.

DESCRIPTION OF THE PROBLEM

The transformation aspect

Like in the work of Bunge et al. [8] the following basic assumption are made:

i) Nucleation of new grains is site saturated, i.e. all nuclei are activated at a very early stage of the transformation, and
ii) the growth rate of new grains is isotropic and constant

It then follows from classical transformation theory that

$$dV = (1-V)\, dV_{ext} \tag{1}$$

where V is the transformed volume and V_{ext} is the extended volume. In the case of a multi-component system, i.e. a system consisting of **i** different recrystallisation texture components growing into the same matrix, with each component satisfying the two basic assumptions, Doherty [9] has suggested that

$$dV_i = (1-V)\, dV_{i,ext} \tag{2}$$

Kolmogorov [10] worked out the mathematical proof of Eq. (1) and Doherty simply assumed that (2) follows from (1). It seems to be a reasonable assumption, however, it is tested here by computer simulation (the "Avrami machine" [11,12]), see below.

Combining Eq. (1) and (2) gives the volume fraction of a recrystallisation component **i**

$$f(i) = \frac{V_i}{V} = \frac{N_i W_i^3}{\sum_{j=1}^{n} N_j W_j^3} \tag{3}$$

where N_i is the number of nuclei, n the total number of components and W_i the growth rate of component i. Note that each fraction is time invariant during the transformation. An independent test of the validity of Doherty's assumption (Eq. 2) can be obtained by computer simulation. A 4-component system was chosen with the starting conditions as given in Table 1. As shown in Fig. 2 the volume fractions are indeed time invariant throughout the transformation in accordance with the prediction of Eq. (2). Further, these computer-generated volume fractions are also identical to those obtained using Eq. (3), which can be regarded as a justification of the assumption by Doherty, Eq. (2).

Component	1	2	3	4
Growth rate W[1]	887	700	640	250
No. of nuclei N[1]	100	120	120	660

1) Arbitrary units

Table 1: Parameters for simulations with the "Avrami-machine"

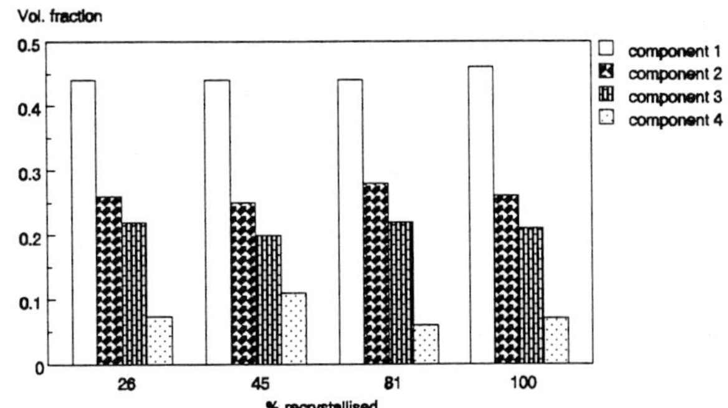

Fig. 2: Volume fractions for the different types of grains calculated by the "Avrami machine"

Bunge et al. [8] used an alternative treatment of the transformation problem by introducing a transformation time t_R. This is however an unphysical and undefined quantity which can only be obtained by applying a normalisation treatment which implies forcing the extended volume at $t=t_R$ to be unity. Despite their unconventional treatment, Bunge and co-workers obtained the same expression for the volume fractions as Eq. (3), but the analysis above is based on the classical treatment of transformation kinetics, and formally correct.

The growth aspect

In the following the growth aspect is treated. Plane strain deformation from an equiaxed grain structure will result in a laminated deformation structure as illustrated schematically in Fig. 1. In the following the growth of a recrystallised grain into such a multicomponent deformation texture will be modelled. While Bunge et al. [8] treated this as an isotropic growth of the nuclei, the growth will in the present work be studied in more detail in two directions; growth parallel ($W_\parallel$) and perpendicular ($W_\perp$) to the deformed grains.

The growing grain is exposed to a matrix of n texture components, each with a volume fraction f_i and mobility M_i. In the perpendicular direction the expression for the growth rate is trivial, and is the same expression as that used by Bunge et al. as the governing growth equation for growth in all directions;

$$W_\perp = \frac{1}{\sum_{i=1}^{n} \frac{f_i}{W_{\perp,i}}} \tag{4}$$

where $W_{\perp,i}=M_i P_D$. Considering now the special case of two deformation texture components with mobilities M_1 and M_2 and equal volume fractions $f_1=f_2=0.5$, one gets:

$$W_\perp = \frac{2P_D}{1/M_1+1/M_2} \tag{5}$$

The parallel direction is, however, non-trivial. A simple model for the migration of the boundary in the $W_\parallel$-direction with two components is seen in Fig. 3. The recrystallised grain is growing into the deformed matrix; direction indicated by the arrow. The two texture components result in two mobilities, M_1 and M_2, and hence two rates of migration. The main effect of the old grain boundary on the migration of the recrystallisation front is to enforce local equilibrium at the triple boundary junction at the ends of the old boundary and the two parts of the migrating boundary with different mobilities. The problem is only solvable if this boundary is assumed to have a very low mobility, M_3. As a further simplification, the boundary is assumed to have only one axis of curvature, i.e. the z-axis in Fig. 1. The boundary velocity of each texture component of the boundary in an x-y coordinate system is given by

$$W_i = M_i\,(P_D - \gamma\kappa) = M_i\,\left(P_D - \gamma\frac{\partial^2 y/\partial x^2}{[1+(\partial y/\partial x)^2]^{3/2}}\right) \qquad (6)$$

where γ is the specific grain boundary energy (assumed to be a constant), κ is the curvature and P_D the driving pressure due to the stored energy in the deformed matrix. Decomposed in the x- and y-direction Eq. (6) yields $W_{x,i}=W_i\cdot[1+(\delta x/\delta y)^2]^{-1/2}$ and $W_{y,i}=W_i\cdot[1+(\delta y/\delta x)^2]^{-1/2}$. For symmetry reasons $W_{\parallel,x,i}(x=d/2)=W_{\parallel,x,i}(x=3d/2)=0$, and it follows that the shape of the boundary is obtained by solving the differential equation for $d/2 \le x \le 3d/2$. However, this equation cannot be solved analytically, but numerically by computer simulations.

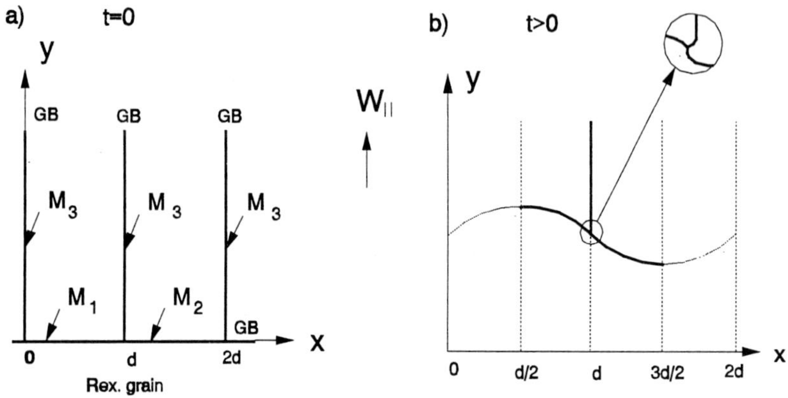

Fig. 3: Schematic model of the growth of a recrystallised grain in the $W_\parallel$-direction: a) starting structure for simulations b) front at time t

COMPUTER SIMULATIONS

The model shown in Fig. 3 was studied by employing a program developed by Sætre [13,14]. The program approximates the boundaries by discrete points and calculates the curvatures and migration rates by using cubic spline mathematics. The equilibrium conditions at the triple line junction (Fig. 3b) are maintained at 3·120° by employing a triangle construction. However, although the equilibrium conditions are met at the junction, the boundaries migrate freely outside the small triangle. This allows a strong curvature to develop at the edges of the triangle which affects the rate of migration. If the part of a boundary with increased curvature cannot reduce the curvature by a higher rate of migration than the average rate, a deviation from equilibrium conditions appears to take place, although 3·120° are maintained inside the triangle.

The driving pressure P_D is assumed to be determined by the stored dislocation substructure, P_D^{def}, and the energy stored in the grain structure, P_D^{GB}. In the $W_\parallel$-direction P_D^{GB} is taken care of through the term $\gamma\kappa$, cf. Eq. (6). However, in the $W_\perp$-direction where the transformation front is planar, the term $P_D^{GB} = \gamma_{GB}/d$ must be added. In pure metals deformed to large rolling reductions (ε>1) the grain boundary term is typically one tenth of P_D^{def}, and was here chosen to be P_D^{GB} =1.2 (while P_D^{def}=20).

The computer simulations were carried out for equal widths of the two texture components, cf. Fig. 4, and with different widths of the two components, cf. Fig. 5. In Fig. 4a) the initial structure is seen, and the following figures b-f) are plotted at identical time revealing the steady state shape of the boundary. The differences in shapes and lengths of the migration front are due to the input parameters (variation in mobility), cf. Table 2. Figure 5 a) and c) show the initial structures where the input parameters of these two cases were identical, cf. Table 2. Fig. 5b) and d) are plotted at identical time.

Figure	M_1	M_2	M_3	Width	P_D	γ
4b)	1	1	0.001	1/2-1/2	20	1
4c)	1.5	1	0.001	1/2-1/2	20	1
4d)	2	1	0.001	1/2-1/2	20	1
4e)	4	1	0.001	1/2-1/2	20	1
4f)	6	1	0.001	1/2-1/2	20	1
5b)	2	1	0.001	1/3-2/3	20	1
5d)	2	1	0.001	2/3-1/3	20	1

Table 2: Input parameters for the simulations of Figs. 4 and 5.

Figure	$W_\parallel$	$W_\perp$	$W_\parallel/W_\perp$
4b)	21.2	21.2	1.0
4c)	29.7	25.4	1.2
4d)	39.2	28.3	1.4
4e)	77.9	33.9	2.3
4f)	117	36.3	3.2
5b)	38.7	31.8	1.2
5d)	39.4	25.4	1.6

Table 3: Rates of migration of Fig. 4 and 5

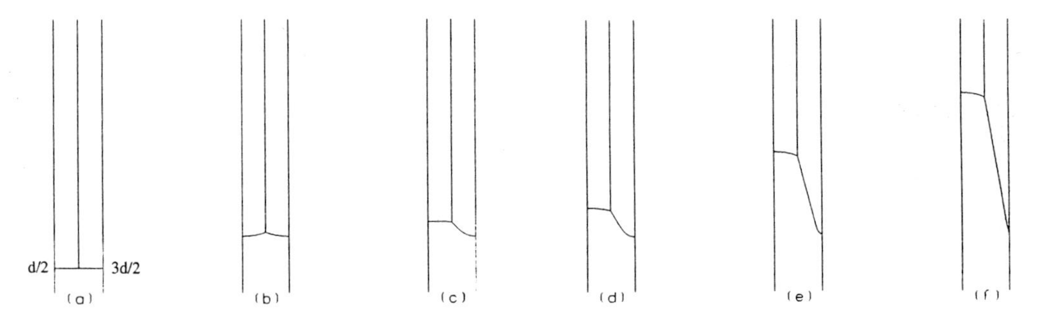

Fig. 4: Computer simulations of grain boundary migration into a deformed matrix with two texture components separated by an old grain boundary. a) initial position, b-f) plotted at identical time, cf. Table 2.

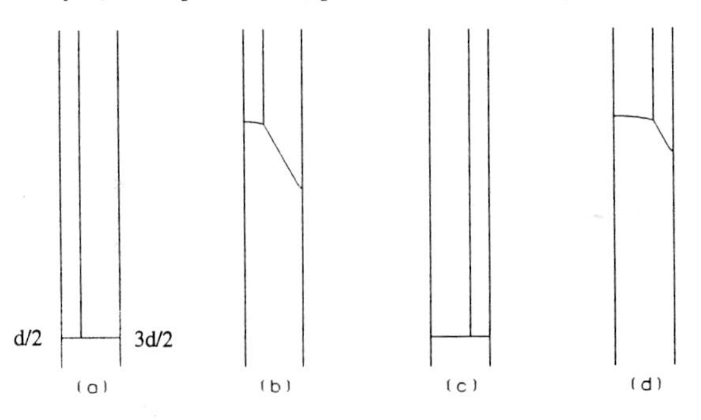

Fig. 5: The effect of variations in the width of the texture components. a) and c): initial positions. b) and d) are plotted at identical time. Input parameters were identical, cf. Table 2.

In all cases, the rates entered gradually into a steady state with constant rates of migration, values given in Table 3. It is seen that the width of the components is of minor importance in the parallel direction when the mobilities are of the same order of magnitude. Table 3 also gives the calculated rates in the perpendicular direction, assuming a planar front. It is seen that the two directions of growth rates, $W_\parallel$ and $W_\perp$, deviate more strongly the larger the difference between the mobilities of the two texture components. It is interesting to note that the steady state migration rate ($W_\parallel$) approaches nearly that of a boundary having the maximum mobility. Physically, this results from the effect of a curvature of the low-mobility part of the front favouring migration in the direction of growth. The results indicate that a recrystallised grain growing in a deformed matrix with two texture components which give rise to significant differences in the mobilities, will obtain an anisotropic (ellipsoidal) shape. The effect should, in significant cases, be observable in micrographs by elliptically shaped grains before impingement. In order to quantify the effect of the ellipsoidal growth, transformation rates were calculated using both an isotropic growth law, $W=W_\perp$ (Eq. 3), and taking the ellipsoidal shape, $W=W(W_\perp,W_\parallel)$, into account. Results of this are plotted in Fig. 6. The transformation can be significantly faster for the elliptical growth. It should be observed that a mobility ratio of only 2 in the elliptic case equals that of ratio 6 in the isotropic case. This demonstrates that care must be emphasised in using an isotropic growth law.

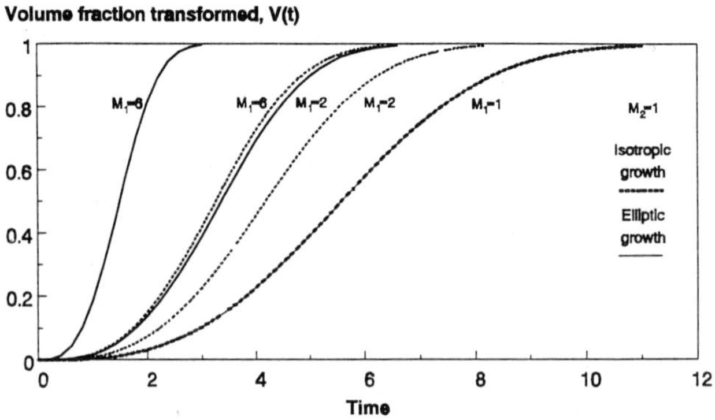

Fig. 6: Effects of isotropic vs. elliptic growth on transformation behaviour for various mobilities

Acknowledgement: The authors want to express their gratitude to NTNF (The Royal Norwegian Council for Scientific and Industrial Research) and NTH/SINTEF through the "Strong Point Center for Light Metals" for financial support

REFERENCES

1) K. Lücke, Can. Metall. Q., **13**, 261, (1974)
2) B. Liebman, K. Lücke, G. Masing, Z. Metallkde., **47**, 57, (1956)
3) P.A. Beck, J. appl. phys., **20**, 633, (1949)
4) P.A. Beck, Advances in Physics, Phil. Mag. Suppl., **3**, 245, (1945)
5) P. A. Beck, H. Hu, Trans AIME, **194**, 83, (1952)
6) H.J. Bunge, Rekristallisation metallischer Werkstoffe, ed. G.E.R. Schulze and H. Ringpfeil, VEB Deutscher Verlag für Grundstoffindustrie, Leipzig, 39, (1966)
7) U. Schmidt, K. Lücke, J. Pospiech, Proc. 4th Int. Conf. on Textures of Materials, London, 147, (1975)
8) U. Köhler, E. Dahlem-Klein, H. Klein, H.J. Bunge, Textures and Microstructures, **19**, 125, (1992)
9) R. Doherty, Scripta metal., **19**, 927, (1985)
10) A.N. Kolmogorov, Izv. Akad. Nauk., USSR Ser. Matemat., **1**, 355, (1937)
11) T.O. Sætre, O. Hunderi, E. Nes, Acta metall., **34**, 981, (1986)
12) T. Furu, K. Marthinsen, E. Nes, Materials Sc. & Tech., **6**, 1093, (1990)
13) T.O. Sætre and N. Ryum, J. Scient. Computing, **3**, 189, (1988)
14) T.O. Sætre and N. Ryum, Proc. TMS Fall Symposium of Coarsening and Grain Growth, Chicago, (1992)

Materials Science Forum Vol. 157-162 (1994) pp. 1901-1908

ANISOTROPIC FINITE-ELEMENT PREDICTION OF TEXTURE EVOLUTION IN MATERIAL FORMING

N. Wang [1], F.R. Hall [1], I. Pillinger [1], P. Hartley [1], C.E.N. Sturgess [1], P. De Wever [2], A. Van Bael [2], J. Winters [2] and P. Van Houtte [2]

[1] School of Manufacturing and Mechanical Engineering, University of Birmingham, Edgbaston, Birmingham, England

[2] Department of Metallurgy and Materials Engineering, Katholieke Universiteit Leuven, de Croylaan 2, B-3001 Leuven, Belgium

Keywords: Anisotropy, Finite Element, Texture Evolution, Cross-Rolling

ABSTRACT

An anisotropic finite-element (FE) program has been developed to predict the texture evolution of materials during forming processes by the incorporation of the Taylor-Bishop-Hill (T-B-H) model. The process of warm cross-rolling of low carbon steel is simulated and the results are presented. Comparisons are made with experimental results.

INTRODUCTION

Anisotropic properties due to crystallographic texture have a substantial effect on the behaviour of materials during some important material forming processes. FE methods have proved valuable in the analysis of material forming processes [1] to calculate the distributions of stress, strain, strain rate and temperature within the workpiece at every stage of the forming process. On the other hand, the T-B-H model is widely used to calculate deformation texture of polycrystalline materials [2-4].

In 1992, the authors have shown the feasibility of predicting texture evolution of material forming process by using an isotropic FE program incorporated with the T-B-H model to simulate a cross-rolling process of low carbon steel plate [8]. To achieve a fully anisotropic analysis of material flow and predicting the final state of material anisotropy, an anisotropic version of the FE program has been developed incorporating the T-B-H model to simulate anisotropic material forming and predict the evolution of crystallographic texture [6-8]. The plastic strain distribution and history, which are used by the T-B-H model, are obtained from the FE calculation on an element-by-element basis and this, in turn, enables a more realistic prediction of the deformation texture in cross-rolling than the previous isotropic simulations in assuming that the workpiece may have inhomogeneous texture and hence material flow characteristics throughout the workpiece, either as a result of the initial treatment or the forming process itself. This inhomogeneity has been ignored by many research workers in deformation texture calculation.

In this paper, a similar cross-rolling process to that presented in [8] is simulated by the anisotropic FE program and the predicted textures are presented and compared with experimental results.

THE EXPERIMENTS

The initial low carbon steel plates had a geometry of 350mm (original transverse direction) x 250mm (original rolling direction) x 25mm (thickness). Six texture samples were cut from one of the plates at chosen positions to guarantee that any inhomogeneity of the texture in the thickness and original transverse direction can be detected. No inhomogeneity is observed from the 6 measurements made at 15% and 50% depths of the 6 samples respectively, which is regarded as a result of hot rolling treatment at the temperature of 870-880°C by the manufacturer. The pole figures of the averaged initial textures are shown in figure 1. A weak BCC texture is observed from the pole figures. This texture was also used to produce the initial yield locus for the FE simulation.

For the cross-rolling tests of the plates, a starting temperature of 500°C was chosen to make sure that recrystallisation will not occur during the forming procedure. The radius of the roller was 230 mm. Three passes with the incremental reduction of 20%, 15% and 15% respectively were required to achieve the final reduction of 50%, all of which were in the original transverse direction of the plates. To obtain the intermediate as well as the final textures, 3 plates were rolled to the total reduction of 20%, 35% and 50% with one, two and three passes applied respectively. Considering the heat loss during the transfer from the furnace to the rolling mill, each plate was overheated to 530°C and reheated before each pass. After the rolling tests, no spread in the transverse direction (TD), which was the original rolling direction of the manufactured plates, was observed apart from at the two ends of the plates. Samples were cut from the centre of all 3 plates and textures were measured at a depth of 13% and 50% respectively through the plate thickness.

Meanwhile, torsion tests of the same material were carried out for a range of temperatures and strain rates to obtain the initial yield stress and the stress-strain curves (figure 2) for the FE simulation.

THE FE SIMULATION AND TEXTURE UPDATING

The mesh used for the FE simulation is shown in figure 3(a). The conditions for the cross-rolling simulation were the same as those in the experiments. Only the top half of the plate is modelled due to symmetry. Plane strain deformation is assumed which means that no spread in TD of the plate is allowed, an assumption validated by the experimental observation.

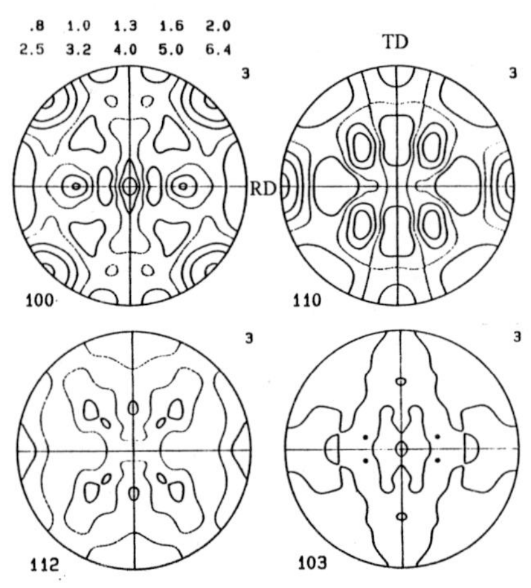

Figure 1 Pole figures of the initial texture of the low carbon steel.

The textures are updated by using the incremental plastic strains calculated at the centroid of each element after every 10 increments to take the complete plastic deformation history into account for the final predicted texture (a total of between 300 to 500 increments is required for each pass). The yield locus of each element, however, is only modified after each pass. The textures calculated for the top and bottom elements at half length of the plate, which are shaded in figure 3, are used to represent the textures at the surface and mid-thickness of the plate respectively. These two elements are later referred to as the top and centre element respectively. Bear in mind that since it is a cross-rolling simulation, the simulated rolling direction (RD) is in the direction of the original transverse direction of the plate, by choosing the corresponding reference systems of the orientation distribution functions (ODFs) and yield loci.

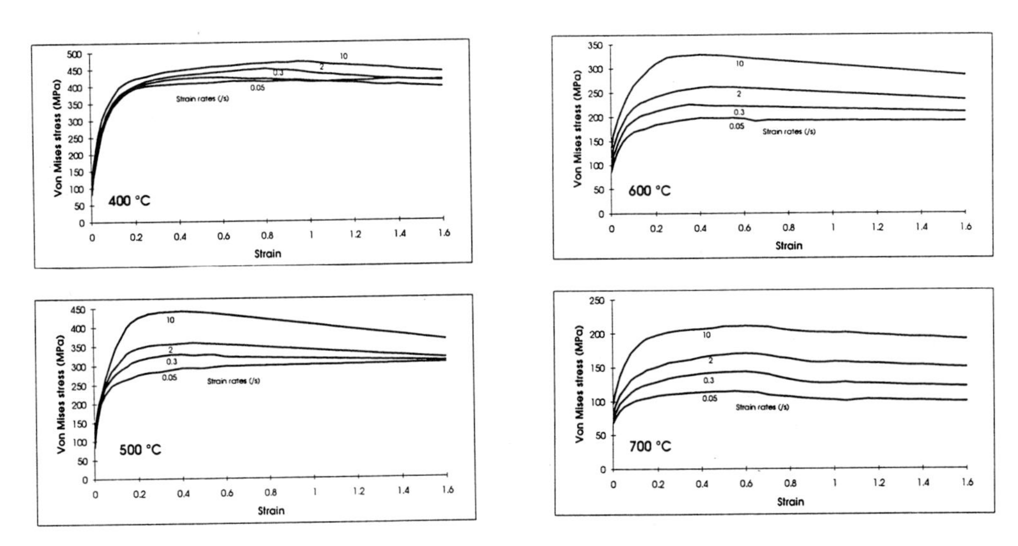

Figure 2 The measured stress-strain curves of the low carbon steel.

As the texture of a rolled product is controlled by many factors [8], here two different deformation models, which are namely the fully constrained model and the pancake model with two relaxations [4], have been applied to the texture updating procedure. After the first pass, however, it was found that the textures predicted by the pancake model were much stronger than the experimental measurement and no inhomogeneity was predicted by this model, which was also contradictory to the experimental results. On the other hand, the textures predicted by the fully constrained model were very close to the experimental measurements as shown in figure 5, and therefore, only this model is used for the texture updating procedure during the second and third passes.

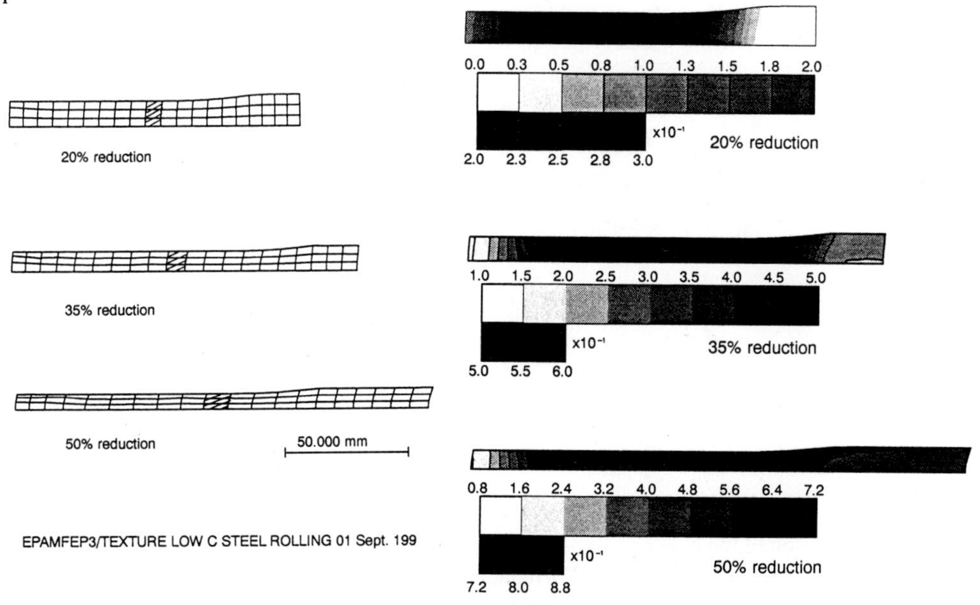

Figure 3 The deformed FE meshes.

Figure 4 The generalised plastic strain.

RESULTS AND DISCUSSIONS

The deformed meshes of the 3 passes and the plastic strain distributions in the plates are shown in figures 3 and 4. The distribution of the generalised plastic strain shows a maximum value at the surface of the plate, which indicates that the plastic deformation on the surface of the plate is bigger than it is at the centre.

The (110) and (100) pole figures shown in figure 5 represent the textures of the plate with 20% reduction. Interesting variations between the predicted textures for the top and centre elements (figure 5(b) and (d)), which represent the surface and the mid-thickness of the plate respectively, are observed. It shows the inhomogeneity as a result of inhomogeneous deformation through the thickness of the plate, as one would expect from the rolling procedure and as observed in the experimental measurements (figure 5(a) and (c)). Compare the texture measured at 50% depth of the plate (figure 5(a)) with that predicted for the centre element by FE simulation (figure 5(b)), similarity is observed both in the distribution of the poles and their magnitude. Even though the $\{100\}\langle 011\rangle$ orientation predicted by the FE simulation for the top element (figure 5(d)) is slightly stronger than measured experimentally at 13% depth of the plate (figure 5(c)), the two sets of pole figures show good agreement and the same tendency of a weaker distribution than their counter parts at the mid-thickness of the plate.

The comparisons of the pole figures and ODFs of the final textures between the experimental results and FE predictions are shown in figures 6 and 7 respectively with the initial texture also shown in each figure for comparison (figures 6(a) and 7(a)). The textures at the surface of the plate become monoclinic (figures 6(d) and (e) and 7(d) and (e)) while those at the centre remain orthorhombic (figures 6(b) and (c) and 7(b) and (c)). Again the measured and predicted textures at the mid-thickness of the plate agree very well with each other (figures 6(b) and (c) and 7(b) and (c)). Both the experimental results and the FE predictions (figures 7(b) and (c)) show the typical BCC rolling texture with the same maximum intensity of ODF at the $\{001\}\langle 110\rangle$ orientations, which is on the α-fibre ($\langle 110\rangle$//RD), and an orientation tube stretching out between the orientations $\{111\}\langle 110\rangle$ and $\{11\ 11\ 8\}\langle 4\ 4\ 11\rangle$, which is near the γ-fibre ($\{111\}$//RD).

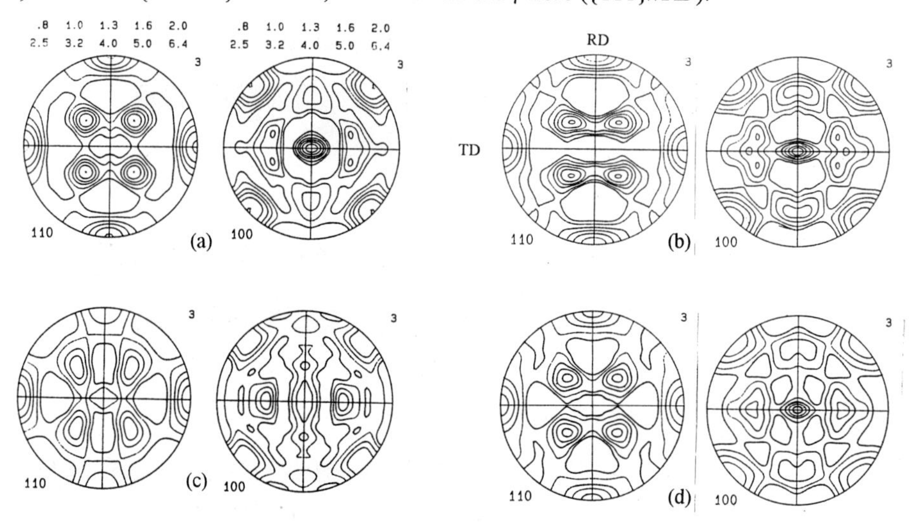

Figure 5 Comparison of the pole figures between the experimental measurements and FE predicted textures with fully constrained model at 20% reduction. (a) Experiment, mid-thickness; (b) FE, centre element; (c) Experiment, 13% depth; (d) FE, top element.

On the other hand, the FE prediction of the textures near the surface of the plate does not compare well with the experimental measurements (figures 6(d) and (e) and 7(d) and (e)), except that they are both monoclinic and they are both weaker than the textures near the mid-thickness of the plate. The FE predictions (figure 7(e)) still show more or less BCC rolling textures, whilst from the experimental measurements (figure 7(d)) one can hardly see any ideal rolling texture components at all. There are two possibilities to explain this observation. Firstly, the area near the surface of the plate is a region of high shear deformation, which is represented by the Goss nuclei (the {110}<001> orientations) in figure 7(d). Since the experimental measurements were made nearer to the surface (13% depth) than the FE predictions (16.7% depth), the shear deformation mode is reflected more completely by the former results. Furthermore, the value of the frictional factor between the rolls and the plate, which is 0.4 for the current simulation, also has influence on the reality of the simulated deformation mode at the surface of the plate. As the anisotropic FE simulation takes much longer CPU time than isotropic simulations (about 20 times with 60 elements on IBM3090 mainframe computer), the limitation of computer resources available for the authors prevents further investigation of the quantitative influence of the frictional factors at present. Secondly, because of technical reasons, microstructural examinations were only taken at the mid-thickness sections, which showed no sign of recrystallisation. It is reasonable to question whether recrystallisation has occurred at the surface of the plate, where plastic deformation is bigger (figure 4) and overheating is easier, and as a result, the ODFs shown in figure 7(d) are very weak.

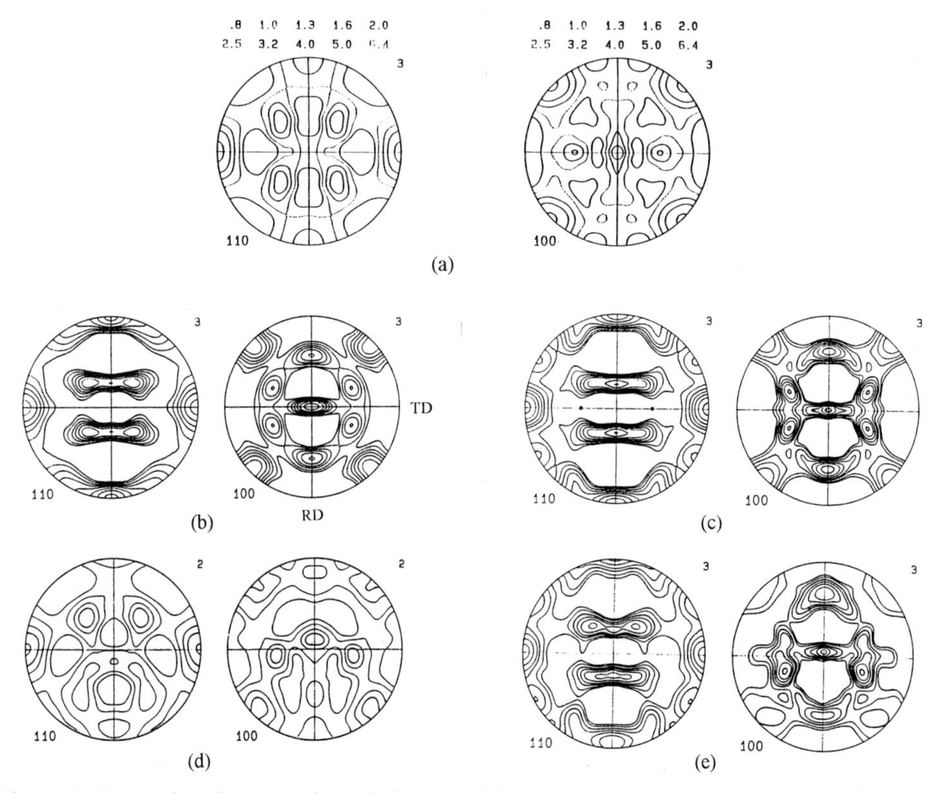

Figure 6 Comparison between the pole figures of the experimental measurements and FE predicted textures with fully constrained model at 50% reduction. (a) The initial texture; (b) Experiment, mid-thickness; (c) FE, centre element; (d) Experiment, 13% depth; (e) FE, top element.

The yield loci calculated from the initial and final predicted textures are shown in figure 8. Deviations between the initial and updated yield loci, as well as between the top and centre elements, are apparent, which shows the evolution and inhomogeneity of the plastic properties of the plate during the forming process.

CONCLUSIONS

The anisotropic FE program incorporated with T-B-H model can predict texture evolution on a much more realistic basis than traditional methods in the sense that the full history and inhomogeneity of plastic deformation during forming processes can be taken into account. Finer tuning of the FE mesh, choosing of the correct relaxation to the T-B-H model and the friction factors as well as the use of powerful computer facilities are all pre-requisite to complete an accurate fully anisotropic FE simulation of material forming with texture evolution.

ACKNOWLEDGEMENT

This work was carried out as part of the Brite/Euram project BREU*0107-C funded by the Commission of the European Communities.

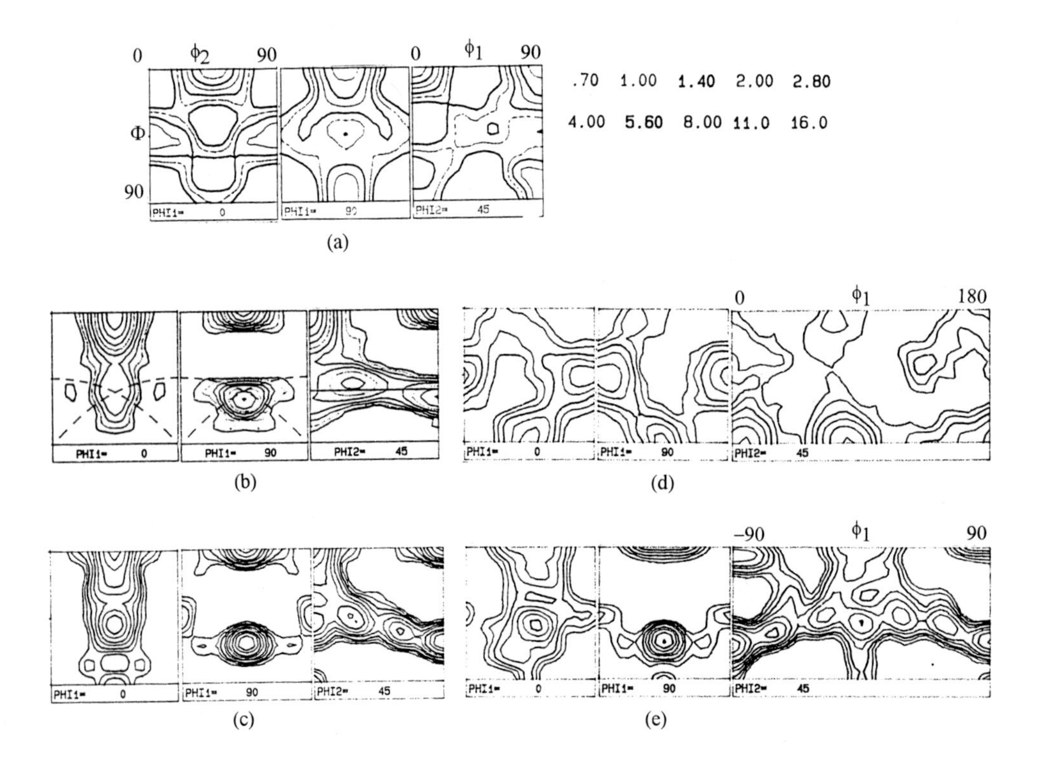

Figure 7 Comparison between the ODFs of the experimental measurements and FE predicted textures with fully constrained model at 50% reduction. (a) The initial texture; (b) Experiment, mid-thickness; (c) FE, centre element; (d) Experiment, 13% depth; (e) FE, top element.

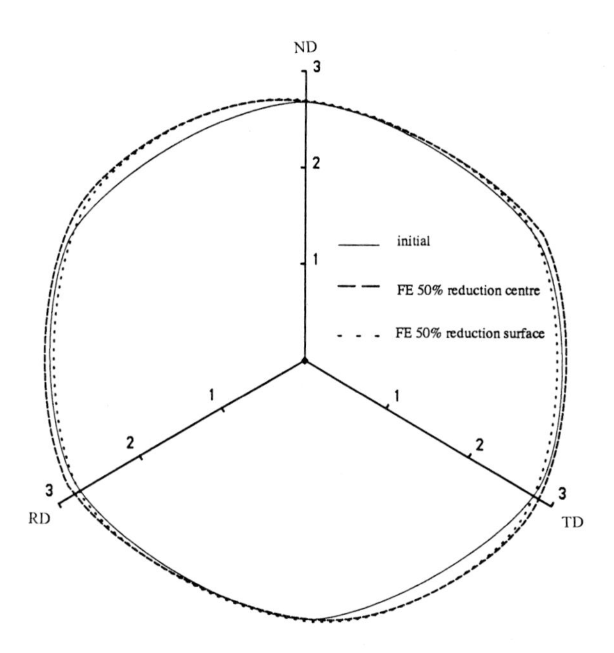

Figure 8 The yield loci calculated from the initial and final FE predicted texture of the low carbon steel plate.

REFERENCES

1). Rowe, G.W., Sturgess, C.E.N., Hartley, P. and Pillinger, I.: *Finite-element plasticity and metal forming analysis*, Cambridge University Press, 1991.
2). Taylor, G.I.: *J. Inst. Met.*, 1938, 62, 307-324.
3). Bishop, J.F.W. and Hill, R.: *Phil. Mag.*, 1951, 42, 414-427, 1298-1307.
4). Van Houtte, P.: *Textures and microstructures*, 1988, 8 & 9, 313-350.
5). Van Houtte, P.: *Directional properties of materials*, DGM Informationsgesellschaft mbH, Overursel, Germany, 1988, 65-76.
6). Van Bael, A., Van Houtte, P., Aernoudt, E., Hall, F.R., Pillinger, I., Hartley, P. and Sturgess, C.E.N.: *ICOTOM9*, Avignon, France, 1990, 1007-1012.
7). Hall, F.R., Pillinger, I., Hartley, P., Sturgess, C.E.N., Van Bael, A., Van Houtte, P. and Aernoudt, E.: *IMA*, Bristol, UK, 1991.
8). Wang, N., Hall, F.R., Pillinger, I., Hartley, P., Sturgess, C.E.N., Van Bael, A., Winters, J., Van Houtte, P. and Aernoudt, E.: *NUMIFORM'92*, Valbonne, France, 1992, 193-198.

Materials Science Forum Vol. 157-162 (1994) pp. 1909-1914

FINITE-ELEMENT PREDICTION OF HETEROGENEOUS MATERIAL FLOW DURING TENSILE TESTING OF ANISOTROPIC MATERIAL

J. Winters [1], A. Van Bael [1], P. Van Houtte [1], N. Wang [2], I. Pillinger [2], P. Hartley [2] and C.E.N. Sturgess [2]

[1] Department of Metallurgy and Materials Engineering, Katholieke Universiteit Leuven, de Croylaan 2, B-3001 Leuven, Belgium

[2] School of Manufacturing and Mechanical Engineering, University of Birmingham, Edgbaston, Birmingham, UK

Keywords: Anisotropic FE-Simulation, Tensile Test, R-Value

ABSTRACT: During tensile testing of anisotropic (orthotropic) sheets, heterogeneous material flow takes place. A simulation of this flow pattern has been performed by means of an elastic-plastic finite element code. Two different anisotropic yield criteria - a yield locus based on crystallographic texture and a Hill-type yield locus - have been used. It is shown that the anisotropic FE-code is capable of predicting the anisotropic behaviour of the material during this test. Due to this heterogeneous material flow, errors might occur when strains are experimentally measured in a classical tensile testing device. This could have an influence on r-values calculated from this strains. It is made clear from the simulations that this influence can be neglected, even for materials with a relatively high planar anisotropy.

INTRODUCTION

The *r-value,* which gives the ratio between the width and the thickness strain in a tensile test, is widely used to characterise anisotropic plastic properties of sheet material. The magnitude of the r-value depends on the orientation of the tensile specimen with respect to the rolling direction of the sheet.

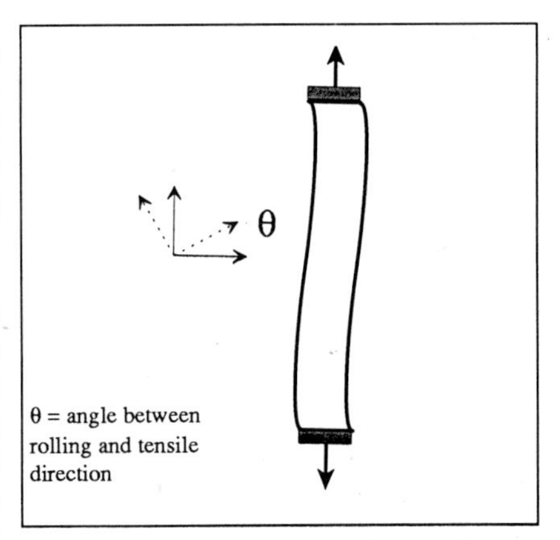

figure 1: S-shaped tensile specimen after deformation (after [1]);

Sheet material is considered to have orthotropic plastic properties. So, in general, only the rolling, the transverse and the normal direction are two-fold rotational symmetry axes for the sheet. It can be shown (see e.g. [1]) that for a specimen for which the tensile direction does not coincide with the rolling nor with the transverse direction, the principal strain axes do not coincide with the principal stress axes. In classical testing devices in which the specimen is clamped, the specimen will therefore tend to get an S-shape. It can be expected that this can have an influence on the obtained *r-values* when using an extensometer or indentation points to measure the width strain.

To investigate this, a finite-element simulation of the heterogeneous material flow during tensile testing has been done. Two different anisotropic yield criteria have been used. The first one is the anisotropic yield criterion that was proposed by Van Houtte et al. [2]. It is derived from the crystallographic texture of the material, in this case a good quality deep drawing steel. The second one, is a Hill-type yield criterion. For both criteria, the evolution of the deformation during the simulation is studied. From the simulations, it was possible to draw useful conclusions concerning the calculation of r-values for anisotropic sheet materials from tensile test data.

FE-MODEL

The material flow is simulated by means of an elastic-plastic implicit FE-code [3]. The anisotropic plastic material behaviour has been studied by means of two different types of yield loci. The first one, very close to a real material, is described by a sixth order series expansion of the strain rate tensor [2]. It is derived from an ODF of a deep drawing steel (see figure 2a). The second one is a simple Hill-type yield locus which is constructed so that the r-values for the rolling (0°), the

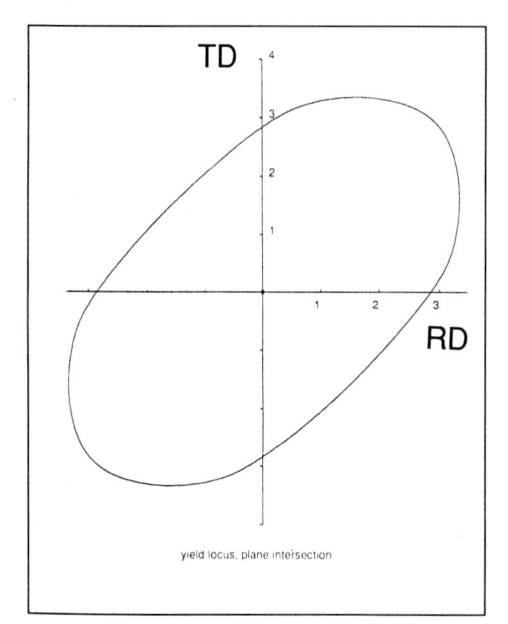

figure 2a: yield locus obtained from an ODF of a deep-drawing steel (material *1*)

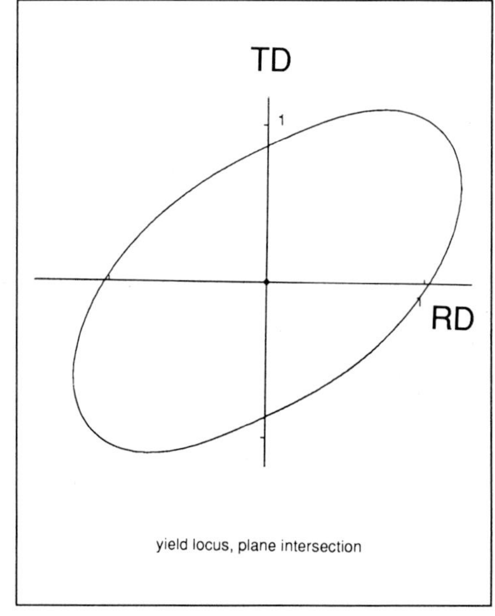

figure 2b: Hill-type yield locus constructed from given r-values (material *2*)

45° and the transverse (90°) direction are 2, 1,5 and 1 respectively (see figure 2b). These are presumed not to change during the test. Both materials exhibit an ideal plastic behaviour. The elastic properties are given in table 1.

material	*1*	*2* (Hill-type)
Young's modulus (MPa)	210000	210000
Poisson ratio	0,33	0,33
yield stress	516 MPa	165 MPa

table 1: elastic properties for both materials

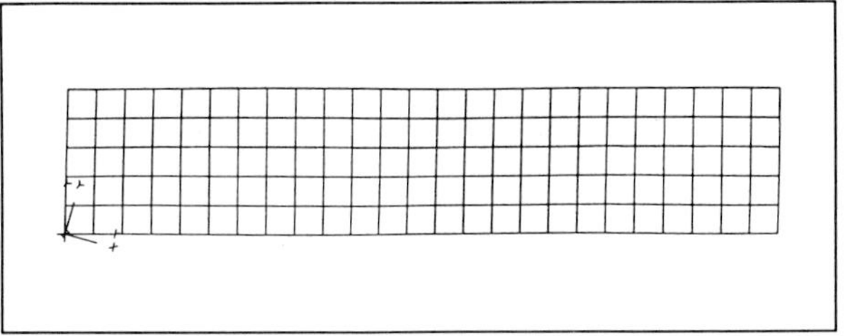

figure 3: initial finite-element mesh

The FE-model consists of 125 linear brick elements (see figure 3). The mid-plane of the sheet is considered to be a symmetry plane, so only half of the specimen needs to be modelled (dimensions of the mesh: length = 50 mm; width = 10 mm, thickness = 0,6 mm). Nodes in the symmetry plane can only move within this plane. A fixed displacement per time-increment along the tensile direction is imposed on the nodes at the two ends of the specimen. Furthermore, these nodes cannot move in transverse direction. For the first material type, the tensile direction was the 30° direction, while for the Hill-type this was the 45° direction.

RESULTS

flow pattern

Figure 4 shows the meshes after an engineering strain of about 10%. For the first material (ODF-yield locus, 30°), the deformation is almost homogeneous throughout the specimen. Therefore the deformed mesh has almost the same form as for a simulation with an isotropic von Mises yield criterion (except, of course, for the different ratio of the width to thickness-strain). For the second material (Hill-type, 45°) this is not the case. Visual examination of the deformed mesh reveals a slight S-shape. The nodal coordinates available from the FE-program confirm this shape.

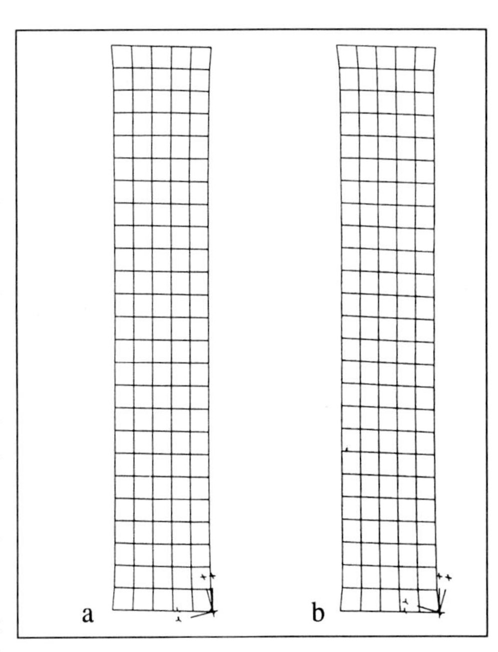

figure 4: deformed FE-meshes for material *1* (a) and for material *2* (b)

As in [1], the flow pattern becomes much clearer when lines of equal displacement are considered (iso-displacement lines). Figure 5 gives the iso-lines of incremental nodal displacement after about 10% engineering strain. Here, for the 45° simulation (material 2), the S-shape is readily visible for the incremental nodal displacement in the transverse direction. The reason why there is no S-shape for material *1*, comes from the fact that the planar anisotropy is too weak to cause noticeable heterogeneous material flow, the r-values being 2.13 at 0°, 1.72 at 45° and 2.08 at 90°.

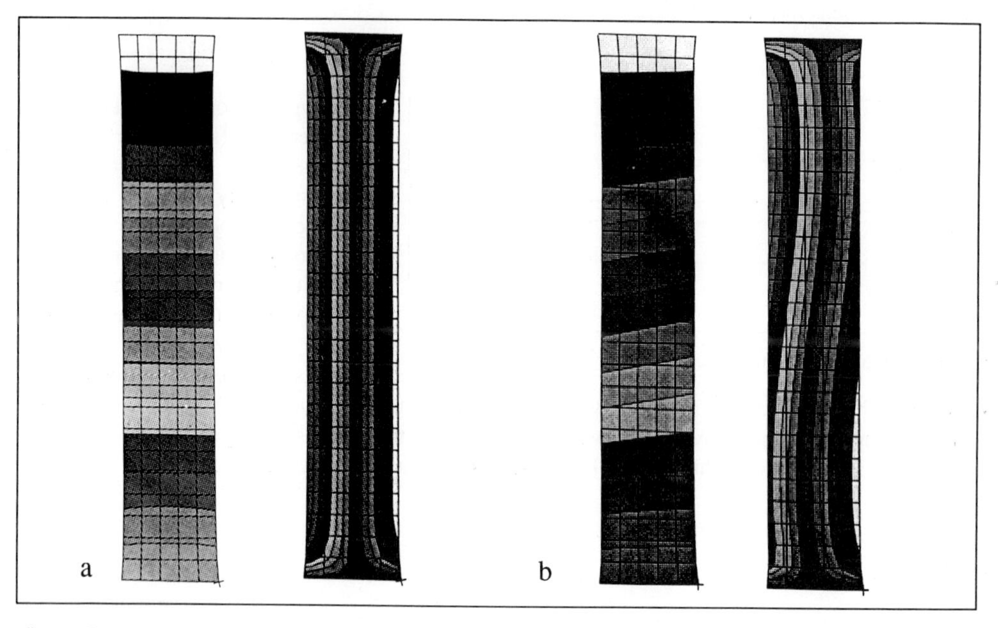

figure 5: iso-lines of incremental nodal displacement for material *1* (a) and for the material *2* (Hill-type yield locus) (b) (left: displacement in tensile direction; right: idem for transverse direction)

strain

A contour plot of the incremental strain in the tensile direction after about 10% engineering strain is given for both materials in figure 6 (it can be interpreted as a plot of the strain rate in tensile direction). From this, it is clear that the deformation is not homogeneous and indeed leads to an S-shape of the specimen. For material *1*, the difference between the lowest and the highest

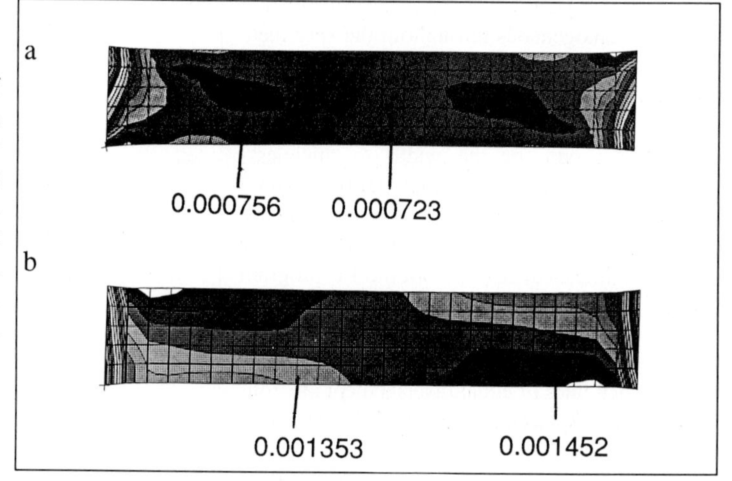

figure 6: incremental strain in tensile direction for material *1* (a) and *2* (b); incremental strain values are indicated.

strain rate in the central part of the specimen (as indicated on the figure) is about 3%. For material 2, this difference goes up to about 10%.

r-values

R-values were calculated from the nodal coordinates in the FE-meshes after an engineering strain of about 10%. The procedure that has been used is as follows: first, from an isotropic simulation, the zone of homogeneous deformation is defined. This is necessary to exclude boundary effects (in this case, this zone starts about 12 mm away from both ends). It is considered as the zone in which the strains are measured during an experiment.

Secondly, the position of 2 nodes at one end of the "homogeneous" zone and 2 at the opposite end of this zone are used to calculate the strains. This is being done by taking the difference between the initial and the final position of the nodes (as in most tensile test experiments, the consequences of anisotropic material behaviour for the shape of the specimen during deformation are neglected during the measuring). In this way, the results of the simulations are treated in very much the same way as results from indentation point measurements. No elastic relaxation is taken into account, though this causes no real extra problem.

The r-values that were obtained after the FE-simulations, were compared with the r-values which were calculated directly from the yield loci. It was seen that the influence of the S-shape on the results was negligible, even for the material with the higher planar anisotropy.

CONCLUSIONS

Simulations of tensile tests revealed that the present anisotropic elastic-plastic FE-code is capable of simulating the heterogeneous material flow that occurs during this test. Comparison with experimental results from literature showed very good agreement. It was shown that the anisotropic material behaviour, which causes in a classical tensile testing equipment the specimen to deform in an S-shape, can be neglected when measuring r-values.

Apart from rather simple anisotropic yield criteria, it is possible - as been shown - to use within this FE-code a yield criterion which is derived from crystallographic texture data. This offers the possibility to do in the near future simulations of e.g. a cup-drawing test for real material data.

ACKNOWLEDGEMENT

Part of this work was carried out with financial support of the EEC. J. Winters also wishes to thank the IWONL for financial support.

REFERENCES

1) J.P. Boehler, L. El Aoufi, J. Raclin: On experimental testing methods for anisotropic materials, Res Mechanica, 1987, 21, 73-95

2) P. Van Houtte, A. Van Bael, J. Winters, E. Aernoudt, F. Hall, N. Wang, I. Pillinger, P. Hartley, C.E.N. Sturgess: The incorporation of an anisotropic yield locus derived from the crystallographic texture in FE modelling of forming, in: Advances in Engineering Plasticity and its Applications, W.B. Lee (ed.), Elsevier Science Publishers B.V., 1993, 93-100

3) G.W. Rowe, C.E.N. Sturgess, P. Hartley, I. Pillinger: Finite-element plasticity and metalforming analysis, Cambridge University Press, 1991

6. Technological Applications of Texture Studies

Materials Science Forum Vol. 157-162 (1994) pp. 1917-1928

PRACTICAL ASPECTS OF TEXTURE CONTROL IN LOW CARBON STEELS

B. Hutchinson

Swedish Institute for Metals Research,
Drottning Kristinas väg 48, S-114 28 Stockholm, Sweden

Keywords: Steel, Texture Processing, Annealing

ABSTRACT

A review is presented of the role of material and process variables in controlling textures and plastic anisotropy in low carbon steels. The separate requirements of batch annealing and continuous annealing are discussed in terms of their underlying physical processes. Consideration is also given to modern ultra-low carbon steels.

INTRODUCTION

Anisotropy, usually defined by the $\bar{r}$-value, is an important requirement of sheet steel intended for press-forming applications. The crystallographic texture which gives rise to anisotropy must therefore be optimised by suitable choice of steel chemistry and processing conditions. However, it is also important to recognise that texture is only one of very many requirements - cost, strength, ductility, stability, surface quality, etc - so the practical control of texture cannot be considered in isolation but rather as part of a complex compromise.

Real textures typically comprise wide spreads of orientations with different frequencies of occurrence and their effect on plastic anisotropy can be calculated with reasonable accuracy, for example using Taylor theory [1]. Using the same principles it is possible to evaluate synthetic textures to identify desirable conditions with respect to anisotropy. Figure 1 shows the predicted variation of $\bar{r}$ for textures having different mixtures of <111> and <100>// ND fibre textures together with random spread. The beneficial effect of <111>// ND texture is very clear as is the detrimental influence of <100>// ND components. This figure indicates that in order to obtain an $\bar{r}$-value of 2 in a texture of mixed components it is necessary to have approximately 60 % of the material close to the <111> orientation.

Fortunately, the <111>// ND (or equivalently, {111} sheet plane) texture components arise naturally during cold rolling of bcc metals and are retained to a greater or lesser degree after recrystallisation. The natural tendency of steel, unlike metals with fcc structures, is to develop a texture in the final annealed condition which is favourable for deep drawability. However, the relative proportions of <111> and other components can vary widely and the key to obtaining the best results lies in maximising the proportion of <111> in the overall texture.

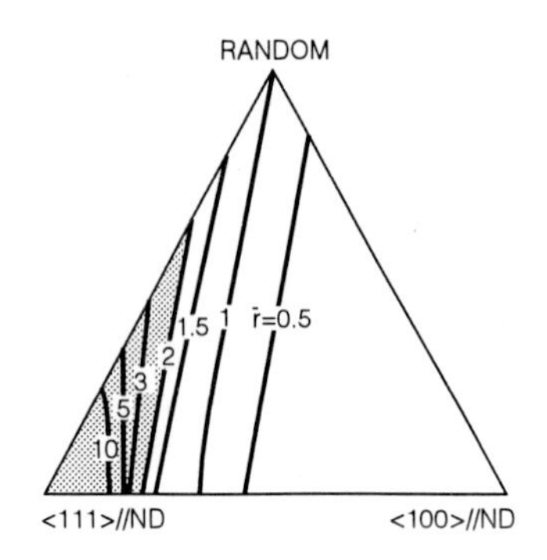

Fig. 1 The effect of different mixtures of <111> and <100> fibre textures together with random spread on calculated r̄-values.

PHYSICAL METALLURGY BACKGROUND

The texture of cold rolled steel sheet can be described to a good approximation as comprising a complete <111>// ND fibre texture (γ-fibre) together with a partial fibre having <110>//RD with a spread of orientations from {001} to {111} parallel to the rolling plane (α-fibre). Intensities of these orientation spreads are affected by the degree of cold rolling as well as the starting texture which is never random in practice, but only to a minor extent by the steel composition. An example of such a texture in the $\phi_2 = 45°$ ODF section is shown in figure 2(a) for a titanium stabilised low carbon steel cold rolled 75 %.

Changes which occur in the overall texture during annealing as recrystallisation proceeds are shown in figures 2(b) - (d). There is relatively little apparent change until the latter stages at which time the α-fibre becomes considerably weakened. The textures of the recrystallised fractions of the microstructures determined from EBSP measurements are shown in figures 2(e) -(g) and the textures of the residual deformed fractions, as differences, are plotted in figures 2(h) -(j). The first formed recrystallised grains have a very sharp γ-fibre texture which consumes preferentially the γ-fibre of the deformation texture. During the later stages of transformation where the α-fibre is consumed, the evolving recrystallisation texture is much more widely spread. Thus, a strong γ-texture in the cold rolled state seems to be a prerequisite. When a partially recrystallised steel sample is examined in the optical microscope, figure 3, it appears that new grains initiate at different locations in the microstructure. Hard second phases such as large carbides, old grain boundaries, and intra-granular sites are all represented. The different types of site are believed to be associated with different orientation spreads of the new grains. There is evidence to suggest, for example, that grains in the <111> orientation spread can nucleate along grain boundaries in the deformation structure and grow into a deformed matrix having some 30° misorientation e.g. {111}<110> grows into {111}<112> and vice versa [2]. Such ~30°<111> relationships have not been identified as corresponding to high mobility in bcc metals so it is inferred that the orientation selectivity is mainly nucleation controlled.

As well as the γ-fibre orientations, Goss texture can appear on recrystallisation, notably in steels having a coarse initial grain size, and where shear banding occurs prominently. The Goss grains nucleate along shear bands and grow, especially in deformed grains in the {111}<112> orientations [3]. Other important annealing textures are the 'near {100}' components, e.g. {411}<148>, which become prominent following heavy cold reductions and which penalise the desired plastic anisotropy. Little seems to be known about their origin. Nucleation at hard second phase particles such as cementite contributes towards a random spread of orientation [4].

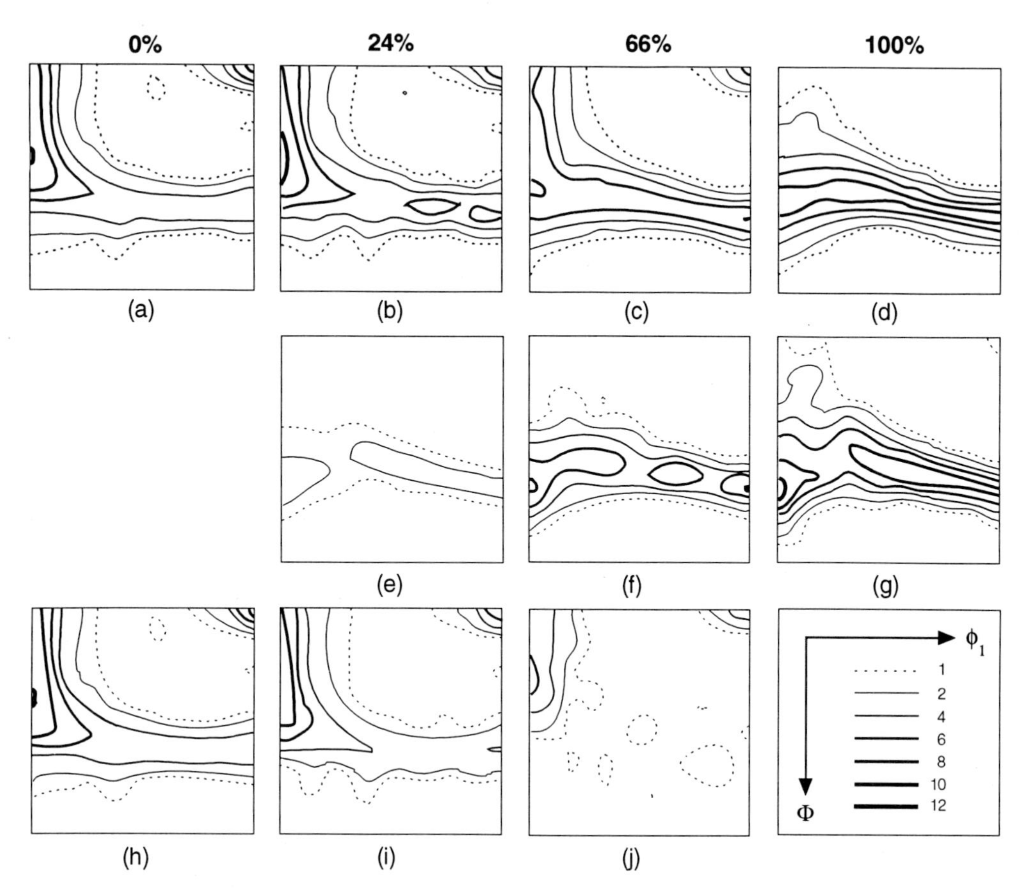

Fig. 2 *ODF sections (ϕ_2 = 45°) for steel at different stages of recrystallisation, (a) - (d) global textures from X-ray measurements, (e) - (g) recrystallised fraction only from EBSP measurements, (h)- (j) textures of the deformed matrix.*

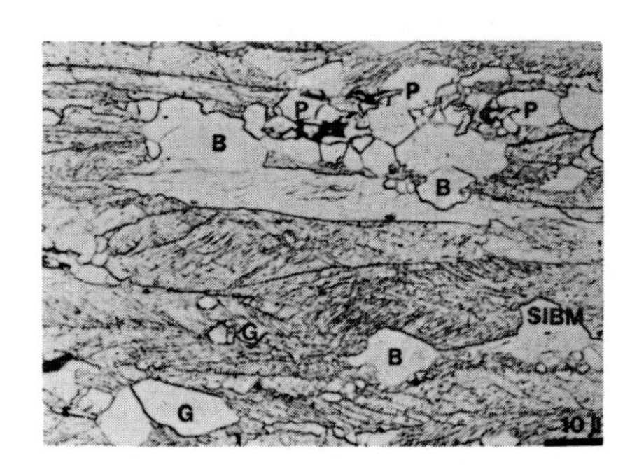

Fig. 3 *Micrograph showing examples of recrystallised grains nucleated at cementite particles (P), along grain boundaries (B) and (SIBM) and within the deformed matrix grains (G) in a cold rolled steel.*

A fine initial grain size is of benefit to the texture after cold rolling and annealing since it enforces a more uniform deformation, reducing the tendency for heterogeneous deformation such as shear banding, and also provides many grain boundary sites giving rise to <111>// ND oriented nuclei [5]. Figure 4(a) and (b) shows how the initial grain size and cold rolling reduction control the $\bar{\mathbf{r}}$-value and $\Delta\mathbf{r}$-value for decarburised steel. The results in figure 4 were calculated from experimental ODFs on the basis of Taylor theory [6]. Rigorous control of the starting grain size (<15 μm) is seen to be necessary to obtain a high $\bar{\mathbf{r}}$-value and, in particular, to prevent excessively large values of $\Delta\mathbf{r}$.

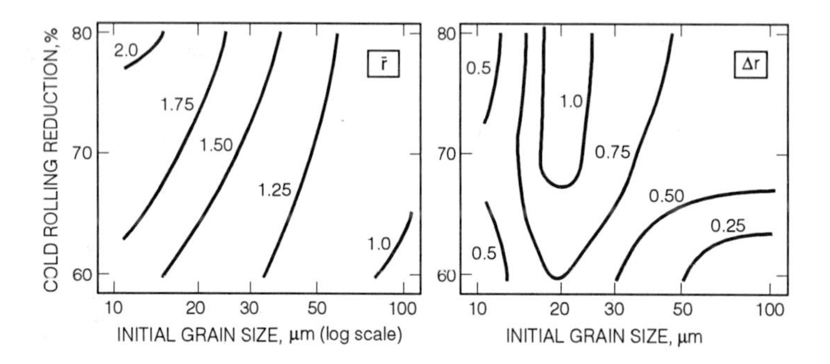

Fig. 4 The effects of initial grain size and cold rolling reduction on $\bar{\mathbf{r}}$*-value and* $\Delta\mathbf{r}$ *for decarburized steel calculated from experimental ODFs on the basis of Taylor theory.*

One of the most significant factors to influence the recrystallisation texture is the carbon content of the steel. Pearlite and cementite as second phase particles can stimulate the nucleation of recrystallised grains with widely spread orientations and these random components act in competition with the <111>// ND orientations. A carbon content of 0.02 % or less is therefore desirable to minimise this tendency. However, carbon dissolved in the ferrite matrix has a further deleterious effect. Where interstitial carbon atoms are present together with manganese atoms they form a type of complex (C-Mn dipole) [7] which obstructs development of the <111>// ND oriented grains and produces a more random spread of texture. A very low level of effective (dissolved) carbon content is therefore desirable. According to results assembled in figure 5 an optimum carbon content exists somewhere in the vicinity of 5 ppm for steels having conventional manganese contents. For Mn-free compositions where dissolved carbon atoms cannot form Mn-C dipoles, the carbon content is much less significant [8].

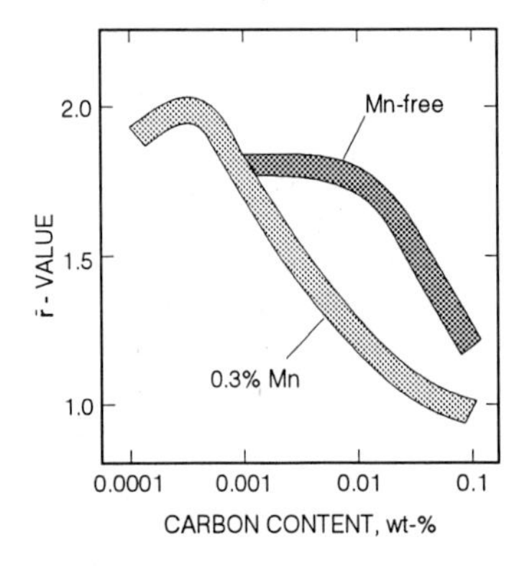

Fig. 5 Effect of carbon content on the $\bar{\mathbf{r}}$*-value of cold rolled and annealed steel.*

Typical annealing treatments produce some degree of grain growth following the completion of primary recrystallisation. Experience shows that this phenomenon is usually associated with strengthening of the <111>// ND fibre texture at the expense of other widely scattered components and accordingly gives rise to enhanced normal anisotropy, figure 6. Grain growth is therefore to be encouraged in general, so control of second phases which may impede the process is an important aspect of practical texture control. The pinning action of the second phase particles should be reduced to the minimum. For this reason it is desirable to convert unavoidable second phases (nitrides, carbides, sulphides etc) into coarse dispersions which hinder grain growth as little as possible.

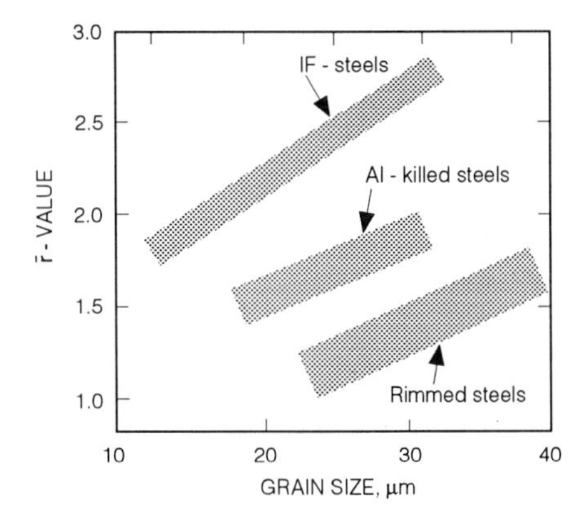

Fig. 6 Beneficial effect of grain growth on r̄-value for three types of steel [9].

INDUSTRIAL PROCESS PARAMETERS

The successful control of texture in practice involves a complex interplay between steel chemistry and process parameters as reviewed in more depth previously [10]. Three different approaches are used in practice for manufacturing sheet steel having high r̄-values. Two of these, namely batch annealing (BA) and continuous annealing (CA) require that the conventional low carbon steel chemistry and prior processing are carefully tailored to the final annealing process. Batch annealing involves very slow heating, ~20 - 50 deg.C/hr, of coiled sheets whereas continuous annealing and metallising lines necessitate rapid heating, ~5 - 20 deg.C/sec, and a very much shorter processing time. The third, interstitial-free (IF) or ultra-low carbon (ULC) steel, is based primarily on the control of steel chemistry and is relatively insensitive to processing conditions. Thus, IF steels can be produced successfully on different types of processing lines.

Table 1 shows an attempt to specify how the various material and process variables are controlled for optimisation of texture in cold rolled low carbon and IF steels, and also gives a ranking of these variables according to their significance.

It is evident from Table 1 that different and conflicting requirements apply according to the particular approach which is adopted. In particular, there are interactions between the coiling temperature after hot rolling and the heating rate after cold rolling as can be seen represented in figure 7. For rapid, continuous annealing the coiling temperature must be high whereas for slow batch annealing it should be low for reasons which are now reasonably well understood and are discussed in the following sections.

Table 1 Control of parameters for optimising the texture/anisotropy of cold rolled and annealed steel sheets

Parameter	Low carbon steels		IF steel
	BA	CA	BA/CA
Carbon content	low (*)	low (**)	low(***)
Manganese content	low (*)	low (**)	( )
Microalloying (Al, Ti or Nb)	Al (***)	( )	Ti/Nb (***)
Soaking temp. for hot rolling	high (***)	low (*)	low (*)
Hot rolling schedule	( )	( )	(**)
Finish rolling temperature	>A3 (**)	>A3 (**)	≲ A3 (*)
Coiling temp after hot rolling	low, <600° (***)	high, >700° (***)	high (*)
Cold rolling reduction	~70 %	~85 %	~90 %
Heating rate of anneal	20 - 50 deg/hr (***)	5 - 20 deg/sec (**)	( )
Max temperature of anneal	~720°C	~850°C	~900°C

( ) not critical (*) significant (**) important (***) vital

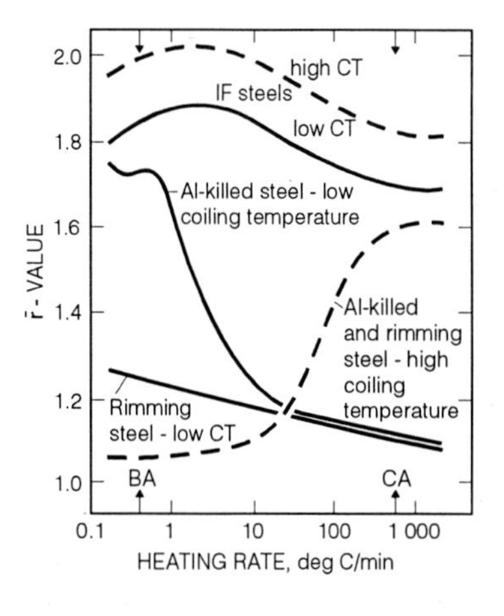

Fig. 7 Effect of heating rate during annealing on $\bar{r}$-value of different steel types with different coiling temperatures after hot rolling.

BATCH ANNEALED ALUMINIUM-KILLED STEELS

This process which has been utilised for many decades relies on a reaction between aluminium and nitrogen atoms to form a compound which interacts with the process of recrystallisation, making it more selective towards the {111} texture components and, in particular, {111}<110>. For the Al and N atoms to be effective they must be initially (after cold rolling) in solid solution. The high soaking temperature, ~1250°C, and low coiling temperature, <600°C, are necessary to ensure that aluminium nitride becomes dissolved and does not precipitate in the hot rolled band.

During the slow heating of the coil the Al and N react together in the cell boundaries and grain boundaries of the cold worked metal in a manner that retards the formation of recrystallisation nuclei. There has been a long discussion as to whether the Al and N form a segregate or whether a true precipitation process (of AlN) is involved. Recent work tends increasingly to support the latter hypothesis and it is quite clear after recrystallisation that rows of AlN particles delineate the positions of the old cell and grain boundaries [11]. A schematic representation of this process may be given in the manner described by Hornbogen [12] where the times for recrystallisation and precipitation are plotted together as functions of temperature, figure 8. At high temperatures recrystallisation precedes precipitation and the resulting {111} texture is only weakly developed. At lower temperatures precipitation occurs first and thus retards the recrystallisation reaction while at the same time rendering the nucleation process more selective towards the {111} oriented grains [13]. The result is a characteristic coarse elongated ('pancake') grain structure with strong {111} texture which is ideally suited for sheet forming purposes.

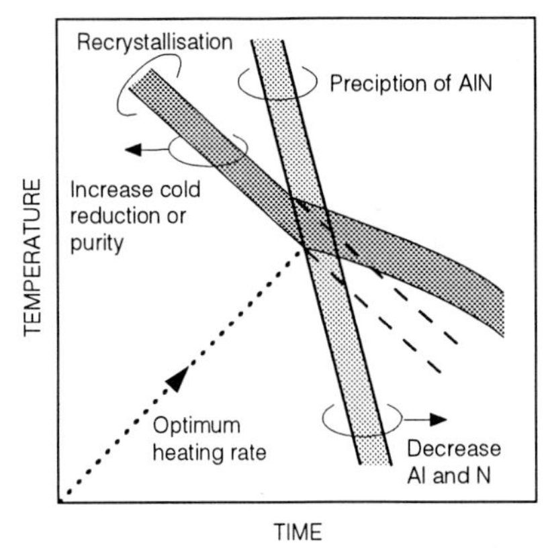

Fig. 8 Schematic diagram showing the kinetics of AlN precipitation and recrystallisation in cold rolled steel.

In practice, the slow heating rate during traditional batch annealing is close to the optimum where a sufficient degree of precipitation precedes the onset of recrystallisation. It is interesting to note that the temperature where the two bands cross in figure 8 which dictates the optimum heating rate, is a function of several parameters. For example, an increased cold rolling reduction or an increased steel purity will accelerate recrystallisation and so move this band towards shorter times. Alternatively, a reduction in the Al and N contents of the steel will reduce the driving force for precipitation and so move that band towards longer times. In either case the optimum condition will be altered in favour of lower heating rates. These effects have been incorporated in a numerical analysis [14] which describes the optimum heating rate (OHR) as:

$$\mathrm{Log\ (OHR)} = 18.3 + 2.7 \log\{[\mathrm{Al}][\mathrm{N}][\mathrm{Mn}]/\mathrm{CR}\}$$

where the heating rate is given in deg.C/hr, the chemical compositions refer to weight % of the elements in solid solution and CR is the cold rolling reduction (%).

While the above discussion is based on long established knowledge it is interesting to observe a renewed interest in the batch annealing process associated with the advent of processing in hydrogen atmospheres. An increase in heating rate is made possible from ~20 deg.C/hr to ~50 deg.C/hr. In this case the material and process parameters need to be adapted to the new circumstances if the optimum texture is to be achieved.

CONTINUOUSLY ANNEALED LOW CARBON STEELS

The high heating rate in continuous annealing lines (~10 deg.C/sec) precludes any use of the AlN precipitation for controlling texture development. A pure iron matrix is sufficient in itself to ensure a strong {111} recrystallisation texture provided that second phases are not too finely dispersed so that grains can grow freely. The problem which can arise, however, is that cementite particles start to dissolve and contaminate the matrix with carbon in solid solution before recrystallisation commences [15]. This carbon, in combination with the manganese which is normally present, weakens the {111} development, producing a more random texture. The total carbon content should therefore be maintained at a low level (~0.02 %) if possible and coiling of the hot rolled band at a high temperature (~730°C) is necessary to produce a sparse dispersion of coarse cementite particles.

Figure 9(a) from the work of Toda et al [16] demonstrates the effect of coiling temperature after hot roling on the $\bar{r}$-values of the fully processed sheet. Micrographs in figures 9(b) and (c) show examples of the cementite structures formed after coiling at low and high temperatures respectively. The spacing of the carbides is evidently very sensitive to coiling temperature.

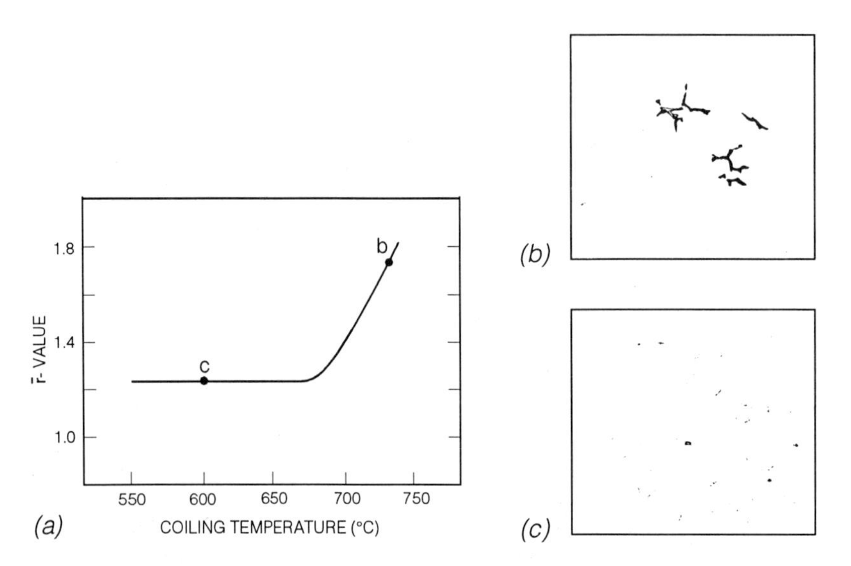

Fig. 9 *(a) Effect of hot band coiling temperature on $\bar{r}$-values of cold rolled and continuously annealed low carbon steel, [17]. Cementite structures produced by (c) low temperature coiling (600°C), and (b) high temperature coiling (730°C).*

The competition between carbide dissolution and recrystallisation is shown schematically in figure 10. Evidently, the carbide dissolution process will be more advanced if the carbide particles are closely spaced. A high rate of heating is also beneficial in this case since carbide dissolution is relatively more favoured at low temperatures. By modelling the two parallel processes of carbide dissolution and recrystallisation it has been possible to calculate the average carbon level reached in the matrix at the temperature where recrystallisation and its associated texture selectively begins [17]. Results in figure 11 show that with finely spaced carbides (λ = 20μm) the average carbon content in the matrix reaches 50 ppm whereas coarse carbides (λ = 80 μm) give rise to less than ~10 ppm under conditions of rapid annealing.

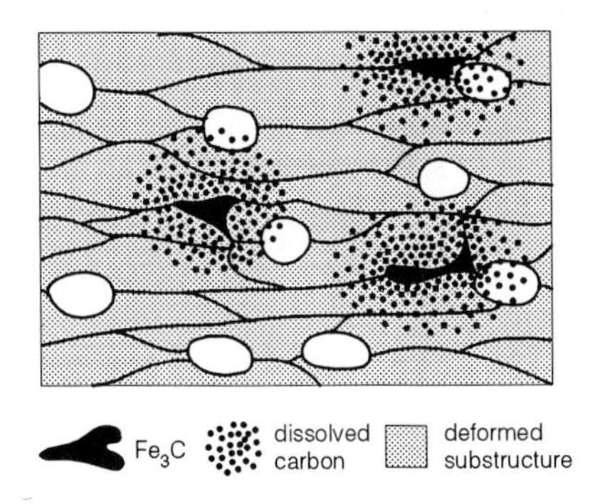

Fig. 10 Schematic microstructure showing the two simultaneous processes of cementite dissolution and recrystallisation.

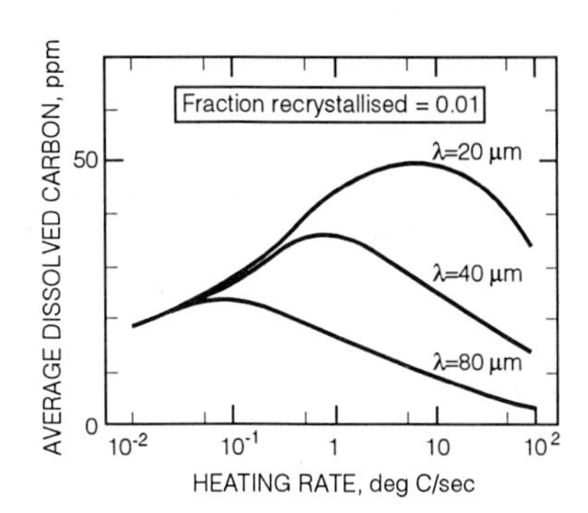

Fig. 11 Calculated dissolved carbon levels for different heating rates and dispersions of cementite particles (λ = inter-particle spacing)

The high coiling temperatures (~730°C) normally employed with steels processed by continuous annealing have a number of practical disadvantages. It has been shown that if the manganese content of the steel is reduced to ~0.1 % then coarse carbide structures can be achieved with lower coiling temperatures (~650°C). This together with the smaller concentration of Mn-C dipoles results in favourable textures and, in particular, better uniformity of properties in the final sheet product [18].

Since the time available for grain growth is very limited during continuous annealing, conditions must be adjusted to encourage grain growth and the associated development of the {111} texture. One way is to increase the annealing temperature although the scope for this is limited to ~850°C since a large fraction of the structure transforms to austenite which weakens the resulting texture. It is also very important to coarsen second phase particles in order to reduce their Zener drag [19]. The most significant second phases are MnS and AlN in normal steels. A low soaking temperature (~1100°C) prior to hot rolling prevents these particles from dissolving and allows them to coarsen by Ostwald ripening. The high coiling temperature also plays another role here in ensuring that such precipitation as does take place on cooling occurs on a relatively coarse scale. A careful control of steel chemistry to balance alloying elements and impurity levels is necessary in order to facilitate growth of the grains and the {111} texture after recrystallisation is complete.

INTERSTITIAL-FREE STEELS

When strong carbide forming elements such as Ti, Nb or Ta are added to steels containing ~0.01%C, there is seen to be a dramatic transition in the recrystallisation texture and plastic anisotropy after cold rolling and annealing. The transition occurs when the alloy element is added in such a quantity as to combine stoichiometrically with the carbon (and nitrogen) in the steel. Some examples of the effect of steel chemistry on $\bar{r}$-values are summarised in figure 12. Several proposals have been put forward to explain the development of strong {111} annealing textures in these micro-alloyed steels, such as:

(i) interference of the recrystallisation process by second phase carbonitride particles,
(ii) modification of the starting texture after hot rolling
(iii) effect of alloy elements in solid solution
(iv) 'scavenging' of C and N to give rise to a pure matrix.

While there is no doubt that particles do affect recrystallisation and texture evolution, most experience shows that the optimum $\bar{r}$-values are obtained in IF steels where the particles are coarse and so have minimum effect. It seems probable that precipitates have only a negative effect on the

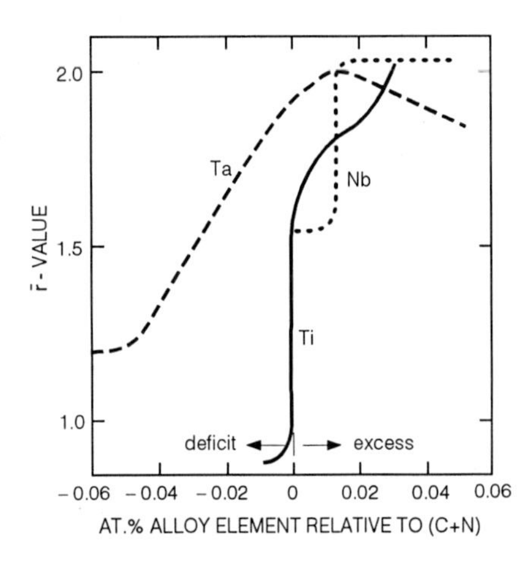

Fig. 12 Effect of stoichiometry on $\bar{r}$-value for various microalloyed IF steels [20, 21, 22].

annealing texture. Starting textures also play a role in influencing the subsequent cold rolling and recrystallisation textures. However, their effects are well documented and small relative to the observed effect on $\bar{r}$-values. Of the last two possibilities, most evidence suggests that it is the absence of C and N rather than the presence of alloy elements in solution which is responsible for the strong {111} recrystallisation texture. Support for this view comes from results on modern ultra-low carbon (ULC) steels which are produced with contents of (C + N) down to 0.005 % or even less. Such steels can give rise to $\bar{r}$-values of ~2 even without the addition of stabilising elements such as titanium or niobium [23].

The role of the alloying elements in such ultra-low carbon steels is to some extent different from that in the previous generation of IF steels. With increasing purity of the steel it becomes less easy to achieve the desired fine grain size in the hot rolled band. Precipitation of carbonitride phases during hot rolling can help to maintain a fine austenite grain size and accordingly refine the ferrite structure. Niobium is claimed to be more effective than titanium in this respect [24] and a current trend is towards ULC steels containing both titanium and niobium. The former element combines preferentially with nitrogen giving relatively coarse precipitates of TiN while niobium reacts with carbon to hinder recrystallisation and grain growth of the hot rolled band both in austenite and ferrite.

The result of having a larger grain size in the hot rolled band, as can be seen from figure 3, is to decrease $\bar{\mathbf{r}}$ and greatly increase $\Delta\mathbf{r}$. Thus, control over microalloying as well as hot rolling and cooling conditions is important for obtaining the most suitable anisotropy in ULC steels. Figures 13 and 14 show how ULC steels are improved by appropriate hot rolling and cooling practice [25, 26] giving a finer grain size which leads to an enhanced $\bar{\mathbf{r}}$-value and reduced $\Delta\mathbf{r}$, principally as a result of the increase in $\mathbf{r}_{45}$.

Unlike plain low carbon steels, the ULC steels do not suffer as a result of hot rolling into the ferrite range. It has even proved possible to produce strong {111} textures in annealed hot rolled sheets provided that recrystallisation does not occur during hot rolling itself. A problem with the practical realisation of this process is the severe through-thickness heterogenity of texture which usually arises due to the high friction conditions during hot rolling and the difficulty of finding suitable lubricants. Nevertheless, there appears to be scope for using hot rolling in the ferrite range, at least as an additive to cold rolling [27]. The 'warm' rolling texture already established means that smaller cold reductions are sufficient in order to obtain a total rolling reduction giving optimum anisotropy after final annealing.

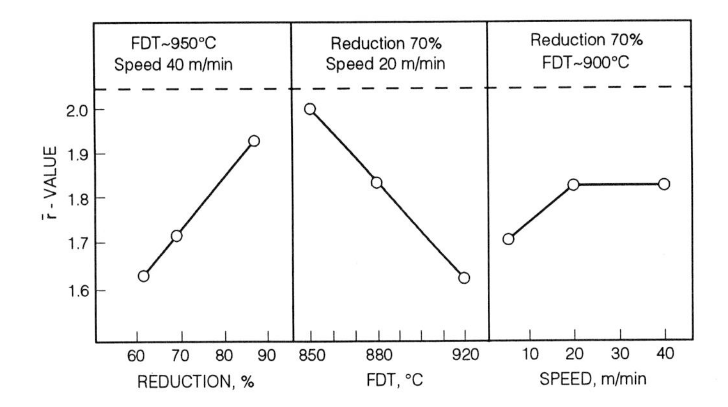

Fig. 13 *Effect of some hot rolling parameters on $\bar{\mathbf{r}}$-value after cold rolling and annealing*

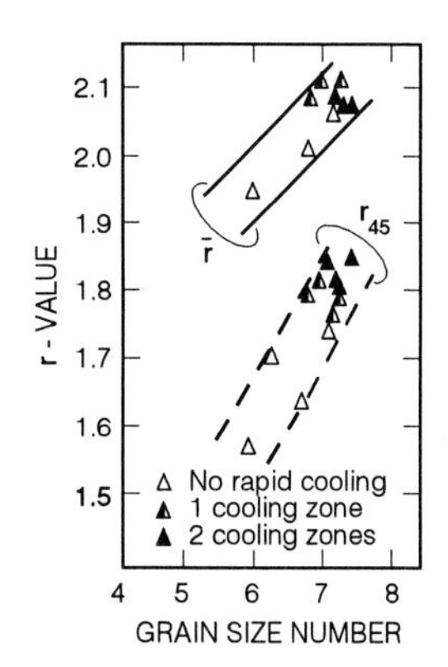

Fig. 14 *Effect of cooling practice on hot band grain size and **r**-values after cold rolling and annealing.*

REFERENCES

1. Bunge, H.J. and Roberts, W.T., J. Appl. Crystallog., 1969, 2, 116.
2. Hutchinson, W.B., Acta Met., 1989, 37, 1047.
3. Ushioda, K., Ohsone, H. and Abe, M., ICOTOM 6, Tokyo, ISIJ, 1981, p. 829.
4. Inagaki, H., Z., Metallkde., 1991, 82, 25.
5. Abe, M., Kokabu, Y., Hayashi, Y. and Hayami, S., Trans. JIM, 1982, 23, 718.
6. Hutchinson, W.B., Nilsson, K-I. and Hirsch, J., Proc. TMS-AIME Conf. Metallurgy of Vacuum Degassed Steel Products, Indianapolis, 1939, p. 109.
7. Abe, H., Scand. J. of Metallurgy, 1984, 13, 226.
8. Osawa, K., et al., Tetsu-to-Hagane,1984, 70, p. 552.
9. Karlyn, D.A., Veith, R.W. and Forand, J.L., Mechanical Working and Steel Processing VII, AIME, 1969, p. 127.

10. Hutchinson, W. B., Int. Met. Rev., 1984, 29, 25.
11. Michalak, J.T. and Schoone, R.D., Trans. AIME, 1968, 242, 1149.
12. Hornbogen, E., Met. Trans., 1979, 10A, 947.
13. Dillamore, I.L., Smith, C.J.E. and Watson, T.W., Met. Sci. J., 1967, 1, 49.
14. Takahashi, M. and Okamoto, A., Sumitomo Met., 1974, 27, 40.
15. Kubotera, H. et al., Trans. ISIJ, 1977, 17, 663.
16 Toda, K. et al., Trans., ISIJ, 1975, 15, 305.
17. Ushioda, K. et al., Mat. Sci. and Techn., 1986, 2, 807.
18. Ushioda, K., Koyama, K. and Takahashi, M., ISIJ International, 1990, 30, 764.
19. Takahashi, N. et al., Metallurgy of Continuous Annealed Sheet Steel, AIME, Dallas, 1982, p. 51.
20. Sudo, M. and Tsukatani, I., Mech. Working and Steel Processing, AIME, 1989, 547.
21. Hayakawa, H. et al., Trans. ISIJ, 1983, 23, B434.
22. Hook, R.E.et al. Met. Trans., 1975, 6A, 1683.
23. Takechi, H., Round Table Conf. on IF Steel, VDE, Düsseldorf, May 1990.
24. Obara, T. et al., Scand J. Metallurgy, 1984, 13, 201.
25. Satoh, S. et al., Tetsu to Hagané, 1980, 66, p. 1243.
26. Kino, N. et al., CAMP-ISIJ, 1990, 3, 785.
27. Hashimoto, S. and Kashima, T., Kobelco Technology Review, 1992, 13, 51.

ACKNOWLEDGEMENT

The reference list above is far too limited to do justice to the vast literature on this subject. The author acknowledges with gratitude his debt to the many researchers in different countries who created the basis for the present review. Thanks are also due to Eva Lindh and Peter Bate who contributed research material and Mineko von Euler for her care and patience in preparing this manuscript.

Materials Science Forum Vol. 157-162 (1994) pp. 1929-1940

INSPECTION AND CONTROL BY ON-LINE TEXTURE MEASUREMENT

H.J. Kopineck

Wildbannweg 36, D-44229 Dortmund, Germany

Keywords: On-Line Measurement, Fixed Angle Texture Analysis, Energy Dispersive Measurement, Ultrasonic Velocity, Magnetic Anisotropy

Abstract

- On-line texture analysis in fast-running metal sheet is described. This has been mainly based on X-ray diffraction and ultrasound velocity. The first industrial application of this method used a fixed-angle texture analyzer equipped with two energy dispersive solid state detectors measuring two diffraction peaks, each. With the short-wave component of the Bremsspectrum, this method can be used in transmission in up to 4mm iron sheet. With empirical calibration this apparatus was successfully used in a continuous annealing line. On the same basis, a universal texture analyzer was developed capable of measuring five diffraction peaks in each of five detectors. With this apparatus, low-order texture coefficients can be obtained without empirical calibration, and hence, anisotropies of various physical properties can be calculated.

  An alternative physical principle is to measure the "flight-time" of ultrasonic pulses in different directions in the metal sheet. In order to separate the influences of texture and stress on sound velocity, this method is backed up by magnetic measurements.

1 Introduction

Crystallographic textures provide information on important technological properties of metallic materials, especially of steel strips [1,2]. Therefore it is desirable to measure texture data on-line during the production processes. Such measuring has great technical and economical significance. For this on-line measuring of metallic strips nondestructive methods are required. The technological data, derived from the texture values are a very important contribution in the field of quality insurance.

The Hoesch Stahl AG, Germany, has developed systems for on-line measurement and study of crystallographic textures of rolled strip in rolling mills and has tested them over many years with excellent results. As far as is known these equipments with Hoesch are the first industrially applied on-line texture measuring instruments [3,4,5].

The first part of this paper informes on the main topics, i.e. on the characteristics of the developed procedure and equipments; next some application results of cold rolled steel strips will be presented; finally further developments will be given.

The texture of steel sheets influences not only the scattering of X-rays but also some other physical testing procedures. That means, there are some other physical testing methods, determining texture values and other technical quality data. Many research groups in the world did interesting research works in the last years, developing further procedures for testing and measuring the texture of steel and strips. Most of them have used ultrasonic waves, i.e. the fact, that the velocity of ultrasonic waves depends on the propagation direction relative to the rolling direction. American, European and Japanese groups studied this phenomene. One of them a research group of the IzfP in Saarbrücken, worked together with a group of Thyssen-Stahl AG, Duisburg [6]. The resulting procedure has great industrial interest. More details of this important research work are given in the last part of this paper.

Metallic strips, particularly wide steel strips, are nowaday produced in computer-controlled production lines. Steel strips - resulting from continuous casting lines, hot and cold strip mills, partly operating in compound - can fulfil the high demands for their quality. An important requirement is: The steel strips - made available for customers - have to have very good and homogeneous technical values; for this reason the production processes have to operate very homogeneously. For low carbon steel the technical values r_m and r are very important. They represent the quality data for the deep drawing property. It is necessary to measure them continuously. The main object of the new determination technique is the nondestructive and contactfree measuring of these values over the whole length of the strips, to get clear informations on the quality and on the technological homogeneity of the strips. The most steel-strip-processing-lines are running at high speed; therefore it is necessary that the components of the measuring equipment should be a sufficient distance away from the strip. This makes the use of X-rays necessary. It means: texture measuring gauges, using X-rays, have a considerable advantage over others using e.g. ultrasonic techniques.

2 Principles of the energy-dispersive measuring technique

The new measuring procedure has two main characteristics:

– The strips are irradiated and investigated by high-energy X-rays in transmission technique. Fig. 1 demonstrates what this means: The transmission technique allows the measuring of the entire thickness of the strips and not just of the near surface layer.

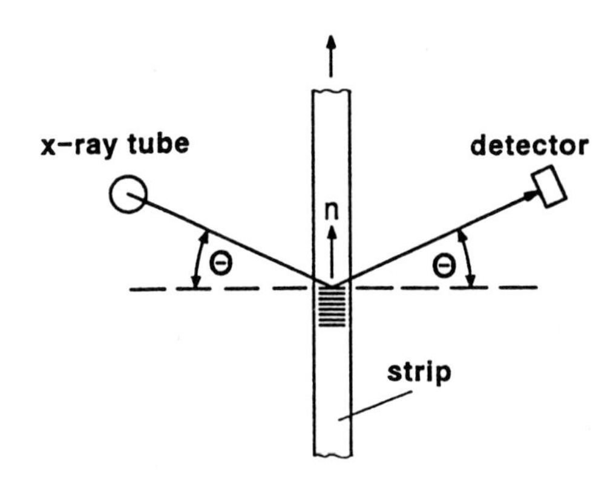

Figure 1: Principle of X-ray texture measurement in transmission technique

This method avoids therefore any requirements for special surface preparation. This is an essential condition for the on-line operation in strip plants. The use of X-rays with 60 up to 100 keV maximum energy allows the transmission through steel strips of thickness up to 2,5 mm or even up to 5 mm. Texture determination of steel strips in X-ray transmission diffraction technique requires X-rays with a nearly white energy spectrum. The bremsspectrum of X-ray tubes fulfils this condition. Physical principles allow only some selected parts of the X-ray bremsspectrum to scatter at the different lattice planes. This selection is given by the Bragg-law [2]:

$$E_{hkl} = \frac{h \cdot c}{2 \cdot \sin \Theta} \cdot \frac{1}{d_{hkl}} \tag{1}$$

written in a form using energy values E instead of the normally used wave length values λ. To get a strict correlation between energy values E_{hkl} and the corresponding lattice plane distances d_{hkl} it is necessary, that the circled part of the equation has to be constant. That means too, that in the new measuring system the diffraction angle has to be constant. From this it follows, that only the mentioned selected parts of the X-ray spectrum will be directed by the diffraction processes in to the 2Θ direction and can be measured there by detectors.

– The second characteristic point is: Energy-dispersive detectors, operating in the 2Θ position, are able to measure simultaneously the different energetic X-ray parts, given by the mentioned scattering process at correlated lattice planes. The conclusion of this is: it is possible to measure simultaneously the whole pole figures or parts of them of many different lattice planes [7].

For the X-ray equipment and the complete detection system it must be required, that they are able to operate reliably and without any problems during 24 hours per day (round the clock). Under this condition the combination of the two characteristic points in a measuring device allows the industrial application of on-line texture measurement under the rough environmental conditions of a production and processing plant for steel strips.

3 The first fixed angle texture analyzer

The first industrial application of this technique was the on-line determination of the anisotropy coefficient r_m , a characteristic parameter for the deep-drawing property of steel strips.

It could be shown, that it is sufficient, to measure only few specific points of the two pole figures (220) and (211) simultaneously and to calculate r_m from the measured intensities.

The technical realization has been done in a "fixed angle texture analyzer". These equipments have been installed as r_m-value measuring devices in the cold rolling plants of the Hoesch Stahl AG, Germany [6]. Fig. 2 shows the structural principle of the equipment: The steel strip under test is irradiated by the collimated Bremsstrahlung of an X-ray tube under a fixed angle. The X-rays are scattered into diffraction cones and are measured in two selected positions by two detectors. They are measuring the specific intensities of two pole figures (220) and (211); it means that 4 pole density values are detected simultaneously. The r_m-values are calculated from these 4 measured values, using a special calibration equation. They are also shown on a monitor and registered.

X-ray tube, detectors, cooling and other additional equipments are installed in a C-frame, guaranteeing stable measuring conditions. The C-frame can move into the measuring position in the production line; e.g. it can be fixed in the center of the line, measuring the middle part of the strip over the length. It is also possible, that the C-frame is scanning over the width of the strip. For maintenance it can be moved out of the line.

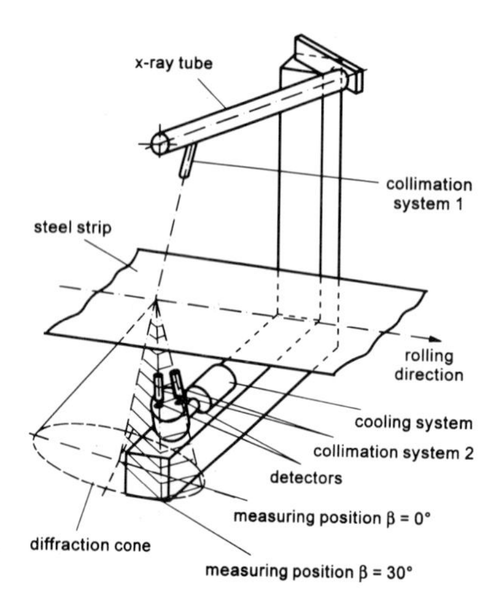

Figure 2: On-line texture analyzer for r_m-value measurement (structural principle) (Löffel-GmbH, Karlsruhe)

Figure 3: On-line measurement equipment at the end of the KRUPP-HOESCH-STAHL AG -CAL in Dortmund (KRUPP-HOESCH Stahl AG, Dortmund)

Fig. 3 shows the device during on-line measuring of steel strips in the inspection area at the end of a Continuos Annealing Line (CAL) for cold rolled steel strips of the Krupp-Hoesch Stahl AG. The strip runs here from left to right between the X-ray tube above the strip and the measuring device concealed by the strip.

In this equipment, different steel grades must be evaluated with different calibration curves. The automatic measuring device has at first to estimate the texture type of the strips from the 4 measured values and then to find out which correlation function has to be used. This is done by a special selection criterion which guarantees the calculation of the correct r_m-values.

This is very important, since logistic rearrangements do not allow clear conclusions from the quality mark of the strip to the given texture type. For an automatic device used in the CAL, this correlation is absolutely necessary. This equipment is, of course, integrated in the computer-aided production-control system of the CAL. It gives the calculated results for each strip to all controlling positions. An immediate reaction is possible. The measured r_m-values of each strip are shown on a monitor as a diagram over the whole strip length, together with additional values of the maximum and minimum and the variance coefficients. The design of the in-plant device permits an easy maintenance.

The C-frame with the X-ray tube etc. can also be used for other applications of this measuring technique; e.g. for application during the production and the processing of hot and cold rolled steel and of non-iron metal strips, especial Al-strips. The basic part of the computer software is identical for all applications.

4 In-plant experience and results

The on-line measurements in two production lines - operating around the clock - have demonstrated that the complete system works accurately and reliably. No components of the equipments had any serious failure during the whole time of more than 9 years. Since disturbances of the fast running strips can not be excluded completely, the above mentioned advantage, the sufficient distance of the equipment components to the running strip has been proved. The measured results, plotted versus strip length, are of high interest. The plots provide in a clear form condensed informations about the homogenity of deep drawing properties of the steel strips. The accuracy of the r_m-data, estimated by X-ray diffraction, is better than ± 0.1. Fig. 4 gives an example of an r_m-value recording. The mean estimated value of r_m over nearly 3 km strip length is 2.09. In this measuring example the variation s is ± 0.0333, whereby a single point represents an integrated value over several meters strip length. Small variations like this characterize steel strips with excellent homogeneity and consistent deep drawing properties. In past years these equipments allowed to learn a lot about the influence of changes of the mill parameters on the r_m-values. Variations in the given rolling parameters of the hot and cold strip mills are important; e.g. the influence of the final rolling temperature, the degree of cold rolling and the change of it by thickness deviations. The investigations gave information on how to control the production process to get the desired uniformity of the deep-drawing properties over the whole strip length. Fig. 5 shows an example with deviation from the uniformity on one end of the strip. The reason was a change in the coiling temperature at this end of the hot rolled strip.

Based on the extensive experiences it is possible to state:

- The on-line r_m-value measuring devices used in the cold rolling plants are well tested and successfully operating industrial instruments.
- They are able to evaluate reliably the deep-drawing quality parameters Πof steel strips.
- TheyΠare important devices for the nondestructive measuring of these technological values under production conditions.
- They allow cold strip mill engineers to offer steel strips to their customers with high homogeneity and quality over the full strip length [8,9,10,11,12].

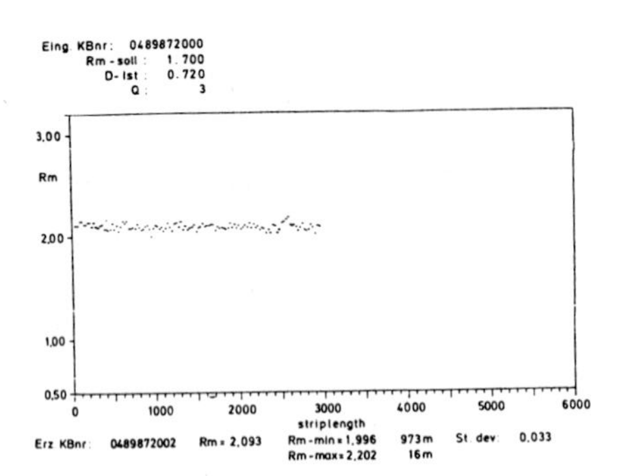

Figure 4: Measurement example: uniform r_m-values along the length of a cold rolled steel strip (KRUPP-HOESCH Stahl AG, Dortmund)

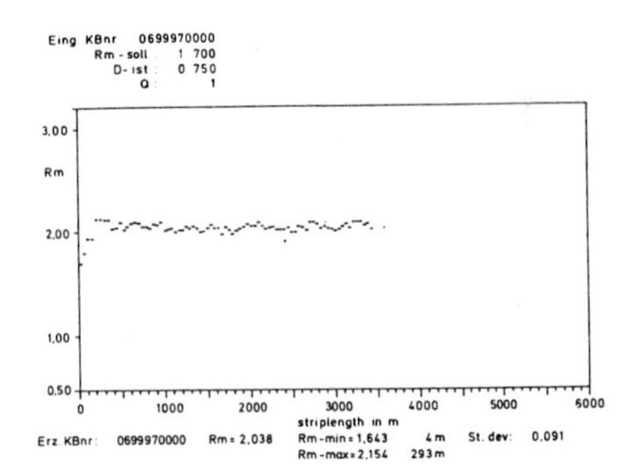

Figure 5: Measurement example: Deviation of the uniformity at one strip end, caused by changing of the coiling temperature of the hot strip (KRUPP-HOESCH Stahl AG, Dortmund)

A second field, using this measuring technique, refers to very thin tinned steel strips or aluminium strips for cans etc. In this material the technological Δr-value has great importance, since it correlates with the value of "earing", a quality value of this material. Investigations with the Hoesch Stahl AG gave the following correlation - see Fig. 6. The ratio of intensities of the two poles of the (220)-reflex are clearly correlated with the ear-values. The figure shows: in the range of the intensity ratio between 2 and 3 there is the domain of minimum ear-values, given by 6 ears at 0°, 60° etc. When the ratio is smaller or larger, the ear-values increase, because there are changes to the 4-fold ears at 0°, 90°, or 45°, 135° [13]. It is possible to supervise the production by a similar on-line texture measurement equipment as mentioned above. It is intended to do so.

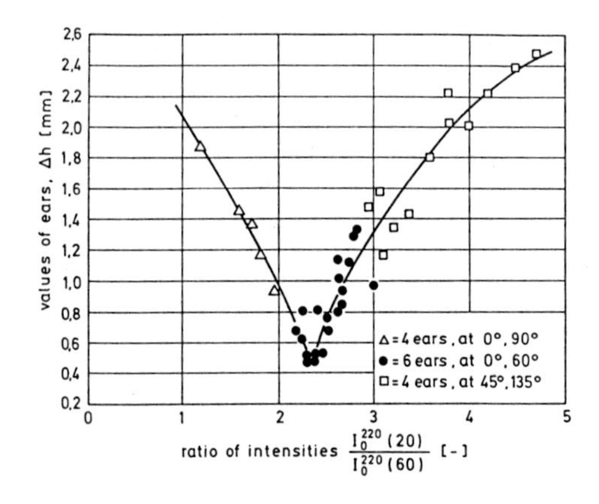

Figure 6: Dependence of the earing values in thin, tinned steel sheets on the ratio of two pole density values (KRUPP-HOESCH Stahl AG, Dortmund)

5 Extension of the measuring technique

The achieved results stimulated an extension of the measuring technique, namely:

- In some cases it is desirable and useful to get not only the r_m-value but also the $r(\beta)$-values, i. e. the dependence of r of the angle to the rolling direction.
- Extensive quality programs furthermore require, to avoid the material dependence of the texture values and the coupled calibration work at the simple measuring technique; it means, it is necessary to get a conditionfree measuring technique.

The two requirements involve an extension of the measuring technique. At first such an extension has to be proved theoretically and than to be put into practice [13-16]. The object of the "conditionfree determination" of the texture-dependent material values is to calculate these values without knowledge of the different quality types and without any calibration. An investigation of this matter with the ODF-method confirmed, that it will be possible to do so [14-17].

It is known that many physical or technological properties of materials depend only on the low-order C-coefficients of the series expansion of the ODF e.g. C_4^{10}, C_4^{12}, C_4^{14}.. It is possible to express the r-value of steel and other material properties in terms of these C-coefficients only. The accuracy of the determination of these coefficients depends on the number of measuring points and on the positions of the points in the used pole figures. It has been shown that this method of conditionfree determination gives very good values, if 20 ore more characteristic measuring points are used. This means, with 4 detectors measuring simultaneously 5 lattice plane intensities at 4 points, a successful measuring by an on-line technique is possible.

An other point is also important: The investigation and texture measuring of hot strips is also very interesting. It could be shown, that an increase of the X-ray maximum energy up to 100 keV still allows texture measurements. This energy value allows to penetrate hot strips up to 5 mm thickness and to measure their texture-values.

The transposition of the theory into praxis gave important results, as shown e. g. in Fig. 7. In this figure the $r(\beta)$ values calculated from X-ray intensities of 20 measured points, are plotted versus mechanically determined values for $\beta = 0°$, 45°, 90°. Also the corresponding measured and calculate r_m-values are plotted. The good agreement can be seen. The result of this investigation is, that it is possible to calculate the $r(\beta)$-value without any precondition from 20 measuring values, i.e. from 5 pole figures with 4 measuring points, each.

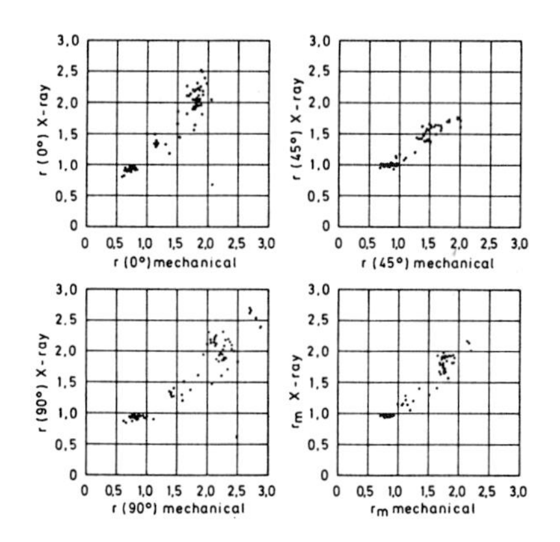

Figure 7: Correlation between X-ray-measured $r(\beta)$-values and the corresponding mechanically-measured values (20 texture measuring points) (KRUPP-HOESCH Stahl AG ,Dortmund)

6 The new universal texture analyzer

The transfer of these laboratory results into an on-line measuring device is given in Fig. 8. The figure shows that the principle has been adopted from the on-line measuring device, described before. Above the steel strip a 100 keV X-ray tube is installed. From the emitted X-rays, 4 single X-ray beams are separated. Below the strip there is the detector system, using 4 energy-dispersive detectors with a closed-loop cooling system. They are measuring simultaneously the intensities of together 20 points. This new equipment has been build. After the usual tests in the workshop it will be installed in the rolling plant. Fig. 9 shows a picture of the measuring part of this device in the workshop (without protection shields).

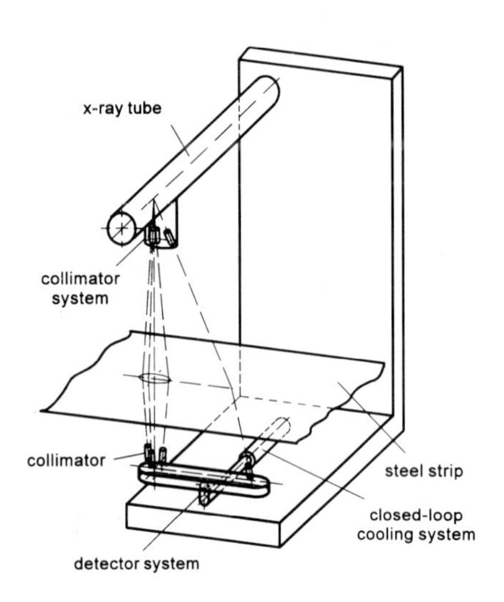

Figure 8: Universal form of the On-line, fixed angle texture analyzer (schematic) (LÖFFEL GmbH, Karlsruhe)

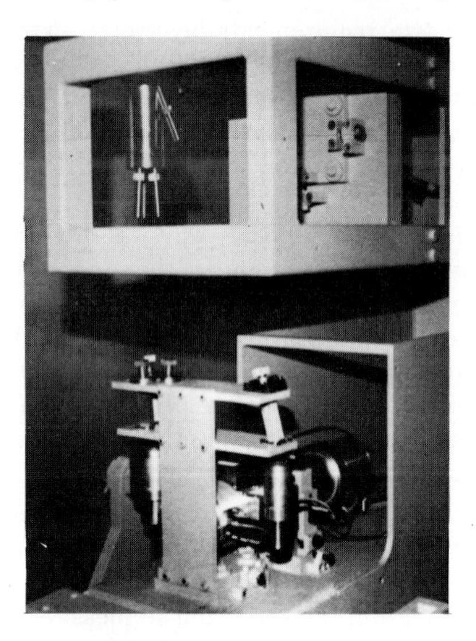

Figure 9: The measurement part of the universal fixed angle texture analyzer (above: X-ray tube, below: detectors and cooling system, without protecting shields) (KRUPP-HOESCH Stahl AG, Dortmund)

This equipment is the universal form of the "fixed angle texture analyzer". With this device it is possible, as mentioned above, to investigate the texture values also of hot strips. It gives the possibility to get information on quality and homogeneity related to the texture over the whole strip length in a very early production step. This is important, since the texture of the hot strip influences the cold strip texture, i.e. the quality of the cold strip. The on-line hot strip texture measurement can be used to give control informations for the following production steps. This point shows the second direction of the extended measuring technique, namely the open-up for additional applications: In table 1 some application fields for the two types of the "fixed angle texture analyzer" are listed.

Table 1: Fixed angle texture analyzer - application fields for On-line evaluation of technological properties of metal strips)

Application fields	Technological properties evaluated by	
	"Normal-type" texture analyzer	"Universal" texture analyzer
Cold rolled steel strips		
Deep drawing qualities	r_m-value Degree of recrystallization	$r(\beta)$-, r_m-, Δr-values Degree of recrystallization Young's modulus E
Packing steel qualities	Δr-value	Δr-value
Electrical steel sheets		Electromagnetical properties
Hot rolled steel strips		
Low carbon steel qualities	r_m-value	$r(\beta)$-, r_m-, Δr-values
Cold and hot rolled aluminum	Δr-value	Δr-value
Other NF-metals (Titanium) ceramics	Test for random	Test for random

The universal texture analyzer allows the determination of $r(\beta)$- and the Δr-values from the C-coefficients. It can be applied at hot and cold steel strips and also at packing material, steel and aluminium, for which the above mentioned Δr-values are characteristic and important. Investigations of the degree of recrystallisation, particularly in the neighbourhood of the complete recrystallisation, are very significant. In this field investigations are running; their object is to measure either with the "simple" or the "universal" analyzer and to use the results to control the annealing lines in an optimized form. Other applications of this proved texture measuring technique seem to be possible in the following fields:

- Determination of electromagnetic properties of electrical steel sheets.
- Supervising the texture of titanium, applied in aircraft structures, with regard to safety questions.
- Ceramics; first investigations allow the conclusion, that a very interesting field of application is given here.

As mentioned above, the adaptation of the equipment to these other application fields is without problems; it is possible by changing only parts of the software.
Finally it can be stated: these texture analyzers, developed and proved by Hoesch are important instruments for quality assurance of rolled strips.

7 A new ultrasonic industrial texture measuring instrument

As mentioned above a research group of the Nondestructive Testing Institute in Saarbrücken (IzfP) developed a new method for determining the texture of steel strips. In this research first the relation between flight-time of ultrasonic waves and the technical values r_m and Δr were investigated. The flight-times are influenced by internal stresses in the strips, too. Therefore additional testing methods, especially several magnetic methods such as measuring the Barkhausen-noise, have been used to eliminate this influence and to obtain measured values in agreement with mechanically measured data e.g. r_m and Δr.

The procedure has been tested in a pilot arrangement , operating in a cold strip mill inspection line of Thyssen-Stahl AG, Duisburg. The results are very good [6]. The pilot equipment has been operating very sufficiently during a long in-plant testing time. The new procedure is helpful for quality insurance and allows the control of production lines. Figure 10 shows, as an example, the correlated data measured mechanically and by the new non-distructive methods [18]. The agreement within one set of sheets of only one quality is very good. It has to be kept in mind, however, that the X-ray gauge has a gap of about 500 mm, whereas the magnetic and ultrasonic gauges use a very small gap of about 1-2 mm which must further be kept constant with only very narrow tolerance. The new testing equipment with Thyssen-Stahl AG guaranties this small gap in a new developed probe. The mentioned report [6,18] gives many details of the procedure and of the pilot equipment as well as of the results. This new procedure allows many further applications. This means, that also this new instrument can be a helpful equipment for quality control and -insurance-.

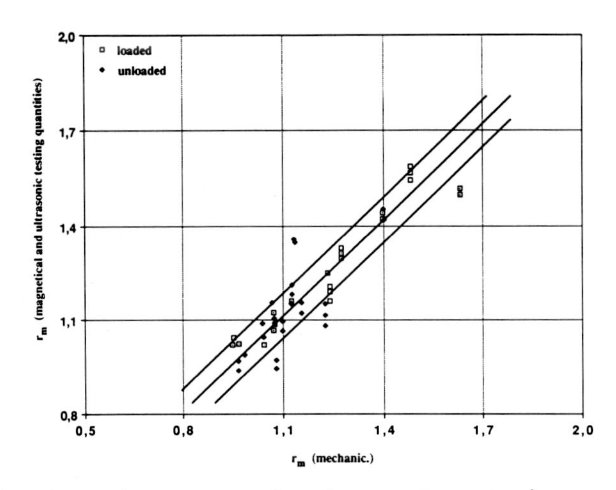

Figure 10: Correlation between r_m-values from mechanical measurements and those obtained from ultrasonic measurements combined with magnetic testing

8 Summary

In the production of metallic strips, especially of steel strips, it is of great importance, that the quality and the technical data should be homogeneous over the entire length of the strips. This can only be monitored using continuous on-line non-destructive methods.

The texture of strips gives interesting informations on their technical behaviour. It influences some physical testing procedures, e.g. the scattering of X-rays. A new X-ray transmission technique has been developed by Hoesch-Stahl AG, Dortmund, for the non-destrucitve determination of texture depended technological data of rolled strips. It can be applied on-line to both hot and cold rolled steel strips and other metallic strips.

Besides the scattering of X-rays, there are also some other physical testing procedures which give information on the texture. Many research groups in the world did a lot of work in the past, developing further procedures for testing and measuring the texture of metallic strips. Some groups use ultrasonic testing respectively the fact that the volocity of ultrasonic waves depends on the propagation direction relative to the rolling direction. The Institute for Non-destructive Testing (IzfP) in Saarbrücken together with Thyssen Stahl AG, Duisburg, has studied this phenomenon. They got good results and found that the ultrasonic wave velocities are additionally influenced by internal stresses. To eliminate this effect, they developed a new procedure, combing ultrasonic and magnetic measurements. They also build a special probe for these two measurements as well as a pilot installation, which has been testet in a cold strip production line. The first results are very good. They have been published in "Stahl und Eisen"[6]. In this paper one may also find many details of the new procedure, and of the measuring pilot installation operating on-line in a cold mill production line. In the present paper only one of these result can be given, published in a report for the C E C [18]. The new procedure has many advantages. So it may be applied and testet in some interestings fields, e. g. at Zinc-coated and annealed steel strips. It can also control the line. The extension for measuring further characteristic values e.g. hardness and tensile strength may be possible, combining the new procedure with eddy current measuring. That means, there are two important texture measuring procedures, which allow to measure one-line and non destructively the important technological data r_m and Δr of steel strips. They also allow to control production processes and they give many informations to the quality-insurance-systems. They are the first industrial texture measuring installations, operating on-line and continuously in steel production lines.

Acknowledgement
The author has to thank the IZfP, Hoesch-Stahl AG, now Krupp-Hoesch-Stahl AG, and Thyssen-Stahl AG for many informations on the newest results of the two measuring procedures.

9 References

1) Bunge, H.J.: Textur und Anisotropie, Z. Metallkunde 70, (1979) 411-418

2) Bunge, H.J.: Texture Analysis in Materials Science, Butterworths Pub., London 1982

3) Maurer, A., Böttcher, W., Kopineck, H.-J.: Vorzugsorientierungen kaltgewalzter Bänder im Verlauf des Walzprozesses,Abschlußbericht zum Forschungsvertrag Nr. 6210-GA04 (August 1978),Herausgegeben von der Kommission der Europäischen Gemeinschaft unter Nr. EUR 6291 DE, 1980

4) Maurer, A., Böttcher, W., Kopineck H.-J.: Bestimmung von Werkstoffkennwerten an kaltgewalzten Blechen mittels Röntgendurchstrahlung und energiedisperiver Meßmethode, Vorträge der 1. Europäischen Tagung für zertörungsfreie Materialprüfung, Bd. 2, 421-426, Deutsche Gesellschaft für zerstörungsfreie Prüfung e.V., Berlin 1978

5) Maurer, A.: Ein neuartiges Untersuchungssystem zur zerstörungsfreien Bestimmung von Werkstoffkennwerten an Feinblechen mittels Röntgenfeinstrukturverfahren, Dissertation Münster 1982

6) Borutzki, M., Thoma, Ch., Bleck, W., Theiner, W.: On-line Bestimmung von Werkstoffeigenschaften in kaltgewalztem Feinblech, Stahl und Eisen 113 (1993) 93-100

7) Böttcher, W., Kopineck, H.-J.: Über ein Röntgentexturmeßverfahren zur zerstörungsfreien On-line-Bestimmung technologischer Kennwerte von kaltgewalzten Stahlbändern, Stahl und Eisen, 105 (1985), 509-516

8) Kopineck, H.-J., Otten, H.: Texture Analyzer for on-line r_m-value Estimation, Textures and Microstructures, 7, (1987), 97-113

9) Kopineck, H.-J., Böttcher, W., Otten, H., Trültzsch, K.L.: The on-line r_m-value measuring of cold rolled steel strips in the exit of a continuous annealing line, 4th International Steel Rolling Conference, Deauville, June 1-3, 1987

10) Kopineck, H.-J., Böttcher, W., Otten, H.: On-line-Meßanlage zur r_m-Wert-Bestimmung mittels Röntgenstrahlbeugung an kaltgewalzten Stahlbändern, Berichte aus Forschung und Entwicklung unserer Gesellschaften der Hoesch AG, Heft 1/87, S. 15-18

11) Bunge, H.-J.: Technological Application of Texture Analysis, Zeitschrift. Metallkunde 76 (1985) 457-470

12) Kopineck, H.J.: Industrial application of on-line texture measurement, 3rd International Symposium on Nondestructive Characterization of Materials, October 2-6, 1988 Saarbrücken

13) Fabian, J.: Texturuntersuchungen zur zerstörungsfreien Ermittlung von Umformeigenschaften an Feinstblechen mittels Röntgenfeinstrukturverfahren, Diplomarbeit - Universität Münster - 1984

14) Kopineck, H.-J., Otten, H., Bunge, H.-J.: On-line determination of technological characteristics of cold and hot rolled steel strips by a fixed angle texture analyzer, 3rd International Symposium on Nondestructive Characterization of Materials, October 2-6, 1988 Saarbrücken

15) Bunge, H.-J., Schulze, M., Grzesik, D.: Calculation of the Yield Locus of Polycrystalline Materials According to the Taylor Theory, Peine + Salzgitter Berichte, Sonderheft 1981

16) Wagner, F., Otten, H., Konineck, H.-J., Bunge, H.-J.: Computer aided optimization of an on-line texture analyzer, 3rd International Symposium on Nondestructive Characterization of Material, October 2-6, 1988 Saarbrücken

17) Otten, H.: Zur Bestimmung von texturabhängigen Werkstoffeigenschaften mittels eines On-line Röntgentexturanalysators, Clausthal, 1988 (Thesis Universität Clausthal)

18) Schneider, E.: Zerstörungsfreie Bestimmung von Texturen in Walzprodukten mit Ultraschall- und magnetischen Verfahren, EGKS-Forschungsprojekt E 1/88 (7210-63111), Abschlußbericht (Entwurf) 1992

Materials Science Forum Vol. 157-162 (1994) pp. 1941-1946

THE INFLUENCE OF TEXTURE ON THE MAGNETIC PROPERTIES OF Co-Cr FILMS

L. Cheng-Zhang [1], J.C. Lodder [2] and J.A. Szpunar [3]

[1] On leave from Institute of Computing Technology, Academia Sinica, PO Box 2704-6, Beijing 100080, China

[2] MESA Research Institute, Faculty of Electrical Engineering, University of Twente, PO Box 217, NL-7500 AE Enschede, The Netherlands

[3] Department of Metallurgical Engineering, McGill University, 3450 University Street, Montreal, Quebec H3A 2A7, Canada

Keywords: Magnetic Properties, Co-Cr Films, Texture, Anisotropy

ABSTRACT

Two Series $Co_{81}Cr_{19}$ films were chosen to investigate the influence of texture on magnetic properties. It was demonstrated that both the texture and the magnetic properties strongly depend on the film thickness. In the thickness range from 12 to 980 nm, the coercivity $Hc\perp$ and the orientation ratio $OR_m\perp$ of the magnetization are mainly determined by the orientation of a hcp crystallites. In the case of Series B films having thickness range of 46-980 nm, there is a critical thickness, at which the $Hc\perp$, the orientation ratio $OR_m\perp$ and the orientation ratio ORc of the crystallite reach their maximum values and the strain ε in film is about zero. The result obtained proves that $Hc\perp$ is mainly controlled by the magnetocrystalline and shape anisotropy, and that the magnetostriction anisotropy exerts a rather small influence on $Hc\perp$. In the thickness range of 5-12 nm, the $Co_{81}Cr_{19}$ films exhibit an anomalous behaviour.

INTRODUCTION

The sputtered Co-Cr film was extensively investigated as a perpendicular recording media where the microstructure can be developed with a columnar grains and a strong texture [1-4]. A vast amount of experimental data shows that the growth direction of the columns and the (0002) direction of a hcp crystallite are oriented along the direction normal to the film. Obviously, such microstructure favours the alignment of the magnetization in the direction normal to the film plane.

As was demonstrated in our previous paper [4], for two Series of $Co_{81}Cr_{19}$ films having the thickness range between 5 and 980 nm, the texture and the internal strains change with the film thickness [Tables 1 and 2]. For Series B films, there is a critical film thickness, below which the ORc value increases with increasing thickness and above which the ORc value decreases with increasing thickness. In the vicinity of 130 nm, the orientation ratio ORc reaches its maximum value. The strain changes with the film thickness as well, as the film thickness increases from 46 to 980nm, the film strain ε along the film normal gradually changes from a tensile to compressive strain, and in the vicinity of 130nm the film is stress-free. Therefore, for Series B films both the highest orientation ratio ORc and the lowest strain ε along the film normal appear at the same thickness . In this case, a high quality perpendicular recording media with strong texture and stress-free conditions can be prepared by optimizing the film thickness. Based on these results, the relationships between film texture and magnetic properties will be investigated.

EXPERIMENTAL PROCEDURE

Two Series of sputtered $Co_{81}Cr_{19}$ films were chosen to investigate the correlation between the texture evolution and the magnetic properties, which is affected by changes in the film thickness. Series A and B films were sputtered on the $Ge/SiO_2/Si$ and Si substrate respectively. The Si substrate is a single crystal of Silicon and the substrate plane coincides with [100] planes. The thickness of the films ranges from 5 to 980 nm. To measure the thickness the SEM, STM (Scanning Tunnel Microscope) and/or AFM (Atomic Force Microscope) were used. The magnetic properties were measured by VSM and a Torque magnetometer. The maximum applied field for measuring the hysteresis loop is about 840 kA/m. The most relevant properties for Series A and B films are summarized in Table 1 and 2 .

Table 1: Properties for Series A $Co_{81}Cr_{19}$ films.

No	Thick. nm	Ms kA/m	Hc⊥ kA/m	K_1 kJ/m³	Rs‖	ORm⊥	ORc	Δα50 degrees
A_1	5	360	4	122	0.08	12.5	3.5	4.75°
A_2	12	460	4.3	116	0.83	0.024	9.6	3.94°
A_3	25	500	15.4	127	0.33	0.18	22	3.25°
A_4	50	485	36.8	112	0.15	0.73	32	2.84°
A_5	100	510	47.2	134	0.09	1.34	40	2.93°
A_6	200	500	69.9	131	0.06	2.84	44	3.03°

Table 2: Properties for Series B $Co_{81}Cr_{19}$ films.

No	Thick. nm	Ms kA/m	Hc⊥ kA/m	K_1 KJ/m³	Rs‖	ORm⊥	ORc	Δα50 degrees
B_1	46	450	79	94	0.11	2.4	31	3.03°
B_2	110	470	96	115	0.034	7.5	41	3.0°
B_3	297	445	65	99	0.031	5.1	36	3.5°
B_4	611	435	46	84	0.034	3.2	27	4.0°
B_5	982	430	40	86	0.036	2.7	19	4.47°

Where the Rs⊥ and Rs‖ are the squareness ratio Mr⊥/Ms and Mr‖/Ms, along perpendicular and in-plane direction respectively. ORm⊥ is so-called orientation ratio of the magnetization and is defined as the ratio of Rs⊥/Rs‖, reflecting the anisotropy of the remanence along the easy and hard direction. Obviously, to achieve a high quality perpendicular recording media, it is desirable to have the ORm⊥ value as high as possible and the Rs‖ value as low as possible.

The pole figures of the specimens was measured using a Siemens D-500 texture goniometer (Mo-Kα radiation). In addition, the precise orientation distribution of (0002) pole density was measured. This pole density F(α) was registered as a function of angular deviation from the film normal. Data were collected in a 12° angular interval as a function of the specimen tilt angle α. The ratio $OR_c(\alpha,\beta)$ of the pole density $F(\alpha,\beta)$ of the film to that of a random specimen was used to evaluate the orientation ratio using the following expression:

$$OR_c(\alpha,\beta) = \frac{\left(\frac{1}{\alpha_2-\alpha_1}\right)\int_{\alpha_1}^{\alpha_2}\int_0^{2\pi} F(\alpha,\beta)d\alpha d\beta}{\frac{2}{\pi}\int_0^{\frac{\pi}{2}}\int_0^{2\pi} F(\alpha,\beta)\, d\alpha d\beta} \tag{1}$$

Where β is the circumferential angle of pole figure. In our case, the value of $F(\alpha,\beta)$ is assumed to be equal to $F(\alpha)$, since the texture in the α-Co phase exhibits high degree of rotational symmetry about the film normal. The difference of (α_2 -α_1) is equal to 0.1°. For the sake of convenience, in this report the ORc value, which represents the maximum $ORc(\alpha,\beta)$ for (0002) planes, will be called the orientation ratio of the crystallite. The $\Delta\alpha 50$ FWHM (full width-half maximum), which is derived from the precise orientation distribution curve for (0002) planes, will be used to characterize the dispersion of (0002) planes.

Based on the model discussed earlier [5], where the magnetic material is considered as an assembly of the single domain fine particles, and we assume the Co-Cr film is composed of the single domain grains, which are completely isolated and separated from each other by paramagnetic Cr-rich boundaries, the coercivity $Hc\perp$ can be calculated using the following formulae:

$$Hc\perp = \frac{2T}{M}\sum_{L=0}^{L} T_L \delta_{CL} \tag{2}$$

$$\frac{2T}{M} = H_a + H_d + gH_\sigma = \frac{2K_1}{Ms} + (N_o - N_t)Ms + g\frac{3\sigma\lambda_s}{Ms} \tag{3}$$

Coefficient δ_{CL} is mathematical constant representing the texture influence on the coercive force. This coefficient can be found out in references [4,5]. The term of 2T/M is, in fact, the effective anisotropy field. The terms Ha, Hd and $H\sigma$ in the formula (3) represent the magnetocrystalline, the shape and the stress anisotropy field. N_o and N_t are the demagnetizing factor along the easy and the hard direction respectively. K is the magnetocrystalline anisotropy constant. In our case, K is replaced by the first order anisotropy constant K_1, because for most of Co-Cr films the second order anisotropy constant is much smaller than K_1. σ and λ_s are the stress and the saturation magnetostriction coefficient($\doteq -12*10^{-6}$) in Co-Cr film respectively. g is a fitting coefficient.

The related texture expansion coefficients T_L in the formulae (2) are calculated using the formula:

$$T_L = \frac{(2L+1)}{2}\int_{\alpha_1}^{\alpha_2} F(\alpha)P_L(\alpha)d\alpha = \frac{(2L+1)}{2}\sum_{L=0}^{L}\sum_{\alpha_1}^{\alpha_2} F(\alpha)P_L(\alpha)\Delta\alpha \tag{4}$$

$F(\alpha)$ is the measured orientation distribution curve of (0002) planes and $P_L(\alpha)$ is the Lth order Legendre polynomial. Considering the fact that this orientation distribution curve $F(\alpha)$ for tested $Co_{81}Cr_{19}$ films exhibits a symmetrical shape in the vicinity of the film normal, only the Legendre polynomials having even L values are involved during the calculation of $Hc\perp$ and $Mr\perp/Ms$. In our case, the maximum L value of the series expansion is taken as 20.

MAGNETIC PROPERTY IN FILMS HAVING DIFFERENT THICKNESSES

For both Series A and B films having the thickness higher than 12 nm, as shown in Tables 1 and 2, as the film thickness increases, a pattern of magnetic properties changes with thickness exhibits the characteristics, which are typical for perpendicular recording media. For example, for Series A films, as the thickness increases from 12 to 200 nm, the coercivity Hc⊥ increases from 4.3 to 69.9 kA/m and the orientation ratio ORm⊥ increases from 0.024 to 2.8. For the same specimens the squareness ratio Rs∥ along the in-plane direction decreases from 0.83 to 0.06. It was demonstrated earlier [9] that in the same thickness range, for Series A films the orientation ratio ORc of the crystallite increases with increasing thickness. Based on these experimental data, it can be argued that for $Co_{81}Cr_{19}$ films having thickness higher than 12 nm, the magnetic properties are mainly determined by the magnetocrystalline anisotropy, because both ORc⊥ and ORm⊥ increase with increasing ORc value. However, if the film thickness is lower than 12 nm, the dependence of magnetic properties on the thickness exhibits anomalous characteristics. For Series A films having thickness higher than 12 nm, the curve representing the normalized hysteresis loss ΔWh⊥ vs. applied field has a single peak only, and the peak position always deviates from the zero field. As the thickness increases, the peak position gradually moves to a higher applied field and the magnitude of these peaks decreases monotonously.

Summing up, for Series B films, as is shown in Table 2, as the film thickness increases, the orientation ratio ORm⊥ and also the coercivity Hc⊥ first increase with increasing the thickness, and then decrease with increasing thickness. It can be concluded from these observations that there is an optimum thickness, where the ORm⊥ and Hc⊥ values reach the maximum values. For example, as the thickness increases from 46 to 110 nm, the ORm⊥ value increases from 2.4 to 7.5 and Hc⊥ from 79 to 96 kA/m. On the contrary, if the thickness continues to increase from 110 to 980 nm, the ORm⊥ and Hc⊥ values monotonously decrease from 7.5 to 2.7 and from 96 to 40 kA/m respectively. In addition, a similar pattern of the change in the dependence of ORc on the thickness was also observed.

COMPARISON OF THE CALCULATED AND MEASURED TEXTURE DEPENDENT MAGNETIC PROPERTIES

In order to calculate the influence of texture on the magnetic properties, a distribution of the crystallographic C-axis orientation I(α) were measured [4]. According to the formula (4), the texture expansion coefficients T_L were calculated. F(α) is the normalized orientation distribution of (0002) planes, which is derived from the measured I(α). Obviously, if we want to calculate the influence of the texture on Hc⊥, the effective anisotropy field must be calculated first. To assess the contribution of the magnetocrystalline anisotropy field (Ha=2K/Ms), the shape anisotropy field ($Hd=(N_o - N_t)Ms$) and the stress anisotropy field ($H\sigma=3\sigma\lambda_s/Ms$) to the effective anisotropy field, the shape anisotropy and stress anisotropy will be discussed.

Based on the experimental data [6,7], it was known that the columnar grains in Co-Cr film are surrounded by the Cr-rich boundaries of paramagnetic phase. It is reasonable to assume that these columnar grains in $Co_{81}Cr_{19}$ film form an assembly of non-interacting single domain columns, whose magnetization reversal mechanism is basically controlled by the rotational reversal. For simplicity, the shape of the columnar grain in our calculation is considered to be a cylinder. For a single-domain column with the cylindrical shape, the demagnetizing factor along the length is a function of the aspect ratio (the ratio of the length to the diameter of the columnar grain), and can be found in references [8].

According to the experimental data of the correlation between the aspect ratio and the thickness [4] as well as the first order magnetocrystalline anisotropy constant K_1, the effective anisotropy field (2T/M =Ha+Hd) can be calculated, if the magnetostriction effect is neglected. Then, according to the formula (2) the Hc⊥ as a function of the thickness is calculated.

For Series A and B films, the calculated Hc⊥ values for different thicknesses calculated without taking into consideration of the magnetostatic interaction between the columns were plotted in Fig. 1. In order to compare the calculated Hc⊥ with the measured Hc⊥ values, the measured Hc⊥ as a function of the thickness was drawn. It can be seen from Fig. 1 that the changes of the calculated Hc⊥ with the thickness are in a good agreement with experimentally observed changes. However, there are a large difference in magnitude between the measured and the calculated Hc⊥ values. This fact might indicate that the influence of the magnetostatic interaction on the calculated Hc⊥ should not be neglected.

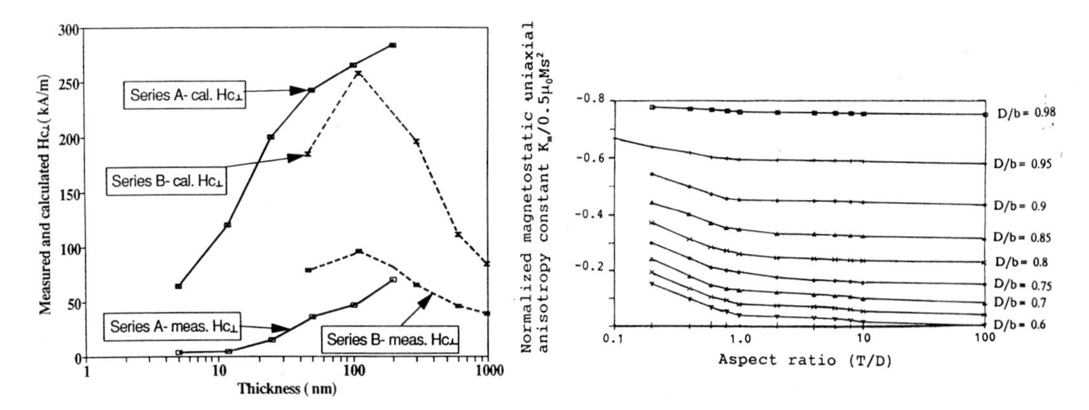

Fig.1 : The measured and calculated Hc⊥ values (without megnetostatic interaction) versus thickness for Series A and B films.

Fig.2: The K_m value as a function of the aspect ratio for different film densities (D/b).

Methods of calculation of the uniaxial anisotropy resulting from the magnetostatic energy in magnetic films with a columnar structure were introduced in reference [9]. The ratio T/b is a parameter of the model, where T and D are the length and diameter of the cylinders respectively, and b is the spacing between neighbouring cylinders. Considering that in Co-Cr films the "packing density" is usually close to 1, for a film having high density the relationship between K_m and the aspect ratio (T/D) can be shown in Fig. 2. It can be seen from Fig. 2 that the larger the film density the larger is the negative magnitude of the magnetostatic uniaxial anisotropy constant. The aspect ratio T/D has only a small influence on the K_m value as compared to the influence of the film density.
In order to fit well the calculated Hc⊥ value to the measured one, using our calculation procedure the film densities are assumed as 0.98 and 0.95 for Series A and B films. According to the formulae (2), (3) and (4) as well as under the prerequisite that $H_d = 2K_m/Ms$, the Hc⊥ values for Series A and B films were calculated. The changes of the measured and calculated Hc⊥ with thickness were plotted in Fig. 3. Comparing the Fig. 1 with the Fig. 3, it can be clearly seen that the Hc⊥ values calculated using the assumption of the influence of the magnetostatic interaction are much closer to the measured Hc⊥ values than those obtained when neglecting the interaction effect, exception are for films thinner than 12 nm.

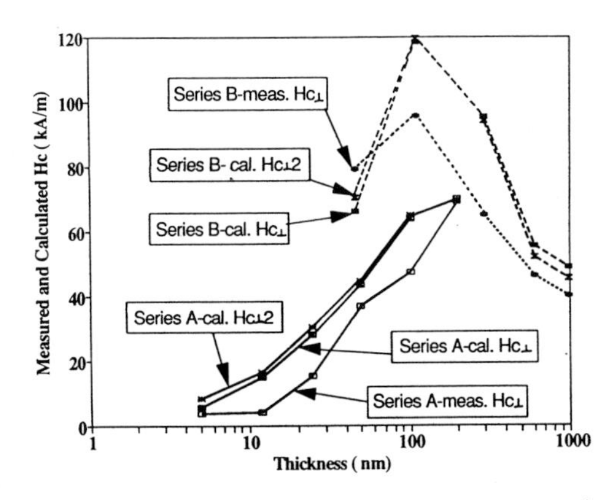

Fig. 3: The measured Hc⊥, calculated Hc⊥ (incl. magnetostatic interaction), and calculated $Hc\perp_2$ (incl. interaction and magnetostriction effects) for Series A and B films.

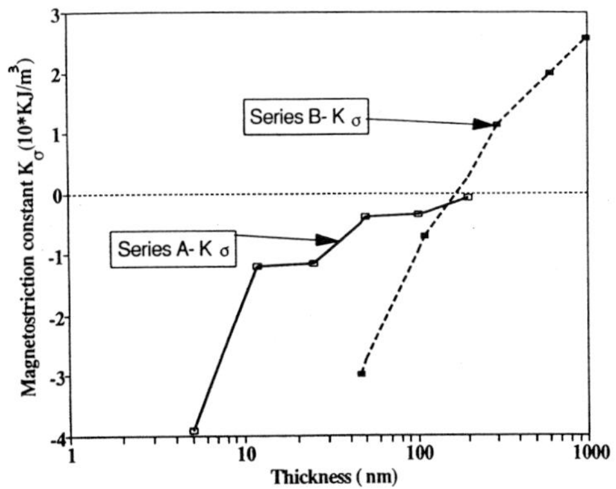

Fig.4 : The magnetostriction constant K_σ for Series A and B films.

Based on the measured changes of the film strain ε as a function of the thickness, for Series A and B films the magnetostriction anisotropy constant K_σ could be calculated, assuming that the Young's modulus is $21*10^{11}$ dynes/cm^2 and the saturation magnetostriction constant is $-12*10^{-6}$ [6]. The calculated magnetostriction constant K_σ is shown as a function of the thickness in Fig. 4. For Series B films, as the film thickness increases from 46 to 980 nm, the film magnetostriction constant K_σ gradually changes from negative to positive value and in the vicinity of 130 nm becomes zero. This fact suggests that the existing magnetostriction anisotropy can strengthen perpendicular anisotropy in thick films. On the contrary, the magnetostriction anisotropy in thinner films (<130 nm) will weaken the perpendicular magnetic anisotropy. For Series A films, the magnitude of negative magnetostriction constant monotonously decreases with increasing thickness. Based on the above-results, if the magnetostriction constant K_σ is used to calculate the magnetostriction anisotropy field $H\sigma$, the $Hc\perp$ value can be calculated. The changes of $Hc\perp$ with the thickness calculated assuming both the effects of the magnetostriction anisotropy and the magnetostatic interaction were compared to experimental results in fig. 3. From this figure the conclusion can be made that the magnetostriction anisotropy has only a minor influence on the coercivity $Hc\perp$ compared to the shape anisotropy.

CONCLUSIONS

1) For $Co_{81}Cr_{19}$ hcp films the textural evolution exerts a tremendous influence on the magnetic properties. In the thickness range of(12-982 nm) the changes of both $Hc\perp$ and the magnetization orientation ratio $ORm\perp$ with the thickness follow the same change in the pattern of the orientation ratio ORc of the crystallite very well, indicating that in this case the orientation of the texture plays a dominant role in the determination of the magnetic properties. It was found out that for Series B films the maximum of $Hc\perp$, $ORm\perp$ and ORc values appear in the vicinity of 130 nm. It means that a high quality media for perpendicular recording can be made by depositing $Co_{81}Cr_{19}$ film having such a thickness. However, in the thickness range of 5-12 nm the $Co_{81}Cr_{19}$ films exhibit an anomalous behaviours.
2) For Series B films as the thickness increases, the film magnetostriction anisotropy constant gradually changes its value from the negative to the positive. For Series A films the magnitude of negative magnetostriction constant monotonously decreases with increasing thickness. In the range of 110-150 nm, the magnetostriction constants for Series A and B films are close to zero.
3) Based on the textural evolution observed as the thickness changes, for films thicker than 12 nm, the calculation of the coercive force were made under the assumption that the magnetization process is dominated by the rotational reversal and that the columnar grains are single-domain fine particles oriented towards the external field. Experimental data show that the changes of the calculated $Hc\perp$ values with the thickness are in much better agreement with the experimental results, if the magnetostatic interaction effects is taken into consideration.

ACKNOWLEDGEMENTS

The authors acknowledge the financial support of the National Sciences and Engineering Research Council of Canada. We would like to thank Dr. W. Geerts and Dr. P. ten Berge, University of Twente, The Netherlands for providing the tested specimens and experimental data of magnetic properties. The authors are also obliged to Mr. S. Poplawski for doing the measurements of X-ray diffraction and texture. We are thankful to Mr. J.D. Baxter for correcting the manuscript in English.

REFERENCES

1) S. Iwasaki, K. Ouchi and N. Honda, IEEE Trans. Vol. Mag-16, pp. 1111, 1980
2) E. Hädicke, A. Werner and H. Hibst, Textures and Microstructure, Vol. 11, pp. 231-248, 1989
3) H.V. Kranenburg, J.C. Lodder, J.J.A. Popma, K. Takei and Y. Maeda,
J. Magn. Soc. Japan, Vol. 15, No. S2, pp. 33-38, 1991
4) Li Cheng-Zhang and J.A. Szpunar "The correlation of texture and magnetic properties in Co-Cr films", 1992
5) B. Szpunar and J.A. Szpunar, J. Magn. Magn. Mater. Vol. 43, pp.317-326, 1984
6) Y. Maeda and M. Takahashi, J. Appl. Phys. Vol. 68, pp. 4751- 4759, 1990
7) H. Cura and A. Lenhart, J. Magn. Magn. Mater. Vol. 83, pp. 72- 74, 1990
8) J.A. Osborn, Phys. Review, Vol. 67, No. 11 and 12, pp. 315-359, 1945
9) M. Masuda, S. Shiomi and M. Shiraki, Japanese J. of Appl. Phys. Vol. 26, No. 10, pp.1680-1689, 1987

Materials Science Forum Vol. 157-162 (1994) pp. 1947-1952

EFFECT OF STRAIN PATH CHANGE ON ANISOTROPY OF YIELD STRESSES OF CUBIC STRUCTURE SHEET METALS

J.H. Chung [2] and D.N. Lee [1]

[1] Department of Metallurgical Engineering and Center for Advanced Materials Research, Seoul National University, Shinrim Dong, 151-742 Kwanak-Ku, Seoul, Korea

[2] R & D Department, Kia Steel Co., Ltd., 3-13 Kuro-dong, Kuro-ku, Seoul 152-050, Korea

Keywords: Strain Path Change, Anisotropy of Yield Stresses, 70-30 Brass, AK Low Carbon Steel, Dislocation Structures, Pole Figures

ABSTRACT

Uniaxial tensile tests in various directions following uniaxial extension, equibiaxial stretching or plane strain rolling have been performed to study effects of changes in strain path on the anisotropy of yield stresses of aluminum-killed low-carbon steel and 70-30 brass sheets. The anisotropy could be predicted from the specimen textures, if dislocation structures were equiaxed as in the case of equibiaxial stretching. However, elongated dislocation cell structures, developed in the steel specimens prestrained in uniaxial tension or plane strain rolling, gave rise to the second-stage yield stresses higher than predicted from textures in the directions different from the maximum prestrain direction. Planar dislocation structures in the brass specimens prestrained in uniaxial tension or plane strain rolling gave the second stage yield stresses lower than predicted from the textures in the directions different from the maximum prestrain direction. The phenomena are discussed based on textures and dislocation structures.

INTRODUCTION

In sheet forming processes, sheet metals are formed to final products through sequential forming processes. It has been found that changes in strain path during plastic deformation of metals give rise to flow stresses and strain hardening rates which differ from those expected from monotonic deformation. Changes in strain hardening, residual ductility, and homogeneity of strain paths are important factors for accurate modeling of stamping processes. Ghosh and Backofen[1] classified the behavior of materials which were subjected to uniaxial tension after biaxial prestraining into ferritic and nonferritic types. The ferritic type (as in aluminum-killed steel) shows increases in yield stresses and decreases in strain hardening rates in the second-stage extensions. On the other hand, the nonferritic type (as in 70-30 brass) shows premature yielding and increase in the strain hardening rate.

The objective of this study was to investigate effects of the texture and dislocation structure on the mechanical properties due to path changes of uniaxial - uniaxial extension, equibiaxial stretching - uniaxial extension and plane strain - uniaxial extension of the AK low carbon steel and 70-30 brass sheets.

EXPERIMENTAL PROCEDURE

Temper-rolled AK low-carbon steel sheet(0.8mm in thickness and 14μm in grain size) and 70-30 brass sheet(0.7mm, 20μm) specimens whose chemical compositions are given in Table 1 were subjected to tensile tests at various angles to the rolling direction at a cross-head speed of 5mm min^{-1}. The strain rate sensitivities were measured by changing the cross-head speed from 5mm min^{-1} to 50mm min^{-1} at 15% strain. The measured tensile properties of the steel and brass specimens are given in Table 2. Three types of strain path changes were carried out. They are three types of prestraining (uniaxial extension, equibiaxial stretching and plane strain rolling) followed by uniaxial extension. The specimens were tensile strained in the rolling direction from which the second-stage tensile specimens were cut at various angles to the prestrain direction. Equibiaxial stretching of sheets was obtained by pushing up the clamped sheets with a cylindrical punch. The plane strain deformation was achieved by rolling. From the equibiaxially stretched sheet and the rolled sheets, tensile specimens were cut at various angles to the rolling direction.

The (110), (200) and (211) pole figures for the AK steel sheets and the (111), (200) and (220) pole figures for 70-30 brass sheets were measured. The pole figure data were used to calculate the orientation distribution functions. The TEM structures of specimens were observed under 200 kV JEOL TEM.

Table 1. Chemical compositions of materials in percent

steel C	Si	Mn	P	S	Ni	Cr	Al	brass Cu	Al	Fe	Pb	Zn
0.062	0.035	0.210	0.011	0.011	0.015	0.040	0.040	69.8	0.003	0.011	0.012	30.1

Table 2. Tensile properties of as-received steel and brass sheet specimens (25 mm gauge length)

	steel YS (MPa)	UTS (MPa)	Uniform Elong.(%)	R	brass YS (MPa)	UTS (MPa)	Uniform Elong.(%)	R
0°	153.9	287.3	28.8	1.964	118.5	335.7	57.2	0.939
22.5°	157.1	191.9	28.2	1.837	117.3	331.9	56.0	1.047
45°	163.4	294.7	28.2	1.491	116.1	329.1	58.8	1.057
67.5°	158.5	289.6	28.4	2.052	119.4	331.3	57.6	1.027
90°	152.0	282.9	29.0	2.661	118.9	333.6	59.2	1.024
$\overline{X}$	158.0	290.3	28.4	1.924	117.9	332.7	57.7	1.028

$\overline{X} = (X_0+2X_{22.5}+2X_{45}+2X_{67.5}+X_{90})/8$ and R is plastic strain ratio.

RESULTS AND DISCUSSION

The anisotropy of yield stresses can be caused by texture. The effect of texture on the yield stress can be calculated based on Taylor's minimum energy theory [2,3,4]. For equibiaxial tension or plane strain, it is necessary to introduce an appropriate effective strain in order for comparison with monotonic flow behavior. The Hosford yield equation[5] for anisotropic materials yields the following effective strain[6].

$$\overline{\varepsilon} = (1+R)^{1/a}\,[1+|\alpha|^a+R|1-\alpha|^a]^{(a-1)/a}\,[1+R\frac{|1-\alpha|^a}{(1-\alpha)}]^{-1}\,\varepsilon_1 \qquad (1)$$

where R is the plastic strain ratio (the ratio of width to thickness strain), α is the ratio of transverse stress to longitudinal stress(σ_2/σ_1), a is a constant (6 for bcc, 8 for fcc) and ε_1 is the longitudinal strain. For equibiaxial stretching(α= 1), the effective strain reduces to

$$\bar{\varepsilon} = 2^{(a-1)/a}(1+R)^{1/a}\varepsilon_1 \quad (2)$$

For plane strain, $d\varepsilon_2/d\varepsilon_1 = [\,|\alpha|^a(1-\alpha) - R|1-\alpha|^a\alpha\,]/[\,\alpha(1-\alpha+R|1-\alpha|^a)\,] = 0$ and

$$\bar{\varepsilon} = (1+R)^{1/a}\left[1+\left[\frac{R^{1/(a-1)}}{1+R^{1/(a-1)}}\right]^{(a-1)}\right]^{-1/a}\varepsilon_1 \quad (3)$$

Figure 1 shows the measured second-stage yield stresses of low carbon steel and 70-30 brass specimens as a function of angle to the rolling direction after equibiaxial stretching as compared with those calculated based on textures (figures 2 and 3). The equibiaxial stretching made the dislocation structure more equiaxed as shown in figures 4(a) and 5(a). Therefore the anisotropy of yield stresses in figure 1 is expected to be caused mainly by the textures of specimens. The excellent agreement between measured data and values calculated using textures reflects that the calculation gives us satisfactory texture effects.

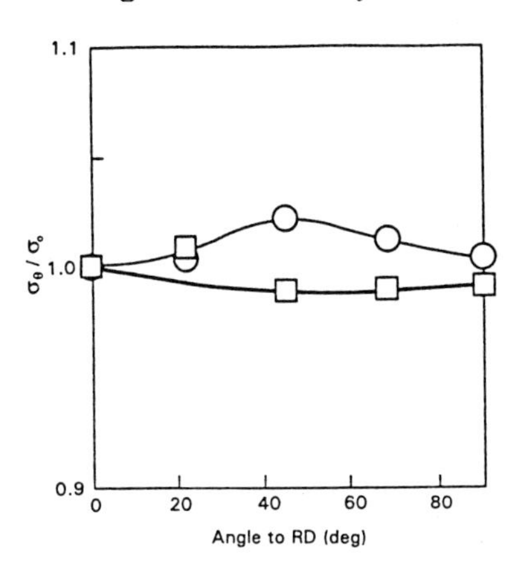

Fig.1 The second-stage yield stresses of (○) the AK steel and (□) the 70-30 brass specimens tested after prestraining in equibiaxial tension ($\bar{\varepsilon}$ = 0.299 for steel and $\bar{\varepsilon}$ = 0.28 for brass) The curves indicate values calculated based on textures.

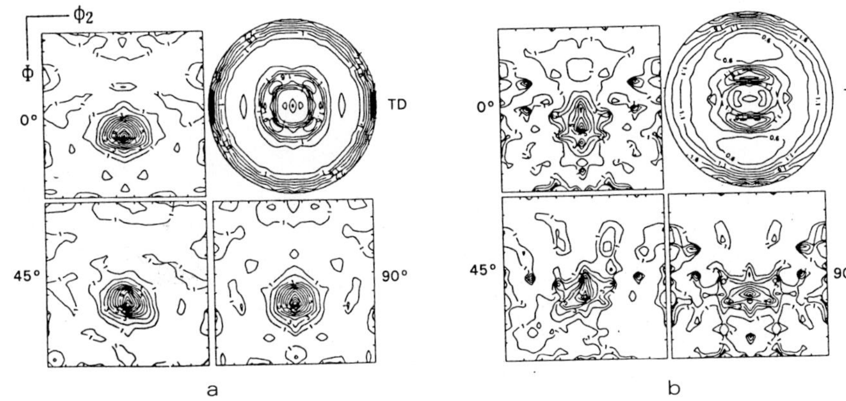

Fig.2 (110) pole figures and ODFs for the AK steel sheets (a) prestrained in equibiaxial tension by effective strain 0.299, and (b) plane strain rolled by effective strain 0.167. The numbers in left and right of ODFs indicate ϕ_1 values.

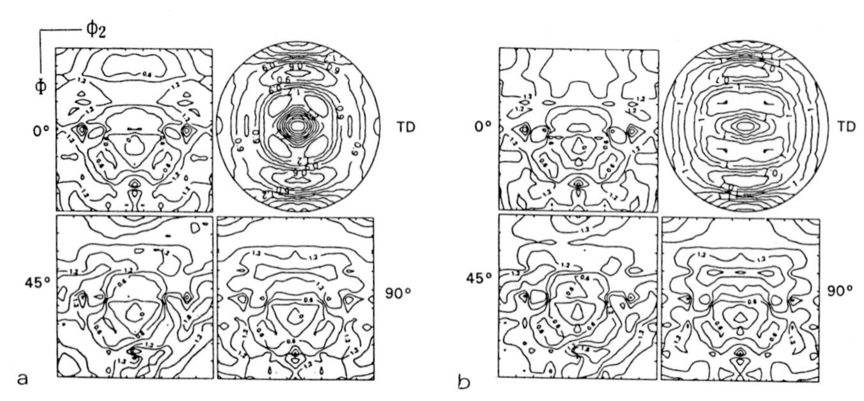

Fig.3 (111) pole figures and ODFs for the 70-30 brass sheets (a) prestrained in equibiaxial tension by effective strain 0.280, and (b) plane strain rolled by effective strain 0.152. The numbers in left and right of ODFs indicate ϕ_1 values.

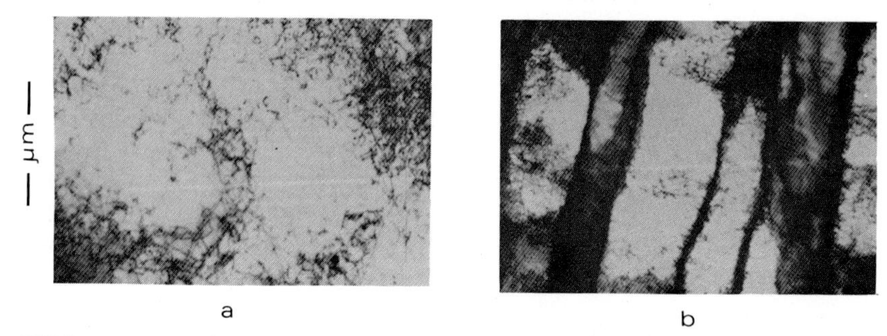

Fig.4 TEM structures of the AK steel specimens (a) after prestraining in equibiaxial tension ($\bar{\varepsilon}$ = 0.126), and (b) after prestraining in uniaxial tension (ε = 0.182)

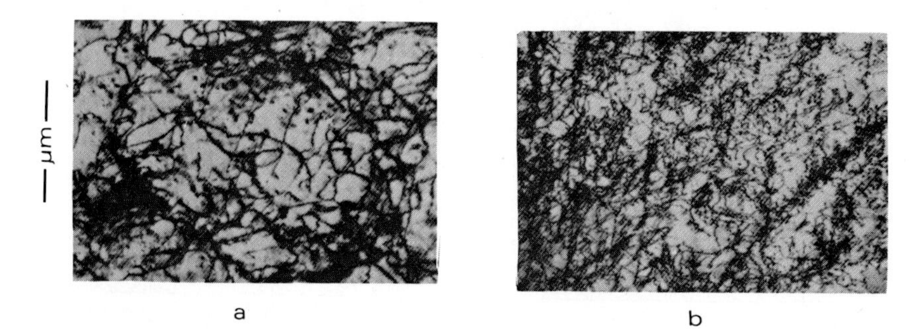

Fig.5 TEM structures of the 70-30 brass specimens (a) after prestraining in equibiaxial tension ($\bar{\varepsilon}$ = 0.28), and (b) after prestraining in uniaxial tension (ε = 0.182)

Figure 6 show anisotropy of yield stresses of the as-received AK steel and 70-30 brass specimens as a function of angle to the rolling direction. For brass, the measured data are in good agreement with values calculated based on textures due to isotropic dislocation structure as shown in figure 5. On the other hand, the steel specimens show a little disagreement between measured and calculated values, possibly due to a slight anisotropic dislocation structure developed during temper rolling.

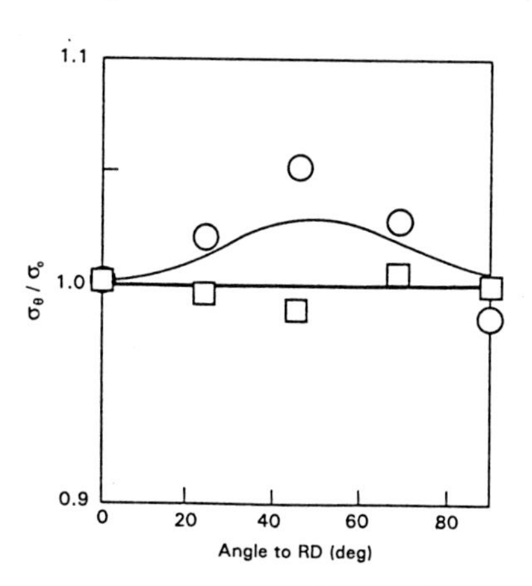

Fig.6 The yield stresses of the AK steel(○) and 70-30 brass (□) specimens in as-received state. The upper and lower curves are values calculated using textures of steel and brass specimens, respectively.

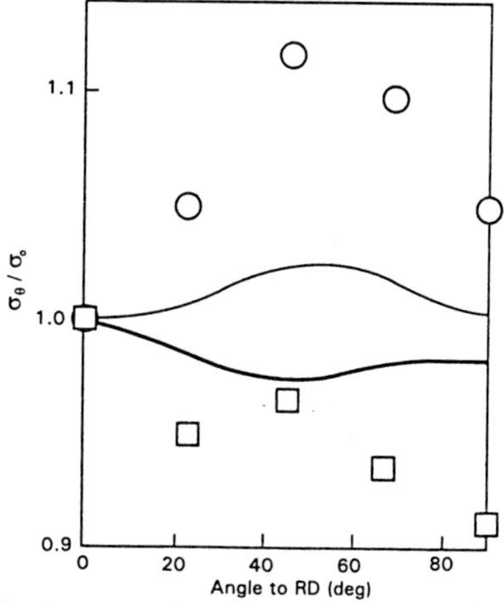

Fig.7 The second-stage yield stresses of the AK steel (○) and 70-30 brass (□) specimens tested after prestraining in uniaxial tension (ε = 0.182). The upper and lower curves are values calculated using textures of steel and brass specimens, respectively.

When tested in different strain paths after prestraining in uniaxial tension in the rolling direction, the yield stress of low-carbon steel specimens deviated positively from the values calculated from their textures, whereas those of 70-30 brass specimens deviated negatively from the calculated values, as shown in figure 7. This behavior can be attributed to the differences in dislocation structures of the two different alloys. The steel specimens had elongated dislocation cell structures as shown in figure 4. If a strain path change takes place, the relatively stable elongated cell walls developed during prestraining will become barriers to dislocation movement due to the new extension. Therefore the second-stage yield stress after the strain path change must be higher than that in the uniaxial prestraining. In this case, a 90° change in strain path will give rise to the highest increase. This may be the case of steel specimens in figure 8, because uniaxial tension and rolling of steel specimens caused elongated cell structure as shown in figure 4. However for the steel specimen in figure 7, the highest increase was observed in a 45° change in strain path, even though they had elongated cell structures (figure 4). The same phenomenon was found in brass specimens.

Brass specimens prestrained in uniaxial tension behave very differently from steel specimens as shown in figure 7. Dislocation structures in brass specimens are planar and relatively homogeneous due to their low stacking fault energy. Therefore, the anisotropic yielding behavior of brass in figure 7 cannot be explained as in low-carbon steel. The planar dislocation structure may be approximated by dislocation pile-ups as shown in figure 9. The figure indicates that uniaxial extension in the rolling direction followed by uniaxial tension in the transverse direction is similar to uniaxial extension followed by uniaxial compression in the rolling direction which causes the Bauschinger effect, unless slip systems in the both cases are different. Therefore the second-stage

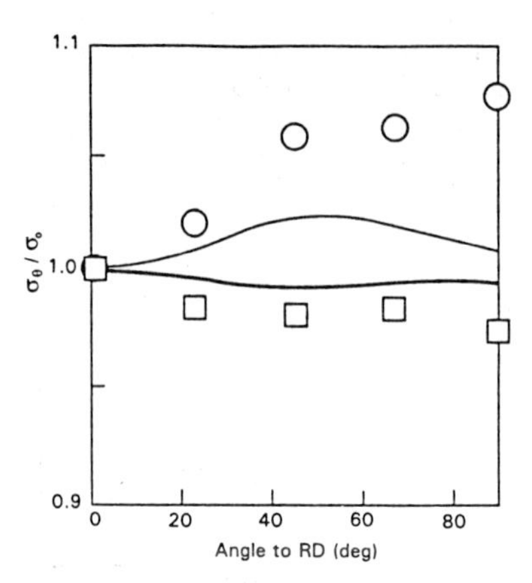

Fig.8 The yield stesses of the AK steel (○) and 70-30 brass (□) specimens after prestraining by rolling ($\bar{\varepsilon}$ = 0.167 for steel and $\bar{\varepsilon}$ = 0.152 for brass). The upper and lower curves are values calculated using textures of steel and brass specimens, respectively.

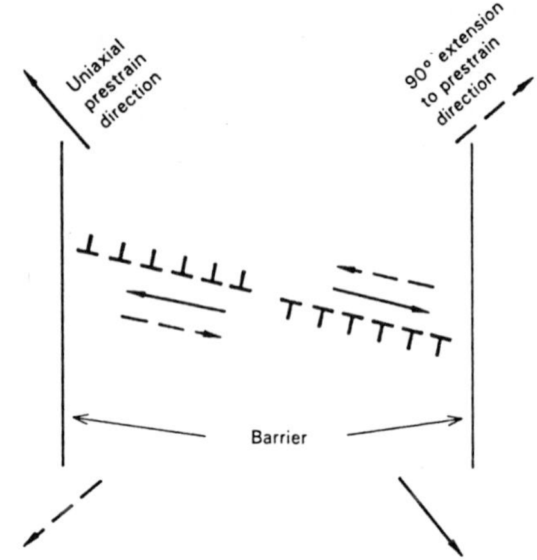

Fig.9 Schematic drawing explaining dislocations generated by uniaxial extension being moved towards the opposite direction by 90° extension to the prior uniaxial extensions.

yield stress of the brass specimen tested after a 90° change in strain path is lower than that expected from its texture. However, when tested after a 45° change in strain path, the dislocations generated during prestraining are in the most difficult position of movement, which in turn gives rise to the higher yield stress as shown in figure7. This explanation could be applied to the low-carbon steel specimens which show the 45° yield stress higher than that expected. The second-stage yield stresses of brass specimens prestrained by rolling (figure8) deviate less from those calculated using textures than specimens prestrained in uniaxial tension. This may be due to the fact that the number of active slip systems in plane strain deformation(as in rolling) is larger than that in uniaxial tension.

ACKNOWLEDGMENTS

This work was financially supported by Korea Science and Engineering Foundation. Authors also want to thank Mr. Jaewook Kwon at Seoul National University for his help with experiment and helpful discussion.

REFERENCES

1. A.K.Ghosh and W.A.Backofen . Metall.Trans., 1973, 4, 1113
2. H.J.Bunge . Texture Analysis in Materials Science, Butterworths, London, 1982, 330.
3. D.N.Lee, K.H.Oh and I.S.Kim . J. Mater. Sci., 1988, 23, 4013
4. G.I.Taylor . J. Inst. Met., 1938, 62, 307
5. W.F.Hosford . Proc. 7th North Am.Metal Working Conf.SME, Dearbon MI, 1979, 191
6. D.N.Lee and Y.K.Kim . Forming Limit Diagrams ; Concepts, Methods and Applications, R.H.Wagoner et al. eds., TMS-AIME, Warrendale, PA, 1989, 153

Materials Science Forum Vol. 157-162 (1994) pp. 1953-1960

A QUANTITATIVE ANALYSIS OF EARING DURING DEEP DRAWING OF TIN PLATE STEEL AND ALUMINUM

A.P. Clarke [1], P. Van Houtte [2] and S. Saimoto [1]

[1] Department of Materials and Metallurgical Engineering,
Queen's University, Kingston, Ontario K7L 3N6, Canada

[2] Department of Metallurgy and Materials Engineering,
Katholieke Universiteit Leuven, de Croylaan 2, B-3001 Leuven, Belgium

Keywords: Earing Profile, ODF Predictions, Cup Height Models

ABSTRACT

An approximate height prediction model has been combined with a Taylor-Bishop-Hill texture evolution model to predict the earing profile on laboratory cups and commercial initial draw products. The ODF of the initial orientation has to be modified by the texture evolution during the flange deformation. A simple additive rule of the original texture to the evolved one results in good agreement between the measured and predicted earing profiles.

INTRODUCTION

The earing phenomenon has long been shown to be a manifestation of material anisotropy[1]. However, quantitative analysis has been most elusive as discussed elsewhere [2]. Our attempts [2,3,4] have been to predict the earing profile by utilizing the orientation distribution function (ODF) of the starting sheet and computer process it using the full constraint Taylor-Bishop-Hill model [5] according to the strain imposed by the cupping or deep draw process.

A simple model for the calculation of the ear height has been described in detail [2] and is not presented here. The blank is assumed to be in a state of plane stress such that the stress in the thickness direction, σ_{zz}, is zero as well as the shear stresses, σ_{zr} and $\sigma_{z\psi}$, where r and ψ are depicted in figure 1.

Although radial flow is assumed, the velocity along any radial path at a given ψ may be different which results in a different thickening rate $-\varepsilon_{zz}$. This variation is due to the crystallographic anisotropy. This radial velocity can be correlated to the punch velocity such that for constant $s=\sigma_{rr}/\sigma_{\psi\psi}$ and a calculated $q=-\varepsilon_{rr}/\varepsilon_{\psi\psi}$ from the ODF at a selected ψ, the height prediction can be derived:

$$h(\psi)=d+(1-\frac{\pi}{4})(2r_{pp}+d)+\frac{r_b^{q+1}-(r_p+d)^{q+1}}{(q+1)(r_p+d/2)^q} \quad (1)$$

where d is the sheet thickness, r_p is the punch radius, r_{pp} is the punch profile radius. A procedure to determine s and q is described in [2] which makes use of the ODF of the material.

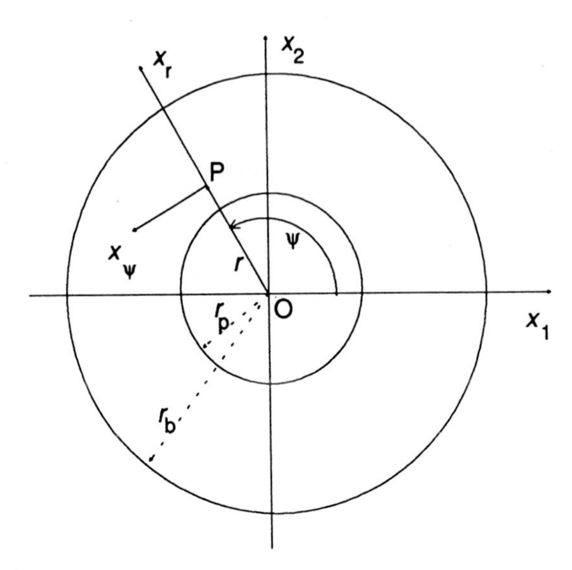

Fig.1 Top view of a circular blank; x_1 is the rolling direction, r_p is the radius of the punch, and r_b the initial radius of the blank. r and Ψ are the polar coordinates of the point P.

Equation 1 has been used to explain the measured profile for the case of laboratory cups of cold-rolled aluminum stock. To achieve this good fit, texture evolution in the flange during the draw had to be incorporated. This was done assuming 70% of the cup flange volume deformed according to the original texture and 30% according to the simulated texture evolution using a full-constraints Taylor model[6] after 24% strain (figure 2). For commercial draw operations, the profile fits were good but the model underpredicted the overall wall height. This underprediction was attributed to ironing caused by close punch-die clearances.

The purpose of this paper is to illustrate that the combined model of the initial and computer processed texture has the capability to predict earing in steel sheets. The one complexity introduced for the case of steel versus aluminum is that the slip plane for the latter is not unique. Thus a selection rule for {011} and {211} slip modes must be invoked. Previous experience [7] in relating measured q values and the computer predicted one to the q value changes with tensile strain indicate that $\tau_{\{011\}}=0.9\tau_{\{211\}}$ in tension and $\tau_{\{011\}}=1.1\tau_{\{211\}}$ in compression. Thus these values were used in the simulation of the texture evolution of steels.

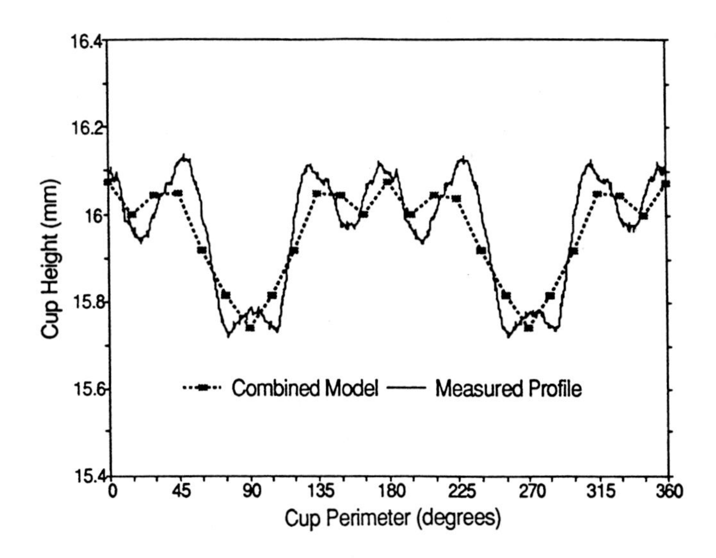

Fig.2 Profile of laboratory rolled Al hot band after cupping test: comparison of measured one (solid line) to that using combined texture (dashed line). After Ref.2.

EXPERIMENTAL PROCEDURE

Cup ear profiles were measured using a profilometer on three types of cups:

a) Commercially rolled and laboratory drawn galvanized autobody sheet steel cup. The blank was cut from 0.8mm galvanized sheet steel for autobody application and had the following chemical composition in weight percent: 0.05 C, 0.35 Mn, 0.054 Cr, 0.003 Mo, 0.010 Si, 0.022 Cu, 0.009 Ni, 0.060 Al, 0.002 P, 0.008. The starting blanks were cut with radii r_b of 27.5mm, the punch radius r_p was 16.5mm and the punch profile radius r_{pp} was 5mm.

b) Commercially rolled and drawn tin plate steel, Manufacturer #1. Cups were obtained after the first draw from a commercial can plant. Blank, punch and punch profile radii were taken to be r_b=70mm, r_p=45mm, r_{pp}=5mm with sheet thickness of 0.3mm.

c) Commercially rolled and drawn tin plate steel, Manufacturer #2. Cups were obtained after the first draw but the original sheet supplier was different. Blank, punch and punch profile radii and thicknesses were the same as above. The typical analyses of tin plate steel are: 0.05/0.07%C, 0.25/0.33Mn, 0.02%P max., 0.012%S max., 0.03%Si max. and 0.02/0.07% ASA max.

The ODF's were determined, as previously described [3,4], from sections near the sheet centre of the samples. From the ODF's determined for each of the different cups, q-values were calculated using combined Taylor-ODF software [5] for seven different ψ values around the cup. The q-values were calculated for s=-0.18 for the commercial cup material and s=-0.20 for the laboratory cupped material and these values of s were in turn derived from the known r_b/r_p ratios.

Using these q-values and substituting into equation 1 an initial profile prediction was obtained based purely on the initial texture. However, previous work [2,4] indicated that the texture changes during the cup draw are significant and that they can be well predicted using the full constraint Taylor deformation software. Therefore, the initial texture was computer drawn in the seven different ψ-directions with 12 incremental steps of ε_o=2% where the strain components are described by the following tensor:

$$\begin{vmatrix} q\cos^2\psi - \sin^2\psi & -(q+1)\sin\psi\cos\psi & 0 \\ -(q+1)\sin\psi\cos\psi & q\sin^2\psi - \cos^2\psi & 0 \\ 0 & 0 & 1-q \end{vmatrix} \epsilon_o \qquad (2)$$

These simulated draws result in a true strain of 24% in the tangential direction in the cup wall which corresponds to the strain roughly halfway up the final cup wall. The starting texture was inherently orthorhombic and remained orthorhombic for draws in the rolling and transverse directions. However, the resultant textures in the other 5 directions produced monoclinic textures. To correctly handle these textures, the starting texture was analyzed assuming monoclinic symmetry before discretization into 4000 representative crystallites to ensure that there were virtual crystals in the correct orientation as the asymmetric draw simulations proceeded. The simulated (110) and (200) pole figures after draws along the RD, TD and 45 directions are shown in figures 3a, 3b and 3c for the commercially rolled sheet steel. The pole figures all have their x_1 axes parallel to the RD direction of the starting sheet and are therefore rotated with respect to the cup radial direction by 0°, 45° and 90° for the RD, 45 and TD draws respectively. This presentation emphasizes the large texture evolution, especially in the 45 draw direction, given the small draw strains involved. It does, however, camouflage the degree to which all the draw directions are evolving toward the *same* end texture with respect to the radial cup frame of reference.

From the simulated draw textures the q-values were again calculated corresponding to the various ψ-directions (one q-value per draw simulation). To incorporate these new values and the effect of the simulated deep drawing in the profile prediction, the blank was modelled as a two texture material with the first 50% of the cup flange volume (compared to 70% for aluminum) having the starting texture and the remaining region having the as-evolved texture.

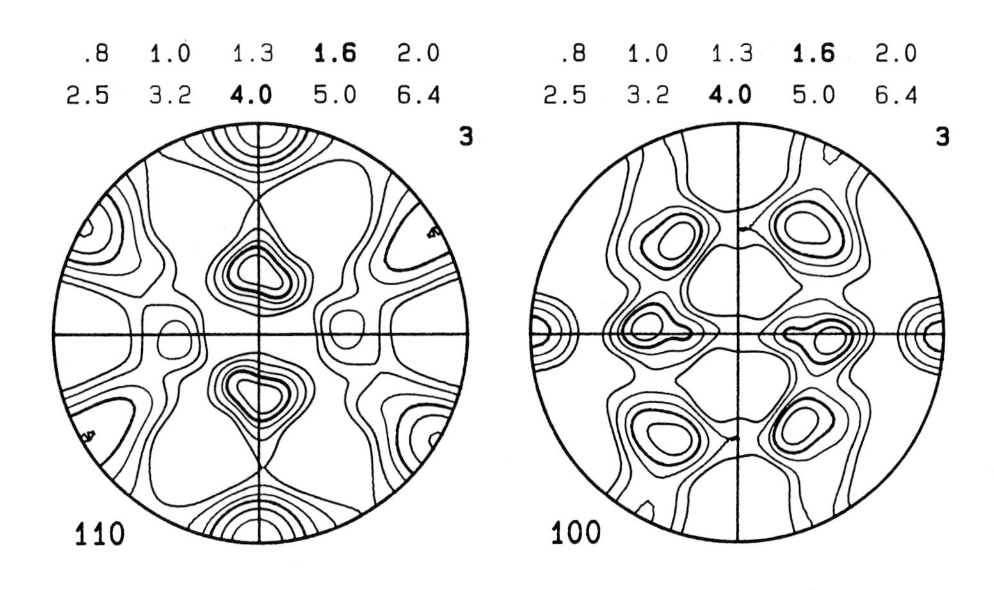

Fig.3a (110) and (200) pole figures after 24% full constraints Taylor-Bishop-Hill draw simulation along the rolling direction for tin plate steel sheet of Manufacturer 1.

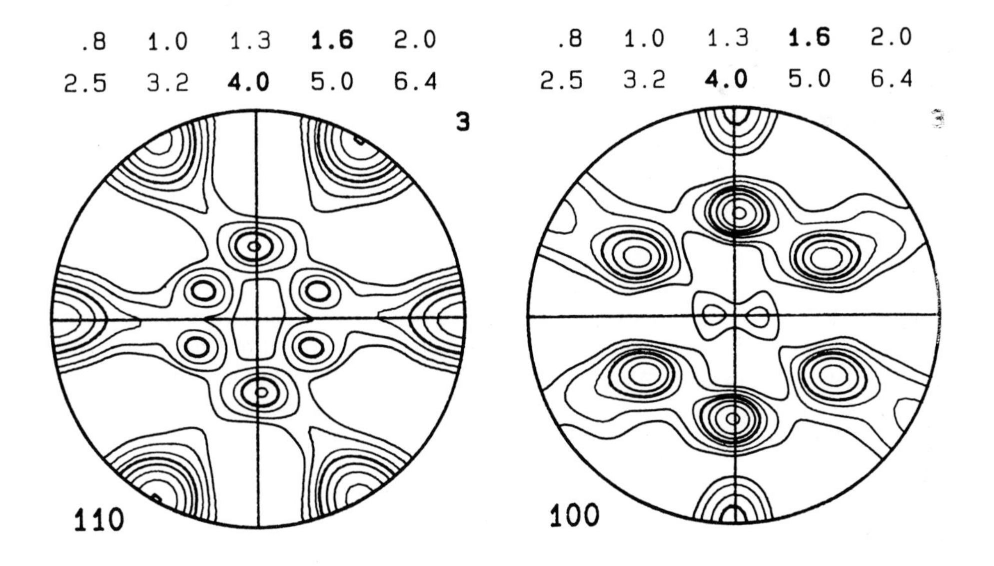

Fig.3b (110) and (200) pole figures after 24% full constraints Taylor-Bishop-Hill draw simulation along the TD direction for tin plate steel sheet of Manufacturer 1.

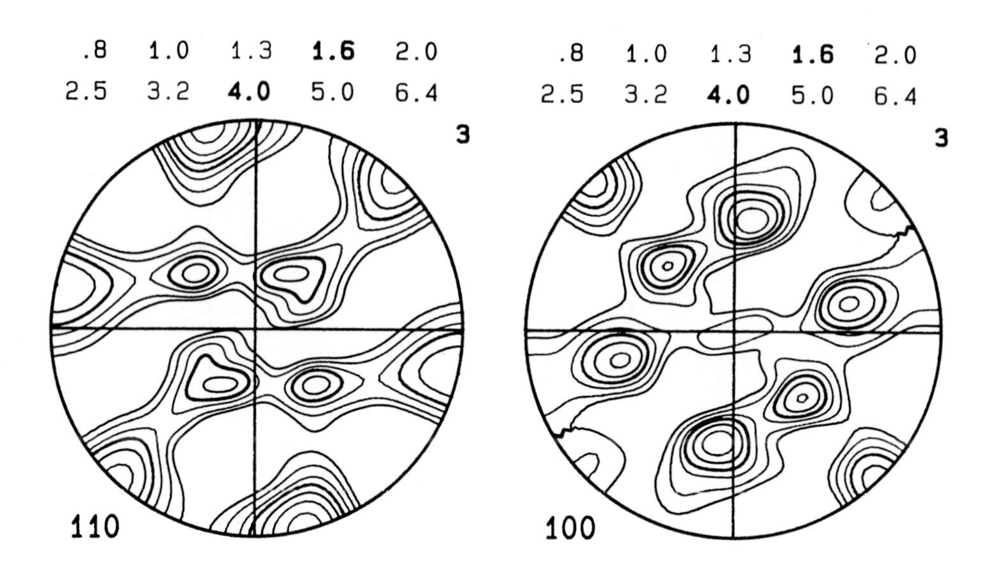

Fig.3c (110) and (200) pole figures after 24% full constraints Taylor-Bishop-Hill draw simulation along the 45° direction for tin plate steel sheet of Manufacturer 1.

RESULTS AND DISCUSSION

Figure 4 shows the cup height predictions of the laboratory steel cup using the original texture and that based on the evolved texture halfway up the cup (that is, 24% strain) and the 50-50 combined model. These predictions seem to underpredict the trough at 45° when compared to the measured profile. On the other hand, the combined model predicts the earing profile very well for the commercially processed cups, but underpredicts the cup height (figures 5 and 6). Moreover, the cup with the higher measured wall height possessed the texture which resulted in a higher predicted wall height. The fact that the relative height changes were well predicted and that the underprediction was consistent for both these materials and for the aluminum commercial cups treated before [2], implies that the underprediction is the result of factors independent of the material or its texture. As mentioned for the case of commercial aluminum cups, this underprediction is attributed to ironing of the cup due to the close tolerances imposed in the commercial machinery.

Although the starting textures are different for the two source materials, the cup earing profile and relative cup height predictions are excellent. This leads credence to the fact that the cup flange ratios to form the combined model is unique to the crystal deformation modes rather than to the starting texture.

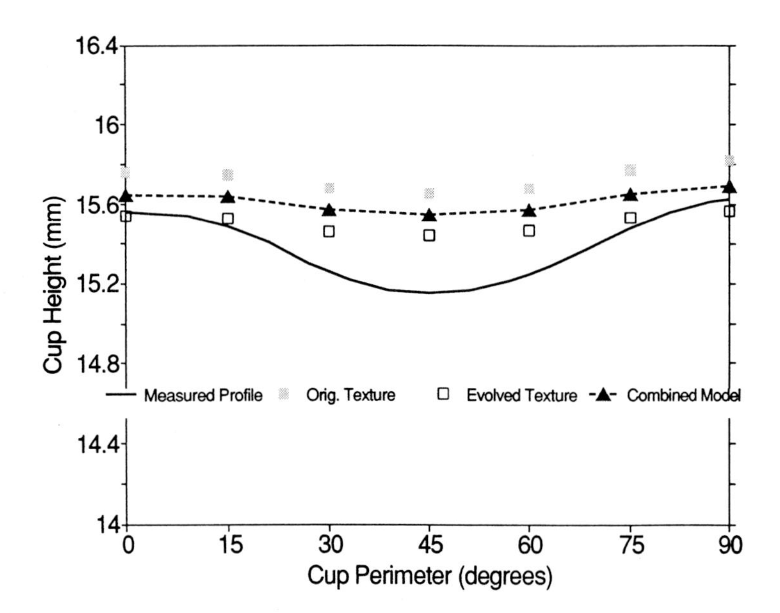

Fig.4 Laboratory cup measured wall height with model predictions.

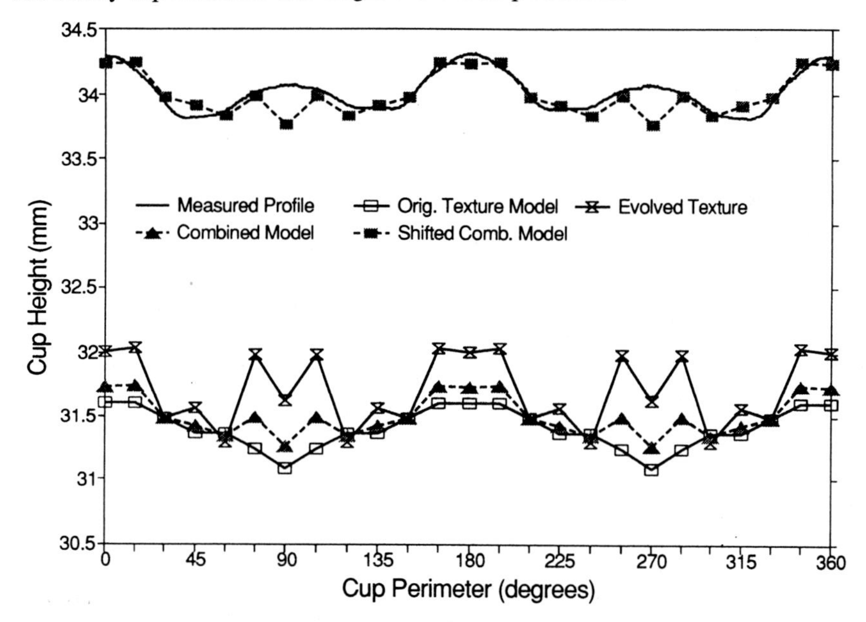

Fig.5 Steel cup wall height predictions and measured profile for Manufacturer 1's commercially cupped tin plate steel.

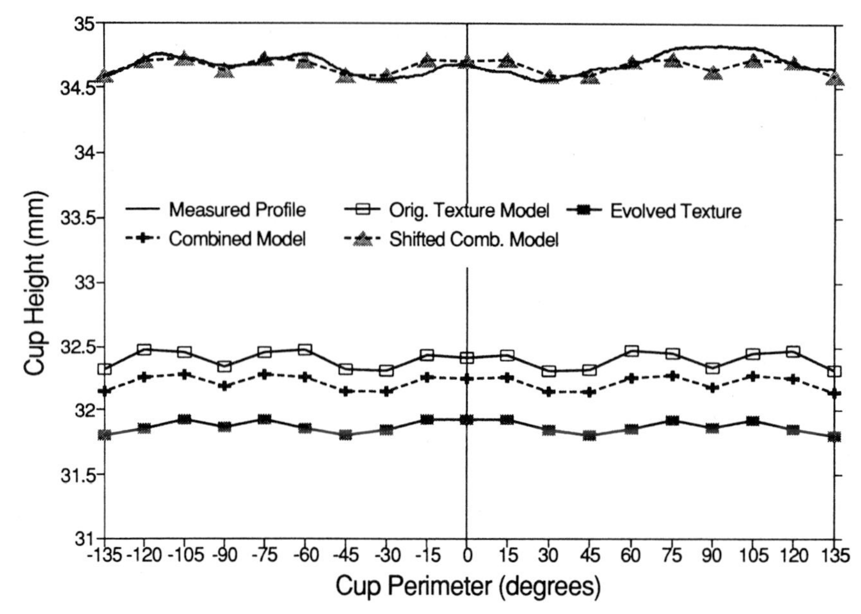

Fig.6 Steel cup wall height predictions and measured profile for Manufacturer 2's commercially cupped tin plate steel.

CONCLUSIONS

The good earing profile fits for the case of commercially produced first draw cups illustrate that, for commercial tin plate steel or aluminum can stock, the relation expressed in equation 1 can be used to predict earing behaviour. Further investigations are required to show that the under predictions of the cup heights are related to specific machinery settings and tolerances.

ACKNOWLEDGMENTS

The authors thank the Ontario Centre for Materials Research, the Natural Sciences and Engineering Council of Canada and Alcan International Inc. for critical support. We are indebted to Mr. J. Akkerman of Alcan R&D, Kingston for profilometric measurements.

REFERENCES

1. Tucker, G.E.G.: Acta Metall., 1961, 9, 275-286.
2. Van Houtte, P., Clarke, P. and Saimoto, S.: in Aluminum Alloys for Packaging, ed. by J.G. Morris, H.D. Merchant, E.J. Westerman and P.L. Morris, Pub. TMS (1993) pp.261-273.
3. Saimoto, S., Van Houtte, P. and Clarke, P.: in International Symposium on Iron and Steel in the Automatic Industry, ed. S. Orr pub. by CIMM (1990) pp.154-167.
4. Saimoto, S., Van Houtte, P., Reesor, D. and Clarke, P.: in Proc. ICOTOM 9, Textures and Microstructures, 1991, 9, 229-244.
5. Van Houtte, P.: Textures and Microstructures, 1988, 8-9, 313-350.
6. Van Houtte, P., Mols, K., Van Bael, A. and Aernoudt, E.: Text. & Microst., 1989, 11, 23-39.
7. Diak, B.J., Saimoto, S. and Van Houtte, P.: to be published.

Materials Science Forum Vol. 157-162 (1994) pp. 1961-1966

EARING PREDICTION FROM EXPERIMENTAL AND TEXTURE DATA

C.S. Da Costa Viana and N.V.V. De Avila

IME-SE/4, Inst. Mil. Eng., Pça. Gen. Tibúrcio,
80 - Urca, 22290-090 Rio de Janeiro (RJ), Brazil

Keywords: Earing, Texture, Prediction

ABSTRACT

In the present work a crystallographic model for the earing of polycrystalline sheet metals is presented. The earing behaviour is both studied experimentally from strain distribution measurements and predicted from upper bound crystallographic yield loci. The cup profiles of some steels, copper and brass are analysed. Cup height and ear profiles are very approximately predicted by the model.

INTRODUCTION

Earing is the most striking manifestation of plastic anisotropy in sheet metals. It is widely recognized that texture is the dominant feature in earing formation. Many have been the attempts to model and quantify the earing behaviour of different materials. The approaches can be classified as continuum - like those of Bourne and Hill [1], Sowerby and Johnson [2], Gotoh and Ishise [3] - Crystallographic - as those due to Tucker [4], Viana et al. [5], Kanetake et al. [6], van Houte et al. [7], Barlat et al. [8], and Rodrigues and Bate [9] - and hybrid, like that of Lin et al. [10].

In the present work a crystallographic model is presented based solely on experimental strain distribution measurements and crystallographic yield loci predicted from texture data.

MATERIALS AND EXPERIMENTS

Four sheet materials were used in this work:

- 0.75 mm-thick cold rolled and annealed IF steel, named IF;
- 0.75 mm-thick cold rolled and annealed low carbon steel, named LC;
- 0.90 mm-thick cold rolled and annealed copper, named CU;
- 1.35 mm-thick cold rolled and annealed α-brass, named BR.

Circular blanks of 102mm in diameter were used in cup drawing all the materials. For the low carbon steel (LC), additional blanks of 80mm, 94mm and 106mm in diameter were also used to study the progress of earing. The blanks were printed with a grid of circles of d_0 = 2.5mm nominal diameter. The strain distribution was obtained from the measurement of the major, d_1, and minor, d_2, axes of the distorted circles (ellipses) and from the position of their centers on the cup wall with respect to the cup bottom. This was done along the ears, troughs, and some intermediate directions. True principal surface strains were calculated as:

$$\varepsilon_i = ln \frac{d_i}{d_0}; \qquad i = 1, 2$$

In addition, blanks with scribed concentric circles and diameters were deep drawn to check the development of shear strains during the operation. Swift 50mm-diameter flat headed punch and corresponding dies were used. The texture was quantified via the crystallite orientation distribution function (CODF) according the method of Roe [11] with a series expansion to the order $l = 20$.

THE EARING MODEL

Figure 1 shows the longitudinal section of an ideal earless drawn cup. The original blank is assumed to have radius R and thickness t_0. The cup wall is considered to have a thickness $t < t_0$.

From volume constancy during cup forming it can be shown that:

$$\varepsilon_l + \varepsilon_t = ln \frac{R^2 - r_b{}^2 - r(2r+\pi r_b)}{2r_0\{R-[r_b{}^2+r(2r+\pi r_b)]^{0.5}\}} = -\varepsilon_w \qquad (1)$$

where $r = r_0 + t_0$ and $r_c = r_0 + t$; $\varepsilon_t = ln(t/t_0)$

In this equation, ε_w is the mean hoop strain in the cup wall of an earless cup. ε_l and ε_t are the mean radial and thickness strains, respectively. ε_w depends only on geometrical parameters related to the original blank and tooling dimensions.

From the experimental strain distributions measured on drawn cups of different materials, local values of the strain ratio, e_w/e_l, could be determined near the wall bottom and top edge, for different radii around the cup. Representative values of these strain ratios were used to bracket the stress states responsible for deep drawing on the upper bound crystallographic yield loci of the materials. From these stress states, the mean hoop strain on a blank radius, at angle α from the rolling direction, RD, is calculated as:

$$\varepsilon_{wm}(\alpha) = \varepsilon_w \frac{\varepsilon_{wc}(\alpha)}{\varepsilon_{wr}} \qquad (2)$$

where:

$$\varepsilon_{wc}(\alpha) = (1/N)\sum_i \frac{\sin\lambda_i}{M_i^n}$$ and ε_{wr} is a normalizing factor obtained

from the random yield locus using the latter expression. N is the number of stress states considered, M is the Taylor factor and λ is the angle shown in figure 2. n is used for work hardening simulation and α is the yield locus rotation angle in the sheet plane. M was obtained from the maximum work principle, for both restricted or pencil glide approaches, as shown elsewhere [12].

Again, from experimental data, the average thickness strain in the wall was estimated as -0.1 which permitted the calculation of ε_{lm}, the mean radial strain, as:

$$\varepsilon_{lm}(\alpha) = -\varepsilon_{wm}(\alpha) + 0.1 \qquad (3)$$

The cup height at angle α to RD can thus be calculated as:

$$H(\alpha) = (R - r^*) \cdot \exp[\varepsilon_{lm}(\alpha)] + r \qquad (4)$$

where r^* is obtained from the equivalence of the curved surface of the arc AB at the cup head to the area of the blank annulus, $R-r^*$, that is conformed to it.

Thus, the ear profile was predicted as a function of the initial blank diameter, initial sheet thickness and the W_{lmn} texture coefficients.

RESULTS AND DISCUSSION

Figure 3 shows a typical strain distribution diagram for one of the cups. It is interesting to point out the behaviour of the thickness strain, almost always below zero and averaging -0.1 for all the materials, independently of the radial direction considered.

The predicted and experimental ear profiles can be seen in figure 4. The agreement can be considered good for both ear position and cup height. For the cube textured copper and the brass, figures 4c and 4d, respectively, the model tends to overestimate both the height and the sharpness of the ears. In figure 4b, the cup profiles of the low carbon steen can be seen for different initial blank diameters. The agreement is good for all of them. This means that the experimental strain ratios used to bracket the stress states occurring in deep drawing are representative and independent of the blank diameter. It is the anisotropy of plastic behaviour brought about in the crystallographic yield locus shape that determines the ear profile.

Based on the present model many earing parameters can be calculated, as the percent earing for instance, once the W_{lmn} are known. This may be useful when the earing behaviour is of prime importance in the development of new materials for deep drawing applications.

ACKNOWLEDGEMENTS

Thanks are due to the Brazilian National Research Council - CNPq for granting support to N. V. V. De Avila, to Cia. Siderurgica Nacional - CSN for help with the experiments and to IBM Brasil for making the presentation of this paper possible.

REFERENCES

1 - Bourne,L. and Hill,R.:Phil. Mag.,1950,41,671
2 - Sowerby,R. and Johnson,W.: J. Strain An.,1974,9,102
3 - Gotoh,M. and Ishise,F.: Int.J. Mech.Sci.,1978,20,423
4 - Tucker,G.E.G.:Acta Met.,1961,9,275
5 - Viana,C.S.C., Kallend,J.S. and Davies,G.J.:Proc.ICOTOM-5, 1978,1,447
6 - Kanetake,N., Tozawa,Y. and Yamamoto,S.: Int.J. Mech.Sci. & Eng., 1985,27,249
7 - Van Houte,P., Cauwenberg,G. and Aernouvat,C.: Mat.Sci. & Eng., 1987,95,115
8 - Barlat,F., Panchanadeeswaran,S. and Richmond,O.: Met.Trans., 1991,22-A,1525
9 - Rodrigues,P.M.B. and Bate,P.S.: Proc. Texture in Non-Ferrous Metals and Alloys, 1985,1,173
10- Lin,D.W., Daniel,D. and Jonas,J.J.: Mat.Sci & Eng.,1991,A131,161
11- Roe,R.J.: j.Appl. Phys.,1965,36,2024
12- Viana,C.S.C.,Kallend,J.S. and Davies,G.J.:Int. J.Mech.Sci.,1979,21,355

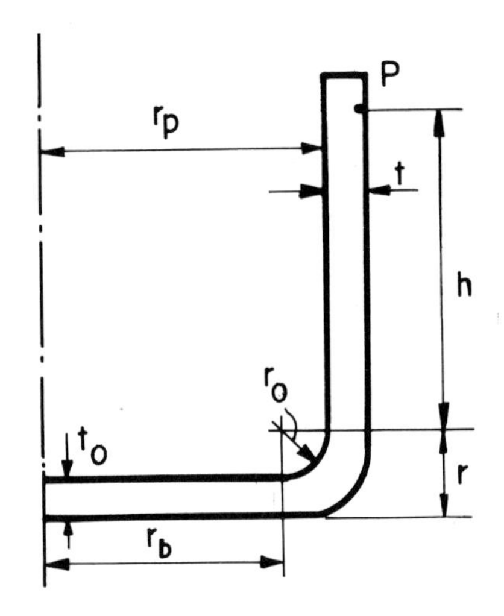

Fig.1. Section of ideal earless cup.

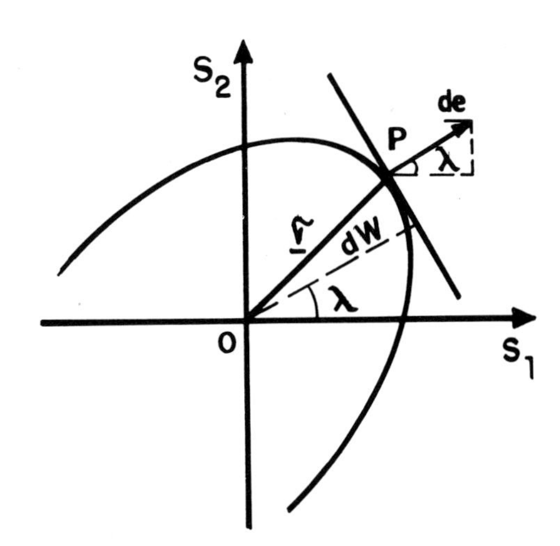

Fig.2. Schematic yield locus diagram.

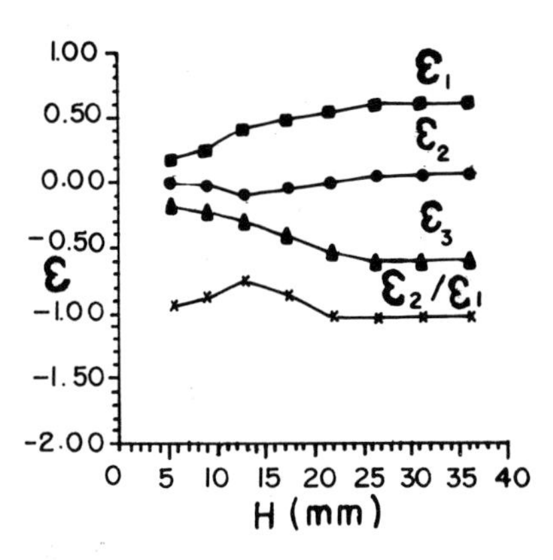

Fig.3. Typical strain distribution.

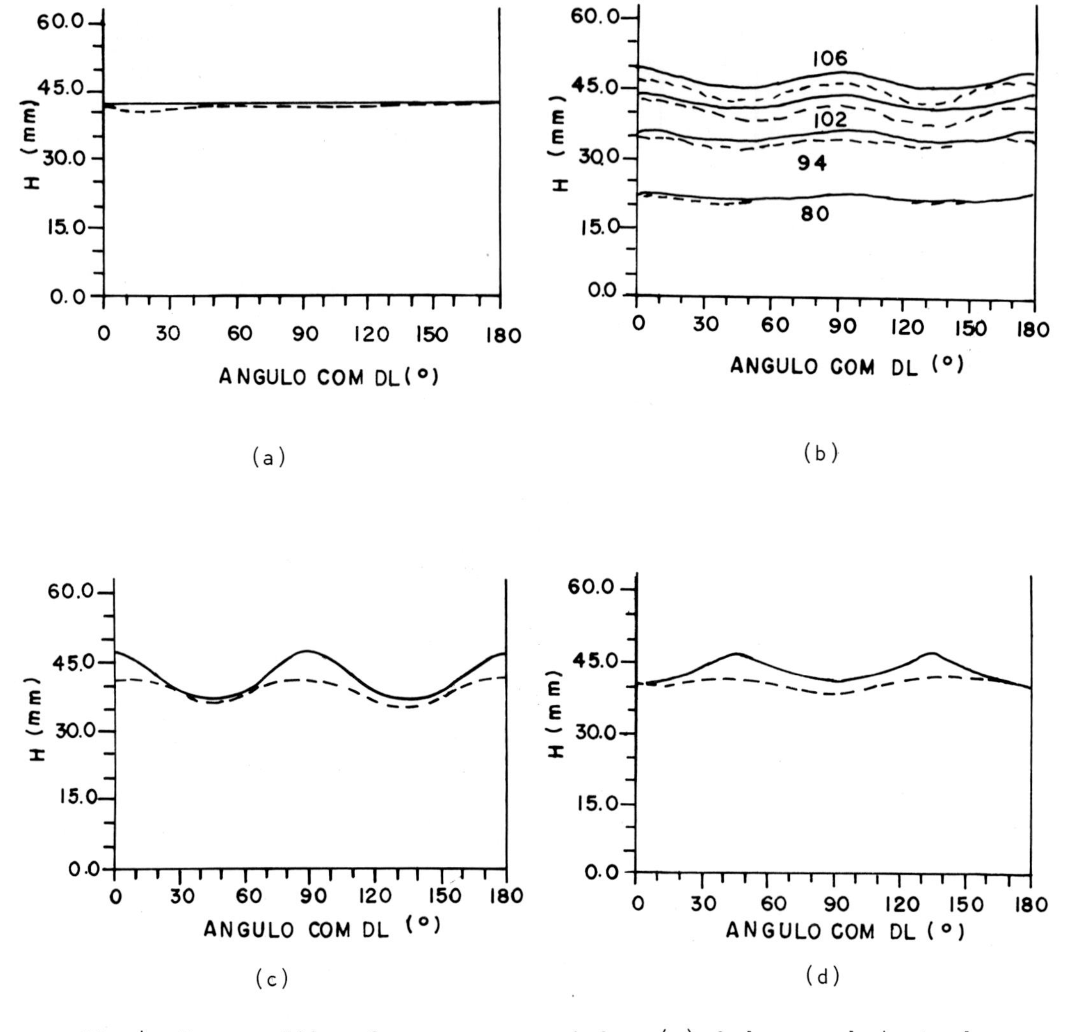

Fig.4. Ear profiles for some materials; (a) Galvannealed steel (b) Low carbon steel; (c) Annealed copper; (d) Annealed brass.

Materials Science Forum Vol. 157-162 (1994) pp. 1967-1970

EARING IN SINGLE CRYSTAL SHEET METALS

C.S. Da Costa Viana

IME-SE/4, Inst. Mil. Eng., Pça. Gen. Tibúrcio,
80 - Urca, 22290-090 Rio de Janeiro (RJ), Brazil

Keywords: Earing, Single Crystal, Prediction

ABSTRACT

A simple upper bound crystallographic model for the earing behaviour of a singly oriented sheet metal is presented. The solutions are based on {111}<110> face centered cubic restricted glide and {hkl}<111> body centered cubic pencil glide approaches. The stress and strain data necessary for the ear profile predictions are taken from predicted crystallographic yield loci of the orientations of interest for different rotations of the locus' axes in the sheet plane. The intention is to investigate the earing contribution of specific orientations embedded in a polycrystalline matrix rather than the earing profile of a drawn single crystal.

INTRODUCTION

Earing is a phenomenon resulting from the anisotropic plastic flow of a sheet metal blank brought about during deep drawing by the stresses reigning in its flange. The control of earing in the drawing of commercial metal sheets depends on the proper balance of the texture components existing in the material. Not only the percent volume fraction of grains with a specific orientation is important but also the magnitude of the earing contribution of that orientation affects the behaviour of the material. Tucker (1) was the first to establish a relationship between predicted lower bound earing profiles and the experimental behaviour of single-crystal aluminium sheets for some selected orientations. His analysis and experimental data are still unique. Da Costa Viana et al. (2) presented a simple purely crystallographic model for the earing behaviour of polycrystalline metals with good agreement between the theoretical and the experimental results for a number of fcc and bcc materials. Many other models have treated the earing phenomenon in polycrystalline materials via both continuum and crystallographic approaches with varying degrees of success(3,4,5). Very few (1,2,6) have, however, tried to relate specific orientations to the type of their contribution to the earing behaviour of metals aiming at giving engineers the possibility of altering their processing routes. The present is an attempt to explore this possibility via a simple crystallographic earing model for singly oriented sheet metals.

THE MODEL

The present model is based on predicted upper bound single crystal yield loci where the stress and strain data pertaining to those that control deep drawing were taken from.

The yield loci were predicted for both {111}<110> face centered cubic restricted glide - using the maximum work principle of Bishop and Hill [7] - and {hkl}<111> body centered cubic pencil glide - using the upper bound solution due to Penning [8]. Earing was considered to result from the relative extension of the blank radii, during cup drawing, enforced by stress states corresponding to those bracketed by plane strain - existing at the die throat - and pure hoop compression - occuring at the blank rim. Each corresponding strain state on the yield locus contributed a radial strain increment - defined as proportional to the reciprocal of its M-value - dW in figure 1 - to the total radial strain. The latter was computed for different rotations of the locus' axes in the sheet plane. The arithmetic mean of the radial strain values was used as the normalization factor. Shear strain components were neglected and no crystal orientation change was considered in the prediction. The definition of strain increment used above amounts to considering dW as the material's resistence to the corresponding imposed strain state. The greater dW the smaller the strain increment.

The present model differs from Tucker's in that only polyslip is assumed to occur. This yields cup profiles that do not generally coincide with Tucker's experimental evidence. However, the intention here is not to reproduce the drawing behaviour of a single crystal sheet but rather to simulate the earing contribution of specific orientations embedded in a hypothetical polycrystalline matrix, and, therefore, subjected to material constraints that enforce that deformation mode.

RESULTS

Figure 2 shows the predicted ear profiles for some important orientations commomly found in FCC and BCC sheet metals. It is easily seen that orientations such as {001}<100> and {110}<112> can be combined so that an earless cup is obtained. This can happen to sheets of materials like copper and brass where these components are found. The same happens to the {112}<111> and {110}<001> orientations normally found in deep drawing steels. Again they can be combined to level out the cup profile. On the other hand, the {111}<110> orientations contribute an undulating low-eared component to the cup profile.

Although the profiles are not quantitative they can be compared in relative terms with respect to their normalized mean value, represented by the number one in the figures, since they are all measured in reciprocal Taylor factor units. For steels, therefore, orientations belonging to the {111}<uvw> fiber should be sought in the development of sheet materials with smaller earing tendency.

ACKNOWLEDGEMENTS

Thanks are due to IBM Brasil for supplying financial support to make the presentation of this paper possible.

REFERENCES

1 - Tucker, G.E.G: Acta Met., 1961,9,275-286
2 - Viana, C.S.C., Kallend,J.S. & Davies, G.J.: Proc. ICOTOM-5, Aachen,1978 1,447-454
3 - Sowerby, R. & Johnson, W.: J. Strain An.,1974,9,102-108
4 - Gotoh, M. & Ishise, F.: Int. J. Mech. Sci., 1978,20,423-435
5 - Kanetake, N. & Tozawa, Y.: Int. J. Mech. Sci., 1985,27,249
6 - Barlat, F., Panchanadeeswaran, S. & Richmond, O.: Met. Trans., 1991, 22-A,1525-1534
7 - Bishop, J.F.W. & Hill, R.: Phil. Mag., 1951,42,414
8 - Penning, P.: Met. Trans., 1976,7-A,1021

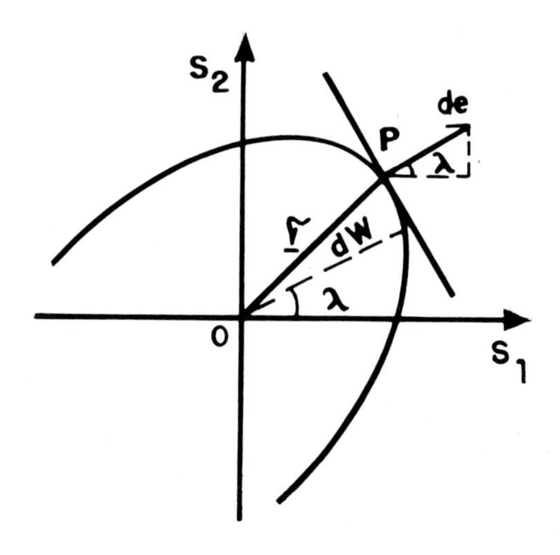

Fig.1. Schematic yield locus diagram.

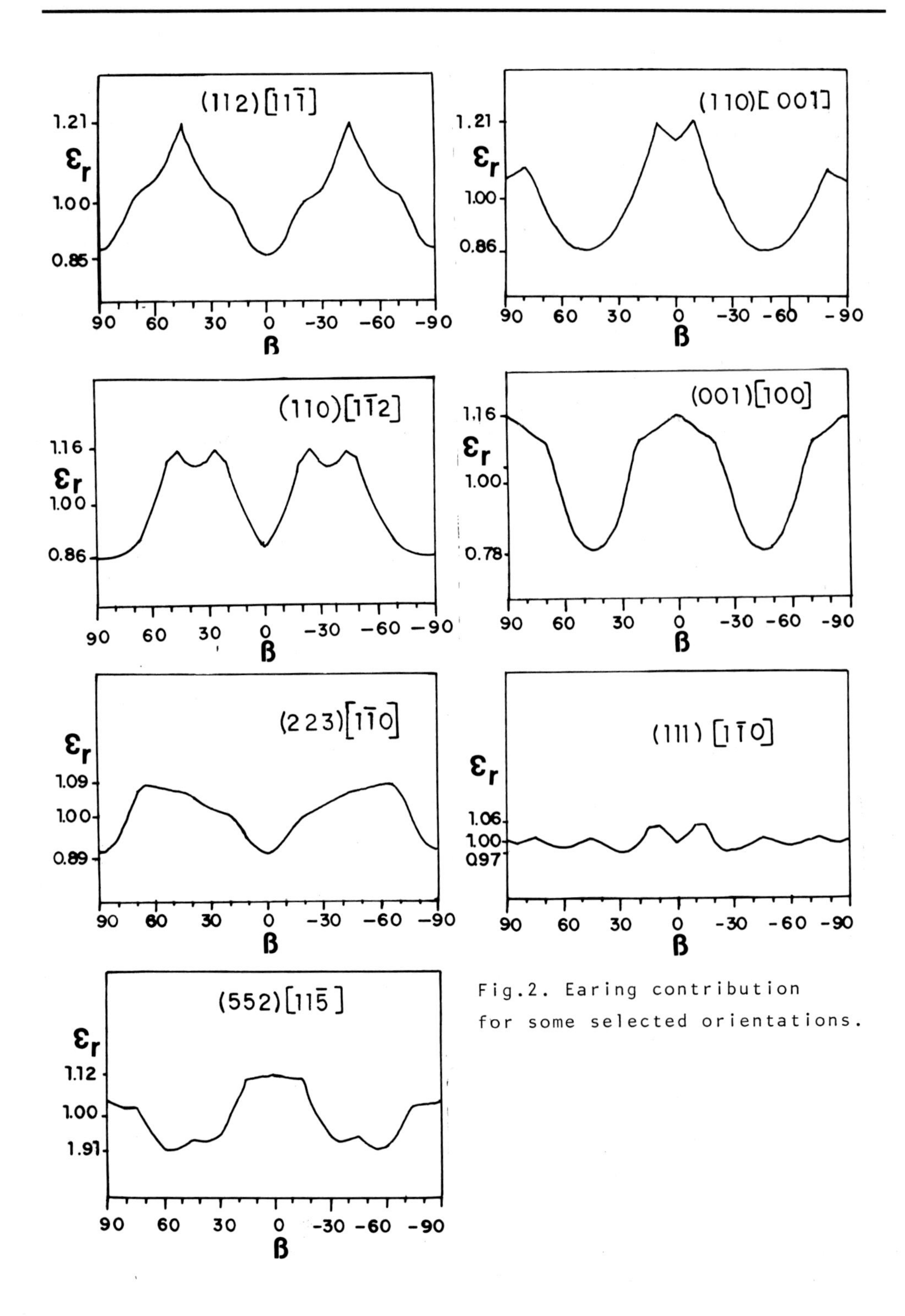

Fig.2. Earing contribution for some selected orientations.

Materials Science Forum Vol. 157-162 (1994) pp. 1971-1978

THE INFLUENCE OF TEXTURE AND MICROSTRUCTURE ON CORROSION-FATIGUE IN Ti-6Al-4V

J.K. Gregory and H.-G. Brokmeier

GKSS Forschungszentrum Geesthacht,
Max-Planck-Str. 1, D-21502 Geesthacht, Germany

Keywords: Titanium Alloys, Corrosion-Fatigue, Stress-Corrosion Cracking

ABSTRACT

Microstructure and texture in the (α+β) titanium alloy Ti-6Al-4V were systematically varied by thermomechanical processing to obtain material suitable for investigating corrosion fatigue in aqueous salt solution. While equiaxed material exhibits higher strength and ductility, coarse lamellar microstructures have superior crack growth characteristics. For equiaxed microstructures, texture is seen to have a dramatic influence on the stress corrosion cracking susceptibility and in turn on corrosion-fatigue. In lamellar microstructures, texture can evidently not be controlled so as to influence on the fracture mechanics properties, but can be used to influence strength. Thus the deliberate control of texture in lamellar microstructures offers a possibility to combine the desirable properties of both conditions. A well-behaved dependence of the relative stress corrosion cracking sensitivity on the basal pole density parallel to the loading axis is observed.

INTRODUCTION

The mechanical properties of (α+β) titanium alloys such as Ti-6Al-4V are sensitive to the microstructure and the preferred crystallographic orientation, or texture [1]. Owing to the inherent anisotropy of the hexagonal α phase, (as opposed to the more isotropic body-centered cubic β phase) it is generally the orientation of the basal planes in the α phase which is responsible for any orientation dependence in mechanical behavior. The most dramatic example is the effect of basal plane orientation on the elastic modulus [2,3].

Fatigue crack growth in (α+β) or near-α titanium alloys in aqueous salt solutions is generally described as having the loading frequency dependence shown schematically in figure 1 [4,5]. At a given frequency, a distinct increase in crack growth rate da/dN is observed at some specific value of the loading parameter ΔK. The lower the frequency, the lower the value of ΔK until a limiting value is reached which is set by the static or quasi-static stress corrosion parameter K_{ISCC}. Previous work on stress corrosion cracking (SCC) susceptibility of (α+β) titanium alloys in aqueous salt solution demonstrated that the degree of sensitivity increases when loading is perpendicular to the basal planes [6]. However, investigations of low-frequency corrosion-fatigue crack growth which focussed on the correlation with stress corrosion cracking phenomena rarely considered texture as a relevant parameter. Furthermore, no attempt was made, even in work on stress corrosion cracking, to control microstructure and texture.

The present work addresses the question of microstructural and texture dependence of corrosion-fatigue of Ti-6Al-4V in aqueous salt solution by applying thermomechanical processing to obtain specific microstructures and textures and subject these well-characterized microstructures to a comprehensive mechanical testing program.

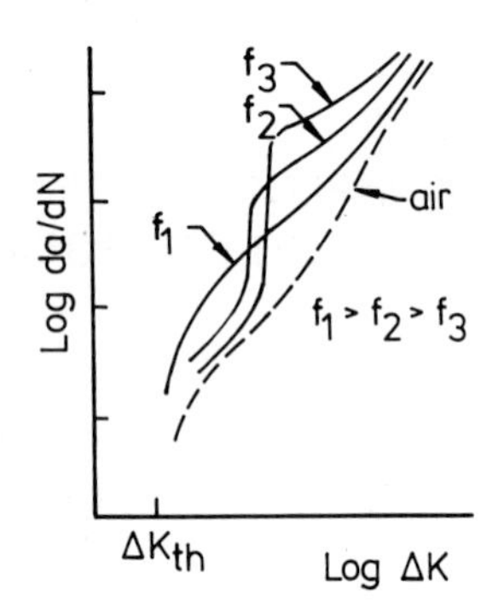

Figure 1: Schematic representation of the frequency dependence of fatigue crack growth of $(\alpha+\beta)$ titanium alloys in salt water.

MATERIAL AND EXPERIMENTAL METHODS

Two distinctly different microstructures, equiaxed and coarse lamellar, were generated by conventional thermomechanical processing procedures [7]. Equiaxed microstructures were obtained by hot-rolling in the $(\alpha+\beta)$ phase field and then recrystallizing, and coarse lamellar microstructures were obtained by slow cooling (1 °C/min) from above the beta transus. Texture was varied in the equiaxed microstructures by varying the rolling temperature. Both a transversal ("T"-basal poles parallel to the transverse direction) and a basal ("B"-basal poles oriented parallel to the sheet normal) texture were produced. In the case of the coarse lamellar microstrucures, the prior processing history (which was unknown) determined the resulting texture. The coarse lamellar microstructure was investigated both with a high number of basal poles parallel to the loading axis ("S"-strong texture) and with a fairly weak ("W") texture. Thermomechanical processing parameters are given in Table I.

Table I: Thermomechanical Processing Parameters

Designation / Microstructure	Thermomechanical Processing Parameters
EQ-T / equiaxed	homogenize 1/2 h 1050 °C, uniaxial rolling at 940 °C/AC, recrystallization 6 h 800 °C/ WQ, age 24 h 500 °C
EQ-B / equiaxed	homogenize 1/2 h 1050 °C, cross rolling 800 °C/WQ, recrystallization 6 h 800 °C/WQ, age 24 h 500 °C
CL-S / coarse lamellar	(rolling conditions unknown), 1/2 h 1050 °C, 1 °C/min cool to 800 °C, 1/2 h 800 °C/WQ, age 24 h 500 °C
CL-W / coarse lamellar	(rolling conditions unknown), 1/2 h 1050 °C, 1 °C/min cool to 800 °C, 1/2 h 800 °C, age 24 h 500 °C

Basal plane pole figures were measured with thermal neutrons. The neutron diffraction method [8] allows on the one hand the determination of the global texture of 1 cm^3 sample cubes and on the other hand the investigation of coarse lamellar as well as fine lamellar microstructures with the same high accuracy. The pole figure measurements were carried out with 1.33 Å neutrons and with an equal area scanning routine of 678 pole figure points, so that complete pole figures of the (0002) reflection were obtained in about 16 hours. Optical micrographs of the microstructures are shown with the corresponding (0002) pole figures in figures 2-5.

Heat treatments of 800 °C and above were performed on specimen blanks in air. A partial aging treatment of 4 h 500 °C was applied prior to machining. After machining, the aging treatment was completed by heat treating an additional 20 h in flowing argon.

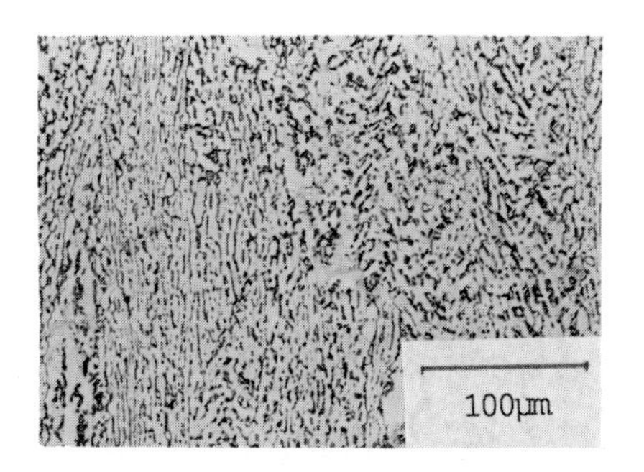

Figure 2a: Microstructure of EQ-T

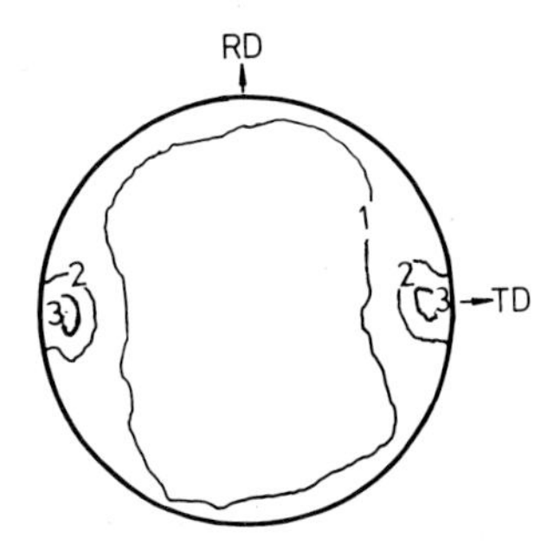

Figure 2b: (0002) Pole figure of EQ-T

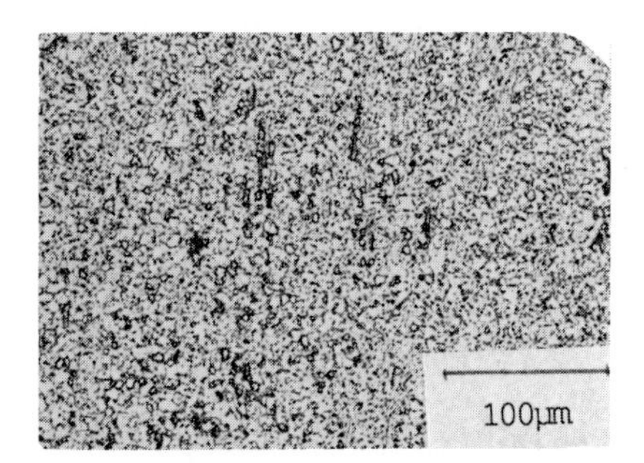

Figure 3a: Microstructure of EQ-B

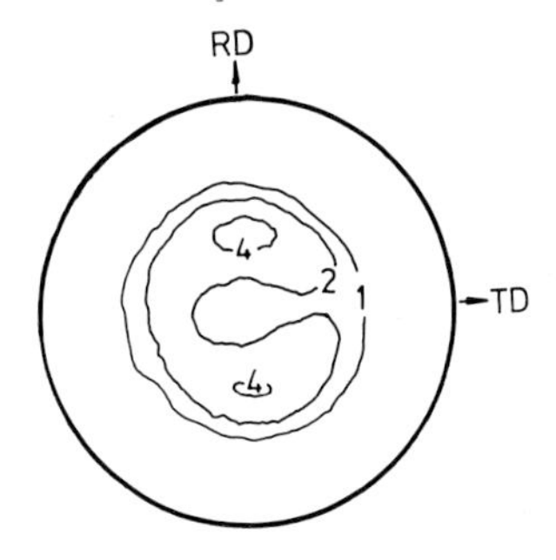

Figure 3b: (0002) Pole figure of EQ-B

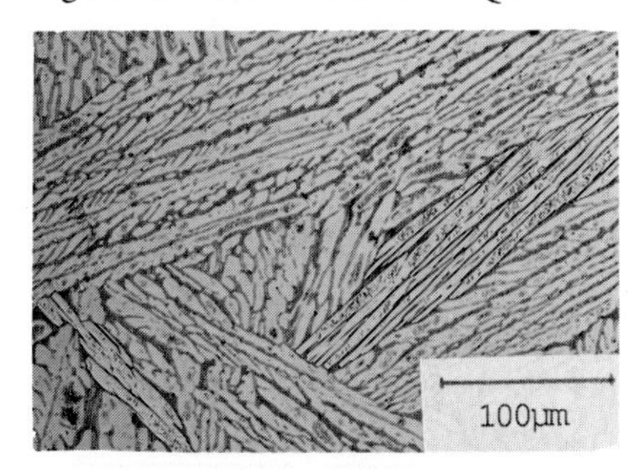

Figure 4a: Microstructure of CL-S

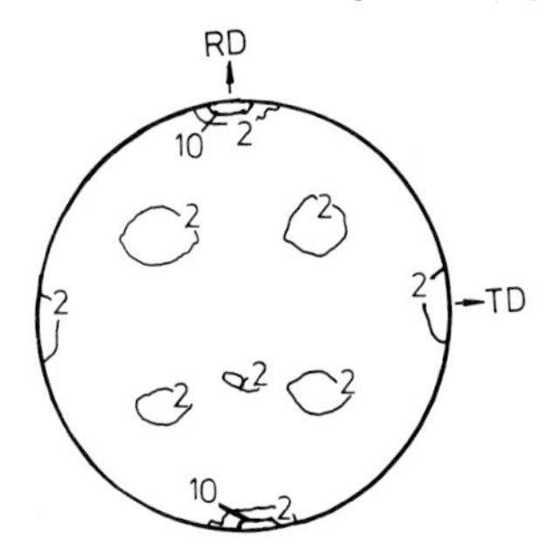

Figure 4b: (0002) Pole figure of CL-S

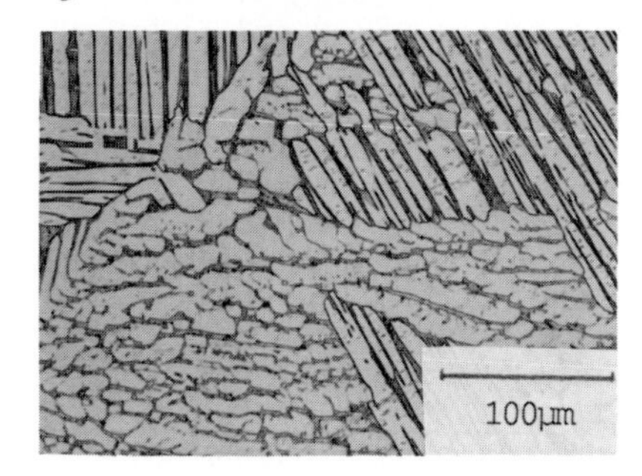

Figure 5a: Microstructure of CL-W

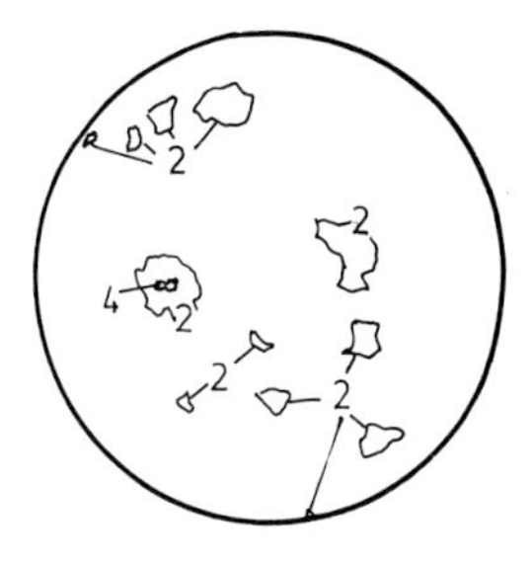

Figure 5b: (0002) Pole figure of CL-W

Tensile properties (strength and ductility) were measured on (German) standard specimens with the gage length equal to five times the specimen diameter, where either d=4 mm or d= 6 mm was used.

Fracture mechanics properties (fracture toughness according to ASTM E 399, K_{ISCC}, and fatigue crack growth in 3.5 % NaCl at R=0.1 according to ASTM E 647) were evaluated using C(T) specimens with W=60 mm and B ranging from 9 to 13 mm. Corrosion-fatigue was measured at loading frequencies of 0.1, 1 and 10 Hz. The frequency dependence of fatigue crack growth in the EQ-T condition was also evaluated in air. The stress corrosion cracking susceptibility K_{ISCC} was evaluated according to the following non-standard procedure: a pre-cracked specimen with $a/W \approx 0.5$ was loaded in the environment of interest to a K-value of roughly 12 MPa√m. If no unstable crack growth occurred after two days, the load was increased roughly 10 % and this process repeated until the onset of unstable crack growth. This last value is reported as K_{ISCC}. This procedure was also used to evaluate the sustained load cracking susceptibility K_{ISLC} of the EQ-T condition in air.

Mechanical testing specimens were oriented so as to obtain a wide variation in number of basal planes perpendicular to the loading axis. The EQ-T condition was tested both in the T-direction (C(T) specimens in the T-L orientation) and 45 ° between the T and L directions. The EQ-B condition was tested in the L-direction (C(T) specimens in the L-T orientation). The CL-S condition was tested in what was presumed to be the L-direction based on the pole figure (L-T orientation). The CL-W condition was tested in an arbitrary orientation.

RESULTS

Tensile and fracture mechanics testing results are summarized in Table II. For simplicity, the density of basal poles coincident to the loading axis is used as a basis for quantifying the texture. This parameter is given as "P_{0002}" in Table II.

Table II: Tensile and fracture mechanics parameters as well as basal pole density for the conditions investigated.

Condition (loading direction)	E	$\sigma_{0.2}$	UTS	el	RA	K_{Ic}	K_{ISCC}	K_{ISCC}/K_{Ic}	P_{0002}
	GPa	MPa	MPa	%	%	MPa√m	MPa√m	-	-
EQ-T (T)	118	1045	1080	12	27	53	28	0.53	3
EQ-T (45 °)	120	1060	1095	13	27	45*	32*	0.71	0.8
EQ-B	102	1035	1040	14	47	51	$\approx K_{Ic}$	1	0.6
CL-S	126	1050	1065	2	9	87	50	0.57	12
CL-W	110	955	980	4	17	87	50	0.57	"1"

* not valid—cracks grew 13 ° out-of-plane

While the strength level exhibits little variation, the well-known dependence of ductility and fracture toughness on microstructure (high tensile ductility with low fracture toughness in equiaxed microstructures, low tensile ductility with high fracture toughness in lamellar microstructures [9,10]) is evident. The CL-S condition exhibited an unusually high strength, as well as a markedly lower ductility than is generally observed in coarse lamellar microstructures.

A significant variation in E, the Young's modulus was found. When E is plotted vs. P_{0002} (figure 6), a dependence in qualitative agreement with that found in [3] is observed in that E tends to increase as more basal planes are subjected to a tensile load.* One exception is EQ-T tested at 45 ° to the rolling direction, which exhibited an unusually high elastic modulus.

* A mathematically more correct correlation between the texture and the elastic modulus can be obtained by integrating the values of the elastic modulus in various single crystal orientations over the inverse pole figure [11]. Since the intent of this work is to correlate texture with fracture mechanics properties as well, this calculation was considered beyond the scope of this paper.

For almost all conditions, the stress corrosion cracking sensitivity as evaluated by K_{ISCC} was significantly lower than the fracture toughness K_{Ic}. The one exception was EQ-B, where these two values were essentially equal. A measure of the relative SCC susceptibility K_{ISCC}/K_{Ic}, i.e., normalized by the fracture toughness, is also provided.

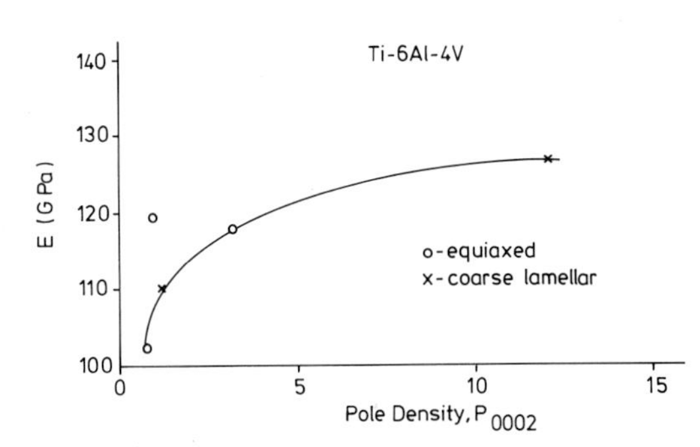

Figure 6: Young's modulus E vs. density of basal poles parallel to the loading axis P_{0002}.

Fatigue crack growth data are shown in figures 7-12. In air, crack growth is independent of loading frequency (for the range of frequency examined) for the EQ-T condition. The sustained load cracking susceptibility in air, K_{ISLC}, was measured to be 63 MPa√m, roughly equal to the K_{max} value measured in the fracture toughness test. In salt water, a marked frequency dependence was observed, as is seen in figure 8. In addition, crack growth rates were higher than in air. EQ-T specimens tested at 45 ° to the L-direction showed no obvious frequency dependence (figure 9). However, all cracks grew out-of-plane by 13 °, making these data difficult to interpret. The EQ-B condition (figure 10) exhibited no frequency dependence. In the coarse lamellar conditions (figures 11 and 12), crack deflection and crack branching occurred, and a slight frequency dependence can be seen in the strongly textured material.

DISCUSSION

The frequency dependence of fatigue crack growth can be predicted qualitatively if both K_{Ic} and K_{ISCC} are known. At low frequencies, i.e., 0.1 Hz, a sharp increase in da/dN occurs at a ΔK

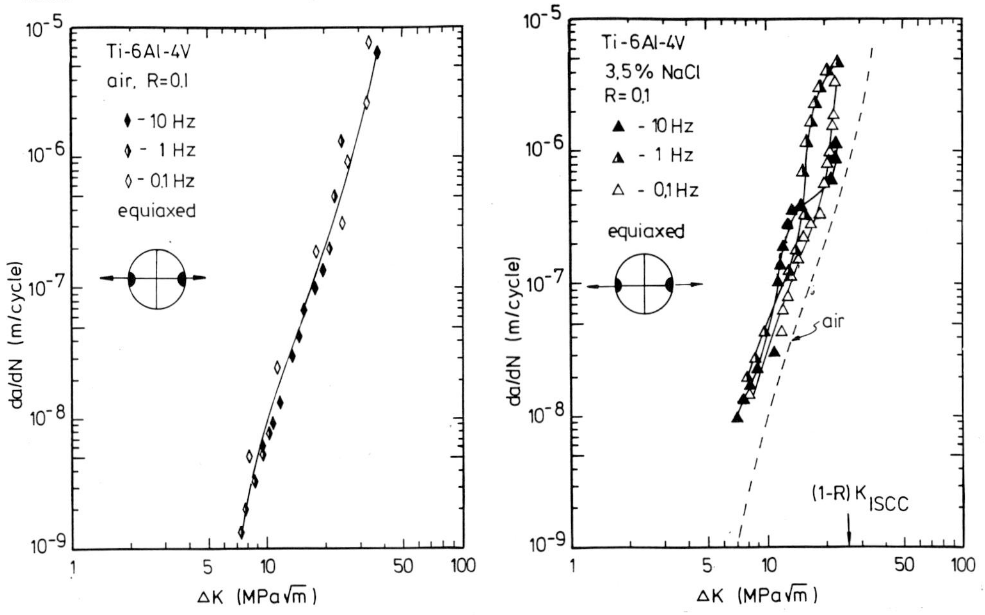

Figure 7: Fatigue crack growth curves for EQ-T in air

Figure 8: Fatigue crack growth curves for EQ-T in 3.5 % NaCl solution

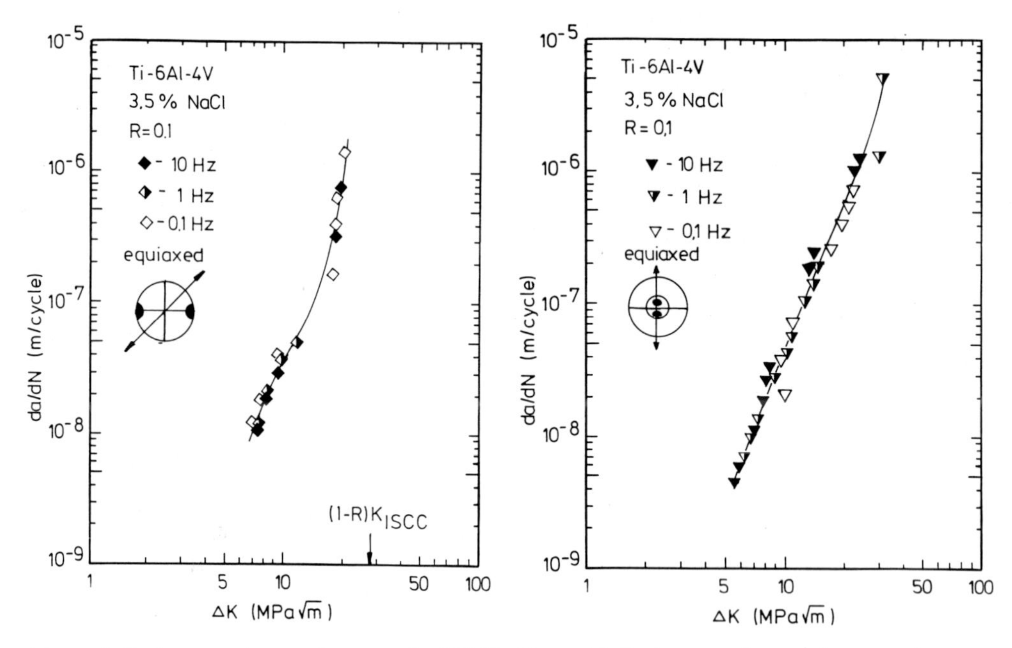

Figure 9: Fatigue crack growth curves for EQ-T (45 ° orientation) in 3.5 % NaCl solution

Figure 10: Fatigue crack growth curves for EQ-B in 3.5 % NaCl solution

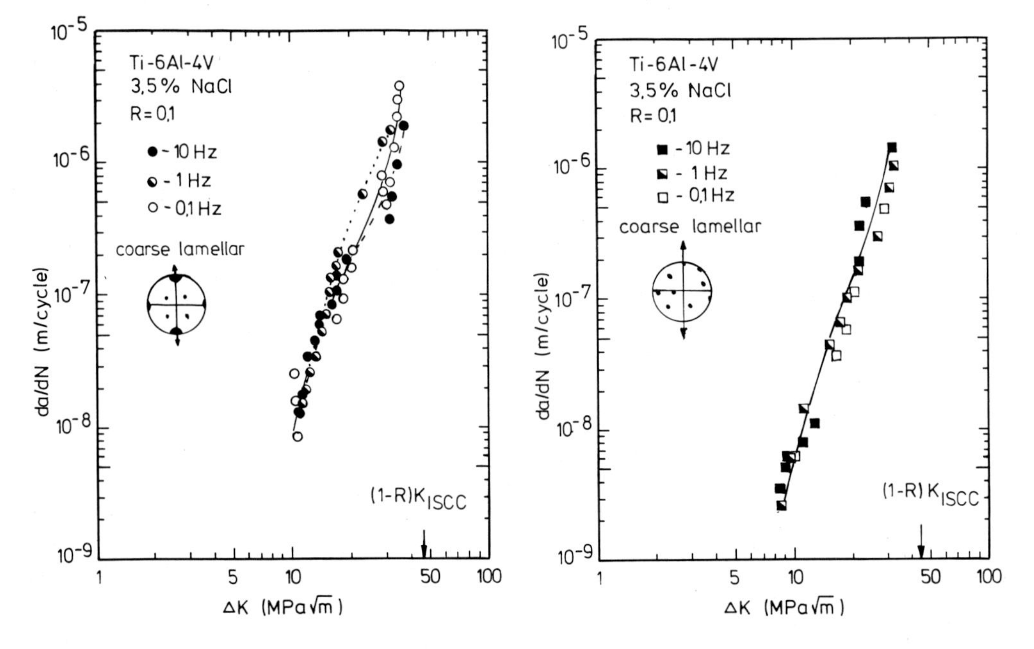

Figure 11: Fatigue crack growth curves for CL-S in 3.5 % NaCl solution

Figure 12: Fatigue crack growth curves for CL-W in 3.5 % NaCl solution

slightly lower than $(1\text{-}R)K_{ISCC}$, that is, when $K_{max}=K_{ISCC}$. When $K_{Ic}=K_{ISCC}$, this "corrosion-fatigue" increase coincides with the normal increase observed when the maximum stress intensity K_{max} approaches the fracture toughness (Region III)[12], and fatigue crack growth is frequency-independent (EQ-T in air, EQ-B).

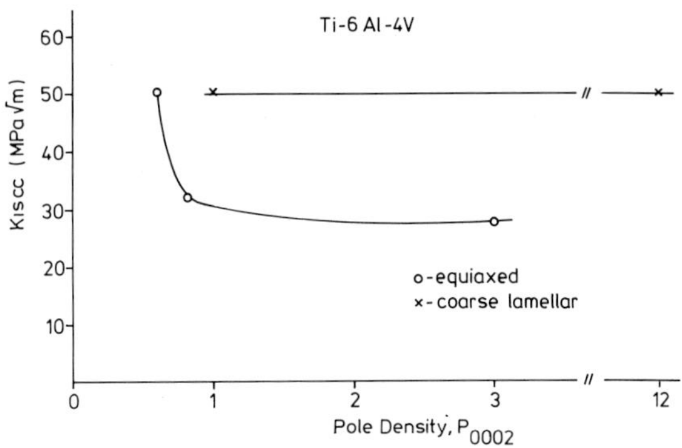

Figure 13: Stress corrosion cracking sensitivity K_{ISCC} vs basal pole density P_{0002}

The issue of fatigue crack growth can thus be reduced to the question of SCC behavior, and it is sufficient to consider the relation between texture and SCC. Taking P_{0002} as a quantitative measure of the texture, it is helpful to graphically depict the dependence found. Figure 13 shows the absolute measure, K_{ISCC} vs. P_{0002}, while figure 14 shows the relative measure, K_{ISCC}/K_{Ic} vs. P_{0002}. Figure 13 thus contains a microstructural dependence (the inherently superior crack growth characteristics of lamellar microstructures) which has been normalized out of figure 14. The relative SCC sensitivity evidently exhibits a well-behaved dependence on the basal pole density, being unity (SCC insensitive) when P_{0002} is significantly less than unity, and decreasing to a value of roughly 0.5 (apparent maximum sensitivity) as P_{0002} increases. This is slightly unusual in that an increase in P_{0002} is associated with an increase in Young's modulus, which generally improves fracture mechanics properties [13].

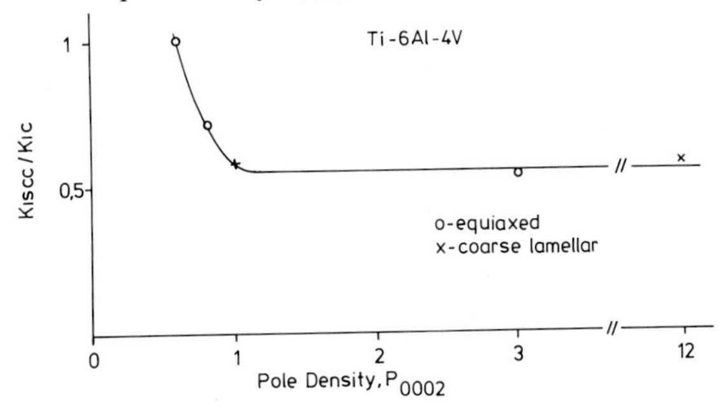

Figure 14: Relative stress corrosion cracking sensitivity K_{ISCC}/K_{Ic} vs basal pole density P_{0002}

That the opposite trend is observed is most likely a result of a tendency for cleavage on basal- or near-basal planes [14], presumably promoted by the formation of similarly oriented embrittling hydrides [15-18]. Equiaxed microstructures can be rendered insensitive to SCC by controlling texture, since single-crystal-like textures with few basal planes being subjected to tensile loading can be achieved by thermomechanical processing. This is not possible in lamellar microstructures, since the nature of their microstructural development is such that several orientation variants of α plates nucleate in the β phase. The results from EQ-T samples tested at 45 ° to the L-direction, where cracks grew out-of-plane at a constant angle, suggest that the strong basal pole located 45 ° to the loading has a significant influence on cracking behavior and possibly on SCC sensitivity. In strongly textured coarse lamellar material, no loading axis at an angle of greater than 45 ° to a pronounced basal pole is possible, and even weakly textured lamellar microstructures exhibit the maximum SCC sensitivity. However, the overall K-values were significantly higher in coarse lamellar than in equiaxed microstructures. In other words, the best EQ condition had a K_{ISCC} equivalent to that of the coarse lamellar microstructures, but with a K_{Ic} of only 50-55 $MPa\sqrt{m}$ as opposed to 87 $MPa\sqrt{m}$. Whereas coarse lamellar microstructures normally have strengths which are 100 MPa lower than equiaxed microstructures, strengths comparable to those of the fine microstructures are apparently possible if an unusually pronounced texture is present. The only inferior property which cannot be improved by texture control is the room temperature ductility.

CONCLUSIONS

The frequency dependence of crack growth in Ti-6Al-4V in 3.5 % aqueous solution (and in air) can be predicted from the stress corrosion cracking (or sustained load cracking) sensitivity when crack growth is essentially in-plane. Pronounced secondary cracking and/or out-of-plane deflection evidently obscures any frequency dependence.

An increased sensitivity to stress corrosion cracking is found with increasing density of basal poles parallel to the loading axis. This is counter to what would be expected considering that the elastic modulus also increases as a greater number of basal planes are subjected to a tensile load.

When normalized by the fracture toughness K_{Ic}, the stress corrosion cracking behavior expressed as K_{ISCC} exhibits a well-behaved, microstructure-independent relation to the pole density P_{0002} parallel to the loading axis. Equiaxed microstructures can be rendered insensitive to stress corrosion cracking by texture control. This is not expected to be possible for coarse lamellar microstuctures.

The generally superior crack growth characteristics of coarse lamellar microstructures as opposed to fine equiaxed microstructures suggest that it may be worthwhile to investigate possibilities for improving strength by deliberate texture control.

ACKNOWLEDGMENTS

The authors thank Fuchs in Meinerzhagen, Robert Zapp Werkstofftechnik in Düsseldorf, and Deutsche Titan GmbH in Essen for providing the Ti-6Al-4V used in this study. Rolling was performed in the laboratory of Prof. Funke, Technical University Clausthal. Technical assistance was provided by Mr. H.-J. Mann (GKSS). Optical microscopy was performed by Mrs. W.-V. Schmitz.

REFERENCES

1) Bowen, A.W.: Scripta Metall., 1977, 11, 17.
2) Fisher, E.S., Renken, D.J.: Phys. Rev., 1964, 135, 482.
3) Zarkades, A., Larson, F.R.: in Int. Conf. on Titanium, ed. R.I. Jaffee, N.E. Promisel, Pergamon Press, London, 1970, 933.
4) Dawson, D.B., Pelloux, R.M.N.: Metall. Trans., 1974, 5, 723.
5) Wanhill, R.J.H.: NLR Report MP 78039 U, Amsterdam, 1978.
6) Fager, D.N., Spurr, W.F.: Trans. ASM, 1968, 61, 283.
7) Peters, M., Lutjering, G.: in Proc. 4th Int. Conf. on Titanium, ed. H. Kimura, O. Izumi, TMS-AIME, Warrendale, PA, 1980, 925.
8) Brokmeier, H.-G.: in Advances and Applications of Quantitative Texture Analysis, ed. H.J. Bunge, C. Esling, DGM-Informationgesellschaft, Oberursel, 1991, 73.
9) Greenfield, M.A., Margolin, H.: Metall. Trans., 1971, 2, 841.
10) Williams, J.C., Chesnutt, J.C., Thompson, A.W.: in Microstructure Fracture Toughness and Fatigue Crack Growth Rate in Titanium Alloys, ed. A.K. Chakrabati and J.C. Chesnutt, TMS-AIME, Warrendale, PA, 1987, 255.
11) Bunge, H.-J.: Texture Analysis in Materials Science, Butterworths, 1982, 304.
12) Gregory, J.K.: in Proc. 4th Int. Conf. on Fatigue and Fatigue Thresholds, ed. H. Kitagawa and T. Tanaka, Materials and Component Engineering Publications, Ltd., Birmingham, 1990, 1845.
13) Clark, W.G. Jr.,: Metals Engg. Quarterly, 1974, 16.
14) Wanhill, R.J.H.: Corrosion, 1973, 29, 435.
15) Wanhill, R.J.H.: Brit. Corr. J., 1975, 10, 69.
16) Wanhill, R.J.H.: in Fracture Mechanics and Technology, ed. G.C. Sih and C.L. Chow, Sijthoff and Noordhoff Inter. Publishers, Alphen aan den Rijn, The Netherlands, 1977, 563.
17) Hall, I.W.: Metall. Trans., 1978, 9A, 815.
18) Boyer, R.R., Spurr, W.F.: Metall. Trans., 1978, 9A, 23.

Materials Science Forum Vol. 157-162 (1994) pp. 1979-1984

TEXTURE EVOLUTION DURING DEEP DRAWING IN ALUMINIUM SHEET

J.R. Hirsch [1] and T.J. Rickert [2]

[1] VAW Aluminium AG, Georg-von-Boeselager-Str. 25, D-53117 Bonn 1, Germany

[2] Allegheny Ludlum Steel, Technical Center, Brackenridge PA, USA

Keywords: Deformation Texture, Deep Drawing, Texture Simulation

ABSTRACT

The texture evolution during deep drawing was analysed in different directions in the cup wall of a deep drawn Al-Mg sheet. The components of the initial rolling texture are strongly affected and major shifts in volume and orientation occur depending on the direction. The ~{112}<111> C-component is strengthened in 0° (= rolling direction) and - less pronounced - in 67°. In the other directions (22°, 45°, 90°) the ~ {011}<211> B-component increases. Experimental results are analysed by theoretical considerations based on the geometry of slip and predicted by Taylor-type calculations.

I. INTRODUCTION

Textures are the main reason for the plastic anisotropy in deep-drawing of sheet material. Commonly the initial texture (usually a cold rolling texture) determines the material properties, e.g., the earing behaviour in deep drawing of aluminium sheet. However, also texture changes occurring during this operation can be important for material properties and surface effects [1, 2]. A considerable amount of deformation occurs, especially in the outer flange of the deep drawn cup which can be affected by the change in strain path from the preceding rolling deformation. Detailed texture data and an analysis of the observed texture effects are still missing.

II. EXPERIMENTAL PROCEDURE AND RESULTS

Samples for texture measurements were taken from an Al-Mg2.5 sheet deep-drawn from a 200 mm blank diameter (drawing ratio = 1.6), completely drawn with a cup wall height of ~ 52 mm with a maximum (circumferential) compressive strain of $\varepsilon_C \approx 0.5$. Samples were taken from the bottom and top of the flange in five different directions - at 0°, 22.5°, 45°, 67.5° and 90° (0, 1/8, 1/4, 3/8, 1/2 · π) angle to the rolling direction. Figure 1 shows the {111} pole figures and Figure 2 the ODFs calculated from four incomplete pole figures after rotating them into a position where best orthorhombic texture symmetry was ensured. This symmetrization significantly simplifies the presentation and the analysis of the results.

The initial texture of the rolled sheet shows the typical aluminium cold rolling texture [3] (Figure 1) with an almost constant density distribution along the β-fibre (Figure 4a). The quantitative ODF analysis shows that about two thirds of the textured volume is accumulated along this fibre, in the ~ {112}<111>"C", ~ {123}<634>"S" and ~ {011}<211>"B" orientation with the maximum intensity of 11.7X random. The rest is located in the two minor fibres: α (= <110> parallel ND) and the $Cube_{RD}$fibre (= <100> parallel RD) including the {011}<100>"G" orientation. About 8 % are random orientations.

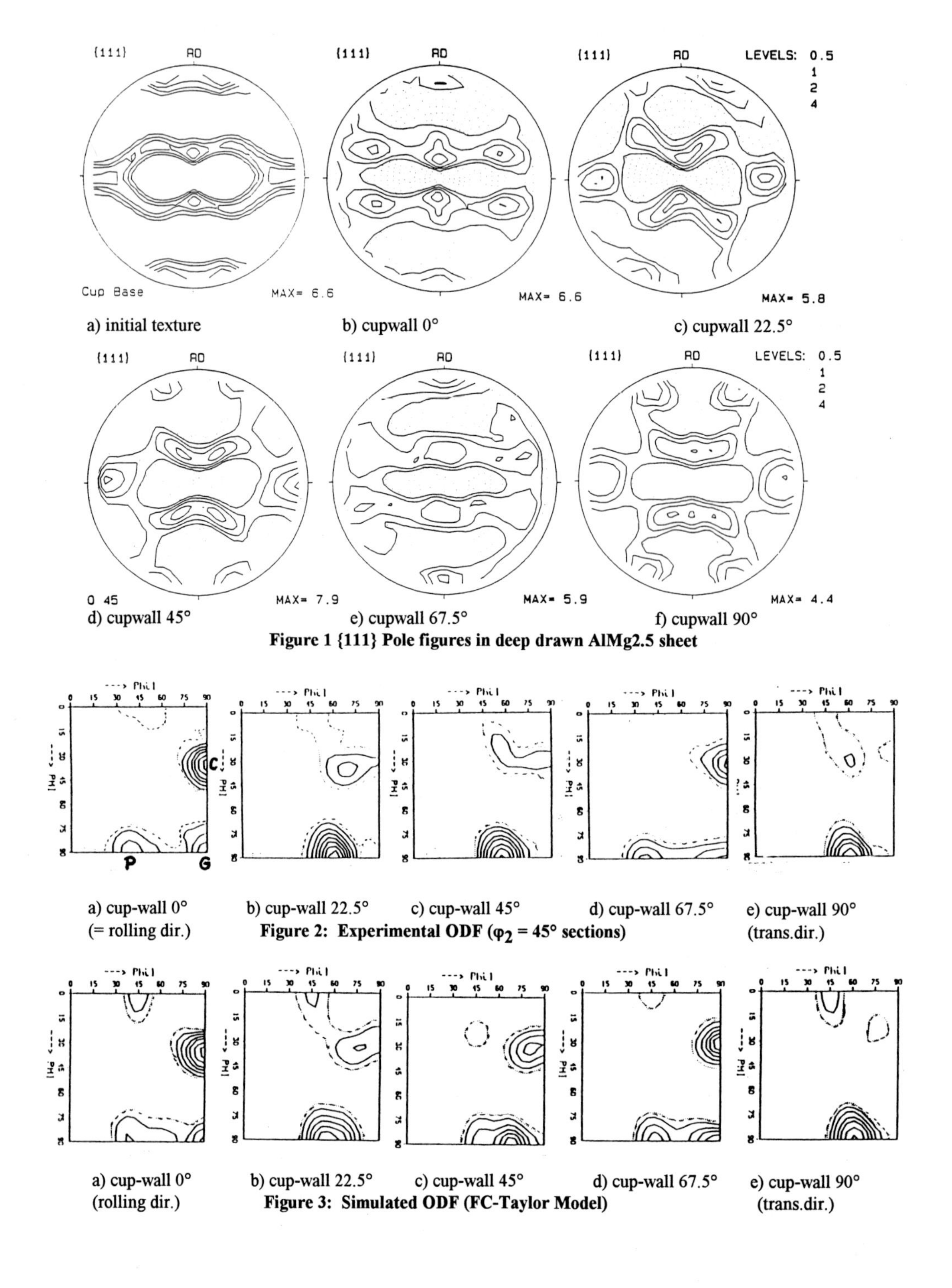

a) initial texture b) cupwall 0° c) cupwall 22.5°

d) cupwall 45° e) cupwall 67.5° f) cupwall 90°

Figure 1 {111} Pole figures in deep drawn AlMg2.5 sheet

a) cup-wall 0° (= rolling dir.) b) cup-wall 22.5° c) cup-wall 45° d) cup-wall 67.5° e) cup-wall 90° (trans.dir.)

Figure 2: Experimental ODF (φ_2 = 45° sections)

a) cup-wall 0° (rolling dir.) b) cup-wall 22.5° c) cup-wall 45° d) cup-wall 67.5° e) cup-wall 90° (trans.dir.)

Figure 3: Simulated ODF (FC-Taylor Model)

The texture evolution observed in the deep drawn cups in the five different directions (in the reference system of the initial rolling texture) can be described as follows [Figures 1, 2, 4]:

- 0° A continuous and relatively strong change of texture towards the C-orientation occurs. A significant shift of the whole β-fibre takes place (Figure 3 b) towards a fibre orientation with <111> almost parallel to RD. Additionally, the depleted B forms a strong G-orientation (~ 20 vol.%).

- 22° Here, in contrast to 0°, the B- and S-orientation increases and the β-fibre orientation shifts in the opposite direction.

- 45° Under 45°, a strong B peak develops and also the β-fibre shift is similar to the one observed under 22°.

- 67° The texture development under 67° shows an almost equal intensity distribution along the β-fibre with a certain preference for the C-orientation. As under 0°, the shift of the main β-fibre leans towards a <111> = RD-fibre and the B-peak splits off and forms a peak at $\varphi_1 = 0°$ (G) and ~ 55° a {110}<111> "P" orientation.

- 90° Here again the B-texture component is dominant. The β-fibre shifts only slightly around ND and is scattering mainly around the rolling direction.

Some scattering and minor peaks around the RD rotated Cube (already present in the initial texture) are present in all textures.

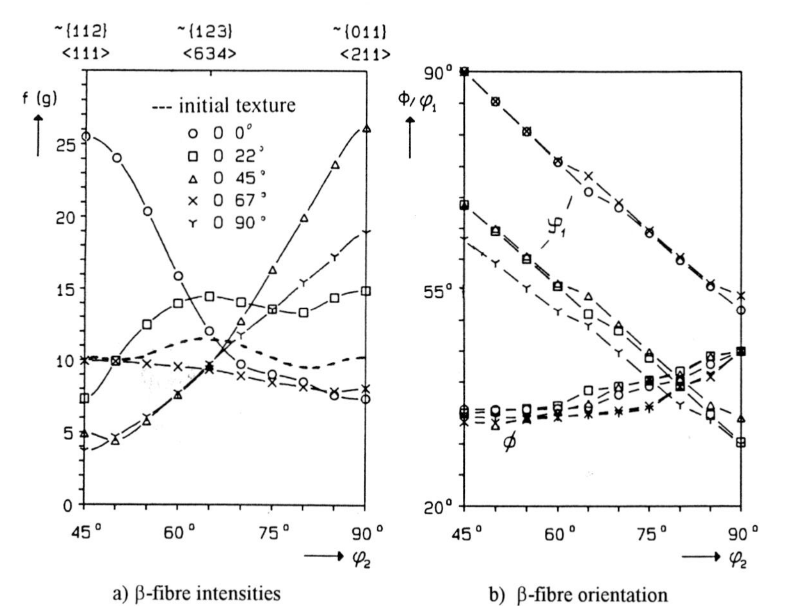

a) β-fibre intensities b) β-fibre orientation

Figure 4: Skeleton line analysis of deep-drawn textures

THEORETICAL PREDICTIONS

In plane strain (rolling) the stable orientations are aligned along the β-fibre [3]. This fibre and its main orientations (C, S, B) are presented in Figure 5 in an inverse pole figure, i.e., plotted in the crystal coordinate system are the three principle directions RD (rolling), ND (normal or short transverse) and TD (long transverse). The range of initial orientations submitted to deep drawing are indicated by changing the angle α form RD towards TD (for one variant).

The deformation of the sheet material in the flange during the deep drawing operation can be described as a plane strain type of deformation for which Tucker's stress state [4] can be applied.

The expected texture changes from basic principles of slip geometry [5], have been derived for the {011} <211> B-orientation. Here the zero strain direction is the highly symmetric <110> = ND direction for which a relatively simple set of slip systems is active which consists of two slip system pairs, a coplanar and a codirectional one. The resulting texture changes and stable orientations of the initial B-orientation in the deep-drawing operation (in its two variants + and -) are listed in Table 1. It can be noted that different parts of the β-fibre have quite different stability orientations. For $\alpha \neq 0°$ or $\neq 90°$, the loss of symmetry must be considered additionally.

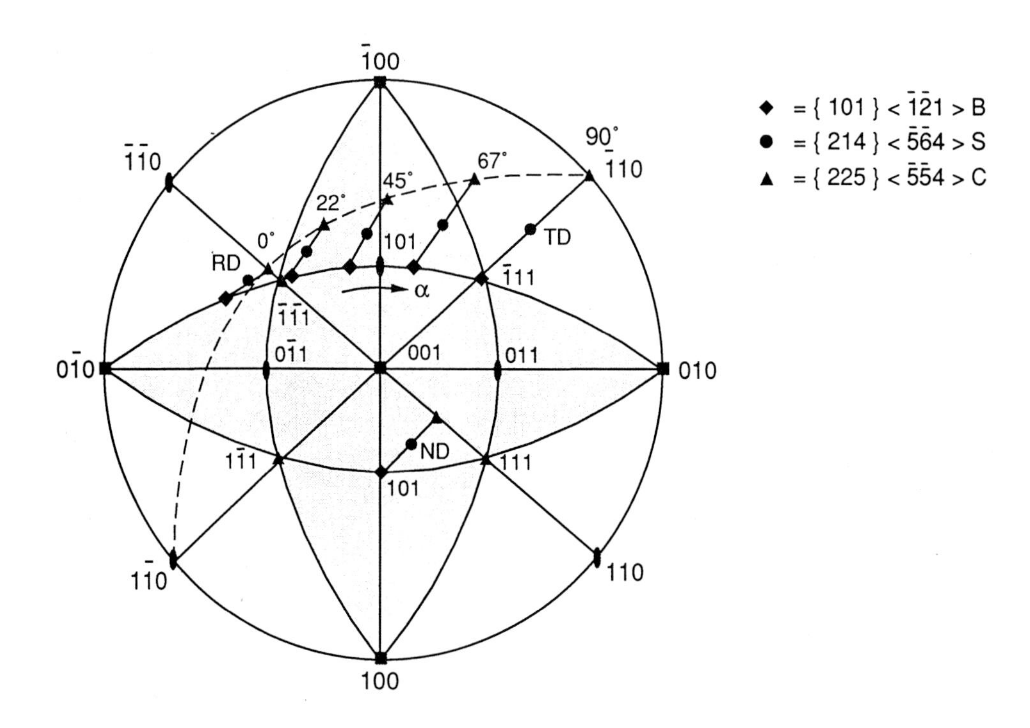

Figure 5: Stereographic projection of the principal sample axes

For a complete prediction of the texture evolution, a multi-slip (Taylor type) model has been applied. For this purpose, the initial rolling texture was decomposed into 218 individual orientations (872 for the calculations with non-symmetric geometry, 22°, 45° and 67°). The orientation change for each grain was calculated for ideal plane strain deformation under full constraints (Taylor) conditions. The results of the simulations have been converted back to continuous ODFs by calculating series expansion coefficients from the individual orientations (assuming a Gaussian distribution with a 5° spread). $\varphi_2 = 45°$ sections of the simulation ODFs are plotted in Figure 3. As can be noted, the principal effects of texture evolution are well predicted (compare to Fig. 2).

DISCUSSION

Usually only 0°, 45° and 90° directions are used to characterise the plastic anisotropy in rolled sheet materials. The results of this investigation show that the texture changes are more complex and quite strong in the flange of the deep drawn cup and the type of texture evolution during deep drawing can be quite different in different directions. In principle, two major types of texture evolution occur. (These can be described in the reference axes of the rolling deformation as the orthorhombic symmetry of the texture is widely maintained.) At 0° and ~67° to RD, a ~{112} <111> "C" (copper) type texture evolves, but with a general <111> RD-fibre component. At the other locations (~ 22°, 45° and 90°) a ~{011} <211> "B" (brass) type texture dominates. Considering the geometry of the main slip systems, the principal texture changes observed during deep drawing can be derived (Table 1) and predicted by a Taylor type model.

The formation of a strong C-peak (as observed in 0° direction in the experimental data) can be explained by the large volume of S-oriented grains which rotate around ND (zero strain direction). However, this is not the truly stable orientation as further slow rotation around TD towards {111} <112> Y must be expected (Table 1). The other two orientations, G and P (at $\varphi_1 = 90°$ and 35°, respectively), found in the 0° deep drawn cup can be derived by a ND rotation of the initial B-component (111) [112].

For the correct interpretation, besides the initial texture also the grain shape is important. The grains which are strongly flattened after the preceding high rolling deformation are now compressed normal to their long transverse axis. For the initial C-orientation, the flat grains lying on top of each other the predicted rotation towards {111} <112> is opposite for both variants, so that it is severely hindered. The B-orientation variants, however, are situated beside each other, and so immediately fulfil the new shear condition which explains the observed fast rotation to either direction.
The difficulty of grains to shear unconstrained, in a subsequent deep drawing operation - with exchanged directions of the principal strains - validates the conditions of full constrained (FC)-Taylor-type deformation that in most cases achieve best results in the simulation.

REFERENCES

1. J. Kusnierz, Archives of Metallurgy, Vol. 33, 2 (1988), pp. 179 - 195.
2. W.A. Anderson, H.C. Stumpf, Metal Progress, 1981, May, pp. 60 - 66.
3. J.R. Hirsch, K. Lücke, Acta Met. Vol. 36, No. 11 (1988), pp. 2863 - 2927.
4. G.E.G. Tucker, J. Inst. Met., Vol. 82, 1953.
5. J.R. Hirsch, (doctoral thesis), RWTH Aachen (1984)

AKNOWLEDGEMENT

Special thanks to B. Kayser and S. Dietrich for careful preparation of the manuscript and to O. Engler from the IMM of the technical university (RWTH) Aachen for texture measurements.

Table 1: **Orientation changes during deep drawing predicted for the two variants of the {011} <211> B rolling texture component in different directions of the sheet.**

Angle α to RD	Rolling Texture Component in Deep Drawing Coordinates				Angle of Rotation	Deep Drawing Texture Components in Rolling Sheet Coordinates			
	nd	rd	cd		<110>=ND	ND	RD	TD	
0°	{011}	<211>	[111]	B±	+20°	{011}	<111>	[211]	P
	(35°,	45°,	0°)			(55°,	45°,	0°)	
	{011}	<211>	[111]	B±	-35°	{010}	<100>	[011]	G
	(35°,	45°,	0°)			(0°,	45°,	0°)	
22.5°	{011}	<111>	[211]	B+	-2.5°	{011}	<211>	[111]	~B
	(57.5°,	45°,	0°)			(32.5°,	45°,	0°)	
	{011}	<611>	[133]	B-	-12.5°	{011}	<722>	[477]	G/B
	(12.5°,	45°,	0°)			(22.5°,	45°,	0°)	
45°	{011}	<144>	[811]	B+	-25°	{011}	<811>	[144]	G/B
	(80°,	45°,	0°)			(10°,	45°,	0°)	
	{011}	<811>	[144]	B-	+10°	{011}	<755>	[1077]	B/P
	(-10°,	45°,	0°)			(45°,	45°,	0°)	
67.5°	{011}	<144>	[811]	B+	+22.5°	{011}	<111>	[211]	~P
	(112.5°,	45°,	0°)			(57.5°,	45°,	0°)	
	{011}	<211>	[111]	B-	-32.5°	{011}	<100>	[011]	~G
	(-32.5°,	45°,	0°)			(2.5°,	45°,	0°)	
					or +22.5°	{011}	<111>	[211]	~P
						(57.5°,	45°,	0°)	
90°	{011}	<111>	[211]	B±	Stable	{011}	<211>	[111]	B
	(125°,	45°,	0°)			(35°,	45°,	0°)	
	(-55°,	45°,	0°)						

ND, RD, TD = normal (short transverse), rolling and (long) transverse direction.
nd, rd, cd = normal (short transverse), elongation and compression direction.

Materials Science Forum Vol. 157-162 (1994) pp. 1985-1990

TEXTURE INHOMOGENEITIES IN HIGH TENSILE STRENGTH STEELS PROCESSED WITH LOW TEMPERATURE CONTROLLED ROLLING

H. Inagaki and K. Inoue

Shonan Institute of Technology,
251 Kanagawaken, Fujisawashi, Tsujidou-Nishikaigan 1-1-25, Japan

Keywords: Surface Texture, Transformation Texture, Controlled Rolling

ABSTRACT

Unusual surface textures and texture inhomogenities in the thickness direction have been found in low temperature control-rolled high tensile strength steel plates with heavy thickness. During hot rolling of such thick materials at intercritical temperatures, temperature gradient developed in the thickness direction is so large that each thickness is rolled in different phase regions. This seems to have resulted in the observed texture anomalies.

1. INTRODUCTION

Recently, demands on very thick high tensile strength steel plates with excellent toughness are increasing. These plates are usually manufactured from very large slabs by heavy, low temperature controlled rolling. Since abnormal surface textures and texture inhomogenities were observed in these plates, their origins were studied in detail.

2. EXPERIMENTALS

Starting materials were 230mm thick slabs of a 0.10%C-0.25%Si-1.27%Mn-0.065% Al steel. According to dilatometer measurements, Ac_3 and Ar_3 of this material were 892 and 790℃, respectively. After soaking at 940℃ for 0.5h, these slabs were hot rolled 50% in the γ range above 800℃, and 70% in the $(\gamma+\alpha)$ range below 790℃ to the final thickness of 41mm. Rolling temperatures were monitored at the plate surface, and finishing temperatures were varied between 770 and 650℃. Reference samples were obtained by finish rolling original slabs at 880℃.

Specimens taken from various depth of final products were subjected to optical and transmission electron microscopies, and to the measurement of {110},{200} and {211} pole figures. From these, ODFs were calculated up to l=22nd order. Also high temperature compression tests were made on cylindrical specimens ($8(\phi)\times12(l)mm^2$) taken from the original slab. After soaking at 1000℃, specimens were cooled to the test temperature, and compressed with the strain rate of $5\times10^{-3}S^{-1}$.

3. RESULTS

3.1 TEXTURE

In reference specimens finished at 880℃, textures were almost random, figure 1. Only at 1/2 thickness, weak textures consisting of {100}<011>, {110}<110> and {110}<001> were observed. They were transformation textures derived from the {100}<001> γ recrystallization texture [1].

In specimens finished at temperatures below Ar_3, textures were strong and inhomogeneous in the thickness direction, figure 1. In the surface layer, <110>//RD fiber textures having their center at {100}<011> were observed. With decreasing finishing temperatures, they developed remarkably. As a weak components, {110}<001> orientations were present in the specimen finished at 770℃. With decreasing finishing temperatures, this component developed into a <110>//TD partial fiber texture lying in the range between {110}<001> and {221}<114>.

Also in the 1/2 thickness, rather strong textures were observed. In the specimens finished at 770℃, {332}<113>, and orientations in the range between {100}<011> and {511}<011> were present. They were transformation textures derived from the rolling textures of γ phase via K-S relationship [1]. With decreasing finishing temperatures, the latter developed into RD//<110> fiber textures. However, their development was much slower than that observed in the surface layer. With decreasing finishing temperature, {332}<113> showed orientation spread toward {554}<225>. Thus, textures observed in the midthickness was α rolling textures, whose initial testures were transformation textures inherited from γ rolling textures.

In the 1/4 thickness, textures were generally weak. Only by lowering the finishing temperature down to 700℃, relatively strong <110>//RD fiber textures having their center at {100}<011>, and <110>//TD fiber textures with their center at {332}<113>, could be observed.

3.2 MICROSTRUCTURES

3.2.1 OPTICAL MICROSCOPY

In the specimen finished at 770℃, figure 2, surface layers consisted of coarse grains. Some of them were equiaxed, but many were deformed and

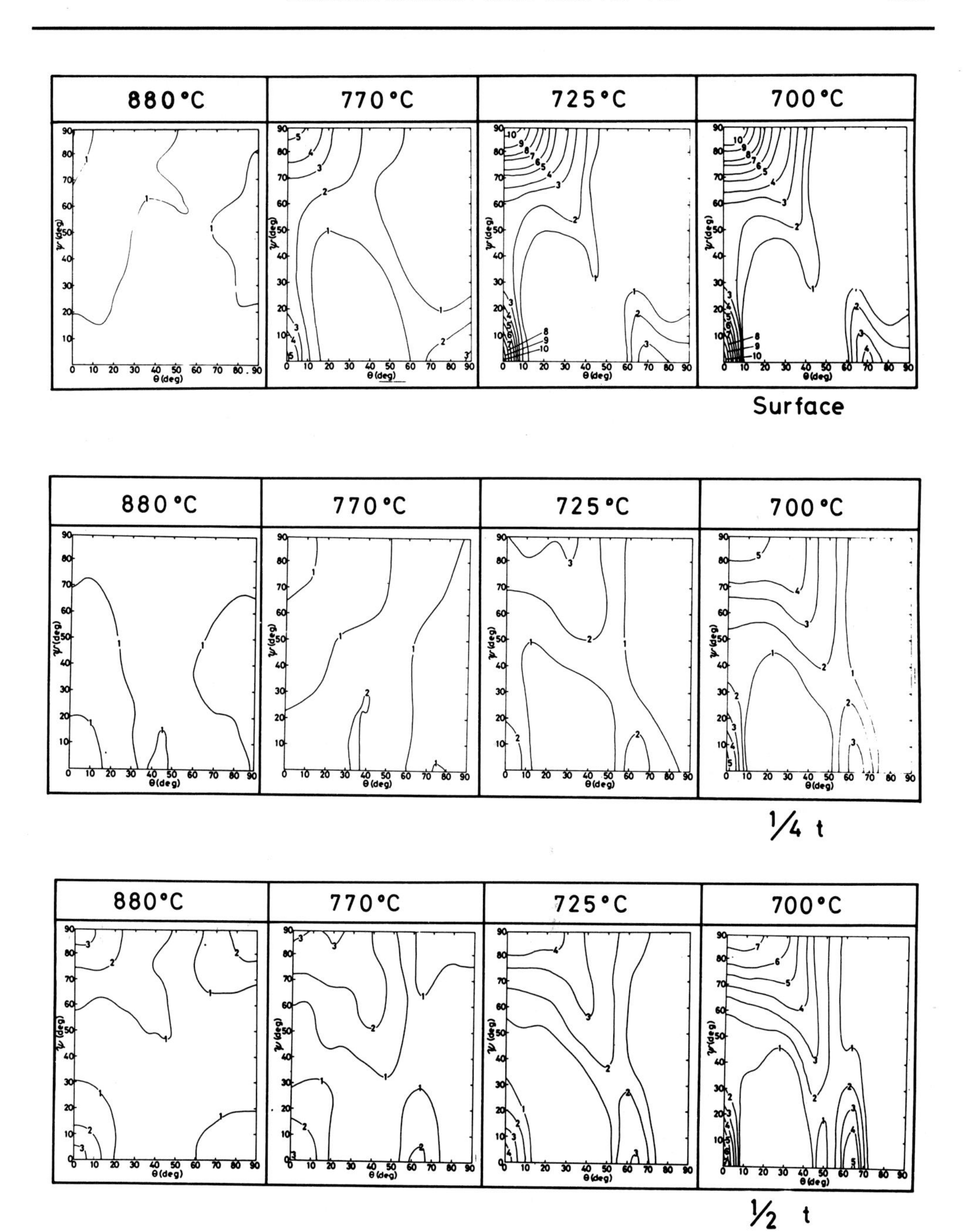

figure 1 ϕ=45° sections of ODF observed in the surface layer, 1/4 thickness and 1/2 thickness of specimens finished at 880, 770, 725 and 700°C.

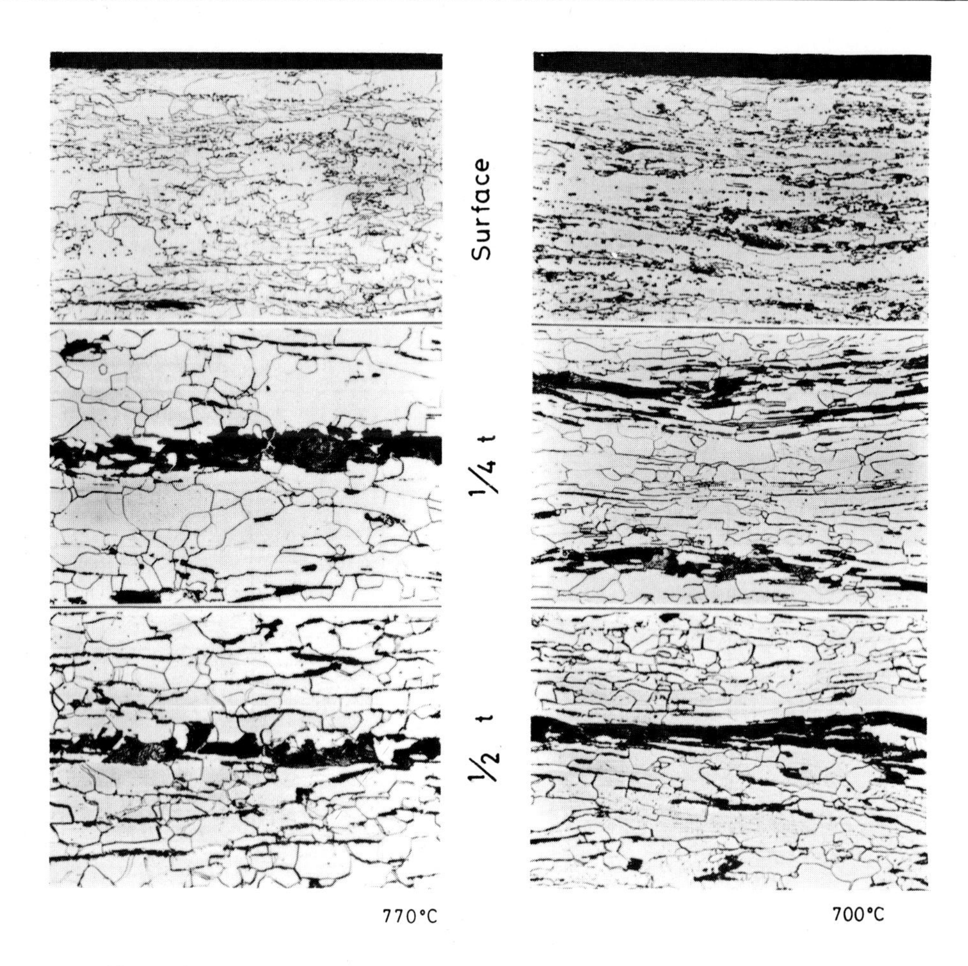

figure 2 Microstructures of specimens finished at 770 and 700℃.

elongated in the rolling direction. Also carbide particles are fragmented. In the 1/4 thickness, coarse, equiaxed grains were observed. In the 1/2 thickness, the microstructure consisted of relatively small, equiaxed grains typically observed in steels finished above Ar_3.

With decreasing finishing temperatures, surface layers were severely deformed and their microstructures consisted of extremely elongated grains and stringers of fragmented cementite particles, figure 2. In the 1/4 thickness, however, ferrite grains and pearlite particles began to be elongated only in the specimens finished below 725℃.

Also in the 1/2 thickness, ferrite grains were elongated, if the finishing temperature was below 725℃. However, the amount of deformation of ferrite grains was much smaller.

3.2.2 TEM

In the specimen finished at 770℃, the surface layer consisted mainly of coarse, dislocation free, recrystallized grains surrounded by small subgrains. In contrast, only large recrystallized grains were observed in the 1/4 thickness. In the 1/2 thickness, slightly dislocated ferrite grains typically formed by the γ to α transformation were observed. In the specimen finished at 725℃, incompletely recovered cell structures were observed in its surface layer. In the 1/4 thickness, recrystallized grains were again observed, but some of them were lightly deformed. Microstructures formed in the 1/2 thickness were thesame as those observed in the same thickness of the specimen finished at 770℃.

4. DISCUSSIONS

In a previous paper[2], textures formed by conventional controlled rolling have already been studied in detail. 80mm thick slabs of a low C steel were soaked at 1150℃ and control-rolled to 12.5mm thickness. In this case, the surface texture has been identified as {110}<001> shear textures.

However, the results described above show that, if much thicker slabs are soaked at lower temperatures just above Ac_3 and subjected to controlled rolling, such surface textures are not formed. Throught the thickness, RD// <110> fiber texture having their centers at {100}<011>, and TD//<110> fiber textures with their center at {332}<113> were always observed. Intensities of these orientations varied significantly in the thickness direction, showing minimum at 1/4 thickness.

Since steel chemistries and frictions between rolls and specimens were similar in both cases, such differences might be ascribed to differences in t/D ratio, and/or temperature gradient in the thickness direction.

During rolling of thick slabs, large temperature gradient is formed in the thickness direction, so that each thickness is rolled at different temperatures. In the temperature range of controlled rolling, deformation resistance of a low C steel depends strongly on deformation temperature[3]. Results of hot compression tests, figure 3, in fact show that, in the γ range above 780℃, both the flow stress at 1% strain, $\sigma_{1\%}$, and peak stress, σ peak, increased linearly with decreasing deformation temperature. At Ar_3, these flow stresses showed sudden decreases of about 30MPa. Also below Ar_3, flow stresses increased with decreasing deformation temperature, but the rate of increase was much rapider than that observed in the γ phase. Such local minimum of the flow stress at Ar_3 can occur, since the flow stress of α phase is lower than that of γ phase at the same temperature in the $(\gamma+\alpha)$ region. In this region, the overall flow stress is determined by the law of mixtures between flow

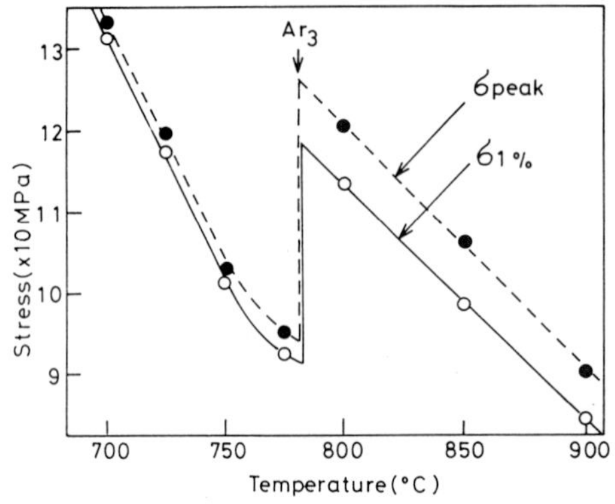

figure 3 Temperature dependence of 1% flow stress and peak stress.

stresses of γ and α phases. In figure 3, comparisons between 1% stress and peak stress further show that work hardening is larger in γ phase. In thick slabs, in which large temperature gradient is established in the thickness dirction, deformation resistance varies therefore in the thickness direction. Deformation may be concentrated in the thickness where the temperature is just below Ar_3, since deformation resistance is minimum. In this layer, recrystallization of α phase would be induced, yielding random textures. In the first stage of controlled rolling, such zones of recrystallized grains are formed in the surface layer, where Ar_3 is first arrived. Inner layers are still in the γ phase with high deformation resistance. With decreasing rolling temperatures such zones of recrystallized grains with random orientation are shifted toward inner thickness. In the surface layers, temperatures are too low to induce recrystallization. Heavily deformed grains are therefore formed, developing strong rolling textures. In the 1/2 thickness, most of rolling reductions are given in the γ phase, so that development of the strongest γ rolling texture, and therefore formation of strong transformation textures is expected in this region. Owing to these processes, textures are strong in the surface layer and in 1/2 thickness, and very weak in the intermediate thickness. However, since t/D is large in thick slabs, large deformation to induce recrystallization cannot be applied to the
inner thickness. Thus, although 1/2 thickness is also rolled in ($\gamma+\alpha$) phase at the lowest rolling temperature, recrystallization of α phase does not occur. In this thickness, α rolling textures are developed from strong transformation textures.

REFEREMCES

1) Inagaki, H: Trans.ISIJ.,1977,17,166
2) Inagaki, H: Z.Metallkde, 1983,14,716
3) Kasper, P., Streissberger, A. and Pawleski, O.: Arch. Eis.,1983,54,195

Materials Science Forum Vol. 157-162 (1994) pp. 1991-1996

ANNEALING TEXTURES IN ALUMINIUM DEFORMED BY HOT PLANE STRAIN COMPRESSION

D. Juul Jensen [1], R.K. Bolingbroke [2], H. Shi [3], R. Shahani [4] and T. Furu [5]

[1] Materials Department, Risø National Laboratory, P.O. Box 49, DK-4000 Roskilde, Denmark

[2] Alcan International Ltd., Banbury Lab., Banbury, Oxon OX16 7SP, UK

[3] Department of Engineering Materials, University of Sheffield, Sheffield SI 3JD, UK

[4] Pechiney, Centre de Recherches de Voreppe, BP 27, F-38340 Voreppe, France

[5] SINTEF, Department of Physical Metallurgy, N-7034 Trondheim Norway

Keywords: Aluminium, Plane Strain Compression, Hot Deformation Texture, Recrystallization Texture, Cube Texture

ABSTRACT

Commercially pure aluminium (AA1050) samples have been hot deformed by plane strain compression and subsequently quenched or annealed at the deformation temperature. The samples were deformed at strain rates 0.25, 2.5 and 25 s^{-1} to strains of 0.5, 1.0 and 2.0 at 300 and 400°C. The deformation and recrystallization textures were measured by neutron diffraction. It was found that during hot deformation typical rolling textures develop, the strength of which is highest at the highest deformation temperature. The recrystallization textures contain cube, Goss and retained rolling components. The strength of the cube texture depends strongly on the strain rate and temperature, being highest at $\dot{\varepsilon} = 0.25\ s^{-1}$ and T = 400°C. It is discussed how the texture variations depend on the deformation parameters.

1. INTRODUCTION

The control of texture during commercial hot rolling is often essential to meet stringent product property requirements specified by the customer. It has been observed that both the material and deformation parameters significantly influence the textures developed after processing. However, on an industrial scale it is not possible to separate the individual effects of the rolling parameters (strain, strain rate and temperature) due to their interdependence. Instead, laboratory simulations using Plane Strain Compression (PSC) can be utilised [1] in which the three parameters can be varied independently. Furthermore PSC allows deformation to high strains in one "pass" without interpass reheats as would be required in the case of hot rolling.

In the present study the hot deformation and subsequent recrystallization textures in aluminium (AA 1050) were studied. The aim is to present a set of texture data which illustrates the effects of each of the individual deformation parameters on the texture development.

2. EXPERIMENTAL

The material used was commercial purity Al containing 0.32% Fe and 0.15% Si. It was prepared by DC casting, homogenization at 600°C, slow cooling, break down rolling and annealing. This starting material contained 0.5 vol.% 1.7 μm eutectic particles. The average grain size was 100 μm. The starting texture was predominantly random with a small contribution from the cube orientation (see Fig. 1). In total ~8 vol.% had orientations near cube and 23 vol.% near the 3 rolling components (C, S, B). (These, and all vol.% given in the following, are calculated by integration over 15° around the corresponding peaks in the ODF).

The samples were deformed by plane strain compression at Pechiney Centre de Recherches de Voreppe. The deformation parameters were temperature (T = 300 and 400°C), strain (equivalent von Mises strain $\varepsilon = 0.5$, 1.0 and 2.0) and strain rate ($\dot{\varepsilon} = 0.25$, 2.5 and 25 s^{-1}). After deformation the samples were either quenched to below 100°C within 3.5 seconds or annealed to complete recrystallization at the deformation temperature. The bulk textures were determined by neutron diffraction [2]. Before these measurements, the samples were scalped so that the texture of only the mid third of the total sample thickness was measured.

3. RESULTS AND DISCUSSION

DEFORMATION TEXTURE. The development of deformation texture in the present material is very similar to that observed for an identical alloy containing 1% Mn [3]. Briefly, this is described below:

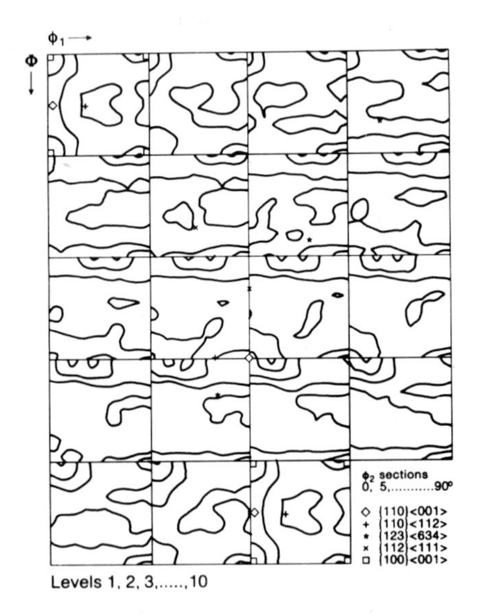

Fig. 1. ODF for the starting material.

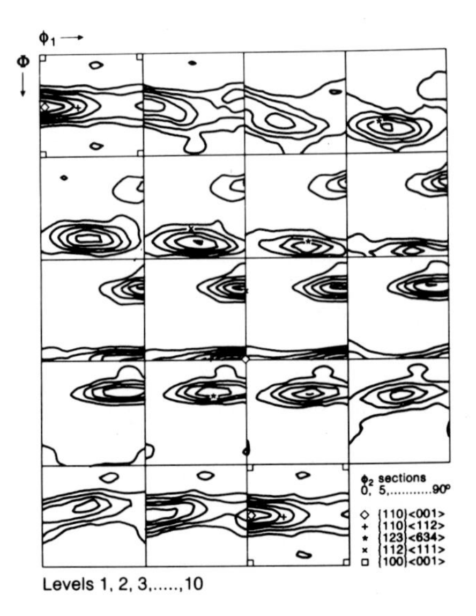

Fig. 2. ODF for a sample deformed by PSC at T = 300°C, $\dot{\varepsilon} = 2.5\ s^{-1}$ to a strain of 2.0 and quenched immediately after deformation.

After PSC the typical copper type rolling texture develops. An example is shown in Fig. 2. It can be seen that it contains orientations lying along the β-fibre running from B {110} <112> over S {123} <634> to C {112} <111>, plus some Goss texture {110} <001>. All three rolling components, B, S and C strengthen with increasing strain and the relative amount of the three components is more or less unchanged during the deformation. A similar behaviour is observed in the Mn containing alloy [3].

In Fig. 3, the total volume concentration of the three rolling components is plotted versus strain for the two deformation temperatures. It can be seen that the strength of the rolling texture increases with strain, initially (in the strain range 0 - ~0.5) at a high rate and later ($\varepsilon \geq 1$) at a lower rate. The strongest texture develops at the highest deformation temperature, T = 400°C. The strain rate, $\dot{\varepsilon}$, does not seem to have any systematic effect on the deformation texture development. The rate at which the texture develops with increasing strain (Fig. 3) is higher in the present material than in the Mn containing alloy. This is most pronounced at the lowest deformation temperature. Studies of the effects of small (non-shearable) particles during cold deformation have shown that they can cause either a weakening or a strengthening of the deformation texture [4]. It appears, that the main effect of small particles on the slip pattern is to promote slip on an increased number of slip systems and to reduce the slip on the individual systems [4]. Such effects may have given the weaker textures observed in the Mn containing alloy compared to the present results.

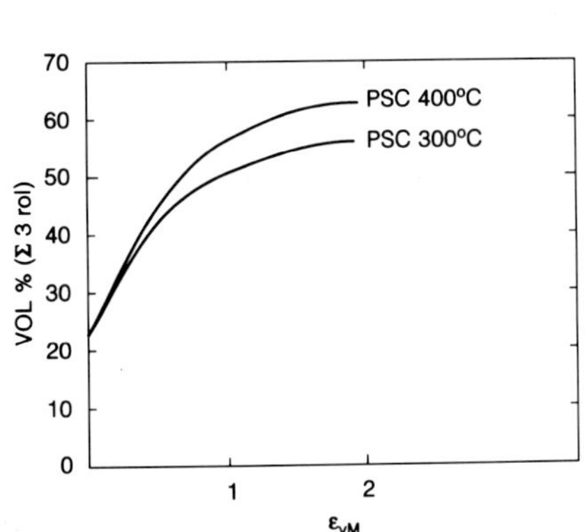

Fig. 3. The development of total volume fraction of the three rolling texture components during PSC.

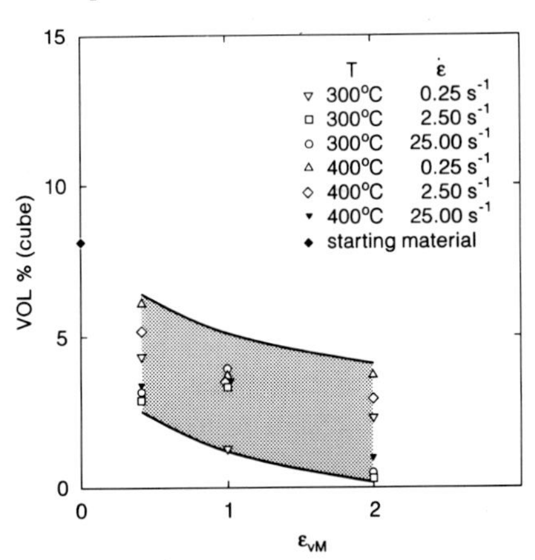

Fig. 4. The development in volume concentration of cube texture during PSC deformation.

During the deformation the initial (weak) cube texture decreases. This is shown in Fig. 4. From the data it is not possible to say if T or $\dot{\varepsilon}$ have any effect on the decrease in cube texture with increasing strain, considering that the experimental error in the texture measurements typically is a few per cent. The decrease in the cube texture is almost identical to that in the Mn containing alloy.

RECRYSTALLIZATION TEXTURE. The recrystallization texture in the PSC and annealed samples contains cube and Goss texture components. However, a large amount (~60 - 90 vol.%) of the material has orientations which do not show up as clear peaks in the ODFs and will be considered as a random component; within this component, ~20 - 30 vol% has orientations near the rolling components. An example of the ODF is shown in Fig. 5. This composition of the recrystallization texture with cube and a relative large volume fraction and random of retained rolling texture agrees well with previous results (for an overview, see [5, 6]). It is generally believed that nuclei formed within deformation zones around large second phase particles are of rolling or random orientations [7]. The deformation conditions under which deformation zones develop have been discussed in [8, 9]. An equation for the critical strain rate above which deformation zones form was derived and at present the constants in the equation are being fine tuned to the present material and deformation parameters [10].

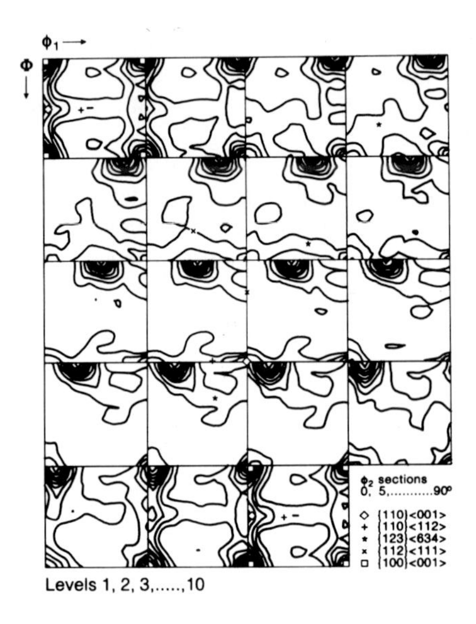

Fig. 5. ODF for a sample deformed at T = 300°C, $\dot{\varepsilon}$ = 2.5 s^{-1} to ε = 2.0 and recrystallized at the deformation temperature.

The strength of the cube texture depends strongly on all three deformation parameters. This can be seen in Fig. 6. Strong cube texture is developed at high strains, low strain rate and high annealing (= deformation) temperature. The strongest cube observed is 26.6 vol.% at ε = 2.0, $\dot{\varepsilon}$ = 0.25 s^{-1} and T = 400°C. The weakest cube textures at both deformation temperatures are ~8 vol.%, i.e. comparable with the strength of the cube in the starting material. That the intensity of the cube increases with increasing strain is well known [eg. 5, 6]. The effect of deformation = annealing temperature has been studied previously for hot rolling [11, 12]. In these studies a more complex behaviour of the cube texture was observed. However, in these papers it was suggested that the observed complex behaviours could be ascribed to effects of the interpass periods between the rolling passes. Effects of strain rate to the authors' knowledge have not been studied before. It is remarkable that the strain rate has such a clear effect on the recrystallization cube texture in the present material, when it has no systematic effect on the deformation texture. In Fig.7 the strength of the cube texture is plotted versus the Zener-Holloman parameter, $Z = \dot{\varepsilon} \exp(Q/RT)$. It can be seen that an increase in Z generally corresponds to a decrease in strength of cube texture. However, for all 3 strains, some discontinuity is observed when the temperature is changed from 400 to 300°C (see Fig.7). A similar effect is observed when the recrystallized grain size is plotted versus Z (see Fig.8). At present a model is being developed for calculation of recrystallization kinetics, grain size and texture [13].

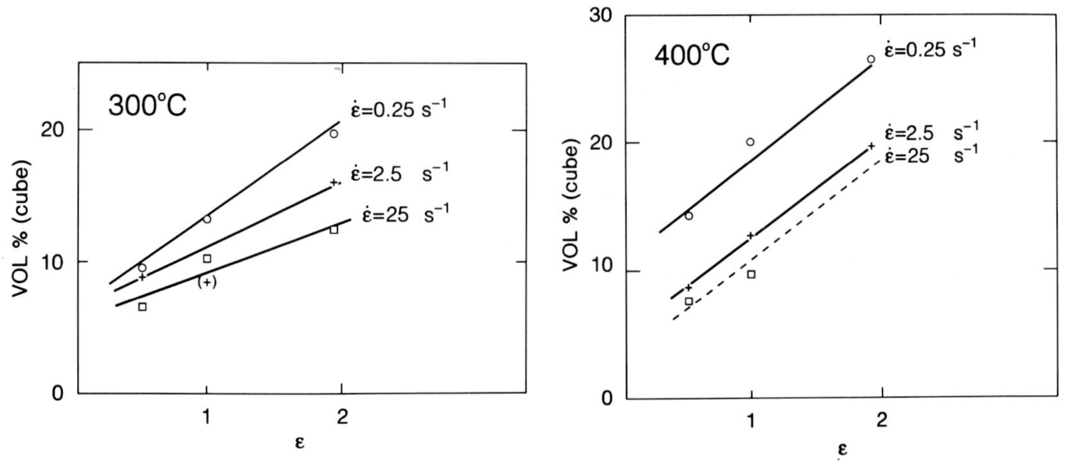

Fig. 6. The development of the cube recrystallization texture component of 300°C, b) 400°C.

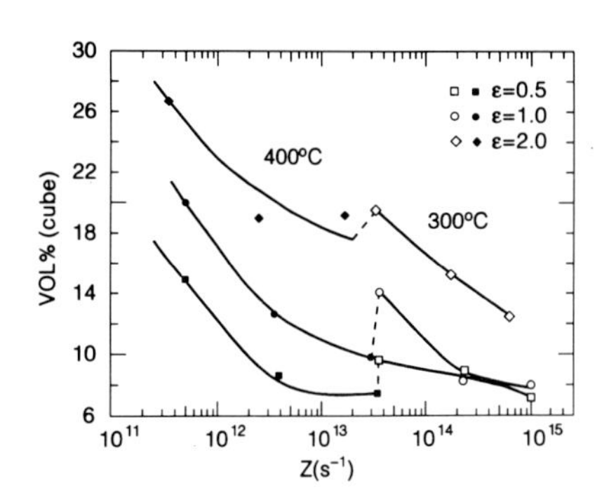

Fig. 7. The volume concentrations of the recrystallized cube component versus the Zener-Holloman parameter (using Q = 156 kJ).

Fig. 8. The recrystallized grain size (linear intercept value) versus the Zener-Holloman parameter (Q = 156 kJ).

The concentration of the Goss recrystallization texture is in the range 4-11 vol.% which is not very different from that in the deformation texture. No systematic variation with ε, $\dot{\varepsilon}$ or T is observed.

4. CONCLUSIONS

The texture development in Al 99.5% (AA 1050) has been studied during hot PSC deformation and subsequent annealing at the deformation temperature. It was found that

- A typical cold rolling type texture develops during the deformation with no particularly strong B ({110} <112>) component. Higher deformation temperature leads to stronger deformation textures whereas the strain rate seems to have no systematic effects.

- The recrystallization texture contains cube, Goss and random components.

- All three deformation parameters - ε, $\dot{\varepsilon}$ and T - affect significantly the strength of the cube component in the recrystallization texture.

ACKNOWLEDGEMENT

The work was carried out as part of a BRITE/EURAM project BREU-CT91-0399. The comments and inspiration from the other partners in the project are gratefully acknowledged.

REFERENCES

1) Bolingbroke, R.K., Creed, E., Marshall, G.J. and Ricks, R.A.: Proc. Sym. Aluminium Alloys for Packaging, TMS Fall Meeting, Chicago, USA, November 1992, (eds.: Morris, J.G. et al.), 215.

2) Juul Jensen, D. and Kjems, J.K.: Textures and Microstructures, 1983, 5, 239.

3) Juul Jensen, D., Shi, H. and Bolingbroke, R.K.: These Proceedings.

4) Hansen, N. and Juul Jensen, D.: Proc. Sym. Formability and Metallurgical Structure, TMS Fall Meeting, Orlando, USA, October 1986 (eds. Sachdev, A.K. and Embury, J.D.), 119.

5) Lücke, K. and Engler, O.: Proc. Aluminium Alloys ICAA3 (eds. L. Arnberg et al.), 1992, III, 439.

6) Hirsch, J.R.: Proc. Recrystallization '90 (ed. T. Chandra, TMS Warrendale), 1990, 759.

7) Humphreys, F.J.: Proc. Recrystallization '90 (ed. T. Chandra, TMS, Warrendale), 1990, 113.

8) Humphreys, F.J. and Kalu, P.N.: Acta metall., 1987, 35, 2815.

9) Lagneborg, R., Zajec, S. and Hutchinson, B.: Scripta metall. and mater., 1993, 29, 159.

10) Vernon-Parry, K.D., Sircar, S. and Humphreys, F.J.: Part of BRITE/EURAM contribution, 1993.

11) Hutschinson, W.B., Oscarson, A. and Karlsson, Å.: Mat. Sci. Tech., 1989, 5, 1118.

12) Merchant, H.D. and Morris, J.G.: Metal. Trans. A, 1990, 21A, 2643.

13) Ørsund, R., Furu, T., Nedreberg, M.L. and Nes, E.: Part of BRITE/EURAM contribution, 1993.

Materials Science Forum Vol. 157-162 (1994) pp. 1997-2004

THE EFFECT OF TEXTURE AND GRAIN BOUNDARY STRUCTURE ON THE INTERGRANULAR CORROSION OF STAINLESS STEEL

H.M. Kim and J.A. Szpunar

Department of Metallurgical Engineering, McGill University,
3450 University Street, Montreal, Quebec H3A 2A7, Canada

Keywords: CSL Boundary, Intergranular Corrosion, Etch Pits

ABSTRACT

Grain boundaries are favored sites for the segregaton and certain grain boundaries can be preferentially attacked. In stainless steels, intergranular corroson occur at the chromium depletion region at the grain boundaries. CSL boundaries, with less segregation, exhibit relatively strong intergranular corrosion resistance. Different CSL statistics and different intergranular corrosion rates are found in this study. Higher frequencies of CSL boundaries may result in strong intergranular corrosion resistance. In addition, etch pit study shows that every single grain shows different pitting potential. Atomic force microscopy is helpful to identify the attacked and unattacked grain boundary as well as to analyze the surface topography.

INTRODUCTION

Grain boundaries are disordered misfit region separating grains of different crystallographic orientaton, so they have relatively high energy and are favored sites for the segregaton of various solute elements or the precipitaton of metal compounds, such as carbides and sigma. Therefore, it is sure that certain corrodent grain boundaries can be preferentially attacked. This type of attack is known as intergranular corrosion and most metals and alloys in specific corrodents can exhibit it.

In stainless steels intergranular corrosion can occur as a result of segregation of certain solute elements to grain bundaries. However, this is known to occur only in highly oxidizing corrodent environment and is not usually encountered in service.[1,2] The usually encountered form of intergranular corrosion is due to sensitizaton. Sensitization is understood in terms of chromium carbide precipitation, and is attributed to chromium depletion in the region of carbides precipitated at grain boundary.[3]

It can be seen that as the chromium content of the alloy is reduced to below approximately 12%, there is a great chance of corrosion attack. Thus, if sensitization reduces the chromium content at the grain boundary to values below 12% while the grains retain a chromium content of about 18%, there will be a significant potential for corrosion at the grain boundaries while the grains remain passive.[4,5]

However, it should be emphasized that even if a given stainless steel is in the sensitized condition, it will not necessarily exhibit intergranular attack. It is noted that certain special boundaries exhibit relatively low energy and less segregation, so they are more corrosion resistant than others. These beneficial grain boundaries, known as CSL boundaries, are believed to act as intergranular corrosion inhibitors. In this study it is aimed to analyze the effect of CSL boundaries on the intergranular corrosion of stainless steel.

EXPERIMENTAL

Materials

The materials used in this study are ferritic stainless steels. Three different 11% ferritic stainless steels are used. However, all three specimens have same chemical compositon and have undergone same heat treatments. Table 1. provides the chemical compositions of given samples.

Table 1. Chemical Compositions (wt.%)

C	Cr	Mn	S	Ti	Si	O_2	N_2
0.008	11.2	0.5	0.002	0.20	0.5	<40ppm	0.008

In addition, one 434 ferritic stainless steel is investigated to study the effect of etch pit characteristics. As etch pit reveals the structural state of materials, it it useful to analyze etch pit distribution(size and depth) in relation to the grain orientaton as well as pitting resistance. This information can later be used to correlate pitting corrosion rate and dependence of microtexture.

Corrosion Tests

Historically, intergranular corrosion was recognized and studied using certain acid immersion tests. These acid tests have remained in use as quality control or acceptance tests in industries because of extensive existing correlations with service experience. Although acid test media bear little relationship to the intended service environment, it is believed to be capable of detecting in relatively short periods of time intergranular attack in some service environments.

Warren test(ASTM Practice D), which uses a mixture of 10%HNO_3 and 3%HF, is performed for the three 11% ferritic stainless steels. After four hours of Warren immersion test, the weight loss per unit area is calculated. This Warren test solution mainly attacks chromium depleted area. For the use of microscopic observation of grain boundaries oxalic etch test, in which 10% $H_2C_2O_4$ solution is used, is performed. This oxalic etch mainly attacks carbides and the various types of attack are analyzed through atomic force microscopy.

For etch pit study a mixture of 100 parts of nitric acid(70% conc.), 30 parts hydrochloric acid(35% conc.), and 30 parts glycerin is used for etching reagent.

Measurement Techniques

For three ferritic stainless steels surface(S=1.0) textures are measured on an automated pole figure goniometer using MoKa radiation. ODFs are obtained from the surface of all three samples, then corrected ODFs are calculated, and then CSL statistics are calcualted from the corrected ODFs. In turn, the CSL statistics are compared with the intergranular corrosion rate.

Atomic force microscopy(AFM) is used to analyze the topography of the surface as well as the types of grain boundary. AFM results give the accurate grain boundary depths, certain grain boundaries like twins, and the location of triple points.

Etch pit test is also performed on the 434 ferritic stainess steel. The number of etch pits per unit area, the depth of etch pits, and the single grain height are measured in order to analyze microtextural pitting potential.

RESULTS AND DISCUSSION

The etch pit results for the 434 ferritic stainless steel produce significant indication of pitting corrosion potential. All the grains are classified into three types; 100 type(CF), 110 type(CE), and 111 type(CC). Each type of a single grain exhibits diffferent resistance against etch pitting. In terms of pit numbers, 100 type grain contains least pits per unit area while 111 type grain contains most pits per unit area as shown in Figure 1. 100 type grain also shows shallow pit depths while 111 type grain shows deep pit depths as shown in Figure 2. However, height of 100 type grain is the lowest while 111 type grain is the highest as indicated in Figure 3. All these results indicate that each grain may exhibit different pitting corrosion resistance.

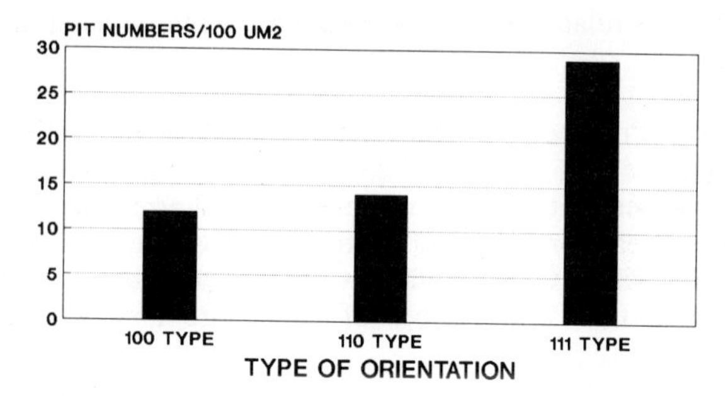

Figure 1. Pit Numbers Vs. Type of Orientation (434 S.S.)

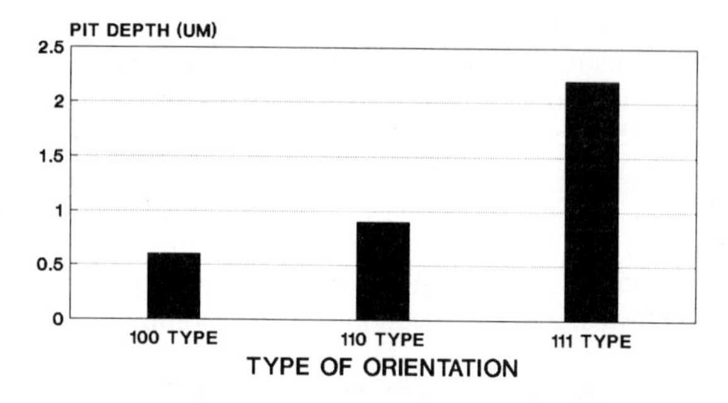

Figure 2. Pit Depths Vs. Type of Orientaton (434 S.S.)

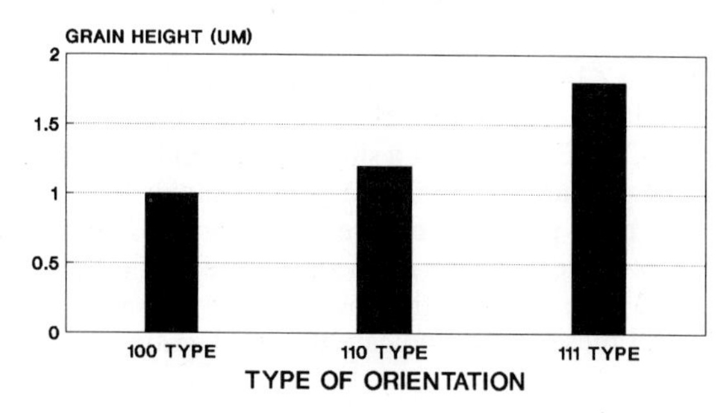

Figure 3. Grain Height Vs. Type of Orientaton (434 S.S.)

The calculated CSL statistics for the three 11%Cr-Fe are located in Figure 4. All three samples have different CSL statistics even though their chemistry and processing variables are the same. The intergranular corrosion rate for these steels are measured after the Warren test and shown in Figure 5. 7500 sample shows the strong resistance against intergranular corrosion while 7315 sample shows relatively weak resistance. It seems that high frequencies of sigma 3(twin) in 7500 sample may be the reason for the strong resistance. However, relatively high frequencies of low angle boundaries(sigma 1) and low frequencies of sigma 5 and 7 in 7500 samplple cannot explain the differences in intergranular corrosion rate. It is believed that up to sigma 29 the special boundaries resist against gain boundary attack while random boundaries do less. Therefore, it is better to consider all the CSLs rather than to consider one specific CSL boundary. The overall intergranular corrosion rate is rather determined by the overall distribution and frequencies of random and CSL boundaries. In order to fully quantify the effect of certain CSL boundaries, EBSP analysis along with the surface analysis equipment such as AFM should be used at the same time.

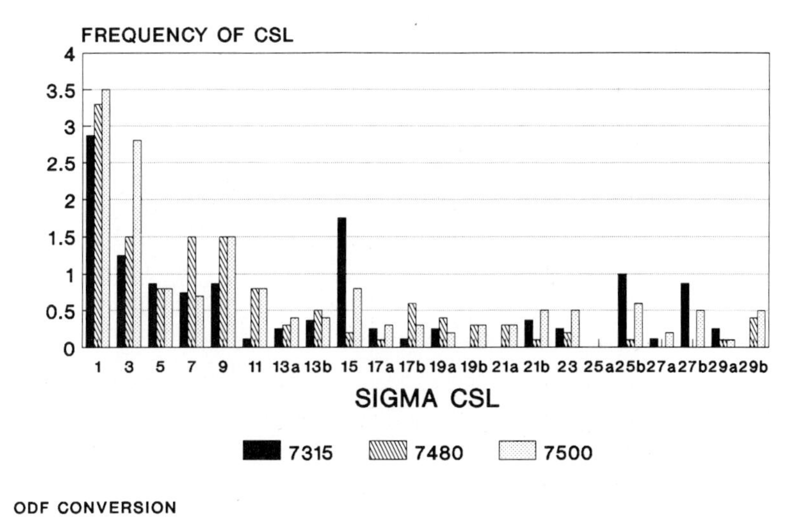

Figure 4. Calculated CSL Statistics for 7315, 7480, and 7500.

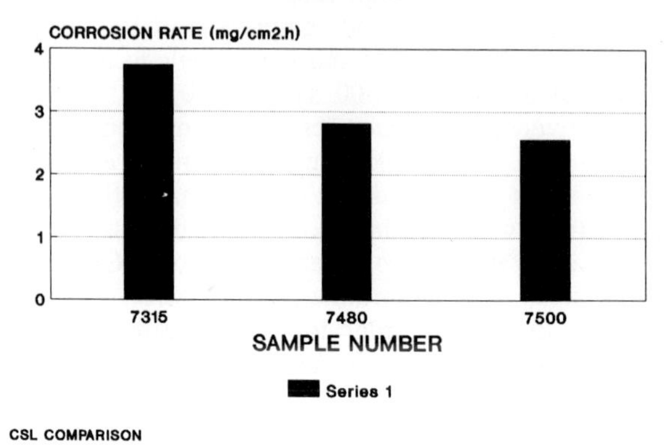

Figure 5. Intergranular Corrosion Rate for 7315, 7480, and 7500.

Atomic force microscopy images of intergranular corrosion attack are shown in Figure 6. While other grain boundaries(probably CSL) remain unattacked or less attacked, certain grain boundary are attacked and corroded to a significant depth. This study show that there are clear relation between the grain boundary structure and the intergranular corrosion rate. However, more work need to be done by EBSP so as to microtexturally analyze the grain boundary and intergranular attack.

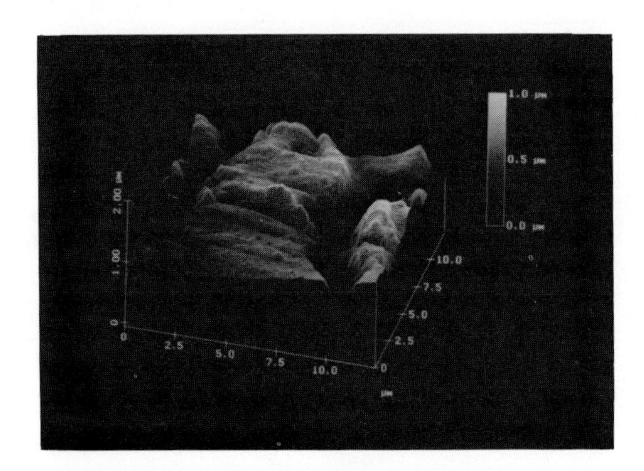

Figure 6. AFM Image of Intergranular Corrosion Attack

CONCLUSION

1. 100(CF) type grain contains least pits, has shallow pits, and has low grain height.
2. 111(CC) type grain contains most pits, has deep pits, and has high grain height.
3. 110(CE) type grain exhibits in between 100(CF) and 110(CE).
4. For three samples(7315, 7480, 7500), overall corrosion rate and CSL statistics are different; there should be correlaton between these.
5. High frequencies of CSL boundaries may enhance the intergranular corroson resistance.

REFERENCES

1. Intergranular Corrosion, Welding Research Council Bulletin No. 138, Feb. 1969.
2. K.Osozawa and H.J. Engell, Corros. Sci., Vol. 6, p.389, 1966.
3. W.H. Hatfield, J.Iron Steel Inst., Vol. 127, p380, 1933.
4. M.Henthorne, Localized Corroson, ASTM STP516, 1972, p.66.
5. R.L. Cowan and C.S. Tedmon, Vol. 3, Plenum Press, New York, 1973, p.293.

Materials Science Forum Vol. 157-162 (1994) pp. 2005-2010

STUDY ON FACTORS AFFECTING R-VALUE OF Cu PRECIPITATION-HARDENING COLD ROLLED STEEL SHEET

M. Morita and Y. Hosoya

Materials and Processing Research Center, NKK Corporation,
1, Kokan-cho, Fukuyama-City, Hiroshima Prefecture, 721, Japan

Keywords: Extra Low Carbon Steel, Interstitial-Free(IF) Steel, Cold-Rolled Steel Sheet, Hot-Coiling Temperature, Cu-Precipitation Hardening, Hot-Band's Microstructure, Ti-Carbonitride, Ti-Carbosulfide, Complex-Precipitate, r-Value, Electrical Resistivity, Recrystallization Texture, γ-Fiber Texture, <100>//ND-Fiber Texture, Orientation Distribution Function, Crystalline Orientation Distribution Function

ABSTRACT

The mechanism of improving r-value of continuously annealed steel sheets by high temperature hot-coiling in Cu-bearing ultra low C interstitial-free(IF) steel has been investigated. By the ODF analysis of recrystallization texture, it was clarified that the γ-fiber texture(<111>//ND) extremely developed by the heat-treatment at the temperature ranging from 720 to 760°C. TEM observations also revealed that the complex-precipitates composed of coarse Ti-carbosulfide and Cu were observed only in the same range of heat-treating temperature. Improvement of r-value by high temperature hot-coiling is caused by the precipitation of Cu combined with coarse Ti-carbosulfides which avoids the detrimental influence of fine pre-precipitation of Cu in matrix.

INTRODUCTION

Hitherto, it has been convinced that a hot-coiling temperature in Cu-bearing ultra low C interstitial-free (IF) steel should be lower than 450°C to avoid a fine pre-precipitation of Cu in hot-band which caused a deterioration of recrystallization texture during continuous annealing[1]. With respect to an improvement of recrystallization texture of IF steel, on the other hand, high temperature hot-coiling treatment was preferable for improving the r-value due to a perfect pre-precipitation of C, N and the coarsening of (Ti,Nb)(C,N), TiS and $Ti_4C_2S_2$ in hot-bands[2]. By optimizing these conflicting conditions in hot-coiling treatments, a new type of deep drawable high strength steel strengthened by precipitation of Cu has been developed by controlling the hot-coiling temperature at 720°C[3].

In this study, mechanism of improving r-value by high temperature hot-coiling has been investigated by observing the precipitates ($Ti_4C_2S_2$, Cu) in hot-bands and measuring the mechanical properties and recrystallization texture.

EXPERIMENTAL PROCEDURE

The chemical composition of steel used in this study is shown in Table 1.

Table 1 Chemical composition of steel used (mass%)

C	Si	Mn	P	S	Sol.Al	N	Cu	Ni	Nb	Ti	B
0.0025	0.16	0.27	0.001	0.002	0.037	0.0036	1.02	0.52	0.016	0.040	0.0004

It was smelted by a 50 ton electric furnace and cast into ingots. These ingots were soaked at 1170°C, hot-rolled into 4.2mm thick with a finishing temperature of 890°C and coiled at 450 and 720°C (Hereinafter, the hot-bands coiled at 450 and 720°C are referred to as LCT and HCT samples, respectively).
After sampling hot-bands, they were subjected to a laboratory heat-treatment according to the sequence shown in Fig. 1 in order to change the size and morphology of Ti-carbosulfides and Cu-precipitates in hot-bands.

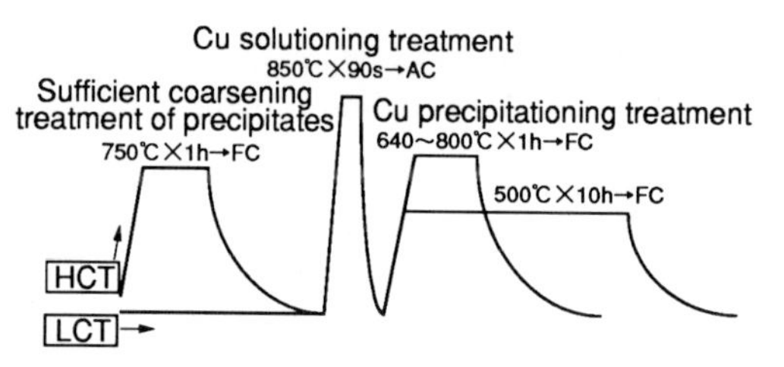

Fig.1 Sequence of heat-treatment of hot-bands.

As a first step, only HCT sample was additionally heat-treated at 750°C for 1h for the purpose of sufficient coarsening of precipitates in hot-bands.
As a second step, every sample was re-heated at 850°C for 90s in order to perfectly dissolve the Cu into ferrite matrix. Finally, they were re-heated at different temperatures ranging from 640 to 800°C for 1h followed by furnace-cooling to simulate the hot-coiling which caused the fractional change in Cu-precipitation and the morphological change of Ti-carbosulfide. Another heat-treatment at 500°C for 10h was carried out to perfectly precipitate the Cu in ferrite matrix. All the hot-bands heat-treated were cold-rolled and annealed at 850°C for 90s followed by air cooling. Annealed sheets were temper-rolled with a reduction of 0.5% and subjected to the mechanical testing. Morphological changes of pre-precipitates in hot-bands were observed by analytical TEM and recrystallization texture of annealed sheets were analyzed by ODFs, which were calculated by Roe`s method by using the data from (200), (110) and (211) pole figures measured at the center layer of the specimen. In order to clarify the behavior of Cu-precipitation during hot-band`s heat-treatment ranging from 500 to 800°C, the electrical resistivity was measured at -196°C by four-terminals method with direct current of 50mA by using the specimens which were precisely machined from the LCT sample into 2mm square in cross-section and 100mm in length.

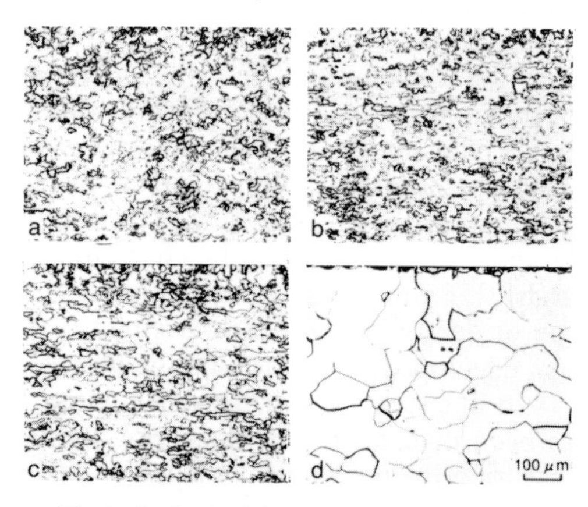

Fig.2 Optical micrographs of hot-band`s microstructure of LCT sample heat-treated by various conditions. (a) 500℃ for 10h, (b) 720℃ for 1h, (c) 760℃ for 1h, (d) 800℃ for 1h

RESULTS

Figure 2 shows the microstructural change of hot-bands of LCT sample heat-treated at 500, 720, 760 and 800°C. The coarsening of ferrite grain is observed in the samples heat-treated at higher than 760°C, and the ferrite structure is perfectly converted into secondary recrystallized grain at 800°C. The microstructure of HCT sample showed the same dependency on the heat-treating temperature observed in the LCT sample.

Figure 3 shows the morphological change of precipitates and the EDS spectra observed in LCT samples heat-treated at the temperature ranging from 500 to 800°C. The size of precipitates increased with elevating the heat-treating temperature up to 760°C. In the sample heat-treated at 800°C, however, the size of precipitates decrease. By EDS analyses, almost all the precipitates were basically composed of Ti and S, and the complex-precipitation of Cu was observed in the precipitates heat-treated higher than 640°C. From the peak-height-ratio of Ti and S, the precipitates observed are presumed to be Ti-carbosulfide. Cu detected is due to complex-precipitates with Ti-carbosulfide which precipitated during heat-treatment. With respect to the morphological change of Cu-precipitates in hot-bands of LCT sample, innumerable Cu-precipitates of which diameter was smaller than 20 nm were observed in the samples heat-treated at 500 and 640°C. Cu-precipitates were hardly observed in that heat-treated at 720°C.

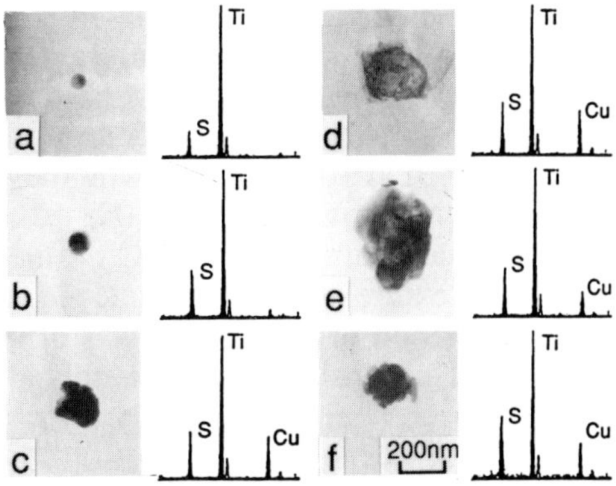

Fig.3 TEM micrographs and EDS spectra of Ti-carbosulphides observed in hot-bands of LCT sample heat-treated by various conditions. (a) as solution-treated, (b) 500℃ for 10h, (c) 640℃ for 1h, (d) 720℃ for 1h, (e) 760℃ for 1h, (f) 800℃ for 1h

Figure 4 shows the change in r-values of annealed sheets of LCT and HCT samples as a function of heat-treating temperature. In the case of LCT sample, the r-value was improved to about 1.9 with elevating the heat-treating temperature up to 760°C and deteriorated by heat-treatment at 800°C. On the other hand, the r-values of HCT samples were sufficiently higher than those of LCT samples because of preliminary coarsening of precipitates. Although the fractional change occurred in Cu-precipitation, the r-values of HCT sample were hardly affected by heat-treating temperatures up to 760°C. The marked deterioration of r-value occurred at 800°C is caused by a coarsening of hot-band`s ferrite structure.

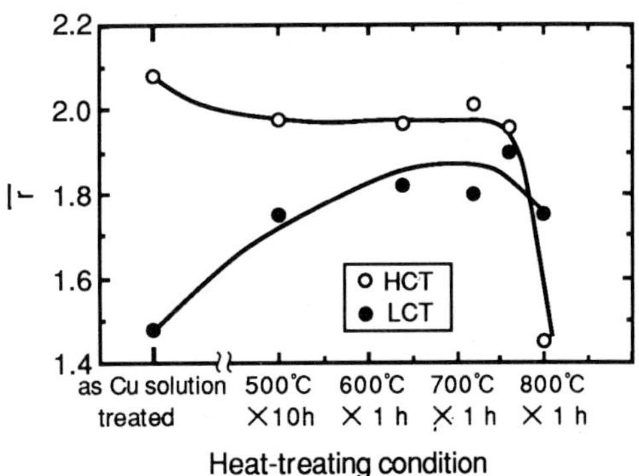

Fig.4 Effect of heat-treating condition on mean r-value of annealed sheet.

The effect of heat-treating condition of hot-bands on the recrystallization texture of annealed sheets of LCT and HCT samples represented by the cross-section of ϕ=45° in ODFs are shown in Figure 5. In the case of LCT samples, the density of γ-fiber texture developed with elevating the heat-treating temperature. This change in the γ-fiber texture agrees with that of r-value as shown in Fig.4. With respect to the <100>//ND-fiber texture, the orientation density of (001)[0$\bar{1}$0] grain (θ=0° and ψ=45°) depended on the heat-treating condition. The orientation density of (001)[0$\bar{1}$0] grain is low in the as-Cu solution-treated condition. In the rest of conditions, however, the (001)[0$\bar{1}$0] texture developed by heat-treatment excepting the treatment at 720°C. In the case of HCT samples, on the other hand, the γ-fiber texture markedly developed except for the heat-treatment at 800°C. The grain orientation of (001)[0$\bar{1}$0] was clearly observed in the as-Cu solution-treated condition. The density of it was hardly affected by other heat-treating conditions.

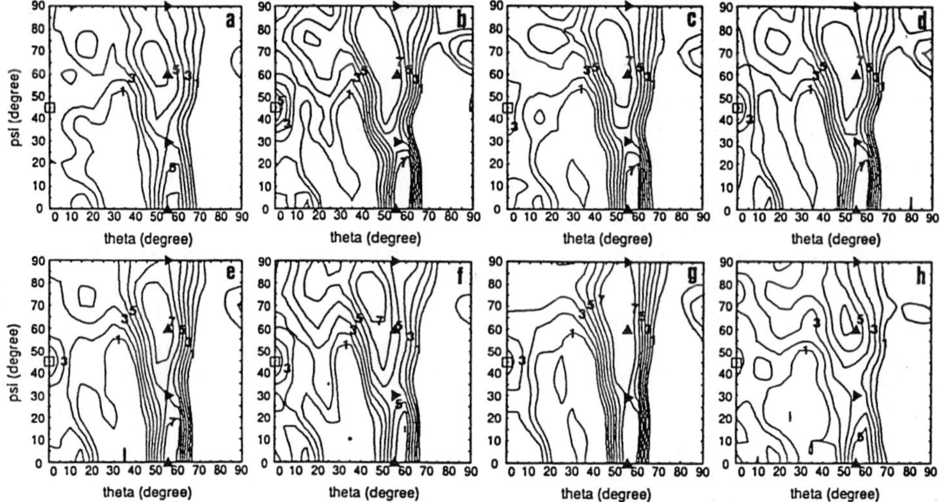

Fig.5 Effect of heat-treating temperature of hot-bands on the recrystallization texture of annealed sheets of LCT and HCT samples represented by the ϕ=45° sections in ODFs.
LCT samples : (a) as-solution treated, (b) 640℃ for 1h, (c) 720℃ for 1h, (d) 800℃ for 1h
HCT samples : (e) as-solution treated, (f) 640℃ for 1h, (g) 720℃ for 1h, (h) 800℃ for 1h
Key; ▶ : {111}<011>, ▲ : {111}<112>, □ : {100}<001>

DISCUSSSION

(1) Behavior of Cu-precipitation during heat-treatment of hot-band

During the heat-treatment of hot-band, the precipitation of Cu is presumed to be dominated by its solubility in ferrite, precipitation sites and diffusion rate in ferrite. In these factors affecting Cu-precipitation, the solubility limit of Cu in ferrite is represented by the following equation[4].

$$\log(\text{mass\%Cu}) = 4.335 - (4499/T) \quad (1)$$

According to the equation-(1), the solubility limit of Cu in ferrite changes from 0.03% to 1.39% during the heat-treatment between 500 and 800°C. In other words, in the steel used in this study, which contained about 1 mass% Cu, the condition of Cu in steel was successively changed from perfect precipitation to perfect solid-solution at the heat-treating temperature from 500 to 800°C.

Then, in order to clarify the behavior of Cu-precipitation during hot-band`s heat-treatment at the temperature ranging from 500 to 800°C for 1h followed by furnace cooling, the electrical resistivity at -196°C was measured on the LCT samples. Figure 6 shows the change in the electrical resistivity (ρ) caused by the different heat-treatment. Under the as-Cu solution-treated condition, ρ was $9.3\times10^{-8}\Omega m$. The decrement in ρ from this value suggests the quantitative change of Cu-precipitation. In the case of heat-treating at 640°C, ρ became the minimum. The increase in ρ at the temperature higher than 640°C is caused by both an quantitative increase of the Cu in solution and an insufficient precipitation of Cu during furnace-cooling after heat-treating. On the other hand, the increase in ρ at 500°C is presumed to be caused by the marked descent of diffusion rate of Cu in ferrite.

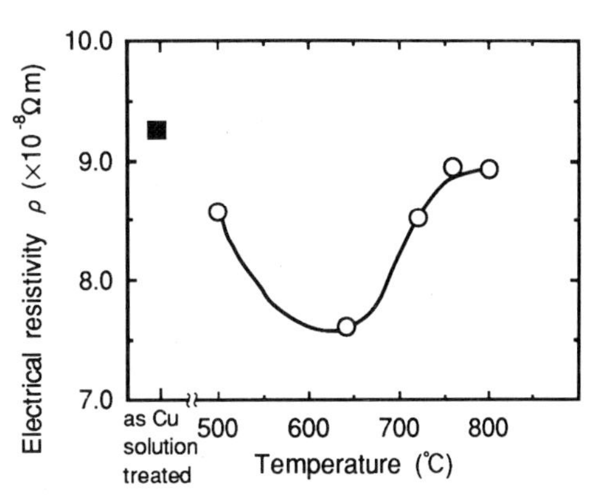

Fig.6 Change in electrical resistivity by heat-treatment after solution-treatment experimented in LCT sample.

(2) Effect of metallurgical factors affecting recrystallization texture

As indicated above, the ferrite grain size, the morphology and size of Ti-carbosulfides and the fraction of Cu-precipitation separately changed during the heat-treatment of hot-band. In order to clarify the each effect of the metallurgical factors on recrystallization texture, the samples of which factors independently changed were selected and subjected to the analyses of γ-fiber and <100>//ND-fiber textures. Table 2 shows the interrelation of heat-treating conditions of hot-bands and metallurgical factors in hot-bands. LCT samples were used for No.1 and 2, and HCT samples were used for No.3 and 4 in Table 2.

Table 2 Interrelation of heat-treating conditions of hot-bands and metallurgical factors in hot-bands

No.	heat-treating condition	G.S. No.	diam. of Ti(CS)*	Condition of Cu in matrix: In solution	Condition of Cu in matrix: Precipitation
1	as solution treated	8.2	70nm	Almost all	few
2	640°Cx1h soaked	8.3	340nm	a few(~10%)	Fine precipitation
3	720°Cx1h soaked	8.3	450nm	Half(~60%)	Coarse precipitation withTi(CS)
4	800°Cx1h soaked	4.8	380nm	Almost all	Coarse precipitation withTi(CS)

* : Ti(CS)=Ti-carbosulfide

(2•1) Change in γ-fiber texture

Figure 7 shows the effects of metallurgical factors on the γ-fiber texture. The γ-fiber texture developed with the growth of Ti-carbosulfides in hot-band. In the case of heat-treatment at 800°C, however, γ-fiber texture markedly declined, it is due to a coarsening of ferrite grains. In the samples heat-treated at 640 and 720°C, with respect to the effect of Cu-precipitation on an improvement of γ-fiber texture, the complex-precipitation with coarse Ti-carbosulfides realized at 720°C is preferable to the fine precipitation of Cu in ferrite at 640°C.

(2•2) Change in <100>//ND-fiber texture

The ODFs obtained from heat-treatment of No.1,2 and 3 shown in Table 2, of which metallurgical factors were different in the size of Ti-carbosulfides and the fraction of Cu-precipitation, were subjected to the analyses of <100>//ND-fiber texture. Figure 8 shows the results of them. The most distinctive change in the <100>//ND-fiber texture is the marked development of the $(001)[0\bar{1}0]$ texture (on ψ=45°) by heat-treating at 640°C. The crystalline orientation distribution function; f(g) of $(001)[0\bar{1}0]$ grain is reduced under both as-Cu solution-treated and heat-treated at 720°C. These results suggest that the existence of the fine Cu-precipitates in ferrite matrix enhances the development of $(001)[0\bar{1}0]$ texture, and the size of Ti-carbosulfides plays little effective roles in the development of <100>//ND-fiber texture.

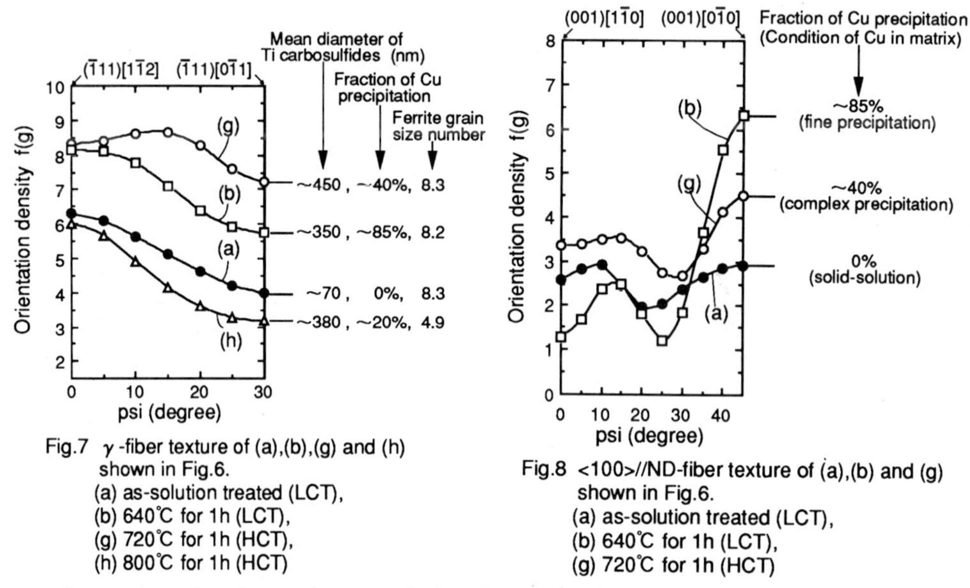

Fig.7 γ-fiber texture of (a),(b),(g) and (h) shown in Fig.6.
(a) as-solution treated (LCT),
(b) 640℃ for 1h (LCT),
(g) 720℃ for 1h (HCT),
(h) 800℃ for 1h (HCT)

Fig.8 <100>//ND-fiber texture of (a),(b) and (g) shown in Fig.6.
(a) as-solution treated (LCT),
(b) 640℃ for 1h (LCT),
(g) 720℃ for 1h (HCT)

Consequently, from the correlation between the metallurgical factors in hot-bands and the recrystallization texture, the coarsening of Ti-carbosulfides, the refining of ferrite grain and the avoidance of fine precipitation of Cu in ferrite matrix are effective to the development of γ-fiber texture, and the amount of fine Cu-precipitates in ferrite matrix dominates the development of $(001)[0\bar{1}0]$ texture.

CONCLUTION

(1) The mean r-value of the Cu-bearing IF-steel sheet is improved to about 2.0 by the heat-treatment at 720°C. TEM observations reveal that complex-precipitates composed of coarse Ti-carbosulfide and Cu are observed in the same heat-treating temperature mentioned above.

(2) With respect to the metallurgical factors affecting recrystallization texture, the coarsening of Ti-carbosulfides, the refining of ferrite grain and the avoidance of fine precipitation of Cu in hot-band play the important roles in development of γ-fiber texture.

REFERENCES

1) K.Kishida et al. : Tetsu-to-Hagané, 76(1990), p.759 2) S.Sanagi et al. : CAMP-ISIJ, 3(1990), p.1768 3) Y.Hosoya et al. : CAMP-ISIJ, 5(1992), p.1823
4) H.A.Wriedt et al. : Trans. Metall. Soc. AIME, 218(1960), p.30

Materials Science Forum Vol. 157-162 (1994) pp. 2011-2016

IMPORTANCE OF PROCESS PARAMETERS FOR TEXTURE AND PROPERTIES OF MICROALLOYED DEEP DRAWING STEELS

C.-P. Reip, W. Bleck, R. Großterlinden and U. Lotter

Forschung und Zentrales Qualitätswesen, Thyssen Stahl AG,
Postfach 11 05 61, D-47161 Duisburg, Germany

Keywords: IF Steel, Bainitic Microstructure, Ferrite Hot Rolling, Cold Rolling, Recrystallization Annealing, r-Value

ABSTRACT

Interstitial-free steels are gaining increasing industrial interest because of their ability to develop deep-drawing properties during continuous annealing. The sequence of texture development from hot strip via cold-rolled to recrystallized cold strip has been investigated. The recrystallization texture is strongly influenced by the initial hot rolling texture. Hot rolling texture results from γ/α-transformation of unrecrystallized austenite, bainitic transformation and/or ferritic hot rolling.

INTRODUCTION

For more than 2 decades IF steels in the form of cold-rolled strip have been used successfully for many applications requiring excellent formability. Because of their non-aging behaviour IF steels are nowadays also used for the in-line production of hot-dip galvanized deep-drawing steels. Current research efforts are directed to develop high strength grades with respect to the demand for a weight reduction of automobiles and to evaluate alternative hot rolling procedures for saving energy and for improving the surface quality of the strip. Strip properties strongly depend on the material texture. This paper reports on the influence of chemical composition and hot rolling procedure on hot strip texture of IF steels. The texture development in the subsequent process steps, i.e. cold rolling and annealing, is described, too.

MATERIALS AND EXPERIMENTAL PROCEDURE

Experiments were carried out on three extra low carbon steels, table 1. Steel A and B are high strength IF steels on a niobium base to which phosphorous is added for solid solution strengthening. To overcome problems with strain-induced embrittlement steel B additionally contains boron. Titanium must be simultaneously alloyed to suppress the formation of boron nitride. Steel C is a conventional Ti-stabilized mild steel for applications requiring excellent formability.

Table 1: Composition, hot rolling process parameters and hot strip microstructure of investigated steels

Steel	C	Mn	P	Nb	Ti	B	FT	CT	Micro-structure
	in mass %						in °C		
A	0.003	0.35	0.050	0.055	-	-	920	800	F
B	0.003	0.35	0.050	0.055	0.020	0.0020	910	710	BF
C	0.003	0.15	0.015	-	0.100	-	<850	630	F-pr

(FT=Finishing temperature, CT=Coiling temperature, F=ferritic, BF=bainitic ferritic, F-pr=ferritic-partially recrystallized)

The various steel grades have commercially been hot rolled with different finishing and coiling temperatures, table 1, subsequent cold-rolled with a thickness reduction of 75 % and then continuously annealed at 840°C in a hot-dip galvanizing line.

ODFs were evaluated from three incomplete pole figures (Co-anode, back reflection mode) by the series expansion method. They are described by means of pole density distributions along the α-fibre (ϕ_1=0°, ϕ_2=45°), ε-fibre (ϕ_1=90°, ϕ_2=45°) and skeleton line. The skeleton line, which often does not follow exactly the path of the γ-fibre (ϕ_2=45°, Φ=55°), connects the points of maximum density in sections ϕ_1=const.

RESULTS

HOT ROLLED STRIP

METALLOGRAPHY. After conventional hot rolling with finishing temperature above A_{r3}-transformation temperature the metallographic structure of hot strip of high strength steels A and B consists of ferrite and bainitic ferrite with parts of ferrite, respectively, table 1. Deformation dilatometry measurements reveal that the formation of ferrite is considerably delayed by boron, figure 1. After ferrite rolling with low coiling temperature metallographic structure of hot strip of steel C is not fully recrystallized, figure 2. With decreasing finishing temperature the percentage of fully recrystallized grains drops. At finishing temperatures of 750°C the whole structure may appear completely unrecrystallized.

TEXTURE. Strong differences exist between the hot strip midplane textures of steel A and steel B, figure 3. Steel A exhibits a weak texture whereas steel B shows a pronounced texture with high densities particularly close to the components (113)[110] and (332)[113]. The unsymmetric path of the density distribution in the three 30° ranges along the skeleton line of steel B is caused by the strong (113)[110] peak which overlaps the skeleton line especially in the case of low ϕ_1-values.

For amounts of recrystallized structure below 50 %, i.e. for finishing temperatures less than 850 °C, steel C reveals sharp midplane textures, figure 4. The plots along the α-fibre display high densities between the components (001)[110] and (112)[110]. With decreasing finishing temperatures, i.e. decreasing amounts of recrystallized structure, a density maximum at (223)[110] is developed. The plot along the skeleton line displays high densities near the component (111)[110] which increase with decreasing amounts of recrystallized structure. The ε-fibre reveals density maxima at (554)[225].

COLD ROLLED STRIP

TEXTURE DEVELOPMENT. The cold rolling textures of the various steels are characterized by high orientation densities along the α-fibre and skeleton line, figure 5. In the case of steel C the corresponding hot strip exhibited a recrystallized structure of 50 %. Prevailing orientations along the α-fibre can be identified between (001)[110] and (111)[110]. Compared to the corresponding hot strip textures these components are intensified, especially in the case of (223)[110]. The skeleton lines of the various steels reveal an enhancement of the component (111)[110] by cold rolling. The density maxima in section ϕ_1=90° near the γ-fibre have been shifted to (111)[112]. This is also obvious from the plot along the ε-fibre.

The recrystallization textures of the various steels are given in figure 6. Along the α-fibre the strong rolling texture components between (001)[110] and (223)[110] are markedly decreased. The plot along the skeleton line displays pronounced densities. The density distributions of steel B and C exhibit a maximum near the component (111)[123], whereas steel A reveals a s-shape density distribution with highest density at ϕ_1=90°. The component (111)[110] remains constant for steel A but has been decreased for steel B and C.

MECHANICAL PROPERTIES. Typical mechanical properties after hot-dip galvanizing are listed in table 2. The plastic anisotropy behaviour of the various steels is graphically shown in figure 7. In this graph the optimum forming behaviour would be represented by a circle with a large radius, i.e. a high perpendicular anisotropy r_m and a low planar anisotropy Δr. With respect to high strength steels boron addition leads to a detoriation of plastic anisotropy parameters. Ferritic hot rolled steel C exhibits reduced r-values compared to conventionally hot rolled mild IF steels.

Table 2: Mechanical properties of hot-dip galvanized steels

Steel	$R_{p0.2}$	R_m	A_{80}	n	r_m	Δr
	in N/mm²		in %			
A	202	364	39	0.22	1.84	0.10
B	204	360	37	0.21	1.68	0.45
C	171	300	42	0.20	1.30	0.56

DISCUSSION

Transformation textures of steels depend on the state of the austenite before transformation and on the transformation mode. Pronounced textures with (113)[110] and (332)[113] preferred orientations or weak textures with predominant component (001)[110] are inherited by ferrite from deformed or recrystallized austenite, respectively [1]. Textures of deformed austenite which has been transformed by a shear or mixed mode (martensite, bainite) are much stronger than those which has been transformed by a diffusion mode (ferrite) [2]. Conventionally hot rolled steel A reveals relative low densities of the major transformation components which indicate a nearly complete recrystallization before transformation. The strong hot strip texture of steel B, which exhibits high densities of the components (113)[110] and (332)[113], may be either a consequence of retarded recrystallization of austenite because of titanium and boron additions or caused by a sharpening of the major transformation texture components due to boron which affects the transformation mode. Ferrite hot rolling of IF steels in the finishing train occurs mainly in the non-recrystallization range. Ti-alloyed IF steels may also exhibit non-recrystallization hot rolling in the austenite because of short interpass times and solute drag effects due to titanium [3]. The hot strip textures of steel C may be thus related to the transformation texture of more or less deformed austenite which has been further sharpened by intensive ferrite rolling depending on the finishing temperature. The densities near the component (112)[110], which is one of the key components of the cold rolling texture, can be taken as a measure of ferrite rolling intensity.

High densities of components along the α- and γ-fibre are typical for cold rolling textures. Ferrite rolling gives rise to a rotation of the transformation texture components (113)[110] and (332)[113]. The former converses to orientations around (112)[110] and the latter via (554)[225] to (111)[112]. For very high strains (111)[112] migrates towards (111)[110]. The component (001)[110] is generally stable and increases in intensity with deformation [4,5]. The recrystallization texture of steel A features a typical deep-drawing texture with low densities of orientation near (001)[110] and rather high densities along the γ-fibre. The higher densities near (111)[112] as well as the development of a density maximum at (111)[123] of steel B and C may be due to the strong cold rolling components near (001)[110] and (112)[110] already present in hot strip texture. High densities of the components (111)[112] and near (111)[123] are normally observed only after high degrees of cold rolling [6,7]. Inspite of the relative weak density along the skeleton line, steel A offers highest r_m- and lowest Δr-values. It can be concluded that for explanation of plastic anisotropy parameters texture inhomogeneities over strip thickness, created by hot rolling, has to be considered as well, especially in steels B and C.

REFERENCES

[1] Ray,R.K.;Jonas,J.J.:Int.Mat.Rev.,1990,35,1
[2] Yutori,T.;Ogawa,R.:in Proc.ICOTOM 6,Vol.1,1981,669,ISIJ,Tokyo
[3] Najafi-Zadeh,A.;Yue,S.;Jonas,J.J.:ISIJ Intern.,1992,32,213
[4] Inagaki,H.:Z.Metallkd.,1983,74,716
[5] Toth,L.S.;Ray,R.K.;Jonas,J.J.:Met.Trans.,1990,21A,2985
[6] Emren,F.;v.Schlippenbach,U.;Lücke,K.:Acta metall.,1986,34,2105
[7] Stickels,C.A.:Trans.Met.Soc.AIME,1965,233,1550

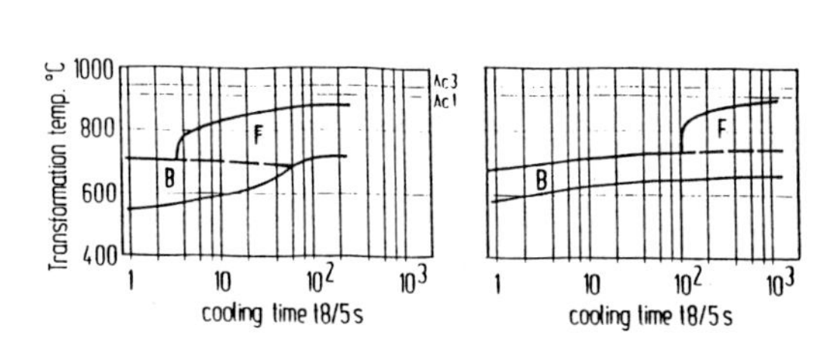

Figure 1: TTT-Diagrams of high strength IF steels A (left) and B (right).

Figure 2: Effect of FT on recrystallized fraction of mild IF steel C.

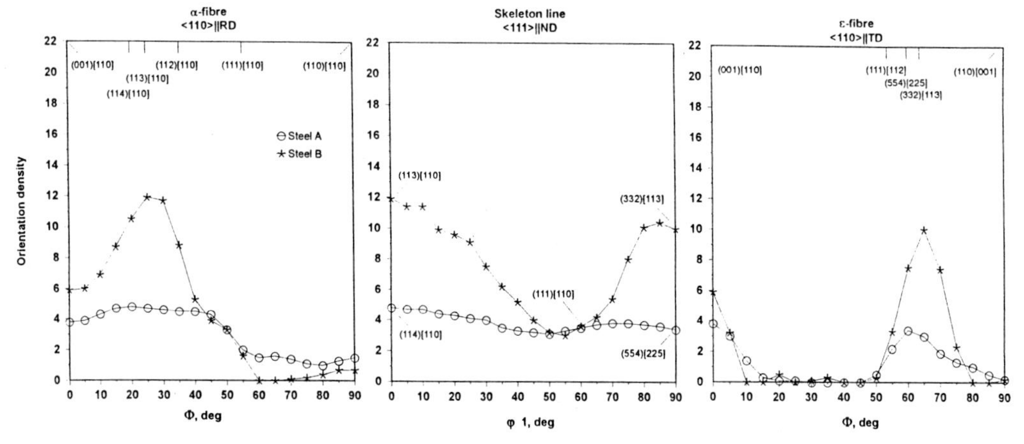

Figure 3: Hot strip textures of high strength IF steels with ferritic (Steel A) and bainitic ferritic (Steel B) microstructure.

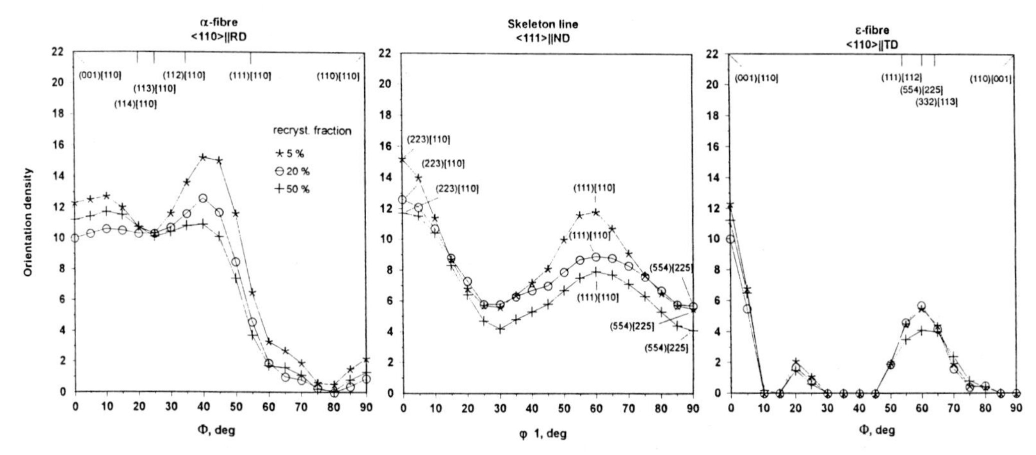

Figure 4: Hot strip textures of mild IF steel C after ferrite hot rolling with different finishing temperatures.

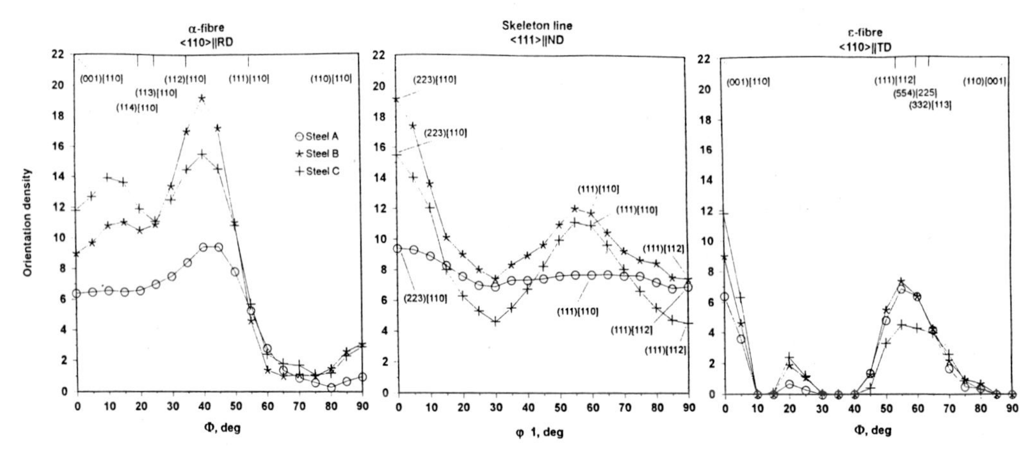

Figure 5: Cold rolling textures of high strength (A,B) and mild (C) IF steels.

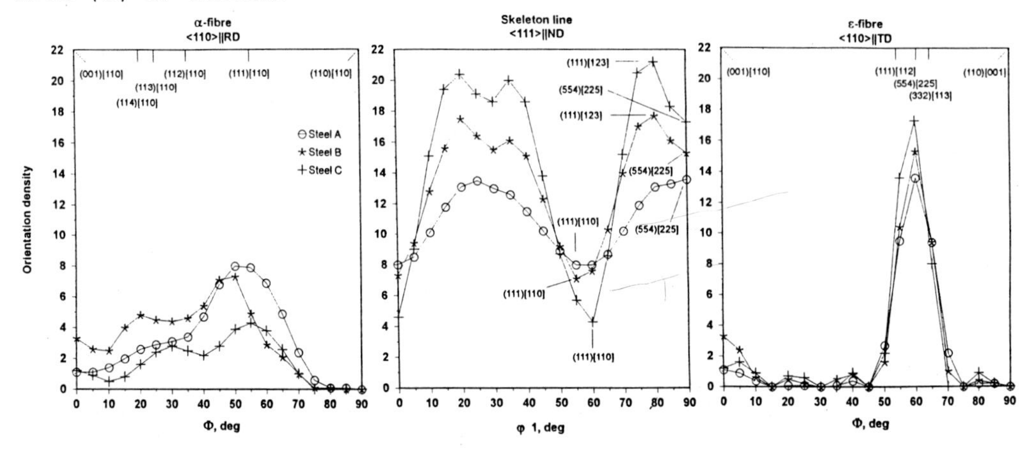

Figure 6: Recrystallization textures of high strength (A,B) and mild (C) IF steels.

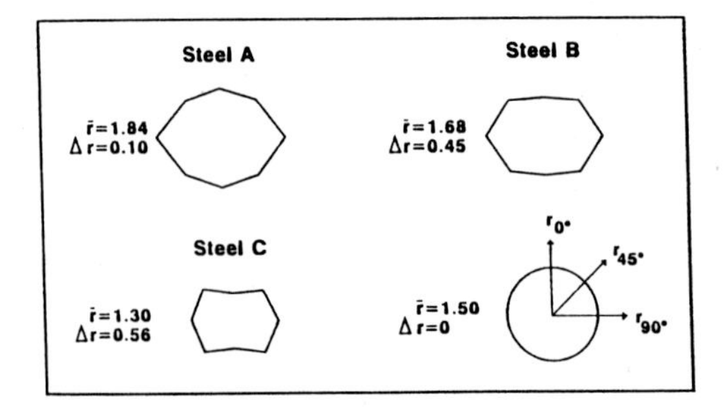

Figure 7: Plastic anisotropy behaviour of high strength (A,B) and mild (C) IF steels.

Materials Science Forum Vol. 157-162 (1994) pp. 2017-2024

EARING AND TEXTURES IN AUSTENITIC STAINLESS STEEL TYPE 305

T.J. Rickert

Allegheny Ludlum Steel Corporation, Technical Center,
Alabama & Pacific Aves., Brackenridge, PA 15014-1597, USA

Keywords: Austenitic Stainless Steel, Type 305, Texture, Earing, Cold Rolling Textures

ABSTRACT

The textures of austenitic stainless steel Type 305 are investigated at different processing stages. The development of the cold rolling textures is found to be very similar to Cu30%Zn, with {112}<111> being stable only at low reductions. Subsequent rolling leads to mechanical twinning and finally the development of {111} orientations through slip. Through-thickness texture homogeneity is found to be rather good. The existence of {112}<111> in the final annealing texture is connected with the earing property. The stronger this texture component, the larger are the 45° ears that develop upon deep drawing.

INTRODUCTION

Austenitic stainless steels are primarily used for their excellent corrosion resistance. Many applications also demand good formability so that intricate parts can easily be manufactured. There is a wide range of products made from these stainless steels, many of them being rather small. However, because of the large numbers produced, the volumes are quite respectable.

Stainless steel of Type 305 contains about 18% chromium and 12% nickel and is normally fully austenitic. It often undergoes severe cold deformation during fabrication of parts. The deep drawing of tube-like shapes from sheet or strip is an example of deformations that necessitate texture control in the material. Earing effects could make the successful forming of such parts impossible.

Textures of F.C.C. materials have been investigated in depth in the past and the texture development during cold rolling and primary recrystallization is understood in general. The work on austenitic stainless steels focused on Type 304 (18%Cr, 8%Ni) since it is the more extensively used austenitic stainless steel. However, Type 304 is not a stable austenitic as Type 305. The formation of martensite influences the texture formation in Type 304 considerably, e.g., [1].

EXPERIMENTAL

The actual chemical composition of the investigated steels is given in Table 1. Samples of hot rolled bands (HRB) were processed to intermediate cold rolled and intermediate annealing stages. Texture samples from these materials were sandwiches made from several transverse cross-sectional pieces which were stacked up to form a sufficiently large measurement area. This way the results cover the full band thickness and can be regarded as "bulk textures." Final gauge samples were investigated as cold rolled and as annealed. These measurements were performed at two different through-thickness levels, at subsurface (s=0.8, 10% below the surface) and at centerline (s=0, mid-thickness). Four polefigures were measured per sample, namely {111}, {200}, {220} and {311}, using MoK_α radiation. ODFs were then calculated by the series expansion method after Bunge [2]. The ODFs are represented in the Euler space using (ϕ_1,Φ,ϕ_2). They were calculated only from even c-coefficients.

Table 1: Chemical composition [wt.%]

Cr	Ni	C	N	Mn	Si	Cu	Mo
18.9	11.9	0.04	0.02	0.87	0.58	0.45	0.44

Earing data were acquired using flat bottom cup tests with 60mm diameter blanks for the final annealed materials. The cup heights in the (four) peaks and valleys were determined and the amount of earing was calculated as the difference between average peak and average valley heights relative to average cup height.

RESULTS

HOT ROLLED BANDS AND INTERMEDIATE GAUGE MATERIAL

Typical HRB, intermediate cold rolling (ICR) and intermediate annealing (IA) textures are presented in Figure 1. The IA and the HRB samples have very weak textures and a finely spaced set of contour lines had to be used to bring out features. The cold rolling texture of the ICR sample, however, is stronger showing the B-, {110}<112> (Brass), and the Goss-, {110}<001>, orientations as the primary components. This is a typical texture for a low stacking fault F.C.C. material [3].

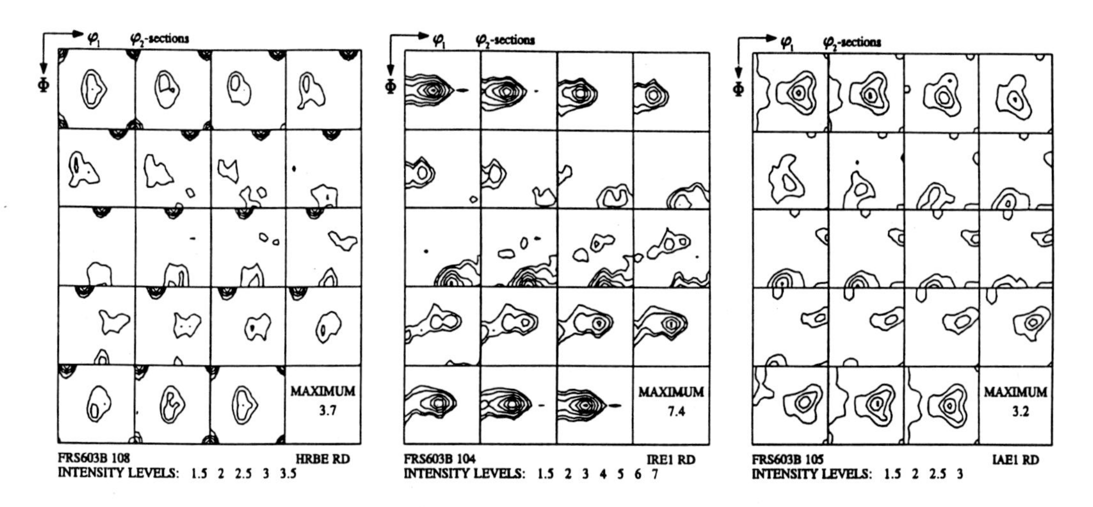

Figure 1: Bulk textures of the hot rolled band (left), intermediate cold rolled (middle) and intermediate annealed band (right).

FINAL COLD ROLLING

Figure 2 shows the final cold rolling textures in the form of recalculated {111} polefigures and the ϕ_2=45° sections of the ODFs. This ODF section includes the τ-fiber, ϕ_1=90° line (<110> in transverse direction), and the α-fiber, Φ=90° line (<110> in normal direction, ND). The {111}//ND orientations lie at ϕ_1=54.7°.

After 30% reduction, the C-component, {112}<111> (Copper), is a prominent feature of the texture. Yet, orientations on the α-fiber are stronger. The maximum is near the B-orientation which has a considerable spread around ND, i.e., along ϕ_1, at this low rolling degree. With increasing reductions, the C-component slowly vanishes and while the intensity of B does not substantially increase, its scattering around ND decreases leading to a more uniform general scattering of this component.

These effects can also be followed in Figure 3 which shows the intensities along selected orientation fibers, namely the β-, τ- and α-fibers of F.C.C. textures. Because the β-fiber tracks the maximum intensities from the C- to the B-orientation along the ϕ_2 angle in the Euler space, i.e., from ϕ_2=45° to ϕ_2=90°, the intensities in this fiber do not necessarily match exactly those in the other plots which include strictly fixed orientations. The positions of the α-fiber maxima are given next to the intensity plot. Besides illustrating the points already made for Figure 2, these plots more clearly show that the position of the maximum on the β-fiber shifts about 10° around ND and the Goss component remains about equally strong with increasing reductions. On the τ-fiber, an intensity increase of orientations between Goss and C-component is clearly visible. First, a component at about Φ=75° grows, but after higher reductions a maximum near 65° develops and also intensities around 55° increase while they strongly decrease in C.

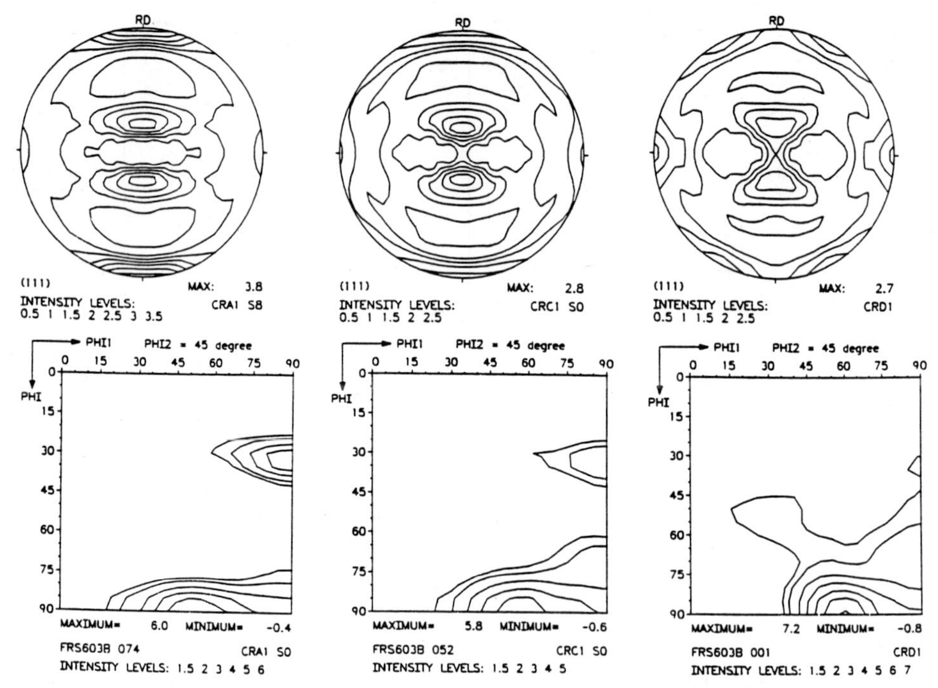

Figure 2: Recalculated {200}-polefigures and ϕ_2=45° sections of the ODFs of textures after 30% (left), 50% (middle) and 70% (right) cold rolling.

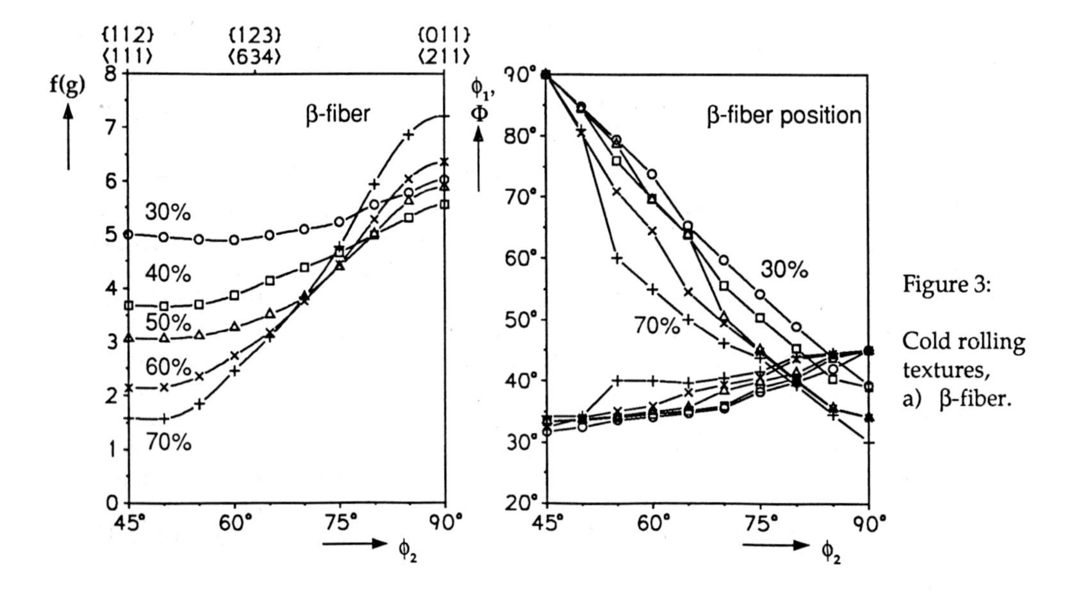

Figure 3: Cold rolling textures, a) β-fiber.

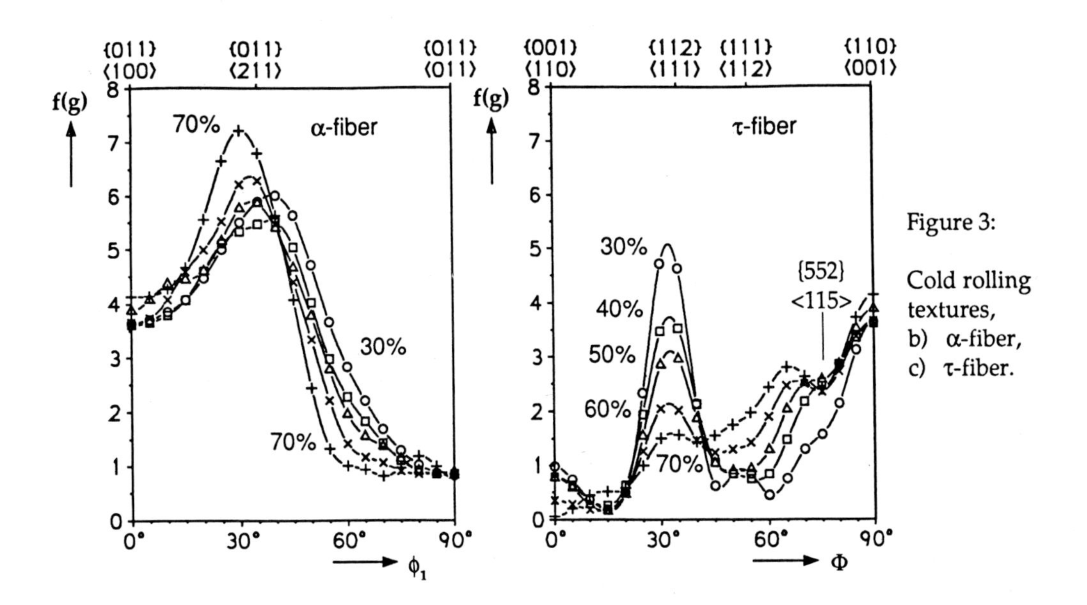

Figure 3:

Cold rolling textures,
b) α-fiber,
c) τ-fiber.

THROUGH-THICKNESS TEXTURE HOMOGENEITY

All final gauge samples were measured at subsurface and mid-thickness levels in order to check for through-thickness texture homogeneity. It was found that there are only very small intensity differences between both levels. But, because of the large number of samples investigated, it can safely be stated that the cold rolling textures are slightly weaker near the surface than at mid-thickness. After the anneal, however, eventual differences cannot be distinguished from intensity variations because of other factors like texture homogeneity over macroscopic distances (processing variations), sample preparation and measurement accuracy.

ANNEALING TEXTURES AND EARING

The final annealing textures are rather weak showing their highest intensities on the α-fiber and around the C-component, {112}<111>. Both these features can be shown favourably in ϕ_2=45° sections of the ODFs. This is done in Figure 4 where these texture results are combined with the results of the earing tests. The figure shows that earing increases as the intensity of the C-component goes up. These intensities are mean values of subsurface and centerline results.

DISCUSSION

The cold rolling textures in this Type 305 austenitic stainless steel can be identified as brass-type which is typical for low stacking fault F.C.C. metals. The C-component, {112}<111>, does develop at very low cold reductions - it is very weak in the intermediate annealed band - but subsequent rolling leads to its continuous weakening and the main texture features become the B-, {110}<112>, and Goss-, {110}<001>, components. As

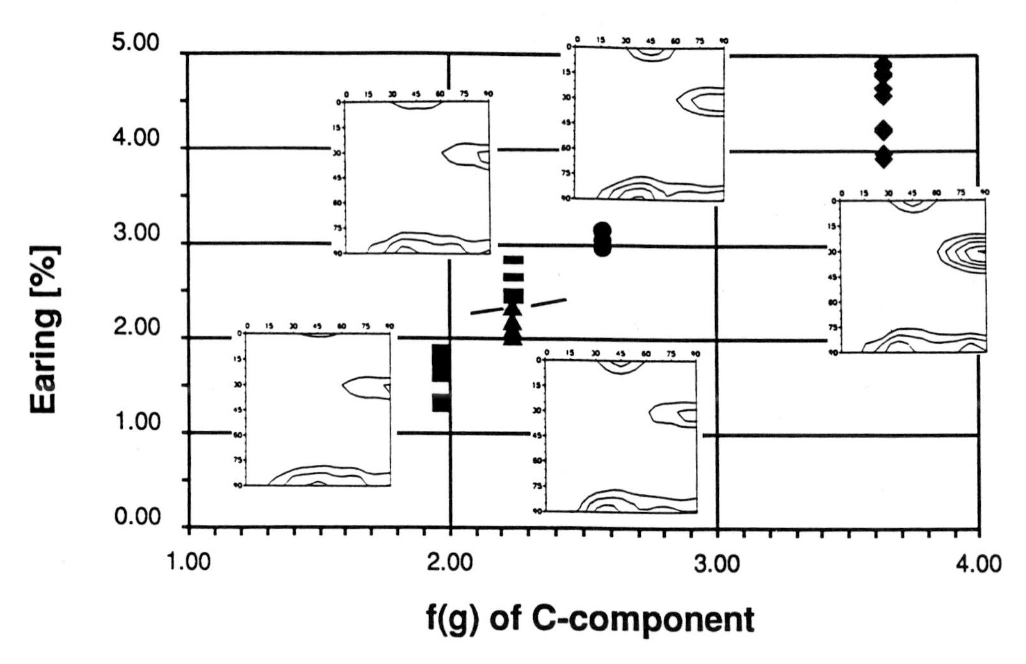

Figure 4: Earing and texture in final annealed samples. The ϕ_2=45° sections of the ODFs are displayed adjacent to the five earing data sets. The five groups represent different processings.

can be followed in Figure 3c, the weakening of the C-component goes hand in hand with the growth of {255}<511> (Φ=75°) and {233}<311> (Φ=65°). The former orientation is a twin of C suggesting the occurence of deformation twinning. Furthermore, it is known that such twins rotate towards {233}<311> and {111}<112> (Φ=54.7°) upon further rolling. All these features have been described before and discussed in detail for Cu30%Zn [4].

The present study is restricted to the industrially feasible cold reduction regime, though, while Hirsch et al. ([4]) have included much higher deformations. For this reason, the emergence of {111}-orientations, i.e., of the γ-fiber, is not as prominently visible. But Figure 5 shows that the whole γ-fiber strengthens above about 50% cold reduction.

There are a number of investigations devoted to cold rolling and annealing textures of austenitic stainless steels to compare the results of the present study to (e.g., [5-11].) However, most of them are on Type 304 and use much higher cold reductions. The majority also does not include ODF analysis which makes more detailed comparisons difficult. The present investigation is in generally good agreement with their

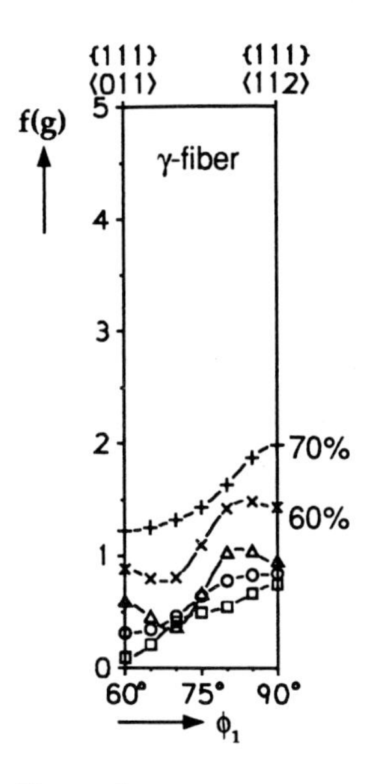

Figure 5:

The development of the {111}-fibers during cold rolling.

rolling texture results but less with their annealing textures. It is assumed that the main reason lies in the usually much higher cold reductions (>90%) used in those investigations. This doesn't change the rolling textures much, but it does change the annealing textures. There is also no resemblance with Cu30%Zn annealing textures which feature a strong brass recrystallization component.

The weak annealing textures are, however, desired for deep drawing applications since they produce low earing. As Figure 4 shows, there is still a considerable band width of earing values for the different textures (and processing routes.) The material that produces around 1.5% earing is substantially better than that on the right side of the Figure with 4% to 5% earing. It is no surprise that the intensity of the C-orientation runs parallel with the earing property since the main texture difference between the samples lies in this component. The ears develop at an angle of 45° to the rolling direction which is the expected type for all rolling orientations (including C.) Cube and Goss orientations which are known to produce 0°/90° ears are present only in small amounts.

REFERENCES

[1] Sumitomo, H. and Yoshimura, H.: Proc. 7th ICOTOM, Netherlands Society for Materials Science, Zwijndrecht, The Netherlands, 1984, 433

[2] Bunge, H.-J.: "Texture Analysis in Material Science", Butterworths, London, 1982

[3] Wassermann, G. and Grewen, J.: "Texturen metallischer Werkstoffe", Springer-Verlag, Berlin, 1962

[4] Hirsch, J., Lücke, K. and Hatherly, M.: Acta metall. 1988, 36, 2905

[5] Goodman, S.R. and Hu, Hsun: Trans. Met. Soc. AIME, 1964, 230, 1413

[6] Goodman, S.R. and Hu, Hsun: Trans. Met. Soc. AIME, 1965, 233, 103

[7] Dickson, M.J. and Green, D.: Mater. Sci. Eng., 1969, 4, 304

[8] Dickson, M.J. and Stratton, R.P.: J. Appl. Cryst., 1972, 5, 107

[9] Donadille, C., Valle, R., Dervin, P. and Penelle, R.: Acta metall., 1989, 37, 1547

[10] Singh, C.D., Ramaswamy, V. and Suryanarayana, C.: Textures and Microstructures, 1992, 19, 101

[11] Sumitomo, H., Yoshimura, H. and Ueda, M.: Tetsu-to-Hagane, 1992, 78, 304

Materials Science Forum Vol. 157-162 (1994) pp. 2025-2030

DIRECT OBERSERVATION OF THE NUCLEATION AND GROWTH RATES OF CUBE AND NON-CUBE GRAINS IN WARM PLANE-STRAIN EXTRUDED COMMERCIAL PURITY ALUMINUM

I. Samajdar [1], R.D. Doherty [1], S. Panchanadeeswaran [2] and K. Kunze [3]

[1] Department of Materials Engineering, Drexel University, Philadelphia, PA 19104, USA

[2] Alcoa Technical Center, PA 15069, USA

[3] Department of Manufacturing Engineering and Engineering Technology, Brigham Young University, 435 CTB, Provo, UT 84602, USA

Keywords: Cube Texture, Commercial Purity Aluminum, Warm Plane Strain Extrusion, Oriented Nucleation, Oriented Growth, OIM, Deformed Cube Bands

ABSTRACT: Commercial purity aluminum was warm plane-strain extruded 83 and 96%, giving recrystallized ODF cube intensities of 50±25 and 200±30 times random. Local orientation studies, including orientation imaging microscopy, showed that grains within 10° and 20° of cube, nucleated much faster than non-cube grains, especially after higher strain. After 83% reduction the cube grains were no larger than the non-cube. At the higher strain, however, the slightly larger cube grain size appeared to come from earlier nucleation. This evidence for oriented nucleation was further supported by TEM evidence of enhanced recovery of near-cube subgrains and from the fall of the spacing of deformed cube bands from 500 to 120μm as the strain increased. The average recrystallized grain thicknesses were 125 and 106 μm respectively. The cube bands rotated towards cube as the reduction increased - matching the decreased mean deviation of the new cube grains.

INTRODUCTION: As discussed in[1], the conflict about the origin of recrystallization texture in terms of the two mechanisms, oriented nucleation (ON) and oriented growth (OG)[2], is likely to be resolved by use of local orientation techniques. This has indeed been shown by studies of heavily cold rolled high purity aluminum[3] which provided clear evidence for ON. Similar results [4,5] were found on warm near plane-strain extruded, commercial purity, aluminum (87% deformed). In these materials, which gave strong cube recrystallization texture, near-cube oriented grains were observed to form at much higher frequencies than expected for random nucleation. Moreover, these near-cube nuclei grew to sizes no larger than non-cube grains. However, in cold rolled commercial purity aluminum, which showed only a very weak cube texture, it was reported that the cube grains did grow larger than non-cube[3,6], giving evidence for a possible role of OG. The present study is a continuation of the past work[4,5] and is aimed at giving more complete data on nucleation and growth and then to begin the investigation of the origin of the apparently dominant ON mechanism in the important case of warm rolled (plane strain extruded) commercial purity aluminum.

EXPERIMENTAL METHODS: The same alloy, Al - 0.15 wt% Fe - 0.07 wt% Si, previously studied[4,5], was used again. Cylindrical, homogenized, DC-cast ingots were warm near plane-strain extruded to 83 and 96% reduction in thickness (measured at the center of the slab) at 320°C at a strain rate of $0.05s^{-1}$. Warm plane strain extrusion was selected, so that the entire deformation process could be undertaken in one pass. This avoided possible complications from recrystallization that might occur during a warm - rolling schedule. The deformation conditions were chosen to be in the regime in which particle stimulated nucleation from the as-cast iron rich constituents was avoided [7], yielding the desired strong cube texture [5]. Two levels of strain were chosen, since preliminary studies [4] had shown an increase in cube texture with larger deformation. Similar observations have been reported for cold rolled high purity copper[7]. Orthogonal orientations were given the equivalent rolling names of "rolling", "transverse" and "normal" directions, RD, TD and ND. Samples from the mid-width and mid-thickness of the extrusions were studied: both as-deformed and after being partially and fully recrystallized in a salt bath for 365°C (96% reduction) and 440°C (83%) for various times to give different degrees of recrystallization. In this study local orientation techniques of manual backscattered Kikuchi diffraction (BKD) and automated BKD giving orientation imaging microscopy (OIM)[4,9] were used to characterize the as-deformed structure and to determine the orientation dependence of the nucleation and growth rates of new grains. Details of the techniques used for sample preparation and local orientation measurements have been given previously[4,5]. Substructural studies, involving TEM and Kikuchi diffraction, were also performed to investigate the faster nucleation of near cube grains. X-ray texture measurements were carried out at Alcoa, using the same techniques as before[5].

RESULTS and DISCUSSION: ODF results derived from both X-ray and BKD/OIM observations report similar values of maximum cube intensities for fully recrystallized 83 and 96% deformed materials at 50(±25) and 200(±30) times random respectively. Figs.1(a) and (b) show the representative microstructures, by OIM, of the two fully recrystallized materials. Grain boundaries have been drawn at all misorientations of 2° or more. Grains within 10°(C_{10}) and 10° to 20°(C_{10}-C_{20}) of exact cube have been identified in the micrographs.

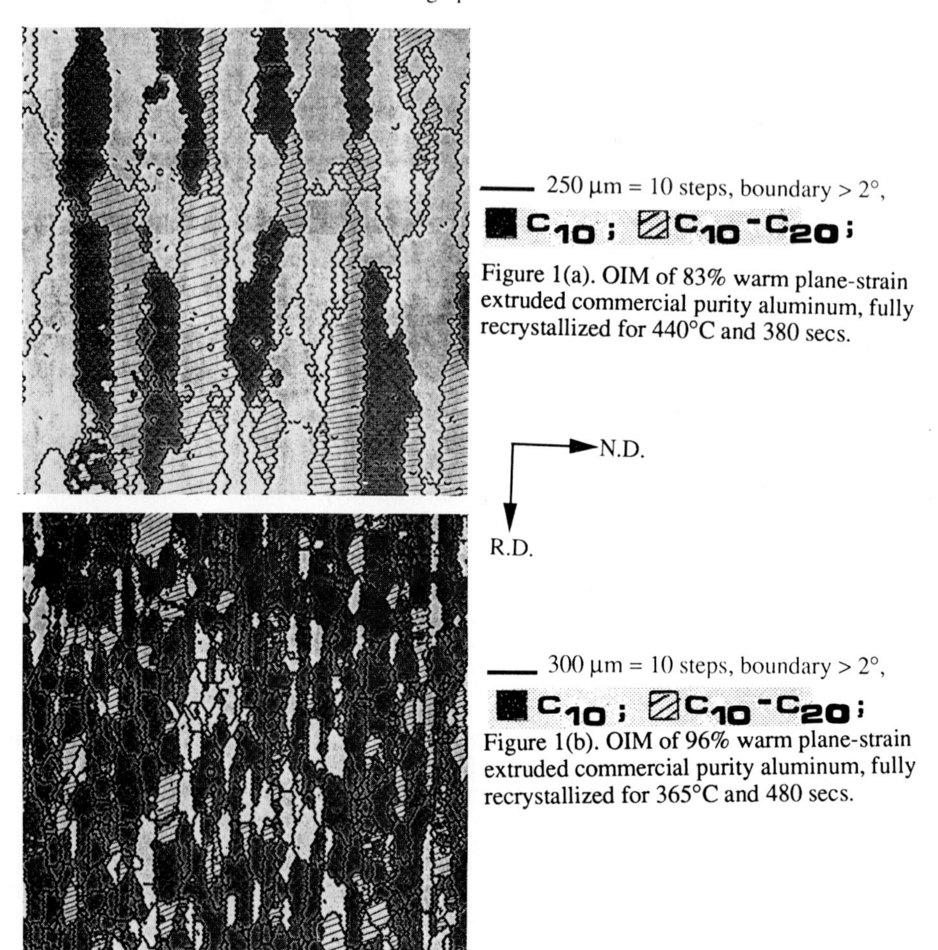

Figure 1(a). OIM of 83% warm plane-strain extruded commercial purity aluminum, fully recrystallized for 440°C and 380 secs.

Figure 1(b). OIM of 96% warm plane-strain extruded commercial purity aluminum, fully recrystallized for 365°C and 480 secs.

Fig.1 shows an overall increase in cube texture by the increase in number and in perfection(closeness to the exact cube) of near cube grains as the strain increased. Although no size advantage in favor of cube was observed in 83% deformed material, grains closest to the exact cube did appear to be somewhat larger at the higher strain. In fully recrystallized samples, the very high fraction of near cube grains in contact with each other (specially in case of 96% deformed material) were difficult to study quantitatively using conventional metallography, because the low angle boundaries between lightly misoriented cube grains were difficult to detect by polarized light optical microscopy, or even by SEM channeling contrast. However, use of images produced from the computer stored OIM data, which are sensitive to small relative misorientations, solved this difficulty and gave the distinct low angle grain boundaries seen in fig.1.

Detailed results on the size and relative frequencies of near-cube and non-cube grains during recrystallization are shown in figs. 2(a) and (b) for 96% deformed material. Fig. 2(a) shows the fraction of near cube((C_{10} and C_{20})) new grains as a function of recrystallization time. Recrystallization times, as shown in the figures, cover the entire range of recrystallization phenomenon(as extruded\0.4% recrystallized, to fully recrystallized samples). Figure 2(b) shows the maximum grain thickness (in the ND) as a function of recrystallization time for both cube(C_{20}) and non-cube.

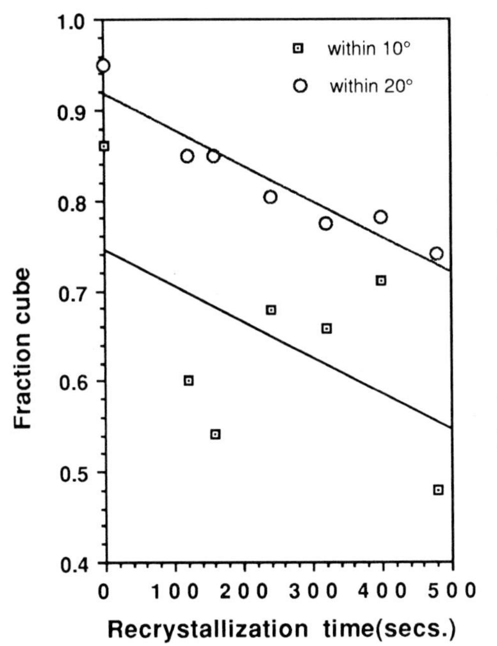

Figure 2(a). Fraction of new grains within 10° and 20° of exact cube as a function of time(secs.) for 96% deformed material recrystallized at 365°C.

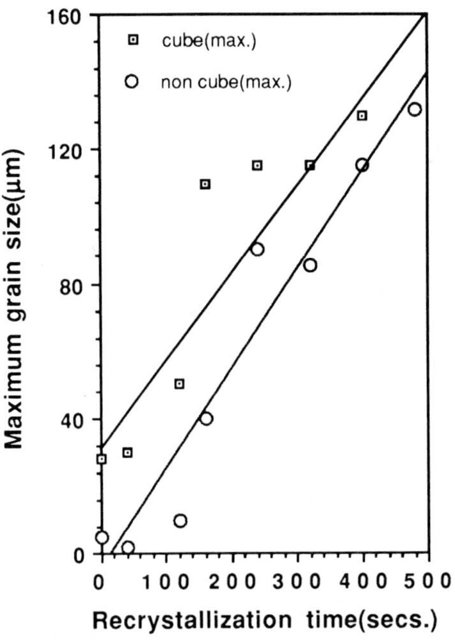

Figure 2(b). Maximum grain size(in normal direction) as a function of time(secs.) for both cube(C_{20})and non-cube in 96% deformed material recrystallized at 365°C.

As shown in fig.2a, at the initial stages of recrystallization, the fractions of C_{10} and C_{20}(within 10° or 20° of exact cube) have the extremely high values of 0.86 and 0.95 respectively. Those cube fractions fell to 0.48 and 0.74 as the recrystallization progressed to completion. These fractions are much larger than the expected random values of 0.006 and 0.04[10], clearly demonstrating the ON mechanism[2,5]. The steady fall of C_{10} and C_{20} fractions at the later stages of recrystallization indicates, as previously reported [5], that cube grains, in general, nucleate first. Preliminary measurements of the initial C_{10} and C_{20} fractions for the 83% deformed material gave values of 0.45 and 0.71; these fell to 0.24 and 0.5 in the fully recrystallized structure. For the same material, the grain sizes of both cube and non cube were indistinguishable at all stages of recrystallization; so showing no sign of OG. However, for the 96% deformed material(fig.2b) the near cube grains, were found to possess, at all times, a size advantage over non-cubes. The as-extruded state had started to recrystallize between extrusion and quenching, giving some cube grains, which were bigger than the very few non-cube grains seen. With further recrystallization time, both near-cube and non-cube grains thickened at the same rate, maintaining the initial size increment of the cube grains. The size advantage, seen here, of the cube grains appears to arise from initial nucleation events rather than from any general growth rate advantage. However if OG is defined[5] by a size advantage, from whatever cause, then such a size advantage is found in the 96% deformed material. The relative size advantage, even after taking volume as the third power of size [2,3], is still clearly very small compared to the nucleation frequency advantage. The ON factor [2,4] in the fully recrystallized 96% material is 80 {48 / 0.6}. This is much larger than the OG factor of $1.95\{(175/140)^3\}$, seen in fig.2b.

TEM studies of cube deformation bands were undertaken to gain some understanding of the nucleation advantage of near-cube grains. Fig. 3 shows bands of cube and non-cube oriented subgrains in the 96% as-extruded material. Convergent beam diffraction patterns outline the position of a cube band in the micrograph. Cube oriented sub-grains, elongated in the RD, were observed to possess, in general, longer cell sizes(approximately 2.5 times) and smaller misorientations(approximately half). Cube sub-grains were found to be more perfect(containing fewer free dislocations in their cell interiors), than their non-cube counterparts. This evidence indicates a lower energy structure for the subgrains in the near-cube oriented bands of deformed material. This lower energy structure, probably arising from more complete recovery, appears to account for the preferential nucleation of near cube grains. Similar observations were made by Ridha and Hutchinson [11] in their study of cold rolled high purity copper. They also suggested that the more extensive recovery of cube oriented cells might be explained by the fact that the dislocations that carry most of the slip in rolled cube oriented fcc metal will have their Burgers vectors at right angles to each other and so have minimal elastic interaction. More detailed studies of the TEM substructure, both in the as-deformed materials and after partial recrystallization, are in progress.

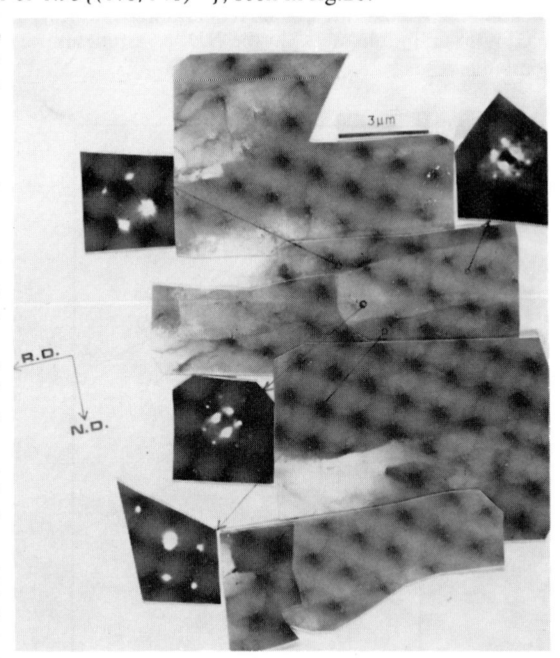

Figure 3. TEM micrograph of cube and non-cube oriented bands in the 96% as deformed material. Convergent beam diffraction patterns indicate the position, but not the accurate orientation, of cube deformed band.

The cube grains were found, as in the previous studies [3-5], to nucleate from the cube oriented sub-grains found in bands lying in the extrusion direction (RD). It is clear that an important feature of the deformed state will be the spacing between these cube-oriented bands, since cube grains only nucleate and grow from these cube oriented bands. The cube band spacings were measured by manual BKD studies in the SEM with the electron beam moved in the ND on cross-sectional planes cut normal to the TD. Figs. 4(a) and (b) show the observed cube band spacings in 83 and 96% deformed material. The band separations are given as average, maxima and minima values for C_{10}(0-10° of exact cube), C_{10} - C_{20}(10-20°), and, C_{20}(within 20°).

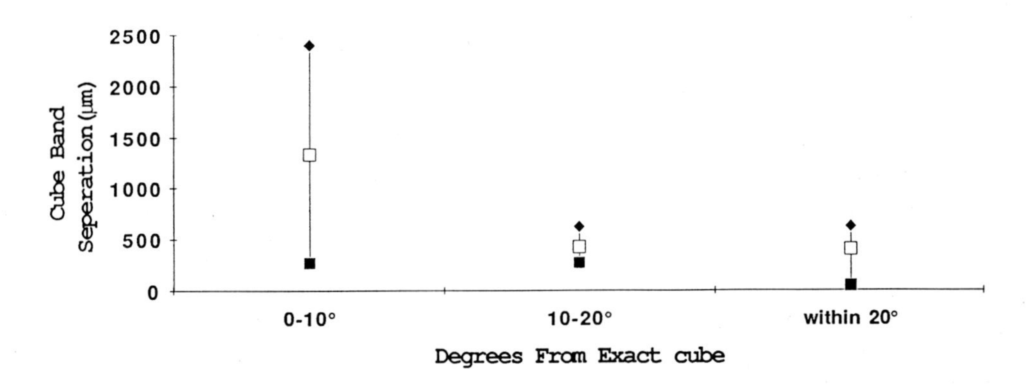

Figure 4(a). Cube band(C_{10}, C_{10}-C_{20}, C_{20}) seperation(in µm) in 83% deformed material.

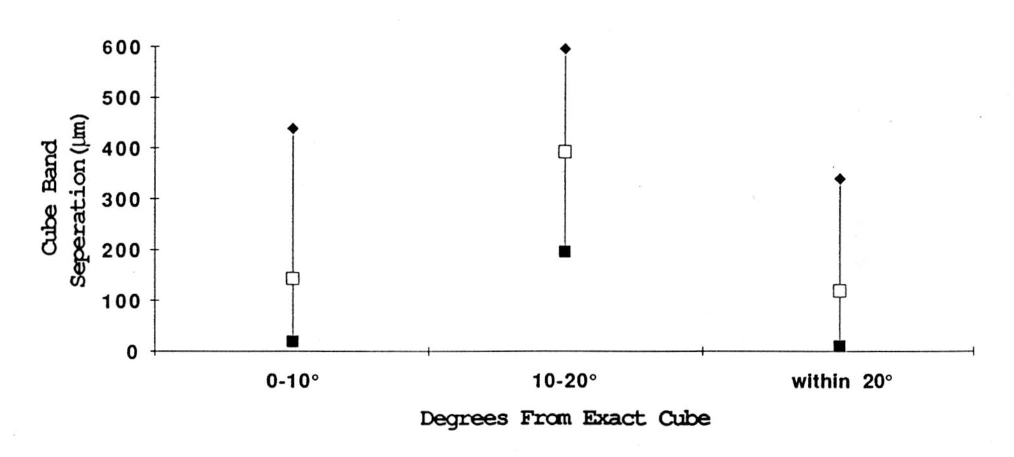

Figure 4(b). Cube band(C_{10}, C_{10}-C_{20}, C_{20}) seperation(in µm) in 96% deformed material.

The average cube band separation for bands within 20° of exact cube are 500 and 120µm for the 83 and 96% reduction respectively. Bands, spaced 500µm apart in material 83% reduced in thickness, will give, at a total reduction of 96 % a spacing of 118 µm if the material is homogeneously deformed with no cube bands being created or destroyed. The close agreement, between observed and predicted spacings at 96%, indicates that, at least between these two strains, the cube bands are stable and the material does appear to deform homogeneously. Deformation band thicknesses also appeared to decrease by the appropriate amount. It should also be noted that, at the higher strain, the cube bands were less deviated, on average, from cube than they were at 83%. So, rotation of the the cube bands towards the exact cube orientation seems to have occurred as the reduction increased from 83 to 96%.

The observed changes in the spacing and mean orientation of the cube bands, fig.4, appears to lead directly to the changes seen in structure and cube recrystallization texture of the two deformed materials - fig.1. The average grain thickness in the recrystallized 83% material was about 125μm, one fourth of the cube band spacing of 500μm, while in the recrystallized 96% material the mean grain thickness of about 106 μm was only slightly smaller than the average cube band spacing. In addition, the fact that the deformed cube bands got closer to the exact cube orientation after the higher strain, may account for the recrystallized cube grains being closer to exact cube orientation. It is clear that since the cube bands in the highly deformed material do remain stable and even rotate towards cube as the strain increase and these cube bands are more recovered, then the observed dramatic rise in cube intensity with increased prior strain is clearly to be expected and this result matches the ON hypothesis very well.

In future work, three major issues are to be addressed: (i) Experimental studies aimed at finding out the origin of the as-deformed cube bands and their stability under plane strain deformation, (ii) Studying, by TEM, the substructure in cube and non-cube bands as a function of strain(iii) Modeling the recrystallization kinetics, texture and microstructure from the experimental characterization of the as deformed state.

CONCLUSIONS:

(1) In warm plane-strain extruded commercial purity aluminum, the cube recrystallization texture was enhanced, both in number of near cube grains and their reduced deviation from the exact cube, by higher deformation.

(2) The cube grains showed much higher than random frequency of nucleation, supporting oriented nucleation.

(3) The cube grains were no larger than the non-cube grains in recrystallized 83% deformed material, but near-cube grains were observed to be slightly larger than non cube in 96% deformed material. This size advantage occurred at the earliest stage of recrystallization and did not increase as recrystallization went to completion.

(4) TEM studies of the as-extruded material showed larger, less misoriented cube oriented sub-grains with fewer free cell-interior dislocations than for non-cube sub-grains. This apparent enhanced recovery may account for the preferential nucleation of the cube grains.

(5) The spacing between cube oriented bands in the as-extruded material decreased from 500 to 120μm as the strain increased. Since the grain thicknesses after recrystallization were of the same order, the greatly increased cube intensity appears to arise directly through the availability of more sites for oriented nucleation of cube.

(6) With the increase in strain the mean deviation from cube of the cube bands in the as deformed material fell. This rotation towards cube corelates with the decreased deviation from exact cube of the recrystallized grains at higher strain(see 1).

ACKNOWLEDGMENTS: The authors are very grateful to Dr. B. Adams of BYU for discussions and for use of the OIM facilities and to Alcoa Technical Center for supply of the deformed material and for X-ray texture evaluation. This work was financed largely by NSF (DMR 9001378).

REFERENCES

1) Doherty, R.D. et al.: ICOTOM 8, Eds. J.S.Kallend and G.Gottstein, AIME Warrendale PA., 1988, p. 563.
2) Doherty, R.D.: Scripta Metall. and Mater., 1985, 19, 927.
3) Hjelen, J., Orsund, R. and Nes, E.: Acta Metall. and Mater., 1991, 39, 137.
4) Doherty, R.D., Samajdar, I. and Kunze, K.: Scripta Metall. and Mater., 1992, 27, 1459.
5) Doherty, R.D., Kashyap, K. and Panchanadeeswaran, S. Acta Metall. and Mater. In Press.
6) Juul-Jensen, D.: Scripta Metall. and Mater., 1992, 27, 533.
7) Kalu, P and Humphreys, F.J.: Acta Metall., 1985, 35, 2815.
8) Necker, C., R.D. and Doherty, R.D. , ICOTOM 10.
9) Wright, S.I., Adams, B.L. and Kunze,K. : Mater. Sc. and Eng., 1993, A160, 229.
10) Mackenzie, M.K. and Thompson, M.J.: Biometrica, 1957, 44, 205.
11) Ridha, A.A. and Hutchinson, W.B.: Acta Metall. and Mater., 1982, 30, 1929.

Materials Science Forum Vol. 157-162 (1994) pp. 2031-2036

GLOBAL TEXTURE DEVELOPMENT IN COLD WORK OF COPPER AND BRASS BY PASS ROLLING

H. Schneider [1], P. Klimanek [1] and H.-G. Brokmeier [2]

[1] Institut für Metallkunde der Berg-Akademie Freiberg,
Gustav-Zeuner-Str. 5, D-09596 Freiberg/Sa., Germany

[2] TU Clausthal und GKSS Forschungszentrum, Institut für Metallkunde und Metallphysik,
Max-Planck-Str. 1, D-21502 Geesthacht, Germany

Keywords: Pass Rolling, Copper, α-Brass, Global Texture, Neutron Diffraction

ABSTRACT

The texture development occurring in copper and α-brass CuZn20 during the first step of pass-rolling in a round-oval and in a square-oval groove sequence at room temperature was investigated by neutron diffraction. The materials were plastically deformed with deformation rates of 1 s^{-1} and 1000 s^{-1} either on technical conditions by means of a pass-rolling mill [1] or in a special compression test [2] which allows to simulate the deformation geometry of pass rolling. In the measurements a global deformation texture with (usually) monoclinic sample symmetry and the main orientations {011}<111>, {011}<100>, {011}<211>, and {112}<111> was found.

INTRODUCTION

Pass rolling of wires and rods is an important technology of metal forming. But though the procedure is well established, hitherto the knowledge about its characteristic influence on the microstructure and especially on the orientation distribution of the rolled metallic materials is rather poor. Former investigations of high-alloy steels (hot working) [3] and nonferrous metals [4] led to the conclusion, that pass rolling textures are very inhomogeneous and cannot be related in a simple manner to other deformation textures. The aim of this paper is to investigate the the deformation texture of pass-rolling more in detail.

EXPERIMENTAL PROCEDURES

The sample materials of the present work were copper and α-brass CuZn20 of technical purity. Table 1 shows the initial state of both materials.

The plastic deformation of the materials was performed with deformation rates of 1 s^{-1} and 1000 s^{-1} at room temperature either on technical conditions by means of a pass-rolling mill [1] or in a special compression test [2], which simulates the pass-rolling. The geometry of the deformation processes is illustrated in figure 1.

Table 1: Initial state

Material	Annealing treatment	Main grain size [μm]	Initial texture {hkl}<uvw>	max. OD
Cu	750°C/20 min/H_2O	34	<111> + <100>	2.8 1.9
CuZn20	700°C/20 min/H_2O	40	<111>	1.3

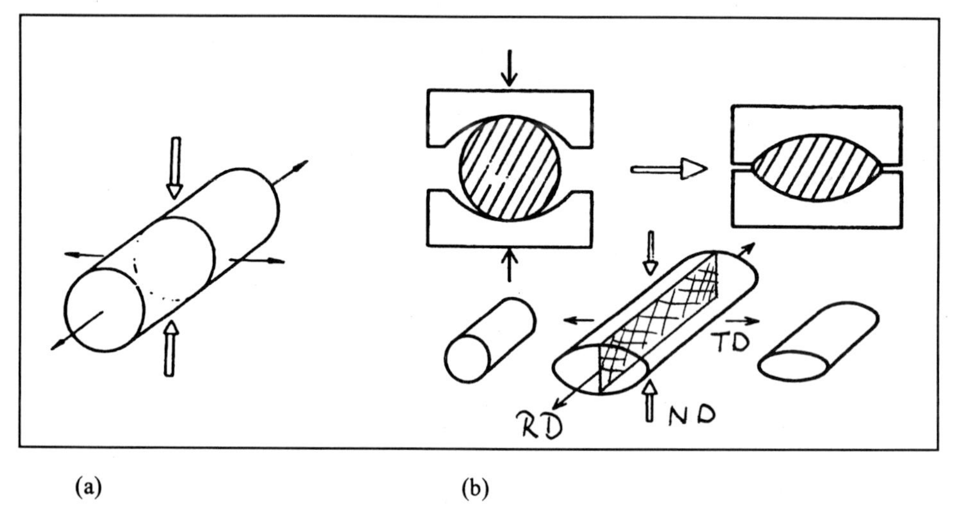

(a) (b)

Figure 1: Deformation by pass-rolling (a) and pass-compressing (b)

The flow of material in pass-compressing is similar to that in pass-rolling. For this reason comparable characteristics and symmetry of the texture can be expected [2, 3]. The obtained values for strains are 45 % - 65 % for pass-rolling and 65 % - 85 % for pass-compressing.

For the texture investigations the neutron diffractometer TEX-2 at the FRG-1 of the GKSS Research Centre Geesthacht was available. The global textures occurring in the central part of the deformed materials were measured (Cu(111) monochromator, λ = 0.1338 nm).
Quantitative texture analysis was based on Odf's calculated from the diffraction pole figures (111), (200) and (220) by Bunge's formalism [5, 6] for triclinic sample symmetry. Moreover, in order to take into account possible ghost phenomena, the calculations were controlled by means of the positivity method [7]. As it should be mentioned, no significant influence of ghosts could be detected.

RESULTS AND DISCUSSION

Sample Symmetry:

Information about the sample symmetry can be taken from the sceleton lines presented in figure 2. From the deformation geometry orthorhombic sample symmetry in the central parts of the rods can be expected. It is approximatly realized in the copper wires, but the α-brass samples show significant deviations from this symmetry. The minimum of the orientation tube is removed to higher phi2- angles. The maxima of the orientation density have equal values.

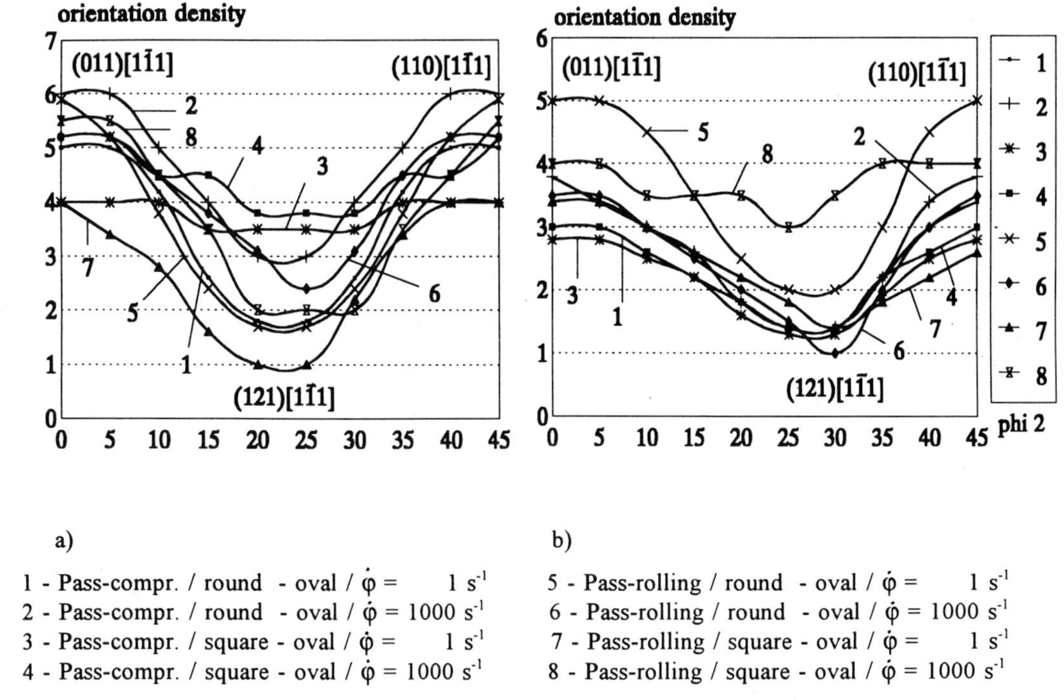

Figure 2: Sceleton lines of copper (a) and α-brass (b)

Representative examples of the triclinic Odf's of the copper and brass wires are shown in figure 3 and 4. They suggest the following interpretation:

- The sample symmetry can be described as monoclinic.
- The symmetry plane of the deformation is defined by the rolling and the normal direction as introduced in figure 1.

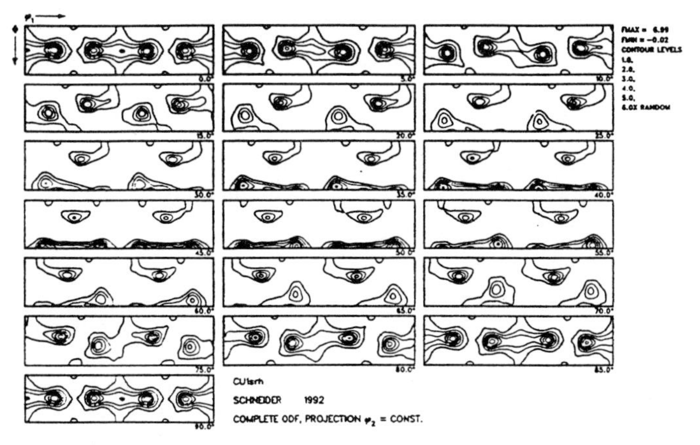

Figure 3: Triclinic Odf of copper (pass-compressing / round - oval/ $\dot{\varphi}$ = 1000 s^{-1}) representative for all deformation states

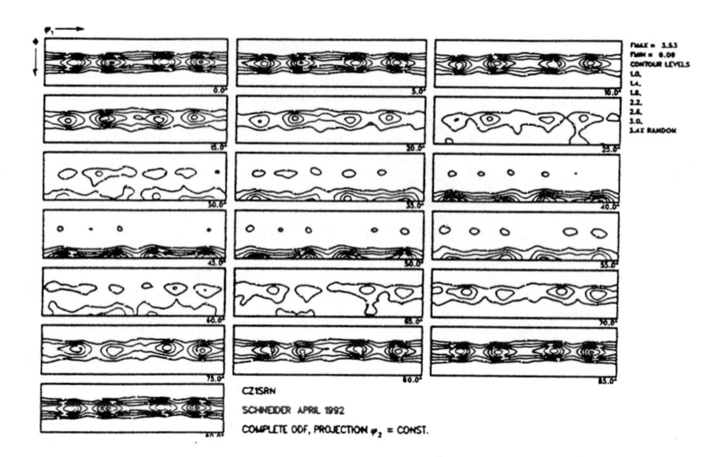

Figure 4: Triclinic Odf of α-brass (pass-compressing / round - oval/ $\dot{\varphi} = 1\ s^{-1}$) representative for all deformation states

Texture-Components:

The pole figures of pass-rolling textures are found to be similar to the well known pole figures of flat-rolling textures (figure 5). However, in the Odf's significant differences become visible.

In figure 3 and 4 an orientation tube between {011}<111> (maximum) and {112}<111> (miminum) can be observed, which obviously represents the pure deformation texture of materials deformed in grooves. Moreover a goss- and a brass-component were found. Obviously, the "deformation - component" {011}<111> is the specific characteristic of pass-rolling textures. Therefore pass-rolling textures form a separate group of deformation textures.

The similar flow of material in pass-rolling and compression leads to similar texture components. Such a conclusion is also true for various groove-sequences (round-oval, square-oval) and strain rates ($1\ s^{-1}$, $1000\ s^{-1}$) because of the similar deformation geometry of the central part of the wires. In general, the pass-rolled samples have a stronger texture than the pass-compressed one's. This can easily be explained by the fail of longitudinal feed in the rolling direction during pass-compressing.

Factors Influencing the Texture:

- The textures received by a round-oval and a square-oval pass sequence correspond qualitatively as a result of the comparable deformation symmetry in the centre of the specimen.
- Cu belongs to materials with intermediate stacking fault energy (SFE) (about $55 \times 10^{-3} Jm^{-2}$ [8]), where the formation of cells or subgrains is the only generally accepted microstructural feature of the deformation structure [9]. In contrast CuZn20 (about $18 \times 10^{-3} Jm^{-2}$ [10]) is a material with low SFE, where planar slip and twinning takes place [9]. Accordingly differences of the texture development should be expected.
  An indication for the influence of the SFE may be that the Odf's show a decrease of the orientation density of the deformation-component {011}<111> and a reduction of the copper- and the goss-position in CuZn20 (table 2). The orientation densities of the brass-position of Cu and CuZn20 are comparable.

Table 2: Texture components and orientation densities of copper and brass

State	{011}<111>		{011}<100>		{112}<111>		{011}<211>	
	Cu	CuZn20	Cu	CuZn20	Cu	CuZn20	Cu	CuZn20
round-oval								
1 s^{-1}	5.7	3.5	2.6	2.2	1.8	1.0	2.6	2.6
1000 s^{-1}	7.0	3.9	4.0	2.6	3.0	1.0	4.0	3.4
square-oval								
1 s^{-1}	4.5	2.9	3.0	2.5	3.0	1.0	3.0	2.5
1000 s^{-1}	5.4	3.3	3.8	3.0	3.8	1.3	3.1	3.0

- Especially at high deformation rates adiabatic overheating should be of importance for the texture formation, e.g. the cross-slip could be easier and the twinning should be reduced. Thence, differences between textures may be smaller.
- The sharpness of the texture is significantly influenced by the strain rate (table 3).

Table 3: Influence of strain rate ($\dot{\varphi}$) (CuZn20, X-Ray)

$\dot{\varphi}$ / s^{-1}	Deformation component	Ms-position	Goss-position	Cu-position
0.06	5.0	5.0	4.0	1.0
1	3.5	2.5	2.5	1.3
1000	3.0	2.3	2.0	2.0

CONCLUDING REMARKS

The experimental results presented here can be summarized as follows:
- A monoclinic sample symmetry of the pass-rolling texture was found.
- Pass-rolling leads to a main texture component which is different from those of flat-rolling (table 4).

Table 4: Distinction between flat-rolling and pass-rolling textures

	F. c. c. flat-rolling texture	F. c. c. pass-rolling texture
Orientation tube	{011}<112>...{112}<111>	{011}<111>...{112}<111>

This result suggest that pass-rolling textures form a separate group of deformation textures [11].

- Similar texture components for pass-rolling and -compressing in a round-oval and in a square-oval groove sequence were found. This allows to conclude, that the results may be transferable to all groove sequences with the same deformation geometry in the central part.
- In order to improve the understanding of the formation mechanism of pass-rolling textures, analysis of the microstructure, local texture analysis and the simulation of flow fields by means of the Thaylor-theory are necessary.

ACKNOWLEDGEMENT

The authors wish to thank Dr.-Ing. M. Dahms for developing the software for the Odf-calculation. Moreover, financial support by the Deutsche Forschungsgemeinschaft (DFG) is grateful acknowledged.

REFERENCES

1) Hensel, A., Oelstöter, G., Lietzmann, K.-D.: Neue Hütte, 1982, 1, 37.
2) Schubert, A., Klimanek, P., Hensger, K.-E.: Neue Hütte, 1987, 11, 423.
3) Schubert, A., Klimanek, P., Hensger, K.-E.: Textures and Microstructures, 1988, 8 & 9, 207.
4) Klimanek, P., Cyrener, K., Mücklich, A., Scholz, U.: Annual Report 1985 of the CINR Rossendorf near Dresden, ZFK-Publ. 584, Rossendorf 1986, p. 70.
5) Bunge, H.-J.: Dreidimensionale Texturanalyse. Neue Hütte, 1973, 18, 742.
6) Bunge H.-J.: Mathematische Methoden der Texturanalyse. Akademie-Verlag Berlin, 1969.
7) Dahms, M., Bunge, H. J.: Texture and Microstructure, 1988, 10, 21.
8) Gallagher, P. C. J.: Metall. Trans., 1970, 1, 2429.
9) Leffers, T., Juul Jensen, D.: Texture and Microstructure, 1991, 14, 933.
10) Thornton, P. R., Mitchell, T. E., Hirsch, P. B.: Phil. Mag., 1962, 7, 1349.
11) Ostwald, D., Schneider, H., Klimanek, P.: DGM Annual Conference 1993, Friedrichshafen, To be published.

(111) (200) (220)

Figure 6: Pole figures of copper (a) and α-brass (b)

Materials Science Forum Vol. 157-162 (1994) pp. 2037-2042

TEXTURE DEVELOPMENT IN FERROUS MATERIALS DURING DEFORMATION BY PASS ROLLING OR COMPRESSION

H. Schneider and P. Klimanek

Institut für Metallkunde der Berg-Akademie Freiberg,
Gustav-Zeuner-Str. 5, D-09596 Freiberg/Sa., Germany

Keywords: Pass Rolling, Pass Compressing, α-Iron, Austenitic High-Alloy Steel, X-Ray Analysis

ABSTRACT

The present work is concerned with the texture formation in α-iron of technical purity and in an austenitic high-alloy steel X8CrNiTi18.10 during the first step of pass-rolling in different groove sequences. The materials were plastically deformed at homologous temperatures $T/T_m \approx 0.4 - 0.5$ with different deformation rates ($1\ s^{-1}$ / $1000\ s^{-1}$) either on technical conditions by means of a pass-rolling mill [1] or in a special compression test [2]. The textures occurring in the central part of the deformed materials were determined by X-ray diffraction. An orientation distribution with nearly orthorhombic or monoclinic sample symmetry was observed. The results are in correspondence with investigations of global pass-rolling textures of non-ferrous f.c.c. materials measured by neutron diffraction [3].

INTRODUCTION

Pass rolling of steel rods and wires is an important technology of hot - working in steel metallurgy and often performed with deformation rates $\dot{\varphi} > 1000\ s^{-1}$. However, hitherto the knowledge of the characteristic relationships between the process parameters of the procedure and the microstructure development in the rolled materials is very limited. In a first study [5], which was performed for a deformation temperature 1173 K and high deformation rates, a mixture of deformation and recrystallization texture components was found. Therefore, in order to get more reliable information about the deformation components of the pass-rolling textures of ferrous materials, further investigations were performed.

EXPERIMENTAL PROCEDURES

The sample materials of the present work were a (b.c.c) α-iron of technical purity (RFe 80: Fe - 99.85 wt.-%, C - 0.02 wt.-%) and a (f.c.c.) high-alloy steel X8CrNiTi18.10 (V2A: C - 0.06 wt.-%, Cr - 17.5 wt.-%, Ni - 9.3 wt.-%), the initial state of which is characterized by table 1.

Table 1: Initial state

Material	Mean grain size [μm]	Initial texture (hkl)<uvw>	OD
RFe80 (b.c.c.)	22	{011}<100>	1.5
X8CrNiTi18.10 (f.c.c.)	36	{001}<110>	1.8

The materials were deformed by pass-rolling and pass-compressing at temperatures $T/T_m = 0.37$ (α-Fe) and 0.48 (γ-Fe) (T_m - melting temperature) by different strain rates ($1s^{-1}$, $1000s^{-1}$) and different pass-sequences (round-oval, square-oval). In order to avoid adiabatic overheating the materials were also deformed at $T/T_m \approx 0.2$ during compression tests with a deformation rate of $0.07\ s^{-1}$. The strains received have a value of 45 % - 70 % for RFe80 and a value of 60 % for X8CrNiTi18.10.

Using an X-ray powder diffractometer HZG4 with the texture unit TZ6 of the Freiberger Präzisionsmechanik GmbH (Co-Kα radiation) the textures occurring in the central part of the pass-rolled and the pass-compressed samples were determined from diffraction pole figures of the lattice planes {110}/{111}, {200} and {211}/{220} by means of the Bunge formalism [6].

RESULTS AND DISCUSSION

Texture Symmetry:

Because of the flow of material and of the existence of 3 mutually perpendicular symmetry planes (figure 1) the texture within the central parts of the rod can be supposed to be of orthorhombic sample symmetry. The sceleton lines in figure 2 show that this is the case in a good approximation for X8CrNiTi18.10. But in the α-iron RFe80 deviations from the orthorhombic sample symmetry, caused by the deformation experiment (use of short steels), are observed (figure 3).

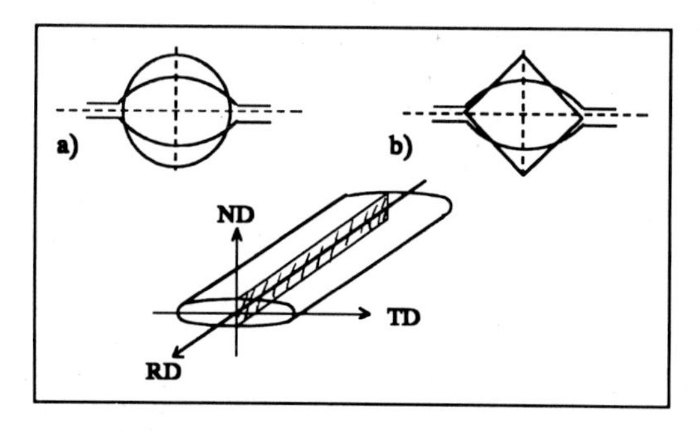

Figure 1: Deformation geometry in a round-oval (a) and in a square-oval (b) groove sequence

1-pass-compr./round -oval/1.00 $s^{-1}/T/T_m = 0.48$
2-pass-compr./square-oval/1.00 $s^{-1}/T/T_m = 0.48$
3-pass-compr./round -oval/0.06 $s^{-1}/T/T_m = 0.22$
4-pass-rolling/round -oval/1.00 $s^{-1}/T/T_m = 0.48$
5-pass-rolling/square-oval/1.00 $s^{-1}/T/T_m = 0.48$

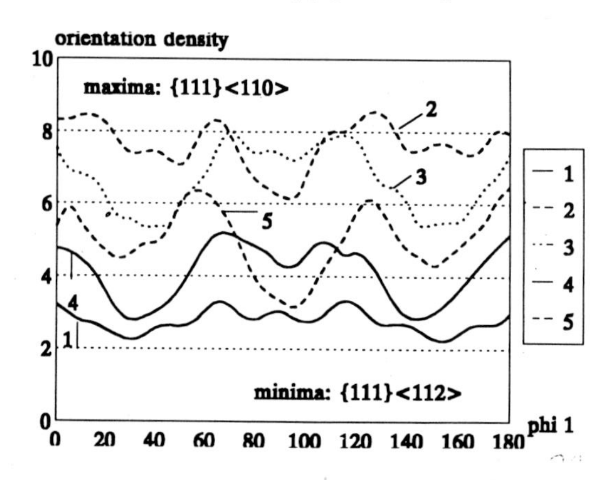

Figure 2: Sceleton lines of X8CrNiTi18.10

1-pass-compr./round-oval/1.00s^{-1}/T/T$_m$=0.16
2-pass-compr./round-oval/1.00s^{-1}/T/T$_m$=0.37
3-pass-compr./round-oval/0.06s^{-1}/T/T$_m$=0.22
4-pass-rolling/round-oval/1.00s^{-1}/T/T$_m$=0.16
5-pass-rolling/round-oval/1.00s^{-1}/T/T$_m$=0.37

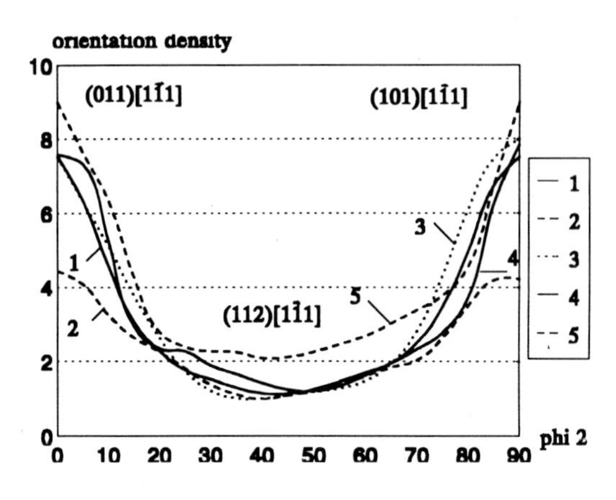

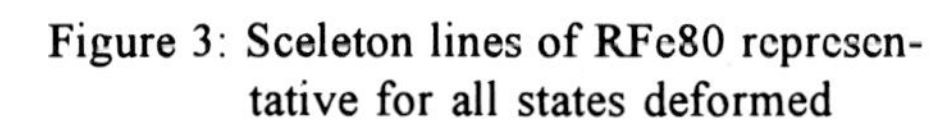

Figure 3: Sceleton lines of RFe80 representative for all states deformed

From the triclinic Odf's in figure 4 follows, that the true sample symmetry can well be described as monoclinic. The symmetry plane of the deformation is defined by the rolling and the normal diction as shown in figure 1.

Texture Components:

The deformation texture of the materials is characterized by occurrence of an orientation tube and ideal positions. A comparison of the texture components of the b.c.c. α-iron and the f.c.c. steel is given in table 2. Figure 5 show selected pole figures of the both materials representative for all states deformed.

Table 2: Texture components of the b.c.c. and the f.c.c. material deformed in grooves

b.c.c. RFe80	f.c.c. X8CrNiTi18.10
{111}<110> - maximum of the tube	{011}<111> - maximum of the tube
{111}<112> - minimum of the tube	{112}<111> - minimum of the tube
{001}<110> - rotated cube-position	{011}<100> - goss-position
{100}<001> - cube-position	{011}<211> - brass-position

According to [4], the comparison shows, that the main texture components of the f.c.c. and b.c.c. material are obtained from each other by exchange of normal and rolling direction. Beyond, as mentioned in [5], the main texture components {011}<111> of the f.c.c. steel and {111}<110> of the α-iron are obtained by combination of the prefered orientations occurring in uniaxial compression and tension.

The similar flow of material during pass-rolling and pass-compressing leads to similar texture components. Such a conclusion is also true for various groove-sequences (round-oval, square-oval) and strain rates (1 s^{-1}, 1000 s^{-1}), because of the similar deformation geometry of the central part of the wires.
In general, the pass-rolled samples of RFe80 have a stronger and more symmetrical texture than the pass-compressed one's. This can easily be explained by the fail of longitudinal feed in the rolling direction in the case of pass-compressing. This effect is overlapped by the differences of strains.

From the experiments follows, that the "deformation - component" {011}<111> is the specific characteristic of f.c.c. and {111}<011> of b.c.c. pass-rolling textures. Therefore pass-rolling textures form a separate group of deformation textures.

Factores Influencing the Texture Development:

- The textures received in round-oval and square-oval pass sequences correspond qualitatively because of comparable symmetry in the centre of the wire.
- A significant influence of the temperature and strain rate on the texture development can be observed (table 3, 4, 5). In the b.c.c. RFe80 with increasing of the deformation temperature the orientation density of all texture components were increased probably due to the activation of additional slip systems. In f.c.c. X8CrNiTi18.10 the main deformation component {011}<111> is decreased with increasing of the temperature and the goss-position becomes stronger. The copper-positon is very weak in both cases.
  The results of X8CrNiTi18.10 are in a good agreement with those of α-brass.[3]
- At high strain rates the texture components of the orientation tubes of RFe80 become stronger, but the components {001}<110> and {100}<001> were reduced. For X8CrNiTi18.10 such a conclusion cannot be drawn because the high flow stress of the material did not allow the deformation experiment.

Table 3: Orientation densities in dependence on homologous temperature (T/T_m) for RFe80

T / T_m	{111}<110>	{100}<110>	{111}<112>	{100}<001>
0.16 (1 s^{-1})	3.3	3.5	2.3	< 1.0
0.22 (0.06 s^{-1})	8.0	4.0	5.4	2.0
0.37 (1 s^{-1})	8.5	6.1	6.2	3.5

Table 4: Orientation densities in dependence on T/T_m for X8CrNiTi18.10

T / T_m	{011}<111>	{011}<100>	{112}<111>	{011}<211>
0.22 (0.06 s^{-1})	8.0	1.0	1.0	6.0
0.48 (1 s^{-1})	7.6	3.9	1.1	5.9

Table 5: Orientation densities of the texture components of RFe80 in dependence on strain rates

Strain rate	{111}<110>	{100}<110>	{111}<112>	{100}<001>
round - oval				
1 s^{-1}	5.2	4.0	2.8	3.5
1000 s^{-1}	6.7	3.0	3.3	2.0

CONCLUDING REMARKS

It can be summerized that

- a monoclinic sample symmetry of the pass-rolling texture was observed.
- the main texture components of pass-rolling are different from those observed in flat-rolling of both f.c.c. and b.c.c. metals. This characterized deformation textures occuring by pass-rolling and pass-compressing as an extra group of textures generally.
- the results are in a good agreement with investigations of copper and α-brass [3].

To understand the formation mechanism of pass-rolling textures, the analysis of microstructure, a simulation of flow fields by means of Thaylor-theory to calculate theoretical texture components of deformation in grooves and textures resulting from several groove sequences are nessesary.

ACKNOWLEDGEMENT

The authors wish to thank Dipl.-Ing. T. Eschner for support in Odf calculation. In addition, financial support of the Deutsche Forschungsgemeinschaft (DFG) is grateful acknowledged.

REFERENCES

1) Hensel, A., Oelstöter, G., Lietzmann, K.-D.: Neue Hütte, 1982, 1, 37.
2) Schubert, A., Klimanek, P., Hensger, K.-E.: Neue Hütte, 1987, 11, 423.
3) Schneider, H., Klimanek, P., Brokmeier, H.-G.: Oral presentation at ICOTOM 10.
4) Dillamore, I. L., Katoh, H.: Met. Sci., 1974, 8, 21.
5) Schubert, A., Klimanek, P., Hensger, K.-E.: Textures and Microstructures, 1988, 8 & 9, 207.
6) Bunge, H.-J.: Neue Hütte, 1973, 18, 742.

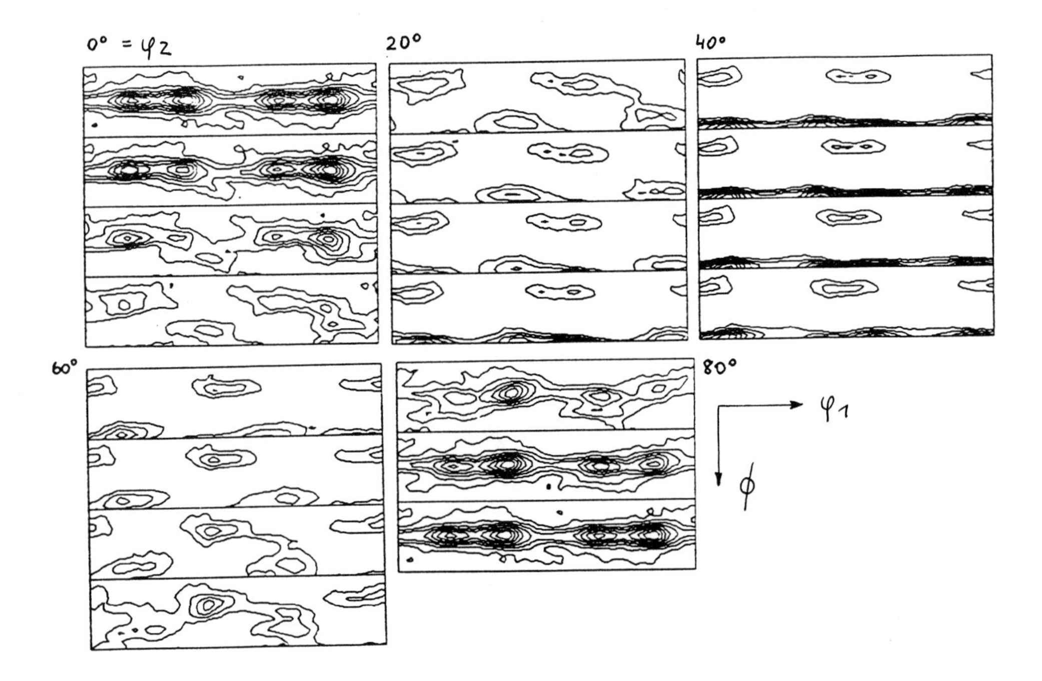

Figure 4a: Triclinic ODF of X8CrNiTi18.10

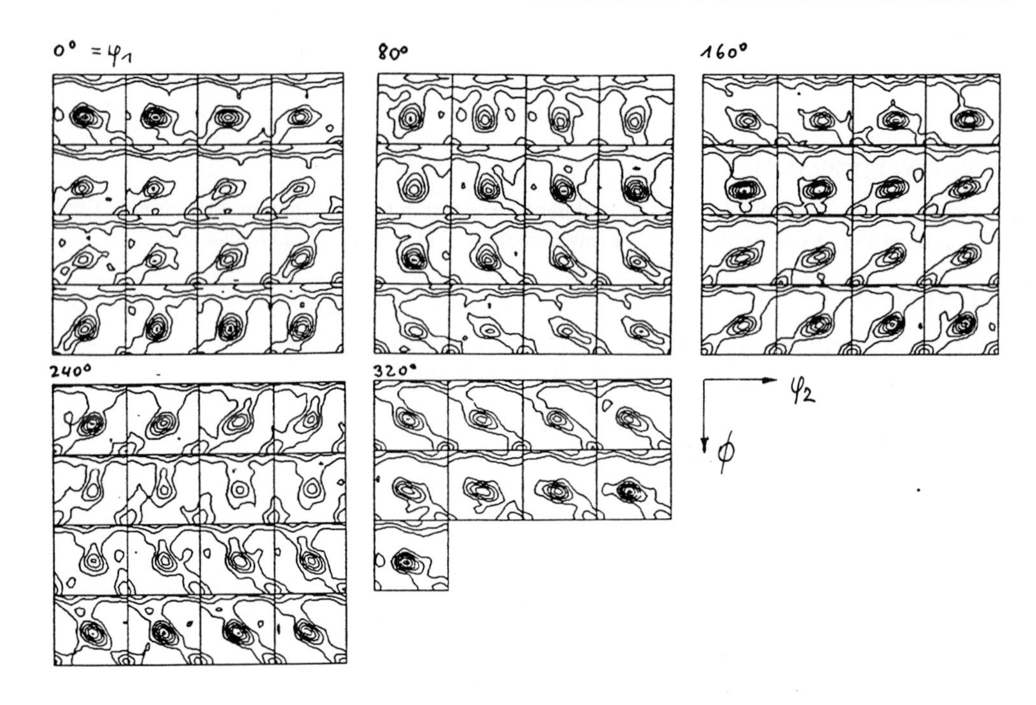

Figure 4b: Triclinic ODF of RFe80

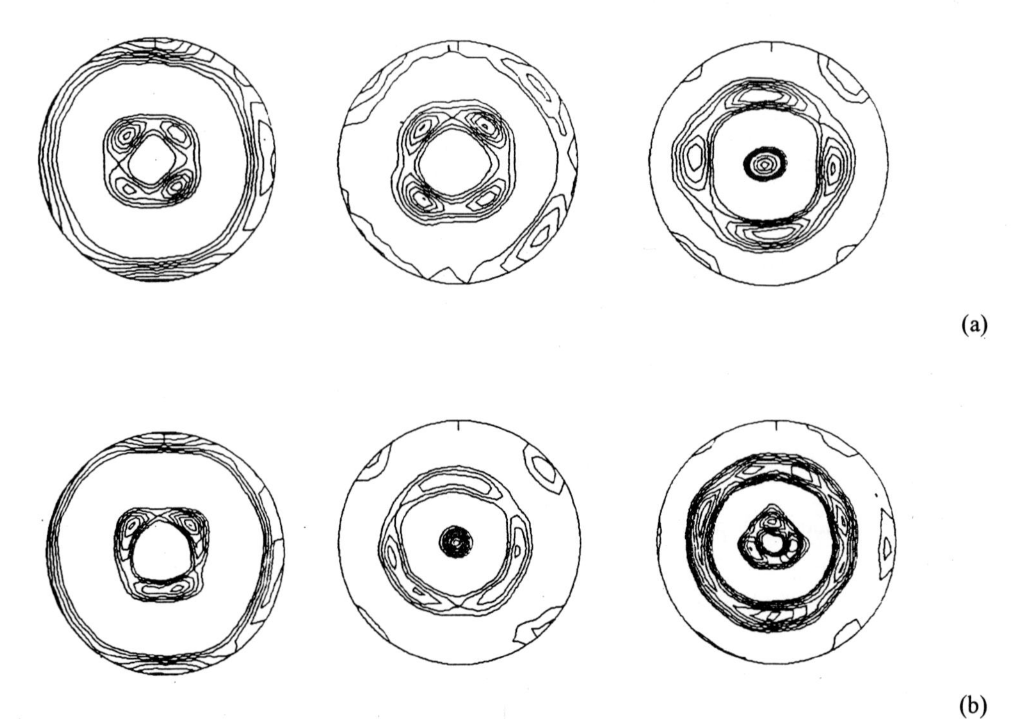

Figure 5: Pole Figures of X8CrNiTi18.10 (a) and RFe80 (b)

Materials Science Forum Vol. 157-162 (1994) pp. 2043-2048

QUANTITATIVE CORRELATION OF TEXTURE AND EARING IN Al-ALLOYS

P. Wagner and K. Lücke

Institut für Metallkunde und Metallphysik,
RWTH Aachen, Kopernikusstr. 14, D-52056 Aachen, Germany

Keywords: Aluminium, Earing, Multiple Linear Regression, Texture

ABSTRACT

In Al-alloys earing is mainly controlled by the texture. While rolling textures produce 45° ears the Cube recrystallization texture leads to the formation of 0°/90° ears. By variation of the manufacturing processes a variety of textures can be obtained leading to different types of earing profiles. In the present work the correlation between earing profiles and the series expansion coefficients (C-coefficients) of the Orientation Distribution Function (ODF) was investigated in differently treated Al samples of the 3xxx and 5xxx series alloys by means of a multiple linear regression. Strong correlations are already obtained with the first 7 C-coefficients. Since the agreement of predicted and measured profiles is very good, the obtained correlations can be used for a quantitative prediction of earing profiles.

INTRODUCTION

Many attempts were made to predict the earing behaviour of fcc metals out of texture data. Based on assumptions of the deformation mode during earing, mostly Taylor factor calculations by means of Full Constraints Taylor models are used [1]. The Taylor factor calculations can be performed in order to calculate the yield locus for a given texture which then can be used for the prediction of earing profiles (see e.g. [2]). These predictions already give a good qualitative information about the expected earing profile but differences occur concerning the quantitative agreement. An improvement may be obtained by integrating the anisotropic yield locii in deformation simulations like the Finite Element Method (FEM) as it has already been done for different deformation modes [3]. Using FEM, however, leads to a strong increase of the required computer time. A further improvement can be expected from modified Taylor models [4], which describe the deformation more realistically. These improved Taylor models, however, also require more computational effort, which for the moment is not suitable for an industrial application. Therefore, in the present work, an investigation of the correlation between texture and earing in Al sheets of the 3xxx and 5xxx series alloys was made by means of a multiple linear regression. This very simple and purely statistical approach allows to make a quantitative prediction of earing profiles without time consuming simulations.

EXPERIMENTAL

By variation of the alloy elements and the processing parameters [1] 24 Al samples of the 3xxx and 5xxx series alloys with the same type of texture (Cube {001}<100>, RD-rot. Cube {hkl}<100>, C {112}<111>, S {123}<634> and B {011}<211>) but different intensities of the main orientations were obtained. The earing profiles of these samples were measured from $\alpha=0°$ to $\alpha=360°$ (α: angle to rolling direction) in steps of 1° and symmetrized with respect to the orthorhombic sample symmetry ($0°\leq\alpha\leq90°$) (figure 1). The strength of earing is characterized by the value

$$Z_{max} = \frac{h_{max} - h_{min}}{h_{min}} \tag{1}$$

with h_{max} and h_{min} being the maximum and minimum heights of the symmetrized earing profile, respectively (figure 1).

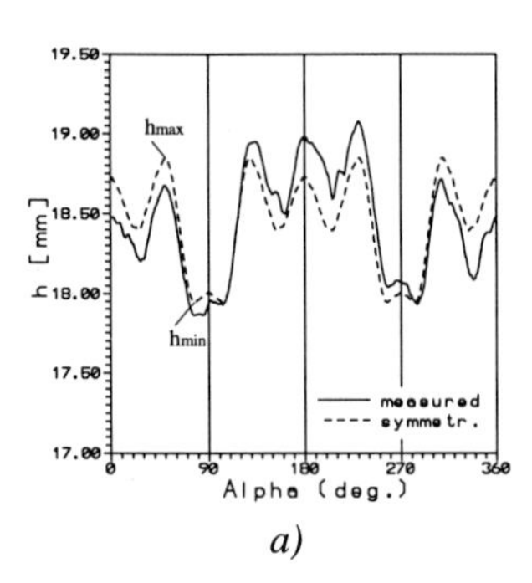

a)

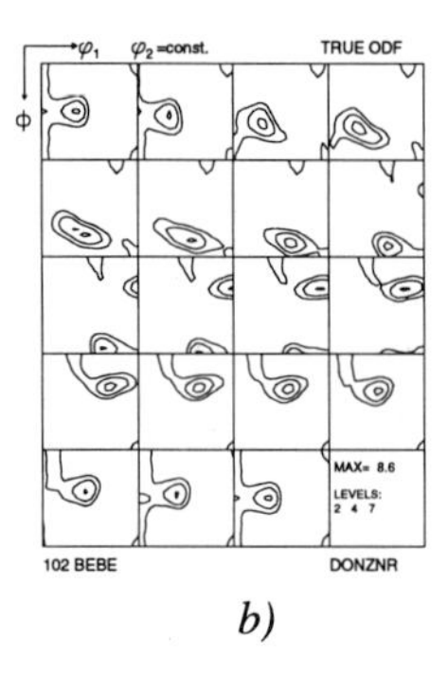

b)

Figure 1: a) measured and symmetrized earing profile and b) ODF of a representative sample

For each sample 4 pole figures were measured by means of an automatic X-ray Siemens-Lücke texture goniometer [5]. From the pole figures the Orientation Distribution Functions (ODFs) were calculated by the series expansion method according to Bunge [6] up to a series expansion degree of $l_{max} = 22$ (figure 2).

MULTIPLE LINEAR REGRESSION

The height h_{exp} of the earing profile is assumed to be only a function of the texture, i.e. of the series expansion coefficients $C_\lambda^{\nu\mu}$ of the ODF, and of the angle α. This leads to the equation

$$h_{exp}(\alpha) = b_0(\alpha) + \sum_{i=1}^{n} (b_i(\alpha) \cdot K_i) + \varepsilon(\alpha) \tag{2}$$

with $b_o(\alpha), b_i(\alpha)$ being the correlation coefficients, K_i the C-coefficients, n the number of C-coefficients and $\varepsilon(\alpha)$ the error. n+1 is then the number of correlation coefficients to be determined. For statistical reasons n was chosen not to exceed N/3-1 with N being the number of samples. Thus, with the 24 samples investigated here n is limited to $n\leq7$, i.e. the correlations could be investigated up to the 7 first C-coefficients C_4^{11}, C_4^{12}, C_4^{13}, C_6^{11}, C_6^{12}, C_6^{13} and C_6^{14}. By means of a multiple linear regression [7] the n+1 coefficients $b_o(\alpha), b_i(\alpha)$ are determined for each angle α, i.e. all in all

91 regressions have to be performed. The correlation coefficients $b_o(\alpha)$,$b_i(\alpha)$ are then used to calculate the theoretical heights $h_{calc}(\alpha)$ according to

$$h_{calc}(\alpha) = b_0(\alpha) + \sum_{i=1}^{n} (b_i(\alpha) \cdot K_i) \tag{3}$$

RESULTS AND DISCUSSION

A first multiple linear regression was carried out for n=3, i.e. up to a series expansion degree of the ODF of l=4 (correlation 1). Figure 2a exhibits the calculated heights as a function of the measured heights for all angles α. The points of each sample (e.g. points inside ellipse) assemble along a straight line nearly parallel to the ideal 45° line, but shifted by Δh to a somewhat higher or lower mean height level. Apart from this shift, which is partly due to an unexact determination of the zero point of the earing profile for some samples, figure 2 shows that a satisfying correlation is already given with the first 3 C-coefficients.

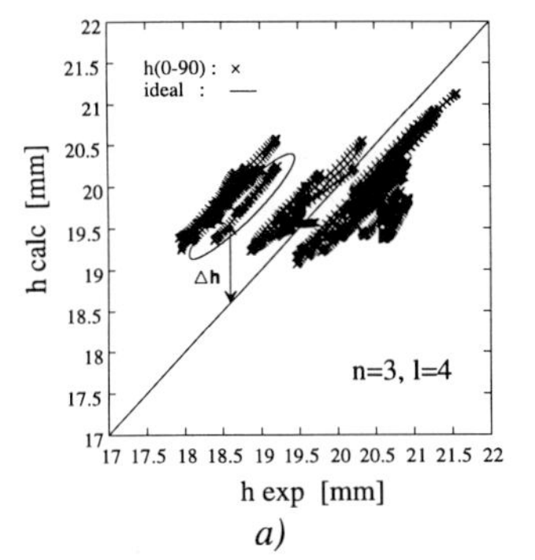

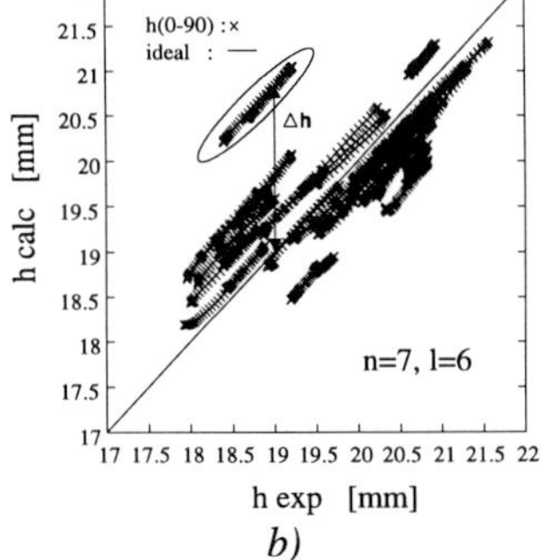

Figure 2: *Calculated heights as a function of measured heights for all angles α*
a) n=3, l=4 (correlation 1)
b) n=7, l=6 (correlation 2)

In a second step the correlation was investigated for n=7, i.e. for a series expansion degree of l=6 (correlation 2). Beside the shift Δh, which also occurs here (figure 2b), the agreement between measured and calculated heights is better than for n=3 (l=4).

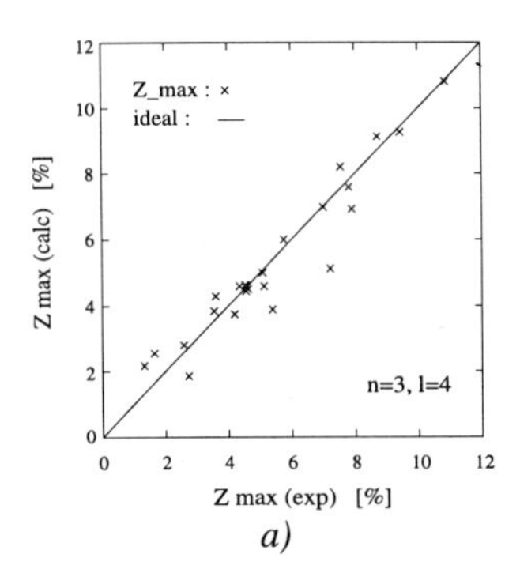

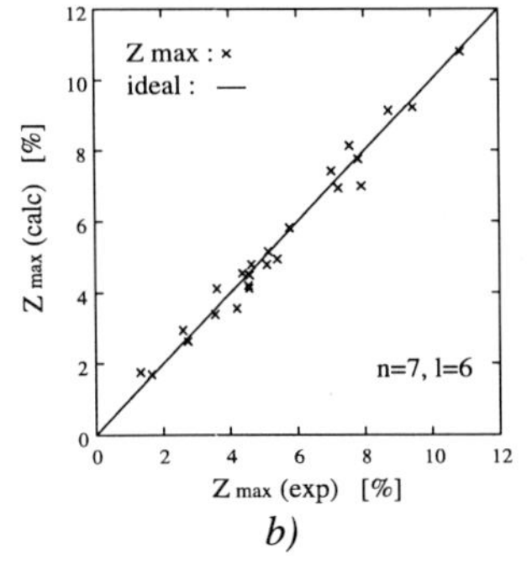

Figure 3: *Calculated Z_{max} values as a function of measured Z_{max} value*
a) n=3, l=4 (correlation 1)
b) n=7, l=6 (correlation 2)

This also becomes evident from figure 3 where the calculated Z_{max} values are plotted as a function of the measured Z_{max} values for both correlations. The agreement is much better for n=7 (figure 3b) than for n=3 (figure 3a).

In order to eliminate the shift Δh from the calculations, a third multiple linear regression was performed (correlation 3) with the mean height level of each sample h_{mean} being substracted from the earing profile. The substraction of h_{mean} allows then to set $b_0=0$ in equation 2. The correlation was investigated for n=7, i.e. l=6. For a better comparison of the results h_{mean} is then again added to the experimental heights h_{exp} and the calculated heights h_{calc}. With the shift Δh now being eliminated, the strong correlation between the heights of the earing profiles and the first 7 C-coefficients is very well illustrated by the excellent agreement between experimental and calculated heights (figure 4a). It should be noted that for all 3 correlations no difference was found between samples of the 3xxx series alloys and samples of the 5xxx series alloys.

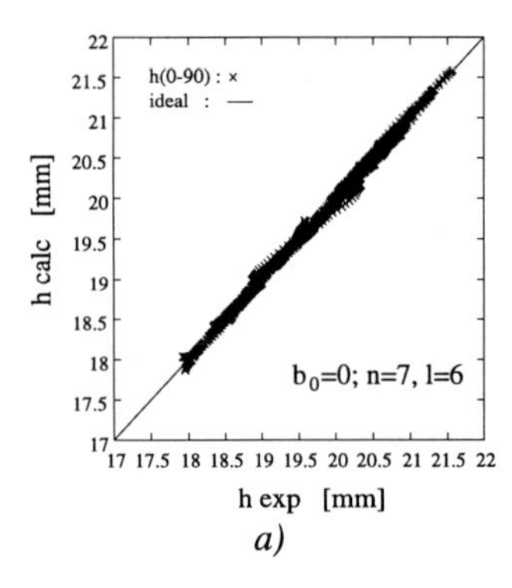

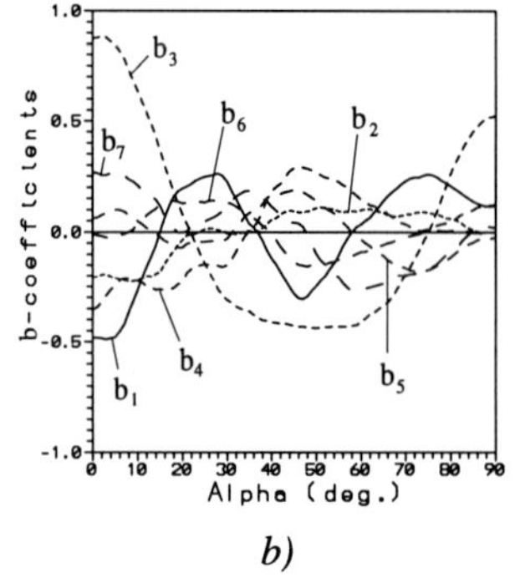

Figure 4: *Correlation 3, $b_0=0$, n=7, l=6*
a) Calculated heights as a function of measured heights for all angles α
b) the 7 coefficients b_i as a function of the angle α

Figure 4b shows the dependency of the 7 correlation coefficients b_i of the angle α according to correlation 3. The coefficient b_3 , which describes the influence of the C-coefficient C_4^{13}, has e.g. a strong influence on the earing profile at 0°, 90° and between 40° and 60°. The sign of the C-coefficient thereby determines wether an ear or a trough is formed. In contrast to b_3 the correlation coefficients b_2 and b_6 seem to be less important.By this sort of examination the influence of the individual C-coefficients can be investigated. This may be important if the number of correlation coefficients n has to be reduced due to a limited number of samples; then less important C-coefficients can be leaved out. In this case the correlation becomes, however, more dependent on the type of texture which then reduces the general validity.

In the following only results obtained with correlation 2 (n=7, l=6 and $b_0 \neq 0$) will be presented since this allows to calculate the absolute height of earing profiles without the need to know the mean height level h_{mean}. As an example figure 5 exhibits the measured and calculated earing profiles of 3 samples with different textures and, thus, different types of earing profiles (4, 6 and 8 ears). The samples were chosen out of the 24 samples used for the determination of the correlation coefficients. They illustrate the excellent agreement between measurement and calculation independently of the type of earing profile. It should, however, be pointed out that a prediction of an earing profile can only be valid for a texture which lies in the range of textures (type and strength of texture) used for the determination of the correlation coefficients.

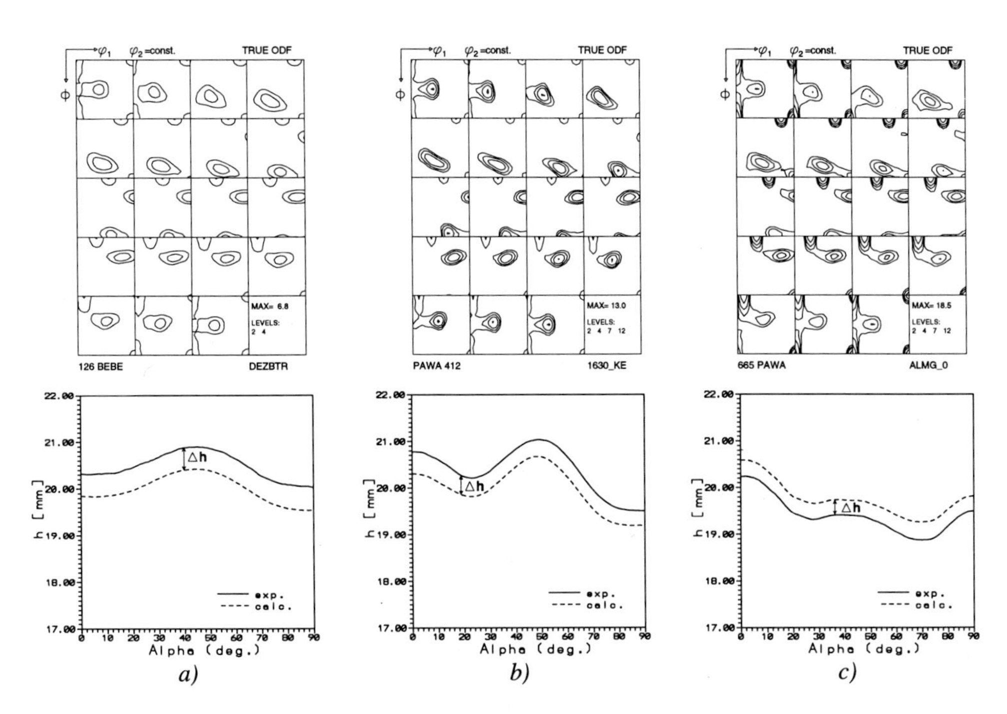

Figure 5: *ODFs and corresponding measured and calculated earing profiles of 3 samples with a) 4ears, b) 6 ears and c) 8 ears (correlation 2)*

In order to check the reliability of the predictions, the earing profiles of two samples which were not used in the determination of the correlation coefficients were calculated and compared to the experimentally measured profiles (figure 6). The textures of both samples lie within the range of textures used for the correlation. Also for these samples a very good agreement between measurement and prediction is obtained.

CONCLUSIONS

The correlation between earing profiles of Al alloys of the 3xxx and 5xxx series was investigated by means of a multiple linear regression. A satisfying correlation was already found with the first 3 C-coefficients, i.e. with a series expansion degree up to l=4. Increasing the series expansion degree to l=6 (7 C-coefficients) leads to a clearly stronger correlation with a very good agreement between measured and calculated earing profiles. It can thus be concluded that the strong correlation between earing profiles and the 7 first C-coefficients can very well be used for the prediction of earing profiles and earing values in Al sheets. From an industrial point of view an application in on-line texture measurements seems to be promising. Furthermore the calculation of the first C-coefficients out of earing profiles is principally possible although one has to keep in mind that the first 7 C-coefficients (l=6) characterize the texture only qualitatively.

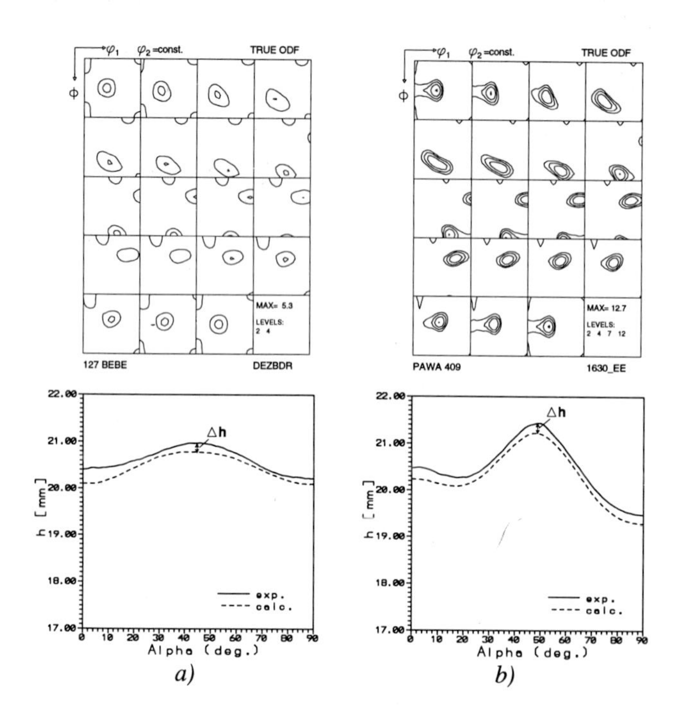

***Figure 6:** ODFs and corresponding measured and calculated earing profiles of 2 samples not used in the correlation (correlation 2)*
a) 4 ears
b) 6 ears

ACKNOWLEGEMENTS

The authors are indebted to *VAW Vereinigte Aluminiumwerke, Bonn* for supplying material and earing profiles and to Dr. J. Hasenclever and Dr. J. Hirsch for helpful discussions.

REFERENCES

1) Hirsch, J., Hasenclever, J.: Proc. ICAA 3, Trondheim 1992, 305
2) Van Houtte, P., Clarke, P., Saimoto, S.: Proc. TMS Symp. "Aluminium Alloys for Packaging", Chicago 1992
3) Van Bael, A., Van Houtte, P., Aernoudt, E., Hall, F.R., Pillinger, I., Hartley, P., Sturgess, C.E.N.: Proc. ICOTOM 9, Avignon 1990, 1007
4) Schmitter, U., Wagner, P., Lücke, K.: Proc. 4th Int. Symp.on Plasticity and Its Current Applications, Baltimore 1993 and
Wagner, P., Schmitter, U., Lücke, K.: to be published in Int. J. of Plasticity
5) Hirsch, J., Burmeister, G., Hoenen, L., Lücke, K.: in "Experimental Techniques of Texture Analysis", ed. H. J. Bunge, DGM Oberursel 1986, 63
6) Bunge, H.J.: in "Mathematische Methoden der Texturanalyse", Akademie Verlag, Berlin1969
7) IMSL Stat/Library Vers. 2.0, 1991, 75

7. Implication of Texture in Powder Diffraction Methods: Internal Stress Analysis, Phase Analysis, Structure Analysis

Materials Science Forum Vol. 157-162 (1994) pp. 2051-2058

DETERMINATION OF RESIDUAL STRESSES IN PLASTICALLY DEFORMED POLYCRYSTALLINE MATERIAL

A. Baczmanski [1], K. Wierzbanowski [1] and P. Lipinski [2]

[1] Wydzial Fizyki i Techniki Jadrowej, Akademia Górniczo-Hutnicza, al. Mickiewicza 30, PL-30-059 Kraków, Poland

[2] Ecole Nationale d'Ingénieurs de Metz, Ile du Saulcy, F-57045 Metz Cédex 1, France

Keywords: Diffraction, Residual Stresses, Elastic Constants, Polycrystalline Material, Texture

ABSTRACT

A new method of the determination of residual stresses in plastically deformed polycrystalline materials is presented. This method is based on standard diffraction measurements of the $\langle d_{hkl} \rangle$ interplanar spacings at different directions. Till now the estimation of the residual stresses was limited to the first order ones. The presented method takes into account both the first and the second order stresses.

1. INTRODUCTION

The internal stresses and strains in a polycrystalline aggregate depend on the history of the material. However, the final state of the stress (or strain) field can be mathematically described.

At the macroscopic scale, the following mean tensors are defined:

$$\sigma^{I}_{ij} = \frac{1}{V} \int_V \sigma_{ij}(\mathbf{r})\, d^3\mathbf{r}, \qquad \varepsilon^{I}_{ij} = \frac{1}{V} \int_V \varepsilon_{ij}(\mathbf{r})\, d^3\mathbf{r} \tag{1}$$

where: σ^I and ε^I are macrostress and macrostrain tensors (first order averages), V is the considered sample volume; $\sigma_{ij}(\mathbf{r})$ and $\varepsilon_{ij}(\mathbf{r})$ local stress and strain tensors at $\mathbf{r}$ position.

The second level of the study concerns the grain size scale. There are two reasons for which the stress and strain for a particular grain differ from the first order averages (see e.g. [1,2]):

- the elastic properties of a crystallite depend on the orientation g of its lattice and also on the coupling with neighboring grains,

- the shape of a crystallite after plastic deformation does not fit, in general, to the surrounding matrix (plastic strain differs from one grain to another). Hence, the grains are subjected to elastic deformation caused by incompatibilities in their boundary regions. This creates additional stresses (and strains) which are "frozen" in the material and they do not depend directly on the macrostress values.

We can conclude that the second order quantities characterize the deviation of the stress and strain for a particular grain from the macroscopic average.

2. EXPERIMENTAL PRINCIPLES

The well known diffraction method of determination of residual (or applied) stresses is based on the measurement of interplanar spacings for various directions characterized by the angles φ and ψ (Fig.1.). In diffraction, only crystallites having the (hkl) planes perpendicular to the scattering vector contribute to the reflected beam intensity. In fact we measure the mean interplanar spacing $\langle d\rangle_{hkl}$ for reflecting grains and the mean value of the strain $\langle\varepsilon'(\phi,\psi)\rangle_{hkl}$ at L_3 direction is equal to:

$$\langle\varepsilon'(\phi,\psi)\rangle_{hkl} = \frac{1}{V_r}\int_{V_r} \varepsilon'_{33}(\mathbf{r})\, d^3\mathbf{r} = \frac{\langle d(\phi,\psi)\rangle_{hkl} - d_0}{d_0} \tag{2}$$

where: d_0 - interplanar spacing for a stress-free specimen,
ε'_{33} - strain at L_3 direction for reflecting grain,
V_r - total volume of reflecting grains.

On the other hand, the measured strain can be expressed through:

$$\langle\varepsilon'(\phi,\psi)\rangle_{hkl} = F_{ij}(hkl,\phi,\psi)\ \sigma^{I}_{ij} + \langle\varepsilon'(\phi,\psi)\rangle^{pi}_{hkl} \tag{3}$$

where $F_{ij}(hkl,\phi,\psi)$ are the diffraction elastic constants for the hkl reflection and $\langle\varepsilon'(\phi,\psi)\rangle^{pi}_{hkl}$ relates to the strain caused by plastic incompatibilities of the grains.

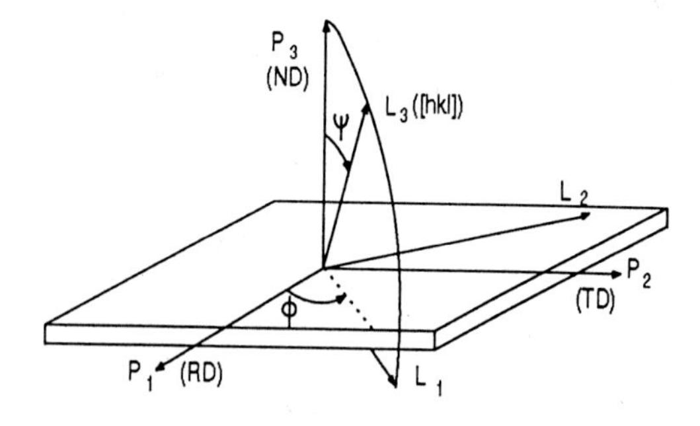

Fig.1.

Definition of ϕ and ψ angles determining the orientation of the scattering vector with respect to the sample system - $\mathcal{P}$. The laboratory system ($\mathcal{L}$) defines the measurement of the lattice distortion $\langle\varepsilon'_{33}(\phi,\psi)\rangle$ along the L_3 axis. The L_2 axis is in the rolling plane ($\mathbf{P}_1 \times \mathbf{P}_2$).

In this work all quantities expressed in the $\mathcal{L}$ coordinate system will be denoted by a prime, while quantities in the sample system ($\mathcal{P}$) are unmarked.

3. MODELS OF PLASTIC DEFORMATION OF POLICRYSTALLINE AGGREGATE

There exists a large spectrum of deformation models for polycrystalline materials. Each model has to describe the interaction between a grain and the surrounding average matrix. Using models, the residual stresses $\tilde{\sigma}_{ij}^{pl}(\mathbf{r})$ caused by the plastic incompatibilities between grains can be predicted.

In this work we have used two models: 1) the Leffers- Wierzbanowski model (L-W model) [3,4] and 2) the self- consistent model developped by M. Berveiller and P. Lipiński [5,6]. Both models take into account the elasto - plastic behaviour of grains in the deformed sample.

4. EVALUATION OF RESIDUAL STRESSES

Let us assume that the second term in the equation (3) can be approximated by:

$$<\varepsilon'(\psi,\phi)>_{hkl}^{pl} = q <\tilde{\varepsilon}'(\psi,\phi)>_{hkl}^{pl} \tag{4}$$

where: $<\tilde{\varepsilon}'(\psi,\phi)>_{hkl}^{lp}$ is **calculated** from a model and q - is a constant factor (it will be a fitting parameter).

The q factor has been introduced in order to find the real magnitude of "plastic" strains assuming that their variation with the ψ and ϕ angles is correctly described by models. Hence, for the hkl reflection, Eq.3 takes the following form:

$$<d(\psi,\phi)>_{hkl} = \left[F_{ij}(hkl,\psi,\phi)\ \sigma_{ij}^{I} + q <\tilde{\varepsilon}'(\psi,\phi)>_{hkl}^{pl} \right] d_o + d_o \tag{5}$$

For **known** diffraction elastic constants $F_{ij}(hkl,\psi,\phi)$, theoretically **predicted** strains $<\tilde{\varepsilon}'(\psi,\phi)>_{hkl}^{pl}$ and **measured** spacings $<d(\psi,\phi)>_{hkl}$, the other quantities from Eq.5 can be determined using a fitting procedure [8]. The unknowns are d_o, q and the first order stresses σ_{ij}^{I}. In general, we have 8 independent parameters, however their number is usually reduced because some components of the macrostress tensor are equal to zero.

Knowing the value of the q parameter from Eq.5 (by applying a fitting procedure) we are able to evaluate the "plastic" second order strain and stress for all grains. Thus, the first and second order stresses may be determined simultaneously.

4.1. DIFFRACTION ELASTIC CONSTANTS

We have theoretically predicted and experimentally verified the elastic diffraction constants F_{ij} for the investigated samples [7]. The calculation of F_{ij} needs some assumptions concerning the elastic behaviour of grains and the texture function. For our samples the Reuss model of F_{ij} calculation fits the best to measured values, thus for the purposes of the present work we have used the F_{ij} predicted by this model.

4.2. DETERMINATION OF RESIDUAL STRESSES FROM EXPERIMENTAL DATA

Two identical samples of cold rolled steel (reduction of 98%) were examinated (we denote them ULC I and ULC II). In both cases the $\langle d(\psi,\phi)\rangle_{211}$ vs $\sin^2\psi$ curves for $\phi = 0^\circ, 30^\circ, 60^\circ$ and 90° were determined using the X-ray diffraction.

Additionally, for the ULC I sample the neutron diffraction experiment has been performed. In this case the $\langle d(\psi,\phi)\rangle_{211}$ spacings have been measured with a wide beam (the sheet was irradiated over the whole cross section).

The theoretical values of the $\langle\tilde{\varepsilon}'(\psi,\phi)\rangle^{pl}_{211}$ strain were predicted by: Leffers-Wierzbanowski model (for 10000 crystallites) and self-consistent model (for 3000 crystallites).

Applying Eq.5 and fitting the results obtained from the models to the experimental data, the first order stresses (σ^I_{ij}), the interplanar spacings for a stress-free material (d_o) and the q- factor have been found. The results are shown in Table 1. In the case of the X-ray experiment (thin surface layer irradiated), we assume the biaxial stress state in the measured region and hence, only the $\sigma^I_{11}, \sigma^I_{22}$ and σ^I_{12} components of the first order stress tensor were determined. On the other hand, for neutron diffraction the whole cross section of the sample is irradiated and consequently we assume: $\sigma^I_{ij} = 0$. The reliability of the q - value is lower for the neutron measurements because of a low number of experimental points. However, its agreement with the same factor determined by X-ray diffraction is surprisingly good.

5. DISCUSSION AND CONCLUSION

The evaluated $\langle d(\psi,\phi)\rangle_{211}$ values (according to Eq.5) have been compared with experimental results. The examplary comparison for X-ray diffraction method is shown in Fig. 2. In order to express the influence of the second order "plastic" strain on the correct interpretation of experimental data the procedure presented above was applied but with the assumption that $\langle\varepsilon'(\psi,\phi)\rangle^{pl}_{hkl} = 0$. The results obtained with this assumption fit worse to the measured values.

In Fig. 3, the results of neutron diffraction are compared with the ones obtained using self-consistent and L-W models.

The behaviour of the interplanar strain q $\langle\tilde{\varepsilon}'(\psi,\phi)\rangle^{pl}_{hkl}$ (determined for 112 reflection) for different directions of the scattering vector is presented on the stereographic projection in Fig 4. Also in Fig.4 , in a similar way, the residual stress q $\langle\tilde{\sigma}'(\psi,\phi)\rangle^{pl}_{hkl}$ in the direction of the scattering vector is shown. We can see that the "plastic" residual stress (or strain) depends on the orientation, however its variation does not directly correspond to the distribution of the {112} planes (Fig.4e).

The new method, proposed in this paper, enables to evaluate magnitude of the first and second order residual stresses using some additional information from theoretical models. The variation of the second order residual stresses in function of the crystal orientation can be determined. The "plastic incompatibility" component plays a key role in a proper interpretation of diffraction measurements of residual stresses.

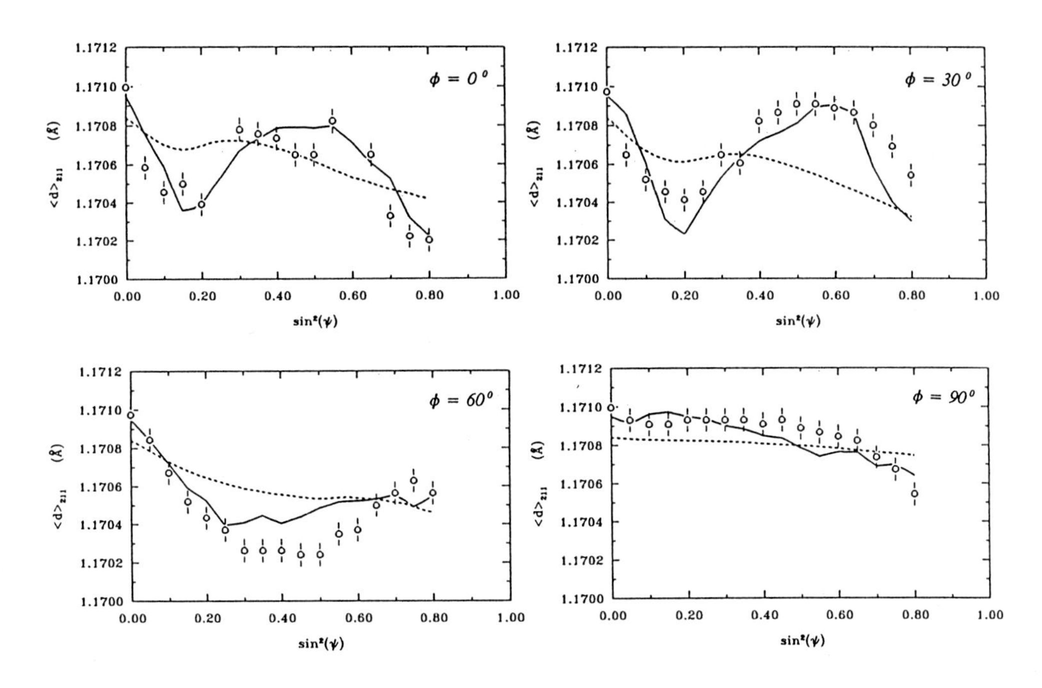

Fig. 2.
The $\langle d(\psi,\phi)\rangle_{211}$ spacings estimated using L-W model with different assumptions, i.e. $\langle \varepsilon'(\psi,\phi)\rangle^{pl}_{hkl} \neq 0$ (solid line) and $\langle \varepsilon'(\psi,\phi)\rangle^{pl}_{hkl} = 0$ (dashed line). The results are compared with the experimental ones. The ULC I sample was investigated using the X-ray technique.

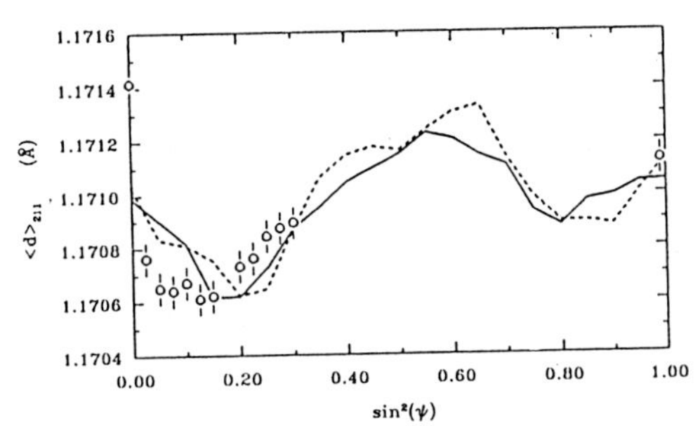

Fig. 3.
Comparison of the calculated $\langle d(\psi,\phi)\rangle_{211}$ spacings (solid line for L-W model and dashed curve for the self consistent model) with the experimental ones. The ULC I sample was investigated using the neutron diffraction.

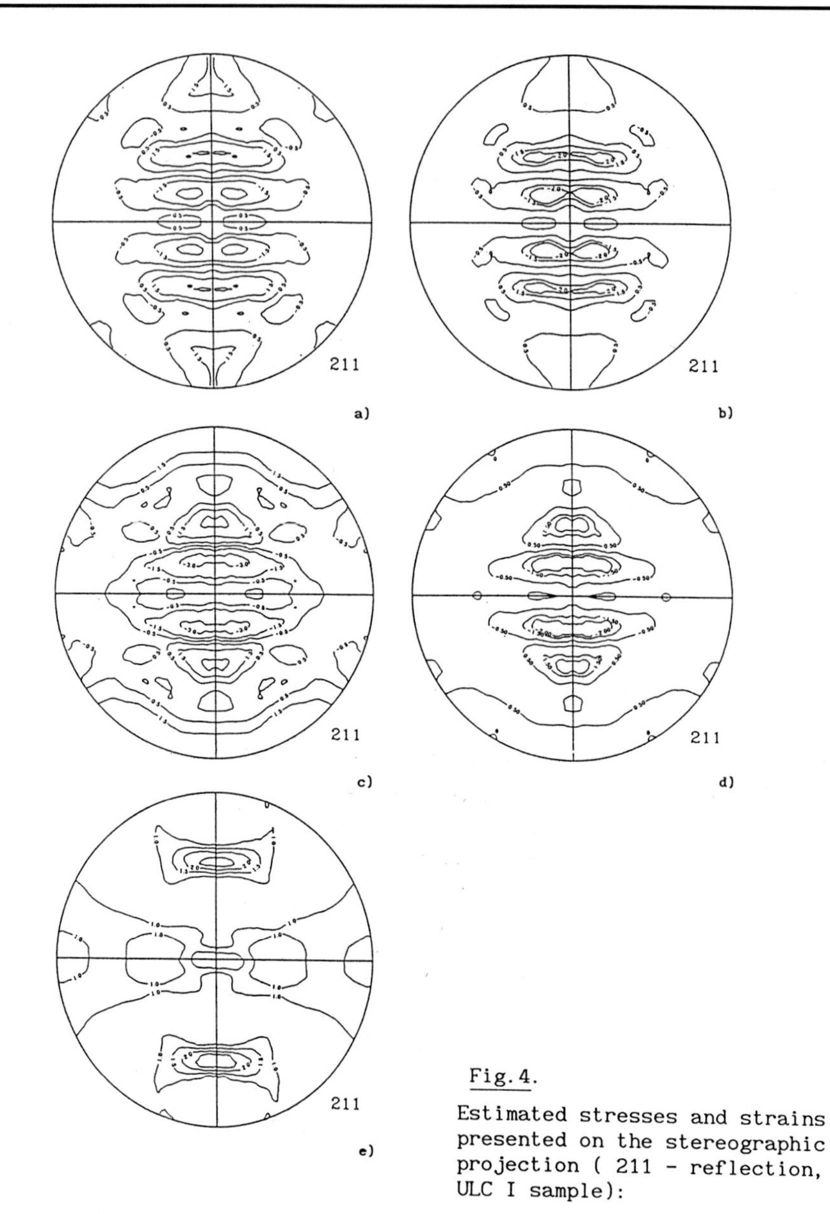

Fig. 4.

Estimated stresses and strains presented on the stereographic projection (211 - reflection, ULC I sample):

a) $q\ \langle\tilde{\sigma}'(\psi,\phi)\rangle^{pi}_{hkl}$ stresses (the levels are in units of 10 MPa), L-W model was used,

c) as above but the self- consistent model was used,

b) $q\ \langle\tilde{\varepsilon}'(\psi,\phi)\rangle^{pi}_{hkl}$ strains (in units of 10^{-4}); L-W model,

d) as above but the self- consistent model was used.

The experimental {211} pole figure for this sample is shown in c).

TABLE 1

Results of evaluation of the σ^{I}_{ij}, q and d_o parameters. The least squares fitting procedure was applied to experimentally measured $<d(\psi,\phi)>_{211}$ spacings (Eq.13). The χ^2 value is defined in [8]. The theoretical $<\tilde{\varepsilon}'(\psi,\phi)>^{pi}_{211}$ values were estimated using:

a) L-W model

	d_o (Å)	q	σ^I_{11} (MPa)	σ^I_{22} (MPa)	σ^I_{12} (MPa)	χ^2
ULC I X-ray	1.1706 ±0.0001	0.321 ±0.015	-104.0 ±6.0	-31.3 ±6.2	-3.9 ±5.9	4.05
ULC II X-ray	1.1703 ±0.0001	0.309 ±0.015	-119.5 ±11.8	-51.3 ±8.7	-24.5 ±7.9	3.62
ULC I neutrons	1.1709 ±0.0001	0.329 ±0.033	-	-	-	10.24

b) self consistent model

	d_o (Å)	q	σ^I_{11} (MPa)	σ^I_{22} (MPa)	σ^I_{12} (MPa)	χ^2
ULC I X-ray	1.1706 ±0.0001	0.405 ±0.020	-101.7 ±5.9	-36.9 ±6.2	-3.8 ±5.2	4.56
ULC II X-ray	1.1704 ±0.0001	0.374 ±0.015	-94.0 ±11.6	-46.9 ±8.7	-35.4 ±7.8	4.76
ULC I neutrons	1.1709 ±0.0001	0.382 ±0.046	-	-	-	9.61

This work has been supported by KBN grant No: 223189203

REFERENCES

1. Hauk,V.M., Adv. X-ray Anal., 1986, 29, 1
2. Noyan,I.C. and Cohen,J.B. Residual Streess - Measurement by Diffraction and Interpretation, Springer-Verlang, 1987
3. Wierzbanowski,K., Jura,J., Haije,W.G. and Helmholdt,R.B., Cryst. Res. Technol., 1992, 27, 513
4. Tarasiuk,J., Wierzbanowski,K., Kuśnierz,J., Arch. of Metall., 1993, 38, No.1
5. Lipiński,P. and Berveiller,M., Int. Journ. of Plasticity, 1989, 5, 149
6. Lipiński,P.,Krier,J. and Berveiller,M.,Revue Phys. Appl., 1990, 25, 361
7. Baczmański,A., Wierzbanowski,K., Haije,W.G., Helmholdt,R.B., Ekambaranathan, G., Pathiraj,B., Cryst. Res. Technol., 1993, 28, 229
8. Press,W.H., Flannery,B.P., Teukolsky,S.A., Vetterling,W.T., Numerical Recipes, The Art of Scientific Computing, Cambridge University Press, 1989.

Materials Science Forum Vol. 157-162 (1994) pp. 2059-2066

X-RAY RESIDUAL MACROSTRESS OF TEXTURIZED POLYCRYSTALS FROM POLE FIGURES

A.R. Gokhman

Technical University, St. Petersburg, Russia

Keywords: Texture, X-Ray Elastic Modulus, Residual Macrostress, Elastic Anisotropy, Plastic Anisotropy

ABSTRACT

It was shown that the crystallographic correlation of properties of the mono- and polycrystals is determined by some integral characteristics of the texture (ICT). For the calculation of the X-ray elastic modulus the ICT are determined by integration not all the pole figure, as in case of mechanical modulus, but only in the region, in which the condition of the coincidence of orientation of the reflex, chosen for determining of the residual deformation, and chosen direction in the polycrystal is full filled. X-ray residual macrostress of the rolling sheets Ti-alloys was carried out.

The correlation between value of the residual deformation ε for the direction (ψ,φ), when ψ - the polar angle and φ - the azimuthal angle, and value of the residual macrostress σ_{ik} for isotropic material has been received in [1,2] with assistance of the transformation law for second rank of a tensor:

$$\varepsilon(\psi,\varphi)=\frac{1+\nu}{E}(\sigma_1\cdot\cos^2\varphi+\sigma_2\cdot\sin^2\varphi)\cdot\sin^2\psi-\frac{\nu}{E}(\sigma_1+\sigma_2), \quad (1)$$

with E is Young's modulus; ν - Poisson's coefficient.

From eq.(1) the linearity of the dependence $\varepsilon=\varepsilon(\sin^2\psi)$, which had been based on the X-ray residual macrostress method - so-called method "$\sin^2\psi$" - had been deduced.
During the last twenty years the essential non-linearity of the dependence $\varepsilon=\varepsilon(\sin^2\psi)$ for a lot of the investigated metals was founded. The causes of evaluating this law can be the following ones:

1. Effect of the material elastic and plastic anisotropy.
2. Effect of the volume stress condition.
3. Effect of the surface relief.

We will discuss the effect of the material's elastic anisotropy.

The set investigations [3-5] have been devoted to account of the elastic modulus anisotropy in the textured specimens. The initial equation in [3-5] is the Hook's law for microvolume:

$$\varepsilon^M_{ik}=S^M_{ijkl}\cdot\sigma^M_{kl}. \quad (2)$$

Equation (2) has been averaged with using of the ideal orientation model or with using of the orientation distribution function (ODF) under the Reuss' approximation as well Foigt's approximation. In these investigations specimen external symmetry isn't

considered in the evident kind although ODF in them satisfies the specimen symmetry as well it satisfies the crystallite symmetry. We suugest the X-ray residual macrostress method, which from the very outset considers the orthotropic external symmetry as well the internal symmetry of the specimen.
In this case specimen elastic modulus have been described by the compliance tensor or stiff tensor with nine independent components and the Hook's law may be written as:

$$\begin{aligned} \varepsilon_{11} &= S^T_{1111}\cdot\sigma_{11} + S^T_{1122}\cdot\sigma_{22} + S^T_{1133}\cdot\sigma_{33}; \quad \varepsilon_{13}=\varepsilon_{31}=2\cdot S^T_{1313}\cdot\sigma_{13} \\ \varepsilon_{22} &= S^T_{1122}\cdot\sigma_{11} + S^T_{2222}\cdot\sigma_{22} + S^T_{2233}\cdot\sigma_{33}; \quad \varepsilon_{23}=\varepsilon_{32}=2\cdot S^T_{2323}\cdot\sigma_{23} \\ \varepsilon_{33} &= S^T_{1133}\cdot\sigma_{11} + S^T_{2233}\cdot\sigma_{22} + S^T_{3333}\cdot\sigma_{33}; \quad \varepsilon_{12}=\varepsilon_{21}=2\cdot S^T_{1212}, \end{aligned} \tag{3}$$

with S^T_{ijkl} are the compliance tensor components of the textured polycrystal.

Let's use the transformation law for second rank of a tensor and neglect the component σ_{i3} , and write the Hook's law for direction (ψ, φ):

$$\begin{aligned} \varepsilon(\psi,\varphi) = \sin^2\psi\,[&\cos^2\varphi\,(\sigma_{11}\cdot S^T_{1111} + \sigma_{22}\cdot S^T_{1122}) + \\ &+ \sin^2\varphi\,(\sigma_{11}\cdot S^T_{1122} + \sigma_{22}\cdot S^T_{2222}) - (\sigma_{11}\cdot S^T_{1133} + \\ &+ \sigma_{22}\cdot S^T_{2233})] + \sigma_{11}\cdot S^T_{1133} + \sigma_{22}\cdot S^T_{2233}. \end{aligned} \tag{4}$$

Correlation (4) may be recommended as Hook's law to carry out the X-ray residual macrostress in the textured specimen with orthotropic symmetry and any internal symmetry.
It is easy to show for random specimen, when $S^T_{1111} = S^T_{2222} = \frac{1}{E}$ and $S^T_{1111} = S^T_{1133} = S^T_{2233} = -\frac{\nu}{E}$, then eq. (4) transformes to eq. (1) from isotropic materials.
The polycrystal is the statistical system of the interaction crystallites with different orientations. Therefore it is necessary to have the ODF, form and size distribution functions to calculate the elastic properties of the polycrystal specimen. Neglecting the crystallites interaction, the connection between crystallites and polycrystal properties by means of direct integration of the orientation dependence of the monocrystal property with using of the ODF or another function as weight function may be determined:

$$S^T_{ijkl} = \langle S'^M_{ijkl} \rangle \tag{5}$$

We found the integral characteristics texture (ICT), which contain the whole information about texture in eq. (5). The number of the ICT is equal to the number of the independent tensor components, which describes the according monocrystal property. It is enough three ICT under studing the elastic modulus of the specimen with external orthotropic symmetry and internal cubic symmetry and five ICT - in case of the internal hexagonal symmetry. In case of the cubic symmetry ICT can be written as:

$$J_1^c = <\alpha_{11}^4 + \alpha_{12}^4 + \alpha_{13}^4>; \quad J_2^c = <\alpha_{21}^4 + \alpha_{22}^4 + \alpha_{23}^4>;$$

$$J_3^c = <\alpha_{11}^2 \cdot \alpha_{21}^2 + \alpha_{12}^2 \cdot \alpha_{22}^2 + \alpha_{13}^2 \cdot \alpha_{23}^2>, \tag{6}$$

with $\alpha_{1i} = \cos \widehat{OX_i\ RD}$; $\alpha_{2i} = \cos \widehat{OX_i\ TD}$;
$OX_i = (OX_1, OX_2, OX_3)$ is the crystal coordinate system;
TD is transvere direction.
Then eq. (5) can be represented as following;

$$\frac{1}{E(\Psi = 90^\circ, \varphi)} = S_{11}^M - K \cdot \Psi^T(\varphi), \tag{7}$$

with $K = S_{11}^M - S_{12}^M - \frac{1}{2} \cdot S_{44}^M$; $\Psi^T(\varphi) = 1 - (J_1^c \cdot \cos^4\varphi + J_2^c \cdot \sin^4\varphi + 1{,}5 J_3^c \cdot \sin^2 2\varphi)$

$$S_{1111}^T = S_{12}^M + S_{44}^M/2 + K \cdot J_1^c \tag{8}$$

$$S_{2222}^T = S_{12}^M + S_{44}^M/2 + K \cdot J_2^c \tag{9}$$

$$S_{3333}^T = S_{12}^M + S_{44}^M/2 + K \cdot (J_1^c + J_2^c + 2J_3^c - 1) \tag{10}$$

$$S_{1122}^T = S_{12}^M + K \cdot J_3^c \tag{11}$$

$$S_{1133}^T = S_{12}^M + K \cdot (1 - J_1^c - J_3^c) \tag{12}$$

$$S_{2233}^T = S_{12}^M + K \cdot (1 - J_2^c - J_3^c) \tag{13}$$

$$S_{1313}^T = S_{44}^M/4 + K \cdot (1 - J_1^c - J_3^c) \tag{14}$$

$$S_{2323}^T = S_{44}^M/4 + K \cdot (1 - J_2^c - J_3^c) \tag{15}$$

$$S_{1212}^T = S_{44}^M/4 + K \cdot J_3^c \tag{16}$$

In case of the hexagonal symmetry of crystallites and orthotropic symmetry of the polycrystal ICT can be written as:

$$J_1^h = <\alpha_{13}^2>; \quad J_2^h = <\alpha_{23}^2>;$$

$$J_4^h = <\alpha_{13}^4>; \quad J_5^h = <\alpha_{23}^4>; \tag{17}$$

$$J_6 = <\alpha_{13}^2 \cdot \alpha_{23}^2>,$$

with $\alpha_{13} = \cos \widehat{RD\ \vec{c}}$; $\alpha_{23} = \cos \widehat{TD\ \vec{c}}$
Then eq. (5) can be represented as following:

$$\frac{1}{E(\Psi = 90^\circ, \varphi)} = S_{11}^M + a \cdot \Psi_2^T(\varphi) + b \cdot \Psi_4^T(\varphi), \tag{18}$$

with $a = 2\cdot(S_{13}^{M} - S_{11}^{M} + \frac{1}{2}S_{44}^{M})$; $b = S_{11}^{M} + S_{33}^{M} - 2S_{13}^{M} - S_{44}^{M}$;

$$\Psi_2^T(\varphi) = J_1^h\cdot\cos^4\varphi + J_2^h\cdot\sin^4\varphi + \frac{1}{4}(J_1^h + J_2^h)\cdot\sin^2 2\varphi;$$

$$\Psi_4^T(\varphi) = J_4^h\cdot\cos^4\varphi + J_5^h\cdot\sin^4\varphi + 1{,}5\cdot J_6^h\cdot\sin^2 2\varphi;$$

$$S_{1111}^T = S_{11}^M + a\cdot J_1^h + b\cdot J_4^h \quad (19)$$

$$S_{2222}^T = S_{11}^M + a\cdot J_2^h + b\cdot J_5^h \quad (20)$$

$$S_{3333}^T = S_{11}^M + a\cdot(1 - J_1^h - J_2^h) + b\cdot(1 + J_4^h + J_5^h + 2(J_6^h - J_1^h - J_2^h)) \quad (21)$$

$$S_{1122}^T = S_{12}^M + b\cdot J_6 + (S_{13}^M - S_{12}^M)\cdot(J_1^h + J_2^h) \quad (22)$$

$$S_{1133}^T = S_{12}^M + b\cdot(J_1^h - J_4^h - J_6^h) + (S_{13}^M - S_{12}^M)\cdot(1 - J_2^h) \quad (23)$$

$$S_{2233}^T = S_{12}^M + b\cdot(J_1^h - J_5^h - J_6^h) + (S_{13}^M - S_{12}^M)\cdot(1 - J_1^h) \quad (24)$$

$$S_{1212}^T = \frac{1}{2}(S_{11}^M - S_{12}^M) + b\cdot J_6^h + c\cdot(J_1^h + J_2^h) \quad (25)$$

$$S_{1313}^T = \frac{1}{2}(S_{11}^M - S_{12}^M) + b\cdot(J_1^h - J_4^h - J_6^h) + c\cdot(1 - J_2^h) \quad (26)$$

$$S_{2323}^T = \frac{1}{2}(S_{11}^M - S_{12}^M) + b\cdot(J_1^h - J_5^h - J_6^h) + c\cdot(1 - J_1^h), \quad (27)$$

with $c = \frac{1}{2}(S_{12}^M - S_{11}^M + \frac{1}{2}S_{44}^M)$.

For random hexagonal specimen $J_1^h = J_2^h = 1/3$; $J_4^h = J_5^h = 1/5$; $J_6^h = 1/15$.
The analogous results had been received by prof. Bunge about thirty years ago, but instead of the ICT the Furie-series coefficients $C_\ell^{\mu\nu}$ of the ODF had been used in [6].
It is easy to deduce the connection between ICT and $C_\ell^{\mu\nu}$ for cubic external symmetry:

$$J_1^c = 0{,}6 + K_1\cdot C_4^{10} - K_2\cdot C_4^{12} + K_3\cdot C_4^{14} \quad (28)$$

$$J_2^c = 0{,}6 + K_1\cdot C_4^{10} + K_2\cdot C_4^{12} + K_3\cdot C_4^{14} \quad (29)$$

$$J_3^c = 0{,}6 + K_4\cdot C_4^{10} - K_3\cdot C_4^{14} \quad (30)$$

with $K_1 = 0{,}021818$; $K_2 = 0{,}03253$; $K_3 = 0{,}043032$; $K_4 = 0{,}007273$
Unlike to $C_\ell^{\mu\nu}$ ICT may be directly found from PF or inverse pole figure (IPF).
The calculation method of the ICT from PF (0002) for hexagonal symmetry is based on the isotropic properties of the basic plane of the hexagonal crystallites.
The calculation method of the ICT from PF (001) and from PF (10$\bar{1}$0) has been carried out with assistance of the probable method.
This method has been based on the consideration of the event to find the crystallite with some orientation in the polycrystal coordinate system as product of the two appropriate dependent

events. First event consists of finding the crystallite with appropriate orientation of the normal $h_1(\psi_1, \varphi_1)$ in the polycrystal coordinate system and second event consists of finding the crystallite with appropriate orientation of the normal $h_2(\psi_2, \varphi_2)$ in the polycrystal coordinate system under the condition that first event took place. From the theorem of multiplication of probable theory for the dependent events we have:

$$f(\psi_1, \varphi_1, \psi_2) = P_{\vec{h}_1}(\psi_1, \varphi_1) \cdot P_{\vec{h}_2/\vec{h}_1}(\psi_2, \varphi_2), \tag{31}$$

with $f(\psi_1, \varphi_1, \psi_2)$ is ODF as function of the $(\psi_1, \varphi_1, \psi_2)$ angles

$$P_{\vec{h}_1}(\psi_1, \varphi_1) = P^{exp}_{\vec{h}_1}(\psi_1, \varphi_1) / \iint P^{exp}_{\vec{h}_1}(\psi_1', \varphi_1') \cdot \sin\varphi_1' \cdot d\psi_1' d\varphi_1' \tag{32}$$

$$P_{\vec{h}_2}(\psi_2, \varphi_2) = P^{exp}_{\vec{h}_2}(\psi_2, \varphi_2) / \oint P^{exp}_{\vec{h}_2}(\psi_2', \varphi_2') \cdot d\ell_2 \tag{33}$$

The integration curve ℓ_2 in the PF $\vec{h}_2$ is satisfies so condition:

$$\vec{h}_1 \cdot \vec{h}_2 = Const \tag{34}$$

or

$$\cos\psi_1 \cdot \cos\psi_1 + \sin\psi_1 \cdot \sin\psi_2 \cdot \cos(\varphi_1 - \varphi_2) = \cos\widehat{\vec{h}_1 \vec{h}_2} \tag{35}$$

Numerous approbations of this method to calculate ICT, including in paper [7] with model PF as initial data, allow to recommend the probability method for calculation ICT in the problems of the texture analysis. Let's spread this method to the calculation of the X-ray elastic modulus.
If the residual elastic deformation has been determined from the X-ray measuring of the crystallographic plane distance (hkl) in the direction (ψ_0, φ_0), for calculation of the X-ray elastic constants it is necessary to consider only crystallites with orientations, which satisfy the condition:

$$\begin{aligned} \alpha_{11} \cdot \gamma_{11} + \alpha_{21} \cdot \gamma_{12} + \alpha_{31} \cdot \gamma_{13} &= \frac{h}{\sqrt{h^2 + k^2 + \ell^2}} \\ \alpha_{12} \cdot \gamma_{11} + \alpha_{22} \cdot \gamma_{12} + \alpha_{32} \cdot \gamma_{23} &= \frac{k}{\sqrt{h^2 + k^2 + \ell^2}} \\ \alpha_{13} \cdot \gamma_{11} + \alpha_{23} \cdot \gamma_{12} + \alpha_{33} \cdot \gamma_{13} &= \frac{\ell}{\sqrt{h^2 + k^2 + \ell^2}}, \end{aligned} \tag{36}$$

with γ_{1i} are the coordinates of the unit normal $\vec{h}$ in the polycrystal's system coordinate: $\gamma_{11} = \sin\psi_0 \cdot \cos\varphi_0$; $\gamma_{12} = \sin\psi_0 \cdot \sin\varphi_0$; $\gamma_{13} = \cos\psi_0$
In case of the hexagonal symmetry for calculation of the X-ray elastic modulus it is necessary to account only crystallites with orientations of the axis "$\vec{c}$", which satisfy condition:

$$\cos\psi \cdot \cos\psi_0 + \sin\psi \cdot \sin\psi_0 \cdot \cos(\varphi - \varphi_0) = \cos\widehat{\vec{c}\,\vec{h}} \tag{37}$$

So ICT and as well X-ray elastic modulus have complicated dependence from values (ψ_0, φ_0) and reflex $\vec{h}$ choice. Therefore the divergence from law "$\sin^2\psi$" may take place for textured materials.

The X-ray residual macrostress in the industrial alloy Ti-3,8Al-1,4V was being carried under suggested method. For determining of the residual elastic deformation the reflex $(10\bar{1}5)$ was used. The dependence $\varepsilon = \varepsilon(\sin^2\psi)$ of the Ti-3,8Al-1,4V alloy's sheets for RD approximately linear, but this dependence is essentially non-linear for TD (fig.1). The non-linearity level of the dependence $\varepsilon = \varepsilon(\sin^2\psi)$ increases after the rolling sheets. The deviation level from linear "$\varepsilon = \varepsilon(\sin^2\psi)$" law depends on the pole density distribution. This level is higher the more texture distribution is variated in this ψ-direction. So far as with increasing of the reduction after the rolling texture shapness increases [8,9] the level of the non-linearity increases too. The X-ray elastic constants and ICT of the investigated specimens essential dependent from $\sin^2\psi$ (tabl.1). RM are found from the condition of the minimal of the following square form:

$$\sum_{i=1}^{N}\{[\varepsilon_{RD}^{exp} - \varepsilon(\psi_i, \varphi=0^\circ)]^2 + [\varepsilon_{TD}^{exp} - \varepsilon(\psi_i, \varphi=90^\circ)]^2\} \quad (38)$$

The sheets after the rolling have RM with value and sign, which change under increasing of the reduction and additional investigations are necessary to explane these results: $\sigma_{11} = -14,8\cdot10^7$ Pa, $\sigma_{22} = 4,0\cdot10^7$ Pa; $\sigma_{11} = -15,1\cdot10^7$ Pa, $\sigma_{22} = -4,7\cdot10^7$ Pa; $\sigma_{11} = 9,8\cdot10^7$ Pa, $\sigma_{22} = -8,9\cdot10^7$ Pa; $\sigma_{11} = 3,3\cdot10^7$ Pa, $\sigma_{22} = -1,4\cdot10^7$ Pa for the cogging degree by thickness 0, 20, 40 and 60% accordingly.

Table 1. ITP and some compliance tensor components $S_{ijkl}^{XT}\cdot10^{-11}$ Pa in the RD and TD of Ti-3,8Al-1,4V initial sheets.

	I_1	I_2	I_4	I_5	I_6	S_{1111}^{XT}	S_{1133}^{XT}	S_{2233}^{XT}
RD								
20°	0.0938	0.3579	0.0667	0.1373	0.0270	0.9938	-0.2039	-0.1470
25°	0.0694	0.3528	0.0153	0.1327	0.0230	1.0247	-0.0805	-0.2832
30°	0.0510	0.3216	0.0088	0.1138	0.0180	1.0493	-0.2293	-0.1450
35°	0.0361	0.2491	0.0048	0.0802	0.0126	1.0695	-0.2551	-0.1283
40°	0.0356	0.1853	0.0040	0.0384	0.0098	1.0699	-0.2732	-0.1103
45°	0.0413	0.1475	0.0042	0.0384	0.0098	1.0615	-0.2819	-0.0974
TD								
20°	0.3428	0.1269	0.1284	0.0400	0.0342	0.6719	-0.1958	-0.0188
25°	0.3385	0.0876	0.1236	0.0214	0.0282	0.6760	-0.2102	-0.0052
30°	0.2697	0.0556	0.1000	0.0107	0.0194	0.7269	-0.2340	-0.0012
35°	0.2087	0.0322	0.0618	0.0046	0.0110	0.8397	-0.2677	-0 0147
40°	0.1584	0.0309	0.0143	0.0035	0.0088	0.9049	-0.2824	-0.0277
45°	0.1342	0.0369	0.0321	0.0037	0.0085	0.9365	-0.2875	-0.0361

References:

1. Vasilyev D.M.: Zavodskaya Laboratoriya, 1959, v.25, 70
2. Macherauch E., Muller P.: Z. Angew.Phys., 1961, Bd.13, 305
3. Houk V., Herlach D., Sesemann H.: Z. Metallkunde, 1975, Bd.66, 734
4. Dolle H., Hauk V.: Z. Metallkunde, 1979, v.70, 682

5. Barral M., Lebrun J.L., Sprauel J.M., Maeder G.: Met.Trans.A., 1987, v.18, 1229
6. Bunge H.J. Mathematische Methoden der Texturanalyse-Berlin: Acad.Verl., 1969, 325s.
7. Gokhman A.R., Divinsky S.V., Dnieprenco V.N.: Metallophysika, 1992, 57
8. Gokhman A.R.: Thin Solid Films, 1993, v.228, 229
9. Bruchanov A.A., Gokhman A.R., Michaylivsky Yu.G.: Phys.Metallov and Metallovedenie, 1991, v.56, 175.

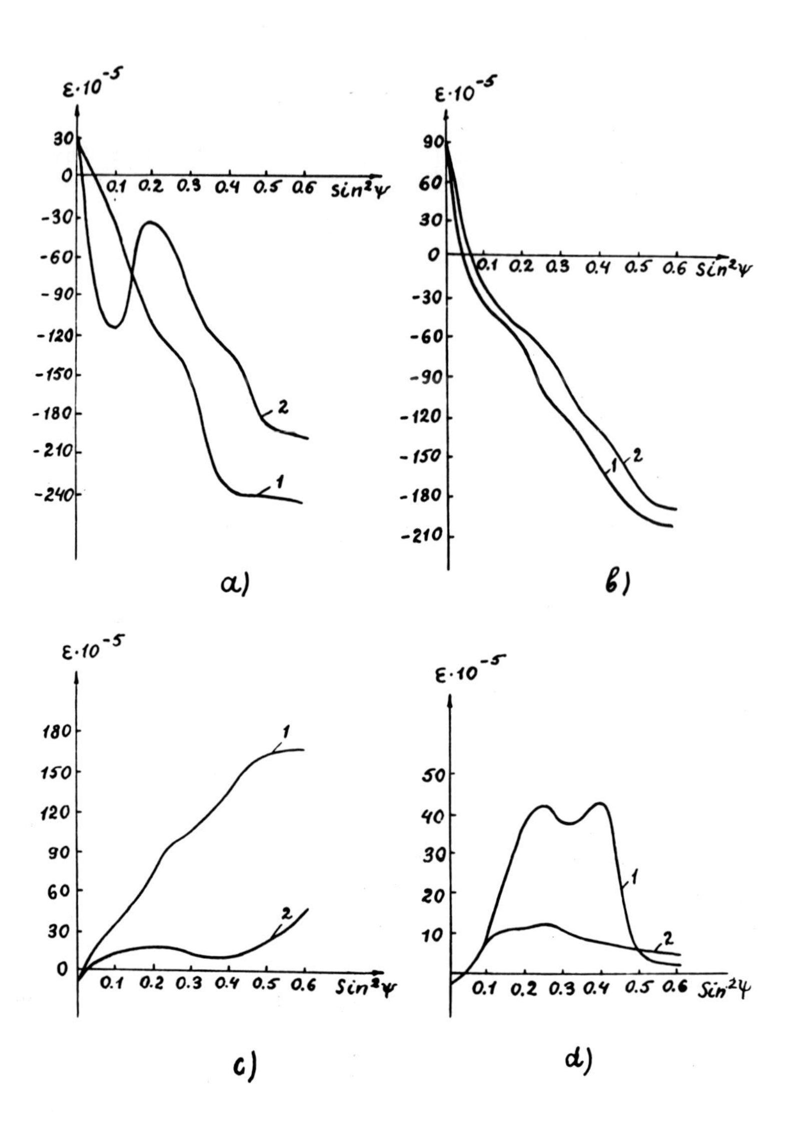

fig.1. The $\varepsilon = \varepsilon(\sin^2\psi)$ dependence of the sheets Ti-3.8Al-1.4V alloy after rolling with cogging 0% (a), 20% (b), 40% (c) and 60% (d) by thickness in the rolling direction (1) and in the transversal direction (2).

Materials Science Forum Vol. 157-162 (1994) pp. 2067-2074

A NEW METHOD FOR CRYSTAL STRUCTURE ANALYSIS OF TEXTURED POWDER SAMPLES

R. Hedel [1], H.J. Bunge [1] and G. Reck [2]

[1] Institut für Metallkunde und Metallphysik, TU Clausthal, Grosser Bruch 23, D-38678 Clausthal-Zellerfeld, Germany

[2] Bundesanstalt für Materialforschung- und prüfung, Berlin, Germany

Keywords: Texture-Structure-Analysis, Orientation Distribution Function, Reflection Separation, Model Textures

ABSTRACT

Crystal structure analysis from powder diffraction patterns has been based on two different strategies, i.e. deconvolution of powder patterns for integrated intensities followed by conventional crystal structure analysis or working with separated diffractions only including however all known structural conditions of the unknown structure. The first strategy can be considerably increased in resolving power by using textured powder samples instead of random ones. In this case, the texture of the sample is an additional unknown quantity which be must determined along with structure. On the other hand, intensity measurement as a function of two sample orientation angles $\{\alpha, \beta\}$ provides a much higher increase of the number of independent experimental input data. A deconvolution method based on a two-step strategy is presented and is tested with some synthetic data.

INTRODUCTION

Crystal structure analysis is mostly based on intensity measurements of single crystal diffraction patterns. In certain cases, single crystals are not available. It is then necessary to determine crystal structures on the basis of polycrystal diffraction patterns. thereby, two different strategies are presently being followed:

- Mathematical deconvolution of the overlapped reflection intensities of powder diffraction diagrams followed by normal crystal structure analysis.
- Reducing the number of unknown structure parameters using already known information about the unknown structure and using only reliable separated reflections in Patterson- or Replacementmethods.

The first strategy was used for instance by Schenk [1] and Giacovazzi [2] where as Baerlocher [3] used the second one. Jansen [1] describes the separation of closely overlapped reflections using Triplet- and negative Quartet relations. This method was successfully tested on known heavy atom structures but an unknown crystal structure was not yet solved. Giacovazzo [2] modified direct method to consider the possible uncertainties of the phase relations. With this method some simple unknown crystal structures could already be solved using powder diffraction data obtained by Synchrotron radiation measurements. Baerlocher [3] determined the crystal structure by cyclic calculation of patterson syntheses followed by Fourier analysis using separated reflections. This method was successfully used to solve the crystal structures of some ceolithes with high crystal symmetry.

In both strategies powder diffraction patterns with random orientation distribution of the crystallites were used. Thereby the reciprocal space is projected onto only one dimension i.e. its radial distance or Bragg angle.

If, instead, powder samples with a high degree of preferred orientation of the crystallites are used the reciprocal space remains essentially three-dimensional with the diffracted intensity depending on two additional angles $\{\alpha, \beta\}$ which describe the sample orientation with respect to the diffraction vector (figure 1). The first strategy of deconvolution can then be based on a three-dimensional intensity distribution instead of only one dimension. In this case, the orientation distribution function of the crystallites in the textured sample is an additional unknown quantity besides the unknown crystal structure. Preliminary investigations have shown however, that the increase of the number of the experimental data, i.e. intensity values $I(\theta, \alpha, \beta)$ is higher than the increase of unknown parameters describing the orientation distribution of the crystallites. Hence, textured powder samples provide a broader basis for the first strategy of deconvolution polycrystal diffraction diagrams.

PRINCIPLES OF THE DECONVOLUTION OF TEXTURED POWDER DIAGRAMS

We assume a textured powder sample, e.g. a flat sample measured in back-reflection technique with the diffraction vector having the orientation $\{\alpha, \beta\}$ with respect to the sample normal direction. Since the tilt angle α leads to a violation of the focusing condition (e.g. Bragg-Brentano condition) the effective diffraction pattern will be broadened. Hence, the diffracted intensity depending on the Bragg angle θ and sample orientation $\{\alpha, \beta\}$ can be expressed in the form:

$$I(\theta, \alpha, \beta) = \sum_{hkl} I_{hkl}^{rand} P_{hkl}(\alpha, \beta) B(\theta - \theta_{hkl}, b(\alpha)) \tag{1}$$

where I_{hkl}^{rand} is the integral intensity of the reflection (hkl) of a random sample, $P_{hkl}(\alpha, \beta)$ is the texture factor which is also called the "pole figure" in texture analysis and $B(\theta - \theta_{hkl}, b(\alpha))$ is the effective peak profile function with the half-width b depending on the sample tilt angle α (It can be assumed to be independent of the rotation β about the sample normal direction). It has been shown that the broadened peak profile can be very well represented by a Gauss function with b(α) depending on the experimental conditions but being at least empirically known.

The texture factor $P_{hkl}(\alpha,\beta)$ can be represented in the form of a series expansions

$$P_{(hkl)}(\alpha,\beta) = \sum_{\lambda=0}^{L} \sum_{\mu=1}^{M(\lambda)} \sum_{\nu=1}^{N(\lambda)} C_{\lambda}^{\mu\nu} k_{\lambda}^{\mu}(\Phi_{hkl},\gamma_{hkl}) k_{\lambda}^{\nu}(\alpha,\beta) \quad (2)$$

where $k_{\lambda}^{\mu}(\Phi_{hkl},\gamma_{hkl})$ and $k_{\lambda}^{\nu}(\alpha,\beta)$ are spherical surface harmonics depending on the crystal direction $\vec{h} \perp$ (hkl) and on the sample direction $\vec{y} = \{\alpha, \beta\}$ respectively [4]. The texture is completely characterised by the coefficients $C_{\lambda}^{\mu\nu}$. Since the texture of the sample is unknown these are additional unknown variables.

The random intensities I_{hkl}^{rand} contain essentially the structure factors

$$I_{hkl}^{rand} = N_{hkl} m_{hkl} |F_{hkl}|^2 \quad (3)$$

with m_{hkl} being the multiplicities of the reflections (hkl) and N_{hkl} summarising all other intensity factors.

REFLECTION SEPARATION WITH THE TWO-STEP METHOD

In order to solve equation 1 for I_{hkl}^{rand}, i.e. for $|F_{hkl}|$ according to equation 3, a sufficient $\{\alpha, \beta\}$-dependence of the texture factor $P_{hkl}(\alpha,\beta)$ is needed. On the other hand the texture must not be to strong such that the series equation 2 converges at a reasonable degree L. In this case the number of unknown $C_{\lambda}^{\mu\nu}$ is not too high. The solution of equation 1 can then be obtained with a two-step method:

In the first step a few non-overlapped reflections (hkl) are chosen. For these the peak profile function $B(\theta - \theta_{hkl}, b(\alpha))$ can be integrated over the whole profile without overlap. Then equation 1 reads

$$I_{hkl}^{int}(\alpha,\beta) = \sum_{hkl} I_{hkl}^{rand} P_{hkl}(\alpha,\beta) \quad (4)$$

which can be solved for $C_{\lambda}^{\mu\nu}$ according to equation 2. For this first step a total of, say, one to ten non-overlapped reflections depending on crystal symmetry are sufficient.

In the second step $P_{hkl}(\alpha,\beta)$ and $B(\theta - \theta_{hkl}, b(\alpha))$ are then known for any (hkl). Hence, equation 1 can be solved for I_{hkl}^{rand} respectively $|F_{hkl}|$ according to equation 3.

VERIFICATION OF THE METHOD WITH MODEL TEXTURES

In order to test the method, synthetic data I (θ, α, β) were generated. For that, crystal structure data of Brannockite [5] and a Polyoxadiazol derivate were used. Synthetic textures were assumed to consist of some preferred orientation $g^i = \{\varphi_1^i, \Phi^i, \varphi_2^i\}$ with

Gaussian spread with the spread widths ω_i about them. In this case, the texture coefficients can easily be expressed in the form [6]

$$C_\lambda^{\mu\nu} = \sum_{i=1}^{I} \nu_i \frac{e^{-\frac{1}{4}\lambda^2\omega_i^2} - e^{-\frac{1}{4}(\lambda+1)^2\omega_i^2}}{1 - e^{-\frac{1}{4}\omega_i^2}} T_\lambda^{\mu\nu}(\varphi_1^i, \Phi^i, \varphi_2^i) \quad (5)$$

The orientation g^i and spread width ω_i were chosen closely to experimentally produced textures of these two materials (table 1, 2).

The generated intensity data I (θ, α, β) of the non-overlapped reflections were used to calculate the texture coefficients $C_\lambda^{\mu\nu}$. They are compared in table 3 with the input values. The separated intensities of a number of overlapped reflections are compared in table 4, 5 with their respective true values.

It is seen that the obtained accuracy is, in most cases, below 3 %. Some larger deviations do, however, occur for some "ill-conditioned" reflections which are not well separated in reciprocal space neither in θ nor in $\{\alpha, \beta\}$. These reflections are, however, known and they may be given a special treatment.

Sections through the intensity distribution I (θ, α, β) in θ, α, β direction respectively are shown in figure 2, 3, 4. The θ-sections figure 2a correspond to "conventional" powder diffraction (with random sample). The sections α and β directions are "new". They are the basis of much more powerful deconvolution provided by the method described here.

ACKNOWLEDGEMENT

This project is supported by the Deutsche Forschungsgemeinschaft.

REFERENCES

/1/ J. Jansen, Doctoral Thesis, "The Determination of Accurate Intensities from Powder Diffraction Data and their Use in Direct Methods Structure Determination", Amsterdam 1991

/2/ C. Giacovazzo, "Direct Method for Powder Diffraction", Veszprem School on Crystallographic Computing, Balatonfüred, Hungary 1992

/3/ Ch. Baerlocher, EXTRACT, a FORTRAN Program for the Extraction of Integrated Intensities from Powder Data, Institute of Crystallography, ETH, Zurich, Switzerland

/4/ H. J. Bunge, "Texture analysis in material science", Butterworth 1982

/5/ T. Armbruster, R. Oberhäusli, American mineralogist, **73** (1988) 595-600

/6/ N. J. Park, H. J.Bunge, Zeitschrift für Metallkunde, **81** (1990) 636-645

ϕ_1 [°]	Φ [°]	ϕ_2 [°]	Fraction [%]	Half Width of the Gaussian Distribution [°]
0,0	07,0	0,0	1,43	12,0
20,0	0,0	0,0	1,43	12,0
40,0	0,0	0,0	1,43	12,0
0,0	45,0	0,0	14,3	12,0
20,0	45,0	0,0	14,3	12,0
40,0	45,0	0,0	14,3	12,0
0,0	75,0	20,0	7,15	12,0
0,0	75,0	40,0	7,15	12,0
20,0	55,0	20,0	4,3	12,0
40,0	55,0	40,0	4,3	12,0
Fraction of the Random Distribution: 30.0 %				

Table 1 Texture components of the generated orientation distribution for the hexagonal Brannockite.

ϕ_1 [°]	Φ [°]	ϕ_2 [°]	Fraction [%]	Half Width of the Gaussian Distribution [°]
0,0	0,0	0,0	14,0	12,0
0,0	25,0	0,0	4,2	12,0
0,0	50,0	0,0	1,4	12,0
30,0	25,0	0,0	14,0	12,0
40,0	40,0	0,0	4,2	12,0
55,0	75,0	30,0	14,0	12,0
40,0	85,0	25,0	4,2	12,0
55,0	25,0	55,0	1,4	12,0
33,0	33,0	45,0	8,4	12,0
45,0	30,0	60,0	4,2	12,0
Fraction of the Random Distribution: 30.0 %				

Table 2 Texture components of the generated orientation distribution for the orthorhombic Polydiaxolderivate.

λ	μ	ν: 1	2	3	4	5	6	7	8
2	1	0,4325 0,47627	-1,2118 -1,1675						
4	1	-1,0396 -0,9703	-1,2562 -1,2968	0,6306 0,6414					
6	1	0,0418 -0,0198	0,0354 -0,0233	-0,1378 -0,0870	-0,5684 -0,4862				
	2	0,0959 0,0705	0,1831 0,1810	0,0783 0,0885	-0,1214 -0,0729				
8	1	1,2258 1,2352	0,7936 0,8908	-0,0407 0,0435	0,1783 0,2300	0,5528 0,2938			
	2	-1,0507 -0,9704	0,9494 0,9491	0,1471 0,1660	-0,2622 -0,2676	-0,2813 -0,2749			
10	1	0,6438 0,6019	0,3482 0,2654	0,3157 0,2353	0,2688 0,2422	0,2287 0,2529	-0,3877 -0,1427		
	2	-1,4793 -1,4507	0,3230 0,4327	0,1086 0,1287	-0,0415 -0,0575	0,3001 0,2625	0,3293 0,1493		
12	1	-0,4120 -0,3465	-0,6585 -0,5735	-0,3089 -0,2904	-0,2899 -0,2209	0,0401 0,0237	-0,1611 -0,2869	0,2324 0,1156	
	2	-0,3137 -0,4199	0,1196 0,1115	-0,1802 -0,1149	-0,0086 -0,0169	0,1053 0,0915	-0,0807 -0,0379	-0,0620 -0,0855	
	3	-0,1141 -0,1175	0,1442 0,0969	-0,0802 -0,0914	-0,0620 -0,0733	0,2752 0,1948	-0,1781 -0,0448	-0,0122 0,0867	
14	1	0,1225 0,1341	0,0585 -0,0182	0,0991 0,0886	0,1213 0,0483	0,2767 0,1917	0,0257 0,0132	0,0520 0,1102	-0,1272 -0,0663
	2	0,3172 0,2015	0,0994 0,1957	-0,0541 -0,0249	0,1463 0,0967	-0,0527 -0,0776	-0,1253 -0,1656	0,0464 -0,0098	-0,0153 -0,1103
	3	0,0169 -0,0857	-0,0809 -0,1366	0,0652 0,0133	0,0464 0,0322	-0,1341 -0,1551	0,1765 0,2339	-0,0734 -0,0706	-0,0277 -0,0055

Table 3 Comparison of the generated texture coefficients (equation 5) with the texture coefficients obtained during calculation of the ODF from non-overlapped reflections (equation 1).

h k l	θ_{hkl} [°]	d [Å]	Intensity [ ] (genereted)	Comment	Intensity [ ] (separated)
1 0 0	5,10	8,662	64,9	non overlapped reflections using to calculate the orientation distribution function	
0 0 2	6,20	7,132	149,5		
1 0 2	8,04	5,506	136,7		
1 1 0	8,86	5,001	326,2		
1 1 2	10,84	4,095	854,1		
2 0 2	12,01	3,702	87,6		
1 0 4	13,51	3,297	128,0	overlapped reflections Region 1	125,8
2 1 0	13,61	3,274	75,9		74,7
2 1 1	13,97	3,191	17,1		17,2
3 1 6	27,12	1,689	155,1	overlapped reflections Region 2	153,8
5 0 2	27,23	1,634	75,4		57,3
1 1 8	27,30	1,679	371,1		374,5
4 1 4	27,47	1,670	931,6		938,3
3 0 2	27,52	1,667	164,2		156,6
2 0 8	27,85	1,649	102,5		101,6
3 2 6	30,35	1,525	143,6	overlapped reflections Region 3	145,7
5 1 2	30,45	1,520	154,6		165,7
3 0 8	30,52	1,517	442,6		445,4
3 3 4	30,67	1,510	496,5		484,0
4 2 4	31,18	1,487	112,7		118,3
4 1 6	31,38	1,479	738,1		740,7
5 4 11	66,06	0,8428	33,4	overlapped reflections Region 4	33,6
6 4 9	66,21	0,8418	33,9		33,0
10 0 4	66,22	0,8417	198.4		206,1
8 3 5	66,42	0,8405	50,5		181,0
7 4 6	66,46	0,8402	646.8		561,5
9 2 3	66,48	0,8401	20,1		-2,1

Table 4 Extract of the generated x-ray spectrum of the hexagonal double ring silicate Brannockite and the texture independed intensities of the reflections after the separation of the overlapped reflections.

h k l	θ_{hkl} [°]	d [Å]	Intensity [ ] (generated)	Comment	Intensity [ ] (separated)
1 1 1	9,8	4,527	365,3	non or in pairs overlapped reflections using to calculate the orientation distribution function	
3 1 1	10,38	4,274	474,6		
5 1 1	11,47	3,875	74,3		
10 0 0	12,11	3,671	90,4		
0 2 0	12,76	3,487	1000,0		
2 2 1	14,99	2,979	170,1		
2 0 2	15,01	2,975	41,2		
4 2 1	15,58	2,867	48,5		
4 0 2	15,60	2,864	82,5		
17 1 1	23,25	1,952	133,0		
5 1 3	24,36	1,868	143,0		
19 1 0	24,44	1,862	1,2		
4 4 1	27,87	1,648	27,6		
1 3 1	20,84	2,165	125,0	overlapped reflections Region 1	123,8
15 1 1	20,93	2,156	274,8		275,9
12 2 1	21,01	2,148	35,5		13,1
12 0 2	21,02	2,147	21,8		27,2
6 2 2	21,13	2,137	10,4		14,4
3 3 1	21,14	2,135	135,1		139,1
9 3 2	27,33	1,678	2,0	overlapped reflections Region 2	-1,6
6 4 0	27,35	1,677	23,5		13,1
0 4 1	27,38	1,675	183,3		178,0
6 2 3	27,38	1,675	13,9		14,4
11 1 3	27,44	1,672	36,3		45,8
22 0 0	27,49	1,669	1,0		-7,8
13 5 1	38,98	1,224	3,0	overlapped reflections Region 3	1,9
14 2 4	39,01	1,223	9,8		11,0
30 0 0	39,01	1,223	10,7		22,4
27 1 2	39,14	1,220	1,2		15,8
29 1 1	39,16	1,219	107,2		76,4
17 1 4	39,26	1,217	2,0		3,5

Table 5 Extract of the generated x-ray spectrum of the orthorhombic Polydiaxolderiva te and the texture independed intensities of the reflections after the separation of the overlapped reflections.

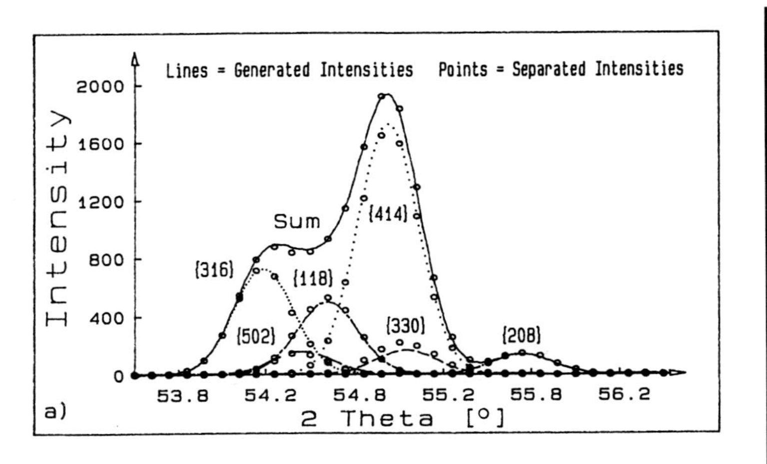

Figure 1 Sample Orientation and diffraction vector of single crystal, textured sample and random sample in the reciprocal space.

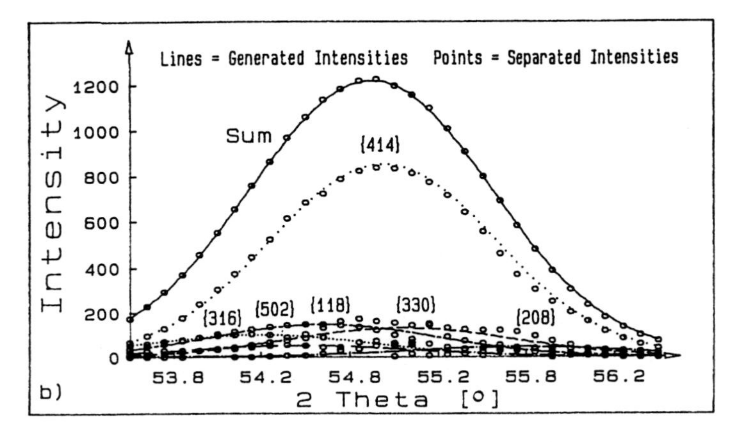

Figure 2 2θ-sections of the generated three-dimensional intensity distribution of the overlapped reflections (316), (502), (118), (414), (330) and (208).
a) $\alpha = 5°$, $\beta = 0°$, b) $\alpha = 70°$, $\beta = 35°$

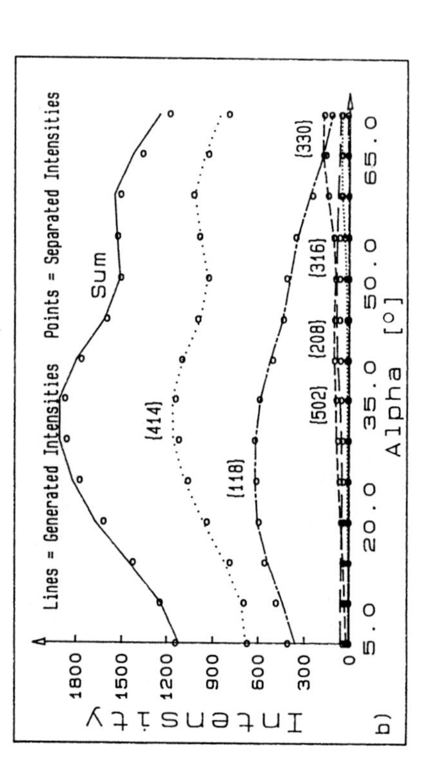

Figure 3 α-sections of the generated three-dimensional intensity distribution of the overlapped reflections (316), (502), (118), (414), (330) and (208).
a) $\theta = 54.7°$, $\beta = 0°$, b) $\theta = 54,7°$, $\beta = 35°$

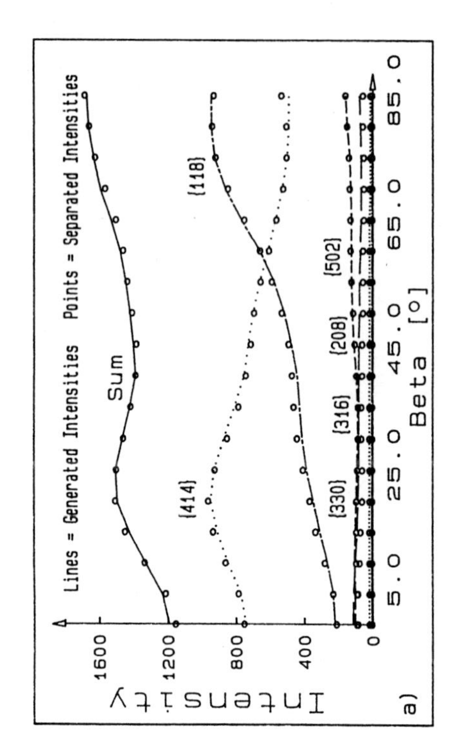

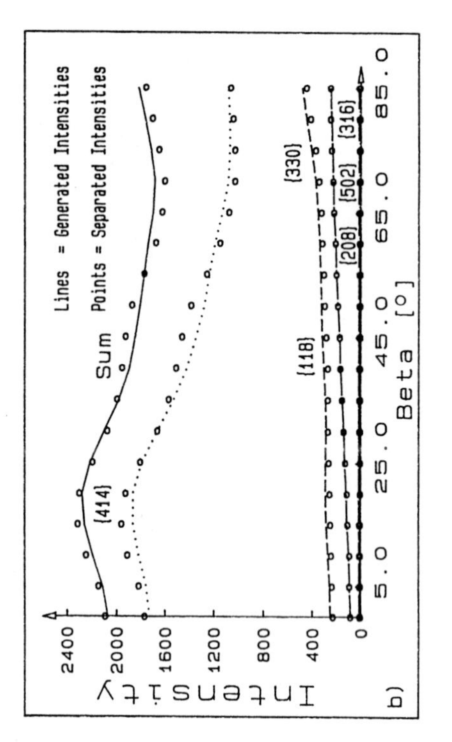

Figure 4 β-sections of the generated three-dimensional intensity distribution of the overlapped reflections (316), (502), (118), (414), (330) and (208).
a) $\theta = 54.7°$, $\alpha = 35°$, b) $\theta = 55.0°$, $\alpha = 70°$

Materials Science Forum Vol. 157-162 (1994) pp. 2075-2082

WIRE GEOMETRY CORRECTION OF TEXTURE AND STRESS MEASUREMENTS ON WIRES

K. Van Acker, P. Van Houtte and E. Aernoudt

Department of Metallurgy and Materials Engineering,
Katholieke Universiteit Leuven, de Croylaan 2, B-3001 Leuven, Belgium

Keywords: Residual Stress, Wire Drawing

ABSTRACT.

The interpretation of texture measurements on wires needs special care. Due to the wire geometry, a screening of incident and diffracted beam occurs and the absorption differs from place to place on the wire. Taking these effects into account together with the defocusing of the X-ray beam, a workable model has been built to correct the experimental pole figures. For this purpose, the model simulates the pole figures of a random textured wire with any diameter. These simulations give an excellent similarity with the measured pole figures of a powder sample with the shape of a wire. A new model has been developed for residual stress measurements on wires. In addition to similar (but different) corrections as for texture measurements, the stress state is assumed to be symmetric with respect to the wire axis. Anisotropic elastic constants for stress calculations could be modelled with the aid of the corrected texture, for both fibre and cyclic textures. The texture and residual stress state was determined in low carbon steel wires (Ø1.65mm) in as-drawn and annealed condition, and in severely cold-drawn high carbon steel wires with Ø0.25mm.

1. INTRODUCTION.

Several attempts have been made to measure textures and residual stresses in drawn wires [1,2,3,4]. A reasonable solution for the measurement of textures at the wire surface could be to drill the core of the wire and subsequently to develop the outer layer. By following such a procedure a flat surface is obtained and can easily be measured and interpreted. However, it only suits for thick wires and such method asks for tedious specimen preparation. For the residual stress measurements, the obvious solution is to use a mechanical method. The stress is then determined by etching or cutting part of the material of the wire or even by splitting the wire. Good quantitative results are hard to obtain. It is particularly difficult to measure circumferential or hoop stresses.

The purpose of the present work is to give the headlines of a method which does not ask any special specimen preparation nor specific instrument equipment other than a common used X-ray goniometer. Both for texture and residual stress measurements, a specimen is prepared which consists of an array of wires lying next to each other so that a square of at least 100 mm^2 is filled with aligned wires. The measurements are done as on an ordinary flat sample. However, the results must be corrected by computer. The first three paragraphs show how these corrections are performed. First, the fundamentals are given for texture correction. Secondly, the adaptations made for the interpretation of residual stress measurements are explained. In the third paragraph the link is made between the texture and residual stress measurements on wires in order to get a good interpretation of the residual stress state in highly textured wires. The validity of all these corrections is assessed on the basis of some experimental results.

2. CORRECTION OF TEXTURE MEASUREMENTS ON WIRE SAMPLES.

Consider now a sample composed of a series of parallel wires. The first correction that has to be made for both texture and residual stress calculations concerns the screening of the incident and diffracted X-ray beams by adjacent wires (see figure 1). Secondly, the measured data have to be corrected for the absorption on the path X-rays followed in the wire. Here, a simple model is assumed. In each point of the wire surface irradiated by the X-ray beam, the curved surface will be replaced by a hypothetical flat facet tangent to the wire in that point. It is assumed that the X-rays are only penetrate in an infinitesimally thin layer at the surface of the wire. More sophisticated models [5] trace the real X-ray path through the wires. Nevertheless, it will be seen that the assumption used in this article even holds for thin metallic wires and that it is not meaningful to use more sophisticated models. The mathematical implications of these corrections are discussed now. For reasons of clearness, the definition of the specimen system S is given in figure 3. In each point of the wire an extra reference system is defined and called the local reference system Lo. This reference system is derived from S by rotating S around the wire axis over an angle κ. In general the diffracted intensity I^r can be written as :

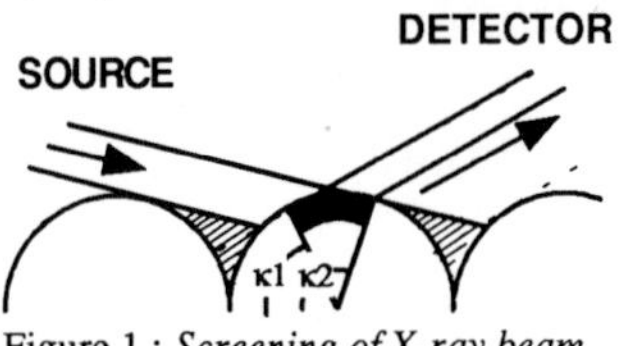

Figure 1 : *Screening of X-ray beam in composed wire sample*

$$I^r = \frac{1}{K} \int_{\kappa=\kappa_1}^{\kappa_2} \int_{r=0}^{R} I^i \, \eta \, P \, e^{-\mu z_{\varphi\psi\kappa r}} \, d\kappa \, dr \qquad \text{with } K = \int_{\kappa=\kappa_1}^{\kappa_2} \int_{r=0}^{R} d\kappa \, dr \tag{1}$$

In equation (1) I^i stands for the intensity of the incident beam, η for the volume fraction of diffracting crystallites, P for the irradiated surface seen from the detector, μ for the linear absorption coefficient and z for the path length of the X-rays in the material. In the most general expression, z is function of φ, ψ, κ and r for a wire. r is the radius of the wire. Figure 1 explains κ_1 and κ_2. See figure 3 for φ and ψ. The integration is done over a section of the wires perpendicular to the wire axis. The limits κ_1 and κ_2 result from the screening correction. In the approach of this article the integral in r is calculated as for flat surfaces and will be disappeared in the following equations.

However, only part of the diffracted beam will be catched by the detector due to defocusing of the diffracted X-ray beam and the malpositioning of the sample. Instrumental defocusing due to the χ tilt (see figure 3) can be corrected for in the usual way, i.e. by measuring a flat powder sample under the same measuring conditions. In the case of wires there is an inherent malpositioning of the sample. Indeed, only the tops of the wires are aligned in the goniometer. The rest of the wire surface is lying further away from the goniometer centre. A variable horizontal shift Δ at the detector is mainly caused by the vertical divergence Ω of the X-ray beam at the source. Δ can be expressed as function of Ω. The shift Δ at the detector is restricted to the interval $\Delta^- \leq \Delta \leq \Delta^+$ by the receiving slits. When an intensity distribution $I(\Omega)$ is assumed at the source and the dimensions of the wires are known, then can be calculated which percentage of the beam I^r is detected.

3. CORRECTIONS FOR RESIDUAL STRESS MEASUREMENTS ON WIRES.

Another stress reference system is necessary for wires (see figure 2). The stresses are respectively Called the longitudinal stress σ_{zz}, the circumferential or hoop stress $\sigma_{\kappa\kappa}$ and the torsional stress $\sigma_{z\kappa}$. They can also be expressed in the local reference system Lo as defined in the previous section.

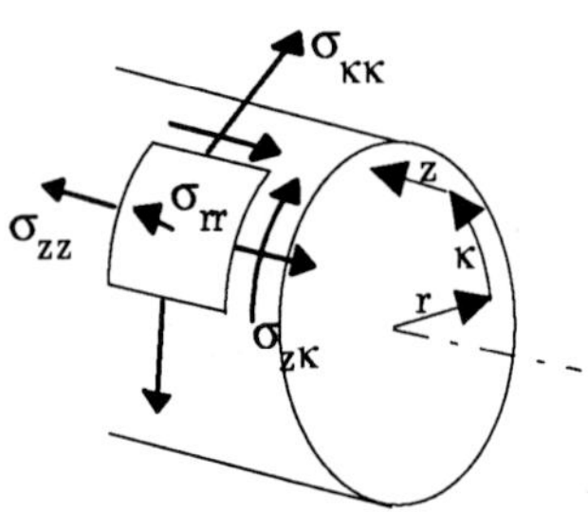

Figure 2 : *Definition of stress components at the wire surface*

Some simplifications for the stress state have been assumed. The stress state after drawing is thought to be rotational symmetric with respect to the wire axis, i.e. the stresses σ_{zz}, $\sigma_{\kappa\kappa}$, $\sigma_{z\kappa}$ will be constant for all points of the surface or at a certain distance of the axis. The stress component σ_{rr} normal to the wire surface is zero. Finally, no stress gradients are taken into account. It is obvious that this assumption doesn't hold for the entire section of a wire. However, stresses determined by X-rays in a common metal are mainly restricted to about the upper 10 μm. So, this layer from which the stress information is originating is relatively small with respect to the wire radius for wires of e.g. 1 mm diameter.

The total strain ε, denoted as ε^L_{33} in the laboratory reference system L (figure 3) and measured in a sample position (φ,ψ) in a ω-goniometer, is

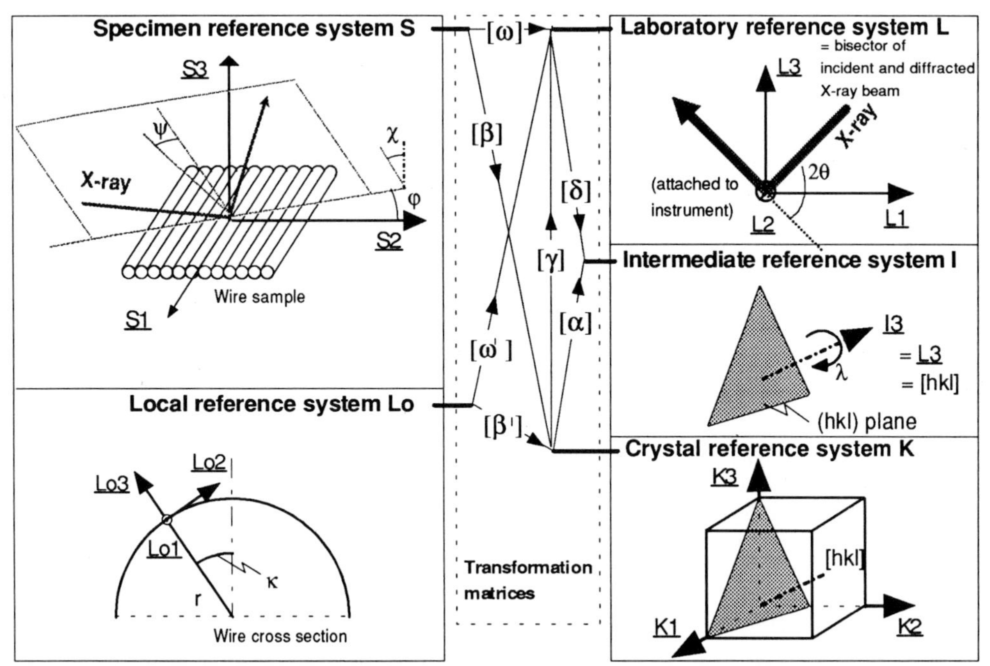

figure 3 : *Definition of used reference systems and the transformation matrices relating these reference systems.*

now related to the stresses $\sigma^{Lo}{}_{ij}$ in the local reference system *Lo* through the elastic compliance's A_{ij} :

$$\varepsilon^{L}_{33} = A_{ij}(\varphi,\psi)\ \sigma^{Lo}_{ij} \tag{2}$$

The total strain ε denoted as $\varepsilon^{L}{}_{33}$ in the laboratory reference system *L* (figure 3) is measured in a sample position (φ,ψ) in a ω-goniometer and is related to the stresses $\sigma^{Lo}{}_{ij}$ through the elastic compliances A_{ij} (equation 3) :

$$\varepsilon^{L}_{33} = A_{ij}(\varphi,\psi)\ \sigma^{Lo}_{ij} \tag{3}$$

The elastic compliances A_{ij} are calculated from the compliances A'_{ij} which give the relationship between contribution $d\varepsilon^{L}{}_{33}$ to the measured strain and $\sigma^{Lo}{}_{ij}$:

$$d\varepsilon^{L}_{33} = A'_{ij}(\varphi,\psi,\kappa)\ d\sigma^{Lo}_{ij} \tag{4}$$

We get after integration over all possible values for κ :

$$A_{ij}(\varphi,\psi) = \frac{1}{I_{\text{tot}}} \int_{\kappa} A'_{ij}(\varphi,\psi,\kappa)\, I(\kappa)\ d\kappa \qquad \text{with } I_{\text{tot}} = \int_{\kappa} I(\kappa)\ d\kappa \tag{5}$$

For isotropic wires equation 5 can be written as follows (The Einstein summation convention is used):

$$\varepsilon^{L}_{33} = r_{33ij} \left(\frac{1}{I_{\text{tot}}} \int_{\kappa} \omega'_{im}\omega'_{jn}\, I(\kappa)\ d\kappa \right) \sigma^{Lo}_{mn} \tag{6}$$

in which r_{33ij} are X-ray elastic constants which depend on the model used : Voigt, Reuss, Kröner, ...[7] The measured strains ε or the measured lattice spacing d_{hkl} can be plotted in a graph as function of $\sin^2\psi$ as is done for classical X-ray calculations.

Other assumptions made for texture measurements have to be implemented here too. However, other rules govern the defocusing correction. The stress goniometer is used in ω-mode and not in χ-mode as for texture goniometers. So, a variable shift Δ at the detector is now mainly caused by the horizontal divergence of the beam at the X-ray source. The question here is not whether or not a diffracted beam is seen by the detector because a whole diffraction profile is scanned during stress measurements. However, the defocusing will cause a supplementary peak shift. It is important to model it as strain is directly related with the exact peak position which acts as an inherent strain gage.

4. TEXTURE CORRECTION FOR RESIDUAL STRESS DETERMINATION.

Consider first the model of Reuss which states that all grains are subjected to the same stress tensor. For flat textured samples can be written [6] :

$$\varepsilon_{33}^{L} = r_{33ij}^{R}\,\omega_{ik}\omega_{jl}\,\sigma_{kl}^{S} = \left(\frac{1}{C}\int_0^{2\pi}\gamma_{3o}\gamma_{3p}\gamma_{iq}\gamma_{jr}\,s_{opqr}^{K}\,f(g(\lambda))\,\mathrm{d}\lambda\right)\omega_{ik}\omega_{jl}\,\sigma_{kl}^{S} \qquad \text{with } C = \int_0^{2\pi} f(g(\lambda))\,\mathrm{d}\lambda \tag{7}$$

The $\omega_{ik}, \gamma_{3o}, \ldots$ are components of the transformation matrices defined in figure 3. s^{K}_{opqr} are the compliancies defined in the crystal reference system K. The path of the integral is taken for all grains for which the normal on the diffracting plane is parallel to the $\underline{L}_3$ axis of the laboratory reference system (i.e. the bisector between incident and diffracted beam). $g(\lambda)$ represents the orientation of these grains and the Euler angles of such an orientation can be extracted of $[\beta] = [\gamma]^{-1}\,[\omega]$ (see figure 3), $f(g(\lambda))$ then is the value of the ODF for orientation $g(\lambda)$.

The real wire surface texture may be a fibre or a cyclic texture. In a cyclic texture is the orientation parameter of the diffraction planes which is free in a fibre texture is not free any more. The intensity of the diffraction planes has a cyclic distribution along the orientation path of the fibre fo the original fibre texture. if the real wire surface texture is known, then the stress state can be calculated with the following equation :

$$\varepsilon_{33}^{L} = \left(\frac{1}{I_{\mathrm{tot}}}\int_{\kappa}\left(\frac{1}{C}\int_0^{2\pi}\gamma_{3o}\gamma_{3p}\gamma_{iq}\gamma_{jr}\,s_{opqr}^{K}\,f(g(\lambda))\,\mathrm{d}\lambda\right)\omega'_{ik}\omega'_{jl}\,I(\kappa)\,\mathrm{d}\kappa\right)\sigma_{kl}^{Lo} \tag{8}$$

The Euler angles of the orientations on the integration path in $g(\lambda)$ are now found in the matrix $[\beta'] = [\gamma]^{-1}\,[\omega']$ (see figure 2). So, the integral in λ need to be recalculated in each point of the curved surface. If no cyclic texture is found and there exist a purely fibre texture with the fibre in the direction of the wire axis, then equation can be simplified as follows :

$$\varepsilon_{33}^{L} = \left(\frac{1}{C}\int_0^{2\pi}\gamma_{3o}\gamma_{3p}\gamma_{iq}\gamma_{jr}\,s_{opqr}^{K}\,f(g(\lambda))\,\mathrm{d}\lambda\right)\left(\frac{1}{I_{\mathrm{tot}}}\int_{\kappa}\omega'_{ik}\omega'_{jl}\,I(\kappa)\,\mathrm{d}\kappa\right)\sigma_{kl}^{Lo} \tag{9}$$

The Voigt case is much simpler. By algorithms given in [6] the Voigt constants r^{V}_{ijkl} can be found and need not to be recalculated in each point of the surface. The following equation is obtained :

$$\varepsilon_{33}^{L} = \omega_{3i}\omega_{3j}\,r_{ijkl}^{V}\,\sigma_{kl}^{S} \tag{10}$$

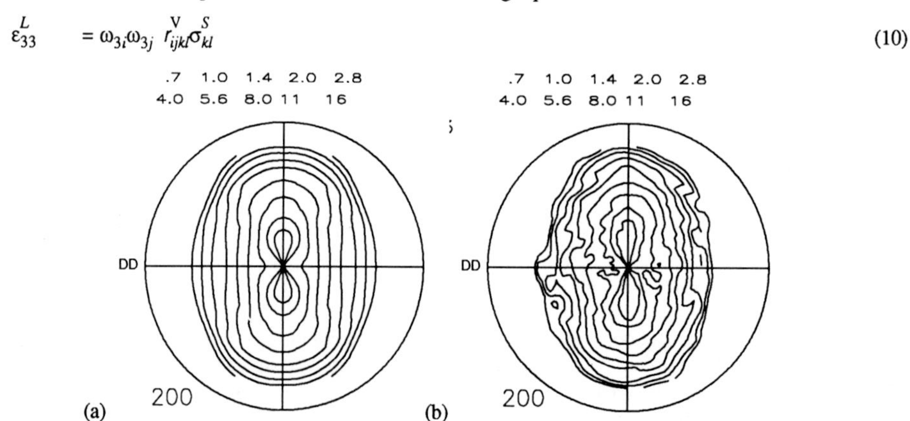

Figure 4 : *(a) Simulated (200) pole figure of a wire with random texture (Ø2mm) (CuKα radiation) and (b) measured (200) pole figures of a powder sample with shape of a wire with Ø2mm (no corrections - CuKα radiation)*

5. EXPERIMENTAL RESULTS.

In figure 4 the simulated pole figure is given for a wire of 1mm diameter. A simulated pole figure of a sample with a random texture can be calculated for any wire diameter and is used in experiments as powder correction. An iron powder sample has been slightly pressed in a stamp to give it the shape of a wire of 1mm diameter. The total surface covered by the sample was 15 by 15 mm. All texture measurements were carried out on a Siemens D500 goniometer with CuKα radiation. If the experimental pole figures of this "wire" powder are compared with the simulated ones then an excellent agreement is seen (figure 4).

Stress measurements were all carried out on a Seifert MZIV goniometer operating in ω-mode. A CrKα X-ray source was mounted on the goniometer. The diffraction peak used to measure residual strains was the (2 1 1) peak. The measurements were performed as on a flat sample. d spacing were measured for $\sin^2\psi$ values varying from 0 to 0.5 in steps of 0.1 for positive and negative ψ angles and for $\varphi = 0°$, 45° and 90°. The stress measurement on the powder sample was only intended to verify if the computed corrections for the wire geometry leave the stress values of an unstressed sample nearly zero. All stress results are given in Table I.

Next, measurements were done on 0.045% C steel wires. The wires were drawn to a diameter of 1.65 mm (total strain 1.0, die angle 3°). Because stresses are low (see Table I) and the texture was not very sharp (maximum of the ODF = 5.3) the texture corrections do not ameliorate the stress values. An annealing procedure (420°C during 2 sec) is successful to lower the residual stress at the surface of the wire.

Finally, measurements were done on severely cold-drawn steel wire with a final diameter of 0.25mm (total strain 3.6, die angle 4.5°). The measured pole figures and the recalculated pole figures after correction for the wire geometry are shown in figure 5. The maximum of the ODF was 9.4. The ODF was calculated enforcing monoclinic sample symmetry with the TD as dyad axis. The pole figures show some cyclic characteristics (lack of rotational symmetry) and look like the simulated texture of severely cold-drawn wires [7]. However, a slight spread of the maxima is still present. The use of this texture for correction in stress measurements gives good results. This is shown by the d-$\sin^2\psi$ curves of figure 6 and in table I. The monoclinic specimen symmetry was not used for stress. Indeed, the stress state which should be monoclinic immediately after drawing can be changed completely by the production steps after drawing : e.g. the straightening, the coiling. There only is a slight difference in stress values corrected with a pure fibre texture method (see equation (8)) or corrected with a cyclic texture method (see equation (9)) because the texture is not much deviating from a fibre texture. Voigt stress values have to be the same in both approaches. An additional surface treatment intended to give higher compressive hoop stresses seems to be successful. This sample as indicated in Table I as treated.

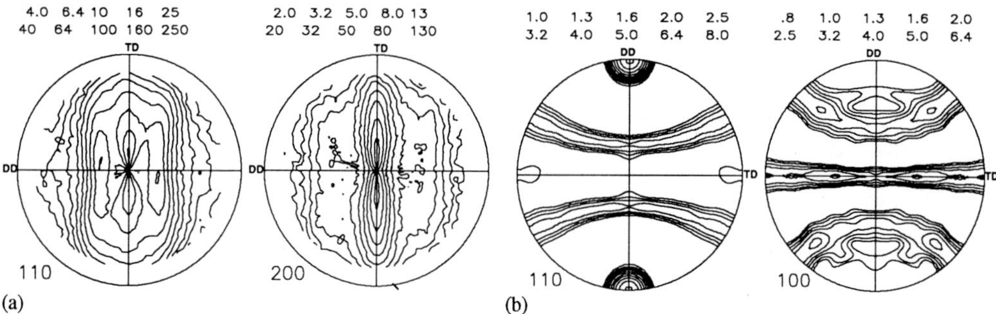

Figure 5 : *(a) Measured pole figures of severely cold-drawn steel wire (Ø0.25mm) and (b) recalculated pole figures of the same wires after correction for wire geometry and ODF calculation.*

6. CONCLUSION.

Screening, absorption and defocusing corrections have been implemented in methods for processing texture and residual stress measurements on wires. Concerning texture, the model is able to simulate the pole figures for wires of any diameter with a random texture. An excellent agreement with measured pole figures of a powder ample is found. The texture correction is implemented on a severely cold-drawn high carbon steel wire. The texture is mainly a fibre texture with some cyclic characteristics. As for the stress measurements, the stresses for low carbon steel wire drawn to a total strain of 1.0 are low and reduce to nearly zero for annealed wires. The stress state in severely cold-drawn high carbon steel wires could also be determined. The accuracy of these calculations is ameliorated when the corrected texture is taken into account. The success of this correction becomes clear in the $d_{(hkl)}$-$\sin^2\psi$ curves.

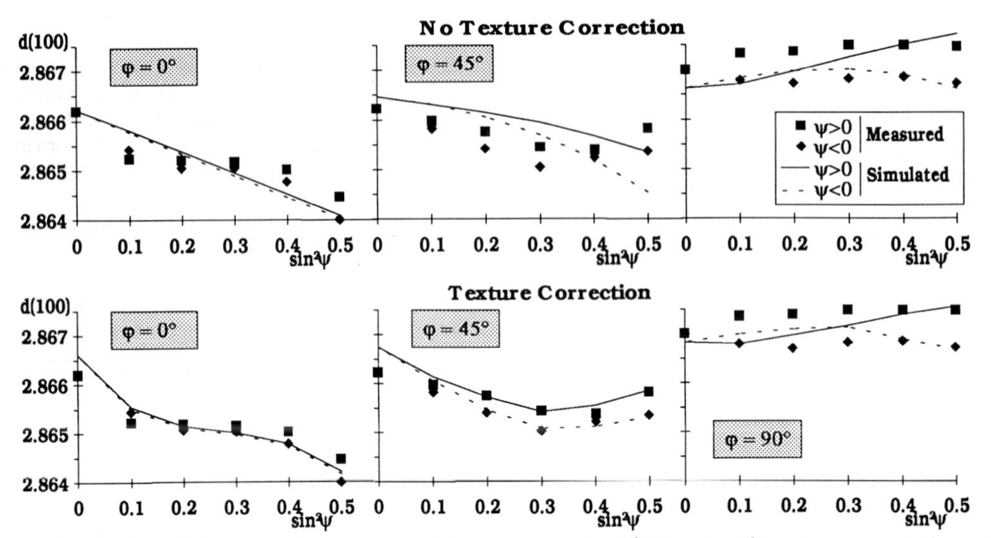

Figure 6 : *(a) d-sin²ψ curves for severely cold-drawn steel wire (Ø0.25mm) without texture correction and (b) d-sin²ψ for the same wire with (cyclic) texture corrections.*

Table I : *Stress values for different wire samples. Δσ is the standard deviation on σ.*

			σ_{ZZ} (Mpa)	$\Delta\sigma_{ZZ}$ (Mpa)	σ_{KK} (Mpa)	$\Delta\sigma_{KK}$ (Mpa)	σ_{ZK} (Mpa)	$\Delta\sigma_{ZK}$ (Mpa)	σ_{ZT} (Mpa)	$\Delta\sigma_{ZT}$ (Mpa)	σ_{KT} (Mpa)	$\Delta\sigma_{KT}$ (Mpa)
Powder Ø2mm	no texture correction	Reuss	-8	-6	-11	-22	20	-14	10	1	9	4
		Voigt	-8	7	-11	23	20	14	11	1	9	4
Steel Wire Ø1.65mm (eps = 1, die 3°)	no texture correction	Reuss	54	18	-44	30	-28	24	6	2	22	6
		Voigt	56	19	-45	31	-29	25	6	2	23	6
Steel Wire Ø1.65mm (eps = 1, die 3°)	(fibre) texture correction	Reuss	39	9	-73	30	-76	25	5	2	24	7
		Voigt	59	21	-42	46	-58	50	13	5	45	12
Steel Wire Ø1.65mm (eps = 1, die 3°),annealed	no texture correction	Reuss	8	3	29	11	32	8	6	1	11	2
		Voigt	8	3	30	11	33	8	6	1	11	2
Steel Wire Ø0.25mm (eps = 3.6, die 4.5°)	no texture correction	Reuss	-339	31	-282	93	-204	132	18	9	80	20
		Voigt	-353	32	-293	97	-212	138	19	9	83	21
Steel Wire Ø0.25mm (eps = 3.6, die 4.5°)	(fibre) texture correction	Reuss	-307	19	-23	63	-164	66	11	5	76	15
		Voigt	-606	64	-539	139	-410	266	37	18	166	42
Steel Wire Ø0.25mm (eps = 3.6, die 4.5°)	(cyclic) texture correction	Reuss	-334	24	-216	81	-37	64	25	9	53	10
		Voigt	-605	64	-538	139	-410	266	37	18	165	42
Steel Wire Ø0.25mm (eps = 3.6, die 4.5°), treated	(cyclic) texture correction	Reuss	-425	15	-337	47	-106	38	3	6	50	6
		Voigt	-887	46	-919	97	-873	211	5	13	175	28

ACKNOWLEDGEMENTS.
One of the authors is indebted to the National Fund for Scientific Research (NFWO,Belgium) for providing a research fellowship. NV Bekaert is gratefully acknowledged for the ready disposal of material.

REFERENCES.

1) Gangli, P., Szpunar, J., Sugondo : Textures & Microstructures, 1991, 13, 243
2) Montesin, T., Heizmann, J. : Textures & Microstructures, 1991, 14-18, 573
3) Willemse, P., Naughton, B., Verbraak, C. : Mat. Sci. Eng., 1982, 56, 25
4) Francois, M., Sprauel, J., Lebrun, J. : Proc. of 3rd Int. Conf. on Residual Stresses (ICRS3), 1991, 933
5) Francois, M., Spraule, J., Lebrun, J., Chalant, G. : Textures & Microstructures, 1991, 14-18, 175
6) Van Houtte, P., De Buyser, L., Acta Met., 1993, 41 (2), 323
7) Van Houtte, P., Watté, P., Aernoudt, E., Gil Sevillano, J., Lefever, I., Van Raemdonck, W., *Taylor simulation of Cyclic Textures at the Surface of Drawn Wires Using a Simple Flow Field Model*, same proceedings.

INDEXES

AUTHOR INDEX

Abdelaoui, A. 611
Adams, B.L. 31, 287, 1243, 1731
Aernoudt, E. 1689, 1783, 1881, 2075
Akdeniz, M.V. 1351
Akdut, N. 865
Anderson, A. 295
Ardakani, M.G. 919

Bacroix, B. 301, 617
Baczmanski, A. 213, 2051
Barthel, M. 1131
Bate, P. 997, 1271
Baudin, T. ... 459, 1027, 1237, 1739, 1847
Beaudoin, A.J. 1695, 1703
Beaujean, I. 1671
Béchade, J.L. 617
Becker, R. 627, 1277
Benay, O. 1357
Benhaddad, S. 1537
Bentdal, J. 143
Benum, S. 913
Bermig, G. 97, 309
Berveiller, M. 1821
Bessières, J. 227, 1371, 1487
Bethke, K. 1137
Bingert, J. 1653
Birsan, M. 1543
Blandford, P. 103, 207
Bleck, W. 2011
Blicharski, M. 627
Bobo, J.F. 227, 1487
Böcker, A. 1411
Böcker, W. 1551
Boeslau, J. 501
Bolingbroke, R.K ... 295, 745, 1145, 1991
Bolmaro, R.E. 1809
Bonarski, J.T. 111
Borchardt, G. 1161
Boulanger, C. 1371
Bouzy, B. 1671, 1861
Bowen, A.W. 315, 919, 1305
Bowman, K.J. 43, 1449, 1577
Brandão, L.P.M. 1365
Brodesser, S. 1153
Brokmeier, H.-G. 59, 119, 251, 515, 633, 685, 733, 821, 1507, 1971, 2031
Brückner, G. 1033, 1153
Brünger, E. 1283
Bucharova, T.I. 323
Bunge, H.J. 13, 71, 111, 167, 251, 333, 481, 507, 563, 685, 767, 827, 1411, 1423, 1551, 1663, 1783, 1791, 1803, 2067
Bunsch, A. 1045

Cai, M.J. 327
Canova, G.R. 645
Ceccaldi, D. 1739
Chang, S.K. 571
Chaoui, M. 235
Chaouni, H. 1371
Chastel, Y.B. 1747
Chateigner, D. 1379
Chauveau, Th. 301
Chavooshi, A. 939
Chen, S.R. 1797
Cheneau-Späth, N. 639
Cheng-Zhang, L. 1387, 1941
Choi, C.S. 1653
Chrusciel, K. 1045
Chung, C.Y. 1765
Chung, J.H. 1947
Ciosmak, D. 521
Clarke, A.P. 1953
Claus, J. 1161
Conradsen, K. 149
Czubayko, U. 125, 181

Da Costa Viana, C.S. ... 1365, 1961, 1967
Daaland, O. 143, 1087
Dahlem-Klein, E. 333
Dahms, M. 341, 507
Dahoun, A. 645
Daniel, D. 1855
Das, S. 555
Dawson, P.R. 1703, 1747, 1797
De Avila, N.V.V. 1961
De Wever, P. 1901
Dendievel, R. 665
Dierking, I. 1393
Dietz, P. 1559
Dingley, D.J. 31

Diot, M. 1671
Divinskii, S.V. 1565, 1753
Dnieprenko, V.N. 653, 1565, 1753
Dörner, B. 927
Doherty, R.D. 1021, 1277, 2025
Dornbusch, J. 1481
Driver, J.H. 585, 639, 807, 1257
Dubke, M. 1039
Dudzinski, D. 1875
Duggan, B.J. 659, 801, 853, 1167, 1513, 1759, 1765
Dunst, D. 665

Ekström, H.-E. 1271
Engler, O. 259, 673, 679, 913, 933, 939, 1109, 1501
Ermrich, M. 119
Ernst, J. 1411
Escher, C. 945
Eschner, T. 633
Esling, C. 493, 847, 1599, 1671, 1861
Etingof, P.I. 287

Faulkner, R.G. 1057, 1063
Fedosseev, A.I. 1771
Field, D.P. 1175, 1181
Fischer-Bühner, J. 1189
Fritsche, L. 1609
Frommeyer, G. 813
Fukutomi, H. 541
Fundenberger, J.J. 349
Furu, T. 1197, 1991

Gangli, P. 953
Gargano, P. 301
Germi, P. 1379
Gerritsen, E. 1299
Gertel-Kloos, H. 515, 685
Gerth, D. 1131, 1205, 1319, 1571
Gervasyeva, I.V. 959
Gieleßen, H. 1559
Gießelmann, F. 1393
Gil Sevillano, J. 1881, 1689
Girlich, I. 357
Göbel, H. 159
Gokhman, A.R. 2059
Gomo, Ø. 137
Gottstein, G. 125, 181, 259, 407, 673, 709, 841, 865, 933, 939, 945, 1109, 1153, 1189, 1283, 1771
Grah, M.D. 1577
Gray III, G.T. 1695
Greene, G.W. 801
Gregory, J.K. 1971
Grewen, J. 3
Großterlinden, R. 71, 2011
Gryziecki, J. 577
Gubernatorov, V.V. 959
Guillen, R. 617
G'sell, C. 645

Haase, A. 71
Haasen, P. 887, 927, 971
Hänel, W. 181
Haessner, F. 1069, 1291
Hall, F.R. 1901
Halley-Demoulin, I. 521
Han, K.Y. 221
Handreg, I. 1405
Hangen, U. 709
Hansen, N. 693, 1211
Hanssen, L. 1197
Harase, J. 899, 1081
Hartig, Ch. 813
Hartley, P. 1901, 1909
Harty, B.D. 1095
Hatakeyama, T. 1585
Hedel, R. 2067
Heidelbach, F. 965, 1243, 1313
Heinicke, F. 1481
Heinitz, J. 131
Heinrich, M. 971
Heizmann, J.J. 227, 235, 459, 465, 521, 611, 701, 1371, 1487
Helming, K. 97, 363, 529, 633, 789, 1219
Hergesheimer, M. 847
Heringhaus, F. 709
Heubeck, C. 1423
Hirano, T. 1103
Hirsch, J. 673, 939, 1777, 1979
Hiwatashi, S. 1585
Hjelen, J. 137
Høier, R. 143
Hölscher, M. 1039, 1137
Hosoya, Y. 2005

Hu, H. 627
Huber, J. 1411
Hufenbach, W. 1593
Hughes, D.A. 693
Humbert, M. 1225, 1599, 1647
Humphreys, F.J. 919
Hutchinson, B. 997, 1271, 1917

Ibe, G. 369
Inagaki, H. 1985
Inakazu, N. 715, 721, 977
Inoue, H. 715, 721, 977
Inoue, K. 1985
Isakov, N.N. 131
Ishida, T. 727
Ishida, Y. 1417
Ishizuki, M. 541
Ito, K. 1417
Ivanova, T.M. 323

Jallon, M. 847
Jansen, E.M. 733, 739
Jasienski, Z. 1231, 1237
Jeong, H.-T. 1603
Jonas, J.J. 879, 1713, 1855
Jung, V. 535
Jura, J. 259, 407, 577, 933
Juul Jensen, D. 149, 693, 745, 1211, 1887, 1991

Kamijo, T. 541
Kaneno, Y. 715, 977
Karhausen, K. 673
Katoh, T. 1417
Kawasaki, K. 1081
Kern, H. 1523
Kestens, L. 1783
Kiewel, H. 1609
Kim, H.M. 753, 1997
Kim, I.S. 1623
Kim, J.H. 761, 983
Kimura, Y. 727
Klein, H. 333, 767, 1423, 1791
Klimanek, P. 275, 357, 1119, 1131, 1405, 2031, 2037
Knorr, D.B. 1327, 1435, 1443
Kocks, U.F. 1797
Köhler, U. 1783, 1803

Kohara, S. 1629
Kopineck, H.J. 1929
Krahn, W. 1411
Krieger Lassen, N.C. 149
Kruger, K.L. 1449
Kruse, R. 529
Kunze, K. 31, 739, 1243, 1277, 2025
Kusnierz, J. 1231, 1635

Lai, J.K.L. 853
Lallemant, M. 521
Laruelle, C. 521
Le Roux, S.D. 1455
Lebensohn, R.A. 783, 835, 1809
Lee, C.S. 659, 1167, 1759
Lee, D.N. 1603, 1947
Lee, H.-C. 761, 983
Lee, K.T. 989
Lee, W.B. 327, 1075, 1251
Lefever, I. 1689, 1881
Leffers, T. 1815
Lefort, A. 847
Leiss, B. 789
Lemoine, X. 1821
Lepper, M. 1593
Li, D.Y. 547, 555, 1827
Li, J. 795
Liang, Z.D. 375, 1839
Lim, E. 801
Lindh, E. 997
Lipinski, P. 2051
Litwora, A. 1231
Liu, W.P. 481
Liu, Y.S. 375, 1839
Lodder, J.C. 1387, 1941
Lohne, O. 1197
Lotter, U. 2011
Lu, J. 1003
Lucas, A.S. 1357
Lücke, K. 413, 571, 597, 679, 761, 841, 865, 983, 1033, 1039, 1137, 1469, 2043
Lutterotti, L. 473

Magri, P. 1371
Mao, W. 1009
Margetan, F.J. 221
Marshall, G.J. 1145

Mason, T.A. 1731
Mathur, K.K. 1703
Matthies, S. 473, 1641, 1647
Maurice, Cl. 807
Mecking, H. 665, 813
Messerschmidt, U. 1131
Michaluk, Ch. 1653
Mizera, J. 1257
Mizui, N. 1015
Molinari, A. 645, 1869, 1875
Molodov, D.A. 125
Montesin, T. 611, 701
Morawiec, A. 1263
Moreau, B. 159
Morita, M. 2005
Moustahfid, H. 1225
Mücklich, A. 275
Mülders, B. 841
Muller, D. 1821
Muller, J. 493, 1599

Nakayama, T. 1081
Neale, K.W. 873
Necker, C.T. 1021
Nes, E. 143, 913, 1087, 1197, 1501, 1895
Nicol, B. 1271
Niederschlag, E. 821
Nikitin, A.N. 131
Nikolayev, D.I. 323, 381, 387, 393
Ning, H. 1513
Novikov, V.Yu. 905, 1063

Obadia, S. 1357
Obata, K. 541
Ochiai, T. 1103
Oertel, C. 309
Oh, K.H. 1603
Oikawa, H. 1103
Oka, M. 727
Olson, T. 1731
Ortega, R. 71
Oscarsson, A. 1271, 1463

Paillard, P. 1027, 1847
Panchanadeeswaran, S. 1277, 2025
Park, N.J. ... 167, 333, 507, 563, 827, 1663
Park, Y.B. 571, 761, 983
Paul, H. 1231
Pawlik, K. 401, 577, 1231
Pelletier, J.B. 611
Penelle, R. ... 375, 459, 1027, 1237, 1537, 1739, 1847
Pernet, M. 1379
Petersen, B.-C. 125
Philippe, M.J. 645, 847, 1225, 1337, 1671, 1861
Piatkowski, A. 1231, 1237
Pillinger, I. 1901, 1909
Pithan, C. 679
Pochettino, A.A. 301, 783, 835
Ponge, D. 1189, 1283
Pospiech, J. 259, 407, 577, 933, 965, 1231

Quade, H. 1675, 1681
Quesne, C. 1537
Qvale, A.H. 137

Raabe, D. .. 413, 501, 571, 597, 709, 841, 1033, 1039, 1469, 1771
Rack, H.J. 1475
Ralph, B. 1075
Randle, V. 175
Ratuszek, W. 1045
Reck, G. 2067
Reher, F. 1039, 181
Reinert, T. 1551, 1675, 1681
Reip, C.-P. 2011
Rickert, T.J. 1979, 2017
Ricks, R.A. 1145
Rocaniere, C. 555
Rodbell, K.P. 1435
Rollett, A.D. 1021
Root, J.H. 1475
Rose, J.H. 221
Royer, F. 1739

Saetre, T.O. 1895
Sahlberg, U. 1463
Saimoto, S. 795, 1953
Sam, D.D. 287, 1731
Samajdar, I. 2025
Sánchez, P. 835
Sang, H. 795
Savoie, J. 879, 1855
Savyolova, T.I. 323, 387, 419

Schacht, J. 1393
Schaeben, H. 349, 423, 431, 453
Scheffzük, C. 309
Scherrer, S. 1161
Schmidt, D. 1675
Schneider, H. 767, 2031, 2037
Schneider, W. 535
Schouwenaars, R. 439
Schuman, C. 847
Schwarzer, R.A. 187, 189, 195, 201, 241, 247, 487, 1131, 1205, 1219, 1319, 1571
Seeger, J. 813
Sekine, K. 727
Senuma, T. 1051
Serghat, M. 1861
Shahani, R. 1991
Shek, C.H. 853
Shen, G.J. 853
Shi, H. 745, 1991
Shvindlerman, L.S. 125, 1057, 1063
Siegesmund, S. 529, 789
Siemes, H. 453, 733, 821, 1507
Skrotzki, W. 1481
Slangen, M.H.J. 1299
Smallman, R.E. 659, 1167, 1759
Sødahl, Ø. 1197
Staszewski, M. 859
Sturgess, C.E.N. 1901, 1909
Suenaga, K. 1417
Suga, Y. 1081
Sukhoparov, W.A. 131
Sun, Z. 1009
Sursaeva, V.G. 1057, 1063
Szpunar, J.A. 103, 207, 547, 555, 753, 953, 989, 1003, 1263, 1387, 1543, 1827, 1941, 1997
Sztwiertnia, K. 1069, 1291

Takahashi, N. 899, 1081
Takayama, Y. 1417
Tarasiuk, J. 213, 447
Thompson, R.B. 295, 221
Tizliouine, A. 227, 1487
To, S. 1075
Tobisch, J. 97, 309
Tomé, C.N. 783, 835
Tomov, I. 1495
Tóth, L.S. 1713, 1869, 1875
Traas, C. 453
Tracy, D.P. 1443
Troost, K.Z. 1299
Truszkowski, W. 577

Ubhi, H.S. 1305
Ullemeyer, K. 131, 1481
Urabe, T. 1713
Ushigami, Y. 899, 1081
Ushioda, K. 1585
Usuda, M. 1585

Vadon, A. 235, 459, 465, 701, 1357
Valot, C. 521
Van Acker, K. 2075
Van Bael, A. 1901, 1909
Van Houtte, P. 439, 1689, 1783, 1881, 1909, 1953, 2043, 2075
Van Raemdonck, W. 1689, 1881
Vandermeer, R.A. 1887
Van der Merwe, D.J. 1455
Vatne, H.E. 1087, 1501, 1895
Vial Edwards, C. 783
Vinel, G.W. 1641
Visser, R.F. 1095

Wagner, F. 159, 1225
Wagner, P. 865, 2043
Walther, K. 131, 381
Wang, C.Q. 827
Wang, F. 375, 1839
Wang, N. 1901, 1909
Wang, Y. 481
Watanabe, T. 1103
Watté, P. 1689, 1881
Wcislak, L. 111
Weber, K. 789
Weber, S. 1161
Wegria, J. 1671
Weiland, H. 1181
Wenk, H.-R. 473, 965, 1243
Wierzbanowski, K. 213, 447, 2051
Wilbrandt, P.-J. 887, 927, 971
Winter, W. 1507
Winters, J. 1901, 1909
Wolff, I.M. 1095
Wood, J.V. 1351

Wright, S.I. 31, 1313, 1695
Wu, L.F. 481

Xia, W. 487
Xiong, L. 1513
Xu, J.Z. 375, 1839

Yala, F. 1739
Yalda-Mooshabad, I. 221
Yang, P. 933, 1109
Yashnikov, V.P. 1057
Ye, D. 1513
Yoshida, Y. 721

Zaefferer, S. 189, 195, 241,
247, 1205, 1319
Zhang, L. 1513
Zhou, Y. 873, 879
Zink, U. 251
Zugenmaier, P. 1393
Zuo, L. 493, 159

SUBJECT INDEX

Absorption 275
- correction 381
Activation energy 125
AgCl 1559
Aggregate structure 13
Alloy
-, amorphous 1003
-, multiphase 665
Aluminium 125, 295, 515, 541, 555, 627, 653, 685, 971, 1057, 1063, 1087, 1095, 1181, 1197, 1211, 1501, 1887, 1991, 2025, 2043
- Alloy 745, 807, 913, 919, 939, 1075, 1145, 1271, 1435, 1443
- Al-Li 1305
- Al-Li-Zr 1257
- Al-Mg 1797
- Al-Mn 933, 1109
- Al-Ni 977
- Al-Si 1205, 1571
- Al-Ti 977
- foil 119
- killed steel 1015, 1947
- oxide 1411, 1507
- sheet 1629
Analysis
-, data 447
-, hyperspherical cluster 431
-, ODF . 363, 493, 529, 633, 1219, 1411
-, on-line 143
-, perturbation 1875
-, profile 473
-, substructure 1119
-, texture 97, 119, 143, 159, 181, 213, 235, 275, 309, 327, 357, 465, 733, 1379
-, x-ray 2037
Angular accuracy 315
Anisotropy 43, 1543, 1565, 1577, 1629, 1753, 1901, 1941, 1947
-, elastic 1585, 2059
-, magnetic 1681, 1929
-, mechanical 1593
-, plastic 939, 1635, 2059
-, seismic 1523
Annealing 439, 841, 853, 959, 1417, 1469, 1917, 2011
-, intermediate 715
- temperature 919, 977
- texture 1365, 1839
- twinning 887
Amorphous alloy 1003
Arbitrary defined cell (ADC) 413
Arithmetic mean 1647
Automation 149

Backscatter Kikuchi pattern (BKP) 31, 137, 149, 259, 315, 501, 795, 927, 933, 965, 997, 1027, 1087, 1109, 1137, 1153, 1161, 1197, 1211, 1277, 1283, 1299, 1305, 1313, 1861
Backscattered noise, ultrasonic 221
Beck-experiment 971
Bendability 1671
Bias 1487
Biaxial 1577
- stretching 873
Bicrystal 627, 865
Bilayer 1487
Bismuth telluride 1371
Boundary
-, coincidence 899
-, CSL (Coincidence Site Lattice) .. 989, 1263, 1997
-, grain 865, 971, 1103
-, low angle 899
Bragg-Brentano geometry 1487
Brass 357, 487, 783, 1947, 2031
Brittle 1577
Bulk texture 633

C60 (Fullerene) 1417
Calculation, numerical 393
Cathodic sputtering 1487
CCD camera 137, 187
Cell block 693
Cellulose 1463
Ceramics 43, 201, 1449, 1507, 1609
Chalcopyrite 733, 739
Channel-die compression 807, 1231, 1237

Channeling pattern 195, 1057, 1063
Chirality 1393
Circular texture 611, 701
Classification 149
Climb 1797
Coalescence 1327
Co-Cr films 1387, 1941
CODF (see ODF)
Cold
- drawn 827
- rolling 439, 617, 693, 853, 939, 977, 1051, 1211, 2005, 2011
Component
- method 309
-, texture 259, 327, 363, 493, 529, 1219, 1481, 1759
Composite 515, 685, 919, 1551
-, metal-matrix 1475
Compression 813, 1417
Compromise texture 1803, 1895
Constraints
-, full 1815
-, relaxed 585, 1815
Controlled rolling 1985
Compatibility 865
Coordinate transformation 369, 1423
Copper 653, 685, 965, 1283, 1291, 1371, 1443, 2031
- Cu-Al 653
- Cu-Mn 679, 927
- Cu-Nb 709
- Cu-Si 1045
- Cu-Zn-Al 167, 507, 563, 827, 1663
- coating 1405
-, phosphorus 1069, 1291
- wire 715
Corona 5 1537
Correction
-, absorption 97
-, defocalization 1379
-, ghost 341
Correlation 1641
- function, two-point 221
Corrosion
- fatigue 1971
-, intergranular 1997
Counting statistic 275
Cross rolling 1075, 1423, 1901
CRSS (Critical Resolved Shear Stress) 639
Crystal
-, cubic 665
-, hexagonal 639, 665
-, liquid 1393
-, monoclinic 507
-, orthorhombic 507
- plasticity 1855
-, single 541, 585, 1237
Crystallization 1003
CSL (Coincidence Site Lattice) 989, 1263, 1997
- boundaries 989, 1263, 1997
Cube/PSN competition 1501
Cube texture 541, 1021, 1087, 1145, 1765, 1803, 1991, 2025
Cubic
- crystal 665
- symmetry 369
Cyclic texture 611, 1881
Cylindrical texture 1299

Data analysis 447
Deep drawing 879, 1051, 1739, 1855, 1979
Deformation 131, 1671, 1875, 865
- banding 585, 1759, 1167, 801
-, channel die 627
-, compressive 721
-, cyclic 1791
-, experimental 733, 739, 821
-, finite 1809
-, hot 309, 1131, 1145, 1257
-, inhomogeneous 767
- pattern 693
-, plastic 1551, 1815
-, polycrystal 1869
-, rolling 501, 665, 1009
- structure 541
- system 241, 1319
-, slip 727
- twin 739
- zone 977
Deformed cube band 2025
Demagnetization 1681
Diamagnetic AMS 1675
Diametral compression test 1507

Differential ODF 439
Diffraction 213, 227, 2051
-, electron 187, 189, 195, 201, 241, 247, 1205, 1571
- multiplet 1119
-, neutron 59, 251, 473, 515, 529, 633, 685, 733, 821, 1119, 1475, 031
-, x-ray 111, 119, 159, 821, 933, 1119, 1495
Diffuse necking 1875
Diffusion bonding 1305
Dislocation structure 653, 1181, 1821, 1947
Domain structure 1543
Doping 1449
Drawing 611, 701, 715, 847, 1417, 1689, 2075
-, wire 701
Dual-phase steel 853
Dynamic
- observation 1069
- recrystallization 721, 1189, 1283, 1449, 1713, 1797

Earing 939, 1739, 1953, 1961, 1967, 2017, 2043
Elastic
- anisotropy 1523, 1585, 2059
- constants 1647, 2051
- modulus 1609, 2059
- property 1599
Elasticity 1731
Electrical resistivity 2005
Electrodeposition 1371, 1405, 1827
Electromigration 1435
Electron diffraction 187, 189, 195, 201, 241, 247, 1205, 1571
Energy dispersive detector 71, 309
Entropy, maximum 349
Epitaxial growth 1405
Epitaxy 535
-, aluminium-iron 227
-, granular 1327
Error
-, ghost 413
- parameters 301
-, truncation 341
Etch pit 753, 1997
Eutectoid pearlite 1689
EVOH (ethylene-vinylalcohol) 727
Experimental deformation 733, 739, 821
Extended nucleation 887
Extinction
-, microstructure induced 275
-, texture induced 275
Extrusion 813, 1197
-, warm plane strain 2025

Fatigue 1791
FCC metals 873, 879, 1167
Fe-Al-Cr alloy 1009
Ferroelectricity 1393
Ferromagnetic material 251
Fe-oligocrystal (s.Oligocrystal) 501
Fe-P alloy 1513
Fe-Si 953, 989, 1027, 1081, 1543, 1847
Fiber texture 535, 611, 685, 715, 1021, 1299, 1435, 1443, 2005
Film
-, Co-Cr 1387, 1941
-, food packaging 727
-, thin 97, 159, 207, 1205, 1379, 1487, 1571
Filtering 1641
Finite
- deformation 1809
- difference method 673
- element model (simulation) 627, 1277, 1585, 1695, 1703, 1901
- - , anisotropic 1909
- strain 1881
Flow field 1881
Food packaging film 727
Forging
-, hot 1449
-, isothermal 721
Formability 1629
Forming limit curve 783, 1629
Fourdimensional unit sphere 369
Fourier transform 149
Fractography 1635
Fracture toughness 1537, 1577
Fullerene (C60) 1417
Function *(continued on next page)*
-, two-point correlation 221

Function *(continued)*
-, Gauss 315, 387
-, indicator 423
-, Kullback-Leibler loss 431
-, spherical harmonic 423
-, standard 413

Gas discharge 201
Gauss
- distribution 323, 387
- function 315, 387
Generalized textural quantities 13
Geophysics 1523
Geometric mean 1647
Ghost errors 413
Goniometer, texture 207
Goss texture 989
Grain
- boundary 865, 971, 1103
- - character distribution 927
- - engineering 175
- - measurement 125
- - migration 125, 1153
- - mobility 1063
- - plane 175
- - special 1175
- - structure 1057
- columnar 1327
- growth 905, 953, 1027, 1057, 1063, 1205, 1411, 1571, 1847
- -, abnormal 899, 1847
-, internal 1277
-, sub 653, 739
- subdivision 1211, 1815
- shape 1675
- - effect 1869
- - fabric 789
- size 315, 801, 1765
- - statistic 275
Granular epitaxy 1327
Grazing incidence 71, 1487
Growth 1021, 1327
-, oriented 1495, 1803, 1895, 2025
- range 1765
- rate 149
- selection 887, 971, 1033, 1087, 1109, 1495
-, selective 571, 1713

Harmonic method 301, 341, 459
Hexagonal
- crystal 639, 665
- material 1337, 1747, 1875
Heating rate 919
Hematite 1681
High
- alloy steel 597, 2037
- pressure device 131
- temperature texture 167, 181, 1225
- - measurement 167, 181
- low cycle fatigue 1153
- resolution texture analysis 71
- T_c superconductor 1161, 1379
Hillocks 1435
Histogram 431
Historical survey 3
Hot
- deformation 309, 1131, 1145, 1257
- forging 1449
- rolling ... 673, 1051, 1095, 1771, 2011
- working 43
Hydrogen co-deposition 1827
Hyperspherical
- cluster analysis 431
- kernel density estimation 431
Hysteresis 251

ICOTOM-review 3
IF steel 103, 1855, 2005, 2011
Image
- processing 149
- analyzer 481
Imaging plate 119
Incidence
-, boundary 899
-, grazing 71, 1487
-, low 1487
-, reduced 159
Indexing 247
Indicator function 423
Ingot 555
Inhomogeneity 1469
Inhomogeneous
- rolling 659
- texture 1423, 1695
Injection moulding 1593

In-situ
- composite 709
- observation 945
- recrystallization 841
Insulators 201
Interconnects 1435
Interface 1305
Interfacial energy 1327
Intergranular
- damage 1175
- fracture 1103
Intermediate annealing 715
Intermetallic phase 515, 721, 945, 1103
Internal stresses 1387
Iron 2037
- quadrangle 1681
Iridium 535
Irreducible spherical basis 1599
Isomorphism 419
Isothermal forging 721
Itacolomite 1675

Jackknife method 431

Kiessig fringes 227
Kikuchi pattern 1131, 1205, 1571
-, backscatter 31, 1277, 1299
-, transmission 143
Kinematics 789
Kullback-Leibler loss function 431
κ-geometry 71

Lankford parameter (see r-value)
Large-scale mathematical programming 423
$La_{2-x}Sr_xCuO_4$ 1161
Lattice
-, coincident site (CSL) 1087, 1175
- image 1513
- rotation 1815
Limited fiber components 1565, 1753
Linear regression 447, 2043
Liquid
- crystal 1393
- crystal polymer (LCP) 1593
Local
- mobility variation 1895
- texture 767, 1069, 1131, 1205, 1257, 1291, 1423, 1571
- orientation 1211
Localized neck 1251, 1635
Location sensitive detector 71
Low
- angle boundary 899
- carbon steel 439, 997, 1051, 2005
- incidence 1487
- - geometry 227

Machining 1251
Macrotexture 259, 709
Magnetic
- anisotropy 1681, 1929
- domain 251
- property 1941
- material, soft - - 1543
Martensitic transformation 167, 563
Material
-, ferromagnetic 251
-, hexagonal 1337, 1747, 1875
-, multiphase 363, 529
Mathematical modelling 873, 879
Mean
- value 1599
- square difference 439
Measurement
-, individual orientation 259
-, on-line 103
Mechanical
- anisotropy 1593
- property 715, 1629, 1731
Melt Spinning 1351, 927
Mesophase 1393
Metal forming 1703
Method
-, component 309
-, finite difference 673
-, harmonic 301, 341, 459
-, jackknife 431
-, Monte Carlo 375, 1847
-, non-destructive 213
-, positivity 341, 507
-, Rietveld 473
-, Schulz 97
-, vector 301, 459
Mica 1675
Micro *(continued on next page)*
- band 1291

Micro *(continued)*
- diffraction 31, 1695
- growth selection 1087
- texture 31, 259, 315, 407, 709, 1109, 1153, 1299, 1305, 1313

Microfibril angle 1463
Micromechanical model 1821
Microscopy
-, scanning electron (SEM) 187, 195, 201, 1299
-, transmission electron (TEM) 189, 201, 241, 571, 945, 1131, 1205, 1219
-, orientation imaging (OIM) 739, 2025

Microstructural evolution 1887
Microstructure 287, 501, 597, 627, 665, 693, 715, 745, 1405, 1537, 1731, 1821, 2005, 2011
- simulation 375, 1839

Minerals 1747
Minimal pole figure range 1219
Misorientation 953, 1271
- correlation index (MCI) 1137
- distribution function (MODF) 259, 965, 1153, 1189, 1243, 1263, 1271

Mobility 125
Model 1167
-, cup height 1953
- data set 315
-, equilibrium-based 1747
-, finite element 627, 1277, 1585, 1695, 1703, 1901
-, mathematical 873, 879
-, micromechanical 1821
-, ODF 413
-, recrystallization 1783
-, structure zone 1327
-, Taylor 617, 847, 1277, 1689, 1777, 1875
- texture 2067
-, viscoplastic self-consistent 645

Modeling 1337, 1803, 1887
-, constitutive 1181
-, texture 585, 673, 953, 1559, 1713

Modulus
-, elastic 1609, 2059
-, Young's 1551, 1663

Monoclinic crystal 507
Monte Carlo method 375, 1847
Moving average 453
Multi-peak pole figure 309
Multiphase material 363, 529
Multilayer 227

Necklace structure 1283
Neutron 131
- diffraction 59, 251, 381, 473, 515, 529, 633, 685, 733, 821, 1119, 1475, 2031
- scattering 357

Ni 515, 653, 693
Ni_3Al 1283, 1103
Noise, ulstrasonic backscattered 221
Non-destructive method 213
Nonlinear elastic theory 1551
Nuclear fuel cladding 1455
Nucleation 945, 1021, 1189, 1765
-, extended 363
-, oriented 887, 1109, 1495, 1803, 2025
-, particle stimulated 977, 1109

Numerical calculation 393

OCF (Orientation Correlation Function) 259
ODDF (Orientation Difference Distribution Function) 259, 1137
ODF (Orientation Distribution Function) 13, 221, 315, 323, 327, 387, 721, 1263, 1271, 1469, 1565, 1753, 2005, 2067
- analysis 363, 493, 529, 633, 1219, 1411
- calculation 401, 473, 487, 1299, 1641
-, differential 439
-, model 413
- prediction 1953
- representation 481
-, true 375

OIM (Orientation Imaging Microscopy) 739, 2025
Oligocrystal 501
On-line
- analysis 143, 71
- texture measurement 103, 1929

Ore, hematite 1681
Orientation 1271, 1291
- chain 645

- change on deformation 1277
- connectivity 175
- coordinate 369
-, crystallite 727
- -, single 1225
- distance 369
- distribution 369
- distribution function (see ODF)
- flow field 1791
- gradient 1257
- grain 801
- imaging 1243
-, - microscopy (OIM) 739, 2025
-, individual (see single)
-, lattice 585, 639
-, local 1211
-, micro 1291
-, preferred (see texture)
- relationship ... 167, 507, 563, 571, 971
-, single 431, 1641
- space 369
-, stable 761
- subspace 369
Oriented
- growth 1495, 1803, 1895, 2025
- nucleation 887, 1109, 1495, 1803, 2025
Orthorhombic crystal 507
Overlapping 235, 465
Oxide texture 521

Pancake grain structure 1015
Parallel
- beam geometry 97
- computing 1703
Particle
- parameter 977
- stimulated nucleation 977, 1109
Pass
- rolling 1131, 2031, 2037
- compressing 2037
Patenting structure 1689
Pattern
-, electron back-scattering (EBSP) (s. Backscatter Kikuchi Pattern)
-, channeling 195, 1057, 1063
-, Kikuchi
-, - backscatter (BKP) 31, 137, 149, 259, 315, 501, 795, 927, 933, 965, 997, 1027, 1087, 1109, 1137, 1153, 1161, 1197, 1211, 1277, 1283, 1299, 1305, 1313, 1861
-, - transmission 143, 187, 189, 201, 241, 247, 1131, 1205, 1319, 1571
- resolution 137
-, spot 189, 247
Pb-Al composites 633
Pearlite 1881
Permeability 1543
Perturbation analysis 1875
Phase
-, intermetallic 515, 721, 945, 1103
-, second 977
- separation 309
- transformation 167, 563, 1225
Photoelasticity 1559
Photoplasticity 1559
Physical property 333
Planar isotropy 1623
Plane strain compression 745, 1991
Plastic
- anisotropy 939, 1635, 2059
- deformation 859, 1815
- strain 585
- strain ratio 1623
Plasticity 1703, 1747, 1809
Platinum 535
Pole figure 251, 381, 401, 453, 627, 733, 1027, 1371, 1487, 1947
-, diffraction 357
- gap 97
-, inverse 795
- measurement 59, 181
-, multi-peak 309
Polycrystal 2051
Polymer 645
-, high barrier 727
-, liquid crystal (LCP) 1593
Polyoxymethylene 645
Polyphase carbonate 789
Positivity method 341, 507
Precipitation 1033, 1501, 2005

Preferred orientation (see texture)
Primary recrystallization 899, 959
Process optimization 1671
Profile
- analysis 473
- rolling 767
Program, mathematical 349
Property
-, elastic 1599
-, magnetic 1941
-, mechanical 715, 1629, 1731
-, physical 333
-, tensorial 1641
-, tensile 1357
Pyrrhotite 821

Quartz 1243, 1675
- mylonite 1481

Rapid solidification 1351
Rate sensitivity 1797
Recovery 715, 801, 1131, 1551
Recrystallization 149, 685, 715, 939, 945, 971, 997, 1021, 1045, 1051, 1069, 1075, 1095, 1109, 1131, 1145, 1319, 1337, 1671, 1803, 2011
-, continous and discontinous 1271
-, dynamic 721, 1189, 1283, 1449, 1713, 1797
- in-situ 841
- model 1783
-, primary 899, 959
-, secondary 899, 905, 953, 959, 989, 1081
-, static 1713
- texture 887, 919, 953, 977, 983, 1009, 1015, 1501, 1713, 1783, 1887, 1991, 2005
Reduced incidence 159
Reflection separation 2067
Refractory metal 597
Relationship, orientation 167, 507, 563, 571, 971
Reliability 1435
Representation theory 287
Residual stress .. 1551, 2051, 2059, 2075
Resolution
-, pattern 137
-, spatial 137
Retarding effect 571
Review (ICOTOM) 3
Ridging 753, 1039, 1137
Rietveld method 473
Robot sample changer 71
Rock 59
Rodrigues vector 431, 1137, 1153
Rolling 309, 795, 835, 841, 1231, 1251, 1417, 1469, 1739, 1861
-, cold 439, 617, 693, 853, 939, 977, 1051, 1211, 2005, 2011
-, controlled 1985
-, cross 1075, 1423, 1901
- gap 859
-, hot 673, 1051, 1095, 1771, 2011
-, inhomogeneous 659
-, profile 767
- reduction 919
Roping 753
Rotation 369, 453
- flow field 1861
- vector 369
r-value 439, 753, 1039, 1653, 1909, 2005, 2011

Sapphire 535
Sample
- geometry 1455
- preparation 201
Scanning electron microscopy (SEM) 187, 195, 201, 1299
Scatter width 315
Schmid's
- factor 721
- law 1603
Schulz method 97
Secant approach 1869
Secondary
- ion mass spectrometry (SIMS) ... 1161
- recrystallization 899, 905, 953, 959, 989, 1081
Segregation of phosphorus 1513
Seismic anisotropy 1523
Selected area electron diffraction 487
Selective growth 571, 887, 971, 1033, 1087, 1109, 1495, 1713

Self-consistent simulation 665, 783, 1809, 1821, 1869
Series expansion 341, 393
Shape memory alloy 167, 235, 563, 827, 1663
Sharp textures 387
Shear 645
- angle 1251
- band 487, 1075, 1219, 1231, 1291
- - fracture 1251
- texture 659, 795, 913
- wave splitting 1523
Sheet metal forming 1251
Short range order 679
Silicon carbide 919, 1475
Silicon steel 989
Simulation 905, 1565, 1753, 1771, 1847
-, microtructure 375, 1839
-, self-consistent 665, 783, 1809, 1821, 1869
-, texture 1777, 1839, 1979
Single
- crystal 541, 585, 1237
- orientation 431, 1641
- - determination 375, 407, 1283
Sintering 43, 1411
Slip 835
-, non-octahedral 807, 1797
-, rate sensitive 1869, 1875
- system 585, 639, 1821
Smectic C* 1393
Smoothing 453, 1641
Soft magnetic material 1543
Solidification 547
- texture 1351
Solid state bonding 1305
Spherical harmonic function 423
Spline 453
Split sample 1277
Spot pattern 189, 247
Stability 1167, 1759
Stable orientation 761
Stainless steel 753, 853, 1033, 1039, 1137, 1469, 1771, 2017
Standard function 413
Stacking fault energy 577, 659, 679, 1045, 1759
Static recrystallization 1713
Statistic 287
Statistical parameter 13, 407
Steel 611, 847, 959, 1263, 1585, 1739, 1881, 1917
-, aluminium killed 1015, 1947
-, dual-phase 853
-, high alloy 597, 2037
-, low alloy 597
-, low carbon 439, 997, 1051, 2005
-, IF 103, 2005, 2011, 1855
-, silicon 989
- sheet 439, 1783
-, stainless 753, 853, 1033, 1039, 1137, 1469, 1771, 2017
Stereographic projection 369
Stiffness 221, 1585
Stochastic process 1137
Strain 1205, 1571
-, inhomogeneous 1277
- localization 1251
- path change 1947
-, plastic 585
- rate sensitivity 673, 807
Stratification 521
Strength 1507
Stress 1205, 1571
- corrosion cracking 1971
-, residual 1551, 2051, 2059, 2075
- voiding 1435
Stretching, biaxial 873
Strip 709
- casting 1039, 1271
Structural inhomogeneity 357
Structure
-, Aggregate 13
-, deformation 541
-, dislocation 653, 1181, 1821, 1947
-, grain boundary 1057
-, micro 287, 501, 597, 627, 665, 693, 715, 745, 1405, 1537, 1731, 1821, 2005, 2011
-, necklace 1283
-, pancake grain 1015
-, patenting 1689
-, subgrain 1189
- zone model 1327
Substitutional alloying element 983

Substrate 535
Superconductors, High-T_c ... 1161, 1379
Superplasticity 1357
Surface
- layer texture 1197
- packing density 547
- texture 1881, 1985
Symmetry 13
-, cubic 369
Synchroton 1081

TA6V 1357
Ta-2.5W 1653
Tangent approach 1869
Tantalum 841, 1423, 1653
Tape casting 43, 1411
Taylor
- model 617, 847, 1277,1689, 1777, 1875
- theory 1791, 1881
Tensile test 1909
Tension 1507
Tensor 287
Tensorial property 1641
Texture 3, 111, 295, 481, 515, 547, 555, 597, 627, 653, 693, 783, 813, 835, 847, 865, 905, 953, 959, 989, 997, 1003, 1027, 1045, 1051, 1357, 1371, 1387, 1405, 1411, 1417, 1455, 1487, 1507, 1513, 1537, 1623, 1629, 1653, 1675, 1681, 1731, 1739, 1809, 1827, 1941, 2017, 2043, 2051, 2059
- analysis 71, 97, 119, 143, 159, 181, 213, 235, 275, 309, 327, 357, 465, 733, 1379
- -, fixed angle 1929
- - program 333
-, annealing 1365, 1839
-, bulk 633
- by irradiation 535
- chain 1463
- change 1551
- comparison 447
- component 259, 327, 363, 493, 529, 1219, 1481, 1759
-, compromise 1803, 1895
-, circular 611, 701
-, cube 541, 1021, 1087, 1145, 1765, 1803, 1991, 2025
-, cyclic 611, 1881
-, cylindrical 1299
-, deformation 577, 585, 745, 821, 887, 919, 1231, 1237, 1337, 1777, 1815, 1881, 1979, 1991
- development 1869, 1875
- diffractometer 71
-, domainal 1481
- evolution 1495, 1901
-, fiber 535, 611, 685, 715, 1021, 1299, 1435, 1443, 2005
- formation 873, 879, 1855
-, global 1423, 2031
- goniometer 207
-, goss 989
- heterogeneity 213
-, high temperature 167, 181, 1225
- index 349
-, inhomogeneous 111, 1423, 1695
-, local 767, 1069, 1131, 1205, 1257, 1291, 1423, 1571
-, macro 259, 709
-, magnetic 251
- mapping 71
-, micro 31, 259, 315, 407, 709, 1109, 1153, 1299, 1305, 1313
-, model 2067
- modeling ... 585, 673, 953, 1559, 1713
-, oxide 521
- processing 1917
-, recrystallization 887, 919, 953, 977, 983, 1009, 1015, 1501, 1713, 1783, 1887, 1991, 2005
-, rolling 679, 761, 1337, 1365, 1635, 1783, 2017
- substructure interrelations 1119
-, sharp 387
-, shear 659, 795, 913
-, solidification 1351
- simulation 1777, 1979, 1839
- spread 327
- stress analysis 71
- structure analysis 2067
-, surface 1881, 1985
-, - layer 1197
-, torsion 1869

- transformation 131, 167, 181, 507, 563, 827, 1759, 1985
Thin film 97, 159, 207, 1205, 1379, 1487, 1571
Titanium 1319, 1875
- Al alloy 309, 721, 813
- alloy 665, 1861, 1971
- oxide 521
- Zr alloy oxide 521
Tomographic inversion problem 423
Transform method of Radon 419
Transformation 453
-, coordinate 369, 1423
-, martensitic 167, 563
-, texture 131, 167, 181, 507, 563, 827, 1759, 1985
- twin 739
Transient liquid phase bonding 1305
Transmission
- electron microscopy (TEM) 189, 201, 241, 487, 945, 1131, 1205, 1219, 1571
- Kikuchi pattern 143, 187, 189, 201, 241, 247, 1131, 1205, 1319, 1571
Transparent metal 1559
Triple junction 175
True ODF 375
Tungsten wire 1299
Two-point correlation function 221
Twin
- roll casting 913
- system 241, 639
- transformation 739
Twinning 659, 835, 927, 1861, 1875

Ultrasonics 221, 295, 1929
Ultrasonic backscattered noise 221
Ultrahyperbolic equation 419
Underlayer 1443
Uniaxial compression 309

Variant selection 507, 563, 1713
Variants of mathematical standard ODFs 423
Variation width of feasible ODFs 423
Vector method 301, 459
Velocity
- anisotropy 1523
- calculation 1523
- ultrasonic 1929
Viscoplastic 665
Visioplasticity 1881
von Mises yield surface 1603
Voigt-Reuss-Hill approximation ... 1663
Volume fraction 315, 327, 493, 919

Wedge-shaped specimen 1095
Whiskers 1475
WIMV 473
Wire 709, 1299, 1689, 2075
- drawing 701
Wood 1463
Workability 1251
Work hardening 1689

X-ray
- analysis 2037
- diffraction 111, 119, 159, 821, 933, 1119, 1495
- penetration depth 111
- topography 1081

Yield 1731
- locus 783, 1603, 1623, 1961, 1967
- potential of polycrystal 1875
- stress 1947
Young's modulus 1551, 1663

Zirconium 835
- oxide 521
Zircalloy 617, 1365, 1455
Zn 859, 1671
- alloy 859, 1351
- layers 1495

ABBREVIATIONS

ADC	Arbitrary Defined Cell
BKP	Backscatter Kikuchi Pattern
CCD camera	Charge Coupled Device camera
CODF	Crystal Orientation Distribution Function (see ODF)
CSL	Coincidence Site Lattice
CRSS	Critical Resolved Shear Stress
EBSP	Electron Back-Scattering Pattern
EVOH	Ethylene-Vinylalcohol
IF	Interstitial Free
LCP	Liquid Crystal Polymer
MCI	Misorientation Correlation Index
MODF	MisOrientation Distribution Function
OCF	Orientation Correlation Function
ODF	Orientation Distribution Function
ODDF	Orientation Difference Distribution Function
OIM	Orientation Imaging Microscopy
PSD	Position Sensitive Detector
PSN	Particle Stimulated Nucleation
SAD	Selected Area Diffraction
SEM	Scanning Electron Microscopy
SIMS	Secondary Ion Mass spectrometry
TEM	Transmission Electron Microscopy